Phytoremediation Potential of Bioenergy Plants

Kuldeep Bauddh • Bhaskar Singh
John Korstad
Editors

Phytoremediation Potential of Bioenergy Plants

Springer

Editors
Kuldeep Bauddh
Centre for Environmental Sciences
Central University of Jharkhand
Ranchi, Jharkhand, India

Bhaskar Singh
Centre for Environmental Sciences
Central University of Jharkhand
Ranchi, Jharkhand, India

John Korstad
Department of Biology
Oral Roberts University
Tulsa, OK, USA

ISBN 978-981-10-9785-0 ISBN 978-981-10-3084-0 (eBook)
DOI 10.1007/978-981-10-3084-0

Printed on acid-free paper

This Springer imprint is published by Springer Nature
The registered company is Springer Nature Singapore Pte Ltd.
The registered company address is: 152 Beach Road, #21-01/04 Gateway East, Singapore 189721, Singapore

Dedicated to my Ph.D. supervisor Prof. Rana Pratap Singh

Kuldeep Bauddh

Dedicated to the memories of my late grandmother Smt. Lalita Devi

Bhaskar Singh

Dedicated to my beautiful and blessed wife Sally

John Korstad

Foreword

INDIAN INSTITUTE OF TECHNOLOGY DELHI
Department of Chemical Engineering
Hauz Khas, New Delhi-110016, India

Dr. K. K. PANT
Petrotech Chair Professor

Tel. : +91-11-2659 6172 (O), 26596 177 (Lab)
: +91-11-2658-1042, 2659 1878 (Resi)
Fax : +91-11-2658 1120
E-mail : kkpant@chemical.iitd.ac.in
: dr.kkpant@gmail.com
Web : http://web.iitd.ac.in/~kkpant/

Phytoremediation Potential of Bioenergy Plants is an important collection of scientific papers that demonstrate practical and feasible ways in remediating degraded habitat through growing plants that sequester unwanted chemicals and can then be harvested for economical byproducts such as biomass for biofuels, pharmaceuticals and other products. Today, there is a great need to develop green technologies that benefit society directly and indirectly and also economically. Phytoremediation using bioenergy plants is a wonderful example of practical sustainability that has a bright future around the world.

Dr. Kuldeep, Dr. Bhaskar and Dr. Korstad deserve praise for editing this book with 19 chapters that deal with various aspects of energy plants bearing potential for phytoremediation. The book has current environmental as well as ecological significance and will be of interest among researchers and industrialists.

(KK PANT)

Preface

The global burgeoning population necessitates production of more goods and services to fulfill the expanding demands of human beings which have resulted in urbanization and industrialization. Uncontrolled industrialization has caused two major problems: an energy crisis and accelerated environmental pollution throughout the world. Presently, there are technologies that have been proposed or shown to tackle both problems. Researchers continue to seek more cost-effective and environmentally beneficial pathways. Plants are considered as a potential feedstock for development of renewable energy through biofuels. Another important aspect of plants is their capacity to sequester carbon dioxide and absorb, degrade and stabilize environmental pollutants such as heavy metals, polyaromatic hydrocarbons, polyaromatic biphenyls, radioactive materials and other chemicals. Thus, plants may be used to provide renewable energy generation and pollution mitigation. An approach that could amalgamate the two aspects can be achieved through phytoremediation (using plants to clean up polluted soil and water) and subsequent generation of energy from the phytoremediator plants. This would be a major advance in achieving sustainability that focuses on optimizing 'people' (social issues), 'planet' (environmental issues) and 'profit' (financial issues). The 'Phytoremediation-Cellulosic Biofuels' (PCB) process will be socially beneficial through lessening pollution impacts on people, ecologically beneficial through pollution abatement and economically viable through providing revenue that supplies an energy source that is renewable and also provides less dependence on importing foreign energy (energy independence). The utilization of green plants for pollution remediation and energy production will also tackle some other important global concerns like global climate change, ocean acidification and land degradation through carbon sequestration, reduced emissions of other greenhouse gases, restoration of degraded lands and waters, and more. This book addresses the overall potential of major plants that have the potential to fulfill the dual purposes of phytoremediation and energy generation. The nonedible bioenergy plants that are explored for this dual objective include *Jatropha curcas*, *Ricinus communis*, *Leucaena leucocephala*, *Millettia pinnata*, *Cannabis sativa*, *Azadirachta indica* and *Acacia nilotica.*

Ranchi, India — Kuldeep Bauddh
Ranchi, India — Bhaskar Singh
Tulsa, OK, USA — John Korstad

Acknowledgement

All the authors made a challenging assignment for me to make this dream a reality by their kind contribution and cooperation. I express my deep sense of gratitude to my Ph.D. supervisor Prof. Rana Pratap Singh and M.Sc. supervisor Dr. Narendra Kumar, Department of Environmental Science, BBAU, Lucknow, and Dr. Manoj Kumar, Head, Centre for Environmental Sciences, CUJ, Ranchi, for their pivotal direction, precious advice and interest which are responsible for the accomplishment of this book.

I would like to thank the coeditors of this book Dr. Bhaskar Singh, CUJ, Ranchi, and Prof. John Korstad, USA, for their wonderful, critical and helpful encouragement to complete this exciting task.

Words are not enough to express my grateful appreciation for my parents (Mr. Suresh Baboo and Mrs. Shyama Devi), elder sister (Mrs. Kalpana Devi), brother-in-law (Mr. Ajay Hans), younger brother (Jagdeep), nieces (Ananya, Komal and Akshay) and all of my relatives. Their dreams and ambition made me strong to complete the target. Their endless inspiration, support, affection and encouragement have made it possible.

I am thankful to my loving wife Sweta. Her cheering praise always prompts me to conclude the dispensed project.

I pay my joyful thanks to all of my friends whose well wishes stay with me.

(**Kuldeep Bauddh**)

I would like to acknowledge and thank each one of the authors who contributed their time and expertise to this book.

I especially thank my fellow editors (Dr. Kuldeep Bauddh and Prof. John Korstad) of the book for their vision and foresight, which inspired me to complete this book along with them.

I acknowledge my family – my parents (Mrs. Usha Kiran and Mr. S.N. Singh), my brother's family (Mrs. Smriti, Mr. Pawan Kumar, Master Jai Kishan), my sister's family (Mrs. Anupama Singh, Mr. S.K. Kashyap, Ms. Raj Nandini and Mr. Gaurav) and my loving, caring and beautiful wife Ragini along with my delightful little Daughter Riyansika. They all kept me going and encouraged me throughout my work. I acknowledge the love and encouragement of my first cousin Mr. Rajesh Kumar in my academic and personal accomplishments.

I earnestly thank my Ph.D. guide and supervisor, Prof. Y.C. Sharma, who is a role model to me and keeps encouraging me to do progressive things.

I take the opportunity to acknowledge the service of the team of Springer Publishing and everyone who collaborated in producing this book.

(Bhaskar Singh)

First and foremost, I honour Jesus Christ, my Lord and Saviour, who has called me to seek to do everything excellently through Him. 'And don't just do the minimum that will get you by. Do your best. Work from the heart for your real Master, for God…' (Colossians 3:23, Message Bible). I am thankful to my wife (Sally) and family for their unending support and encouragement. You give me great joy and fulfillment! To my colleagues, current and former students and friends, thank you for helping me grow as a servant-leader.

(John Korstad)

Contents

Contributors

O.P. Abioye Department of Microbiology, Federal University of Technology, Minna, Nigeria

Sangeeta Anand Department of Environmental SciencesI, Babasaheb Bhimrao Ambedkar University, Lucknow, India

Faiz Ahmad Ansari Institute for Water and Wastewater Technologies, Durban University of Technology, Durban, South Africa

S.A. Aransiola Bioresources Development Centre, National Biotechnology Development Agency, Ogbomoso, Nigeria

Ingvar Backeus School of Natural Sciences, Technology and Environmental Studies, Sodertorn University, Huddinge, Sweden

Department of Plant Ecology and Evolution, Uppsala University, Uppsala, Sweden

S.C. Barman Environmental Monitoring Division, Indian Institute of Toxicology Research, Lucknow, India

Kuldeep Bauddh Centre for Environmental Sciences, Central University of Jharkhand, Ranchi, Jharkhand, India

Priyanka Bharti Centre for Environmental Sciences, Central University of Jharkhand, Ranchi, Jharkhand, India

Sushil Kumar Bharti Department of Environmental Sciences, Babasaheb Bhimrao Ambedkar University, Lucknow, India

Nitesh Bhatia Centre for Business Studies, Central University of Jharkhand, Ranchi, Jharkhand, India

Faizal Bux Institute for Water and Wastewater Technology, Durban University of Technology, Durban, South Africa

Poulomi Chakravarty Centre for Environmental Sciences, Central University of Jharkhand, Ranchi, Jharkhand, India

R.K. Dey Centre for Applied Chemistry, Central University of Jharkhand, Ranchi, Jharkhand, India

Neetu Dviwedi Department of Environmental Studies, Sri Aurobindo College, Delhi University, Delhi, India

Abhishek Guldhe Institute for Water and Wastewater Technology, Durban University of Technology, Durban, South Africa

Sanjay Kumar Gupta Environmental Engineering, Department of Civil Engineering, Indian Institute of Technology, Delhi, India

U.J.J. Ijah Department of Microbiology, Federal University of Technology, Minna, Nigeria

Rajeev Jain Forensic Toxicology Division, Central Forensic Science Laboratory, Ministry of Home Affairs, Government of India, Guwahati, Assam, India

Ambuj Bhushan Jha Crop Development Centre/Department of Plant Sciences, University of Saskatchewan, Saskatoon, SK, Canada

C.K. Jisha Centre for Tribal Folklore, Language and Literature, Central University of Jharkhand, Ranchi, Jharkhand, India

John Korstad Department of Biology and Renewable Energy, Oral Roberts University, Tulsa, OK, USA

Atul Kumar Department of Electrical and Electronic Engineering, Universiti Teknologi Petronas, Perak, Malaysia

Dhananjay Kumar Department of Environmental Sciences, Babasaheb Bhimrao Ambedkar University, Lucknow, India

Dipesh Kumar Centre for Environmental Sciences, Central University of Jharkhand, Ranchi, Jharkhand, India

Manoj Kumar Center for Environmental Sciences, Central University of Jharkhand, Ranchi, Jharkhand, India

Narendra Kumar Department of Environmental Sciences, Babasaheb Bhimrao Ambedkar University, Lucknow, India

Sanjeev Kumar Department of Environmental Science, Central University of Rajasthan, Ajmer, Rajasthan, India

Santhosh Kumar Department of Biotechnology and Food Technology, Durban University of Technology, Durban, South Africa

Virendra Kumar Centre for Environmental Sciences, Central University of Jharkhand, Brambe, Ranchi, India

Sheena Kumari Institute for Water and Wastewater Technology, Durban University of Technology, Durban, South Africa

Amarendra Narayan Misra Centre for Life Sciences, Central University of Jharkhand, Ranchi, Jharkhand, India

Rohit Misra Institute for Water and Wastewater Technology, Durban University of Technology, Durban, South Africa

Mahmoud Nasr Sanitary Engineering Department, Faculty of Engineering, Alexandria University, Alexandria, Egypt

Arvind Kumar Nema Environmental Engineering, Department of Civil Engineering, Indian Institute of Technology, Delhi, India

Muhammad Ntale Department of Chemistry, College of Natural Sciences, Makerere University, Kampala, Uganda

H. Oryem-Origa Department of Plant Sciences, Microbiology and Biotechnology, College of Natural Sciences, Makerere University, Kampala, Uganda

Vimal Chandra Pandey Plant Ecology and Environmental Science Division, CSIR-National Botanical Research Institute, Lucknow, India

Anju Patel School of Earth Sciences, Banasthali University, Vanasthali, Rajasthan, India

D.D. Patra Bidhan Chandra Krishi Viswavidyalaya, Nadia, West Bengal, India

Poonam Department of Environmental Science, Babasaheb Bhimrao Ambedkar University, Lucknow, India

Anita Rani Department of Botany, Dyal Singh College, University of Delhi, New Delhi, India

Nirmal Renuka Institute for Water and Wastewater Technology, Durban University of Technology, Durban, South Africa

Pallavi Sharma Centre for Life Sciences, Central University of Jharkhand, Brambe, Ranchi, Jharkhand, India

Yogesh Chandra Sharma Department of Chemistry, Indian Institute of Technology (BHU), Varanasi, India

Rashwet Shrinkhal Centre for Tribal and Customary Law, Central University of Jharkhand, Ranchi, India

Sushil Kumar Shukla Centre for Environmental Sciences, Central University of Jharkhand, Ranchi, Jharkhand, India

Bhaskar Singh Centre for Environmental Sciences, Central University of Jharkhand, Ranchi, Jharkhand, India

D.P. Singh Department of Environmental Sciences, Babasaheb Bhimrao Ambedkar University, Lucknow, India

Gulshan Singh Institute for Water and Wastewater Technology, Durban University of Technology, Durban, South Africa

Poonam Singh Institute for Water and Wastewater Technology, Durban University of Technology, Durban, South Africa

Ritu Singh Department of Environmental Science, Central University of Rajasthan, Ajmer, Rajasthan, India

Amritanshu Sriwastav Environmental Engineering Department of Civil Engineering, Indian Institute of Technology, Kanpur, India

Jamilu Edrisa Ssenku Department of Plant Sciences, Microbiology and Biotechnology, College of Natural Sciences, Makerere University, Kampala, Uganda

Thor Axel Stenström Institute for Water and Wastewater Technology, Durban University of Technology, Durban, South Africa

Jaya Tiwari Department of Environmental Sciences, Asian Institute of Public Health, Bhubaneswar, Odisha, India

Neha Vishnoi Department of Environmental Sciences, Babasaheb Bhimrao Ambedkar University, Lucknow, India

About the Editors

Kuldeep Bauddh is currently an assistant professor of environmental science at the Centre for Environmental Sciences, Central University of Jharkhand, Ranchi, India. He received his Ph.D. in environmental science from the Babasaheb Bhimrao Ambedkar University, Lucknow, UP, India, in 2014. Dr. Bauddh is currently engaged in teaching the thrust areas of environmental sciences like environmental pollution and management, environmental chemistry, environmental toxicology, soil science and hazardous waste management. He is also working as associate editor of the journal *Climate Change and Environmental Sustainability* and coeditor of the magazine *Kahaar*. Dr. Bauddh is serving as secretary of the Ranchi chapter of the society 'Professor HS Srivastava Foundation for Science and Society', joint secretary of the 'Society for Science of Climate Change and Sustainable Environment' and joint secretary of 'Professor HS Srivastava Foundation for Science and Society'. His main research areas are phytoremediation, restoration of contaminated sites and slow-release fertilizers. He has published one book (edited) with Springer titled *Algae and Environmental Sustainability*, more than 28 research papers in national and international journals, 6 book chapters and 6 popular articles in magazines. Dr. Bauddh is reviewer of several international research journals like *International Journal of Phytoremediation*; *Chemosphere*; *Ecological Engineering*; *Sustainability of Water Quality and Ecology*; *CLEAN – Soil, Air, Water*; *Land Development and Degradation; Acta Physiologiae Plantarum*; and *Climate Change and Environmental Sustainability*.

Bhaskar Singh received his Ph.D. from the Indian Institute of Technology (BHU), Varanasi, UP, India, in 2010 and M.Phil. from Pondicherry (Central) University, India, with a gold medal (2006). Dr. Bhaskar did his postdoctoral fellowship in Durban University of Technology, Durban, South Africa. He is currently serving as an assistant professor at the Centre for Environmental Sciences, Central University of Jharkhand, Ranchi, India. Dr. Singh is currently engaged in teaching in the thrust area of environmental sciences like environmental earth science, natural resources management, environmental chemistry, environmental pollution monitoring and control technologies, environmental management and sustainable development and environmental law and legislation. He is reviewer of several international journals such as *Fuel*, *Energy*, *Renewable Energy*, *Renewable & Sustainable Energy Reviews*, *Energy & Fuels*, *Journal of Cleaner Production*, *Materials Chemistry and*

Physics, *Biofuel Research Journal*, *Oceanologia* and *Journal of Agricultural Biotechnology and Sustainable Development*. Dr. Singh is serving as treasurer of the Ranchi chapter of the society 'Professor HS Srivastava Foundation for Science and Society'. His current interests lie in the application of biodiesel synthesis from plant oil and algal biomass, development of heterogeneous and green catalysts and corrosion aspects of biodiesel fuel. Dr. Bhaskar has coedited one book with Springer titled *Algae and Environmental Sustainability* along with Dr. Kuldeep Bauddh and Prof. Faizal Bux. His publications include 35 research and review papers in peer-reviewed and high-impacted international journals along with 9 book chapters with reputed international publishers. As per Scopus, the current *h*-index of publications of Dr. Bhaskar Singh is 16.

John Korstad is a professor of biology and past director of the honours programme at Oral Roberts University in Tulsa, Oklahoma (USA). His educational background is a B.A. in geology and B.S. in biology from California Lutheran University; an M.S. in environmental science from California State University, Hayward; and an M.S. and Ph.D. in zoology from the University of Michigan. His specialty is limnology (lake ecology) with specific expertise in nutrient-phytoplankton (microalgae)-zooplankton interactions. He has vast experience in aquaculture, including two sabbatical years doing research at the Foundation for Scientific and Industrial Research (SINTEF) in Trondheim, Norway, resulting in three published and numerous technical publications. He started doing algae to biofuels research during his Fall 2009 sabbatical leave. His desire is to connect various industries like local oil refineries, electrical power plants and cement factories that produce carbon dioxide waste and local industries like waste management landfills and the city's sewage treatment plants with high nutrient loads in their wastewater and cultivate algae for biofuels. The algae would improve the air and water quality by taking up the waste CO_2 and nutrients and then be of benefit by providing a feedstock for biodiesel and/or 'high-end' products like omega-3 fatty acids. This would essentially be a win-win situation and would hopefully develop into a vibrant industry that brings added jobs and revenue to the local area.

1 Phytoremediation: A Multidimensional and Ecologically Viable Practice for the Cleanup of Environmental Contaminants

Poulomi Chakravarty, Kuldeep Bauddh, and Manoj Kumar

Abstract

The humungous load of pollutants added to the environment every day by the human activities is one of the major menaces facing by the world. Toxic substances released into the ecosystems are said to create imbalance to the equilibrium of the environment. Phytoremediation is a set of processes which have been considered as one of the most sustainable approaches to combat the problem of contaminants. Phytoremediation is considered to be more effective in comparison with traditional techniques because of the added benefits provided by the plants. The mechanisms adapted by the plants for extraction, accumulation, stabilization and degradation of contaminants from the polluted sites have been explored in this chapter. Various floral species which have been reported by several researchers that have the potential to remediate contaminated sites are listed in this report. The bioenergy crops, medicinal plants, trees and weeds have been found to be the best options for phytoremediation. Phytoremediation has proven to have a holistic approach which can help in restoration of contaminated sites with production timber, essential oils, energy, and employment to the rural peoples and with several other ecosystem services.

Keywords

Bioenergy • Electrokinesis • Heavy metals • Phytoremediation • Pollution • Transgenic plants

P. Chakravarty • K. Bauddh (✉) • M. Kumar
Centre for Environmental Sciences, Central University of Jharkhand, Ranchi-835205, Jharkhand, India
e-mail: kuldeep.bauddh@cuj.ac.in

K. Bauddh et al. (eds.), *Phytoremediation Potential of Bioenergy Plants*,
DOI 10.1007/978-981-10-3084-0_1

1.1 Introduction

The world population has exceeded seven billion and is rapidly approaching eight billion. This ever-increasing population has exerted tremendous chaos on the existing natural resources and has created immeasurable amount of wastes across the globe. When pollution is in manageable amount, the terrestrial, aquatic and atmospheric ecosystems can dilute, degrade or absorb the contaminants naturally. The rising burden of pollutants requires additional measures to curb the detrimental effects of pollution (Glick 2003; Glick 2010). Contaminants pose a threat to the environment because of their abundance and recalcitrant nature. Rampant industrialization and urbanization are the main culprits for the gradual degradation in environmental quality. The release of natural and anthropogenic contaminants is a major concern in the last few decades. There are numerous contaminants that continuously cause problems, some of which are easily curable but many are not. Plants act as Green Livers for the ecosystem clarifying any ill effects caused by contaminants and toxicants in the ambient environment (Sanderman 1994).

1.1.1 Contaminants: Sources, Types and Effects

A pollutant is anything that is present in the environment in excess to its original concentration. Waste generation by anthropogenic activities is so diverse in nature that it is difficult to categorize them effectively. Contaminants that create nuisance in soil and water are usually industrial wastes, municipal solid wastes, agricultural runoffs and leachates (organic pollutants) and radioactive wastes. The organic pollutants, heavy metals and radioactive wastes are dealt here as they are potentially the most problematic pollutants in terms of soil and water. They cause adverse effects directly to the plants as well as animals including human beings and sometimes indirectly by changing the natural composition of ecosystems (Fig. 1.1).

1.1.2 Heavy Metals

Heavy metals have been reported as one of the major nemeses for the environment. Apart from natural processes, maximum number of anthropogenic activities releases heavy metals (Tangahu et al. 2011). The problem lies when contaminants migrate to pristine areas in the form of metal dust or leachates as in the case of soil and also as sewage sludge (Gaur and Adholeya 2004). Heavy metals are those elements which have an atomic number more than 20. Metals are also present naturally in soil. Many of them are essential for growth and sustenance of soil flora and fauna. Zinc, copper, manganese, nickel and cobalt are imperative for survival of the plants. The importance of some metals such as cadmium, lead and mercury is unknown in respect to plants (Lasat 2000; Gaur and Adholeya 2004). Heavy metals are non-biodegradable, therefore creating problems in the overall biological systems. Heavy metals such as lead, cobalt and cadmium are more deleterious in nature because of their high bioaccumulation rate even at lower concentration

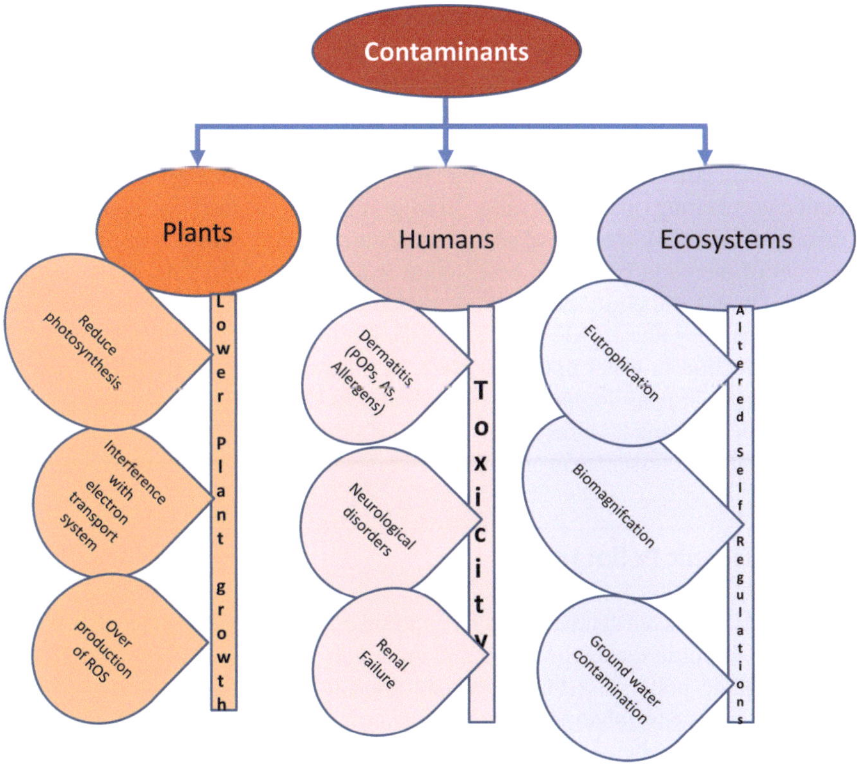

Fig. 1.1 Adverse impacts of contaminants on the environment

(Pehlivan et al. 2009; Tangahu et al. 2011). Heavy metals may cause negative impact on plant growth and soil microflora (Roy et al. 2005). Arsenic is one major environmental pollutant which falls under the category of heavy metal having atomic number 33. Arsenic is found in the environment as organic arsenic species, inorganic arsenic compounds and arsine gas. Arsenic is a very toxic element, and its toxicity is usually dependents on the species. The inorganic compounds of arsenic are usually more toxic than its organic counterparts. Arsenites are more toxic in nature than arsenates as they are more prone to cause DNA breakdown (Ampiah-Bonney et al. 2007; Vaclavikova et al. 2008). Arsenates are found to be more stable thermodynamically than arsenites; therefore, they cause groundwater contamination (Chutia et al. 2009). Arsenic compounds are carcinogenic in nature and cause dermatitis where the groundwater is contaminated. Lead with atomic number 82 is a highly toxic element which is non-biodegradable and remains in the environment for a very long time and accumulates in the first 8 in. of the soil and remains immobile. Sources of lead include natural sources, industrial sites, leaded fuels and orchards where the use of lead arsenate takes place (Traunfeld and Clement 2001; Tangahu et al. 2011). The harmful effects of lead are spread across a wide range of organisms such as humans, animals, plants and microbes. In terms of human health, lead causes major adverse impacts such as mental retardation and brain damage (Cho-Ruk et al. 2006). Mercury is another heavy metal that is notoriously toxic and is available in soil in three soluble forms. It is a toxic element with a high bioaccumulation potential in living organisms such as human beings, fish and other animals. Mercury is found in naturally as well as by anthropogenic activities in the environment. Mercury pollution in the environment is caused by mining, petrochemical, painting industries, also from fertilizers, medical instruments, etc. (Resaee et al. 2005). Usually terrestrial plants are not very sensitive to the adverse impacts of mercury, but it has been found that mercury interferes with electron transport in mitochondria and chloroplasts and adversely affects oxidative metabolism and photosynthesis. Mercury acts as an inhibitor of aquaporin activities and causes reduction in water uptake in plant. In human beings, the toxic impacts of mercury include neurological and renal disorders (Resaee et al. 2005). As toxic metallic species cannot be degraded, there is a requirement of physical removal or transformation to lesser toxic or non-toxic compounds.

1.1.3 Organic Pollutants

Organic pollutants are synthetic and recalcitrant in nature. These organic xenobiotics are persistent in the environment and are highly toxic. They are known as persistent organic pollutants (POPs) as they are not easily degradable. Pesticides, petroleum products, pharmaceuticals, polyaromatic hydrocarbons (PAHs) and polychlorinated biphenyls (PCBs) are some of the existing organic pollutants (Abhilash and Singh 2009). Twelve major POPs are known as the 'dirty dozen' which have been called for elimination and phasing out by the United Nations Environmental Program (US EPA 2005). Aldrin, dieldrin, chlordane, DDT, endrin, heptachlor, mirex, toxaphene, PCBs, HCBs, dibenzodioxins and dibenzofurans are the twelve most dangerous pollutants in respect to organic contaminants. Organic pollutants are a real menace for the ecosystem because of their persistence in the environment, lipophilic nature and high bioconcentration potential. These pollutants tend to get deposited in the adipose tissues of organisms (POPs, WHO Report 2008). Over a period of time, the pollutants reach a high level of toxicity because of their high bioconcentration potential, even though the exposure is limited. The pollutants move up the food chain as a result of biomagnification. Therefore, it is reported that the apex consumers reveal the maximum amount of organic pollutant concentration in their tissues. Marine mammals are known to have the highest concentration of these pollutants which caused reproductive disorders and higher susceptibility to infections resulting from microbes. The soil that is contaminated by organic pollutants causes death of soil microflora and reduction in plant growth and yield. Leaching of these pollutants causes groundwater contamination. Fertilizers when reaching the surface water bodies cause eutrophication by nutrient enrichment. The algal bloom caused by this nutrient enrichment reduces the dissolved oxygen level of the water bodies culminating in the death of aquatic flora and fauna. These are just few of the impacts of organic pollutants; there can be numerous direct and indirect effects of these contaminants. It is essential to remove these harmful toxicants from the environment to continue the balanced functioning of the ecosystems.

1.1.4 Radioactive Contaminants

Radioactive contaminants are introduced into the environment mostly by anthropogenic activities. Although radioactive elements are present in the environment naturally, they are not as harmful as contamination caused by anthropogenic causes because in nature, they are in a very low concentration. The environment is contaminated by radionuclides by nuclear weapon testing, disposal of nuclear wastes, emissions from nuclear power plants and also from spillage from plant operations such as nuclear fuel mining, milling and nuclear testing fallout, etc. In the process of oil drilling, sometimes radionuclides that occur naturally are brought up to the surface of the Earth (Fulekar et al. 2010). Chernobyl disaster in 1986 was one of the first nuclear power plant disasters which exposed the devastating effects of nuclear accidents to the world. The most recent nuclear accident occurred at Fukushima, Japan, during an earthquake at 2011; at Fukushima, an explosion was caused by failure of emergency cooling. Radionuclides are highly unstable nuclei possessing additional energy. There is a constant radioactive decay experienced by the radionuclides which forms alpha, beta and gamma particles as a result (Ghosh and Singh 2005; Fulekar et al. 2010). Consumption of food crops and water contaminated by radionuclides is one of the major causes of exposures to humans. The persistence of radiation in the environment can be over billions of years; therefore, it can cause irreparable damage to organisms as well as the ecosystem (Fig. 1.2) (Malhotra et al. 2014). Generally, the radiation released by the radionuclides can be carcinogenic and mutagenic in nature and is also known to cause birth defects and abnormalities in humans over a long period of exposure. Uranium-238 the most common natural isotope of uranium has a half-life of 4.46 billion years that is used in nuclear weapons and nuclear fuel. It is known to cause birth defects, cancer and mutations in the genes of humans (Jadia and Fulekar 2008). Thorium-232 is the most stable isotope with a half-life of 14 billion years, is used in nuclear fuel and alloying agent and is found to be carcinogenic in nature. Spinks and Woods (1990) state that radium-226 has a half-life of 1600 years and is used in an abundant fashion in our daily lives in the form of luminous paints and in dials of watches. An exposure for a long duration may cause fatal diseases like bone cancer, lymphoma, aplastic anaemia and leukaemia. During the Chernobyl accident, several radionuclides were released into the atmosphere; among them were isotopes of caesium-134 and caesium-137. These isotopes are retained by the soil and not washed away even by the heaviest rainfall. Isotopes of caesium are taken up by the plants, and they easily enter the food chain; also adverse effects are caused when there is an exposure to the contaminated soil surface (Westhoff 1999). The beta and gamma radiations of the radionuclides are highly dangerous and can cause ulcers, erythema or tissue necrosis in humans.

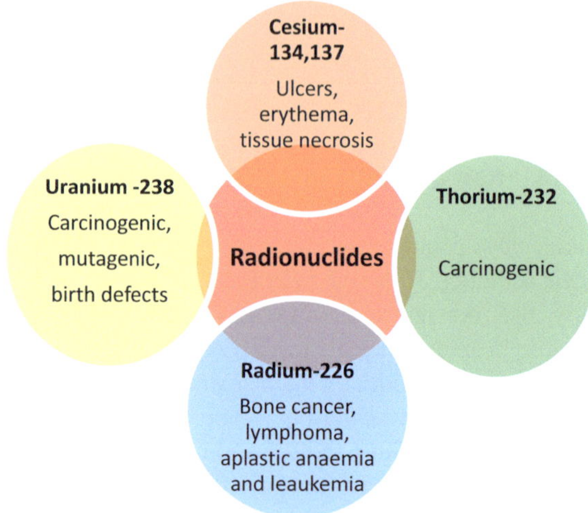

Fig. 1.2 Adverse impacts of radionuclides on humans

1.2 Contaminant Remediation Techniques

The above-mentioned problems are just the tip of the iceberg, and there are several underlying issues related to these contaminants that can cause direct or indirect impact on the environment. It is highly imperative to remediate the contaminated spheres of the environment. There are several conventional methods and techniques applied for the remediation of the contaminated areas. Some of the traditional methods to combat the problem of contaminated soil include:

1. *Soil excavation*: Treatment or removal of contaminants in the case of soil is done by onsite management or by excavation of the contaminated soil and by its disposal at a landfill site. This method of disposal is not a real solution of the problem as it merely dislocates the contaminants from one area to another (Tangahu et al. 2011).
2. *Soil washing*: As an alternative to the dislocation of contamination from the source to a landfill area, an onsite management method is applied. Soil washing is carried out by two processes: first of them is by dissolution or suspension of contaminated soil in a wash solution which is chemical in nature and the second process concentrates the contaminants into a smaller volume of soil by techniques such as gravity separation, particle size separation and attrition scrubbing. Heavy metals, organic xenobiotics and radionuclides can be removed by this process. This method is not cost-effective, and residues rich in contaminants require additional treatment. Therefore, this process is not extensively used (Tangahu et al. 2011).
3. *Stabilization/solidification*: In this process, the contaminants present in the soil are stabilized or solidified either by physical or chemical interactions between the contaminant and a stabilizing agent (Gomes 2012).

4. *Vitrification*: In the process of vitrification, heat is used for melting and subsequently solidifying the contaminants in a solid material which is glasslike in nature. Vitrification can be carried out onsite (in situ vitrification) and also aboveground in a separate treatment unit (ex situ vitrification).
5. *Electrokinetic treatment*: The electrokinetic remediation technique is solely in situ-based where direct electric potential is applied using cathodes and anodes. According to Cameselle et al. (2013), various reactions take place in the contaminated soil due to the electric potential; as a result, the contaminants move towards the cathode or anode. The mobilization or transport mechanisms in electrokinetic treatment are of two types, electro-osmosis and electromigration. When there is a combined effect of electric charge and electric field on soil particle surface, it results in an electro-osmotic flux which causes the movement of negatively charged particles towards the cathode. In the electromigration mechanism, movement of ionic species takes place in the electric field towards the oppositely charged electrode (Cameselle and Reddy 2012).

   Some other methods such as incineration and chemical oxidation/reduction are also used for the remediation of contaminated soil, but most of these traditional methods are not feasible because of high cost and problems regarding disposal of contamination-rich residues. Some of these techniques also destroy the soil biota causing the area to become devoid of life. Hence, it is essential for the sake of the environment to find alternative technologies that are environment-friendly and green in approach. These technologies must be cost-effective and reduce the pollutant load in the environment, and at the same time, the technique should have features which help them to resolve other major concerns like fuel crisis, emission of greenhouse gases, etc.

1.3 Phytoremediation: A Successful and Environment-Friendly Approach

Everyday new technologies are being developed by humans to vanquish the evil effects of pollution created by humans themselves. The solution lies in the hands of nature itself; plants are the nature's best defence against all man-made pollution. The word phytoremediation originates by combining two words *Phyto* (Greek) meaning plants and *remedium* (Latin) meaning removal or correction of evil. In general words, phytoremediation means removal, degradation or stabilization of pollutants using plants. At current time, plants have regained their former status of importance because of their multifaceted applications. The contaminants are removed from soil, water and sediments using plants. Certain plant root systems have special uptake capabilities, and also the shoot systems are capable in translocation, accumulation and degradation of the contaminants. These features allow efficient uptake and removal of harmful toxicants from the environment. Phytoremediation is a solar energy-driven process and does not require external energy, so it is cost-effective and less (zero) polluting in comparison with traditional methods. There are several definitions of phytoremediation given by various researchers; few have been compiled in Table 1.1.

Table 1.1 Definitions of phytoremediation

No.	Definition	Reference
1	Phytoremediation is a set of techniques or processes where plants are used for extracting, containing, degrading/destroying or immobilizing contaminants from the medium (soil, water or sediments)	EPA (2000)
2	The usage of plants for remediation of toxicants found in groundwater, contaminated soil, sludge, wastewater, surface water and sediments	Rodriguez et al. (2005)
3	Phytoremediation is a technology that makes use of plants to purify contamination from water, sediments or soil	Tangahu et al. (2011)
4	The application of plants for extraction and sequestration followed by detoxification of the contaminants	Ismail (2012)
5	A sustainable and green process in which live plants are used for removing or degrading contaminants from the environment	Cameselle et al. (2013)

1.3.1 Types of Phytoremediation

1.3.1.1 Phytoextraction

In terms of economic opportunities, phytoextraction presents the largest benefits (Raskin et al. 1997; Ismail 2012). Phytoextraction is considered as the most efficient method for removal of an isolation of contaminants from the polluted medium that is the soil where the fertility and structure of the soil is retained (EPA 2000). In the process of phytoextraction, the plant absorbs contaminants from the soil/water through roots and transfers or translocates them to the aerial parts of the plants. The aerial parts can be burnt to gain energy, and the metal can be recycled from the ash (Liu et al. 2000; Prasad and Freitas 2003 Erakhrumen and Agbontalor 2007; Moreno et al. 2008). Phytoextraction is most effective in large areas which have a contamination level of low to medium range, and the depth is also shallow (Kumar et al. 1995a, b; Blaylock and Huang 2000). The plant must possess some special characteristics to be efficient in the process of phytoextraction. These characteristics include tolerance towards the specific contaminant, efficient translocation of contaminants to aerial and harvestable parts of the plant and ability of plant to survive in stress conditions like soil pH, salinity, soil structure, water content and resistance to pests (Brooks 1994; Ismail 2012).

1.3.1.2 Phytostabilization

There are certain plant species that specialize in immobilizing contaminants in the soil or groundwater itself. These plants absorb and accumulate the contaminants in plant tissues, adsorb on the root surface or precipitate them within the root zone thereby preventing migration of contaminants in the soil and their movement by erosion (Liu et al. 2000; Prasad and Freitas 2003; Erakhrumen and Agbontalor 2007; Moreno et al. 2008). This method of phytoremediation is also known as phytorestoration. The plants used for phytostabilization must be weak in translocating the contaminants from the root to the aerial parts; must grow fast, having developed root systems and canopies, and must be tolerant towards abiotic and biotic stresses (Ismail 2012).

1.3.1.3 Phytofiltration

The process of phytofiltration can be of two types, one through the roots that is known as rhizofiltration and another one by seedlings that is known as blastofiltration. The roots or seedlings of the plant accumulate the contaminants from the effluents when grown in water that is aerated (Raskin et al. 1997). In this technique, plants are grown hydroponically; then they are transplanted in polluted water where they accumulate the contaminants (Dushenkov et al. 1995; Salt et al. 1995; Flathman and Hannza 1998). The phytoremediation of effluent or domestic wastewater is carried out using rhizofiltration. The contaminants are adsorbed or precipitated onto the plant roots and also in some cases absorbed and sequestered in the roots of plants present in constructed wetland for purification of effluent and wastewater (Liu et al. 2000; Prasad and Freitas 2003 Erakhrumen and Agbontalor 2007; Moreno et al. 2008). Ideally for rhizofiltration, plants must have roots that are fast growing and have higher efficiency in accumulation of contaminants over a longer time period. The toxic contaminants form a precipitate over the root surface which is then harvested and disposed (Flathman and Hannza 1998). The process of blastofiltration belongs to the second generation of water treatment technology which is plant based. After germination as there is an immense increase in the surface and volume ratio, the seedlings more effectively absorb or adsorb larger amounts of contaminants in ionic form making it more efficient than rhizofiltration (Raskin et al. 1997).

1.3.1.4 Phytovolatilization

In the process the contaminant is taken up by the plant and released by the process of transpiration either in the same form or in a modified form. In the process of phytovolatilization, the plant uptakes water which includes the contaminants, and the contaminants when reaching the aerial parts of the plants move out by transpiration (Liu et al. 2000; Prasad and Freitas 2003 Erakhrumen and Agbontalor 2007; Moreno et al. 2008). Some toxic contaminants exist in the atmosphere in gaseous form, for example, metallic species-like arsenic, mercury and selenium. In case of heavy metals, the plants adsorb metals in their elemental form, and then they are biologically converted into gaseous species which is known as biomethylation to create volatile molecules that are released into the atmosphere. There is a major disadvantage of this process in that volatile gaseous species may return to the ecosystem by precipitation thus creating havoc by spreading the toxic metals to a wider range of area (Henry 2000).

1.3.2 Mechanism of Phytoremediation

The basic steps involved in metal detoxification include metal ion binding on the cell wall of roots, metal ion transportation to the shoots and chelation of contaminants in cytosol (Fig. 1.5). The first step of mechanism of contaminant accumulation is the adsorption of metals on the root surface of the plants. Numerous metal transporters are located in the cell wall which allows metal ions to move inside the cell. Metal transporters can be grouped into ZIP family, NRAMP family and CTR family. IRT1 was found in *Arabidopsis thaliana* that belongs to the ZIP family expressed to accumulate higher amount of Fe at the time of Fe deficiency (Eide et al. 1996; Zaal et al. 1999; Guerinot 2000; Vert et al. 2002). This element has also been found to be characterized in *A. thaliana* and responsible for the accumulation and transport of Mn, Zn and Cd (Cohen et al. 1998; Korshunova et al. 1999; Zaal et al. 1999). Nishida et al. (2011) reported that expression of AtIRT1 enhances Ni accumulation in *Saccharomyces cerevisiae*. NRAMP is another metal transporter family which helps the plants to transport a number of metals like Cd, Ni, Zn, Fe, Cu, etc. (Nevo and Nelson 2006; Krämer et al. 2007).

In metal accumulator and hyperaccumulator plants, there are several defence mechanisms involved like (1) production of antioxidative components, e.g. ascorbate peroxidase (ASP), catalase (CAT), superoxide dismutase (SOD), glutathione S-transferase (GST), glutathione reductase (GR), proline, etc. (Ni et al. 2013; **Shanmugaraj et al.** 2013**; Yu et al.** 2013**;** Bauddh and Singh 2012a, b, 2015a, b), (2) production of phytochelatins (Cobbett 2000; Lee et al. 2003; Manara 2012), (3) production of metallothioneins (Nordberg 2004; Zimeri et al. 2005; Zhigang et al. 2006), (4) production of ferritins (Ravet et al. 2009; Liu et al. 2010; Yin et al. 2008; Rastgoo and Alemzadeh 2011), etc. These systems make a plant tolerant and enhance the metal-accumulating ability of plants at an even higher contamination level.

The production of metallothioneins in metal accumulator plants has been reported, and it is found that this component has the ability to detoxify the metal ion (Cobbett and Goldsbrough 2002; Papoyan and Kochian 2004; Zhigang et al. 2006; Mijovilovich et al. 2009). Many studies showed a substantial role of MTs in detoxification of Cu in many plants like *Nicotiana tobacum*, *N. caerulescens*, *Thlaspi caerulescens*, etc. (Kägi 1991; Maiti et al. 1991; Roosens et al. 2004; Papoyan and Kochian 2004; Mijovilovich et al. 2009; Leitenmaier and Küpper 2013).

It has been observed that during exposure to a biotic stresses like heavy metals, drought, salinity, etc. plants experience the overproduction of reactive oxygen species (ROS), e.g. superoxide radical (O_2^-), hydroxyl radical ($OH^\bullet$), hydrogen peroxide (H_2O_2), singlet oxygen ($^1O_2$), etc. (Fig. 1.3) which can lead to a number of abnormalities like peroxidation of lipids and damage of proteins, enzymes, cell wall, etc. (Mittler 2002; Sharma and Dubey 2005; Asada 2006; Vanderauwera et al. 2011; Sharma et al. 2012; Noctor et al. 2014; Arora et al. 2016).

To overcome these adverse changes caused by ROS, plants produce antioxidative defence system which comprises of both enzymatic components like superoxide dismutase (SOD), catalase (CAT), peroxidase, ascorbate peroxidase (APX), glutathione reductase (GR), guaiacol peroxidase (GPX), etc. and several non-enzymatic components like ascorbate, carotenoids, glutathione (GSH), phenolics, tocopherols, etc. (Fig. 1.4) (Asada 2006; Slater et al. 2008; Sharma et al. 2012; Sewelam et al. 2016).

Phytochelatins are low molecular weight cysteine-rich proteins synthesized from glutathione by an enzyme phytochelatin synthase during prolonged exposure of

Fig. 1.3 Overproduction of reactive oxygen species (ROS) under biotic and abiotic stresses (Ashry and Mohamed 2012; Sharma et al. 2012; Noctor et al. 2014)

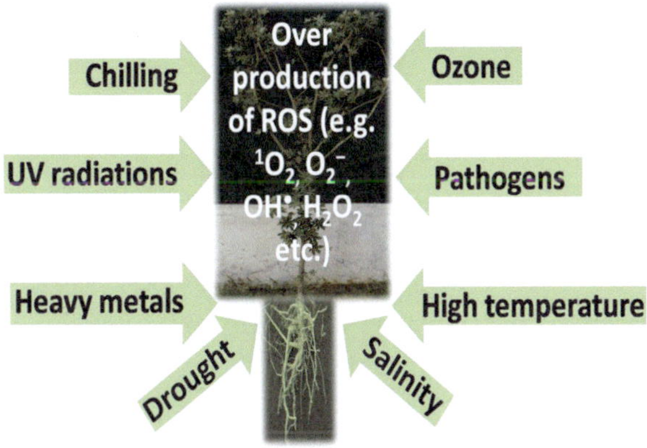

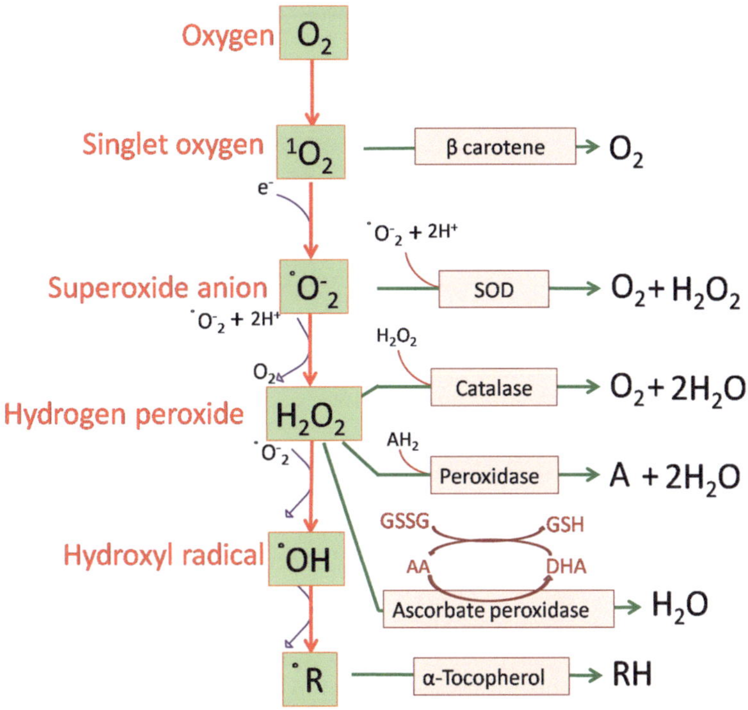

Fig. 1.4 The mechanism of formation of reactive oxygen species and their removal by antioxidants and antioxidative enzymes. *AA* ascorbic acid, *DHA* dehydroascorbic acid, *GHS* glutathione, *GSSG* oxidized glutathione, *SOD* superoxide dismutase (Adopted from Slater et al. 2008; Page No. 230)

heavy metals (Tommasini et al. 1998; Cobbett 2000; Clemens 2001; Schützendübel and Polle 2002; Harada et al. 2002; Gao et al. 2013). Phytochelatins contain gamma glutamylcystein and glycine in its structure ($(\gamma\text{-Glu-Cys})_n$-Gly) (Kondo et al. 1984;

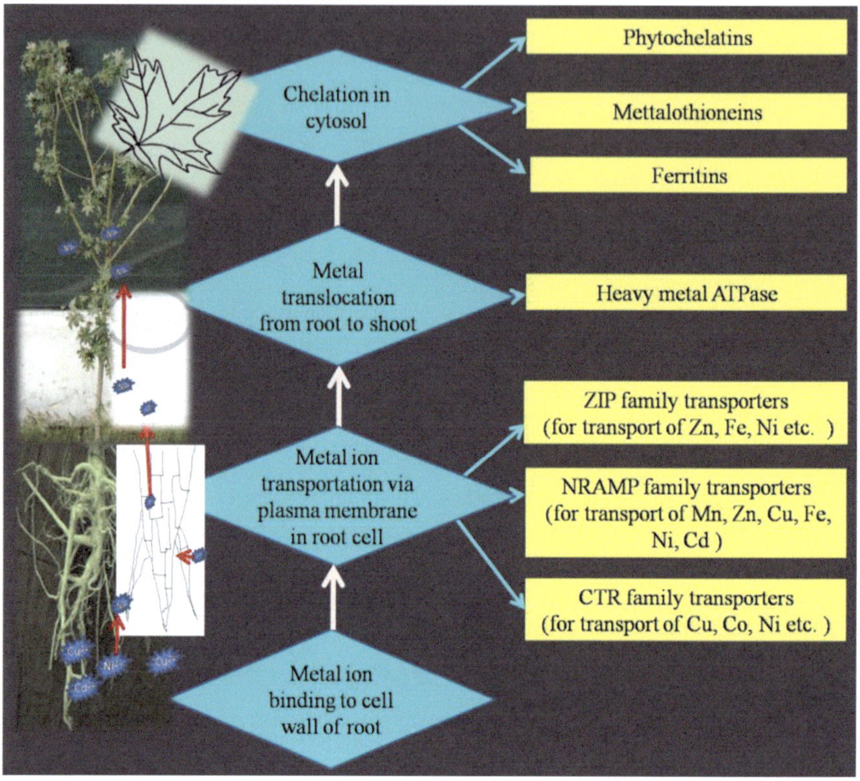

Fig. 1.5 The properties which make a plant metal accumulator/hyperaccumulator (Harada et al. 2002; Vert et al. 2002; Manara 2012; Gao et al. 2013)

Grill et al. 1986). An enhanced transcription of genes which synthesizes the precursor (glutathione reductase) of PCs was reported by Xiang and Oliver (1998) which confirmed the role of PCs as metal detoxifier (Hartley-Whitaker et al. 2001; Andresen et al. 2013). Further, Gao et al. (2013) demonstrated that the synthesis of PCs in plant *Phytolacca americana* is Cd dose dependent.

Ferritins are the proteins which have the ability to bind excess content of Fe in plants (Briat 1996; Fabisiak et al. 1999; Briat et al. 2006; Ravet et al. 2009; Briat et al. 2010). Phytoferritins are basically found in the mitochondria (Zancani et al. 2004, 2007) and non-photosynthetic plastids such as chromoplasts, proplastids, etioplasts, etc. (Seckback 1982; Ragland et al. 1990). Deák et al. (1999) proposed that ferritin can protect the plant from oxidative damage persuaded by a number of abiotic as well as biotic stresses (Fig. 1.5).

On the other hand, many plants secrete exudates from their roots which can chelate the metals and in soil only and prevent metal uptake inside the cell (Fig. 1.6) (Marschner 1995; Salt et al. 2000; Jung et al. 2003; Liao and Xie 2004, Schwab et al. 2005; Bais et al. 2006; Dong et al. 2007). The production of several organic acids as root secretion like malate, citrate, succinic, malonic, oxalate, etc. have been

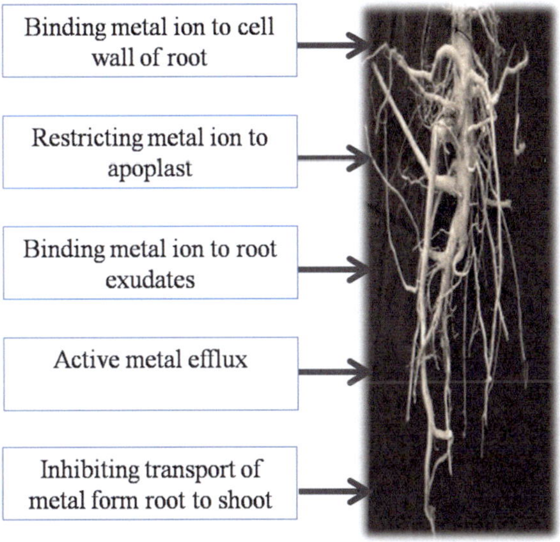

Fig. 1.6 Mechanism of metal excluder plants (Marschner 1995; Salt et al. 2000; Jung et al. 2003, Liao and Xie 2004, Schwab et al. 2005; Bais et al. 2006; Dong et al. 2007)

also reported to serve as a line of defence against toxic metals (Bidwell et al. 2002; Hall 2002; Pittman 2005; Hinsinger et al. 2006; Sun et al. 2006; Verbruggen et al. 2009; Gao et al. 2013). Verbruggen et al. (2009) suggested that these organic acids help in vacuolar transportation of heavy metals especially for Cd. Phytosiderophores have been reported by many authors that they produced specially by roots of leguminous crops during exposure of several heavy metals like Cu, Zn, Cd, etc. and play an important role in restricting the entry of metal ions inside the cell (Awad and Römheld, 2000; Shenker et al. 2001 Chaignon et al. 2002; Xu et al. 2005; Phytotechnology Mechanism 2005). Active metal efflux system in metal excluder plants also helps to restrict the entry of toxic metals (Baker 1981; van Hoof et al. 2001; Tong et al. 2004; Yang et al. 2005; Kushwaha et al. 2016).

1.3.2.1 Factors That Affect Uptake Mechanisms

The uptake mechanisms of plants used in phytoremediation are affected by several factors. The knowledge of these factors can be used to increase the efficiency of the phytoremediation potential of the plants.

1.3.2.1.1 Plant Species

Certain species of plants have superior remediation properties than other species; therefore, more efficient species must be selected for phytoremediation of contaminants. The plants that are most suitable must be hyperaccumulators and must produce more amounts of biomass (Rodriguez et al. 2005).

1.3.2.1.2 Properties of Growing Medium

Development of agronomical practices is carried out for enhancement of phytoremediation; factors such as pH, chelators and fertilizers are adjusted to increase the phytoremediation efficiency (Prasad and Freitas 2003).

1.3.2.1.3 Root Zone

Root zone is the main site for extraction, accumulation and stabilization of the contaminants. Therefore, the root zone must be well developed with high extraction, accumulation and stabilization efficiency. Sometimes, the degradation of contaminants takes place by enzymes that are exuded by the plant roots (Merkl et al. 2005).

1.3.2.1.4 Uptake Mechanism by Vegetative Parts

The environmental factors play a critical role in the uptake mechanism by vegetative parts. The growth enzymes are affected by the temperature which in turn affects the root length. The fate of metabolic activities of the contaminants inside the plants is very important in deciding the phytoremediation potential and efficacy (Mwegoha 2008).

1.3.2.1.5 Chelating Agents

The addition of chelating agents can enhance the capacity of the plants to extract and accumulate contaminants from the soil. Even micronutrients can be added along with the chelators to increase uptake. Chelating agents like EDTA are added in case of heavy metal contaminants. There is a chance of leaching in case of addition of chelators which are synthetic in nature (Van Ginneken et al. 2007; Tangahu 2011).

1.3.3 Indices Used for Assessment of Phytoremediation Potential

The suitability of plant for the purpose of phytoremediation depends on several factors: some of them are intrinsic plant characteristics; others are dependent on the environment or the contaminants. It is of utmost importance for the plants to accumulate a large amount of contaminants from the site. Also the ability of the plant to translocate the contaminants from the roots to shoots is of concern. Enrichment coefficient and translocation factor are two methods to measure the amount of contaminant accumulated and translocated by the plant. The amount or degree of heavy metal concentration/accumulation in the plants which are grown on contaminated sites is determined by enrichment coefficient (Kisku et al. 2000).

$$EC = \frac{\text{Metal concentration in roots or shoots}}{\text{Metal concentration } \textit{at} \text{ the site}} \quad (1.1)$$

Translocation factor (TF) is the ratio which defines the movement or mobilization of metal from roots to shoots of any plant. Equation 1.2 gives the formula for calculation of TF (Barman et al. 2000; Gupta et al. 2008; Shi et al. 2011).

$$TF = \frac{\text{Metal concentration in plant shoots}}{\text{Metal concentration in plant roots}} \quad (1.2)$$

Tolerance index is another major index which determines the suitability of any plant for the purpose of phytoremediation. It is imperative for a plant to exhibit healthy growth for its own survival and for extraction and accumulation of toxicants. The TI of any plant is based on the biomass produced by the plant. Equation 1.3 states the formula for the calculation of tolerance index (de Souza et al. 2012).

$$TI = \frac{\text{Biomass of plants cultivated in contaminanted soil}}{\text{Biomass of plants cultivated in control conditions}} \tag{1.3}$$

These indices are used by the researchers to test the potential of the desired plants for phytoremediation.

1.3.4 Different Aspects of Phytoremediation

1.3.4.1 Application of Edible Crops

In the present world, the availability of land as a resource is a major cause of concern due to the exponential population rise. It is imperative that land usage should be judicious and serve multidimensional benefits. Therefore, researchers have tried hitting two birds with one stone and have developed phytoremediation techniques using edible crops. Application of edible crops for remediation will serve several benefits such as decontamination of the land, food production, and efficient land usage. The edible crops studied for phytoremediation potential by various researchers include wheat (Khan et al. 2011), maize (Mojiri 2011), sunflower (Liphadzi et al. 2003), Indian mustard (Sainger et al. 2014), *Amaranthus* (Shevyakova et al. (2011), tobacco (Chitra et al. 2011), tomato (Uera et al. 2007), *Trapa* (Sweta et al. 2015), etc. Tabulation of these examples has been done in Table 1.2. These are just few examples of the edible crop plants utilized for phytoremediation; there are plenty of literatures available on many other plants as well. Albeit, numerous studies have been carried out testing the phytoremediation potential of edible crops; there are some major demerits associated with them. According to Bauddh et al. (2015a, b), the first obvious demerit is the bioaccumulation of toxicants in the edible plant which can further lead to biomagnification and move up the food chain causing toxicity to animals and humans. Other negative traits of edible crops regarding phytoremediation include short life span, low biomass production and high palatability. For efficient remediation of contaminants, the plants ideally must have a long life span, should be unpalatable and must produce larger amount of biomass for higher accumulation of contaminants (Pandey and Singh 2011). The abovementioned problems of edible crops reduce the overall feasibility of phytoremediation by using these crops. If these problems can be solved like containing the contaminants in the unpalatable portions of the plant and increasing the biomass by technological interventions (biotechnological), only then edible crops may also be effectively used for the remediation purposes.

Table 1.2 Edible crops used in phytoremediation

Plant species	Family	Contaminants	Remarks	References
Triticum aestivum (wheat)	Poaceae	Cu, Cr, Zn, Fe, Ni, Cd, Pb and Mn	Soil used as substrate, maximum accumulation of iron succeeded by manganese and zinc	Chandra et al. (2009)
Triticum vulgare	Poaceae	Methyl parathion, *p*-nitrophenol and hydroquinone	Experiments were carried out using wheat plants to remediate soil contaminated with methyl parathion which after hydrolysis forms *p*-nitrophenol which in turn forms hydroquinone with release of nitrate after metabolism. The uptake and degradation of methyl parathion, *p*-nitrophenol and hydroquinone were increased by 64.85 %, 94.7 % and 55.8 % when wheat plants were used in unsterilized soil	Khan et al. (2011)
Oryza sativa (rice)	Poaceae	As, Hg, Pb, Cr and Cd	Maximum amount stored in roots and least in grains; translocation of As lowest and Hg highest from roots to other parts of the plant	Liu et al. (2007)
Hordeum vulgare (barley)	Poaceae	Hg	In the field experiment carried on for 3 years, mercury was phytoextracted by the barley plant up to 719 mg/ha in a soil where Hg concentration was 29.17 μg/g at 0–10 cm soil depth. The concentration of Hg in the plants was equivalent or higher than Hg present in the soil	Rodriguez et al. (2005)
Solanum tuberosum (potato)	Solanaceae	Ni, Zn, Cu, Cr, Cd, Pb and Mo	Potato plants were planted in sewage sludge and yard manure compost, and it was found that concentration of heavy metals Zn, Cu and Mo were significantly high in the tubers, whereas Ni, Cd, Cr and Pb did not exhibit significant accumulation in the plants	Antonious and Snyder (2007)
Zea mays (maize)	Poaceae	Cd and Pb	Roots of maize were more efficient in phytoextraction and accumulation of lead and cadmium than the shoots. Phytoremediation potential of the plant decreased when the levels of dosage of Cd were increased from 8 to 16 ppm	Mojiri (2011)
Helianthus annuus (sunflower)	Asteraceae	Cd, Pb and Ni	EDTA $Na_4.H_2O$ added 1 g/kg to the soil for enhancing the phytoremediation potential of the sunflower plant. It was observed in the field experiment that leaves of the plants grown in the soil with the chelating agent accumulated more Cd, Ni and Pb than the soil without the chelator	Liphadzi et al. (2003)

Brassica juncea (*cv. T-59*) (Indian mustard)	Brassicaceae	Ni	Out of four cultivars, cv. T-59 variety of mustard plant exhibited high-tolerance index accumulation potential for nickel. The plants accumulated (roots and shoots) 6–6.51 μg Ni g^{-1} from the soil dosed with 8 mM of Ni	Sainger et al. (2014)
Amaranthus spp.	Amaranthaceae	Ni and Fe	Three hybrids of the *Amaranthus* plants were experimented on with varying concentration of $Nicl_2$ and 2 μM Fe^{3+}; EDTA was also added. It was observed that *Amaranthus paniculatus* f. *cruentus* that is red *Amaranthus* (Vishneviy dzhem) accumulated maximum amount of Ni in the shoots and caused Fe reduction in the roots. Ni and Fe exhibit antagonistic behaviour towards each other. Accumulation of Ni in the shoots inhibited Fe accumulation	Shevyakova et al. (2011)
Nicotiana tabacum (tobacco)	Solanaceae	Cd	The author cultivated tobacco, corn and wheat plants in a pot experiment to their study of phytoremediation potential against Cd at varied concentrations. It was found that the tobacco plants recorded maximum accumulation of Cd in their roots and shoots in comparison with corn and wheat. Thus, tobacco was considered as the hyperaccumulator species among the three	Chitra et al. (2011)
Solanum lycopersicum (tomato)	Solanaceae	Ethidium bromide	The study found that tropical plants have a high uptake potential for EtBr. 1.0±0.23 μg $^{kg-1}$ EtBr was taken up by the tomato plants in this experiment	Uera et al. (2007)

1.3.4.2 Application of Weeds

Phytoremediation can effectively curb the toxic impacts of the environmental contaminants. Researchers have tried developing methods of phytoremediation using weeds. Terrestrial as well as aquatic weeds have been experimented with, and encouraging results have been recorded. If aquatic weeds efficiently remove contaminants from effluents, it will prove to be a boon as they are fast growing, and surface water can be easily treated by using them. Several researchers have used aquatic weeds such as alligator weed, duckweed, water lettuce, water hyacinth and *Azolla spp.* for the remediation of several toxicants from water (Cho-Ruk et al. 2006; Skinner et al. 2007; Zhang et al. 2008; Rahman et al. 2008). Terrestrial weeds such as *Parthenium hysterophorus*, *Tridax procumbens*, *Cyperus procera*, *Euphorbia hirta* and *Datura stramonium* are just few of the examples which have been studied for their phytoremediation potential against heavy metals (Kumar et al. 2013). Table 1.3 describes some of the successful experiments on remediation of contaminants by terrestrial as well as aquatic weeds.

In the studies conducted by Kumar et al. (2012 and 2013), emphasis has been given on EC and TF of the contaminants in the plant bodies. Enrichment coefficient gives an accurate estimation of the total contaminant (heavy metal accumulated by the plants from a contaminated site). If the EC is high, the plant is considered to be suitable for phytoremediation. *Eichhornia crassipes* showed high values of EC and TF when tested with heavy metals such as Cr, Pb, Ni and Cd making this aquatic weed most suitable among other weeds for phytoremediation of heavy metals (Kumar et al. 2012). Among the terrestrial weeds, *Tridax procumbens*, *Cyperus procera*, *Euphorbia hirta*, *Parthenium hysterophorus and Datura stramonium* exhibited higher EC in that order and were found suitable for phytoremediation purpose by Kumar et al. (2013). According to Baker (1981) if the translocation factor is more than one, then the plant is termed as metal accumulator, and if it is below one, the plant is known as metal excluder. Kumar et al. (2013) in their study found that TF of the terrestrial weeds ranged between 0.119 for Cd in *T. procumbens* and 3.86 for lead in *S. oleracea* (described in Fig. 1.7). *P. hysterophorus* and *S. oleracea* exhibited TF more than one for all the heavy metals studied (Cu, Pb, Cd and Ni) which made them ideal metal accumulators. The *Cyprus spp.* (*C. procera and C. rotundus*) recorded all the TF values less than one making the weeds unsuitable for phytoremediation.

The aquatic weeds studied by Kumar et al. (2012) presented impressive results regarding TF. It was found that most of the aquatic weeds had TF above one. The study of the average TF (Fig. 1.8) for all these aquatic weeds disclosed the fact that *Marsilea minuta* (2.82), *Bacopa monnieri* (1.84) and *Hydrilla verticillata* (1.69) were most efficient in translocation of heavy metals from the roots to shoots and thus can be used for remediation of heavy metal-contaminated sites.

1.3.4.3 Application of Trees

Trees are considered as one of the most important entities in terms of phytoremediation. Trees have higher biomass and extensive root system which enable them to accumulate more contaminants from the surrounding soil. Many authors have

Table 1.3 Terrestrial and aquatic weeds used in phytoremediation

Plant species	Family	Contaminant	Remarks	References
Alternanthera philoxeroides (Alligator weed)	Amaranthaceae	Pb	The uptake of Pb in the plant was by phytoextraction, and accumulation efficiency was about 30–80%. High accumulation efficiency was attributed to very long stolons and largely spread fibrous roots.	Cho-Ruk et al. (2006)
Pistia stratiotes (water lettuce)	Araceae	Hg	Different concentration of mercury in the form of $HgSO_4$ was tested for extraction in the laboratory experiment. It was observed that the highest concentration of Hg used showed maximum accumulation by the plant	Skinner et al. (2007)
Azolla spp.	Azollaceae	As	Two species of Azolla, namely, *A. caroliniana* and *A. filiculoides* were treated in the Arsenic nutrient solution. It was found that *A. caroliniana* released more arsenate and arsenite in the form of efflux than *A. filiculoides*. The efflux in the form of arsenate was released nine times higher than arsenite	Zhang et al. (2008)
Spirodela polyrhiza (duckweed)	Araceae	Arsenate and Dimethylarsinic acid (DMAA)	In the hydroponic media, arsenic was absorbed as well as adsorbed by the plant. In the plant by phosphate uptake, pathway arsenic was taken up; also by physiochemical adsorption on Fe plaques on the plant surfaces, arsenic was adsorbed in the form of arsenate	Rahman et al. (2008)
Eichhornia crassipes (water hyacinth)	Pontederiaceae	Cr, Cu, Ni and Pb	*Eichhornia* was found to be a very good accumulator of Cu, Ni and Pb. Other macrophytes were more efficient in accumulating Cr. It was also observed that translocation factor (TF) and enrichment coefficient (EC) of Eichhornia were also high making it suitable for phytoremediation	Kumar et al. (2012)
Cyperus procera	Cyperaceae	Cr, Cu, Ni, Pb and Cd	The enrichment coefficient in the roots and shoots was studied along with translocation factor. It was observed that for the heavy metals Cr and Pb, the EC was highest 318.960 and 318.876 in the roots, and similar values were found for shoots also. In case of Cu and Cd, the ECs in roots were 7.305 and 6.665; shoots were 5.062 and 5.445, respectively	Kumar et al. (2013)

(continued)

Table 1.3 (continued)

Plant species	Family	Contaminant	Remarks	References
Euphorbia hirta	Euphorbiaceae	Cr, Cu, Ni, Pb and Cd	The EC in the roots and shoots of *E. hirta* was highest for Cr and Pb. Lowest values were obtained in case of Ni and Cu	Kumar et al. (2013)
Datura stramonium	Solanaceae	Cr, Cu, Ni, Pb and Cd	For Cr the EC in shoot was recorded as 133.418 and 68.840 in roots. For Pb EC was 135.905 in shoots and 161.031 in roots. The values of EC for Cu were as low as 2.885 and 2.254 in roots and shoots. Similarly low values were recorded for Ni which was 6.690 in roots and 12.147 in shoots	Kumar et al. (2013)
Parthenium hysterophorus (congress grass)	Asteraceae	Cr, Cu, Ni, Pb and Cd	The highest EC was found for Pb, i.e. 227.606 in roots and 242.394 in shoots followed by Cr for which EC was 174.567 in roots and 206.153 in shoots; for the other heavy metals, EC was below 20	Kumar et al. (2013)
Tridax procumbens (coat buttons)	Asteraceae	Cr, Cu, Ni, Pb and Cd	For Cr the ECs in roots and shoots were very high, i.e. 609.157 and 1130.372, respectively. For the rest of the heavy metals, the EC varied. Among them for Pb higher, EC was observed than Ni and Cd	Kumar et al. (2013)

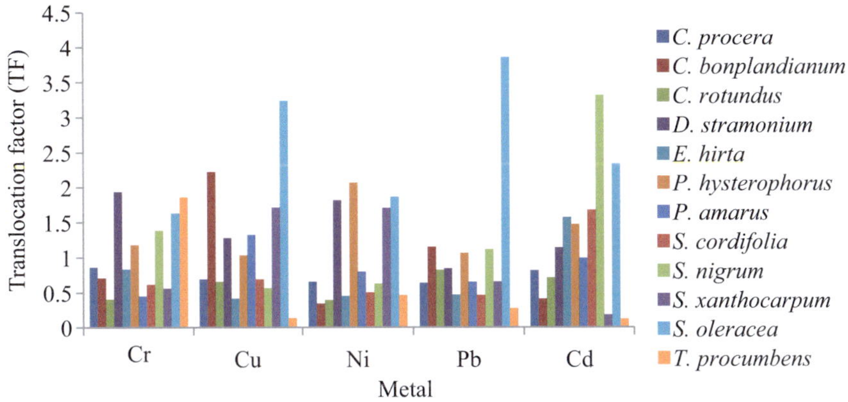

Fig. 1.7 Translocation factor (TF) of the terrestrial weeds grown naturally in the metal-contaminated sites (Kumar et al. 2013)

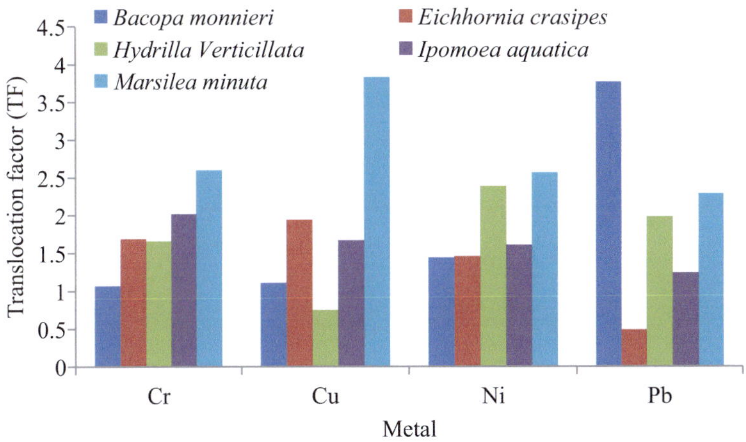

Fig. 1.8 Translocation factor (TF) of the aquatic weeds (macrophytes) naturally growing in the drain receiving tannery effluent (Kumar et al. 2012)

studied the phytoremediation potential of the trees extensively, and few of the studies have been compiled in Table 1.4. The species of trees from the Salicaceae family (willow, poplar) were found to be most appropriate for the phytoremediation purpose of contaminants. More research needs to be carried out using multipurpose trees that would help in remediation of contaminants in addition to carbon sequestration and employment generation.

De Souza et al. (2012) studied three species of leguminous plants *Erythrina speciosa*, *Schizolobium parahyba* and *Mimosa caesalpiniaefolia* for their lead tolerance at seedling stage. The indices studied by the author were TF, BCF and TI. The tolerance index is calculated on the basis of the biomass yield of the plant (Shi et al.

Table 1.4 Trees used for phytoremediation

Plant species	Family	Contaminants	Remarks	References
Salix viminalis (willow)	Salicaceae	Cd and Zn	The field study was conducted in two different contaminated sites one with calcareous soil and another with acidic soil. The phytoextraction of Cd and Zn was found to be more efficient in the acidic soil. There was larger biomass production and more metal accumulation in shoots in the acidic soil; this attributed to more extraction of Cd and Zn	Hammer et al. (2003)
Populus spp. (hybrid poplar)	Salicaceae	Zinc	Different concentrations of Zn were applied as $ZnSo_4$ solution. The root tissues accumulated more amounts of heavy metal than other parts of the plants	Hamon et al. (1999)
Populus spp. (poplar)	Salicaceae	As, Co, Cu, Pb and Zn	Three species *P. alba*, *P. nigra* and *P. tremula* were studied in a pot and field experiment. The uptake mechanisms by the plants were phytoextraction and phytostabilization. The fine roots accumulated the highest amount of trace metals than other aerial parts. *P. nigra* reported the highest accumulation of the contaminants	Vamerali et al. (2009)
Salix spp. (willow)	Salicaceae	Cd	Two clone species of willow used for the study and uptake mechanism by the plants were phytoextraction and phytostabilization. High levels of accumulation were found in the aerial parts of the plant	Vandecasteele et al. (2005)
Schizolobium parahyba (Brazilian fire tree)	Fabaceae	Lead	Different concentrations of lead were added to the soil (250, 500 and 1000 mg Kg^{-1}), and seedlings of *S. parahyba* were tested with tolerance factor, bioconcentration factor and tolerance index indices. It was found that there is maximum lead accumulated in the roots followed by leaves and stem. The accumulation of Pb in the leaves reached toxic levels	de Souza et al. (2012)

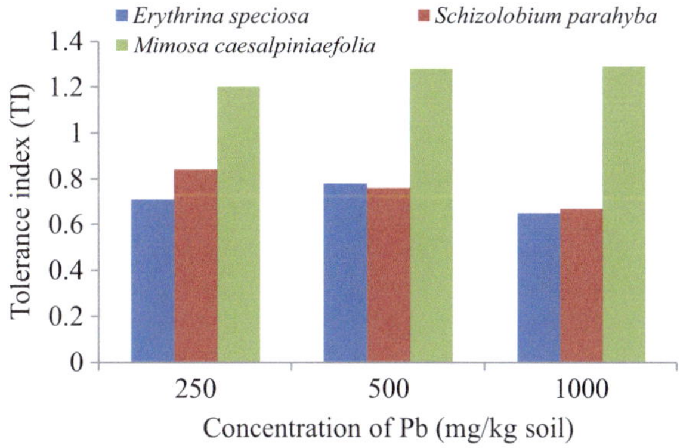

Fig. 1.9 Tolerance index (TI) of some tree species growing in lead-contaminated soil (Source: de Souza et al. 2012)

2011). The biomass yield of the controlled plant and plants grown in a contaminated site are compared. The author found that the tolerance index of *Mimosa caesalpiniaefolia* recorded the highest readings of 1.20, 1.28 and 1.29 for 250, 500 and 1000 mg Kg^{-1} of lead; *Erythrina speciosa* recorded 0.71, 0.78 and 0.65 for 250, 500 and 1000 mg Kg^{-1} of lead; and *Schizolobium parahyba* recorded 0.84, 0.76and 0.67 for the similar Pb concentrations. Therefore, it can be concluded that *Mimosa caesalpiniaefolia* was the most tolerant species followed by *Schizolobium parahyba* and *Erythrina speciosa.* The TI of the three species is represented in Fig. 1.9.

1.3.4.4 Application of Bioenergy Crops

Holistic approach should be applied for remediation of toxicants from the environment. It is of utmost importance to detoxify the contaminants using sustainable means. Amalgamation of phytoremediation techniques with sustainable approach would provide multidimensional benefits for the entire Earth. Using bioenergy crops or trees is one such measure that is sustainable in approach and can be effectively tapped for phytoremediation. Several bioenergy crops have been tested for phytoremediation potential by the researchers in the recent past. If bioenergy crops are used for phytoremediation, it would save contaminated sites from being discarded; also it would generate employment and increase the interest of the people in plantation of such crops. Both edible and nonedible energy crops have been tested for their phytoremediation potential by researchers with encouraging results (Rowe et al. 2009; Shi and Cai 2009; Meers et al. 2010; Bauddh and Singh 2012a, b, 2015a, b; Bauddh et al. 2015a, b, 2016a, b). The use of edible crops for phytoremediation poses a bit of a concern because it is assumed that toxicants might enter the food chain. The study conducted by Meers et al. (2010) showed that the grains, the edible part of maize, accumulated the lowest amount of heavy metals. The researcher attributed this result to the defence mechanism of the plant to restrict toxicity from reaching the reproductive parts and seeds and constraining them within the vegetative parts of the plants. More research needs to be carried out to test the phytoremediation potential of the bioenergy crops as it would help in detoxifying the environment along with the generation of clean fuel and lower the carbon emission into the atmosphere. Using bioenergy crops would provide the most wholesome results in comparison with all other plants combined (Table 1.5).

1.3.4.5 Aromatic Plants Used in Phytoremediation

It is of preference to use nonedible crops for the purpose of phytoremediation because of the obvious reasons of avoiding bioaccumulation and biomagnifications of toxicants. Very recently few aromatic plants have been tested for their potential to remediate contaminants. This will serve the dual purpose of providing essential oils derived from the plant along with cleansing the environment. The plants such as *Ocimum basilicum* (basil), *Cymbopogon martinii* (palmarosa), *Vetiveria zizanioides* (vetiver), *Cymbopogon flexuosus* (lemon grass), *Mentha sp.* (geranium mint) (citronella) and *Cymbopogon winterianus* have been considered for their phytoremediation potential. Gupta et al. (2013) suggest that the likes of basil are viable and feasible for phytoremediation, and other aromatic grasses (lemon grass, citronella,

Table 1.5 Bioenergy crops used for phytoremediation

Plant species	Family	Contaminants	Remarks	References
Ricinus communis (castor)	Euphorbiaceae	Cadmium and nickel	*Ricinus communis* extracted large amounts of Ni from the soil because of its high above- and belowground biomass.	Bauddh and Singh (2015b)
			In a comparative study between two plants, *Ricinus communis* and *Brassica juncea* for Cd, drought and salinity tolerance, it was found that *Ricinus communis* was more tolerant to the stresses applied singly or in a combination than *Brassica juncea*	Bauddh and Singh (2012b)
Linum usitatissimum	Linaceae	Cadmium	The plant showed high bioconcentration factor values and highest values for translocation factor of 54–66% and was overall tolerant in Cd-contaminated soil. Flax accumulated a high amount of Cd from soil; the values were >100 mg/kg	Shi and Cai (2009)

(continued)

Table 1.5 (continued)

Plant species	Family	Contaminants	Remarks	References
Cannabis sativa (hemp)	Cannabaceae	Cadmium	In the study, BCF was found to be highest for hemp roots and low in shoots. Due to higher biomass, Cd accumulation was high in hemp. TI was moderate for hemp	Shi and Cai (2009)
Zea mays (maize)	Poaceae	Zinc, cadmium and lead	Overall accumulation and tolerance for heavy metals by the plants showed promising results. The accumulation of heavy metals in the grains was found to be the lowest that was attributed to the defence mechanism of the plant. If energy was produced at the site from maize, renewable energy of 33,000–46,000 kW h ha^{-1} y^{-1} could be generated	Meers et al. (2010)
Arachis hypogaea (peanut)	Fabaceae	Cadmium	Maximum amount of Cd was accumulated by peanut plant in the shoots making it efficient for phytoremediation of Cd. The highest values for Cd uptake were recorded in peanut; the values were 56.0–68.9 μg/ plant	Shi and Cai (2009)
Salix calodendron (willow)	Salicaceae	Cadmium and zinc	The Zn and Cd present in the contaminated soil could be reduced at the rate of 96 and 5.6 mg Kg^{-1}, respectively, over the period of 20 years	Rowe et al. (2009)

palmarosa and vetiver) are perennial in nature as well as stress tolerant. These qualities make them appropriate for removal of toxicants from the environment. These perennial herbs can be planted at the contaminated sites, and they can accumulate contaminants in the biomass. The plants can be harvested, and their essential oil can be extracted by steam distillation. In this process, the essential oil forms a separate layer on the top, and the water containing the contaminants is left in the lower layer; the essential oil can be separated and used after its quality assessment (Pandey et al. 2015).

1.3.4.6 Plants as Hyperaccumulators

Certain plants have the tendency to accumulate larger amount of contaminants from the environment without showing adverse effects. These plants can be considered to be most ideal in terms of suitability for removal of toxicants. Different researchers have put forth several definitions for hyperaccumulator plant species; the plant species that can accumulate contaminants (metals, metalloids, etc.) at levels 50–500 times higher than their concentrations in soil are considered as hyperaccumulators (Clemens 2006, Kotrba et al. 2009). Another variation is mentioned by Brooks et al. (1998) who state that hyperaccumulators are those plant species which accumulate any element from the substrate at concentration 100 times higher than the substrate or medium. There are certain standards set for considering any plant as a hyperaccumulator; specifically for metals the concentration must be 0.1 weight % as dry weight; for Cd it is variable up to 0.01 weight % and 1 % for Zn (Reeves and Baker 2000). More than 45 families of plants are known to belong to hyperaccumulating species and over 450 plants. The number of hyperaccumulating plants is less in context to the problem of pollution because of their biomass which is low, their slower growth rate and being specific in contaminant accumulation (Chaney et al. 2005). Few examples of hyperaccumulating plant species have been tabulated in Table 1.5. It is seen that the plant family Brassicaceae is dominant in producing hyperaccumulators; other families such as Fabaceae and Crassulaceae also contain hyperaccumulators. Certain plants that are hyperaccumulators can be made more efficient with genetic engineering and with biological amendments (Table 1.6).

1.3.5 Application of Chemical and Biological Amendments to Enhance Phytoremediation

Although phytoremediation is an excellent option for the effective removal of contaminants from the environment, there are few drawbacks of this technique too. One of the major drawbacks is the time taken for complete remediation of a particular site which could be as long as 15–20 years, even if hyperaccumulating species are used. At the recent past, certain amendments in the process of phytoremediation are applied to make it more effective in terms of time and efficiency. Even highly efficient plants exhibit deleterious effects of heavy dosage of contaminants. There is usually a reduction in growth and yield of plants due to over accumulation of the contaminants. The phytoremediation potential of plant species as well as other organisms is being thoroughly studied to find methods to eliminate the risk of

Table 1.6 Example of hyperaccumulator plants used for phytoremediation

Plant species	Family	Contaminants	Reference
Brassica juncea	Brassicaceae	Ni, Cd, Pb and Zn	Sainger et al. (2014)
Astragalus racemosus	Fabaceae	Heavy metals and metalloids	Reeves and Baker (2000)
Sedum alfredii	Crassulaceae	Zn^{2+}	Yang et al. (2006)
Thlaspi caerulescens	Brassicaceae	Zn^{2+}, Ni^{2+} and Cd^{2+}	Milner and Kochian (2008)
Alyssum sp.	Brassicaceae	Heavy metals and metalloids	Reeves and Baker (2000)

ever-increasing contaminant load. Algae, fungi and bacteria are few organisms which have the ability to speed up the process of phytoremediation. Since the past few years, researchers are working on making biological amendments to plant species to increase their efficiency in remediation of toxicants. The importance of bacteria and fungi in increasing plant efficiency for phytoremediation has been dealt in this chapter.

1.3.6 Role of Bacteria in Enhancement of Phytoremediation Potential of Plants

According to Glick (2010), there are rich population of bacteria near the rhizosphere because of the release of nutrient-rich exudates; these bacteria can degrade organic contaminants by phytostimulation or rhizodegradation (Kuiper et al. 2004). In context to phytoremediation, the biodegradative bacteria and bacteria that promote plant growth are very useful. Bacterial species such as *Pseudomonas* spp. are capable of degrading organic xenobiotics with the help of several enzymes produced on its plasmids (Cork and Krueger 1991; Glick 2010). The bacteria that are degradative in nature are capable of converting nonhalogenated compounds in easily metabolizable compounds catechol or protocatechuate. Halogen-based aromatic compounds which are the main constituents of biocides are very slowly degraded by plasmid-encoded enzymes (Glick 2010). Growth-promoting bacterial species releases phytohormones such as auxin which have a direct effect on the plant (Brown 1974; Patten and Glick 1996). A higher concentration of the heavy metals in the plant body causes synthesis of stress ethylene and deficiency in iron content (Glick 2010). A few bacteria release an enzyme ACC deaminase that is capable of lowering the phytohormone ethylene in a plant that is subjected to stress (Glick 2010). Another such enzyme IAA is released by IAA bacteria which helps in adventitious and lateral root elongation and prevent environmental stress-related adverse effects (Lindberg et al. 1985; Frankenberger and Arshad 1995). Table 1.5 represents few examples of bacteria and associated plants used for phytoremediation (Table 1.7).

Table 1.7 Bacteria and plant combination for enhanced phytoremediation

Plant species	Family	Bacteria	Contaminants	Remarks	References
Brassica juncea	Brassicaceae	*Azotobacter chroococcum* HKN-5 + *B. mucilaginosus* HKK-1 + *B. megaterium* HKP-1	Zn, Pb, Cu and Cd	There was a distinct increase in metal bioavailability and biomass of the plant when the bacterial strains were incorporated with the mustard plant	Wu et al. (2006)
Sedum alfredii	Crassulaceae	*Burkholderia cepacia*	Zn and Cd	The results of the study indicated increased metal uptake by the plant; also there was an increase in translocation of metals from root to shoot. The plant also exhibited increased biomass.	Li et al. (2007)
Brassica juncea	Brassicaceae	*Bacillus sp.* 32, *Pseudomonas* sp. A4	Cr	The presence of IAA enzyme, siderophores and phosphate solubilization caused increase in length of roots and shoots	Rajkumar et al. (2006)
Thlaspi caerulescens	Brassicaceae	Rhizosphere bacteria	Zn	The uptake efficiency of the plant for the metal zinc increased in presence of rhizosphere bacteria	Whiting et al. (2001)
Vigna radiata (green gram)	Fabaceae	*Bradyrhizobium* sp. RM8	Zn and Ni	Due to the presence of IAA and siderophores, there was a positive addition in nodules and plant nutrition	Wani et al. (2007)
Ricinus communis	Euphorbiaceae	*Pseudomonas* sp. M6, *Pseudomonas jessenii* M15	Ni, Cu and Zn	Biomass increase was observed due to IAA, ACC enzyme increase and phosphate solubilization	Rajkumar and Freitas (2008)
Solanum lycopersicum	Solanaceae	*Pseudomonas* sp. RJ10, *Bacillus* sp. RJ16	Cd and Pb	There was a distinct increase in aboveground biomass, length of the roots and metal uptake capacity of the plant	He et al. (2009)
Helianthus annuus	Asteraceae	*Streptomyces tendae* F4	Cd	Decreased metal uptake and increased iron content exhibited due to siderophores	Dimpa et al. (2009)
Cajanus cajan (pigeon pea)	Fabaceae	*Glomus mosseae*	Cd and Pb	There was an enhancement of biomass, nodules, nitrogenase activity and leghemoglobin content when the microbes were incorporated with the metal-stressed plant. The process of biosorption was the mechanism of contaminant removal adapted by the plants	Garg and Aggarwal (2011)

1.3.7 Role of Fungi in Enhancement of Phytoremediation Potential of Plants

According to Glick (2010) almost 90 % of plants that are terrestrial have mycorrhizal association. Therefore it is prudently suggested that the beneficial impacts of fungi in regard to phytoremediation must also be taken into account to increase the efficiency of the plants for the remediation of harmful toxicants. The species of fungi that form mycorrhizal association with the plants have proven to increase the accumulation and tolerance of contaminants from the soil or water. Few examples of fungi and plant association that remediates contaminants have been listed in Tables 1.7 and 1.8.

1.3.8 Technological Interventions in Plants Used for Phytoremediation

It is said that plants have intrinsic qualities that enable them to detoxify contaminants, but there is a lacuna in terms of catabolic pathway which they lack, inhibiting complete degradation of the contaminants. Microbes are efficient in this matter and can completely degrade xenobiotics (Abhilash et al. 2009). Genetic engineering plays a pivotal role in enhancement of the plants' ability to accumulate and detoxify contaminants. Transgenic plants as well as electrokinetic techniques have been employed to enhance the phytoremediation potential, and it has been successfully implemented. The role of transgenic crops and electrokinetic process in enhancement of phytoremediation potential has been briefly described in this section. For the enhancement of phytoremediation potential, another approach has been followed by Bauddh and Singh (2015a). The authors have used inorganic fertilizers, biofertilizers (*Bacillus subtilis* and *Azotobacter chrocoocum*), slow-release fertilizers and vermicompost to study their effects on accumulation and partitioning capacity of *Brassica juncea* and *Ricinus communis* for cadmium. It was found that protein content that decreased due to Cd stress was recovered by using biofertilizers. The use of biofertilizers increased metal accumulation, whereas vermicompost decreased bioaccumulation by the plants. The biofertilizers and vermicompost increased the overall health of the plants. *Ricinus communis* was found to be more tolerant and accumulated more Cd than *Brassica juncea.*

1.3.8.1 Transgenic Plants and Phytoremediation

Earlier applied only for inorganic pollutants; gradually, transgenic plants have progressed towards remediation of organic pollutants such as explosives, chlorinated solvents and hydrocarbons (Salt et al. 1998; Pilon-Smits 2005). Heavy metals were the first contaminants to be remediated by transgenic plants using tobacco plant which expressed a metallothionein gene to create higher tolerance for cadmium and *Arabidopsis thaliana* plant which overexpressed a reductase gene mercuric ion for creating more tolerance to Hg (Misra and Gedamu 1989; Rugh et al. 1996). The plants that have been developed with transgenes are used in two ways for

Table 1.8 Mycorrhizal fungi used for enhancement of phytoremediation

Plant species	Family	Fungi	Contaminants	Remarks	References
Apium graveolens (celery)	Apiaceae	*Glomus macrocarpum*	Cd	The Cd^{2+} accumulation increased in the roots, but it was found that the translocation of the metal to shoots was diminished	Kapoor and Bhatnagar (2007)
Oryza sativa	Poaceae	*Glomus versiforme*, *G. mosseae* and *G. diaphanum*	Zn, Pb, Cu and Cd	The root and shoot biomass increased when the plants were inoculated with the arbuscular mycorrhizal fungi, and the accumulation of the heavy metals in the roots was found to be high, but lower translocation was found in the aerial parts	Zhang et al. (2005)
Zea mays	Poaceae	*Glomus mosseae*	Cd	The Cd uptake by the plants increased by inoculation of *G. mosseae*	Malekzadeh et al. (2011)
Pteris vittata	Pteridaceae	*Glomus intraradices*, *G. geosporum* and *G. mosseae*	As	The study found that plant growth increased, and nitrogen, phosphorus and chlorophyll contents also were found to be significantly higher	Leung et al. (2010)
Calopogonium mucunoides	Fabaceae	*Glomus etunicatum*	Pb	The inoculated plants showed higher concentration of metals in the roots but lower concentration in the shoots	de Souza et al. (2012)

phytoremediation purpose: first is the use of transgenes for metabolizing the contaminants and second is the use of transgenes to increase the resistance of the plants towards the toxicants (Abhilash et al. 2009). Some examples of transgenic plants used for remediation of contaminants have been listed below in Tables 1.7 and 1.9.

1.3.8.2 Role of Electrokinesis for Enhanced Phytoremediation

In situ treatment of contaminated sites can be done by the techniques associated with electrokinetic remediation (Reddy and Cameselle 2009). In this technique the contaminated soil is subjected to electric potential directly by inserting electrodes into the ground. Various transport processes and reactions are induced by the electric potential; this causes the movement of contaminants towards the oppositely charged electrodes. The mobilization of the toxicants occurs by two processes: (a) electromigration is a process in which the contaminants move towards the electrodes of opposite charge and (b) electro-osmosis is a process in which the net flux of water is induced by electric field through structure of soil that is porous in nature. Usually, the particles of soil are charged negatively; thus they move towards the cathode (Cameselle and Reddy 2012). Phytoremediation coupled with electrokinetic techniques have a promising future and need to be researched further for contaminants like heavy metals and others as well. Several researchers imply electrokinetics during cultivation of plants in contaminated sites and have been found that the application of electrokinetics enhanced the bioaccumulation of contaminants (Tables 1.8 and 1.10).

1.3.9 Multitasking Approach of Phytoremediation

It is known that all plants provide innumerable benefits to the ecosystem. We are aware of only a small fraction of ecosystem services that is provided by the plants. Hence, the preference of plants over traditional techniques for remediation of contaminants is understandable. The traditional methods would only address the problem of the contaminants, but when plants are applied for the same purpose, several added advantages would be achieved (Fig. 1.10). The first and the foremost advantage of phytoremediation is the release of oxygen by the plant which would be a major boon. The second merit would be the carbon sequestration by the plants. It is well known that plants are the major storehouses of carbon. If trees are used for phytoremediation, a large amount of CO_2 can be fixed by the plants which would help in curbing the greenhouse effect. The use of bioenergy crops for phytoremediation would remove the contaminant along with energy generation; this would be a very major advantage for the people as well as the environment. As phytoremediation is a solar energy-driven process, using plants, the energy may be used up in application of the traditional methods. If the plants used for the remediation of contaminants are cash crops, they would provide employment for the masses. This is the most important merit for the humans especially the ones living in the developing countries. Employment generation would boost the application of plants for

Table 1.9 Transgenic crops used in phytoremediation

Plant species	Family	Gene	Contaminants	Remarks	References
Nicotiana tabacum	Solanaceae	Nfsl	Trinitrotoluene	The gene released the nitroreductase enzyme which removed a large amount of TNT from the solution and reduced TNT to form 4-hydroxylamino-2, 6-dinitrotoluene	Hannink et al. (2001), (2007)
Oryza sativa	Poaceae	CYP1A1	Simazine and atrazine	Cytochrome P450 monooxygenase enzyme was successfully incorporated to remediate xenobiotics simazine and atrazine	Kawahigashi et al. (2005)
Solanum tuberosum	Solanaceae	CYP1A1, CYP2B6 and CYP2C19	Herbicides and sulphonylurea	The enzyme cytochrome P450 monooxygenase increased resistance of the plants towards sulphonylurea and other herbicides	Inui and Ohkawa (2005)
Brassica juncea	Brassicaceae	γ-ECS, GS γ-	1-chloro-2,4-dinitrobenzen, phenanthrene, metolachlor and atrazine	The two enzymes glutamylcysteine synthetase and glutathione synthetase derived from the gene sourced by the plant itself resulted in increased tolerance towards 1-chloro-2,4-dinitrobenzene, phenanthrene, metolachlor and atrazine	Flocco et al. (2004)
Nicotiana tabacum	Solanaceae	LAC	Bisphenol A and PCP	The enzyme laccase was secreted into the rhizosphere, and this helped in removal of the pollutants bisphenol A and PCP	Sonoki et al. (2005)

Adapted and modified from Abhilash et al. (2009)

Table 1.10 Plants used in phytoremediation treated with electric current

Plant species	Family	Contaminants	Remarks	References
Lactuca sativa (lettuce)	Asteraceae	Cd	In the hydroponic culture, the nutrient solution and Cd were added; the plant was subjected to 1 V cm^{-1} AC current for 60 days for remediation of Cd	Bi et al. (2010)
Lolium sp. (ryegrass)	Poaceae	Cu, Cd and As	The DC of 30 V was applied after 5 days of germination for 90 days or remediation of As, Cu and Cd	O'Connor et al. (2003)
Solanum tuberosum	Solanaceae	Zn, Pb, Cd and Cu	For remediation of the heavy metals, AC or DC 500 mA for 90 days after 30 days of plantation was applied	Aboughalma et al. (2008)
Brassica juncea	Brassicaceae	Zn, Pb, Cd and Cu	For a period of 16 days,8 h a day each direct current was applied for remediation of the heavy metals	Lim et al. (2004)
Poa pratensis (Kentucky bluegrass)	Poaceae	Pb	Remediation of Pb was done after adding urea to the plants and applying DC continuously for 15 days at 500 mA intensity	Putra et al. (2013)

Adapted and modified from Cameselle et al. (2013)

phytoremediation as the plants can be harvested for their parts, and the pollutants can be removed at the same time. It would help in the overall societal development and improve the ambient environment.

1.3.10 Economic Feasibility of Phytoremediation Over Conventional Methods

Any technology or process needs to be economically feasible to be practically applied. It is same in the case of the phytoremediation also as the process needs to be beneficial in terms of monetary gains as well. It has been found that using plants for remediating pollutants has indeed been superior to traditional techniques in

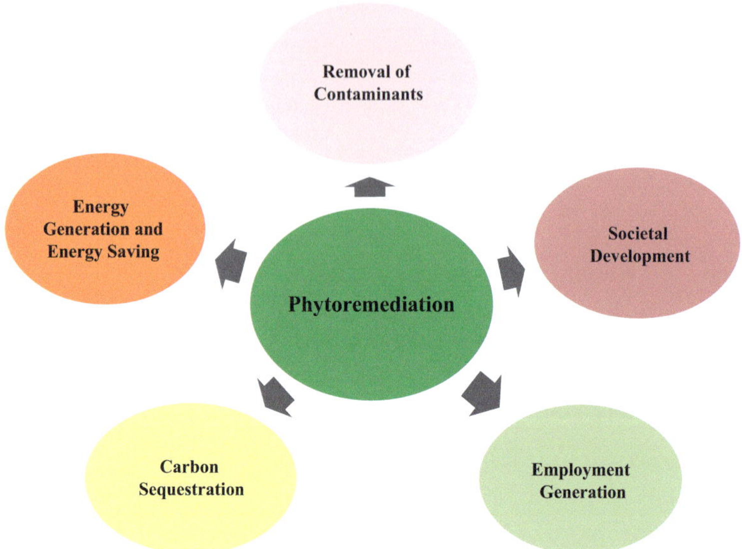

Fig. 1.10 Additional merits of phytoremediation

Table 1.11 Cost comparison between phytoremediation and traditional methods for contaminant removal

Contaminants	Estimated cost of traditional techniques	Estimated cost of phytoremediation	Reference
Petroleum hydrocarbons	$850,000	$70,000	Jipson (1996)
Heavy metals	$250 per cubic yard	$80 per cubic yard	Black (1995)
Lead (10 acres)	$12 million	$500,000	Plummer (1997)
Nitrogen and phosphorous (present in water causing eutrophication and algal bloom)	–	121.1 Yuan/ton of water hyacinth (shadow price) 1,332,581 Yuan (annual cost)	Wang and Wan (2013)

monetary terms. Table 1.9 is a compilation of comparison of the cost between phytoremediation and traditional techniques of studies conducted by several authors like Black (1995), Jipson (1996), Plummer (1997), Wang and Wan (2013), etc. Traditional techniques have cost more than the phytoremediation processes making phytoremediation feasible for implementation. Phytoremediation is more economically beneficial than traditional techniques because of the additional merits such as energy generation, food production, essential oil production, timber production and several other ecosystem and societal services (Table 1.11).

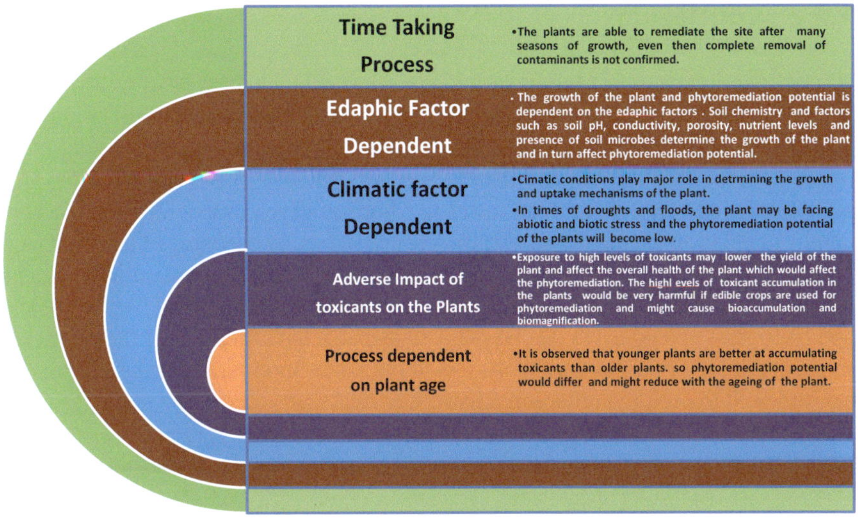

Fig. 1.11 Constraints of phytoremediation (Tu et al. 2004; Mwegoha 2008)

1.3.11 Constraints of Phytoremediation

All technologies and processes comprise of some pros and cons, and this is also applicable in the case of phytoremediation. Phytoremediation is a time-taking process as a long time taken for maximum removal of contaminants from the site; even then complete removal of the contaminants is not guaranteed. After excavation, incineration or disposal might take maximum time in months to accomplish the task, whereas phytodegradation or phytoextraction might take several years (Mwegoha 2008). The phytoremediation process is dependent on edaphic factors and soil chemistry where the soil pH, conductivity, porosity, nutrient levels and presence of soil microbes are instrumental in deciding the uptake mechanisms of the plants. Climatic factors are also very essential in determining the uptake mechanisms, and climatic stress can cause lower phytoremediation potential of the plants. Toxicants are known to have detrimental effects on the plant bodies; even hyperaccumulators exhibit negative impacts after prolonged exposure to the toxicants. Therefore, over a period of time, the efficiency of the plants for phytoremediation reduces making the process unfeasible. Another factor that might hamper the phytoremediation potential of the plant is the age of the plant. Younger plants are said to accumulate more contaminants that the older plants. Some studies suggest that older plant having more biomass accumulates more toxicants in total which can compensate for their lower physiological activities (Tu et al. 2004). Overall despite several constraints, phytoremediation proves to be an environment-friendly and sustainable approach which can be implemented effectively.

1.4 Conclusions

At present era, phytoremediation provides a solution to the most disastrous problem of pollution that is faced by mankind. Phytoremediation not only addresses the problem of pollution but also provides several ecosystem services along with making it a viable and feasible approach. Especially the use of bioenergy crops, aromatic plants and tree species can result in a holistic development of the ecosystem and its population. Being economically feasible, it can be encouraged to be adapted by the masses for decontamination of the sites. A wide range of contaminants can be remediated by plants at a lower cost which is a commendable feat. Technological and biological amendments can be made to increase the efficiency of the plants for the remediation of the contaminants. It is of immense importance for the sake of our environment to promote phytoremediation.

Acknowledgement Dr. Kuldeep Bauddh is thankful to UGC for the award of UGC Start-up grant (3(B):2202.03.789.03.01.31).

References

Abhilash PC, Singh N (2009) Pesticide use and application: an Indian scenario. J Hazard Mater 165:1–12

Abhilash PC, Jamil S, Singh N (2009) Transgenic plants for enhanced biodegradation and phytoremediation of organic xenobiotics. Biotechnol Adv 27:474–488

Aboughalma H, Bi R, Schlaak M (2008) Electrokinetic enhancement on phytoremediation in Zn, Pb, Cu and Cd contaminated soil using potato plants. J Environ Sci Health Part A 43:926–933

Ampiah-Bonney RJ, Tyson JF, Lanza GR (2007) Phytoextraction of arsenic from soil by *Leersia oryzoides*. Int J Phytoremed 9(1):31–40

Andresen E, Mattusch J, Wellenreuther G, Thomas G, Arroyo Abad U, Küpper H (2013) Different strategies of cadmium detoxification in the submerged macrophytes Ceratophyllum demersum L. Metallomics 5(10):1377–1386

Antonious GF, Snyder JC (2007) Accumulation of heavy metals in plants and potential phytoremediation of lead by potato, *Solanum tuberosum* L. J Environ Sci Health A Tox Hazard Subst Environ Eng 42(6):811–816

Arora D, Jain P, Singh N, Kaur HC (2016) Mechanisms of nitric oxide crosstalk with reactive oxygen species scavenging enzymes during abiotic stress tolerance in plants. Free Radic Res 50(3):291–303

Asada K (2006) Production and scavenging of reactive oxygen species in chloroplasts and their functions. Plant Physiol 141(2):391–396

Ashry NA, Mohamed HI (2012) Impact of secondary metabolites and related enzymes in flax resistance and/or susceptibility to powdery mildew. Afr J Biotechnol 11(5):1073–1077

Awad F, Römheld V (2000) Mobilization of heavy metals from contaminated calcareous soils by plant born, microbial and synthetic chelators and their uptake by wheat plants. J Plant Nutr 23:1847–1855

Bais HP, Weir TL, Perry LG, Gilroy S, Vivanco JM (2006) The role of root exudates in rhizosphere interactions with plants and other organisms. Annu Rev Plant Biol 57:233–266

Baker AJM (1981) Accumulators and excluders-strategies in the response of plants to heavy metals. J. Plant Nutr 3:643–654

Barman SC, Sahu RK, Bhargava SK, Chatterjee C (2000) Distribution of heavy metals in wheat, mustard and weed grains irrigated with industrial effluents. Bull Environ Contam Toxicol 64:489–496

Bauddh K, Singh RP (2012a) Cadmium tolerance and its phytoremediation by two oil yielding plants *Ricinus communis* (L.) and *Brassica juncea* (L.) from the contaminated soil. Int J Phytoremed 14:772–785

Bauddh K, Singh RP (2012b) Growth, tolerance efficiency and phytoremediation potential of *Ricinus communis* (L.) and *Brassica juncea* (L.) in salinity and drought affected cadmium contaminated soil. Ecotoxicol Environ Saf 85:13–22

Bauddh K, Singh RP (2015a) Effects of organic and inorganic amendments on bio-accumulation and partitioning of Cd in *Brassica juncea* and *Ricinus communis*. Ecol Eng 74:93–100

Bauddh K, Singh RP (2015b) Assessment of metal uptake capacity of castor bean and mustard for phytoremediation of nickel from contaminated soil. Biom J 19(2):124–138

Bauddh K, Singh K, Singh B, Singh RP (2015a) *Ricinus communis*: a robust plant for bio-energy and phytoremediation of toxic metals from contaminated soil. Ecol Eng 84:640–652

Bauddh K, Singh K, Singh RP (2015b) *Ricinus communis* L. a value added crop for remediation of cadmium contaminated soil. Bull Environ Contam Toxicol 96(2):265–269

Bauddh K, Kumar A, Srivastava S, Singh RP, Tripathi RD (2016a) A study on the effect of cadmium on the antioxidative defense system and alteration in different functional groups in castor bean and Indian mustard. Arch Agron Soil Sci 62(6):877–891

Bauddh K, Singh B, Singh RP (2016b) *Ricinus communis* L. as a value added alternative for restoration of cadmium contaminated and degraded agricultural ecosystem. Bull Environ Contam Toxicol 96(2):265–269

Bi R, Schlaak M, Siefert E, Lord R, Connolly H (2010) Alternating current electrical field effects on lettuce (Lactuca sativa) growing in hydroponic culture with and without cadmium contamination. J Appl Electrochem 40:1217–1223

Bidwell SD, Batianoff GN, Sommer-Knudsen J, Woodrow IE (2002) Hyperaccumulation of manganese in the rainforest tree *Austromyrtus bidwillii* (Myrtaceae) from Queensland, Australia. Funct Plant Biol 29:899–905

Black H (1995) Absorbing possibilities: phytoremediation. Environ. Health Perspect 103:1106–1108

Blaylock MJ, Huang JW (2000) Phytoextraction of metals. In: Raskin I, Ensley BD (eds) Phytoremediation of toxic metals: using plants to clean up the environment. Wiley, New York, pp 53–70

Briat JF (1996) Roles of ferritin in plants. J Plant Nutr 19:8-9

Briat JF, Cellier F, Gaymard F (2006) Ferritins and iron accumulation in plant tissues. In: Barton LL, Abadía J (eds) Iron nutrition in plants and rhizospheric microorganisms, Chapter 17. Springer, New York, pp 341–357

Briat J, Ravet K, Arnaud N, Duc C, Boucherez J, Touraine B, Cellier F, Gaymard F (2010) New insights into ferritin synthesis and function highlight a link between iron homeostasis and oxidative stress in plants. Ann Bot 105(5):811–822

Brooks RR (1994) In: Gargo ME (ed) VCH Verlagsgesellsschaft. Plants and chemical elements: biochemistry, uptake, tolerance and toxicity. Weinheim, 88–105

Brooks RR, Chambers MF, Nicks LJ, Robinson BH (1998) Phytomining. Trends Plant Sci 3:359–362

Brown ME (1974) Seed and root Bacterization. Annu Rev Phytopathol 12:181–197

Cameselle C, Reddy KR (2012) Development and enhancement of electro-osmotic flow for the removal of contaminants from soils. Electrochim Acta 86:10–22

Cameselle C, Chirakkara RA, Reddy KR (2013) Electrokinetic-enhanced phytoremediation of soils: status and opportunities. Chemosphere 93(4):626–636

Chaignon V, Di Malta D, Hinsinger P (2002) Fe-deficiency increases Cu acquisition by wheat cropped in a Cu-contaminated, vineyard soil. New Phytol 154:121–130

Chandra R, Bharagava RN, Yadav S, Mohan D (2009) Accumulation and distribution of toxic metals in wheat (*Triticum aestivum* L.) and Indian mustard (*Brassica campestris* L.) irrigated with distillery and tannery effluents. J Hazard Mater 162(2–3):1514–1521

Chaney RL, Angle JS, McIntosh MS, Reeves RD, Li YM, Brewer EP, Chen KY, Roseberg RJ, Perner H, Synkowski EC, Broadhurst CL, Wang S, Baker AJM (2005) Using hyperaccumulator plants to phytoextract soil Ni and Cd. Z Naturforsch C 60(3–4):190–198

Chitra K, Sharavanan S, Vijayaragavan M (2011) Tobacco, corn and wheat for phytoremediation of cadmium polluted soil. Recent Res Sci Technol 3(2):148–151

Cho-Ruk K, Kurukote J, Supprung P, Vetayasuporn S (2006) Perennial plants in the phytoremediation of lead contaminated soils. Biotechnol 5(1):1–4

Chutia P, Kato S, Kojima T, Satokawa S (2009) Arsenic adsorption from aqueous solution on synthetic zeolites. J Hazard Mater 162(1):440–447

Clemens S (2001) Molecular mechanisms of plant metal tolerance and homeostasis. Planta 212:475–486

Clemens S (2006) Toxic metal accumulation, responses to exposure and mechanisms of tolerance in plants. Biochimie 88:1707–1719

Cobbett CS (2000) Phytochelatin biosynthesis and function in heavy metal detoxification. Curr Opin Plant Biol 3:211–216

Cobbett C, Goldsbrough P (2002) Phytochelatins and metallothioneins: roles in heavy metal detoxification and homeostasis. Annu Rev Plant Biol 53:159–182

Cohen CK, Fox TC, Garvin DF, Kochian LV (1998) The role of Iron-deficiency stress responses in stimulating heavy-metal transport in plants. Plant Physiol 116:1063–1072

Cork DJ, Krueger JP (1991) Microbial transformations of herbicides and pesticides. Adv Appl Microbiol 36:1–66

de Souza SCR, de Andrade SAL, de Souza LA, Schiavinato MA (2012) Lead tolerance and phytoremediation potential of Brazilian leguminous tree species at the seedling stage. J Environ Manag 110:299–307

Deák M, Horváth GV, Davletova S, Török K, Sass L, Vass I, Barna B, Király Z, Dudits D (1999) Plants ectopically expressing the ironbinding protein, ferritin, are tolerant to oxidative damage and pathogens. Nat Biotechnol 17:192–196

Dimkpa CO, Merten D, Svatos A, Büchel G, Kothe E (2009) Siderophores mediate reduced and increased uptake of cadmium by Streptomyces tendae F4 and sunflower (*Helianthus annuus*), respectively. J Appl Microbiol 107:1687–1696

Dong J, Mao WH, Zhang GP, Wu FB, Cai Y (2007) Root excretion and plant tolerance to cadmium toxicity – a review. Plant Soil Environ 53(5):193–200

Dushenkov V, Kumar P, Motto H, Raskin I (1995) Rhizofiltration: the use of plants to remove heavy metals from arduous streams. Environ Sci Technol 29:1239–1245

Eide D, Broderius M, Fett J, Guerinot ML (1996) A novel iron-regulated metal transporter from plants identified by functional expression in yeast. Proc Natl Acad Sci U S A 93:5624–5628

EPA (2000) (U. S. Environmental Protection Agency), "Introduction to Phytoremediation," National Risk Management Research Laboratory, EPA/600/R-99/107, 2000, http://www.clu-in.org/download/remed/introphyto.pdf

Erakhrumen A, Agbontalor A (2007) Review Phytoremediation: an environmentally sound technology for pollution prevention, control and remediation in developing countries. Educ Res Rev 2(7):151–156

Fabisiak JP, Pearce LL, Borisenko GG, Tyhurina YY, Tyurin VA, Razzack J, Lazo JS, Pitt BR, Kagan VE (1999) Bifunctional anti/pro-oxidant potential of metallothionein: redox signaling of copper binding and release. Antioxid Redox Signal 1:349–364

Flathman PE, Hannza GR (1998) Phytoremediation: current view on an emerging green technology. Soil Contamin 7(4):415–432

Flocco CG, Lindblom SD, Smits E (2004) Over expression of enzymes involved in glutathione synthesis enhances tolerance to organic pollutants in *Brassica juncea*. Int J Phytoremed 6:289–304

Frankenberger JWT, Arshad M (1995) Microbial synthesis of auxins. In: Frankenberger WT, Arshad M (eds) Phytohormones in soils. Marcel Dekker, New York, pp 35–71

Fulekar MH, Singh A, Thorat V, Kaushik CP, Eapen S (2010) Phytoremediation of 137Cs from low level nuclear waste using *Catharanthus roseus*. Indian J Pure Appl Phys 48(7):516–519

Gao L, Peng K, Xia Y, Wang G, Niu L, Lian C, Shen Z (2013) Cadmium and manganese accumulation in Phytolacca americana L and the roles of non-protein thiols and organic acids. Int J Phytoremed 15(4):307–319

Garg N, Aggarwal N (2011) Effects of interactions between cadmium and lead on growth, nitrogen fixation, phytochelatin, and glutathione production in mycorrhizal *Cajanus cajan* (L.) Millsp. J Plant Growth Regul 30(3):286–300

Gaur A, Adholeya A (2004) Prospects of arbuscular mycorrhizal fungi in phytoremediation of heavy metal contaminated soils. Curr Sci 86(4):528–534

Ghosh M, Singh SP (2005) A review on phytoremediation of heavy metals and utilization of its by products. Asian J Energy Environ 6(4):18

Glick BR (2003) Phytoremediation: synergistic use of plants and bacteria to clean up the environment. Biotechnol Adv 21(5):383–393

Glick BR (2010) Using soil bacteria to facilitate phytoremediation bacteria. Biotechnol Adv 28:367–374

Gomes HI (2012) Phytoremediation for bioenergy: challenges and opportunities. Environ Technol Rev 1(1):59–66

Grill E, Gekeler W, Winnacker EL, Zenk HH (1986) Homo-phytochelatins are heavy metal-binding peptides of homo-glutathione containing Fabales. FEBS Lett 205(1):47–50

Guerinot ML (2000) The ZIP family of metal transporters. Biochim Biophys Acta 1465(1–2):190–198

Gupta S, Nayek S, Saha RN, Satpati S (2008) Assessment of heavy metal accumulation in macrophyte, agricultural soil and crop plants adjacent to discharge zone of sponge iron factory. Environ Geol 55:731–739

Gupta AK, Verma SK, Khan K, Verma RK (2013) Phytoremediation using aromatic plants: a sustainable approach for remediation of heavy metals polluted sites. Environ Sci Technol 47(18):10115–10116

Hall JL (2002) Cellular mechanisms for heavy metal detoxification and tolerance. J Exp Bot 53(366):1–11

Hammer D, Kayser A, Keller C (2003) Phytoextraction of Cd and Zn with Salix viminalis in field trial. Soil Use Manag 19(3):187–192

Hamon RE, Holm PE, Lorenz SE, McGrath SP, Christensen TH (1999) Metal uptake by plants from sludge-amended soils: caution is required in the plateau interpretation. Plant Soil 216(1–2):53–64

Hannink N, Rosser SJ, French CE, Basran A, Murray JAH, Nicklin S, Bruce NC (2001) Phytodetoxification of TNT by transgenic plants expressing a bacterial nitroreductase. Nat Biotechnol 19:1168–1172

Hannink NK, Subramanian M, Rosser SJ, Basran A, Murray JA, Shanks JV, Bruce NC (2007) Enhanced transformation of TNT by tobacco plants expressing a bacterial nitroreductase. Int J Phytorem 9(5):385–401

Harada E, Yamaguchi Y, Koizumi N, Sano H (2002) Cadmium stress induces production of thiol compounds and transcripts for enzymes involved in sulfur assimilation pathways in Arabidopsis. J Plant Physiol 159:445–448

Hartley-Whitaker J, Ainsworth G, Vooijs R, Ten Bookum W, Schat H, Meharg AA (2001) Phytochelatins are involved in differential arsenate tolerance in *Holcus lanatus*. Plant Physiol 126:299–306

He LY, Chen ZJ, Ren GD, Zhang YF, Qian M, Sheng XF (2009) Increased cadmium and lead uptake of a cadmium hyperaccumulator tomato by cadmium resistant bacteria. Exotoxicol Environ Safety 72:1343–1348

Henry JR (2000) Overview of the phytoremediation of lead and mercury. In: Overview of the phytoremediation of lead and mercury. EPA, Washington, DC
Hinsinger P, Plassard C, Jaillard B (2006) Rhizosphere: a new frontier for soil biogeochemistry. J Geochem Explor 88:210–213
Inui H, Ohkawa H (2005) Herbicide resistance in transgenic plants with mammalian P450 monooxygenase genes. Pest Manag Sci 61:286–291
Ismail S (2012) Phytoremediation: a green technology. Iranian J Plant Physiol 3(1):567–576
Jadia CD, Fulekar MH (2008) Phytoremediation: the application of vermicompost to remove zinc, cadmium, copper, nickel and lead by sunflower plant. Environ Eng Manag J 7(5):547–558
Jipson E (1996) Chevron grows a new remediation technology: Alfalfa and Poplars. Envirobiz News and Press Release Archive, Available from http://www.envirobiz.com
Jung C, Maeder V, Funk F, Frey B, Sticher H, Frossard E (2003) Release of phenols from *Lupinus albus* L. roots exposed to Cu and their possible role in Cu detoxification. Plant Soil 252:301–312
Kägi JHR (1991) Overview of metalothioneins. Methods Enzymol 205:613–623
Kapoor R, Bhatnagar AK (2007) Attenuation of cadmium toxicity in mycorrhizal celery (*Apium graveolens* L.). World J Microbiol Biotechnol 23:1083–1089
Kawahigashi H, Hirose S, Inui H, Ohkawa H, Ohkawa Y (2005) Enhanced herbicide cross-tolerance in transgenic rice plants co-expressing human CYP1A1, CYP2B6, and CYP2C19. Plant Sci 168:773–781
Khan NU, Varma B, Imrana N, Shetty PK (2011) Phytoremediation using an indigenous crop plant (wheat): the uptake of methyl parathion and metabolism of p-nitrophenol. Ind J Sci Technol 4(12):1661–1667
Kisku GC, Barman SC, Bhargava SK (2000) Contamination of soil and plants with potentially toxic elements irrigated with mixed industrial effluent and its impact on the environment. Water Air Soil Pollut 120(1–2):121–137
Kondo N, Imani K, Isobe M, Goto T, Murasugi A, Wada-Nakagawa C, Hayashi Y (1984) Cadystin A and B, major unit peptides comprising cadmium binding peptides induced in a fission yeast separation, revision of structures and synthesis. Tetrahedron Lett 25:3869–3872
Korshunova YO, Eide D, Clark WG, Guerinot ML, Pakrasi HB (1999) The IRT1 protein from Arabidopsis thaliana is a metal transporter with a broad substrate range. Plant Mol Biol 40:37–44
Kotrba P, Najmanova J, Macek T, Ruml T, Mackova M (2009) Genetically modified plants in phytoremediation of heavy metal and metalloid soil and sediment pollution. Biotechnol Adv 27:799–810
Krämer U, Talke IN, Hanikenne M (2007) Transition metal transport. FEBS Lett 581:2263–2272
Kuiper I, Lagendijk EL, Bloemberg GV, Lugtenberg BJ (2004) Rhizoremediation: a beneficial plant-microbe interaction. Mol Plant-Microbe Interact 17(1):6–15
Kumar PB, Dushenkov ANV, Motto H, Raskin I (1995a) Phytoextraction: the use of plants to remove heavy metals from soils. Environ Sci Technol 29(5):1232–1238
Kumar PBAN, Dushenkov V, Motto H, Raskin I (1995b) Phytoextration-the use of plants to remove heavy metals from soils. Environ Sci Technol 29:1232–1238
Kumar N, Bauddh K, Barman SC, Singh DP (2012) Accumulation of metals in selected macrophytes grown in mixture of drain water and tannery effluent and their phytoremediation potential. J Environ Biol 33:323–327
Kumar N, Bauddh K, Kumar S, Dwivedi N, Singh DP, Barman SC (2013) Heavy metal uptake by plants naturally grown on industrially contaminated soil and their phytoremediation potential. Ecol Eng 61:491–495
Kushwaha A, Rani R, Kumar S, Gautam A (2016) Heavy metal detoxification and tolerance mechanisms in plants: implications for phytoremediation. Environ Rev 24(1):39–51
Lasat MM (2000) Phytoextraction of metals from contaminated soil: a review of plant/soil/metal interaction and assessment of pertinent agronomic issues. J Hazard Sub Res 2(5):1–25

Lee S, Moon JS, Ko TS, Petros D, Goldsbrough PB, Korban SS (2003) Over expression of Arabidopsis phytochelatin synthase paradoxically leads to hypersensitivity to cadmium stress. Plant Physiol 131:656–663

Leitenmaier B, Küpper H (2013) Compartmentation and complexation of metals in hyperaccumulator plants. Front Plant Sci 4:374

Leung HM, Wu FY, Cheung KC, Ye ZH, Wong MH (2010) The effect of arbuscular mycorrhizal fungi and phosphate amendement on arsenic uptake, accumulation and growth of *Pteris vittata*in as-contaminated soil. Int J Phytoremed 12(4):384–403

Li WC, Ye ZH, Wong MH (2007) Effects of bacteria on enhanced metal uptake of the Cd/Zn hyperaccumulating plant, *Sedum alfredii*. J Exp Bot 58:4173–4182

Liao M, Xie XM (2004) Cadmium release in contaminated soils due to organic acids. Pedosphere 14:223–228

Lim JM, Salido AL, Butcher DJ (2004) Phytoremediation of lead using Indian mustard (*Brassica juncea*) with EDTA and electrodics. Microchem J 76:3–9

Lindberg T, Granhall U, Tomenius H (1985) Infectivity and acetylene reduction of diazotrophic rhizosphere bacteria in wheat (*Triticum aestivum*) seedlings under gnotobiotic conditions. Biol Fertil Soils 1:123–129

Liphadzi MS, Kirkham MB, Mankin KR, Paulsen GM (2003) EDTA-assisted heavy metal uptake by poplar and sunflower grown at a long-term sewage sludge farm. Plant Soil 257:171–182

Liu D, Jiang W, Liu C, Xin C, Hou W (2000) Uptake and accumulation of lead by roots, hypocotyls and shoots of Indian mustard [*Brassica juncea* (L.)]. Bioresour Technol 71(3):273–277

Liu WX, Shen LF, Liu JW, Wang YW, Li SR (2007) Uptake of toxic heavy metals by rice (*Oryza sativa* L.) cultivated in the agricultural soil near Zhengzhou city, People's Republic of China. Bull Environ Contam Toxicol 79(2):209–213

Liu X, Peng K, Wang A, Lian C, Shen Z (2010) Cadmium accumulation and distribution in populations of *Phytolacca americana* L. and the role of transpiration. Chemosphere 78:1136–1141

Maiti IB, Wagner GJ, Hunt AG (1991) Light inducible and tissue specific expression of a chimeric mouse metallothionein cDNA gene in tobacco. Plant Sci 76:99–107

Malekzadeh E, Alikhani HA, Savaghebi-Firoozabadi GR, Zarei M (2011) Influence of arbuscular mycorrhizal fungi and an improving growth bacterium on Cd uptake and maize growth in Cd-polluted soils. Span J Agric Res 9(4):1213–1223

Malhotra R, Agarwal S, Gupta P (2014) Phytoremediation of radioactive metals. J Civil Eng Environ Technol 1:75–79

Manara A (2012) Plant responses to heavy metal toxicity. In: Furini A (ed) Plants and heavy metals, Springer briefs in biometals. Springer, Dordrecht

Marschner H (1995) Mineral nutrition of higher plants, 2nd edn. Academic Press, London

Meers E, Van Slycken S, Adriaensen K, Ruttens A, Vangronsveld J, Du Laing G, Witters N, Thewys T, Tack FMG (2010) The use of bio-energy crops (*Zea mays*) for 'phytoattenuation' of heavy metals on moderately contaminated soils: a field experiment. Chemosphere 78(1):35–41

Merkl N, Schultze-Kraft R, Infante C (2005) Phytoremediation in the tropics—influence of heavy crude oil on root morphological characteristics of graminoids. Environ Pollut 138(1):86–91

Mijovilovich A, Leitenmaier B, Meyer-Klaucke W, Kroneck PM, Götz B, Küpper H (2009) Complexation and toxicity of copper in higher plants. II. Different mechanisms for copper versus cadmium detoxification in the copper-sensitive cadmium/zinc hyperaccumulator Thlaspi caerulescens (Ganges ecotype). Plant Physiol 151:715–731

Milner MJ, Kochian LV (2008) Investigating heavy-metal hyperaccumulation using *Thlaspi caerulescens* as a model system. Ann Bot 102:3–13

Misra S, Gedamu L (1989) Heavy metal tolerant transgenic *Brassica napus* L. and *Nicotiana tabacum* L. plants. Theor Appl Genet 78:161–168

Mittler R (2002) Oxidative stress, antioxidants and stress tolerance. Trends Plant Sci 7(9):405–410

Mojiri A (2011) The potential of corn (*Zea mays*) for phytoremediation of soil contaminated with cadmium and lead. J Biol Environ Sci 5(13):17–22
Moreno FN, Anderson CWN, Stewart RB, Robinson BH (2008) Phytofiltration of mercury-contaminated water: volatilisation and plant-accumulation aspects. Environ Exp Bot 62(1):78–85
Mwegoha WJS (2008) The use of phytoremediation technology for abatement soil and groundwater pollution in Tanzania: opportunities and challenges. J Sust Dev Africa 10(1):140–156
Nevo Y, Nelson N (2006) The NRAMP family of metal-ion transporters. Biochim Biophys Acta 1763:609–620
Ni LX, Acharya K, Hao XY, Li SY (2013) Antioxidant and metabolism responses to polyphenol stress in cyanobacterium *Microcystis aeruginosa*. J Environ Sci Health, Part B 48:153–161
Nishida S, Tsuzuki C, Kato A, Aisu A, Yoshida J, Mizuno T (2011) AtIRT1, the primary iron uptake transporter in the root, mediates excess nickel accumulation in *Arabidopsis thaliana*. Plant Cell Physiol 52:1433–1442
Noctor G, Mhamdi A, Foyer CH (2014) The roles of reactive oxygen metabolism in drought: not so cut and dried. Plant Physiol 164:1636–1648
Nordberg GF (2004) Cadmium and health in the 21st century—historical remarks and trends for the future. Biometals 17:485–489
O'Connor CS, Lepp NW, Edwards R, Sunderland G (2003) The combined use of electrokinetic remediation and phytoremediation to decontaminate metal polluted soils: a laboratory-scale feasibility study. Environ Monit Assess 84:141–158
Pandey VC, Singh K (2011) Is *Vigna radiata* suitable for the re-vegetation of fly ashlandfills? Ecol Eng 37:2105–2106
Pandey VC, Pandey DN, Singh N (2015) Sustainable phytoremediation based on naturally colonizing and economically valuable plants. J Clean Prod 86:37–39
Papoyan A, Kochian LV (2004) Identification of *Thlaspi caerulescens* genes that may be involved in heavy metal hyperaccumulation and tolerance. Characterization of a novel heavy metal transporting ATPase. Plant Physiol 136(3):3814–38223
Patten C, Glick BR (1996) Bacterial biosynthesis of indole-3- acetic acid. Can J Microbiol 42:207–220
Pehlivan E, Özkan AM, Dinç S, Parlayici S (2009) Adsorption of Cu^{2+} and Pb^{2+} ion on dolomite powder. J Hazard Mater 167(1–3):1044–1049
Persistent Organic Pollutants (POPs) (2008) Children's health and the environment WHO training package for the health sector world health organization www.who.int/ceh
Phytotechnology Mechanism (2005) Solid waste and emergency response (5102G), EPA 542-R-05-002, 2005, http://www.clu-in.org/download/remed/542-r-05-002.pdf
Pilon-Smits E (2005) Phytoremediation. Annu Rev Plant Biol 56:15–39
Pittman JK (2005) Managing the manganese: molecular mechanisms of manganese transport and homeostasis. New Phytol 167:733–742
Plummer C (1997) Interest increases in using plants for environmental remediation. In: Glaser L (ed) Industrial uses of agricultural materials situation and outlook. Economic Research Service, US Department of Agriculture, Washington, DC, USA, Available from http://www.ers.usda.gov/publications/IUS6/ius6g 002.pdf
Prasad MNV, Freitas De HMO (2003) Metal hyperaccumulation in plants biodiversity prospecting for phytoremediation technology. Electron J Biotechnol 6(3):110–146
Putra RS, Ohkawa Y, Tanaka S (2013) Application of EAPR system on the removal of lead from sandy soil and uptake by Kentucky bluegrass (Poa pratensis L.). Sep Purif Technol 102:34–42
Ragland M, Briat JF, Gagnon J, Laulhere JP, Massenet O, Theil EC (1990) Evidence for conservation of ferritin sequences among plants and animals and for a transit peptide in soybean. J Biol Chem 265:18339–18344
Rahman MA, Hasegawa H, Ueda K, Maki T, Rahman MM (2008) Arsenic uptake by aquatic macrophyte Spirodela polyrhiza L.: interactions with phosphate and iron. J Hazard Mater 160(2–3):356–361

Rajkumar M, Freitas H (2008) Influence of metal resistant-plant growth-promoting bacteria on the growth of Ricinus communis in soil contaminated with heavy metals. Chemosphere 71:834–842

Rajkumar M, Nagendran R, Lee KJ, Lee WH, Kim SZ (2006) Influence of plant growth promoting bacteria and Cr6þ on the growth of Indian mustard. Chemosphere 62:741–748

Raskin I, Smith RD, Salt DE (1997) Phytoremediation of metals using plants to remove pollutants from the environment. Curr Opin Biotechnol 8:221–226

Rastgoo L, Alemzadeh A (2011) Biochemical responses of Gouan (*Aeluropus littoralis*) to heavy metals stress. AJCS 5:375–383

Ravet K, Touraine B, Boucherez J, Briat J-F, Gaymard F, Cellier F (2009) Ferritins control interaction between iron homeostasis and oxidative stress in Arabidopsis. Plant J 57:400–412

Reddy KR, Cameselle C (2009) Electrochemical remediation technologies for polluted soils, sediments and groundwater. Wiley, Hoboken

Reeves RD, Baker AJM (2000) Metal-accumulating plants. In: Raskin I, Ensley BD (eds) Phytoremediation of toxic metals, using plants to clean up the environment. Wiley-Interscience, New York, pp 193–229

Resaee A, Derayat J, Mortazavi SB, Yamini Y, Jafarzadeh MT (2005) Removal of mercury from chlor-alkali industry wastewater using *Acetobacter xylinum* cellulose. American J Environ Sci 1(2):102–105

Rodriguez L, Lopez-Bellido FJ, Carnicer A, Recreo F, Tallos A, Monteagudo JM (2005) Mercury recovery from soils by phytoremediation. In: Book of environmental chemistry. Springer, Berlin, pp 197–204

Roosens NH, Bernard C, Leplae R, Verbruggen N (2004) Evidence for copper homeostasis function of metallothionein (MT3) in the hyperaccumulator Thlaspi caerulescens. FEBS Lett 577:9-16

Rowe RL, Street NR, Taylor G (2009) Identifying potential environmental impacts of large- scale deployment of dedicated bioenergy crops in the UK. Renew Sust Energ Rev 13(1):271–290

Roy S, Labelle S, Mehta P et al (2005) Phytoremediation of heavy metal and PAH-contaminated brown field sites. Plant Soil 272(1–2):277–290

Rugh CL, Wilde D, Stack NM, Thompson DM, Summer AO, Meagher RB (1996) Mercuric ion reduction and resistance in transgenic Arabidopsis thaliana plants expressing a modified bacterial merA gene. Proc Natl Acad Sci U S A 93:3182–3187

Sainger M, Sharma A, Bauddh K, Sainger PA, Singh RP (2014) Remediation of nickel-contaminated soil by Brassica juncea L. cv. T-59 and effect of the metal on some metabolic aspects of the plant. Biom J 18:100–110

Salt DE, Blaylock M, Kumar NPBA, Duskenkov V, Eustry D, Chet D, Raskin I (1995) Phytoremediation: a novel strategy for the removal of toxic metals from the environment using plants. Nat Biotechnol 13:468–474

Salt DE, Smith RD, Raskin I (1998) Phytoremediation. Annu Rev Plant Physiol Plant Mol Biol 49:643–668

Salt DE, Kato N, Krämer U, Smith RD, Raskin I (2000) The role of root exudates in nickel hyperaccumulation and tolerance in accumulator and nonaccumulator species of Thlaspi. In: Terry N, Banuelos G (eds) Phytoremediation of contaminated soil and water. CRC Press LLC, Boca Raton

Sanderman H (1994) Higher plant metabolism of xenobiotics: The 'green liver' concept. Pharmacogenetics 4:225–241

Schützendübel A, Polle A (2002) Plant responses to abiotic stresses: heavy metal-induced oxidative stress and protection by mycorrhization. J Exp Bot 53(372):1351–1365

Schwab AP, He YH, Banks MK (2005) The influence of organic ligands on the retention of lead in soil. Chemosphere 61:856–866

Seckback JJ (1982) Ferreting out the secret of plant ferritin—a review. J Plant Nutr 5:369–394

Sewelam N, Kazan K, Schenk PM (2016) Global plant stress signaling: Reactive oxygen species at the cross-road. Front Plant Sci, 23 February 2016 http://dx.doi.org/10.3389/fpls.2016.00187

Shanmugaraj BM, Chandra HM, Srinivasan B, Ramalingam S (2013) Cadmium induced physio-biochemical and molecular response in *Brassica juncea*. Int J Phytoremed 15:206–218

Sharma P, Dubey RS (2005) Drought induces oxidative stress and enhances the activities of antioxidant enzymes in growing rice seedlings. Plant Growth Regul 46(3):209–221

Sharma P, Jha AB, Dubey RS, Pessarakli M (2012) Reactive oxygen species, oxidative damage, and antioxidative defense mechanism in plants under stressful conditions. J Bot 2012:1–26 Article ID 217037, 26 pages

Shenker M, Fan TWM, Crowley DE (2001) Phytosiderophores influence on cadmium mobilization and uptake by wheat and barley plants. J Environ Qual 30:2091–2098

Shevyakova N, Cheremisina A, Kuznetsov V (2011) Phytoremediation potential of Amaranthus hybrids: antagonism between nickel and iron and chelating role of polyamines. Russ J Plant Physiol 58(4):634–642

Shi G, Cai Q (2009) Cadmium tolerance and accumulation in eight potential energy crops. Biotechnol Adv 27(5):555–561

Shi X, Zhang X, Chen G, Chen Y, Wang L, Shan X (2011) Seedling growth and metal accumulation of selected woody species in copper and lead/zinc mine tailings. J Environ Sci 23:266–274

Skinner K, Wright N, Porter-Goff E (2007) Mercury uptake and accumulation by four species of aquatic plants. Environ Pollut 145(1):234–237

Slater A, Scott NW, Fowler MR (2008) Plant Biotechnology; the genetic manipulation of plants 2ndnd edn; Oxford University Press, Oxford

Sonoki T, Kajita S, Ikeda S, Uesugi M, Tatsumi K, Katayama Y, Iimura Y (2005) Transgenic tobacco expressing fungal laccase promotes the detoxification of environmental pollutants. App Microbiol Biotechnol 67:138–142

Spinks JWT, Woods RJ (1990) An introduction to radiation chemistry, 3rd edn. Wiley-Interscience, New York

Sun R, Zhou Q, Jin C (2006) Cadmium accumulation in relation to organic acids in leaves of *Solanum nigrum* L. as a newly found cadmium hyperaccumulator. Plant Soil 285:125–134

Sweta BK, Singh R, Singh RP (2015) *Trapa natans* cultivated in contaminated water is found hazardous and not suitable for phytoremediation of heavy metals. Ecol Eng 83:39–42

Tangahu BV, Sheikh Abdullah SR, Basri H, Idris M, Anuar N, Mukhlisin M (2011) A review on heavy metals (As, Pb, and Hg) uptake by plants through phytoremediation. Int J Chem Eng: http://dx.doi.org/10.1155/2011/939161

Tommasini R, Vogt E, Fromenteau M, Hoertensteiner S, Matile P, Amrhein N, Martinoia E (1998) An ABC transporter of Arabidopsis thaliana has both glutathione conjugate and chlorophyll catabolite transport activity. Plant J 13:773–780

Tong Y, Kneer R, Zhu Y (2004) Vacuolar compartmentalization: a second-generation approach to engineering plants for phytoremediation. Trends Plant Sci 9(1):7–9

Traunfeld JH and Clement DL (2001) Lead in garden soils. Home and garden. Maryland Cooperative Extention, University of Maryland, 2001, http://www.hgic.umd.edu/media/documents/hg18.pdf

Tu S, Ma LQ, Fayiga AO, Zillioux EJ (2004) Phytoremediation of arsenic-contaminated groundwater by the arsenic hyperaccumulating fern Pteris vittata L. Int J Phytoremed 6(1):35–47

U. S. Environmental Protection Agency (2005) Use of field-scale phytotechnology, for chlorinated solvents, metals, explosives, and propellants, and pesticides, vol 9. U.S. Environmental Protection Agency, Solid Waste and Emergency Response, Washington, DC, pp. 7–9

Uera RB, Paz-Alberto AM, Sigua GC (2007) Phytoremediation potentials ofselected tropical plants for ethidium bromide. Environ Sci Pollut Res 14(7):505–509

Vaclavikova M, Gallios GP, Hredzak S, Jakabsky S (2008) Removal of arsenic from water streams: an overview of available techniques. Clean Techn Environ Policy 10(1):89–95

Vamerali T, Bandiera M, Coletto L, Zanetti F, Dickinson NM, Mosca G (2009) Phytoremediation trials on metal- and arsenic-contaminated pyrite wastes (Torviscosa, Italy). Environ Pollut 157(3):887–894

Van Ginneken L, Meers E, Guisson R, Ruttens A, Elst K, Tack FM, Vangronsveld J, Diels L, Dejonghe W (2007) Phytoremediation for heavy metal-contaminated soils combined with bioenergy production. J Environ Eng Landsc 15(4):227–236

Van Hoof NA, Koevoets PL, Hakvoort HW, Ten Bookum WM, Schat H, Verkleij JA, Ernst WH (2001) Enhanced ATP dependent copper efflux across the root cell plasma membrane in copper tolerant *Silene vulgaris*. Physiol Plant 113(2):225–232

Vandecasteele B, Meers E, Vervaeke P, Vos BD, Quataert P, Tack FMG (2005) Growth and trace metal accumulation of two Salix clones on sediment-derived soils with increasing contamination levels. Chemosphere 58(8):995–1002

Vanderauwera S, Suzuki N, Miller G, van de Cotte B, Morsa S, Ravanat JL (2011) Extranuclear protection of chromosomal DNA from oxidative stress. Proc Natl Acad Sci U S A 108:1711–1716

Verbruggen N, Hermans C, Schat H (2009) Molecular mechanisms of metal hyperaccumulation in plants. New Phytol 181:759–776

Vert G, Grotz N, Dédaldéchamp F, Gaymard F, Guerinot ML, Briat J-F, Curie C (2002) IRT1, an Arabidopsis transporter essential for iron uptake from the soil and for plant growth. Plant Cell 14:1223–1233

Wang Z, Wan J (2013) An economic analysis of the use of water hyacinth for phytoremediation and biogas production in Dianchi Lake, China, No. rr2013025. Economy and Environment Program for Southeast Asia (EEPSEA)

Wani PA, Khan MS, Zaidi A (2007) Effect of metal tolerant plant growth promoting Bradyrhizobium sp. (vigna) on growth, symbiosis, seed yield and metal uptake by greengram plants. Chemosphere 70:36–45

Westhoff Alex (1999) Mycorrhizal plants for phytoremediation of soils contaminated with radionuclides. University of Minnesota, Department of Horticultural Science. Retrieved from the University of Minnesota Digital Conservancy, http://hdl.handle.net/11299/59501

Whiting SN, de Souza SP, Terry N (2001) Rhizosphere bacteria mobilize Zn for hyperaccumulation by Thlaspi caerulescens. Environ Sci Technol 35:3144–3150

Wu SC, Cheung KC, Luo YM, Wong MH (2006) Effects of inoculation of plant growth-promoting rhizobacteria on metal uptake by *Brassica juncea*. Environ Pollut 140:124–135

Xiang C, Oliver DJ (1998) Glutathione metabolic genes coordinately respond to heavy metals and jasmonic acid in Arabidopsis. Plant Cell 10:1539–1550

Xu JK, Yang LX, Wang YL, Wang ZQ (2005) Advances in the study uptake and accumulation of heavy metal in rice (*Oryza sativa*) and its mechanisms. Chinese Bull Bot 22:614–622

Yang X, Feng Y, He Z, Stoffella P (2005) Molecular mechanisms of heavy metal hyperaccumulation and phytoremediation. J Trace Elem Med Biol 18:339–353

Yang XE, Li TQ, Yang JC, He ZL, Lu LL, Meng FH (2006) Zinc compartmentation in root, transport into xylem, and absorption into leaf cells in the hyperaccumulating species of *Sedum alfredii* Hance. Planta 224(1):185–195

Yin H, Chen Q, Yi M (2008) Effects of short-term heat stress on oxidative damage and responses of antioxidant system in Lilium longiflorum. Plant Growth Regul 54:45–54

Yu F, Liu K, Li M, Zhou Z, Deng H, Chen B (2013) Effects of cadmium on enzymatic and non-enzymatic antioxidative defences of rice (*Oryza sativa* L.). Int J Phytoremed 15(6):513–521

Zaal BJ, Neuteboom LW, Pinas JE, Chardonnes AN, Scjat H, Verleij JAC, Hooykass PJJ (1999) Over expression of a novel Arabidopsis gene related to putative Zn transporter genes from animals lead to enhanced zinc resistance and accumulation. Plant Physiol 19:1047–1055

Zancani M, Peresson C, Biroccio A, Federici G, Urbani A, Murgia I, Soave C, Micali F, Vianello A, Macri F (2004) Evidence for the presence of ferritin in plant mitochondria. Eur J Biochem 271:3657–3664

Zancani M, Peresson C, Patui S, Tubaro F, Vianello A, Macri F (2007) Mitochondrial ferritin distribution among plant organs and its involvement in ascorbate-mediated iron uptake and release. Plant Sci 173:182–189

Zhang XH, Zhu YG, Chen BD, Lin AJ, Smith SE, Smith FA (2005) Arbuscular mycorrhizal fungi contribute to resistance of upland rice to combined metal. J Plant Nutr 28:2065–2077
Zhang X, Lin AJ, Zhao FJ, Xu GZ, Duan GL, Zhu YG (2008) Arsenic accumulation by the aquatic fern Azolla: comparison of arsenate uptake, speciation and efflux by *A. caroliniana* and *A. filiculoides*. Environ Pollut 156(3):1149–1155
Zhigang A, Cuijie L, Yuangang Z, Yejie D, Wachter A, Gromes R, Rausch T (2006) Expression of BjMT2, a metallothionein 2 from *Brassica juncea*, increases copper and cadmium tolerance in Escherichia coli and *Arabidopsis thaliana*, but inhibits root elongation in *Arabidopsis thaliana* seedlings. J Exp Bot 57:3575–3582
Zimeri AM, Dhankher OP, McCaig B, Meagher RB (2005) The plant MT1 metallothioneins are stabilized by binding cadmium and are required for cadmium tolerance and accumulation. Plant Mol Biol 58:839–855

Bioenergy: A Sustainable Approach for Cleaner Environment

2

Abhishek Guldhe, Bhaskar Singh, Nirmal Renuka, Poonam Singh, Rohit Misra, and Faizal Bux

Abstract

Bioenergy is a renewable form of energy generated from the biomass. It holds great potential to provide energy security as it can be produced from locally available biomass resource into various forms. This chapter discusses the various bioenergy forms, feedstocks such as energy crops and waste biomass, used for the bioenergy generation. The chapter also deals with the socio-economic and environmental impacts of the bioenergy usage. The integrated approach of coupling phytoremediation and bioenergy production is also discussed.

Keywords

Bacteria • Bioenergy • Biorefinary • Fungi • Microalgae

2.1 Bioenergy

Bioenergy is the sustainable solution for ever-increasing energy demands. Currently fossil fuels are catering to the global energy demands. Fossil fuels are limited and non-renewable resources and also significantly contribute towards the increased carbon dioxide in the atmosphere. In quest for the renewable energy supply and to mitigate the climate change constraints globally, bioenergy is gaining interest. Currently 14% of the global energy demand is catered by the bioenergy from biomass. Biomass can be converted to the electricity, heat or liquid biofuels with

A. Guldhe (✉) • N. Renuka • P. Singh • R. Misra • F. Bux
Institute for Water and Wastewater Technology, Durban University of Technology, P. O Box 1334, Durban 4000, South Africa
e-mail: abhiguldhe@gmail.com

B. Singh
Centre for Environmental Sciences, Central University of Jharkhand, Ranchi-835205, Jharkhand, India

K. Bauddh et al. (eds.), *Phytoremediation Potential of Bioenergy Plants*,
DOI 10.1007/978-981-10-3084-0_2

advanced technologies. The International Energy Agency has set its goal to achieve 25% of the global transportation fuel demand through biofuels by 2050 (Pandey et al. 2016). Many countries have set the specifications for blending the biofuels, in order to achieve energy security, reduce dependence on fossil fuel and mitigate carbon dioxide emission.

Biomass is a broad term which includes energy crops, residues and other biological materials which can be used to produce different renewable energy forms. During generation of biomass, carbon dioxide is utilized, while use of bioenergy emits the carbon dioxide which results in carbon neutral cycle. Biomass can be categorized in three types depending upon its origin. Primary biomass includes photosynthetically generated biomass which is directly taken from the land, e.g. short rotation woody crops, oilseeds, herbaceous plants and agriculture and forestry residues (McKendry 2002). Secondary biomass includes chemically, physically or biologically treated primary biomass. Tertiary biomass includes the waste material after use such as used cooking oil, animal fats and grease and solid waste material.

Biomass can be directed towards various forms of bioenergy depending upon its composition. The bioenergy includes liquid transportation biofuels such as bioethanol and biodiesel, electricity and heat generated from woodchips and biomass and gaseous energy forms such as biomethane and biohydrogen. Liquid biofuels are the major section of the bioenergy. Biomass rich in sugars and starch is most suitable for bioethanol production. Biomass rich in oil is directed towards the biodiesel production. Solid waste biomass can be primarily utilized for biomethane generation via anaerobic digestion and also for liquid biofuels.

Energy from biomass is directly associated with the crops, forest and ecosystem unlike other renewable energy sources such as wind and solar. Because of this the use of biomass has immediate effect on the environment and society both positive and negative. The major negative impact is the 'food vs. fuel' debate. First-generation biomass includes food crops such as corn, vegetable oils, etc. Limited arable land is also a major constraint. Thus, growing the energy crops for biomass on nonagricultural land is imperative to mitigate this challenge. Phytoremediation is used for contaminated lands. Bioenergy plantation can serve both purposes of biomass generation as well as phytoremediation. This integrated approach is gathering interest for providing energy security and cleaner environment.

2.2 Bioenergy Forms

2.2.1 Combustion: Heat and Power

Heat is a form of energy that is due to the random motion of atoms and molecules, whereas power is the rate of energy use. Power is calculated as energy divided by time. Biomass on combustion generates heat which can be used for cooking and electricity production. In power plants biomass is used to generate high-pressure steam via combustion and used to run the turbines. There are two types of engines used in transport sector: spark ignition and compression ignition (CI). While the former runs on gasoline, the latter runs on mineral diesel. High fuel conversion efficiency is obtained upon combustion of mineral diesel in CI engine as compared to combustion of gasoline in spark ignition engine. However, high emission of soot and NO_X occurs due to rapid auto-ignition of mineral diesel upon combustion when compressed inside the cylinder along with heterogeneous air.

Bioenergy is an alternate for the non-renewable and polluting fossil fuel. It is reported that 95–97% of the world's bioenergy is being produced by direct combustion of biomass. However, direct combustion results in generation of high amount of ash (on ground and in ambient air) that restricts its application (Vassilev and Vassileva 2016). Combustion of fuels containing carbon results in the emission of carbon dioxide. The carbon-containing fuel includes biomass, wood, coal, oil, petroleum and its subordinates. The oxygen content in biofuels (viz. biodiesel) ensures a better combustion as compared to fossil fuels. The oxygen reacts with the unburned carbon atoms during combustion, and thereby, a higher concentration of CO_2 and a lower concentration of CO are produced (Imdadul et al. 2016). The two prominent sectors responsible for CO_2 emissions are electricity generation and transport. Combustion of biomass as fuel retards the net emission of CO_2 in the atmosphere making it a green fuel (as a subsequent amount of carbon has already been sequestered by the plant during photosynthesis when it was alive) (Johnston and van Kooten 2016). Schönnenbeck et al. (2016) report that prior to the combustion process, moisture content from the biomass needs to be reduced to abate the release of high concentrations of CO, volatile organic compounds and particulates. Hence, to enhance the efficiency of biomass as fuel, it needs to be converted or transformed to a higher grade (viz. bioethanol, biomethane, biohydrogen, biodiesel) by various processes and techniques.

2.2.2 Gaseous Energy Forms

The gaseous energy forms include natural gas and biogas (both of which have methane as its major constituent), butane and propane. While some of the energy forms are non-renewable, others belong to the renewable energy category. Globally, oil and natural gas provide 70–80% of primary energy. However, the percentage consumption of oil is quite high in Middle East countries and the coal is the dominant fuel in Asian countries. Natural gas exists in trapped form in between shale rocks named as 'shale gas'. It has recently been trapped in the USA and constitutes a significant fraction of natural gas supply. Methane is also found in frozen form trapped in oceans as methane hydrates. These methane hydrates are found in deep ocean waters that are cold enough to provide high pressure and low temperature. Methane is also being tried to recover from coal bed where it is present in substantial amount. Butane is the main component of petroleum gas that is liquefied and used as fuel for heating, cooking and other purposes. Another gaseous fuel is synthetic gas, i.e. a mixture of carbon monoxide and water, and is commonly produced by gasification of a lower-grade coal.

Biomass if converted to gaseous energy forms can be used easily for various purposes. Various gaseous energy forms can be produced from biomass, viz. biomethane, biohydrogen, etc. Among the renewable, biogas, is a mixture of methane and carbon dioxide. The common mode of production of biogas is gasification, a thermochemical process. Methane is also produced by anaerobic digestion (the other name being anaerobic fermentation) of biomass (Botkin and Keller 1996). Hydrogen is another renewable gaseous fuel that is produced either from natural gas (through steam reforming) or water (through electrolysis, thermolysis or photolysis). Biohydrogen can be produced from the biomass and it is considered as clean energy source. The advantage with hydrogen is its being a clean fuel as it gives water as a product during combustion. Its major disadvantage is its high manufacturing cost and safety issues during transportation. Hence, it is mostly used near the site of its production.

2.2.3 Liquid Biofuels

Transportation fuel demands (up to 98%) are mainly supplied by petroleum. Use of petroleum fuels is associated with energy security and emission concerns. Biofuels such as biodiesel and bioethanol can be locally produced and considered as carbon neutral. Liquid biofuels such as biodiesel and bioethanol have the advantage of usage in existing vehicle engines without any major modification required in the engine. The advantage of liquid biofuels (biodiesel and bioethanol) over gaseous biofuels (biomethane and biohydrogen) is the ease of storage and transportation of the former over the latter. The liquid biofuels also have high hydrogen content as compared to gaseous biofuels (Singh et al. 2014). Among the liquid biofuels, bioethanol production has gained attention specifically in Europe, the USA, Brazil and few Asian countries. Bioethanol is produced by fermentation process involving feedstocks, such as sugarcane, or any grain (viz. corn, wheat). A litre of ethanol is comprised of 66% of energy compared to a litre of petrol (Nigam and Singh 2011). Ethanol has higher octane level; thus, if blended with petrol, it improves the performance and also reduces the emission. The other liquid biofuel, biodiesel, is obtained from a variety of feedstock oil (edible and nonedible, waste cooking and frying oil, waste animal and fish oil and microalgae). Biodiesel can also be blended with diesel fuel to be used as a transport fuel. It is prepared by the process called transesterification using chemical or enzymatic catalysts.

Biofuels can be classified based on the feedstock used for their production. First-generation biofuels include bioethanol produced from crops such as sugarcane, maize and corn and biodiesel produced from oilseeds. Second-generation biofuels are produced from nonedible lignocellulosic biomass or nonedible oil plants. Third-generation biofuels are produced from the microorganisms and waste materials. First-generation biofuels are currently produced at commercial scale in many countries. First-generation biofuels are directly associated with the food security concerns. Second-generation biofuels address the food security issue and, however, are associated with land requirement and complex conversion process. Third-generation biofuels are still at its early stages and need to be investigated thoroughly for its commercial-scale production. Liquid biofuels hold the promising potential to cater the energy demand of transportation sector.

2.3 Plant-Based Feedstocks for Bioenergy

2.3.1 Oil Crops

The biodiesel is mono alkyl esters of long-chain fatty acids derived from vegetable and nonedible oils in form of acylglycerol via transesterification process. The oilseeds contain droplets of lipids which can be extracted for biodiesel generation. The lipids mainly include triacylglycerol. The fatty acid chain lengths generally range from 10 to 24 carbon atoms. The oil for biodiesel production can be derived from vegetable oilseeds such as soybean and rapeseed or nonedible oils such as *Jatropha* oil (*Jatropha curcas*) and karanja oil (*Pongamia pinnata*). Currently 85% of the world biodiesel production comes from the rapeseed oil, 13% from sunflower oil, 1% from sunflower oil and 2% from soybean and other oils (Gui et al. 2008). Use of edible oil resources for biodiesel generation is not sustainable because of food security issue. Various nonedible feedstocks such as jatropha, castor, karanja and rubber seed have been gaining interest to resolve the food security concern. Nonedible oil crops can be grown on unproductive nonagricultural lands and degraded forests. Suitability of oil crop for biodiesel production depends upon a number of factors such as oil content, adaptability to the local climate, fatty acid profile, etc.

2.3.2 Woody Feedstock

Wood is the predominant biomass used in developing countries for energy generation. It is mostly used for non-commercial purposes in heating and cooking stoves. Woody biomass is also composed of lignocellulose which can be converted to bioethanol. Cellulosic content of woody biomass can be broken down into fermentable sugars which can eventually be converted to bioethanol. Woody biomass can be obtained from various sources. These sources include short rotation woody crops, forest wood, residues from timber and milling process and sawdust. Short rotation woody crops most commonly grown include poplars (*Populus* spp.) and willow (*Salix* spp.) which are grown for a short duration of time with intensive crop management. Short rotation woody crops along with biofuel production are also used for paper and pulp industry and saw timber. Energy from woody biomass contributes to only 3–4% of total energy in industrialized countries, while it accounts for more than 30% of the total energy in developing countries (Rockwood et al. 2004). In many parts of the world, wood is the only locally available energy source. Woody biomass can be easily collected and converted to usable forms. Increasing energy demands are causing deforestation to supply woody biomass. There is a need to produce the woody biomass with proper integration and management to cater the energy needs of developing countries. Agroforestry is the possible solution for supply concerns of the woody biomass. Woody biomass has several challenges for its use as a bioenergy feedstock. Woody biomass is not cost competitive compared to fossil fuels.

2.3.3 Energy Crops

Producing biomass for energy using agricultural land is increasing to meet the renewable energy production demands. Energy crops are classified into two groups: annual crops, i.e. crops grown and ready for harvest in one season such as wheat and maize, and perennial crops, i.e. crops which grow over a period of more than 2 years such as *Miscanthus* and *Salix*. Annual crops require higher input of fertilizers especially N source than the perennial crops. Biomass generated from energy crops can be used for heat and power generation as well as liquid biofuels. First-generation biodiesel and bioethanol are mainly produced from the annual energy crops such as oilseeds, corn, maize, etc. Second-generation biofuels are mostly synthesized from lignocellulosic biomass of perennial crops. Energy crops can also be classified depending upon their major constituent that can be used as a bioenergy feedstock. Starch crops such as wheat and corn, sugar crops such as sugarcane and beet, forage crops such as grasses and alfalfa and oilseed crops such as rapeseed and sunflower are the different types of energy crops (Demirbas 2001). Starchy cop corn is being used as a major feedstock for bioethanol production with 60 billion litres produced in 2012 followed by sugarcane with 20 billion litres globally (Ho et al. 2014). Bioethanol production from sugar crops is an easy process where sugars (mostly C6) are fermented using yeast. Starch crops need to be first hydrolysed to fermentable sugars and then fermented to bioethanol and are thus more energy intensive. The common oil crop in Europe is rapeseed; in the USA and Latin America, soybean; and in Asian countries, palm and coconut oil (Ho et al. 2014). Selection of energy crops depends upon the quality of biomass, adaptability towards local soil and climate conditions. Competition of energy crops for land with the food crops is the major sustainability concern associated with socio-economic and environmental impact.

2.4 Microorganisms for Bioenergy

2.4.1 Microalgae

Microalgae have emerged as the promising third-generation feedstock for biofuel production. Microalgae provide several benefits over the conventional first- and second-generation energy crops. Microalgae have faster growth rates, it can be grown in different climatic conditions (wastewater can be used as a growth medium) and have minimum land requirements due to high productivities (Singh et al. 2015). Biodiesel has been the primary focus while considering microalgal biomass for bioenergy production. Microalgae can accumulate up to 20–80% of lipids based on their dry weight (Singh et al. 2016). Lipid accumulation capabilities can be enhanced by subjecting it to stress conditions or adapting specific cultivation strategies. Microalgae can also be utilized for the generation of biomethane, biohydrogen and bioethanol. Microalgal biomass can be used to produce more than one biofuel in an integrated biorefinery concept (Singh et al. 2015). Oil can be extracted from microalgal biomass for biodiesel production, while lipid-extracted biomass can be used for bioethanol or biomethane generation. However, commercial-scale cultivation of microalgae is still a bottleneck due to high production cost. Fundamental research is needed for suitable strain selection and engineering, reducing the technological challenges and improving the economics of microalgal mass cultivation.

2.4.2 Bacteria

Bacteria are associated with bioenergy production directly as a feedstock or indirectly as a conversion tool for other feedstock. Microbial fuel cell generates electrical energy by converting energy stored in chemical bonds by catalytic reactions of bacteria. Bacteria as a feedstock for bioenergy generation are scarcely exploited area. However, its use as a conversion agent for production of different bioenergy forms such as biomethane, biohydrogen and bioethanol has been extensively studied. Biohydrogen can be produced from dark fermentation of organic substrates by anaerobic bacteria. However, this technology is still not mature enough to reach commercial-scale production. Biomethane is produced from various organic waste materials and also from some energy crops. Various bacteria play a role in the process of biomethane generation. Bacteria are responsible for hydrolysis of complex compounds into simpler ones; in acetogenesis acetic acid and carbon dioxide are produced, and finally methanogens produce methane, carbon dioxide and some amounts of NH_3 and H_2S. Bacteria can be engineered genetically to perform efficiently in bioenergy production process. More research and development work is needed to exploit the potential of bacteria for bioenergy production.

2.4.3 Fungus

Fungus and yeasts can accumulate lipids as high as 70% of their dry cell biomass. *Rhodotorula* sp., *Candida* sp., *Lipomyces* sp., *Aspergillus* sp. and *Mortierella* sp. are some of the efficient oleaginous fungus and yeasts used for biodiesel production. Unlike microalgae, biomass of fungus and yeast can be produced in conventional microbial bioreactor at much lower cost. This gives advantage of higher biomass and lipid yields at minimum input cost. Oleaginous fungus has not been studied thoroughly for its application as a biodiesel feedstock. Vicente et al. (2009) studied

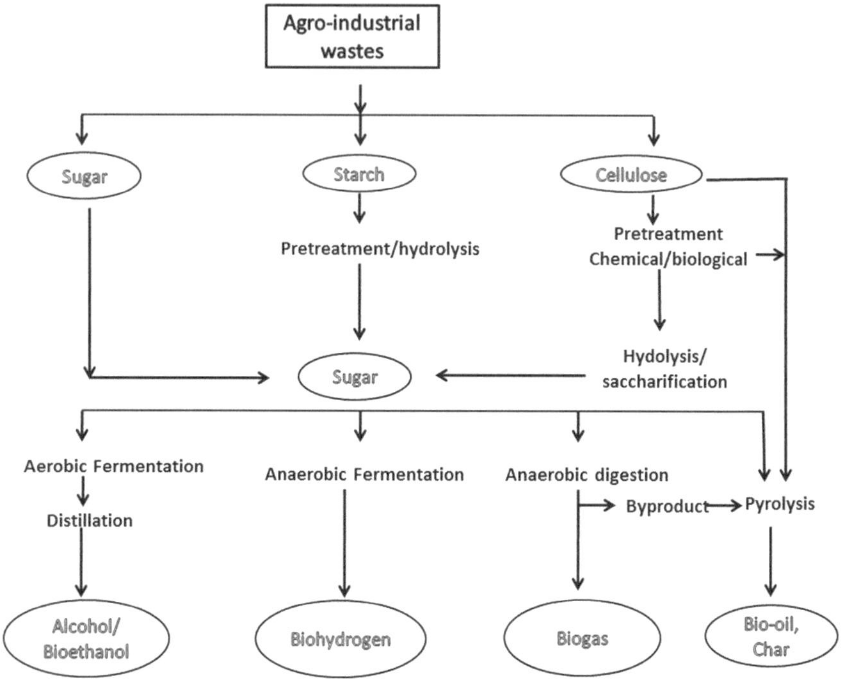

Fig. 2.1 Schematic representation of various processes involved for bioenergy production from agro-industrial wastes

filamentous fungus *Mucor circinelloides* for production of biodiesel. This fungus showed lipid accumulation of around 19% based on the dry cell weight. The fungal lipids in this study were composed of more than 85% of the saponifiable lipids and showed suitable lipid profile for biodiesel production. Further investigations are needed to explore high lipid accumulating fungus, lipid accumulation enhancement strategies and extraction techniques to make this technology sustainable and economically viable.

2.5 Bioenergy from Waste

2.5.1 Agro-industrial Waste Biomass

The use of agro-industrial residues for bioenergy production has attracted the great attention in the recent years. Many types of agro-industrial wastes are available, which include crop residues, pulp from juice industry, pomace or residues from oil industry, etc. These wastes can be converted into three types of potential energy forms, viz. bioethanol, biohydrogen and biogas. Figure 2.1 depicts the various energy forms and processes involved in their production. The use of agro-industrial waste biomass for bioenergy is important due to environmental concerns related to their disposal. It has been estimated that ethanol demand could surpass 125 billion litres by 2020, and current agricultural status could not support the increasing future ethanol demand (Bohlmann 2006). Bioethanol production from agro-industrial wastes could be a beneficial approach producing energy alternate to fossil fuel, as it does not involve food substrates.

There is huge generation of agro-industrial wastes throughout the world every year. These agro-wastes mainly consist of lignocellulose, starch, simple sugars (glucose, fructose, etc.) disaccharide (sucrose), polysaccharides (pectin, inulin, xylans and galactoxylans), nitrogen, minerals, lipids or a mixture of them in different proportions into their biomass. The residues from oil agro-industries are of special interest for fermentative processes, due to the presence of large amounts of carbohydrates, nitrogen, residual oil and minerals. Lignocellulosic biomass from agro-wastes is the most difficult biomass for bioenergy conversion, since it consists of three types of polymers – cellulose, hemicellulose and lignin (Lim and Matu 2014). The conversion of lignocellulosic biomass to bioethanol requires pretreatment for efficient utilization of starch- and hemicellulose-derived carbohydrates. There are two types of methods involved in the pretreatment (saccharification) of lignocellulosic wastes – chemical and biological treatment. The chemical pretreatment involves hydrolysis of biomass using acids and alkali, ozonolysis, etc. The most crucial factors affecting the efficiency of acid hydrolysis are temperature, retention time and acid concentration. Dilute acid hydrolysis is considered to be a low-cost technology with high reaction rates for the conversion of hemicellulose to fermentable sugars (Chandel et al. 2012). Appropriate acid concentrations are very crucial for industrial scale, because sometimes acid decomposes the hemicellulose, degrades the sugars into inhibitors and causes damage to the equipment. Therefore, a conventional method of starch and hemicellulose hydrolysis using acid has now been replaced by biological saccharifying enzymes (Maiti et al. 2016). Biological treatment involves the use of fungi (such as actinomycetes and white rot fungi) and bacteria, producing enzymes capable of degradation of complex polysaccharides lignin and hemicellulose into simpler sugars or compounds (Saritha et al. 2012; Sarkar et al. 2012). Biological pretreatment is a time-consuming process; however, it has advantages such as the process is more efficient and consumes less energy. One of the efficient biological pretreatment methods is enzymatic hydrolysis via the use of cellulase and hemicellulose enzymes (derived from certain bacterial and fungal species) with advantages such as substrate specificity, no inhibitory product formation and low toxicity and cost (Ferreira et al. 2009). The bacterial species reported to produce cellulase enzyme are *Bacillus*, *Bacteroides*, *Erwinia*, *Streptomyces*, *Clostridium*, Thermomonospora, *Cellulomonas*, etc. Various fungi capable of cellulase production include *Trichoderma*, *Fusarium*, *Humicola*, *Aspergillus* and *Schizophyllum* sp. (Saritha et al. 2012).

Agro-industrial wastes are converted into many valuable bioproducts, like ethanol via solid state fermentation. The pretreated saccharified agro-waste biomass is followed by fermentation by different microorganisms for the production of bioethanol. The industrial utilization of lignocelluloses is constrained by the lack of ideal microorganisms capable of efficient fermentation of pentose and hexose sugars (Talebnia et al. 2010). However, a wide range of microorganisms such as *Aspergillus oryzae*, *A. niger*, *Penicillium* sp., *Trichosporon fermentans*, *Pseudomonas aeruginosa*, *Streptomyces*, etc., are used for fermentation of biomass for alcohol production, where *Saccharomyces cerevisiae* is the most commonly used organism (Domínguez-Bocanegra et al. 2015; Orozco et al. 2008). Rodríguez et al. (2010) conducted a laboratory-scale study to evaluate the efficiency of grape and sugar beet pomace for alcohol production. More than 82% of yield was obtained by the treatment of pomace with *Saccharomyces cerevisiae* yeasts at the rate of 10^8 cells/g, incubated in anaerobic conditions at 28 °C for 48 h. The production of bioethanol through direct fermentation of apple pomace with coculturing of *Trichoderma harzianum*, *Aspergillus sojae* and *Saccharomyces cerevisiae* was also investigated (Evcan and Tari 2015). The highest bioethanol yield of 0.945 g/g could be obtained from apple pomace with inoculation rates of 6% (w/v) for *T. harzianum* and *A. sojae* and 4% (v/v) for *S. cerevisiae* with agitation (200 rpm) and vented aeration (Evcan and Tari 2015). Kouteu Nanssou et al. (2016) studied the ethanol production from cassava plant stem and peelings by using *Rhizopus* spp. and *Saccharomyces cerevisiae*. The hydrolysates of cassava stems and peelings could produce 5.27 g/100 g and 2.6 g/100 g of ethanol from stems and peel biomass, respectively.

Anaerobic fermentation of agro-industrial residues can be a promising process for biohydrogen production with low cost and higher yields (Robledo-Narvaez et al. 2013). Hydrogen is a promising energy source as an alternative fuel. The most commonly used bacteria for anaerobic fermentation of agro-industrial wastes for biohydrogen production are anaerobe *Clostridium* spp. (Al-Shorgani et al. 2013; Collet 2004). Various studies revealed higher yields of biohydrogen in pretreated biomass as compared to untreated biomass. Further, the biologically treated biomass leads to higher biohydrogen production (Al-Shorgani et al. 2013; Reginatto and Antonio 2015).

Agro-industrial residues can also be used as substrates for biogas generation. Anaerobic digestion of these wastes is a beneficial process for biogas generation and substitute for energy crops. However, some residues need pretreatments before anaerobic digestion due to their low biodegradability (Schievano et al. 2009). Recently, several pretreatment methods of these residues have been developed to enhance their anaerobic biodegradability and to exploit them for biogas production. There are some obstacles such as pretreatment procedures, production costs and ideal technology development that need to be overcome in the production of biofuel through these wastes.

2.5.2 Sewage Sludge

Sewage sludge generated from wastewater treatment usually contains organic matter (cellulose, hemicellulose and lignin) and is rich in nitrogen and phosphorus; therefore, it is considered as a good source of energy (Adar et al. 2016). Recently, more attention is being given for the energy recovery from sewage, due to the environmental concerns related to its disposal into the landfills and land such as emission of landfill gases and heavy metal contamination. Annual production of sewage sludge surpasses ten million tonnes and is likely to be increased gradually due to increase in population and industrialization (Cao and Pawłowski 2012). A number of methods are available for the energy recovery from sewage sludge, viz. anaerobic digestion, incineration, gasification, pyrolysis, supercritical water gasification, liquefaction, etc. (Anjum et al. 2016). Anaerobic digestion of sludge is considered as one of the most cost-effective methods to convert sludge to biogas and can be utilized as a fuel to generate heat and electricity. However, the risks associated with hazardous heavy metals in sludge cannot be eliminated by this process and can impact on the environment adversely if necessary treatments are not executed. Further, digestion process of sewage sludge cannot sufficiently extract the energy; therefore, the potential use of different pretreatment methods, viz. thermal treatment, ultrasonication, photocatalysis, microwave, wet oxidation, Fenton, etc., can improve the anaerobic treatment of sludge and biogas production, with low energy demand (Anjum et al. 2016). These pretreatment methods can be employed as single procedure or in combination.

Digested sewage sludge is further considered an energy source due to enriched organic matter but is poor in biodegradability. Sludge can also be converted into useful bioenergy forms through pyrolysis such as bio-oil or gas, leading to the formation of stabilized residue termed as biochar (Alvarez et al. 2016; Hwang et al. 2007). Sludge pyrolysis can be carried out with raw and digested sludge and indicates the possibilities of the use of two parallel processes (anaerobic digestion followed by pyrolysis) for efficient energy conversion (Cao and Pawłowski 2012). However, emphasis has been given for the use of pretreated or digested sludge for energy conversion, due to poor degradability of raw sludge. Various studies have indicated the formation of bio-oil from fast pyrolysis of sewage sludge at temperature of 500–600 °C range with high calorific value and maximized liquid yields (Alvarez et al. 2016). Sludge pyrolysis oil corresponded to more than 70% similarity to gas oil, revealing the potential to substitute conventional fuel. However, further refinements were emphasized to increase the heating value and decrease the nitrogen content in the bio-oil (Alvarez et al. 2016).

2.5.3 Animal Waste

There is a huge generation of biomass feedstock due to animal production worldwide which can be exploited as energy source (Cantrell et al. 2008). Animal wastes (wastes from poultry, livestock, animal fat, etc.) are usually rich in organic matter and fats; and their use as energy feedstock can generate new marketplace for farmers (Baladincz and Hancsók 2015). These wastes can be converted into biogas and bio-oil, and methods employed include biological treatment and thermochemical process (Abdeshahian et al. 2016). It has been estimated that biological treatment and anaerobic digestion of animal wastes can generate 0.1–0.3 $m^3m^{-3}d^{-1}$ of methane-rich biogas (Cantrell et al. 2008). However, methanogenic biogas production rate is dependent on factors such as changes in influent, pH, temperature, organic loading rate and hydraulic retention time and needs to be controlled to maximize the biogas production. Other efficient processes for energy recovery from animal wastes include pyrolysis and thermal liquefaction. Little has been known for the bio-oil formation from the animal wastes; however, recent studies have explored the possibilities of conversion of waste animal fat into biofuel (Baladincz and Hancsók 2015). It has been found that slow pyrolysis of animal wastes leads to the formation of char, which can be used as alternate fuel to coal combustion and gasification and source of activated carbon (Koutcheiko et al. 2007).

2.6 Environmental and Socio-economic Significance

The development and usage of biofuels bear environmental and economic significance. There is considerable reduction in the emission of CO_2 from the combustion of biofuels. Other contaminants, viz. particulate matter (PM), carbon monoxide (CO), dust and carcinogens, also get reduced or eliminated to a large extent. A report of National Renewable Energy Laboratory, USA states that replacing biodiesel with mineral diesel leads to reduction in 90% of air toxins (viz. PM, CO and hydrocarbons). It also reports that even the usage of B20 (i.e. 20% biodiesel and remaining mineral diesel) reduces air toxins by 20–40%. Bioenergy generation increases CO_2 levels in the atmosphere at the instant it undergoes combustion. When left on the forest ground, it still will be released into the atmosphere, albeit after a long time (years or decades) (Repo et al. 2015). Various strategies have been suggested to improve the climatic impacts of bioenergy. The strategies reported are an increase in forest carbon stocks that could be achieved by extending rotation lengths, forest fertilization and forest thinning. Extending rotation period in forest allows the trees to grow longer and in turn more forest litter is generated that incorporates organic matter in the soil. Forest fertilization and forest thinning enhance the growth of tree which in turn incorporates more litter in the soil. The decay rate of the litter decreases with the increasing amount of litter in the forest. This reduces the carbon loss from the forest and thus reduces the climatic impacts arising from bioenergy produced from forest residue (Repo et al. 2015).

Increased food, feed and fibre demands are exerting tremendous pressure on existing land resources. Increasing renewable energy requirements are also escalating the need of biomass for production of bioenergy. The use of limited land resources for food, feed, fibre and energy need is a major challenge, which needs to be addressed by integrated policies for sustainable future. Rotation of crop's use of nonagricultural land for biomass production could be a possible solution to overcome the land usage challenges. Biodiversity at a particular site maintains the interrelations between vegetation, soil, air and water. Introducing a bioenergy plantation could hamper the ecosystem. Thus, it is imperative to assess the long-term problems of bioenergy plantation on ecosystem.

Bioenergy can be produced at various scales catering the need of small communities to large industrial-scale productions. Various bioenergy forms can be produced from locally available biomass throughout the year. Bioenergy thus provides energy security to the community and thus aids in economic development of the country. A rise in awareness for enhancing the degrading environment among various nations has led to emergence of biofuels. Several nations now have emphasized on the use of biofuels for the protection of the environment. In Germany, several terms have been coined to lay emphasis on renewable energy, viz. 'bioenergy villages', 'community energy' and 'energy self-sufficient villages'. As per the German Federal Ministry of Food and Agriculture, a 'bioenergy village' satisfies at least 50% of its total energy demand from indigenous biomass (Grundmann and Ehlers 2016). It has been observed worldwide that there has been a steady increase in the consumption of biofuels. The target of production of renewable energy in Europe is 27% of the total energy share by the year 2035. It has also been reported that between the year 2005 and 2010, a significant portion of electricity was generated from biofuels of which the largest contribution was from biomass (comprising mainly of wood and wood wastes) production major. In India, as per the National Biofuel Policy, an indicative target of a 20% blending of biofuels by the year 2017 (for biodiesel and bioethanol) is proposed. Ethanol produced from the fermentation of molasses (by-product of sugar industry) is being blended with gasoline (at 5% blend) which is likely to increase in coming years. Similarly, other nations also target to supplement their energy requirement with biofuels in coming years.

2.7 Coupling Phytoremediation with Bioenergy: An Integrated Biorefinery Approach

Phytoremediation is a sustainable approach to address the issue of contaminated land. Integrated generation of biomass and phytoremediation is a biorefinery approach towards the greener future. The limited land resource is the major challenge for bioenergy production. The use of contaminated land for biomass generation where agricultural production is a challenge can address the issue of food versus fuel debate. Phytoremediation can clear number of pollutants such as heavy metals, persistent organic pollutants, metalloids and other organic pollutants. These pollutants are added to the soil by various urban and industrial activities. Energy crops such as *Miscanthus*, *Ricinus*, *Salix*, *Jatropha* and *Populus* can be used for phytoremediation purpose. This integrated approach has several indirect advantages. These advantages include aesthetically pleasing landscape, CO_2 sequestration, decreased leaching of pollutants into water bodies and reduced soil erosion. Using contaminated land can also improve the economics of the bioenergy generation as the capital cost for required land is minimal in case of contaminated land sites. However, this approach is associated with challenges like low yield, contaminated biomass, imbalance in biodiversity, etc. This integrated approach can cater to the increasing demand for renewable energy as well as mitigate the challenge of the contaminated land.

2.8 Conclusion

Bioenergy is promising means to cater the world renewable energy demand. Various forms of bioenergy are available which can be used for heating, cooking, transportation fuel and electricity. Various feedstocks which are used for bioenergy generation across the world depending upon climate conditions are available bioresources. Several types of waste biomass can be used for bioenergy generation. There are challenges of limited agricultural land, loss of biodiversity and high production cost associated with the bioenergy. Use of energy crops for phytoremediation of contaminated land sites can address the challenges such as land usage and high cost associated with bioenergy production. Integrated phytoremediation and bioenergy generation are sustainable approach for greener future.

References

Abdeshahian P, Lim JS, Ho WS, Hashim H, Lee CT (2016) Potential of biogas production from farm animal waste in Malaysia. Renew Sust Energ Rev 60:714–723

Adar E, Karatop B, İnce M, Bilgili MS (2016) Comparison of methods for sustainable energy management with sewage sludge in Turkey based on SWOT-FAHP analysis. Renew Sust Energ Rev 62:429–440

Al-Shorgani NKN, Tibin E-M, Ali E, Hamid AA, Yusoff WMW, Kalil MS (2013) Biohydrogen production from agroindustrial wastes via Clostridium saccharoperbutylacetonicum N1-4 (ATCC 13564). Clean Techn Environ Policy 16(1):11–21

Alvarez J, Lopez G, Amutio M, Artetxe M, Barbarias I, Arregi A, Bilbao J, Olazar M (2016) Characterization of the bio-oil obtained by fast pyrolysis of sewage sludge in a conical spouted bed reactor. Fuel Process Technol 149:169–175

Anjum M, Al-Makishah NH, Barakat MA (2016) Wastewater sludge stabilization using pre-treatment methods. Process Saf Environ Prot 102:615–632

Baladincz P, Hancsók J (2015) Fuel from waste animal fats. Chem Eng J 282:152–160

Bohlmann GM (2006) Process economic considerations for production of ethanol from biomass feedstocks. Ind Biotechnol 2:14–20

Botkin DB, Keller EA (1996) Environmental science: Earth as a living planet. Cambridge University Press, Cambridge

Cantrell KB, Ducey T, Ro KS, Hunt PG (2008) Livestock waste-to-bioenergy generation opportunities. Bioresour Technol 99(17):7941–7953

Cao Y, Pawłowski A (2012) Sewage sludge-to-energy approaches based on anaerobic digestion and pyrolysis: Brief overview and energy efficiency assessment. Renew Sust Energ Rev 16(3):1657–1665

Chandel AK, Antunes FAF, de Arruda PV, Milessi TSS, da Silva SS, de Almeida Felipe MDG (2012) Dilute acid hydrolysis of agro-residues for the depolymerization of hemicellulose: state-of-the-art. Springer, Dordrecht, pp. 39–61

Collet C (2004) Hydrogen production by Clostridium thermolacticum during continuous fermentation of lactose. Int J Hydrog Energy 29(14):1479–1485

Demirbas A (2001) Biomass resource facilities and biomass conversion processing for fuels and chemicals. Energy Convers Manag 42:1357–1378

Domínguez-Bocanegra AR, Torres-Muñoz JA, López RA (2015) Production of Bioethanol from agro-industrial wastes. Fuel 149:85–89

Evcan E, Tari C (2015) Production of bioethanol from apple pomace by using cocultures: Conversion of agro-industrial waste to value added product. Energy 88:775–782

Ferreira S, Duarte AP, Ribeiro MHL, Queiroz JA, Domingues FC (2009) Response surface optimization of enzymatic hydrolysis of *Cistus ladanifer* and *Cytisus striatus* for bioethanol production. Biochem Eng J 45(3):192–200

Grundmann P, Ehlers M-H (2016) Determinants of courses of action in bioenergy villages responding to changes in renewable heat utilization policy. Util Policy 41:183–192

Gui MM, Lee KT, Bhatia S (2008) Feasibility of edible oil vs. non-edible oil vs. waste edible oil as biodiesel feedstock. Energy 33(11):1646–1653

Ho SH, Chan MC, Liu CC, Chen CY, Lee WL, Lee DJ, Chang JS (2014) Enhancing lutein productivity of an indigenous microalga Scenedesmus obliquus FSP-3 using light-related strategies. Bioresour Technol 152:275–282

Hwang IH, Ouchi Y, Matsuto T (2007) Characteristics of leachate from pyrolysis residue of sewage sludge. Chemosphere 68(10):1913–1919

Imdadul HK, Masjuki HH, Kalam MA, Zulkifli NWM, Alabdulkarem A, Kamruzzaman M, Rashed MM (2016) A comparative study of C4 and C5 alcohol treated diesel–biodiesel blends in terms of diesel engine performance and exhaust emission. Fuel 179:281–288

Johnston CMT, van Kooten GC (2016) Global trade impacts of increasing Europe's bioenergy demand. J For Econ 23:27–44

Koutcheiko S, Monreal CM, Kodama H, McCracken T, Kotlyar L (2007) Preparation and characterization of activated carbon derived from the thermo-chemical conversion of chicken manure. Bioresour Technol 98(13):2459–2464

Kouteu Nanssou PA, Jiokap Nono Y, Kapseu C (2016) Pretreatment of cassava stems and peelings by thermohydrolysis to enhance hydrolysis yield of cellulose in bioethanol production process. Renew Energy 97:252–265

Lim S-F, Matu SU (2014) Utilization of agro-wastes to produce biofertilizer. Int J Energy and Environ Eng 6(1):31–35

Maiti S, Sarma SJ, Brar SK, Le Bihan Y, Drogui P, Buelna G, Verma M (2016) Agro-industrial wastes as feedstock for sustainable bio-production of butanol by Clostridium beijerinckii. Food Bioprod Process 98:217–226

McKendry P (2002) Energy production from biomass (part 1): overview of biomass. Bioresour Technol 82:37–46

Nigam PS, Singh A (2011) Production of liquid biofuels from renewable resources. Prog Energy Combust Sci 37(1):52–68

Orozco AL, Pérez MI, Guevara O, Rodríguez J, Hernández M, González-Vila FJ, Polvillo O, Arias ME (2008) Biotechnological enhancement of coffee pulp residues by solid-state fermentation with Streptomyces. Py–GC/MS analysis. J Anal Appl Pyrolysis 81(2):247–252

Pandey VC, Bajpai O, Singh N (2016) Energy crops in sustainable phytoremediation. Renew Sust Energ Rev 54:58–73

Rajesh Kumar B, Saravanan S (2016) Partially premixed low temperature combustion using dimethyl carbonate (DMC) in a DI diesel engine for favorable smoke/NOx emissions. Fuel 180:396–406

Reginatto V, Antonio RV (2015) Fermentative hydrogen production from agroindustrial lignocellulosic substrates. Braz J Microbiol 46(2):323–335

Repo A, Ahtikoski A, Liski J (2015) Cost of turning forest residue bioenergy to carbon neutral. Forest Policy Econ 57:12–21

Robledo-Narvaez PN, Munoz-Paez KM, Poggi-Varaldo HM, Rios-Leal E, Calva-Calva G, Ortega-Clemente LA, Rinderknecht-Seijas N, Estrada-Vazquez C, Ponce-Noyola MT, Salazar-Montoya JA (2013) The influence of total solids content and initial pH on batch biohydrogen production by solid substrate fermentation of agroindustrial wastes. J Environ Manag 128:126–137

Rockwood DL, Naidu CV, Carter DR, Rahmani M, Spriggs TA, Lin C, Alker GR, Isebrands JG, Segrest SA (2004) Short-rotation woody crops and phytoremediation: opportunities for agroforestry. Agrofor Syst 61:51–63

Rodríguez LA, Toro ME, Vazquez F, Correa-Daneri ML, Gouiric SC, Vallejo MD (2010) Bioethanol production from grape and sugar beet pomaces by solid-state fermentation. Int J Hydrog Energy 35(11):5914–5917
Saritha M, Arora A, Lata (2012) Biological pretreatment of lignocellulosic substrates for enhanced delignification and enzymatic digestibility. Indian J Microbiol 52(2):122–130
Sarkar N, Ghosh SK, Bannerjee S, Aikat K (2012) Bioethanol production from agricultural wastes: an overview. Renew Energy 37(1):19–27
Schievano A, D'Imporzano G, Adani F (2009) Substituting energy crops with organic wastes and agro-industrial residues for biogas production. J Environ Manag 90(8):2537–2541
Schönnenbeck C, Trouvé G, Valente M, Garra P, Brilhac JF (2016) Combustion tests of grape marc in a multi-fuel domestic boiler. Fuel 180:324–331
Singh B, Guldhe A, Rawat I, Bux F (2014) Towards a sustainable approach for development of biodiesel from plant and microalgae. Renew Sust Energ Rev 29:216–245
Singh B, Guldhe A, Singh P, Singh A, Rawat I, Bux F (2015) Sustainable production of biofuels from microalgae using a biorefinery approach. In: Kaushik G (ed) Applied environmental biotechnology: present scenario and future trends. Springer, New Delhi, pp 115–128
Singh P, Kumari S, Guldhe A, Misra R, Rawat I, Bux F (2016) Trends and novel strategies for enhancing lipid accumulation and quality in microalgae. Renew Sust Energ Rev 55:1–16
Talebnia F, Karakashev D, Angelidaki I (2010) Production of bioethanol from wheat straw: an overview on pretreatment, hydrolysis and fermentation. Bioresour Technol 101(13):4744–4753
Vassilev SV, Vassileva CG (2016) Composition, properties and challenges of algae biomass for biofuel application: an overview. Fuel 181:1–33
Vicente G, Bautista LF, Rodríguez R, Gutiérrez FJ, Sádaba I, Ruiz-Vázquez RM, Torres-Martínez S, Garre V (2009) Biodiesel production from biomass of an oleaginous fungus. Biochem Eng J 48(1):22–27

Phytoremediation of Heavy Metal-Contaminated Soil Using Bioenergy Crops

3

Ambuj Bhushan Jha, Amarendra Narayan Misra, and Pallavi Sharma

Abstract

Heavy metal contamination of soils affects large areas worldwide. Excessive amount of metals, whether essential or nonessential, adversely affects the health of wildlife, humans, and plants and makes the land unusable for agricultural production. Phytoremediation, a sustainable, environment-friendly, and potentially cost-effective technology, can be used to decontaminate heavy metal-contaminated land. Use of nonfood, dedicated bioenergy crops for remediation of heavy metal-polluted sites has the advantage that biomass produced can be used to generate bioenergy, a cheaper, safer, sustainable, and renewable energy source compared to fossil fuels, avoids direct competition with food, and uses land unsuitable for growing food crops. Identifying dedicated bioenergy crops suitable for a particular metal-contaminated land and strategies to increase their phytoremediation potential are important for the success of this approach. Some dedicated bioenergy crops including poplars (*Populus* spp.), willows (*Salix* spp.), elephant grass (*Miscanthus* × *giganteus*), castor bean (*Ricinus communis*), and switchgrass (*Panicum virgatum*) can tolerate high concentrations of heavy metal, accumulate metal, and grow well on contaminated lands. Phytoremediation potential of these crops can be further improved by the effective use of metal solubilizing agents, endophytic bacteria, and genetic engineering. A better understanding of the mechanisms of heavy metal uptake, translocation, accumu-

A.B. Jha
Crop Development Centre/Department of Plant Sciences, University of Saskatchewan, 51 Campus Drive, Saskatoon, SK S7N 5A8, Canada

A.N. Misra • P. Sharma (✉)
Centre for Life Sciences, Central University of Jharkhand, Brambe, Ranchi, Jharkhand 835205, India
e-mail: pallavi.sharma@cuj.ac.in

K. Bauddh et al. (eds.), *Phytoremediation Potential of Bioenergy Plants*, DOI 10.1007/978-981-10-3084-0_3 lation, and tolerance in normal and metal hyperaccumulator plants will help scientists to develop effective and economic transgenic bioenergy crops for remediation of heavy metals in soil.

Keywords
Bioenergy crops • Biomass • Heavy metals • Phytoremediation • Renewable energy

3.1 Introduction

Heavy metal-contaminated soil, a major problem worldwide, affects the health of humans, animals, and plants (Taiz and Zeiger 1998; Shah and Dubey 1998; Sharma and Dubey 2007). Among heavy metals, Pb, Cd, Cu, Hg, Ni, Al, Se, Zn, Cr, and As are most commonly found in soil and are known to exert their toxic symptoms (Scott and Smith 1981). Concentrations of heavy metals in soil vary from traces to 100,000 mg kg^{-1} (Blaylock and Huang 2000). Although, high concentrations of heavy metals are naturally present in some areas, discharge of heavy metals in the environment due to human activities is mainly responsible for continuous increase in soil metal contamination. Soil can get contaminated with excessive levels of heavy metals due to various processes including exploitation of natural resources, mining, metal industries, and farming including wastewater irrigation, power plants, urban traffic, garbage dumping, etc. In humans and animals, direct ingestion or contact with contaminated soil can create health problems. Also, heavy metals from contaminated soil can enter various domain of plant system through root absorption processes. Some metals might be essential nutrients in small doses, but in higher doses these are toxic (Claire et al. 1991; Fernandes and Henriques 1991; Maheshwari and Dubey 2007). Plants can accumulate metals from contaminated soil and pass toxic concentrations of these to the next trophic levels of the food chain (Harrison and Chirgawi 1989; Welch et al. 1993). Remediation of heavy metal-contaminated soils is required to minimize the metal-related health hazards, increase the land resource accessible for food production, and increase food security.

Conventional remediation technologies which are based on biological, physical, and chemical methods are complex. Although competent, these techniques are environmentally destructive, costly, time taking, produce high quantities of waste, and usually damaging to the natural soil milieu (Cunningham et al. 1995; Danh et al. 2009; Ahmadpour et al. 2012). Phytoremediation, the use of plants to decontaminate soils, sediments, and water, should be used for remediation of heavy metals as it is a sustainable, cost-effective, environmentally beneficial, and aesthetically advantageous approach (Cluis et al. 2004). Mechanisms by which plant remediate heavy metal-contaminated soils are phytoextraction, phytostabilization, phytovolatilization, and rhizofiltration. Among these phytoextraction and phytostabilization are the most reliable. Choice of the suitable phytoremediation strategy for heavy metal-contaminated sites depends on levels and characteristics of heavy metals, plant species, physical and chemical properties of soil, and climate (Salt et al. 1998; Laghlimi et al. 2015).

If bioenergy crops are used for phytoremediation, produced biomass could be used to generate bioenergy (biogas and biofuels) which is cheaper, safer, sustainable, and renewable energy source compared to fossil fuels. In addition, they can provide other advantages including control of erosion and establishment of wildlife habitats. Bioenergy based on food crops or using land and water that would otherwise go for food production will lead to higher prices and are risk to food security. These problems can be solved by using nonedible, dedicated energy crops like poplars (*Populus* spp.), willows (*Salix* spp.), elephant grass (*Miscanthus* × *giganteus*), and switchgrass (*Panicum virgatum*) that can accumulate and tolerate high levels of heavy metals and grow well on contaminated lands.

Research is going on to improve the capability of bioenergy crops to remove heavy metals from contaminated land. Metal solubilizing agent and symbiotic endophytic bacteria have been successfully used for enhanced removal of toxic metals from polluted sites (Schmidt 2003; Weyens et al. 2010; Khan and Dotty 2011; Li et al. 2012). Genes involved in the uptake, transport, and detoxification of heavy metals from microbes, plants, and animals have been identified and utilized successfully to develop transgenic plants with improved capability to decontaminate metal-polluted sites (Song et al. 2007; Park et al. 2012). This review focuses on bioenergy crops, their use in phytoremediation of heavy metals and strategies to increase their phytoremediation potential. A greater understanding of the mechanism by which these bioenergy crops tolerate and hyperaccumulate heavy metal in their parts will lead to development of bioenergy crops with high biomass and increased phytoremediation potential.

3.2 Bioenergy Crops

Dwindling fossil fuels, increasing energy prices, and consumption and climatic changes due to emission of greenhouse gas have focused the world attention on the need to find low-cost, safer, and more renewable energy source which can substitute petroleum-based fuel. Bioenergy, a form of renewable energy derived from recently living biological sources such as wood, crops, or animal waste, is a promising alternative. Bioenergy crops include dedicated energy crops and agricultural coproducts that can be used to produce bioenergy. These crops act as thermochemical energy storage systems by collecting solar energy. They can reduce greenhouse gas emissions and thus play an important role as environmentally safe renewable energy supplier globally (Karp and Shield 2008). At present, biomass produced from bioenergy crops is the largest potential source of renewable energy that currently provides around 10% of world's primary energy supply. It has been estimated that by 2050, bioenergy crops can produce up to 273–1381 EJ/yr (Smeets et al. 2007).

Characteristics that make a particular plant species desirable as potential bioenergy crop are fast growth, large volume of biomass, high planting density and harvest frequency, high biotic and abiotic stresses tolerance, and low input requirements.

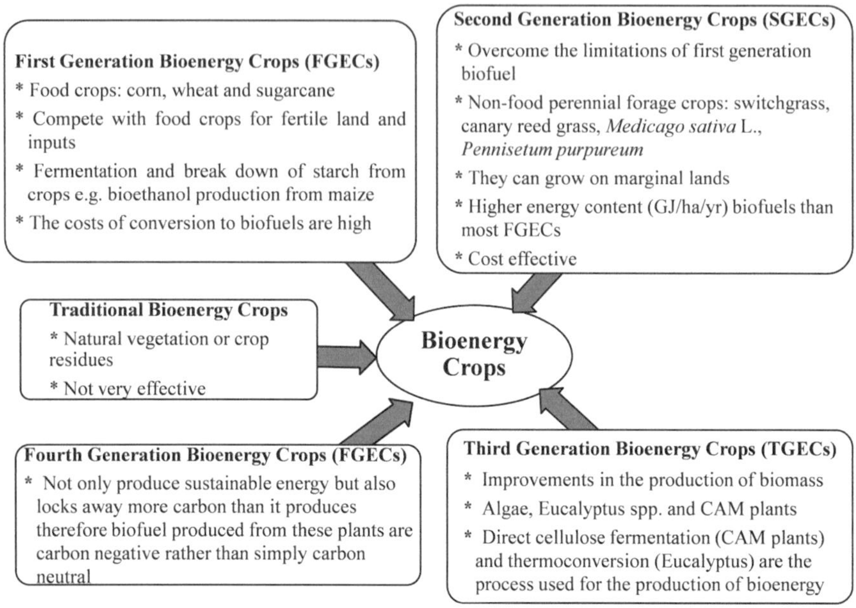

Fig. 3.1 Classification of bioenergy crops

Bioenergy crops can be grouped into annuals and perennials. Annual energy crops include rapeseed (*Brassica napus* L. and *B. carinata*), fiber sorghum (*S. bicolor* L. Moench), sweet sorghum (*Sorghum bicolor* L. Moench), and kenaf (*Hibiscus cannabinus* L.), whereas perennial energy crops include *Miscanthus*, switchgrass, reed canary grass (*Phalaris arundinacea*), cordgrass (*Spartina pectinata*), wheat (*Triticum aestivum*), maize (*Zea mays*), sugar beet (*Beta vulgaris*), cardoon (*Cynara carduncul*us), giant reed (*Arundo donax* L.), poplars (*Populus*), willows (*Salix*), and eucalyptus (*Eucalyptus camaldulensis* Dehnh. and *E. globulus* Labill.) (Zabaniotou et al. 2008; Simpson et al. 2009; Gomes 2012).

Bioenergy crops are further classified on the source of biomass used for bioenergy production (Fig. 3.1). Traditional bioenergy crops include natural vegetation and plant debris which are not very efficient. First-generation bioenergy crops (FGECs) include food crops such as wheat, corn, sugarcane, and rapeseed. As they can also be used for food, they compete with food crops for fertile soil and resources creating problems of high food prices and threat to food security. In crops such as maize, fermentation and breakdown of starch lead to production of bioethanol, whereas transesterification of plant oils present in soybean produces biodiesel. The costs of converting some of these energy crops to biofuels are high. Second-generation bioenergy crops (SGECs) such as switchgrass, canary reed grass, *Medicago sativa* L., and *Pennisetum purpureum* were developed to overcome the limitations of first-generation biofuels. They can grow on marginal lands and produced biofuels from lignocellulosic biomass. SGECs are cost effective, environment friendly, and generate more energy than FGECs. Third-generation bioenergy crops (TGECs) include algae, *Eucalyptus* spp., and crassulacean acid metabolism (CAM) plants which are based on improvements in the production of biomass. Algae can be used to produce different kinds of fuels including biodiesel, gasoline, petrol, and jet fuel. Direct cellulose fermentation (CAM plants) and thermoconversion (*Eucalyptus*) are the processes used for the production of bioenergy. Fourth-generation bioenergy crops not only produce sustainable energy but also fix more carbon than it produce; therefore, biofuels produced from these plants are carbon negative rather than simply carbon neutral. They also reduce CO_2 emissions by replacing fossil fuels. Due to limited land and water availability, raising biomass for energy production conflicts with food production leading to higher prices and food insecurity. These problems can be resolved by growing nonedible dedicated bioenergy crops (DECs) on wastelands and contaminated lands (Tandon et al. 2013). DEC crops include cellulosic biomass producing short-rotation trees and shrubs such as poplar, willow, birch (*Betula* spp.), eucalyptus, and perennial grasses such as giant reed, reed canary grass, switchgrass, elephant grass, and nonedible oil crops such as castor (*Ricinus communis*), physic nut (*Jatropha curcas*), and pongamia (*Pongamia* spp.). Nonfood dedicated bioenergy crops that can grow on marginal and contaminated lands and provide high yields of reliable and sustainable biomass with low production costs are highly desirable in order to fulfill the growing bioenergy needs. Among these crops, switchgrass, elephant grass, poplar, willow, mesquite, etc. are regarded most important.

Willow shrubs or trees are found mostly in north-temperate regions. They are a good candidate for bioenergy crop as well as phytoremediation. Globally, around 450 willow species are found. *Salix viminalis* is one of the most widely used species. Erect stems, ability to grow rapidly, and extensive root development are the traits that make bush *Salix* suitable for biomass coppice (Ahman and Larsson 1994). Willow can be easily propagated vegetatively from dormant cuttings. It resprouts after harvests which are frequently done by coppicing. Willow yield up to 10–15 dry t ha^{-1} $year^{-1}$ has been reported (Riddell-Black 1993). Willow grown for 21–22 years under nitrogen fertilization produced 140 t ha^{-1} dry biomass, whereas under nonfertilized condition, 86 t ha^{-1} yields were observed (González-García et al. 2012). Biomass yield of willow can be increased using genetic engineering, traditional breeding, and hybridization. Various forms of biofuel including solid, liquid, and gaseous ones are efficiently generated from the lignocellulosic biomass of willow. The net energy ratio associated with willow varies from 10 to 13 (Keoleian and Volk 2005). Greenhouse gas emissions are much less in comparison to carbon dioxide produced. Further, it also reduces NOx, SO_2, and particulate emissions (Keoleian and Volk 2005).

Poplar, another promising bioenergy crop, is the general term for trees such as cottonwoods (poplars) and aspens. Most species of poplars are native to the temperate and cold northern hemisphere. Different species of poplars are cross-fertilized to form hybrids. In areas where two crossable species overlap, hybrid formation occurs naturally. Poplars, due to large mass yield, fast growth, extended and deep root apparatus, short-rotation coppice (SRC) cultures, easy propagation, and high adaptability to a variety of soils and climate, are promising candidates for bioenergy crop as well as phytoremediation (Rockwood et al. 2004). It is typically harvested after 5–10 years. The lignocellulosic biomass present in poplar may be converted into biofuels which is used to generate heat energy, electric energy, and transport fuel (Gross et al. 2003; Vaezi et al. 2012). In vitro methods can be used to propagate poplar. As whole genome sequence data is available for poplar (Tuskan et al. 2006), they can be conveniently used in genetic engineering programs (Confalonieri et al. 2003).

Jatropha curcas (L.), also known as physic nut, belongs to Euphorbiaceae family which originated in Central America. It is now found dispersed in tropical regions of Asia as well as Africa (Openshaw 2000). It is a nonedible oilseed crop and is one the most promising bioenergy crops. It is a small tree and can grow between 6.0 and 18.0 ft. It has high growth rates and biomass yields. It is drought tolerant and grows well in low to high rainfall and different types of soils including infertile, gravelly, and sandy soils (Dagar et al. 2006) but is susceptible to freezes. It can grow in sands to clay loams with pH range of 5.5–9.0 (Foidl et al. 1996). However, irrigation and fertilization are required for commercial yields of *Jatropha*. Oil yields of *Jatropha* are similar or more than other oilseed crops such as soybean and rapeseed but lower than castor. It yields 25–40% oil by weight (Deng et al. 2010; Heller 1996). *Jatropha* seeds contain 21% saturated and 79% unsaturated fatty acids with oleic acid, linoleic acid, palmitic acid, and stearic acid as major fatty acids (Gubitz et al. 1999). *Jatropha* oil is transesterified to generate biodiesel which is quite similar to diesel fuel. Therefore it can directly replace/supplement conventional fuel needed in cooking, machines, and small diesel engines (Cerrate et al. 2006; Ndong et al. 2009). As *Jatropha* oil is a little less viscous than the diesel, it passes easily via the injector (Brittaine and Lutaladio 2010). Twenty percent mixing of biodiesel (B20) by 2012 was targeted by the National Mission on Biofuels (India). *Jatropha* was aimed to be cultivated on 400,000 ha of marginal land (Rajagopal 2007). The growth behavior, metal accumulation, translocation, and bioconcentration indicated that *Jatropha* could be a good candidate to remediate metal-polluted sites (Mangkoedihardjo and Surahmaida 2008; Marrugo-Negrete 2015). Biodiesel obtained from bioenergy crop *Jatropha* shows higher flash point and emits 100% lower SO_2 and 80% less CO_2 in comparison to petroleum diesel (Tiwari et al. 2007).

Castor, which is a perennial C3 oilseed plant, is now considered promising alternative source of bioenergy (Olivares et al. 2013). It belongs to Euphorbiaceae family and is native of tropical Asia and Africa. Castor, with large biomass, faster growth, and good yields, is a good candidate for crop rotation and companion planting. New castor hybrids and varieties have been developed which show early maturity (150–210 days) and yield (1500–1800 kg ha^{-1}) even in rainfed situations (Rao et al. 2003; Lavanya and Mukta 2008). Castor has well-developed root system and can be grown in a variety of environments across vast geographical regions including marginal lands which are considered unsuitable for production of food crops. High amounts of oil and biomass were obtained when castor was grown on marginal lands and irrigated with wastewater (Berman et al. 2011; Chatzakis et al. 2011). Compared to *Jatropha*, oil content (47–49%) in castor seed is much higher (Weiss 2000; Kumar et al. 2008). Although nonedible for humans or animals, castor oil has many applications in industry (Olivares et al. 2013). Castor oil can be used to produce environment-friendly biodiesel which is biodegradable, nontoxic, and has low emission properties (Meher et al. 2006). Crude castor oil is approximately 100-fold more viscous than diesel fuel (Conceicao et al. 2007; Bello and Makanju 2011). Transesterification significantly reduces the viscosity. It consists of about 90% ricinoleic acid which is responsible for the solubility of castor oil in alcohol at 30 °C facilitating transesterification without heating and thus lowers the cost of production (Da Silva et al. 2006). Castor fatty acid methyl esters should be blended 10 or 20% in volume (B10 or B20) with petrodiesel (Berman et al. 2011).

Miscanthus belongs to Poaceae family and is native to eastern Asia (Hastings et al. 2008). *M. sacchariflorus*, *M. sinensis*, and *M. × giganteus*, the sterile triploid hybrid generated from a cross between these species are the most important *Miscanthus* species for bioenergy (Karp and Shield 2008). It is considered a good candidate for bioenergy crop due to high production of biomass under low input conditions (Clifton-Brown et al. 2004). High yield of *Miscanthus* biomass is largely due to its efficient C4 photosynthesis pathway which is retained even at low temperature. In *M. × giganteus*, maintenance of pyruvate phosphate dikinase (PPDK) and RuBisCO large subunit amounts was reported to be critical to retain enhanced rates of C4 photosynthesis at low temperature (Naidu et al. 2003). *Miscanthus* spreads naturally via rhizomes, but some species can also propagate by seeds. It can be harvested annually and has a lifetime of 20–25 years (Lewandowski et al. 2003). Extended plantations of *Miscanthus* sequester carbon in soil (Hansen et al. 2004). *Miscanthus* produces lignocellulosic biomass. In an experiment performed in Europe, *Miscanthus* provided one of the highest yields of energy and one of the lowest energy costs of production (Sims et al. 2006). *Miscanthus* species with high lignin content such as *M. × giganteus* and *M. sacchariflorus* show properties appropriate for thermochemical conversion to biofuels, whereas species with low lignin content such as *M. sinensis* were suitable for biochemical conversion (Arnoult and Brancourt-Hulmel 2015). Various experiments were done in order to identify and specify metal remediation properties of *Miscanthus* (Arduini et al. 2006; Bang et al. 2015).

Growing interest in biomass crops for energy production has focused attention on switchgrass, a high-yielding perennial C4 grass native to North America (Mehdi 2000; Casler and Boe 2003). Switchgrass is now regarded as one of the most promising bioenergy crops. It is a contender for ethanol making and fuel supply for power generation due to its high cellulosic content. The benefits of using switchgrass include high biomass yield, perennial life form, tremendous soil conservation traits, suitability to conventional agricultural practices, and adaptability to marginal soils (McLaughlin et al. 1999, 2005). The plant is an erect, immense biomass producer and its height ranges between 1.5 and 8 ft. It has a highly efficient photosynthesis pathway and typically produces 10–12 dry t ha^{-1} on agricultural soils in average growing conditions. Switchgrass is drought resistant and can grow well in sands to clay loams with pH values of 4.5–7.6. Roots of these grasses are capable of reaching 10 ft under the soil. Two general morphological types are recognized (Newman et al. 2008). The lowland type adapted to poorly drained soils has tall, coarse stems, whereas upland type is more drought tolerant and has short, fine stems. It can grow on soils of moderate fertility without fertilizing (Parrish and Fike 2005). Even with limited additions of fertilizer, switchgrass can maintain productivity. In the establishment year, switchgrass requires little or no nitrogen as it can efficiently utilize native nitrogen supply of soil. However, nitrogen fertilizers may be needed to sustain high yields under multiple-harvest systems (Pedroso et al. 2011). Switchgrass is established from seeds. Switchgrass shows tolerance to metals and also has the potential for extraction of these contaminants. Therefore, growth of switchgrass on metal-contaminated land could provide biomass and make such lands suitable for agricultural production.

3.3 Heavy Metals and Their Remediation Using Bioenergy Crops

Heavy metals including Cd^{2+}, Pb^{2+}, Al^{3+}, Zn^{2+}, Ni^{2+}, Cu^{2+}, Zn^{2+}, As^{3+}, and Hg^{2+} are common environmental pollutants and are released in the atmosphere by industrial effluents, mining, repeated use of metal-containing pesticides or fertilizers, and urban runoff (Foy et al. 1978; di Toppi and Gabbrielli 1999; Morera et al. 2001). Heavy metals generally have densities > 5 g cm^{-3}. Arsenic and Al with partial metal characteristics and densities <5 g cm^{-3} are also considered in this context because of the toxic effects these metals have on plants (Sharma and Dietz 2006).

Seventeen metals with importance for both the organism and ecosystems have been recognized. They are generally available in the vicinity of living cells (Weast 1984). Heavy metals, Fe, Mo, and Mn, are regarded as micronutrients, whereas Zn, Ni, Cu, V, Co, W, and Cr are considered as trace elements. These metals are toxic at high concentrations. As, Hg, Ag, Sb, Cd, Pb, and U are toxic to all plants and microorganisms and have no known nutrient function (Breckle 1991; Nies 1999; Jha et al. 2004 a, b, c; Jha et al. 2005; Sharma and Dubey 2007; Mishra et al. 2011).

Heavy metals present in higher concentrations in the soil exert adverse effects on soil biota, plants, animals, and human health. Various mechanisms have been proposed for heavy metal lead damage. These include enhanced production of ROS and accumulation of methylglyoxal production, direct interaction with proteins because of their affinities for thionyl, histidyl, and carboxyl groups, and substitution of essential metal ions from specific binding sites, making it nonfunctional (Schützendübel and Polle 2002; Sharma and Dietz 2006). Heavy metals can be categorized as redox active such as Cu, Cr, Co, and Fe and redox inactive such as Ni, Cd, Zn, Al, etc. As redox-active heavy metals directly take part in redox reactions, they lead to production of $O_2^{\bullet-}$ and afterward H_2O_2 and $^{\bullet}OH$ via Haber-Weiss and Fenton reactions in plants (Dietz et al. 1999, Schützendübel and Polle 2002). Heavy metals which are redox inactive also cause oxidative damage indirectly by disturbing electron transport chain, enhancing lipoxygenase (LOX) activity which in turn increases lipid peroxidation and interaction with the antioxidant defense system (Hossain et al. 2012). Strong binding of heavy metals with O, N, and S atoms (Nieboer and Richardson 1980) and substitution of one heavy metal ion (cofactor) by another cause decrease in enzyme activities. Heavy metals cause oxidation and cross-linking of –SH groups in proteins leading to inhibition of important membrane protein including H^{+}-ATPase. Cd has been reported to bind to –SH group of structural proteins and enzymes leading to misfolding and decline in activity (DalCorso 2008, Hall 2002). In radish, calmodulin-dependent phosphodiesterase activity was inhibited when Ca^{2+} was displaced by Cd^{2+} in calmodulin, a key protein in cellular signaling (Rivetta et al. 1997). Similarly, displacement of Mg^{2+} by divalent cations such as Co^{2+}, Ni^{2+}, and Zn^{2+} resulted in a loss of activity of ribulose-1,5-bisphosphate-carboxylase/oxygenase (RuBisCO) (Wildner and Henkel 1979; Van Assche and Clijsters 1986). In plants, impaired nutrient uptake, changes in cell ultrastructure, reduced growth, and productivity have been reported (Pahlsson 1989; Trivedi and Erdei 1992; Sresty and Rao 1999).

Categorization of metals on the basis of their affinity with biological molecules was provided by Nieboer and Richardson (1980). They divided metals into three classes. Metals with affinity for O-bearing ligands were placed in class A, whereas metals with affinity for N- or S-bearing ligands were placed in class B. Metals with affinity for O-, N-, as well as S-bearing ligands with apparent likings were placed in class C (Nieboer and Richardson, 1980). Cadmium, lead, and mercury show a strong affinity for ligands such as phosphates, cysteinyl and histidyl side chains of proteins, purines, pteridines, and porphyrins. Therefore, these metals can interact with various proteins or enzymes containing –SH groups. Metals can bind and change the conformation of nucleic acids and affect the oxidative phosphorylation pathway. Specific reaction relies on the characteristic of the particular metal (Fodor 2002). Cadmium is known to specifically inhibit chlorophyll biosynthesis by its interference with the sulfhydryl site (Prasad and Strzalka 1999).

Heavy metals do not undergo biodegradation. Once a soil becomes contaminated with metal, it continues to be a long-term source of heavy metal toxicity. Therefore, remediation of heavy metal-polluted land is required to decrease the health hazards, make the land accessible for food production, and increase food security. Phytoremediation which utilizes plants for remediating heavy metal from contaminated soil could be a sustainable environment-friendly and potentially cost-effective technology. Use of nonedible dedicated bioenergy crops for heavy metal phytoremediation programs will not only generate energy but will also reduce the pressure to produce energy crops on land that would otherwise be used to produce food crops and make contaminated sites available for agriculture (Fig. 3.2). As detrimental effect of heavy metal varies with the properties of metal as well as plant species, several studies have been focused on identifying the bioenergy crops suitable for a particular metal-contaminated land which is important for the success of this approach (Table 3.1).

3.3.1 Willow

Willow is a rapidly growing woody plant that generates high biomass in short time. Its biomass is used as a source for biofuels including bioethanol (Punshon and

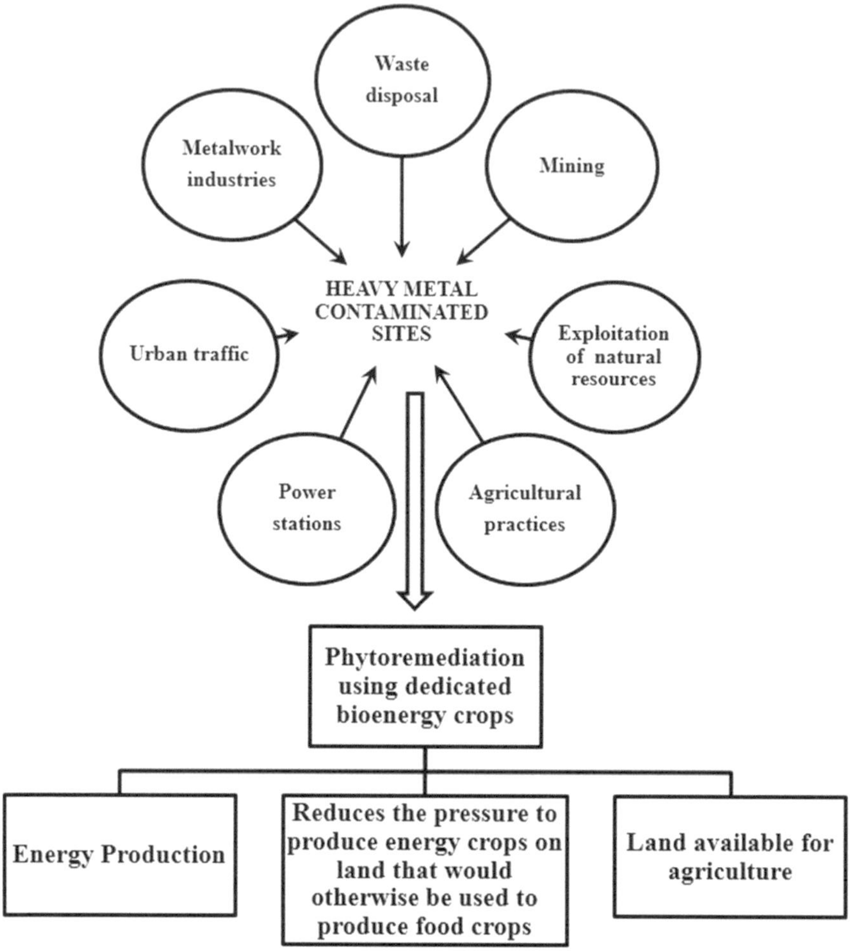

Fig. 3.2 Sources of heavy metal contamination and use of bioenergy crops for phytoremediation of contaminated sites

Dickinson 1997; Pulford and Watson 2003; Brereton et al. 2010). Besides high biomass productivity, willow trees have capacity for Cd, Cr, Cu, Ni, Pb, Mn, Fe, Zn, As, and Al uptake and are suitable for phytoremediation purposes. In a 90-day pot trial, around 30% of the bioavailable Cd was successfully removed by willow (Greger 1999). Many other studies have demonstrated support to the study that willow can clean up heavy metal-polluted soil within a few years. To estimate the phytoremediation efficiency of five willow species, Kuzovkina et al. (2004) grew willow plants under Cu and Cd. They observed that willow species were highly resistant to Cu and Cd, and among tested species, *S. nigra* was most resistant to both metals. Substantial disparity in survival, biomass yield, and metal uptake was reported in 20 different willow species planted on soils contaminated with Cd, Zn,

Table 3.1 Phytoremediation of heavy metals using bioenergy crops

Bioenergy crops	Metal(s)	Observations	Reference
Willow (*Salix*) and poplar (*Populus*)			
5 *Salix* spp.	Cu and Cd	*S. nigra* showed highest tolerance to both Cu and Cd and also exhibited highest accumulation levels of metals	Kuzovkina et al. (2004)
20 varieties of *Salix* spp.	Cd, Zn, Cu, and Ni	Eleven *Salix* varieties found promising for phytoremediation	Pulford et al. (2002)
16 *Salix* clones	Cu	Selected *Salix* clones had potential for reclamation and restoration of contaminated soils	Punshon et al. (1995)
8 *Salix* clones	Cd and Zn	Zwarte Driebast showed high phytoremediation potential	Slycken et al. (2013)
2 *Salix* spp.	Cd, Cr, Cu, Ni, Pb, Mn, Fe, and Zn	*S. viminalis* as well as *S. fragilis* showed high concentrations of Cd and Zn in aerial parts and Cu, Cr, Pb, Fe, Mn, and Ni mostly in the roots	Vandecasteele et al. (2005)
5 *Salix* clones	As	Clone 99202–011 showed highest As-tolerance and aboveground As accumulation	Purdy and Smart (2008)
9 *Salix* spp.	Cd, Cu, and Zn	*S. caprea* was most amenable to resistance acclimation	Punshon and Dickinson (1997)
3 *Salix* spp.	Fe, Zn, and Pb	High metal removing potential of *Salix* spp.	Ali et al. (1999)
S. viminalis	Cu, Zn, and Cr	Extraordinary tolerance of *S. viminalis* to even relatively high levels of heavy metal ions (Cu, Zn, Cr) in water solution was observed	Łukaszewicz et al. (2009)
5 *Salix* clones and 5 *Populus* clones	Cu, Zn, B, Fe, and Al	*Populus* exhibited the greatest uptake of Cu, whereas *Salix* showed the highest uptake of Zn, B, Fe, and Al	Zalesny and Bauer (2007)
6 clones of *Salix* spp.	Hg	*Salix* has potential role in removing metal contaminants	Wang and Greger (2004)
Populus and *Salix*	Cd, Zn, As, Pb, Cu, and Ni	*Populus* along with *Salix* and *Alnus* were useful to reduce Cd and Zn contamination	French et al. (2006)
SRC of *Populus* and *Salix*	Cd	*Populus* and *Salix* can be used on metal-contaminated soils	Ruttens et al. (2011)
Two Italian *Populus* clones	Zn, Cu, Cr, and Cd	Increasing plant growth and metal accumulation in both clones	Sebastiani et al. (2004)

(continued)

Table 3.1 (continued)

Bioenergy crops	Metal(s)	Observations	Reference
Three *Populus* species along with *Salix*	As, Co, Cu, Pb, and Zn	Phytostabilization and effective immobilization can be attained for these metals	Vamerali et al. (2009)
Hybrid willow and hybrid poplar	Pb and As	Promising phytoremediation potential	Hinchman et al. (1996)
Jatropha			
J. curcas	Cd	*J. curcas*, tolerant to Cd, can be used for phytoremediation of Cd-polluted soils	Marques et al. (2013)
J. curcas	Fe, Al, Cr, Cu, and Mn	Useful in extraction of heavy metals (Fe, Al, Cr, Cu, and Mn) from fly ash	Jamil et al. (2009)
J. curcas	Cd and Pb	Useful for phytoremediation of Pb- and Cd-contaminated sites	Mangkoedihardjo and Surahmaida (2008)
J. curcas, *Pterocarpus indicus*	Cr	Jatropha was more effective than *Pterocarpus* to remediate hexavalent Cr-polluted soil	Mangkoedihardjo et al. (2008)
J. curcas	As, Cr, and Zn	Potential use in phytoremediation	Yadav et al. (2009)
Castor bean (*Ricinus communis*)			
23 genotypes of *R. communis*	Cd	Can be used for eliminating Cd from polluted soils	Huang et al. (2011)
R. communis	Cu, Zn, Mn, Pb, and Cd	Potential role of this plant for phytostabilization of metals	Olivares et al. (2013)
R. communis	Pb	Good candidate for phytoextraction of Pb	Romeiro et al. (2006)
R. communis	Cd	Potential role in Cd phytoextraction	Hadi et al. (2015)
R. communis	Cd	*R. communis* can remove greater amount of Cd from contaminated soil	Bauddh and Singh (2012)
R. communis	Ni	Ni accumulator and promising for remediation of Ni-polluted soils	Adhikari and Kumar (2012)
R. communis	Cu	Strong tolerance of castor bean for Cu with high accumulation in root cell	Kang et al. (2015)
R. communis	Cu	Potential plant for Cu remediation of polluted soil	Andreazza et al. (2013)
R. communis cv. Guarany	As	Could not qualify for commercial phytoextraction; however, it could be suitable to revegetate As-polluted areas with its usefulness for oil production	Melo et al. (2009)

(continued)

Table 3.1 (continued)

Bioenergy crops	Metal(s)	Observations	Reference
R. communis	Zn, Pb, and Cd	Can be used to remediate metal-polluted sites	Angelova et al. (2016)
Grasses			
Arundo donax	Zn, Cr, and Pb	Suitable for Zn phytoextraction	Barbosa et al. (2015)
Miscanthus× *giganteus*, *M. sinensis*, and *M. floridulus*	Zn	*M.* × *giganteus* could be most suitable for Zn phytoextraction	Barbosa et al. (2015)
Miscanthus sp. Goedae-Uksae 1	As, Cu, Pb, Ni, Cd, and Zn	Goedae-Uksae 1 removed all the metals from contaminated soil as well as water except As from contaminated water	Bang et al. (2015)
M. x giganteus	Pb and Zn	Potential crop for Pb and Zn phytoremediation	Pogrzeba et al. (2011, 2013)
M. sinensis x giganteus	Pb and Cd	Can be used profitably in metal-polluted soils	Barbu et al. (2010)
A. donax and *M. sacchariflorus*	Zn and Cr	Both exhibited high tolerance and accumulation ability for Zn/Cr, and hence, they can be used for phytoremediation of Zn-/Cr-polluted sites	Li et al. (2014)
M. × *giganteus* and *Sida hermaphrodita*	Zn and Pb	*Miscanthus* showed greater ability to phytoextract Zn and Pb from the contaminated soil than mallow	Kocoń and Matyka (2012)
Spartina pectinata Link ("Savoy") and *Panicum virgatum* L. ("Cave-In-Rock" (CIR))	Cd	Savoy has the higher tolerance, translocation, and Cd accumulation capabilities than Cave-In-Rock	Zhang et al. (2015)

Cu, and Ni (Pulford et al. 2002). Results indicated that 11 out of these 20 species had high survival, biomass yield, and metal uptake and thus could be a potential candidate for phytoremediation. The enhanced uptake and transfer of Cd and Zn from root to shoot by willow from the soil contaminated with these metals suggested phytoextraction technology, whereas high level of Cu and Pb in roots and low translocation rate indicated the phytostabilization method for remediation of these heavy metals. Similarly, Punshon et al. (1995) also reported significant differences in root length increase, lateral roots, and metal uptake with improved resistance to Cu toxicity in various British willows. van Slycken et al. (2013) evaluated short-rotation coppice of willow in metal-polluted agricultural soils and observed that even at low biomass production this plant could remove 72 g Cd and 2.0 kg Zn ha^{-1} $year^{-1}$, and this can be enhanced by 40% if leaves are used too. Two willow clones, *S. fragilis* "Belgisch Rood" and *S. viminalis* "Aage," grown in six sediment-derived soils were evaluated for growth and metal (Cd, Cr, Cu, Ni, Pb, Mn, Fe, and Zn) uptake in a greenhouse experiment (Vandecasteele et al. 2005). No inhibition in growth of both clones makes them suitable for phytoextraction. Cd and Zn accumulated in aboveground plant parts, whereas Cu, Cr, Pb, Fe, Mn, and Ni were observed typically in the roots. Hydroponic culture of four willow clones showed large differences in As uptake and tolerance and potential use of *Salix* spp. for As phytoremediation (Purdy and Smart 2008). Several other experiments also showed higher uptake, accumulation, and tolerance of various species of *Salix* for Cd, Cu, and Zn (Punshon and Dickinson 1997); Fe, Zn, and Pb (Ali et al. 1999); Cu, Zn, and Cr (Łukaszewicz et al. 2009); Zn, Fe, and Al (Zalesny and Bauer 2007); and Hg (Wang and Greger, 2004), suggesting potential role of *Salix* in removing metal contaminants. Annual uptake of Cd content of willow stems in 12 different stands on 5 locations in central Sweden showed that willow can decrease the soil Cd content far less than the natural level (Ostman 1994). Eriksson and Ledin (1999) also suggested that *Salix* plantation can decrease the quantity of soil Cd accessible to plant. They estimated a period of time of 12–20 years to decrease Cd to half of its original value of 600 g ha^{-1} Cd in topsoil, if stem-wood production is 10 t ha^{-1} $year^{-1}$.

3.3.2 Poplar

Poplars are considered good candidates for bioenergy crop as well as phytoremediation due to their large mass yield, fast growth, extended and deep root apparatus, and short-rotation coppice cultures (Rockwood et al. 2004; Pulford and Watson 2003). They are being used to clean up various contaminants from contaminated soil and groundwater. In comparison to hyperaccumulators, metal uptake by trees is less efficient. However, due to greater biomass yield, trees can lead to more effective removal of metals from soil (Fischerová et al. 2006; Mench et al. 2009; Vangronsveld et al. 2009). Short-rotation coppices (SRC) of *Populus* along with *Salix* and *Alnus* were suggested to be useful to decrease Cd and Zn pollution in the region of Liverpool and St. Helens in NW England (French et al. 2006). Further, it reduced the exposure of less mobile contaminants, As, Pb, Cu, and Ni, as lability of these elements was not augmented 3 years after planting the coppice. Ruttens et al. (2011) suggested possible use of SRC of poplars and willows on an acid- and metal-polluted soil in an experimental field in the Campine, Belgium, which is not only advantageous for soil remediation but also for biomass production for energy purpose. As it would take long time to attain the legal threshold values for metals, it was suggested that this sort of land utilization can provide economical support to local farmers. The effects of heavy metals (Zn, Cu, Cr, and Cd) on two poplar clones generally planted in Italy showed increased growth and accumulation of metal in these plants (Sebastiani et al. 2004). Further, in both clones phytoextraction and phytostabilization strategies were witnessed.

The growth of three poplar species, *P. alba*, *P. nigra*, and *P. tremula*, along with *Salix* was monitored at a site in Udine, Italy, contaminated with As, Co, Cu, Pb, and Zn from wastes obtained from mineral roasting for sulfur extraction (Vamerali et al. 2009). It was suggested that phytostabilization and effective immobilization can be attained for these metals by soil improvement along with growth of woody biomass species. Growth of poplar along with gray willow and common alder as short-rotation biomass on two reclaimed coal tips in the Cynon Valley of South Wales showed very high rate of survival (Steer and Baker 1997) suggesting that these crops could be useful for biomass production and betterment of the environment. To evaluate the Pb and As remediation potential of hybrid poplar and hybrid willow, Hinchman et al. (1996) grew these plants in greenhouse in contaminated soil obtained from a field as well as in quartz sand supplements with solutions of Pb and As. Both plants exhibited uptake and partitioning of Pb and As in the sand media as well as in the polluted soil. The willow plants had removed about 9.5% Pb and 1% As from the metal-polluted soil, whereas the poplars removed about 1% Pb and 0.1% As in 1 month. In the sand experiment, the willows removed about 40% and 30–40% of supplied Pb and As, respectively.

3.3.3 *Jatropha*

Jatropha, one of the most promising oilseed crops, has superior biomass yields for bioenergy production. It is also now being used for remediation of metal-polluted sites. In order to assess their remediation capability, effect of Cd on photosynthetic pigments, the mineral composition of plants, soluble proteins, and enzymes related to defense mechanism was studied in *Jatropha* (Marques and do Nascimento 2013). Results indicated that *J. curcas* can be used in phytoremediation of Cd-polluted soils as this species can grow on contaminated sites due to its relative tolerance to Cd. A greenhouse experiment performed by Jamil et al. (2009) indicated usefulness of *J. curcas* in extraction of heavy metals (Fe, Al, Cr, Cu, and Mn) from fly ash. It was observed that accumulation of Fe and Mn was greater in roots, whereas Cu, Al, and Cr were translocated mostly to the shoot. In another study, *J. curcas* demonstrated the best remediation ability for Cd, Cr, Ni, and Zn (Chang et al. 2014). Cd, Ni, and Zn were easily absorbed and transferred to aerial parts of *Jatropha*, whereas Cu, Cr, and Pb remained mostly in roots (Chang et al. 2014). Several other researchers have also indicated role of *J. curcas* for phytoremediation of As-, Cr-, Zn-, Cd-, and Pb-contaminated soil (Mangkoedihardjo and Surahmaida 2008; Mangkoedihardjo et al. 2008; Yadav et al. 2009).

3.3.4 Castor

Castor with large biomass and shorter growing period is a good candidate for bioenergy production. Phytoremediation capability of castor has been reported in several studies (Prasad and Freitas 2003; Huang et al. 2011; Bauddh and Singh 2012; de Souza Costa et al. 2012). Castor can grow in metal-polluted soils and accumulate metals; thus, it can be planted on metal-polluted site for phytoremediation and bioenergy production (Bauddh and Singh 2012; de Souza Costa et al. 2012). Experiment conducted by Huang et al. (2011) on 23 genotypes of *R. communis* showed huge differences in the uptake and accumulation of Cd in leaf, stem, and root in these genotypes in the Cd-contaminated soil. Results suggested that castor can be used for eliminating Cd from polluted soils due to rapid growth, large biomass, and strong absorption and accumulation of Cd. Similar experiment conducted in hydroponic condition also indicated potential role of this plant in the phytoextraction of Cd (Hadi et al. 2015). Further, Bauddh and Singh (2012) reported that castor can grow in wasteland soils and can remove greater amount of Cd from contaminated soil due to high biomass yield. The potential use of castor for energy crop and metal phytoremediation capability was studied in mine tailings in the arid region of Zimapan, Hidalgo state, Mexico. Mining started there in 1632, and tailings have elevated levels of Cu, Zn, Mn, Pb, and Cd (Olivares et al. 2013). Results indicated potential role of this plant for phytostabilization of metals. Further, plants grown in mine tailings had high oil yields (41–64%) and reduced human health risks in these areas. In greenhouse experiments, Romeiro et al. (2006) observed hyperaccumulation of Pb by castor and suggested that this plant might be a good candidate for phytoextraction of this metal. For phytoremediation purposes, castor (*R. communis* cv. Guarany) plant's ability to tolerate and grow in Cd- and Pb-polluted solutions was assessed (de Souza Costa et al. 2012). Cd contamination led to toxicity symptoms in root and shoot of plants, whereas Pb exerted no undesirable effect (de Souza Costa et al. 2012). It was proposed that castor can be utilized as indicator plant for Cd and tolerant for Pb and hence can be employed for phytoremediation of polluted soils (de Souza Costa et al. 2012). In a separate pot culture experiment, Adhikari and Kumar (2012) indicated that castor plant was Ni accumulator and promising for remediation of Ni-polluted soils. Similarly, Kang et al. (2015) reported a strong tolerance of castor for Cu with high accumulation in root cell. High growth and biomass with efficient Cu hyperaccumulation in polluted soils in Tonglu Mountain, Daye City of Hubei Province, China, established castor as a potential plant for Cu remediation of polluted soil (Andreazza et al. 2013). To evaluate phytoextraction capability for As, castor plants were grown in the presence of increasing concentration of As (0–5000 $\mu g\ L^{-1}$). Although decreased shoot and root biomass was observed, no severe toxicity symptom was detected (Melo et al. 2009). It was suggested that castor plant could not qualify for commercial phytoextraction; however, it could be suitable to revegetate As-polluted areas with its usefulness for oil production. Growth of castor on industrial sites polluted with Zn, Pb, and Cd near the nonferrous metalworks in Plovdiv, Bulgaria, showed potential use of this plant to remediate metal-polluted sites (Angelova et al. 2016).

3.3.5 Grasses

Grasses such as giant reed, *Miscanthus*, cordgrass, and switchgrass have been found useful both for bioenergy production and phytoremediation. *A. donax* was grown in soil under increasing concentrations of Zn, Cr, and Pb, whereas *Miscanthus* spp., *M.* × *giganteus*, *M. sinensis*, and *M. floridulus* were grown under Zn to see heavy metal tolerance and phytoremediation capacity of these plants (Barbosa et al. 2015). Metal accumulation and translocation were highest for Zn followed by Cr and Pb in giant reed; however, different phytoremediation potential was obtained in *Miscanthus* genotypes. Results indicated that *A. donax* and *M.* × *giganteus* could be most suitable for Zn phytoextraction. Further, *A. donax* and *Miscanthus* genotypes are important for phytostabilization of heavy metal contamination. Similar experiment was conducted by Bang et al. (2015) to assess the remediation capability of *Miscanthus* sp. Goedae-Uksae 1, hybrid, for six heavy metals (As, Cu, Pb, Ni, Cd, and Zn). Results indicated that Goedae-Uksae 1 removed all the metals from polluted soil as well as water except As from polluted water. In fact, Cd, Pb, and Zn were entirely removed from polluted water samples, whereas in case of soil, maximum removal (97.7%) was reported for As. A higher amount of Pb and Zn was accumulated in the tissues of *M.* × *giganteus* grown in polluted soils in Silesia, Poland, where 5–10% of agricultural lands are polluted with Pb, Cd, and Zn (Pogrzeba et al. 2011, 2013). *M.* × *giganteus* in contaminated soils can accumulate up to 700 mg Zn kg^{-1}, 200 mg Pb kg^{-1}, and 5 mg Cd kg^{-1} (Pogrzeba et al. 2011). These results suggested potential role of this crop for Pb and Zn phytoremediation. Similarly, Barbu et al. (2010) reported profitable use of *M. sinensis* × *giganteus* in the soils of Copsa Mica region, Romania, polluted with Pb and Cd.

Pot experiments conducted on two grasses, *A. donax* and *M. sacchariflorus*, grown under increasing concentrations of Zn (0–2000 mg kg^{-1}) and Cr (0–1000 mg kg^{-1}) showed strong tolerance of these grasses to Zn and Cr (Li et al. 2014). *A. donax* accumulated 17.5 mg Zn $plant^{-1}$ and 3.9 mg Cr $plant^{-1}$, whereas *M. sacchariflorus* accumulated 12.1 mg Zn $plant^{-1}$ and 2.9 mg Cr $plant^{-1}$. These grasses could be promising bioenergy crop for the phytoremediation because of strong tolerance and high capacity to accumulate metals. Similar experiment conducted by Kocoń and Matyka (2012) in Puławy, Poland, showed higher uptake and phytoextraction potential of *M.* × *giganteus* for Zn and Pb. They suggested growth of energy crop is economical and pro-ecological to clean metal-polluted environment. Studies of growth rate, Cd accumulation, translocation, and tolerance under different Cd concentrations on two warm season perennials, prairie cordgrass (*Spartina pectinata* Link, "Savoy") and switchgrass (*Panicum virgatum* L., "Cave-In-Rock" (CIR)), indicated their usefulness in phytoremediation under 10 mg L^{-1} concentration (Zhang et al. 2015). Among these two species, Savoy showed higher tolerance, translocation, and accumulation abilities than Cave-In-Rock and could be useful for phytoremediation. Switchgrass has also shown great potential for phytoremediation of Pb- and Cr-contaminated soil. Cr concentration of 872.5 mg kg^{-1} was found in roots of switchgrass when grown on heavy Cr-polluted soil (600 mg kg^{-1}).

3.4 Strategies to Increase Phytoremediation Potential of Bioenergy Crops

Phytoremediation by bioenergy crops can suffer from several limitations including low biomass yields, depth of soil that plant can remediate, low uptake and translocation rate of metal from root to shoot, and low availability of metal at particular time. A high biomass plant with an ability to accumulate high content of toxic

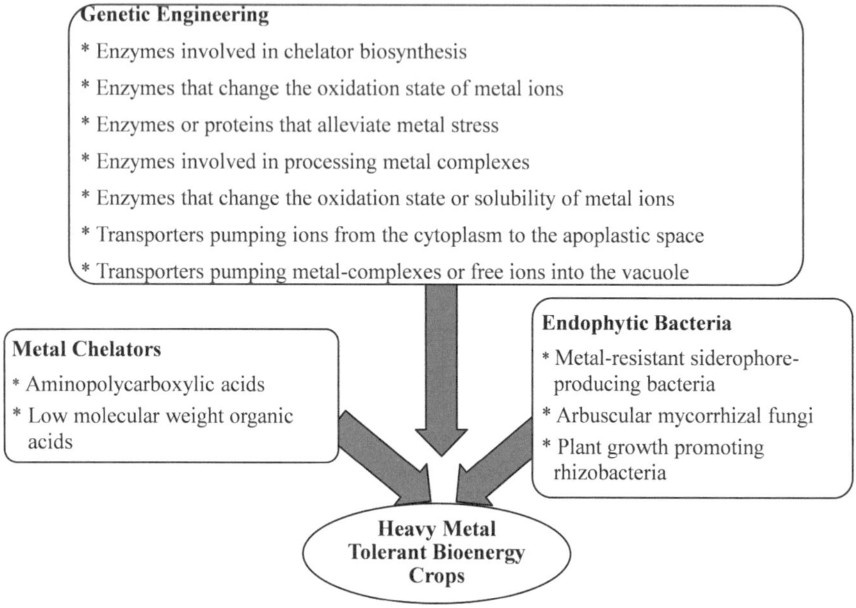

Fig. 3.3 Strategies to increase phytoremediation potential of bioenergy crops

metals is desirable for the success of phytoremediation. Effective use of metal solubilizing agents, prevention of leaching, use of endophytic bacteria, and production of new plant materials having high biomass and capacity to accumulate high levels of toxic metals, through breeding and transgenic approach, can be used to increase the phytoremediation potential of bioenergy crops (Fig. 3.3). Various strategies that can be promising when used alone or in combination in bioenergy crops are discussed in this section.

3.4.1 Metal Solubilizing Agent

The crucial factor regulating the effectiveness of phytoextraction is the availability of heavy metals to plant roots (Felix 1997). Many soil characteristics including cation exchange capacity, pH, organic matter content, as well as type of metal species present in soil affect the phytoavailability of metal for uptake by plant. Application of chelating agents to soil is one of the strategies to improve the metal accumulation capacities and uptake efficiency of plant. Persistent aminopolycarboxylic acids (APCAs) such as ethylenediaminetetraacetic acid (EDTA) as well as biodegradable APCAs, ethylenediamine disuccinate (EDDS), and nitrilotriacetic acid (NTA) have been successfully employed in many phytoextraction experiments (reviewed in Evangelou et al. 2007). Low molecular weight organic acids (LMWOA) such as citric and tartaric acid also increase phytoextraction capacity of plants by enhancing the solubility of heavy metals in soil. Addition of EDTA to polluted soils resulted in more tolerance and up to 100 times increase of Pb concentration in bioenergy crops like willow, Indian mustard, corn, and sunflower (Schmidt 2003). Citric acid led to increased concentration of uranium, whereas inorganic agents like elemental sulfur or ammonium sulfate led to increased concentrations of cadmium and zinc (Schmidt 2003). However, addition of these agents can result in increased mobility of heavy metals in soils. Therefore, there is possibility of leaching of these metals in the soil. Reduction of leaching will be critical for the success of this strategy (Evangelou et al. 2007). As rate of degradation of these agents has a direct effect on the leaching possibility, it should be studied in detail. Further the fate of these agents and its toxicity to plants and soil microorganisms should also be studied in detail for success of this strategy.

3.4.2 Symbiotic Endophytic Microorganisms

The success of the heavy metal phytoremediation using bioenergy crops relies on sufficient yield of crops and on the efficient uptake and translocation of metals from the roots to shoots. Growing evidence suggests that in addition to weather and soil characteristics, plant-microbe interactions also regulate the metal phytoextraction competence of host plants (Khan and Dotty 2011; Li et al. 2012). Symbiotic endophytic microorganisms, which play an important role in host plant adjustment to contaminated environments, have been used to increase phytoremediation potential of plant by improving growth, producing highly branched root systems, enhancing metal tolerance, and mobilizing/degrading or immobilizing pollutants in the soil (Chen et al. 2010; Weyens et al. 2010). Various heavy metal tolerance mechanisms such as exclusion, active removal, biosorption, precipitation, or accumulation both in external and intracellular spaces have been identified in a variety of microbes (Haferburg and Kothe 2007; Harrison et al. 2007). These methods affect the solubility and the accessibility of the metal to the plant and hence modify the harmful effects of the metal.

Among the rhizosphere microorganisms, arbuscular mycorrhizal fungi (AMF) and plant growth-promoting rhizobacteria (PGPR) have been studied for their use in phytoremediation process. These microorganisms increased plant growth, protected it from harmful effects of metals, and enhanced plant uptake of trace elements (Weller and Thomashow 1994; Glick 1995; Sheng and Xia 2006). However, in some studies they reduced or had no effect on the metal uptake (Rajkumar et al. 2006; Madhaiyan et al. 2007). Arora et al. (2015) examined the bioremediation capability of bioenergy crop switchgrass together with plant-associated microbes (AMF and *Azospirillum*) for Pb and found that these microbes enhanced the length, number of branches, surface area, and biomass of roots. Shoot biomass also increased. The lower values of bioconcentration factor and tolerance index were obtained revealing high retention of Pb and Cd in the roots and limited transfer of Pb and Cd to shoots.

Metal-resistant endophytes which are present in a variety of hyperaccumulator plants contribute in successful survival and development of these plants on heavy metal-contaminated soils. They stimulate plant development by a variety of mechanisms including nitrogen fixation, solubilization of minerals, phytohormone and siderophore synthesis, use of 1-aminocyclopropane-1-carboxylic acid as the only N source, and alteration of supplement components (Glick et al. 1999; RajKumar et al. 2009). Inoculation of bioenergy crop *S. viminalis* plants with a *Rahnella* sp. strain increased extraction potential due to a considerably enhanced twig biomass (Janssen et al. 2015). Rajkumar and Freitas (2008) reported enhanced Zn accumulation in shoot tissues of *R. communis* when *Pseudomonas jessenii* was added to the surface-sterilized root of this crop. Further, the plant-associated bacteria *Pseudomonas koreensis* AGB-1, isolated from roots of bioenergy plant *M. sinensis* growing in mine tailing soil, demonstrated high tolerance to heavy metals and plant growth-promoting characters. Extracellular sequestration of heavy metals by AGB-1 was revealed using transmission electron microscope analysis. It was observed as dark metal complexes on AGB-1 surfaces and exterior of the cells. *M. sinensis* biomass, chlorophyll, and protein content were increased by 54%, 27%, and 28%, respectively, when mining site soil was inoculated with AGB-1. Metal(loid) levels in roots and shoots of plants grown in inoculated soil were also more compared to controls indicating that AGB-1 in combination with *M. sinensis* can be used to help phytostabilization and remediation of heavy metal-polluted sites (Babu et al. 2015).

Plants having extensive network of fine roots with large surface area are desirable for effective absorption and accumulation of heavy metals. A soil bacterium, *Agrobacterium rhizogenes*, can stimulate the root growth proliferation (hairy roots) via plasmid (Ri T-DNA) insertion into plant genome. It has been shown to enhance the metal uptake and root biomass in various hyperaccumulator plants (Metzger et al. 1992; Hughes et al. 1997; Nedelkoska and Doran 2000). As hairy roots can grow in vitro without addition of exogenous growth regulators, the problem of possible inhibitory activity of their phytohormones can be solved. Some endophytes can increase the growth of host plants and thus enhance phytoremediation by producing indole-3-acetic acid, cytokinins, and/or gibberellins (Feng et al. 2006; Oelmüller et al. 2009; Hamayun et al. 2010). Ethylene is a key phytohormone which regulates the plant growth (Ping and Boland 2004). Overproduction of ethylene promoted by stresses has been reported to reduce plant developmental processes, such as root elongation, lateral root growth, and root hair formation (Mayak et al. 2004). A number of endophytes can enhance plant growth by producing 1-aminocy clopropane-1-carboxylate (ACC) deaminase which hydrolyzes ACC, an ethylene precursor to modulate the ethylene levels in plants (Hardoim et al. 2008; Chen et al. 2010; Ma et al. 2011; Zhang et al. 2011).

Besides changing the phytohormone concentrations, a number of endophytes can enhance plant growth by nitrogen (N) fixation, supply of mineral nutrition, and enhancing tolerance to biotic and abiotic stresses (Ryan et al. 2008). Many bioenergy crops including maize, wheat, poplar, and grass are colonized by nitrogen fixation endophytes ensuring sufficient N to the plant and enhance plant growth (Hurek and Reinhold-Hurek 2003; Doty et al. 2009; Rajkumar et al. 2009). Kuklinsky-Sobral et al. (2004) found that mineral phosphate can be solubilized by 52% of the endophytic bacteria isolated from soybean plants. Plant-associated bacteria directly enhance phytoextraction by producing organic acids, chelators, and/or changing redox status thus varying the solubility, accessibility, and transport of heavy metal and nutrients. Siderophore-producing bacteria (SPB) which are metal resistant contribute in the successful survival and development of plants in polluted soils by removing metal from polluted sites. They mitigate the metal toxicity and provide plant with nutrients, specifically Fe. Siderophores, low-molecular-mass iron chelators, have high association constants for binding Fe. These chelators can bind with other metals including Cd, Cu, Ga, Al, In, Pb, and Zn (Miethke and Marahiel 2007). The formation of complex between siderophore and metal enhances its efficient uptake by increasing the concentration of soluble metal (Rajkumar et al. 2010). Under iron limitation, siderophore makes iron available to plants. Siderophore-producing bacterial strain led to enhanced growth of bioenergy crop *B. juncea* growing under Cr stress but Cr uptake capability of plants was not influenced (Rajkumar et al. 2005). Some endophytes secrete low-molecular-mass organic acids and thus increase heavy metal mobilization. For instance, significant increase in the water-soluble Pb and decrease in pH of the solution were observed with endophyte growth (Sheng et al. 2008). It was suggested that organic acids released by the endophytes could be responsible for this (Sheng et al. 2008). Endophytic diazotroph *Gluconacetobacter* releases 5-ketogluconic acid which solubilizes a variety of Zn sources, such as ZnO, $ZnCO_3$, or $Zn_3(PO_4)_2$, and hence makes it accessible for plant uptake (Saravanan et al. 2007). Actinobacterium endophytes also release some metal-mobilizing metabolites that can mobilize Zn and/or Cd and hence increased the metal content in *S. caprea* leaves (Kuffner et al. 2010). Interestingly, some studies revealed that metal-resistant endophytes reduce metal uptake and accumulation (Lodewyckx et al. 2001; Huo et al. 2012). Whether endophytes enhance or decrease metals uptake and accumulation depends principally on the levels of the metals in the soil.

3.4.3 Genetic Engineering

Plants survive in presence of high levels of metals probably by various detoxifying mechanisms including exclusion, sequestration, chelation, and compartmentalization of metals. Mechanism for uptake, transport, and sequestration is specific for each metal in plants (Eapen and D'Souza 2005; Mello-Farias and Chaves 2008). Genetic and molecular techniques are being used to identify gene families and candidate genes in plants that are involved in these processes. Genes reported to be involved in these processes and have been utilized to develop transgenic plants with high phytoremediation capability are *gshI*, *gshII*, *CAX-2*, *AtNramps*, *AtPcrs*, *CAD1*, *PCS*, *gst*, and *Nt CBP4* (Ha et al. 1999; Zhu et al. 1999; Thomine et al. 2000; Heiss et al. 2003; Song et al. 2004). These genes can also be expressed/overexpressed in bioenergy crops to enhance their phytoremediation capability.

An understanding of uptake, translocation, and regulatory control mechanism in common and metal hyperaccumulators plants will help scientists to develop effective and economic transgenic plants for remediation of metals in soil (Chaney et al. 2000; Fulekar et al. 2008). Different engineering approaches have been proposed to increase metal uptake in plants (Clemens et al. 2002). These comprise increasing the number of uptake and intracellular binding sites and change of specificity of metal uptake system in order to decrease competition by other cations. Using functional complementation in yeast, it was demonstrated that ZNT1 from *T. caerulescens* roots (Zn transporter gene) participate in high-affinity uptake of Zn^{2+} and low-affinity uptake of Cd^{2+} (Pence et al. 2000). Therefore, ZNT1 can be expressed in bioenergy crops to enhance the uptake of these metals. Several classes of proteins including natural resistance-associated macrophage protein (Nramp) family, cysteine-proline-serine (CPx)-type ATPases, and cation diffusion facilitator (CDF) family proteins have been shown to be involved in transport of metals in plants (Williams et al. 2000). Conserved intramembranous motif cysteine-proline-serine, cysteine-proline-cysteine, or cysteine-proline-histidine present in CPx-type metal ATPases are known to be involved in the transport of essential as well as potentially toxic metals like Pb, Cu, Zn, and Cd across cell membranes (Williams et al. 2000). Candidate genes for improvement of heavy metal resistance such as *ScYcf1* and *ScYHL035C* (MRP) subfamily of ABC transporter, *ScPdr13*, a heat shock protein 70 (*HSP70*), and *AtPcr1* have been identified by yeast mutant screening and *Arabidopsis* cDNA library screening (Song et al. 2007). In budding yeast, *ScYCF1* (yeast cadmium factor 1) encodes a transporter that sequesters toxic metals/metalloids into the vacuoles. Compared to the non-transgenic plants, the poplar plants overexpressing YCF1 showed increased growth, higher Cd level in the aerial tissue, and reduced toxicity symptoms. Higher levels of Cd, Zn, and Pb were also observed to be accumulated in the root. These results suggest that in highly polluted sites, where survival of wild plant is not possible, YCF1-expressing transgenic poplar may be useful for phytostabilization and phytoattenuation (Shim et al. 2013). In *Arabidopsis thaliana*, two ABCC-type transporters, AtABCC1 and AtABCC2, that participate in As detoxification are known. Increased Cd(II) tolerance and accumulation were also observed in Arabidopsis plant overexpressing AtABCC1, whereas hypersensitive phenotypes were observed in *atabcc1* single or *atabcc1 atabcc2* double knockout mutants in the presence of Cd(II) and Hg(II). These results revealed that these vacuolar transporters participate in Cd and Hg tolerance along with their involvement in As detoxification. These transporters offer important tools for developing plants with desirable characteristics for phytoremediation (enhanced metal tolerance and accumulation) using genetic engineering (Park et al. 2012).

AtPcr1 is a plant protein which reduces Cd(II) in the cell. Comparison of growth of the transgenic poplar plant harboring AtPcr1 and wild-type plant in Cd- and Pb-contaminated media led to conclusion that expression of this gene enhances heavy metal resistance and has potential for phytoremediation of Cd- and Pb-polluted sites (Song et al. 2007). Phytochelatins (PCs) are metal-binding thiol-reactive peptides that are synthesized posttranslationally in plants and constitute a major pathway of metal detoxification (Gasic and Korban 2007; Mishra et al. 2011). Indian mustard plant expressing an *Arabidopsis thaliana AtPCS1* gene showed higher tolerance to Cd and As. Shoots of PCS plants with Cd showed considerably higher levels of PCs and thiols and lower concentration of Cd compared to wild-type plants. Accumulation of As in transgenic plants was comparable to that in wild-type plants. In spite of the fact that the capacity of Indian mustard to tolerate Cd and As was enhanced by PCs, accumulation capability of these metals in the aerial tissues couldn't be increased (Gasic and Korban 2007). Metallothionein (MT), a small cysteine-rich protein, participates in cellular detoxification of metals by sequester ing metal ions. JcMT2a from *J. curcas* has been shown to participate in Cu and Cd homeostasis and therefore might be a potential source to be utilized in genetic engineering approaches for phytoremediation. For instance, expressing a gene coding for a metallothionein-like protein in transgenic white poplar enhanced its tolerance to $CuCl_2$ in in vitro culture (Balestrazzi et al. 2009).

Modification of oxidative stress-related enzymes glutathione-S-transferase and peroxidase could also result in altered metal tolerance (Ezaki et al. 2000; Eapen and D'Souza 2005). Phytoremediation capabilities of four poplar lines including *P. nigra*, *P. canescens*, and two transgenic *P. canescens* clones were examined using in vitro leaf disk cultures (Bittsánszky et al. 2005). An elevated level of glutathione was observed in the transgenic poplars overexpressing a bacterial gene encoding gamma-glutamylcysteine synthetase (γ-ECS) in the cytosol or in the chloroplasts. Different concentrations of $ZnSO_4$ applied on leaf disks of poplar clones for 21 days revealed that Zn (2+) was phytotoxic to transgenic *P. canescens* lines only at high concentrations (0.1–1 M) of Zn, whereas *P. nigra* was sensitive line. Elevated heavy metal uptake was observed in transgenic poplars than non-transformed clones. These observations revealed that transgenic poplars are more appropriate for phytoremediation of Zn-polluted sites (Bittsánszky et al. 2005). Similarly, overexpression of *γ-ECS* gene in Indian mustard plants (Zhu et al. 1999) led to accumulation of 2.8 times more Se in leaves in comparison to wild type, whereas accumulation was 2.3-fold in transgenic plant overexpressing glutathione synthetase gene (GS; Liang et al. 1999). Under field conditions, transgenic Indian mustard lines overexpressing genes encoding the enzymes adenosine triphosphate sulfurylase (APS), ECS, and GS were tested for their ability to remediate Se from Se- and B-polluted saline sediment. Plants overexpressing APS, ECS, and GS accumulated 4.3, 2.8, and 2.3 times more Se in their leaves compared to wild type, respectively. These observations suggested that these genetically engineered plants can be successfully utilized for phytoremediation under field conditions (Bañuelos et al. 2005). GS plants significantly tolerated the polluted soil better and attained an aboveground biomass/area almost 80% compared to 50% for wild-type plants (Bañuelos et al. 2005). In Indian mustard, overexpression of selenocysteine methyltransferase (SMT) or selenocysteine lyase led to increased capability for Se phytoremediation under field conditions (Bañuelos et al. 2007). Up to nine times more Se accumulation was observed in Indian mustard plants in which both *APS* and the *SMT* gene from the Se hyperaccumulator, *Astragalus bisulcatus*, were co-expressed (LeDuc et al. 2006). Similarly, overexpression of ACC deaminase resulted in an increased level of various metals (Grichko et al. 2000; Eapen and D'Souza 2005) in plant and hence can be introduced in bioenergy crops to increase their metal remediation potential.

Using transgenic approach, genes associated with hyperaccumulation or phytovolatilization can be expressed in those plants in which these are not originally present. Introduction of mercuric ion reductase (*merA*) and organomercury lyase (*merB*) genes led to Hg tolerance in plants (Bizily et al. 2000; Dhankher et al. 2002; Eapen and D'Souza 2005). The bacterial *merA* and *merB* genes were simultaneously expressed in eastern cottonwood *P. deltoides*. The *merA*/*merB*-expressing plants displayed enhanced Hg tolerance and could detoxify organic Hg compounds three- to fourfold faster compared to controls (Lyyra et al. 2007). In the presence of 40 ppm Hg(II), *P. deltoids* trees, overexpressing modified *merA9* and *merA18* genes, showed higher biomass compared to control plants (Che et al. 2003). Within 2 weeks, control plants died but transgenic cottonwoods were still alive.

In addition to high metal-accumulating capability, candidate bioenergy crops for metal phytoremediation should have high biomass. Various molecular biology techniques can be used to identify genetic variation, explore canopy architecture, and study carbon allocation patterns that affect biomass yield. Overexpression of sucrose synthase (SUS) PvSUS1 in switchgrass enhances height, biomass, and tiller number up to 37, 13.6, and 79% compared to control plants. The overexpression of this gene in switchgrass and other bioenergy feedstocks might be a possible approach to enhance biomass and stacking with other genes that take part in uptake, translocation, and tolerance of metal, might lead to increased metal remediation potential (Poovaiah et al. 2015). Poplar plants which overexpress a cytosolic glutamine synthetase (GS1) under elevated N levels displayed enhanced biomass (6%); increased amount of chlorophylls, proteins, and total sugars; and higher N use efficiency mainly in young leaves in comparison to non-transformed controls (Castro-Rodríguez et al. 2016). In conclusion, transgenic plants carrying genes implicated in metal uptake, translocation, and tolerance will lead to increased metal remediation potential which will improve the plant's capacity to tolerate and accumulate heavy metals. Further, genes used for increasing growth and biomass production are also key resource for increasing metal remediation potential of plants (Castro-Rodríguez et al. 2016).

3.5 Concluding Remarks and Future Prospects

Use of biomass as a source of renewable energy can decrease reliance on petroleum fuels. Bioenergy based on food crops is a risk to food security through higher prices and use of land and water that would otherwise go for food production. Pressures on food supply can be reduced through technologies that make use of nonedible bioenergy crops growing on marginal lands. Dedicated energy crops that can provide reliable and sustainable biomass with high yields, low production costs, and competence to tolerate and accumulate high levels of heavy metals are highly desirable for this purpose. Among dedicated bioenergy crops, switchgrass, elephant grass, poplar, willow, mesquite, etc. appear to be the most promising. Biomass yield and capacity to accumulate and tolerate heavy metals can be enhanced using transgenic and breeding approaches. Use of metal-tolerant endophytic symbiont and metal chelating agents are also effective in increasing heavy metal extraction efficiency of these plants. More work needs to be done for effective prevention of leaching when using metal chelator and efficient conversion of heavy metal-contaminated biomass as biofuels. Lignocellulosic biomass which provides a natural resistance to its degradation is one of the main challenges for the second-generation biofuel industry. Simultaneous development of bioenergy crops with high yield, heavy metal phytoremediation potential, and bioconversion technologies that efficiently convert biomass into bioenergy are required for making marginal lands reusable for agricultural production and reducing dependency on fossil fuels.

Acknowledgments Financial Support from DBT Builder programme project No. BT/PR-9028/INF/22/193/2013 is gratefully acknowledged (ANM and PS). PS is thankful to UGC-FRP.

References

Adhikari T, Kumar A (2012) Phytoaccumulation and tolerance of *Ricinus communis* L. to nickel. Int J Phytoremediation 14:481–492

Ahmadpour P, Ahmadpour F, Mahmud TMM et al (2012) Phytoremediation of heavy metals: a green technology. Afr J Biotechnol 11:14036–14043

Åhman I, Larsson S (1994) Genetic improvement of willow (*Salix*) as a source of bioenergy. Nor J Agric Sci 18:47–56

Ali MB, Tripathi RD, Rai UN et al (1999) Physico-chemical characteristics and pollution level of Lake Nainital (U.P., India): role of macrophytes and phytoplankton in biomonitoring and phytoremediation of toxic metal ions. Chemosphere 39:2171–2182

Andreazza R, BortolonL PS et al (2013) Use of high-yielding bioenergy plant castor (*Ricinus communis* L.) as a potential phytoremediator for copper-contaminated soils. Pedosphere 23:651–661

Angelova V, Perifanova-Nemska M, Ivanov K (2016) Potential of castor (*Ricinus communis* L.) for phytoremediation of soils contaminated with heavy metals. World Academy of Science, Engineering and Technology, Int J Environ Ecol Eng 3:2016

Arduini I, Masoni A, Ercoli L (2006) Effects of high chromium applications on miscanthus during the period of maximum growth. Environ Exp Bot 58:234–243

Arnoult S, Brancourt-Hulmel M (2015) A review on Miscanthus biomass production and composition for bioenergy use: genotypic and environmental variability and implications for breeding. Biol Res 8:502–526

Arora K, Sharma S, Monti A (2015) Bio-remediation of Pb and Cd polluted soils by switchgrass: a case study in India. Int J Phytoremediation. doi:10.1080/15226514.2015.1131232

Babu AG, Shea PJ, Sudhakar D et al (2015) Potential use of *Pseudomonas koreensis* AGB-1 in association with *Miscanthus sinensis* to remediate heavy metal(loid)-contaminated mining site soil. J Environ Manag 151:160–166

Balestrazzi A, Botti S, Zelasco S et al (2009) Expression of the PsMTA1 gene in white poplar engineered with the MAT system is associated with heavy metal tolerance and protection against 8-hydroxy-2′-deoxyguanosine mediated-DNA damage. Plant Cell Rep 28:1179–1192

Bang J, Kamala-Kannan S, Lee KJ et al (2015) Phytoremediation of heavy metals in contaminated water and soil using *Miscanthus* sp. Goedae-Uksae 1. IntJ Phytoremediation 17:515–520

Banuelos G, Terry N, LeDuc DL et al (2005) Field trial of transgenic Indian mustard plants shows enhanced phytoremediation of selenium-contaminated sediment. Environ Sci Technol 39:1771–1777

Bañuelos G, LeDuc DL, Pilon-Smits EAH et al (2007) Transgenic Indian mustard overexpressing selenocysteine lyase or selenocysteine methyltransferase exhibit enhanced potential for selenium phytoremediation under field conditions. Environ Sci Technol 41:599–605

Barbosa B, Boléo S, Sidella S et al (2015) Phytoremediation of heavy metal-contaminated soils using the perennial energy crops *Miscanthus* spp. and *Arundo donax* L. Biol Res 8:1500–1511
Barbu CH, Pavel BP, Sand C et al. (2010) *Miscanthus sinensis x giganteus* cultivated on soils polluted with heavy metals- a valuable replacement for coal. Papers presented at Green Remediation Conference, University of Massachusetts Amherst, pp. 2–5, 15–17 June 2010
Bauddh K, Singh RP (2012) Cadmium tolerance and its phytoremediation by two oil yielding plants *Ricinus communis* (L.) and *Brassica juncea* (L.) from the contaminated soil. Int J Phytoremediation 14:772–785
Bello EI, Makanju A (2011) Production characterization and evaluation of castor oil biodiesel as alternative fuel for diesel engines. J Emerg Trends Eng Appl Sci 2:525–530
Berman P, Nisri S, Wiesman Z (2011) Castor oil biodiesel and its blends as alternative fuel. Biomass Bioenergy 35:2861–2866
Bittsánszky A, Kömives T, Gullner G et al (2005) Ability of transgenic poplars with elevated glutathione content to tolerate zinc (2+) stress. Environ Int 31:251–254
Bizily SJ, Rugh CL, Meagher RB (2000) Phytodetoxification of hazardous organomercurials by genetically engineered plants. Nat Biotechnol 18:213–217
Blaylock MJ, Huang JW (2000) Phytoextraction of metals. In: Raskin I, Ensley BD (eds) Phytoremediation of toxic metals: using plants to clean up the environment. Wiley, New York, pp 53–70
Breckle CW (1991) Growth under heavy metals. In: Waisel Y, Eshel A, Kafkafi U (eds) Plant roots: the hidden half. Marcel Dekker, New York, pp 351–373
Brereton NJ, Pitre FE, Hanley SJ et al (2010) Mapping of enzymatic saccharification in short rotation coppice willow and its independence from biomass yield. Biol Res 3:251–261
Brittaine R, Lutaladio N (2010) Jatropha: a smallholder bioenergy crop- the potential for pro-poor development. Integrated crop management, vol 8, Rome
Casler MD, Boe AR (2003) Cultivar x environment interactions in switchgrass. Crop Sci 43:2226–2233
Castro-Rodríguez V, García-Gutiérrez A, Canales J et al (2016) Poplar trees for phytoremediation of high levels of nitrate and applications in bioenergy. Plant Biotechnol J 14:299–312
Cerrate S, Yan F, Wang Z et al (2006) Evaluation of glycerine from biodiesel production as a feed ingredient for broilers. Int J Poult Sci 5:1001–1007
Chaney RL, Li YM, Brown SL et al (2000) Improving metal hyperaccumulator wild plants to develop phytoextraction systems: approaches and progress. In: Terry N, Banuelos G (eds) Phytoremediation of contaminated soil and water. Lewis Publishers, Boca Raton, pp 129–158
Chang FC, Ko CH, Tsai MJ et al (2014) Phytoremediation of heavy metal contaminated soil by *Jatropha curcas*. Ecotoxicology 23:1969–1978
Chatzakis MK, Tzanakakis VA, Mara DD et al (2011) Irrigation of castor (*Ricinus communis* L.) and sunflower (*Helianthus annus* L.) plant species with municipal wastewater effluent: Impacts on soil properties and seed yield. Water 3:1112–1127
Che D, Meagher RB, Heaton ACP et al (2003) Expression of mercuric ion reductase in Eastern cottonwood (*Populus deltoides*) confers mercuric ion reduction and resistance. Plant Biotechnol J 1:311–319
Chen L, Luo S, Xiao X et al (2010) Application of plant growth-promoting endophytes (PGPE) isolated from *Solanum nigrum* L. for phytoextraction of Cd-polluted soils. Appl Soil Ecol 46:383–389
Claire LC, Adriano DC, Sajwan KS et al (1991) Effects of selected trace metals on germinating seeds of six plant species. Water Air Soil Pollut 59:231–240
Clemens S, Palmgren MG, Krämer U (2002) A long way ahead: understanding and engineering plant metal accumulation. Trends Plant Sci 7:309–314
Clifton-Brown JC, Stampfl PF, Jones MB (2004) *Miscanthus* biomass production for energy in Europe and its potential contribution to decreasing fossil fuel carbon emissions. Glob Chang Biol 10:509–518

Cluis C (2004) Junk-Greedy Greens: phytoremediation as a new option for soil decontamination. BioTeach J 2:61–67

Conceicao MM, Candeia RA, Silva FC et al (2007) Thermochemical characterization of castor oil biodiesel. Renew Sust Energ Rev 11:964–975

Confalonieri M, Balestrazzi A, Bisoffi S et al (2003) *In vitro* culture and genetic engineering of *Populus* spp.: synergy for forest tree improvement. Plant Cell Tissue Organ Cult 72:109–138

Cunningham SD, Berti WR, Huang JW (1995) Phytoremediation of the contaminated soils. Trends Biotechnol 13:393–397

Da Silva NDL, Maciel MRW, Batistella CB et al (2006) Optimization of biodiesel production from castor oil. Appl Biochem Biotechnol 130:405–414

Dagar JC, Tomar OS, Kumar Y et al (2006) Performance of some under-explored crops under saline irrigation in a semiarid climate in northwest India. Land Degrad Dev 17:285–299

DalCorso G, Farinati S, Maistri S et al (2008) How plants cope with cadmium: staking all on metabolism and gene expression. J Integr Plant Biol 50:1268–1280

Danh LT, Truong P, Mammucari R et al (2009) Vetiver Grass, *Vetiveria zizanioides*: a choice plant for phytoremediation of heavy metals and organic wastes. Int J Phytoremediation 11:664–691

de Souza Costa ET, Guilherme LR, de Melo EE et al (2012) Assessing the tolerance of castor to Cd and Pb for phytoremediation purposes. Biol Trace Elem Res 145:93–100

Deng X, Fang Z, Liu YH (2010) Ultrasonic transesterification of *Jatropha curcas* L. oil to biodiesel by a two-step process. Energy Convers Manag 51:2802–2807

Dhankher OP, Li Y, Rosen BP et al (2002) Engineering tolerance and hyperaccumulation of arsenic in plants by combining arsenate reductase and gamma-glutamylcysteine synthetase expression. Nat Biotechnol 20:1140–1145

di Toppi LS, Gabbrielli R (1999) Response to cadmium in higher plants. Environ Exp Bot 41:105–130

Dietz KJ, Bair M, Krämer U (1999) Free radical and reactive oxygen species as mediators of heavy metal toxicity in plants. In: Prasad MNV, Hagemeyer J (eds) Heavy metal stress in plants from molecules to ecosystems. Spinger-Verlag, Berlin, pp 73–79

Doty SL, Oakley B, Xin G et al (2009) Diazotrophic endophytes of native black cottonwood and willow. Symbiosis 47:23–33

Eapen S, D'Souza SF (2005) Prospects of genetic engineering of plants for phytoremediation of toxic metals. Biotechnol Adv 23:97–114

Eriksson J, Ledin S (1999) Changes in phytoavailability and concentration of cadmium in soil following long term *Salix* cropping. Water Air Soil Pollut 11:171–184

Evangelou MWH, Ebel M, Schaeffer A (2007) Chelate assisted phytoextraction of heavy metals from soil. Effect, mechanism, toxicity, and fate of chelating agents. Chemosphere 68:989–1003

Ezaki B, Gardner RC, Ezaki Y et al (2000) Expression of aluminium induced genes in transgenic Arabidopsis plants can ameliorate aluminium stress and/or oxidative stress. Plant Physiol 122:657–665

Felix HRZ (1997) Field trials for in situ decontamination of heavy metal polluted soils using crops of metal-accumulating plants. Z Pflanzenernähr Bodenk 160:525–529

Feng Y, Shen D, Song W (2006) Rice endophyte *Pantoea agglomerans* YS19 promotes host plant growth and affects allocations of host photosynthates. J Appl Microbiol 100:938–945

Fernandes JC, Henriques FS (1991) Biochemical, physiological, and structural effects of excess copper in plants. Bot Rev 57:246–273

Fischerová Z, Tlustoš P, Száková J et al (2006) A comparison of phytoremediation capability of selected plant species for given trace elements. Environ Pollut 144:93–100

Fodor F (2002) Physiological responses of vascular plants to heavy metals. In: Prasad MNV, Strzalka K (eds) Physiology and biochemistry of metal toxicity and tolerance in plants. Kluwer Academic Publishers, Dordrecht, pp 149–177

Foidl N, Foidl G, Sánchez M et al (1996) *Jatropha curcas* L. as a source for the production of biofuel in Nicaragua. Bioresour Technol 58:77–82

Foy CD, Chaney RL, White MC (1978) The physiology of metal toxicity in plants. Annu Rev Plant Physiol 29:511–566
French CJ, Dickinson NM, Putwain PD (2006) Woody biomass phytoremediation of contaminated brownfield land. Environ Pollut 141:387–395
Fulekar MH, Singh A, Bhaduri AM (2008) Genetic engineering strategies for enhancing phytoremediation of heavy metals. Afr J Biotechnol 8:529–535
Gasic K, Korban SS (2007) Expression of *Arabidopsis* phytochelatin synthase in Indian mustard (*Brassica juncea*) plants enhances tolerance for Cd and Zn. Planta 225:1277–1285
Glick BR (1995) The enhancement of plant growth by free-living bacteria. Can J Microbiol 41:109–117
Glick BR, Patten CL, Holguin G et al (1999) Biochemical and genetic mechanisms used by plant growth promoting bacteria. Imperial College Press, London
Gomes HI (2012) Phytoremediation for bioenergy: challenges and opportunities. Environ Technol Rev 1:59–66
González-García S, Mola-Yudego B, Dimitriou I et al (2012) Environmental assessment of energy production based on long term commercial willow plantations in Sweden. Sci Total Environ 421-422:210–219
Greger M (1999) Metal availability and bioconcentration in plants. In: Prasad MNV, Hagemeyer J (eds) Heavy metal stress in plants. Springer-Verlag, Berlin, pp 1–27
Grichko VP, Filby B, Glick BR (2000) Increased ability of transgenic plants expressing the bacterial enzyme ACC deaminase to accumulate Cd, Co, Cu, Ni, Pb and Zn. J Biotechnol 81:45–53
Gross R, Leach M, Bauen A (2003) Progress in renewable energy. Environ Int 29:105–122
Gubitz GM, Mittelbach M, Trabi M (1999) Exploitation of the tropical oil seed plant *Jatropha curcas* L. Bioresour Technol 67:73–82
Ha SB, Smith AP, Howden R et al (1999) Phytochelatin synthase genes from Arabidopsis and the yeast *Schizosaccharomyces pombe*. Plant Cell 11:1153–1163
Hadi F, Ul Arifeen MZ, Aziz T et al (2015) Phytoremediation of cadmium by *Ricinus communis* L. in hydrophonic condition. Am-Eurasian J Agric Environ Sci 15:1155–1162
Haferburg G, Kothe E (2007) Microbes and metals: interactions in the environment. J Basic Microbiol 47:453–467
Hall JL (2002) Cellular mechanisms for heavy metal detoxification and tolerance. J Exp Bot 53:1–11
Hamayun M, Khan SA, Khan AL et al (2010) Exogenous gibberellic acid reprograms soybean to higher growth and salt stress tolerance. J Agric Food Chem 58:7226–7232
Hansen EM, Christensen BT, Jensen LS et al (2004) Carbon sequestration in soil beneath long-term Miscanthus plantations as determined by ^{13}C abundance. Biomass Bioenergy 26:97–105
Hardoim PR, van Overbeek LS, Elsas JD (2008) Properties of bacterial endophytes and their proposed role in plant growth. Trends Microbiol 16:463–471
Harrison RM, Chirgawi MB (1989) The assessment of air and soil as contributors of some trace metals to vegetable plants. Use of a filtered air growth cabinet. Sci Total Environ 83:13–34
Harrison JJ, Ceri H, Turner RJ (2007) Multimetal resistance and tolerance in microbial biofilms. Nat Rev Microbiol 5:928–938
Hastings A, Clifton-Brown JC, Wattenbach M et al (2008) Potential of Miscanthus grasses to provide energy and hence reduce greenhouse gas emissions. Agron Sustain Dev 28:465–472
Heiss S, Wachter A, Bogs J et al (2003) Phytochelatin synthase (PCS) protein is induced in *Brassica juncea* leaves after prolonged Cd exposure. J Exp Bot 54:1833–1839
Heller J (1996) Physic nut. *Jatropha curcas* L. Promoting the conservation and use of underutilized and neglected crops. 1. Institute of Plant Genetics and Crop Plant Research, Gatersleben/Intemational Plant Genetic Resources Institute, Rome
Hinchman RR, Negri MC, Gatliff EG (1996) Phytoremediation: using green plants to clean up contaminated soil, groundwater, and wastewater. In: Proceedings, International Topical Meeting on Nuclear and Hazardous Waste Management, Spectrum, 96, Seattle, WA, p 1–13, August 1996

Hossain MA, Piyatida P, da Silva JAT and Fujita M (2012) Molecular mechanism of heavy metal toxicity and tolerance in plants: central role of glutathione in detoxification of reactive oxygen species and methylglyoxal and in heavy metal chelation. J Bot Article ID 872875, 37 pages http://dx.doi.org/10.1155/2012/872875

Huang H, Yu N, Wang L et al (2011) The phytoremediation potential of bioenergy crop *Ricinus communis* for DDTs and cadmium co-contaminated soil. Bioresour Technol 102:11034–11038

Hughes JB, Shanks J, Vanderford M et al (1997) Transformation of TNT by aquatic plants and plant tissue cultures. Environ Sci Technol 31:266–271

Huo W, Zhuang C, Cao Y et al (2012) Paclobutrazol and plant-growth promoting bacterial endophyte *Pantoea* sp. enhance copper tolerance of guinea grass (*Panicum maximum*) in hydroponic culture. Acta Physiol Plant 34:139–150

Hurek T, Reinhold-Hurek B (2003) *Azoarcus* sp. strain BH72 as a model for nitrogen-fixing grass endophytes. J Biotechnol 106:169–178

Jamil S, Abhilash PC, Singh N et al (2009) *Jatropha curcas*: a potential crop for phytoremediation of coal fly ash. J Hazard Mater 172:269–275

Janssen J, Weyens N, Croes S et al (2015) Phytoremediation of metal contaminated soil using willow: exploiting plant-associated bacteria to improve biomass production and metal uptake. Int J Phytoremediation 17:1123–1136

Jha AB, Dubey RS (2004a) Arsenic exposure alters the activities of key nitrogen assimilatory enzymes in growing rice seedlings. Plant Growth Regul 43:259–268

Jha AB, Dubey RS (2004b) Carbohydrate metabolism in growing rice seedlings under arsenic toxicity. J Plant Physiol 161:867–872

Jha AB, Dubey RS (2004c) Effect of arsenic on nitrogen assimilatory enzymes in germinating rice seeds. Indian J Plant Physiol 9:438–441

Jha AB, Dubey RS (2005) Effect of arsenic on behaviour of enzymes of sugar metabolism in germinating rice seeds. Acta Physiol Plant 27:341–347

Kang W, Bao J, Zheng J et al (2015) Distribution and chemical forms of copper in the root cells of castor seedlings and their tolerance to copper phytotoxicity in hydroponic culture. Environ Sci Pollut Res 22:7726–7734

Karp A, Shield I (2008) Bioenergy from plants and the sustainable yield challenge. New Phytol 179:15–32

Keoleian GA, Volk TA (2005) Renewable energy from willow biomass crops: life cycle energy, environmental and economic performance. Crit Rev Plant Sci 24:385–406

Khan Z, Dotty S (2011) Endophyte-assisted phytoremediation. Curr Top Plant Biol 12:97–105

Kocon A, Matyka M (2012) Phytoextractive potential of *Miscanthus giganteus* and *Sida hermaphrodita* growing under moderate pollution of soil with Zn and Pb. J Food Agric Environ 10:1253–1256

Kuffner M, De Maria S, Puschenreiter M et al (2010) Bacteria associated with Zn and Cd-accumulating *Salix caprea* with differential effects on plant growth and heavy metal availability. J Appl Microbiol 108:1471–1484

Kuklinsky-Sobral J, Araujo WL, Mendes R et al (2004) Isolation and characterization of soybean-associated bacteria and their potential for plant growth promotion. Environ Microbiol 6:1244–1251

Kumar RV, Tripathi YK, Yadav V et al (2008) Oil percentage in *Jatropha curcas* L. germplasm of national agroforestry repository. Indian J Gen Plant Breed 68:463–466

Kuzovkina YA, Knee M, Quigley MF (2004) Cadmium and copper uptake and translocation in five willow (*Salix* L.) species. Int J Phytoremediation 6:269–287

Laghlimi M, Baghdad B, El Hadi H et al (2015) Phytoremediation mechanisms of heavy metal contaminated soils: a review. Open J Ecol 5:375–388

Lavanya C, Mukta N (2008) Varieties and hybrids of castor. Directorate of Oilseeds Research, Hyderabad, p. 88

LeDuc DL, Abdel Samie M, Montes-Bayon M et al (2006) Overexpressing both ATP sulfurylase and selenocysteine methyltransferase enhances selenium phytoremediation traits in Indian mustard. Environ Pollut 144:70–76

Lewandowski I, Scurlock J, Lindvall E et al (2003) The development and current status of perennial rhizomatous grasses as energy crops in the US and in Europe. Biomass Bioenergy 25:335–361
Li HY, Wei DQ, Shen M et al (2012) Endophytes and their role in phytoremediation. Fungal Divers 54:11–18
Li C, Xiao B, Wang QH, Yao SH, Wu JY (2014) Phytoremediation of Zn- and Cr-contaminated soil using two promising energy grasses. Water Air Soil Pollut 225:2027
Liang ZY, Pilon-Smits EA, Jouanin L et al (1999) Overexpression of glutathione synthetase in Indian mustard enhances cadmium accumulation and tolerance. Plant Physiol 119:73–80
Lodewyckx C, Taghavi S, Mergeay M et al (2001) The effect of recombinant heavy metal resistant endophytic bacteria in heavy metal uptake by their host plant. Int J Phytoremediat 3:173–187
Łukaszewicz JP, Wesołowski RP, Cyganiuk A (2009) Enrichment of *Salix viminalis* wood in metal ions by phytoextraction. Pol J Environ Stud 18:507–511
Lyyra S, Meagher RB, Kim T et al (2007) Coupling two mercury resistance genes in Eastern cottonwood enhances the processing of organomercury. Plant Biotechnol J 5:254–262
Ma Y, Prasad MNV, Rajkumar M et al (2011) Plant growth promoting rhizobacteria and endophytes accelerate phytoremediation of metalliferous soils. Biotechnol Adv 29:248–258
Madhaiyan M, Poonguzhali S, Sa TM (2007) Metal tolerating methylotrophic bacteria reduces nickel and cadmium toxicity and promotes plant growth of tomato (*Lycopersicon esculentum* L.). Chemosphere 69:220–228
Maheshwari R, Dubey RS (2007) Nickel toxicity inhibits ribonuclease and protease activities in rice seedlings: protective effects of proline. Plant Growth Regul 51:231–243
Mangkoedihardjo S, Surahmaida A (2008) *Jatropha curcas* L. for phytoremediation of lead and cadmium polluted soil. World Appl Sci J4:519–522
Mangkoedihardjo S, Ratnawati R, Alfianti N (2008) Phytoremediation of hexavalent chromium polluted soil using *Pterocarpus indicus* and *Jatropha curcas* L. World Appl Sci J 4:338–342
Marques MC, do Nascimento CWA (2013) Analysis of chlorophyll fluorescence spectra for the monitoring of Cd toxicity in a bio-energy crop (*Jatropha curcas*). J Photochem Photobiol B Biol 127:88–93
Marrugo-Negrete J, Durango-Hernández J, Pinedo-Hernández J et al (2015) Phytoremediation of mercury-contaminated soils by *Jatropha curcas*. Chemosphere 127:58–63
Mayak S, Tirosh T, Glick BR (2004) Plant growth-promoting bacteria confer resistance in tomato plants to salt stress. Plant Physiol Biochem 42:565–572
McLaughlin SB, Kszos LA (2005) Development of switchgrass (*Panicum virgatum*) as a bioenergy feedstock in the United States. Biomass Bioenergy 28:515–535
McLaughlin S, Bouton J, Bransby D et al (1999) Developing switchgrass as a bioenergy crop. In: Janick J (ed) Perspectives on new crops and uses. ASHS Press, Alexandria, pp 282–299
Mehdi B, Duxbury P, Samson R (2000) Development of bioenergy feedstocks: agronomy data from eastern Canada, final report. Resource Efficient Agricultural Production (REAP), Quebec
Meher LC, Vidya Sagar D, Naik SN (2006) Technical aspects of biodiesel production by transesterification- a review. Renew Sust Energ Rev 10:248–268
Mello-Farias PC, Chaves ALS (2008) Biochemical and molecular aspects of toxic metals phytoremediation using transgenic plants. In: Tiznado-Hernandez ME, Troncoso-Rojas R, Rivera Domínguez MA (eds) Transgenic approach in plant biochemistry and physiology. Research Signpost, Kerala, pp 253–266
Melo EEC, Costa ETS, Guilherme LRG et al (2009) Accumulation of arsenic and nutrients by castor plants grown on an As-enriched nutrient solution. J Hazard Mater 168:479–483
Mench M, Schwitzguebel JP, Schroeder P et al (2009) Assessment of successful experiments and limitations of phytotechnologies: contaminant uptake, detoxification and sequestration, and consequences for food safety. Environ Sci Pollut Res 16:876–900
Metzger L, Fouchault I, Glad C et al (1992) Estimation of cadmium availability using transformed roots. Plant Soil 143:249–257
Miethke M, Marahiel MA (2007) Siderophore-based iron acquisition and pathogen control. Microbiol Mol Biol Rev 71:413–451

Mishra S, Jha AB, Dubey RS (2011) Arsenite treatment induces oxidative stress, upregulates antioxidant system and causes phytochelatin synthesis in rice seedlings. Protoplasma 248:565–577

Morera MT, Echeverria JC, Garrido JJ (2001) Mobility of heavy metals in soils amended with sewage sludge. Can J Soil Sci 81:405–414

Naidu SL, Moose SP, AL-Shoaibi AK et al (2003) Cold tolerance of C4 photosynthesis in *Miscanthus x giganteus*: adaptation in amounts and sequence of C4 photosynthetic enzymes. Plant Physiol 132:1688–1697

Ndong R, Montrejaud-Vignolesk M, Saint Girons O et al (2009) Life cycle assessment of biofuels from *Jatropha curcas* in West Africa: a field study. GCB Bioenergy 1:197–210

Nedelkoska TV, Doran PM (2000) Hyperaccumulation of cadmium by hairy roots of *Thlaspi caerulescens*. Biotechnol Bioeng 67:607–615

Newman Y, Williams MJ, Helsel Z and Vendramini J (2008) Production of biofuel crops in Florida: Switchgrass. Publication #SS AGR 291, Series of the Agronomy Department, UF/IFAS Extension. http://edis.ifas.ufl.edu

Nieboer E, Richardson DHS (1980) The replacement of the nondescript term % heavy metals' by a biologically and chemically significant classification of metal ions. Environ Pollut B 1:3–26

Nies DH (1999) Microbial heavy-metal resistance. Appl Microbiol Biotechnol 51:730–750

Oelmuller R, Sherameti I, Tripathi S, Varma A (2009) *Piriformospora indica*, a cultivable root endophyte with multiple biotechnological applications. Symbiosis 49:1–17

Olivares AR, Carrillo-González R, González-Chávez MC et al (2013) Potential of castor (*Ricinus communis* L.) for phytoremediation of mine tailings and oil production. J Environ Manag 114:316–323

Openshaw K (2000) A review of *Jatropha curcas* L.: an oil plant of unfulfilled promise. Biomass Bioenergy 19:1–15

Ostman G (1994) Cadmium in *Salix* - a study of the capacity of *Salix* to remove cadmium from arable soils. In: Aronsson P, Perttu K (eds) Proceedings of a study tour, conference and workshop in Sweden, Swedish University of Agricultural Sciences Report 50, p 153–155 June 1994

Pahlsson AMB (1989) Toxicity of heavy metals (Zn, Cu, Cd, Pb) to vascular plants. Water Air Soil Pollut 47:287–319

Park J, Song WY, Ko D et al (2012) The phytochelatin transporters AtABCC1 and AtABCC2 mediate tolerance to cadmium and mercury. Plant J 69:278–288

Parrish DJ, Fike JH (2005) The biology and agronomy of switchgrass for biofuels. Crit Rev Plant Sci 24:423–459

Pedroso G, De Ben C, Hutmacher R et al (2011) Switchgrass is a promising, high-yielding crop for California biofuel. Cal Ag 65:168–173

Pence NS, Larsen PB, Ebbs SD et al (2000) The molecular physiology of heavy metal transport in the Zn/Cd hyperaccumulator *Thlaspi caerulescens*. Proc Natl Acad Sci U S A 97:4956–4960

Ping L, Boland W (2004) Signals from the underground: bacterial volatiles promote growth in *Arabidopsis*. Trends Plant Sci 9:263–266

Pogrzeba M, Krzyżak J, Sas-Nowosielska A et al. (2011) A heavy metal environmental threat resulting from combustion of biofuels of plant origin. In: Simeonov LI, Kochubovski M, Simeonova B (eds) Environmental heavy metal pollution and effects on child mental development- risk assessment and prevention strategies. Springer Science + Business Media BV, Dordrecht

Pogrzeba M, Krzyżak J, Sas-Nowosielska A (2013) Environmental hazards related to *Miscanthus x giganteus* cultivation on heavy metal contaminated soil. E3S Web Conf 1, 29006, p 1–4

Poovaiah CR, Mazarei M, Decker SR et al (2015) Transgenic switchgrass (*Panicum virgatum* L.) biomass is increased by overexpression of switchgrass sucrose synthase (PvSUS1). Biotechnol J 10:552–563

Prasad MNV, Freitas HMO (2003) Metal hyperaccumulation in plants- biodiversity prospecting for phytoremediation technology. Electron J Biotechnol 6:284–321

Prasad MNV, Strzalka K (1999) Impact of heavy metals on photosynthesis. In: Prasad MNV, Hagemeyer J (eds) Heavy metal stress in plants: from molecules to ecosystems. Spinger-Verlag, Berlin, pp 117–138

Pulford I, Watson C (2003) Phytoremediation of heavy metal-contaminated land by trees-a review. Environ Int 29:529–540
Pulford ID, Riddell-Black D, Stewart C (2002) Heavy metal uptake by willow clones from sewage sludge-treated soil: the potential for phytoremediation. Int J Phytoremediation 4:59–72
Punshon T, Dickinson NM (1997) Acclimation of *Salix* to metal stress. New Phytol 137:303–314
Punshon T, Lepp NW, Dickinson NM (1995) Resistance to copper toxicity in some British willows. J Geochem Explor 52:259–266
Purdy JJ, Smart LB (2008) Hydroponic screening of shrub willow (*Salix* spp.) for arsenic tolerance and uptake. Int J Phytoremediation 10:515–528
Rajagopal D (2007) Rethinking current strategies for biofuel production in India. Paper presented at the international conference on linkages in water and energy in developing countries, ICRISAT, Hyderabad, 29–30 January
Rajkumar M, Freitas H (2008) Influence of metal resistant-plant growth-promoting bacteria on the growth of *Ricinus communis* in soil contaminated with heavy metals. Chemosphere 71:834–842
Rajkumar M, Lee KJ, Lee WH et al (2005) Growth of *Brassica juncea* under chromium stress: influence of siderophores and indole-3-acetic acid producing rhizosphere bacteria. J Environ Biol 26:693–699
Rajkumar M, Nagendran R, Lee KJ et al (2006) Influence of plant growth promoting bacteria and Cr^{6+}on the growth of Indian mustard. Chemosphere 62:741–748
Rajkumar M, Ae N, Freitas H (2009) Endophytic bacteria and their potential to enhance heavy metal phytoextraction. Chemosphere 77:153–160
Rajkumar M, Ae N, Prasad MNV et al (2010) Potential of siderophore-producing bacteria for improving heavy metal phytoextraction. Trends Biotechnol 28:142–149
Ram SG, Parthiban KT, Kumar RS et al (2008) Genetic diversity among *Jatropha* species as revealed by RAPD markers. Genet Resour Crop Evol 55:803–809
Rao CH, Lavanya C, Anjani K et al (2003) Crop improvement in castor. In: Hegde DM, Sujatha M, Singh NB (eds) Castor in India. Directorate of Oilseeds Research, Hyderabad, pp 14–35
Riddell-Black DA (1993) Review of the potential for the use of trees in the rehabilitation of contaminated land, WRc report CO 3467. Water Research Centre, Medmenham
Rivetta A, Negrini N, Cocucci M (1997) Involvement of Ca^{2+}-calmodulin in Cd^{2+} toxicity during the early phases of radish (*Raphanus sativus* L.) seed germination. Plant Cell Environ 20:600–608
Rockwood DL, Naidu CV, Carter DR et al (2004) Short-rotation woody crops and phytoremediation: opportunities for agroforestry? Agrofor Syst 61:51–63
Romeiro S, Lagôa AMMA, Furlani PR et al (2006) Lead uptake and tolerance of *Ricinus communis* L. Braz J Plant Physiol 18:483–489
Ruttens A, Boulet J, Weyens N et al (2011) Short rotation coppice culture of willows and poplars as energy crops on metal contaminated agricultural soils. Int J Phytoremediation 13:194–207
Ryan RP, Germaine K, Franks A et al (2008) Bacterial endophytes: recent developments and applications. FEMS Microbiol Lett 278:1–9
Salt DE, Smith RD, Raskin I (1998) Phytoremediation. Ann Rev Plant Physiol Plant Mol Biol 49:643–668
Saravanan VS, Madhaiyan M, Thangaraju M (2007) Solubilization of zinc compounds by the diazotrophic, plant growth promoting bacterium *Gluconacetobacter diazotrophicus*. Chemosphere 66:1794–1798
Schmidt U (2003) Enhancing phytoremediation: the effect of chemical soil manipulation on mobility, plant accumulation, and leaching of heavy metals. J Environ Qual 32:1939–1954
Schützendübel A, Polle A (2002) Plant responses to abiotic stresses: heavy metal-induced oxidative stress and protection by mycorrhization. J Exp Bot 53:1351–1365
Scott JS, Smith PG (1981) Dictionary of waste and water treatment. Butterworths, London
Sebastiani L, Scebba F, Tognetti R (2004) Heavy metal accumulation and growth responses in poplar clones Eridano (*Populus deltoides* x *maximowiczii*) and I-214 (*P.* x *euramericana*) exposed to industrial waste. Environ Exp Bot 52:79–88

Shah K, Dubey RS (1998) A 18 kDa Cd inducible protein complex: its isolation and characterization from rice (*Oryza sativa L.*) seedlings. J Plant Physiol 152:448–454
Sharma SS, Dietz KJ (2006) The significance of amino acids and amino acid-derived molecules in plant responses and adaptation to heavy metal stress. J Exp Bot 57:711–726
Sharma P, Dubey RS (2007) Involvement of oxidative stress and role of antioxidative defense system in growing rice seedlings exposed to toxic concentrations of aluminum. Plant Cell Rep 26:2027–2038
Sheng XF, Xia JJ (2006) Improvement of rape (*Brassica napus*) plant growth and cadmium uptake by cadmium-resistant bacteria. Chemosphere 64:1036–1042
Sheng XF, Xia JJ, Jiang CY et al (2008) Characterization of metal-resistant endophytic bacteria from rape (*Brassica napus*) roots and their potential in promoting the growth and lead accumulation of rape. Environ Pollut 156:1164–1170
Shim D, Kim S, Choi YI et al (2013) Transgenic poplar trees expressing yeast cadmium factor 1 exhibit the characteristics necessary for the phytoremediation of mine tailing soil. Chemosphere 90.1478–1486
Simpson JA, Picchi G, Gordon AM et al (2009) Short rotation crops for bioenergy systems: environmental benefits associated with short-rotation woody crops, Task 30, Technical review no 3. IEA Bioenergy secretariat, Rotorua
Sims REH, Hastings A, Schlamadinger B et al (2006) Energy Crops: current status and future prospects. Glob Chang Biol 12:1–23
Smeets EMW, Faaij APC, Lewandowski IM et al (2007) A bottom up assessment and review of global bio-energy potentials to 2050. Prog Energy Combust Sci 33:56–106
Song WY, Martinoia E, Lee J et al (2004) A novel family of cys-rich membrane proteins mediates cadmium resistance in Arabidopsis. Plant Physiol 135:1027–1039
Song WY, Choi YI, Shim D et al (2007) Transgenic poplar for phytoremediation. In: Xu Z, Li J, Xue Y, Yang W (eds) Biotechnology and Sustainable Agriculture 2006 and Beyond, Proceedings of the 11th IAPTC&B Congress, August 31–18, 2006 Beijing. China, Springer science & business media, pp 265–271
Sresty TVS, Rao KVM (1999) Ultrastructural alterations in response of zinc and nickel stress in the root cells of pigeonpea. Environ Exp Bot 41:3–13
Steer P, Baker RM (1997) Colliery spoil, sewage and biomass- potential for renewable energy from wastes. Paper presented at Biomass and energy crops, meeting of the association of applied biologists, Royal Agricultural College, Cirencester, UK, 7–8 April 1997,Aspects Appl Biol 49:300–305
Taiz L, Zeiger E (1998) Plant Physiology, 2nd edn. Sinauer Associates Inc, Publishers, Sunderland
Tandon M, Vasudevan P, Naik SN et al (2013) Oil bearing seasonal crops in India: energy and phytoremediation potential. Int J Energ Sect Manage 7:338–354
Thomine S, Wang R, Ward JM et al (2000) Cadmium and iron transport by members of a plant metal transporter family in Arabidopsis with homology to Nramp genes. Proc Natl Acad Sci U S A 97:4991–4996
Tiwari AK, Kumar A, Raheman H (2007) Biodiesel production from Jatropha oil (*Jatropha curcas* L.) with high free fatty acid: an optimized process. Biomass Bioenergy 31:569–575
Trivedi S, Erdei L (1992) Effects of cadmium and lead on the accumulation of Ca^{2+} and K^+ and on the influx and translocation of K^+ in wheat of low and high K^+ status. Physiol Plant 84:94–100
Tuskan GA, DiFazio S, Jansson S et al (2006) The genome of black cottonwood, *Populus trichocarpa* (Torr. and Gray). Science 313:1596–1604
Vaezi M, Passandideh-Fard M, Moghiman M et al (2012) On a methodology for selecting biomass materials for gasification purposes. Fuel Process Technol 98:74–81
Vamerali T, Bandiera M, Coletto L et al (2009) Phytoremediation trials on metal and arsenic-contaminated pyrite wastes (Torviscosa, Italy). Environ Pollut 157:887–894

Van Assche F, Clijsters H (1986) Inhibition of photosynthesis in *Phaseolus vulgaris* by treatment with toxic concentration of zinc: effect on ribulose-1,5-bisphosphate carboxylase/oxygenase. J Plant Physiol 125:355–360
van Slycken S, Witters N, Meiresonne L et al (2013) Field evaluation of willow under short rotation coppice for phytomanagement of metal-polluted agricultural soils. Int J Phytoremediation 15:677–689
Vandecasteele B, Meers E, Vervaeke P et al (2005) Growth and trace metal accumulation of two *Salix* clones on sediment-derived soils with increasing contamination levels. Chemosphere 58:995–1002
Vangronsveld J, Herzig R, Weyens N et al (2009) Phytoremediation of contaminated soils and groundwater: lessons from the field. Environ Sci Pollut Res Int 16:765–794
Wang Y, Greger M (2004) Clonal differences in mercury tolerance, accumulation and distribution in willow. J Environ Qual 33:1779–1785
Weast RC (1984) CRC handbook of chemistry and physics, 64th edn. CRC Press, Boca Raton
Weiss EA (2000) Oil seed crops, 2nd edn. Blackwell Science, Oxford, p. 364
Welch RM, Norvell WA, Schaefer SC et al (1993) Induction of iron (III) and copper (II) reduction in pea (*Pisum sativum* L.) roots by Fe and Cu status: Does the root-cell plasmalemma Fe(III)-chelate reductase perform a general role in regulating cation uptake? Planta 190:555–561
Weller DM, Thomashow LS (1994) Current challenges in introducing beneficial microorganisms into the rhizosphere. In: O'Gara F, Dowling DN, Boesten B (eds) Microbial ecology of rhizosphere microorganisms. Weinheim, VCH, pp 1–18
Weyens N, Croes S, Dupae J et al (2010) Endophytic bacteria improve phytoremediation of Ni and TCE co-contamination. Environ Pollut 158:2422–2427
Wildner GF, Henkel J (1979) The effect of divalent metal ions on the activity of Mg^{2+} depleted ribulose-1,5-bisphosphate oxygenase. Planta 146:223–228
Williams LE, Pittman JK, Hall JL (2000) Emerging mechanisms for heavy metal transport in plants. Biochim Biophys Acta 1465:104–126
Yadav SK, Juwarkar AA, Kumar GP et al (2009) Bioaccumulation and phyto-translocation of arsenic, chromium and zinc by *Jatropha curcas* L.: impact of dairy sludge and biofertilizer. Bioresour Technol 100:4616–4622
Zabaniotou AA, Kantarelis EK, Theodoropoulos DC (2008) Sunflower shells utilization for energetic purposes in an integrated approach of energy crops: laboratory study pyrolysis and kinetics. Bioresour Technol 99:3174–3181
Zalesny RS Jr, Bauer EO (2007) Selecting and utilizing Populus and Salix for landfill covers: implications for leachate irrigation. Int J Phytoremediation 9:497–511
Zhang YF, He LY, Chen ZJ et al (2011) Characterization of ACC deaminase-producing endophytic bacteria isolated from copper-tolerant plants and their potential in promoting the growth and copper accumulation of *Brassica napus*. Chemosphere 83:57–62
Zhang K, Johnson L, Prasad PVV et al (2015) Comparison of big bluestem with other native grasses: chemical composition and biofuel yield. Energy 83:358–365
Zhu YL, Pilon-Smits EA, Tarun AS et al (1999) Cadmium tolerance and accumulation in Indian mustard is enhanced by overexpressing gamma-glutamylcysteine synthetase. Plant Physiol 121:1169–1178

Phytoremediation of Soil Contaminants by the Biodiesel Plant *Jatropha curcas*

4

O.P. Abioye, U.J.J. Ijah, and S.A. Aransiola

Abstract

Contamination of soil is prominent among the most critical environmental issues throughout the world, and it has a hugely harmful impact on people, animals, microorganisms, and plants. Organic pollution pollutes the environment because of its innate characteristics—for example, solubility, instability, and biodegradability. Correspondingly, heavy metals pollute the earth and have injurious impacts on the environment and its sectors. Current technology involves soil excavation and either landfilling or soil washing for physical or substance detachment of the contaminants. Although they are highly variable and depend on the contaminants of concern, soil properties, site conditions, etc., the generally huge expenses connected with the removal of metals from soils by conventional physicochemical methods explain why most organizations have a tendency to overlook the problem. The removal of these metals from the environment is thus imperative. Phytoremediation involves treatment of ecological problems (bioremediation) using florae that reduce ecological contamination, avoiding the need to uncover the polluted substances and dispose of them elsewhere. Remediation using plants (phytoremediation) consists of attenuating contaminant concentrations in polluted water, air, or soil with plants that are able to reduce or eliminate metals, pesticides, solvents, crude oil and its derivatives, and heavy metals. *Jatropha curcas* is a widely used plant species, with many advantages for remediation and biodiesel production. In this chapter, the phytoremediation method using the biodiesel plant *Jatropha curcas* for remediation of contaminated soil is examined, along with its effectiveness and efficiency.

O.P. Abioye (✉) • U.J.J. Ijah
Department of Microbiology, Federal University of Technology, P. M. B 65, Minna, Nigeria
e-mail: bisyem2603@yahoo.com

S.A. Aransiola
Bioresources Development Centre, National Biotechnology Development Agency, KM 5 Ogbomoso/Iresapa Road, Onipaanu, Ogbomoso, Nigeria

K. Bauddh et al. (eds.), *Phytoremediation Potential of Bioenergy Plants*,
DOI 10.1007/978-981-10-3084-0_4

Keywords
Biodiesel • Contamination • Environment • *Jatropha curcas* • Phytoremediation • Soil

4.1 Introduction

An inexorably industrialized worldwide economy has significantly elevated the release of anthropogenic chemicals into the environment in the course of the most recent century and brought about contamination of numerous areas of the earth. Pollution can be a result of improper chemical production (e.g., oil slicks from drilling, explosive from assembling plants), transport (e.g., oil slicks from tankers or pipelines), storage (e.g., chemicals from leaking storage tanks), utilization (e.g., pesticides and manure from agribusiness, explosives from weapons firing), or disposal processes (e.g., explosives from neutralization facilities). Simultaneously with increasing contamination levels, energetic enthusiasm for creating techniques for remediation of environmental contaminants—utilizing physical, chemical, and biological procedures—has risen.

As great "suck and truck" methodologies involving off-site treatments are costly, in situ bioremediation forms—such as monitored natural attenuation, biostimulation, bioaugmentation, and phytoremediation—have turned into an appealing approach to restore contaminated sites (Ayoub et al. 2010). The site conditions, the contaminant type and source, the source control measures, and the potential effect of the conceivable remedial measure influence the decision on a remediation methodology and technology. Heavy metal contamination in soil is one of the major ecological issues throughout the world. Heavy metals have a critical lethal impact on humans, animals, microorganisms, and plants. Besides, heavy metals are not subject to degradation and, along these lines, remain uncertainly in the environment (Raskin and Ensley 2000). Heavy metals are discharged by different anthropogenic activities into the environment—for example, manufacturing processes in industry, household refuse, and waste materials, especially sawdust muck, sewage sludge, textile industry sludge, and slaughterhouse sludge. Polluted soils need to be remediated. There are some current techniques—for example, physical and chemical treatments, or removal and entombment at a risk waste site—that are expensive, tedious, and inadequate, as the treated soils remain incompletely cleaned. Tree plantation can be utilized as a technique for remediation of polluted soil and to restore soil fertility and efficiency. For viable phytoremediation, the plant species ought to be inedible (Abioye et al. 2015) and able to grow copiously on a huge scale in polluted soil, yet some edible plants, such as *Glycine max* L. (Aransiola et al. 2013) and *Arachis hypogaea* (Ijah et al. 2015), have turned out to be excellent for phytoremediation. Such plants can endure harsh conditions, such as heavy metal contamination, low nutrient content, and dry seasons (Kumar et al. 2008). Considering all of the aforementioned choices, among inedible trees, *Jatropha curcas* is a species of choice (Jussieu 1789). *Jatropha* is native to Central America and has become naturalized in numerous tropical and subtropical regions, including India, Africa, and North America. It develops essentially all over India under an assortment of agro-climatic conditions. *Jatropha curcas* is a multipurpose plant with numerous properties and impressive potential. In recent decades, phytoremediation, which is inexpensive and offers the most naturally benign innovation and methodology (Aransiola et al. 2013), has been utilized effectively for the remediation of soils polluted with different contaminants. Furthermore, phytoremediation is being progressively utilized as a technological complement to treatment of contaminated water in various sorts of treatment wetlands (Horne 1999; Zhang et al. 2010).

Jatropha curcas is a perennial oil-yielding plant developed in the tropical and subtropical areas of the globe. It has been broadly utilized for energy plantation and wetland recovery because of its capacity to endure harsh environmental conditions—for example, dry seasons and infertile and heavily contaminated soils. In light of the accumulation of different trace elements (Cu, Cr, Mn, Ni, and Zn) and use of supplements (Ca, K, Mg, Na, and P), *Jatropha curcas* could be utilized for revegetation programs in mining areas in mica belt regions (Nagaraju and Karimulla 2002). Studies have, for the most part, centered around changes in development execution and metal tolerance of *Jatropha curcas* in metal-polluted calcareous soils with the introduction of organic waste (biosludge and dairy sludge) and biofertilizer (*Azotobacter chroococcum*) (Kumar et al. 2008; Yadav et al. 2009, 2010). Jamil et al. (2009) recommended *Jatropha curcas* plantation for phytoremediation in coal fly ash landfills with use of crucial plant supplements and suitable chelating agents.

In addition to generation of biodiesel, by-products of *Jatropha curcas* can be utilized to make an extensive variety of products, including quality paper, health supplements, soap, beautifying agents, toothpaste, embalming fluid, pipe joint cement, hard drug, and a moistening agent in tobacco. *Jatropha curcas* (part of the Euphorbiaceae family) develops well under harsh climatic conditions because of its low moisture demands, low fertility requirements, and resilience to high temperatures (Augustus et al. 2002). Hence, the present review was conducted to assess the plant's potential in remediation of pollutants in the soil.

4.2 Soil Contaminants

Environmental contamination has been a serious concern globally. Water, soil, and air are being contaminated in various ways. As an "all-inclusive sink," soil is subject to heavy contamination. Soil can be polluted in different ways. There is a desperate need in monitoring soil contamination, keeping in mind the end goal of preserving soil fertility and increasing productivity (Muhammad et al. 2014). Soil is the very foundation of farming. All harvests for human sustenance and animal survival rely on it. This essential regular asset is being lost to quickened disintegration. What is more, huge amounts of man-made waste items, ooze, and different items from waste treatment plants and contaminated water are likewise bringing about or promoting soil contamination (Muhammad et al. 2014). Soil contamination is characterized as the build-up in soils of persistent toxic compounds, chemicals, salts, radioactive materials, or disease-causing agents, which have effects on plant development and animal well-being (Okrent 1999). Soil is the thin layer of organic and inorganic materials that covers the earth's surface; the former come from the decayed remains of plants and animals, and the latter have formed over thousands of years by physical and chemical weathering of bedrock. Productive soils are important for farming to supply the world with adequate nourishment (Belluck et al. 2003).

There is a wide range of ways in which soil can become contaminated. These include seepage from a landfill; release of mechanical waste into the soil; permeation of sullied water into the soil; leakage from underground storage tanks; abundant utilization of pesticides, herbicides, or compost; and strong waste drainage.

The most widely recognized chemicals involved in causing soil contamination are solvents, pesticides, heavy metals, and petroleum hydrocarbons.

4.2.1 Inorganic Poisonous Mixtures

Inorganic buildups in modern waste cause difficult issues in respect of their disposal. They include metals, which have high potential for poisonous qualities. Modern transportation likewise releases a lot of arsenic fluorides and sulfur dioxide (SO_2) (Richardson et al. 2006). Fluorides are found in the environment from superphosphate, phosphoric acid corrosive, aluminum, steel, and earthenware commercial ventures. Sulfur dioxide emitted by industrial facilities and power plants may make soils extremely acidic. These metals cause leaf harm and annihilate plantations. Cadmium, lead, nickel, copper, mercury, and arsenic are components that can aggregate in the soil if they pass into it through sewage, modern waste, or mine washings. A portion of fungicides containing copper and mercury likewise add to soil contamination. Vehicle emissions contain lead, which gets adsorbed by soil particles and is poisonous to plants. The harmfulness can be minimized by working up soil natural matter, adding lime to soils, and keeping it basic (Van Zorge 1996).

4.2.2 Organic Waste

Natural squanders (organic wastes) of different sorts lead to contamination hazards. Residential junk, civil sewage, and modern squanders—when left in stacks or dumped as refuse—undoubtedly influence the health of people, plants, and animals (Crane and Giddings 2004). Natural squanders contain borates, phosphates, and cleansers in huge amounts. In the event that they remain unprocessed, they influence the development of plants. The principal natural pollutants are coal and phenol. Asbestos; ignitable tools; gases such as carbon dioxide, hydrogen sulfide, methane, carbon monoxide, and sulfur dioxide; and petroleum are additional pollutants. Harmful materials such as thorium, strontium, and uranium additionally cause soil pollution. Aftermath of strontium for the most part remains in the soil and is moved with the residue (Nathanail et al. 2005). Different fluid wastes such as sewage and sewage sludge are additionally leading causes of soil problems.

4.2.3 Sewage and Sewage Sludge

Contamination of soil, frequently brought about by an unrestrained transfer of sewage and different fluid squanders, comes about because of household use of water, modern squanders containing an assortment of toxins, agrarian effluent from animal farming, wastage of watering system water, and urban spillover (Tarazona et al. 2005; Evans et al. 2006). Watering systems with sewage water cause significant changes in flooded soils. Among the different changes that occur in the soil because of sewage watering systems are physical changes in draining; changes in humus substance, porosity, and so forth; compound changes in the soil response, base exchange status, and saltiness; and changes in the amounts and accessibility of nutrients such as nitrogen, potash, and phosphorus. Sewage oozes contaminate the soil by aggregating metals such as lead, nickel, zinc, and cadmium. This may cause phytoxicity in plants.

4.2.4 Heavy Metal Contamination

Soil contamination by metals has become a major theme of the examination of ecological issues nowadays. Substantial metals exist in ionic, particulate, colloidal, and broken-up forms. Heavy metals are available in the soil as soluble metal compounds, free metal particles, replaceable metal particles, naturally bound metals, Precipitated or insoluble mixtures such as oxides, carbonates, and a portion of silicate materials or hydroxides (Leyval et al. 1997).

Metals are common components of soil, and they endure in soils and have a moderate draining rate; consequently, they have a tendency to aggregate in soils. Trace amounts of some substantial metals are required by living creatures, yet in overabundance they are harmful. The eco-toxicological dangers of metal pollution result in possible damage to plants, animals, people, and microorganisms. Heavy metal contamination can smother or even kill plants and soil microbial groups or alter their qualities and structures. When they accumulate in the food chain, they have unfavorable impacts at various trophic levels because of biomagnification. Then again, substantial metals such as Fe, Mn, Ni, Cu, and Zn are vital for plant development and are vital constituents of numerous chemicals. Heavy metals such as As, Cd, Cr, Hg, Pb, Sb, Al, and Se are superfluous and harmful over certain thresholds (Panda and Choudhury 2005). Mostly, urban and modern pressurized canned products, ignition of fills, fluids and strong waste from animals and people, mining squanders, mechanical and farming chemicals, etc., are contributing substantial metal contamination. Overwhelming metals are available in all uncontaminated soils as a consequence of weathering from their materials of origin. The concentrations of substantial metals in the lithosphere, soil, and plants are given in Table 4.1.

The fate of heavy metals in the soil can be influenced by physical and organic processes within the soil. Metal particles penetrate the soil structure from these different types of mixtures at various rates, and they may either stay within the

Table 4.1 Concentrations of heavy metals in the lithosphere, soil, and plants (μg/g dry matter)

Heavy metal	Lithosphere	Soil range	Plants
Cadmium (Cd)	0.2	0.01–0.7	0.2–0.8
Cobalt (Co)	40	1–40	0.05–0.5
Chromium (Cr)	200	5–3000	0.02–1.0
Copper (Cu)	70	2–100	4–15
Iron (Fe)	50,000	7000–5,50,000	140
Mercury (Hg)	0.5	0.01–0.3	0.015
Manganese (Mn)	1000	100–4000	15–100
Molybdenum (Mo)	2.3	0.2–5	1–10
Nickel (Ni)	100	10–1000	1
Lead (Pb)	16	2–200	0.1–10
Tin (Sn)	40	2–100	0.3
Zinc (Zn)	80	10–300	8–100

Source: Muhammad et al. (2014)

structure, go into the seepage water, be taken up by plants growing in the soil, or be held by the soil in sparingly soluble or insoluble structures. Natural materials in the soil have an incredible proclivity for substantial metal cations, which form stable compounds, in this way promoting diminished nutrient content (Huinink 1998; Urzelai et al. 2000).

4.2.5 Organic Pesticides

Pesticides are mostly employed to resolve certain nuisances. Pesticides may have destructive impacts on microorganisms and, as a consequence, plant development may be influenced. Pesticides that do not deteriorate quickly may create such issues. Collection of pesticide deposits in higher fixations is harmful. Persistence of pesticides in soil and their introduction into water streams may likewise promote their introduction into foods and endanger health. Pesticides, especially aromatic natural mixtures, are not debased quickly and can therefore remain active for a long time, as shown in Table 4.2.

4.3 Reasons for Soil Contamination

Contamination of soil is brought about by the presence of artificial chemicals or other adjustments in the regular soil environment. This sort of pollution normally emerges from leakage from underground storage tanks, the use of pesticides, permeation of sullied surface water into the subsurface strata, oil and fuel dumping, draining of squanders from landfills, or direct release of mechanical squanders into the soil. The most widely recognized chemicals involved are solvents, pesticides,

Table 4.2 Persistence times of some selected pesticides

Pesticide	Persistence time
Benzene hexachloride (BHC)	11 years
Dichlorodiphenyltrichloroethane (DDT)	10 years
2,4-Dichlorophenoxy acetic acid (2,4 D)	2–8 weeks
Aldrin	9 years
Diuron	16 months
Atrazine	18 months
Siwazine	17 months
Chlordan	12 years
2,3,6-Trichlorobenzene (TBA)	2–5 years

Source: Muhammad et al. (2014)

lead, petroleum hydrocarbons, and other substantial metals. These events can be connected to the level of industrialization and the intensity of concoction utilization. A soil toxin is any variable that degrades the quality, surface, or mineral substance of the soil or that disturbs the natural equalization of the life-forms in the soil. Contamination in soil has antagonistic impacts on plant development.

4.3.1 Causes of Soil Contamination

4.3.1.1 Indiscriminate Utilization of Fertilizers

Oxygen is obtained from water and air; however, other vital nutrients such as nitrogen, potassium, calcium, magnesium, phosphorus, and sulfur need to be acquired from the soil. Farmers, for the most part, utilize manure to rectify soil inadequacies. Fertilizer contaminates the soil with impurities, which originate from the crude materials utilized for their production. Blended composts regularly include phosphorus as P_2O_5, potassium as K_2O, and ammonium nitrate (NH_4NO_3). As a case in point, As, Pb, and Cd present in rock phosphate minerals are incorporated into superphosphate manure. Since the metals are not degradable, their amassing in the soil beyond harmful levels (because of inordinate utilization of phosphate composts) turns them into indestructible substances that are toxic to crops. Overutilization of NPK fertilizer decreases the yields of vegetables and crops grown in the soil over periods of years. It additionally diminishes the protein content of wheat, maize, grams, and so forth grown in that soil. The carbohydrate quality of such crops additionally becomes degraded (Provoost et al. 2008). An abundance of potassium content in the soil diminishes vitamin C and carotene content in vegetables and fruits. Vegetables and fruits grown in overfertilized soil are more vulnerable to assaults by insects and disease.

4.3.1.2 Indiscriminate Utilization of Herbicides, Pesticides, and Insecticides

Plants on which humans depend for sustenance are often attacked by insects, fungi, bacteria, viruses, rodents, and other animals, and must compete with weeds for nutrients. To destroy undesirable populations living in or on their yields, farmers use pesticides. The primary widespread insecticide use started toward the end of World War II and included DDT (dichlorodiphenyltrichloroethane) and gammaxene. Insects soon became resistant to DDT, and as the chemical did not break down promptly, it persisted in the environment. Since it was soluble in fat as opposed to water, it biomagnified up the food chain and disturbed the calcium digestion system in birds, making eggshells thin and fragile (Toccalino and Norman 2006).

4.3.1.3 Discarding of Large Amounts of Solid Waste

By and large, solid waste incorporates garbage, residential refuse, and disposed-of solid materials—for example, those from commercial, industrial, and agricultural operations. Solid waste contains expanding measures of paper, cardboard, plastics, glass, old building materials, packaging materials, and harmful or generally hazardous substances. Since a lot of urban solid waste has a tendency to be paper and food waste, the majority is recyclable or biodegradable in landfills. Likewise, most agricultural waste is reused, and mining waste is left at the site. The bit of solid waste that is dangerous—for example, oils, battery metals, and heavy metals from purifying commercial enterprises and organic solvents—are the ones that require specific consideration. These can, over the long haul, accumulate in the soils in the surrounding area and pollute them, altering their chemical and biological properties (Patterson et al. 2007).

4.3.1.4 Deforestation and Erosion of Soil

Soil erosion happens when weathered soil particles are disturbed and removed by wind or water. Deforestation, agricultural advancement, temperature extremes, precipitation (including acid rain), and human activities add to this erosion. Humans speed up this process by development, mining, cutting of timber, overcropping, and overgrazing, resulting in flooding and causing soil erosion. Forests and grasslands are incredible binding systems, which keep the soil in place and solid. They support numerous habitats and ecosystems, which give endless nourishing pathways or food chains of life to all involved species. Their misfortune undermines the food chain of life and the survival of numerous species. In recent years, considerable expanses of green areas have been changed into deserts. The valuable rain forest habitats of South America, tropical Asia, and Africa are bearing the brunt of their populations' developments and improvements (particularly timber logging, building development, and agriculture). Numerous researchers believe that an abundance of therapeutic substances, including cures for cancer and AIDS, lie in these forests. Deforestation is gradually annihilating the most bountiful flora and fauna areas in the world, which also form vast tracts of a very valuable sink for CO_2 (Leon Paumen 2008).

4.3.1.5 Pollution Due to Urbanization

Surface soils are becoming contaminated by both biodegradable materials (such as vegetables, animal waste, paper, wood, corpses, plant twigs, leaves, fabric waste, and additional sweepings) and numerous nonbiodegradable materials (for example, plastic bags, plastic containers, plastic waste, glass bottles, glass pieces, and stone/cement pieces) (Simcox et al. 1995; Nawrot et al. 2006). One alarming assessment reported that Indian urban communities are consistently outputting strong city squanders to the tune of 50,000–80,000 metric tons "everyday". In the event that these are left uncollected and decay, they cause a few issues, for example:

1. Stopping up of channels: Causing genuine waste issues including bursting/spillage of seepage lines, causing health hazards
2. Hindrance to movement of water: Solid squanders genuinely harm the ordinary movement of water, thereby causing issues of immersion, harm to establishment of structures and, in addition, general health hazards
3. Foul smells: these are generated by dumping of squanders at particular sites
4. Expanded microbial activity: microbial decomposition of natural squanders produces huge amounts of methane and numerous other chemicals, contaminating the soil and water streaming on its surface
5. Health hazards from medical waste: at sites where such strong squanders include medical waste, they can cause numerous well-being issues, as they may contain unsafe pathogens, hazardous medications, and infusion substances.

4.3.1.6 Contamination of Subsurface Soil

Subsurface soil in urban communities is liable to be contaminated by chemicals discharged in industrial waste and by decomposed and partly deteriorated materials in sanitary waste.

Numerous hazardous chemicals—such as chromium, lead, arsenic, cadmium, and selenium—are prone to being placed in subsurface soil. Furthermore, underground soils contaminated by sanitary waste produce numerous harmful chemicals. These can harm the typical processes and biological balance in the underground soil.

4.4 Phytoremediation Techniques

Phytoremediation is defined as the utilization of green plants to remove, degrade, or sequester hazardous environmental contaminants (Cunningham and Berti 1993; Aransiola et al. 2013). In this procedure, specially chosen or engineered plants are utilized, which are able to directly take up contaminants from the environment (Macek et al. 2000). Phytoremediation can be applied to both inorganic and organic pollutants present in solid and fluid substrates (Salt et al. 1998). For the most part, phytoremediation of contaminants by a plant includes the following steps: uptake, translocation, transformation, compartmentalization, and sometimes mineralization (Schnoor et al. 1995). Factors influencing the uptake, circulation, and transformation of organic compounds by a plant are, for the most part, identified with the physical and chemical properties of the compound (e.g., its water solubility, subatomic weight, and octanol–water segment coefficient) and, in addition, environmental conditions (e.g., temperature, pH, organic matter, and soil moisture) and plant attributes (e.g., the root system and enzymes) (Susarla et al. 2002; Suresh and Ravishankar 2004). Although the designations of various phytoremediation techniques differ in their descriptions, the basic plan is given in Fig. 4.1.

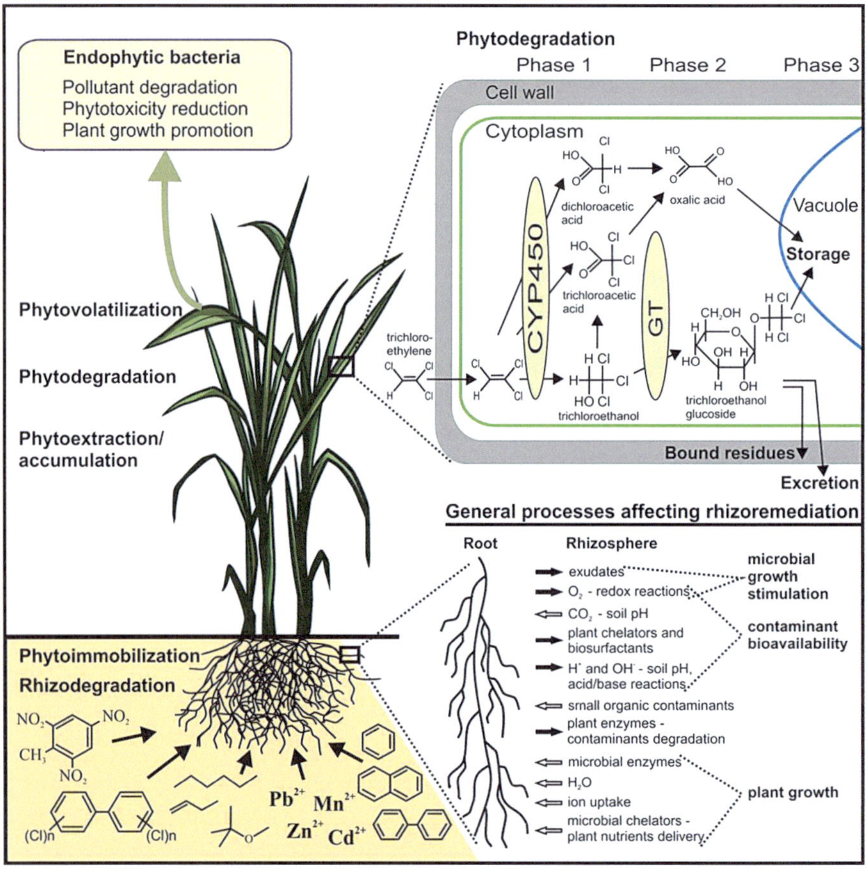

Fig. 4.1 Techniques of phytoremediation

4.4.1 Rhizoremediation and Microbe–Plant Interactions in Phytoremediation

Rhizoremediation (rhizodegradation, organism-assisted phytoremediation, rhizosphere bioremediation) involves remarkable interactions between roots, root exudates, rhizosphere soil, and microorganisms to degrade contaminants into harmless or less dangerous compounds.

Plant roots invigorate rhizosphere microbial groups by circulating air through the soil and by discharging exoenzymes and nutrients through root exudates while likewise providing surfaces for colonization and niches to insure microscopic organisms against dehydration and other abiotic and biotic stresses (Kuiper et al. 2004). Rhizospheric microorganisms thus advance plant development by nitrogen fixation, nutrient (e.g., phosphorus) activation, creation of plant development regulators, diminishing plant stress hormone levels, giving security against plant pathogens, and degradation of contaminations before they adversely affect the plant (Fig. 4.1) (Chaudhry et al. 2005; Segura et al. 2009). Consequently, these common associations, also called rhizosphere effects, result in increased numbers, varieties, and metabolic movements of microorganisms that are able to degrade contaminants or bolster plant development in close proximity to the roots, contrasted with the bulk soil (Ramos et al. 2000; Kent and Triplett 2002).

The sum and composition of root exudates that make a nutrient-rich environment in the region of roots is particular to the plant family or species. Root exudates contain organic acids (lactate, acetic acid derivative, oxalate, succinate, fumarate, malate, and citrate), sugars, and amino acids as the primary parts and, additionally, secondary metabolites (isoprenoids, alkaloids, and flavonoids), which are discharged into the soil as rhizodeposits (Singer et al. 2003). The fundamental parts of exuded organic acids are available in the soil as separated anions (carboxylates) (Martin et al. 2014). It has been suggested that 10–44% of the settled carbon is discharged by rhizodeposition (Baiš et al. 2006; Kumar et al. 2006). Root exudates can be utilized as a nutrient source by microorganisms. In expansion, the structures of numerous auxiliary metabolites take after those of the contaminants, in this way instigating the outflow of particular catabolic qualities of smaller-scale life-forms essential for degradation of the contaminant. As a case in point, the plant optional metabolite salicylate has been connected to the microbial degradation of polycyclic aromatic hydrocarbons (PAHs; naphthalene, fluoranthene, pyrene, and chrysene) and PCB (Master and Mohn 2001), while terpenes can impel the microbial degradation of toluene, phenol, and TCE (Kim et al. 2002). Effectively, degradable root-radiated mixtures can likewise serve as cometabolites in procedures where contaminants cannot be utilized as a sole carbon source (e.g., vigorous degradation of trichloroethylene (Hyman et al. 1995) because of a negative energy balance) (Reichenauer and Germida 2008). This is vital under numerous circumstances where microorganisms cannot depend on vitality gain from the contaminant and cometabolism is the main mechanism of the debasement of the contaminant. Plant roots alongside some rhizospheric microorganisms might likewise discharge biosurfactants, in this manner expanding the bioavailability and uptake of toxins

(Wenzel 2008). This perspective can be particularly helpful in dealing with aged soils with low contaminant bioavailability, which, by and large, have all of the earmarks of being a great deal less receptive to rhizodegradation than freshly spiked soil (Dams et al. 2007; Gunderson et al. 2007).

A recent technology to enhance phytoremediation and detoxification of contaminants is the utilization of endophytic bacteria. Endophytic bacteria are portrayed as nonpathogenic microbes, and they appear to have a ubiquitous presence in most, if not all, higher plant species. They regularly have a place in genera ordinarily found in soil, including *Pseudomonas, Burkholderia, Bacillus*, and *Azospirillum* (Yrjala et al. 2010). Endophytic bacteria are additionally known for their capacity to advance plant development and control pathogens (Berg et al. 2005; Ryan et al. 2008). A noteworthy point regarding the preference for utilizing endophytic bacteria over rhizospheric microbes in phytoremediation is that while a rhizospheric bacterial population is hard to control, and rivalry between rhizospheric bacterial strains regularly reduces the quantity of the desired strains (unless metabolism of the contamination is specific), utilization of endophytes that actually inhabit the internal tissues of plants decreases the issue of rivalry between bacterial strains (McGuinness and Dowling 2009). Studies have recommended that these microbes be utilized to supplement the metabolic capability of their host plants through direct degradation (Philips et al. 2008, 2009) and also exchange of degradative plasmids between different endophytes (Wang et al. 2007). Notwithstanding the contamination degradation pathway, endophytic bacteria might likewise have the capacity to improve plant development and adaptation by 1-aminocyclopropane-1-carboxylate (ACC) deaminase movement, siderophore creation, and nutrient solubilization. The presence of such bacteria in plants prompts more effective phytoremediation action and reduces the requirement for extra treatment (Afzal et al. 2014).

Indeed, even enhanced rhizoremediation may be extensively slower than ex situ treatments because of environmental restrictions at field locales—for example, competition by weed species that are better adjusted to the site (Nedunuri et al. 2000), constrained plant development in intensely and unevenly contaminated soil, and the presence of plant pathogens and other biotic and abiotic stressors (Gerhardt et al. 2009). Besides, rhizoremediation is viable just in the rooting zone and is unsatisfactory for utilization in deeper subsurface layers. Some toxic contaminant metabolites can likewise bioaccumulate in plants, making strict regulations for plant material treatment important. Be that as it may, regardless of the previously stated weaknesses, rhizoremediation is rising as a standout among the best means by which plants can influence the remediation of organic contaminants, especially large recalcitrant compounds (Gerhardt et al. 2009; Azaizeh et al. 2011). Other than its moderately low upkeep costs, no size limitations for the range, and environmental friendliness, the quality and "texture of soil" is also improved by additions of organic materials, supplements, and oxygen by means of plant and microbial metabolic procedures. Regardless of the difficulties of upscaling phytoremediation from the laboratory and greenhouse, rhizoremediation has been utilized to treat field sites contaminated with petroleum hydrocarbons (Siciliano et al. 2003) and PAHs (White et al. 2006). As with other bioremediation systems, point-by-point and persistent monitoring of synthetic and organic markers is key to ensuring phytoremediation process proficiency and environmental safety.

4.4.2 Uses of Phytoremediation in Treatment Wetlands

Treatment wetlands are successful and require minimal effort to operate different options for customary conventional technology for the disposal of an extensive variety of contaminants from wastewater and contaminated groundwater (Braeckevelt et al. 2007; Calheiros et al. 2007; Matamoros et al. 2007; Calheiros et al. 2010). Recent years have seen increasing use of treatment wetlands to treat city or industrial wastewater (Vymazal 2011) and for the removal of excessive nitrogen and phosphorous from contaminated surface and subsurface waters to safeguard sea-going biological systems (Mitsch et al. 2012; Ligi et al. 2014). An integration of phytoremediation has been proposed to enhance the execution of existing wastewater treatment in built wetlands, particularly to deal with micropollutants, e.g., organic chemicals, personal care items, and pharmaceuticals (including antimicrobials) (Hijosa-Valsero et al. 2011). In any case, the biogeochemical forms connected with the change of the organic chemicals in vegetated treatment wetlands have so far seldom been assessed, probably because of the highly complex and synergistic nature of the processes involved. In a complex treatment wetland system, a few disposal pathways for organic compounds (volatilization, photochemical oxidation, sedimentation, sorption, and biodegradation) may happen at the same time, while plants may contribute either by direct contaminant uptake and accumulation, phytovolatilization, and metabolic change, or by making the conditions ideal for contaminant removal inside the treatment system. The last includes providing a suitable surface for biofilm anchorage, advancing the improvement and development of various microbial species inside the systems by discharging root exudates, pumping and discharging oxygen to the deeper layers of the wetland media, holding suspended solid particles, and protecting against low temperatures (Shelef et al. 2013). The relative significance of a specific procedure differs depending upon the organic or inorganic contaminant to be dealt with, the treatment wetland type (free water, subsurface stream, level stream, or vertical stream; and the kind of vegetation), and the operational configuration (the wastewater stacking rate and maintenance time, and the soil matrix type). Zhang and collaborators examined how very different procedures—for example, microbial degradation, photodegradation, and plant uptake—add to the expulsion of pharmaceutical compounds from wastewater in an aquatic plant–based system (Zhang et al. 2014). They found that plant uptake accounted for an overwhelming proportion of the disposal of clofibric acid and caffeine, and likewise largely accounted for the disposal of ibuprofen. In any case, the effect of the plants' surroundings and the capacity of specific species to enhance the efficiency of the evacuation of certain pharmaceutical mixtures and personal care items remain unclear. This is on account of numerous different components, such as the structure of the rhizosphere microbial groups and the properties of the wastewater. In addition, the ecological conditions (e.g., the temperature and accessibility of electron acceptors) and the operational conditions (e.g., the water-driven maintenance time, particular surface zone, and stacking mode) may all demonstrate effects (Verlicchi and Zambello 2014). For instance, in a study of a surface stream developed wetland, planted and unplanted mesocosms were not altogether diverse in their capacities to remove pharmaceuticals (Cardinal et al. 2014).

4.5 *Jatropha curcas*

Jatropha curcas is a versatile plant species. Its seeds contain, by and large, 38% oil, with extraordinary qualities for the generation of biodiesel. This plant is an enduring bush with an expected beneficial life cycle of up to 40 years. Its yield potential is 5 t/ha of grains, which converts to 1.9 t/ha of oil, and its productive peak begins from the fourth year after planting. These features make *Jatropha curcas* a promising oil plant for business development, particularly for family agribusinesses (Foidl et al. 1996; Heller 1996; Openshaw 2000; Dias et al. 2007; Jongschaap et al. 2007; Achten et al. 2008; Dias 2011).

Scientists with various claims to fame and extensive enterprises from the vitality sector have promoted *Jatropha curcas* as a standout among the most encouraging oil plant species for biodiesel generation. Governments with their research and development agencies are creating a developing bulk of assets accessible for exploration, and substantial partnerships have made impressive arrangements for widespread planting. India and China, for instance, as of now have 2.6 million hectares planted in *Jatropha curcas* (Fairless 2007). In Latin America, there is altogether less enthusiasm for *Jatropha curcas* (Fig. 4.2).

4.5.1 History and Domestication of *Jatropha curcas*

Common names: Physic nut, Barbados nut, curcas bean, purge nut, purging nut (Parsons and Cuthbertson 2001). *Jatropha curcas* is known by numerous different names in various dialects (Sunil et al. 2013; National Germplasm Resources Laboratory (NGRP) 2014). The focal point of its source, the center of its domestication, and the presence of conceivable secondary focuses on its differing qualities are still open inquiries for *Jatropha curcas*. The center of origin, or the essential focus on its differing qualities, is the waterfront district of the Gulf of Mexico. There are records from times long past of the use and knowledge of *Jatropha curcas* by the Olmeca people in Mexico. It is now known that this specific development thrived somewhere around 1500–3000 years BCE. This is the oldest known use of *Jatropha curcas*. All things considered, until a history preceding that of the Olmecas is discovered, *Jatropha curcas* is thought to have initially come from Mexico. McVaugh (1945), Wilbur (1954), and Dehgan and Schutzman (1994) have substantiated its Mexican source, since *Jatropha curcas*—the most primitive plant of its type—and related taxa occur there. Basha et al. (2009), examining the biochemical and subatomic DNA markers of the RAPD, SCAR, and ISSR types in 72 accessions of

Fig. 4.2 *Top left*: Inflorescence of *Jatropha curcas*. *Top right*: Seeds and capsule. *Bottom left*: Habit. *Bottom right*: Leaves and fruit (Source: Starr and Starr 2009–2012)

determined *Jatropha curcas* from 13 nations, exhibited that the Mexican varieties are interesting regarding phorbol ester content (trunk borer and stinkbugs) and the subatomic profile. These rich assorted qualities of the Mexican *Jatropha curcas* germplasm should be fused in hereditary change projects of the species.

4.5.2 Cultivation and Distribution of *Jatropha curcas*

Jatropha curcas is an overwhelmingly drought-tolerant and pest-tolerant plant and is unpalatable to animals. It is planted in tropical nations chiefly as a fence, shielding cropland from dairy cattle, sheep, and goats (Openshaw 2000; Francis et al. 2005). *Jatropha curcas* is a diploid plant species type with 2n = 22 chromosomes. The size of the *Jatropha* genome (416 Mb) is about equivalent to those of the rice genome (400 Mb) and the castor genome (323 Mb). Customarily, *Jatropha* seed and plant branches have been utilized for cleansers, restorative mixtures, and oil (Kohli et al. 2009). *Jatropha* has been advanced as a "one of a kind" competitor among renewable vitality sources because of its exceptional qualities, such as dry season resilience (Openshaw 2000), quick development and simple breeding, greater oil content than other oil plants (Achten et al. 2008), a short development period, an extensive variety of environmental adjustments, and an ideal plant size, and architecture make it as a sole candidate for further consideration (Sujatha et al. 2008). Its development requires basic innovation and relatively little capital investment.

4.5.2.1 Climate

Jatropha develops in tropical and subtropical districts, with development limits of 30°N and 35°S. It develops at lesser altitudes of 0–500 m above sea level. *Jatropha* is not sensitive to day length (blossoming is autonomous of scope) and might flowers whenever of the year (Heller 1996). The plant is a succulent bush, which sheds its leaves in times of drought, with profound roots that make it appropriate for semi-bone-dry circumstances. While *Jatropha* can get by with as little as 250–300 mm of yearly precipitation, no less than 600 mm is needed for blossoming and fruit setting. The ideal precipitation for seed creation is considered to be 1000–1500 mm, which relates to subhumid ecosystems. But *Jatropha* has been seen to develop with 3000 mm of precipitation (Foidl et al. 1996, referred to by Achten et al. 2008), and higher rainfall is liable to bring about fungi assault and confine root development in all but the most free-draining soils. *Jatropha curcas* is not found in the more humid parts of its territory of origin, Mexico and Central America. Rainfall instigates blooming and, in ranges of unimodal precipitation, blossoming is consistent all throughout the vast majority of the year. Ideal temperatures are somewhere around 20–28 °C. High temperatures can discourage yields (Gour 2006). *Jatropha* has been seen to be intolerant of frost. The plant is best suited to states of high light force (Jongschaap et al. 2007) and is unsuited to growth in shade (Fig. 4.3).

4.5.2.2 Soils

Jatropha curcas thrives well in an aerated soil containing sand and loam of no less than 45 cm profundity (Gour 2006). Overwhelming mud soils are less suitable and

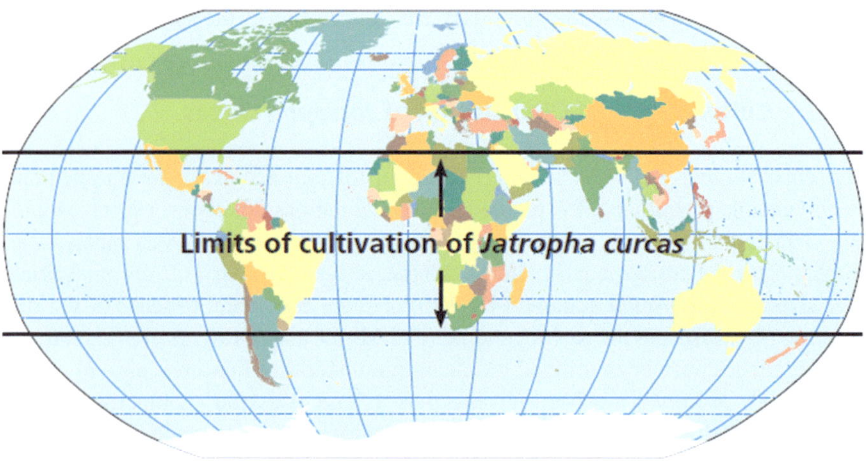

Fig. 4.3 Cultivation limits for *Jatropha curcas*

ought to be kept away from the plants, especially where seepage is weakened, as the plant is intolerant of waterlogged circumstances. Capacity for the plant to develop in basic soils has been broadly reported; however, the soil pH should be between 6.0 and 8.0/8.5.

4.5.2.3 Propagation and Crop Establishment

Plant propagation should be done from cuttings or seeds that have demonstrated, over more than a few seasons, high returns and seed oil content under the same watering system and treatment conditions that are proposed for the new plantation. Seed from high-yielding *jatropha* plants is not generally available; due to the fact that the out-crossing seed selected from productive plants may or may not result in high-yielding and high-quality *Jatropha* plants. Trees fit for creating more than two tons of dry seed per hectare with 30% seed oil substance ought to be chosen as the source material (Achten et al. 2008). Opinion is divided on the decision of choosing seed or cuttings. Heller (1996) has pointed out that the capacity of seedlings to create taproots is essential. The difficulty of acquiring adequate cuttings is a consideration. Heller (1996) found that cuttings of no less than 30 mm diameter gave earlier and higher introductory yields than plants raised from seed, albeit that practically no difference in yield was seen for later collections. Be that as it may, cuttings taken from precisely picked plants with higher yield potential would most likely outperform seed-raised plants. Raising plantlets from tissue cultivation is being examined and protocols have been developed but, that as it may, as it is a latex-delivering plant, the method is not clear. There have been no reports of tissue cultivation of *Jatropha* being conducted on a huge scale. Heller (1996) observed that nursery-raised plants from seed and cuttings and directly planted cuttings have a higher survival rate (more than 80%) than straightforward seed sowing in the field.

4.5.2.4 Vegetative Propagation Using Cuttings

The absolute benefits of utilizing cuttings are hereditary consistency, early yield, and quick foundation. The disadvantages are lack of material and the expenses of collecting, transporting, arranging, and establishing the stems, contrasted with seeds. A further downside is that this cutting method does not produce a taproot, which means the plant needs to reach soil water and nutrients at a shallower depth and this may correspondingly bring down potential yields, although the impact of this, in various situations, has not yet been established. The lack of a taproot makes for less dependability in unsheltered, breezy locations, and cuttings compete more for water and nutrients with intercrops.

4.5.2.5 Propagation from Seed

Seeds ought to be sown (nursed) in nylon bags or tubes 3 months prior to the start of the rainy season. The packs ought to be sufficiently long to avoid unduly confining taproot development in the field, in the least expensive technique. The sacks ought to be loaded with free-depleting developing media comprising natural matter, and should be watered all around before planting (Achten et al. 2008). Seeds ought to be taken from full-grown fruit. Just the biggest ought to be chosen for sowing, and any that do not sink in water ought to be disposed of. Pretreatment to soften the seed coat will improve sprouting. Soaking in cow compost slurry for 12 h gave 96% sprouting in comparison with soaking in ice water or scratching the seed coat, which yielded around 72% sprouting (Achten et al. 2008). Only a single seed should be put in each sack, which ought to be placed at 2–4 cm depth with the caruncle pointing downward (Srimathi and Paramathma 2006). The seeds ought to be kept well watered and will then develop within 6–15 days. Seedlings may be planted out after a few months, after achieving a stature of 30–40 cm and before taproot growth becomes excessively confined.

4.5.2.6 Planting

Jatropha curcas is planted at densities of 1100–2500 plants per hectare. The yield per tree tends to be greater with wider spacing between trees but with a decrease in yield per hectare (Achten et al. 2008). Spacing choices ought to be based on the site conditions, e.g., how they influence rivalry among trees for water, light, and nutrients. Semi-dry, low-density systems ought to utilize more extensive spacing—for example, 3.0 × 2.0 m, 3.0 × 2.5 m, or 3.0 × 3.0 m. Exchange planting in succeeding columns will minimize common shading. What is more, thought ought to be given to access. A space of no less than 2.5 m between trees permits less demanding access for organic product pickers, while a 5-m back road at each fourth column allows access for trucks. Planting pits 30–45 cm wide and deep ought to be arranged, with natural matter added before planting. A bug spray may be incorporated as a precautionary measure against termites. The seedlings may require a watering system for the first few months after planting. Where the plant is being planted as a living fence, cuttings of 60–120 cm in length ought to be embedded 15–25 cm apart and 20 cm into the ground. This ought to be done a few months before the onset of the rainy season.

4.6 Uses of *Jatropha curcas*

4.6.1 Potential Phytoremediators

Specialists everywhere throughout the globe are seeking novel plant varieties appropriate to be utilized in phytoremediation. When choosing a plant species category for phytoremediation, a few variables are taken into consideration. The plant species ought to be one that develops quickly, delivers high biomass, has a plentiful root system, tolerates unfavorable environment conditions, is unpalatable, and is monetarily productive (Garbisu et al. 2002; Alkorta et al. 2004). Thinking about every one of these variables, *Jatropha curcas* seems suitable for remediation. It is viewed as a possible biofuel plant for the future because of its low moisture requirements, immaculate strength, and stress-tolerant capacity (Kaushik et al. 2007). It develops fast with little support and can achieve a stature of 3–8 m (Kaushik et al. 2007; Gunaseelan 2009). It has been identified in India and different nations as the most suitable oil-bearing plant and has been suggested for plantation on wasteland, as it requires minimal input for its establishment (Gunaseelan 2009). Thus, phytoremediation of contaminated soil with an inedible biodiesel crop such as *Jatropha curcas* offers an eco-friendly and financially savvy strategy for cleaning contaminated soil. Jamil et al. (2009) reported that *Jatropha curcas* is capable of extracting heavy metal from fly ash (FA), and the extraction is improved by numerous folds in the presence of synthetic chelants such as EDTA. Recently, some scientists have additionally reported that *Jatropha curcas* has the potential for remediation of metalloid and metal–polluted soil systems. *Jatropha curcas* has been involved in remediation of soil polluted by heavy metals (Al, Fe, Cr, Mn, Ar, Zn, Cd, and Pb) because of its bioaccumulation potential (Juwarkar et al. 2008; Yadav et al. 2009). Alternate types of *Jatropha*—for example, *Jatropha dioica*—accrued Zn (6249 mg/kg) at concentrations close to the criteria for hyperaccumulator plants (Gonzalez and Gonzalez 2006). Agamuthu et al. (2010) demonstrated that *Jatropha curcas* with the addition of organic matter has potential in remediating hydrocarbon-polluted soil. Abhilash et al. (2009) suggested that remediation of lindane (a pesticide) is likewise possible with *Jatropha curcas*. Each of these endeavors is in progress to develop a suitable technique to remove oil contaminants/heavy metal contaminants from soil while promoting development of a financially viable plant such as *Jatropha curcas*, which has a seed useful for biodiesel production (Abhilash and Yunus 2011).

4.6.2 Soil Carbon Sequestration

From the perspective of environmental considerations, biodiesel is considered carbon neutral in light of the fact that all of the CO_2 discharged in its utilization has been requisitioned from the atmosphere for the development of crops. As a clean, renewable resource, it has no outflow of CO_2 to cause unnatural contamination (El-Bassam 1998). Establishment of energy plants is helpful for easing CO_2 levels in the environment, plus the outflow of CO_2 by burning of biodiesel will dependably be lower when contrasted with fossil fuel (petro-diesel), thus creating a general decrease of CO_2 emissions into the open atmosphere (Khan and el Dessouky 2009). Optimistic forecasts for the diminishment of greenhouse gases (GHGs) are estimated on the basis that the harmful global warming capability of the creation and utilization of *Jatropha* biodiesel is 23% of the global warming boost capability of fossil diesel (Tobin and Fulford 2005). According to research based on an appraisal of the biomass capability of marginal lands in Northern China (Yan and Chen 2001), the reduction in CO_2 emissions by utilizing biovitality can be relied upon to be around 75 million tons of carbon equivalents in 2020, which would represent 4% of the aggregate 1.8 billion tons of carbon-equivalent CO_2 discharge in China. In 2050, the decrease in CO_2 outflow because of bioenergy is predicted to be 150 million tons of carbon equivalents, representing 5% of aggregate CO_2 discharges. The carbon sink could be sold on the worldwide carbon exchange market, which could decrease the expense of biovitality generation and support bioenergy improvement (Dayal 2004). A few authors have likewise reported that biodiesel emissions depend on feedstock, engine innovation, engine power, and motor operating conditions (Knothe and Steidley 2005). Feedstock and the injection system play an important role because they influence the fuel spray and consequently the burning attributes (McCormick et al. 2001; Agarwal 2007). Along these lines, utilization of *Jatropha* biodiesel reduces CO_2 emissions and brings down the carbon footprint (Gan et al. 2011). All things considered, the *Jatropha* yield might likewise earn carbon credits, but distinctive soil conditions have not received due consideration in life cycle evaluation investigations of *Jatropha* for carbon sequestration to date (Romijn 2011).

4.6.3 Reduction of Environmental Pollutants

Besides the significant diminishment in net CO_2 releases resulting from use of bioenergy instead of coal, airborne contaminations—for example, lethal heavy metals, ozone-forming chemicals, and sulfur dioxide adding to acid rain—will likewise be decreased. These atmospheric compounds (CO_2, toxic heavy metals, ozone-forming chemicals, SO_2, air contaminant gases, etc.) interact with agricultural systems and impact crop performance, either straightforwardly by influencing development and quality or in an indirect way by changing the plant's capacity to adapt to other abiotic and biotic burdens (Bender and Weigel 2011). A few scientists have compared fossil diesel with biodiesel and found that biodiesel by and large caused a decline in unburned HC, carbon monoxide (CO), and particulate (PM) discharges alongside an expansion in NOx emissions (Kalligeros et al. 2003; Klein-Douwel et al. 2009). Numerous scientific papers have reported that the engine operation on biodiesel blended with diesel gave lower discharges than diesel fuel aside from a possible increase in NOx. In the event of NOx emission, there is a 2% increase in NOx with B20 mixing and 10% with B100. Other natural contaminations were decreased to different degrees by utilizing biodiesel (Fig. 4.4). The mixing of methanol or ethanol with fossil diesel or biodiesel obtained from *Jatropha curcas* has been broadly examined as an approach to diminishing smoke and NOx (Senthil et al. 2003).

4.6.4 Soil Erosion Control

Worldwide agro-biological zoning has estimated that somewhere in the range of 16% of the aggregate worldwide territory is in danger of soil disintegration (Koohafkan 2000). The proportions of regions under potential threat from soil disintegration differ from Europe (19%) to North Africa and the Near East (10%) (Koohafkan 2000); in this sense, relief of soil disintegration utilizing *Jatropha curcas* plantations is vital for sustainability of these areas. *Jatropha curcas* builds up a deep taproot and, at first, four shallow parallel roots (Reubens et al. 2011). The taproot may stabilize the soil against landslips, while the shallow roots are thought to help prevent and control soil disintegration brought about by wind or water, yet this potential has not been researched systematically (Achten et al. 2008). *Jatropha*

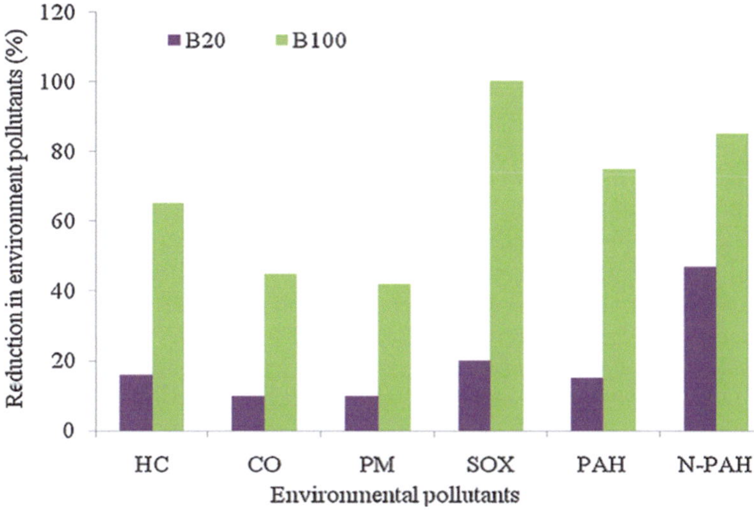

Fig. 4.4 Using biodiesel for lessening environmental contaminants (Khan and el Dessouky 2009) such as unburned hydrocarbons (HC), carbon monoxide (CO), particulate matter (PM), sulfur (SOx), polycyclic aromatic hydrocarbons (PAH), and nitrated polycyclic aromatic hydrocarbons (N-PAH). (Source: Pandey et al. 2012)

curcas shows high initial establishment success and survival (Zahawi 2005). Planting in hedges can be set up rapidly (plantation proliferation) and straightforwardly in the field, but taproots do not form in cutting-propagated *Jatropha curcas*, which makes the florae more vulnerable to dislodgement by wind. Reubens et al. (2011) showed that parallel rows of *Jatropha curcas* could diminish soil erodibility through extra soil attachment by its taproots, and its sinker roots may be able to exploit subsurface soil dampness and therefore encourage vegetative spread, even in exceptionally dry situations. Other than biodiesel generation, *Jatropha* can be planted to diminish soil disintegration and to stabilize embankments (Reubens et al. 2011). Plantation establishment with *Jatropha curcas* in various nations has had different objectives: (i) soil erosion control with fences and oil production (biodiesel) in Mali (Lutz 1992); (ii) reforestation of arid areas for soil erosion control in Cape Verde (Spaak 1990); (iii) utilization as a support plant for vanilla in Madagascar; (iv) biodiesel production in marginal area of India; and (v) use as an energy plantation for generation of methyl esters in Nicaragua. Wind erosion and sand rises could likewise be settled incredibly by the environment reproduction of corrupted area, especially in dry districts.

4.6.5 Utilizing Marginal Land for *Jatropha* Agro-forestry

Marginal lands can be managed for the generation of a renewable vitality source, as their soil quality and impoverished fertility would not have the capacity to sustain field crops, and their recovery, especially in dry-land polluted biological systems, could be conceivable through establishment of *Jatropha* plantations. Ogunwole et al. (2008) studied the effect of *Jatropha* development, with or without soil amendments, on the structural stability, carbon content, and nitrogen content of a degraded entisol under restoration in western India. *Jatropha curcas* sheds its leaves, giving abundant organic matter around the root zone of the plants and encouraging microbial action and earthworms, which enhance the fertility of the soil, thereby improving the ecology of the site. The plant itself is known to be useful for recovery of wetland (Openshaw 2000; Francis et al. 2005). However, no information is available to date on nutrient cycles and the impact on soil biological activities. The growing concern on these issues must be validated by focused research. Utilizing *Jatropha curcas* on wasteland encourages the auxiliary progression of local species to advance neighborhood biodiversity. A significant part of the enthusiasm for *Jatropha curcas* has emerged because of its capacity to grow in "marginal areas"; in this respect, it does not compete with food crops for arable areas. It is estimated that there are currently 2.5 million hectares of area planted with *Jatropha curcas* in India and China alone, with arrangements for an extra 93078.7 ha by 2010 (Fairless 2007). Indian Railways utilizes 2 million kiloliters of diesel each year. The Indian government made it an objective to utilize biodiesel mixing at a 5% level in the normal diesel supply by 2005–2007; however, it was not feasible because of a lack of adequate biodiesel creation, despite the fact that biofuel crops had been planted along both sides of railroad tracks covering a zone of 2500 km. Additionally, a few state governments have begun to sponsor agriculturists to develop *Jatropha* on communal badlands in villages. Around 63.85 million hectares, or roughly 20.17% of the aggregate geological area in India, has been designated as debased areas or "no-man's-land" (Pandey et al. 2010), which directly requires revegetation to avert further degradation. *Jatropha* plantations for supply of biodiesel (conceivably with the extra advantage of carbon sequestration in soils and standing biomass) could assume an imperative part in the rebuilding of these lands. Some areas might be focused on for *Jatropha curcas* suitability—for example, those with degraded and eroded soil, those with tolerably sodic or saline soil, group wastelands, mine spoils, gorges, rainfed lands (low-rain zones/rain shadow regions), water shortage areas, those needing supplanting of uneconomical crops, railroad tracks, roadsides, river sides, Jhum fallows in sloping regions, embankments, erosion-prone watershed territories, fly ash ponds, and heavy metal–polluted areas. Marginal land usage has extraordinary prospects for increasing bioenergy assets on the planet, with concomitant advantages—for example, carbon sequestration, water/soil preservation, and wind erosion protection. Marginal land is characterized as an area that has a delicate eco-environment and is unsatisfactory for agriculture (Wu and Gao 1998; Hamdar 1999). Bioenergy plant species for marginal land ought to have a few qualities—for example, the properties of low water utilization, dry spell tolerance, saltiness and sodicity resistance, high net productivity, and high energy value, and in this way they have huge potential for being widely planted and for usage as energy crops. Attendant eco-environmental advantages can be accomplished by utilizing marginal land for planting bioenergy shrubs rather than relinquished pasture with low acceptable value. The capacity for soil/water preservation could come about fundamentally from the protection of the soil surface and from changing the soil structure through root penetration and expansion of organic matter by breakdown of leaves, roots, and wood (Li et al. 2007). Marginal lands could be more healthily utilized for *Jatropha* agro-ranger services with intercropping of occasional harvests to produce income between crops of *Jatropha curcas*. A few shade-loving plants, short duration pulses, vegetables, and shade-loving fragrant herbs can be gainfully developed under *Jatropha* plantations for the initial 2 years.

4.6.6 Medicinal Value

The genus name *Jatropha* comes from the Greek word *giatros* (doctor), and *trophe* (nourishment) implies restorative properties. As indicated by Correll and Correll (Coelho-Ferreira 2009), *curcas* is the common name for the physic nut in Malabar, India. Therapeutic properties are mainly found in *Jatropha curcas*, *Jatropha multifida, Jatropha gossypifolia, Jatropha macrorhiza* and *Jatropha cinerea* (Coelho-Ferreira 2009). *Jatropha curcas* has numerous restorative qualities for human and veterinary usefulness and potentially has other unrecognized helpful qualities for utilization in other ways. Some therapeutic uses of *Jatropha curcas* are given in Table 4.3. A decoction of leaves is utilized against cough and as a antiseptic after birth. Branches are utilized as a chewing stick in Nigeria (Isawumi 1978).

Table 4.3 Uses of different parts of *Jatropha curcas*

S. no.	Usable plant parts	Diseases curing
1	Seeds	To treat gout, arthritis and jaundice, wound-healing, fractures, burns, purge
2	Seed oil	Eczema, skin diseases, soothe rheumatic pain, purgative action
3	Stem	Toothache, gum inflammation, gum bleeding, pyorrhea
4	Stem bark	Infectious diseases, including sexually transmitted diseases
5	Plant sap	Dermatomucosal diseases
6	Water extract of branches	HIV, tumour
7	Plant extract	Wound healing, allergies, burns, cuts and wounds, inflammation, leprosy
8	Leaves and latex	Refractory ulcers, septic gums, styptic in cuts and bruises
9	Latex	Reduced the clotting time of human blood, sore mouth, oral thrush, fish barb wounds, snake-bites, infected sores, trating newborns' umbilical cords, coughs, mouth and throat sores
10	Root powder	In the treatment of inflammation
11	Leaf	Scabies, Eczema, Syphilis, blood cleansing, headache, flu, cough, congestion, evil eye, cleansing house
12	Fruit	Stroke, toothache, numbness after bug sting, to clean mother's and baby blood during the pregnancy

Source: Samy et al. (1998), Osoniyi and Onajobi (2003), Mujumdar and Misar (2004), and Kaushik et al. (2007)

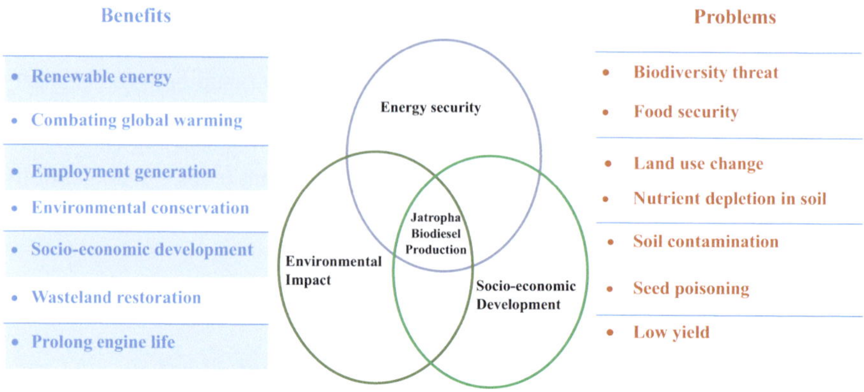

Fig. 4.5 Summary of *Jatropha* uses (Source: Pandey et al. 2012)

The sap spilling out of the stem is utilized to stem bleeding from wounds. Nath and Dutta (1992) showed the healing properties of curcain, a proteolytic catalyst isolated from latex. Latex has antimicrobial properties against *Staphylococcus aureus*, *Escherichia coli*, *Klebsiella pneumoniae*, *Streptococcus pyogenes*, and *Candida albicans* (Thomas 1989). The latex itself has been observed to be a solid inhibitor of watermelon mosaic virus (Tewari and Shukla 1982). HIV inhibitor properties have been found in *Jatropha* (Matsuse et al. 1998; Wender et al. 2008). Goonasekera et al. (1995) found that concentrates of *Jatropha* organic products had an abortive effect. A methanol concentrate of *Jatropha curcas* roots showed systemic and significant anti-inflammatory activity in intense carrageenan-instigated rodent paw edema (Mujumdar and Misar 2004), and it was found that a methanol concentrate of physic nut leaves moderately safeguarded refined human lymphoblastoid cells against the cytopathic impacts of human immunodeficiency infection. A concentrate of the leaves demonstrated intense cardiovascular activity in guinea pigs and may be a conceivable wellspring of beta-blocker agents (Fojas et al. 1986). All parts of the plant are utilized as a part of conventional pharmaceuticals, and dynamic segments are being researched in experimental trials. A few parts seem to have promising applications both in prescription medications and as plant protectants (Table 4.3 and Fig. 4.5).

4.7 Remediation of Soil Contaminants by *Jatropha curcas*

4.7.1 Heavy Metals

Soil contamination has recently been attracting a lot of open consideration since the scale of this problem in soils calls for quick action (Garbisu and Alkorta 2003). As a consequence of human activities—for example, mining and smelting of metalliferous substances, electroplating, gas fumes, energy and fuel production, and manure and pesticide application—metal contamination has ended up as a standout among the most genuine environmental issues today. Because of their permanent nature, metals are a group of contaminants of much concern. Truth be told, although several metals are fundamental for biological systems and must be available within a specific concentration range (Garbisu and Alkorta 2003), at high concentrations, metals can act in a deleterious manner by blocking essential functional groups, dislodging other metal particles, or changing the dynamic adaptation of biological molecules. Phytoremediation could provide a solution for such pollution, and *Jatropha curcas* has great potential for this. The plant belongs to the *Euphorbiaceae* family, which may be compelling in removing Pb and Cd, as exhibited by *Euphorbia cheirandenia* (Eapen and D'Souza 2005). In Indonesia, the plant is grown under different soil conditions; it is planted as a green corridor along streets and outside houses, and it has been proposed as a wellspring of possible biofuel. What is more, *Jatropha* has been observed to be capable of remediating hexavalent chromium (Mangkoedihardjo et al. 2008). Accordingly, planting of *Jatropha curcas* could have a few uses, e.g., remediation of soil contaminated with Pb and Cd, procurement of green space, and as a wellspring of alternative biofuel.

4.7.2 Organic Contaminants (Petroleum, Spent Lubricating Oil, Trinitrotoluene, Pesticides)

The principal objective in phytoremediation is to discover a plant variety that is impervious to or can endure a specific pollutant, with a view to amplifying its latent potential for remediation. Resistant crops are normally found developing on earth with hidden minerals or on the edges of contaminated sites. When a tolerant plant variety has been chosen, conventional reproducing techniques are utilized to advance the resistance of the plant variety to a specific pollutant. Agrarian strategies—for example, the use of chelators, pH adjusters, and fertilizer—can be used to enhance the possibility of phytoremediation.

For effective remediation using plants, microorganisms and the plants must subsist and develop in unrefined petroleum–polluted soil, which is frequently low in nutrients. Microorganisms require nitrogen, phosphorus, potassium, and micronutrients to degrade natural soil pollutants (Rangzan 2006). Among all of these, nitrogen is required in the highest concentrations (Besalatpour et al. 2008). Compost containing nitrogen is regularly added to encourage crop development in petroleum-polluted soils (Ahmadi and Mousavi 2005). Although nitrogen prerequisites for normal agronomic systems are well documented (Anderson et al. 1993), these necessities often do not coincide with the nitrogen requirements in raw petroleum–polluted systems. Suitable agronomic practices—for example, culturing and lime additions—can likewise be utilized to enhance the physical and chemical conditions of the soil for improved plant and microorganism development. Contamination of groundwater resources and soil with organic compounds around oil refineries and in the region of fuel transportation facilities are significant signs of an environmental contamination problem. Organic hydrocarbons (e.g.,. petroleum) are often found in different chemicals utilized for human activities. Spilled and released crude oil fills soil pores by moving vertically with gravity and capillary force into unsaturated soil. A substantial spillage can reach the water level, accumulate, and move into the groundwater (as its particular gravity is less than that of water) and will drift in the water. Total petroleum hydrocarbons (TPHs) for the most part defile the environment because of their innate capacity for solubility, instability, and biodegradability. Petroleum that has already been discharged from an underground tank or a refinery will probably be found in soils and perhaps at the same time in shallow groundwater and air, threatening the health of humans and the environment.

Idowu and Fayinminnu (2015) utilized *Jatropha curcas* seedlings to phytoremediate organically amended soil with 0%, 3%, and 6% (w/w) spent lubricating oil for 84 days. They found that at the end of 12 weeks, the 0% contamination level recorded the greatest mean height (56.15 cm), mean stem diameter (7.42 cm), and mean number of leaves (49.28). The mean height for the 3% contamination level was 22.34 cm, while the mean diameter and mean leaf number were 3.86 cm and 10.23, respectively. The lowest values of mean height (15.83 cm), mean stem diameter (2.81 cm), and mean number of leaves (5.73) were seen at the 6% level of contamination. In altered soil, higher values of mean height (43.04 cm), mean stem diameter (5.66 cm), and mean number of leaves (34.96) were observed in comparison with values of 19.84 cm, 3.74 cm, and 8.54 for mean height, stem diameter, and number of leaves in unamended soil. The outcomes indicated huge contrasts ($P<0.05$) in height, stem diameter, and leaf generation over the 0% (control), 3%, and 6% levels of contamination in the altered and unamended soil types that were researched. In summary, the observed growth parameters demonstrated that soil amendment brought about better performance in terms of the mean height, mean diameter, and mean number of leaves. Then again, increased contamination of the soil with spent lubricating oil (3% and 6% levels of contamination) had recognizable antagonistic consequences for the development parameters.

Phytoremediation of soil polluted with 2.5% and 1% (w/w) spent lubricating oil utilizing *Jatropha curcas* and improvement with organic wastes [banana skin (BS), brewery spent grain (BSG), and spent mushroom compost (SMC)] was performed for a time of 180 days under room conditions (Agamuthu et al. 2010), and 56.6% and 67.3% losses of spent lubricating oil were recorded in *Jatropha*-remediated soil without organic amendment at the 2.5% and 1% contamination levels, respectively. However, addition of organic waste (BSG) to *Jatropha* remediation quickly increased the removal of spent lubricating oil to 89.6% and 96.6% in soil polluted with 2.5% and 1% oil, respectively. *Jatropha* roots did not accumulate hydrocarbons from the soil; rather, the quantity of hydrocarbon-using bacteria was high in the rhizosphere of the *Jatropha* plant, accordingly suggesting that the mechanism of the oil degradation was rhizodegradation (Fig. 4.6).

4.7.3 Radionuclides

Radionuclides (organic contaminants) can be either taken up from the soil and immobilized by the roots (phytoimmobilization) or transported to the plant shoot

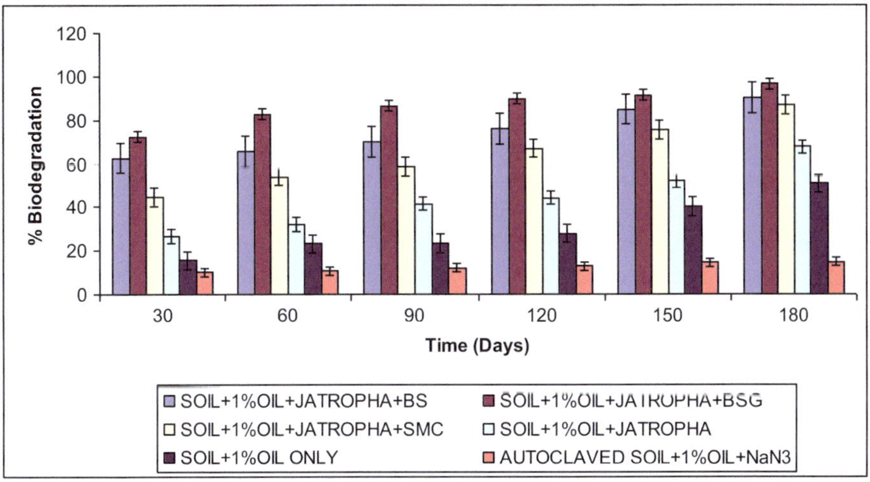

Fig. 4.6 Percentage biodegradation of waste lubricating oil in soil contaminated with 1% oil. *Bars* indicate standard errors ($n = 3$) (Agamuthu et al. 2010)

(phytoextraction) (Reichenauer and Germida 2008). Since, under most circumstances, the bioavailability of metals (including a few metals fundamental to life) in soil is fairly low, plants (including *Jatropha curcas*) have exceptionally viable metal uptake systems utilizing transporter molecules—for example, zinc-regulated transporter protein and copper transporter protein (Krämer et al. 2007). Furthermore, plants are fit for discharging metal-chelating particles such as siderophores and organic acids (malate, citrate) and biosurfactants (e.g., rhamnolipids) to the surrounding soil, furthermore expelling protons from the roots to acidify the soil and activate soil-bound metals (Garbisu and Alkorta 2001). Organic pollutants, radionuclides, and heavy metals cannot be biodegraded inside the plant; they are just changed from one oxidation state or organic complex into another (Garbisu and Alkorta 2001). Accordingly, metals have a tendency to accumulate in the plant. Almost 450 hyperaccumulator plants, ranging from annual herbs to perennial shrubs and trees (e.g., tobacco, sunflower, mustard, maize, pennycress, brake fern, Russian thorn, rattlebush, python tree, willow, poplar, and *Jatropha*), have been described as accumulating and detoxifying extraordinary amounts of metal particles—for example, Ni, Co, Pb, Zn, Mn, and Cd—in their above-ground tissues (Meagher 2000; Padmavathiamma and Li 2007; Shah and Nongkynrih 2007; Sheoran et al. 2009; Abioye et al. 2010). For some inorganic components (Hg, As, and Se), uptake by roots followed by transport to the shoot and transpiration to the environment through the leaf stomata (phytovolatilization) have been observed (Padmavathiamma and Li 2007). Since the volatile forms of Hg and Se are dangerous, it is unclear whether the volatilization of these components into the atmosphere is desirable or safe (Watanabe 1997; Padmavathiamma and Li 2007).

4.8 Microbial-Assisted Remediation of Soil Contaminants by *Jatropha curcas*

By rhizospheric association, plants up take most mineral nutrients through interactions between microorganisms and products from the plant in exudates of the root comprising numerous natural corrosive anions, vitamins, inorganic particles, amino acids, purines, nucleosides, root fringe cells, sugars, vaporous atoms, chemicals, and phytosiderophores (Dakora and Phillips 2002). The exudation rate is increased by the presence of microbes in the rhizospheric zone (Gardner et al. 1983). A few exudations from the roots act as metal chelators and improve the availability of metallic soil mini-nutrients. Metal chelators bind to metals in the soil, in this way diffusing the metals through the soil and enhancing their solubility and portability. Binding of substantial metals by cysteine-rich peptides is a noteworthy mechanism utilized by plants for limiting heavy metal danger. Phytochelatins have been observed to bind substantial metals with high affinity (Dakora and Phillips 2002). Furthermore, plant siderophores and microbes (bacteria) additionally act as chelating operators to dissolve the iron (Fe) attached to soil particles. With the completion of solubilization, iron (Fe) is taken up by the living cells through particular layer transporters and is absorbed. The roots of the plant can influence microbial activities in the rhizosphere, support advantageous interactions, influence the synthetic and physical properties of soil, and repress the development of competitive plant species. Plant growth–promoting rhizobacteria (PGPR) advance plant development straightforwardly or in a roundabout way by delivering plant development promoters (e.g., cytokinin, gibberellin, auxin, and indole acidic corrosives), siderophores, chelating operators, phytohormones, anti-infection agents (Shanahan et al. 1992), cyanide (Flaishman et al. 1996), nonsymbiotic nitrogen fixers (Boddey and Dobereiner 1995), solubilizing mineral nutrients, and phosphate (de Freitas et al. 1997), and serving in a couple of different capacities (Kamnev and van der Lelie 2000; Joseph et al. 2007). These vital plant growth–promoting rhizobacteria include *Azotobacter, Azospirillum, Acetobacter, Pseudomonas, Rhizobium*, and *Achromobacter*. An expansive cluster of microbes—including species of *Gluconacetobacter*, *Alcaligenes*, *Arthrobacter*, *Burkholderia*, *Bacillus*, *Klebsiella*, *Seratia*, and *Enterobacter*—have additionally been observed to improve plant development (Okon and Labandera-Gonzalez 1994; Glick 1995; Saravanan et al. 2008).

Agamuthu et al. (2010) discovered positive effects of some of these microorganisms in remediation of hydrocarbon contamination in the soil with 2.5% and 1% spent lubricating oil. Polluted soil treated with BSG and *Jatropha* remediation indicated high counts of hydrocarbon-utilizing bacteria (HUBs; 240×10^5 CFU/g and 193×10^5 CFU/g) in soil contaminated with 2.5% and 1% oil, respectively, while treatment with just *Jatropha* plants without organic waste amendment resulted in low HUB counts (48×10^5 and 45×10^5 CFU/g) with 2.5% and 1% contamination, respectively. The explanation behind the increase in HUB counts in polluted soil amended with organic waste may be that the presence of nutrients in the organic waste, particularly nitrogen and phosphorus, enhanced the augmentation of bacteria in the soil. The HUBs isolated from the polluted soil were identified as species of

Pseudomonas, *Bacillus megaterium*, *Micrococcus*, and *Corynebacterium.* These bacterial species, together with root exudates of *Jatropha* plants, potentially help in the removal of spent lubricating oil from the soil. The rhizosphere of *Jatropha* harbors metabolically various bacteria measured as HUBs. In this manner, it is proposed that oil removal from the soil may occur through the mechanism of rhizodegradation. Addition of organic waste to the polluted soil further upgraded the development of *Jatropha* and the proliferation of bacteria in the soil, along these lines representing increases in oil removal of 33% and 29.3% in soil contaminated with 2.5% and 1% oil, respectively, compared with treatment with *Jatropha* alone. The study in this way demonstrated the reasonability of utilizing *Jatropha curcas* with BSG amendment in remediating hydrocarbon-contaminated soil. Thus strategies could be employed to remove oil contaminants from soil while promoting growth of economically viable plants, such as *Jatropha*, whose seed can be used for production of biodiesel (Agamuthu et al. 2010).

4.9 Mechanisms of Soil Contaminant Remediation by *Jatropha curcas*

Jatropha curcas has significant advantages for phytoremediation because of its potential to endure natural environmental stresses. New research has suggested that expression of the aldehyde dehydrogenase gene *JcBD1* in *Jatropha curcas* releases the JcBD1 enzyme, which helps the plant to survive environmental stresses caused by heat, salt, and dry spells (Zhang et al. 2008). In *Escherichia coli*, the release of the JcBD1 enzyme resulting from expression of this gene makes it impervious to abiotic stressors such as salt. Use of *Jatropha curcas* to produce a petroleum-substitute biodiesel crop for renewable energy holds promise, given the depletion of fossil fuel sources. Some plants amass fundamental and insignificant metals in their roots and shoots in higher concentrations than the levels present in the soil (Raskin et al. 1994). Plants that can ingest large amounts of pollutants by concentrating them in roots and shoots are called hyperaccumulators.

Phytoaccumulation (phytoextraction) utilizes plants or green growth (algae) to expel pollutants from sediments, soils, or water into harvestable plant biomass. Phytoaccumulation has rapidly become widespread globally throughout the last quarter century. By and large, this procedure has been attempted more frequently to extract heavy metals than for organics. By the time of harvest, pollutants are commonly amassed in the much smaller volume of the plant matter rather than in the initially tainted soil or dregs. This crop ingests pollutants through the root system and stores them in the root biomass and/or transports them up into the stems and/or leaves. A living plant may keep on accumulating pollutants until it is harvested. When the harvest is done, a lower level of the contaminant will stay in the soil, so the development cycle should typically be repeated for a few yields to accomplish thorough remediation. After the procedure, the remediated soil can support other vegetation (Ahmadi and Mousavi 2005).

Rhizofiltration is fundamentally the same as phytoaccumulation because it evacuates pollutants by capturing them into harvestable plant biomass. To start with, plants are placed in contact with the pollution. They assimilate contaminants through their root systems and store them in their root biomass and/or transport them up into the stems and/or leaves. The plants are then replaced to proceed with the development/harvest cycle until acceptable levels of contaminants are reached (Ahmadi and Mousavi 2005). Phytoremediation is an innovation taking into account synergistic participation of plant roots and soil microorganisms for disintegrating, exchanging, deactivating, and rendering inert contaminated mixtures of soil and underground water (Cunningham et al. 1996). In phytoremediation, the rhizosphere stimulates microbial activities, which provide a suitable domain for development and expansion, helping to diminish the petroleum defilement of the soil. Microorganisms in the rhizosphere benefit from the root exudates, and the plants benefit from the metabolic detoxification of conceivably dangerous mixtures achieved by those microbial groups.

Restriction of plant development parameters (germination, plant length, and biomass) can be brought on by dangerous compounds of petroleum hydrocarbons (e.g., low atomic weight hydrocarbons). They can enter and pass through cell layers, leading to reduced membrane integrity and/or to death of the plant cell. Phytoremediation is a site-particular remediation technique, which is why some conflicting results have been reported with respect to the effectiveness of this innovation in expelling contaminants from soil. Utilizing local plant species that are tolerant of high concentrations of TPHs in soil can be a key variable in the accomplishment of phytoremediation. Phytoremediation is the utilization of green plants and root-related soil microorganisms, soil amendments, and agronomic procedures to uplift, contain, or render innocuous natural contaminants (Shahriari et al. 2006). Phytoremediation involves the rhizosphere, or the zone of soil nearest to and specifically impacted by plant roots (Besalatpour et al. 2011). Phytoremediation, like bioremediation, is moderately nonintrusive and involves minimal effort for a therapeutic choice appropriate to numerous locales.

4.10 Advantages of Using *Jatropha curcas* in Soil Remediation

Jatropha roots of various treatments were Soxhlet extracted to figure out whether there was phytoaccumulation of hydrocarbons in the plant root. Gas chromatography/mass spectrometry (GC/MS) investigation of the concentrate did not demonstrate the presence of hydrocarbons with any of the treatments. This is in sharp contrast to the results reported by Palmroth et al. (2002), who observed an uptake of diesel oil by grass roots. *Jatropha curcas* removes carbon from the atmosphere (carbon sequestration), stores it in woody tissues, and helps with the building of soil carbon (Agbogidi and Ekeke 2011). Industrialists, miners, environmentalists, and others have endorsed the idea of regular remediation by *Jatropha curcas*, and

Fig. 4.7 Growth of *Jatropha curcas* with different treatment levels of refinery sludge (Source: Jyoti and Smita 2012)

numerous nations have consequently shown incredible enthusiasm toward the development of this species for bioremediation purposes.

Studies have demonstrated that *Jatropha curcas* could remediate soils polluted with heavy metals and hydrocarbons (Kumar et al. 2008; Agbogidi and Eruotor 2012; Agbogidi et al. 2013). Jones and Miller (1993) reported that *Jatropha curcas* is a multipurpose species for problematic sites. The growth parameters of *Jatropha* plants, as elucidated by Jyoti and Smita (2012), were significantly influenced by the harmfulness of refinery sludge and the multi-metal contamination in it. Be that as it may, at lower levels of contamination up to S3 treatment (Fig. 4.7), the observed impact was unimportant. The plants in the control soil had the greatest normal shoot length (62 ± 2.0 cm), followed by those in the S2 and S1 treatments, while those in the S6 treatment had the shortest shoot length (31 ± 1.0 cm). The plants in the control treatment and S1 treatment likewise had the greatest numbers of leaves, followed by those in the S2 treatment, and the plants in the S5 and S6 treatments appeared to have a 40–60% reduction in the numbers of leaves. *Jatropha curcas* was observed to be fit for proficient removal of all observed heavy metals—for example, Cd, Cr, Cu, and Ni—from the soil, and the percentage removal was highest in the case of cadmium. Similar patterns were seen in the cases of chromium and nickel removal (Fig. 4.7) (Jyoti and Smita 2012).

4.11 Constraints Associated with Use of *Jatropha curcas* in Soil Remediation

Prolonged treatment times for recovery and control of ecological elements, where possible, keeps phytoremediation under criticism. A significant disadvantage in phytoremediation is that plants experience more numerous stressors in field trials than they do in greenhouse systems and laboratories. Among the many constraints

Fig. 4.8 *Jatropha curcas:* Healthy leaf (**a**), necrosis (**b**, **c**), chlorosis (**d**, **e**), and complete destruction (**f**) with initial lead and cadmium concentrations in soil of more than 80 mg/kg

are nutrients, precipitation (rainfall), herbivory, plant pathogens, variations in temperature, antagonistic impacts of pesticides, weeds, and weedicide applications, though completely controlled conditions are maintained in ex situ strategies (Gerhardt et al. 2009). Moreover, the root structure, soil surface (texture) and quality, and bioavailability of nutrients, among other things, can change after some time and take an undesirable turn simultaneously. Regardless of a few impediments, phytoremediation is generally acknowledged for being a characteristic method for heavy metal remediation of the earth. Some of the bases for this are its in situ process of bioaugmentation, which increases the value of the procedure; alleviation of soil disintegration and dangerous atmospheric deviation; and creation of biofuels and natural gas from plant biomass as a distinct alternative to fossil fuels.

The appearance of *Jatropha* plants in response to various concentrations of oil was checked throughout the 180 days of the trial; no plant deaths were recorded in any of the soil treatments polluted with 1% spent lubricating oil; however, some of the plants showed signs of phytotoxicity (e.g., yellowing of leaves or stunted growth compared with the controls). Vouillamoz and Mike (2009) described comparable objective findings. Plants in soil polluted with 2.5% lubricating oil indicated high manifestations of phytotoxicity, with the death of at least one *Jatropha* plant being recorded in every treatment. Thus, *Jatropha* plants can endure minor exposure to hydrocarbons (Agamuthu et al. 2010). Sarwoko and Surahmaida (2008) examined phytoremediation of cadmium and lead by *Jatropha* and observed week by week advancement of the toxins' impact on leaf necrosis and chlorosis (Fig. 4.8) at underlying lead and cadmium concentrations of 90 mg/kg. These results were observed in conditions with a maximum starting lead concentration of around 80 mg/kg and a cadmium concentration of around 50 mg/kg for phytoremediation using *Jatropha*.

Sarwoko and Surahmaida (2008) concluded that *Jatropha* was not an accumulator of lead and cadmium. Be that as it may, the plant could be utilized for Pb and Cd phytoremediation of contaminated soil provided that the most extreme underlying concentrations of each metal were around 50 mg/kg. Cadmium was more poisonous to plant dry matter than lead, but no adverse impact on plant dry matter was found in the presence of both contaminants. Lead was a synergic poison with cadmium in terms of effects on plant dry matter. Furthermore, Agamuthu et al. (2010) found that *Jatropha* could become extremely successful in remediation of pollutants if further enhanced.

4.12 Conclusion

Phytoremediation is an innovation that depends on the combined activity of plants and their related microbial communities to degrade, remove, transform, or immobilize toxic compounds situated in soils, sediments, groundwater and surface water. Phytoremediation has been utilized to treat numerous classes of contaminants, including petroleum hydrocarbons, chlorinated solvents, pesticides, explosives, heavy metals, and radionuclides in soil and polluted water. This strategy will be more productive if the right kinds of plants are chosen. Consequently, *Jatropha* has an incredible capacity to reduce and degrade different soil contaminants through its ruggedness and resilience in wide atmospheric conditions.

References

Abhilash PC, Yunus M (2011) Can we use biomass produced from phytoremediation. Biomass Bioenergy 35:1371–1372

Abhilash PC, Jamil S, Singh N (2009) Transgenic plants for enhanced biodegradation and phytoremediation of organic xenobiotics. Biotechnol Adv 27:474–488

Abioye OP, Agamuthu P, Abdul Aziz A (2010) Phytoaccumulation of zinc and iron by *Jatropha curcas* grown in used lubricating oil–contaminated soil. Malays J Sci 29(3):207–213

Abioye OP, Ekundayo PO, Oyeleke SB, Aransiola SA (2015) Remediation of zinc contaminated soil using *Acalypha wilkesiana*. J Glob Ecol Environ 2(2):82–89

Achten WMJ, Verchot L, Franken YJ, Mathijs E (2008) Jatropha bio-diesel production and use. Biomass Bioenergy 32:1063–1084

Afzal M, Khan QM, Sessitsch A (2014) Endophytic bacteria: prospects and applications for the phytoremediation of organic pollutants. Chemosphere 117:232–242

Agamuthu P, Abioye OP, Aziz AA (2010) Phytoremediation of soil contaminated with used lubricating oil using *Jatropha curcas*. J Hazard Mater 179:891–894

Agarwal AK (2007) Biofuels (alcohols and biodiesel) applications as fuels for internal combustion engines. Prog Energy Combust Sci 33:233–271

Agbogidi OM, Ekeke EA (2011) *Jatropha curcas*: Linn an important but neglected plant species in Nigeria. J Biol Chem Res 28(1):52–62

Agbogidi OM, Eruotor PG (2012) Morphological changes due to spent engine oil contamination and its heavy metal components of *Jatropha curcas*. In: Baby S, Sandhu PS (eds) Proceedings of the International Conference on Bioscience, Biotechnology and Health Sciences ICBBHs

2012 organized by Planetary Science Centre Research December 14 and 15, in Singapore, pp 88–93

Agbogidi OM, Mariere AE, Ohwo A (2013) Metal concentration in plant tissues of *Jatropha curcas* L. grown in crude oil contaminated soil. J Bioinnovation 2(3):137–145

Ahmadi, A.M. and Mousavi, B.A. (2005). Phytoremediation, a novel method for correcting the contaminated soil and water. International Center for Advanced Science, Technology and Environmental Sciences, 4th National Biotechnology Congress of Iran, Kerman

Alkorta I, Herna AJ, Becerril JM, Amezaga I, Albizu I, Garbizu C (2004) Recent findings on the phytoremediation of soil contaminated with environmentally toxic heavy metals and metalloids such as zinc, cadmium, lead and arsenic. Environ Sci Biotechnol 3:71–90

Anderson TA, Guthrie EA, Walton BT (1993) Bioremediation in the rhizosphere. Environ Sci Technol 27:2630–2636. Int J Farm Alli Sci 2(21):909–913, 2013 913

Aransiola SA, Ijah UJJ, Abioye OP (2013) Phytoremediation of lead polluted soil by *Glycine max* L. Appl Environ Soil Sci, 2013, Article ID 631619, doi:10.1155/2013/631619

Augustus GDPS, Jayabalan M, Seiler GJ (2002) Evaluation and bioinduction of energy components of *Jatropha curcas*. Biomass Bioenergy 23:161–164

Ayoub K, Van Hullebusch ED, Cassir M, Bermond A (2010) Application of advanced oxidation processes for TNT removal: a review. J Hazard Mater 178(1–3):10–28

Azaizeh H, Castro PML, Kidd P (2011) Biodegradation of organic xenobiotic pollutants in the rhizosphere. In: Schröder P, Collins CD (eds) Organic xenobiotics and plants. Springer, Dordrecht, pp 191–215

Bais HP, Weir TL, Perry LG, Gilroy S, Vivanco JM (2006) The role of root exudates in rhizosphere interactions with plants and other organisms. Annu Rev Plant Biol 57:233–266

Basha SD, Franis G, Makkar HPS, Becker K et al (2009) A comparative study of biochemical traits and molecular markers for assessment of genetic relationships between *Jatropha curcas* germplasm from different countries. Plant Sci 176:812–823

Belluck DA, Benjamin SL, Baveye P, Sampson J, Johnson B (2003) Widespread arsenic contamination of soils in residential areas and public spaces: an emerging regulatory or medical crisis? Int J Toxicol 22:109–128

Bender J, Weigel HJ (2011) Changes in atmospheric chemistry and crop health: a review. Agron Sustain Dev 31:81–89

Berg G, Krechel A, Ditz M, Sikora RA, Ulrich A, Hallmann J (2005) Endophytic and ectophytic potato-associated bacterial communities differ in structure and antagonistic function against plant pathogenic fungi. FEMS Microbiol Ecol 51:215–229

Besalatpour A, Melali A, Hajabbasi M, Afyouni M,,Khoshgoftarmanesh A (2008) Responses of plants to soil pollution around refineries in Tehran. J Sci Technol Agric Nat Resour. (SoilWater Sci), 12th Year, No.44

Besalatpour A, Hajabbasi M, Afyouni M,, Khoshgoftarmanesh A (2011) Decisions based on genetic galgorithm to select the right plants for phytoremediation of soils scontaminated with petroleum. Jpn Soil Res (SoilWater Sci), A, No.25

Boddey RM, Dobereiner J (1995) Nitrogen fixation associated with grasses and cereals: recent progress and perspectives for the future. Nutr Cycl Agrosys 42:241–250

Braeckevelt M, Rokadia H, Imfeld G et al (2007) Assessment of in situ biodegradation of monochlorobenzene in contaminated groundwater treated in a constructed wetland. Environ Pollut 148:428–437

Calheiros CSC, Rangel A, Castro PML (2007) Constructed wetland systems vegetated with different plants applied to the treatment of tannery wastewater. Water Res 41:1790–1798

Calheiros CSC, Teixeira A, Pires C et al (2010) Bacterial community dynamics in horizontal flow constructed wetlands with different plants for high salinity industrial wastewater polishing. Water Res 44(17):5032–5038

Cardinal P, Anderson JC, J.C. C, Low JE, Challis JK, Beattie SA, Wong CS (2014) Macrophytes may not contribute significantly to removal of nutrients, pharmaceuticals, and antibiotic resistance in model surface constructed wetlands. Sci Total Environ 482–483:294–304

Chaudhry Q, Blom-Zandstra M, Gupta S, Joner EJ (2005) Utilising the synergy between plants and rhizosphere microorganisms to enhance breakdown of organic pollutants in the environment. Environ Sci Pollut Res 12(1):34–48

Coelho-Ferreira M (2009) Medicinal knowledge and plant utilization in an Amazonian coastal community of Maruda, Para State (Brazil). J Ethnopharmacol 126:159–175

Crane M, Giddings JM (2004) 'Ecologically acceptable concentrations' when assessing the environmental risks of pesticides under European Directive 91 414/EEC. Hum Ecol Risk Assess 10:733–747

Cunningham SD, Berti WR (1993) Remediation of contaminated soils with green plants: an overview. In Vitro Cell Dev Biol 29P:207–212

Cunningham SD, Anderson TA, Schwab AP, Hsu FC (1996) Phytoremediation of soils contaminated with organic pollutants. Adv Agron 56:55–114

Dakora FD, Phillips DA (2002) Root exudates as mediators of mineral acquisition in low-nutrient environments. Plant Soil 245:35–47

Dams RI, Paton GI, Killham K (2007) Rhizoremediation of pentachlorophenol by *Sphingobiumchlorophenolicum* ATCC 39723. Chemosphere 68(5):864–870

Dayal P (2004) Rural electrification in developing countries using renewable energy and carbon credits. Air Water Manage Assoc 11:25–29

de Freitas JR, Banerjee MR, Germida JJ (1997) Phosphate solubilizing rhizobacteria enhance the growth and yield but not phosphorus uptake of canola (*Brassica napus* L.). Biol Fertil Soils 24:358–364

Dehgan B, Schutzman B (1994) Contributions toward a monograph of neotropical *Jatropha*: phenetic and phylogenetic analyses. Ann Mo Bot Gard 81:349–367

Dias LAS (2011) Biofuel plant species and the contribution of genetic improvement. Crop Breed Appl Biotechnol S1:16–26

Dias LAS, Leme LP, Laviola BG, Pallini A, et al (2007) Cultivo de Pinhão Manso (*Jatropha curcas*) para Produção de Óleo Combustível. UFV, Viçosa

Eapen S, D'Souza SF (2005) Prospects of genetic engineering of plants for phytoremediation of toxic metals. Biotechnol Adv 23:97–114

El-Bassam N (1998) Energy plant species: their use and impact on environment and development. James and James Science Publishers, London

Evans J, Wood G, Miller A (2006) The risk assessment–policy gap: an example from the UK contaminated land regime. Environ Int 32:1066–1071

Fairless D (2007) Biofuel: the little shrub that could—maybe. Nature 449:652–655

Flaishman MA, Eyal ZA, Zilberstein A, Voisard C, Haas D (1996) Suppression of *Septoria tritci* blotch and leaf rust of wheat by recombinant cyanide producing strains of *Pseudomonas putida*. Mol Plant-Microbe Interact 9:642–645

Foidl N, Foidl G, Sanchez M, Mittelbach M et al (1996) Jatropha curcas L. as a source for the production of biofuel in Nicaragua. Biores Technol 58:77–82

Fojas FR, Garcia LL, Venzon EL, Sison FM, Villanueva BA, Fojas AJ et al (1986) Pharmacological studies on *Jatropha curcas* as a possible source of anti-arrhythmic (beta-blocker) agent. Philipp J Sci 115:317–328

Francis G, Edinger R, Becker K (2005) A concept for simultaneous wasteland reclamation, fuel production, and socio-economic development in degraded areas in India: need, potential and perspectives of *Jatropha* plantations. Nat Resour Forum 29:12–24

Gan Y, Liang C, Hamel C, Cutforth H, Wang H (2011) Strategies for reducing the carbon footprint of field crops for semiarid areas. A review. Agron Sustain Dev 31:643–656

Garbisu C, Alkorta I (2001) Phytoextraction: a cost-effective plant-based technology for the removal of metals from the environment. Bioresour Technol 77:229–236

Garbisu C, Alkorta I (2003) Basic concepts on heavy metal soil bioremediation. Eur J Miner Process Environ Prot 3:58–66

Garbisu C, Hernandez-Allica J, Barrutia O, Alkorta I, Becerril JM (2002) Phytoremediation: a technology using green plants to remove contaminants from polluted areas. Rev Environ Health 17:75–90
Gardner WK, Barber DA, Parbery DG (1983) The acquisition of phosphorus by *Lupinus albus* L. III. The probable mechanism by which phosphorus movement in the soil/root interface is enhanced. Plant Soil 70:107–124
Gerhardt KE, Huang XD, Glick BR, Greenberg BM (2009) Phytoremediation and rhizoremediation of organic soil contaminants: potential and challenges. Plant Sci 176:20–30
Glick BR (1995) The enhancement of plant growth by free living bacteria. Can J Microbiol 41:109–114
Gonzalez RC, Gonzalez-Chavez MCA (2006) Metal accumulation in wild plants surrounding mining wastes. Environ Pollut 144:84–92
Goonasekera MM, Gunawardana VK, Jayasena K, Mohammed SG, Bala-subramaniam S (1995) Pregnancy terminating effect of *Jatropha curcas* in rats. J Ethnopharmacol 47:117–123
Gour VK (2006) Production practices including post-harvest of *Jatropha curcas*. In: Singh B, Swaminathan R, Ponraj V (eds) Biodiesel conference towards energy independence—focus on *Jatropha*, Rashtrapati Nilayam, Hyderabad, 9–10 June 2006, pp 223–251
Gunaseelan VN (2009) Biomass estimates, characteristics, biochemical methane potential, kinetics and energy flow from *Jatropha curcas* on dry lands. Biomass Bioenergy 33:589–596
Gunderson JJ, Knight JD, Van Rees KCJ (2007) Impact of ectomycorrhizal colonization of hybrid poplar on the remediation of diesel-contaminated soil. J Environ Qual 36(4):927–934
Hamdar B (1999) An efficiency approach to managing Mississippi's marginal land based on the conservation reserve program (CRP). *Resour Conserv Recycl* 26:15–24
Heller J (1996) Physic nut (*Jatropha curcas*): promoting the conservation and use of underutilized and neglected crops. Institute of Plant Genetics and Crop Plant Research, Rome
Hijosa-Valsero M, Fink G, Schlüsener MP et al (2011) Removal of antibiotics from urban wastewater by constructed wetland optimization. Chemosphere 83(5):713–719
Horne A (1999) Phytoremediation by constructed wetlands. In: Terry N, Banuelos GS (eds) Phytoremediation of contaminated soil and water. CRC Press, Florida, pp 13–40
Huinink JTM (1998) Soil quality requirements of use in urban environments. Soil Tillage Res 47:157–162
Hyman MR, Russell SA, Ely RL, Williamson KJ, Arp DJ (1995) Inhibition, inactivation, and recovery of ammonia-oxidizing activity in cometabolism of trichloroethylene by *Nitrosomonaseuropaea*. Appl Environ Microbiol 61(4):1480–1487
Idowu OD, Fayinminnu OO (2015) Phytotoxicity effect of spent soil on *Jatropha curcas* seedlings used in soil phytoremediation. Ethiop J Environ Stud Manag 8(2):906–915
Ijah UJJ, Aransiola SA, Abioye OP (2015) Restoration of lead contaminated soil using *Arachis hypogaea*. Int J Environ Pollut Control Res 1(1)
Isawumi MA (1978) Nigerian chewing sticks. Niger Field 43:111–121
Jamil S, Abilash PC, Singh N, Sharma PN (2009) *Jatropha curcas*: a potential crop for phytoremediation of coal fly ash. J Hazard Mater 172:269–275
Jones N, Miller JH (1993) *Jatropha curcas*: a multipurpose species for problematic sites. Washington DC. A World Bank paper.
Jongschaap REE, Corré WJ, Bindraban PS, Brandenburg WA (2007) Claims and facts on Jatropha curcas L. Wageningen, Plant Resource International
Joseph B, Patra RR, Lawrence R (2007) Characterization of plant growth promoting rhizobacteria associated with chickpea (*Cicer arietinum* L.). Int J Plant Prod 1:141–151
Jussieu AL (1789) Cucurbitaceae classis XV, OrdoII. G E N P1:393–399
Juwarkar AA, Yadav SK, Kumar GP, Singh SK (2008) Effect of biosludge and biofertilizer amendment on growth of *Jatropha curcas* in heavy metal contaminated soils. Environ Monit Assess 145:7–15
Jyoti L, Smita C (2012) Phytoremediation potential of *Jatropha curcas* for removal of heavy metals from refinery sludge. Int J Sci Eng Res 3(10):1–5

Kalligeros S, Zannikos F, Stournas S, Lois E, Anastopoulos G, Teas C et al (2003) An investigation of using biodiesel/marine diesel blends on the performance of a stationary diesel engine. Biomass Bioenergy 24:141–149

Kamnev AA, van der Lelie D (2000) Chemical and biological parameters as tools to evaluate and improve heavy metal phytoremediation. Biosci Rep 20:239–258

Kaushik N, Kumar K, Kumar S, Kaushik N, Ray S (2007) Genetic variability and divergence studies in seed traits and oil content of *Jatropha* (*Jatropha curcas* L.) accessions. Biomass Bioenergy 31:497–502

Kent AD, Triplett EW (2002) Microbial communities and their interactions in soil and rhizosphere ecosystems. Annu Rev Microbiol 56:211–236

Khan NA, el Dessouky H (2009) Prospect of biodiesel in Pakistan. Renew Sust Energ Rev 13:1576–1583

Kim D, Park MJ, Koh SC, So JS, Kim E (2002) Three separate pathways for the initial oxidation of limonene, biphenyl, and phenol by *Rhodococcus* sp strain t104. J Microbiol 40:86–89

Klein-Douwel RJH, Donkerbroek AJ, van Vliet AP, Boot MD, Somers LMT, Baert RSG et al (2009) Soot and chemiluminescence in diesel combustion of bio-derived, oxygenated and reference fuels. Proc Combust Inst 32:2817–2825

Knothe G, Steidley KR (2005) Kinematic viscosity of biodiesel fuel components and related compounds, influence of compound structure and comparison to petrodiesel fuel components. Fuel 84:1059–1065

Kohli A, Raorane M, Popluechai S, Kannan U, Syers JK, O'Donnell AG (2009) *Jatropha* as a novel non-edible oilseed plant for biodiesel. In: Ferry N, Gatehouse AMR (eds) Environmental impact of genetically modified novel crops. CAB International, London, pp 294–322 Chapter 14

Koohafkan AP (2000). Land resources potential and sustainable land management: an overview. Nat Res Forum 24:69–81

Krämer U, Talke IN, Hanikenne M (2007) Transition metal transport. FEBS Lett 581:2263–2272

Kuiper I, Lagendijk EL, Bloemberg GV, Lugtenberg BJ (2004) Rhizoremediation: a beneficial plant–microbe interaction. Mol Plant-Microbe Interact 17:6–15

Kumar R, Pandey S, Pandey A (2006) Plant roots and carbon sequestration. Curr Sci 91:885–890

Kumar GP, Yadav SK, Thawale PR, Singh SK, Juwarkar AA (2008) Growth of *Jatropha curcas* on heavy metal contaminated soil amended with industrial wastes and *Azotobacter*—a greenhouse study. Bioresour Technol 99:2078–2082

Leon PM (2008) Invertebrate life cycle responses to PAC exposure. PhD thesis. Amsterdam: University of Amsterdam

Leyval C, Turnau K, Haselwandter K (1997) Effect of heavy metal pollution on mycorrhizal colonization and function: physiological, ecological and applied aspects. Mycorrhiza 7:139–153

Li NY, Zhi L, Wang GS, Li HJ, Hong JY (2007) Status and industrialization development of wood biomass energy in the north regions. Soil Water Conserv 5:70–74

Ligi T, Oopkaup K, Truu M et al (2014) Characterisation of bacterial communities in soil and sediment of a created riverine wetland complex using high-throughput 16S rRNA amplicon sequencing. Ecol Eng. in press

Lutz A (1992) Vegetable oil as fuel. An environmentally and socially compatible concept for Mali. Gate 4:38–46

Macek T, Mackova M, Kas J (2000) Exploitation of plants for the removal of organics in environmental remediation. Biotechnol Adv 18:23–34

Mangkoedihardjo S, Ratnawati R, Alfianti N (2008) Phytoremediation of hexavalent chromium polluted soil using *Pterocarpus indicus* and *Jatropha curcas*. World Appl Sci J 4(3):338–342

Martin BC, George SJ, Price CA, Ryan MH, Tibbett M (2014) The role of root exuded low molecular weight organic anions in facilitating petroleum hydrocarbon degradation: current knowledge and future directions. Sci Total Environ 472:642–653

Master ER, Mohn WW (2001) Induction of *bphA*, encoding biphenyl dioxygenase, in two polychlorinated biphenyl-degrading bacteria, psychrotolerant *Pseudomonas* strain cam-1 and mesophilic *Burkholderia* strain lb400. Appl Environ Microbiol 67:2669–2676

Matamoros V, Mujeriego R, Bayona JM (2007) Trihalomethane occurrence in chlorinated reclaimed water at full-scale wastewater treatment plants in NE Spain. Water Res 41:3337–3344
Matsuse IT, Lim YA, Hattori M, Correa M, Gupta MP (1998) A search for anti-viral properties in Panamanian medicinal plants. The effects on HIV and its essential enzymes. *J Ethnopharmacol* 64:15–22
McCormick RL, Graboski MS, Alleman TL, Herring AM (2001) Impact of biodiesel source material and chemical structure on emissions of criteria pollutants from a heavy-duty engine. Environ Sci Technol 35:1742–1747
McGuinness M, Dowling D (2009) Plant-associated bacterial degradation of toxic organic compounds in soil. Int J Environ Res Public Health 6:2226–2247
McVaugh R (1945) The genus *Jatropha* in America: principal intergeneric groups. Bull Torrey Bot Club 72:271–294
Meagher RB (2000) Phytoremediation of toxic elemental and organic pollutants. Curr Opin Plant Biol 3:153–162
Mitsch WJ, Zhang L, Stefanik KC et al (2012) Creating wetlands: primary succession, water quality changes, and self-design over 15 years. Bioscience 62(3):237–250
Muhammad AA, Jamil MM, Ismail Y (2014) Soil Contamination, Risk Assessment and Remediation. Licensee InTech. (http://creativecommons.org/licenses/by/3.0), http://dx.doi.org/10.5772/57287
Mujumdar AM, Misar AV (2004) Anti-inflammatory activity of *Jatropha curcas* roots in mice and rats. J Ethnopharmacol 90:11–15
Nagaraju A,, Karimulla S (2002) Accumulation of elements in plants and soils in and around Nellore mica belt, Andhra Pradesh, India: a biogeochemical study. Environ Geol 41:852–860. Downloaded By: [Sun Yat-Sen University] At: 12:09 4 June 2011
Nath LK, Dutta SK (1992) Wound healing response of the proteolytic enzyme curcain. Indian J Pharmacol 24:114–115
Nathanail P, McCaffrey C, Earl N, Forster ND, Gillett AG, Ogden R (2005) A deterministic method for deriving site-specific human health assessment criteria for contaminants in soil. Hum Ecol Risk Assess 11:389–410
National Germplasm Resources Laboratory, NGRP (2014) World economic plants in GRIN (Germplasm Resources Weed Risk Assessment for *Jatropha curcas*)
Nawrot T, Plusquin M, Hogervorst J, Roels HA, Celis H, Thijs L, Vangronsveld J, Van Hecke E, Staessen J (2006) Environmental exposure to cadmium and risk of cancer: a prospective population-based study. Lancet Oncol 7:119–126
Nedunuri KV, Govindaraju RS, Banks MK, Schwab AP, Chen Z (2000) Evaluation of phytoremediation for field-scale degradation of total petroleum hydrocarbons. J Environ Eng 126(6):483–490
Ogunwole JO, Chaudhary DR, Gosh A, Daudu CK, Chikara J, Patolia S (2008) Contribution of *Jatropha curcas* to soil quality improvement in a degraded Indian entisol. Acta Agric Scand Sect B 58:245–251
Okon Y, Labandera-Gonzalez CA (1994) Agronomic applications of *Azospirillum*. In: Ryder MH, Stephens PM, Bowen GD (eds) Improving plant pproductivity with rhizosphere bacteria. CSIRO, Glen Osmond, p 274
Okrent D (1999) On intergenerational equity and its clash with intragenerational equity and on the need for policies to guide the regulation of disposal of wastes and other activities posing very long time risks. Risk Anal 19:877–901
Openshaw K (2000) A review of *Jatropha curcas*: an oil plant of unfulfilled promise. Biomass Bioenergy 19:1–15
Osoniyi O, Onajobi F (2003) Coagulant and anticoagulant activities in *Jatropha curcas* latex. J Ethnopharmacol 89:101–105
Padmavathiamma PK, Li LY (2007) Phytoremediation technology: hyper-accumulation metals in plants. Water Air Soil Pollut 184:105–126

Palmroth MRT, Pichtel J, Puhakka JA (2002) Phytoremediation of subarctic soil contaminated with diesel fuel. Bioresour Technol 84:221–228

Panda SK, Choudhury S (2005) Chromium stress in plants. Braz J Plant Physiol 17:95–102

Pandey AK, Bhargava P, Gupta N, Sharma D (2010) Performance of *Jatropha curcas*: a biofuel crop in wasteland of Madhya Pradesh, India. Int J Energ Environ 1:1017–1026

Pandey VC, Jay KS, Singhc S, Akhilesh K, Bajrang S, Rana PS (2012) *Jatropha curcas*: a potential biofuel plant for sustainable environmental development. Renew Sust Energ Rev 16(2012):2870–2883

Parsons WT, Cuthbertson EG (2001) Noxious weeds of Australia, Second edn. CSIRO Publishing, Collingwood, 698 pp

Patterson MM, Cohen E, Prommer H, Thomas DG, Rhodes S, McKinley AI (2007) Origin of mixed brominated ethane groundwater plume: contaminant degradation pathways and reactions. Environ Sci Technol 41:1352–1138

Philips LA, Germida JJ, Farrell RE, Greer CW (2008) Hydrocarbon degradation potential and activity of endophytic bacteria associated with prairie plants. Soil Biol Biochem 40:3054–3064

Philips LA, Greer CW, Farrell RE, Germida JJ (2009) Field-scale assessment of weathered hydrocarbon degradation by mixed and single plant treatments. Appl Soil Ecol 42:9–17

Provoost J, Reijnders L, Swartjes F, Bronders J, Carlon C, D'Allessandro M, Cornelis C (2008) Parameters causing variation between soil screening values and the effect of harmonization. J Soils Sediments 8:298–311

Ramos C, Mølbak L, Molin S (2000) Bacterial activity in the rhizosphere analyzed at the single-cell level by monitoring ribosome contents and synthesis rates. Appl Environ Microbiol 66(2):801–809

Rangzan N (2006) The role of plants in the remediation of gasoline-contaminated soils, M.Sc. thesis, Department of Agriculture, Shahid Chamran University of Ahvaz

Raskin I, Ensley BD (2000) Phytoremediation of toxic metals: using plants to clean up the environment. Inter Science Publication, Princeton

Raskin I, Nanda Kumar PBA, Dushenkov S, Salt DE (1994) Bioconcentration of heavy metals by plants. Curr Opin Biotechnol 5:285–290

Reichenauer TG, Germida JJ (2008) Phytoremediation of organic contaminants in soil and groundwater. ChemSusChem 1:708–717

Reubens B, Achten WMJ, Maes WH, Danjon F, Aerts R, Poesen J et al (2011) More than biofuel? *Jatropha curcas* root system symmetry and potential for soil erosion control. J Arid Environ 75:201–205

Richardson GM, Bright DA, Dodd M (2006) Do current standards of practice in Canada measure what is relevant to human exposure at contaminated sites? II: oral bioaccessibility of contaminants in soil. Hum Ecol Risk Assess 12:606–618

Romijn HA (2011) Land clearing and greenhouse gas emissions from *Jatropha* bio-fuels on African Miombo woodlands. Energ Policy 39:5751–5762

Ryan RP, Germaine K, Franks A, Ryan DJ, Dowling DN (2008) Bacterial endophytes: recent developments and applications. FEMS Microbiol Lett 278:1–9

Salt E, Smith RD, Raskin I (1998) Phytoremediation. Annu Rev Plant Physiol Plant Mol Biol 49:643–668

Samy RP, Lgnacimuthu S, Sen A (1998) Screening of 34 Indian medicinal plants for antibacterial properties. *J Ethnopharmacol* 62:173–181

Saravanan VS, Madhaiyan M, Osborne J, Thangaraju M, Sa TM (2008) Ecological occurrence of *Gluconacetobacter diazotrophicus* and nitrogen-fixing *Acetobacteraceae* members: their possible role in plant growth promotion. Microb Ecol 55:130–140

Sarwoko M, Surahmaida A (2008) *Jatropha curcas* L. for phytoremediation of lead and cadmium polluted soil. World Applied Sciences Journal 4(4):519–522

Schnoor JL, Licht LA, McCutcheon SC, Wolfe NL, Carreira LH (1995) Phytoremediation of organic and nutrient contaminants. Environ Sci Technol 29:318–323

Segura A, Rodríguez-Conde S, Ramos C, Ramos JL (2009) Bacterial responses and interactions with plants during rhizoremediation. Microb Biotechnol 2(4):452–464
Senthil KM, Ramesh A, Nagalingam B (2003) An experimental comparison of methods to use methanol and *Jatropha* oil in a compression ignition engine. Biomass Bioenergy 25:309–318
Shah K, Nongkynrih JM (2007) Metal hyperaccumulation and bioremediation. Biol Plant 51:618–634
Shahriari M, Savaghebi GH, Minaee D,,Padidaran M (2006) Effects of *Medicago sativa* and *Festuca arundinacea* mix in phytoremediation soil polluted with light crude oil. Environ Sci. 13
Shanahan P, ÓSullivan DJ, Simpson P, Glennon JD, ÓGara F (1992) Isolation of 2,4-diacetylphlorogucinol from a fluorescent pseudomonad and investigation of physiological parameters influencing its production. Appl Environ Microbiol 58:353–358
Shelef O, Gross A, Rachmilevitch S (2013) Role of plants in a constructed wetland: current and new perspectives. Water 5(2):405–419
Sheoran V, Sheoran AS, Poonia P (2009) Phytomining: a review. Miner Eng 22:1007–1019
Siciliano SD, Germida JJ, Banks K, Greer CW (2003) Changes in microbial community composition and function during a polyaromatic hydrocarbon phytoremediation field trial. Appl Environ Microbiol 69(1):483–489
Simcox NJ, Fenske RA, Wolz SA, Lee IW, Kalman DA (1995) Pesticides in household dust and soil: exposure pathways for children of agricultural families. Environ Health Perspect 103:1126–1134
Singer AC, Crowley DE, Thompson IP (2003) Secondary plant metabolites in phytoremediation and biotransformation. Trends Biotechnol 21:123–130
Spaak JD (1990) Boiser les lles du Cap-Vert—pourquoi, comment, pour qui? Bois For Trop 225:47–54
Srimathi P, Paramathma M (2006) Influence of seed management techniques for production of quality seedlings in Jatropha. In: Singh B, Swaminathan R, Ponraj V (eds) Proceedings of the biodiesel conference toward energy independence – focus of Jatropha, Hyderabad, India, June 9–10. New Delhi, Rashtrapati Bhawan, pp 158–178
Starr F, Starr K(2009—2012) Plants of Hawaii: *Jatropha curcas*. Starr Environmental. http://www.starrenvironmental.com/images/search/?q=Jatropha+curcas&o=plants. (Archived at PERAL)
Sujatha M, Reddy TP, Mahasi MJ (2008) Role of biotechnological interventions in the improvement of castor (*Ricinus communis* L.) and *Jatropha curcas* L. Biotechnol Adv 26:424–435
Sunil N, Kumar V, Varaprasad K (2013) Origin, domestication, distribution and diversity of *Jatropha curcas* L. In: Bahadur B, Sujatha M, Carels N (eds) *Jatropha,* challenges for a new energy crop. Volume 2: Genetic improvement and biotechnology. Springer, New York, pp 137–151
Suresh B, Ravishankar GA (2004) Phytoremediation—a novel and promising approach for environmental clean-up. Crit Rev Biotechnol 24:97–124
Susarla S, Medina VF, McCutcheon SC (2002) Phytoremediation: an ecological solution to organic chemical contamination. Ecol Eng 18:647–658
Tarazona JV, Fernandez MD, Vega MM (2005) Regulation of contaminated soils in Spain. J Soils Sediments 5:121–124
Tewari JP, Shukla IK (1982) Inhibition of infectivity of two strains of watermelon mosaic virus by latex of some angiosperms. Geobios 9:124–126
Thomas OO (1989) Re-examination of the antimicrobial activities of *Xylopia aethiopica*, *Carica papaya*, *Ocimum gratissimum* and *Jatropha curcas*. Fitoterapia 60:147–155
Tobin J, Fulford DJ (2005) Life cycle assessment of the production of biodiesel from *Jatropha*. MSc dissertation, UK: University of Reading. http://www.extra.rdg.ac.uk/eng/energy/book/msc05b.html
Toccalino PL, Norman JE (2006) Health-based screening levels to evaluate US geological survey groundwater quality data. Risk Anal 26:1339–1348
Urzelai A, Vega M, Angulo E (2000) Deriving ecological risk-based soil quality values in the Basque country. Sci Total Environ 247:279–284

Van Zorge JA (1996) Exposure to mixtures of chemical substances: is there a need for regulations? Food Chem Toxicol 34:1033–1036
Verlicchi P, Zambello E (2014) How efficient are constructed wetlands in removing pharmaceuticals from untreated and treated urban wastewaters? A review. Sci Total Environ 470-471:1281–1306
Vouillamoz J, Milke MW (2009) Effect of compost in phytoremediation of diesel contaminated soils. Water Sci Technol 43(2):291–295
Vymazal J (2011) Constructed wetlands for wastewater treatment: five decades of experience. Environ SciTechnol 45(1):61–69
Wang Y, Xiao M, Geng X, Liu J, Chen J (2007) Horizontal transfer of genetic determinants for degradation of phenol between the bacteria living in plant and its rhizosphere. Appl Microbiol Biotechnol 77:733–739
Watanabe ME (1997) Phytoremediation on the brink of commercialization. Environ SciTechnol 31:182–186
Wender PA, Kee JM, Warrington JM (2008) Practical synthesis of prostratin, DPP, and their analogs, adjuvant leads against latent HIV. Science 320:649–652
Wenzel WW (2008) Rhizosphere processes and management in plant-assisted bioremediation (phytoremediation) of soils. Plant Soil 321(1–2):385–408
White PM Jr, Wolf DC, Thoma GJ, Reynolds CM (2006) Phytoremediation of alkylated polycyclic aromatic hydrocarbons in a crude oil–contaminated soil. Water Air Soil Pollut 169(1–4):207–220
Wilbur RL (1954) A synopsis of *Jatropha*, subsection *Eucurcas*, with the description of two new species from Mexico. J Elisha Mitchell Sci Soc 70:92–101
Wu G, Gao L (1998) Research on exploitation and sustainable development counter-measure of marginal lands in the Three Gorge reservoir areas. Chin J Environ Sci 19:89–93
Yadav SK, Juwarkar AA, Kumar GP, Thawale PR, Singh SK, Chakrabarti T (2009) Bioaccumulation and phyto-translocation of arsenic, chromium and zinc by *Jatropha curcas* L.: impact of dairy sludge and biofertilizer. Bioresour Technol 100:4616–4622
Yadav SK, Dhote M, Kumar P, Sharma J, Chakrabarti T, Juwarkar AA (2010) Differential antioxidative enzyme responses of *Jatropha curcas* to chromium stress. J Hazard Mater 180:609–615
Yan LG, Chen JW (2001) Key issues on energy sustainable development in China. Science Press, Beijing, pp 354–378
Yrjala K, Mancano G, Fortelius C, Akerman ML, Sipila TP (2010) The incidence of *Burkholderia* in epiphytic and endophytic bacterial cenoses in hybrid aspen grown on sandy peat. Boreal Environ Res 15:81–96
Zahawi RA (2005) Establishment and growth of living fence species: an overlooked tool for the restoration of degraded areas in the tropics. Restor Ecol 13:92–102
Zhang FL, Niu B, Wang YC, Chen F, Wang SH, Xu Y, Jiang LD, Gao S, Wu J, Tang L, Jia YJ (2008) A novel betaine aldehyde dehydrogenase gene from Jatropha curcas, encoding an enzyme implicated in adaptation to environmental stress. Plant Sci 174:510–518
Zhang BY, Zheng JS, Sharp RG (2010) Phytoremediation in engineered wetlands: mechanisms and applications. Procedia Environ Sci 2:1315–1325
Zhang D, Gersberg RM, Ng WJ, Tan SK (2014) Removal of pharmaceuticals and personal care products in aquatic plant–based systems: a review. Environ Pollut 184:620–639

Ricinus communis: An Ecological Engineer and a Biofuel Resource

5

Dhananjay Kumar, Poonam, Kuldeep Bauddh, Jaya Tiwari, D.P. Singh, and Narendra Kumar

Abstract

Degradation of soil quality due to industrial and agricultural activities has caused worldwide concerns. In order to remediate the soil contamination, several physicochemical techniques like soil washing, soil vapor extraction, solidification, stabilization, vitrification, electrokinetic, etc. had been applied. These modes of remediation may irretrievably affect the soil quality, destroy biodiversity, and render the soil useless for the plant growth. Hence, there is need of suitable, cost-effective, and eco-friendly techniques for remediation of soil without interfering with soil fertility. Phytoremediation being interdisciplinary in nature validated with a series of enthralling scientific research has emerged as a most promising, cost-effective, eco-friendly, and aesthetically acceptable technique for soil restoration. During phytoremediation, whole life span of a plant produces profound effects on the chemical, physical, and biological processes that occur in its instant vicinity. Processes like water and mineral acquisition, senescence, and biomass decay can greatly influence the rhizospheric and subsequently result in land restoration. Further, in the search of a suitable phytoremediator, *Ricinus communis* has proven its potential as a good phytoremediator for several organic and inorganic toxic chemicals particularly heavy metals and polyaromatic hydrocarbons along with other associated benefits being a medicinal and oil-producing plant. Unlike

D. Kumar • Poonam • D.P. Singh • N. Kumar (✉)
Department of Environmental Sciences, Babasaheb Bhimrao Ambedkar University, Lucknow 226025, India,
e-mail: narendrakumar_lko@yahoo.co.in

K. Bauddh
Centre for Environmental Sciences, Central University of Jharkhand, Ranchi-835205, Jharkhand, India

J. Tiwari
Department of Environmental Sciences, Asian Institute of Public Health, Bhubaneswar, Odisha, India

K. Bauddh et al. (eds.), *Phytoremediation Potential of Bioenergy Plants*,
DOI 10.1007/978-981-10-3084-0_5

the other plant, it can grow throughout the year even in drought and saline condition which makes it better than other plants. Being a hyperaccumulator plant for several metals, it can be sustainably used for the reclamation of sodic soil.

Keywords
Bioenergy • Heavy metals • Phytominer • Phytoremediation • *Ricinus*

5.1 Introduction

Over the past several epochs, human being depends upon the plants to satisfy their basic needs as food, clothing, shelter, and health care. Plants are appreciated for aesthetic reasons and more altruistically for providing habitations for other species. Plants have been also documented for their consumption of CO_2 and several industrial gaseous by-products. More recently, plants have been further appreciated scientifically for their role in slackening the rate of global warming. With the increasing interest of corporate sectors, the things have gradually changed. The life support needs of mankind are growing rapidly because of a growing world population and urbanization. The growing human demands have also forced to change the cultivation and utilization of goods and services provided by the plant species. Today, the proficiency of plants to satisfy these growing necessities is not a new concern. Human beings have altered the planet's ecosystem by clearing forests, replacing the riotous diversity of nature with uniform monocultures, industrialization, and urbanization, although, to some extent, these changes have improved the quality of lives but also led to degradation of environment. Alteration in environmental quality is a natural process, but the human activities like improper waste management practices, landfill operations, mining, application of chemical fertilizers, sewage sludge, etc. have accelerated the level and rate of pollution. The anthropogenic happenings like mining, urban sewage, industrial discharge, agriculture, constructions, etc. release various organic and inorganic toxicants in biosphere, and consequently it has degraded the quality of air, water, and soil (Kumar et al. 2012). Today, soil pollution due to industrial and agricultural activities has caused worldwide concerns (Kumar et al. 2013a, b; Ghavri et al. 2013a, b). As a consequence, many life forms including human being have to face serious threats to their life-supporting resources (Fig. 5.1).

5.2 Soil Metal Contamination and Conventional Treatment Techniques

The rising rate of anthropogenic activities in the biosphere poses an unprecedented threat that leads to a disturbance in natural ecosystem. Degradation of soil quality resulting from industrial and agricultural happenings had caused a great concern in recent era (Ha et al. 2014; Ghavri et al. 2013a, b). Domestic and industrial activities

Fig. 5.1 Sources of soil contamination

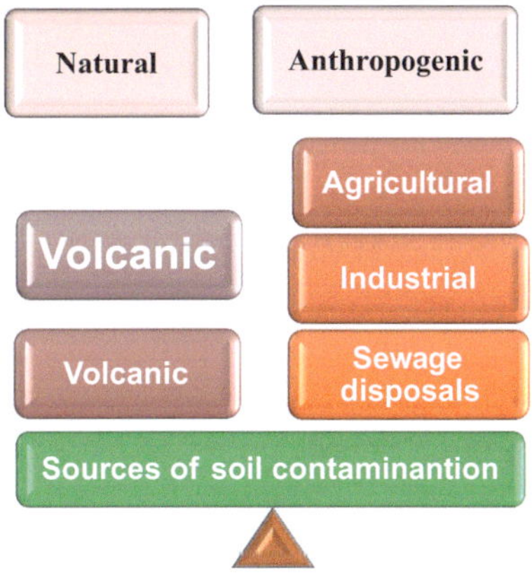

such as mining, paper mills, chemical works, metal fabrication shops, textile plants, and metal refineries are particularly guilty for the contamination of soil (Wong 2003; Freitas et al. 2004). A large number of organic (polycyclic aromatic hydrocarbon, pesticides, chlorophenols, petroleum and related products, etc.) and inorganic (radionuclides, heavy metals and metalloids, etc.) materials enter in the soil, causing threats to human health and natural ecosystem. Today, the contamination of soil had become a great concern in the industrial- and agricultural-intensified areas. The contaminants of soil, with several toxic organic and inorganic chemicals, affect crop yields, soil biomass, and fertility (Gratao et al. 2005; Rajkumar et al. 2009). Inorganic contaminants like heavy metals (e.g., Cr, Cu, Cd, As, Fe, Ni, Pb, etc.) are an important class of environmental pollutants, and many among these are very toxic to plants, animals, and humans in both elemental and soluble forms. Being nonbiodegradable and persistent in nature, they remain in the environment for a long duration and also get accumulated in human and animal body through food chain and pose serious health hazards to humans and animals (Sakakibara et al. 2011; Kumar and Kumar 2016). Therefore, removal of the metal contaminants from the polluted environment is essential for the safety of environment worldwide. In order to remediate the soil contaminated with heavy metals, several chemical, physical, or biological techniques had been applied (Fig. 5.2) (McEldowney et al. 1993). A summary of these techniques is presented in Table 5.1.

Further, chemical and physical modes of remediation irretrievably affect the soil quality, destroy biodiversity, and may render the soil useless for the plant growth. It was also reported that these modes of remediation are not cost-effective (Glass 1999). Hence, there is a need of suitable, cost-effective, and eco-friendly techniques for remediation of soil without affecting soil fertility.

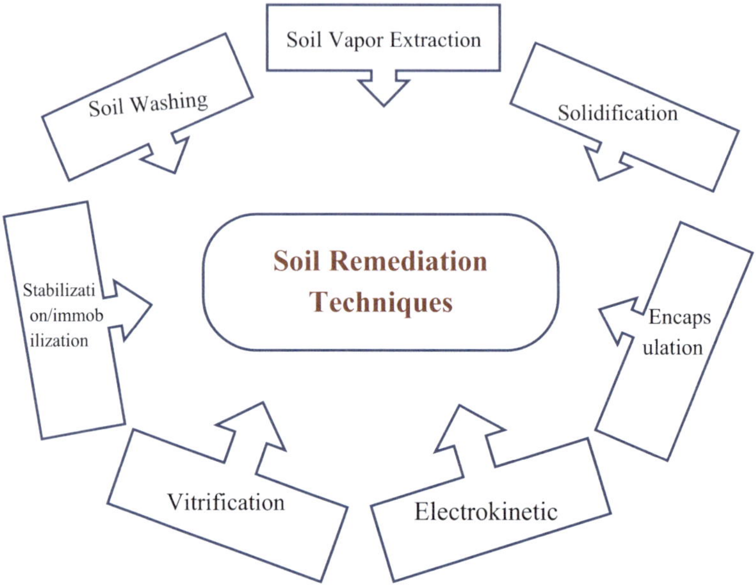

Fig. 5.2 Conventional soil remediation techniques

5.3 Phytoremediation: An Emerging, Economical, and Eco-friendly Technique for Environmental Cleanup

Application of plants for the soil remediation is not a new concept. Through an interdisciplinary approach associated with a series of enthralling scientific research has endorsed the phytoremediation to develop into a most promising, cost-effective, eco-friendly, and aesthetically acceptable technique (Padmavathiamma and Li 2007; Kumar et al. 2013a, b; Sainger et al. 2014; Sweta et al. 2015). This technique also involves the association of plants with the rhizospheric microorganism to remove, degrade, or immobilize various soil contaminants. The idea of using plants to remediate metal from contaminated soil was reintroduced and developed by Utsunamyia (1980) and Chaney (1983), while the first field experiment for phytoextraction of Cu and Zn was conducted by Baker et al. (1991).

The term "phytoremediation" is derived from the Greek word *φυτο* (phyto) that means plant and a Latin *remedium* that means "to restore" or "clean." It deals with mitigation of pollutants from contaminated environment with naturally occurring or genetically engineered plants that have potential to accumulate, degrade, or eliminate the contaminants like metals, pesticides, solvents, explosive, crude oil, its derivatives, etc. from the growing medium (Flathman and Lanza 1998; Prasad and Freitas 2003). The whole life span of a plant produces profound effects on the chemical, physical, and biological processes that occur in its instant vicinity.

Table 5.1 Conventional soil remediation techniques

Techniques	Descriptions
Soil washing	Water-based method for scrubbing soils to remove contaminants. It removes contaminants from soils in two ways:
	By dissolving or suspending contaminated soil in the wash solution (which can be sustained by chemical manipulation of pH for a period of time)
	By concentrating them into a smaller volume through particle-size separation, gravity separation, and attrition scrubbing (similar to those techniques used in sand and gravel operations)
Soil vapor extraction	Vacuum is applied through a well to evaporate the volatile components of contaminated soil, which are subsequently withdrawn in an extraction well
Solidification	Techniques that encapsulate the waste in a monolithic solid of high structural integrity. Encapsulation may be of fine waste particles (microencapsulation) or of a large block or container of wastes (macroencapsulation)
Stabilization/immobilization	Reduce the toxicity of a waste by transforming into less soluble, mobile, or toxic forms. Physical nature and handling characteristics of waste do not necessarily change
Vitrification	During this process, high temperature is applied which may volatilize and/or destroy organic contaminants or volatile metal species (such as Hg) that must be collected for treatment or disposal. Soils contaminated with a wide range of inorganic and organic contaminants may be treated using vitrification
Electrokinetic	During this process, soil contamination removal is achieved by application of an electric field
Thermal deposition	Polluted soil is excavated, screened, and heated to temperatures up to the boiling point of the contaminants
Encapsulation	It involves physical isolation and control of contaminant. The impacted soils are isolated by low-permeability caps or wall to confine the infiltration of precipitation

Conner and Hoeffner (1998), Mulligan et al. (2001), Khan et al. (2004), Ottosen and Jensen (2005), and Marques et al. (2009)

5.4 *Ricinus communis:* A Multipurpose Plant

Ricinus is a fast-growing, perennial shrub growing up to 6 meters or more. It is an economical plant cultivated mainly for oil production which is obtained from its seed. After extraction of oil from caster seeds, it is heated at high temperature for destruction of toxins and used as a fertilizer and also incorporated into animal feeds. Oil obtained from caster seeds are mainly used as a purgative and laxative. Other commercial application of castor oil includes its use as a lubricant, lamp fuel, and a component of cosmetics. It is also used in the manufacture of several valuable products like soaps, printer's ink, plastic, fibers, hydraulic fluid, brake fluid, varnishes, paints, embalming fluid, textile dyes, leather finishes, adhesive, waxes, and fungicides (Jena and Gupta 2012). The leaves of castor are used as a food for seri silkworms and stalks for fuel purpose (Kadambi and Dabral 1955). Castor oil is extensively used in Ayurvedic, allopathic, homeopathic, and Unani system of medicines. Further, the castor plant is well reported for its antioxidant, anti-implantation, anti-inflammatory, antidiabetic, central analgesic, antitumor, larvicidal and adult emergence inhibition, antinociceptive, and antiasthmatic properties. All these properties are due to the presence of certain phytochemical constituents, viz., methyl ricinoleate, ricinoleic acid, 12-octadecadienoic acid, and methyl ester for antioxidant; saponins, steroids, and alkaloids for antinociceptive; saponin, flavonoids, apigenin, and luteolin for antiasthmatic; ethanolic extract of roots for antidiabetic and antihistaminic; flavonoids and tannins for hepatoprotective activity; and ricinine for molluscicidal, insecticidal, and larvicidal activity (Shukla et al. 1992; Ilavarasan et al. 2006; Gupta et al. 2006; Singh et al. 2010; Taur et al. 2011; Jena and Gupta 2012).

5.4.1 Geographical Distribution of *Ricinus communis*

Ricinus communis most commonly known as "castor oil plant" and "Arandi" is an evergreen, fast-growing, herbaceous shrub, C3 plant, belonging to the family Euphorbiaceae (Kirtikar and Basu 1975; Huang et al. 2011; Aserse et al. 2012; Baishya and Kalita 2015). It is the native of northeastern Africa but has naturalized in middle east, with a wide distribution in the tropical and subtropical regions of the world (Vwioko and Fashemi 2005). Some of the species also have their distribution in the Mediterranean Basin, Ethiopia, Africa, Brazil, and India (Gana et al. 2013).

5.4.2 *Ricinus communis*: An Ecological Engineer

5.4.2.1 Heavy Metal Extractor

Heavy metals are the elements found to be present in the environment with the atomic number in between 63 and 200 and density greater than 5.0 gm/cm^3, e.g., Cr, Cd, Ni, As, Pb, Mn, Al, Co, etc. (Garbarino et al. 1995; Jjemba 2004). In recent years, environmental pollution by different heavy metals has crossed the limit of the safe and pure environment, especially in the developing countries (Zhang et al. 2014). Some of these metals like Cu, Mn, Fe, and Zn are essential micronutrients required by the plants and animals in low doses for proper growth and development, but higher doses may be extremely toxic due to their association with the metabolic malfunctioning, growth inhibition, and adverse health effects (Sapci and Ustun 2012). Moreover, these metals also have the potential to get bioaccumulated and thus can also contaminate the food chain (Dirilgen 2011). In urban areas, activities like industrial and municipal effluent discharge without proper treatment, sewage disposal, mining, smelting processes, etc. are the main source of the metal contamination in the water bodies and soil. A number of conventional methods have been identified for treating the metal-contaminated water and soil, e.g., ion exchange/chelation, coagulation, flocculation, chemical precipitation, adsorption, liquid membrane separation, size exclusion, etc. (Khellaf and Zerdaoui 2009). However, these technologies have different constraints, as they are generally expensive, not applicable at mild contamination levels, generate by-products, and require intensive energy (Kumar et al. 2013a, b; Garg et al. 2008; Priya and Selvan 2014). Thus, exploration of techniques for efficient removal of pollutants, particularly from the mildly contaminated water with no or little amount of waste generation, cost-effectiveness, and less energy demands are urgently required.

Phytoremediation, i.e., application of green plants for the accumulation and remediation of different heavy metals, has emerged as a cost-effective, adorable, and eco-friendly technology for the extraction and removal of heavy metals from the environment (Tangahu et al. 2011; Bosiacki et al. 2013; Kumar et al. 2014). Plants that accumulate very high concentrations of heavy metals in their organs, and are capable of growing under extreme adverse conditions and toxic for other species, are referred to as "hyperaccumulators," and the whole process is known as "phytoextraction" (Brooks et al. 1997). Hyperaccumulators are conventionally defined as species capable of accumulating metals at levels 100-fold greater than those typically measured in common non-accumulator plants (Lasat 2000). These hyperaccumulators take the advantages of their well-developed root canal system, which enable them to uptake different metals from the surrounding soil, sediment, and water, and shoot system for the translocation, accumulation, and degradation abilities for the purpose of removing the toxic metals from the environment (Hinchman et al. 1995).

Phytoextraction can be described as the uptake, translocation, and accumulation of the heavy metals and other inorganic and organic contaminants from the soil, sediments, and water by the means of root, shoot, and leaves of the green plants for the purpose of harvesting and restoration of heavy metals (Bhattacharya et al., 2006; Ginneken et al. 2007; Moreno et al. 2008; Ekwumemgbo et al. 2013). The accumulated metals can be recovered from the ash, thus contributing toward the clean and pure environment.

5.4.2.2 Mechanism of Phytoextraction

Heavy metal extraction in plants is a multistep process that includes mobilization from soil into the soil solution and then into the plants. Several steps involved during the process of phytoremediation are outlined below (Clemens et al. 2002):

1. Dissolution
2. Uptake by roots
3. Transport to shoot
4. Distribution in aerial sinks
5. Sequestration and storage in the leaf tissue

1. **Dissolution** – Heavy metal ions directly present in the system cannot be absorbed by the plant until they are in mobile condition, i.e., dissolved in the medium and available for root uptake. When ion is soluble in the soil solution, it comes into mobile form and can be either taken directly by the cell wall of the root or carried up by specific metal transporters. Root secretes some chemicals which enable the process of capturing the metal in the rhizosphere and making it available to the roots, e.g., organic acids and carboxylates. Ananthi et al. (2012) studied the Pb phytoextraction potential of *Ricinus communis L.* and *Brassica juncea* (L.) Czern. They pointed out that solubility and bioavailability of Pb in the soil are limited due to its complexation with organic matter, sorption on oxides and clays, and precipitation as hydroxide, carbonates, and phosphates. Thus, for making the metal bioavailable, different chelating agents such as EDTA and citric acid are used. When EDTA solution was applied in the soil, significant increase in the concentration of Pb was observed which in turn increase the Pb uptake by the plants, i.e., phytoremediation.
2. **Uptake by roots** – Cellular membranes are lipophilic in nature and as metal ions are charged; they cannot move freely across the cellular membranes. Therefore, ion transport into cells must be mediated by membrane proteins with transport functions, generically known as transporters (Lasat 2000). Heavy metal ions enter inside the root through the process of absorption on the root hairs and root cell wall via the plasma membrane, probably involving cationic channels such as calcium channel. Roots are also capable of accumulating significant quantities of these heavy metals and simultaneously restrict its translocation to the shoot (Lane and Martin 1977). This occurs through either active (symplastic) or passive (apoplastic) pathway (Lu et al. 2009). Symplastic pathway is dependent upon the formation of complexes and chelates, and apoplastic pathway is energy dependent.
3. **Transport to shoot** – As compared to non-hyperaccumulator plants, hyperaccumulator plants efficiently transport heavy metals from roots to shoots. Movement of metal-containing sap from the root to the shoot, termed translocation, is primarily controlled by two processes: root pressure and leaf transpiration (Lasat 2000). Heavy metals are transported by means of xylem, and xylem loading is mediated by membrane transport proteins (Clemens et al. 2002). This is the most complex process of the phytoextraction, as the translocation to shoot is facilitated by complexing of metal with low-molecular-weight chelators, gene expression, etc. Ethylenediaminetetraacetic acid (EDTA), 1,2-cyclohexylenedinitrilotetraacetic acid (CDTA), ethylenediamine-N-N′-bis(2-hydroxyphenylacetic acid) (EDDHA), etc. are the chelators, as reported by many researchers, used for enhancing the potential of the plants to phytoextract heavy metals by the process of chelation in the cytoplasm or storing them into vacuoles (Shen et al. 2002; Lin et al. 2009; Rascio and Navari-Izzo 2011). These chemicals have the ability to change the physiochemical and biological properties of the soil, which may facilitate the phytoextraction of the heavy metals (Ultra et al. 2005).
4. **Distribution to aerial sinks** – Complexing with organic ligands, which may occur at any point along the transport pathway, converts the metal into less toxic

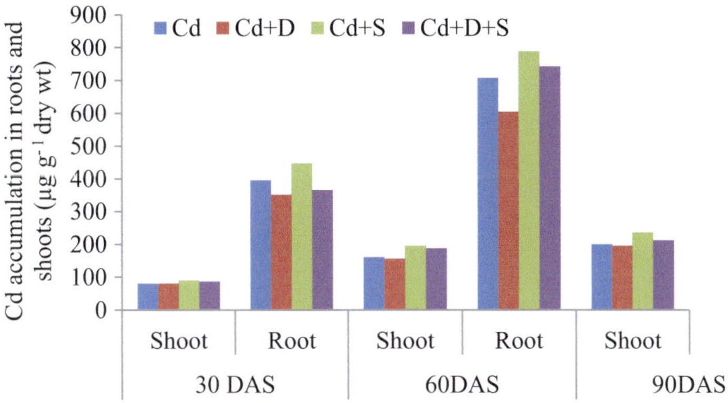

Fig. 5.3 Effects of drought and salinity stress on Cd accumulation in the roots and shoots of *R. communis*. Cd = 100 mg Cd/kg soil, Cd + D = 100 mg Cd/kg soil + drought, Cd + S = 100 mg Cd/kg soil + salinity, Cd + D + S = 100 mg Cd/kg soil + drought + salinity (Bauddh and Singh 2012b)

form, thus conferring high metal tolerance in hyperaccumulators (Peer et al. 2005).

5. **Sequestration and storage in the leaf tissue** – In the leaf tissues, metals are sequestered in extracellular or subcellular compartments, mainly in vacuoles. Metal is transported (via xylem sap) to the apoplast of leaves from where it is distributed within the leaf tissue via apoplast or transporters-mediated uptake by symplast (Mahmood 2010).

Ricinus communis, an evergreen herb, can grow even in very harsh environmental conditions and have been studied by different researchers and scientists for their potential to phytoextract various heavy metals from the soil because of its ability to grow in heavily polluted soil together with its capacity for metal ion accumulation and fast growth rate (Rajkumar and Freitas 2008; Shi and Cai 2009). Researchers have reported that the plant has the ability to absorb the toxic metals, translocate, and accumulate in their upper parts (Babita et al. 2010; Bauddh and Singh 2012a, b). *R. communis* has been found to have substantial tolerance against several abiotic stresses, e.g., salinity and drought, and have also accumulated significant amount of metal in its tissues (Fig. 5.3) (Bauddh and Singh 2012b).

5.4.2.3 The Phytoextraction of the Heavy Metals Depends on the Following Factors

5.4.2.3.1 Tolerance to Higher Concentrations of Metals

The amount of heavy metals phytoextracted from the soil and water is affected mainly by the tolerance tendency of the plants (Schmidt 2003). Shi and Cai (2009) compared Cd tolerance of eight energy crops and reported *R. communis* as a moderately tolerant species toward higher levels of Cd contamination in comparison to rapeseed, sunflower, soybean, hemp, safflower, flax, and peanut. Coscione and

Berton (2009) and Pandey (2013) reported that *R. communis* has the ability to extract Ba from the contaminated soil. Pandey (2013) and Coscione and Berton (2009) have reported that the plants grow vigorously and extract a number of heavy metals (Cd, Zn, Cr, Cu, Pb, Mn, and Fe) when cultivated on the soil contaminated with fly ash.

In two different studies conducted by Bauddh and Singh (2012a and 2015a), *R. communis* was found to be a tolerant species during its cultivation in Ni- and Cd-contaminated soil (up to 150 mg metal Kg^{-1} soil) (Figs. 5.4 and 5.5).

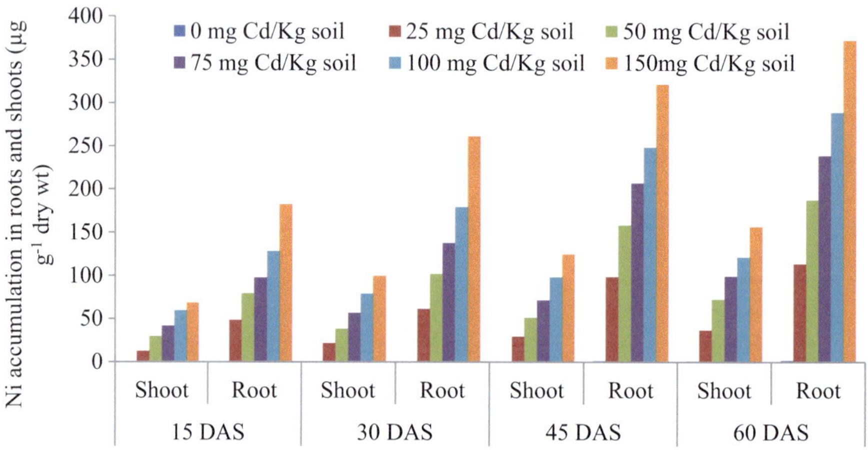

Fig. 5.4 Ni accumulation potential of *R. communis* during cultivation in soil having different levels of the metal (Bauddh and Singh 2015a)

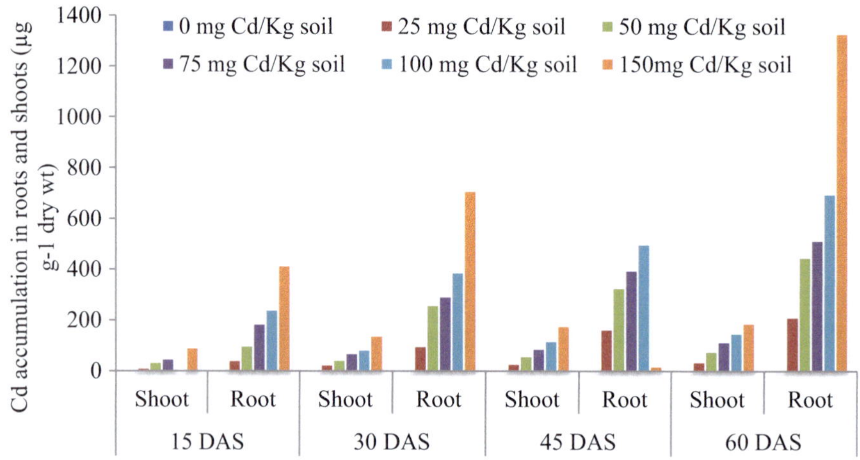

Fig. 5.5 Cd accumulation potential of *R. communis* during cultivation in soil having different levels of the metal (Bauddh and Singh 2012a)

Romeiro et al. (2006) have reported *Ricinus communis* to be a hyperaccumulator of Pb and showed tolerance properties in lead at concentration of 100, 200, and 400 $\mu molL^{-1}$. Huang et al. (2011) demonstrated that *R. communis* L. has high potential for removing Cd from contaminated soils because of its fast growth, high biomass, strong absorption, and accumulation of Cd. Andreazza et al. (2013) studied the effect of the use of castor bean (*Ricinus communis* L.) as a potential phytoremediator for copper-contaminated soils. They concluded it to be highly resistant to soil contaminated with heavy metals like Cu and demonstrated a strong potential application in phytoremediation purposes as castor bean plants did not present any visual symptoms of Cu toxicity when the plants were cultivated in the Cu-contaminated soil, even in the Cu mining waste, which contains a Cu concentration of 575 mg/kg of soil.

5.4.2.3.2 Accumulation Ability

This factor influence most of the processes of phytoextraction, as the higher is the capacity of the plant to accumulate concentration of the heavy metals, the higher is the probability of phytoextraction (Bosiacki et al. 2013). Shi and Cai (2009) observed the metal accumulation tendency of eight energy crops and reported that Cd accumulation by this energy crop is not much significant as compared to other energy crops like hemp, flax, and peanut. Bosiacki et al. (2013) studied the evaluation of suitability of *Amaranthus caudatus* L. and *Ricinus communis* L. in phytoextraction of cadmium and lead from contaminated substrates. It has been reported that the highest concentration of Cd and Pb was observed in stem and lowest in the inflorescence of *R. communis*. According to Shi and Cai (2009), *R. communis* is moderately tolerant species toward Cd toxicity. Likewise, Hadi et al. (2015) reported maximum accumulation of Cd in the shoot of *Ricinus communis* when treatment of 25 mg/L of Cd was given to the plant in comparison to other treatments of 0, 5, 10, 15, 20, and 25 mg/L. Bauddh and Singh (2015b) studied the effects of amendment of organic and inorganic fertilizers on bioaccumulation and partitioning of Cd in *R. communis*. They reported that the Cd accumulation increased with application of inorganic fertilizers. However, organic fertilizers (viz., vermicompost) that are considered slow-release fertilizers reduced the metal accumulation in the plant tissues (Fig. 5.6).

Romeiro et al. (2006) reported *R. communis* to hyperaccumulate Pb at 10.54–24.61 g Pb kg^{-1} dry weight of the plants. Melo et al. (2009) and Romeiro et al. (2006) have reported *R. communis* to be a better accumulator of Cd and Pb. Pandey (2013) has reported suitability of *Ricinus communis* L. for phytoremediation of fly ash disposal sites. The plant showed the tendency to accumulate different heavy metals in their parts. The metal accumulation trend was found to be in the following order: Ni>Cd>Zn>Cu>Pb at fly ash-polluted site except in the leaves which mainly depend upon some factors like pH, presence of humic substances, metal concentration, presence of other metals, type of plant species, age of vegetation, and sampling season (Maiti and Jaiswal 2008; Pandey 2013). Pandey (2013) had also reported the trend of bioaccumulation in different parts of the plant in the following order: root>stem>leaf for the fly ash-polluted site. Mahmud et al. (2008) studied the

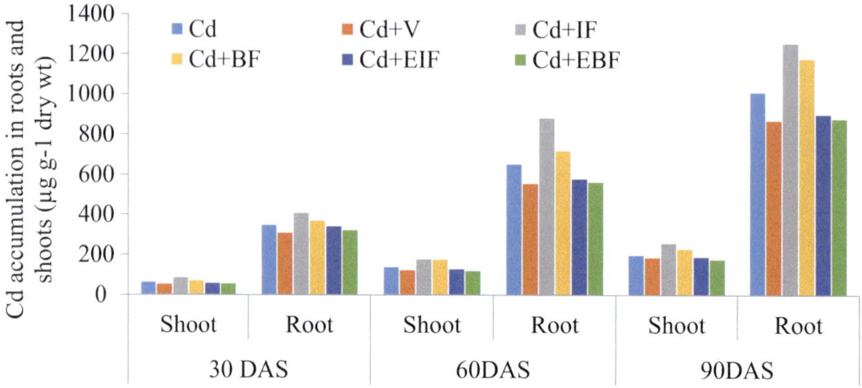

Fig. 5.6 Effects of amendments of different fertilizers on Cd accumulation in the roots and shoots of *R. communis*. Cd = 100 mg Cd/kg soil, Cd + V = 100 mg Cd/kg soil + vermicompost, Cd +IF =100 mg Cd/kg soil + inorganic fertilizer, Cd + BF = 100 mg Cd/kg soil + biofertilizer, Cd + EIF = 100 mg Cd/kg soil + entrapped inorganic fertilizer, and Cd + EBF = 100 mg Cd/kg soil + entrapped biofertilizers (Bauddh and Singh 2015b)

phytoremediation potential of 49 plant species and reported that among them, *R. communis* showed a high ability to accumulate As from the soil with bioconcentration factors (BCFs) and translocation factors (TFs) of 6.79 and 1.47, respectively. Further, Bauddh et al. (2016) studied the Cd accumulation potential of *R. communis* and its accumulation, translocation, and partitioning in different plant parts and reported that about 51% Cd was stabilized in its roots and the rest of the metal was transferred to stem and leaves. Thus, the study indicates this plant to be highly tolerant to Cd contamination that could be used for the remediation of polluted sites.

5.4.2.3.3 Heavy Metal Concentration in the Medium (Soil)

The concentration of the toxic metal in the medium influences the accumulation potential of the plants. The higher is the concentration of the metal in the medium (soil, water), the higher is the probability of phytoextraction depending upon the extent of tolerance of the plants. *R. communis* has shown distinct accumulation at different concentration of heavy metals in their tissues, which affects the growth and development of the plant. Lu and He (2005) reported that *R. communis* can bioaccumulate 10–400 mg/kg of Cd which affected the growth of plant. At the beginning, during low concentration of Cd, there was hardly any damage caused to the plant, but as the concentration increased, the growth of the plant slowed down and was totally inhibited beyond the concentration of 40 mg/kg. The maximum bioaccumulation was 4460.3 mg/kg when the concentration of the metal in the soil was 360 mg/kg suggesting that higher metal concentration (i.e., >360 mg/kg) adversely affected the metabolism of *R. communis*. Mathiyazhagan and Natarajan (2013) studied metal extraction competence of plants on waste dumps of magnesite mine. It was reported that *R. communis* phytoextracted 85 mg/kg of Cd from the soil. Mahmud et al. (2008) studied the potential of indigenous plant species for the phytoremediation of arsenic-contaminated areas of Bangladesh. The results indicated that among 49 examined plant species belonging to 29 families, only one species of fern (*Dryopteris filix-mas*), three herbs (*Blumea lacera*, *Mikania cordata*, and *Ageratum conyzoides*), and two shrubs (*Clerodendrum trichotomum* and *Ricinus communis*) were found to be suitable for phytoremediation on the basis of high bioconcentration factors (BCFs) and translocation factors (TFs) of As.

5.4.2.3.4 Rate of Metal Uptake by the Roots and Translocation

The metal uptake by roots is initiated by the transport across the root cellular membrane. Translocation of heavy metals from root to upper aerial parts is likely to occur via the xylem and is driven by the transpiration of leaves (Hart et al. 1998), whereas translocation factor can be defined as the transfer of metals from growing medium to other parts (root and shoot) of the plant species (Barman et al. 2000; Gupta et al. 2008; Andreazza et al. 2013). Both of these phenomena influence the potential of the plants to phytoextract the heavy metals. Huang et al. (2011) studied the phytoremediation potential of bioenergy crop *Ricinus communis* for DDTs and cadmium co-contaminated soil and reported that the total uptake of Cd varied from 66 to 155.1 μg per pot, indicating that *R. communis* has great potential for removing Cd from contaminated soils which may be attributed to its fast growth, high biomass, strong absorption, and accumulation. Hadi et al. (2015) have also studied the phytoremediation of cadmium by *R. communis* L. in hydroponic condition and reported that Cd contents in plant tissues were found increasing with increasing dose of Cd. The maximum Cd uptake was found at 25 mg/L of Cd. The pattern of Cd accumulation in different parts of plant was found as root>leaf>stem, and maximum bioaccumulation factor (0.455) was found in treatment of 10 mg/L of Cd. TF value ranged from 0.645 to 0.857 and 0.626 to 0.837 for root–shoot and shoot–leaves, respectively, at the treatments of increasing doses of Cd (0, 5, 10, 15, 20, and 25 mg/L Cd for 28 days.

Andreazza et al. (2013) reported *R. communis* to be hyperaccumulator of Cu as the bioconcentration factors (BCF) were found to be between 4.1 and 5.5 in the vineyard soils. Contrary to the other studies, Olivares et al. (2013) worked on the potential of *R. communis* for phytoremediation of mine tailings and oil production and analyzed accumulation of Cu, Zn, Mn, Pb, and Cd in the root and shoot of *R. communis*. They concluded that *R. communis* accumulated low concentrations of metals in aboveground plant parts and these were below phytotoxic values, and root metal concentrations were higher than those found in aerial plant parts except for Mn. Broadly, *Ricinus* had low metal translocation factors, except for Mn, and according to its metal concentrations in shoot and translocation factors, *Ricinus* was concluded not to be a hyperaccumulator plant. Mathiyazhagan and Natarajan (2013) studied metal extraction competence of plants on waste dumps of magnesite mine in South India and reported that *R. communis* had a higher value of translocation factor which indicated that metals (Pb and Cd) are accumulated by the plants and are equally retained in roots and shoots.

5.4.2.3.5 Biomass Production

Biomass can also influence the tolerance level of the plants toward the toxic metals. According to Papoyan and Kochain (2004), plant species with higher biomass are better suited for the purpose of phytoremediation. Zhi-Xin et al. (2007) have reported that the biomass of the plants decreased with increase in the concentration of the toxic metals, as higher concentration of the toxic metals may be harmful to the tissues of the plants than lower. However, some of the researchers have reported that most of the hyperaccumulators were found to be small and slow growing (Lasat 2002). Ananthi et al. (2012) studied the potential of *Ricinus communis* L. and *Brassica juncea* (L.) Czern. in natural and induced Pb phytoextraction. They reported that there was no significant reduction in biomass and root and shoot length of the *Ricinus communis* under different levels (0–800 mg/kg) of Pb indicating high tolerance of *Ricinus communis* to Pb. They reported that when soil Pb content was only up to 800 mg/kg, the plant was able to accumulate 190 mg/kg of Pb in its shoots within short duration of exposure. The translocation factor was also found to be >2, concluding it to be a hyperaccumulator. Rajkumar and Freitas (2008) and Ma et al. (2001) reported that plant growth-promoting bacterium has the potential to increase phytoextraction efficiency of *R. communis* L. directly by enhancing the Zn and Ni metal accumulation in plant tissues and indirectly by promoting the shoot and root biomass. Andreazza et al. (2013) have also recorded high levels of biomass production in both fresh mass and the dry mass of the shoots and roots of *R. communis* after 57 days of growth in Cu-contaminated soil, indicating its high tolerance to Cu and great potential of phytoextraction of Cu from the soil. Further, it has also been reported that *R. communis* produces higher biomass and lower reduction in fresh and dry weights on Ni contamination than *B. juncea*. Antioxidant production such as proline and malondialdehyde in *R. communis* leaves also increases with the increase of concentration of Ni in soil (Baudhha and Singh 2015a, b). Melo et al. (2009) reported that application of As at higher concentrations decreased the growth of castor bean plant's shoots and roots, contrary to this, Mahmud et al. (2006) reported less biomass reduction and high As accumulation from As-spiked soil by *Ricinus communis* in a greenhouse experiment.

5.4.2.3.6 Metabolic Activity of the Plants

Toxic metals interfere with different metabolic activities of the plants like photosynthesis and respiratory carbohydrate metabolism in plant cells, probably by substituting another micronutrient in critical enzymes (Zhi-Xin et al. 2007). Heavy metals also inhibit the formation of chlorophyll and synthesis of amino Valeric acid which is required for photosynthesis (Stobart et al. 1985). According to Habash et al. (1995), heavy metals interfere with different steps of Calvin cycle, resulting in the inhibition of photosynthetic CO_2 fixation. However, Lu and He (2005) reported that the low concentration of Cd may improve the growth of *R. communis*. Andreazza et al. (2013) studied the use of high-yielding bioenergy plant castor bean (*Ricinus communis* L.) as a potential phytoremediator for vineyard soils (inceptisol and mollisol) contaminated with Cu and Cu mining waste. Plants exhibited a high growth potential in Cu-contaminated areas and attained the significant plant height after

57 days of growth in the mollisol soil. The plant attained an average height of 28.3 cm which was higher than the non-contaminated soil (native soil), i.e., 24.7 cm. In the inceptisol, the castor bean plants exhibited an average plant height of 20 cm after 57 days of growth. Copper mining waste showed the worst growth of castor bean plants; however, their growth was still considered to be high, with an average height of 15 cm. Thus, the results showed the increased growth rate of plant, even at the higher concentration of toxic metal, claiming it to be a potential phytoextractor of Cu. Further it has been reported that the *R. communis* has a high accumulation potential and also possesses a good antioxidant defense system against Cd stress and may be used for the phytoremediation of metal-contaminated soils in place of edible crops, which enhance the risk of contaminating the food chain (Baudhha et al. 2015a, b).

5.4.2.3.7 Physicochemical Properties of Medium

The uptake of heavy metals largely depends upon the distribution of heavy metals in different phases of the medium which in turn is greatly influenced by pH, Eh, organic matter, and other properties (Zhi-xin et al., 2007). Mobility of heavy metals is mainly dependent upon the pH of medium in which plants are grown (Lim et al. 2002). Increase in the pH of the medium increases the content of exchangeable ion's activity and makes it more available for phytoextraction by plants. Naturally, the plants also secrete some diverse array of organic compounds which forms complexes with metals, changing the pH of the medium, making them more bioavailable for the purpose of phytoextraction (Prasad 1995). Likewise, Eh could change the forms of heavy metals or immobilize them, thereby reducing the toxicity (Shuaman and Wang 1997).

5.4.2.3.8 Other Interactions

This category includes different kinds of interactions like ion–ion interaction, effect of presence of other chemicals, and organic substances. The presence of one metal ion can interfere with the extraction of other metals by forming or breaking different kind of bonds, complexes, etc. Bosiacki et al. (2013) have reported that *R. communis* have a preferential removal tendency toward Cd than Pb as higher mean weights of leaves and stems as well as total weight of plants were obtained in substrate polluted with Cd in comparison to Pb. Zhang et al. (2015) have studied the effects of citric acid in increasing the phytoextraction of Cd/Pb by *Ricinus* and reported that citric acid positively interferes with the phytoextraction of Cd and Pb. The concentration of Cd and Pb in plant shoots was increased by the addition of citric acid in all treatments by 78% and 18–45%, respectively, at the dosage of 10 mM kg^{-1} soil without affecting shoot biomass production. This was due to the reduction of cation exchange capacity, weakening of heavy metal adsorption, and activation of Cd and Pb in soil on addition of citric acid. They observed that addition of citric acid has enhanced the transportation of Cd and Pb from roots to shoots, which was beneficial for phytoextraction. Chaves et al. (2010) studied the content and distribution of copper and zinc in castor bean (*R. communis*) cultivar. They reported that Zn translocation was higher than Cu in the castor bean plants. Thus, they concluded that castor bean plants could not be considered as heavy metal hyperaccumulator; nevertheless, they can be considered as zinc accumulator. Hence, it may be concluded that *R. communis* is selective toward the accumulation of Zn than Cu, or the presence of Zn influences the availability of Cu for phytoextraction.

5.4.2.4 Biotransformation of Polyaromatic Hydrocarbons

Polyaromatic hydrocarbons(PAHs) are a group of over 100 different organic compounds that are formed during the incomplete combustion of coal, oil, gas, garbage, or other organic substances like tobacco or meat cooked at very high temperature (Ravindra et al. 2008). These organic compounds contain carbon and hydrogen bond with multiple fused aromatic rings, e.g., naphthalene, anthracene, phenanthrene, benzopyrene, fluorine, etc. These compounds are generally white or light yellow crystals with high melting temperature and are neither soluble in water nor volatile. These are found naturally in fire and grass fire, oil seeps, and volcanoes and in chlorophyllous plants, fungi, and bacteria and anthropogenically in petroleum and electric power plants, during incineration, home heating, and production of coke, black carbon, coal tar, and asphalt (Wilson and Jones 1993; Henner et al. 1997; Soclo et al. 2000; Paria and Yuet 2006; Okon and Mbong 2013). Today, PAH contamination is a great environmental concern due to its ubiquitous occurrence, recalcitrance, toxicity, mutagenicity, carcinogenic properties, high retention time, and bioaccumulation properties (Juhasz and Naidu 2000; Mrozik et al. 2003; Kot-Wasik et al. 2004; Haritash and Kaushik 2009 Al-Sbani et al. 2016). The polycyclic aromatic hydrocarbon transformations in the natural environment include biotic and abiotic processes, such as volatilization, adsorption, photolysis, chemical oxidation, microbial degradation, phytoremediation, and biodegradation (Lors et al. 2012; Al-Sbani et al. 2016). Plant density and root parameters, such as diameter, length, and surface area, are important features involved in nutrient uptake as faster plant growth and larger total root surface area may facilitate the dissipation of PAHs (Casper and Jackson 1997; Søndergaard 1988; Liu et al. 2014). Rentz et al. (2005), Spriggs et al. (2005), and Hamdi et al. (2007) have reported that when soils were subjected to phytoremediation, there was a significant decrease in low- and high-molecular-weight PAH content of soils (Rentz et al. 2005; Spriggs et al. 2005; Hamdi et al. 2012). In this context, the rhizospheric region of plants plays a significant role in the bioremediation of soils contaminated with PAHs and other organic chemicals (Spriggs et al. 2005; Hamdi et al. 2012). Rhizosphere which has high microbial activity region in relation to soil and root influences the dissipation of PAHs mainly because of high microbial growth stimulation and selection of specific microbial communities (Reilley et al. 1996; Kirk et al. 2005). Plant roots secrete different organic acids and sugars to facilitate and stimulate the activity of the rhizospheric microorganisms and also favor the aerobic metabolism of PAHs (Schwitzguébel 2015). Thus, interactions between rhizospheric microorganisms and plant roots are the most important factors to phytoremediate the organic contaminants like PAHs, PCBs, and total petroleum hydrocarbons (TPHs).

In addition, the root proliferation improves oxygen flux, permeability, and breakdown of soil contents, mechanically, which causes an increase in bioavailability of

PAH for plant (Hamdi et al. 2007). Thus, the plants like *R. communis*, with deep, taprooted, dense secondary roots, have been extensively accepted for the phytoremediation of PAH-contaminated soils (Reilley et al. 1996; Fan et al. 2008).

Liu et al. (2014) have suggested that lower plant density (e.g., 260 plants m^{-2}) is better suited for phytoremediation of PAHs when considering sustainability of the ecosystem. Wang et al. (2013) performed a pot experiment to explore the potential of phytoextraction of heavy metals and rhizoremediation of PAHs in co-contaminated soil by co-planting a cadmium/zinc hyperaccumulator and lead (Pb) accumulator *Sedum alfredii* with either *Lolium perenne* or *Ricinus communis*. They reported that the co-planting of *S. alfredii* with *R. communis* decreased the shoot biomass as compared to that in monoculture. Further, Cd concentration in shoot of *S. alfredii* decreased, whereas, there was no reduction of Zn or Pb concentration observed, when grown with ryegrass or *R. communis* as compared to that in monoculture or co-planting treatments. The study reported that removal of Cd, Zn, and Pb by plants was similar across *S. alfredii* monoculture or co-planting with ryegrass or *R. communis*, except higher Pb removal in *S. alfredii* and *L. perenne* co-planting treatment. Co-planting of *S. alfredii* with *L. perenne* or *R. communis* significantly increased the dissipation of pyrene and anthracene as compared to that in the bare soil or monoculture *of S. alfredii*. Dissipation rates were in the range of 98.6–99.1, 91.4–96.1, and 65.0–74.9% for phenolphthalein, pyrene, and anthracene, respectively. This dissipation was because of the enhanced soil microbial population and activities in both co-planting treatments making it a relevant technology for the degradation of PAHs. According to Wang et al. (2013), *Ricinus communis* has extensive root biomass which helps to increase the potential of the plant to accumulate organic pollutants like PAHs, DDT, and other pesticides, including Cd and other heavy metals, whereas co-planting of the hyperaccumulator plants like *R. communis*, *B. juncea*, and others may also enhance the chances of decontamination of soil co-contaminated with PAHs (Wang et al. 2013).

5.4.2.5 Sodic Soil and Degraded Land Reclamation

Soil is the most important part of the environment which fulfills the need of food and shelter of the living beings (Dominati 2013). Sodicity is defined by the amount of efficient exchangeable sodium (Na) which can adversely affect the growth of the plants. As the amount of Na increases in the soil, it tends to become more dispersed which results in the breakdown of the soil aggregates and thus lower the permeability of soil for air and water. Similar to sodicity, land degradation is also one of the major factors which affect the growth and health of the soil. It may be because of the use of various chemical fertilizers and other industrial pollutants and has emerged as major problem in the developing countries like India (Bauddh et al. 2015a, b).

The restoration of such degraded lands by the means of chemical and mechanical engineering has been widely used. Nevertheless, these techniques are expensive and not suitable for the structure and fertility of the soil (Gaur and Adholeya 2004). To overcome with all these constrains, phytoremediation has emerged as a promising method to restore degraded lands (Ma et al. 2011; Rajkumar et al. 2012). Many edible and nonedible plants like *Oryza sativa* (Liu et al. 2007; Jamil et al. 2013), *Brassica* sp. (Gupta et al. 2009), *Raphanus sativus* (Hamon et al. 1999; Shevyakova et al. 2011), *Triticum aestivum* L. (Shtangeeva et al. 2004; Chandra et al. 2009), *J. curcas*, *Salix viminalis* (Vyslouzilova et al. 2003; Vervaeke et al. 2004; Hammer et al. 2003), *Chilopsis linearis* (Rodríguez et al. 2009), *Calotropis procera*, *Parthenium hysterophorus*, *Milletti*a *pinnata* (Tulod et al. 2012), and *Portulaca grandiflora* (Cho-Ruk et al. 2006) have been used for phytoremediation of contaminated sites. *Ricinus communis* is a nonedible, energy crop and a potential phytoremediator of heavy metals, PAHs, and POPs. Further, it is a salt-tolerant plant species (Li et al. 2010). The application of *R. communis* for phytoremediation purpose in place of edible as well as nonedible stress-sensitive crops/herbs may be a good alternative for the remediation of contaminated, salt-stressed, and sodic soils. These properties make the plant a promising and viable substrate for amelioration of saline soils as well as biofuel production (Barnes et al. 2009). The crop growth and productivity is greatly reduced by soil salinity and sodicity as about 560 × 106 hectares of sodic and saline–sodic soils are present all over the world which requires efficient and economic reclamation (Tanji 1990; Abro et al. 2015).

Reclamation and restoration of sodic soil require the removal of most of the exchangeable Na and its replacement by more favorable other ions like Ca^{2+}. Joshi et al. (2012) studied the effect of Ca^{2+} on NaCl-stressed *R. communis* with respect to the growth response and seed germination. They reported that the salinity and sodicity of the soil adversely affected the seed germination and growth of the plant. However, when Ca^{2+} was supplied 1:025 Na^+/Ca^{2+} ratio, the deleterious effect of NaCl on seed germination and plant growth was restored. They also suggested that salt stress also reduced N, P, K, and Ca content in the tissue of R. *communis*, but, by the addition of Ca^{2+} at critical levels to the saline soil, all the nutrients were restored. Sodicity of the soil may also increase with NaCl, but supply of Ca^{2+} significantly decreases Na content in the plant. The reclamation of such soil is dependent on the displacement of Na^+ from productive soil horizons by Ca. Thus, gypsum, as source of soluble Ca, has been used as amendment by several workers for soil reclamation (Amezketa et al. 2005). However, many researchers have used inorganic conditioners and organic manures successfully for the reclamation of saline–sodic soils (de Mesquita et al. 2015; Li and Keren 2009).

De Mesquita et al. (2015) studied the effect of chemical and organic conditioners on the chemical attributes of saline–sodic soil on the initial growth of two castor bean (*Ricinus communis*) cultivars. Five treatments of saline–sodic soil were given without conditioners i.e. saline–sodic soil + biofertilizer, saline–sodic soil + chalk, saline–sodic soil + chalk + biofertilizer, and nonsaline soil. It was reported that the corrective effect of chalk on sodicity and on the nutrient availability for the plants as the application of chalk combined with biofertilizer promoted a better initial development of the *Ricinus communis* plant in comparison with other recovering treatments. Abro et al. (2015) studied the growth of *Ricinus communis* L. on newly reclaimed saline–sodic soil to assess the quality improvement of the sodicity–salinity of the soil. After 2 months of the reclamation process by leaching (control), gypsum, farmyard manure (FYM), and FYM + gypsum along with the non-decomposed, the castor bean was grown on post-reclaimed soil. The results indicated that the combination of FYM + gypsum and the longest decomposition period, i.e., 12 weeks of decomposition, produced the highest mean and relative growth of *R. communis*. The authors conclude that the reduction in electronic conductivity (EC) and sodium adsorption ratio (SAR) does not ensure the productivity of reclaimed soil.

Wu et al. (2012) studied ameliorative effect of castor bean (*Ricinus communis* L.) planting on physicochemical and biological properties of seashore saline soil. They reported that after planting castor bean for two growing seasons, soil salinity lowered down than that under the control scenario, and the value of EC is also decreased significantly.

5.4.2.6 Radionuclide Accumulator

Nuclear power plants are operating in around 31 countries of the world. Nuclear power plants are polluting the environment by generating various nuclear wastes, anthropogenically, from different activities (viz., reactor operations, mining, fuel fabrication, fuel reprocessing, military operations, research laboratories) with radionuclide production and applications of radioisotopes in medicine and industry, power plants, accidents, and disasters (Sharma et al. 2015). Naturally, earthquakes and tsunamis also cause the leakage of a considerable amount of radioactive pollutants (Yoshida and Kanda 2012). Further, the presence of a large number of fission products along with multiple oxidation states and long-lived radionuclides such as neptunium (^{237}Np), plutonium (^{239}Pu), americium ($^{241/243}Am$), curium (^{245}Cm), cesium (^{137}Cs), and strontium (^{90}Sr) makes the waste streams a potential radiological threat to the environment (Sharma et al. 2015). These radionuclides are capable of producing potential health threat due to their long half-lives, bioaccumulation tendency, and effortless translocation into the human body. Other than physiochemical techniques for the removal of these contaminants from the environment, phytoremediation has emerged as an eco-friendly, cost-effective, and viable technology to cope with the problems. This technique offers a green chemistry-based route to remediate contaminated sites containing radionuclides (Eapen et al. 2006; Sharma et al. 2015). Among different hyperaccumulators, castor (*Ricinus communis*) is also becoming popular as a value-added plant for the phytoremediation of contaminated sites along with economic and ecological services (Bauddh et al. 2015a, b).

5.4.2.7 Potential Phytominer

Phytomining is bio-harvesting of metals from plants with higher biomass grown in substrate contaminated by mine sites along with recovery of economic metals from the plants (Chaney et al. 1998; Anderson et al. 1999). This is among one of the recent branches of phytoremediation producing low-value, sulfide-free bio-ore which either can be safely disposed off or, if the targeted metal is economically valuable, may be smelted or recovered. All over the world, a number of sites are available for the phytomining process as these sites are enriched with various valuable metals. Further, this technology can also be good mediator among research institutes, industries, and communities to develop good understanding to target industrial waste dumping sites with a goal to mitigate hazard without causing any threat to the environment. But, as compared to other phytoremediation techniques, phytomining is in its infancy and more research is required to be done to harvest this technique efficiently in order to cope with the soil pollution.

Phytomining is a subbranch of phytoextraction which is used for the metal recovery from different metal-contaminated sites (Hunt et al. 2014). A commercial metal mining is usually performed from ores having a high concentration of target metals and requires huge capital investment. However, hyperaccumulators are known to efficiently extract metals from the metalliferous soils and translocate it to shoot tissues. After sufficient growth, plant is harvested, dried, and reduced to ash. This ash is further treated by roasting, sintering, or smelting methods, which allows the recovery of the metals present in the ash (Sheoran et al. 2009). Phytoextraction takes place in the plants which have higher growth rates, higher biomass composition, deep roots, tolerance to metal toxicity, and higher transportation rates (Cunningham and Ow 1996; Watanabe 1997). The most efficient plants are those which can readily translocate metals from the roots and accumulate them in their shoots. The shoot biomass, in turn, is harvested and processed for the recovery of the metals.

5.5 *Ricinus communis:* Potential Bioenergy Plant

In the future, the global energy demand is expected to increase considerably as a result of increased population growth and economic development, although fossil fuels like coal and petroleum will play an important role to fulfill the increased energy demand over the next 20 years. While fossil fuels will remain the conventional fuel, renewable energy sources will be in great demands over the next decades. In the search for a renewable, biodegradable, and eco-friendly fuel, the castor oil has proven to be technically, economically, and ecologically beneficial and also offers as prospects for agricultural development (Gressel 2008; Mehmood et al. 2011). Seeds from *Ricinus* contain around 45% oil which is used as a lubricant in high-speed engines and airplanes. *Ricinus* oil has an ash content of about 0.02% and sulfur content less than 0.04% with a higher cetane number (CN). The CN of *Ricinus* oil biodiesel is in good range for diesel engines with desirable properties, i.e., very low cloud and pour point values, which displays that the fuel is suitable for use in extreme winter temperature. Castor oil contains over 85% of the hydroxylate fatty acid, ricinoleic acid (D-12-hydroxyoctadec-*cis*-9-enoic acid), which is the source of an 18-carbon hydroxylated fatty acid with one double bond (Caupin 1997). Further, properties like unsaturated bond, high molecular weight (298), low melting point (5 °C), and very low solidification point (−12 °C to −18 °C) make it industrially useful. In addition since it is soluble in alcohol, it can be used in synthesis of biodiesel.

5.6 Conclusion

R. communis is a robust and fast-growing economical crop with multiple applications. *R. communis* or castor plant is widely explored for its pharmacological uses due to the presence of certain specific chemicals, viz., ricins A, B, and C, which are well recognized for their antitumor property. The alkaloid (ricinine) and glycoside are being used in several herbal formulations for anti-inflammatory, analgesic, antipyretic, cardiac tonic, and antiasthmatic effects. Although *R. communis* is a noble medicinal plant, it requires more investigation for maximum utilization of its pharmacological properties. Further, *R. communis* is an evergreen herb which can be cultivated even in drastic environmental conditions. It has ability to grow in heavily polluted soil together with high metal accumulation and fast growth rate (Rajkumar and Freitas 2008; Shi and Cai 2009). Several researchers have reported about its potential to absorb the toxic metals, translocate, and accumulate in their upper parts (Babita et al. 2010; Bauddh and Singh 2012a, b). Generally, rhizospheric activity and root proliferation improve oxygen flux, permeability, and breakdown of soil content mechanically which causes an increase in dissipation and bioavailability of PAH in soil. *R. communis* with a deep, tap, and dense secondary root system has been widely adopted and successfully reported for the remediation of PAH-contaminated soil. Further, being a perennial plant, it can achieve the remediation of metal from contaminated soil throughout the year.

References

Abro SA, Rajput MT, Mangrio SM, Abbasi AR, Bughio FA, Jakhar AM (2015) Growth of castor (*Ricinus communis* L.) on newly reclaimed saline-sodic soil as an indicator of soil quality improvement. Sindh Univ Res Jour (Sci Ser) 47(3):549–554

Al-Sbani NH, Abdullah SRS, Idris M, Hasana HA, Jehawia OH, Ismail N (2016) Sub-surface flow system for PAHs removal in water using *Lepironia articulate* under greenhouse conditions. Ecol Eng 87:1–8

Amezketa E, Aragüés R, Gazol R (2005) Efficiency of sulfuric acid, mined gypsum, and two gypsum by-products in soil crusting prevention and sodic soil reclamation. Agron J 97(3):641–653

Ananthi TAS, Meerabai RS, Krishnasamy R (2012) Potential of *Ricinus communis* L. and *Brassica juncea* (L.) Czern. under natural and induced Pb phytoextraction. Univers J Environ Res Technol 2(5):429–438

Anderson CWN, Brooks R, Stewart R, Simcock R, Robinson B (1999) The phytoremediation and phytomining of heavy metals. Pacrim 99, Ball, pp 127–135

Andreazza R, Bortolon L, Pieniz S, Camargo FAO (2013) Use of high-yielding bioenergy plant castor bean (*Ricinus communis L.*) as a potential phytoremediator for copper-contaminated soils. Pedo 23(5):651–661

Aserse AA, Räsänen LA, Assefa F, Hailemariam A, Lindström K (2012) Phylogeny and genetic diversity of native rhizobia nodulating common bean (*Phaseolus vulgaris L.*) in Ethiopia. Syst Appl Microbiol 35(2):120–131

Babita M, Maheswari M, Rao LM, Shanker AK, Rao DG (2010) Osmotic adjustment, drought tolerance and yield in castor (*Ricinus communis* L.) hybrids. Environ Exp Bot 69:243–249

Baishya M, Kalita MC (2015) Phytoremediation of crude oil contaminated soil using two local varieties of castor oil plant (*Ricinus communis*) of Assam. Int J Pharm Bio Sci 6(4):1173–1182

Baker AJM, Reeves RD, McGrath SP (1991) In situ decontamination of heavy metal polluted soil using crops of metal accumulating plants-A feasibility study. In: Hinchee RE, Olfenbuttel RF (eds) In situ bioremediation. CRC, Boca Raton, pp 155–177

Barman SC, Sahu RK, Bhargava SK, Chaterjee C (2000) Distribution of heavy metals in wheat, mustard and weed grown in field irrigated with industrial effluents. Bull Environ Contam Toxicol 64:489–496

Barnes D, Baldwin BS, Braasch DA (2009) Degradation of ricin in castor seed meal by temperature and chemical treatment. Ind Crop Prod 29:509–515

Bauddh K, Singh RP (2012a) Cdmium tolerance and its phytoremediation by two oil yielding plants *Ricinus communis* (L.) and *Brassica juncea* (L.) from the contaminated soil. Int J Phytoremediation 14:772–785

Bauddh K, Singh RP (2012b) Growth, tolerance efficiency and phytoremediation potential of *Ricinus communis* (L.) and *Brassica juncea* (L.) in salinity and drought affected cadmium contaminated soil. Ecotox Environ Safety 85:13–22

Bauddh K, Singh RP (2015a) Assessment of metal uptake capacity of castor bean and mustard for phytoremediation of nickel from contaminated soil. Biore J 19(02):124–138

Bauddh K, Singh RP (2015b) Effects of organic and inorganic amendments on bio-accumulation and partitioning of Cd in *Brassica juncea* and *Ricinus communis*. Ecol Eng 74:93–100

Bauddh K, Singh K, Singh RP (2016) *Ricinus communis* L. a value added crop for remediation of cadmium contaminated soil. Bull Environ Contam Toxicol 96:265–269

Bauddh K, Kumar A, Srivastava S, Singh RP, Tripathi RD (2015a) A study on the effect of cadmium on the antioxidative defense system and alteration in different functional groups in castor bean and Indian mustard. Arch Agro Soil Sci. doi:10.1080/03650340_2015_1083554

Bauddh K, Singh K, Singh B, Singh RP (2015b) *Ricinus communis*: a robust plant for bio-energy and phytoremediation of toxic metals from contaminated soil. Ecol Eng 84:640–652

Bhattacharya T, Banerjee DK, Gopal B (2006) Heavy metal uptake by *Scirpus littoralis* Schrad from fly ash dosed and metal spiked soils. Environ Monit Assess 121(1–3):363–380

Bosiacki M, Kleiber T, Kaczmarek J (2013) Evaluation of suitability of *Amaranthus caudatus* L. and *Ricinus communis* L. in phytoextraction of cadmium and lead from contaminated substrates. Arch Environ Prot 39(3):47–59

Brooks RR, Lee J, Reeves RD, Jaffre TJ (1997) Geol Explor 7:49–57

Casper BB, Jackson RB (1997) Plant competition underground. Annu Rev Ecol Syst 28:545–570

Caupin HJ (1997) Products from castor oil: past, present, and future. In: Gunstone FD, Padley FB (eds) Lipid technologies and applications. Marcel Dekker, New York, pp 787–795

Chandra R, Bharagava RN, Yadav S, Mohan D (2009) Accumulation and distribution of toxic metals in wheat (*Triticum aestivum* L.) and Indian mustard (*Brassica campestris* L.) irrigated with distillery and tannery effluents. J Hazard Mater 162(2–3):1514–1521

Chaney RL (1983) Plant uptake of inorganic waste constitutes. In: Parr JF, Marsh PB, Kla JM (eds) Land treatment of hazardous wastes. Noyes Data Corporation, Park Bridge, pp 50–76

Chaney RL, Angle JS, Baker AJM, Li JM (1998) Method for phytomining of nickel, cobalt, and other metal from soil. US Patent 5(711):784

Chaves LHG, de Mesquita EF, de Araujo DL, de França CP (2010) Acúmulo e distribuição de cobre e zinco em mamoneira cultivar brs paraguaçu e crescimento da planta. Engenharia Ambiental - Espír ito Santo do Pinhal 7(3):263–277

Cho-Ruk K, Kurukote J, Supprung P, Vetayasuporn S (2006) Perennial plants in the phytoremediation of lead contaminated soils. Biotech 5(1):1–4

Clemens S, Palmgren MG, Krämer U (2002) A long way ahead: understanding and engineering plant metal accumulation. Trends Plant Sci 7:309–315

Conner JR, Hoeffner SL (1998) A critical review of stabilization/solidification technology. Crit Rev Environ Sci Technol 28(4):97–462

Coscione AR, Berton RS (2009) Barium extraction potential by mustard, sunflower and castor bean. Scientia Agricola (Piracicaba, Braz) 66(1):59–63

Cunningham SD, Ow DW (1996) Promises and prospects of phytoremediation. Plant Physiol 110:715–719

de Mesquita EF, Sá FVD, Bertino AMP, Cavalcante LF, Paiva EPD, Ferreira NM (2015) Effect of soil conditioners on the chemical attributes of a saline-sodic soil and on the initial growth of the castor bean plant. Semina: Ciências Agrárias. doi:10.5433/1679-0359.2015_36_4_2527

Dirilgen N (2011) Mercury and lead: assessing the toxic effects on growth and metal accumulation by *Lemna minor*. Ecotoxicol Environ Saf 74:48–54

Dominati EJ (2013) Natural capital and ecosystem services of soils. In: Dymond JR (ed) Ecosystem services in New Zealand—conditions and trends. Manaaki Whenua Press, Lincoln

Eapen S, Singh S, Thorat V, Kaushik CP, Raj K, D'Souza SF (2006) Phytoremediation of radiostrontium (^{90}Sr) and radiocesium (^{137}Cs) using giant milky weed (*Calotropis gigantea* R.Br.) plants. Chem 65:2071–2073

Ekwumemgbo, PA, Eddy NO, Omoniyi IK (2013) Decontamination of Heavy metals in polluted soil by phytoremediation using *Bryophyllum pinnatum.* E3S Web of Conferences, 1, 13004

Fan S, Li P, Gong Z, Ren W, He N (2008) Promotion of pyrene degradation in rhizosphere of alfalfa (*Medicago sativa* L.). Chemosphere 71:1593–1598

Flathman PE, Lanza GR (1998) Phytoremediation: current views on an emerging green technology. J Soil Contam 7(4):415–435

Freitas H, Prasad MNV, Pratas J (2004) Plant community tolerant to trace elements growing on the degraded soils of Sao Domingos mine in the south east of Portugal: environmental implications. Environ Int 30:65–72

Gana AK, Yusuf AF, Apuyor B (2013) Castor oil plant and its potentials in transformation and industrialization of under developing nations in the world. Adv J Agric Res 1(5):72–79

Garbarino JR, Hayes H, Roth D, Antweider R, Brinton TI, Taylor H (1995) Contaminants in the Mississippi river. U. S. Geological Survey Circular, Virginia, U.S.A. 1133

Garg KU, Kaur MP, Garg VK, Sud D (2008) Removal of Ni(II) from aqueous solution by adsorption on agriculture waste biomass using a response surface methodological approach. Bioresour Technol 99:1325–1331

Gaur A, Adholeya A (2004) Prospects of arbuscular mycorrhizal fungi in phytoremediation of heavy metal contaminated soils. Curr Sci 86(4):528–534

Ghavri SV, Bauddh K, Kumar S, Singh RP (2013a) Bioaccumulation and translocation potential of Na+ and K+ in native weeds grown on industrially contaminated soil. Int J ChemTech Res 5(4):1869–1875

Ghavri SV, Kumar S, Bauddh K, Singh RP (2013b) Enrichment coefficient and translocation factors of Fe and Cu in weeds growing in Sandila Industrial area, Hardoi District, Uttar Pradesh, India. Geophytol 43(2):153–161

Ginneken LV, Meers E, Guisson R (2007) Phytoremediation for heavy metal-contaminated soils combined with bioenergy production. J Environ Eng Landsc Manag 15(4):227–236

Glass DJ (1999) U.S. and international markets for phytoremediation, 1999–2000. D. Glass Associates, Needham, p 266

Gratao PL, Prasad MNV, Cardoso PF, Lea PJ, Azevedo RA (2005) Phytoremediation: green technology for the cleanup of toxic metals in the environment. Braz J Plant Physiol 17:53–64

Gressel J (2008) Transgenics are imperative for biofuel crops. Plant Sci 174:246–263

Gupta MK, Sharma PK, Ansari SH (2006) In-vitro antioxidant activity of the successive extracts of *Ricinus communis* leaves. Int J Plant Sci 1(2):229–231

Gupta N, Khan DK, Santra SC (2008) An assessment of heavy metal contamination in vegetable grown in waste water irrigated areas of Titagarh, West Bengal, India. Bull Environ Contam Toxicol 80:115–118

Gupta M, Sharma P, Sarin NB, Sinha AK (2009) Differential response of arsenic stress in two varieties of *Brassica juncea* L. Chem 74(9):1201–1208

Ha H, Olson J, Bian L, Rogerson PA (2014) Analysis of heavy metals sources in soil using kriging interpolation on principal components. Environ Sci Technol 48:4999–5007

Habash DZ, Paul MJ, Parry MAJ (1995) Increased capacity for photosynthesis in wheat grown at elevated CO2: the relationship between electron transport and carbon metabolism. Planta 197:482–489
Hadi F, Arifeen MZU, Aziz T, Nawab S, Nabi G (2015) Phytoremediation of Cadmium by *Ricinus communis* L. in hydrophonic condition. Am Eurasian J Agric Environ Sci 15(6):1155–1162
Hamdi H, Benzart S, Manusadzianas L, Aoyama I, Jedidi N (2007) Solid-phase bioassays and soil microbial activities to evaluate PAH-spiked soil ecotoxicity after a long-term bioremediation process simulating land farming. Chem 70:135–143
Hamdi H, Benzarti S, Aoyama I, Jedidi N (2012) Rehabilitation of degraded soils containing aged PAHs based on phytoremediation with alfalfa (*Medicago sativa L.*). Int Biodeterior Biodegrad 67:40–47
Hammer D, Kayser A, Keller C (2003) Phytoextraction of Cd and Zn with *Salix viminalis* in field trials. Soil Use Manag 19(3):187–192
Hamon RE, Holm PE, Lorenz SE, McGrath SP, Christensenm TH (1999) Metal uptake by plants from sludge-amended soils: caution is required in the plateau interpretation. Plant Soil 216(1–2):53–64
Haritash A, Kaushik C (2009) Biodegradation aspects of polycyclic aromatic hydrocarbons (PAHs): a review. J Hazard Mater 169:1–15
Hart JJ, Welch RM, Norvell WA (1998) Characterization of cadmium binding, uptake, and translocation in intact seedlings of bread and durum wheat cultivars. Plant Physiol 116(44):1413–1420
Henner P, Schiavon M, Morel JL, Lichtfouse E (1997) Polycyclic aromatic hydrocarbon (PAH) occurrence and remediation methods. Analusis 25:56–59
Hinchman RR, Negri MC, Gatliff EG (1995) Phytoremediation: using green plants to clean up contaminated soil, groundwater, and wastewater, Argonne National Laboratory Hinchman, Applied Natural Sciences, Inc. Available http://www.treemediation.com/Technical/Phytoremediation 1998.pdf
Huang H, Yu N, Wang L, Gupta DK, He Z, Wang K, Zhu Z, Yan X, Li T, Yang X (2011) The phytoremediation potential of bioenergy crop *Ricinus communis* for DDTs and cadmium co-contaminated soil. Bioresour Technol 102:11034–11038
Hunt AJ, Anderson CWN, Bruce N, García AM, Graedel TE, Hodson M, Meech JA, Nassar NT, Parker HL, Rylott EL, Sotiriou K, Zhang Q, Clark JH (2014) Phytoextraction as a tool for green chemistry. Green Processes Synth 3:3–22
Ilavarasan R, Mallika M, Venkataraman S (2006) Anti-inflammatory and free radical scavenging activity of *Ricinus communis* root extract. J Ethnopharmacol 103:478–480
Jamil M, Zeb S, Anees M, Roohi A, Ahmed I, ur Rehman S, Rha ES (2013) Role of Bacillus licheniformis in phytoremediation of nickel contaminated soil cultivated with rice. Int J Phytorem 16(6):554–571
Jena J, Gupta AK (2012) *Ricinus communis* Linn: a phytopharmacological review. Int J Pharm Pharm Sci 4(4):25–29
Jjemba PK (2004) Interaction of metals and metalloids with microorganisms in the environment. In: Jjemba PK (ed) Environmental microbiology–principles and applications. Science Publishers, Enfield, pp 257–270
Joshi SV, Patel NT, Pandey IB, Pandey AN (2012) Effect of supplemental Ca^{2+} on NaCl-stressed castor plants (*Ricinus communis* L.). Acta Bot Croat 71(1):13–29
Juhasz AL, Naidu R (2000) Bioremediation of high molecular weight polycyclic aromatic hydrocarbons: a review of the microbial degradation of benzo[a]pyrene. Int Biodeterior Biodegrad 45:57–88
Kadambi K, Dabral SN (1955) The silviculture of *Ricinus communis* Linn. Indian Forester 81(1):53–58
Khan FI, Husain T, Hejazi R (2004) An overview and analysis of site remediation technologies. J Environ Manag 71:95–112
Khellaf N, Zerdaoui M (2009) Phytoaccumulation of zinc by the aquatic plant, *Lemna gibba* L. Biores tech 100:6137–6140

Kirk JL, Klironomos JN, Lee H, Trevors JT (2005) The effects of perennial ryegrass and alfalfa on microbial abundance and diversity in petroleum contaminated soil. Environ Pollut 133:455–465

Kirtikar KR, Basu BD (1975) Indian medicinal plants, 2, vol 2. International Book Distributor, Dehradun, pp 894–895

Kot-Wasik A, Dănbrowska D, Namiésnik J (2004) Photodegradation and biodegradation study of benzo(a)pyrene in different liquid media. J Photochem Photobiol A 168:109–115

Kumar D, Kumar N (2016) Tannery effluent toxicity assessment on the growth and Germination of *Phaseolus vulgaris* L. (Bean). Int J Green Herbal Chem 5(2):139–144

Kumar N, Bauddh K, Barman SC, Singh DP (2012) Accumulation of metals in selected macrophytes grown in mixture of drain water and tannery effluent and their phytoremediation potential. J Environ Biol 33:323–327

Kumar N, Bauddh K, Kumar S, Dwivedi N, Singh DP, Barman SC (2013a) Accumulation of metals in weed species grown on the soil contaminated with industrial waste and their phytoremediation potential. Ecol Eng 61:491–495

Kumar N, Bauddh K, Kumar S, Dwivedi N, Singh DP, Barman SC (2013b) Extractability and phytotoxicity of heavy metals present in petrochemical industry sludge. Clean Techn Environ Policy 15:1033–1039

Kumar N, Bauddh K, Kumar S, Dwivedi N, Singh DP, Barman SC (2014) Heavy metal uptake by plants naturally grown on industrially contaminated soil and their phytoremediation potential. Ecol Eng 61:491–495

Lane SD, Martin ES (1977) A histochemical investigation of lead uptake in *Raphanus sativus*. New Phytol 79:281–286

Lasat MM (2000) Phytoextraction of metals from contaminated Soil: a review of plant/soil/metal Interaction and assessment of pertinent Agronomic issues. J Hazard Sub Res 2:5–1

Lasat MM (2002) Phytoextraction of toxic metals: a review of biological mechanisms. Environ Qual 31:109–120

Li FH, Keren R (2009) Calcareous sodic soil reclamation as affected by corn stalk application and incubation: a laboratory study. Pedos 19(4):465–475

Li G, Wan SW, Zhou J, Yang ZY, Qin P (2010) Leaf chlorophyll fluorescence, hyperspectral reflectance, pigments content, malondialdehyde and proline accumulation responses of castor bean (*Ricinus communis* L.) seedlings to salt stress levels. Ind Crop Prod 31(1):13–19

Lim TT, Tay JH, Teh CI (2002) Contamination time effect on lead and cadmium fractionation in a tropical coastal clay. Environ Qual 31:806–812

Lin C, Liu J, Liu L, Zhu T, Sheng L, Wang D (2009) Soil amendment application frequency contributes to phytoextraction of lead by sunflower at different nutrient levels. Environ Exp Bot 65(2–3):410–416

Liu WX, Shen LF, Liu JW, Wang YW, Li SR (2007) Uptake of toxic heavy metals by rice (*Oryza sativa* L.) cultivated in the agricultural soil near Zhengzhou City, People's Republic of China. Bull Environ Contam Toxicol 79(2):209–213

Liu H, Meng F, Tong Y, Chi J (2014) Effect of plant density on phytoremediation of polycyclic aromatic hydrocarbons contaminated sediments with *Vallisneria spiralis*. Ecol Eng 73:380–385

Lors C, Damidot D, Ponge JF, Périé F (2012) Comparison of a bioremediation process of PAHs in a PAH-contaminated soil at field and laboratory scales. Environ Pollut 165:11–17

Lu XY, He CQ (2005) Tolerance, uptake and accumulation of cadmium by *Ricinus communis L.* J Agro Environ Sci 24(4):674–677

Lu L, Tian S, Yang X, Li T, He Z (2009) Cadmium uptake and xylem loading are active processes in the hyperaccumulator *Sedum alfredii*. J Plant Physiol 166:579–587

Ma LQ, Komar KM, Tu C, Zhang W, Cai Y (2001) A fern that hyper accumulates arsenic. Nature:409–579

Ma Y, Prasad MNV, Rajkumar M, Freitas H (2011) Plant growth promoting rhizobacteria and endophytes accelerate phytoremediation of metalliferous soils. Biotechnol Adv 29:248–258

Mahmood T (2010) Phytoextraction of heavy metals – the process and scope for remediation of contaminated soils. Soil Environ 29(2):91–109

Mahmud R, Inoue N, Kasajima S, Kato M, Shaheen R, Miah MAM, Rahman MS (2006) Response of common buckwheat and castor oil plant against different levels of soil arsenic concentration: a comparative study. Fagopyrum 23:45–51

Mahmud R, Inoue N, Kasajima S, Shaheen R (2008) Assessment of potential indigenous plant species for the phytoremediation of arsenic- contaminated areas of Bangladesh. Inter J Phytorem 10:119–132

Maiti SK, Jaiswal S (2008) Bioaccumulation and translocation of metals in the natural vegetation growing on fly ash lagoons: a field study from Santaldih thermal power plant, West Bengal. India Environ Monit Assess 136:355–370

Marques APGC, Rangel AOSS, Castro PML (2009) Remediation of heavy metal contaminated soils: phytoremediation as a potentially promising clean-up technology. Crit Rev Environ Sci Technol 39:622–654

Mathiyazhagan N, Natarajan D (2013) Metal extraction competence of plants on waste dumps of magnesite mine, Salem District, South India. J Min Environ 4(2):113–124

McEldowney S, Hardman DJ, Waite S (1993) Treatment technologies. In: McEldowney S, Hardman DJ, Waite S (eds) Pollution, ecology and biotreatment. Longman Singapore Publishers Pvt. Ltd, Singapore, pp 48–58

Mehmood F, Khan AUH, Khan ZUD (2011) Appraisal of ecological significance of *Ricinus communis Linn.* in the wasteland of Lahore, Pakistan. Biologia 57(1&2):97–103

Melo EEC, Costa ETS, Guilherme LRG, Faquin V, Nascimento CWA (2009) Accumulation of arsenic and nutrients by castor bean plants grown on an As-enriched nutrient solution. J Hazard Mater 168:479–483

Mrozik A, Piotrowska-Seget Z, Labuzek S (2003) Bacterial degradation and bioremediation of polycyclic aromatic hydrocarbons. Pol J Environ Stud 12:15–25

Moreno FN, Anderson CWN, Stewart RB, Robinson BH (2008) Phytofiltration of mercury-contaminated water: volatilisation and plant-accumulation aspects. Environ Exp Bot 62(1):78–85

Mulligan CN, Yong RN, Gibbs BF (2001) Remediation technologies for metal contaminated soils and groundwater: an evaluation. Eng Econ 60:193–207

Okon JE, Mbong EO (2013) Effects of nutrient amendments of spent engine oil polluted soil on some growth parameters of *Abelmoschus esculentus* (L.) Moench in South-South Nigeria. Bull Environ Pharmacol Life Sci 2(5):75–78

Olivares AR, Carrillo-González R, MadelCarmen A, González-Chávez MCA, Hernández RMS (2013) Potential of castor bean (*Ricinus communis* L.) for phytoremediation of mine tailings and oil production. J Environ Manag 114:316–323

Ottosen LM, Jensen PE (2005) Electro-remediation of contaminated soil. In: Lens P, Grotenhuis T, Malina G, Tabak H (eds) Soil and sediment remediation. IWA Publishing, London, pp 264–287

Padmavathiamma PK, Li LY (2007) Phytoremediation technology: hyperaccumulation metals in plants. Plant Soil 184:105–126

Pandey VC (2013) Suitability of *Ricinus communis* L. cultivation for phytoremediation of fly ash disposal sites. Ecol Eng 57:336–341

Papoyan A, Kochian LV (2004) Identification of Thlaspi caerulescens genes that may be involved in heavy metal hyperaccumulation and tolerance. Characterization of a novel heavy metal transporting ATPase. Plant Physiol 136(3):3814–23

Paria S, Yuet PK (2006) Solubilization of naphthalene by pure and mixed surfactants. Ind Eng Chem Res 45:3552–3558

Peer WA, Baxter IR, Richards EL, Freeman JL, Murphy AS (2005) Phytoremediation and hyperaccumulator plants. In: Tamás MJ, Martinoia E (eds) Molecular biology of metal homeostasis and detixification. Springer-Verlag, Heidelberg, pp 299–340

Prasad MNV (1995) Cadmium toxicity and tolerance in vascular plants. Environ Exp Bot 35(4):25–545

Prasad MNV, Freitas H (2003) Metal hyperaccumulation in plants-Biodiversity prospecting for phytoremediation technology. Electron J Biotechnol 6:285–321
Priya ES, Selvan PS (2014) Water hyacinth (Eichhornia crassipes)- An efficient and economic adsorbent for textile effluent treatment– a review. Arab J Chem. doi:http://dx.doi.org/10.1016/j.arabjc.2014.03.002
Rajkumar M, Freitas H (2008) Influence of metal resistant-plant growth promoting bacteria on the growth of *Ricinus communis* in soil contaminated with heavy metals. Chemo 71(5):834–842
Rajkumar M, Prasad MNV, Freitas H, Ae N (2009) Biotechnological applications of serpentine soil bacteria for phytoremediation of trace metals. Crit Rev Biotechnol 29:120–130
Rajkumar M, Sandhya S, Prasad MNV, Freitas H (2012) Perspectives of plant-associated microbes in heavy metal phytoremediation. Biotechnol Adv 30:1562–1574
Rascio N, Navari-Izzob F (2011) Heavy metal hyperaccumulating plants: how and why do they do it? And what makes them so interesting? Plant Sci 180:169–181
Ravindra K, Sokhi R, Van Grieken R (2008) Atmospheric polycyclic aromatic hydrocarbons: source attribution: emission factors and regulation. Atmos Environ 42:2895–2921
Reilley KA, Banks MK, Schwab AP (1996) Dissipation of polycyclic aromatic hydrocarbons in the rhizosphere. J Environ Qual 25:212–219
Rentz AJ, Alvarez PJJ, Schnoor JL (2005) Benzo[a]pyrene co-metabolism in the presence of plant root extracts and exudates: implications for phytoremediation. Environ Pollut 136:477–484
Rodríguez E, Peralta-Videa JR, Israr M (2009) Effect of mercury and gold on growth, nutrient uptake, and anatomical changes in *Chilopsis linearis*. Environ Exp Bot 65(2–3):253–262
Romeiro S, Lagôa AMMA, Furlani PR, de Abreu CA, de Abreu MF, Erismann NM (2006) Lead uptake and tolerance of *Ricinus communis* L. Braz J Plant Physiol 18:483–489
Sainger M, Sharma A, Bauddh K, Sainger PA, Singh RP (2014) Remediation of nickel contamination from the sub-tropical soil of northern India using Indian mustard (*Brassica juncea* L.) and effect of the metal on growth, nitrate assimilation, diamine metabolism and proline accumulation in the plant. Biore J 18:100–110
Sakakibara M, Ohmoril Y, Ha NTH, Sano S, Sera K (2011) Phytoremediation of heavy metal-contaminated water and sediment by *Eleocharis acicularis*. Clean Soil Air Water 39(8):735–741
Sapci Z, Ustun EB (2012) Interactions between contaminated aquatic environments and element uptake by *Echinodorus amazinocus* and *Cryptocoryne undulata*. Ecotox Environ Saf 76:114–125
Schmidt U (2003) Enhancing phytoextraction: the effect of chemical soil manipulation on mobility, plant accumulation, and leaching of heavy metals. J Environ Qual 32:1939–1954
Schwitzguébel J (2015) Phytoremediation of soils contaminated by organic compounds: hype, hope and facts. J Soils Sediments. doi:10.1007/s11368-2015_1253_9
Sharma S, Singh B, Manchanda VK (2015) Phytoremediation: role of terrestrial plants and aquatic macrophytes in the remediation of radionuclides and heavy metal contaminated soil and water. Environ Sci Pollut Res 22:946–962
Shen ZG, Li XD, Wang CC, Chen HM, Chua H (2002) Lead phytoextraction from contaminated soil with high-bio-mass plant species. J Environ Qual 31:1893–1900
Sheoran V, Sheoran AS, Poonia P (2009) Phytomining: a review. Miner Eng 22:1007–1019
Shevyakova N, Cheremisina A, Kuznetsov V (2011) Phytoremediation potential of amaranthus hybrids: antagonism between nickel and iron and chelating role of polyamines. Russ J Plant Physiol 58(4)
Shi G, Cai Q (2009) Cadmium tolerance and accumulation in eight potential energy crops. Biotechnol Adv 27:555–561
Shtangeeva I, Laiho JVP, Kahelin H, Gobran GR (2004) Phytoremediation of metal-contaminated soils. symposia papers presented before the division of environmental chemistry. American Chemical Society, Anaheim, Calif, USA, http://ersdprojects.science.doe.gov/workshop pdfs/california 2004/p050.pdf

Shukla B, Visen PKS, Patnaik GK, Kapoor NK, Dhawan BN (1992) Hepatoprotective effect of an active constituent isolated from the leaves of *Ricinus communis* Linn. Drug Dev Res 26(2):183–193
Shuman LM, Wang J (1997) Effect of rice variety on zinc, cadmium iron and manganese content in rhizosphere and non-rhizosphere soil fractions. Comm Soil Sci Plant Ann 28:23–26
Singh RK, Gupta MK, Katiyar D, Srivastava A, Singh P (2010) In-vitro antioxidant activity of the successive extracts of *Ricinus communis* stems. IJPSR 1(8)
Soclo HH, Garriges P, Ewald M (2000) Origin of polycyclic aromatic hydrocarbons (PAHs) in coastal marine sediments: case studies in Cotonou (Benin) and Aquitaine (France) areas. Mar Pollut Bull 40:387–396
Søndergaard M (1988) Seasonal variations in the loosely sorbed phosphorus fraction of the sediment of a shallow and hypertrophic lake. Environ Geol Water Sci 11:115–121
Spriggs T, Banks MK, Schwab P (2005) Phytoremediation of polycyclic aromatic hydrocarbons in manufactured gas plant-impacted soil. J Environ Qual 34:1755–1762
Stobart AK, Griths WI, Ameen B (1985) The effect of Cd^{2+} on the biosynthesis of chlorophyll in leaves of barley. Physiol Plant 63:293–298
Sweta, Bauddh K, Singh R, Singh RP (2015) *Trapa natans* cultivated in contaminated water is found hazardous and not suitable for phytoremediation of heavy metals. Ecol Eng 83:39–42
Tangahu BV, Abdullah SRS, Basri H, Idris M, Anuar N, Mukhlisin M (2011) A review on heavy metals (As, Pb, and Hg) uptake by plants through Phytoremediation. Inter JChem Eng. doi:10.1155_2011_939161
Tanji KK (1990) Nature and extent of agricultural salinity. In: Tanji KK (ed) Agricultural salinity assessment and management. American Society of Civil Engineers, New York, pp 1–17
Taur DJ, Waghmare MG, Bandal RG, Patil RY (2011) Antinociceptive activity of *Ricinus communis* L. leaves. Asian Pac J Trop Biomed 1(2):139–141
Tulod AM, Castillo AS, Carandang WM, Pampolina NM (2012) Growth performance and phytoremediation potential of Pongamia pinnata (L.) Pierre, Samanea saman (Jacq.) Merr. and Vitex parviflora Juss. in copper contaminated soil amended with zeolite and VAM. Asia Life Sci 21(2):499–522
Ultra VU Jr, Yano A, Iwasaki K, Tanaka S, Kang Y, Sakurai K (2005) Influence of chelating agent addition on copper distribution and microbial activity in soil and copper uptake by brown mustard (*Brassica juncea*). Soil Sci Plant Nutr 51(2):193–202
Utsunamyia T (1980) Japanese patent application no. 55–72959
Vervaeke P, Tack FMG, Lust N, Verloo M (2004) Short and longer-term effects of the willow root system on metal extractability in contaminated dredged sediment. J Environ Qual 33(3):976–983
Vwioko DE, Fashemi DS (2005) Growth response of *Ricinus communis* L (Castor oil) in spent lubricating oil polluted soil. J Appl Sci Environ Manag 9(2):73–79
Vyslouzilova M, Tlustos P, Szakova J, Pavlikova D (2003) As, Cd, Pb and Zn uptake by Salix spp clones grown in soil enrich by high load of this elements. Plant Soil Environ 49(5):191–196
Wang K, Huang H, Zhu Z, Li T, He Z, Yang X, Alva A (2013) Phytoextraction of metals and rhizoremediation of pahs in co-contaminated soil by co-planting of sedum alfredii with ryegrass (*Lolium perenne*) or castor (*Ricinus communis*). Intern J Phytoremediation 15:283–298
Watanabe ME (1997) Plants ranging from pennycress to poplar trees are proving their worth as cleanup tools. Environ Sci Technol 31:182A–186A
Wilson SC, Jones KC (1993) Bioremediation of soil contaminated with poly nuclear aromatic hydrocarbons (PAHs): a review. Environ Pollut 81(3):229–249
Wong MW (2003) Ecological restoration of mine degraded soils, with emphasis on metal contaminated soils. Chemosphere 50:775–780
Wu X, Zhang H, Li G, Liu X, Qin P (2012) Ameliorative effect of castor bean (*Ricinus communis* L.) planting on physico-chemical and biological properties of seashore saline soil. Ecol Eng 38:97–100
Yoshida N, Kanda J (2012) Tracking the fukushima radionuclides. Science 336:1115–1116

Zhang X, Gao B, Xia H (2014) Effect of cadmium on growth, photosynthesis, mineral nutrition & metal accumulation of bana grass & vetiver grass. Ecotoxicol Environ Saf 106:102–108

Zhang H, Chen X, He C, Liang X, Oh K, Liu X, Lei Y (2015) Use of energy crop (*Ricinus communis L.*) for phytoextraction of heavy metals assisted with citric acid. Inter J Phytoremediation 17:1549–7879

Zhi-xin N, Li-na S, Tie-heng S, Yu-shuang L, Hong W (2007) Evaluation of phytoextracting cadmium and lead by sunflower, ricinus, alfalfa and mustard in hydroponic culture. J Environ Sci 19:961–967

Bioenergy and Phytoremediation Potential of *Millettia pinnata*

6

Dipesh Kumar, Bhaskar Singh, and Yogesh Chandra Sharma

Abstract

Phytoremediation is currently being envisaged as an ecologically sustainable and economical decontamination technique for polluted sites. Phytoremediation being a slow process limits the utility of the site to an extent for other activities, and therefore valorisation of the biomass produced by some means is very important. Biomass-based energy (bioenergy) is increasingly gaining popularity as a relatively clean, renewable and carbon-neutral alternative to fossil fuels. Coupling phytoremediation with bioenergy production by using fast-growing hyperaccumulating species having bioenergy potential appears to be a rather attractive opportunity. *Millettia pinnata* is a hardy tree species with well-established bioenergy potential. The use of *Millettia pinnata* for simultaneous decontamination (of heavy metal contaminated sites) and bioenergy production has been discussed in this chapter.

Keywords

Biomass • Bioenergy • Fatty acids • *Millettia pinnata* • Phytoremediation

---

D. Kumar • B. Singh (✉)
Centre for Environmental Sciences, Central University of Jharkhand, Ranchi-835205, Jharkhand, India
e-mail: bhaskar.singh@cuj.ac.in

Y.C. Sharma
Department of Chemistry, Indian Institute of Technology (BHU), Varanasi-221005, India

K. Bauddh et al. (eds.), *Phytoremediation Potential of Bioenergy Plants*,
DOI 10.1007/978-981-10-3084-0_6

6.1 Introduction

Soil is one of the most important natural resources bestowed upon humankind that forms through a very slow natural process of rock weathering. Maintaining the fertility and integrity of soil is invaluable for the survival of our socio-economic structure. Anthropogenic activities have been generating humungous quantity of wastes that are being dumped in the environment with or without proper treatment. Soil has been the recipient of several types of waste originating from several industrial, municipal, commercial and agricultural activities. Among various types of waste, those containing heavy metals are of great environmental concern. Since the dawn of industrial revolution, human beings have been extensively mining and using heavy metals for several economic purposes. Heavy metals have been defined in several ways, but generally they include those elements having specific density >5gm/cm^3 and atomic number >20 (Järup 2003). Although there can be ambiguity in defining what a heavy metal is, there is certainly no doubt about their hazardous nature. Heavy metals include several metals, transition metals, and metalloids having relatively high density such as Pb, Hg, Ni, Cd, Cr, Zn, Co, Cu, As, Mn, Fe and Se. Some of the major sources of heavy metal pollution in soil include those arising from mine tailings, the use of paints and leaded gasoline, application of pesticides and fertilisers containing heavy metals, sewage sludge, combustion residue of coal, sewage sludge disposal, irrigation using contaminated wastewater, atmospheric deposition, etc. (Chander et al. 2001).

The presence of heavy metals in soil or groundwater is potentially hazardous due to their inherent toxicity to biota in elemental as well as in the combined form. Some of these heavy metals such as Fe, Mn, Co, Cu, Mo, Zn and Ni are essential for normal metabolism and growth of plants, while others such as Se, As, Cd, Pb or Hg have no known physiological role in plants. The former metals cause toxicity when their concentration exceeds supra optimal levels, while the latter elements are inherently toxic to plants. Heavy metal toxicity arises when biota is exposed to bioavailable forms of heavy metals beyond their threshold tolerance levels for individual organisms. The phytotoxic effects may arise from alteration of several physiological processes operating at cellular and/or molecular levels. These include inactivation of enzymes, blocking of functional groups present in several metabolically important molecules, displacement or substitution of essential elements by heavy metals, disruption of plasma membrane integrity and generation of reactive oxygen species (Rascio and Navari-Izzo 2011). These reactive oxygen species give rise to oxidative stress leading to membrane dismantling, cleavage of DNA strand, ion leakage, lipid peroxidation and deterioration of biomolecules (Quartacci et al. 2001). Since heavy metals cannot be degraded by any metabolic means, they tend to accumulate at successive trophic levels and pose risk of exposure to other organisms at the higher trophic levels. Leaching of heavy metals from contaminated sites can lead to pollution transfer to other non-contaminated sites and also to the groundwater. Traditional approach of soil decontamination is based on soil excavation followed by its dumping to landfill sites. In the present era, there are several competitive demands for land and excavation and dumping are merely methods of waste containment in the

Table 6.1 Different techniques involved in phytoremediation and their respective mechanism

Phytoremediation technique	Mechanism
Phyto-extraction	Uptake of contaminants by plant roots and translocation to above-ground biomass
Phyto-stabilisation	Accumulation of contaminants in plant roots or their immobilisation (transformation to bio-unavailable forms) in soil matrix by root exudates
Rhizo-filtration	Removes contaminants from aqueous media of groundwater, soil moisture, surface runoff or water used for irrigation
Phyto-degradation	Enzymatic degradation of contaminants to less toxic forms by root exudates and cellular enzymes
Phyto-volatilisation	Contaminants are released through the stomata to the atmosphere
Rhizo-degradation/ phyto-stimulation	Plants stimulate rhizosphere-based microbes for enhanced microbial enzymatic degradation of contaminants

Tangahu et al. (2011)

landfill sites. Besides, land filling is not a permanent solution as the availability of land is limited, and there always exists possibility of release of contaminants through leaching or by other means. Stabilisation or immobilisations of contaminants by using chemical additives have also been investigated (Alkorta et al. 2004), but it is essentially a non-decontamination technique. Therefore, other alternative methods have been explored, and research is being carried out to develop environment-friendly techniques. Phytoremediation involves the use of plant species capable of tolerating low to moderate levels of contaminants and having ability to accumulate contaminants in their biomass for bioremediation of pollution. Plants have also been used as environmental sentinel for determining whether metals are close to the surface for ascertaining the feasibility for metal mining at such locations. Phytoremediation is a group of mechanisms employed by plants for uptake, accumulation, degradation, translocation, filtration, sequestration and stabilisation/immobilisation of contaminants present in soil or groundwater. Six major types of phytoremediation mechanisms have been identified (Table 6.1). These include (1) phyto-extraction, (2) phyto-stabilisation, (3) rhizo-filtration, (4) phyto-volatilisation, (5) rhizo-degradation and (6) phyto-degradation (Tangahu et al. 2011). Out of these six, only four types of mechanisms (phyto-extraction, phyto-stabilisation, phyto-volatilisation and rhizo-filtration) are recognised for phytoremediation of heavy metals from contaminated sites, while the remainders are involved in remediation of organic contaminants.

Plants have a series of defence mechanisms that control the uptake, storage and translocation of heavy metals. Plants are commonly known to entrap heavy metals in the apoplastic environment by binding them to exudated organic acids or to the anionic groups present in the cell wall and thus prevent the entry of heavy metals in the root cell. When these contaminants find their way into the root cell through the plasma membrane, detoxification is achieved by their complexation with intracellular organic acids, amino acids, metal-binding peptides and by storage in vacuoles or by release of free ionic forms from the cytoplasm. These mechanisms somewhat limit the translocation of heavy metals to above-ground biomass, and most of the metals remain entrapped in below-ground biomass (root). Plants tackle the oxidative stress that may arise due to exposure to high levels of heavy metals by enhancing its antioxidant system. Although a relatively new development, phytoremediation has been attracting tremendous scientific attention due to the potential benefit that it might offer over other decontamination techniques. Phytoremediation is an appealing decontamination technique as it is relatively inexpensive and aesthetically pleasant when compared to other techniques. Several species of plants have been identified which can extract, accumulate, transform, stabilise and store significant quantity of heavy metals in their biomass without showing any signs of phytotoxicity. Certain plants which grow naturally on metalliferous soils have developed tolerance to presence of excessive levels of heavy metals in soil and an ability to accumulate massive amounts of metals (indigenously present in the soil) in their tissues without suffering from phytotoxic effect (Prasad and Freitas 2003). Tolerance to very high levels of heavy metals by certain plants (hypertolerant plants) has been explained by two different mechanisms. Majority of hypertolerant plants retain and detoxify most of the heavy metals in root tissues with minimum translocation to above-ground biomass (excluders), while few hypertolerant plants actively take up very high levels of heavy metals, and most of it is translocated and retained in above-ground biomass (hyperaccumulators). Hyperaccumulators have been defined on the basis of threshold concentrations of different heavy metals beyond which they are known to have phytotoxic effects on congener non-hyperaccumulator plants. In comparison to congener non-hyperaccumulators, hyperaccumulators can accumulate up to 100–1000-fold higher levels of heavy metals in their shoot without suffering from phytotoxic damage. About 450 species of hyperaccumulator angiosperms have been identified which take up one or more types of heavy metals from the soil, and the concentration of heavy metals in their foliar biomass can be 100–1000-fold higher than in non-hyperaccumulators. Hyperaccumulators are known to accumulate exceedingly high levels of heavy metals in their foliar biomass without showing any signs of phytotoxicity which far exceeds the threshold toxicity levels for most other plants; therefore, hyperaccumulation is an important trait that can be used for phyto-extraction of heavy metals from soil. Hyperaccumulation is a valuable trait which can also be used for phytomining of heavy metals from heavily contaminated soil. Harvestable biomass of hyperaccumulators can be incinerated, and the metal contained in ash can be reused. Several hypotheses have been proposed to explain the tendency of some plants to accumulate exceedingly high levels of heavy metals. Currently the most widely tested and accepted hypothesis is the *elemental defence hypothesis* (Jhee et al. 2006 and Jiang et al. 2005). The elemental defence hypothesis states that the hyperaccumulation trait in plants has evolved as a means to avoid herbivory and parasitism by animals. Several evidence based on research have supported this hypothesis including some recent studies on role of Ni (Jhee et al. 2006), Cd (Jiang et al. 2005), As (Rathinasabapathi, et al. 2007), Zn (Behmer et al. 2005) and Se (Galeas et al. 2008) in defence against herbivory. However, some of the studies such as those conducted by Noret et al. (2007) and Huitson and Macnair (2003) seem to suggest that heavy metals in hyperaccumulators are not involved in any such defence mechanism. More research is required in order to test the elemental defence hypothesis by studying more and more hyperaccumulators; by analysing additional heavy metals as most of the studies have focused on Ni, Zn, Cd, Se and As; and by exposing hyperaccumulators to an array of herbivores under field conditions taking into account the level of tolerance to heavy metals in such animals (Rascio and Navari-Izzo 2011). Plants make use of secondary organic metabolites in defence against several types of enemies. These secondary metabolites (organic defence) and heavy metals (elemental defence) seem to function jointly and additively against herbivore enemies (*joint effect hypothesis*) in Ni hyperaccumulator *S. polygaloides* (Jhee et al. 2006). Plants can only take up bioavailable forms of heavy metal, and depending on their bioavailability, metals have been categorised as readily bioavailable metals (Cu, Cd, Ni, Zn, Se, As), moderately bioavailable metals (Fe, Mn, Cu) and marginally bioavailable metals (Pb, Cr) (Prasad 2003). The metals remain bio-unavailable under alkaline conditions and when they are strongly bonded to mineral particles present in the soil or precipitated. Thus, the physicochemical characteristics of soil are critical factors that affect the efficiency of phyto-extraction. Plants are known to secrete certain substances (phytosiderophores) for improving the mobility of metals (Salt et al. 1998). Plants secrete certain organic acids in the rhizosphere for effectively altering the pH of the soil and thereby improve the bioavailability of certain metals. H^+ released by plant root can displace the heavy metal adsorbed on mineral particles of soil and thus renders them bioavailable (Alford et al. 2010). Chelating agents such as EDTA, elemental sulphur, citric acid and ammonium sulphate can be added to the soil matrix for improving the bioavailability of heavy metals (Lai and Chen 2004, Lone et al. 2008). Care should be taken in using synthetic chelators like EDTA as when present in excess it may have toxic effects on the plants, and use of biodegradable and nontoxic chelators such as citric acid should be promoted (Ali et al. 2013, Smolinska and Krol 2012).

One significant disadvantage of phytoremediation is that most of the identified hyperaccumulators are slow-growing and low biomass-producing species, and thus phytoremediation of contaminated soils is a rather slow affair (Huang et al. 2004). This potentially impairs their phyto-extraction efficiency which is a function of harvestable biomass and concentration of heavy metals in the biomass. Therefore, more research is required for determining the phytoremediation efficiency of alternative fast-growing and high biomass-producing species. Since, phytoremediation is a slow process, the contaminated land cannot be used for any sort of economic activity unless an agronomic crop is employed. Valorisation of produced biomass by some means is required for adding an economic value to phytoremediation (Ghosh and Singh 2005). Edible crops cannot be used for heavy metal phytoremediation as heavy metals tend to bioaccumulate at successive trophic levels and doing so might expose the biota at higher trophic levels to hazardous concentrations of heavy metals.

Since the negative consequences of using fossil fuels are well known, prospects of using other renewable and clean alternatives forms of energy are growing day by day. Growing bioenergy plants as bioremediation agent for decontamination of polluted sites can have bright prospects as it offers the possibility of simultaneous reclamation of contaminated sites and bioenergy production (Pandey et al. 2016). Bioenergy production from nonedible feedstocks eliminates the famous food versus fuel dilemma associated with edible bioenergy feedstocks. Bioenergy refers to energy derived from biomass of recent origin in the form of heat, electricity or biofuel. Several biochemical and physicochemical conversion routes for biomass to bioenergy are well established. Gasification, direct combustion, hydrothermal liquefaction, pyrolysis, anaerobic digestion, alcoholic fermentation and transesterification are among the most commonly employed biomass to bioenergy conversion routes. Some of the biomass to bioenergy conversion routes can be more suitable than others for a given biomass depending on its biochemical composition. The presence of heavy metals in biomass does not have any detrimental effect on biomass to bioenergy conversion efficiency; instead, heavy metals might catalyse some of the conversion techniques (Lievens et al. 2008). Among biomolecules, lipids have the greatest energy density (Evershed 1993), and plants which produce seeds rich in oil can serve as feedstock for lipid-based biofuels such as biodiesel. Most of the work on hyperaccumulators has revolved around species of Brassicaceae which account for as much as 25% of all discovered hyperaccumulators. Some of the species of Brassicaceae such as *Brassica napus* (rapeseed/canola), *Brassica juncea* (Indian mustard) and *Brassica rapa* (field mustard) are reportedly good accumulators of metals and also have a fast growth rate (Ebbs and Kochian 1997). These plants produce seeds rich in oil which can be transformed to biodiesel via a catalytic or noncatalytic reaction of transesterification. Biodiesel is a relatively clean alternative to mineral diesel, besides being renewable, nontoxic, carbon neutral and biodegradable. The residual biomass after lipid extraction from seed can be subjected to other conversion routes for extraction of energy or for the production of energy carriers (biofuels). Pyrolysis of heavy metal-infested biomass has also been investigated, and the results obtained from these studies are promising. Lievens et al. (2008) and Liu et al. (2012) employed flash pyrolysis as a biomass to biofuel conversion technology using heavy metal-infested lignocellulosic biomass and reported pyrolysis of such biomass to be a viable route for biofuel production. Some of the heavy metals (particularly Cu) served as a catalyst during pyrolytic degradation of lignocellulosic biomass to bio-oil. According to these and similar studies, the bio-oil and other non-condensable fractions produced as a result of biomass pyrolysis remain virtually devoid of heavy metals (which mostly remain enriched in residual char). Hydrothermal liquefaction (HTL) of biomass can also be performed under conditions of high temperature and pressure for production of bio-oil. Unlike pyrolysis which requires prior drying of biomass, HTL can process wet biomass. Yield of bio-oil is reportedly enhanced in the presence of several transition metals such as Ni, Pt, Pd and Ru which are known to catalyse the process of HTL. Bio-oil derived from biomass via pyrolysis and HTL is inherently acidic, viscous and unstable and needs to be upgraded prior to its use as biofuel for purposes other than serving as heating oil. Several types of fossil fuel alternatives such as biodiesel and biogasoline can be produced from upgraded bio-oil. Alcoholic fermentation of cellulosic biomass produces bioethanol which can be used in place of or in addition to gasoline/petrol. Poplar, willow, *Miscanthus*, etc. are being extensively studied as bioethanol feedstock due to their fast rate of growth and cellulose rich biomass. In addition to the above-mentioned biomass to bioenergy conversion techniques, other techniques such as biomass gasification, direct combustion, co-combustion, hydrotreatment, etc. are also being investigated.

As phytoremediation efficiency is a function of metal uptake rate and rate of biomass production, a fast-growing species is likely to have better metal removal efficiency. Very limited number of species have been investigated for their phytoremediation coupled with bioenergy production potential. In this chapter, we propose the utilisation of *M. pinnata* for simultaneous phytoremediation and bioenergy production.

Phytoremediation coupled with bioenergy energy production is a relatively new development; thus, most of the studies have either not considered heavy metal accumulator plant/crop for bioenergy production or they have not analysed the fate of metals in biomass or bioenergy carriers. For instance, the concentration of metals in oil of oilseed crops grown on metal-infested soil has not been analysed extensively, and the concentration of metals in oil-based biofuels has not been reported. Compared to other plant parts, the concentration of metals in oil seeds is relatively low but, their fate when metal infested biomass based biofuels are combusted is not clearly understood. Since phytoremediation is still an emerging decontamination technique and the fate of heavy metals in infested biomass or bioenergy carriers derived from such biomass has not been studied extensively more research dedicated towards these issues will be invaluable for better understanding of simultaneous phytoremediation and bioenergy production.

6.2 Introduction to *Millettia pinnata*

M. pinnata (L.) Pierre (previously known as *Pongamia pinnata*) is a medium-sized relatively fast-growing tree/shrub. It has been described as a briefly deciduous or evergreen tree having a broad crown, and branching is of either drooping or spreading type (Orwa et al. 2009). It has the capability to fix atmospheric N_2 symbiotically in association with several species of *Rhizobium* and *Bradyrhizobium* (Arpiwi et al. 2013 and Samuel et al. 2013). The trunk is 50–80 cm or more in diameter, and it can be crooked or straight. The bark is either greyish or greyish brown in colour with smooth or faint fissures (vertical). Branchlets are devoid of hair and the pale scars of stipule are prominent. The arrangement of leaves (imparipinnate) is of alternate type, and each leaf has five to nine leaflets, and the terminal leaflet is the largest. When young, leaflets are pinkish red which mature to become glossy dark green in colour with prominent veins and dull green colour underneath. Inflorescence is raceme-like bearing two to four strongly fragrant flowers which grow up to 15–18 mm in length. Flowering in general starts after 3–4 years with raceme-like inflorescence bearing pealike white, pink and purple flowers in small clusters which often blossom throughout the year. Pollination is achieved by several bee species including *Apis dorsata*, *A. cerana indica*, *Amegilla* sp., *Megachile* sp. and *Xylocopa* sp. *M. pinnata* serves as an excellent source of nectar for production of honey by honeybees during summer (Raju and Rao 2006). The brown pods which are smooth, thick walled, flattened and elliptical appear immediately after pollination and take up to 10–11 months in maturation. The pods contain one or two bean-shaped brownish-red seeds which are about 1.5–2.5 cm long. The seeds are rich in oil and contain about 30–35% (w/w) oil. Figure 6.1 depicts the *M. pinnata* tree, its leaves, pods and inflorescence.

M. pinnata is a perennial tree growing natively in littoral zones of Southeast Asia and Australia. It symbiotically fixes atmospheric N_2 in association with several strains of *Bradyrhizobium* and *Rhizobium* and thus might not need significant input of synthetic fertiliser or manure. It can survive varying degrees of aridity and is known to flourish in areas receiving 500–2500 mm of annual rainfall and can even survive in waterlogged soils. It tolerates slight frost and can resist and overcome scorching heat and drought. It is also tolerant to alkaline and saline soil. *M. pinnata* produces a dense network of lateral roots, and therefore it is preferred for checking soil erosion and for binding of sand dunes. It has widespread distribution in India and has been used extensively in *Ayurveda* (Indian traditional system of medicine) for the treatment of bronchitis, tumours, piles, whooping cough, wounds, rheumatic arthritis, ulcer and diabetes (Tanaka et al. 1992, Pandeya et al. 2013). In traditional systems of medicine several parts of *M. pinnata* are being used due to their well-known antiulcer, anti-plasmodial, anti-hyperglycaemic, anti-nociceptive, anti-lipid peroxidative, anti-hyperammonemic, anti-inflammatory, antidiarrhoeal and antioxidant activity (Chopade et al. 2008).

6.3 Bioenergy Potential of *M. pinnata*

M. pinnata is among the most popular second-generation biodiesel feedstocks known for its superior characteristics in terms of oil content, fatty acid profile, agronomic performance and tolerance towards several types of abiotic stress particularly for tropical nations like India where it is present indigenously. The national biofuel policy of India mandates blending of 20% biodiesel derived from nonedible feedstocks (particularly *J. curcas* and *M. pinnata*) grown on wasteland in high-speed diesel by 2017 (Government of India 2009). Since the yield of oil from *J. curcas* is reportedly highly variable depending on the growth conditions, *M. pinnata* appears to be a viable alternative. The seeds of *M. pinnata* contain about 30–35% (w/w) of oil which can be used for the production of biodiesel. Biodiesel is a renewable, cleaner, nontoxic and biodegradable alternative to mineral diesel. It can be used in place of or in addition to mineral diesel virtually for every purpose the latter is used for. Since vegetable oil in its natural form is usually very viscous, it cannot supplement or substitute fuel used in compression ignition engines which are designed to run on a fuel having low viscosity (mineral diesel). Therefore, the viscosity of vegetable oil must be lowered to levels comparable to that of mineral diesel prior to its use in internal combustion compression-ignition (CI) engines. Four techniques for using vegetable oil as fuel in CI engines have been reported, and these include (1)

Fig. 6.1 (**a**) *M. pinnata* tree, (**b**) leaves and pods of *M. pinnata* and (**c**) *M. pinnata* inflorescence

direct use and blending, (2) transesterification reaction, (3) pyrolysis or thermal cracking and (4) microemulsion (Ma and Hanna 1999). Out of these techniques, transesterification appears to be the most promising opportunity for utilisation of vegetable oil as fuel in CI engines. Transesterification involves reaction between

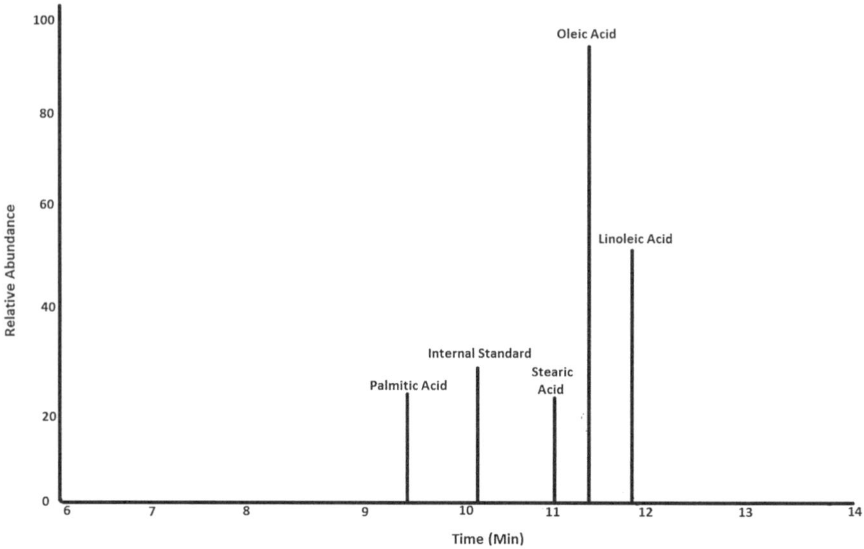

Fig. 6.2 Gas chromatogram of biodiesel derived from *M. pinnata* oil (Scott et al. 2008)

vegetable oil and a monohydric alcohol usually in the presence of an alkaline or basic catalyst. Transesterification reaction converts each triacylglycerol molecule in vegetable oil to three lower molecular weight molecules of fatty acid alkyl ester (biodiesel) and produces one molecule of glycerol as a by-product (Meher et al. 2006). Several properties of biodiesel such as cetane number, viscosity, oxidative stability and cold flow properties (cloud point and pour point) are inherently dependent on the fatty acid profile of the vegetable oil used as feedstock for transesterification. Figure 6.2 depicts the gas chromatogram of *M. pinnata* oil. Fatty acid profile of vegetable oil in turn is dependent on several biotic and abiotic factors.

Cetane number indicates the ignition quality of a fuel to be used in CI engines relative to cetane ($C_{16}H_{34}$), and higher value of cetane number for mineral diesel alternatives is desirable. Oxidative stability in biodiesel depicts the susceptibility of the fuel to oxidation under variable conditions of storage and use, and oxidised biodiesel produces several undesirable effects on fuel quality, engine performance and exhaust composition. Cold flow properties such as cloud point and pour point indicate the temperature at which biodiesel starts gelling and can no longer flow, respectively. Lower values of cloud point and pour point are desirable for the use of biodiesel under cold climate conditions. In general the cetane number and viscosity increase with increase in degree of saturation in fatty acids, while the values of oxidative stability and cold flow properties increase with increase in degree of unsaturation. The fatty acid profile of *M. pinnata* oil is reported in Table 6.2.

Table 6.3 reports several properties of biodiesel derived from *M. pinnata* oil and number 2 mineral diesel

As biodiesel contains about 11% oxygen (w/w), it facilitates completion of combustion reaction between the fuel (biodiesel) and the oxidant (oxygen), and thus the use of biodiesel reduces the emission of unburned hydrocarbons and carbon

Table 6.2 Fatty acid profile of *M. pinnata* oil

Fatty acid	Percentage in oil
Palmitic acid (C16:0)	3.7–7.9
Stearic acid (C18:0)	2.4–8.9
Oleic acid (C18:1)	44.5–71.3
Linoleic acid (C18:2)	10.8–18.3
Linolenic acid (C18:3)	1.8–2.6
Arachidic acid (C20:0)	2.2–4.7
Eicosenoic acid (C20:1)	9.5–12.4
Behenic acid (C22:0)	4.2–5.3
Lignoceric acid (C24:0)	1.1–3.5

Bobade and Khyade (2012)

Table 6.3 Comparison of no. 2 diesel, biodiesel standards and *M. pinnata* biodiesel

Fuel property	No. 2 diesel[a]	*Biodiesel limits*	*M. pinnata* biodiesel[b]
Fuel standard	ASTM D975	ASTM D6751	–
Higher heating value, (Btu/gal)	~137,640	–	–
Lower heating value, (Btu/gal)	~129,050	–	–
Kinematic viscosity, (at 40 °C)	1.3–4.1	1.9–6.0	5.44
Oxygen, (wt %)	0	-	11
Sulphur, (wt %)	0.0015 max	0.05 (max.)	<0.005
Flash point, (°C)	60–80	93.0 (min.)	158
Cloud point, (°C)	−35 to 5	Report to customer	5
Pour point, (°C)	−35 to −15	Report to customer	−2
Cetane number	40–55	47 (min.)	57
Acid value, (mg of KOH/g)	–	0.50 (max.)	0.44
Oxidative stability, (h)	–	3 (min.)	2.33

[a]Tyson (2009), [b] Sharma et al. (2010)

monoxide (EPA 2002) (Fig. 6.3). Biodiesel contains negligible amounts of sulphur, and hence the emission of oxides of sulphur is also checked. However, the use of biodiesel reportedly enhances the emission of oxides of nitrogen marginally, which is possibly due to the generation of high temperature in the combustion chamber facilitated by the presence of relatively higher levels of oxygen in biodiesel. The use of biodiesel in CI engines along with mineral diesel in proportions up to 20% biodiesel is being practiced without any need of engine modification, but the use of neat (100%) biodiesel is hampered by the design of the CI engines, and the use of neat biodiesel requires engine modifications. The fuel properties and emission characteristics of biodiesel blend are greatly dependent on the proportion of biodiesel in the blend.

In addition to transesterification, other routes of biomass to bioenergy conversion are available. But unfortunately, most of the research interest on plants like *M. pinnata* as a source of bioenergy has traditionally revolved around the production of

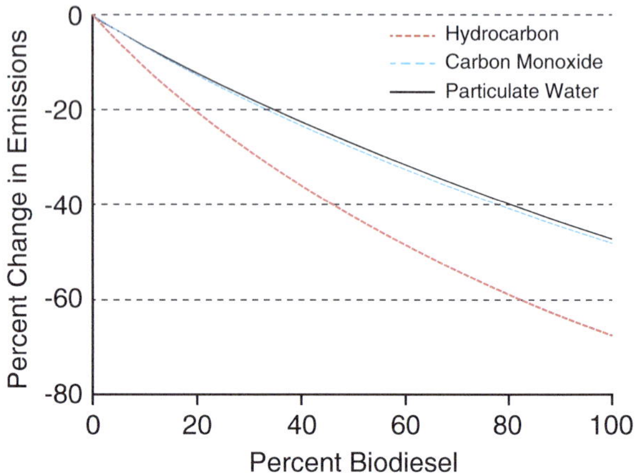

Fig. 6.3 Emission profile of biodiesel relative to content of biodiesel in diesel-biodiesel blend (EPA 2002)

Table 6.4 Physical properties of *M. pinnata* seed bio-oil

Properties	Bio-oil *from M. pinnata* seed
Density at 15 °C (kg/m^3)	0.9384
Kinetic viscosity at 40 °C (cSt)	27.9
Pour point (°C)	16
Cloud point (°C)	28
Gross calorific value (kcal/kg)	8113
Sulphur content (wt %)	0.05

Nayan et al. (2012)

biodiesel from its seeds rich in oil via transesterification. Unlike transesterification which employs only the lipids present in seed, other routes also facilitate the use of lignocellulosic biomass for production of biofuels or bioenergy. In this regard, pyrolysis of biomass deserves attention. Pyrolysis involves thermal cracking of biomass at high temperature in the absence of air, which results in production of bio-oil, char and non-condensable gases having higher energy density than the parent biomass. Flash or fast pyrolysis technique in which biomass is heated rapidly with very short vapour residence time has been developed as a means to enhance the production of bio-oil. The energy density of bio-oil is low (15–20 MJ/kg) attributed to the presence of high moisture and oxygen content. Pyrolysis involves a series of fragmentation and depolymerisation reactions of lignocellulosic biomass constituents. Bio-oil obtained from fast pyrolysis of biomass is a highly complex mixture of over 400 compounds. Bio-oil is a dark-brown colour, highly viscous and often corrosive liquid having high oxygen (35–40 wt %.) and water content (15–30 wt %.). Nayan et al. (2012) in their study on pyrolysis of *M. pinnata* seeds reported maximum yield of bio-oil (57 wt %) at 500 °C having physical properties similar to that of bio-oil obtained from other feedstocks (Table 6.4). The bio-oil was reportedly useable as fossil fuel substitute and as a feedstock for valuable chemicals after proper upgradation/treatment, while the suggested use of solid char was as an adsorbent or as a solid fuel. Bio-oil as such, cannot be used as a fuel for internal combustion engines as its characteristics are very different than that of petroleum and need to be upgraded catalytically for moisture and oxygen removal in order to produce several types of biofuels such as biogasoline and biodiesel along with several other bioproducts. Bio-oil can also be obtained via hydrothermal liquefaction of biomass in which moist biomass is subjected to temperature of about 300–350 °C and high pressure (5–20 Mpa) in the presence of a catalyst for breaking down the biomass components into smaller fragments. Gasification, on the other hand, is a partial oxidation of biomass at high temperature for the production of syn-gas (a combustible mixture of gases consisting of H_2, CH_4, CO, CO_2, etc.). Biomass has been one of the most important sources of primary energy in which biomass is directly combusted as fuel wood for cooking and also as a source of heat for several industrial and commercial purposes.

Triacylglycerol present in oil can be converted to simpler hydrocarbons by its hydrocracking or hydrotreatment under conditions of elevated temperature and pressure in the presence of a catalyst which involves addition of hydrogen at the points of unsaturation and removal of oxygen. The product of vegetable oil hydrocracking (usually referred to as green diesel or renewable diesel) can be used as a diesel fuel substitute having properties very similar to that of mineral diesel (Liu et al. 2009). The carbohydrate polymers (cellulose and hemicellulose) present in lignocellulosic biomass can be broken down via biomass pretreatment followed by chemical or enzymatic hydrolysis to their respective monomers which can either serve as substrate for alcoholic fermentation for the production of bioethanol or as growth substrate for the production of single cell oil or biomass by microorganisms (Kumar et al. 2008). Clearly, multiple approaches for biomass to bioproduct conversion are possible, and using these conversion techniques, multiple bio-based products from a given plant can be obtained. Recently the concept of *biorefinery* (an integrated facility analogous to petroleum refinery) has gained tremendous scientific interest. It uses different biomass conversion routes for the effective utilisation of virtually every component of biomass for the production of different bio-based products including biofuels and several other value-added products (Hasunuma et al. 2013). Carbohydrates (cellulose, hemicelluloses and starch) can be used for the production of bioethanol, lipids can be used for the production of lipid-based fuels (particularly biodiesel or green diesel) and the residual fraction (e.g. lignin) of limited utility can be combusted for delivering heat and energy for biorefinery unit operations. Likewise, there are several possible routes for the conversion of different biomolecules in the biomass and even for the bulk biomass, and biorefinery is based on the concept of industrial symbiosis which involves synergy of waste material/energy between different unit operations in order to derive maximum value out of the biomass in an environmentally sustainable manner. Different biomass to bioenergy conversion routes are illustrated in Fig. 6.4.

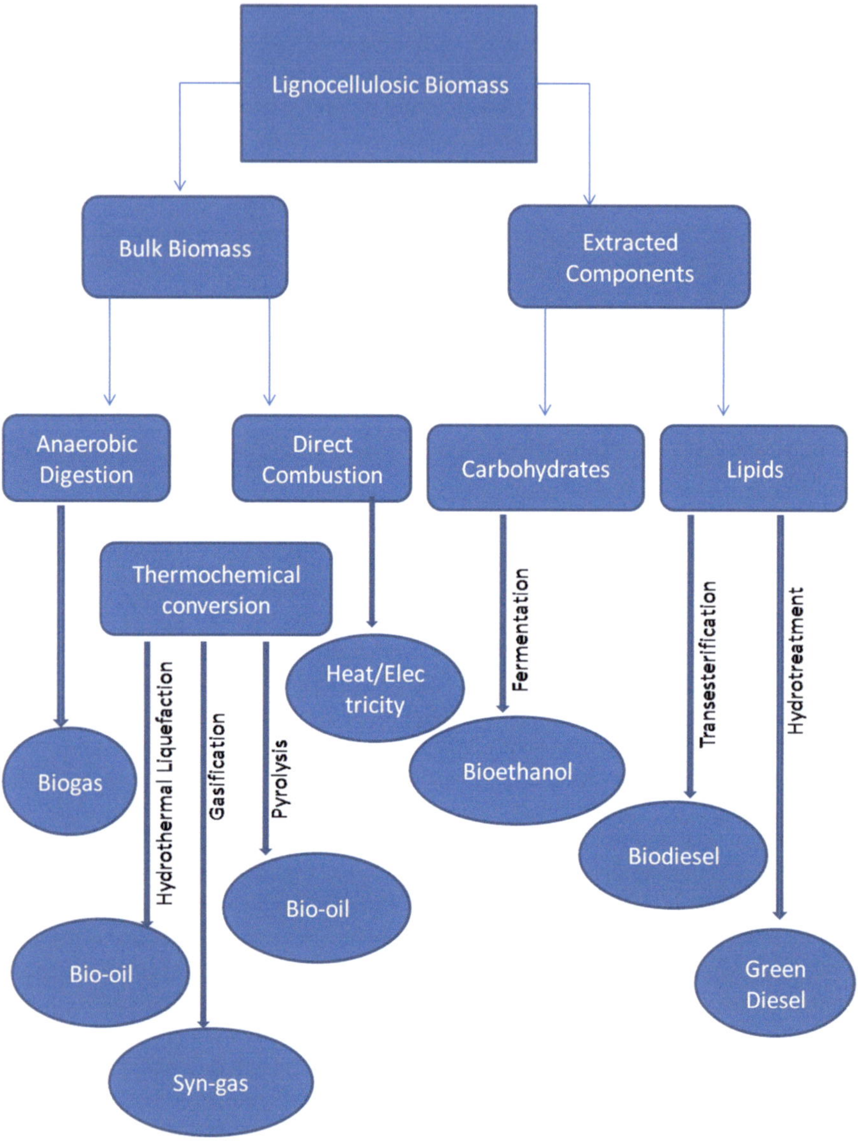

Fig. 6.4 Lignocellulosic biomass to bioenergy conversion routes

6.4 Phytoremediation Potential of *M. pinnata*

Employing agronomic crops for phytoremediating contaminated sites is highly advantageous. Phytoremediation using plants having high bioenergy potential can offer several ecological and economic benefits to the society in terms of decontamination of environment, oxygen production, wasteland reclamation, erosion control, sink for air pollutants and timber in addition to biomass for bioenergy production. In a study on remediation of coal overburden (coal mine spoil dump) using ten phytoremediator species (including *M. pinnata*) assisted by addition of biofertilisers (VAM and nitrogen-fixing bacteria) and sugar industry sludge, Singh and Juwarkar (2014) reported an economic return (including ecological benefits) to the tune of 36,562 US $ 20 years after plantation. The total investment required was reported to be 1250 US $ per hectare, and the reported payback period was only 2.5 years.

Although bioenergy potential (particularly as biodiesel feedstock) of *M. pinnata* has been explored extensively, its phytoremediation potential has only been examined in few preliminary comparative studies.

A decrease in concentration of metals (Cd, Zn, Fe, Pb, Ni, Cr, Mn and Cu) in coal mine spoil dump by 50–60% within 3 years of plantation was reported (Singh and Juwarkar 2014). Plant species when supplemented with biofertilisers and sugar industry sludge were able to tolerate high metal content in spoil dump and produced biomass at a rapid rate (Table 6.5).

M. pinnata when grown on landfill site waste (municipal origin) containing various heavy metals was able to substantially reduce the concentration of metals (except Fe) in waste soil within 2 months of plantation (Table 6.6).

In comparison to control soil, the plantations were able to produce higher biomass when grown on waste soil at a rapid rate. The concentration of chlorophyll *a*, chlorophyll *b* in leaves and Na^+ and K^+ in plant tissues was reportedly relatively higher in plantations grown in waste soil. The authors also examined the T. S of

Table 6.5 Comparative growth characteristics of different plants grown on metal-infested soil dump

Plant species	Total biomass (T ha^{-1})	Total carbon (T ha^{-1})
Ficus religiosa	12.54	6.27
Eugenia jambolana	10.50	5.25
M. pinnata	28.86	14.43
Gmelina arborea	21.62	10.81

Singh and Juwarkar (2014)

Table 6.6 Reduction in concentration of heavy metals after 60 days of *M. pinnata* plantation

Heavy metal	Initial concentration of metals in contaminated soil (in mg $0.1gm^{-1}$ soil)	Metal concentration in soil after 60 days of *M. pinnata* plantation (in mg $0.1gm^{-1}$soil)
Cu	8.83	6.06
Zn	14.490	5.372
Cr	4.97	0.564
Ni	1.12	0.251
Fe	21.07	22.175
Mn	14.464	5.372

Shirbhate and Malode (2012)

M. pinnata grown in control and waste soil for comparison and several differences were reported. The accumulation of pollutants by plants grown on waste soil was reportedly in the form of blackish deposits in vascular and cortical tissues (Shirbhate and Malode 2012). Kumar et al. (2009) in their in vitro study on effect of Cu and Cr on germination of *M. pinnata* seeds, its growth, metal tolerance and uptake reported that the seedlings of *M. pinnata* can tolerate the presence of Cu (50–400 μM) and Cr (100–800 μM) in the growth medium. According to Kumar et al. (2009), Cu is more toxic to *M. pinnata* than Cr, the accumulation of metals in the stem was restricted and seed coats has an excellent property to hold metals, and it can be employed as an absorbent for decontamination of liquid waste. Overall the presence of heavy metals in the growth medium did not seemingly affect the growth characteristics of *M. pinnata* seedling. The growth performance of 1-year-old seedlings of *M. pinnata* irrigated with industrial effluent (in various proportions of 0, 25, 50, 75 and 100% with tap water) containing heavy metals was evaluated (Manzoor et al. 2015). They reported an increase of 7.4% in shoot length and 33% increase in number of leaves compared to control when plantations were grown on 10% effluent. Leaf area and number of branches reportedly grew by 8.5 and 3%, respectively, when plantations were irrigated with 50% effluent. However, the value of all the estimated parameters for growth and tolerance analysis decreased with any further increase in proportion of effluent. Three tree species (*V. parviflora*, *S. saman* and *M. pinnata)* were compared for their growth and phytoremediation potential when grown on Cu-infested soil amended with or without VAM and/or zeolite for 170 days, and the most efficient combination was found to be that of *M. pinnata* grown on Cu-infested soil amended by both VAM and zeolite. When *M. pinnata* was grown on copper-contaminated soil (amended by addition of VAM and zeolite), it was able to accumulate 1219.75 and 26 μg g^{-1} dry matter as copper in its root and shoot, respectively (Tulod et al. 2012).

6.5 Discussion

Phytoremediation is among the most promising and perhaps the most sustainable set of techniques available till date for the decontamination of polluted sites. Since it is a slow process, utilisation of fast-growing species capable of producing high biomass having some sort of economic value is critical. Valorisation of biomass produced for production of bioenergy seems to be an attractive opportunity. For this purpose, identification of ideal hyperaccumulator plants having high bioenergy potential is of utmost importance. *M. pinnata* being a hardy legume tree in terms of tolerance to a wide range of abiotic stress such as endurance to variable degree of aridity, water logging, salinity, alkalinity, slight frost, etc. is increasingly being examined for its bioenergy potential owing to its relatively rapid growth rate and 30–35% of oil in its seed. Therefore, exploration of its ability to tolerate and flourish in growth media contaminated with hazardous heavy metals with significant translocation of heavy metals to above-ground biomass without suffering from phytotoxicity is invaluable. *M. pinnata* a plant native to tropical and subtropical Asia is among one of the most commonly researched nonedible oil-producing biodiesel feedstock. Two *M. pinnata* biomass to bioenergy conversion techniques of transesterification of its oil and pyrolysis of its oil and/or biomass have been reported in the literature. Extensive amount of research on transesterification of its oil is available, and the findings are very promising particularly for tropical countries like India where it has found its place in national biofuel policy as potential biodiesel feedstock. It can grow even on marginal lands or lands of limited utility (wasteland) and reportedly has the potential to bring about wasteland reclamation and employment opportunity. However, only few preliminary comparative studies have been carried out on soil decontamination using *M. pinnata*. The results are promising, but since phytoremediation is still a new and relatively less examined technique of decontamination, more and more objective-based research is required. The future course of research should be dedicated to decipher the mechanism of tolerance in *M. pinnata* to heavy metal(s), its hyperaccumulation ability, mechanism and regulation of metal uptake, accumulation, translocation and sequestration. Further the fate of accumulated metal(s) in biomass when it is used for the production of bioenergy is not fully known except for simple combustion of biomass in air as firewood. Although in comparison to other plant parts seeds sequester relatively less amount of heavy metal, the fate of metal upon transesterification, subsequent use of biodiesel in internal combustion engine and exhaust quality are not clearly understood. Pb, Zn, Cu, Sn, brass and bronze may catalyse the oxidation and polymerisation of biodiesel components (Tyson 2009). Bio-oil derived from fast pyrolysis of lignocellulosic biomass grown of metal-infested media usually contains relatively very trace amounts of heavy metal with major fraction remaining entrapped in solid charcoal. The metals contained in charcoal can be effectively recycled for utility purposes. Phytoremediation in combination with bioenergy production has been reported to have economic and environmental feasibility (Singh and Juwarkar 2014). Genetic engineering of plants for enhanced Cd, As, Hg and Se accumulation has been carried out by introduction of foreign metal transporter and metal chelator genes. Extensive research leading to discovery of genes regulating metal metabolism and genome sequencing projects are likely to open new vistas for development of efficient specialist/generalist hyperaccumulators (Eapen and D'souza 2005).

6.6 Conclusion

Valorisation of biomass produced by hyperaccumulating plant species by bioenergy production appears to be a promising opportunity. *M. pinnata* being a hardy plant with well-established bioenergy potential has not been examined extensively for its phytoremediation potential. Some of the preliminary comparative studies on phytoremediation potential of *M. pinnata* have yielded promising results. But several research gaps in terms of tolerance and ability of *M. pinnata* to accumulate different heavy metals or combination thereof without suffering from phytotoxicity, regulatory mechanism governing phytoremediation, effect of different soil amendments, rhizospheric interaction with microorganisms, and fate of accumulated metals and their effect on combustion process and exhaust quality when contaminated biomass is used for bioenergy production still persist.

References

Alford ER, Pilon-Smits EAH, Paschke MW (2010) Metallophytes – a view from the rhizosphere. Plant Soil 337:33–50

Alkorta I, Hernández-Allica J, Becerril JM, Amezaga I, Albizu I, Onaindia M, Garbisu C (2004) Chelate-enhanced phytoremediation of soils polluted with heavy metals. Rev Environ Sci Biotechnol 3(1):55–70

Ali H, Khan E, Sajad MA (2013) Phytoremediation of heavy metals – concepts and applications. Chemos 91(7):869–881

Arpiwi NL, Yan G, Barbour EL, Plummer JA, Watkin E (2013) Phenotypic and genotypic characterisation of root nodule bacteria nodulating *Millettia pinnata* (L.) Panigrahi, a biodiesel tree. Plant soil 367(1–2):363–377

Behmer ST, Lloyd CM, Raubenheimer D, Stewart-Clark J, Knight J, Leighton RS, Harper FA, Smith JAC (2005) Metal hyperaccumulation in plants: mechanisms of defence against insect herbivores. Funct Ecol 19(1):55–66

Bobade SN, Khyade VB (2012) Detail study on the properties of *Millettia pinnata* (Karanja) for the production of biofuel. Res J Chem Sci 2(7):16–20

Chander K, Dyckmans J, Joergensen R, Meyer B, Raubuch M (2001) Different sources of heavy metals and their long-term effects on soil microbial properties. Biol Fertil Soils 34(4):241–247

Chopade V, Tankar A, Pande V, Tekade A, Gowekar N, Bhandari S, Khandake S (2008) *Millettia pinnata*: phytochemical constituents, traditional uses and pharmacological properties: a review. Int J Green Pharm 2(2):72

Eapen S, D'souza SF (2005) Prospects of genetic engineering of plants for phytoremediation of toxic metals. Biotechnol Adv 23(2):97–114

Ebbs SD, Kochian LV (1997) Toxicity of zinc and copper to Brassica species: implications for phytoremediation. J Environ Qual 26(3):776–781

EPA (Environmental Protection Agency) Draft Technical Report, A Comprehensive Analysis of Biodiesel Impacts on Exhaust Emissions, EPA420-P-02-001, 2002; www.epa.gov/OMS/models/biodsl.htm.

Evershed RP (1993) Biomolecular archaeology and lipids. World Archaeol 25(1):74–93

Galeas ML, Klamper EM, Bennett LE, Freeman JL, Kondratieff BC, Quinn CF, Pilon-Smits EA (2008) Selenium hyperaccumulation reduces plant arthropod loads in the field. New Phytol 177(3):715–724

Ghosh M, Singh SP (2005) A review on phytoremediation of heavy metals and utilization of it's by products. Asian J Energy Environ 6(4):18

Government of India (2009) National policy on biofuels. Ministry of New & Renewable Energy, New Delhi

Hasunuma T, Okazaki F, Okai N, Hara KY, Ishii J, Kondo A (2013) A review of enzymes and microbes for lignocellulosic biorefinery and the possibility of their application to consolidated bioprocessing technology. Bioresource Technol 135:513–522

Huang XD, El-Alawi Y, Penrose DM, Glick BR, Greenberg BM (2004) A multi-process phytoremediation system for removal of polycyclic aromatic hydrocarbons from contaminated soils. Environ Pollut 130(3):465–476

Huitson SB, Macnair MR (2003) Does zinc protect the zinc hyperaccumulator Arabidopsis halleri from herbivory by snails? New Phytol 159(2):453–459

Järup L (2003) Hazards of heavy metal contamination. Brit Med Bull 68(1):167–182

Jhee EM, Boyd RS, Eubanks MD, Davis MA (2006) Nickel hyperaccumulation by *Streptanthus polygaloides* protects against the folivore *Plutella xylostella* (Lepidoptera: Plutellidae). Plant Ecol 183:91–104

Jiang RF, Ma DY, Zhao FJ, McGrath SP (2005) Cadmium hyperaccumulation protects Thlaspi caerulescens from leaf feeding damage by thrips (*Frankliniella occidentalis*). New Phytol 167:805–814

Kumar S, Mehta JU, Hazra S (2009) In vitro studies on chromium and copper accumulation potential of *Millettia pinnata* (L.) Pierre seedlings. Biorem Biodiv Bioavail 3(1):43–48

Kumar R, Singh S, Singh OV (2008) Bioconversion of lignocellulosic biomass: biochemical and molecular perspectives. J Ind Microbiol Biotechnol 35(5):377–391

Lai HY, Chen ZS (2004) Effects of EDTA on solubility of cadmium, zinc, and lead and their uptake by rainbow pink and vetiver grass. Chemos 55:421–430

Lievens C, Yperman J, Vangronsveld J, Carleer R (2008) Study of the potential valorisation of heavy metal contaminated biomass via phytoremediation by fast pyrolysis: Part I. Influence of temperature, biomass species and solid heat carrier on the behaviour of heavy metals. Fuel 87(10):1894–1905

Liu WJ, Tian K, Jiang H, Zhang XS, Ding HS, Yu HQ (2012) Selectively improving the bio-oil quality by catalytic fast pyrolysis of heavy-metal-polluted biomass: take copper (Cu) as an example. Environ Sci Technol 46(14):7849–7856

Liu Y, Sotelo-Boyás R, Murata K, Minowa T, Sakanishi K (2009) Hydrotreatment of Jatropha oil to produce green diesel over trifunctional Ni-Mo/SiO2-Al2O3 catalyst. Chem Lett 38(6):552–553

Lone MI, He ZL, Stoffella PJ, Yang XE (2008) Phytoremediation of heavy metal polluted soils and water: progresses and perspectives. J Zhejiang University Sci B 9(3):210–220

Ma F, Hanna MA (1999) Biodiesel production: a review. Bioresour Technol 70(1):1–15

Manzoor SA, Mirza SN, Zubair M, Nouman W, Hussain SB, Mehmood S, Irshad A, Sarwar N, Ammar A, Iqbal MF, Asim A (2015) Estimating genetic potential of biofuel forest hardwoods to withstand metal toxicity in industrial effluent under dry tropical conditions. Genet Mol Res 14(3):9543–9554

Meher LC, Sagar DV, Naik SN (2006) Technical aspects of biodiesel production by transesterification—a review. Renew Sust Energ Rev 10(3):248–268

Nayan NK, Kumar S, Singh RK (2012) Characterization of the liquid product obtained by pyrolysis of karanja seed. Bioresour Technol 124:186–189

Noret N, Meerts P, Vanhaelen M, Dos Santos A, Escarré J (2007) Do metal-rich plants deter herbivores? A field test of the defence hypothesis. Oecol 152(1):92–100

Orwa C, Mutua A, Kindt R, Jamnadass R, Simons A (2009) Agroforestree database: a tree reference and selection guide version 4.0. World Agroforestry Centre

Pandeya KB, Tripathi IP, Mishra MK, Dwivedi N, Pardhi Y, Kamal A, Gupta P, Dwivedi N, Mishra C (2013) A critical review on traditional herbal drugs: an emerging alternative drug for diabetes. International Journal of Organic Chemistry 3(1):1–22

Pandey VC, Bajpai O, Singh N (2016) Energy crops in sustainable phytoremediation. Renew Sust Energ Rev 54:58–73

Prasad MNV (2003) Phytoremediation of metal-polluted ecosystems: hype for commercialization. Russ J Plant Physiol 50(5):686–701

Prasad MNV, de Oliveira Freitas HM (2003) Metal hyperaccumulation in plants: biodiversity prospecting for phytoremediation technology. Electron J Biotechnol 6(3):285–321

Quartacci MF, Cosi E, Navari-Izzo F (2001) Lipids and NADPH-dependent superoxide production in plasma membrane vesicles from roots of wheat grown under copper deficiency and excess. J Exp Bot 152:67–75

Raju AS, Rao SP (2006) Explosive pollen release and pollination as a function of nectar-feeding activity of certain bees in the biodiesel plant, *Millettia pinnata* (L.) Pierre (Fabaceae). Curr Sci 90(7):960

Rascio N, Navari-Izzo F (2011) Heavy metal hyperaccumulating plants: how and why do they do it? And what makes them so interesting? Plant Sci 180(2):169–181

Rathinasabapathi B, Rangasamy M, Froeba J, Cherry RH, McAuslane HJ, Capinera JL, Srivastava M, Ma LQ (2007) Arsenic hyperaccumulation in the Chinese brake fern (*Pteris vittata*) deters grasshopper (*Schistocerca americana*) herbivory. New Phytol 175(2):363–369
Salt DE, Smith RD, Raskin I (1998) Phytoremediation. Annu Rev Plant Biol 49(1):643–668
Scott PT, Pregelj L, Chen N, Hadler JS, Djordjevic MA, Gresshoff PM (2008) *Millettia pinnata*: an untapped resource for the biofuels industry of the future. Bioenergy Res 1(1):2–11
Sharma YC, Singh B, Korstad J (2010) Application of an efficient nonconventional heterogeneous catalyst for biodiesel synthesis from *Millettia pinnata* oil. Energy & Fuels 24(5):3223–3231
Shirbhate N, Malode SN (2012) Heavy metals phytoremediation by *Millettia pinnata* (L) growing in contaminated soil from municipal solid waste landfills and compost Sukali Depot, Amravati (Ms). Int J Adv Biol Res 2(1):147–152
Singh SK, Juwarkar AA (2014) Sustainable reclamation of coal mine spoil dump using microbe assisted phytoremediation technology. Int J Env Sci Tox 2(3):43–54
Samuel S, Scott PT, Gresshoff PM (2013) Nodulation in the legume biofuel feedstock tree *Millettia pinnata*. Agric Res 2(3):207–214
Smolinska B, Krol K (2012) Leaching of mercury during phytoextraction assisted by EDTA, KI and citric acid. J Chem Technol Biotechnol 87:1360–1365
Tanaka T, Iinuma M, Yuki K, Fujii Y, Mizuno M (1992) Flavonoids in root bark of *Millettia pinnata*. Phytochemistry 31(3):993–998
Tangahu BV, Sheikh Abdullah SR, Basri H, Idris M, Anuar N, Mukhlisin M (2011) A review on heavy metals (As, Pb, and Hg) uptake by plants through phytoremediation. Int J Chem Eng 2011
Tulod AM, Castillo AS, Carandang WM, Pampolina NM (2012) Growth performance and phytoremediation potential of *Millettia pinnata* (L.) Pierre, *Samanea saman* (Jacq.) Merr. and *Vitex parviflora* Juss. in copper-contaminated soil amended with zeolite and VAM. Asia Life Sci 21(2):499–522
Tyson KS (2009) Biodiesel handling and use guidelines. DIANE Publishing, Golden

7 Phytoremediation Potential of *Leucaena leucocephala* (Lam.) de Wit. for Heavy Metal-Polluted and Heavy Metal-Degraded Environments

Jamilu Edrisa Ssenku, Muhammad Ntale, Ingvar Backeus, and H. Oryem-Origa

Abstract

Regeneration of heavy metal-polluted and heavy metal-degraded sites has remained a global challenge despite the existence of numerous conventional physico-chemical techniques that can be applied. In view of the large size of the degraded areas and the cost implications, the application of the inexpensive "green" and sustainable technique of phytoremediation is unrivalled by any other possible alternative techniques. However, its effectiveness is largely dependent on judicious selection of the plant to be used. We thus assessed the suitability of *Leucaena leucocephala* for phytoremediation of heavy metal-polluted and heavy metal-degraded sites. *L. leucocephala* has numerous inherent characteristics that can be exploited to augment phytoremediation and lower the cost of regeneration. The species can survive in harsh environmental conditions with the exception of heavily frosted conditions and occurs in a wide range of ecological settings. The species is fast growing, capable of reaching maturity in 6 to 7 months to produce a vast amount of seeds that can germinate into numerous seedlings to carry on further remediation of the polluted site. It can produce large quantities of phytomass that can accumulate heavy metals and can repeatedly be

J.E. Ssenku (✉) • H. Oryem-Origa
Department of Plant Sciences, Microbiology and Biotechnology, College of Natural Sciences, Makerere University, Kampala, Uganda
e-mail: jssenku@gmail.com

M. Ntale
Department of Chemistry, College of Natural Sciences, Makerere University, Kampala, Uganda

I. Backeus
School of Natural Sciences, Technology and Environmental Studies, Sodertorn University, 141 89 Huddinge, Sweden

Department of Plant Ecology and Evolution, Uppsala University, Norbyv. 18D, 752 36 Uppsala, Sweden

K. Bauddh et al. (eds.), *Phytoremediation Potential of Bioenergy Plants*,
DOI 10.1007/978-981-10-3084-0_7

harvested to regenerate a polluted area through phytoextraction. Heavy metal-laden phytomass of *L. leucocephala* moulds into furniture and is used for construction to preclude contamination at the site of use. It is excellent on coppicing, thus eliminating the costs of replanting during the phytoremediation programme. The species is endowed with high proficiency for nitrogen fixation through nodule formation and can substantially revitalize microbial mass and micro-bioactivities to pave way for re-establishment of self-sustaining plant communities over the polluted sites. Its flexibility to nodulate with rhizobia of other legumes and its rhizobia to nodulate with other legumes could optimize nitrogen content revitalization of the polluted soils. However, the species is invasive and could be adopted under stringent measures to avoid its spread. It is also very palatable to animals and may thus be of limited application in the phytoremediation of areas accessible to animals. Suitability of the species in heavily polluted areas is minimal as many of the inherent characteristics may not fully be expressed.

Keywords
Bioenergy plants • Phytoremediation • Restoration • Heavy metals • Phytosiderephore • Hyper-accumulation

7.1 Brief Biology and Ecology of *Leucaena leucocephala*

Leucaena leucocephala (Lam.) de Wit. is a tree species belonging to the family Fabaceae and subfamily Mimosoideae, which is a host to many species capable of prolific vegetative growth even under extremely unfavourable environmental conditions (Dias et al. 2008). It is native to southern Mexico and northern Central America (Hill 1971). Being a free-seeding and colonizing plant, it has over the years managed to establish itself in areas with significantly varied ecological settings but free of frost. Currently, the species has naturalized itself in many areas and in some cases outside the tropics (Hughes 1998; Jayanthy et al. 2014), where climatic conditions are detrimental to its growth. According to Suttie (2005), the species occurs in a wide range of geographical areas including those that are semiarid provided it can find some soil moisture. In highly frosted periods, aerial growth of the species ceases, but the crowns survive and subsequently grow robustly in the following summer with numerous branches (Shelton and Brewbaker 1994). It is a deciduous plant species of varied forms that range from shrubby varieties to medium-sized trees (National Research Council 1984; Parrotta 1992).

L. leucocephala tolerates a wide range of soil conditions, from skeletal and stony soils to heavy clays, although best growth occurs on well-drained soils with pH ranging between 6.0 and 7.5 (National Research Council 1984, Parrotta 1992). According to Shelton and Brewbaker (1994), it can grow on mildly acidic soils with pH >5.2. It does best on deep, well-drained, neutral to calcareous soils and is often found naturalized on the rocky coralline terraces of Pacific Island countries (Suttie 2005). It is well adapted to clay soils and requires good levels of phosphorus and calcium for best growth (Shelton and Brewbaker 1994). Generally the species has always failed to grow on soils with low pH, low phosphorous, low calcium and high aluminium concentration (Orwa et al. 2009), which are characteristic of most heavy metal-polluted sites. Further still, high soil salinity and water logging are detrimental to the growth of the species.

Although *L. leucocephala* can survive on sites receiving less than 600 mm to more than 2000 mm annual rainfall, it grows best on sites receiving about 1500 mm annual rainfall with a dry season lasting approximately 4 months (Parrotta 1992). It is not tolerant of poorly drained soils, especially during seedling growth, and production can be substantially reduced during periods of water logging, but once established, it can survive short periods of excess moisture (Suttie 2005). It grows well in areas with annual temperatures ranging from 20 to 30 °C with mean monthly minimum and maximum temperatures of 16 and 32 °C, respectively (National Research Council 1984; Van den Beldt and Brewbaker 1985). For optimal growth, it requires warm temperatures of 25–30 °C (Jayanthy et al. 2014). Brewbaker et al. (1985) suggest that temperature limitations occur above 1000 m elevation within 10° latitude of the equator and above 500 m elevation within the 10–25° latitude zone. Under conditions of adequate moisture content, *L. leucocephala* flowers and produces fruits throughout the year with associated suppression of vegetative growth (Orwa et al. 2009).

7.2 Global Mine Waste Disposal and Remediation Challenges

Mining activities generate enormous volumes of wastes which are usually deposited on the surface and abandoned after definitive closure of the mines. Due to pollution by the wastes, the soils close to the wastes are usually degraded to magnitudes that lead to complete loss of vegetation cover and impediment of revegetation for a prolonged period of time. Heavy metal pollution of the lithosphere and hydrosphere ecosystems by abandoned mine wastes is still a global challenge as the wastes from which it arises are generated in large quantities and have a long-term persistence in the environment (Ssenku et al. 2014d). The exposure of the pyrite materials to water and oxygen usually leads to the generation of acid mine drainage (AMD) through oxidation and reduction reactions below:

1. $FeS + 7O_2 + 2H_2O \rightarrow 2FeSO_4 + 2H_2SO_4$
2. $4Fe^{2+} + 3\ O_2 + 4H^+ \rightarrow 4Fe^{3+} + 2H_2O$
3. $Fe^{3+} + 3H_2O \rightarrow Fe(OH)_3 + 3H^+$
4. $4FeS\ (s) + 15O_2 + 14H_2O \leftrightarrow 16H^+ + 8SO_4^- + 4Fe(OH)_3\ (s)$

The AMD characterized with extremely low pH and elevated concentrations of heavy metals pollutes and degrades sites close and far away from the wastes. Remediation of such AMD-polluted and AMD-degraded soils is a hard task (Tandy et al. 2009), due to the excessive costs incurred when conventional techniques are applied. The conventional techniques usually involve pneumatic fracturing, solidification/stabilization, vitrification, excavation and removal of contaminated soil layers, physical stabilization or washing of contaminated soils with strong acids or heavy metal chelators (Bhargava et al. 2012). These rely heavily on "dig and dump" or encapsulation, neither of which decontaminates the soil (Pulford and Watson 2003). Phytoremediation has in the recent years proved to be economical, eco-friendly and aesthetically acceptable (Santana et al. 2012; Stingu et al. 2012; Witters et al. 2012). Given the size of most of the degraded and polluted sites and the cost implications, the application of the "green" and sustainable technique of phytoremediation is unrivalled by any other possible alternative techniques. *L. leucocephala* is known to be suitable for effective and yet inexpensive restoration of degraded mining sites through afforestation (Maleko and Mtupile 2012). There are numerous bioenergy plants with varying suitability for phytoremediation. Thus effectiveness of any of the above phytoremediation technique adopted to clean up the environment can greatly be enhanced by judicious selection of the bioenergy plant to use.

7.3 What Is Phytoremediation?

Phytoremediation is a biotechnological approach involving the use of green plants to clean up contaminated environment. It is a technique that is increasingly becoming popular among scientists, environmentalists and remediation engineers. It comprises of plant-based remediation technologies such as phytoextraction, phytostabilization, phytovolatilization, phytofiltration and phytodegradation (Fulekar et al. 2009; Marques et al. 2009). Restoration of the normal pre-mining physico-chemical characteristics of heavy metal-polluted sites through phytoremediation majorly involves extraction of the heavy metals from the polluted soil (phytoextraction) and/or inactivation of heavy metals (phytostabilization) in the soil. Through phytoextraction, the plant takes up heavy metals from the soil or water and translocates them into the shoot or any other harvestable parts (Sas-Nowosielska et al. 2008). In the long run, repeated cycles of harvesting metal loaded shoots that are transported away followed by sprouting may reduce the concentration of heavy metals in the polluted medium. On the other hand, phytostabilization involves the establishment of a plant cover on the surface of the polluted site with the ultimate aim of reducing the mobility of pollutants within the vadose zone through accumulation by roots or immobilization within the rhizosphere (Nanthi et al. 2011). Additionally, the established plant cover reduces both wind and water erosion of fine soil particles laden with heavy metals and through production of root exudates that may bind heavy metals restricting their leaching to the water table.

7.4 Criteria for Selection of Bioenergy Plants for Phytoremediation

Various parameters have been devised to gauge the suitability of a plant species for phytoremediation. According to USEPA (2012), gauging the suitability of the plant species should be based on its ability to extract or degrade the pollutant targeted, adaptation of the species to the local climate, biomass productivity, depth of root penetration, compatibility with soils to be remediated, plant species growth rate and ease of planting and maintenance. Generally, tree and shrubby bioenergy plants like *L. leucocephala* have high potential to grow deeper roots that can scavenge for heavy metals resident in deeper layers of the polluted substratum. On a large scale, the deeper rooting by these plants coupled with their inherent high phytomass productivity can boost heavy metal uptake from the polluted substratum leading to its regeneration through phytoextraction.

In some instances, indices like bioconcentration factor (BCF) – proportion of metal concentration of the dry weight of the plant to the metal concentration of the dry weight of the soil (Ashraf et al. 2011) – and translocation factor (TF), proportion of metal concentration in above-ground dry matter to the metal concentration in the dry weight of the roots (Deng et al. 2004; Santillan et al. 2010), have been used. Translocation factor value greater than 1 is indicative of the plant's high accumulation efficiency and potential to phytoextract a particular heavy metal (Zhang et al. 2002; Santillan et al. 2010), while values less than 1 are indicative of the species potential for phytostabilization of a particular heavy metal (Li et al. 2007; Zacchini et al. 2008; Saraswat and Rai 2009). In this chapter, the suitability of *L. leucocephala* for phytoremediation of heavy metal-polluted and heavy metal-degraded sites with respect to the above factors and other inherent characters has been evaluated.

7.5 Suitability of *Leucaena leucocephala* for Phytoremediation

L. leucocephala is a multipurpose tree species currently used to improve the fertility of agricultural soils and serves as a fodder tree if grown on non-contaminated soil (Daimon 2006; Franzel et al. 2014) and as a hedge plant, timber and bioenergy crop plant. Its innumerable roles has led to its worldwide recognition as a miracle tree (Brewbaker 1987) and has attracted a lot of interests from environmentalists and scientists to explore its potential for phytoremediation of polluted sites. *L. leucocephala* is now considered as one of the tree species that has potentials for phytoremediation in many tropical and subtropical areas (Ma et al. 2009). The species can grow on poor soils most probably due to its nitrogen-fixing capabilities. *L. leucocephala* has been reported to also grow on chromite overburdens in India by Rout et al. (1999) to provide a plant cover that has effectively delimited the dispersal of heavy metals by erosion from the mine tailings to the neighbouring environments. Adoption of the species could eliminate maintenance costs in the form of fertilizer application and replanting. Specifically, the species suitability for phytoremediation is ascribed to its attributes as discussed below.

7.5.1 Enhancement of Secondary Succession by Native Species on Heavy Metal-Polluted Sites

Heavy metal-polluted sites remain devoid of any vegetation cover due to the persistence of heavy metals and impoverished nutritional status. Nitrogen deficiency of mine wastes is one of the factors that preclude plant species re-establishment and maintenance of healthy plant growth (Yang et al. 2003, Song et al. 2004). Adoption of *L. leucocephala* for phytoremediation improves soil conditions paving way for secondary succession by the native species that leads to phytostabilization of the heavy metal-polluted sites. The leaf litter derived from the species can improve soil organic matter content (Majule 2006) and also enhance the nutritional status over time through its decomposition and mineralization (Brewbaker et al. 1985) to create conducive environmental conditions that attract other native plant species to grow on sites under regeneration. Phytoremediation of amended acidic pyrite and copper tailings by *L. leucocephala* has been shown to lead to re-establishment of understorey plant communities with higher species diversity, richness and cover than *E. grandis* (Sssenku et al. 2014a). This highlights the species higher potential for phytostabilization of sites with similar pollution problems and climatic conditions.

7.5.2 Growth Rate and Phytomass Accumulation

Besides heavy metal accumulation and tolerance, plant species adopted in phytoremediation strategy should be fast growing, with deep rooting system, and producing abundant phytomass. *L. leucocephala* is a well-known fast-growing leguminous tree species, with a high potential for producing large quantities of seeds and phytomass within a short interval of time that can repeatedly be harvested, leading to regeneration of the affected area through phytoextraction. Under optimum conditions, stands of the species have been reported by Balasundaran and Ali (1987) to yield annually the highest quantities of wood ever recorded. *Leucaena* is extensively cultivated mainly for phytomass production due to its high regenerative capacity, which is more than 50 t ha^{-1} year. *L. leucocephala* yields higher quantities of wood than species; it has been compared with most short-duration trials ranging between 3 and 5 years (Brewbaker 1987). Wood may be harvested from the species on cycles ranging from 6 months to 10 years. Upon harvests of the wood during phytoremediation, the species coppices readily with rare loss of a tree, eliminating costs of replanting.

The heavy metal-laden wood of the species harvested from the site under regeneration has several uses that eliminate pollution of new sites where it is put to use. The wood has average heating values on oven-dry basis of about 19.4 MJ kg^{-1} (4640 kcal kg^{-1}) and thus could be used as fuel wood, generating less than 1 percent volume of ash (Brewbaker 1987) that could safely be kept or have its heavy metals extracted for commercial purposes. Wood from the species can be used to make excellent charcoal, with heating values of 29 MJ kg^{-1} (Brewbaker 1987). The wood from *L. leucocephala* has fibre contents that are similar to that of other tropical hardwoods and could produce paper with good printability and low tearing and folding strengths (Brewbaker 1987). The wood could also be used to make furniture, craftwood items and other items used in daily life. All the aforementioned economic values provide safe means of disposing of heavy metal-laden wood or immobilizing heavy metals after phytoextraction to preclude contamination at the site of use.

The inherent faster growth of the species could result in increased nutrient exudation and secretion of mucigel from roots and decay of roots to create an organic carbon-rich environment in which microbes can thrive and act optimally to restore the polluted area (Hernandez-Allica et al. 2006; Khan et al. 2009). Further still, by growing very fast, its new thick canopy could diminish erosion and inflow of heavy metals to surface of water bodies and promote water conservation within the rhizospheres through reduced evaporation. The faster growth could also enhance production of phytosiderophore that can sequester heavy metals within the rhizospheres of *L. leucocephala*, precluding their leaching to underground water. Alongside its modulation of soil condition during phytoremediation, by growing faster on larger polluted sites, it can trap and sequester large quantities of carbon dioxide into its biomass, thereby helping to mitigate climate change impacts.

However, its growth and survival are adversely affected when grown in highly heavy metal-polluted unamended soils for remediation (Plate 7.1). Thus, like any other plant species, amelioration of the harsh conditions is a prerequisite to ensure its growth on highly heavy metal-polluted soils (Haoab et al. 2013) for remediation. Successful pretreatment of soils heavily polluted with pyrite and copper tailings to pave way for establishment of *L. leucocephala* has been demonstrated by Ssenku et al. (2014a) under field conditions through application of limestone and compost.

7.5.3 Heavy Metal Uptake and Hyperaccumulation

In addition to high phytomass productivity, proficiency of in situ restoration of heavy metal-polluted sites through phytoremediation by vascular plants is dependent on heavy metal uptake and translocation to the harvestable phytomass. The roots of *L. leucocephala* have been reported to accumulate higher concentrations Fe, Zn, Cu and Mn than the leaves (Gupta et al. 2000). Similarly, less quantities of As are translocated from the roots to the shoot (Schneider et al. 2013). In pot experiments conducted by Shanker et al. (2006), *L. leucocephala* seedlings accumulated exceedingly high concentrations of Cr in roots that were ninefold higher than the concentrations which accumulated in shoots after 1 year of establishment. The reduced translocation of the above heavy metals from the roots to the shoot by *L. leucocephala* is an interesting mechanism not only for using the species for regeneration of sites polluted with the heavy metals cited but also for protection of

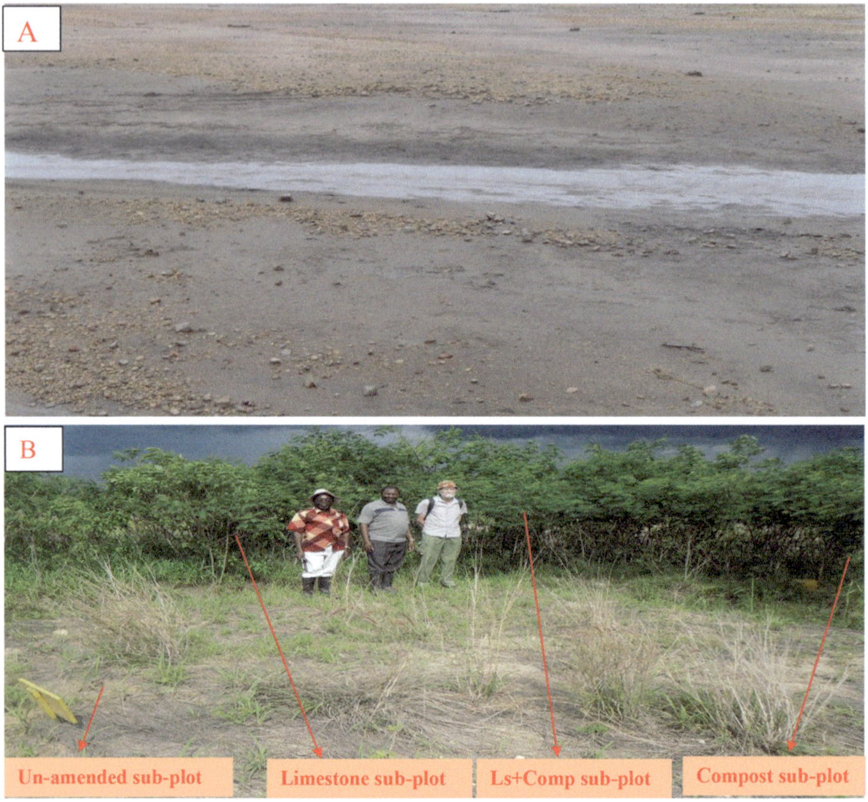

Plate 7.1 A completely bare heavy metal-polluted site before amendment application (**a**) and the establishment and growth of *L. leucocephala* after amendment application (**b**). Note the failure of growth on unamended subplot and the paltry growth of grasses due to leaching of pollutants by fresh water after delimiting fresh inflow of polluted storm water. *Ls* = limestone; *Comp* = compost

the species (Smith et al. 2010a, b; Wong 2003) against phytotoxicity. Furthermore, the translocation pattern is suggestive of the species suitability for remediation of sites polluted with these heavy metals through phytostabilization rather than phytoextraction.

L. leucocephala has been shown to accumulate Pb in shoot to concentrations >1000 mg kg^{-1} (Meeinkuirt et al. 2012). Studies by Yitao et al. 2014 have revealed that *L. leucocephala* accumulates Pb and Cd in the above-ground parts than in the roots. This is reminiscent of its high phytoextraction potential of Pb and Cd. However, it should be avoided for remediation of Pb polluted site accessed by both wild and domestic animals to avoid consumption of Pb and Cd intoxicated foliage. In a study to evaluate the suitability of *L. leucocephala* for phytoremediation of Zn and Cd polluted soils by Saraswat and Rai (2011), the species attained maximum remediation efficiency in moderately contaminated soils definitely due to high phytomass production and/or high metal accumulation. This is in conformity with earlier observations on the species made by Vyslouzilova et al. (2003) and Tlustos et al. (2006). More than 80% of the total Pb absorbed by the entire *L. leucocephala* plant accumulates in its parts comprising the root, branch, stem, bark and xylem (Quan et al. 2001), in which it can permanently be restrained from returning to the environment. This is indicative of the species superior phytoremediation potential for Pb as a large quantity of it was accumulated in the plant parts for its phytostabilization (root) and phytoextraction (harvestable parts – branch, stem, bark and xylem) leaving a small portion of it in the leaves that would be taken back to the soil under regeneration as litter.

7.5.4 Susceptibility to Pathogens and Insect Folivores

The health of any plant adopted in a phytoremediation strategy is paramount to ensure optimal production of phytomass, uptake and accumulation of heavy metals and cover of the polluted soils to preclude heavy metal dispersal. Although *L. leucocephala* is known for its high resistance to diseases, it is susceptible to some pathogens (Pound and Martinez-Cairo 1983) and entomological folivores that may compromise its suitability for phytoremediation of a heavy metal-polluted and heavy metal-degraded site. The species is prone to attacks by *Camptomeris leucaenae* (Stev. and Dalbey) Syd. that causes leaf spot (Pound and Martinez Cairo 1983). It may also be affected by *Colletotrichum* spp. pathogens to cause leaf spot.

The species is also susceptible to fungal infections that may cause undesirable effects to the plant and ultimately its phytoremediation potential. Stem gummosis owing to *Fusarium semitectum* (Singh 1981; Singh et al. 1983) and *F. acuminatum* Ell. & Ev. occurs to the species. Some *Fusariums* have also been reported to cause stem and root rots. Infections by *Phytophthora drechsleri* have led to the development of necrotic patches or cankers and death of the entire tree of *L. leucocephala* in some parts of the world notably Hawaii. A collar rot of young trees of *L. leucocephala* due to infection by *Sclerotium rolfsii* has been reported in Florida by Brewbaker (1987). The pods and seeds may be affected by different fungi and bacteria such as *Pseudomonas fluorescens* with severe effects on the fecundity of the species during phytoremediation.

Apart from pods and seeds, the damage to other parts of the species by insect is not common (Pound and Martinez-Cairo 1983). Thus, insects can adversely affect the self-seeding potential of the species to provide a plant cover over the heavy metal-polluted soils by inflicting damage on the pods and fruits. Two insects, ants and psyllids (jumping plant lice), are known to cause severe damage to *Leucaena* trees including *L. leucocephala* (Brewbaker 1987). According to Brewbaker (1987), the establishment of the species in the East African region without application of insecticides to eliminate termites or harvester ants is virtually impossible. Even though termites may initially be absent in a heavy metal-polluted site, subsequent changes to the soils by *L. leucocephala* and amendments applied may lead to recolonization of the site by termites and ultimately causing severe damage to the species. The *Leucaena* psyllid, *Heteropsylla cubana* Crawford, is one of the most severe pests of *Leucaena* on record, capable of causing severe setback to fodder production in some parts of the world and significant effects on production of wood (Brewbaker 1987). Leaf feeding by the beetles, *Apogonia rouca* and *Adoretus sinicus*, has been reported in India and Hawaii, all of which can greatly compromise the suitability of the species for phytoremediation. In view of the potential pathological and entomological constraints, optimal exploitation of the traits of *L. leucocephala* to optimize phytoremediation would require at some instances proper management of the infections. Under severe infestation and attacks, treatment against these pathogens and pests may escalate the cost of cleanup, thus undermining one of the phytoremediation cardinal goals of cost minimization.

7.5.5 Reproduction Potential

High reproduction rate by early colonizers on heavy metal-polluted sites is vital for the species to perpetuate their existence and occupy more bare areas of the site targeted for phytoremediation. *L. leucocephala* grows very fast within 6 months to reach reproductive maturity (National Research Council 1984; Van den Beldt and Brewbaker 1985) and produce numerous seeds that can germinate to carry on further remediation of the site. During field trials of the species on heavily polluted soils in Queen Elizabeth Conservation Area (QECA) in Uganda by Ssenku et al. (2014a), it grew very fast only on soils amended with limestone and compost to reach reproductive maturity 7 months after planting before *Eucalyptus grandis* and *Senna siamea*. Subsequent pod formation and dispersal of numerous seeds led to seedling emergence that was directly proportional to pod formation (Table 7.1). The

Table 7.1 Pod formation and seedling emergence of *L. leucocephala* in unamended and amended heavy metal-polluted plots in QECA

Site	Amendment application	Average number of pods/ plant (n=4)	Average seedling emergency/m^2
KTDS	Unamended	0	0
	Limestone	79	139
	Compost	42	188
	Limestone+compost	36	111
LPPTS	Unamended	0	0
	Limestone	71	216
	Compost	50	196
	Limestone+compost	39	106
HPPTS	Unamended	0	0.
	Limestone	0	0
	Compost	20	24
	Limestone+compost	9	16

From Ssenku (2015)

KTDS Kilembe tailings dam site, *LPPTS* low polluted pyrite trail site, *HPPTS* highly polluted pyrite trail site

Plate 7.2 Ground of a *Leucaena leucocephala* amended plot covered by seedlings that emerged after dispersal of seeds from the pods

numerous seedlings covered a significant proportion of the floor of the plots (Plate 7.2), thus limiting any further dispersal of the heavy metals. In highly polluted zones, seedling emergence and establishment are low due to poor pod setting, low seed production and high concentration of heavy metals that according to Shafiq et al. (2008), adversely affect germination of seeds and subsequent seedling establishment and growth in *L. leucocephala*.

This trend has also been demonstrated by Ssenku et al. (2014a). Albeit prolific emergence of seedlings, there may not be a recognizable new generation of *L. leucocephala* trees establishing themselves under the canopy of the mature trees. This is attributed to the vulnerability of seedlings to numerous environmental factors. The seedlings are intolerant of frosts which are known to cause leaf shedding (Isarasenee et al. 1984). Shading from the mature trees reduces the growth of the seedlings of the species. However, the species possesses a superior tolerance to reduce light when compared with other tree legumes (Benjamin et al. 1991). Proper growth, development and establishment of the seedlings to enhance phytoremediation by *L. leucocephala* require full sunlight (Parrotta 1992) and proper management of other emerging species since the seedlings are more susceptible to suppression by competing vegetation (Parrotta 1992).

7.5.6 Rooting Pattern and Volume

Root pattern and volume characteristics are salient traits that need to be put into consideration when selecting plants for phytoremediation. The two root traits are related to higher capacity for absorption and accumulation of essential nutrients and heavy metals during phytoremediation. Roots are essential for the growth and

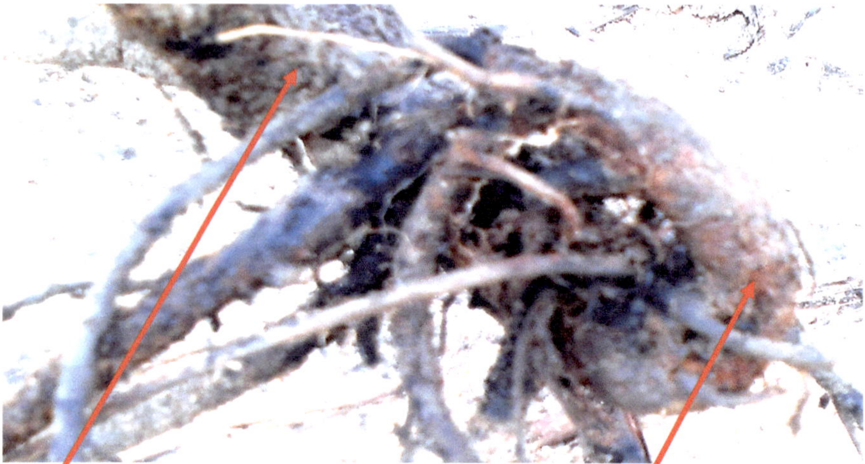

Plate 7.3 Curved root of *L. leucocephala* in response to amendment application

development of a plant, and their modifications have consequences on other plant parts (Biermacki and Lovett-Doust 2002). Rooting in *L. leucocephala* is heavily dependent on the magnitude of heavy metal pollution. It forms very thin roots and ultimately smaller root volume on heavily polluted pyrite and copper tailings leading to low phytomass production. The elevated concentrations of heavy metals inhibit root elongation through their adverse effects on cell division, including inducement of chromosomal aberrations and abnormal mitosis (Huillier et al. 1996; Jiang et al. 2001; Liu et al. 2003; Radha et al. 2010) that culminates into disorders of root cell metabolism and depression of root growth, which severely undermines the potential of the species for phytoremediation.

Rooting in heavy metal-polluted soils by *L. leucocephala* can significantly be enhanced through application of both organic and inorganic amendment materials. However, the much desired potential for forming deeper roots to scavenge and regenerate polluted deeper layers is lost, as roots curve (Plate 7.3) and grow superficially avoiding deeper layers of the polluted soil profiles having concentrations that are prohibitive to growth. The response of root growth by *L. leucocephala* upon amendment application greatly limits its potential for phytoextraction of the deeper layers in heavily polluted sites.

7.5.7 Nodulation and Nutrient Replenishment

Optimization of growth and survival of any plant selected for phytoremediation on the stressful polluted site is obligatory and greatly challenging. According to Al-menaie et al. (2010), growth and survival of trees introduced for phytoremediation are dependent on plant growth requirements, of which mineral nutrients are very important. However, heavy metal-polluted sites are nutritionally very poor (King 1988) characterized with very low content of the key macronutrient phosphorous and total nitrogen due to low organic matter content and low mineralization rates. Thus, remediation of heavy metal-polluted sites would require nutritional enhancement either through addition of artificial fertilizers or use of plants that can boost their fertility. Adoption of nitrogen-fixing plant species in a phytoremediation strategy could substantially lower the costs as opposed to the use of fertilizers. The use of nitrogen-fixing *L. leucocephala* has been commended for this task (Schneider et al. 2013). *L. leucocephala* is a perennial bioenergy plant that can fix more than 500 kg N/ha a year (Aganga and Tshwenyane 2003) that could substantially ameliorate nitrogen deficiency which delimits re-establishment of native plant species in heavy metal-polluted sites. The species has dual characteristics of heavy metal sequestration and nitrogen fixation and is thus suitable for metal decontamination (Tawfik 2008) and fertility improvement (Ma et al. 2006).

L. leucocephala forms symbiotic relationship with the nitrogen-fixing bacterium *Rhizobium loti* (Halliday and Somasegaran 1983), which is generally fast growing (3–5 h generation time) and acid producing, probably responsible for the species intolerance to acid soils. The seedlings of *L. leucocephala* nodulate with the rhizobia which is not exceptionally species specific. Its rhizobium can nodulate a number of leguminous plant species and most especially from the genus *Acacia* (Dommergues 1982). Analogous to the above, rhizobia from a number of species of *Acacia, Mimosa, Dalea* spp., *Neptunia* spp., *Piptadenia* spp., *Sabinea* sp. and *Sesbania grandiflora* can nodulate *L. leucocephala*. Fluidity of the rhizobia of *L. leucocephala* to nodulate other legumes and the rhizobia of other legumes to nodulate it may lead to elevated nodulation, nitrogen enrichment and establishment of plant cover that can substantially control heavy metal pollution. *L. leucocephala* roots have poorly developed root hairs and appear to rely heavily on mycorrhiza for nutrient uptake. Several species of *Glomus* and *Gigaspora* can infect the species. Huang et al. (1985) have reported *Glomus fasciculatus* to significantly improve uptake of minerals by *L. leucocephala*, notably P, K and Ca. *L. leucocephala* growth improves microbial mass C and N in the heavy metal soils (Saraswat and Rai 2011), which can also be released upon the death of the microorganisms in forms that can be used by *L. leucocephala* and other successional species during phytoremediation.

Growth and nodulation of *L. leucocephala* in heavy metal-polluted soils improve upon fertility attributes of heavy metal-polluted soils. Improvement of soil fertility by the species has been shown to improve with increasing duration of phytoremediation and to decrease with increasing concentration of heavy metals (Saraswat and Rai 2011). The potential of the species to nodulate and improve the nutritional status of highly heavy metal-polluted soils has been reported to be severely diminished. Nodule formation (number) by *L. leucocephala* has been shown to fall with increasing free aluminium in acid soils. Similarly, field trials of the species on heavily polluted pyrite and copper tailings revealed complete failure of nodulation and growth (Ssenku et al. 2014c). Failure of nodulation of *L. leucocephala* in untreated soils heavily polluted by acidic pyrite and copper tailings was due to extremely low pH that was in the range of 2.27–3.48 (Ssenku et al. 2014d). In such acidic soils,

Balasundaran and Ali (1987) have reported abnormal formation of roots, impairment of nodulation, dismal growth of the whole plant and low survival of *Rhizobium* to occur in *L. leucocephala*. These adverse effects could be ameliorated to enhance phytoremediation by the species through application of organic and/or inorganic amendments (Ssenku et al. 2014a). The findings by various authors are suggestive of the species suitability for phytoremediation of moderately heavy metal-polluted sites as opposed to extremely polluted ones.

7.5.8 Restoration of Microbial Activity

The exposure of soil to heavy metals results in reduction of microbial diversity and microbial activities (Wang et al. 2007), thus deterioration of soil quality. Such soils are characterized by low microbial biomass (Clemente et al. 2006), poor structure and poor plant growth potential. Restoration of such heavy metal-polluted and heavy metal-degraded soils is strongly hinged on the revitalization of the microbial activity. During restoration, the microorganisms play a cardinal role in soil redevelopment and biogeochemical cycling to improve and maintain soil fertility. Phytoremediation of heavy metal-polluted ecosystems would thus require adoption of a plant that can substantially revitalize microbial mass and activities to pave way for re-establishment of self-sustaining plant communities. *L. leucocephala* can be employed as the phytoremediator, especially because it is a host plant for arbuscular mycorrhizal fungi (Smith and Read 1997) and other microfloral species. These fungi play a fundamental role in the absorption of nutrients (Gardezi et al. 2000; Gardezi et al. 2004) by *L. leucocephala* to boost its establishment and growth on the soil under remediation. It is known to form a mutualistic relationship with the nitrogen-fixing bacteria *Rhizobium loti* (Halliday and Somasegaran 1983). According to González-Chavez (2000), the mutualistic association occurring between *L. leucocephala* and Rhizobium and mycorrhizal fungi seems to tolerate the presence of heavy metals and could be exploited for phytoremediation of soils polluted with heavy metals.

In some instances, faster regeneration of the polluted site would be desirable. However, phytoremediation is rather too slow but could be accelerated by exploiting the wide range of microfloral species that inhabit its rhizospheres. Several bacterial species associated with plants can play a major role in accelerating phytoremediation of heavy metal-contaminated soils by promoting plant growth and health (Compant et al. 2010; Dary et al. 2010). The microfloral species of *L. leucocephala* may enhance phytoextraction of heavy metals by acidification that increases bioavailable fractions of heavy metals, phosphate solubilization and uptake which could improve phytomass production and promote the release of phytosiderophores and biosurfactants that have a synergistic effect on phytoremediation.

Studies by Saraswat and Rai (2011) have shown that phytoremediation of Cd and Zn polluted soils for duration of 1 year by *L. leucocephala* significantly increased dehydrogenase and urease activities by 70% and 50%, respectively, over the control soils. According to Vasquez-Murrieta et al. (2006), the species enhancement of enzymatic activities during phytoremediation is attributed to its ability to reduce heavy metal content through sequestration thus minimizing their detrimental effects on the enzymes in polluted soils under regeneration. Comparative studies by Ssenku et al. (2014b) on soils polluted by pyrite soils and copper tailings revealed a higher efficacy for revitalization of bacterial functional diversity by *L. leucocephala* than *Eucalyptus grandis*. At a rhizosphere and symbiotic level, the species has been reported to improve environmental conditions for bacterial communities (Vázquez-Luna 2015). The species is now considered as a good resource for development of bioinoculants (Gopalakrishnan et al. 2014) that may be applied to augment phytoremediation efficiency.

7.5.9 Palatability to Ruminants and Invasiveness

Phytoremediation is very slow and would thus require long-time survival of the species adopted. In addition to the pathological problems, the species should be free of attacks from animals to which it is palatable and guarded against anthropogenic attacks due to its economic values to the surrounding community. Due to its palatability, its survival during field phytoremediation trials at sites within Queen Elizabeth Conservation Area was minimal partly due to recurrent consumption of its foliage by *Kobus kob thomasi* (Uganda kob), despite deterrent measures that were instituted to limit their accessibility to the plant (Ssenku et al. 2014a). Thus, for phytoremediation of heavy metal-polluted sites within wildlife protected areas and sites accessible to domestic animals, *L. leucocephala* may not be suitable since it may in turn serve as link through which heavy metals may be channelled into food chains of wildlife and man.

L. leucocephala is a belligerent colonizing species mainly due to its continuous flowering and fruiting throughout the year, self-fertility and prolific seed production. The species is also a renowned resprouter capable of resuming growth after exposure to fires and chopping off of the aerial parts. According to Orwa et al. (2009), *L. leucocephala* is an ubiquitous invasive species capable of growing in a wide array of areas including those that are disturbed and degraded. Usually the ultimate goal of mine land rehabilitation is to reinstate the pre-disturbed land use or ecosystem (DEHP 2012). Due to its invasiveness, it may, once adopted in a phytoremediation strategy, establish itself on the polluted and degraded sites leading to the emergence of a plant community that is totally different from the pre-disturbed one and above all spread over to the untargeted areas. This is undesirable more so if the remediation is to be carried out within a protected area such as the heavily heavy metal-polluted pyrite areas in Queen Elizabeth Conservation Area. However, *L. leucocephala* has numerous advantages that comes along with its adoption in a phytoremediation programme and thus be used under stringent measures to control its invasiveness.

7.6 Conclusions and Recommendations

L. leucocephala has multifaceted inherent properties that can optimize the effectiveness of phytoremediation of heavy metal-polluted sites and substantially lower the costs involved. Its environmental plasticity that has enabled it naturalizes itself in a wide range of geographical regions, and to grow on poor and degraded soils makes it probably a more suitable candidate for phytoremediation than most of the other bioenergy plant species. The species has a high proficiency for nitrogen fixation through nodule formation and revitalization of microbial activity that can improve the conditions of heavy metal-polluted soils for secondary succession of the native plant species and ultimately the establishment of self-sustaining plant communities. The high reproduction potential due to its prolific seed production, high growth rate and phytomass accumulation can substantially enhance phytostabilization and phytoextraction of heavy metal-polluted sites. Suitability of *L. leucocephala* for regeneration of areas heavily polluted with heavy metals is minimal, rendering pretreatment with amendment materials a prerequisite. The unsuitability is exacerbated by low soil pH that increases the available fractions and uptake of heavy metals to levels that are phytotoxic to the species. The species is also susceptible to insect and mammal folivores and bacterial and fungal pathogens that may undermine the suitability of the species in areas where they are prevalent.

L. leucocephala should be adopted for phytoremediation of sites with climatic conditions to which it is well adapted or can grow optimally. The species potential to cope with the magnitude of pollution should be tested under laboratory and/or greenhouse conditions at planning stage before full-scale application. For sites that can easily be accessed by animals, strategies to preclude browsing and destruction of *L. leucocephala* should be devised to avoid channelling of heavy metals into their food chains and costs that come along with replanting. In circumstances where conservation of the native flora is much treasured, the species may be used but under stringent measures to control its invasiveness.

Acknowledgement The authors are deeply grateful to the Swedish International Development Agency (Sida) for the invaluable financial support that was extended through Makerere University to carry out research from which part of the materials for this chapter was sourced.

References

Aganga AA, Tshwenyane SO (2003) *Lucerne, Lablab* and *Leucaena leucocephala* forages: Production and utilization for livestock production. Pak J Nutr 2:46–53

Al-menaie HS, Al-Ragon O, Al-Shatti A, Mathew M, Suresh N (2010) Evaluating the growth performance of *Cassia nodosa* and *Cassia fistula* L. Seedlings using different potting mixtures. Academic Journal of Plant Sciences 3:33–36

Ang LH, Tang LK, Ho WM, Hui TF, Theseira GW (2010) Phytoremediation of Cd and Pb by four tropical timber species grown on an Ex-tin mine in peninsular Malaysia. World Acad Sci Eng Technol 62:459–463

Ashraf MA, Maah MJ, Yusoff I (2011) Heavy metal accumulation in plants growing in ex-tin mining catchment. Int J Environ Sci Technol 8:401–416

Balasundaran M, Ali MI (1987) Root Nodulation Potentialities of *Leucaena leucocephala* in KERALA. Kerala Forest Resarch Institute, Peechi

Benjamin A, Shelton HM, Gutteridge RC (1991) Shade tolerance of some tree legumes. In: Shelton HM, Stür WW (eds), Forages for Plantation Crops. ACIAR Proceedings No. 32, Canberra, pp. 75–76.

Bhargava A, Carmona FF, Bhargava M, Strivastava S (2012) Approaches for enhanced phytoextraction of heavy metals. J Environ Manag 105:103–120

Biermacki M, Lovett-Doust J (2002) Developmental shifts in watermelon growth and reproduction caused by the squash bug, *Anasa tristis*. New Physiologist 155:265–273

BIO-WISE (2000) Contaminated land remediation: a review of biological technology. DTI, London

Brewbaker JL (1987) *Leucaena*: a multipurpose tree genus for agroforestry. In: Steppler HA, Nair PKR (eds) Agroforestry, a decade of development. ICRAF, Nairobi

Brewbaker JL, Hegde N, Hutton EM, Jones RJ, Lowry JB, Moog F, van den Beldt R (1985) Leucaena – Forage Production and Use. NFTA, Hawaii, 39 pp

Careillo-Gonzalez R, Gonzalez-Chavez MCA (2006) Metal accumulation in wild plants surrounding mining wastes. Environ Pollution 144:84–92

Chaney RL, Angle JS, Broadhurst CL, Peters CA, Tappero RV, Sparks DL (2007) Improved understanding of hyperaccumulation yields, commercial phytoextraction and phytomining technologies. J Environ Qual 36:1429–4143

Clemente RC, Almela MP, Bernal MP (2006) A remediation strategy based on active phytoremediation followed by natural attenuation in a soil contaminated by pyrite waste. Environ Pollut 143:397–406

Compant S, Clemente C, Sessitsch A (2010) Plant growth-promoting bacteria in the rhizo- and endosphere of plants: their role, colonization, mechanisms involved and prospects for utilization. Soil Biol Biochem 42:669–678

Daimon H (2006) Traits of genus Crotolaria used as green manure legume on sustainable cropping systems. Jpn Agric Res Q 40(4):299–305

Dary M, Chamber-Perez MA, Palomares AJ, Pajuelo E (2010) "*In situ*" Phytostabilisation of heavy metal polluted soils Using *Lupinus luteus* inoculated with metal resistant plant-growth promoting rhizobacteria. J Hazard Mater 177:323–330

DEHP (2012) Rehabilitation requirements for mining projects. EM1122 version 1. Queensland Department of Environment and Heritage Protection, Brisbane.

Deng H, Ye ZH, Wong MH (2004) Accumulation of lead, zinc, copper and cadmium by 12 wetland plant species thriving in metal-contaminated sites in China. Environ Pollut 132(1):29–40

Dias LE, Franco AA, Campello EFC, Faria SM, Castilho A, Henriques JC, Daniels WL (2008) A two-phase process for revegetation of acidic bauxite tailings in the Amazon Region, Brazil. In: National meeting of the America society of mining and reclamation, Richmond, Proceedings. Richmond, New Opportunities to Apply Our Science, ASMR, 2008.CD-ROM

Dommergues Y (1982) Ensuring effective symbiosis in nitrogen-fixing trees. In: Graham PH, Harris SC (eds) Biological nitrogen fixation technology for tropical agriculture. CIAT, Colombia

Franzel S, Carsan S, Lukuyu B, Sinja J, Wambugu C (2014) Fodder tress for improving livestock productivity and smallholder livelihoods in Africa. Curr Opin Environ Sustain 6:98–103

Fulekar HM, Singh A, Bhaduri AM (2009) Genetic engineering strategies for enhancing phytoremediation of heavy metals. Afr J Biotechnol 8:529–535

Gardezi AK, Cetina VM, Talavera D, Ferrera-Cerrato R, Rodríguez F, Larque M (2000) Efecto de inoculación con endomicorrhiza arbuscular y dosis creciente de fertilización fosfatada en el crecimiento de chapulixtle (*Dodonaea viscosa*). Terra 18:153–159

Gardezi AK, Víctor MC, Alcalá., Enrique M. Sáenz., Joel Pérez N., Erasmo Rubio G., Juán Velásquez M, Mario L Saavedra (2004) Growth response of *Sesbania emerus* (Aubl) Urban associated with Arbuscular Endomycorrhizal fungi and fertilized with inorganic phosphorus. In: Proceedings of the 8th World Multiconference of Systemics, Cybernetics and Informatics, Orlando Florida, USA XV: 41–46

González-Chavez MC (2000) Biorremediación de suelos contaminados con metales pesados. La edafología y sus perspectivas al siglo XXI. Tomo II. UNAM, UACH, SEMARNAT pp 714–731

Gopalakrishnan S, Sathya A, Vijayabharathi R, Varshney RK, Gowda CL, Krishnamurthy L (2014) Plant growth promoting rhizobia: challenges and opportunities. Biotech 4(18):1–23

Gupta M, Kumari A, Yunus M (2000) Effect of fly-ash on metal composition and physiological responses in *Leucaena leucocephala* (lamk.) de. wit. Environ Monit Assess 61:399–406

Halliday J, Somesagaran P (1983) Nodulation, nitrogen fixation and Rhizobium strain affinities in the genus *Leucaena*. In: *Leucaena* research in the Asia-Pacific region. In . Ottawa, IDRC

Haoab X, Taghavib S, Xiea P, Orbachc MJ, Alwathnanid HA, Rensinge C, Weia G (2013) Phytoremediation of Heavy and Transition Metals Aided by Legume-Rhizobia Symbiosis. Int J Phytoremediat 16:179–202

Hernandez-Allica J, Becerril JM, Zarate O, Garbisu C (2006) Assessment of the efficiency of a metal phytoextraction process with biological indicators of soil health. Plant Soil 281:147–158

Hill GD (1971) Herbae Abstracts 4:111–119

Huang RS, Smith WK, Yost RS (1985) Influence of vesicular-arbuscular mycorrhiza on growth, water relations, and leaf orientation in *Leucaena leucocephala* (Lam.) de Wit. New Phytol 99:229–243

Hughes CE (1998) Systematic botany monograph; pp. 55

Huillier LL, Auzac JD, Durand M, Michaud-Ferriere N (1996) Nickel effects on two maize (*Zea mays*) cultivars: Growth, structure, nickel concentration, and localization. Can J Bot 74:1547–1554

Isarasenee A, Shelton HM, Jones RM (1984) Accumulation of edible forage of *Leucaena leucocephala cv.* Peru over late summer and autumn for use as dry season feed. Leucaena Research Reports: 3–4.

Jayanthy V, Geetha R, Rajendran R, Prabhavathi P, KarthikSundaram S, Dinesh Kumar S, Santhanam P (2014) Phytoremediation of dye contaminated soil by *Leucaena leucocephala* (subabul) seed and growth assessment of Vigna radiata in the remediated soil. Saudi J Biol Sci 21(4):324–333

Jiang W, Liu D, Liu X (2001) Effects of copper on root growth, cell division, and nucleolus of *Zea mays*. Biol Plantarum 44:105–109

Khan MS, Zaidi A, Wani PA, Oves M (2009) Role of plant growth promoting rhizobacteria in the remediation of metal contaminated soils. Environ Chem Lett 7:1–19

King JA (1988) Some physical features of soil after opencast mining. Soil Use Manag 4:23–30

Lasat MM (2002) Phytoextraction of toxic metals: a review of biological mechanisms. J Environ Qual 31:109–120

Li MS, Luo YP, Su ZY (2007) Heavy metal concentrations in soils and plant accumulation in a restored manganese mine land in Guangxi, South, China. Environ Pollut 147:168–175

Liu D, Jiang W, Gao X (2003) Effects of cadmium on root growth, cell division and nucleoli in root tip cells of garlic. Biologia Plantarum 47:79–83

Ma Y, Dickinson NM, Wong MH (2006) Beneficial effects of earthworms and arbuscular mycorrhizal fungi on establishment of leguminous shrubs on Pb/Zn mine tailings. Soil Biol Biochem 38:1403–1412

Ma Y, Rajkumar M, Freitas H (2009) Inoculation of plant growth promoting bacterium *Achromobacter xylosoxidans* strain Ax10 for the improvement of copper phytoextraction by *Brassica juncea*. J Environ Manag 90:831–837

Majule AE (2006) Evaluating the performance of *Leucaena accessions* for agroforestry in sub-humid environment, southern Tanzania

Maleko D, Mtupile E (2012) The potential of *Leucaena leucocephala* pioneer trees and *Cenchrus ciliaris* understory grass species in soil improvement and forage production at Wazo Hill Quarry. Quarry life award report 2012

Marques APGC, Rangel AOSS, Castrol PML (2009) Remediation of metal contaminated soils: Phytoremediation as a potentially promising clean-up technology. Crit Rev Environ Sci Technol 39:622–654

Meeinkuirt W, Pokethitiyook P, Kruatrachue M, Tanhan P, Chaiyarat R (2012) Phytostabilization of a pb-contaminated mine tailing by various tree species in pot and field trial experiments. Int J Phytoremediat 14(9):925–938. doi:10.1080/15226514.2011.636403

Nanthi S, Bolan NS, Park JH, Robinson B, Naidu R, Huh KY (2011) Phytostabilization: a green approach to contaminant containment. Adv Agron 112:145–204

National Research Council (1984) *Leucaena*: Promising forage tree crop for tropics, 2nd edn. National Academy Press, Washington D.C, 100pp

Orwa C, Mutua A Kindt R, Jamnadass R, Anthony S (2009) Agroforestree database: a tree reference and selection guide version 4.0. http://www.worldagroforestry.org/sites/treedbs/treedatabases.asp

Parrotta JA (1992) Fabaceae (Bean family): *Leucaena leucocephala* (Lam.) de wit. International Institute of Tropical Forestry, USDA Forest Service

Pound B, Martinez-Cairo L (1983) *Leucaena*: Its cultivation and use. Overseas Development Administration, London

Pulford ID, Watson C (2003) Phytoremediation of heavy metal contaminated land by trees-a review. Environ Int 29:529–540

Pulford ID, Dickson NM (2005) Phytoremediation technologies using trees. In: MNV P, Saiwan KS, Naidu R (eds) Trace elements in the environment, pp 375–395

Quan ZZ, Wensheng S, Chongyu L, Ong MWI (2001) Uptake and translocation of heavy metals in dominant plants of soil seed banks introduced to a lead/zinc mine tailings pond. Acta Phytoecological Sinica 25(3):306–311

Radha J, Srivastava S, Solmon S, Srivastava AK, Chandra A (2010) Impact of excess zinc on growth parameters, cell division, nutrient accumulation, photosynthetic pigments and oxidative stress of sugar cane (*Saccharum* spp.). Acta Physiologiae Plantarum 32:979–986

Rout GR, Samantaray S, Das P (1999) Chromium, nickel and zinc tolerance in *Leucaena leucocephala* (K8). Sivae Genetica 48:3–4

Saraswat S, Rai JPN (2009) Phytoextraction potential of six plant species grown in multimetal contaminated soil. Chem Ecol 25:1–11

Saraswat S, Rai JPN (2011) Prospective application of *Leucaena leucocephala* for phytoextraction of Cd and Zn and nitrogen fixation in metal polluted soils. Int J Phytoremediat 13(3):271–288

Santana KB, de Almeida AF, Souza VL, Mangabeira PAO, Silva D, da Gomes FP, Dutruch L, Loguercio LL (2012) Physiological Analyses of *Genipa americana* L

Santillan LFJ, Constantino CAL, Rodriguez GAV, Ubilla NMC, Henandez RIB (2010) Manganese accumulation in plants of the mining zone of Hidalgo. Mexico Biores Tech 110:5836–5841

Sas-Nowosielska A, Kucharski R, Pogrzeba M, Krzyzak J, Kuperberg JM, Japeng J (2008) Phytoremediation techniques used to reduce environmental threat posed by metal contaminated soils: theory and reality. In: Barnes I, Kharytonov MM (eds) Simulation and assessment of chemical processes in a multiphase environment. Springer, New York, pp 285–297

Shafiq M, Zafar MI, Athar M (2008) Effect of lead and cadmium on germination and seedling growth of *Leucaena leucocephala*. J Appl Sci Environ Manag 12(2):61–66

Shanker AK, Ravichandran V, Pathmanabhan G (2006) Phytoaccumulation of chromium by some multipurpose-tree seedlings. Agrofor Syst 64(1):83–87

Schneider J, Labory CR, Rangel WM, Alves E, Guilherme LRG (2013) Anatomy and ultrastructure alterations of *Leucaena leucocephala* (Lam.) inoculated with mycorrhizal fungi in response to arsenic-contaminated soil. J Hazard Mater 262:1245–1258

Shelton HM, Brewbaker JL (1994) *Leucaena leucocephala* – the most widely used forage tree legume. In: Gutteridge RC, Shelton HM, Mathur NK (eds) Forage tree. Legumes in tropical agriculture. CAB International, Wallingford

Singh S (1981) Gummosis and canker in *Leucaena leucocephala*. In: Kaul RN, Gogate MG, Mathur NK (eds) *Leucaena leucocephala* in India. Vanguard Press, Dehra Dun

Singh S, Khan SN, Misra BN (1983) Gummosis, brown spot and seedling mortality in Subabul: Disease incidence and pathology of the host. Indian Forester 109:185–192

Smith SE, Read DJ (1997) Mycorrhizal symbiosis.structure and development of ectomycorrhizal roots. Academic Press, New York, pp. 161–290

Smith SE, Christophersen HM, Pope S, Smith FA (2010a) Arsenic uptake and toxicity in plants: integrating mycorrhizal influences. Plant Soil 327:1–21

Smith SE, Facelli E, Pope S, Smith FA (2010b) Plant performance in stressful environments: interpreting new and established knowledge of the roles of arbuscular mycorrhizas. Plant Soil 326:3–20

Song SQ, Zhou X, Wu H, Zhou YZ (2004) Application of municipal garbage compost on revegetation of tin tailing dams. Rural Eco-Environ 20:59–61

Ssenku EJ, Ntale M, Backéus I, Oryem-Origa H (2014a) Assessment of seedling establishment and growth performance of *Leucaena leucocephala* (Lam.) De Wit., *Senna siamea* (Lam.) and *Eucalyptus grandis* W. Hill ex Maid. in amended and untreated Pyrite and copper tailings. J Biosci Med 2:33–50

Ssenku EJ, Ntale M, Oryem-Origa H (2014b) The efficacy of compost, limestone and growth of *Leucaena leucocephala* (Lam.) de wit, *Senna siamea* (Lam.) and *Eucalyptus grandis* W. Hill ex Maid. for the restoration of bacterial functional diversity in the rhizosphere in copper tailings and pyrite soils. J Nat Sci Res 4(10):39–54

Ssenku EJ, Ntale M, Backéus I, Lehtilä K, Oryem-Origa H (2014c) Dynamics of plant species during phytostabilisation of copper mine tailings and pyrite soils, Western Uganda. J Environ Eng Ecol Sci 3(4):1–12

Ssenku EJ, Ntale M, Oryem-Origa H (2014d) Physico-chemical characteristics of copper tailings and pyrite soils in western Uganda: implication for phytoremediation. Int J Environ Monit Anal 2:191–198

Ssenku EJ (2015) Phytoremediation of acid mine waste polluted and degraded habitats in Kasese District by selected tree species. PhD thesis, Makerere University, Kampala

Stingu A, Volf I, Popa VI, Gostin I (2012) New approaches concerning the utilisation of natural amendments in cadmium phytoremediation. Ind Crops Prod 35:53–60

Suttie JM (2005) *Leucaena leucocephala* (Lam.) de Wit Leguminosae Mimosaceae. Online at: http://www.fao.org/ag/AGP/AGPC/doc/Gbase/data/pf000158.htm. Accessed on 10 Apr 2015

Tandy S, Healey JR, Nasson MA, Williamson JC, Lones DL (2009) Remediation of metal polluted mine soil with compost: composting versus incorporation. Environ Pollut 157:690–697

Tawfik KM (2008) A monitory field study at El Saaf-Helwan faba bean farms irrigated by industrial waste water and polluted water with sewage. J Appl Sci Res 4:492–499

Tlustos P, Szakova J, Hruby J, Hartman I, Najmanova J, Nedelnik J, Pavlíkova D, Batysta M (2006) Removal of As, Cd, Pb, and Zn from contaminated soil by high biomass producing plants. Plant Soil Environ 52:413–423

USEPA (2012). Using phytoremediation to clean up sites. www.epa.gov/superfund/accomp/news/phyto.htm. 12/03/2015

Van Den Beldt RJ, Brewer JL (1985) *Leucaena* wood production and use. Nitrogen Fixing Tree Association, Holualoa, 50pp

Vasquez-Murrieta MS, Migueles-Garduio I, Franco-Hernandez O, Govaerts B, Dendooven L (2006) C and N mineralization and microbial biomass in heavy-metal contaminated soil. Eur J Soil Biol 42:89–98

Vázquez-Luna D (2015) Biological indices of toxicity in tropical legumes grown in oil-contaminated soil. Ecol Indic 53:43–48

Vyslouzilova M, Tlustos P, Szakova J (2003) Zn and Cd phytoextraction potential of seven clones of Salix spp. planted on heavy metal contaminated soils. Plant Soil Environ 49:542–547

Wang Y, Shi J, Wang H, Lin Q, Chen X, Chen Y (2007) The influence of soil heavy metals pollution on soil microbial biomass, enzyme activity, and community composition near a copper smelter. Ecotoxicol Environ Saf 67(1):75–81

Witters N, Mendelsohn RO, Slycken SV, Weyens N, Schreurs E, Meers E, Tack F, Carleer R, Vangronsveld J (2012) Phytoremediation, a sustainable remediation technology? Conclusions from a case study. I: Energy production and carbon dioxide abatement. Biomass Bioenergy 39:454–469

Wong MH (2003) Ecological restoration of mine degraded soils, with emphasis on metal contaminated soils. Chemosphere 50:775–780

Yang B, Shu WS, Ye ZH, Lan CY, Wong MH (2003) Growth and metal accumulation in vetiver and two Sesbania species on lead/zinc mine tailings. Chemosphere 52:1593–1600

Yitao XU, Qixian F, Jixiu W, Mingrui LI, Fangdong Z, Xiao Y (2014) Heavy metal uptake characteristics of three plants grown in lead/zinc mine tailings. Environ Sci Technol 4(2):118–138

Zacchini M, Pietrini F, Mugnozza G, Lori V (2008) Metal tolerance, accumulation and translocation in poplar and willow clones treated with cadmium in hydroponics. Water Air Soil Pollut 197:23–34

Zhang W, Cai Y, Ma LQ (2002) Arsenic speciation and distribution in an arsenic hyper-accumulating plant. Sci Total Environ 300:167–177

Phytoremediation Potential of Industrially Important and Biofuel Plants: *Azadirachta indica* and *Acacia nilotica*

8

Jaya Tiwari, Atul Kumar, and Narendra Kumar

Abstract

A huge gap between demand and supply of energy, soil contamination with heavy metals, radionuclides, pesticides and other harmful chemicals, degradation of soil due to salinity/sodicity, acidity, soil erosion and decrease in soil microbial diversity, etc. are some great challenges of the twenty-first century. Subsequently, there is an undisputable need to create viable option for economical and sustainable remediation of the polluted soils. Phytoremediation as a strategy to solve environmental defilement has been quickly accepted on the grounds because it is energy efficient, financially savvy and satisfying procedure for the remediation of polluted sites; furthermore it is relevant to a wide scope of contaminants, including heavy metals and radionuclides, and additionally organic solvents like chlorinated solvents, polycyclic aromatic hydrocarbons, pesticides/insecticide, explosives and surfactants. Phytoremediation of a contaminated soil becomes a 'win–win' strategy if it is done with such plant species which are not only efficient in removing the pollutants but also have industrial application and generate large amount of biomass in form of food, fodder and biofuel. Therefore, the present study comprehensively reviewed the above-mentioned quality of *Azadirachta indica* and *Acacia nilotica* and also compared the fuel efficiency of both the plants. It was found that *Azadirachta indica* and *Acacia nilotica* both

J. Tiwari (✉)
Department of Environmental Sciences, Asian Institute of Public Health, Bhubaneswar, Odisha, India
e-mail: jaya.tiwari@gmail.com

A. Kumar
Department of Electrical and Electronic Engineering, Universiti Teknologi Petronas, Tronoh, Perak, Malaysia

N. Kumar
Department of Environmental Sciences, Babasaheb Bhimrao Ambedkar University, Lucknow 226025, India

K. Bauddh et al. (eds.), *Phytoremediation Potential of Bioenergy Plants*,
DOI 10.1007/978-981-10-3084-0_8

have phytoremediation capability in soil and water medium along with tremendous medicinal and therapeutic properties. The biodiesel derived from both the tree species is energy efficient and meets the requirement of international standards.

Keywords
Azadirachta • *Acacia* • Bioenergy • Heavy metals • Phytoremediation • Sustainability

8.1 Introduction

Soil is the fundamental requirement of our agricultural framework, nourishment security, worldwide economy and ecological sustainability. Fast rate of urbanization and industrialization rendered the soil contaminated with heavy metals and organic pollutants, which debilitate biological systems, surface and groundwater quality, sustenance and human wellbeing. Consequently, it is the need of hour to create successful advances for sustainable management and remediation of the contaminated soils (Oh et al. 2014). Worldwide researchers have shown their interest in accepting phytoremediation as a technique to tackle environmental contamination. In this green innovation, 'tolerant plants' have been used to clean up soil and groundwater from heavy metals and other harmful organic compounds. Phytoremediation includes growing plants in a defiled media to eradicate environmental contaminants by encouraging sequestration and/or degradation (detoxification) of the contaminants. Plants are unique organism furnished with astounding metabolic and assimilation abilities and in addition transport frameworks that can take up supplements or contaminants selectively from the polluted soil or water media (Alberto and Gilbert 2013).

Phytoremediation has developed as an innovative device with extraordinary potential that is most valuable when pollutants are restricted close to the root zone of the plants (top 3–6 ft). Moreover, phytoremediation needs very less energy; it is cost-effective, aesthetically pleasing technique of remediation of polluted sites with low to medium levels of pollutants (Mojiri et al. 2013). The strategy for phytoremediation makes the utilization of either natural metal hyperaccumulator plants or genetically designed plants (Setia et al. 2008). A variety of contaminated waters can be phytoremediated, whether it is sewage or municipal wastewater, agricultural runoff/drainage water, industrial wastewater, coal pile runoff, landfill leachate, mine waste and groundwater plume (Olguin and Galvan 2010).

Many extensive research studies and small-scale demonstrations on phytoremediation have been carried out, but large-scale applications are still limited to a very small number of projects (Cunningham and Ow 1996). The selection criteria for the appropriate plant species could be one of the reasons. The uptake of contaminants by plants is governed by several factors. These factors are:

- *Plant species:* The physical characteristics, genetic makeup and biological processes including absorption/adsorption, translocation and transformation of pollutant play important role in the uptake of pollutants. Higher biomass production and climatic adaptability of species (seeds or plants should match the climate of the phytoremediation site) and soil conditions are required for effective phytoremediation. Selection of the plant species whether annuals or perennials, monoculture or deciduous is an important consideration.
- *Physical properties:* Physicochemical parameters such as pH, electrical conductivity and organic matter, concentration of contaminants, texture and mineralogy of the soil affect the removal/degradation of the contaminant.
- *Root zone:* Root length and diameter affect uptake and degradation of contaminants. Plant enzymes and root exudates also affect the degradation of contaminants. These plant root exudates organic acids such as citrate and oxalate that affect the bioavailability of metals. The type, amount and effectiveness of root exudates and enzymes produced vary between species and even within subspecies or varieties of one species.
- *Plant biomass:* For high contaminant removal, the plant species should produce high biomass (Schnoor et al. 1995).

The following steps are included in the phytoremediation process:

Step 1. Problem area identification
Step 2. Physicochemical analysis of the soil before use of the phytoremediation
Step 3. Plantation of accumulators
Step 4. Use of agricultural and specialized measures and assessment of vegetative development
Step 5. Picking and drying the plants
Step 6. Physicochemical analysis of soil near the root after completed phytoremedia-tion
Step 7. Chemical composition analysis of green leaves of plants
Step 8. Determination of coefficient factors of plants

The key criteria determined for potential phytoremediators incorporate seed germination, seedling growth, plant development and their reproduction (Jonathan et al. 2004; Krishna and Govil 2004). Consequently in light of its low cost, ease and minimum technical problems, phytoremediation has turned out to be exceptionally prominent technology worldwide. In India, serious effort is being given to identify indigenous plant species that can be utilized to remediate contaminations, e.g. pharmaceutical wastes, arsenic, fly ash, debris and metals (Fig. 8.1) (Willey 2007).

Furthermore, for better experimental setup of phytoremediation by using a species, it is important to have complete information about the geological and geographical conditions of the study area. Also, it is necessary to collect the right information about the various kinds of species existing in the degraded lands in India (Table 8.1).

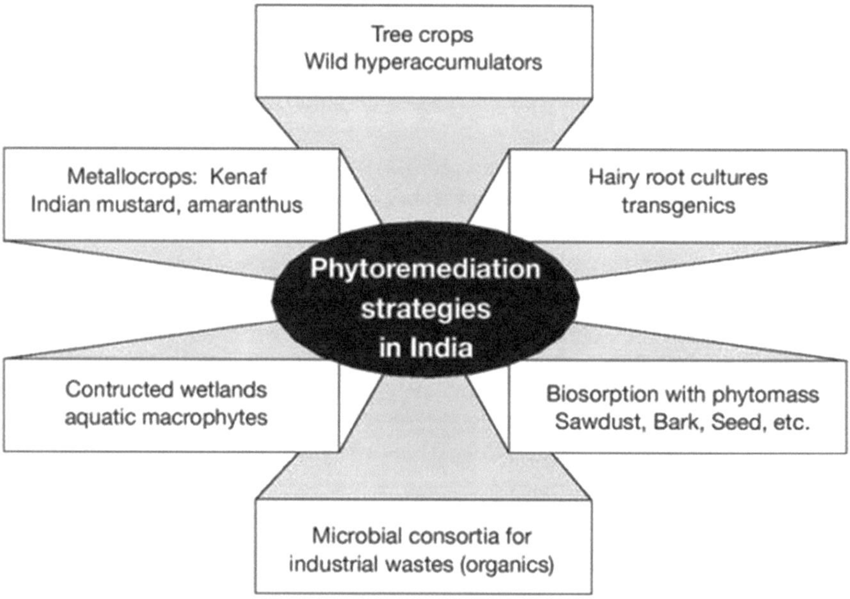

Fig. 8.1 Feasible phytoremediation studies in India (Tangahu et al. 2011)

Table 8.1 Species used for phytoremediation and rehabilitation of degraded lands in India

Salt-effected ravine areas	Forage and fuel species for eroded areas	Vegetative barriers for runoff and soil loss from Doon Valley
Trees and shrubs	**Trees**	**Grasses**
Acacia tortilis	*Prosopis juliflora*	*Panicum maximum*
Albizia amara	*Prosopis cineraria*	(Guinea grass)
Dichrostachys cinerea	*Acacia tortilis*	*Eulaliopsis binata*
Leucaena leucocephala	*Acacia radianoa*	(Bhabar grass)
Acacia nilotica	*Ziziphus mauritiana*	*Vetivera zizinoides*
D. sissoo		(Khus grass)
Albizia lebbeck	**Shrubs**	
Prosopis juliflora	*Calligonum polygonoides*	
Terminalia arjuna	*Crotolaria burhia*	
Azadirachta indica	*Aerva javanica*	
	Ziziphus nummularia	
Grasses and legumes		
Chrysopogon fulvus	**Grasses**	
C. ciliaris	*Lasiurus sindicus*	
Pennisetum pedicellatum	*Panicum turgidum*	
Saccharum spontaneum	*P. antidotale*	
D. annulatum		
Phaseolus atropurpureus		
Stylosanthes species		

Tangahu et al. (2011)

Hence, the overall analysis signifies that the selection of plant variety is critical to ensure effective remediation and the scientific world is looking for such plants species which are not only suitable for phytoremediation but also fast growing, produce high biomass with well-developed and profuse root system, stress tolerant, cheap to cultivate, industrially and commercially important and having potential to produce biofuels (biodiesel). The present study attempted to comprehensively review various qualities of *A. indica* and *A. nilotica* and also compare the fuel efficiency of both the plants.

8.2 Problems Related to Soil

Soil is considered as 'the biogeochemical engine of Earth's life support system' (Ang et al. 2012) and asset that provides us food, fodder, fibre and fuel. In addition to these important agricultural service products, soils provide such biological community benefits that cannot be computed in terms of money. These life-supporting services incorporate, for instance, reuse of carbon and essential nutrients for all living organism, filter and store water, regulate environment and control insects. The soil health has been overlooked after green revolution. This has affected plant growth and has led to decline in crop productivity. More importance has been given to nutrition of crops than 'nutrition of soils' (Lee et al. 2006). At times, physicochemical parameters of soil are altered in such a way that it makes the cultivation on such soil very difficult and it requires proper management.

Soil is not just the source of nutrients for plants but also acts as an assimilation site for the removal of contaminants which enter the soil due to industrial and other anthropogenic activities. These activities may introduce various toxic metals, viz. arsenic, cadmium, cobalt, chromium, cooper, mercury, molybdenum, nickel, lead, vanadium, zinc, etc. If accumulation of metals in soil becomes excessive, it will accumulate in the plant tissues and causes serious impacts on the health of human beings and animals (Saraswat and Chaudhary 2014). There is another problem associated with soil worldwide, i.e. contamination with radionuclide elements. It goes into soil fundamentally from atomic weapons-related exercises and various nuclear accidents like nuclear disasters at Chelyabinsk and Chernobyl (Karavaeva et al. 1994). The concentration of radionuclides in contaminated soil varies greatly. Sometimes even their low concentration affects the agricultural production and causes potential hazards to human health via food chain and other pathways. In this way, there is an extraordinary pressure to make huge volumes of soil free from radionuclides and other chemical contaminations (Willey 2007).

8.3 Remediation Techniques

Soil contamination is a noteworthy environmental issue around the world, as an aftereffect of mining, manufacturing and urban exercises over the past several years. The chemical composition of soil, especially its metal concentration, has environmental significance as its high concentration can decrease soil fertility responsible for biological magnification which eventually can jeopardize human health. The human health risk associated with metal-contaminated soil has attracted attention worldwide (Arantzazu et al. 2000; Denti et al. 1998; Krzyztof et al. 2004; Sandaa et al. 1999). Various physical and chemical methods have been utilized as a part of soil remediation (Barceló and Poschenrieder 2003; Cunningham et al. 1995). Some of these methods are:

1. Washing of soil
2. Physical inclusion or chemical interactions between the stabilizing agent and the pollutant for solidification/stabilization
3. Vitrification
4. Electrokinetic treatment
5. Chemical oxidation or reduction
6. Excavation and off-site treatment or storage at a more appropriate site ('dig and dump')
7. Incineration

All these remediation methods have some negative environmental impacts. Therefore, researchers have offered the adoption of less invasive, alternative remediation options ('gentle' remediation technologies), the so-called green remediation, based on life cycle analysis (LCA) in order to conserve resources and minimize environmental impacts (Gomes 2012; Glass 2000). Phytoremediation is a word derived from the Greek prefix 'phyto' which implies plant and suffix 'remedium' which means to clean (or) restore (Vasavi et al. 2010). It has been used as a natural option for the ecologically harmful physical remediation strategies currently in practice, which depends on the utilization of green plants to separate, sequester and/or detoxify toxins (Facchinelli et al. 2001; Govil et al. 2001). The constituents which can be removed by phytoremediation incorporate Pb^{2+}, Sr, Cd^{2+}, Cu^{2+}, Ni^{2+}, $Zn^{2+,}$ Cr^{6+} and U. The plant-influenced soil environment can change metals from a soluble to an insoluble oxidation state As, Cd, Cr+, Pb and Zn.

Phytoremediation of contaminated soils could be achieved by one or more of the following mechanisms: phytoextraction, phytostabilization, phytodegradation, phytovolatilization, rhizofiltration and rhizodegradation as shown in Fig. 8.2. Phytoremediation can be applied to a broad range of contaminants, including heavy metals and radionuclides, as well as xenobiotic compounds (Abbas et al. 2016; Ang et al. 2012). The success of phytoremediation processes is highly dependent on the potential of plants to extract and/or transform pollutants to less toxic substances since uptake, accumulation and degradation of contaminants shift from plant to plant. The plants utilized in phytoremediation are by and large chosen on the basis of their growth rate and biomass, their capacity to endure and immobilized contaminants, the depth of their root zone and their capability to transpire groundwater (Basavaraja et al. 2011). The plants utilized in phytoremediation should not only immobilize, break down or volatilize the contaminants but should also grow rapidly in different conditions. As a promising innovation, phytoremediation in management

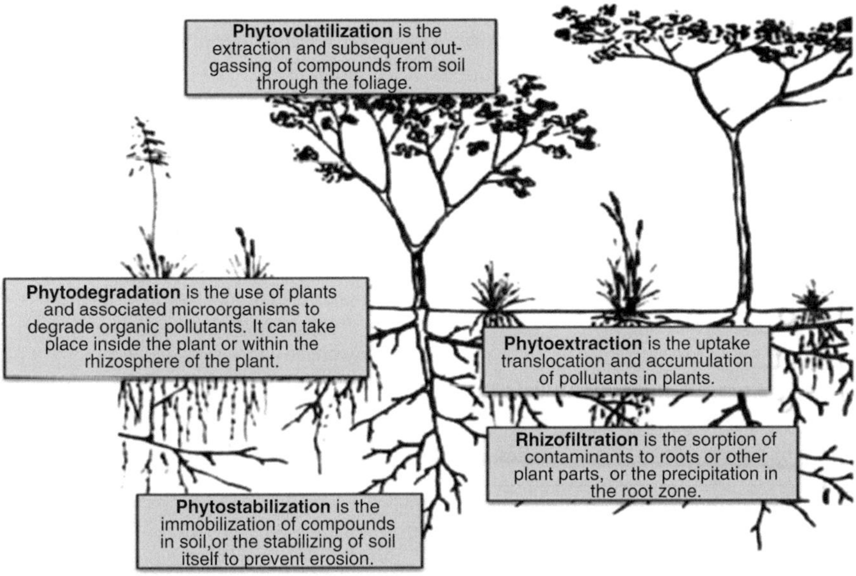

Fig. 8.2 Mechanisms involved in phytoremediation (Singh et al. 2003; Fischerová et al. 2012; Krishna and Govil. 2007; Meagher 2000)

and remediation of contaminated soils has its own merits and demerits. The advantages are that phytoremediation is a natural and in situ remediation framework driven by sun-powered and green plants. It is quicker than natural attenuation and can preserve the soil resources. It is cheap and does not induce the secondary pollutants. Successful phytoremediation immobilizes the toxins towards groundwater and prevents it from being contaminated, maintains the soil structure and improves the soil quality and productivity. Soils treated with phytoremediation are still more suitable for its original purpose, i.e. agricultural application, and this way prevents the loss of soil resources (Abbas et al. 2016; Ang et al. 2012).

8.3.1 Phytoremediation: Main Mechanisms and Strengths, Weaknesses, Opportunities and Threats (SWOT) Analysis

The uptake and aggregation of contaminants differ from plant to plant and species to species of same genus (Singh et al. 2003). Selection of appropriate plant species for phytoremediation plays a vital part in the advancement of remediation strategies (decontamination or stabilization), particularly on low-or-medium-contaminated soils (Singh et al. 2003). Several distinct groups of plant species according to their trace element accumulation capability are excluders with avoidance (or limitations) system for element uptake, exceptionally sensitive markers lacking a protection mechanism and accumulators with mechanism of metal resilience and accumulation ability in the above-ground biomass. Among various accumulators, there is a particular subgroup which is represented as hyperaccumulators (Singh et al. 2003). Hyperaccumulator plants are commonly grown on metalliferous soils and do not show any symptoms of metal phytotoxicity in their entire life cycle. Unfortunately, less biomass is produced from these plants as compared to other plant species. Hyperaccumulation limits were defined with respect to individual metal as 100 mg kg^{-1} of Cd, 1000 mg kg^{-1} of Pb and 10,000 mg kg^{-1} of Zn of plant dry weight (Fischerová et al. 2012).

Soil contaminants can be remediated by plants through several biophysical and biochemical processes (Fig. 8.2). In case of removal of heavy metals and radionuclides, plants extract and translocate harmful oxyanion to above-ground tissues for later reap, convert more toxic substances into less toxic substances or at least sequester the metals in roots to prevent leaching from the site (Krishna and Govil 2007; Meagher 2000). For organic pollutants, there are four mechanisms, which are included in phytoremediation (Trapp and Karlson 2001):

1. Direct uptake and accumulation of contaminants and transformation in plant tissues
2. Transpiration of volatile organic compounds (VOCs) from leaves
3. Release of exudates that stimulate microbial movement and biochemical changes in the soil
4. Enhancement of mineralization into less toxic constituents, e.g. carbon dioxide, nitrate, chlorine and ammonia at the root–soil interface – attributed to mycorrhizal fungi and the microbial consortia connected with that surface

A strengths, weaknesses, opportunities and threats (SWOT) examination of phytoremediation technique is presented in Table 8.2. This table focuses on the internal and external factors that are favourable and unfavourable to its execution.

8.3.1.1 Advantages

1. The cost in phytoremediation is very minimal than that of conventional process both in situ and ex situ.
2. Plants can be easily maintained and monitored.
3. The possibility of the regeneration and reuse of valuable metals.
4. It causes very less harm to ecosystem as it uses naturally occurring organisms and conserves the natural state of the environment.

8.3.1.2 Limitations

1. With phytoremediation, it is impractical to fully prevent the draining of contaminants into the groundwater
2. The survival of the plant is influenced by the toxicity-polluted land and general states of the soil.

Table 8.2 SWOT analysis for phytoremediation

Strengths	*Weaknesses*
Natural pumping and filtering systems derived energy from sun. Since this method is ecofriendly, it has received more public acceptance	Plants suitable for phytoremediation should have various characteristics: rapid growth, high biomass and deep roots, simple to harvest and should tolerate and accumulate a wide range of heavy metals in their aerial and harvestable parts
Provides various ecosystem services by controlling soil erosion, carbon sequestration and wildlife habitat. It additionally makes socioeconomic advantages by broadening local manufacturing into new items that engage local labours, therefore develop value-added industry	Phytoremediation works at a low pace than physicochemical processes and should be considered as a consistent and sustainable remediation process
Cost-effective option as compared to other physical remediation systems	The limited number of target metals that can be extracted by phytoremediation, the roots can reach up to limited depth, the difficulty of generating a high-biomass crop of wanted species and the paucity of enough knowledge on the agronomic practices and management. This technology is applicable to moderately polluted land and cannot work as an equivalent on more intensely polluted sites
Generation of a recyclable metal-rich plant deposit, applicable to wide range of toxic metals and radionuclides; removal of secondary air or waterborne wastes. A few metals can be recovered from the biomass, which assist in reducing generation of hazardous waste and create reusing and recycling revenues. Less site destruction and destabilization, low ecological effect and aesthetically pleasant benefits compared with biosorption incorporate persistent in situ recovery of the biomass and the capacity of living plant cells to supplement passive sorption of metals with metabolic mechanism of metal uptake and detoxification	Significance of choosing the appropriate plant for the phytoremediation application, whether it is local to that ecosystem where the phytoremediation is practising Limited successful experience; not applicable for all chemical compounds
Opportunities	*Threats*
Can be clubbed with different strategies. The coupling of phytoextraction with other soil reclaiming method (e.g. soil cleaning, soil vapour extraction) in so-called 'treatment trains' is picking up interest; this might be particularly helpful in situations where mixed type of pollutants requires the utilization of more than one strategy to adequately remediate site	Soil composition and texture, pH, salinity and concentration of pollutant and other toxins must be within the range of plant tolerance. In most polluted soils, the quantity of microorganisms is declined so that there are insufficient microbes either to encourage contaminant degradation or to support plant development

(continued)

Table 8.2 (continued)

Optimization of normal procedures through cultivation, air circulation, seepage, the utilization of soil added substances and ameliorants (e.g. lime, manure, zeolites and phosphates), the encouragement of soil microflora (including mycorrhizae), earthworm inoculation and specific planting	Contaminants that are exceptionally hydrophilic may drain outside the root zone and require control and should also include the identification of organic contaminant metabolites to guarantee that there are no harmful products
Perspectives on amelioration of the method like chelate-enhanced phytoextraction and genetic engineering, metagenomics, etc.	Successful phytodegradation requires biodegradable pollutants to be biologically accessible for retention to, or uptake and transformation by, plant or plant-related microbial system. This relies upon the relative lipophilicity of the compound, the soil properties and the age of the contaminant
Soil bioaugmentation positively affects metal bioavailability by increasing the concentration of the most available fractions of metals and subsequently phytoextraction	Disposal of crop after phytoextraction mechanism
If generated biomass from phytoremediation method could be valorized, then the primary downside of phytoextraction (slow rate of remediation process) gets to be invalid and slower working phytoremediation plans based on gradual attenuation of the pollutants instead of short-term forced extraction might be envisaged	In addition there is concern about metal-accumulating plants giving an exposure pathway to toxic pollutants to enter the ecological food chain if nearby herbivores consume these plants

Meagher (2000), Trapp and Karlson (2001), Sas-Nowosielska et al. (2004), Kim and Owens (2010), Lebeau et al. (2008), and Lokeshwari and Chandrappa (2006))

3. Possible bioaccumulation of contaminants which then went into the food chain from primary level consumers to higher level of consumers.

Phytoremediation is applicable to a considerable number of organic and inorganic chemical compounds and can be applied either in situ or ex situ. It is an innovative technology and also an intricate cleanup method, which provides a cost-effective, environment-friendly alternative to conventional physical and chemical cleanup methods (Reeves and Baker 2000; Bañuelos et al. 1999). Therefore, in this study phytoremediation technology has been reviewed on two species of nonedible trees (*A. indica* and *A. nilotica*) assessing their remediation efficiency and determining which plant is more efficient in accumulating pollutants and better in producing biofuel (Pierzynski et al. 1994).

8.4 *Azadirachta indica (A. indica)* and *Acacia nilotica (A. nilotica)*

8.4.1 Overview of *A. indica* and *A. nilotica*

8.4.1.1 *A. indica* (Neem)

Neem tree is predominantly cultivated in the Indian subcontinent and is considered as supreme tree and a divine gift of nature. Neem is a member from the mahogany family, Meliaceae. It is known by the botanical name *A. indica* A. Juss. Neem has been utilized widely by mankind to treat different ailments before the accessibility of written records which recorded the start of history. Since ancient times, neem has been utilized by mankind. The Latin name of neem, '*A. indica*', originated from the Persian. Azad signifies 'free'; dirakht signifies 'tree'; i-Hind signifies 'of Indian origin'. Henceforth it actually signifies 'the free tree of India' (Abbas et al. 2016; Ang et al. 2012). The neem tree is considered as an incredible plant that has been pronounced as the 'Tree of the 21st century' by the United Nations (Kumar and Navaratnam 2013). *A. indica* A. Juss and *M. azedarach* are two closely related species of Meliaceae. The former is prominently known as Indian neem (margosa tree) or Indian lilac and the latter as the Persian lilac. Neem is an evergreen tree, grown in different parts of the Indian subcontinent. All parts of the tree have been utilized as conventional medication for homemade cure against different human sicknesses. Neem has been widely utilized as a part of Ayurveda, Unani and homoeopathic medicine and has turned into a cynosure of modern medicine. The Sanskrit name of the neem tree is 'Arishtha' signifying 'reliever of ailments' and thus is considered as 'Sarbaroganibarini'. The tree is still viewed as 'village dispensary' in India. The significance of the neem tree has been noticed by the US National Academy of Sciences, which published a report in 1992 entitled *Neem: A Tree for Solving Global Problems* (Biswas et al. 2002).

8.4.1.2 Taxonomical Classification

The taxonomic grouping of neem is as follows: Kingdom, Plantae; Order, Rutales; Suborder, Rutinae; Family, Meliaceae; Subfamily, Melioideae; Tribe, Melieae; Genus, *Azadirachta*; Species, *indica* (Hashmat et al. 2012).

8.4.1.3 Botanical Description

The height of neem tree varies between 40 and 50 ft or higher with a straight stem and long spreading branches which give them wide round crown; barks are dark brown and rough with wide longitudinal gaps separated by flat ridges. The leaves are compound, imparipinnate, each comprising 5–15 leaflets (Parotta 2001). The compound leaves are themselves alternative to one another. It bears numerous flowered panicles, mostly in the leaf axils. The seeds are ovate and about 1 cm long with sweet aroma and white oblanceolate petals. It produces yellow drupes that are ellipsoid and glabrous, 12–20 mm long. Fruits are green, becomes yellow on maturity, scented with garlic like odour. Young leaves and flowers come in spring season

(March–April). Ripening of fruits occurs between April and August depending upon region (Parotta 2001; Ross 2001).

8.4.1.4 *A. nilotica*

Acacia nilotica (L.) Willd. ex Del, commonly known as babul, Kikar or Indian gum Arabic tree, is worldwide known as a multipurpose tree (National Academy of Sciences 1980). According to Bargali and Bargali (2009), it is considered as one of the worst weeds in Australia because of its invasiveness, tendency to spread very fast and economic and environmental impacts. It is widely distributed throughout arid and semiarid regions of the world. *Acacia* species are commonly known as 'babool' in India and ethnomedicinally used for the treatment of different ailments.

8.4.1.5 Taxonomical Classification

The taxonomic division of babool is as follows: Kingdom, Plantae; Order, Fabales; Family, Fabaceae; Subfamily, Mimosoideae; Genus, *Acacia*; Species, *nilotica*.

8.4.1.6 Botanical Description

The height of babool tree varies from 7 to 13 m, with a trunk diameter of 20–30 cm. The spread branches form a broad crown at medium height which is almost symmetrical. The bark is very dark brown to black, rough with deep regular vertical fissures in older tress. The thistles are almost straight, paired at the nodes of the stem and usually pointing slightly towards backside. The leaves are bipinnate, 4.5–7 cm long, with 2–14 pairs of pinnae (Spicer et al. 2007). The bright yellow flowers are borne on globe-shaped flower heads. The flowers have sweet aroma and appear near the start of the rainy season. The nutritious pods retain their seeds at maturity and are dispersed by animals. The pods are compressed, slightly curved and vary from slightly constricted to almost rosary like (string of beads). The pods are smooth or covered with fine hairs (Orwa et al. 2009).

8.4.2 Geographical Distribution and Cultivation

8.4.2.1 *A. indica*

It is a native of East India and Burma and extensively grown in Southeast Asia and West Africa and very recently in the Caribbean and South and Central America. It is planted and often naturalized throughout the dry regions of tropical and subtropical India, Pakistan, Sri Lanka, Thailand and Indonesia. Similarly planted and frequently naturalized in Peninsular Malaysia, Singapore, the Philippines, Australia, Saudi Arabia, Tropical Africa, the Caribbean, and Focal and South America (Parotta 2001). In the South Pacific, neem is found in the Fiji Islands. In Australia, it was initially introduced around 60–70 years back. In China, *A. indica* was planted on subtropical island of Hainan and Southern China. In Nepal, Neem trees are found in the southern, low-lying areas (Terai region). In Sri Lanka, it is widely distributed in northern parts of the island. In Iran, neem trees are grown near the coast up to the Chat el Arab in Iraq on the Arabian Peninsula. In Qatar and Abu Dhabi, neem was planted under proper watering system utilizing desalted seawater in streets and public gardens. A huge neem tree plantation was set up on the Arafat fields near to Makkah to give shade to pilgrims (http://www.neemfoundation.org/). *A. indica* can be successfully grown on all types of soil. Its root system is highly developed and has unique ability to extract nutrients and moisture even from highly leached soils. Thereby, it plays significant role in preventing soil erosion and desertification. The species has vast climatic adaptability. It thrives well at very high temperature where maximum temperatures are as high as 50 °C and tolerates cold up to 0 °C, on altitudes up to 1500 m. Two species of *Azadirachta* have been identified, *A. indica* A. Juss – native to Indian subcontinent – and *Azadirachta excelsa* Kack. confined to the Philippines and Indonesia (Jattan et al. 1995; Hegde 1995; Ogbuewu et al. 2011). The former grows as a wild tree in India, Bangladesh, Burma, Pakistan, Sri Lanka, Malaysia, Thailand and Indonesia. Presently neem trees are successfully grown worldwide in about 72 countries, in Asia, Africa, Australia and North, Central and South America (Ahmed et al. 1989; Sidhu 1995; Sateesh 1998; Fathima 2004).

It has been assessed that there are around 25 million trees cultivating all over India, out of which 55.7% are found in Uttar Pradesh alone followed by Tamil Nadu (17.8%) and Karnataka (5.5%). Other states of India where neem tree are grown include Andhra Pradesh, Assam, Bihar, Delhi, Gujarat, Haryana, Himachal Pradesh, Kerala, Madhya Pradesh, Maharashtra, Meghalaya, Orissa, Punjab, Rajasthan, West Bengal alongside Andaman and Nicobar Islands, the Union domain (Sindhuveerendra 1995; Chakraborthy and Konger 1995; Bahuguna 1997; Fathima 2004). India stands first in neem seed generation, and around 442,300 tons of seeds are generated yearly yielding 88,400 tons of neem oil and 353,800 tons of neem cake (Ogbuewu et al. 2011).

8.4.2.2 *A. nilotica*

A. nilotica has been grown widely for generations in arid and semiarid regions of India and Pakistan. It is a plant of arid regions and does not grow in areas receiving annual rainfall more than 1250 mm or in regions susceptible to frost or very low temperature (Puri et al. 1994). This species has capability to tolerate salt and dry environments (Minhas et al. 1997), grazing, drought and fire. Acacia often forms the significant fraction of the woody vegetation in the semiarid areas on African savannas (Tybirk 1989).

A. nilotica is naturally occurring widespread in the drier areas of Africa, from Senegal to Egypt and down to South Africa and from Arabia eastward to India, Burma and Sri Lanka in Asia. The biggest tracts are found in Sindh. It is widely distributed in Indian forests, roadsides, farmlands, tank foreshores, crop fields, village pasture lands, wastelands, bunds, along the national roadways and railway tracts. Most of the time it occurs as an isolated tree and rarely found in patches in forests (Bargali and Bargali 2009).

8.4.3 Industrial Application of *A. indica* and *A. nilotica*

8.4.3.1 *A. indica*

In 2002, at the World Neem Conference, the thought of advancing neem as an 'Industrial Plant' was proposed (Kumar 2003). Several industries including drugs, cosmetics and textile industries utilize neem oil (Jattan et al. 1995). In India, neem is widely utilized by many Ayurvedic drug industries. Neem oil and powdered neem leaves are used in different cosmetic products such as face creams, nail enamel, nail oils, shampoos and conditioners (Jattan et al. 1995). A new hair cleanser (shampoo), using seed extract of neem, was found very effective, more than permethrin-based product, against head lice under in vitro conditions neem cake, a byproduct of neem oil industry, is utilized as animal food, manure and natural pesticide. Neem oil is often utilized in production of soap. Medicated neem soaps have become very popular in India (Ogbuewu et al. 2011). Uses of neem and its derivatives are as follows:

- Neem-based toothpaste is commonly used in India and European countries.
- Neem acts as a source for many oral hygiene preparations and dental care items.
- Neem bark generates gum and tannins which are utilized in tanning, dyeing, etc.
- Neem seed pulp is a rich source of carbohydrate, which is used in fermentation industries and for methane gas production.
- Plantation of neem and manufacturing of neem products generate employment and income.

Collection of neem seeds further supplied to the industries to create important means of employment and income for the poor households, especially the women from rural area. About 540,000 tons of seeds are produced annually in India and yield 107,000 tons of neem oil and 425,000 tons of neem cake. The amount of azadirachtin available is calculated to be about 1600 tons for every year, generating huge amount of crude material for pesticide industry. In the production sector, the yearly calculated turnover is about 1000–1200 crores INR. Small-scale industries have a key role to play in harnessing the potential.

8.4.3.1.1 Phytochemistry

Biologically active compounds extracted from various parts of the plant include: azadirachtin, meliacin, gedunin, nimbidin, nimbolide, salannin, nimbin and valassin. Meliacin forms the bitter component of neem oil; the neem seed also possesses tignic acid, which is responsible for the peculiar smell of the oil (Sharma et al. 2011). Neem kernels contain 30–50% of oil mostly used in soap, pesticide and pharmaceutical industries and contain many active compounds which are together called as triterpene or limonoids (Djenontin et al. 2012). The four best limonoids compounds of neem are azadirachtin, salannin, meliantriol and nimbin. Limonoids have insecticidal and pesticidal properties (Mondal and Mondal 2012).

8.4.3.1.2 Medicinal Use

The whole tree has been used medicinally for centuries. It has been utilized in Ayurvedic medicine for more than 4000 years because of its medicinal values. The ancient Sanskrit medical literatures reveal several advantages from neem's fruits, seeds, oil, leaves, roots and bark; each part has been utilized in the Indian Ayurvedic and Unani medicine and is currently used in drug and cosmetics industries (Brototi and Kaplay 2010).

8.4.3.1.3 Pharmacological Actions

Various parts of neem act as abortifacient (drug which causes miscarriage), analgesic (pain killer), anthelminthic (drugs that remove parasitic helminths from human body), antibacterial (kills harmful bacteria), antiyeast, antiulcer, antifertility, antifilarial, antifungal, antihyperglycaemic (counteracting high levels of blood glucose level), anti-inflammatory, antiviral, antimalarial, diuretic (these are medicines that help in reducing the amount of water from the body) , antinematodal, antipyretic, antispasmodic (drug which is used to relieve spasm of involuntary muscles, insecticidal, antispermatogenic, antitumor, hypercholesterolemic (effective against elevated serum cholesterol levels), immunomodulator (chemical agent that modifies the immune response (Parotta 2001; Ross 2001).

8.4.3.1.4 Health and Personal Care Products

Many fast-moving consumer goods like personal health and hygiene products manufactured from seed, oil and leaf of neem tree are skin-related products (eczema ointment, antiseptic lotion), hair care products (hair cleanser, conditioner and hair oils), oral cleanliness (tooth powder and paste and neem twigs), curative (loose neem leaves, tea, green capsules, powders), products used for household purposes (bathing soap bars, detergents, mosquito repellent (spray and cream) and candles) (Sharma et al. 2011; Srivastava and Rupainwar 2010).

8.4.3.1.5 Therapeutic Uses

Different parts of neem are also utilized for various types of cure. Bark extract with hot water is consumed orally by the adult female as a tonic to improve menstrual release (emmenagogue). Anthraquinone portion of dried flowers, fruits and leaves is taken orally for leprosy. Hot water extract of the flowers and leaves are consumed orally to cure hysteria and are utilized to cure external injuries. The dried flower is effective for the diabetic patients. Hot water extract of dried fruit is utilized against piles and external skin problems and ulcers. Hot water concentrate of the whole plant is utilized as anthelmintic, an insect-killing spray and laxative. Juices of bark of *Andrographis puniculata*, *A. indica*, *Tinospora cordifolia* are taken orally for curing filariasis (Hasmat et al. 2012). The warm water extract is taken for fever, diabetes and as a tonic, refrigerant and anthelmintic. Fruit leaf and root, ground and blended with dried ginger and 'triphala', are brought orally with tepid water to treat common fever. Leaves because of insecticidal properties are kept with woollen and different materials for a long time. Leaf juice is given to cure gonorrhoea and leucorrhoea. Washing with neem leaves is useful for curing itching and other skin illnesses. Leaf juice is utilized as nasal drop to treat worm infestation in the nose. Inhalation of bark steams is valuable in treatment of throat infection. Extract of flower is given in dyspepsia and general debility (Chatterjee and Pakrashi 2008). The delicate twigs of the tree are utilized as toothbrush (Kabeeruddin and Mufradat 2007; Tandon and Sirohi 2010). The seed oil of neem is utilized in leprosy, syphilis, skin inflammation and chronic ulcer (Hasmat et al. 2012; Kabeeruddin and Mufradat 2007).

8.4.3.2 *A. nilotica*

A. nilotica is a multipurpose tree in true sense and widely used in conventional agroforestry system. In the current scenario of climate change, agricultural activities, cultivation of *A. nilotica* has emerged as a viable option for controlling the harmful impacts of climate change (Singh et al. 2013). It is found to be well nodulated with *Rhizobium* species (Dreyfus and Dommergues 1981).

The wood is a common fuel in Sindh region in Pakistan, and huge quantities are used as a firewood and charcoal. The calorific value of sapwood is 4800 kcal per kg, while that of heartwood is 4950 kcal per kg. The wood is bulky (specific gravity, 0.67–0.68) and the trees occasionally cut back to ground level periodically to stimulate growth rate. The hard, tough wood is termites resistant, waterproof, and is commonly used for railroad sleepers, tool handles and carts. It is an alluring wood, is useful for cutting and turnery and is still utilized for boat building. It is one of the best mining timbers in Pakistan. It is commonly utilized for furniture. The leaves and pods are generally utilized as grain and, in dry areas of Sindh, constitute the main food for goats and sheep. Pods contain much as 15% unrefined protein. The bark and pods are broadly utilized in tanneries; their tannin content differs from 12 to 20% gum. *A. nilotica* is most likely the first commercial source of gum Arabic; however, this variable item now comes basically from *Acacia senegal* (*A. senegal*). The gum is still utilized in production of matches, inks, paints and sweet shop (Firewood crops 1980; Natural vegetation assessment report. 2008).

8.5 Application of *A. indica* and *A. nilotica* in Soil Reclamation

8.5.1 Amelioration of Sodic Soil by *A. indica* and *A. nilotica*

Large portion of Indo-Gangetic alluvial plain is sodic land. In past few decades, their reclamation has gained much attention. From 1980 onwards, several workers initiated afforestation on sodic soils with salt-tolerant species (Sissay 1986; Totey et al. 1987; Sharma 1988). It has been observed that the growth of trees depends on the variation in the sodicity levels (Singh et al. 2011). Salt-tolerant species were identified according to the characteristics of sodic soil which may vary from place to place and from low to high exchangeable sodium percent (ESP) (Yadav and Singh 1970; Toth 1981). Later on, focus had been shifted from mere growth observation of salt-tolerant trees to biomass and productivity of few trees (Chaturvedi

1985; Dogra 1989; Singh 1991; Singh 1998; Chaturvedi et al. 1991; Chaturvedi and Behl 1996; Jain and Singh 1998). Keeping this in view, various plant species like *A. indica*, *A. nilotica*, *Eucalyptus*, etc. were grown which not only ameliorate the sodic/alkali soil but also provide biomass in terms of fuelwood, fodder, fibre, timber, etc.

Furthermore, the survival and growth performance of *Prosopis juliflora, Terminalia arjuna, Pongamia pinnata, Acacia nilotica, Azadirachta indica, Eucalyptus tereticornis* has been evaluated and studied (Singh et al. 2008), and it has been observed that survival of *A. nilotica* was >95%. After 10 years of planting, diameter at breast height was also maximum in *A. nilotica*, whereas crown diameter, lopped biomass and litter fall were minimum in *A. indica*. When the data on soil amelioration was evaluated, it was found that *A. indica* reduced the pH from 10.6 to 9.8, electrical conductivity (EC) from 1.43 to 0.33(ds/m), organic carbon from 0.8 to 2.7 and ESP from 85–92 to 56. *A. nilotica* showed more significant results which lowered the pH of same soil up to 9.7, EC to 0.77(ds/m), organic carbon to 3.5 and ESP to 56.The study also compared the rhizosphere acidification and phytoremediation in *A. nilotica* grown on saline sodic soil and *A. ampliceps* and observed that both the species of *Acacia* have shown higher improvement in the soil properties like pH, sodium absorption ratio (SAR), bulk density (BD) and infiltration rate (Abbas et al. 2016). *A. ampliceps* has been identified as more suitable species than *A. nilotica*. Furthermore, *A. nilotica* was also evaluated for determining its potential to improve sodic soil in Hosadurga Taluk of Chitradurga district, central dry zone of Karnataka (Basavaraja et al. 2011). On evaluation, it has been found that this species had showed a noticeable reduction in pH and EC, SAR, ESP and increased concentration of Ca, Mg and K, organic carbon, cation exchange capacity (CEC) and other major nutrients. Suwalka and Qureshi (1995) reported study of nine tree species on sodic soil (Typic Natrustalf) in Rajasthan, India. Out of nine tree species, *A. nilotica* survived better. However, *A. indica* failed to survive. It was also reported that higher ameliorative effect was showed by *A. nilotica* than *A. indica*.

Thus, based on above studies, it can be concluded that *A. nilotica* has more capability to reclaim sodic soil. However, *A. indica* has less potential to reclaim it; in some of the studies it did not even survived on the sodic soil.

8.5.2 Anticlastogenic Activity of *A. indica* and *A. nilotica* by Removing Organic Contaminants

Anticlastogenic activity is the mutation which occurs when the base pair sequence of DNA or RNA is altered. It may occur because of several factors, e.g. exposure to ultraviolet or ionizing radiation, chemicals, during cell division or DNA replication. Mutation can be minimized by using anticlastogens which interfere with DNA repair, mutagen metabolites or with free radicals. There are several natural components-originated anticlastogens (e.g. flavonoids, tannins, carotenoids, phenolics, pigments, etc.). It has been reported that these chemicals also help in prevention of cancer (Vishwanatha et al. 2010).

Several research findings reported that the extracts of different parts of various plants have anticlastogenic potential. *A. indica* and *A. nilotica* are among them. Vinod et al. (2011) evaluated the mutagenic and antimutagenic property of neem oil and its DMSO extract in in vitro Ames Salmonella/microsome assay and in vivo mouse bone marrow micronucleus test and reported anticlastogenic activity of neem oil in the micronucleus test. However, in the Ames assay neem oil showed lesser antimutagenicity against MMC, but it has been proven that the reduction of clastogenic effect of cyclophosphamide and mitomycin C can be extracted by neem leaf (Mukhopadhaya and Mukherjee 1998).

On the basis of results, it was concluded that methanol extracts of neem flowers (MENF) at 100 and 500 mg kg^{-1} body weight showed no clastogenic effects. But, it possesses anticlastogenic potential in the rat liver, particularly at high doses (Kupradinun et al. 2013).

A. arabica Wild or *A. nilotica* Wild is now known for several biological activities, viz. anti-free radical, anti-quorum sensing, chemopreventive, anthelmintic, antimutagenic, cytotoxic, antimicrobial, anti-inflammatory, etc. These properties of *A. nilotica* are mainly attributed to various metabolic active compounds in different parts of the plant, e.g. polyphenolic compound in bark, polysaccharides, calcium, magnesium in gums, proteins and amino acids, tannins, gallic acid, chlorogenic acid, etc. (Alambayam et al. 2014).

Furthermore, a correlation has been established between the antimutagenic and chemopreventive property of the bark of *A. auriculiformis* and *A. nilotica* by using Ames antimutagenicity assay and mouse mammary gland organ culture (MMOC) model in two different strains using both direct-acting and indirect-acting mutagens (Kaur et al. 2002). Anticarcinogenic activity test was done by observing the effect of chemical carcinogen 7,12-dimethylbenz[a]anthracene (DMBA). A good correlation between antimutagenic assay and MMOC model suggested the antimutagenic and chemopreventive properties of both the species of *Acacia*.

Moreover, a leaf extract of *A. nilotica* showed significant anticlastogenic and chemopreventive properties as compared to other parts of this tree species (Kalavani and Mathew 2010). Extract of flower and gum are next to leaf in showing such activities (Meena et al. 2006). Kaur et al. (2005) reported that the anticlastogenic and cytotoxic activities against direct-acting mutagens (NPD or sodium azide) and S9-dependent mutagens 2AF are mainly due to the presence of gallic acid and polyphenols in acetone extract.

8.5.3 Biosorption

Biosorption is a process that allows certain biomass to accumulate and adsorb pollutants onto its cellular structure. It is an economical method for the remedy from heavy metals or any other environmental contaminants from wastewater. Low cost, less generation of sludge, recovery of metal and regeneration of biosorbent make it successful over other conventional methods (Volesky and Holan 1995; Kratochvil and Volesky 1998). Bacteria, fungi, algae, agricultural waste, saw dust, etc. are different categories of biosorbent (Vijayaraghvan and Yun 2008). Besides this, there are several other low-cost biomaterials that can be used as biosorbent of heavy metals. Several important phytochemicals, azadirachtin, salanin, meliantriol, nimbin and nimbidin of *A. indica*, have a great metal-binding capacity (Ang et al. 2012). Several scientific studies have been conducted on the utility of *A. indica* leaf as biosorbent material for the sorption of cadmium(II) (Sharma and Bhattacharya (2005), chromium(VI) (Babu and Gupta 2008), lead(II) (Bhattacharya and Sharma 2004)) and copper (Febriana et al. 2010). Furthermore, the adsorption of *A. indica* bark for iron removal from aqueous solution was found to be 80% within few minutes of contact time (Rajjak and Bapurao 2013). In another study of adsorptive potentiality of *A. indica*, leaf powder and activated charcoal were compared for the removal of chromium(VI) ions from aqueous solution. Neem leaf powder was found to be more efficient and economical than the activated charcoal in the chromium removal (Tawde and Bhalerao 2010). The successful attempts have been made to remove basic dye called Rhodamine B from aqueous solution by *A. indica* leaf powder in acidic pH range (Sharma et al. 2008). Bhattacharya and Sharma (2004) found 52–99% removal of dye Congo red from its aqueous solution by using neem leaf powder as biosorbent. Biosorbent prepared from mature dried neem leaves was also effective in 80% removal of fluoride from aqueous solutions at pH range of 5–7 using batch adsorption process (Bharali and Bhattachrya 2015). Furthermore, a study was conducted to evaluate the efficiency of neem leaf, stem and bark powder in the removal of zinc from wastewater. It was found that the biosorbent was efficient in zinc uptake from wastewater. The bark powder of *A. indica* could also remove dye (Arshad et al. 2008; Srivastava and Rupainwar 2010).

Besides *A. indica* there are other plants also whose various parts could be considered as very effective biosorbent. In this context *A. nilotica* is a very good example. Thenmozhi and Santhi (2015) characterized activated *A. nilotica* seed pods for the adsorption of nickel from water solution. *A. nilotica* was found as a nontraditional low-cost biosorbent for the removal of Pb(II) and Cd(II) ions from aqueous solution (Waseem et al. 2014). The investigation of the biomass derived from the stem of *A. nilotica* had been done for the removal of As ions from surface water samples and observed 95% sorption efficiency (Baig et al. 2010). Thus powder of stem of *A. nilotica* can be used as low-cost biosorbent for As. Few researchers have used *A. nilotica* seed shell ash-supported $Ni_{0.5}Zn_{0.5}Fe_2O_4$ magnetic nanoparticles for the removal of Pb(II) from aqueous solution and reported it to be effective in controlling water pollution due to Pb(II) ions (Hosseini and Moeinpour 2016). The improved activated char of *A. nilotica* branches (CANBI) has been utilized to see the adsorption efficiency. This process not only efficiently removed the phenol, but CANBI was also successfully regenerated (Dass and Jha 2015). Hence, it has been investigated that the physical characteristics of *A. nilotica* leaves can be utilized as an adsorbent for the removal of hazardous Cr(VI) ions from wastewater (Prasad and Thirumalisamy 2013).

8.5.4 Removal of Radionuclides

Radionuclides in the soil pose significant risk to human and ecological wellbeing. These radionuclides go into the soil because of periodic nuclear accidents (e.g. Chernobyl disaster of 1986 and Fukushima Daiichi nuclear disaster in 2011) and utilization of radioactive material in different research, and formative tasks, bio-medical, mechanical destinations, mining, atomic power plant and small fraction of radionuclides may even show genuine human wellbeing impacts. The commonly known radionuclides include cobalt-60 (^{60}Co), plutonium-239 (^{239}Pu), radium-226 (^{226}Ra), radon-222 (^{222}Rn), technetium-99 (^{99}Tc), thorium-232 (^{232}Th) and uranium-238 (^{238}U). However, the common radionuclides created through atomic reactors by means of the part of essential elemental atoms are thallium-201 (^{201}Tl), iridium-192 (^{192}Ir), caesium-137 (^{137}Cs) and strontium-90 (^{90}Sr), which takes longer time to decay (Kurnaz et al. 2007). Additionally, ^{238}U decays to form ^{226}Ra, which has a half-life of 1600 years (Prakash et al. 2013). During the process of photosynthesis, both the stable inorganic components and the unstable ones (radioactive component) discover their way into these plants. Natural radionuclides are exchanged and cycle through characteristic natural procedures and between the different ecological compartment by going into the biological community and food chain through direct or indirect contamination of natural radionuclides (Adewumi 2011; Elujoba et al. 2005). It is possible for plants to absorb radionuclides during nutrient absorption through their roots and transport such nutrients via their phloem to their active portions. The rate of natural radionuclide uptake is very much dependent on the activity concentration in the soil. The root uptake relies upon soil properties, e.g. pH, mineralogical organization, organic matter substance and nutrient status, and also metabolic and physiological attributes of the plant species (IAEA 2006). Plant uptake of radionuclides is one of numerous vectors for the movement of characteristic radionuclides into people from nature by means of food chain.

In recent years interest has been developed for usage of plants for the removal of radionuclides from soil (Entry et al. 1997; Zhu and Shaw 2000). Phytoremediation has been practised to a variety of radionuclide-affected areas. Different plants have different ability to accumulate radionuclides and decontaminate the soil (Dushenkov 2003).

8.5.4.1 Estimation of Transfer Factor (TF)

The uptake of radionuclides or various elements by plants from the soil is regularly called as transfer factor (TF), the ratio of concentration of radionuclide/heavy metal in plant tissue and soil (in Bq.kg^{-1} or mg.kg^{-1}) (Rao and Sudhakar 2013):

$$TF = \frac{\text{Metal concentration in Plant tissue}\,(Dry\ \text{weight})}{\text{Metal concentration in Soil}\,(Dry\ \text{weight})\,\text{from where the plant was grown.}}$$

An experiment was carried out for uranium uptake from sandy soil treated with various concentration of uranium by utilizing two types of *Acacia* (*A. albida* and *A. nilotica*). Results revealed that there is a distinction in the capacity of the *Acacia* seedlings examined to digest different concentration of uranium through their roots. *A. nilotica* showed the largest amounts of absorption and storage of uranium in dry weight of roots in various concentrations (202, 339, 1175 and 1477 $\mu g.g^{-1}$, respectively) of the concentrations (50, 100, 200 and 500 $mgkg^{-1}$). In contrast with the root of *A. albida*, the absorption of uranium was (60, 54, 133 and 526 $\mu g\ g^{-1}$) in the concentration of the same samples. The capacity of *A. nilotica* is superior to that of *A. albida* to uptake uranium from the soil, where 80–90% of the uranium is absorbed by the seedlings, contrasted with 44–85% in *A. albida*.

8.5.4.2 Case Study

Emphasis on plant research has increased in recent times, and a large body of evidence has been collected to show the substantial potential of medicinal plants used in different traditional systems all over the world. About 70–80% of the world population, particularly in the developing countries, relies on nonconventional medicine in their primary healthcare (Chan 2003; Desideri et al. 2010). Natural radionuclides are present in every domain of the environment: air, water, soil, food and humans (WNA 2014).

As indicated by the International Food Safety Authorities Network (INFOSAN 2011), plants utilize as nourishment regularly have ^{40}K, ^{232}Th and ^{238}U and their progenies. It is normal that resemblance would be found in plants utilized for therapeutic purposes since plants are the essential pathway of natural radionuclides going into the human body through the food chain. In a variety of concentrations, Naturally Occurring Radioactive Materials (NORMs) have always been present in every part of the earth and in the tissue of all living beings. Natural radionuclides such as ^{40}K, ^{232}Th and ^{238}U can be found almost everywhere, in soil, public water supplies, oil and atmosphere, thereby subjecting human beings to reasonable exposure (Ali 2008; Varier 2009).

8.5.4.2.1 Materials

Tree barks of neem (*A. indica*) have been collected from the campus of Banaras Hindu University. Barks were grinded and washed over and over by double distilled water and dried at room temperature before being utilized as adsorbents.

Stock solutions of Cd^{2+}, Hg^{2+} and Cr^{3+} (1.0 mol dm^{-3}) were prepared by dissolving suitable measures of their salts (cadmium nitrate, mercuric nitrate and chromic chloride) in double distilled water and were standardized according to EDTA titration method (Flaschka 1964). The stock solutions have been diluted in the desired experimental sportive concentration as and when required and after that marked with their respective radionuclides (115mCd, t1/2 = 44.8 d, specific activity = 177.6 ± 10% MBq/g in dilute HNO_3; 203Hg, t 1/2 = 47.0 d, specific activity = 151 ± 10% MBq/g in dilute HNO_3 and 51Cr, t 1/2 = 27.8 d, specific activity = 222.7 ± 10% MBq/g in dilute HNO_3) procured from the Board of Radiation and Isotope Technology (BRIT), Mumbai (India).

8.5.4.2.2 Sorption Measurement

The sorption estimations were done by mixing at regular interval and equilibrating 0.1 g of tree bark test with 10.0 cm^3 of named sorptive solution, centrifuged for phase division, and after that supernatant solution was investigated for its β-activity estimation utilizing an end window GM counter. Radioactivities of a few specimens were additionally checked for their g-activity utilizing either a single channel γ-spectrometer (ECIL, Nucleonix, Hyderabad) or an MCA coupled gamma spectrometer (Canberra Series 35 Plus or Nucleonix, Hyderabad). Methods for the estimation of quantity adsorbed and different parameters were similar to those given in earlier works (Mishra et al. 1996).

8.5.4.2.3 Irradiation of Adsorbents

Irradiation of sorbent (neem barks) at various time intervals was done utilizing a 11.1 GBq (Ra-Be) neutron source having an integral neutron flux of 3.8×10^6 n/cm^2/s and associated with a nominal γ-dose of ca. 1.72 Gy/h furthermore utilizing a high dose γ-cell (^{60}Co source) unit GC-900 (BRIT, Mumbai, India) having an action of ca. 94.4 ATBq (mean dose rate of 4.66 kGy/h). The irradiated biomasses were then utilized along with unirradiated materials for the uptake investigation of three ions (viz. Cd^{2+}, Hg^{2+} and Cr^{3+}) from aqueous solutions.

8.5.4.2.4 Results and Discussion

All results reported are an average of, at least, four independent measurements. The effect of concentration (10^{-2} to 10^{-8} mol dm^{-3}) of Hg^{2+} and Cr^{3+} on the uptake behaviour of neem bark was studied. The resultants of such analysis were displayed as a graph for the amount sorbed against time for uptake of Hg^{2+} on neem bark (Fig. 8.3a) and Cr^{3+} on neem bark (Fig. 8.3b).

It is evident from these time-rate curves that at first the uptake of metal particles was very fast, turning out to be slower with the time and afterwards an evident balance between the two stages was reached inside of ca. 4 h of contact time. No further

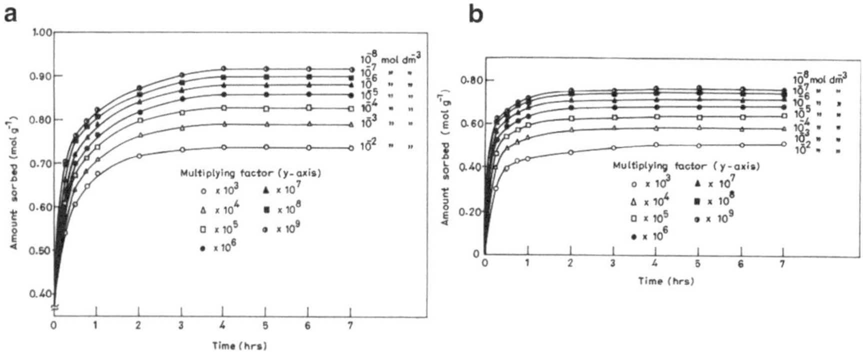

Fig. 8.3 Time variation of sorption of (**a**) Hg^{2+} and (**b**) Cr^{3+} ions by neem (*A. indica*) bark at different concentrations of sorptive (Tiwari et al. 1999)

change in sorption occurs even after 24 h of contact of the bulk solution with solid sorbents. It was observed that most metal uptake happens during the initial time of contact which is comparable with prior reports (Mishra et al. 1997a, b; Scott et al. 1995) on dead biomasses. Moreover, on living biomasses, *Aspergillus niger*, *Penicillium spinulosum* and *Trichoderma viride* (Townsley et al. 1986), the sorption of heavy metal particles was observed to follow the same pattern, and it was recommended that the fast accumulation is because of increased pH during the uptake process.

Furthermore, the concentration data has been investigated for Langmuir and Freundlich adsorption isotherms. It was observed that the data fit well to the established Freundlich equation, instead of the Langmuir mathematical equation for the studied systems:

$$\log a_e = 1/n \log c_e + \log k$$

where *ae* and *ce* are the amounts sorbed (mol g^{-1}) and bulk concentration (mol dm^{-3}) at equilibrium and K and 1/n are the Freundlich constants referring to sorption capacity and intensity of sorption, respectively. The straight lines were found for these systems on plotting log *ae* vs log *ce* (Fig. 8.4) at 293 K, which gives the value of K and 1/*n* by intercept and slope of these lines, respectively, and are returned in Table 8.3.

The estimation of 1/*n* being less than 1 demonstrates that the surface of sorbent is heterogeneous in nature and demonstrates the energy of the adsorption sites to have an exponential dispersion (Benes and Majer 1980).

In addition of the two heavy metal toxic ions (Hg^{2+} and Cr^{3+}), an attempt was made for the sorption of Cd^{2+} ion on the surface of neem bark tests, and results demonstrated that this ion did not adsorb considerably (just up to 3–5%), and

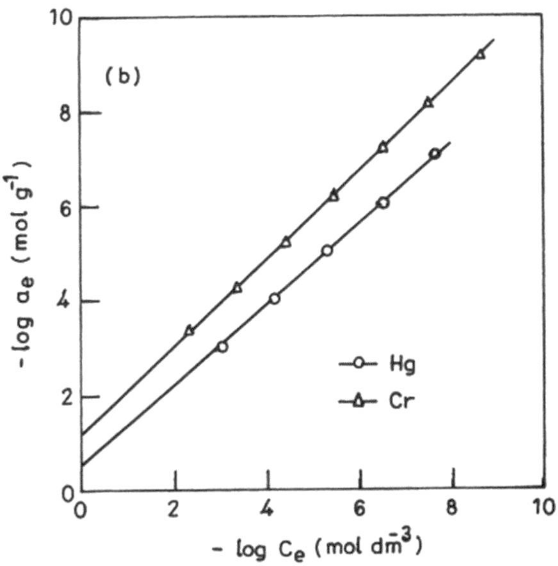

Fig. 8.4 Freundlich isotherms of Hg^{2+} and Cr^{3+} on neem (*A. indica*) bark (temperature: 293 K) (Tiwari et al. 1999)

Table 8.3 Freundlich parameters for the sorption of Hg^{2+} and Cr^{3+} on neem (*A. indica*) bark (temperature: 293 K)

System	$1/n$	K (mol g^{-1}) × 10^2
Neem bark, Hg^{2+}	0.846 (±0.007)	33.1 (±0.97)
Neem bark, Cr^{3+}	0.936 (±0.003)	6.91 (±0.79)
The uncertainty quoted is $\pm 3\sigma$		

Tiwari et al. (1999)

consequently further tests were not done with this ion. Similar results were also obtained for Cd^{2+} on rice husk (Mishra et al. 1997a)

8.6 Comparative Analysis of Phytoremediation Potential and Biofuel Efficiency of *A. indica* and *A. nilotica*

8.6.1 Biofuel Efficiency of *A. indica*

Biodiesel has advantages that include properties such as biodegradability, renewability, less toxicity and few emissions of gaseous and particulate pollutants with larger cetane number than normal fossil fuels. The overall process for extracting biodiesel from natural oil from the plant species is described in Fig. 8.5.

Charcoal which is made up from *A. indica* wood is of very good quality, and the wood has been used as firewood since ages. Oil from *A. nilotica* is used for burning the lamps throughout India. *A. indica* seeds as nonedible plant oil sources can also be used for biodiesel and bioethanol production. To determine the catalytic conversion of nonedible oil sources for the generation of biodiesel and maximizing the use of biomass, two case studies have been discussed here.

8.6.1.1 Case I: Biofuel Production from *A. indica* Seeds Using Transesterification

Transesterification utilizes alcohols in the presence of acid, a base or enzyme as a catalyst relying on the free fatty acid content of the crude materials. This chemical process breaks the triglyceride molecules into alkyl esters as biodiesel fuels and glycerol as a byproduct (Khandelwal and Chauhan 2012).

8.6.1.1.1 Materials

Seeds of *A. indica* were collected, washed and dried in an oven (37 °C) for the oil extraction. Some chemical content such as sodium hydroxide (99% purity, analytical grade, Acros Organics) and methanol (99.8%, analytical grade, Fluka) were added before applying yeast strains to culture the *A. indica* in the biological laboratory.

8.6.1.1.2 Methods

A. indica seeds were mechanically pressed while adding warm water (10–20 ml) to press out the oil. Solvent extraction using n-hexane has been performed to enhance the recovery of oil from the seeds and then the n-hexane has been evaporated using

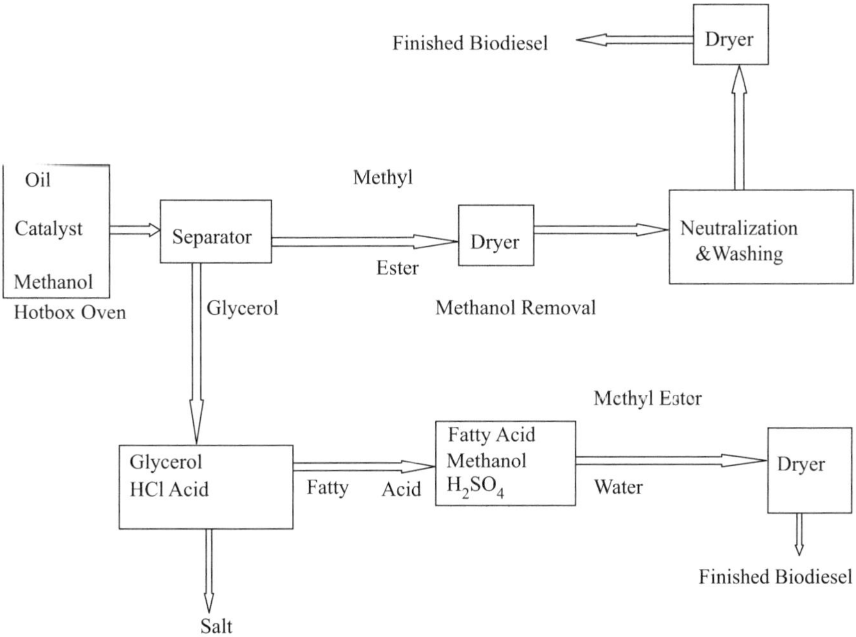

Fig. 8.5 Schematic biodiesel process (Khandelwal and Chauhan 2012)

Table 8.4 Viscosity and specific gravity of different fuels (Lin and Tanaka 2006)

Fuel	Viscosity at 40 °C (mm^2/s)	Specific gravity at 25 °C
Standard for diesel fuel (ASTM Standard D 975)	1.3–4.1	0.85
Standard for biodiesel fuel (ASTM D 6751)	1.6–6.0	0.88
A. indica oil	15.29	0.93
A. indica biodiesel	5.246	0.85

a rotary evaporator. The collected oils from *A. indica* seeds have been stored to perform the transesterification process on it. Transesterification process of recovered oil from the neem seeds was performed.

8.6.1.1.3 Ethanol Fermentation from Biomass

At the final stage of producing ethanol from the biomass, 30 g of *A. indica* was mixed with 100 ml of distilled water. In sterilized mixture, cultured yeast (3.5 g) was added in 50 ml of water to perform its fermentation. Finally, ethanol was collected through distillation (1–3% yield) (Lin and Tanaka 2006).

The viscosity and specific gravity of *A. indica* oil and *A. indica* biodiesel were compared with standard for diesel fuel (ASTM Standard D 975) and standard for biodiesel fuel (ASTM D 6751) in Table 8.4.

8.6.1.2 Case II: Production of Biodiesel via Neem Seeds Oil by Utilizing Their Base-Catalysed Transesterification and Its Mixing with Diesel

8.6.1.2.1 Plant Material

The organic substances consisted of some neem oil acquired by solvent extraction from *A. indica*'s fully developed fruit seeds. Such collected seeds were then washed and first dried naturally and followed by oven drying at 40 °C for a week.

8.6.1.2.2 Methods

The process for extracting biofuels from the *A. indica* involved extraction of oil, physiochemical characterization, fatty acid analysis, transesterification of neem seed oil and blending of diesel and neem seeds oil.

8.6.1.2.3 The Extraction of Oil

The oil utilized for the transesterification and mixing with diesel was separated with the Soxhlet technique by taking cyclohexane as solvent.

8.6.1.2.4 Transesterification of the Neem Seeds Oil

Generally biodiesel is produced by the process of transesterification. A reaction between oil and an alcohol (usually methanol or ethanol) produced alkyl esters (often methyl or ethyl) and glycerol. These reactions are reversible, so an excess of alcohol is used to shift the balance in the direction of the product (esters and glycerol) synthesis. In the industrial process, a molar ratio (alcohol/oil) of 6:1 is mostly used to obtain greater conversion rate to 98% (Kaura et al. 1998).

Most frequently methanol is used as alcohol because it is cheaper, but it is more toxic. Hence, ethanol is considered as a better choice. The kinetics of reaction transesterification process could be enhanced by using a base, an acid or enzymes as catalyst. The transesterification was carried out by taking ethanol/oil in a molar ratio 6:1, and the catalyst (sodium hydroxide) was fixed at 1% of the oil (w/w). The reaction took place in a protected reactor manufactured by PIGNAT Company (France), equipped with a stirrer EUROSTAR of Ika-Werke type at 75 °C. The stirring speed was fixed at 1200 rpm. Towards the completion of the transesterification process, the reaction mixture was neutralized by adding citric acid at 6%. The partition of the two phases (glycerol and ester) was done by utilizing a separator funnel after 24 h of decantation (Kaura et al. 1998).

8.6.1.2.5 Blending of Diesel and Neem Seeds Oil

Mixing or blending of neem seeds oil with diesel is a process called as biodiesel synthesis, in which vegetable oils are diluted with diesel fuel or with an ethanol solvent. In this study the NSO–diesel blend (NSODB) was prepared by taking 1:5 ratio of oil–diesel as the ratios in this range, i.e. 1:10 to 1:5, were found to be effective. Dilution is then followed by mixing in which mixture was homogenized by stirring in the reactor at room temperature before being analysed (Djibril et al. 2015; Hashmat et al. 2012).

8.6.1.2.6 Physicochemical Properties of the Neem Seeds Oil

To characterize neem seed oil, various physicochemical parameters were analysed and furnished in Table 8.5. The calorific value of neem seeds oil was calculated by utilizing the following empirical relationship: calorific value = 11,380 – iodine value – 9.15 × saponification value (Djibril et al. 2015).

Neem seed oil showed high saponification index (200.54 mg g^{-1}) which indicated its high degree of saponification. In this case, more water or little alcohol may cause soap formation during the transesterification. But the less water content of neem seed oil (<0.05%) did not cause saponification during the transesterification and subsequent process of separation and cleaning. The ash (0.02%) content was found within the maximum value (0.1%); it would not affect the functioning of the injection pump and the cylinder. The sulphur content (0.11%) was lower than its maximum limit. However, the carbon residue content was found noticeably very high (1.45%), which is beyond permissible limit for diesel (0.2%), which increases the chances of their deposition after combustion. NSO has relatively high net calorific value (39.53 MJ/kg), which renders it as good source for biodiesel production. The high flashpoint of NSO (227 °C) ensures safety from risk of catching fire during working. It cannot be used as fuel in the colder condition because it has freezing point (10 °C) and the pour point (12 °C), which are well below those of diesel. The viscosity of the NSO was reported 49.79 mm^2/s much higher than that of diesel and for fuel application; therefore it should be reduced by transformation of the triglycerides into fatty acids esters or by dilution with diesel.

Table 8.5 Physicochemical properties of NSO

Characteristics	Value	Methods
Acid value ($mg.g^{-1}$)	9.10	AFNOR T60–204
Saponification value ($mg.g^{-1}$)	200.54	AFNOR T60–206
Iodine value ($g.100\ g^{-1}$)	74.82	AFNOR T60–203
INS	126	-
Peroxide value (meq $O_2.kg^{-1}$)	1.49	AFNOR T60–220
Density at 25 °C	0.912	ASTM D 4052
Refraction index at 25 °C	1.465	Direct reading
Viscosity at 37.8 °C ($mm^2.s^{-1}$)	49.79	ASTM D 445
Calorific value ($MJ.kg^{-1}$)	39.53	-
Flashpoint (°C)	227 °C	ASTM D 93
Freezing point (°C)	10 °C	ASTM D 97
Pour point (°C)	12 °C	ASTM D 97
Ash (wt%)	0.02	ASTM D 482
Carbon residue (wt%)	1.45	ASTM D 189
Sediments (wt%)	0.01	ASTM D 4052
Water content (% v)	< 0.05	ASTM D 9590

Djibril et al. (2015), Kaura et al. (1998), and Hashmat et al. (2012)

8.6.1.2.7 Determination of Fatty Acid Composition of Neem Seeds Oil

The fatty acid profile was found by analysis of its fatty acid methyl esters (FAME) in gas chromatography (GC) using the French Standard NF ISO 5508. The four primary fatty acids (Table 8.6) found in neem seed oil were oleic acid (41.91 ± 0.69%), linoleic acid (19.59 ± 0.44%), stearic acid (18.71 ± 0.46%) and palmitic acid (15.59 ± 0.27%). The high proportion of linoleic and oleic acid proportion was also reported in Jatropha oil and *Pongamia pinnata* (Karanja). In neem seed oil, these fatty acids constitute major proportion (95.80%) of the total fatty acids. The ratio of unsaturated/saturated fatty acids was around 60%, which renders the neem seed oil as unsaturated in nature.

8.6.1.2.8 Impact of Reaction Time on Ester Yield

The impact of reaction time on the synthesis of the ethyl esters was studied at 75 °C with 1% (w/w oil) of catalyst and a molar ratio 6:1. The analysis conducted at fixed time intervals, i.e. after 30 min, 60 min, 90 min, 120 min, 150 min 180 min, 210 min and 240 min (Figs. 8.6 and 8.7), showed the fast production of esters. After 30 min, there was 58% of ester yield, and after 60 min it was 92%. However, the maximum

Table 8.6 Fatty acid composition of neem seeds oil

Composition	%
C16:0 Palmitic acid	15.59±0.27
C16:1 Palmitoleic acid	0.12±0.00
C18:0 Stearic acid	18.71±0.46
C18:1 Oleic acid	41.91±0.69
C18:2 Linoleic acid	19.59±0.44
C20:0 Arachidic acid	1.33±0.01
C18:3 Linolenic acid	0.44±0.01
C20:1 Gadoleic acid	0.08±0.00
C22:0 Behenic acid	0.86±0.38
Saturated fatty acids	37.00
Unsaturated fatty acids	63.00

Djibril et al. (2015), Kaura et al. (1998), Hashmat et al. (2012)

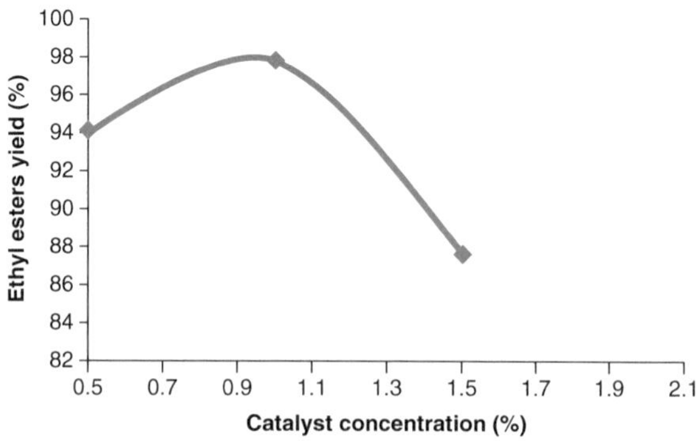

Fig. 8.6 Impact of concentration of catalyst on ethyl ester yield at 75 °C, molar ratio 6:1 (Djibril et al. 2015; Kaura et al. 1998; Hashmat et al. 2012)

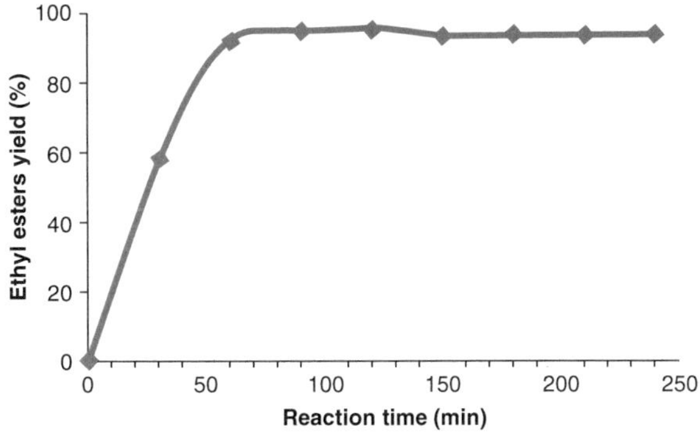

Fig. 8.7 Impact of reaction time on ethyl ester yield at 75 °C, molar ratio 6:1 and catalyst 1% (w/w) (Djibril et al. 2015; Kaura et al. 1998; Hashmat et al. 2012)

yield (93–95%) was observed after passing of 90 min of reaction. These results were similar to those researchers who used Jatropha oil with KOH catalysis, on the Thumba oil with NaOH and KOH catalysis and on the palm oil in alkali catalysis.

After the investigation of above study, it is evident that *Azadirachta* (neem) seeds oil could be considered as a renewable source of fuel. The physical and thermal properties of the ethyl esters (NSOB) acquired by transesterification of NSO with NaOH catalysis are very close to those of diesel. The NSO–diesel mixture (NSODB) provides even closer properties. Thus, the utilization of the NSOB and the NSODB as biofuel in the diesel engines can be envisaged.

Furthermore, the analysis and comparative study of *Azadirachta* biodiesel (ABD) with petroleum diesel (PD) indicates the results as (Djibril et al. 2015):

Properties	Result (ABD)	PD	ASTM
Density at 15 °C kg/m^3	874	838	890
Viscosity at 40 °C (mm^2/s)	4.4	5	6.0
Specific gravity at 20° (0 °C)	0.876	0.816 0.862	0.862
Calorific value (MJ/kg)	39.25	44.21	–
Flashpoint (0 °C)	138	53	130–173
Fire point (0 °C)	145	59	–
Cetane number	59	40–55	51
Cloud point (0 °C)	11	7	–1
Pour point	–20	–20	–
Acid value (mg KOH/g)	0.88	0.22	0.50
Carbon residue (%)	0.3	0.17	0.30
Water content	0.04	< 0.03	< 0.03
Sulphur content (%)	0.36	0.39	–
Sediment	Nil	–	–

Note: *ABD Azadirachta* biodiesel, *PD* petroleum diesel, *ASTM* American Society for Testing and Material

8.6.1.2.9 Study of Engine to Justify the Biodiesel Efficiency

This work was done to analyse the efficiency of diesel engine utilizing *Azadirachta* biodiesel mix (10–50%) in a single cylinder diesel engine and compared to petroleum diesel. The engine specification was given below in Table 8.7.

8.6.1.2.10 Engine Performance

The mixes of ABD and PD were utilized to test a single cylinder, four stroke and water-cooled motor at 1500 rpm. The motor was linked with electrical dynamometer and DC generator. The efficiency [specific fuel utilization (SFC) and brake warm productivity (BTE)] of the motor was examined at various motor loads utilizing different ABD mixes at steady speed which were contrasted with PD. Figures 8.8 and 8.9 demonstrated the summary of the trials attempted and their outcomes (Balat and Balat 2010).

Table 8.7 Engine efficiency for experimental analysis Djibril et al. (2015), Kaura et al. (1998), and Hashmat et al. (2012)

Name	Kirloskar
BHP	5 HP at 1500 rpm
Speed	1500 rpm
Bore	80 mm
Stroke	110 mm
Fuel	High-speed diesel
Loading	EDDY current dynamometer
Cooling	Water cooling system
Lubrication	Forced type

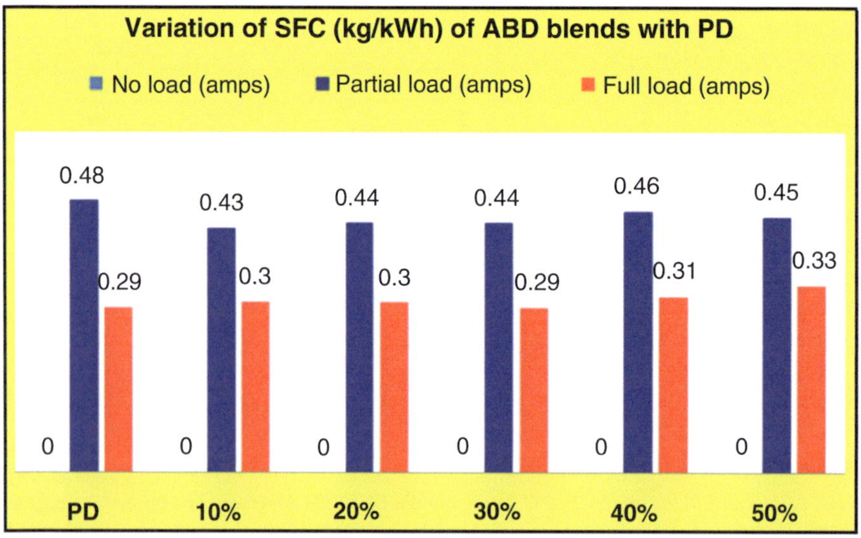

Fig. 8.8 Variation of SFC of ABD blends with PD (Balat and Balat 2010; Mathiyazhagan et al. 2013)

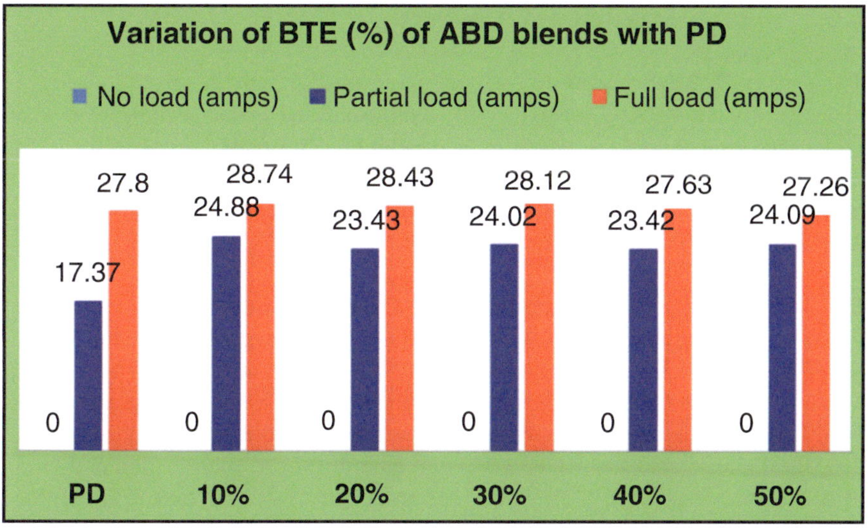

Fig. 8.9 Variation of BTE of ABD blends with PD (Balat and Balat 2010; Mathiyazhagan et al. 2013).Hence, the above investigation and analysis recommend the application of ABD as an alternative fuel to diesel engines

8.6.1.2.11 Specific Fuel Consumption

Variation of SFC utilizing various ABD mixes at different motor loads was illustrated in Fig. 8.8. Information extracted from this figure that the SFC diminishes with increasing engine load. Usually SFC utilization of biodiesel increases with increase in the concentration of biodiesel, but the study demonstrated lower SFC utilization; this is because of more oxygen content and cetane number. The presence of this higher oxygen content leads to efficient ignition of ABD fuel ended with formation of lower SFC.

8.6.1.2.12 Brake Thermal Efficiency

The variations of BTE utilizing PD and ABD mixtures at various loads were shown in Fig. 8.9. In all the mixtures (10–50%), the BTE efficiency was enhanced when the load increases. This is attributed to increase in power output with the increase of load. The maximum BTE efficiency (28.74%) was achieved from 10% ABD which was higher than the PD (27.80%). The lowest BTE performance was noticed from 50% ABD (27.26%). This is because of reduction in calorific value of ABD when increasing the concentration of BD with PD. However, the BTE of ABD blended up to 50% did not show significant difference. But further increase of BD level (above 50%) showed reduction of BTE. Hence, this study recommended ABD mixture up to 50% as an alternative fuel to diesel engines (Balat and Balat. 2010; Mathiyazhagan et al. 2013).

8.6.2 Biofuel Efficiency of *A. nilotica*

The pods of *A. nilotica* (family Leguminosae) are generally 7–15 cm long, green and covered with densely matted woolly hairs which are called as tomentose when young and greenish black when fully developed. The pods are indehiscent and tightly constricted between the seeds giving a neck jewellery appearance 8–12. Seeds are available per pod. The seeds are tightly packed, ovoid, dark brown shinning with hard testa (Fig. 8.10c) (Adhikesavan et al. 2015).

8.6.2.1 Case I: Biodiesel from *A. nilotica* (Karuvel) Seed Oil

For study analysis a plant seed named Karuvel is considered which belongs to family of *A. nilotica*. The Karuvel oil contains high content of free fatty acid (FFA); hence, a two-step stage process (first esterification of FFA and subsequently transesterification) was carried out for the production of biodiesel (Adhikesavan et al. 2015). Sulphuric acid and KOH were used as catalyst for esterification and transesterification, respectively. The biodiesel produced was characterized by physiochemical properties and using gas chromatography, and high amount of unsaturated fatty acid methyl esters was found.

Fig. 8.10 (**a**–**c**) Karuvel tree and seeds (Adhikesavan et al. 2015)

Table 8.8 Fatty acid composition of Karuvel oil

Fatty acids	Percentage
Myristic acid (C14:0)	0.4
Palmitic acid (C16:0)	15.7
Stearic acid (C18:0)	9.0
Arachidic acid (C20:0)	1.2
Vaccenic acid (C18:1)	29.0
Linoleic acid (C18:2)	44.5

Adhikesavan et al. (2015)

8.6.2.1.1 Materials

The Karuvel pods were collected from the Karuvel trees in and around Government College of Engineering, Tirunelveli, Tamil Nadu. The seeds were then removed from the pods for further processing. AR grade of methanol, n-hexane, potassium hydroxide and sodium sulphate were used. The fatty acid composition of Karuvel seed oil taken from the literature is given in Table 8.8 (Adhikesavan et al. 2015).

In CXX:Y, XX indicates the length of the carbon chain, and Y indicates the number of double bonds.

8.6.2.1.2 Oil Production by Solvent Extraction

Solvent extraction process was adopted to extract the oil from Karuvel seeds. 0.5 kg of seed powder was used for the extraction each time. The seeds were finely powdered by a dry grinder and added to the extraction chamber. Anhydrous n-hexane (organic solvent) was added to the flask, and the temperature of the flask was maintained at 50 °C by using a thermostat. At this temperature, the hexane gets vaporized and reaches the condenser where it gets condensed and trickles in to the extraction chamber. The powdered seeds percolate in n-hexane solvent; thereby, the oil content from the seeds gets dissolved in to the solvent. This process was repeated till most of the oil from the seeds was extracted by the solvent. Then, distillation has been carried out to separate oil from n-hexane. The oil thus produced was filtered using Whatman No. 1 filter paper. Subsequently degumming process was carried out in a centrifuge with water to remove water-soluble phospholipids. The degummed oil was then treated with sodium sulphate to remove any traces of moisture. The oil yield was about 20% by weight (Hoekman et al. 2012).

8.6.2.1.3 Biodiesel Production by Transesterification

The Karuvel oil produced had a high acid number of about 18.2 mg KOH/g, respectively (Adhikesavan et al. 2015).Therefore esterification of free fatty acids was done first. Methanol to oil molar ratio used for esterification was 9:1. Sulphuric acid was used as catalyst at 1 % w/w of oil. The esterification was carried out for about 2 h. The reactant mixture was allowed to stand in a separating funnel for 24 h. The oil layer that got settled at the bottom has been separated and used for transesterification. Transesterification process was done in a same experimental setup. The transesterification variables such as methanol to oil ratio, reaction temperature and speed of the stirrer were maintained as constant for all experiments at 6:1, 60 °C and 600 rpm, respectively (Adhikesavan et al. 2015).

8.6.2.1.4 Quantification of Biodiesel

The quantification of methyl esters in Karuvel biodiesel was done using gas chromatography (GC). Fatty acid profile is given in Table 8.9. The percentage of saturated and unsaturated fatty acid methyl esters in the biodiesel is 19.83% and 79.55%, respectively (Adhikesavan et al. 2015).

8.6.2.1.5 Estimation of Properties of Biodiesel

The estimated values of density, viscosity and ultrasonic speed were $\pm 2.8\times10^{-3}$ g/cm^{-3}, $\pm 1.9\times10^{-3}$ mm^2/s^{-1} and 0.03%, respectively (Adhikesavan et al. 2015). The flashpoint of the biodiesel was measured using the Cleveland open cup apparatus. The cetane number of the biodiesel was estimated using equation. The measured physiochemical properties of biodiesel were given in Table 8.10 and compared with the biodiesel standards. The biodiesel produced from Karuvel oil had density, viscosity, acid number, flashpoint and cetane number values well within the limits specified by ASTM D6751 and EN 14214 standards.

Table 8.9 Composition of fatty acid methyl ester in biodiesel

Fatty acids	Percentage
Palmitic acid methyl ester	13.8
Behenic acid methyl ester	0.26
Arachidic acid methyl ester	0.74
Isostearic acid methyl ester	4.88
Lignoceric acid methyl ester	0.15
Elaidic acid methyl ester	11.22
Linoleic acid methyl ester	68.33

Adhikesavan et al. (2015)

Table 8.10 Physical and chemical properties of biodiesel

		Standards	
Properties	Karuvel biodiesel	EN14214	ASTM D6751
Density[a] (kg/m^3)	890.1	860–900	–
Kinematic viscosity[b] (mm^2/s)	4.2045	3.5–5	1.9–6
Acid number (mg KOH/g)	0.1923	0.5 max	0.80 max
Flashpoint (°C)	160	>101	130 min
Cetane number	51	51 min	47 min

Adhikesavan et al. (2015)
[a] measured at 288.15 K
[b] measured at 313.15 K

8.6.2.1.6 Impact of Catalyst Concentration on Biodiesel Yield

The influence of catalyst concentration on the methyl ester conversion of the vegetable oil has been studied. From Fig. 8.11 it was observed that methyl ester yield was less when the catalyst concentration was lower at 0.5% w/w (Adhikesavan et al. 2015). The optimal yield was achieved with 1% w/w catalytic concentration. Any further increase in the catalytic concentration had adversely affected the methyl ester yield that may be probably due to saponification process.

8.6.2.1.7 Influence of Reaction Time on Biodiesel Yield

It was observed (Fig. 8.12) that the conversion of methyl ester was lower at 30 min reaction time. However when the reaction time was increased to 1 h, the yield was found to be maximum. Any further increase in reaction time has not drastically yielded more methyl ester. Adhikesavan et al. (2015) have reported the similar observations.

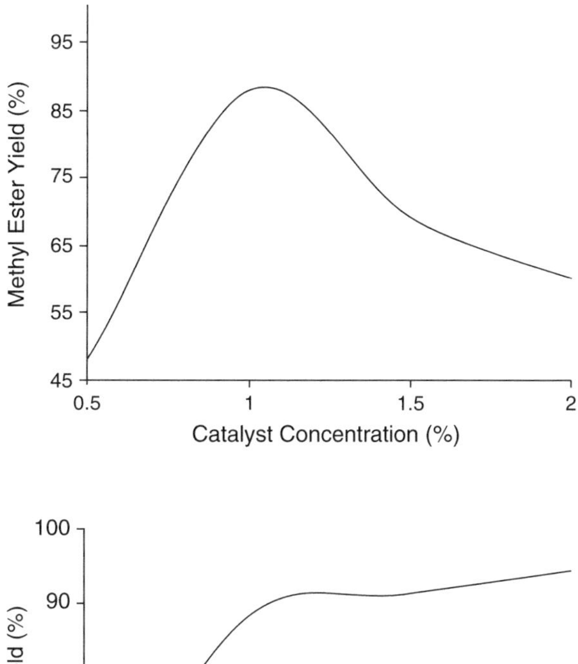

Fig. 8.11 Effect of catalyst concentration on biodiesel yield (Adhikesavan et al. 2015)

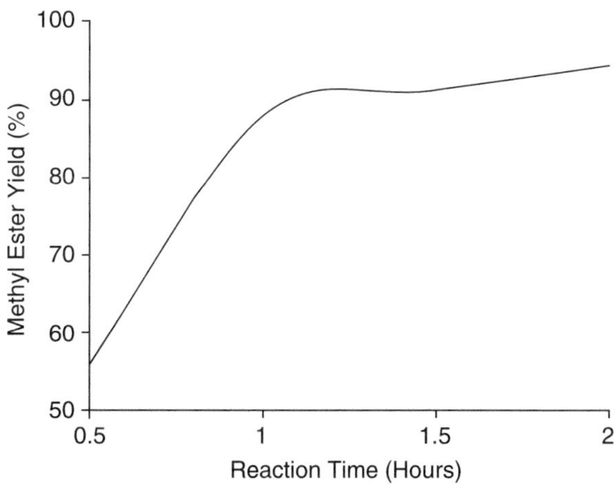

Fig. 8.12 Effect of reaction time on biodiesel yield (Adhikesavan et al. 2015)

Hence, from the experiments conducted, it can be concluded that biodiesel has been successfully produced from nonedible Karuvel seeds. From GC analysis, it was found that the amount of unsaturated fatty acids was high (79.55 %). The physicochemical properties of the biodiesel were found well within the limits specified by international standards, thereby indicating that this biodiesel may be used in diesel engine with proper blending with petrodiesel (Adhikesavan et al. 2015).

8.7 Challenges and Opportunities

A review of several pilot-scale applications reveals that phytoremediation is at least 50% less costly than excavation and it is also cheaper than bioremediation, in accordance with other data (Adhikesavan et al. 2015; Hoekman et al. 2012). Some traditional remediation methods usually have several demerits such as generation of atmospheric emissions and huge quantity of additional wastes that need to be disposed of and not being suited for the treatment of soils that are to be reused for agricultural or similar purposes of plant/biomass production (Hashmat et al. 2012).

A number of authors have argued that the accomplishment of phytoremediation at commercial level relies on the generation of profitable or valuable biomass on contaminated land rather than as a pure remediation technique that may not compare positively with the costs of inaction or alternative technologies (Balat and Balat 2010; Mathiyazhagan et al. 2013). Valuable biomass includes timber, biofuel, feedstock for pyrolysis and biofortified products (enriched in Fe, Zn or Se) for food nutrients or ecologically important species.

The main challenge regarding the use of biomass for bioenergy is the issue of pollutant emission and heavy metal present in the biomass. Pollutants in the crops may cause problems in later phases of biofuel production, and the decision on whether crop uptake should be encouraged or not varies from case to case. There must be a risk management of the crops or crop choices, and clones can be made that prevent take-up of contaminants using excluders instead of hyperaccumulators (Basavaraja et al. 2011). There is insufficient knowledge regarding the emissions that may be generated in the use of plants and wood material for bioenergy. More research is needed about the uptake mechanism of different crops, rate of combustion and emissions, ingredients present in biofuel under different conditions and different techniques in order to confident in using end products of phytoremediation for bioenergy with minimum ecological impacts.

Another important challenge is that current legislation and practice in soil remediation are based on the total concentrations of the contaminants present in the soil and not on functionality of soil or risk-associated land management, which can be a barrier to the use of phytoremediation which is a slow remediation process (Qadir et al. 2006). Regarding the opportunities, there are various lands in the world where growing of food and feedstock crops is not possible because as a result of non-point source contaminations, excessive amounts of pollutants are present, and repeated applications of chemical fertilizers and pesticides, along with atmospheric deposition, cause economic losses and adverse impacts on the human food chain and health. In Germany 1999, approximately 10,000 ha of cultivable field have been taken out of food production due to heavy metal contamination (Abbas et al. 2016). In Sweden 2009, it was found that 75,000 ha of contaminated land could be suitable for phytoremediation with bioenergy crops (Trapp and Karlson 2001). In England 1993, there are some 39,000 ha of 'derelict land', defined as 'Irreparable damage of land by industrial or other developmental activities, which cannot be used without treatment' (Pierzynski et al. 1994).

Now much attention is paid to the ideas of the disposal of agricultural and municipal solid wastes (to remove nitrates and other nutrients from municipal wastewater 'polishing', farmland drainage water and sewage sludge) on energy crops (Adhikesavan et al. 2015). This practice will meet the requirement of organic matter and nutrients needed for plant growth at a low cost, whilst enabling controlled disposal of wastes on a nonedible crop (Pierzynski et al. 1994). Other authors have argued that the irrigation with leachate from municipal solid waste would have negative impacts on soil micro- and mesofauna, which are known to be essential in healthy ecosystem functioning (Hoekman et al. 2012).

The profound and vast root systems of short rotation coppice (SRC) are used as a purifier during the process of wastewater treatment with an excessive biological oxygen demand (BOD) and eutrophication condition. An extensive distribution of SCR in the landscape (particularly along watercourses) could minimize non-point source pollution of water with plant nutrients (Reeves and Baker 2000). A plant habitat and the community along the river banks are the key filter that moderates and slows down runoff flow, accumulating sediments, manures, plant waste and humic substances that would be otherwise deposited in the stream flow (Reeves and Baker 2000; Hoekman et al. 2012).

Other authors have suggested that phytoremediation is the ideal technology for mitigating landfill environmental problems including soil and groundwater pollution, leachate formation and release of various greenhouse gases (especially if post-closure treatment of landfills is not proper or the conventional clay landfill capping is deteriorated) (Adhikesavan et al. 2015; Hoekman et al. 2012).

It could also be presumed that the growing of short rotation coppice (SRC) and perennial energy grasses (PEG) for heat and power production would become more significant when new technologies introduced in market and bio-heat options, even for domestic use, are further developed (Hoekman et al. 2012).

Other benefit of phytoremediation is that crops with high concentration of metal pollutants could be regenerated for further use (Hoekman et al. 2012).The technique used for the recovery of these metal pollutants is electrokinetics, which takes out pollutants from the media by electric current and transfers them to electrolyte solutions, where they are collected and treated (Djibril et al. 2015; Adhikesavan et al. 2015). Several workers have successfully removed heavy metal from fly ash obtained from the combustion of municipal solid waste and waste wood (Adhikesavan et al. 2015; Hoekman et al. 2012).

Recent development of transgenic plants and bacteria that enhance phytodegradation of contaminants could also be an exciting opportunity for bioenergy production through phytoremediation (Hoekman et al. 2012). Phytoremediation may provide the solution, given the low costs involved. Bioenergy crop cultivation on contaminated land could be a cost-effective alternative of phytoremediation of these sites, simultaneously providing soil that does not compete for food production, reducing the combustion of fossil fuel, generating employment options and providing efficient treatment of contaminated sites (Djibril et al. 2015; Adhikesavan et al. 2015; Hoekman et al. 2012). However, the use of bioenergy resulting from phytoremediation is constrained due to paucity of knowledge regarding the emissions that may be generated and the issues associated with pollution transfer, especially for heavy metals. There is certainly an opportunity for soils contaminated with organics and nutrients (N and P) where that problem does not exist.

8.8 Conclusion

The above study clearly showed that *A. indica* and *A. nilotica* are two tree species which not only possess immense industrial, commercial and therapeutic properties, but their various parts (seeds, leaves, root, stems, barks, etc.) can also be used in phytoremediation to remove various contaminants (heavy metals, dye, radionuclides, pesticides, etc.) from the soil. Both these nonedible tree species can effectively amend the sodic/alkaline soil. Anticlastogenic and chemopreventive activity of both the plants have also been reported in various studies. Application of phytoremediation technique on some wasteland can be converted into a success story if we use such plants which possess above said quality and whose biomass can be used as biofuel. The above study revealed that *A. indica* and *A. nilotica* are two tree species which can produce biodiesels. Their density, viscosity, specific gravity, calorific values, cetane number, flashpoint, etc. meet the requirement specified by American Society for Testing and Material (ASTM) with slight variation from each other. Their comparative study revealed that *A. indica*-based biodiesel (ABD) showed density at 15 °C kg/m^3 = 874, viscosity at 40 °C (mm^2/s) = 4.4, specific gravity at 20 °C = 0.876 and calorific value (MJ/kg) = 39.25), cetane no. = 59 and flashpoint (0 °C) 138, whereas *A. nilotica* (Karuvel) biodiesel showed density at 15^0 C kg/m^3= 890.1, viscosity at 40 °C (mm^2/s) = 4.2, flashpoint (0 °C) =160 and cetane no. = 51. Thus by growing these tree species, we can maintain good health of various components of environment (soil, water, air, etc.) and solve the problem of energy crisis by increasing the efficiency of biodiesels derived from both the trees.

References

Abbas G, Saqib M, Akhbar J et al (2016) Relationship between rhizosphere acidification and phytoremediation in two *Acacia* species. J Soils Sediments 4:1392–1399

Adewumi AA (2011) Assessment of NORM-containing food crops/stuffs in OML 58 & OML 61 within the Niger delta region of Nigeria. Proceedings of the 1st international technology, education and environment conference. Afr Soc Sci Res (ASSR), pp 594–603

Adhikesavan C, Rajagopal K, Rajaduari JS (2015) Production and characterization of biodiesel from *Acacia nilotica* seeds. Int J Chem Tech Res 8:854–859

Ahmed S, Bamofleh M, Munshi A (1989) Cultivation of neem (*Azadirachta indica*) in Saudi Arabia. Econ Bot 43:35–38

Alambayan J, Vats M, Sardana S, Chopra AK (2014) Phytochemical and Hptlc analysis of roots of *Acacia arabica* willd. Sch Acad J Pharm 32:133–139

Alberto AMP, Gilbert C and Sigua (2013) Phytoremediation: a green technology to remove environmental pollutants. Am J Clim Chang 2:71-86

Ali FA (2008) Measurements of naturally occurring radioactive materials (NORMs) in environmental samples. http://personal.ph.surrey.ac.uk/~phs1pr/msc_dissertations/msc-diss-2008/fuad-ali-msc-diss-NORM-2008.pdf

Ang XW, Sethu VS, Andresen JM, Sivakumar M (2012) Copper (II) ion removal from aqueous solutions using biosorption technology: thermodynamic and SEM–EDX studies. Clean Technol Environ 15:401–407

Arantzazu U, Vega M, Angul E (2000) Deriving ecological risk based soil quality values in the Barque country. Sci Total Environ 247:279–284

Arshad M, Zafar MN, Younnis S, Nadeem R (2008) The use of Neem biomass for the biosorption of zinc from aqueous solutions. J Hazard Mater 157:534–540

Babu BV, Gupta S (2008) Adsorption of Cr (VI) using activated neem leaves: kinetic studies. Chem Mat Sci 14(1):85–92

Bahuguna VK (1997) Silviculture and management practices for cultivation of *Azadirachta indica* (Neem). Indian For 123:379–386

Baig JA, Kazi TG, Shah AQ et al (2010) Biosorption studies on powder of stem of *Acacia nilotica*: removal of arsenic from surface water. J Hazard Mater 178(1–3):941–948

Balat M, Balat M (2010) Progress in biodiesel processing. Appl Energ 87:1815–1837

Bañuelos GS, Shannon MC, Ajwa H et al (1999) Phytoextraction and accumulation of boron and selenium by poplar (*populus*) hybrid clones. Int J Phyt 1:81–96

Barceló J, Poschenrieder C (2003) Phytoremediation: principles and perspectives. Contrib Sci 2:333–344

Bargali K, Bargali SS (2009) *Acacia nilotica*: a multipurpose leguminous plant. Nat Sci 7(4):11–19

Basavaraja PK, Sharma SD, Dhananjaya BN et al (2011) *Acacia nilotica*: a tree species for amelioration of sodic soils in central dry zone of Karnataka India. Res J Agric Sci 2:311–316

Benes P, Majer V (1980) Trace chemistry of aqueous solutions, vol 11. Elsevier, Amsterdam, p 200

Bharali RKand Bhattacharya KG (2015) Biosorption of fluoride on Neem (*Azadirachta indica*) leaf powder. J Environ Chem Eng 3(2):662–669

Bhattacharyya KG, Sharma A (2004) Adsorption of Pb (II) from aqueous solution by *Azadirachta indica* (Neem) leaf powder. J Hazard Mater 113:97–109

Biswas K, Chattopadhyay I, Banerjee RK et al (2002) Biological activities and medicinal properties of Neem (*Azadirachta indica*). Curr Sci 82(11):136–1345

Brototi B, Kaplay RD (2010) *Azadirachta indica* (Neem): its economic utility and chances for commercial planned plantation in Nanded district. Int J Pharm 1(2):100–104

Chakraborthy R, Konger G (1995) Root-rot of Neem (*Azadirachta indica*) caused by Ganoderma applanatum. Indian For 121:1081–1083

Chan K (2003) Some aspects of toxic contaminants in herbal medicines. Chemosphere 52:1361–1371

Chatterjee A, Pakrashi SC (2008) The treatise on Indian medicinal plants. Nat Inst Sci Commun (CSIR) 3:75–78

Chaturvedi AN (1985) Firewood farming on degraded lands in the Gangetic plain, Uttar Pradesh. For Bullet 50:50

Chaturvedi AN, Behl HM (1996) Biomass production trials on sodic site. Indian For 122:439–455

Chaturvedi AN, Bhatia S, Behl HM (1991) Biomass assessment for shrubs. Indian For 117:1032–1035

Cunningham SD, Ow DW (1996) Promises and prospects of phytoremediation. Plant physiol 110:715–719

Cunningham SD, Berti WR, Huang JW (1995) Phytoremediation of contaminated soils. Trends Biotechnol 13:393–397

Dass B, Jha P (2015) Batch adsorption of phenol by improved activated *Acacia nilotica* branches char: equilibrium, kinetic and thermodynamic studies. Int J Chem Tech Res 8(12):269–279

Denti B, Cocucci SM, Di Givolamo F (1998) Environmental pollution and forest stress: a multidisciplinary approach study on alpine forest ecosystems. Chemosphere 36:1049–1054

Desideri D, Meli MA, Roselli C (2010) Natural and artificial radioactivity determination of some medicinal plants. J Environ Radioact 101:751–756

Djenontin TS, Amusant N, Dangou J et al (2012) Screening of repellent, termiticidal and preventive activities on wood, of *Azadirachta indica* and *Carapa procera* (Meliaceae) seeds oils ISCA. J Biol Sci 1(3):25–29

Djibril D, Lamine DM, Mamadou F et al (2015) Biodiesel production from Neem seeds (*Azadirachta indica* A. Juss) oil by its base-catalyzed transesterification and its blending with diesel. Res J Chem Sci 5:13–19

Dogra PD (1989) Fuel wood production potential in India through short rotation intensive culture (SRIC) system applied to tree growing for fuel wood on reduced land area or on wasteland sites. Indian For 115:11–16

Dreyfus BL, Dommergues YR (1981) Nodulation of Acacia species by fast- and slow-growing tropical strains of Rhizobium. Appl Environ Microb 41:97–99

Dushenkov D (2003) Trends in phytoremediation of radionuclides. Plant Soil 249:167–175

Elujoba AA, Odeleye OM, Ogunyemi CM (2005) Traditional medicine development for medical and dental primary health care delivery system in Africa. Afr J Tradit Complement Altern Med 2(1):46–61

Entry JA, Watrud LS, Manasse RS et al (1997) Phytoremediation and reclamation of soils contaminated with radionuclides. In: Kruger EL, Anderson TA, Coats JR (eds) Phytoremediation of soil and water contaminants. American Chemical Society, Washington, DC, pp 299–306

Facchinelli A, Sacchi E, Malleri L (2001) Multivariate statistical and GIS-based approach to identify heavy metal sources in soils. Environ Pollut 114:313–324

Fathima SK (2004) Investigations on the biology and management *of Phomopsis azadirachtae* on Neem. Ph.D. thesis, University of Mysore, Mysore, India

Febriana N, Lesmana SO, Soetaredjo FE et al (2010) Neem leaf utilization for copper ions removal from aqueous solution. J Tai Inst Chem Eng 41(1):111–114

Firewood crops 1980. National Academy of Sciences, Washington, DC

Fischerová Z, Tlustoš P, Száková J, Šichorová K (2012) A comparison of phytoremediation capability of selected plant species for given trace elements. Environ Pollut 144:93–100

Flaschka H (1964) EDTA-titrations: an introduction to theory and practice, 2nd edn. Pergamon Press, Oxford, p 138

Glass DJ (2000) Phytoremediation's economic potential. In: Raskin I, Ensley BD (eds) Phytoremediation toxic met. Using plants to clean up environment. Wiley, New York, pp 15–31

Gomes HI (2012) Phytoremediation for bioenergy: challenges and opportunities. Environ Technol Rev 1:59–66

Govil PK, Reddy GLN, Krishna AK (2001) Contamination of soil due to heavy metals in the Patancheru industrial development area, Andhra Pradesh, India. Environ Geol 41:461–469

Hasmat I, Hussain A, Ahmed A (2012) Neem (*Azadirachta indica* A. Juss) – a nature's drugstore: an overview. Int Res J Biol Sci 1(6):76–79

Hegde NG (1995) Neem and small farmers constraints at grass root level. Indian For 121:1040–1048

Hoekman SK, Broch A, Robbins C et al (2012) Review of biodiesel composition, properties, and specifications. Renew Sustainable Energy Rev 16:143–169

Hosseini O and Moeinpour F (2016) Removal of Pb (II) from aqueous solutions using *Acacia nilotica* seed shell ash supported Ni0.5 Zn0.5Fe2O3 magnetic nanoparticles. J Water Reuse Desal 6:1

IAEA (2006) Classification of soil systems on the basis of transfer factors or radionuclides from soil to reference plants IAEA-TECDOC 1497. Vienna

INFOSAN (2011) Information on nuclear accidents and radioactive contamination of foods International Food Safety Authorities Network. World Health Organization (WHO), Geneva
Jain RK, Singh B (1998) Biomass production and soil amelioration in a high density *Terminalia arjuna* plantation in sodic soils. Biomass Bioenerg 15(2):187–192
Jattan SS, Shashikumar M, Pujar G (1995) Perspectives in intensive management of Neem plantations. Indian For 121:981–988
Jonathan MP, Ram Mohan V, Srinivasalu S (2004) Geochemical variation of major and trace elements in recent sediments of the Gulf of Mannar the southeast coast of India. Environ Geol 45:466–480
Kabeeruddin H and Mufradat M (2007) Aijaz Publishing House, New Delhi, pp 400–411
Kalaivani T, Mathew L (2010) Free radical scavenging activity from leaves of *Acacia nilotica* (L.) Wild. ex Del. An Indian medicinal tree. Food Chem Toxicol 48:298–305
Karavaeva YN, Kulinov NV, Molchanova IV et al (1994) Accumulation and distribution of long-lived radionuclides in the forest ecosystems of the Kyshtym accident zone. Sci Total Environ 157.47–151
Kaur K, Arora S, Hawthorne ME et al (2002) A correlative study on antimutagenic and chemopreventive activity of *Acacia auriculiformis* A. Cunn. and *Acacia nilotica* (L.) Wild.ex Del. Drug Chem Toxico 25(1):39–64
Kaur K, Michael H, Arora S et al (2005) In vitro bioactivity-guided fractionation and characterization of polyphenolic inhibitory fractions from *Acacia nilotica* (L.) Willd. ex Del. J Ethnopharmacol 99:353–630
Kaura SK, Gupta SK, Chowdhury JB (1998) Morphological and oil content variation in seeds of *Azadirachta indica* A. Juss. (Neem) from northern and western provenances of India. Plant Foods Hum Nutr 52:293–298
Khandelwal S, Chauhan R (2012) Biodiesel production from nonedible-oils: a review. J Chem Pharm Res 4:4219–4230
Kim KR, Owens G (2010) Potential for enhanced phytoremediation of landfills using biosolids: a review. J Environ Manage 91:791–797
Krishna AK, Govil PK (2004) Heavy metal contamination of soil around Pali Industrial Area, Rajasthan, India. Environ Geol 47:38–44
Krishna AK, Govil PK (2007) Soil contamination due to heavy metals from an industrial area of Surat, Gujrat, Western India. Environ Monit Assess 124:263–275
Krzyztof L, Wiechua D, Korus I (2004) Metal contamination of farming soils affected by industry. Environ Int 30:159–165
Kratochvil D, Volesky B (1998) Advances in the biosorption of heavy metals. Trends Biotechnol 16:291–300
Kumar ARV (2003) Neem for the industry or for the common man: where does India stand. Curr Sci 84:265–267
Kumar VS, Navaratnam V (2013) Neem (*Azadirachta indica*): prehistory to contemporary medicinal uses to humankind. Asian Pac J Trop Biomed 3(7):505–514
Kupradinun P, Jarratwisarutporn Y, Wongpoomchai R et al (2013) Clastogenic and anticlastogenic potential of Neem flower extract evaluated by rat liver micronucleus assay. Thai J Vet Med 43(4):589–594
Kurnaz A, Kucukomeroglu B, Keser R et al (2007) Determination of radioactivity levels and hazards of soil and sediment samples in Firtina Valley (Rize, Turkey). Appl Radiat Isot 65:1281–1289
Lebeau T, Braud A, Jézéquel K (2008) Performance of bioaugmentation-assisted phytoextraction applied to metal contaminated soils: a review. Environ Pollut 153:497–522
Lee H, Ho SH, Lee CH et al (2006) Fly ash effect on improving soil properties and rice productivity in Korean paddy soils. Bioresour Technol 97:1490–1497
Lin Y, Tanaka S (2006) Ethanol fermentation from biomass resources: current state and prospects. Appl Microbiol Biotechnol 69:627–642
Lokeshwari H, Chandrappa GT (2006) Impact of heavy metal contamination of Bellandur Lake on soil and cultivated vegetation. Curr Sci 91:622–628

Mathiyazhagan M, Elango T, Senthilkumar T et al (2013) Assessment of fuel efficiency of Neem biodiesel (*Azadirachta indica*) in a single cylinder diesel engine. Int J Energ Technol 9:279–285

Meagher RB (2000) Phytoremediation of toxic elemental and organic pollutants. Curr Opin Plant Biol 3:153–162

Meena PD, Kaushik P, Shukla S et al (2006) Anticancer and antimutagenic properties of *Acacia nilotica* (L.) on 7, 12-dimethylbenz (a) anthracene-induced skin papillomagenesis in Swiss albino mice. Asian Pac J Can Prev 7:627–632

Minhas PS, Singh YP, Tomar OS et al (1997) Effect of saline irrigation and its schedules on growth, biomass production and water use by *Acacia nilotica* and *Dalbergia sissoo* in a highly calcareous soil. J Arid Environ 36:181–192

Mishra SP, Singh VK, Tiwari D (1996) Inorganic particulates in removal of toxic heavy metal ions. Efficient removal of cadmium ions from aqueous solution by hydrous zirconium oxide. Radiochim Acta 73:49

Mishra SP, Tiwari D, Dubey RS (1997a) The uptake behaviour of rice (Jaya) husk in the removal of Zn (II) ions: a radiotracer study. Appl Radiat Isotop 48:877–882

Mishra SP, Tiwari D, Dubey RS et al (1997b) Removal behaviour of rice *Oryza sativa (*L.) hulls for submicro-concentrations of mercury and chromium from aqueous solutions: a radiotracer study. Radiochim Acta 29(22):47–52

Mojiri A, Aziz HA, Zahed MA et al (2013) Phytoremediation of heavy metals from urban waste Leachate by Southern Cattail (*Typha domingensis*). Int J Scienti Res Environ Sci 1(4):63–70

Mondal D, Mondal T (2012) A Review on efficacy of *Azadirachta indica* A. Juss based biopesticides: An Indian perspective. Res J Recent Sci 1(3):94–99

Mukhopadhyay MJ, Mukherjee A (1998) Inhibition of clastogenic effect of cyclophosphamide and mitomycin C by Neem leaf-extract in mice. Phytother Res 12:409–412

National Academy of Sciences (1980) Firewood crops; shrubs and tree species for energy production (Vol. I & II). National Academy Press, Washington, DC, p 237

Natural vegetation assessment report (2008) Indus for All Programme, WWF – Pakistan

Ogbuewu IP, Odoemenam VU, Obikaonu HO et al (2011) The growing importance of Neem (*Azadirachta indica* A. Juss) in Agriculture, Industry, medicine and environment. Rev Res J Med Plants 5:230–245

Oh K, Cao T, Li T et al (2014) Study on application of phytoremediation technology in management and remediation of contaminated soils. J Clean Energ Technol 2(3):216–220

Olguín EJ, Galván GS (2010) Aquatic phytoremediation: novel insights in tropical and subtropical regions. Pure Appl Chem 82(1):27–38

Orwa C, Mutua A, Kindta R et al. (2009) Agroforestry database: a tree reference and selection. http://www.worldagroforestry.org/af/treedb/

Parotta JA (2001) Healing plants of Peninsular India. CABI Publishing, New York, pp 495–496

Pierzynski GM, Schnoor JL, Banks MK et al (1994) Vegetative remediation at superfund sites. In: Hester RE, Harrison RM (eds) Issues in environmental science and technology, Royal Society of Chemistry, vol 1. Cambridge, England, pp 49–69

Prakash D, Gabani P, Chandel AK et al (2013) Bioremediation: a genuine technology to remediate radionuclides from the environment. Microb Biotechnol 6(4):349–360

Prasad AL, Thirumalisamy S (2013) Evaluation of the use of *Acacia nilotica* leaf as an ecofriendly adsorbent for Cr (VI) and its suitability in real waste water: study of residual errors. J Chem:1–7

Puri S, Singh S, Kumar A (1994) Growth and productivity of crops in association with an *Acacia nilotica* tree belt. J Arid Environ 27:37–48

Qadir M, Noble AD, Schubert S et al (2006) Sodicityinduced land degradation and its sustainable management: problems and prospects. Land Degrad. Dev 17:661–676

Rajjak SP, Bapurao BA (2013) Biosorptive behavior of Neem (*Azadirachta indica*) bark for iron removal from aqueous solution. J Appl Phytotechnol Environ Sanit 2(1):31–36

Rao CM, Sudhakar G (2013) Removal of heavy metal radionulcide contaminants by plants from Uranium mill tailings. Int J Adv Eng Glob Technol 1(5):288–302

Reeves RD, Baker AJM (2000) Metal- Accumulating Plants. In: Raskin I, Ensley BD (eds) Phytoremediation of toxic metals: using plants to clean-up the environment. Wiley, New York, pp 193–230

Ross IA (2001) Medicinal plants of the world: chemical constituents, In: Traditional and modern medicinal uses, vol 2.Totowa, New Jersy, pp 81–85

Sandaa RA, Enger O, Torsvik V (1999) Abundance and diversity of archae in heavy-metal-contaminated soils. Appl Environ Microb 65:3293–3297

Saraswat PK, Chaudhary K (2014) Effect of Fly Ash (FA) improving soil quality and increase the efficiency of crop productivity. Eur J Biotechnol Biosci 2(6):72–78

Sas-Nowosielska A, Kucharski R, Malkowski E, Pogrzeba M, Kuperberg JM, Krynski K (2004) Phytoextraction crop disposal – an unsolved problem. Environ Pollut 128:373–379

Sateesh MK (1998) Microbiological investigations on die-back disease of Neem (*Azadirachta Indica* A. Juss). Ph.D. thesis, University of Mysore, Mysore, India

Schnoor JL, Licht LA, McCutcheon SC et al (1995) Phytoremediation of organic and nutrient contaminants. J Environ Sci Technol 29:318A–323A

Scott JA, Karanjkar AM, Rowe DL (1995) Biofilm covered granular activated carbon for decontamination of streams containing heavy metals and organic chemicals. Miner Eng 8(1):221–230

Setia RC, Kuar N, Setia N, Nayyar H (2008) Heavy metal toxicity in plants and phytoremediation. Crop improvement: strategies and applications. New Delhi, I.K. International Publishing House Pvt. Ltd., pp 206–218

Sharma SK (1988) Recent advances in afforestation of salt affected soils in India. Adv For Res 2:17–31

Sharma A, Bhattacharyya KG (2005) *Azadirachta indica* (Neem) leaf powder as a biosorbent for removal of Cd(II) from aqueous medium. J Hazard Mater 125:102–112

Sharma J, Sharma A, Bhattacharya KG (2008) Biosorption of commercial dyes on *Azadirachta indica* leaf powder: a case study with a basic dye rhodamine. Ind Eng Chem Res 47(15):5433–5440

Sharma P, Tomar L, Bachwani M et al (2011) Review on Neem (*Azadirechta indica*): thousand problems one solution. Int Res J Pharm 2(12):97–102

Sidhu DS (1995) Neem in agro forestry as a source of plant derived chemicals for pest management. Indian For 121:1012–1021

Sindhuveerendra HC (1995) Variation studies in provenances of *Azadirachta indica* (The Neem tree). Indian For12:1053-1056

Singh P (1991) Management strategies for wasteland development. In: Singh RK, Saxena AK, Singh IS (eds) Agroforestry— present status and scope for future development in farming systems. Narendra Dev University of Agriculture and Technology, Kumarganj, Faizabad, pp 133–142

Singh B (1998) Biomass potentials of some firewood shrubs of North India: short communication. Biomass 16:199–203

Singh OV, Labana S, Pandey G, Budhiraja R (2003) Phytoremediation: an overview of metallic ion decontamination from soil. Appl Microbiol Biotechnol 61:405–412

Singh YP, Sharma DK, Singh G et al. (2008) Alternate land use management for sodic soils. Tech Bull No 2/2008. Central Soil Salinity Research Institute, Regional Research Station, Lucknow-226005 (INDIA), p 16

Singh B, Tripathi KP, Singh K (2011) Community structure, diversity, biomass and net production in a rehabilitated subtropical forest in North India. Open J For 1(2):11–26

Singh NR, Jhariya MK, Raj A (2013) Tree crop interaction in agroforestry system. Readers Shelf 10(3):15–16

Sissay B (1986) Salt wasteland in ethiopia: potential for production of forage and fuel. Recl Reveget Res 5:59–64

Spicer N, Barnes R, Timberlake J (2007) Acacia handbook. DFID Forestry Research Programme, London
Srivastava R, Rupainwar DC (2010) A comparative evaluation for the adsorption of dye on Neem Bark and Mango Bark powder. Indian J Chem Technol 18:67–75
Suwalka RL, Qureshi FM (1995) Amelioration of some sodic soils in arid regions of Rajasthan. J Indian Soc of Soil Sci 4:660–662
Tandon P, Sirohi A (2010) Assessment of larvicidal properties of aqueous extracts of four plants against *Culex quinquefasciatus* Larvae. Jordan J Biol Sci 3(1):1–6
Tangahu VB, Abdullah SRS, Basri H et al (2011) A review on heavy metals (As Pb and Hg) uptake by plants through phytoremediation. Int J Chem Eng:1–31
Tawde SP, Bhalerao SA (2010) Biosorption of Chromium (VI) from an aqueous solution using *Azadirachta indica*. A. Juss. (Neem) and activated charcoal: a comparative study. Biol Forum — Int J 2(2):4–10
Thenmozhi R, Santhi T (2015) Characterization of activated *Acacia nilotica* seed pods for adsorption of Nickel from aqueous solution. IntJ Environ Sci Technol 12:1677–1686
Tiwari D, Mishra SP, Mishra P et al (1999) Biosorptive behaviour of Mango (*Mangifera indica*) and Neem (*Azadirachta indica*) bark for Hg2+, Cr3+ and Cd2+ toxic ions from aqueous solutions: a radiotracer study. Appl Radiat Isot 50:631–642
Totey NG, Kulkarni R, Showmik AK et al (1987) Afforestation of salt affected wasteland I—screening of forest tree species of Madhya Pradesh for salt tolerance. Indian For 113:805–815
Toth B (1981) Afforestation on salt affected soils in Hungary. Agro- kemia Es Talajtan Tom 30:205–212
Townsley CC, Ross IS, Atkins AS (1986) In: Lawrence RW, Branion RMR, Enner HG (eds) Fundamental and applied biohydrometallurgy. Elsevier, Amsterdam, pp 279–289
Trapp S, Karlson U (2001) Aspects of phytoremediation of organic pollutants. J Soil Sediments 1:1–7
Tybirk K (1989) Flowering, pollination and seed production of *Acacia nilotica*. Nordic J Bot 9(4):375–381
Varier MK (2009) Nuclear radiation detection: measurements and analysis. Narosa Publishing House, New Delhi
Vasavi A, Usha R, Swamy PM (2010) Phytoremediation – an overview review. JIndust Pollut Cont 26:83–88
Vijayaraghavan K, Yun YS (2008) Bacterial biosorbents and biosorption. Biotechnol Adv 26:266–291
Vinod V, Tiwari PK, Meshram GP (2011) Evaluation of mutagenic and antimutagenic activities of Neem (*Azadirachta indica*) seed oil in the invitro Ames Salmonella/Microsome assay and invivo mouse bone marrow micronucleus test. J Ethno pharmco 12:931–937
Vishwanatha GLS, Vaidya SK, Ramesh C et al (2010) Antioxidant and antimutagenic activities of bark extract of *Terminalia arjuna*. Asian Pacific J trop med 3(12):965–970
Volesky B, Holan ZR (1995) Biosorption of heavy metals. Biotechnol Prog 11:235–250
Waseem S, Din MI, Nasir S, Rasool A (2014) Evaluation of *Acacia nilotica* as a non- conventional low cost biosorbent for the elimination of Pb (II) and Cd (II) ions from aqueous solution. Arab J Chem 7(6):1091–1098
World Nuclear Association (2014) http://www.world-nuclear.org/info/Safety-and-Security/Radiation-and-Health/Naturally-Occurring-Radioactive-Materials-NORM/
Willey N (2007) Soils contaminated with radionuclides. Some insights for phytoextraction of inorganic contaminants. In: Willey N (ed) Phytoremediation Methods and Reviews. Humana Press Inc, Totowa, pp 305–317
Yadav JSP, Singh K (1970) Tolerance of certain forest species to varying degree of salinity and alkalinity. Indian For 96:587–599
Zhu YG, Shaw G (2000) Soil contamination with radionuclides and potential remediation. Chemosphere 41:121–128

Efficiency of an Industrially Important Crop *Hibiscus cannabinus* for Phytoremediation and Bioenergy Production

9

Neha Vishnoi and D.P. Singh

Abstract

The present chapter discusses the geographical distribution of *Hibiscus cannabinus* and its versat ile applications. It is an annual, short-day, low input requirement, high biomass-yielding non-food crop which is cultivated since 4000 BC in Africa. It belongs to family Malvaceae and section Furcaria. It is an important fibre crop which has numerous industrial applications. It is mainly cultivated for bast and core fibres which are used in making of ropes, twine, composites, bedding materials, absorbents, highly efficient paper, fabrics and building materials. It is also considered as a potent candidate for phytoremediation of heavy metals and oils. Seed oil can be used for treatment of various health disorders like cholesterol level, blood pressure, etc. Numerous studies have been conducted on *H. cannabinus* (kenaf) which proved that it is also a green alternative for the production of eco-friendly and reliable bioenergy. It also sequesters carbon dioxide which is considered as a greenhouse gas and thus contributes in reducing the consequences of global warming.

Keywords

Bioenergy • Biomass • Heavy metals • *Hibiscus cannabinus* • Phytoremediation

N. Vishnoi • D.P. Singh (✉)
Department of Environmental Sciences, Babasaheb Bhimrao Ambedkar University, Lucknow 226025, India
e-mail: dpsingh_lko@yahoo.com

K. Bauddh et al. (eds.), *Phytoremediation Potential of Bioenergy Plants*,
DOI 10.1007/978-981-10-3084-0_9

9.1 Introduction

Kenaf (*Hibiscus cannabinus*) is an annual, spring, herbaceous, dicotyledonous non-food crop belonging to family Malvaceae section Furcaria (Dempsey 1975; Saba et al. 2015). It is found in tropical and temperate climates and distributed between 45°N and 30°S (Muchow and Wood 1983; Duke 1983). It is a fibre, non-food crop originated in Africa about 6000 years ago (Dempsey 1975). Its cultivation for the production of fibre came into light from the beginning of the twentieth century in countries like India, Sri Lanka, Thailand, Malaysia and Java (FAO 2003). Kenaf has wide industrial applications as its bast, and core fibres of stem are used as a raw material in the production of ropes, twines, composites, absorbents, paper, building materials, etc. (Hamid 2008; Lips et al. 2009). Workers like Le Roux (2007); Zwane and Masarirambi (2009) suggested that the natural, low-cost, high-quality and low-density fibres of Kenaf have been used as an eco-friendly alternative to plastics and thus contributes to environmental sustainability. The industrial product such as paper made from kenaf can be recycled and thus provides an alternative to overcome the economic, social and environmental challenges of the coming future. It also acts as a suitable phytoremediator for removal of heavy metals like copper, chromium, cadmium, lead, zinc, etc. and hydrocarbons from the environment as it produces high biomass (Arthur et al. 2003; Nabulo et al. 2011; Wong et al. 2013). Many medicinal properties are exhibited by the oil extracted from the seed of this plant (Agbor et al. 2005; James et al. 2013). The oil contains antioxidants and polyunsaturated fatty acids which cure many health disorders like cholesterol level, blood pressure, etc. Nutritional value of oil is high due to presence of lipid-soluble bioactives and linoleic, oleic and palmitic acid (Coetzee et al. 2008). This plant also sequesters CO_2 gas and thus reduces the level of this greenhouse gas in the environment which ultimately contributes to global warming which is considered as a global environmental issue (Mazumder et al. 2005; Bakhtiari et al. 2011; Hossain et al. 2016). Energy crisis is one of the other major problems faced by the population due to continuous consumption of non-renewable energy resources (Langan et al. 2011). Kenaf biomass has been proved to be a cost-effective and renewable resource for the production of bioenergy (Conde-Mejía et al. 2012; Azadi et al. 2013; Muhammad et al. 2013; Kojima et al. 2014; Saba et al. 2015). Heat and electricity are being produced by the solid fuel, solid residue produced by the pretreatment and hydrolysis of kenaf biomass (Demirbas 2004). Kenaf has proven to be an economically important plant as it produces eco-friendly biofuels which are important for sustainable development of ecosystem. This chapter lays emphasis on the geographical distribution and economical importance of kenaf.

9.2 Geographical Distribution of *Hibiscus cannabinus*

Hibiscus cannabinus is commonly known with several names like kenaf, Bimli jute, teel, mesta, palungi, stokroos, Java jute, dah, gambo, rama, papoula de São Francisco, Deccan hemp, vegetable kenaf, Guinea hemp, etc. More than 400 species of *Hibiscus* have been reported. Kenaf resembles closely with cotton, okra and hollyhock. Lateral root system of this plant makes it drought tolerant. Kenaf stem consists of a core region and the outer bark, both containing fibres. The main components of fibres are alpha cellulose (58–63%), hemicelluloses (21–24%) and lignin (12–14%). Under appropriate conditions, kenaf can grow up to 4–5.6 m height with stem diameter 24–52 mm in about 4–6 months. Kenaf is a fibre plant originated in Africa about 6000 years ago (Kaldor et al. 1990; LeMahieu et al. 2003; Cheng et al. 2004). It is found in tropical and temperate climates and distributed between 45°N and 30°S (Duke 1983; Muchow and Wood 1983). It is found mainly in the savanna region of West and Central Africa. Dempsey (1975) reported the existence of kenaf in Southern Asia in the beginning of the eighteenth century and was cultivated in India for the first time. From India, it was then disseminated in other Asian countries. Literature about cultural relationship between the Indus Valley and Egypt has given the evidence of kenaf distribution from Africa to other Asian countries. Kenaf got expanded in mainland China from Taiwan in the beginning of 1900 (Charles 2002; Li 2002). Russia started to produce kenaf in the year 1902, and its cultivation at commercial level started in USSR and Asia in the 1930s. During the Second World War, it was introduced in the USA. Its cultivation for the production of fibre came into light from the beginning of the twentieth century in countries like India, Sri Lanka, Thailand, Malaysia and Java. It is cultivated as a commercial fibre crop in India, Indo-China, Togo, Benin, Niger, Kenya, Tanzania, Malawi, Taiwan, Indonesia and North and South America. Kenaf is being cultivated in more than 20 countries as evident from the reports of FAO (2003). Countries like India, China and Thailand are the major kenaf producers at commercial level.

Hibiscus cannabinus (kenaf) (Source: Google Images)

9.3 Industrial Importance of *H. cannabinus*

9.3.1 Medicinal Importance

Various parts of *H. cannabinus*, viz. stem, leaves and seed, exhibit medicinal properties. The presence of carotenoids, flavonoids, alkaloids, vitamins and phenolics in kenaf confers it with numerous antioxidative properties (Thaiponga et al. 2006; Makari et al. 2008; Mukherjee et al. 2010). It has been reported from the study of Mishra et al. (2007) that intake of these antioxidants recovers disorders in lipid metabolism, prevents erythrocyte haemolysis, improves dyslipidaemia and also prevents the body from various diseases like cataract, cancer, nerve disorders, high blood pressure and cholesterol level, arthritis, cold and cough (Anila and Vijayalakshmi 2002). The oil extract of this plant also shows anti-inflammatory and antibacterial effect (Russo et al. 2001; Pawar et al. 2008). The seeds are used as a food additive due to the presence of protein, fibre and calcium (Webber and Bledsoe 1991; Mosquera et al. 2007; Alexopoulou et al. 2015). Various workers reported that kenaf stem parts serve as therapy for anaemia, fever, bruises, fatigue puerperium, aperitif, stomachic, etc. and as antitoxic agent against numerous toxicants like acids, alkali and pesticides (Mohamed et al. 1995; Russo et al. 2001; Coetzee et al. 2008; Nyam et al. 2009). An investigation was carried out by Agbor et al. (2005) and James et al. (2013) in which they found that the plant leaf extract contains some natural products which act as antioxidants.

Oil is an important dietary component as it reflects many nutritional benefits. Oil is mainly produced as a by-product from seeds of the plant such as legumes, cotton, kenaf, etc. *Hibiscus cannabinus* produces high yields of edible seed oil as cottonseed oil. Seeds also contain fatty acids, sterols and phospholipids. Presence of high amount of monounsaturated and polyunsaturated fatty acids in kenaf seed oil contributes to many health benefits such as lowering of cholesterol level in the blood and preventing coronary and heart diseases (Yusri et al. 2012). Tocopherols, sterols and phenolics present in seed oil behave as antioxidants and thus play a significant role in defence mechanism.

Some acids like caffeic, feluric and vanillic behave as antioxidants and thus alter the taste of the oil. Different types of phenols and tocopherols are important biological antioxidants present in seed oil of kenaf. Phytosterols lower the cholesterol level in the blood. Tocopherol reduces the chances of occurrence of Alzheimer's disease, cancer and heart diseases. They have wide applications in food, pharmaceuticals and cosmetics. Alpha-tocopherols/vitamin E prevents oxidation of fatty acids and lipid components of cells. Nutritional value of oil is high due to presence of lipid-soluble bioactives and linoleic, oleic and palmitic acid. A study was conducted by Coetzee et al. (2008) in which it was found that among the tested eight cultivars of kenaf (El Salvador, Endora, Dowling, Gregg, Cuba, SF 459, Tainung, Everglades 41, Endora), mainly linoleic, oleic and palmitic acid were the dominant fatty acids. Phospholipid content in oil extracted from seed of kenaf is high in comparison to soya bean and cottonseed oil. Phospholipid and oil quality varies among different genotypes, and the yield could be increased by genetic improvement (Mohamed et al. 1995). Thus, kenaf seeds serve as a by-product for the production of oil which is an important component of food (Coetzee et al. 2008).

9.3.2 *H. cannabinus* as a Fibre Crop

Kenaf is a fast-growing, industrially important biomass crop and thus used globally due to its applications. It is mainly cultivated for fibres in countries like India, Bangladesh, the USA, Indonesia, Malaysia, South Africa, Vietnam and Thailand and Europe. One of the major uses of this crop is that it contains high fibre or lignocellulosic material and thus used as a good source of fibre for paper and textile industries (Neto et al. 1996; Ardente et al. 2008). Kenaf pulping is accepted worldwide due to consumption of less amount of energy, minimal uses of chemicals and higher rate of production over wood pulping (Nelson et al. 1962). Since 1962, the USA was using writing and printing paper made from kenaf fibres. The industrial product such as paper made from kenaf can be recycled. This provides an alternative to overcome the economic, social and environmental challenges of the coming future. The stem of this plant yields two kinds of fibres: the core and bast fibres. The former is derived from the core of the stem and latter from the bark of stem. The bast fibres are more efficient and longer than core fibres (Marisol et al. 2013). Various industrial applications have been observed by core fibres as they can be used as a good absorbent agent. This fibre-yielding plant is also used as a bioremediator in the remediation of oil contamination, wastewater treatment and soil contamination (Benson and George 2005; Borazjani and Diehl 2010; Hernández et al. 2012). Results from the findings of Neto et al. (1996) suggested that the core of kenaf contains higher amount of lignin (14.5–16.1%) than bark (9.4–10.1%) but less than as compared to wood. The fibres can be used in the production of various fabrics, adsorbents, nanofibres, ropes, twine, canvas, cordage and building materials such as particle board, wallpaper backing, low-density panels, synthetic fibres for thermal resistance and sound absorption, automotive components and bedding material for poultry (Kugler 1988; Sellers et al. 1993; Kaldor et al. 1990; Kulger 1996; Hamid 2008; Lips et al. 2009). Le Roux (2007) suggested that the natural, low-cost, high-quality and low-density fibres of kenaf have been used as an eco-friendly alternative to plastics and thus contribute to environmental sustainability.

9.3.3 Efficacy of *H. cannabinus* in Carbon Sequestration

Global warming is an environmental consequence caused by the emission of greenhouse gases in the atmosphere due to rapid industrialization and some natural activities. It is an important global problem which has received more attention in the present scenario because it alters the composition of ecosystem. Carbon dioxide is one of the major greenhouse gases and is present in amount more than 60% in the atmosphere (Zaini and Kamarudin 2014). Kenaf can sequester carbon dioxide from the atmosphere and thus plays a crucial role in reducing the carbon dioxide concentration from the atmosphere through the process of photosynthesis (Hossain et al. 2016). Kenaf is a C3 plant and accumulates carbon dioxide more than a tree (Mazumder et al. 2005; Bakhtiari et al. 2011). Reports suggested that one acre of kenaf can sequester about 10–20 tons of carbon dioxide from the air (Cosentino and Copani 2003).

9.4 *H. cannabinus* as an Important Phytoremediator

Soil and water pollution is increasing at an alarming rate due to various natural and anthropogenic activities. Toxic contaminants enter the soil, water and air mainly through many industrial processes like burning of coal, discharge of untreated industrial effluents, mining process, etc. These contaminants which are mainly inorganic or organic in nature enter the food chain and through biomagnification and cause toxicity in human beings. Various physical and chemical methods have been employed to decontaminate the environment but have not become successful over biological methods (Cunningham et al. 1995; Kumar et al. 1995; Barcelo and Poschenrieder 2003). These physical and chemical methods are not cost-effective and disturb the ecology of soil. So there is an urgent need to use some cheaper alternatives to clean the ecosystem. Various biological methods have been employed to remediate the problem of environmental pollution. In biological methods, biological agents like plants, bacteria, fungi, algae, etc. are used to minimize the toxicity caused by different contaminants. Industrial discharge mainly contains heavy metals like cadmium, chromium, copper, lead, zinc, nickel, mercury, etc. and some organic compounds like pesticides, dye, oil, etc. Some of the metals like zinc, copper, nickel, iron, etc. did not show toxicity symptoms in plants at their low concentration. They generally appeared to be toxic when usually present in high concentration. Heavy metals interfere with the metabolism process and thus disturb the normal functioning of biological systems. They also cause alterations in the natural biota of the ecosystem.

Several studies have been conducted which showed the removal of inorganic and organic pollutants from soil and water by the use of green plants through extraction, sequestration and detoxification (Raskin et al. 1997; Meagher 2000; Arthur et al. 2003; Duggan 2005; Jankaite and Vasarevisius 2007). Since heavy metals are nondegradable, they can be remediated by using plants through stabilization or extraction in harvestable plant parts (Ho et al. 2008; Meera and Agamuthu 2011).

Phytoremediation is the process by which plants remove the pollutants from environment via different mechanisms like phytoextraction, phytostabilization, phytovolatilization, rhizofiltration and phytodegradation (Burken and Schnoor 1997; Raskin et al. 1997; Maestri et al. 2010; Wu et al. 2010). Accumulation and distribution of metals in plant organs are the two important factors that determine the potential of the plant to remediate contaminated sites. Phytoremediation has gained more attention in the current scenario due to several reasons such as:

1. Minimal site destruction thus maintaining the aesthetic value of the site
2. Controls soil erosion and causes carbon sequestration
3. Cost-effective process
4. Production of recyclable metal-rich residue from plant

Plants producing high biomass are considered good remediators of heavy metals. One of the best examples of plants producing high biomass is *Hibiscus cannabinus* which accumulates metals in their roots, shoot and leaves from soil and water. Several studies have been conducted that proved *Hibiscus cannabinus* to be a suitable candidate for removal of chromium, zinc, cadmium, lead, arsenic, iron, oil, etc. (Ho et al. 2008). Arsenic, a carcinogen in soil and water, is geogenic in origin. It also occurs in the environment through mining operations, use of pesticides, wood preservatives, etc. and causes many abnormalities in animals and human beings (Dudka and Miller 1999). Iron, a non-biodegradable metal, enters the water mainly through discharge of untreated wastewater and causes toxicity to aquatic fauna (Wittbrodt and Palmer 1995). Kenaf roots have proved to be a good bioavailable sink for heavy metals like arsenic and iron from soil contaminated with landfill leachate. These results were confirmed by the studies of Meera and Agamuthu (2011)which showed that kenaf can sequester 0.06–0.58 mg arsenic and 66.82–461.71 mg iron per gram plant dry weight from landfill leachates. Results from the findings of Schmoger et al. (2000), Ho et al. (2008) and Meera and Agamuthu (2011) suggested that translocation factor < 1 in case of kenaf which indicates that fibrous roots translocate the major amount of toxic metal by phytoextraction. Thus, it minimizes the metal mobilization to aerial parts of the plants.

Kenaf roots showed high metal uptake ability and could be used for phytoremediation of lead-, chromium-, cadmium-, zinc-, copper-, nickel- and zinc-contaminated sites (Ho et al. 2008; Arbaoui et al. 2013). Maximum absorption of lead was observed up to 20 mg/plant in roots of kenaf, and aerial parts showed only a maximum of 1 mg Pb plant accumulation as evident from the studies of Ho et al. (2008).

Since cadmium is also a toxicant, its removal from the soil and water is necessary by some cost-effective process. Zinc is considered to be an essential metal required by the plants at low levels, but it causes toxicity at high levels. Some plants like brassica, corn, kenaf, etc. have proven to be good remediators of cadmium and zinc. Kenaf also accumulates high concentration of metals like cadmium and zinc in their stem (Cartoga et al. 2005; Nabulo et al. 2011). An investigation was carried out by Arbaoui et al. (2013) in which they concluded that kenaf shoots accumulate 2.49 mg/kg zinc and 82.5 mg/kg cadmium, respectively, from dredging sludge. Since kenaf cultivation requires least field management practices, it can also be grown in sites which are less fertile and contaminated with trace metals. Accumulation potential of toxicant also varied among different varieties of same cultivars (Santos et al. 2010; Wuana and Okieimen 2010). This was demonstrated by Cartoga et al. (2005) who reported that Tainung 2 variety of *H. cannabinus* accumulated high levels of cadmium in shoots among the tested five varieties of the same cultivar when cultivated in site contaminated with trace metals. Nabulo et al. (2011) tested metal accumulation capacity of 20 tropical species when cultivated in metal-polluted soil with respect to five trace metals and concluded that kenaf ranked second in accumulating cadmium and zinc next to *Gynandropsis gynandra*. Phytoaccumulation of metals also depends on several physical and chemical factors of the soil like amount of metal present in soil, pH, soil composition, etc. This was confirmed by the studies of Arthur et al. (2003) and Guo et al. (2011) who reported that higher amount of cadmium was absorbed and phytoextracted by kenaf when grown in soil containing high levels of cadmium. Bioavailability index determines the absorption potential of plants (Rotkittikhun et al. 2006). Bada and Raji (2010) reported maximum cadmium absorption of 2.84 mg/kg soil at 6 mg/kg cadmium concentration in soil, while the absorption decreases further when the concentration of cadmium in soil decreases.

The core fibres of kenaf in powdered form have proved to be a better absorbent of waxes. These results were confirmed by the experiments of Wong et al. (2013) who reported that kenaf core fibres absorb waxes released from textile industry wastewater. This plant also showed removal of hydrocarbons from the amended soil (Abioye et al. 2010).

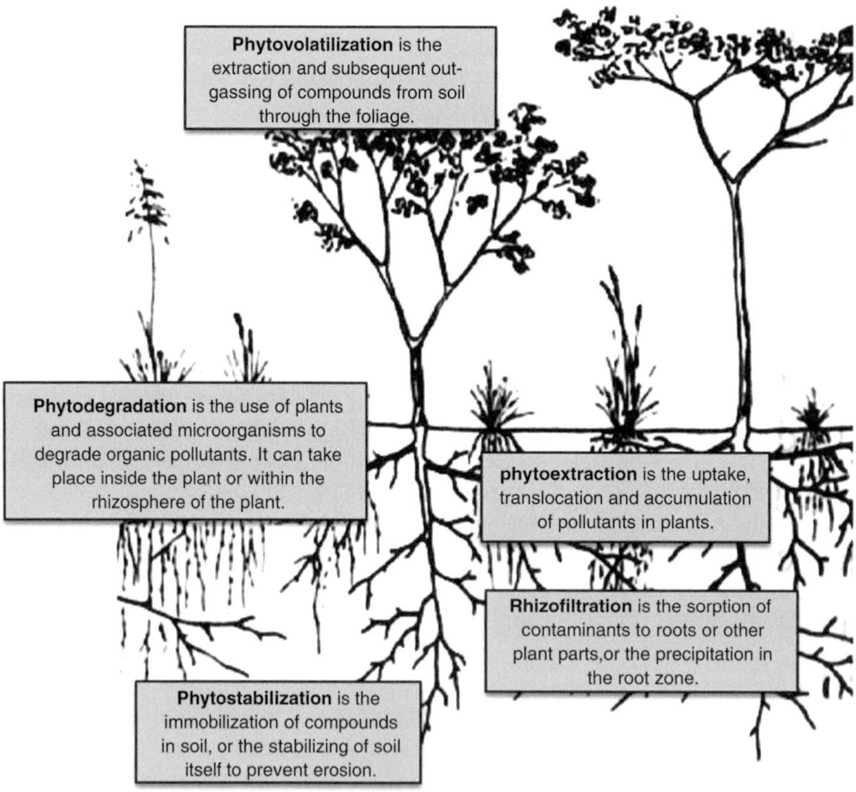

Mehanism involved in Phytoremediation. Source: Google

Table 1.1 Metal accumulation by kenaf from contaminated soil

Metals	Soil (mg/kg)	Root (mg/kg)	Stem (mg/kg)	Leaves (mg/kg)	References
Cd	3.54	2.67 ± 0.41	0.37 ± 0.07	1.95 ± 0.32	Bettaieb et al. (2013)
Zn	1157	65.33 ± 4.45	53.90 ± 1.74	60.98 ± 1.55	
Cd	3.54	–	1.50 ± 0.14	0.99 ± 0.29	Arbaoui et al. (2013)
Zn	1157	–	59.09 ± 7.12	23.41 ± 5.14	
Cd	6.0	2.78 ± 0.51			Bada and Raji (2010)
As	12.31 ± 2.65	7.13 ± 0.78	0.42 ± 0.09	1.31 ± 0.22	Meera and Agamuthu (2011)
Fe	2230.78 ± 24.21	13709.61 ± 153.21	511.58 ± 25.32	920.43 ± 16.54	
Pb	5.0	1.50	1.21	1.17	Bada (2015)
Cr	5.5	0.15	0.13	0.12	
Zn	27.5	3.18	13.10	23.48	
Cd	0.5	0.14	0.12	0.13	
Zn	4500 μM	285.15 ± 20.23	162.9 ± 35.56	272.7 ± 25.84	Arbaoui et al. (2014)
Cd	50 μM	40.65 ± 2.56	14.09 ± 1.36	35.34 ± 1.88	

9.5 Bioenergy Production Efficacy of *H. cannabinus*

Overexploitation of the population causes an increase in energy requirements. Conventional fossil fuels are used to produce energy in order to meet the energy demands of current population. Burning of fossil fuels causes emission of greenhouse gases which ultimately give rise to an alarming environmental problem, i.e. 'global warming' (Langan et al. 2011). It causes a disturbance in ecosystem. Fossil fuels are not considered as sustainable and reliable source for the production of energy due to continuous emission of greenhouse gases (Muhammad et al. 2013). So, alternative eco-friendly and cost-effective bioresources have been developed by the researchers for the energy production (Cetin et al. 2005; Umeki et al. 2012).

Biomass of certain plants like soya bean, rice, jatropha, ricinus, etc. are used to produce biofuels which are considered as an alternative green source of energy. Biofuels like biogas, bioethanol, biohydrogen and biodiesel can be extracted from numerous plant parts like stem, grain, seed, etc. (Bhutto et al. 2011). Lignocellulosic biomass is considered as the potent resource for bioenergy production with an annual growth of 170–200 billion tons per hectare (Gonzalez-Garcia et al. 2010).

Kenaf is considered as one of the best candidate for the annual source of bioenergy production due to presence of lignocellulosic biomass. The biomass is used as the feedstock to produce biofuels like biodiesel, biogas, biohydrogen and bioethanol. Due to its high yield and more ecological adaptations, it is considered as an alternative source of energy. Current research unfolds the maximum utilization of this plant for the production of biofuels. A pilot study was conducted, and it was estimated that 1 kg of kenaf can produce 18,000 kJ of energy (Saba et al. 2015).

Kenaf core undergoes carbonization and formed wood gas which can be used as a fuel. After the carbonization process of kenaf core, it gets converted into char that further undergoes gasification and forms water gas (Kojima et al. 2014). Production of bioethanol from lignocellulosic material involves several steps like pretreatment, enzymatic hydrolysis, separation and concentrating (Conde-Mejía et al. 2012). Kenaf biomass undergoes gasification and forms syngas which finally produces electricity, heat and methanol (Azadi et al. 2013). Anaerobic digestion of kenaf biomass converts biohydrogen, biogas into biofertilizer and bioelectricity (Muhammad et al. 2013). Kenaf oil undergoes transesterification and produces biodiesel (Rathana et al. 2010). Heat and electricity are being produced by the solid fuel, solid residue produced by the pretreatment and hydrolysis of kenaf biomass (Demirbas 2004). It has been estimated that 1 kg of lignin with steam yields 62 mol hydrogen and 53 mol CO (Azadi et al. 2013). Hence, based on the above reports, it can be concluded that kenaf is an economically important plant as it produces highly efficient and eco-friendly biofuels which are necessary for sustainable development of ecosystem.

9.6 Conclusion

It can be concluded that *H. cannabinus* is an economically and industrially important non-food crop. It is considered as an alternative, eco-friendly, sustainable and cost-effective green feedstock for the production of numerous commercial-based products like fibres and bioenergy. Kenaf is utilized in the paper and textile industry and production of building materials, composites, bedding materials, etc. It is also used in the remediation of toxic metals and oil spills from the environment and thus acts as a potent phytoremediator. It sequesters carbon dioxide gas which is considered as a major greenhouse gas and thus reduces global warming. Kenaf oil from seeds possesses nutritional and medicinal value due to the presence of antioxidants, sterols, polyunsaturated fatty acids, etc. and helps in the treatment of various disorders of the body. Many economic and environmental benefits are shown by this plant as it produces recycled and efficient quality of paper, reduces use of synthetic fibres, reduces soil erosion and consumes less chemicals and energy in paper production and thus plays a crucial role in the sustainable development of ecosystem.

References

Abioye OP, Abdul Aziz A, Agamuthu P (2010) Enhanced biodegradation of used engine oil in soil amended with organic wastes. Water Air Soil Pollut 209:173–179

Agbor GA, Oben JE, Ngogang JY (2005) Antioxidant activity of *Hibiscus cannabinus* leaf extract. www.foodafrica.nri.org/nutrient/nutrition proceeding, August 2009

Alexopouloua E, Lib D, Papatheoharic Y, Siqib H, Scordiad D, Testa G (2015) How kenaf (*Hibiscus cannabinus L.*) can achieve high yields in Europe and China. Ind Crop Prod 68:131–140

Anila L, Vijayalakshmi NR (2002) Flavonoids from Emblica officinalis and Mangifera indica-effectiveness for dyslipidemia. J Ethnopharmacol 79:81–87

Arbaoui S, Evlard A, El Wafi MM, Campanella B, Roger P, Bettaieb T (2013) Potential of kenaf (*Hibiscus cannabinus* L.) and corn (*Zea mays* L.) for phytoremediation of dredging sludge contaminated by trace metals. Biodegradation J24:563–567

Arbaoui S, Campanella B, Rezgui S, Paul R, Bettaieb T (2014) Bioaccumulation and photosynthetic activity response of Kenaf (*Hibiscus cannabinus L.*) to cadmium and zinc. Greener J Agri Sci 4:091–100

Ardente F, Beccali M, Cellura M, Mistretta M (2008) Building energy performance: ALCA case study of kenaf-fibers insulation board. Energ Build 40:1–10

Arthur LS, Kelly LH, Jae Min L, David JB (2003) Phytoremediation of arsenic and lead in contaminated soil using Chinese Brake Ferns and Indian Mustard. Int J Phytoremed 5:89–103

Azadi P, Inderwildi OR, Farnood R, King DA (2013) Liquid fuels, hydrogen and chemicals from lignin: a critical review. Renew Sust Energ Rev 21:506–523

Bada BS (2015) Bioremediation of textile effluent polluted soil using kenaf *(Hibiscus cannabinus Linn.*) and composted market waste. J Appl Sci Environ Manag 19:773–776

Bada SB, Raji KA (2010) Phytoremediation potential of kenaf (*Hibiscus cannabinus* L.) grown in different soil textures and cadmium concentrations. Afr J Environ Sci Technol 4:250–255

Bakhtiari MR, Ahmad D, Othman J, Ismail N (2011) Physical and mechanical properties of kenaf-seed. Trans ASABE 27:263–268

Barcelo J, Poschenrieder C (2003) Phytoremediation: principles and perspectives. Contrib Sci 2:333–344

Benson TJ, George CE (2005) Cellulose based adsorbent materials for the dehydration of ethanol using thermal swing adsorption. Adsorption 11:697–670

Bettaieb T, Arbaoui S, Campanella B, Paul R (2013) Potential of Kenaf (Hibiscus cannabinus L.) for phytoremediation of land irrigated by waste water. In: Proceedings of the Ramiran 2013. 15th International Conference, Versailles, France, June. S2.27 ref.12

Bhutto AW, Bazmi AA, Zahedi G (2011) Greener energy: issues and challenges for Pakistan—biomass energy prospective. Renew Sust Energ Rev 15:3207–3219

Borazjani A, Diehl S (2010) Kenaf core as an enhancer of bioremediation. http://journeytoforever.org/biofuel.html
Burken JG, Schnoor JL (1997) Uptake and metabolism o fatrazine by poplar trees. Environ Sci Technol 31:1399–1406
Cartoga A, Fernando A, Oliveira JS (2005) Effects on growth, productivity and biomass quality of kenaf of soils contaminated with heavy metals. Sci Ref Cosmos 4:1–4
Cetin E, Gupta B, Moghtaderi B (2005) Effect of pyrolysis and heating rate on radiate pine char structure and apparent gasification reactivity. Fuel 84:1328–1334
Charles I (2002) Trends in new crops and new use. ASHS Press, Alexandria
Cheng Z, Lu BR, Sameshima K, Fu DX, Chen JK (2004) Identification and genetic relationships of kenaf (*Hibiscus cannabinus L.*) germplasm revealed by AFLP analysis. Genet Resour Crop Evol 51:393–401
Coetzee R, Labuschagne MT, Hugo A (2008) Fatty acid and oil variation in seed from kenaf (*Hibiscus cannabinus* L.). Ind Crop Prod 27:104–109
Conde-Mejía C, Jiménez-Gutiérrez A, El-Halwagi M (2012) A comparison of pretreatment methods for bioethanol production from lignocellulosic materials. Spec Issue Energy Waste 90:189–202
Cosentino SL, Copani V (2003) Leaf photosynthesis in kenaf (*Hibiscus cannabinus* L.) in response to water stress. Agro Ind 2:137–145
Cunningham SD, Berti WR, Huang JW (1995) Phytoremediation of contaminated soils. Trends Biotechnol 13:393–397
Demirbas A (2004) Combustion characteristics of different biomass fuels. Prog Energy Combust Sci 30:219–230
Dempsey JM (1975) Fiber crops. The University Press of Florida, Gainesville, pp 457
Dudka S, Miller WP (1999) Permissible concentrations of Arsenic and Lead in soils based on risk assessment. Water Air Soil Pollut 113:127–132
Duggan J (2005) The potential for landfill leachate treatment using willows in the UK—a critical review. Resour Conserv Recycl 45:97–113
Duke JA (1983) Handbook of energy crops (unpublished). http://www.hort.purdue.edu/newcrop/duke_energy/Hibiscus_cannabinus.html. Accessed 11 Nov 2004
FAO (2003) Consultation on natural fibres, The production and consumption of Kenaf in China. ESC-Fibres consultation No: 03/6
Gonzalez-Garcia S, Moreira MT, Feijoo G (2010) Comparative environmental performance of lignocellulosic ethanol from different feedstocks. Renew Sust Energ Rev 14:2077–2085
Guo XF, Wei ZB, Wu QT, Qiu JR, Zhou JL (2011) Cadmium and zinc accumulation in maize grain as affected by cultivars and chemical fixation amendments. Pedosphere 21:650–656
Hamid MRA (2008) Kenaf to replace tobacco: farmers in Kedah to go for plants attributed to various items. Beriata Harians news update
Hernández R, Fernando S, French T (2012) Development of a bioadsorbent for the biodiesel industry. http://www.technologyalliance.ms/strategic-biomass-solu-tions/ms-projects/university-research.php
Ho W, Ang L, Lee D (2008) Assessment of Pb uptake, translocation and immobilization in kenaf (*Hibiscus cannabinus* L.) for phytoremediation of sand tailings. J Environ Sci 20:1341–1347
Hossain MD, Musa MH, Nuruddin AA, Jol H, Hamid HA (2016) Physio-agronomic performance of kenaf as influenced by different carbon levels. Plant OMICS 9:61–72
James SA, Ladan MJ, Goje DJ (2013) Antioxidant potential of *Hibiscus cannabinus* methanolic leaf extract. Sci World J 8:8–12
Jankaite A, Vasarevisius S (2007) Use of *Poaceae f.* species to decontaminated soil from heavy metals. Ekologija 53:84–89
Kaldor AF, Karlgren C, Verwest H (1990) Kenaf-a fast growing fiber source for papermaking. TAPPI J 73:205–208
Kojima Y, Kato Y, Yoon S-L, Lee M-K (2014) Kenaf as a bioresource for production of hydrogen-rich gas. Agrotechnology 3:1–8

Kugler DE (1988) Kenaf newsprint: realizing commercialization of a new crop after four decades of research and development. A report on the Kenaf Demonstration Project. United States Cooperative State research service, special projects and programme systems, Washington, DC, pp 13

Kulger DE (1996) Kenaf commercialization 1986–1995. In: Janick J (ed) Progress in new crops. ASHS Press, Alexandria, pp 129–132

Kumar PBAN, Dushenkov V, Motto H, Raskin I (1995) Phytoextraction: the use of plants to remove heavy metals from soils. Environ Sci Technol 29:1232–1238

Langan P, Gnanakaran S, Rector KD, Pawley N, Fox DT, Chof DW et al (2011) Exploring new strategies for cellulosic biofuels production. Energy Environ Sci 4:3820–3833

Le Roux M (2007) South Africa: Seardel hopes new plant will oust plastics. http://allafrica.com/stories/200702060312.html

LeMahieu PJ, Oplinger ES, Putnam DH (2003) Kenaf in: alternative field crops manual. http://www.corn.agronomy.wisc.edu/FISC/Alternatives/Kenaf.htm

Li D (2002) Kenaf production research and development in China. In: International Kenaf symposium, TN, USA

Lips SJJ, Iniguez de Heredia GM, Op den Kamp RGM, van Dam JEG (2009) Water absorption characteristics of kenaf core to use as bedding material. Ind Crop Prod 29:73–79

Maestri E, Marmiroli M, Visioli G, Marmiroli N (2010) Metal tolerance and hyperaccumulation: costs and trade offs between traits and environment. Environ Exp Bot 68: 1–13

Makari HK, Haraprasad N, Patil HSR (2008) In vitro antioxidant activity of the hexane and methanolic extracts of Cordiawallichii and celastruspaniculata. Internet J Aesthetic Antiaging Med 1:1–6

Marisol Berti T, Kamireddy SR, Ji Y (2013) Row spacing affects biomass yield and composition of Kenaf (*Hibiscus cannabinus* L.) as a lignocellulosic feedstock for bioenergy. J Sust Bioenerg Syst 3:68–73

Mazumder BB, Nakgawa-izumi A, Kenichi K, Ohtani Y, Sameshima K (2005) Evaluation of harvesting time effects on kenafbast lignin by pyrolysis-gas chromatography. Ind Crop Prod 21:17–24

Meagher RB (2000) Phytoremediation of toxic elemental and organic pollutants. Curr Opin Plant Biol 3:153–162

Meera M, Agamuthu P (2011) Phytoextraction of As and Fe using *Hibiscus cannabinus* from soil polluted with landfill leachate. Int J Phytoremed 14:186–199

Mishra J, Srivastava RK, Shukla SV, Raghav CS (2007) Antioxidants in aromatic and medicinal plants. Science Tech Entrepreneur:1–16

Mohamed A, Bhardwaj H, Hamama A, Webber C III (1995) Chemical composition of kenaf (*Hibiscus cannabinus* L.) seed oil. Ind Crop Prod 4:157–165

Mosquera OM, Correa YM, Buitrago DC, Nino J (2007) Antioxidant activity of twenty five plants from Colombian biodiversity. Mem Inst Oswaldo Cruz Rio de Janerio 102:631–634

Muchow RC, Wood IM (1983) Effect of sowing date on the growth and yield of kenaf (*Hibiscus cannabinus*) grown under irrigation in tropical Australia I. Phenology and seed production. Field Crop Res 7:91–102

Muhammad SR, Naim R, Ameena S, Tariq M, Jong-In H (2013) Potential of bioenergy production from industrial hemp (*Cannabis sativa*): Pakistan perspective. Renew Sust Energ Rev 18:154–164

Mukherjee S, Jagtap SD, Kuvalekar AA, Kale YB, Kulkarni OP, Harsulkar AM, Pawar PK (2010) Demonstration of the potential of *H. cannabinus* Linn. flowers to manage oxidative stress, bone related disorders and free- radical induced DNA damage. Ind J Nat Prod Res 1:322–327

Nabulo G, Black CR, Young SD (2011) Trace metal uptake by tropical vegetables grown on soil amended with urban sewage sludge. Environ Pollut 15:368–376

Nelson GH, Nieschlag HJ, Wolff IA (1962) A search for new fiber crops, V. Pulping studies on kenaf. TAPPI J 45:780–786

Nyam KL, Tan CP, Lai OM, Long K, Che Man YB (2009) Physicochemical properties and bioactive compounds of selected seed oils. LWT Food Sci Technol 42:1396–1403

Pascoal Neto C, Seca A, Fradinho D, Coimbra MA, Domingues F, Evtuguin D, Silvestre A, Cavaleiro JAS (1996) Chemical composition and structural features of the macro molecular components of *Hibiscus cannabinus* grown in Portugal. Ind Crop Prod 5:189–196

Pawar PK, Borse TP, Pinjari RZ, Maheshwari VL (2008) A simple technique for rapid quantitative determination of solasodine from cultured hairy roots of *Solanum surattense*. J Herbal Med Toxicol 2:7–10

Raskin II, Smith RD, Salt DE (1997) Phytoremediation of metals: using plants to remove pollutants from the environment. Curr Opin Biotechnol 8:221–226

Rathana Y, Roces SA, Bacani FT, Tan RR, Kubouchi M, Yimsiri P (2010) Microwave enhanced alkali catalyzed trans esterification of kenaf seed oil. Int J Chem React Eng 8:132–141

Rotkittikhun R, Kruatrachue M, Chaiyarat R, Ngernsansaruay C, Pokethitiyook P, Paijitprapaporn A, Baker AJM (2006) Uptake and accumulation of lead by plants from the Bongam lead mine area in Thailand. Environ Pollut 144:681–688

Russo A, Izzo AA, Cardile V, Borrelli F, Vanella A (2001) Indian medicinal plants as antiradicals and DNA cleavage protectors. Phytomedicine 8:125–132

Saba N, Jawaid M, Hakeem KR, Paridah MT, Khalina A, Alothman O.Y (2015) Potential of bioenergy production from industrial kenaf (*Hibiscus cannabinus L.*) based on Malaysian perspective. Renew Sust Energ Rev 42:446–459

Santos GCG, Rodella AA, Abreu CA, Coscione AR (2010) Vegetable species for phytoextraction of boron, copper, lead, manganese and zinc from contaminated soil. Sci ELO Brasil 67:713–719

Schmoger MEV, Oven M, Grill E (2000) Detoxification of arsenic by phytochelatins in plants. Plant Physiol 122:793–802

Sellers T, Miller GD, Fuller MJ (1993) Kenaf core as a board raw material. For Prod J 43:69–71

Thaiponga K, Boonpra Koba U, Crosbyb K, Cisneros-Zevalllosc L, Byrne DH (2006) Comparison of ABTS, DPPH, FRAP and ORAC assays of estimating antioxidant activity from guava fruit extracts. J Food Compos Anal 19:669–675

Umeki K, Namioka T, Yosikawa K (2012) The effect of steam on pyrolysis and char reactions behavior during rice straw gasification. Fuel Process Technol 94:53–64

Webber CL, Bledsoe RE (1991) Kenaf production, harvesting, processing and products. In: Janick J, Simon JE (eds) New crops. Wiley, New York, pp 416–421

Wittbrodt PR, Palmer CD (1995) Reduction of Cr-VI in the presence of excess soil fulvic acid. Environ Sci Technol 29:255–265

Wong YC, Lim SH, Atiqah NA (2013) Remediation of industry wastewater effluent by using Kenaf as waxes. Present Environ Sustain Dev 7:290–295

Wu G, Kang H, Zhang X, Shao H, Chu L, Ruan C (2010) A critical review on the bio-removal of hazardous heavy metals from contaminated soils: issues, progress, eco environmental concerns and opportunities. J Hazard Mater 174:1–8

Wuana RA, Okieimen FE (2010) Phytoremediation potential of Maize (*Zea mays* L.). A review. Afr J Gen Agric 6:275–287

Yusri NM, Chan KW, Iqbal S, Ismail M (2012) Phenolic content and antioxidantactivity of *Hibiscus cannabinus* L. Seed extracts after sequential solvent extraction. Molecules 17:12612–12621

Zaini N, Kamarudin KSN (2014) Adsorption of carbon dioxide on monoethanolamine (MEA)–impregnated Kenaf core fiber by pressure swing adsorption system (PSA). J Teknologi Sci Eng 68:11–16

Zwane PE, Masarirambi MT (2009) Kenaf (*Hibiscus cannabinus*) and allied fibres for sustainable development in Swaziland. J Agric Soc Sci 5:35–39

Cannabis sativa: A Plant Suitable for Phytoremediation and Bioenergy Production

10

Sanjeev Kumar, Ritu Singh, Virendra Kumar, Anita Rani, and Rajeev Jain

Abstract

Phytoremediation has been gaining interest as a sustainable approach for management of the sites contaminated with toxic/hazardous substances. Several plants have been reported in the literature that have the potential for simultaneous execution of phytoremediation and production of useful products like biogas, bioethanol, biodiesel, charcoals, fibres, etc. during the process. Among several plant species, the present chapter has been focused on *Cannabis sativa* L., a multipurpose annual herbaceous plant species which has wide range of application as seed oil, industrial fibre, food, livestock feed, medicine as well as significant place in recreation, religious and spiritual practices. *Cannabis sativa* L. has the potential to serve as phytoremedial agent for removal of toxic metals from contaminated sites as well as yields high biomass which could be used for production of bioenergy. The energy yield of *Cannabis sativa* L. for biofuel and biogas production has been reported comparable to most of the energy crops. The present chapter provides an overview of the phytoremediation capacity of *Cannabis sativa* L. for resolving environmental issues of contaminated soil along

S. Kumar (✉) • R. Singh (✉)
Department of Environmental Science, Central University of Rajasthan, NH8, Bandarsindri, Kishangarh, Ajmer 305817, Rajasthan, India
e-mail: sanjivk845@gmail.com; ritu2735@gmail.com; ritu_ens@curaj.ac.in

V. Kumar
Centre for Environmental Sciences, Central University of Jharkhand, Brambe, Ranchi 835205, India

A. Rani
Department of Botany, Dyal Singh College, University of Delhi, New Delhi 110003, India

R. Jain
Forensic Toxicology Division, Central Forensic Science Laboratory, Ministry of Home Affairs, Government of India, Tetelia, Gotanagar, Guwahati 781033, Assam, India

K. Bauddh et al. (eds.), *Phytoremediation Potential of Bioenergy Plants*,
DOI 10.1007/978-981-10-3084-0_10

with its potential abilities to generate bioenergy to meet energy demand of future generations.

Keywords
Bioenergy • *Cannabis sativa* L. • Phytoremediation • Biogas • Biomethanol • Bioethanol • Biodiesel

10.1 Introduction

Soil, as the most basic natural resource of our planet, plays an important role in the maintenance of the ecological balance of the communities. Soil functions as a vital living system that contains several biological elements that are key to ecosystem functions. It is by virtue of these functions the nature sustains the biological productivity of soil, maintains air and water quality and upholds plant, animal and human health (Nathan 2009). Owing to the rapid growth of industrialization and urbanization, the soil is being contaminated increasingly by various kinds of toxic, hazardous pollutants affecting large areas of agricultural land. The persistent and recalcitrant nature of these pollutants is a serious threat to everyone's health and the environment. Consequently, the management of contaminated soil has become a great priority in developed and developing countries for sustainable development (Enell et al. 2016). Table 10.1 represents different causes of contamination in the environment. To get rid of the burden of toxic wastes, several remediation technologies have been proposed for removal, elimination or sequestration of toxic/hazardous pollutants from contaminated sites such as stabilization/solidification, incineration, solvent extraction, soil washing, thermal treatment, chemical treatment, air sparging, verification, etc. However, high cost, requirement of skilled manpower, labour-intensiveness, toxic by-products, technical difficulties, etc. are proven setbacks to these technologies. Moreover, in case of ex situ methods, risks of secondary contamination pose further problems (Surriya et al. 2015).

In recent years, phytoremediation has gained interest as a sustainable approach to the management of the contaminated sites. Basically, phytoremediation refers to the use of living plants to mitigate the concentrations of both inorganic and organic pollutants present in soil. It represents a simple, economic, environmentally friendly technology for eradicating toxic heavy metals and other soil contaminants. Worldwide, more than 400 plant species have been recognized that have the potential to clean up the contaminated sites (Surriya et al. 2015). This method encompasses the use of plant to extract, sequester and detoxify the pollutants (Jiang et al. 2015). It employs methodologies like phytoextraction, phytostabilization, phytostimulation, phytofiltration, phytodegradation and phytovolatilization (Koptsik 2014). Being an in situ method, phytoremediation provides low cost, environmentally benign and aesthetically pleasing alternative with minimal disruption to the environment. The level of cleanup achieved by phytoremediation is dependent on the plant species, contaminants and the land conditions.

Table 10.1 Source of environmental contamination

Contaminant	Source of contamination in soil and water reservoir	References
Heavy metals		
Pb	Batteries, paints, e-waste, smelting operations, ceramics, coal-based thermal power plants, bangle industry	Cai et al. (2015), Cheng et al. (2015), George et al. (2015), Goix et al. (2015), Ono et al. (2015) and Quan et al. (2015)
Cr	Mining, coolants, chromium salts manufacturing, leather tanning	Liu et al. (2011), Foldi et al. (2013), Kumar et al. (2014a), Ma et al. (2014) and Candeias et al. (2015)
Hg	Chlor-alkali plants, thermal power plants, fluorescent lamps, hospital waste, electrical appliances, etc.	Raposo and Roeser (2001), Dufault et al. (2009), Hu and Cheng (2012), Wei-Hua et al. (2012) and George et al. (2015)
As	Geogenic, natural processes, smelting operations, thermal power plants, fuel burning	Reilly et al. (2010), George et al. (2015) and Rodriguez-Iruretagoiena et al. (2015)
Cu	Mining, electroplating, smelting operations	Liu et al. (2011), Konecny et al. (2014), Kumar et al. (2014b, 2015)
Ni	Smelting operations, thermal power plants, battery industry	George et al. (2015) and Rodriguez-Iruretagoiena et al. (2015)
Cd	Zinc smelting, waste batteries, e-waste, paint sludge, incinerations and fuel combustion	Zhang et al. (2012), Hong et al. (2009), Kumar et al. (2012, 2014b; 2015), Ono et al. (2015) and Quan et al. (2015)
Mo	Spent catalyst	Marafi and Stanislaus (2008)
Zn	Smelting, electroplating	Kumar et al. (2009), Liu et al. (2011) and Kumar et al. (2014b, 2015)
Pesticides		
Organochlorine, polybrominated diphenyl ethers, phthalate esters	Pesticides manufacture factory, pesticide runoff to agricultural soil and water	Zhang et al. (2009), Chen et al. (2012), Jennings and Li (2014), Niu et al. (2014), Sun et al. (2015) and Sun et al. (2016)
Polycyclic aromatic hydrocarbons	Combustion of fuels in vehicular engines, coke production, incineration of industrial and domestic waste, oil refinery	Ahel and Tepic (2000), Wang et al. (2004), Cachada et al. (2016) and Jiang et al. (2016)

10.2 Phytoremediation Coupling to Bioenergy Production

In recent years, research has been focused on investigating synergism possibilities between phytoremediation to remove contaminants from contaminated sites and bioenergy production. This integrated approach could simultaneously provide environmental benefits such as contaminant cleanup, improved soil quality, erosion control, improved habitat, etc. and a good alternative to conventional energy resources (Helena 2012). The biomass produced during phytoremediation could be used for gaining economic benefits such as production of bioenergy (biogas, biohydrogen, biofuels, bioethanol, biomethanol, combustion for energy production and heating), charcoal, alkaloids, fibres, etc. In present time, bioenergy is the sustainable source of renewable energy. Bioenergy primarily refers to energy derived from living material such as crops, wood or animal waste. Bioenergy crops refer to any plant material which is used to generate bioenergy. These energy crops have large capacity to produce high volume of biomass. They can be grown in marginal soils. The produced biomass could be used as solid material, i.e. biomass for combustion or as liquid products, i.e. biofuels. Both the biomass and biofuels can be derived from energy crops, agricultural coproducts or waste materials. Switchgrass (*Panicum virgatum* L.), elephant grass (*Pennisetum purpureum* Schum.), poplar (*Populus* spp.), willow (*Salix* spp.), mesquite (*Prosopis* spp.), etc. have been touted as crops with the most widespread promise. Cultivation of energy crops on arable land can decrease our dependency on depleting reservoirs of fossil fuels and can also mitigate climate change. Table 10.2 represents list of plant species which have been used for phytoremediation as well as bioenergy production.

10.3 *Cannabis sativa*

Cannabis sativa (genus, *Cannabis*; family, Cannabaceae) is an annual dioecious herb with an erect stem (1–5 m height), palmate leaves and long tap roots. It is also known as hemp in different regions of the world. It has been grown since ancient times for deriving different products like seeds, wooden core, fibres, cannabinoids, etc. *Cannabis sativa* is considered as an ancient non-food crop, used for manufacturing rope, fabric, oil, varnish, insulating board and paper, etc. worldwide. In the start of the ninetieth century, hemp seed oil is used for human nutrition, cosmetic, paints and most importantly in biodiesel production. Figure 10.1 shows multiple uses of *C. sativa*.

10.3.1 Ecology and Geographical Distribution of *Cannabis sativa*

C. sativa is adaptive to various climatic conditions and wide ranges of soil. It requires mild temperate (14–21 °C) climate with minimum of 67 cm annual rainfall; abundant rainfall is required when the seeds are germinating till young plantlets are established. It can tolerate annual temperature of 6–27 °C, precipitation of 3–40 dm

Table 10.2 Plant species used in phytoremediation process and bioenergy process

Plant	Contaminants	Process of removal	Sustainable bioenergy approach	References
Jatropha curcas	Cd	Phytoremediation	Bioenergy production	Marques and Nascimento (2013)
Canola, oat, wheat	Cd	Phytoremediation	Biogas production	Zhang et al. (2013)
King grass (*Pennisetum americanum, Pennisetum purpureum*)	Cd	Phytoremediation	Bioenergy (biomass) production	Zhang et al. (2014)
Water hyacinth	Inorganic nutrients	Phytoremediation	Biogas production	Wang and Calderon (2012)
Sweet potatoes	Pb	Phytoremediation	Bioenergy (bioethanol) production	Cheng et al. (2015)
Willow	Contaminants area	Phytoremediation	Bioenergy (willow-based ethanol and willow chips in an industrial furnace	Gonzalez-Garcia et al. (2013)
Poplars (*Populus* spp.) and willows (*Salix* spp.)	Pollutants including fertilizers, inorganic metals and metalloids, petrochemical compounds and soluble radionuclides	Phytoremediation	Bioenergy (biomass) production	Licht and Isebrands (2005)
Peanut (*Arachis hypogaea* L.)	Cd phytoextraction	Phytoremediation	Bioenergy (biofuel) production	Su et al. (2013)
Miscanthus spp. and giant reed (*Arundo donax* L.)	Zn, Cr and PB contamination	Phytoremediation	Bioenergy (biomass) production	Barbosa et al. (2015)
Energy grasses (*Arundo donax* and *Miscanthus sacchariflorus*)	Zn, Cr	Phytoremediation	Bioenergy (biomass) production	Li et al. (2014)

(continued)

Table 10.2 (continued)

Plant	Contaminants	Process of removal	Sustainable bioenergy approach	References
Poplar clones	Zn	Phytoremediation	Bioenergy (biomass) production	Romeo et al. (2014)
Castor bean (*Ricinus communis* L.)	Cu, Zn, Mn, Pb and Cd	Phytoremediation	Bioenergy (oil production)	Olivares et al. (2013), Bauddh and Singh (2012, 2015) and Bauddh et al. (2016a, b)
Zea mays	Cu	Phytoremediation	Bioenergy (biomass) production	Sheng et al. (2012)
Willow (*Salix klara* and *Salix inger*)	Cr, Cu, Ni, Pb, Zn	Phytoremediation	Bioenergy (biomass) production	Enell et al. (2016)
Elephant grass (*Pennisetum purpureum* Schum.) and sugarcane (*Saccharum* spp.)	Soil P	Phytoremediation	Bioenergy (biomass) production	Silveira et al. (2013)
Ricinus communis L.	DDT and Cd	Phytoremediation	Bioenergy (biomass and oil) production	Huang et al. (2011)
Sunflowers (*Helianthus annuus*)	Pb, Zn and Cd	Phytoremediation	Oil yielding	Angelova et al. (2012)

and pH ranging from 4.5 to 8.2. It grows well on nitrogen-rich, fertile, neutral to slightly alkaline, well-drained silt or clay loam soil. It has around 1 m-deep roots that grow fast and easily form dense stands.

10.3.2 Phytoremediation Potential of *Cannabis sativa*

The basic requirement of phytoremediation is the ability of the plant to store heavy metal in its body parts, but selecting those plants which store large quantities of metal and are much resistant to the contaminated soil and other environmental pollution gives ecological as well as economic benefits (Angelova et al. 2004). Hemp is considered as a good candidate for soil phytoremediation as it has high metal tolerance and yields high biomass (Table 10.3). The high biomass production, the

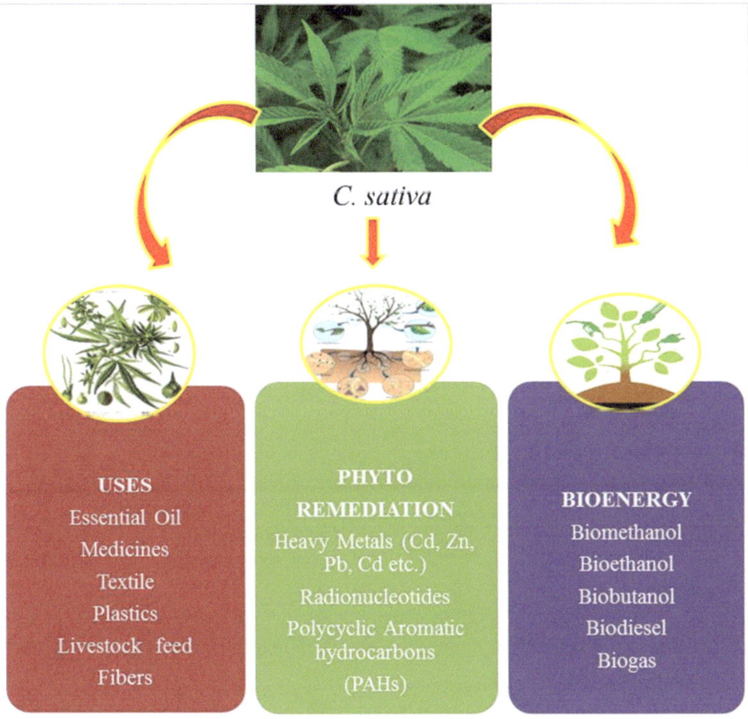

Fig. 10.1 *Cannabis sativa*: a multipurpose plant species

large plasticity that allows hemp to be grown under a varied range of agro-ecological conditions and the possibility to use its biomass in non-food industries make this species attractive for phytoremediation (Linger et al. 2002; Citterio et al. 2003; Arru et al. 2004).

Shi and Cai (2009) studied cadmium accumulation and tolerance in eight energy plants (glycine, *Arachis*, *Brassica*, *Cannabis*, *Ricinus*, *Carthamus*, *Helianthus*, *Linum*). Cd accumulation and growth parameter in all the plants were studied for 28 days in substrates containing 0, 50, 100 and 200 mg Cd/kg. Among all, *Cannabis*, *Linum*, *Ricinus* and *Arachis* were found more tolerant for Cd. In comparison to shoot, the roots of *Cannabis* showed high accumulation with bioconcentration factor ranging from 2026 to 3099%. Such a high BCF showed that hemp has higher capacities for Cd phytostabilization. The translocation factor was reported in the range of 2.5–3.7%. As the capacity for metal accumulation highly differs between different species and varieties, Shi et al. in 2012 investigated cadmium tolerance and its accumulation capacity, both, in 18 cultivars of hemp. They also screened cultivars that can be planted in Cd-contaminated sites for bioenergy production as well as phytoremediation. The hemp cultivars were found to cope with 25 mg Cd/ kg^{-1} soil stress with approximately 50 % reduction of biomass. They reported most of the Cd adsorbed by the plants was retained in the roots of the plants. As the

Table 10.3 Hemp plant used in phytoremediation process and bioenergy process

Plants	Removal of environmental contamination by phytoremediation	Bioenergy approach	References
Flax, cotton and hemp	Pb, Cu, Zn, Cd	Biomass used in textile industry, paper industry, chemical factories, building and furniture industry	Angelova et al. (2004)
Cannabis sativa L.	Cd, Cr, Ni	Used biomass in non-food industries and other economic activities	Citterio et al. (2003)
Cannabis sativa L.	Cu	High biomass	Arru et al. (2004)
Cannabis sativa L.	Ni, Pb, Cd	Used fibre as bioenergy and other economic activities	Linger et al. (2002)
Brassica rapa, *Cannabis sativa*, *Helianthus annuus*, *Zea mays*	Zn, Cu, Cd, Pb, Ni	Used biomass for bioenergy production	Meers et al. (2005)
Cannabis sativa L.	Cd, Zn, Fe	Used both fibre and oil	Mihoc et al. (2012)

translocation factor (TF) for cultivars were found in the range of 3.15–12.7 %, the authors suggested that hemp could be considered as an excluder for Cd-contaminated sites (Shi and Cai 2009; Shi et al. 2009, 2012). Metal exclusion is defined as the capacity of plant to avoid absorption and restrict its translocation to the shoots. Besides Cd, hemp was also studied for removal of Cr and Ni from contaminated sites (Citterio et al. 2003). For Cr, hemp did not show any significant result, but for Ni, accumulation was much higher in roots with accumulation factor ranging from 0.1 to 0.4. Here the accumulation factor was calculated using equation:

$$AF = \frac{\text{mean plant metal conc.}}{\text{mean total metal conc.}}$$

In comparison to other species, *C. sativa* has been reported to have a high translocation rate for Zn. Thus, *C. sativa* could be used as a potential hyperaccumulator for Zn. Experimental findings showed TF values higher than 1, which indicate that the accumulation of Zn in shoots is more in comparison to other heavy metals (Malik et al. 2010). The influence of metal accumulation on hemp fibre quality was explored by Linger et al. (2002). Their study revealed that accumulation of heavy metals in all the parts of *C. sativa* restricts its use as a raw material in clothes, but the quality of fibres remains unaffected which allows them to be used in special products like combined material. As per the results observed by Arru et al. (2004), primary bast fibres are not involved in Cu accumulation. In contrast, Linger et al.

(2002) observed heavy metal (Cd, Ni and Pb) accumulation in almost all parts of hemp, with maximum accumulation in the leaves part of the plant. These results indicate that different mechanisms exist for uptake and accumulation of different species of metal. Similar kind of research was done by Angelova et al. (2004) displaying metal concentration in roots, stems, leaves, seeds and flowers of hemp plant using dry mineralization process. The metal distribution in the plant axis of hemp was found to decrease in the following pattern: root > stems > leaves > seeds while the pattern in cotton followed leaves > seeds > root > stems. This suggests that the heavy metal distribution in the plant axis is highly selective.

The potential of *C. sativa* for phytoextraction of heavy metals from calcareous dredged sediment-derived surface soil in presence of amendments, viz. ethylenediamine tetraacetate (EDTA) and ethylenediamine disuccinate (EDDS), was demonstrated by Meers et al. (2005). The application of amendments showed significant effect on heavy metal accumulation. EDTA treatment displayed more accumulation of lead and cadmium, while EDDS induced higher concentrations of Cu and Ni in shoot. No significant difference was noticed in case of Zn. The authors supported the fact that preharvest application of amendments gives better extraction results.

As the hemp seed varieties could form the base of a diet containing lots of oil and non-fatty acid, Mihoc et al. (2012) determined the nutritional value of hemp seed in terms of their oil content and metal concentration (Zn, Cd, Mn, Fe, Ca, Mg) for five varieties of hemp seeds in Romanian soils. Significant amount of Fe, Mn, Zn and Cd was noticed in seeds of all varieties, restricting their use for food purposes.

Vandenhove and Hees (2005) tested feasibility of the production of *C. sativa* on radionuclide-contaminated areas. The experiments were performed over a well-known radionuclide radiocaesium in sandy soil under greenhouse and lysimeter conditions for studying the transfer of radiocaesium from contaminated soil to the hemp. The distribution of radiocaesium during crop conversion was also studied. The translocation factor (TF) to the hemp was found to be 0.6×10^{-3} m^2/kg in stem, 0.5×10^{-3} m^2/kg in straw and 1.0×10^{-3} m^2/kg in fibres, respectively. Low translocation factor permits hemp for production of clean end products such as biofuel, fibre and seed oil even on degraded lands. The use of hemp stem as biofuel is restricted to areas with contamination levels <1050 k Bq/m^2.

Industrial hemp (*Cannabis sativa*) has been used efficiently for remediation of two polycyclic aromatic hydrocarbons, namely, benzo[α]pyrene and chrysene (Campbell et al. 2006). Experiments were conducted with the plant grown for 45 days in soil spiked with PAHs as concentration of 25, 50 and 75 μg/g. The concentration of PAHs was found reduced in all the experiments. The weight growth rate obtained for the above-mentioned concentration of PAHs was found above 150%, and the height growth rate was found above 130% for all the three concentrations. The authors suggested that metabolites of benzo[α]pyrene and chrysene stimulate the growth of hemp.

10.3.3 Bioenergy Production Efficiency of *Cannabis sativa*

C. sativa also known as industrial hemp is a real promising source of bioenergy production in present scenario. It is an economic option which also has low environmental impacts (Li et al. 2010). In comparison to agricultural lignocellulosic materials like corn stover and wheat straw, which have around 37% (dry mass) cellulose, cellulose content of hemp is relatively high (~44%) (Ohgren et al. 2005; Linde et al. 2008). The presence of high cellulosic content as well as high biomass yield makes hemp a very potential crop for bioethanol production. Hemp stem mainly consists of two major constituents: (1) bast fibres which have high cellulose and low lignin and (2) a woody core which contains significantly higher lignin content (Garcia-Jaldon et al. 1998). Postponing of hemp harvesting by 1–2 months could help in achieving high biomass yield per hectare months (Keller et al. 2001).

The physico-chemical properties of fuel obtained from energy crop affect its suitability and applicability as solid biofuel. The physical properties of the fuel such as its bulk density, the size of the particle, the tendency of bridging and angle of repose can be adjusted by various physical treatments such as milling, grinding and compaction (Mattsson and Briere 1984). The chemical fuel properties include content of sodium, potassium, magnesium, calcium, silicon chlorine, aluminium, sulphur and phosphorus which are of significant importance in the combustion of hemp fuel (Baxter et al. 1998). The oil content in seeds of hemp, which ranges from 26 to 38%, contains a larger amount of polyunsaturated fatty acids (Oomah et al. 2002; Kriese et al. 2004; Matthäus and Brühl 2008; Latif and Anwar 2009). Hemp also serves as a source of fuel as it yields high biomass that can be transformed efficiently into low-carbon fuels like bioethanol and biobutanol (Moxley et al. 2008). Li et al. (2010) investigated hemp's seed oil potential and efficacy as a biodiesel source. By using a base-catalysed transesterification method, they produced biodiesel from hemp seed oil. The product recovery yield (PRY, actual biodiesel recovered/theoretical amount of biodiesel) of 97% indicated that the biodiesel loss is not much significant. The author further studied certain other fuel properties such as free glycerine content, total glycerine, kinematic viscosity, flash point, cloud point, density, acid number (mg KOH/g oil) and its sulphur content. The hemp biodiesel meets all the standards required for biodiesel fuels as given by ASTM 6751-09. The authors concluded that various properties of hemp biodiesel such as cold flow make it a good competitor to petroleum-based diesel fuel. Another study reported that 96% of the glucose could be recovered from the cellulosic hurds of hemps, which enables further recycling of the fibre wastage (Moxley et al. 2008).

Steam pretreatment of hemp fibres has been done by many researchers for separating the fibres from other components (Garcia-Jaldon et al. 1998). To obtain highest glucose content in enzymatic hydrolysis during ethanol conversion, Sipos et al. (2010) optimized steam pretreatment parameters for hemp. The dry hemp stems and its silage were subjected to steam pretreatment under various conditions. The experiments were also carried with and without sulphur dioxide impregnation. The steam pretreatment efficiency was assessed with the help of sugar recovery and polysaccharide conversion in slurry hydrolysis. The overall glucose yield, hemicellulose solubilization efficiency and sugar degradation, all of these, were calculated on the basis of slurry enzymatic hydrolysis and the mass balance of the pretreatment. Apart from this, simultaneous saccharification and fermentation (SSF) of both, i.e. the whole pretreatment slurry and separated solid fraction of pretreated material, at optimized conditions was also investigated. It was noticed that steam pretreatment with a catalyst such as SO_2 is a competent method for production of ethanol from both materials, dry hemp and silage. For both of these materials, i.e. dry hemp and its silage, impregnation with SO_2 (2%) followed by 5 min steam pretreatment at 210 °C showed highest overall yield of glucose. The effect of the ensiling process remains insignificant to the conversion to ethanol at its optimized environment. SSF experiments under optimized conditions yielded ethanol 163 g/kg ensiled hemp and 171 g/kg dry hemp, which corresponds to 206–216 lMg^{-1} ethanol based on initial dry material. In order to increase the energy recovery, further investigation is required to utilize the liquid fraction and the residues of SSF for production of biogas.

The esterification process parameters (molar ratio, catalyst concentration, reaction time) for generation of methyl ester of raw hemp oil were standardized by Ragit et al. (2012). For obtaining lowest kinematic viscosity, hemp oil (6:1 M ratio) was treated with methanol for 45 min at 60 °C in presence of potassium hydroxide (0.01 w/w). Afterwards, it was kept undisturbed for 24 h. The ester recovery with this method was reported to be 90.62 %. The authors also tested different fuel properties such as pour point, density, flash point, fire point, free fatty acid content, cloud point, kinematic viscosity and calorific value of the hemp methyl ester and hemp oil. The calorific value, flash point and fire point of hemp oil were found higher in comparison to diesel oil. For hemp methyl ester, the pour point was reported to be greater than that of diesel fuel, whereas cloud point, flash point, fire point and the viscosity were found lesser than diesel. Since, the properties of biodiesel derived from hemp were found closer to diesel fuel; thus, hemp oil could serve as a substitute for diesel oil. The authors recommended methyl esters of hemp seed oil as a prospective fuel and a performance-improving additive in compression-ignition engines.

The high biomass of hemp is utilized for biogas production also, as an alternative source of bioenergy. Several studies have been carried out over commercial use of hemp biomass for energy production (Castleman 2006; Burczyk et al. 2008; Rice 2008). The growth stage of the plants influences the biomass yield, hence the energy production. As the plant ages, the amount of structural carbohydrates and lignin increases, but the carbohydrates (cellulose and hemicellulose) start degradation (Jones 1970). Pakarinen et al. (2011) evaluated the suitability of biomass crop hemp along with maize and faba bean under boreal conditions for biogas production. They also tested their potential efficacy to serve as a source for sugar raw material for bioethanol production. The methane yield was found to be 379 $Ndm^3/kg/VS$ from maize, 387 $Ndm^3/kg/VS$ from faba bean and 239 $Ndm^3/kg/VS$ from hemp. For methane production, maize was reported to be the most potential raw material on the basis of yield per hectare. Maize was also observed as the most efficient raw material which has a conversion yield of about 80% of the theoretical sugars. As far as the total amount of carbohydrates was concerned, the highest theoretical yield per hectare was displayed by hemp. On the basis of result obtained, which includes both the requirement for weeding and fertilizers, it was concluded that all the three crops (maize, faba bean and hemp) may serve as an attractive option for cultivation and energy production in boreal conditions.

A comprehensive life cycle assessment was done for ethanol produced from hemp hurds (a by-product of fibre crops) in an enzymatic hydrolysis process (González-Garcıá et al. 2012). This process converts lignocellulosic biomass into ethanol. The authors analysed and compared the environmental performance of two ethanol fuel blends, viz. E10 and E85, in a flexi-fuel vehicle. They selected two allocation approaches for checking environmental performance. These are mass allocation and economic allocation. They also carried out a sensitivity analysis which involved different allocation coefficients. For this, the market price and production volume of each agricultural product were taken into account. It was revealed that ethanol-based fuels offer improved environmental performance and a decrease in fossil fuel-based energy resources. The authors suggested that petrol is the best choice if impact categories like acidification, eutrophication, photochemical oxidants formation, etc. are concerned. They concluded that the selection of the allocation approach considerably influences the environmental performance, thus the question: Which one, E10 or E85, has smaller environmental impacts?

The fuel chain greenhouse gas (GHG) balance along with farm economics of hemp which was grown for the purpose of bioenergy was examined and compared with bioenergy plants. They selected *Miscanthus* and willow as perennial and oilseed rape and sugar beet as traditional annual bioenergy crops (Finnan and Styles 2013). The greenhouse gases load of hemp cultivation was found in between perennial and annual energy crop GHG loads. The net fuel chain GHG abatement potential of 11 t/CO_2 eq/ha/year in the mid-yield estimate was found comparable to perennial crops and 140–540% greater than for annual crop fuel chains. As far as gross margins were concerned, it was considerably lower for hemp than that of oilseed rape and sugar beet, but for *Miscanthus* it exceeded when organic fertilizers were used. The authors suggested that replacement of 25% of oilseed rape or sugar beet with hemp could increase greenhouse gases abatement by 21 Mt./CO_2eq/year. They further suggested that integration of hemp into food crop rotations could become a good alternate for developing more sustainable non-transport bioenergy supply chains.

Prade et al. (2012) assessed the net energy yields and energy output to input ratios (RO/I) for production of power, heat and vehicle fuel from industrial hemp. They compared four scenarios for hemp biomass which are as follows: (a) combined heat and power (CHP) from spring-harvested baled hemp, (b) heat from spring-harvested briquetted hemp, (c) CHP and (d) vehicle fuel from autumn-harvested ensiled hemp processed to biogas. The authors recorded above average net energy yields for combined heat and power and heat production from hemp biomass. Further, the hemp combustion showed high conversion efficiency and RO/I. In terms of biogas production, hemp competes with sugar beets and maize, whereas it gives competition to perennial crops like reed canary grass, *Miscanthus* and willow if production of solid biofuel is concerned.

10.4 Conclusion

Phytoremediation is a eco-friendly process to remediate heavy metal and other forms of contamination in the contaminated land. The feasibility of this approach must be supported by those crops which have the potential to produce economic benefits. In current scenario, the fastest growing source of renewable energy is bio energy. The cultivation of energy crops on arable land can decrease our dependency on continuously depleting fossil resources. Further, it could also mitigate climate change. Hemp is an annual herbaceous crop that has been used by human civilization for several years for its seeds and fibres. Hemp (*C. sativa*) has also been proved as an efficient phytoremediator for heavy metal-contaminated sites. Apart from phytoremediation, hemp produces high biomass which have been used for production of bioenergy such as bioethanol, biodiesel, biogas, biomethanol, etc. Using hemp as a remediation agent, the sustainability and applicability of the phytoremediation approach could be managed well by the profitable income generated through the production of bioenergy.

References

Ahel M, Tepic N (2000) Distribution of polycyclic aromatic hydrocarbons in a municipal solid waste landfill and underlying soil. Bull Environ Contam Toxicol 65:236–243

Angelova V, Ivanova R, Delibaltova V, Ivanov K (2004) Bio-accumulation and distribution of heavy metals in fibre crops (flax, cotton and hemp). Ind Crop Prod 19:197–205

Angelova V, Ivanova R, Ivanov, K, Perifanova-Nemska MN, Uzunova GI (2012) Potential of sunflower (*Helianthus annuus* L.) for phytoremediation of soils contaminated with heavy metals. In: BALWOIS-Ohrid, Republic of Macedonia, pp 1–11

Arru L, Rognoni S, Baroncini M, Bonatti PM, Perata P (2004) Copper localization in *Cannabis sativa* L. grown in a copper-rich solution. Euphytica 140:33–38

Barbosa B, Boleo S, Sidella S, Costa J, Duarte MP, Mendes B, Cosentino SL, Fernando AL (2015) Phytoremediation of heavy metal-contaminated soils using the perennial energy crops *Miscanthus* spp. and *Arundo donax* L. Bioenerg Res 8:1500–1511

Bauddh K, Singh RP (2012) Cadmium tolerance and its phytoremediation by two oil yielding plants *Ricinus communis* (L.) and *Brassica juncea* (L.) from the contaminated soil. Int J Theor Phys 14:772–785

Bauddh K, Singh RP (2015) Assessment of metal uptake capacity of castor bean and mustard for phytoremediation of nickel from contaminated soil. Bioremed J 19(2):124–138

Bauddh K, Kumar A, Srivastava S, Singh RP, Tripathi RD (2016a) A study on the effect of cadmium on the antioxidative defense system and alteration in different functional groups in castor bean and Indian mustard. Arch Agron Soil Sci 62(6):877–891

Bauddh K, Singh B, Singh RP (2016b) *Ricinus communis* L. as a value added alternative for restoration of cadmium contaminated and degraded agricultural ecosystem. Bull Environ Contam Toxicol 96(2):265–269

Baxter LL, Miles TR, Miles TR, Jenkins BM, Milne T, Dayton D, Bryers RW, Oden LL (1998) The behavior of inorganic material in biomass-fired power boilers: field and laboratory experiences. Fuel Process Technol 54(1–3):47–78

Burczyk H, Grabowska L, Kolodziej J, Strybe M (2008) Industrial hemp as a raw material for energy production. J Indust Hemp 13(1):37–48

Cachada A, Silva EF, Duarte AC, Pereira R (2016) Risk assessment of urban soils contamination: the particular case of polycyclic aromatic hydrocarbons. Sci Total Environ 551–552:271–284

Cai ZP, Doren JV, Fang ZG, Li WS (2015) Improvement in electrokinetic remediation of Pb-contaminated soil near lead acid battery factory. Trans Nonferrous Metals Soc China 25(9):3088–3095
Campbell S, Paquin D, Awaya JD, Li QX (2006) Remediation of Benzo[a]pyrene and chrysene-contaminated soil with industrial hemp (*Cannabis sativa*). Int J Theor Phys 4(2):157–168
Candeias C, Avila PF, Silva EF, Teixeira JP (2015) Integrated approach to assess the environmental impact of mining activities: estimation of the spatial distribution of soil contamination (Panasqueira mining area, Central Portugal). Environ Monit Assess 187:135
Castleman T (2006) Hemp biomass for energy. Fuel and Fiber Company, Sacramento
Chen Y, Li J, Liu L, Zhao N (2012) Polybrominated diphenyl ethers fate in China: a review with an emphasis on environmental contamination levels, human exposure and regulation. J Environ Manag 113:22–30
Cheng SF, Huang CH, Chen KL, Lin SC, Lin YC (2015) Exploring the benefits of growing bioenergy crops to activate lead-contaminated agricultural land: a case study on sweet potatoes. Environ Monit Assess 187:144
Citterio S, Santagostino A, Fumagalli P, Prato N, Ranalli P, Sgorbati S (2003) Heavy metal tolerance and accumulation of Cd, Cr and Ni by *Cannabis sativa* L. Plant Soil 256:243–252
Dufault R, LeBlance B, Schnoll R, Cornett C, Schweitzer L, Wallinga D, Hightower J, Patrick L, Lukiw WJ (2009) Mercury from chlor-alkali plants: measured concentrations in food product sugar. Environ Health 8:2. doi:10.1186/1476-069X-8-2
Enell A, Andersson Y, Vestin J, Wagelmans M (2016) Risk management and regeneration of brownfields using bioenergy crops. J Soils Sediments 16(3):987–1000
Finnan J, Styles D (2013) Hemp: a more sustainable annual energy crop for climate and energy policy. Energ Policy 58:152–162
Foldi C, Dohrmann R, Matern K, Mansfeldt T (2013) Characterization of chromium-containing wastes and soils affected by the production of chromium tanning agents. J Soils Sediments 13:1170–1179
Garcia-Jaldon C, Dupeyre D, Vignon MR (1998) Fibers from semiretted hemp bundles by steam explosion treatment. Biotechnol Bioeng 14:251–260
George J, Masto RE, Ram LC, Das TB, Rout TK, Mohan M (2015) Human exposure risks for metals in soil near a coal-fired power-generating plant. Arch Environ Contam Toxicol 68:451–461
Goix S, Mombo S, Schreck E, Pierart A, Leveque T, Deola F, Dumat C (2015) Field isotopic study of lead fate and compartmentalization in earthworm-soil-metal particle systems for highly polluted soil near Pb recycling factory. Chemosphere 138:10–17
González-Garcıá S, Luo L, Moreira MT, Feijoo G, Huppes G (2012) Life cycle assessment of hemp hurds use in second generation ethanol production. Biomass Bioenergy 36:268–279
Gonzalez-Garcia S, Yudego BM, Murpphy RJ (2013) Life cycle assessment of potential energy uses for short rotation willow biomass in Sweden. Int J Life Cycle Assess 18:783–795
Hong CO, Gutierrez J, Yun SW, Lee YB, Yu C, Kim PJ (2009) Heavy metal contamination of arable soil and corn plant in the vicinity of a zinc smelting factory and stabilization by liming. Arch Environ Contam Toxicol 56:190–200
Hu Y, Cheng H (2012) Mercury risk from fluorescent lamps in China: current status and future perspective. Environ Int 44:141–150
Huang H, Yu N, Wang L, Gupta DK, He Z, Wang K, Zhu Z, Yan X, Li T, Yang X (2011) The phytoremediation potential of bioenergy crop Ricinus communis for DDTs and cadmium co-contaminated soil. Bioresour Technol 102:11034–11038
Jennings AA, Li Z (2014) Scope of the worldwide effort to regulate pesticide contamination in surface soils. J Environ Manag 146:420–443
Jiang Y, Lei M, Duan L, Longhurst P (2015) Integrating phytoremediation with biomass valorisation and critical element recovery: a UK contaminated land perspective. Biomass Bioenergy 83:328–339

Jiang Y, Yves UJ, Sun H, Hu X, Zhan H, Wu Y (2016) Distribution, compositional pattern and sources of polycyclic aromatic hydrocarbons in urban soils of an industrial city, Lanzhou, China. Ecotoxicol Environ Saf 126:154–162

Jones DIH (1970) Cell-wall constituents of some grass species and varieties. J Sci Food Agric 21(11):559–562

Keller A, Leupin M, Mediavilla V, Wintermantel E (2001) Influence of the growth stage of industrial hemp on chemical and physical properties of the fibres. Ind Crop Prod 13:35–48

Konecny L, Ettler V, Kristiansen SM, Amorim MJB, Kribek B, Mihaljevic M, Sebek O, Nyambe I, Fordsmand JJ (2014) Response of Enchytraeus crypticus worms to high metal levels in tropical soils polluted by copper smelting. J Geochem Explor 144:427–432

Koptsik GN (2014) Problems and prospects concerning the phytoremediation of heavy metal polluted soils: a review. Eurasian Soil Sci 47(9):923–939

Kriese U, Schumann E, Weber WE, Beyer M, Bruhl L, Matthaus et al (2004) Oil content, tocopherol composition and fatty acid patterns of the seeds of 51 *Cannabis sativa* L. genotypes. Euphytica 137:339–351

Kumar N, Kumar S, Bauddh K, Dwivedi N, Singh KA, Singh DP (2009) Phytotoxicity of industrial effluent to *Raphanus sativus* L. and *Trigonella foenumgracum* L. J Ecophysiol Occup Health 9:163–169

Kumar N, Bauddh K, Kumar S, Dwivedi N, Singh DP, Barman SC (2012) Extractability and phytotoxicity of heavy metals present in petrochemical industry sludge. Clean Techn Environ Policy 15:1033–1039

Kumar R, Alamelu D, Acharya R, Rai AK (2014a) Determination of concentrations of chromium and other elements in soil and plant samples from leather tanning area by Instrumental Neutron Activation Analysis. J Radioanal Nucl Chem 300:213–218

Kumar N, Kumar S, Bauddh K, Dwivedi N, Singh DP, Barman SC (2014b) Toxicity assessment of effluent from flash light manufacturing industry by bioassays tests in methi (Trigonella foenumgracum). J Environ Biol 35:1107–1113

Kumar N, Kumar S, Bauddh K, Dwivedi N, Singh DP, Barman SC (2015) Toxicity assessment and accumulation of metals in Raphanus sativus L. irrigated with battery manufacturing industry effluent. Int J Veg Sci 21(4):373–385

Latif S, Anwar F (2009) Physicochemical studies of hemp (Cannabis sativa) seed oil using enzyme-assisted cold-pressing. Eur J Lipid Sci Technol 111:1042–1048

Licht L, Isebrands JG (2005) Linking phytoremediated pollutant removal to biomass economic opportunities. Biotechnol Bioeng 28:203–218

Li S, Jd S, Richard SP (2010) The feasibility of converting *Cannabis sativa* L. oil into biodiesel. Bioresour Technol 101:8457–8460

Li C, Xiao B, Wang H, Yao SH, Wu JY (2014) Phytoremediation of Zn- and Cr-contaminated soil using two promising energy grasses. Water Air Soil Pollut 225:20–27

Linde M, Jakobsson EL, Galbe M, Zacchi G (2008) Steam pretreatment of dilute $H_2SO_4^-$ impregnated wheat straw and SSF with low yeast and enzyme loadings for bioethanol production. Biotechnol Bioeng 32:326–332

Linger P, Mussig J, Fischer H, Kobert J (2002) Industrial hemp (*Cannabis sativa* L.) growing on heavy metal contaminated soil: fibre quality and phytoremediation potential. Ind Crop Prod 16:33–42

Liu J, Zhang X, Tran H, Wang DQ, Zhu YN (2011) Heavy metal contamination and risk assessment in water, paddy soil, and rice around an electroplating plant. Environ Sci Pollut Res 18:1623–1632

Ma Z, Li T, Qu C, Bi J, Huang L (2014) Evaluation and source identification of trace element contamination of soils in the Qixia lead-zinc mining area, Jiangsu, China. J Soils Sediments 14:1703–1712

Malik RN, Husain SZ, Nazir I (2010) Heavy metal contamination and accumulation in soil and wild plant species from industrial area of Islamabad. Pak J Bot 42(1):291–301

Marafi M, Stanislaus A (2008) Spent hydroprocessing catalyst management: a review Part II. Advances in metal recovery and safe disposal methods. Resour Conserv Recycl 53:1–26

Marques MC, Nascimento A (2013) Analysis of chlorophyll fluorescence spectra for the monitoring of Cd toxicity in a bio-energy crop (*Jatropha curcas*). J Photochem Photobiol B Biol 127:88–93

Matthäus B, Brühl L (2008) Why it is so difficult to produce high quality virgin rapeseed oil for human consumption? Eur J Lipid Sci Technol 110:611–617

Mattsson JE, Briere DM (1984) External biomass characteristics for efficient energy conversion-some considerations. Biomass characteristics for efficient energy conversion. In: Gambles RL et al (eds) Proceedings of joint workshop of IEA/FE programme Groups B, C and D, Uppsala, Sweden, Canadian Forestry Service, Ottawa, Canada

Meers E, Ruttens A, Hopgood M, Lesage E, Tack FMG (2005) Potential of *Brassica rapa*, *Cannabis sativa*, *Helianthus annuus* and *Zea mays* for phytoextraction of heavy metals from calcareous dredged sediment derived soils. Chemosphere 61:561–572

Mihoc M, Pop G, Alexa E, Radulov I (2012) Nutritive quality of romanian hemp varieties (*Cannabis sativa* L.) with special focus on oil and metal contents of seeds. Chem Central J 6:122

Moxley G, Zhu Z, Zhang YP (2008) Efficient sugar release by the cellulose solvent-based lignocellulose fractionation technology and enzymatic cellulose hydrolysis. J Agric Food Chem 56:7885–7890

Nathan M (2009) Soil testing for lead for garden and landscape soils. Missouri environment and garden Columbia, MO: University of Missouri, College of Agriculture, Food and Natural Resources, Division of Plant Sciences, Plant Protection Programs 15(4):23–32

Niu L, Xu Y, Xu C, Yun L, Liu W (2014) Status of phthalate esters contamination in agricultural soils across China and associated health risks. Environ Pollut 195:16–23

Öhgren K, Galbe M, Zacchi G (2005) Optimization of steam pretreatment of SO_2 impregnated corn stover for fuel ethanol production. Appl Biochem Biotechnol 121-1241055-1068

Olivares AR, Carrillo-Gonzalez R, Gonzalez-Chavez MC, Hernandez RMS (2013) Potential of castor bean (*Ricinus communis* L.) for phytoremediation of mine tailings and oil production. J Environ Mngmnt 114:316–323

Ono FB, Penido ES, Tappero R, Sparks D, Guilherme LRG (2015) Bioaccessibility of Cd and Pb in tailings from a zinc smelting in Brazil: implications for human health. Environ Geochem Health. doi:10.1007/s10653-015-9774-0

Oomah BD, Busson M, Godfrey DV, Drover JCG (2002) Characteristics of hemp (Cannabis sativa L.) seed oil. Food Chem 76:33–43

Pakarinen A, Maijala P, Stoddard FL, Santanen A, Tuomainen P (2011) Evaluation of annual bioenergy crops in the boreal zone for biogas and ethanol production. Biotechnol Bioeng 35:3071–3078

Prade T, Svensson SE, Mattsson JE (2012) Energy balances for biogas and solid biofuel production from industrial hemp. Biotechnol Bioeng 40:36–52

Quan SX, Yan B, Yang F, Li N, Xiao X, Fu J (2015) Spatial distribution of heavy metal contamination in soils near a primitive e-waste recycling site. Environ Sci Pollut Res 22:1290–1129

Ragit SS, Mohapatra SK, Gill P, Kundu K (2012) Brown hemp methyl ester: transesterification process and evaluation of fuel properties. Biotechnol Bioeng 41:14–20

Raposo C, Roeser HM (2001) Contamination of the environment by the current disposal methods of mercury-containing lamps in the state of Minas Gerais, Brazil. Waste Manag 21(7):661–670

Reilly JO, Watts MJ, Shaw RA, MArcilla AL, Ward NI (2010) Arsenic contamination of natural waters in San Juan and La Pampa, Argentina. Environ Geochem Health 32:491–515

Rice B (2008) Hemp as a feedstock for biomass-to-energy conversion. J Ind Hemp 13(2):145–156

Rockwood DL, Naidu CV, Carter DR, Rahmani M, Spriggs TA, Lin C, Alker GR, Isebrands JG, Segrest SA (2004) Short-rotation woody crops and phytoremediation. Opportunities for agroforestry? Agrofor Syst 61(1):51–63

Rodriguez-Iruretagoiena A, Vallejuelo SF, Gredilla A, Ramos CG, Oliveira MLS, Arana G, Diego A, Madarianga A, Silva LFO (2015) Fate of hazardous elements in agricultural soils surrounding a coal power plant complex from Santa Catarina (Brazil). Sci Total Environ 508:374–382

Romeo S, Francini A, Ariani A, Sebastiani L (2014) Phytoremediation of Zn: identify the diverging resistance, uptake and biomass production behaviours of poplar clones under high zinc stress. Water Air Soil Pollut 225:1813

Sheng X, Sun L, Huang Z, He L, Zhang W, Chen Z (2012) Promotion of growth and Cu accumulation of bio-energy crop (*Zea mays*) by bacteria: Implications for energy plant biomass production and phytoremediation. J Environ Manag 103:58–64

Shi G, Cai Q (2009) Cadmium tolerance and accumulation in eight potential energy crops. Biotechnol Adv 27:555–561

Shi G, Cai Q, Liu Q, Wu L (2009) Salicylic acid-mediated alleviation of cadmium toxicity in hemp plants in relation to cadmium uptake, photosynthesis, and antioxidant enzymes. Acta Physiol Plant 31:969–977

Shi G, Liu C, Cui M, Ma Y, Cai Q (2012) Cadmium tolerance and bioaccumulation of 18 hemp accessions. Appl Biochem Biotechnol 168:163–173

Silveira ML, Vendramini JMB, Sui X, Sollenberger L, Connor GA (2013) Screening perennial warm-season bioenergy crops as an alternative for phytoremediation of excess soil P. BioEnerg Res 6:469–475

Sipos B, Kreuger E, Svensson SE, Réczey K, Björnsson L (2010) Steam pretreatment of dry and ensiled industrial hemp for ethanol production. Biomass Bioenergy 34(12):1721–1731

Sun Y, Yuan G, Li J, Li J, Wang G (2015) Polybrominated diphenyl ethers in surface soils near the Changwengluozha Glacier of Central Tibetan Plateau. China Sci Tot Environ 511:399–406

Sun J, Pan L, Zhan Y, Lu H, Tsang DCW, Liu W, Wang X, Li X, Zhu L (2016) Contamination of phthalate esters, organochlorine pesticides and polybrominated diphenyl ethers in agricultural soils from the Yangtze River Delta of China. Sci Total Environ 544:670–676

Su G, Li F, Lin J, Liu C, Shi G (2013) Peanut as a potential crop for bioenergy production via Cd-phytoextraction: a life-cycle pot experiment. Plant Soil 365:337–345

Surriya O, Saleem SS, Wagar K, Kazi AG (2015) Phytoremediation of soils: prospects and challenges, Ch 1. In: Hakeem K, Sabir M, Ozturk M, Murmet A (eds) Soil remediation and plants, prospects and challenges. Academic, Amsterdam, pp 1–36

Vandenhove H, Hees MV (2005) Fibre crops as alternative land use for radioactively contaminated arable land. J Environ Radioact 81:131–141

Wang Z, Calderon MM (2012) Environmental and economic analysis of application of water hyacinth for eutrophic water treatment coupled with biogas production. J Environ Manag 110:246–253

Wang XJ, Chen J, Zhang ZH, Piao XY, Hu JD, Tao S (2004) Distribution and sources of polycyclic aromatic hydrocarbons in soil profiles of Tianjin area, People's Republic of China. Bull Environ Contam Toxicol 73:739–748

Wei-Hua Z, Ying-Xin W, Simonnot MO (2012) Soil contamination due to e-waste disposal and recycling activities: a review with special focus on China. Pedosphere 22(4):434–455

Zhang L, Dong L, Shi S, Zhou L, Zhang T, Huang Y (2009) Organochlorine pesticides contamination in surface soils from two pesticide factories in Southeast China. Chemosphere 77(5):628–633

Zhang X, Yang L, Li Y, Li H, Wang W, Ye B (2012) Impacts of lead/zinc mining and smelting on the environment and human health in China. Environ Monit Assess 184:2261–2273

Zhang H, Tian Y, Wang L, Zhang L, Dai L (2013) Ecophysiological characteristics and biogas production of cadmium-contaminated crops. Bioresour Technol 146:628–636

Zhang X, Zhang X, Gao B, Xia H, Li H, Li J (2014) Effect of cadmium on growth, photosynthesis, mineral nutrition and metal accumulation of an energy crop, king grass (*Pennisetum americanum, P. purpureum*). Biomass Bioenergy 67:179–187

11 Phytoremediation and Bioenergy Production Efficiency of Medicinal and Aromatic Plants

C.K. Jisha, Kuldeep Bauddh, and Sushil Kumar Shukla

Abstract

The cultivation of aromatic and medicinal plants for their direct as well as indirect uses is a common practice from ancient time. There are several medicinal plants which have not only the tolerance ability against the environmental contaminants but also may extract them from the polluted sites. Essential oil-bearing crops like peppermint (*Mentha* sps.), tulsi (*Ocimum basilicum* L.), industrial hemp (*Cannabis sativa* L.), *Cymbopogon citratus* etc., have been found to bear substantial efficiency to accumulate toxic metals e.g., Cd, As, Ni, Cu, Fe, etc. Generally the process used to extract the essential oil is steam distillation which has the least chance to allow the contaminants to move in oil. After harvesting the oil, residual biomass may be utilized for energy production. This energy may be produced by direct burning of biomass or production of biogas through the gasification of biomass. This integrated approach will not only reduce the cost of petroleum oil but also will help to develop a sustainable model which will help in mitigation of many environmental issues like reduction of greenhouse gases, pollution alleviation etc.

Keywords

Biomass • Bioenergy • Medicinal plants • Phytoremediation

C.K. Jisha
Centre for Tribal Folklore, Language and Literature, Central University of Jharkhand, Ranchi-835205, Jharkhand, India

K. Bauddh (✉) • S.K. Shukla
Centre for Environmental Sciences, Central University of Jharkhand, Ranchi-835205, Jharkhand, India
e-mail: kuldeepenvir0811@gmail.com

K. Bauddh et al. (eds.), *Phytoremediation Potential of Bioenergy Plants*,
DOI 10.1007/978-981-10-3084-0_11

11.1 Introduction

Generally, the aromatic plants have several unique features like social and environmental benefits, industrial/commercial crops, multiple harvests within a year, tolerance against both biotic and abiotic stress conditions, unpalatable nature etc. (Verma et al. 2014). Aromatic and medicinal plants are cultivated from ancient time for extracting the essential oils. The oil of these plants is commonly used as an important ingredient in detergents, biological pesticides, soaps, cosmetic etc. The biomass (as by-product) may be utilized for energy production or manufacturing of biomanure/green fertilizer which may reduce the over extraction of crude oil and excessive utilization of synthetic fertilizers, respectively. Utilization of biomass for bioenergy is an old practice which is a cheap alternative and renewable energy source (Mckendry 2002). The biomass of nonedible plants, grasses (both terrestrial and aquatic), etc., may be utilized for the purpose of energy generation. The basic features of interest of material of biomass for processing as energy source are described in Fig. 11.1 (McKendry 2002).

The waste/by-products of some industrial plants may also be used for the purpose of energy production. The biomass derived from medicinal plants after extraction of essential oils or other important component may become a good alternative (Gupta et al. 2013; Tamari et al. 2014). Numerous medicinal plants have been reported that have the ability to accumulate the environmental contaminants (both organic and inorganic) (Gupta et al. 2013). These plants may be applied for the remediation of environmental contaminants simultaneously can fulfill several objectives like production of pharmaceutical component, energy production, carbon sequestration, and pollution remediation (Fig. 11.2).

Aromatic and medicinal plants may be used to remediate the contaminated sites because they have least chances of metal transfer from soil to essential oil or alterations in its composition (Zheljazkov et al. 2006) which is due to the process used for oil extraction (Scora and Chang 1997; Bernstein et al. 2009; Pandey and Singh 2015). Further, it is confirmed that metals linger in residues of plants (at bottom of pot) during oil extraction by steam distillation (Zheljazkov and Nielsen 1996a, b; Scora and Chang 1997; Zheljazkov et al. 2006). In the present chapter, efforts have been made to gather and interpret the available information which shall extend the knowledge of this area one step ahead. The medicinal plants that have phytoremediation ability and may also be utilized for energy production have been discussed.

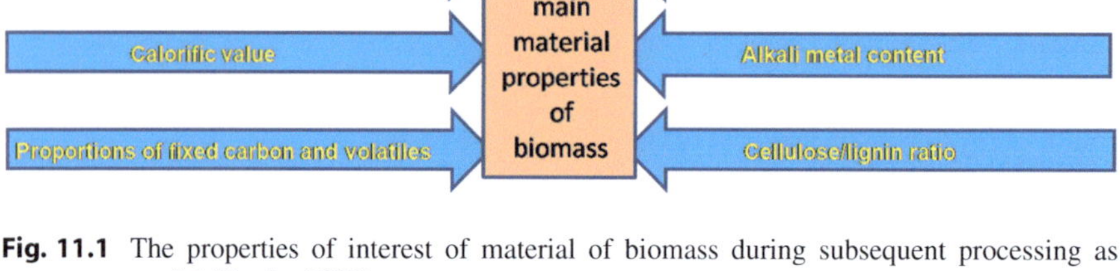

Fig. 11.1 The properties of interest of material of biomass during subsequent processing as energy source (McKendry 2002)

Fig. 11.2 An integrated approach and multi-aspects of medicinal plants which may also be utilized for the remediation of contaminated sites and the by-product (biomass) may be utilized for biofuel production (Image source: Google)

11.2 Important Medicinal Plants and Their Uses

Humans have a close relation with nature which started along with them on earth. Without nature it does not have any existence. They were used to collect their food from nature and gradually they realized medicinal values of the plants too. These realizations lead to the origin of medicine from nature and commonly popularized as folk medicinal practices. Generally, folk medicinal practices can be divided into preventive and therapeutic medicine. All human beings are practicing the preventive medicinal practices in their daily life. The traditional foods, dress, and the hygiene systems are part and partial of the preventive medicinal practices. Human beings are conscious and aware about their dress and food by considering the weather condition. The vegetables and fruits which they are having have a prominent role in preventive medicinal practice. Our daily practices are protecting our body and health from diseases. Therapeutic medicinal practice is for curing the diseases directly. It needs proper care along with medicines. Therapeutic medicine can be classified in two categories as home medicinal practices and traditional medicinal practices. Home medicinal practice is related with the family. It is done in the premises of home by using the available medicinal plants. Traditional medicinal practices are done by some particular communities. Medicinal plants have a prominent role in folk medicinal practices. There are plenty of medicinal plants in the world with multipurpose uses that simultaneously protect the human beings and nature. Several medicinal plants which are commonly used by traditional medicinal practitioners are listed in Table 11.1 This is a field-based report done by the author (Jisha CK) at Kerala state of India.

Table 11.1 Important medicinal plants and their uses

Plant	Family	Features	Image
Hibiscus Botanical name: *Hibiscus rosa-sinensis*	Malvaceae	It is normally seen in India and China. It is included in bush type of plants with various colors. It is useful for medicine of menstrual problems and as antidandruff shampoo	
Indian borage Botanical name: *Plectranthus amboinicus* (Lour) Spreng	Lamiaceae	Its stem is fleshy and leaves are soft with light green color in oval shape. The extract of leaves is used to prepare oils which protect small children from cold and fever	
Tulsi Botanical name: *Ocimum tenuiflorum*	Lamiaceae	Tulsi or holy basil is considered a holy plant in India and is used in traditional medicines. There are different varieties of tulsi. It is considered good for the treatment of cough and cold	
Brahmi Botanical name: *Bacopa monnieri*	Plantaginaceae	Brahmi is a creeper with small leaves and bitter taste. It is used to enhance memory. It is also useful in the treatments of asthma and leprosy	
Biophytum Botanical name: *Biophytum sensitivum*	Oxalidaceae	Biophytum is an annual plant with small leaf bunch and yellow flower. It is good for wound healing. It is used to cure loose motion and used to clean the womb and cure the wounds inside the womb	
Peepal Botanical name: *Ficus religiosa*	Moraceae	Peepal is a sacred tree in India with many medicinal values. It is known as a natural air freshener. Peepal has several components which are good to prepare the medicines to cure asthma, loose motion, and other gastric problems	
Purple fleabane Botanical name: *Cyanthillium cinereum*	Asteraceae	Purple fleabane is an annual plant. It is mainly used to cure kidney problems as well as fever. Fleabane is the excellent medicine to cure malaria	

Pepper Botanical name: *Piper nigrum*	Piperaceae	Pepper is a woody vine with dark-green leaves with oval shape. Pepper is an important spice of many countries especially of India. It is good for toothache and is used as an ingredient for most of the tooth powders	
Aloe Botanical name: *Aloe vera*	Liliaceae	Aloe is an Ayurvedic plant seen in almost all geographic conditions. The juice of the aloe leaves is used as traditional shampoo to remove dandruff. It can be directly used as a medicine to heal the wounds and burn on the skin. It is a good medicine for stomach disorders	
Malabar nut Botanical name: *Adhatoda beddomei Clarke*	Acanthaceae	Malabar nut is a medicinal plant used in both Ayurveda and traditional medicine. It is good for blood purification and it can cure vomiting, cough, and menstrual problems. Only raw Malabar nut is used as medicine because it becomes poisonous once it dries	
Butterfly pea (blue) Botanical name: *Clitoria ternatea*	Fabaceae	Butterfly pea (Asian pigeonwings) is a woody vine with blue and light yellow color flowers. It is a useful medicine for increasing memory power. It is also used to treat infertility problems	
Creeping tick trefoil Botanical name: *Desmodium triflorum*	Fabaceae	Creeping tick trefoil is a branched creeper and commonly known as the creeping herb. It is a good medicine to cure skin infections	
Betel Botanical name: *Piper betle*	Piperaceae	Betel is a wood vine having an important role in traditional medicine used for poisonous infection of the skin. Betel is used to prepare oil for hair growth. It is used along with areca nut to prepare betel nut and used for chewing as a mouth freshener	
Nut grass Botanical name: *Cyperus rotundus*	Cyperaceae	Nut grass has thin and long leaves and nut-shaped white flowers. It is good for fever and stomachache. It has been proved as a good medicine for leprosy	

(continued)

Table 11.1 (continued)

Plant	Family	Features	Image
Leucas aspera Botanical name: *Leucas aspera*	Lamiaceae	*Leucas aspera* is an important medicinal plant commonly used in traditional medicines. It is generally used for the treatment of indigestion and to heal wounds and kid's stomach problems. Its juice is used after delivery for curing the internal wound in the womb and to clean the womb	
Dhub grass Botanical name: *Cynodon dactylon*	Poaceae	Dhub grass is used for both medicinal and sacred purposes. It bears thin green leaves and flat stems. It is used to cure wounds, headache, stomach problems, and skin diseases	
Black musli Botanical name: *Curculigo orchioides*	Amaryllidaceae	Black musli is a good medicine for strength and body rejuvenation. It can cure headache, earache, and gynecology problems in female. It is considered good in curing jaundice	
Arrowroot Botanical name: *Maranta arundinacea* Linn.	Marantaceae	Arrowroot is used in the treatment of stomach disorders like dysentery	
Turmeric Botanical name: *Curcuma longa*	Zingiberaceae	Turmeric is a good germ fighter. It is considered good for the skin	
Bael Botanical name: *Aegle marmelos*	Rutaceae	Bael is a long-lived bushy tree. Bael has both medicinal and religious values. It is good for stomach and sugar problems. Consumption of the juice of bael in summers prevents loo (wind of extreme hot and dry summers)	
Tamarind Botanical name: *Tamarindus indica*	Fabaceae	It is found in almost all geographic areas. The fruit is fleshy and juicy. It is used for cooking to add flavor. Its leaves are good for treatment of pain and swelling	

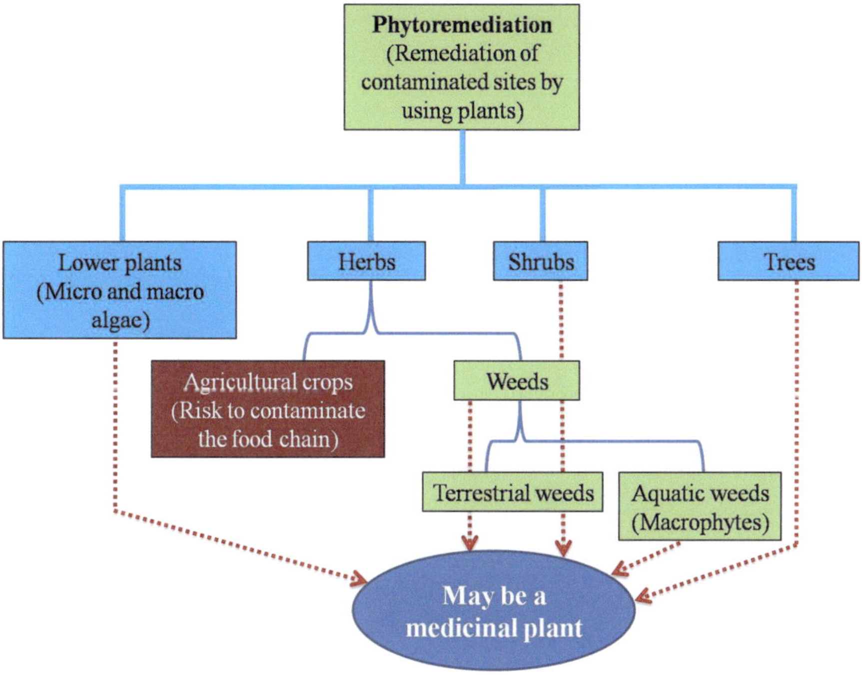

Fig. 11.3 Selection of medicinal plants for phytoremediation

11.3 Phytoremediation Potential of Medicinal Plants

The utilization of plants or plant products for the removal of environmental contaminants is known as phytoremediation. Albeit this technique is almost 2–3 decades old, the recent developments made it more attractive and efficient. At the beginning when the technique emerged, more emphasis was given to search plants that have the ability to accumulate contaminants especially toxic metals. Numerous plants have been investigated as phytoremediator and many of them are listed as hyperaccumulators (plants which can accumulate high amount of metals in ground parts and transfer them into their aerial parts). The selection of plant for the removal of contaminants is an important step of phytoremediation. At present the researchers are looking to find plants which have the contaminant-accumulating efficiency and simultaneously could provide some value-added products (either at the same time or after harvesting the crop). Several medicinal and aromatic plants (herb, shrub, as well as tree Fig. 11.3) are found to be tolerant to both biotic and abiotic stresses and have the ability to accumulate contaminants that may be used as phytoremediator plant (Table 11.2).

Chrysopogon zizanioides commonly known as vetiver grass has been found suitable for the removal of organic (e.g., 2, 4, 6-trinitrotoluene, phenol, and petroleum hydrocarbon) as well as inorganic (specially toxic metals like lead, cadmium,

Table. 11.2 Medicinal plants studied for their phytoremediation potential of different heavy metals

Plant	Family	Metal	Tissue	Reference
Bacopa monnieri	Plantaginaceae	Hg, Cd, Cu, Cr, Mn, Pb	Roots and shoots	Sinha and Chandra (1990); Sinha et al. (1996); Sinha (1999); and Hussain et al. (2011)
Ocimum basilicum	Lamiaceae	Cd	Roots, shoots, and leaves	Putwattana et al. (2010)
Catharanthus roseus	Apocynaceae	Cd, Pb, Ni, Cr	Leaves, stem, and roots	Pandey et al. (2007); Subhashini and Swamy (2013); and Ahmad and Misra (2014)
Mentha sps.	Lamiaceae	Cr, Ni, Cd, Pb	Roots and shoots	Zurayk et al. (2001, 2002) and Sa et al. (2014)
Aloe vera	Xanthorrhoeaceae	Cr		Sharma and Adholeya (2011)
Cannabis sativa	Cannabaceae	Ni, Pb, Cr, Cd	Seeds, leaves, fibers, and hurds	Linger et al. (2002); Citterio et al. (2003); Shi and Cai (2009)
Cymbopogon citratus	Poaceae	Cr, Ni, Pb, Cd	Roots and shoots	Pandey et al. (2015)
Ricinus communis	Euphorbiaceae	Cd, Ni	Root and shoots	Shi and Cai (2009); Bauddh and Singh (2012a, b, 2015a, b); Bauddh et al. (2016a, b)
Azadirachta indica	Meliaceae	Tannery effluent-contaminated soils (Cd, Cu, Zn, Cr, Pb)	Root and shoot	Thangaswamy et al. (2015)
Withania somnifera	Solanaceae	Cu	Root and shoot	Varun et al. (2012)
Euphorbia hirta		Cd	Root and shoot	Hamzah et al. (2016)
Vetiver grass (*Chrysopogon zizanioides*)		As 45–450 mg As/kg soil	Root and shoot	Datta et al. (2011)

copper, zinc, and arsenic) contaminants (Chen et al. 2004; Brandt et al. 2006; Makris et al. 2007; Singh et al. 2008; Singhakant et al. 2009; Datta et al. 2011; Balasankar et al. 2013; Ho et al. 2013). Due to high root biomass and ability to penetrate deep layers of the soil, the plant vetiver may be used for the remediation

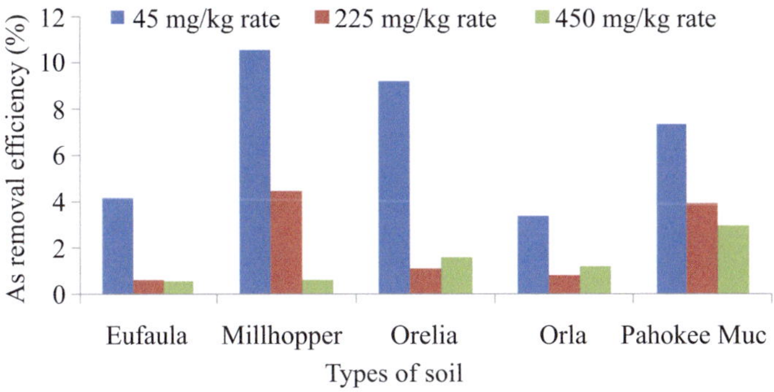

Fig. 11.4 Arsenic removal efficiency of vetiver grass when cultivated in different types and different level of As contamination (Datta et al. 2011)

of deep contaminated soil (Truong 2000; Pichai et al. 2001). Truong (2000) reported that this plant has the ability to restore the soil loaded with mining waste. In an experiment conducted by Datta et al. (2011), vetiver grass was found to be a potential phytoremediator of As when cultivated in different types of soils. The plant was found to have significant As removal efficiency, i.e., up to 10.6% when grown in 45 mg As/kg of soil (Millhopper soil) (Fig. 11.4).

Hamzah et al. (2016) have conducted an experiment to assess the cadmium accumulation efficiency of some indigenous plants. They found that *Euphorbia hirta*, an industrially important plant (especially in medical and petro industries), has significant Cd bioaccumulation efficacy. *Euphorbia hirta* has been also found suitable for the remediation of radioactive waste (Hu et al. 2014). *Hypericum perforatum* L. is an important medicinal plant commonly used as an antidepressive agent and has excellent ability to accumulate substantial amount of Cd (Schneider and Marquard 1996; Kim et al. 1999; Müller et al. 1999; Malko 2002; Verotta 2003). Workers have suggested that *Hypericum* sp. can accumulate good content of Cd in their aerial parts without significant negative effects on the growth and dry biomass. Hypericin, an important component of essential oil, was not found adversely affected with Cr when cultivated in Cr-contaminated media (Tirillini et al. 2006).

Ricinus communis commonly known as castor is used as a medicinal plant from very ancient times as anti-implantation activity, anti-inflammatory activity, antitumor activity, antiasthmatic activity, etc., and has been recently popularized as a vigorous phytoremediator for the cleanup of numerous pollutants both organic and inorganic types (Adhikari and Kumar 2012; Abreu et al. 2012; Bauddh and Singh 2012a, b, 2015a, b; Bauddh et al. 2015a, b, 2016a, b). According to Zahir et al. (2010), the extracts of castor have acaricidal and insecticide properties which work against *Haemaphysalis bispinosa* and hematophagous fly, *Hippobosca maculata*. The plant has excellent ability to grow easily up to 150 mg/kg Cd and also accumulate substantial amount of the metal in its roots and shoots (Bauddh and Singh 2012a; Fig. 11.5). The plant has good resistance capacity toward biotic and abiotic stresses. Moreover, *R. communis* is an annual herb that produces high biomass

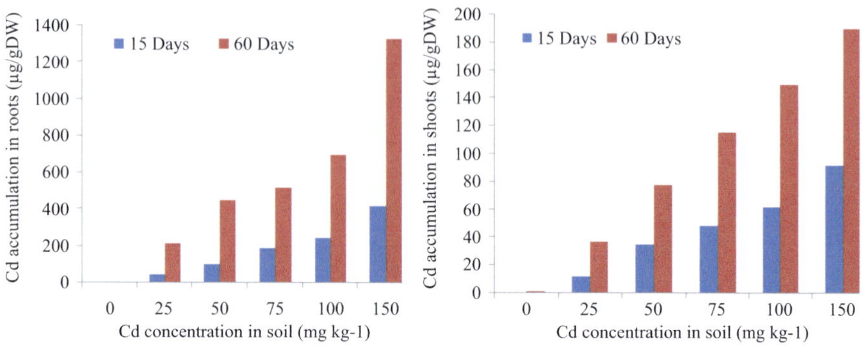

Fig. 11.5 Cadmium accumulation in roots and shoots of *Ricinus communis* (Bauddh and Singh 2012a)

which can be utilized for energy production as biomass. The husk and cake produced during oil extraction may also be applied as manure because it has significant amount of nutrients (Zahir et al. 2010). The plant has been found to accumulate higher amount of heavy metals like Cd, Ni, Pb, Cu, As, Cr, Zn, Ba, etc. (Romeiro et al. 2006; Zhi-Xin et al. 2007; Mahmud et al. 2008; Bauddh and Singh 2012a, b, 2015a, b; Bauddh et al. 2015a, b, 2016a, b; Adhikari and Kumar 2012; Abreu et al. 2012). The plantation of *R. communis* has also been reported to improve the soil physicochemical properties when cultivated in wasteland soils (Wu et al. 2012).

Hypericum perforatum L., a medicinally important plant especially used in depression (Verotta 2003), has been found to be effective for remediation of Cd from soil (Malko 2002; Schneider and Marquard 1996). The plant did not express adverse effects on growth as well as dry biomass production. Bishehkolaei et al. (2011) found that *Ocimum basilicum* L. has good efficiency to accumulate chromium (Cr) in its tissues. The plant developed defense mechanism by restricting the metal in their roots. In a recent study, *Ocimum basilicum* has expressed its suitability for the phytoremediation of Cd-contaminated soil which was further improved when the plants were supplied with different types of fertilizers (Zahedifara et al. 2016). Rai et al. (2004) reported that *Ocimum tenuiflorum* L. has the ability to bear up phytotoxicity of Cr by changing different metabolic processes.

Six wild plants, viz., *Malva parviflora*, *Datura stramonium*, *Citrullus colocynthis*, *Rhazya stricta*, *Phragmites australis*, and *Lycium shawii*, were assessed for their phytoremediation potential of Cd, Zn, Cu, Ni, and Pb by Ibrahim et al. (2013). They found *Datura stramonium*, a medicinal plant, suitable for phytostabilization of soil contaminated with Ni and Cu (Fig. 11.6).

Phytoremediation potential of *Cymbopogon martinii*, *Cymbopogon flexuosus*, and *Vetiveria zizanoides* was assessed for Cd by Lal et al. (2008). *Vetiveria zizanoides* was found to have high Cd tolerance. Metal (viz., Ni, Cr, Cd, Al, etc.) tolerance and bioaccumulation potential of several *Mentha* species have been reported by many authors (Zurayk et al. 2001, 2002; Ali et al. 2002; Manikandan et al. 2015). Zurayk et al. (2001) cultivated 12 different hydrophyte species along with four

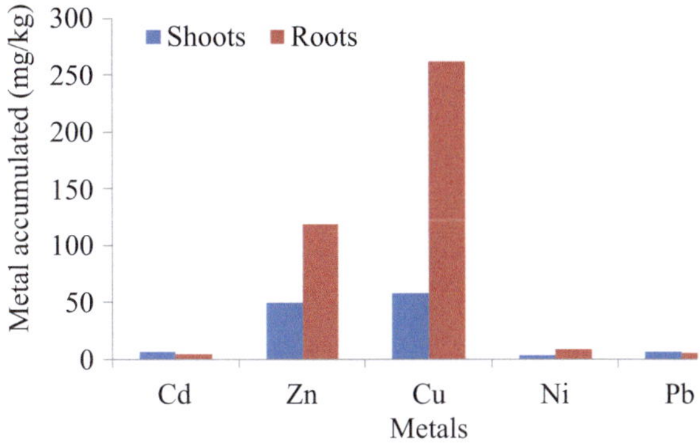

Fig. 11.6 Metal accumulation in roots and shoots of *Datura stramonium* cultivated in a site polluted by industrial wastes (Ibrahim et al. 2013)

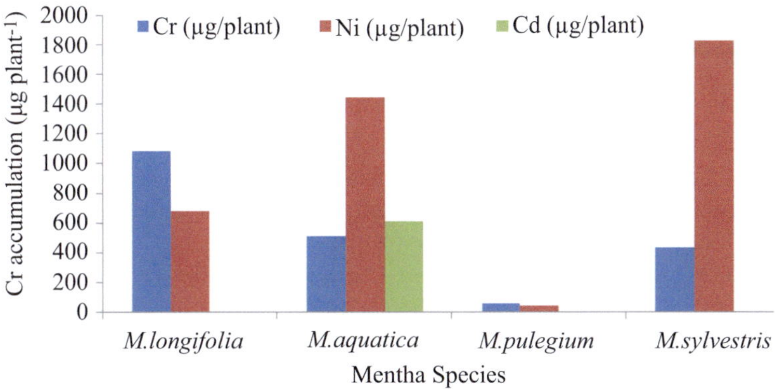

Fig. 11.7 Accumulation of Cr, Ni, and Cd in four *Mentha* species grown for 21 days in solution with 1.0 ppm metal levels (Zurayk et al. 2001)

Mentha species, viz., *Mentha longifolia*, *Mentha aquatica*, *Mentha pulegium*, and *Mentha sylvestris*, in 1.0 ppm Cr, Ni, and Cd contamination. They found that all *Mentha* species accumulated a substantial amount of all studied metals (Fig. 11.7). *M. longifolia* was found to accumulate Cr (1076.8 µg Cr plant^{-1}), whereas *M. sylvestris* accumulated Ni (1822 µg Cr plant^{-1}) which was the highest level of the metals accumulated among all (12) studied plants.

Two species of *Mentha*, i.e., *Mentha aquatica* L. and *Mentha sylvestris* L., were cultivated in the solutions having 1.0, 2.0, 4.0, and 8.0 mg Ni L^{-1} for 14 days (Zurayk et al. 2002). Both the species accumulated high amount of Ni in their roots (8327 mg Ni kg^{-1} dry weight in *M. aquatica* and 6762 mg/kg dry weight in *M. sylvestris*) which indicated this plant as a vigorous phytoremediator. Two medicinally important plants (*Centella asiatica* and *Orthosiphon stamineus*) have been analyzed for their phytoremediation potential of heavy metals by several researchers (Abdu et al. 2011; Salim et al. 2013; Manan et al. 2015). In a study conducted by Manan et al. (2015), the comparison was made between these two medicinal plants (*C. asiatica*

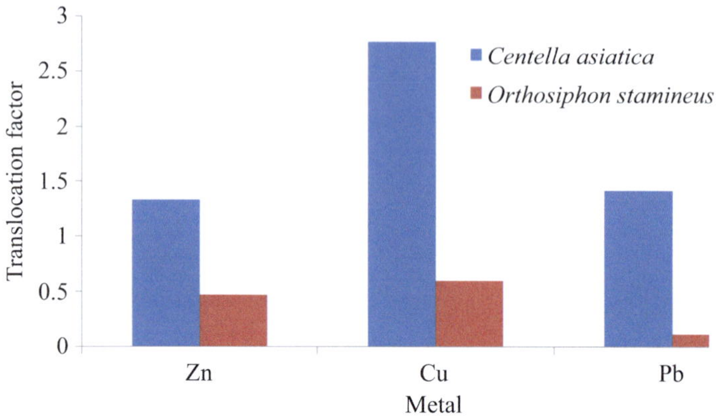

Fig. 11.8 Translocation factor of Zn, Cu, and Pb in *Centella asiatica* and *Orthosiphon stamineus* (Manan et al. 2015)

and *O. stamineus*) for their Zn, Cu, and Pb accumulation and translocation and found that *C. asiatica* has translocation factor of more than one for all the studied metals (Fig. 11.8; Zn = 1.34, Cu = 2.77, and Pb = 1.42). Translocation factor more than one for all the metals indicates that *C. asiatica* may be used for phytoextraction. However, another plant *O. stamineus* has translocation factor of less than one but was found to accumulate substantial amount of metals and may be suitable for phytostabilization.

Allium sativum (garlic) was found to have good efficiency to accumulate cadmium in its roots which was 1826 times higher than the control (Jiang et al. 2001). The plant showed enough tolerance toward Cd and did not show toxic effects at lower concentrations. *Cannabis sativa* L. (hemp), an industrial (for fiber) and medicinal important plant cultivated in Ni-, Pb-, and Cd-contaminated soil to assess their phytoremediation potential and effects of these metals on quality of fiber, was investigated by Linger et al. (2002). They found that the highest amount of all studied metals was estimated in the leaves of the plant. Moreover, they found that none of the studied metal affects the quality of fiber. Citterio et al. (2003) grew *Cannabis sativa* in Cd-, Ni-, and Cr-contaminated soil and after 2 months of seed sowing found no substantial changes in plant growth (morphology). Metals caused enhanced production of phytochelatins which showed a strong defense mechanism against metal toxicity.

Cymbopogon martinii commonly known as palmarosa was found as a suitable candidate for phytostabilization of metals when cultivated in tannery sludge mixed soil (Pandey et al. 2015). The amendment of sludge was found to enhance the growth and productivity of *C. martini*. The translocation factor for all the studied metals, viz., Cr, Ni, Pb, and Cd, was found <1; however, bioconcentration factor was assessed >1 which indicates the suitability of this plant as an excellent phytostabilizer.

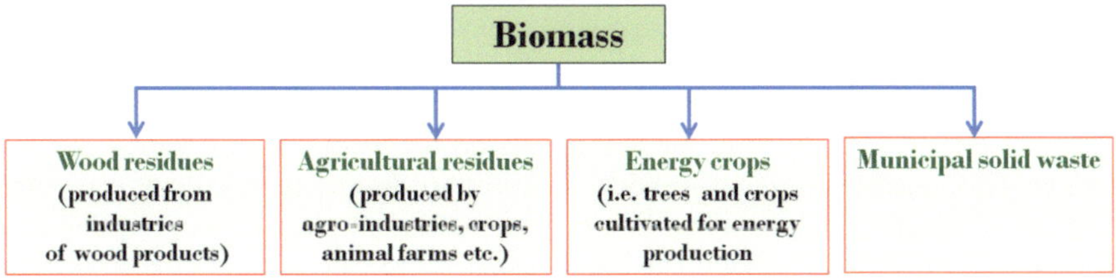

Fig. 11.9 Different sources of biomass residues (Easterly and Burnham 1996)

11.4 Bioenergy Production Efficiency of Medicinal Plants

Due to rocketing price of crude/petroleum oil and concern of environmental pollution, the entire world is searching new alternatives of energy. In the last few decades, bioenergy emerged as a potential solution which also found economically feasible and environmentally safe alternative of energy. Concerns of food versus fuel draw a boundary line to use the edible crops/plants, e.g., maize, sugarcane, soybeans, etc., as energy source (Sagar and Kartha 2007). Many other options have been investigated in the last few decades; many have been adopted commercially like blends of oil extracted from seeds of petro crops (*Jatropha curcas*, *Ricinus communis*, *Millettia pinnata*, etc.), solar energy, biogas, etc. Production of bioethanol, biobutanol, lipids, lignin, starch, etc., from the plants is a very common practice these days.

Proper utilization of waste material derived from the plants as biomass after harvesting the desired part may also become an integrated approach for many industries. Different sources of biomass residues are described in Fig. 11.9 (Easterly and Burnham 1996). The direct use of plant and plant parts (biomass) is one of the processes by which the consumption of petroleum oil may be reduced. It has been reported by several researchers (Karekezi and Kithyoma 2006; Hashiramoto 2007) that the biomass is a potential source of bioenergy. However, this practice will require a huge agricultural area to cultivate these plants which indirectly affect the agricultural productivity. Generally, woody crops, grasses, or herbaceous plants are considered potential energy crops. These crops may require low input, low water for irrigation, and low nutrients and provide high yield of biomass (McKendry 2002). The biomass of industrial crops/plants as a by-product may be directly or indirectly utilized for the energy production (Boyle 1996; Wereko-Broddy and Hagen 1996).

The biomass of medicinal crops (as a by-product) may also be used as an efficient energy source. This practice may initiate an innovative and sustainable business especially in pharmaceutical industries. Numerous medicinal plants like *Ricinus communis*, *Azadirachta indica*, *Syzygium cumini*, *Phyllanthus emblica*, *Cannabis sativa*, *Moringa oleifera*, *Madhuca longifolia*, etc., are high biomass-producing plants, and their biomass may be directly used for energy purposes. The oil of *Ricinus communis* has numerous components (e.g., ricinoleic acid, linoleic acid, oleic acid, stearic acid, palmitic acid, dihydroxystearic acid, linolenic acid, and eicosanoic acid) which make it a robust biodiesel plant (Maleki et al. 2013; Mejia et al. 2013; Ustra et al. 2013).

11.5 Conclusions

Due to high cost and unavailability of chemical drugs in outlying areas in ancient time, the total dependency for medicines was on the natural plants. Cultivation of aromatic and medicinal plants provides not only medicines but also helps in other ways like reduction in level of carbon dioxide, pollution remediation, biofuel production, etc. Many medicinal plants like vetiver, neem, peepal, and others produce huge amount of biomass and can be used for energy generation. The medicinal plants like *Bacopa monnieri*, *Ocimum basilicum*, *Catharanthus roseus*, *Mentha* sps., *Ricinus communis*, *Cannabis sativa*, *Chrysopogon zizanioides*, etc., have substantial potential to extract the contaminants when cultivated in polluted sites and may be adopted for pollution remediation.

Acknowledgment Dr. Kuldeep Bauddh is thankful to UGC for the award of UGC Start-Up Grant (3(B):2202.03.789.03.01.31).

References

Abdu A, Aderis N, Abdul-Hamid H, Majid NM, Jusop S, Karam DS, Ahmad K (2011) Using *Orthosiphon stamineus* B. for phytoremediation of heavy metals in soils amended with sewage sludge. Am J Appl Sci 8(4):323–331

Abreu CA, Cantoni M, Coscione AR, Paz-Ferreiro J (2012) Organic matter and barium absorption by plant species grown in an area polluted with scrap metal residue. Appl Environ Soil Sci: 1–7. http://dx.doi.org/10.1155/2012/476821

Adhikari T, Kumar A (2012) Phytoaccumulation and tolerance of *Ricinus communis* L. to nickel. Int J Theor Phys 14:481–492

Ahmad R, Misra N (2014) Evaluation of phytoremediation potential of *Catharanthus roseus* with respect to chromium contamination. Am J Plant Sci 5:2378–2388

Ali MS, Saleem M, Ahmad W, Parvez M, Yamdagni R (2002) A chlorinated monoterpene ketone, acylated β-sitosterol glycosides and a flavanone glycoside from *Mentha longifolia* (Lamiaceae). Phytochemistry 59(8):889–895

Balasankar D, Vanilarasu K, Preetha PS, Rajeswari S, Umadevi M, Bhowmik D (2013) Traditional and medicinal uses of vetiver. J Med Plant Stud 2(3):191–200

Bauddh K, Singh RP (2012a) Cadmium tolerance and its phytoremediation by two oil yielding plants *Ricinus communis* (L.) and *Brassica juncea* (L.) from the contaminated soil. Int J Theor Phys 14:772–785

Bauddh K, Singh RP (2012b) Growth, tolerance efficiency and phytoremediation potential of *Ricinus communis* (L.) and *Brassica juncea* (L.) in salinity and drought affected cadmium contaminated soil. Ecotoxicol Environ Saf 85:13–22

Bauddh K, Singh RP (2015a) Effects of organic and inorganic amendments on bio-accumulation and partitioning of Cd in *Brassica juncea* and *Ricinus communis*. Ecol Eng 74:93–100

Bauddh K, Singh RP (2015b) Assessment of metal uptake capacity of castor bean and mustard for phytoremediation of nickel from contaminated soil. Bioremed J 19(2):124–138

Bauddh K, Singh K, Singh B, Singh RP (2015a) *Ricinus communis*: a robust plant for bio-energy and phytoremediation of toxic metals from contaminated soil. Ecol Eng 84:640–652

Bauddh K, Singh K, Singh RP (2015b) *Ricinus communis* L. a value added crop for remediation of cadmium contaminated soil. Bull Environ Contam Toxicol 96(2):265–269

Bauddh K, Kumar A, Srivastava S, Singh RP, Tripathi RD (2016a) A study on the effect of cadmium on the antioxidative defense system and alteration in different functional groups in castor bean and Indian mustard. Arch Agron Soil Sci 62(6):877–891

Bauddh K, Singh B, Singh RP (2016b) *Ricinus communis* L. as a value added alternative for restoration of cadmium contaminated and degraded agricultural ecosystem. Bull Environ Contam Toxicol 96(2):265–269

Bernstein N, Chaimovitch D, Dudai N (2009) Effect of irrigation with secondary treated effluent on essential oil, antioxidant activity, and phenolic compounds in oregano and rosemary. Agron J 101:1–10

Bishehkolaei R, Fahimi H, Saadatmand S, Nejadsattari T (2011) Ultrastructural localisation of chromium in Ocimum basilicum. Turk J Bot 35:261–268

Boyle G (1996) Renewable energy power for a sustainable future. Oxford University Press, Oxford

Brandt R, Merkl N, Schultze-Kraft R, Infante C, Broll G (2006) Potential of vetiver (*Vetiveria zizanioides* (L.) Nash) for phytoremediation of petroleum hydrocarbon-contaminated soils in Venezuela. Int J Theor Phys 8:273–284

Chen Y, Shen Z, Li X (2004) The use of vetiver grass (*Vetiveria zizanioides*) in the phytoremediation of soils contaminated with heavy metals. Appl Geochem 19:1553–1565

Citterio S, Santagostino A, Fumagalli P, Prato N, Ranalli P, Sgorbati S (2003) Heavy metal tolerance and accumulation of Cd, Cr and Ni by *Cannabis sativa* L. Plant Soil 256(2):243–252

Datta R, Quispe MA, Sarkar D (2011) Greenhouse study on the phytoremediation potential of vetiver grass, *Chrysopogon zizanioides* L., in arsenic-contaminated soils. Bull Environ Contam Toxicol 86:124–128

Easterly JL, Burnham M (1996) Overview of biomass and waste fuel resources for power production. Biomass Bioenergy 11(2–3):72–92

Gupta AK, Verma SK, Khan K, Verma RK (2013) Phytoremediation using aromatic plants: a sustainable approach for remediation of heavy metals polluted sites. Environ Sci Technol 47:10115–10116

Hamzah A, Hapsari RI, Wisnubroto EI (2016) Phytoremediation of cadmium-contaminated agricultural land using indigenous plants. Int J Environ Agric Res (IJOEAR) 2(1):8–14

Hashiramoto O (2007) Cross-sectoral policy developments in forestry. In: Dubé YC, Schmithüsen F (eds) Wood-product trade and policy issues. CABI, Wallingford.

Ho Y, Hsieh J, Huang C (2013) Construction of a plant–microbe phytoremediation system: combination of vetiver grass with a functional endophytic bacterium, Achromobacter xylosoxidans F3B, for aromatic pollutants removal. Bioresour Technol 145:43–47

Hu N, Ding D, Li G (2014) Natural plant selection for radioactive waste remediation. In: Gupta DK, Walther C (eds) Radionuclide contamination and remediation through plants. pp 33–53. doi:10.1007/978–3-319-07665-2_2

Hussain K, Abdussalam AK, Chandra PR, Salim N (2011) Heavy metal accumulation potential and medicinal property of Bacopa monnieri – a paradox. J Stress Physiol Biochem 7(4):39–50

Ibrahim MM, Alsahli AA, El-Gaaly G (2013) Evaluation of phytoremediation potential of six wild plants for metal in a site polluted by industrial wastes: a field study in Riyadh, Saudi Arabia. Pak J Bot 42(2):571–576

Jiang W, Liu D, Hou W (2001) Hyperaccumulation of cadmium by roots, bulbs and shoots of garlic (*Allium sativum* L.). Bioresour Technol 76(1:9–13

Karekezi S, Kithyoma W (2006) Bioenergy and agriculture: promises and challenges. Bioenergy and the poor. In: 2020 vision for food, agriculture, and the environment. International Food Policy Research Institute, Washington, DC

Kim HL, Streltzer J, Goebert D (1999) St. John's wort for depression: a metal-analysis of well-defined clinical trials. J Nerv Ment Dis 187:532–538

Lal K, Minhas PS, Chaturvedi RK, Yadav RK (2008) Cd uptake and tolerance of three aromatic grasses on the Cd-rich soil. J Indian Soc Soil Sci 56:290–294

Linger P, Müssig J, Fischer H, Kobert J (2002) Industrial hemp (*Cannabis sativa* L.) growing on heavy metal contaminated soil: fibre quality and phytoremediation potential. Ind Crop Prod 16(1):33–42

Mahmud R, Inoue N, Kasajima S, Shaheen R (2008) Assessment of potential indigenous plant species for the phytoremediation of arsenic-contaminated areas of Bangladesh. Int J Theor Phys 10(2):117–130

Makris KC, Shakya KM, Datta R, Sarkar D, Pachanoor D (2007) Chemically catalyzed uptake of 2,4,6-trinitrotoluene by *Vetiveria zizanioides*. Environ Pollut 148:101–106

Maleki E, Aroua MK, Sulaiman NMN (2013) Improved yield of solvent free enzymatic methanolysis of palm and Jatropha oils blended with castor oil. Appl Energy 104:905–909

Malko A (2002) Untersuchung zum Wirkstoffgehalt, zur Cadmiumaufnahme und Rotwelkeanfälligkeit von *Hypericum perforatum* L. Dissertation Justus-Liebig-Universität Gieβen, Germany, 163 pp

Manan FA, Tsun-Thai C, Samad AA, Mamat DD (2015) Evaluation of the phytoremediation potential of two medicinal plants. Sains Malaysiana 44(4):503–509

Manikandan R, Sahi SV, Venkatachalam P (2015) Impact assessment of mercury accumulation and biochemical and molecular response of *Mentha arvensis*: a potential hyperaccumulator plant. World Sci World J. http://dx.doi.org/10.1155/2015/715217

McKendry P (2002) Energy production from biomass (part 1): overview of biomass. Bioresour Technol 83:37–46

Mejia JD, Salgado N, Orrego CE (2013) Effect of blends of diesel and palm-castor biodiesels on viscosity, cloud point and flash point. Ind Crop Prod 43:791–797

Müller WE, Singer A, Wonnemann M (1999) Johanniskraut. Vom Nerventee zum Modernen Antidepressivum. Dt Apoth Ztg 139:1741–1750

Pandey VC, Singh N (2015) Aromatic plants versus arsenic hazards in soils. J Geochem Explor 157:77–80

Pandey S, Gupta K, Mukherjee AK (2007) Impact of cadmium and lead on *Catharanthus roseus* a phytoremediation study. J Environ Biol 28(3):655–662

Pandey J, Chand S, Pandey S, Rajkumari PDD (2015) Palmarosa [*Cymbopogon martinii* (Roxb.) Wats.] as a putative crop for phytoremediation, in tannery sludge polluted soil. Ecotoxicol Environ Saf 122:296–302

Pichai NMR, Samjiamjiaras R, Thammanoon H (2001) The wonders of a grass, Vetivers and its multifold application. Asian Infrastruct Res Rev 3:1–4

Putwattana N, Kruatrachue M, Pokethitiyook P, Chaiyarat R (2010) Immobilization of cadmium in soil by cow manure and silicate fertilizer, and reduced accumulation of cadmium in sweet basil (*Ocimum basilicum*). Sci Asia 36:349–354

Rai UN, Vajpayee P, Singh SN, Mehrotra S (2004) Effect of chromium accumulation on photosynthetic pigments, oxidative stress defense system, nitrate reduction, proline level and eugenol content of Ocimum tenuiflorum L. Plant Sci 167:1159–1169

Romeiro S, Lagôa AMMA, Furlani PR, de Abreu CA, de Abreu MF, Erismann NM (2006) Lead uptake and tolerance of *Ricinus communis* L. Braz J Plant Physiol 18(4):483–489

Sá RA, Sá RA, Alberton O, Gazim ZC, Laverde A Jr, Caetano J, Amorin AC, Dragunski DC (2014) Phytoaccumulation and effect of lead on yield and chemical composition of *Mentha crispa* essential oil. Desalin Water Treat 53(11):3007–3017

Sagar AD, Kartha S (2007) Bioenergy and sustainable development? Annu Rev Environ Resour 32:131–167

Salim MRJ, Adenan MI, Amid A, Jauri MH, Sued AS (2013) Statistical analysis of metal chelating activity of *Centella asiatica* and *Erythroxylum cuneatum* using response surface methodology. Biotechnol Res Int, Article ID 137851

Schneider M, Marquard R (1996) Investigation on the uptake of cadmium in *Hypericum perforatum* L. (St. John's wort). Acta Hortic 426:435–442

Scora RW, Chang AC (1997) Essential oil quality and heavy metal concentrations of peppermint grown on a municipal sludge-amended soil. J Environ Qual 26:975–979

Sharma S, Adholeya A (2011) Phytoremediation of Cr-contaminated soil using *Aloe vera* and *Chrysopogon zizanioides* along with AM fungi and filamentous saprobe fungi: a research study towards possible practical application. Mycorrhiza News 22(4):16–20

Shi G, Cai Q (2009) Cadmium tolerance and accumulation in eight potential energy crops. Biotechnol Adv 27:555–561

Singh S, Melo J, Eapen S, D'Souza S (2008) Potential of vetiver (*Vetiveria zizanoides* L. Nash) for phytoremediation of phenol. Ecotoxicol Environ Saf 71:671–676

Singhakant C, Koottatep T, Satayavivad J (2009) Enhanced arsenic removals through plant interactions in subsurface-flow constructed wetlands. J Environ Sci Health, Part A 44:163–169

Sinha S (1999) Accumulation of Cu, Cd, Cr, Mn and Pb from artificially contaminated soil by *Bacopa monnieri*. Environ Monit Assess 57:253–264

Sinha S, Chandra P (1990) Removal of Cu and Cd from water by *Bacopa monnieri* (L.). Water Air Soil Pollut 51:271–276

Sinha S, Gupta M, Chandra P (1996) Bioaccumulation and biochemical effects of mercury in the plant *Bacopa monnieri* (L.). Environ Toxicol Water Qual 11:105–112

Subhashini V, Swamy AVVS (2013) Phytoremediation of Pb and Ni contaminated soils using *Catharanthus roseus* (L.). Univers J Environ Res Technol 3(4):465–472

Tamari N, Mine A, Sako A, Tamagawa S, Tabira Y, Kitamura Y (2014) Possible application of the medicinal plant *Hyoscyamus albus* in phytoremediation: excess copper compensates for iron deficiency, depending on the light conditions. Am J Plant Sci 5:3812–3822

Thangaswamy S, Manjunatha B, Suarez JP, Babu M, Ponnuswamy G, Reddy GVS, Mariadoss S (2015) Phytoremediation of tannery effluent polluted soils of Dindigul, Tamil Nadu, using *Arbuscular mycorrhizal* fungi inoculated *Azadirachta indica*. J Chem Pharm Res 7(7):905–914

Tirillini B, Ricci A, Pintore G, Chessa M, Sighinolfi S (2006) Induction of hypericins in *Hypericum perforatum* in response to chromium. Fitoterapia 77:164–170

Truong P (2000) Vetiver grass technology for environmental protection. In: The 2nd international. Vetiver conference: Vetiver and the environment. Cha Am, Thailand

Ustra MK, Silva JRF, Ansolin M, Balen M, Cantelli K, Alkimim IP, Mazutti MA, Voll FAP, Cabral VF, Cardozo-Filho L, Corazza ML, Oliveira JV (2013) Effect of temperature and composition on density, viscosity and thermal conductivity of fatty acid methyl esters from soybean, castor and *Jatropha curcas* oils. J Chem Thermod 58:460–466

Varun M, D'Souza R, Pratas J, Paul MS (2012) Metal contamination of soils and plants associated with the glass industry in North Central India: prospects of phytoremediation. Environ Sci Pollut Res 9(1):269–281

Verma SK, Singh K, Gupta AK, Pandey VC, Trevedi P, Verma SK, Patra DD (2014) Aromatic grasses for phyto management of coal fly ash hazards. Ecol Eng 73:425–428

Verotta L (2003) *Hypericum perforatum*, a source of neuroactive lead structures. Curr Top Med Chem 3:87–201

Wereko-Broddy C, Hagen EB (1996) Biomass conversion and technology. Wiley, Chichester

Wu XH, Zhang HS, Gang L, Liu XC, Qin P (2012) Ameliorative effect of castor bean (*Ricinus communis* L.) planting on physicochemical and biological proper-ties of seashore saline soil. Ecol Eng 38:97–100

Zahedifara M, Moosavib AA, Shafigh M, Zareib Z, Karimian F (2016) Cadmium accumulation and partitioning in *Ocimum basilicum* as influenced by the application of various potassium fertilizers. Arch Agron Soil Sci 62(5):663–673

Zahir AA, Rahuman AA, Bagavan A, Santhoshkumar T, Mohamed RR, Kamaraj C, Rajakumar G, Elango G, Jayaseelan C, Marimuthu S (2010) Evaluation of botanical extracts against Haemaphysalis bispinosa Neumann and Hippobosca maculata Leach. Parasitol Res 107(3):585–592

Zheljazkov VD, Nielsen NE (1996a) Studies on the effect of heavy metals (Cd, Pb, Cu, Mn, Zn and Fe) upon the growth, productivity and quality of lavender (*Lavandula angustifolia* Mill) production. J Essent Oil Res 8:259–274

Zheljazkov VD, Nielsen NE (1996b) Effect of heavy metals on peppermint and cornmint. Plant Soil 178:59–66

Zheljazkov VD, Craker LE, Xing B (2006) Effects of Cd, Pb, and Cu on growth and essential oil contents in dill, peppermint, and basil. Environ Exp Bot 58:9–16

Zhi-Xin N, Li-Na S, Tie-Heng S, Yu-Shuang L, Hing W (2007) Evaluation of phytoextracting cadmium and lead by sunflower, ricinus, alfalfa and mustard in hydroponic culture. J Environ Sci 19:961–967

Zurayk R, Sukkariyah B, Baalbaki R (2001) Common hydrophytes as bioindicators of nickel, chromium and cadmium pollution. Water Air Soil Pollut 127(1–4):373–388

Zurayk R, Sukkariyah B, Baalbaki R, Ghanem DA (2002) Ni phytoaccumulation in *Mentha aquatica* L. and *Mentha sylvestris* L. Water Air Soil Pollut 139(1):355–364

A Sustainable Approach to Clean Contaminated Land Using Terrestrial Grasses

12

Anju Patel and D.D. Patra

Abstract

After the industrial revolution, due to various activities like mining, sewage disposal, pesticides, etc., an enormous increase in pollution load in the soil environment has been observed. Mitigation of soil pollution in a sustainable manner can be achieved by the help of grasses. Grasses are fast-growing plants that hold the soil surface and check the soil degradation. It increases soil microbial population and helps in the removal of various contaminants like heavy metals, pesticides, polyaromatic hydrocarbon, and various other pollutants from the soil. Grasses also help in carbon sequestration and prevent desertification of land. Grasses are tolerant to several abiotic stresses because of their strong antioxidant defense system. Various grasses like vetiver, *Miscanthus*, switchgrass, lemongrass, etc. have been used as an ideal phytoextractant as they produce high biomass in less time and accumulate high concentration of metals mostly in root. These grasses can also be utilized for the production of biofuels. Thus, grasses help in meeting the needs of increasing energy demands without affecting food crop prices.

Keywords

Phytoremediation • Grasses • Biofuel • Antioxidant • Carbon sequestration • Microbial activity

A. Patel
School of Earth Sciences, Banasthali University, Vanasthali, Rajasthan 304022, India
e-mail: patel7anju@gmail.com

D.D. Patra (✉)
Bidhan Chandra Krishi Viswavidyalaya, PO: Krishi Viswavidyalaya, Mohanpur, Nadia 741252, West Bengal, India
e-mail: ddpatra@rediffmail.com

K. Bauddh et al. (eds.), *Phytoremediation Potential of Bioenergy Plants*,
DOI 10.1007/978-981-10-3084-0_12

12.1 Introduction

Owing to staggering rise in human and animal population and expanding agricultural, mining, and industrial activities, heavy metal pollution of soil and water has become a global issue all around the world. Colossal disposal of industrial effluents, mainly in the developing countries, is causing soil pollution. The use of contaminated grounds for the agricultural and urban development needs a secure and proficient purification system. With the enhancement of the use of agrochemicals, unwanted elements such as heavy metals and many other compounds find their way in agricultural soils, which pose a potential threat to the human and livestock (Chaney et al. 1998). Restoration of contaminated lands by forming a grass/plant cover is crucial before the contaminants can cause major health hazard by relocating the metals into the neighboring areas. Grasses have shown to be most effective in phytoextraction of many toxic substances and degradation of organic contaminants present in the soils (Anderson et al. 1993). Their fibrous root system offers an increased root surface for microbial growth and activity. Furthermore, grasses can effectively stabilize surface soils by improving physical and chemical properties of the soil and retarding the degradation of soils (Aprill and Sims 1990).

12.2 Geographical Distributions of Grasses

The grass family is one of the most widely distributed and abundant groups of plants on Earth. Grasses belong to the family Poaceae. The cereal grasses, bamboos, natural grasslands, and pastures come under Poaceae family. Grasslands such as savannah and prairie where grasses are dominant are estimated to constitute 20% of the vegetation cover of the Earth. Grasses are also an important part of the vegetation in many other habitats, including wetlands, forests, and tundra. Various grass species are found around the world. There are some essential oil-bearing aromatic grasses, viz., lemongrass (*Cymbopogon flexuosus*), palmarosa (*Cymbopogon martinii*), citronella (*Cymbopogon winterianus*), and vetiver (*Vetiveria zizanioides*). Out of these four, high-value essential oil of *C. martinii*, *C. winterianus*, and *C. flexuosus* is extracted from shoot portion, while that of vetiver is extracted from the root.

Vetiver grass is native to South and Southeast Asia where it has been grown for centuries for roof thatching, fodder for livestock, and perfumery and cosmetic industries. Vetiver grass (*Vetiveria zizanioides* or *Chrysopogon zizanioides*) is a tall (2–3 m), fast-growing, perennial tussock grass. It has a deep elongated (3–4 m), extensive, and complex root system that usually diffuses in the deeper layers of the soil (Dalton et al. 1996; Truong 2000; Pichai et al. 2001). Vetiver was first used for soil conservation and land stabilization purposes in Fiji in the early 1950s and promoted by the World Bank for soil and water conservation in India in the 1980s (Dalton et al. 1996).

Switchgrass (*Panicum virgatum*) is a warm-season perennial grass and has been native to North America. Switchgrass has broad adaptability, high growth rates, and tolerance to a wide variety of climatic and edaphic conditions. It has earned attention because it has a great demand in biofuel industry. Timothy grass (*Phleum pratense*) is also perennial grass native to Central Europe which has been introduced to New England in the early eighteenth century. It is tolerant to cold as well as drought conditions (Balsamo et al. 2015).

Miscanthus sinensis is of tetraploid origin consisting of two subgenomes. Several species occur naturally in Eastern and Southeastern Asia (Clifton-Brown et al. 2008), Africa, and the Himalayas (Hodkinson et al. 2002). *Miscanthus*, a C_4 grass similar to sugarcane, develops well in moderately polluted soil and also on degraded land where soil quality is poor, specifically in nitrogen. Being a high-biomass-producing grass, it has gained a high interest among researchers as it can be used as a phytoremediator as well as energy crop.

Cymbopogon citratus is a tall perennial grass that grows in warm temperate and tropical regions of the world. Its common names are lemongrass, barbed wire grass, and silky heads (Kazembe and Chauruka 2012). The insecticidal (Arias et al. 1992), antimicrobial (Syed et al. 1995), and therapeutic properties (Akendengue 1992) of its oil and extracts have been reported. Lemongrass is one of the uttermost extensively cultivated essential oil-bearing grasses of the tropics and subtropics of India, Indonesia, and Madagascar and also in Africa and South America (Anaruma et al. 2010; Ganjewala and Luthra 2010; Weiss 1997). Lemongrass is usually found in wild habitats like in dense/mixed forest and also along the banks of rivers and canals (Singh and Sharma 2001; Weiss 1997). Lately, its agrotechnology is now available, and it is commercially cultivated in several parts of the world.

The C_3 grass species are mainly found at high latitude and high elevation and grow in low-temperature season. C_4 grasses mainly occur in warm seasons. The mechanism interpreting the poor performance of C_4 plant species in colder climates is not known. C_4 grasses and C_3 grasses differ immensely in their capability to stimulate rhizodegradation and metabolize contaminants (Ehleringer 1978; Sage and Kubien 2007). Phylogenetic literature showed that all C_4 lineages originated from C_3 ancestors adapted to warm climates. Various C_4 species have independently evolved cold tolerance, and these C_4 species survive well without showing complications from chilling injury. Cold-adapted C_4 species show a high level of antioxidant and carotenoid pigments and also have amplified photoprotective ability at low temperature. Rubisco capacity is similar to gross CO_2 assimilation rate below 20 °C in cold-adapted C_4 species, indicating that it is an important limitation on C_4 photosynthesis at cool temperature (Raghavendra and Sage 2010).

12.3 Utilization of Grasses for Different Purposes

12.3.1 Effectiveness of Grasses in Remediation of Contaminated Soil

Procedures of phytoremediation are common to many genera of plants. Foremost rules in selecting grass species for phytoremediation of a contaminated land are resistance to contaminant; degree of agglomeration, translocation, and uptake potential of various contaminants; climatic acclimatization to abiotic stress like heat, wind, cold, waterlogging, and drought; tolerance against extreme pH or salinity; and economic value for marketable product (Dickinson et al. 2009; Hemen 2011).

Grasses usually accumulate metals in intensified amount inside their tissues (hyperaccumulators) and showing low biomass (Chaudhry et al. 1998) possesses restricted root system (Ernst 1996), although some grasses which are tolerant and show high biomass production have some adaptation mechanisms that concede them to survive in contaminated land better than others (Palazzo and Lee 1997). Keller et al. (2003) reported that resistance of grass species and its suitability for particular contaminants at the root level must be considered while choosing plants for phytoremediation. Root system morphology empowered some plants to be more proficient than others in nutrient uptake from stressed soil (Fitter and Stickland 1991).

Cynodon dactylon played an essential role in phytoremediation of chromite–asbestos mine waste and agriculturally contaminated sites when compared with *Sorghastrum nutans*, *A. concinna*, and *Cajanus cajan*. Cr content in grasses is high and more than the critical plant total metal content (5–30 mg kg^{-1}) (Kabata-Pendias 2011). However, the soil contains very high amount of Cr, but its accumulation inside the cereals is low (<10 mg kg^{-1}). This may be because of immobile nature of Cr and grass–legumes are acting as a potential barrier for metal uptake by cereals (Alloway 2013). Addition of organic matter (OM) in terms of farmyard manure retards the Cr mobility in soil, resulting in lower Cr accumulation in crops (Singh et al. 2007a, b). The translocation factor (TF) of *C. dactylon*, *S. nutans*, and *C. cajan* is found to be greater than 1, which means a high agglomeration of Cr in the leaves as compared to the roots (Brunetti et al. 2009).

Grass cover has the potential to agglomerate metals and facilitates phytoremediation of chromite–asbestos mine waste (Hoflich et al. 1994; Zong et al. 2015). When grasses (*Cymbopogon* sp., *Chrysopogon* sp.) are added with biochar, farmyard manure, and chicken manure along with *Pseudomonas* sp. and *Bacillus* sp., this not only enhances the growth of plant but also enhances the remediation of Cr from polluted site (Masarovicova et al. 2009). The use of various grasses for removal of different contaminants is presented in Table 12.1.

12.3.2 Phytoremediation of Various Metals by Vetiver

Vetiver can sustain in soil environment that contains various heavy metals, and it agglomerates these heavy metals inside the roots and shoots. Numerous studies revealed the ability of *Chrysopogon zizanioides* (CZ) to extract metals such as arsenic, chromium, lead, zinc, cadmium, aluminum, copper, mercury, and so on from contaminated land (Truong 1999).

CZ is tolerant against high levels of arsenic (As^{5+}) in the soil. However, arsenic mainly agglomerates in the root region (Srisatit et al. 2003). When CZ was cultivated from the arsenic-contaminated soil (125 mg As kg^{-1}), roots contained about

Table 12.1 Removal of contaminant by various grasses

Grasses	Contaminants	References
Cynodon dactylon	Pb	Wong and Lau (1985)
	Fe, Cu, and Pb	Maric et al. (2013)
Festuca rubra	Cu, Pb, and Zn	Wong (1982)
Phragmites australis	Fly ash, Pb/Zn mine soils	Ye et al. (1998)
Vetiver zizanioides	Pb/Zn mine tailings, Pb, Cd, Zn, and Pb	Shu et al. (2000), Chen et al. (2004), and Lai and Chen (2004)
Panicum virgatum	Polyaromatic hydrocarbon (PAH)	Pradhan et al. (1998)
Lolium perenne	Gulf war-contaminated soils	Yateem et al. (1999)
	p,p′-DDE	White (2000)
Saccharum officinarum	Cd, Cu, Pb, and Zn (mixture of wastewater and sewage sludge)	Muchuweti et al. (2006)
Miscanthus sp.	Arsenic	Hartley et al. (2009)
Forage grass (*Bromus inermis*, *Elymus dauricus*)	2-Chlorobenzoic acid	Siciliano and Germida (1997)
Reed canary grass	4-Chlorophenol, naphthalic anhydride	Urbanek et al. (2005)
Grass rhizosphere	Parathion and diazinon; 2,4-D and 2,4,5-T	Hsu and Bartha (1979) and Shann and Boyle (1994)
Dactylis glomerata	Fe, Cu, and Pb	Maric et al. (2013)
Anthoxanthum odoratum		
Agrostis gigantea		

Abbreviation: *2,4-D* 2,4-Dichlorophenoxyacetic acid, *2,4,5-T* 2,4,5 Trichlorophenoxyacetic acid, *DDE* Dichlorodiphenyldichloroethylene

9.78 mg kg^{-1} dry weight (DW) of arsenic as compared to 0.53 mg kg^{-1} DW in shoots after 45 days of experiment (Truong and Baker 1998a, b). However, arsenic concentration decreases with the growth of the grass (Srisatit et al. 2003). Chiu et al. (2005) reported that when CZ was grown in soils containing 100 mg kg^{-1} of arsenic for 3 months, around 3 mg kg^{-1} of arsenic got agglomerated in shoots, while roots enclosed around 13 mg kg^{-1} of arsenic. Arsenic uptake by CZ can be enhanced by using chelating agents like organic matter, chelates enhance the uptake capacity of grass to accumulate more metal in less time. Singh et al. (2007a) observed that CZ grows well in soil contaminated with dairy waste and about 500 mg kg^{-1} of NaH_2AsO_4. After 6 months, CZ contained about 185.4 mg kg^{-1} of arsenic in root and about 100.6 mg kg^{-1} in leaves. Summing up, CZ can absorb 62% of arsenic from the contaminated land. However, CZ cannot be considered as a hyperaccumulator of arsenic; it can extract heavy metals from contaminated sites and thus cuts down the contaminant level present in the soil (Truong 1999).

Angin et al. (2008) revealed that CZ can be used to remove boron from contaminated soil. CZ has been successfully established on boron-contaminated soils, and the agglomeration of boron is higher in shoots (about 28 mg kg^{-1} DW) than in roots (about 17 mg kg^{-1} DW) (Angin et al. 2008).

Cadmium is extremely harmful to plant growth (Baker and Eldershaw 1993). Truong (1999) revealed that CZ has tolerance to 60 mg kg^{-1} of Cd. CZ also sustains in mine tailing soils which contained about 32 mg kg^{-1} of Cd in the soil (Yang et al. 2003). Vo (2007) found that when CZ was grown in 10, 20, and 40 mg kg^{-1} of Cd-incorporated soils, it showed acceptable growth and development. Cd concentration in CZ grasses enhanced with the increasing dose of Cd in the soil. Overall, Cd concentration inside root got increased from 0.14 to 0.23 mg kg^{-1}, and in shoot it increased from 0.11 to 0.15 mg kg^{-1} after 70 days of exposure (Vo 2007). Zhuang et al. (2005) found that CZ grown on shale oil tailings (Xia 2004) and contaminated farm lands has the Cd content ratio in root and shoot varied between 2 and 3, respectively. Major amount of Cd got accumulated in CZ shoots, which is approximately 25 mg kg^{-1} (Lai and Chen 2004). Various studies (Zhuang et al. 2005; Lai and Chen 2004) revealed that no enhancement in Cd uptake has been observed after the addition of chelating agent like EDTA by the root and shoot of CZ.

Chiu et al. (2006) reported that CZ also sustains well on Cu mine tailing soils that contained 1084 mg kg^{-1} of Cu (Antiochia et al. 2007). Castillo et al. (2007) also revealed that CZ successfully established in Cu tailings (La Africana mine, Chile) which contained about 3921 mg $kg^{-1}$of Cu. The agglomeration of Cu by CZ is low. Yang et al. (2003) revealed that CZ also survived well in Pb/Zn mine soils which contained 35 mg kg^{-1} of Cu. The root accumulated more Cu than the shoot. The accumulation was 58.7 and 3.7 mg kg^{-1} of Cu in root and shoot, respectively, after 20 weeks of establishment of CZ grass. Another study (Chiu et al. 2006) reported that CZ accumulated 330 mg $kg^{-1}$of Cu in roots and around 10 mg kg^{-1} in shoots when grown in Cu mine soils. Various studies (Chiu et al. 2006; Wilde et al. 2005) revealed that the translocation ratio (TR), i.e., transfer from root to shoot of Cu in CZ, is generally less than 10%. Cu uptake by CZ roots and shoots can enhance by the addition of chelating agents, although the effect of different chelating agents (cyclohexane-1,2-diaminetetraacetic acid, ethylenediaminetetraacetic acid, citric acid, and malic acid) has different results on Cu uptake by CZ (Lou et al. 2007; Chiu et al. 2005). Organic matter, viz., sewage sludge and manure compost amended in the Pb/Zn-contaminated soil, retards the uptake of Cu by roots and shoots of CZ (Chiu et al. 2006).

CZ is also tolerant against high amount of Cr (about 600 mg kg^{-1}) in contaminated soils (Truong 1999). Hoang et al. (2007) revealed that CZ can sustain on the canal sludge that contained about 2290 mg kg^{-1} of Cr. Antiochia et al.(2007) also reported that CZ has grown well in Cr-contaminated water that contained about 623 mg kg^{-1} of Cr. Chromium is mainly absorbed by the roots of CZ. CZ has easily grown in soils containing 50, 200, and 600 mg kg^{-1} of Cr and agglomerates about 404, 1170, and 1750 mg kg^{-1} in roots, respectively, while shoot accumulated low amount of Cr which is 4, 5, and 18 mg kg^{-1}, respectively (Truong 1999). The accumulation of Cr inside root and shoot is also dependent on exposure time. Cr accumulation by the CZ roots enhanced as the Cr exposure time increased, while Cr accumulation by the leaves remained constant (Antiochia et al. 2007).

CZ showed its tolerance against high range of lead (Pb) in the soil. CZ showed 100% survival rate when grown in high levels of Pb-contaminated soils (9020–10,750 mg kg^{-1} of Pb) (Rotkittikhun et al. 2007). Chen et al. (2004) revealed that CZ displayed visible growth in Pb-contaminated soils (about 5000 mg kg^{-1} of Pb) (Wilde et al. 2005). CZ growth has been enhanced with the use of organic matter and arbuscular mycorrhizal fungi (Rotkittikhun et al. 2007) in Pb-contaminated soil. The application of organic matter (OM) into the polluted soils lowered the extractable metal from the soil. Thus, CZ exhibits reduced Pb uptake when the polluted soils are incorporated with organic matter, such as pig manure, sewage sludge, and domestic refuse (Chiu et al. 2006; Yang et al. 2003). Majority of Pb is absorbed by the roots only, and a small portion is translocated to the shoots (Cunningham and Berti 2000).

CZ is also useful in remediating soil contaminated with 2,4,6-trinitrotoluene (TNT). TNT is a potent mutagen and classified as a group C human carcinogen (USEPA, 1991; Cenas et al. 2001). It is mainly used in the production of munitions and explosives. Makrisa et al. (2007a) observed that CZ sustained well in 40 mg L^{-1} of TNT in hydroponic system, whereas it does not show any toxic effects after 8 days of exposure. However, Makrisa et al. (2007b) reported that CZ accumulated 1026 mg of TNT. Although the absorption kinetics was slow, CZ also demonstrates the ability to clean radionuclide from soils as well as water (Shaw and Bell 1991). Singh et al. (2008a) showed the potential of CZ to extract 61% of 137Cs and 94% of 90Sr from the Cs and Sr solution after 168 h of exposure (Singh et al. 2008b).

12.3.3 *Miscanthus* sp. Role in Remediation of Metals

Miscanthus usually takes small amounts of the toxic heavy metals (Pogrzeba et al. 2010), whereas *M. sacchariflorus* contains remarkable concentrations of toxic heavy metals, viz., Zn, Cd, and Pb. The role of *Miscanthus* on heavy metal uptake behavior when grown in contaminated soil has been studied by Iqbal et al. (2013). Ollivier et al. (2012) and Wanat et al. (2013) reported that *Miscanthus* was tolerant to long-term contaminated soil near a gold mine.

Studies (Kao et al. 1998; Sun et al. 2006) revealed that *M. floridulus* showed a high potential for remediation in nonferrous mining sites, which contained about 760 mg kg^{-1} of Pb, Cd (> 4 mg kg^{-1}), Zn (approximately 370 mg kg^{-1}), and Cu (>95 mg kg^{-1}). The agglomeration of metals was higher in stems and leaves than in the roots. Li et al. (2011) demonstrated that *M. floridulus* is uranium tolerant when cultivated on shale of uranium mill tailings (in South China). In Cd- and Pb-polluted soils, *M. x giganteus* agglomerated about 35–55 g/ha/year of heavy metals (Barbu et al. 2009; Iordache et al. 2010; Barbu 2010). Iordache et al. (2010) reported that more than 680 mg kg^{-1} of Pb and more than 13 mg kg of Cd got accumulated by the *Miscanthus*. *M. x giganteus* displayed good adaptation against acidic soil. *Miscanthus* has been evaluated for its potential to grow on heavy metal-polluted land due to tin mining (Visser et al. 2001). Kerr et al. (1998) revealed that *Miscanthus* sustain easily in soils incorporated with sewage sludge, regardless of high amount of phytotoxic metals present in the growing medium. Though metal accumulation was low, it sustained well in contaminated soil.

12.3.4 Remediation of Polyaromatic Hydrocarbon and Pesticides

Polycyclic aromatic hydrocarbons (PAHs) are the major organic pollutants, and most of PAHs show mutagenic and carcinogenic characteristics. Li et al. (2006) found that CZ has a good adaptation potential when grown in soil contaminated with benzo[a]pyrene (B[a]P). The CZ grass tolerated up to 100 mg kg^{-1} of B[a]P. The removal of B[a]P from the soils is accompanied by an enhancement in soil microbial biomass.

Aprill and Sims (1990) revealed that for phytoremediation of PAH, the species of grasses are most suitable. Grasses like fescue (*Festuca* sp.), ryegrass (*Lolium* sp.), *Panicum* sp., and prairie grasses (e.g., *Buchloe dactyloides*, *Bouteloua* sp.) are popular because they possess very dense and deep root system which in turn gives a large root surface area (Aprill and Sims 1990). Mulberry trees were very useful in degradation of PCB and PAH in the soil by triggering the expression of microbial genes involved in PCB and PAH degradation (Cajthaml et al. 2002; Renoldi et al. 2003; Mueller and Shann 2006).

Brandt et al. (2006) observed that CZ has grown well in land contained about 5% (w/w) of crude oil. Crude oil inhibits the growth and biomass of CZ grass, and after 6 months of CZ cultivation in crude oil, no further uptake enhancement has been observed in the field trials. Hence, CZ can be successfully established on slightly polluted sites with PAH. Merkl et al. (2005) established that *Brachiaria brizantha* can remediate the crude oil-contaminated soil and also enhances the microbial oil degradation. Although this grass is native to tropical Africa, it has been introduced into many countries of the world. Organic fertilizer catalyzes the process of remediation. Physicochemical and structural properties of contaminants determine the uptake by grasses from the soil. Moderately hydrophobic organic compounds (characterized with the octanol–water partition coefficient, log K_{ow}, with values between 1 and 3) available in diesel fuel have been quickly taken up by roots. Hydrophilic (log $K_{ow} < 1$) and strongly hydrophobic compounds (log $K_{ow} > 3$) are unavailable for uptake because of their strong bonding to soil particles or to plant roots (Kamath et al. 2004). The growth of *B. brizantha* is not restricted by the diesel-contaminated grass in the soil. This grass species is notably tolerant to the diesel concentrations and is effective in remediating the contaminated soil surface of diesel spills (Mezzari et al. 2011).

CZ can also control the agrochemical pollutions like atrazine, and this grass sustains about 1000–2000 μg L^{-1} of atrazine (Winters 1999; Cull et al. 2000). Truong et al. (2000) also found that CZ has effectively removed agrochemicals like carbofuran, atrazine, and monocrotophos from cotton and sugarcane farms (Pinthong et al. 1998; Sripen et al. 1996). Vetiver enhanced the biodegradation of organic contaminants such as phenol, benzo[a]pyrene, organophosphate pesticides, and atrazine (Truong et al. 2000; Boonsaner et al. 2002; Li et al. 2006; Singh et al. 2008a).

Studies (Techer et al. 2011, 2012c) showed that *M. x giganteus* removed various PAH from contaminated soils. Biodegradation of pyrene, pyrene + phenanthrene, pyrene + salicylate, and pyrene + diesel fuel was accelerated by root exudates. Another study by Maliszewska-Kordybach (2005) revealed that quercetin (as a root exudate) is responsible for biodegradation of PAH when grown in long-term PAH-contaminated soils (Techer et al. 2012a, b).

Xia (2004) showed that various grasses like vetiver grass (*Vetiveria zizanioides*), bahia grass (*Paspalum notatum*), St. Augustine grass (*Stenotaphrum secundatum*), and bana grass (*Pennisetum glaucum* and *P. purpureum*) have the potential to remediate the oil shale mined land in the Southwest of Guangdong Province, China (Xia 2004). Considering all above grasses, vetiver has the highest survival rate, i.e., 99%, followed by bahia and St. Augustine, and their survival rates are 96% and 91%, respectively, although bana grass has the lowest survival rate of 62%.

A study by Truong (1999), Xia, and Shu (2000) revealed that vetiver grass has a strong resistance against despicable environment and sustains high concentrations of heavy metals. In Australia, *C. zizanioides* has been used successfully to stabilize in saline, alkaline (pH 9.5) coal mined land, and highly acidic (pH 2.7) gold mined land (Truong and Baker 1998a, b; Truong 1999). *C. zizanioides* repeated a similar pattern in China (Xia and Shu 2000). Other studies (Kalmbacher and Martin 1998; Reynolds et al. 1999; Xia and Shu 2000) suggest that bahia grass (*Paspalum notatum*) is also an excellent grass that helps in ecological restoration and turf and forage application. *Stenotaphrum secundatum*, a lesser known grass, is generally used as turfgrass in Guangdong and as forage in few countries. This grass possesses tolerance against drought and salinity (Marcum and Murdoch 1994; Carrow 1996; Qian and Engelke 1999). Bana grass is similar to elephant grass (*Pennisetum purpureum*), and they are tolerant to high metal concentration (Koester et al. 1992a, b; Van De et al. 1999).

12.3.5 Other Useful Aspects of Grasses

Saline area is expanding in both dry and irrigated lands in low rainfall and semiarid regions of the world. This is the supreme concern because salinity has been increased in agricultural soil. Truong and Baker (1997) revealed that vetiver grass not only redeems the saline soil but also prevents soil erosion. Many tropical countries in Asia (viz., Thailand and Vietnam) and Africa face the problem of acid sulfate soils. The leachates from these soils cause various diseases in fish and finally cause their death (Truong and Baker 1998a). Vetiver has been easily cultivated on these areas and it rehabilitates the acidic soil.

Kilpatrick (2012) reported that *Miscanthus* can be cultivated on contaminated soils as well as on marginal land. Its productivity can be enhanced by wastewater treatment and by organic amendments coming from distillery effluents (Galballey et al. 2012).

Seashore paspalum and Bermuda grass are mainly used as a vegetative cap for a brine landfarm, whereas weeping love grass does not sustain in contaminated soil. *Seashore paspalum* agglomerate large concentration of Cl than the other turf species. This result showed that *Seashore paspalum* act as a phytoremediator of Cl in brine-impacted soils and also sustain in contaminated soils (Fontenot et al. 2015).

Some grasses are rich in essential oil content, and they were widely used in food or drinks, mosquito repellents, perfumery, and health and care products (Jantan et al. 1999). Citronella (*Cymbopogon nardus*) essential oil has been utilized for insect-repellent purpose; however, combined dose of citronella, lemon (*Citrus limon*), rose (*Rosa damascena*), lavender, and basil essential oils with 1 liter of distilled water is very efficient to prevent indoor pests (Koul et al. 2008). But, these natural oils do not last as long as synthetic repellents (Debboun et al. 2000; Barnard and Xue 2004). Biomass of another aromatic grass, namely, lemongrass, has a wide application. Its essential oil has a great demand in cosmetic and pharmaceutical industries and also in production of eco-friendly pesticides (Ganjewala and Luthra 2010; Weiss 1997). Lemongrass oil possesses refreshing aroma and also has antifungal and antibacterial properties (Anaruma et al. 2010; Guynot et al. 2003; Pandey et al. 2003; Kumar et al. 2007, 2009; Inouye et al. 2001; Pattnaik et al. 1996). Cultivation of lemongrass is usually found in the tropics and subtropics, and major importers of lemongrass oil are the United States and European countries.

Grasses protect the soil from soil erosion because of its extensive root system, which grow very fast unlike trees which develop deep root system after several years. Soil erosion occurs due to various reasons like high rainfall, high wind, or flood. Vetiver grass technology (VGT) gives a very efficient protection against high wind and floods. Civil construction like roads, dams, etc. as well as farmlands remained stable if they were protected by VGT. Miller (1999) revealed that many countries like Central and South America and speakers from El Salvador, Mexico, Venezuela, and Costa Rica found vetiver grass technology effective in protecting and stabilizing the infrastructure of a region. Similarly this technology protects us from landslides as well as deforestation. The deep roots of grasses holds the soil efficiently from being washed away; on the other hand, its thick top growth retards the flow velocity and its erosive power (Truong 1999).

12.3.6 Carbon Sequestration by Grasses

Miscanthus can be easily grown in nutrient-poor marginal lands. However over the time, *Miscanthus* sequestered the C; as a result of this, soil C increased (Clifton-Brown et al. 2007). The C capture differs during summer and winter. Studies revealed that *M. x giganteus* grown for 15 years in Ireland captured around 0.6 t/ha/year; however, *M. sinensis* showed 2 t/ha/year soil C (Stewart et al. 2009).

A comparative study in Poland by Borzecka-Walker et al. (2008) revealed that *Miscanthus* captured about 0.6 t/ha^{-1}/year^{-1} of C, whereas willow sequestered around 0.3 t/ha^{-1}/year^{-1} of C. Thus, *Miscanthus* perform better in terms of C sequestration. Zimmermann et al. (2012) established that captured C amounts by

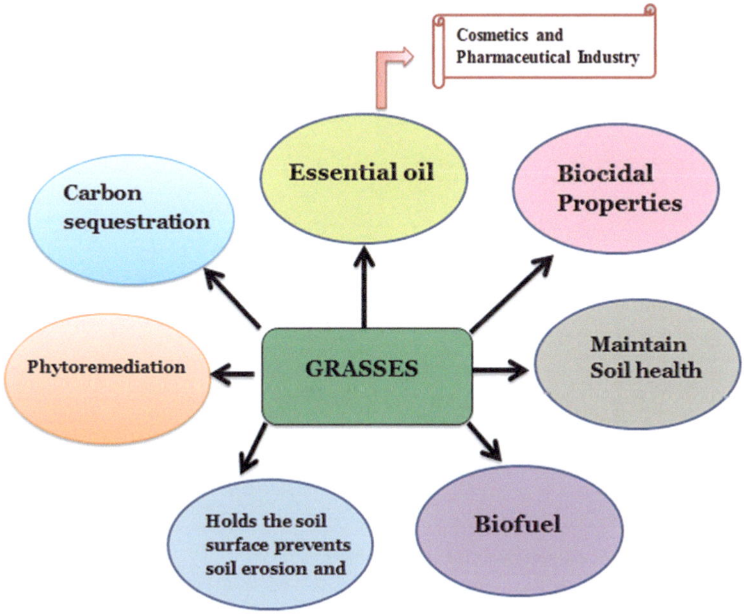

Fig. 12.1 Various uses of grasses

Miscanthus varied and it depends on the soil type. Captured carbon range varied between 0.2 and 1 t/ha^{-1}/year^{-1}. Schneckenberger and Kuzyanov (2007) studied the long-term carbon sequestration by the *Miscanthus* in the field condition. They reported that *Miscanthus* grown in a loamy soil for 9 years showed inclusion of around 0.23 t/ha^{-1}/year^{-1} soil organic C in the top 1 m of soil, whereas cultivation in a sandy soil for 12 years showed the half of incorporation as above. Hence, the soil texture also contributes much in carbon fixation by the plant in the soil. Singh et al. (2011) reported that carbon sequestered is found to be the highest by vetiver in comparison to two other aromatic grasses, viz., lemongrass (*Cymbopogon flexuosus*) and palmarosa (*Cymbopogon martinii* var. motia). Figure 12.1 depicts the various uses of grasses.

12.4 Mechanism Adopted by the Grasses Against Metal/ Xenobiotic Toxicity

Mainly, all the plants/grasses usually show similar response against environmental stress like drought, chilling, metal toxicity, and UV radiation as well as pathogen attack which causes the generation of ROS in plants due to disturbance in cellular homeostasis (Mittler 2002; Hu et al. 2008; Han et al. 2009). ROS are highly harmful to organisms at high concentrations. When the amount of ROS exceeds the defense mechanisms, a cell is said to be in a state of "oxidative stress." The elevated level of ROS during environmental stresses causes threat to cells by causing peroxidation of

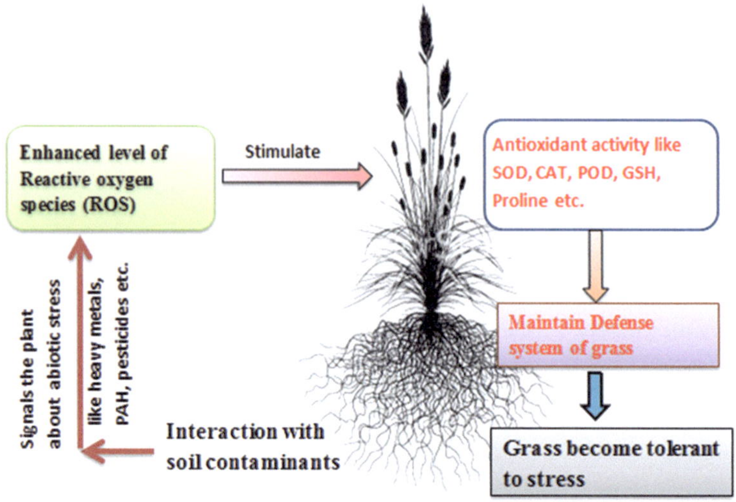

Fig. 12.2 Mechanism adopted by grass against abiotic stress

lipids, oxidation of proteins, and damage to nucleic acids ultimately leading to cell death (Verma and Dubey 2003; Meriga et al. 2004; Sharma and Dubey 2005; Maheshwari and Dubey 2009; Tanou et al. 2009). Scavenging or detoxification of excess ROS is achieved by an efficient antioxidative system comprising of the non-enzymic as well as enzymic antioxidants (Noctor and Foyer 1998). The enzymic antioxidants include superoxide dismutase (SOD), catalase (CAT), guaiacol peroxidase (GPX), and enzymes of ascorbate–glutathione (AsA-GSH) cycle such as ascorbate peroxidase (APX), monodehydroascorbate reductase (MDHAR), dehydroascorbate reductase (DHAR), and glutathione reductase (GR) (Noctor and Foyer 1998). Ascorbate (AsA), glutathione (GSH), carotenoids, tocopherols, and phenolics serve as potent nonenzymic antioxidants within the cell. Maintenance of a high antioxidant capacity to scavenge the toxic ROS has been linked to increased tolerance of grasses to these environmental stresses (Urbanek et al. 2005; Zaefyzadeh et al. 2009; Chen et al. 2010; Patel et al. 2016). Figure 12.2 shows the coping mechanism of the grasses against abiotic stress.

Dawes and Bowler (1959) found that the root systems of grasses also supply organic matter (OM) all through the soil environment. Roots secrete a gel-like material which has been a notable source of OM. The extent of gel formation by root depends on various factors like soil type, type of plant species, and age of root (Foster 1982). Mainly, gel is secreted by the epidermal root cells and comprises carbohydrates having low molecular weight (Floyd and Ohirogge 1970). Generally, gel is colonized and availed by bacteria of soil (Foster and Rovira 1976, 1978). Studies revealed that bacterial populations are higher in the rhizosphere than outside the rhizospheric soil environment. Furthermore, growth of actinomycetes and fungi is also improved with enhancement in root growth (Bowen and Rovira 1976).

The most intense fibrous root arrangement belongs to the grass family. Based on their growth tendency, grasses can be differentiated into two main groups, (1) bunch grasses which are recognized by erect growth of all stems and (2) sod-forming grasses which produce horizontal stems that may grow above (stolons) or below (rhizomes) the surface of the soil. Sod-forming grasses are known for their immense capacity to stabilize soil surfaces against erosive forces (Aprill and Sims 1990). Due to their fibrous root arrangement, they form vegetative cover on the polluted soil surface thus provide in situ removal of contaminants. Sod-forming grasses are perennials. Because of having well-developed root system, grasses provide appropriate environment for the growth of microorganism. These microorganisms in turn speed up the nitrogen uptake and metal binding and degrade the hydrocarbon as well as organic pollutants (Das and Adholeya 2012). There are three suggested mechanisms for the stimulation of soil contaminant disappearance by plants: (1) nonspecific enhancement of microbial communities by the plant (Haby and Crowley 1996), (2) enhancement of microbial activity in the rhizosphere that protects the plant from contaminants and results in disappearance (Walton et al. 1994), or (3) the development of specific degradative microbial communities in the rhizosphere (Ferro et al. 1994; Siciliano and Germida 1997).

12.5 Applications of Grasses for Bioenergy Production

Biofuel production with the help of grasses comes under second-generation biofuel crops. The second generation chiefly comprises nonfood crops, and they are preferred over food crops because these nonfood crops are not in an immediate competition with the food crops. Thus, food crop prices have not been directly affected by the second-generation crop (Lewandowski et al. 2000; Fowler et al. 2003; Rosillo-Calle et al. 2006). Biofuels are derived from cellulose, hemicellulose, or lignin, i.e., biomass obtained from the forests, grassland, or aquatic ecosystems, or by cultivation of bioenergy crops (Corma et al. 2007; Pu et al. 2008; Naik et al. 2010). Furthermore, lignocellulosic waste could also be used for the biofuel purpose (Rosillo-Calle et al. 2006; Clifton-Brown et al. 2007). Many biofuel perennial grasses belong to family Poaceae and are represented by *Miscanthus* (*M. sinensis*, *M. sacchariflorus*, *M. hybrida x giganteus*), switchgrass (*Panicum virgatum* L.), reed canary grass (*Phalaris arundinacea* L.), common reed (*Phragmites australis*), Johnson grass (*Sorghum halepense* Cav.), and sugarcane (*Saccharum* spp.) (Hoogwijk et al. 2003; Pidlisnyuk et al. 2014).

Collura et al.(2006) found that *Miscanthus* annual harvestable energy production is about 17 MJ/kg dry matter and more than 10,000 kg/ha^{-1}/yield (total 170,000 MJ/ha/year) (Dahl and Obernberger 2004; Dorge et al. 2011). Studies revealed that *Miscanthus* generates high biomass which in turn provides more energy per hectare (Heaton et al. 2010). This may be due to higher photosynthetic rates at low temperatures (Dohleman et al. 2009). However, *Miscanthus* is a C_4 grass and very effectively uses water at high temperatures and light intensity. It is adapted in a C_4 mode even at relatively low temperature, compared to switchgrass.

Zheljazkov et al. (2011) found that lemongrass biomass promptly converted to ethanol in comparison to the biomass obtained from other species which has not received pretreatment with steam. Biomass obtained from lemongrass is invariable because it has long leaves and did not form a stem. Thus, it facilitates more bioethanol production from the biomass. Tilman et al. (2006) proposed that approximately 235 million hectares of grassland can be converted into cellulosic ethanol. They predicted that about 1032 liters of ethanol can be produced through the conversion of the 4 t/ha/year of grasses. However, this disturbs the natural ecosystem as grasses play a crucial role in maintaining the natural ecosystem. Pimentel and Patzek (2008) revealed that switchgrass can also produce good amount of ethanol and the cost of producing a liter of ethanol using switchgrass was 93¢ (Ghana Cedi).

Various research has been carried out on diverse grass species for biofuel production and also determining impacts on nutrient cycle (Masarovicova et al. 2009; Rakhmetov 2007; Stefanovska et al. 2011; Gubisova et al. 2011; Pidlisnyuk 2012; Behnke et al. 2012).

12.6 Effectiveness of Grasses in Soil Microbial Growth

Soil enzymes mainly originate from plant, animal, and microbial sources. There is a prevalent postulation that soil enzymes are derived predominantly from soil microorganisms (Ladd 1978; Marinari et al. 2014). Enzyme activities are higher in rhizosphere soils than those of the bulk soil. The higher enzyme activity of the rhizosphere mainly lies on the stimulation of root-associated microbial activity by rhizo-deposition and also on the release of enzymes by roots or by lysis of root cells (Gramss et al. 1999; Chroma et al. 2002; Harvey et al. 2002). Dick (1994, 1997) established that soil enzyme activities (including rhizosphere ones) can be useful indexes of changes occurring in the microbial functioning in soil, as affected by various and different factors like input of excretory waste by herbivores (Ruess and McNaughton 1987), changes in soil porosity due to trampling of soil by animals (Abdel-Magid et al. 1987), and changes in competition for N between microorganisms and the defoliated sward (Busso et al. 2001). Moreover, grazing/mowing by animals also affects the identity of major plant species (Collins et al. 1998; Olff and Ritchie 1998), which can influence the structure of soil microbial communities (Ibekwe and Kennedy 1998; Priha et al. 1999; Briones et al. 2002; Söderberg et al. 2002).

Microorganisms in soil are also influenced by moisture content in the soil, temperature, pH, and organic matter present in the vicinity of the rhizospheres of plants (Lynch 1982). Phytoremediation practices also affect the microbial communities present in the remediated sites, and they are very crucial for revegetation as well as the extraction of metals from the contaminated soils (Pawlowska et al. 2000; Khan 2003).

Various studies (Wardle and Nicholson 1996; Wardle et al. 1998, 1999; Bardgett and Shine 1999) revealed that the perennial grasslands are an important component of plant species and played a determining role in soil microbial population. They help in maintaining soil ecosystem and nutrient cycling. However, the dominant plant species are important determinants of ecosystem function (Hopper and Vitousek 1997; Wardle et al. 1997; Grime 1998). Urease activity was higher in the soil from management system, and increased urease activity is observed in low levels of tannery waste (Patel and Patra 2014). Urease activity is more in rainy season as comparison to summer and winter (McGarity and Myers 1967). Dehydrogenase enzymes appear to be linked with microbial activity associated with initial breakdown of organic matter (Ross 1971); they are known to maintain redox potential of soil by transferring electrons and protons from substrate to acceptors. Ross (1970) stated that dehydrogenase activity appeared to be more dependent on the metabolic state of the soil or on the biological activity of the microbial population than on any free enzyme present. Enzyme activity has been enhanced in rhizospheric zone of grasses, when soil has amended with organic matter (Anderson et al. 1993).

The well-developed root provides an effective means of distributing and enhancing organic matter in the soil. The expansion of roots also helps in distribution of soil microorganism. Hence, better microbial growth is observed, and it also removes the toxic compound because interaction between microbes and toxic compound is enhanced. The fibrous root arrangement is more proficient than the tap root arrangement because the former holds the high microbial population growth than the later due to more profuse and increased surface area in fibrous root system (Anderson et al. 1993). The genetic diversity and dense, fibrous rooting system of prairie grasses facilitates the growth of microbial population and an enhanced microbial biomass (Aprill and Sims 1990).

12.7 Futures of Grasses After Remediation and Conclusions

After phytoremediation, grasses can be disposed in the incinerator, where its biomass turned into ash, which further can be converted in bricks. Another method is compositing and compaction (Kumar et al. 1995; Raskin et al. 1997; Garbisu and Alkorta 2001). Hetland et al. (2001) exhibit that composting remarkably reduces the volume of harvested biomass, although metal-contaminated biomass further requires treatment before disposal. The advantage of this method is that it lowers the transportation cost. Compaction was suggested by Blaylock and Huang (2000). The leachate collected will treat separately as it contains heavy metals. Overall schematic representation is explained in Fig. 12.3.

After phytoremediation, plant biomass can be used to extract profitable heavy metal, and the remaining is converted into bio-ore (Brooks et al. 1998), an important use for hyper-accumulator grasses for economic gain in the mining industry (Nicks and Chambers 1994).

Combustion and gasification are other routes of disposal of plant after remediation. Recovery of energy from biomass by burning or gasification could help make phytoextraction more cost-effective. Bio-fumigation is a new concept of utilizing plants insecticidal properties. Bio-fumigation is usually applied to plants, which can stimulate synthesis of glucosinolates (GLS), which are organic compounds

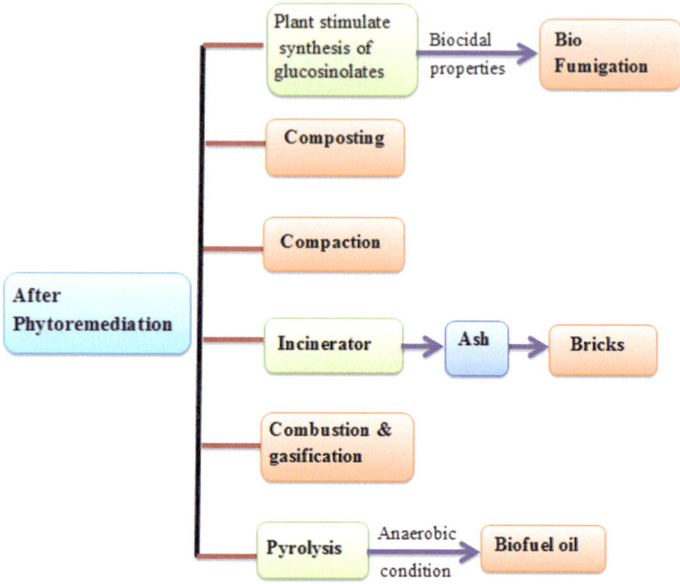

Fig. 12.3 Schematic representation of plant process after phytoremediation

containing sulfur. The products of their enzymatic degradation, mainly isothiocyanates, exhibit biocidal properties, which are used in bio-fumigation.

Pyrolysis decomposes material under anaerobic conditions. The final products are pyrolytic fluid oil and coke; heavy metals will remain in the coke, which could be used in smelter. Koppolu et al. (2003) reported that 99% of the metal recovered in the product stream was concentrated in the char formed by pyrolyzing the synthetic hyper-accumulator biomass used in the pilot-scale reactor. The metal component is concentrated by 3.2–6 times in the char, compared to feed. Indeed, what makes phytoextraction even more interesting is the fact that interest in using biofuels as an alternative energy source is high on the agenda of policy makers in many countries (Pimentel and Patzek 2007). The impact of metals on conversion efficiency, as well as the energy needed to properly use or dispose the rest product after conversion, should be considered. This has not yet been calculated extensively.

Grasses seem to be very beneficial in extracting contaminants from the soil. They not only provide a vegetative cover but its fibrous root helps in the growth of soil microbes. Grasses capture the carbon and reduce the greenhouse load of the atmosphere. Various essential oils that are obtained from aromatic grasses are used in cosmetic and pharmaceutical industry. After phytoremediation grasses can be used as biopesticides or can be utilized in producing biofuels.

References

Abdel-Magid AH, Trlica MJ, Hart RH (1987) Soil and vegetation responses to simulated trampling. J Range Manag 40:303–306

Akendengue B (1992) Medicinal plants used by the Fang traditional healers in Equatorial Guinea. J Ethnopharmacol 37:165–173

Alloway BJ (2013) Heavy metals in soils: trace metals and metalloids in soils and their bioavailability. Environ Pollut 22. Springer, Dordrecht

Anaruma ND, Schmidt FL, Duarte MCT, Figueira GM, Delarmelina C, Benato EA, Sartoratto A (2010) Control of *Colletotrichum gloeosporioides* (Penz.)Sacc.in yellow passion fruit using *Cymbopogon citratus* essential oil. Braz J Microbiol 41:66–73. doi:10.1590/S1517-83822010000100012

Anderson TA, Guthrie EA, Walton BT (1993) Bioremediation in the rhizosphere – plant roots and associated microbes clean contaminated soils. Environ Sci Technol 27(13):2630–2636

Angin I, Turan M, Ketterings QM, Cakici A (2008) Humic acid addition enhances B and Pb phytoextraction by Vetiver grass (*Vetiveria zizanioides* (L.). Nash). Water Air Soil Pollut 188:335–343

Antiochia R, Campanella L, Ghezzi P, Movassaghi K (2007) The use of vetiver for remediation of heavy metal soil contamination. Anal Biochem 388:947–956

Aprill W, Sims RC (1990) Evaluation of the use of prairie grasses for stimulating polycyclic aromatic hydrocarbon treatment in soil. Chemosphere 20:253–265

Arias RJ, Schmeda-Hirschmann G, Falcao A (1992) Feeding deterrency and insecticidal effects of plant extracts on Lutzomyia longipalpis. Phytother Res 6:64–67

Baker DE, Eldershaw VJ (1993) Interpreting soil analyses for agricultural land use in Queensland. Project Report Series Q093014, QDPI, Brisbane, Australia

Balsamo Ronald A et al (2015) Utilization of grasses for potential biofuel production and phytoremediation of heavy metal contaminated soils. Int J Phytoremed 17(5):448–455

Barbu CH (2010) *Miscanthus sinensis x giganteus* cultivated on soils polluted with heavy metals – a valuable replacement for coal. In: International conference on green remediation, conference summary papers. University of Massachusetts Amherst, pp 2–5

Barbu CH, Pavel BP, Sand C, Pop MR (2009) *Miscanthus sinensis x giganteus'* behavior on soils polluted with heavy metals. Metal Elements in Environment, Medicine and Biology, Tome 9. Cluj University Press, pp 21–24. (Proceedings of 9th international symposium of Romanian Academy – Branch Cluj- Napoca, Cluj-Napoca, Romania, 16–17 Oct 2009). Available at web: http://www.rutsolmeg.ro/raport02.pdf

Bardgett RD, Shine A (1999) Linkages between litter diversity, soil microbial biomass and ecosystem function in temperate grasslands. Soil Biol Biochem 31:317–321

Barnard DR, Xue RD (2004) Laboratory evaluation of mosquito repellents against *Aedes albopictus, Culex nigripalpus, and Ochlerotatus triseriatus* (Diptera: Culicidae). J Med Entomol 41(4):726–730

Behnke GD, Davis MB, Voight NB (2012) Greenhouse gas emissions, nitrate leaching, and biomass yields from production of *Miscanthus x giganteus* in Illinois, U.S.A. Bioenergy Res 1:801–813

Blaylock MJ, Huang JW (2000) Phytoextraction of metals. In: Raskin I, Ensley BD (eds) Phytoremediation of toxic metals: using plants to clean up the environment. Wiley, New York, pp 53–70

Boonsaner M, Kaewumput T, Boonsaner A (2002) Bioaccmulation of organophosphate pesticides in vetiver grass (*Vetiveria zizanioides* Nash.). Thammasat Int J Sci Tech January–April: 30–40

Borzecka-Walker M, Faber A, Borek R (2008) Evaluation of carbon sequestration in energetic crops (Miscanthus and coppice willow). Int Agrophys 22:185–190

Bowen GD, Rovira AD (1976) Microbial colonization of plant roots. Annu Rev Phytopathol 14:121–144

Brandt R, Merkl N, Kraft SR, Infante C, Gabriele B (2006) Potential of vetiver (*Vetiveria zizanioides* (L) Nash) for phytoremediation of petroleum hydrocarbon-contaminated soils inVenezuela. Int J Phytorem 8:273–284
Briones AM, Okabe S, Umemiya Y, Ramsig NB, Reichardt W, Okuyama H (2002) Influence of different cultivars on populations of ammonia-oxidizing bacteria in the root environment of rice. Appl Environ Microbiol 68:3067–3075
Brooks RR, Chambers MF, Nicks LJ, Robinson BH (1998) Phytomining. Trends Plant Sci 1:359–362
Brunetti G, Soler-Rovira P, Farrag K, Senesi N (2009) Tolerance and accumulation of heavy metals by wild plant species grown in contaminated soils in Apulia region, Southern Italy. Plant Soil 318:285–298
Busso CA, Briske DD, Olalde-Portugal V (2001) Root traits associated with nutrient exploitation following defoliation in three coexisting perennial grasses in a semiarid savanna. Oikos 93:332–342
Cajthaml T, Bhatt M, Săsek V, Matĕju V (2002) Bioremediation of PAH contaminated soil by composting: a case study. Folia Microbiol 47(6):696–700
Carrow RN (1996) Drought resistance aspects of turf grasses in the Southeast: root-shoot responses. Crop Sci 36:687–694
Castillo M, Fonseca R, Candia JR (2007) Report on the pilot study on the use of Vetiver grass for Cu mine tailings phytostabilisation at Anglo American El Solado mine, Chile. Fundacion Chile (report in Spanish)
Cenas N, Ceniene AN, Sergediene E, Nivinskas H, Anusevicius Z, Sarlauskas J (2001) Quantitative structure-activity relationships in enzymatic single-electron reduction of nitroaromatic explosives: implications for their cytotoxicity. Biochim Biophys Acta 1528:31–38
Chaney RL, Ryan AA, Li YM, Brown SL (1998) Soil cadmium as a threat to human health. In: McLaughlin MJ, Singh BR (eds) Cadmium in soils, plants and the food chain. Kluwer, Dordrecht, pp 219–256
Chaudhry TM, Hayes WJ, Khan AG, Khoo CS (1998) Phytoremediation – focusing on accumulator plants that remediate metal-contaminated soils. Australas J Ecotoxicol 4(1):37–51
Chen Y, Shena Z, Lib XD (2004) The use of vetiver grass (*Vetiveria zizanioides*) in the phytoremediation of soils contaminated with heavy metals. Appl Geochem 19(10):1553–1565
Chen Q, Zhang M, Shen S (2010) Effect of salt on malondialdehyde and antioxidant enzymes in seedling roots of Jerusalem artichoke (*Helianthus tuberosus* L.). Acta Physiol Plant 33(2):273–278
Chiu KK, Ye ZH, Wong MH (2005) Enhanced uptake of As, Zn, and Cu by *Vetiveria zizanioides* and Zea mays using chelating agents. Chemosphere 60:1365–1375
Chiu KK, Ye ZH, Wong MH (2006) Growth of *Vetiveria zizanioides* and *Phragmities australis* on Pb/Zn and Cu mine tailings amended with manure compost and sewage sludge: a greenhouse study. Bioresour Technol 97:158–170
Chroma L, Mackovam M, Kucerovam P, Der Wiesche C, Burkhard J, Macek T (2002) Enzymes in plant metabolism of PCBs and PAHs. Acta Biotechnol 22:35–41
Clifton-Brown J, Joern B, Jones M (2007) Carbon mitigation by the energy crop, miscanthus. Glob Chang Biol 13:2296–2307
Clifton-Brown JOHN, Robson PAUL, Allison GORDON, Lister S, Sanderson R, Hodgson E et al (2008) Miscanthus: breeding our way to a better future. Asp Appl Biol 90:199–206
Collins SL, Knapp AK, Briggs JM, Blair JM, Steinauer EM (1998) Modulation of diversity by grazing and mowing in native tall grass prairie. Science 280:745–747
Collura S, Azambre B, Finqueneisel G, Zimny T, Weber JW (2006) *Miscanthus x giganteus* straw and pellets as sustainable fuels. Combustion and emission tests. Environ Chem Lett 4:75–78
Corma A, Iborra S, Velty A (2007) Chemical routes for the transformation of biomass into chemicals. Chem Rev 107:4044–4098
Cull RH, Hunter H, Hunter M, Truong P (2000) Application of vetiver grass technology in off-site pollution control. II. Tolerance of vetiver grass towards high levels of herbicides under wetland

conditions. In: Proceedings of second international Vetiver conference (ICV-2), Phetchaburi, Thailand, 18–22 January

Cunningham SD, Berti WR (2000) Phytoextraction and phytostabilization: technical, economic and regulatory considerations of the soil-lead issue. In: Terry N, Banuelos GS (eds) Phytoremediation of contaminated soil and water. CRC Press, Boca Raton, pp 359–376

Dahl J, Obernberger I (2004) Evaluation of the combustion characteristics of four perennial energy crops (*Arundo donax, Cynara cardunculus, Miscanthus x Giganteus,* and *Panicum virgatum*). In: Proceedings 2nd world conference on biomass for energy, industry, and climate protection, Rome, Italy, May 2004

Dalton PA, Smith RJ, Truong PNV (1996) Vetiver grass hedges for erosion control on a cropped flood plain: hedge hydraulics. Agric Water Manag 31:91–104

Das M, Adholeya A (2012) Role of microorganisms in remediation of contaminated soil. In: Microorganisms in environmental management. Springer, Dordrecht, pp 81–111

Dawes CJ, Bowler E (1959) Light and electron microscope studies of the cell wall structure of the root hairs of *Raphanus sativus*. Am J Bot 46:561–565

Debboun M, Strickman D, Solberg VB, Wilkerson RC, McPherson KR, Golenda C et al (2000) Field evaluation of deet and a piperidine repellent against Aedes communis (Diptera: Culicidae) and Simulium venustum (Diptera: Simuliidae) in the Adirondack mountains of New York. J Med Entomol 37(6):919–923

Dick RP (1994) Soil enzymes activities as indicators of soil quality. In: Doran JW, Coleman DC, Bezdicek DF, Stewart BA (eds) Defining soil quality for a sustainable environment, vol 35. SSSA Special Publication, Madison, pp 107–124

Dick RP (1997) Soil enzyme activities as integrative indicators of soil health. In: Pankhurst CE, Doube BM, Gupta VVSR (eds) Biological indicators of soil health. CAB International, Wallingford, pp 121–156

Dickinson NM, Baker AJ, Doronila A, Laidlaw S, Reeves RD (2009) Phytoremediation of inorganics: realism and synergies. Int J Phytoremed 11(2):97–114

Dohleman FG, Heaton EA, Leakey A, Long SP (2009) Does greater leaf-level photosynthesis explain the larger solar energy conversion efficiency of miscanthus relative to switchgrass? Plant Cell Environ 32:1525–1537

Dorge S, Jeguirim M, Trouve G (2011) Thermal degradation of Miscanthus pellets: kinetics and aerosol characterization. Waste Biomass Valoriz 2:149–155

Ehleringer JR (1978) Implications of quantum yield differences on the distributions of C3 and C4 grasses. Oecologia 31(3):255–267

Ernst WHO (1996) Bioavailability of heavy metals and decontamination of soils by plants. Appl Geochem 11(1):163–167

Ferro AM, Sims RC, Bugbee B (1994) Hycrest crested wheatgrass accelerates the degradation of pentachlorophenol in soil. J Environ Qual 23:272–279

Fitter AH, Stickland TR (1991) Architectural analysis of plant root system. 2. Influence of nutrient supply on architecture in contrasting plant species. New Phytol 118:383–389

Floyd RA, Ohlrogge AJ (1970) Gel formation on nodal root surfaces of *Zea mays*. I. Investigation of the gel's composition. Plant Soil 33:331–343

Fontenot D, Bush E, Beasley J, Fontenot K (2015) Evaluating Bermudagrass (Cynodon dactylon), seashore Paspalum (Paspalum vaginatum), and weeping lovegrass (*Eragrostis curvula*), as a vegetative cap for industrial brine landform stabilization and phytoremediation. J Plant Nutr 38(2):237–245

Foster RC (1982) The fine structure of epidermal cell mucilages of roots. New Phytol 91:727–740

Foster RC, Rovira AD (1976) Ultrastructure of wheat rhizosphere. New Phytol 76:343–352

Foster RC, Rovira AD (1978) The ultrastructure of the rhizosphere of *Trifolium subterraneurn* L. In: Loutit MW, Miles JAR (eds) Microbial ecology. Springer, New York

Fowler PA, McLauchlin AR, Hall LM (2003) The potential industrial uses of forage grasses including Miscanthus, report published by biocomposites centre, University of Wales, Wales, 40 pages. Available at: http://www.bc.bangor.ac.uk

Galbally F, Fagan C, Ryan D, Finnan J, Grant J, McDonnell K (2012) Biosolids and distillery effluent amendment to Irish *Miscanthus x giganteus* plantations: impacts on groundwater and soil. J Environ Qual 41:114–123
Ganjewala D, Luthra R (2010) Essential oil biosynthesis and regulation in the genus Cymbopogon. Nat Prod Commun 5(1):163–172
Garbisu C, Alkorta I (2001) Phytoextraction: a cost-effective plant-based technology for the removal of metals from the environment. Bioresour Technol 77:229–236
Gramss G, Voigt KD, Kirsche B (1999) Degradation of polycyclic aromatics hydrocarbons with three to seven aromatic rings by higher fungi in sterile and unsterile soils. Biodegradation 10:51–62
Grime JP (1998) Benefits of plant diversity to ecosystems: immediate, filter and founder effects. J Ecol 86:902–910
Gubisova M, Porvaz P, Gubis J, Zofajova A, Mihalik D, Kraic J (2011) Growing technology and biotechnology of miscanthus in Slovakia. In: Proceeding of Pannonian plant biotechnology workshops. The bioenergy question: reality or wishful thinking, Vienna, Austria, 16–18 May 2011. Available at: http://www.pannonbiotech.hu/user/browser/File/PPBATulln2011Book.pdf.
Guynot ME, Ramos AJ, Setó L, Purroy P, Sanchis V, Marín S (2003) Antifungal activity of volatile compounds generated by essential oils against fungi commonly causing deterioration of bakery products. J Appl Microbiol 94:893–899. doi:10.1046/j.1365-2672.2003.01927.x
Haby PA, Crowley DE (1996) Biodegradation of 3-chlorobenzoate as affected by rhizodeposition and selected carbon substrates. J Environ Qual 25:304–310
Han C, Liu Q, Yang Y (2009) Short-term effects of experimental warming and enhanced ultraviolet-B radiation on photosynthesis and antioxidant defense of *Picea asperata* seedlings. Plant Growth Regul 58(2):153–162
Hartley W, Dickinson NM, Riby P, Lepp NW (2009) Arsenic mobility in brownfield soils amended with green waste compost or biochar and planted with Miscanthus. Environ Pollut 157(10):2654–2662
Harvey PJ, Xiang M, Palmer JM (2002) Extracellular enzymes in the rhizosphere. In: Proceedings of inter-cost workshop on soil-microbe-root interactions: maximising phytoremediation/bioremediation, Grainau, Germany, pp 23–25
Heaton EA, Dohleman FG, Miguez AF, Juvik JA, Lozovaya V, Widholm J, Zabotina OA, McIsaac GF, Davis MB, Voigt TB, Boersma NN, Long SP (2010) Miscanthus. A promising biomass crop. Adv Bot Res 56:76–137
Hemen S (2011) Metal hyperaccumulation in plants: a review focusing on phytoremediation technology. J Environ Sci Technol 4(2):118–138
Hetland MD, Gallagher JR, Daly DJ, Hassett DJ, Heebink LV (2001) Processing of plants used to phytoremediate lead-contaminated sites. In: Leeson A, Foote EA, Banks MK, Magar VS (eds) Phytoremediation. Wetlands, and sediments, – the sixth international in situ and on-site bioremediation symposium, San Diego, California, 4–7 June. Battelle Press, Columbus, pp 129–136
Hoang TTT, Tu TCL, Dao PQ (2007) Progress and results of trials using vetiver for phytoremediation of contaminated canal sludge around Ho Chi Minh City. In: Proceedings of Vetiver workshop, Hanoi, Vietnam, May 2007 (in Vietnamese)
Hodkinson TR, Chase MW, Renvoize SA (2002) Characterization of a genetic resource collection for Miscanthus (Saccharinae, Andropogonae, Poaceae) using AFLP and ISSR PCR. Ann Bot 89:627–636
Hoflich G, Wiehe W, Kuhn G (1994) Plant growth stimulation with symbiotic and associative rhizosphere microorganisms. Experientia 50:897–905
Hoogwijk MM, Faaij A, van den Broek R, Berndes G, Gielen D, Turkenburg W (2003) Exploration of the ranges of the global potential of biomass for energy. Biomass Bioenergy 25:119–133
Hooper D, Vitousek PM (1997) The effects of plant composition and diversity on ecosystem processes. Science 277:1302–1305
Hsu TS, Bartha R (1979) Accelerated mineralization of two organophosphate insecticides in the rhizosphere. Appl Environ Microbiol 37:36–41

Hu WH, Song XS, Shi K, Xia XJ, Zhou YH, Yu JQ (2008) Changes in electron transport, superoxide dismutase and ascorbate peroxidase isoenzymes in chloroplasts and mitochondria of cucumber leaves as influenced by chilling. Photosynthetica 46(4):581–588

Ibekwe AM, Kennedy AC (1998) Phospholipid fatty acid profiles and carbon utilization patterns for analysis of microbial community structure under field and greenhouse conditions. FEMS Microbiol Ecol 26:151–163

Inouye S, Takizawa T, Yamaguchi H (2001) Antibacterial activity of essential oils and their major constituents against respiratory tract pathogens by gaseous contact. J Antimicrob Chemother 47:565–573. doi:10.1093/jac/47.5.565

Iordache V, Mihail D, Neagoe A., Sandu R, Bodescu F, Marinoiu L (2010) Decision support for the management of agrosystems contaminated with metals: research program and preliminary results. In: Barbu CH, Sand C (eds) Modern technologies and biotechnologies for environmental protection. "Lucian Blaga". University of Sibiu, pp 83–93

Iqbal M, Bermond A, Lamy I (2013) Impact of miscanthus cultivation on trace metal availability in contaminated agricultural soils: complementary insights from kinetic extraction and physical fractionation. Chemosphere 91:287–294

Jantan I, Zaki ZM, Ahmad AR, Ahmad R (1999) Evaluation of smoke from mosquito coils containing Malaysian plants against *Aedes aegypti*. Fitoterapia 70(3):237–243

Kabata-Pendias A (2011) Trace elements in soils and plant, 4th edn. Taylor and Francis Group, LLC, Boca Raton

Kalmbacher RS, Martin FG (1998) Effect of flooding on seed germination and emergence of three pasture grasses. Soil Crop Sci Soc Fla Proc 57:73–77

Kamath R, Rentz JA, Schnoor JL, Alvarez PJJ (2004) Phytoremediation of hydrocarbon-contaminated soils: principles and applications. In: Vazquez-Duhalt R, Quintero-Ramirez R (eds) Studies in surface science and catalysis. Elsevier, Amsterdam, pp 447–478

Kao WY, Tsai TT, Chen WH (1998) Acomparative study of *Miscanthus floridulus* (Labill) Warb. and *M.transmorrisonensis* Hayata: photosyntheticgas exchange, leaf characteristics and growth in controlled environments. Ann Bot 81:295–299

Kazembe T, Chauruka D (2012) Mosquito repellence of Astrolochii hepii, Cymbopogon citratus and Ocimum gratissimum extracts and mixtures. Bull Environ Pharmacol Life Sci 1:60–64

Keller C, Hammer D, Kayser A, Richner W, Brodbeck M, Sennhauser M (2003) Root development and heavy metal phytoextraction comparison of different species in the field. Plant Soil 249:67–81

Kerr J, Pulford I, Duncan H, Wheeler C (1998) Phytoremediation of heavy metal contaminated sites by potential fiber crops. In: Proceedings of the 6th international FZK/TNO conference contaminated soil, vol 98. Telford, London, pp 1119–1120

Khan AG (2003) Vetiver grass as an ideal phytosymbiont for Glomalian fungi for ecological restoration of derelict land. In: Truong P, Hanping X (eds) Proceedings of the third international conference on vetiver and exhibition: vetiver and water, Guangzou, China, October 2003. China Agricultural Press, Beijing, pp 466–474

Kilpatrick LA (2012) Sustainable growth of miscanthus on marginal lands amended with flue gas desulfurization gypsum and sewage biosolids. Paper No. 12–1337624, ASABE. www.asabe.org, 37 pp

Koester HH, Meissner HH, Coertze RJ et al (1992a) Variation in the production and quality of bana grass over the growing season using hand-clipped samples. Suid Afrikaanse Tydskrif Vir Veekunde 22:31–34

Koester HH, Meissner HH, Coertze RJ et al (1992b) Voluntary intake and quality of diet selected by cattle grazing bana grass, kikuyu and forage sorghum. Suid Afrikaanse Tydskrif Vir Veekunde 22:35–41

Koppolua L, Agblover FA, Clements LD (2003) Pyrolysis as a technique for separating heavy metals from hyperaccumulators. Part II: Lab-scale pyrolysis of synthetic hyperaccumulator biomass. Biomass Bioenerg 25:651–663

Koul O, Walia S, Dhaliwal GS (2008) Essential oils as green pesticides: potential and constraints. Biopestic Int 4:63–84

Kumar PBAN, Dushenkov V, Motto H, Raskin I (1995) Phytoextraction: the use of plants to remove heavy metals from soils. Environ Sci Technol 29:1232–1238
Kumar R, Dubey NK, Tiwari OP, Tripathi YB, Sinha KK (2007) Evaluation of some essential oils as botanical fungitoxicants for the protection of stored food commodities from fungal infestation. J Sci Food Agric 87:1737–1742. doi:10.1002/jsfa.2906
Kumar A, Shukla R, Singh P, Dubey NK (2009) Biodeterioration of some herbal raw materials by storage fungi and aflatoxin and assessment of *Cymbopogon flexuosus* essential oil and its components as antifungal. Int Biodeterior Biodegrad 63:712–716. doi: 10.1016/j
Ladd JN (1978) Origin and range of enzymes in soil. In: Burns RG (ed) Soil enzymes. Academic, New York, pp 51–96
Lai HY, Chen ZS (2004) Effect of EDTA on solubility of cadmium, zinc and lead and their uptake by rainbow pink and vetiver grass. Chemosphere 55:421–430
Lewandowski I, Clifton-Brown J, Scurlock J, Huisman W (2000) Miscanthus: European experience with a novel energy crop. Biomass Bioenergy 19:210–227
Li H, Luo YM, Song J, Wu LH, Christie P (2006) Degradation of benzo[a]pyrene in an experimentally contaminated paddy soil by vetiver grass (*Vetiveria zizanioides*). Environ Geochem Health 28:183–188
Li GY, Hu N, Ding DX, Zheng JF, Liu YL, Wang YD, Nie XQ (2011) Screening of plant species for phytoremediation of uranium, thorium, barium, nickel, strontium and lead contaminated soils from a uranium mill tailing repository in South China. Bull Environ Contam Toxicol 86:646–652
Lou LQ, Ye ZH, Wong MH (2007) Solubility and accumulation of metals in Chinese brake fern, Vetiver and Rostrate Sesbania using chelating agents. Int J Phytoremed 9:325–343
Lynch JM (1982) Limits to microbial growth in soil. Microbiology 128(2):405–410
Maheshwari R, Dubey RS (2009) Nickel-induced oxidative stress and the role of antioxidant defense in rice seedlings. Plant Growth Regul 59(1):37–49
Makrisa KC, Shakyaa KM, Dattaa R, Sarkar D, Pachanoora D (2007a) High uptake of 2,4,6-trinitrotoluene by vetiver grass – potential for phytoremediation? Environ Pollut 146:1–4
Makrisa KC, Shakyaa KM, Dattaa R, Sarkar D, Pachanoora D (2007b) Chemically catalyzed uptake of 2,4,6-trinitrotoluene by *Vetiveria zizanioides*. Environ Pollut 148:101–106
Maliszewska-Kordybach B (2005) Dissipation of polycyclic aromatic hydrocarbons in freshly contaminated soils—the effect of soil physicochemical properties and aging. Water Air Soil Pollut 168(1–4):113–128
Marcum KB, Murdoch CL (1994) Salinity tolerance mechanisms of six C-4 turf grasses. J Am Soc Hortic Sci 119:779–784
Maric M, Antonijevic M, Alagic S (2013) The investigation of the possibility for using some wild and cultivated plants as hyperaccumulators of heavy metals from contaminated soil. Environ Sci Pollut Res 20(2):1181–1188
Marinari S, Moscatelli C, Grego S (2014) Enzymes at plant-soil interface. In: Gianfreda L, Rao MA (eds) Enzymes in agricultural sciences. OMICS eBooks Group, 94–109
Masarovicova E, Kralova K, Pesko M (2009) Energetic plants- cost and benefit. Ecol Chem Eng 16:263–276
McGarity JW, Myers MG (1967) A survey of urease activity in soils of northern New South Wales. Plant Soil 27(2):217–238
Meriga B, Reddy BK, Rao KR, Reddy LA, Kishor PBK (2004) Aluminium-induced production of oxygen radicals, lipid peroxidation and DNA damage in seedlings of rice (*Oryza sativa*). J Plant Physiol 161(1):63–68
Merkl N, Schultze-Kraft R, Arias M (2005) Influence of fertilizer levels on phytoremediation of crude oil-contaminated soils with the tropical pasture grass *Brachiaria brizantha* (Hochst. ex a. rich.) stapf. Int J Phytoremed 7(3):217–230
Mezzari MP, Zimermann DMH, Corseuil HX, Nogueira AV (2011) Potential of grasses and rhizosphere bacteria for bioremediation of diesel-contaminated soils. Revista Brasileira de Ciência do Solo 35(6):2227–2236

Miller J (1999) Report on bioengineering workshop for post-Hurricane Mitch construction – experience with the use of vetiver for the protection and stabilization of infrastructure, Latin America Vetiver Network, San Salvador, May 1999

Mittler R (2002) Oxidative stress, antioxidants and stress tolerance. Trends Plant Sci 7(9):405–410

Muchuweti M, Birkett JW, Chinyanga E, Zvauya R, Scrimshaw MD, Lester JN (2006) Heavy metal content of vegetables irrigated with mixtures of wastewater and sewage sludge in Zimbabwe: implications for human health. Agric Ecosyst Environ 112(1):41–48

Mueller KE, Shann JR (2006) PAH dissipation in spiked soil: impacts of bioavailability, microbial activity and trees. Chemosphere 64:1006–1014

Naik SN, Goud VV, Rout PK, Dalai AK (2010) Production of first and second generation biofuels: a comprehensive review. Renew Sust Energ Rev 14:578–597

Nicks L, Chambers MF (1994) Nickel farm. Discover, September, p 19

Noctor G, Foyer CH (1998) Ascorbate and glutathione: keeping active oxygen under control. Annu Rev Plant Biol 49:249–279

Olff H, Ritchie ME (1998) Effects of herbivores on grassland plant diversity. Trees 13:261–265

Ollivier J, Wanat N, Austruy A, Hitme A, Joussein E, Welzl G, Munch JC, Schloter M (2012) Abundance and diversity of ammonia oxidizing prokaryotes in the root-rhizosphere complex of *Miscanthus x giganteus* grown in heavy metal-contaminated soils. Microb Ecol 64:1038–1046

Palazzo AJ, Lee CR (1997) Root growth and metal uptake of plants grown on zinc-contaminated soils as influenced by soil treatment and plant species. In: Extended abstracts of the 4th international conference on the biogeochemistry of trace. Elements, Berkeley, CA, USA, 23–26 June 1997, pp 441–442

Pandey AK, Rai MK, Acharya D (2003) Chemical composition and antimycotic activity of the essential oils of corn mint (*Mentha arvensis*) and lemon grass (*Cymbopogon flexuosus*) against human pathogenic fungi. Pharm Biol 41:421–425. doi:10.1076/phbi.41.6.421.17825

Patel A, Patra DD (2014) Influence of heavy metal rich tannery sludge on soil enzymes vis-à-vis growth of Tagetes minuta, an essential oil bearing crop. Chemosphere 112:323–332

Patel A, Pandey V, Patra DD (2016) Metal absorption properties of *Mentha spicata* grown under tannery sludge amended soil- its effect on antioxidant system and oil quality. Chemosphere 147:67–73

Pattnaik S, Subramanyam VR, Kole C (1996) Antibacterial and antifungal activity of ten essential oils in vitro. Microbios 86:237–246

Pawlowska TE, Chaney RL, Chin M, Charvat I (2000) Effect of metal extraction practices on the indigenous community of arbuscular mycorrhizal fungi at a metal contaminated landfill. Appl Environ Microbiol 66:2526–2530

Pichai NMR, Samjiamjiaras R, Thammanoon H (2001) The wonders of a grass, Vetiver and its multifold applications. Asian Infrastruct Res Rev 3:1–4

Pidlisnyuk V (2012) Expanding the potential of second generation biofuels crops by using for phytoremediation of sites contaminated by heavy metals: laboratory stage. Sci Bull Kremenchuk Nat Univ 74(4):104–108 (in Ukrainian)

Pidlisnyuk V, Stefanovska T, Lewis EE, Erickson LE, Davis LC (2014) Miscanthus as a productive biofuel crop for phytoremediation. Crit Rev Plant Sci 33(1):1–19

Pimentel D, Patzek T (2007) Ethanol production: energy and economic issues related to U.S. and Brazilian sugarcane. Nat Resour Res 16(3):235–242

Pimentel D, Patzek T (2008) Chapter 15: Ethanol production using corn, switchgrass and wood; biodiesel production using soybean. In: Pimentel D (ed) Biofuels, solar and wind as renewable energy systems: benefits and risks. Springer, Dordrecht, pp 375–396

Pinthong J, Impithuksa S, Ramlee A (1998) The capability of vetiver hedgerows in decontamination of agrochemical residues: a case study on the production of cabbage at Nong Hoi Development Center. In: Proceedings of the first international conference on vetiver, Chiang Rai, Thailand, pp 91–98

Pogrzeba M, Kryzyzak J, Sas-Nowosielska A, Majtkowski W, Malkowski E, Kita A (2010) A heavy metals environmental threat resulting from combustion of biofuels of plant origin. In: Simenov LI, Kochubovski MV, Simenova BG (eds) Environmental heavy metal pollution and effects on child mental development: risk assessment and prevention strategies, NATO Science for Peace and Security Series C: Environmental Security, pp 213–225

Pradhan SP, Conrad JR, Paterek JR, Srivastava VJ (1998) Potential of phytoremediation for treatment of PAHs in soil at MGP Sites. J Soil Contam 7:467–480

Priha O, Grayston SJ, Pennanen T, Smolander A (1999) Microbial activities related to C and N cycling and microbial community structure in the rhizospheres of *Pinus sylvestris, Picea abies* and *Betula pendula* seedlings in an organic and mineral soil. FEMS Microbiol Ecol 30:187–199

Pu Y, Zhang D, Singh PM, Ragauskas AJ (2008) The new forestry biofuels sector. Biofuels Bioprod Bioref 2:58–73

Qian YL, Engelke MC (1999) Performance of five turfgrasses under linear gradient irrigation. Hortscience 34:893–896

Raghavendra AS, Sage RF (eds) (2010) C4 photosynthesis and related CO_2 concentrating mechanisms, vol 32. Springer, Dordrecht

Rakhmetov DB (2007) Role of new cultures in bioenergy. Sci Release Nat Agric Univ 116:13–20 (in Ukrainian)

Raskin I, Smith RD, Salt DE (1997) Phytoremediation of metals: using plants to remove pollutants from the environment. Curr Opin Biotechnol 8(2):221–226

Renoldi F, Lietti L, Saponaro S, Bonomo L, Forzatti P (2003) Thermal desorption of a PAH-contaminated soil: a case study. In: Brebbia CA (ed) Ecosystems and sustainable development IV, vol 2. WIT Press, Southampton, pp 1123–1132

Reynolds CM, Wolf DC, Gentry TJ et al (1999) Plant enhancement of indigenous soil microorganisms: a low-cost treatment of contaminated soils. Polar Rec 35(192):33–40

Rosillo-Calle F, de Groot P, Hemstock SL, Woods J (eds) (2006) The biomass assessment handbook: bioenergy for a sustainable environment. Earthscan Publications Ltd., London

Ross DJ (1970) Effects of storage on dehydrogenase activities of soils. Soil Biol Biochem 2(1):55–61

Ross DJ (1971) Some factors influencing the estimation of dehydrogenase activities of some soils under pasture. Soil Biol Biochem 3(2):97–110

Rotkittikhun P, Chaiyarat R, Kruatrachue M, Pokethitiyook P, Baker AJM (2007) Growth and lead accumulation by the grasses *Vetiveria zizanioides* and *Thysanolaena maxima* in lead contaminated soil amended with pig manure and fertilizer: a glasshouse study. Chemosphere 66:45–53

Ruess RW, McNaughton SJ (1987) Grazing and the dynamics of nutrient and energy regulated microbial processes in the Serengeti grasslands. Oikos 49:101–110

Sage RF, Kubien DS (2007) The temperature response of C3 and C4 photosynthesis. Plant Cell Environ 30(9):1086–1106

Schneckenberger K, Kuzyanov Y (2007) Carbon sequestration under miscanthus. In: Second international Vetiver conference (ICV-2), Phetchaburi, Thailand, 18–22 January

Shann J R, Boyle JJ (1994) Influence of plant species on in situ rhizosphere degradation. In: Anderson TA, Coats JR (eds) Bioremediation through rhizosphere technology. ACS Symposium Series 563. ACS, Washington, DC, pp 70–81

Sharma P, Dubey RS (2005) Drought induces oxidative stress and enhances the activities of antioxidant enzymes in growing rice seedlings. Plant Growth Regul 46(3):209–221

Shaw G, Bell JNB (1991) Competitive effects of potassium and ammonium on cesium uptake kinetics in wheat. J Environ Radioact 13:283–296

Shu WS, Lan CY, Zhang ZQ, Wong MH (2000) Use of vetiver and other three grasses for revegetation of Pb/Zn mine tailings at Lechang, Guangdong Province: field experiment. In: 2nd international Vetiver conference, Bangkok Thailand, January

Siciliano SD, Germida JJ (1997) Bacterial inoculants of forage grasses that enhance degradation of 2-chlorobenzoic acid in soil. Environ Toxicol Chem 16(6):1098–1104

Singh M, Sharma S (2001) Influence of irrigation and nitrogen on herbage and oil yield of palmarosa (*Cymbopogon martinii*) under semi-arid tropical conditions. Eur J Agron 14:157–159. doi:10.1016/S1161-0301(00)00083-6

Singh G, Brar MS, Malhi SS (2007a) Decontamination of chromium by farmyard manure application in spinach grown in two texturally different Cr-contaminated soils. J Plant Nutr 30:289–308

Singh SK, Juwarkar AA, Kumar S, Meshram J, Fan M (2007b) Effect of amendment on phytoextraction of arsenic by *Vetiveria zizanioides* from soil. Int J Environ Sci Technol 4:339–344

Singh S, Eapen S, Thorat V, Kaushik CP, Raj K, D'Souzaa SF (2008a) Phytoremediation of 137cesium and 90strontium from solutions and low-level nuclear waste by *Vetiveria zizanioides*. Ecotoxicol Environ Saf 69:306–311

Singh S, Melo JS, Eapen S, D'Souza SF (2008b) Potential of vetiver (*Vetiveria zizanioides* L. Nash) for phytoremediation of phenol. Ecotoxicol Environ Saf 71:671–676

Singh M, Guleria N, Prakesa Rao EVS, Goswami P (2011) A strategy for sustainable carbon sequesteration using Vetiver (*Vetiveria zizanioides* (L.): a quantitative assessment over India. CSIR Network Progect: Integrated Analysis for Impact, Mitigation and Sustainability (IAIMS)

Söderberg KH, Olsson PA, Baath E (2002) Structure and activity of the bacterial community in the rhizosphere of different plant species and the effect of arbuscular mycorrhizal colonization. FEMS Microbiol Ecol 40:223–231

Sripen S, Komkris T, Techapinyawat S, Chantawat S, Masuthon S, Rungsuk M (1996) Growth potential of vetiver grass in relation to nutrients in wastewater of Changwat Phetchaburi. Paper presented at ICV-1. In: Abstracts of papers presented at ICV-1. ORDPB, Bangkok, p 44

Srisatit T, Kosakul T, Dhitivara D (2003) Efficiency of arsenic removal from soil by *Vetiveria zizanioides* and *Vetiveria nemoralis*. Sci Asia 29:291–296

Stefanovska T, Lewis E, Pidlisnyuk V (2011) Evaluation of potential risk for agricultural landscapes from second generation biofuel production in Ukraine: the role of pests. Aspects of applied biology, 109, Agricultural ecology research: its role in delivering sustainable farm systems, pp 165–169

Stewart JR, Toma T, Fernandez FG, Nishiwaki A, Yamada T, Bellero G (2009) The ecology and agronomy of *Miscanthus sinensis*, a species important to bioenergy crop development, in its native range in Japan: a review. GCB Bioenergy 1:126–153

Sun J, Tie BQ, Qin PF, Isao A, Luo R (2006) Investigation of contaminated soil and plants by heavy metals in Pb-Zn mining area. J Plant Resour Environ 15:63–67

Syed M, Qumar S, Riaz M, Choudhary FM (1995) Essential oil of the family gramineae with antibacterial activity. Part 2 the antibacterial activity of a local variety of *Cymbopogon citratus*. Pak J Sci Ind Res 38:146–148

Tanou G, Molassiotis A, Diamantidis G (2009) Induction of reactive oxygen species and necrotic death-like destruction in strawberry leaves by salinity. Environ Exp Bot 65(2–3):270–281

Techer D, Laval-Gilly P, Henry S, Bennasroune A, Formanek P, Martinez-Chois C, D'Innocenzo M, Muanda F, Dicko A, Rejsek K, Falla J (2011) Contribution of *Miscanthus x giganteus* root exudates to the biostimulation of PAH degradation: an in vitro study. Sci Total Environ 409(20):4489–4495

Techer D, Martinez-Chois C, Laval-Gilly P, Henry S, Bennarroune A, D'Innocenzo M, Falla J (2012a) Assessment of *Miscanthus x giganteus* for rhizoremediation of long term PAH contaminated soils. Appl Soil Ecol 62:42–49. Thailand

Techer D, Laval-Gilly P, Henry S, Bennasroune A, Martinez-Chois C, D'Innocenzo M, Falla J (2012b) Prospects of *Miscanthus x giganteus* for PAH phytoremediation: a microcosm study. Ind Crop Prod 36:276–281

Techer D, D'Innocenzo M, Laval-Gilly P, Henry S, Bennasroune A, Martinez-Chois C, Falla J (2012c) Assessment of *Miscanthus x giganteus* secondary root metabolites for the biostimulation of PAH-utilizing soil bacteria. Appl Soil Ecol 62:142–146

Tilman D, Hill J, Lehman C (2006) Carbon-negative biofuels from low-input high-diversity grassland biomass. Science 314(5805):1598–1600

Truong P (1999) Vetiver grass technology for mine rehabilitation. Pac Rim Vetiver Netw Tech Bull 2:1–19

Truong P (2000) Vetiver grass for mine site rehabilitation and reclamation. Extended Abstract. In: Proceedings of remade lands international conference, Fremantle, Australia

Truong P, Baker D (1997) The role of vetiver grass in the rehabilitation of toxic and contaminated lands in Australia. In: Proceedings of international Vetiver workshop, Fuzhou, China, 21–26 October

Truong PNV, Baker DE (1998a) Vetiver grass for the stabilization and rehabilitation of acid sulfate soils. Proceedings of second conference on acid sulfate soils, Coffs Harbour, Australia, pp 196–198

Truong P, Baker D (1998b) Vetiver grass system for environmental protection. Pac Rim Vetiver Netw Tech Bull 1

Truong P, Mason F, Waters D, Moody P (2000) Application of vetiver grass technology in off-site pollution control: trapping agrochemicals and nutrients in agricultural lands. In: Proceedings of second international Vetiver conference (ICV-2), Phetchaburi, Thailand, 18–22 January

USEPA (1991) Risk assessment guidance for superfund, vol 1, Human OSWEF. Directive Publ, 25 March, Health Evaluation Manual, Supplemental Guidance 9285, 6-03

Urbanek H, Majorowicz H, Zalewski M, Saniewski M (2005) Induction of glutathione S-transferase and glutathione by toxic compounds and elicitors in reed canary grass. Biotechnol Lett 27(13):911–914

Van De WM, Hanson J, Luethi I (1999) Morphological and agronomic characterisation of a collection of napier grass (Pennisetum purpureum) and P. purpureum X P. glaucum. Trop Grassl 33:150–158

Verma S, Dubey RS (2003) Lead toxicity induces lipid peroxidation and alters the activities of antioxidant enzymes in growing rice plants. Plant Sci 164(4):645–655

Visser P, Pignatelli V, Jørgensen U, Santos Oliveira JF (2001) Utilisation of miscanthus. In: Jones MB, Walsh M (eds) Miscanthus for energy and fibre. James and James, London, p 150

Vo VM (2007) Uptake potential of cadmium from soil by vetiver grass (*Vetiveria zizanioides* (L.) Nash). J Sci. Da Nang University, 21: 17–21 (in Vietnamsese)

Walton BT, Hoylman AM, Perez MM, Anderson TA, Johnson TR, Guthrie EA, Christman RF (1994) Rhizospheremicrobial communities as a plant defense against toxic substancesin soils. In: Anderson TA, Coats JR (eds) Bioremediation through rhizosphere technology, ACS symposium series 563, 7th edn. American Chemical Society, Washington, DC, pp 82–92

Wanat N, Austruy A, Joussein E, Soubrand M, Hitmi A, Gautier-Moussard C, Lenain JF, Vernay P, Munch JC, Pichon M (2013) Potentials of *Miscanthus x giganteus* grown on highly contaminated technisols. J Geochem Explor 126:78–84

Wardle DA, Nicholson KS (1996) Synergistic effects of grassland species on soil microbial biomass and activity: implications for ecosystem-level effects of enriched plant diversity. Funct Ecol 10:410–416

Wardle DA, Zackrisson O, Hornberg G, Gallet C (1997) The influence of island size area on ecosystem properties. Science 277:1296–1299

Wardle DA, Barker GM, Bonner KI, Nicholson KS (1998) Can comparative approaches based on plant ecophysiological traits predict the nature of biotic interactions and individual plant species effects in ecosystems? J Ecol 86:405–420

Wardle DA, Bonner KI, Barker GM, Yeates GW, Nicholson KS, Bardgett RD et al (1999) Plant removals in perennial grassland: vegetation dynamics, decomposers, soil biodiversity, and ecosystem properties. Ecol Monogr 69(4):535–568

Weiss EA (1997) Essential oil crops. In: Weiss EA (ed). Cambridge University Press, Cambridge, pp 86–103

White JC (2000) Phytoremediation of weathered p,p-DDE residues in soil. Int J Phytoremed 2(2):133–144

Wilde EW, Brigmon RL, Dunn DL (2005) Phytoextraction of lead from firing range soil by Vetiver grass. Chemosphere 61:1451–1457

Winters S (1999) Plants reduce atrazine levels in wetlands. Fourth year report. School of Land and Food, University of Queensland, Brisbane

Wong MH (1982) Metal co-tolerance to copper, lead and zinc in Festuca rubra. Environ Res 29:42–47

Wong MH, Lau WM (1985) Root growth of *Cynodon dactylon* and *Eleusine indica* collected from motorways at different concentrations of lead. Environ Res 36:257–267

Xia HP (2004) Ecological rehabilitation and phytoremediation with four grasses in oil shale mined land. Chemosphere 54:345–353

Xia H, Shu W (2000) Resistance to and uptake of heavy metals by Vetiveria zizanioides and paspalum notatum from lead/zinc mine tailings. Acta Ecol Sin 21(7):1121–1129

Yang B, Shu WS, Ye ZH, Lan CY, Wong MH (2003) Growth and metal accumulation in vetiver and two Sesbania species on lead/zinc mine tailings. Chemosphere 52:1593–1600

Yateem A, Balba MT, El-Nawawy AS, Al-Awadhi N (1999) Experiments in phytoremediation of Gulf War contaminated soil. Soil Groundw Cleanup 2:31–33

Ye ZH, Wong MH, Baker AJM, Willis AJ (1998) Comparison of biomass and metal uptake between two populations of Phragmites australis grown in flooded and dry conditions. Ann Bot 80:363–370

Zaefyzadeh M, Quliyev RA, Babayeva SM, Abbasov MA (2009) The effect of the interaction between genotypes and drought stress on the superoxide dismutase and chlorophyll content in durum wheat landraces. Turk J Biol 33(1):1–7

Zheljazkov VD, Cantrell CL, Astatkie T, Cannon JB (2011) Lemongrass productivity, oil content, and composition as a function of nitrogen sulfur, and harvest time. Agron J 103:805–881

Zhuang P, Ye ZH, Lan CY, Xie ZW, Shu WS (2005) Chemically assisted phytoextraction of heavy metal contaminated soils using three plant species. Plant Soil 276:153–162

Zimmermann J, Daubert J, Jones MB (2012) Soil carbon sequestration under *Miscanthus x giganteus*. A large scale survey. In: Conference proceedings SIMBIOSYS, 28 June 2012, Integrating biodiversity research, policy, and practice. Sectorial Impacts on Biodiversity and Ecosystem Services: www.tcd.ie/research/simbiosys/images/zimmermann.pdf

Zong K, Huang J, Nara K, Chen Y, Shen Z, Lian C (2015) Inoculation of ectomycorrhizal fungi contributes to the survival of tree seedlings in a copper mine tailing. J For Res 20:493–500

Macrophytes for the Reclamation of Degraded Waterbodies with Potential for Bioenergy Production

13

Sangeeta Anand, Sushil Kumar Bharti, Neetu Dviwedi, S.C. Barman, and Narendra Kumar

Abstract

Macrophytes have excessive efficiency to remove various inorganic and organic contaminants including heavy metals, nutrients, pesticides, POPs, oils from wastewater. The removal of contaminants depends upon the concentration of contaminant, duration of exposure and others factors including environmental characteristics (pH, temperature etc.), physicochemical properties of pollutants (solubility, pressure etc.) and plants characteristics (species, root system etc.). To ascertain large scale execution of this process, management of phytoremediating macrophytes will be a chief concern bioenergy production is effective and low cost practices for the optimum utilization and eco-friendly management of these macrophytes. Due to high photosynthetic efficiency and higher biomass production, macrophytes can produce useful quantities of carbohydrate and cellulose; raw material forbio-gas, bioethanol and lipids which are non-polluting and renewable sources of energy. Several aquatic macrophytes, such as *Eichhornia crassipes, Trapa natans, Typha latifolia, Pistiastratiotes, Phragmites australis, Lemna gibba* can easily degraded, and produce high bioenergy yield. In addition, macrophytes can be used for several other purposes such as recreational, household, flowers, fodder, fertilizers, mulch etc. Macrophytes are also capable for sequestering carbon through photosynthesis and accumulation of organic matter

S. Anand • S.K. Bharti • N. Kumar (✉)
Department of Environmental Sciences, Babasaheb Bhimrao Ambedkar University, Lucknow 226025, India
e-mail: narendrakumar_lko@yahoo.co.in

N. Dviwedi
Department of Environmental Studies, Sri Aurovindo College, Delhi University, Delhi 110017, India

S.C. Barman
Environmental Monitoring Division, Indian Institute of Toxicology Research, Lucknow 226001, India

K. Bauddh et al. (eds.), *Phytoremediation Potential of Bioenergy Plants*,
DOI 10.1007/978-981-10-3084-0_13

in sediments and plant biomass. This chapter highlights the macrophytes potential for removal of inorganic and organic contaminants and their subsequent use for bioenergy production.

Keywords
Bioenergy • Heavy metals • Macrophytes • Phytoremediation • Wastewater

13.1 Introduction

Water is a precious natural resource that supports life on earth. Unfortunately, it has been continuously subjected to over-exploitation due to ever-increasing population, industrialization and urbanization. Huge amount of organic and inorganic pollutants have been added in the environment due to organic contaminants such as pesticides, fertilizers, detergents, petroleum hydrocarbons, food additive, human antibiotics, etc., and inorganic contaminants, e.g. heavy metal, nitrate, ammonium, phosphate, radioisotopes, etc., are known to be carcinogenic, endocrine disruptor and neurotoxins. In addition, they harm nervous system, cause irritation in respiratory, damage reproductive system and also cause immunological disorder in children. Considering the effects of these pollutants on human beings, animals, aquatic organisms and plants, appropriate treatment techniques are required (Kumar et al. 2013a). Various conventional treatment techniques such as ion exchange, chemical precipitation, adsorption, reverse osmosis, electrochemical treatment, coagulation-flocculation, filtration, etc., are used for the removal of these organic and inorganic contaminants. However, weak implementation and limited financial resource pose serious challenges to the environment.

Hence, sustainable management of aquatic ecosystem needs cost-effective and eco-friendly remediation technologies. Many scientists demonstrated the potential of aquatic macrophytes for inorganic and organic pollutant removal. Phytoremediation is an innovative technique to handle the surface water pollution problem because plants are solar-driven, their relative growth rate is high, and they are also natural accumulators of elements, cost-effective, less destructive and environmentally friendly. Phytoremediation technique is more efficient as compared with conventional techniques and can be applied at low concentration of pollutants.

13.1.1 Macrophytes

Macrophytes are a diverse group of photosynthetic organism found in waterbodies. They include bryophytes (mosses, liverwort, etc.), pteridophytes (ferns) and spermatophytes (flowering plants). Chamber et al. (2008) represented macrophytes into seven different plant divisions: Spermatophyta, Pteridophyta, Bryophyta, Xanthophyta, Rhodophyta, Chlorophyta and Cyanobacteria. Arber (1920) and Sculthorpe (1967) classified macrophytes into four different categories (Fig. 13.1) depending on their growth forms:

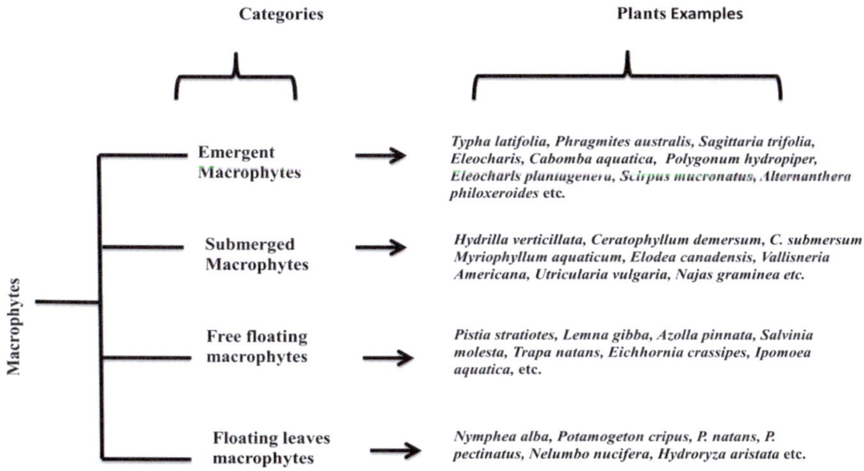

Fig. 13.1 Classification of macrophytes and their example

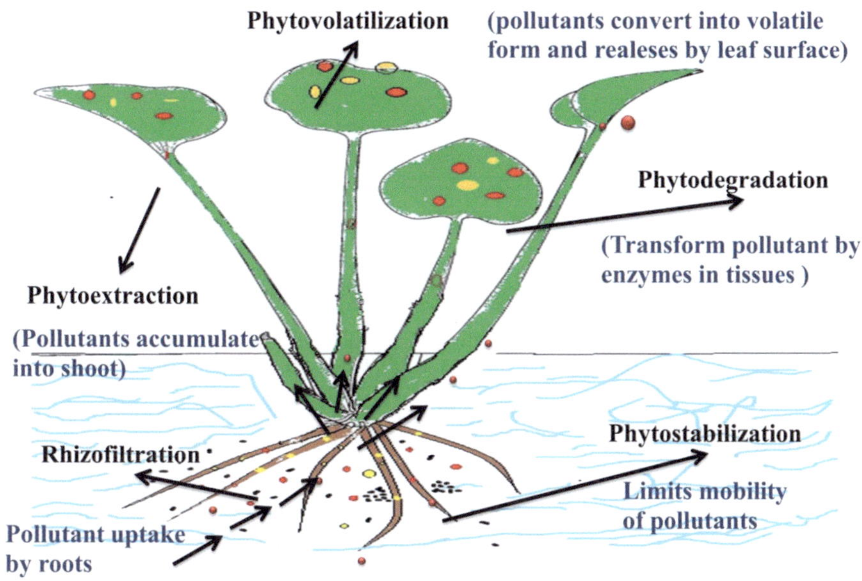

Fig. 13.2 Phytoremediation mechanism and its techniques (Rai 2009)

1. Emergent macrophytes: Plants rooted in soil and also emerging to a significant height above water.
2. Submerged macrophytes: Plants grow below the surface water including few ferns, numerous mosses and some angiosperm.
3. Free-floating macrophytes: Plants that are non-rooted to the substratum and float on the surface of the water.

4. Floating leaf macrophytes: Plants occur in submerged sediment, and leaves are floating with long flexible petiole on the surface, which mainly include angiosperm.

The aquatic macrophytes are a major part of the aquatic ecosystem and show high efficiency in the removal of organic and inorganic contaminants. They absorb and concentrate the contaminants in their roots and shoots from polluted wastewater (Zhu et al. 1999; Rai 2009). Root exudates and pH of rhizosphere also may cause inorganic and organic pollutants to precipitate on root surfaces.

13.1.2 Phytoremediation

Phytoremediation chiefly refers to diverse collection of plant-based technology that uses plants to reduce the concentrations or toxic effects of contaminants in the environments (Greipsson 2011). This technology can be used for the removal of inorganic pollutants (heavy metals, radionuclides and nutrients) and organic pollutants (PAHs, POPs and pesticides). It is a simple, cost-effective, eco-friendly, novel remediation strategy also called as green technology (Kalve et al. 2011; Sarma 2011; Singh and Prasad 2011; Vithanage et al. 2012). Day by day, phytoremediation is gaining interest because of effective and inexpensive cleanup of hazardous waste sites. There are several techniques by which plants can clean up or remediate contaminated sites (Table 13.1).

Table 13.1 Techniques of phytoremediation and their advantages

S. no.	Techniques	Mechanism	Advantages
1.	Rhizofiltration	Utilization of roots to absorb, concentrate and precipitate toxic contaminants from water. It reduces the mobility of contaminant and prevents migration to the groundwater	In situ and ex situ application retains contaminants in the root
2.	Phytoextraction	Contaminant absorption by roots and subsequent translocation to shoots	Inexpensive and permanent contaminant removal
3.	Phytostabilization	Immobilization of heavy metals in soil through sorption by root, precipitation and complexation	Disposal of toxic waste is not required, reduce soil erosion
4.	Phytovolatilization	Plants uptake pollutants and convert them to volatile form and release them in the atmosphere. It is used for remediation of organic pollutants and some heavy metals like Se and Hg	Contaminants directly released to the atmosphere
5.	Phytodegradation	Degradation of organic pollutants by plants in the presence of enzyme like oxygenase and dehalogenase. It is independent of rhizospheric microorganisms	Reduce waste and also are used to produce bioenergy

Rhizofiltration is a process applicable for aquatic macrophytes. The process involves raising plants hydroponically and then transplanting them into polluted waters, where macrophytes absorb and concentrate the pollutants in their roots and shoots (Rai 2009). Macrophytes accumulate significant amounts of pollutants in combination with low maintenance cost, easy handling and least secondary waste disposal and also produce large amounts of root biomass. Roots play an important role in element uptake from water (Sharma and Gaur 1995). Uptake mechanism is controlled by several factors such as pH, temperature, physiochemical properties of pollutants (solubility, pressure, etc.) and plant characteristics (species, root system, etc.) (Greger 1999; Blaylock and Huang 2000).

13.2 Phytoremediation Potential of Aquatic Macrophytes

Aquatic macrophytes are known for accumulating and concentrating contaminants from their environment. Several studies have shown that aquatic macrophytes are very effective in removing inorganic and organic chemicals from polluted water.

13.2.1 Remediation of Inorganic Contaminants

Aquatic macrophytes possess the ability to remediate inorganic contaminants including nutrients like nitrate, nitrite, ammonium, phosphorus, heavy metals and radioactive isotopes from wastewaters. The process efficiency varies with plant species and is also dependent on concentration of the contaminant and duration of exposure.

13.2.1.1 Remediation of Heavy Metals

Heavy metals are the largest part of inorganic contaminants and their common sources are mining, vehicle emission, batteries, fertilizers, paints, etc. Heavy metals enter the human body through skin contact, respiration or ingestion and pose risk to human health. Many of these heavy metals are carcinogenic, endocrine disruptor and mutagenic, and some cause neurological and behavioural changes especially in children (Kumar et al. 2013b). Several aquatic macrophytes have the capacity to remove heavy metals from water, e.g. *Eichhornia crassipes* (Kay et al. 1984; Zhu et al. 1999), *Hydrocotyle umbellata* (Dierberg et al. 1987), *Lemna minor* L. (Mo et al. 1989; Rai 2007a) and *Azolla pinnata* (Mo et al. 1989; Rai 2007a, b). Sasmaz and Obek (2009) reported that *L. gibba* is an accumulator of As, U and B from effluents. Several species of *Salvinia* including *S. herzogii*, *S. minima*, *S. natans* and *S. rotundifolia* have shown potential to remove various heavy metals, viz. Pb, Cr, Ni and Cu from wastewaters (Nichols et al. 2000; Olguin et al. 2005; Sune et al. 2007; Sanchez-Galvan et al. 2008; Xu et al. 2009). Odjegba and Fasidi (2004) reported that *Pistia* is a potential candidate for the removal of Zn, Cr, Cu, Cd, Pb and Hg. It accumulates Zn and Cd up to high concentration. Mercury is moderately accumulated by the plant and showed poor accumulation of Ni. Four aquatic plants

Eichornia crassipes, *Pistia stratiotes*, *Scirpus tabernaemontani* and *Colocasia esculenta* were evaluated for their efficiency in the removal of Hg from wastewater and were found to be hyperaccumulator of Hg (Skinner et al. 2007; Kumar et al. 2012, Rezania et al. 2015). Potential of various macrophytes for heavy metal accumulation is presented in Table 13.2.

Table 13.2 Phytoremediation potential of aquatic various macrophytes for heavy metal

S. no	Macrophyte species	Heavy metals	Accumulation in macrophytes	References
1.	*Eichhornia crassipes*	Cu	6000–7000 mg kg^{-1}	Molisani et al. (2006)
		Cr	4000–6000 mg kg^{-1}.	Hu et al. (2007)
		Ni	1200 mg kg^{-1}	Low et al. (1994)
		Cd	2200 μg kg^{-1}	Zhu et al. (1999)
		Zn	1677 mg g^{-1}	Kamel (2013)
		As	909.58 mg kg^{-1}	Delgado et al. (1993)
		Hg	119 ng g^{-1}	Molisani et al. (2006)
		Mn	300 mg kg^{-1}	Dixit et al. (2011)
2.	*Azolla Pinnata*	Ni	16,252 l μg g^{-1}	Arora et al. (2004)
		Cd	740 μg g^{-1}	Rai (2008)
		Cr	1095 μg g^{-1}	Rai (2010)
		Hg	940 μg g^{-1}	Rai and Tripathi (2009)
		Pb	1383 mg kg^{-1}	Iqbal et al. (2013)
3.	*Azolla filiculoides*	Cd	2608 μg g^{-1}	Arora et al. (2004)
		Cu	6013 μg g^{-1}	Valderrama et al. (2012)
		As	>60 μg g^{-1}	Zhang et al. (2008)
		Cr	12,383 μg g^{-1}	Arora et al. (2006)
		Pb	1607 mg kg^{-1}	Vesely et al. (2011)
4.	*Azolla caroliniana*	Cr	356 mg kg^{-1}	Bennicelli et al. (2004), Zhang et al. (2008)
		Hg	578 mg kg^{-1}	
		As	284 mg kg^{-1}	
5.	*Hydrilla verticillata*	As	231 mg kg^{-1}	Srivastava et al. (2010)
		Cu	771–3,0831 mg kg^{-1}	Srivastava et al. (2011)
6.	*Typha latifolia*	Ni	295.6 mg kg^{-1}	Afrous et al. (2011)
		Cu	1157.7 mg kg^{-1}	Nguyen et al. (2009)
7.	*Pistia stratiotes*	Pb	519 mg kg^{-1}	Vesely et al. (2011)
8.	*Salvinia minima*	Pb	5469 mg kg^{-1}	Vesely et al. (2011)
9.	*Salvinia natans*	Cr	7.40 mg g^{-1}	Dhir et al. (2009)
10.	*Lemna gibba*	U	896.9 mg kg^{-1}	Mkandawire et al. (2004)
		As	1021.7 mg kg^{-1}	
11.	*Lemna minor*	Pb	561 mg g^{-1}	Leblebici and Aksoy 2011
		Cu	400 μg g^{-1}	Boule et al. 2009
12.	*Typha angustifolia*	Mn	860mgkg^{-1}	Sasmaz et al. (2008)
		Cu	50mgkg^{-1}	

(continued)

Table 13.2 (continued)

S. no	Macrophyte species	Heavy metals	Accumulation in macrophytes	References
13.	*Myriophyllum spicatum*	Pb	8.94 mg g^{-1}	Kamel (2013)
		Zn	2.66 mg g^{-1}	
14.	*Ceratophyllum submersum*	Pb	258.62 mg kg^{-1}	Kamel (2013)
		Zn	1172.8 mg kg^{-1}	
15.	*Wolffia globosa*	As	> 1000 mg kg^{-1}	Zhang et al. (2009)
16.	*Phragmites communis*	Fe	2813 μg g^{-1}	Chandra and Yadav (2011)
		Mn	814.40 μg g^{-1}	
		Zn	265.80 μg g^{-1}	
		Pb	92.80 μg g^{-1}	
17.	*Trapa natans*	Cd	0.015 μg g^{-1}	Sweta et al. (2015)

13.2.1.2 Remediation of Nitrate, Nitrite, Ammonium and Phosphate

Nitrite, nitrate, ammonia and phosphate are the main nutrient ions present in contaminated water and cause threat to water quality and increase eutrophication. Mostly nutrients are derivative of the decomposed organic wastes (nitrogenous) and are easily taken up by rooted green plants or floating plants in the water. Various aquatic macrophytes showed potential for the proficient removal of inorganic contaminants such as nitrogen and phosphorus from water (Dhir et al. 2009).

Ammonia is the by-product of decomposed nitrogenous organic wastes and ordinary constituent of groundwater and sometimes is also added to drinking water to eliminate odour and taste of free chlorine (Dhir 2009). Ismail et al. (2015) reported that ammonia removal efficiency of *E. crassipes* and *P. stratiotes* was 91.1 and 95.7%, respectively. This indicated that *P. stratiotes* is slightly superior to *E. crassipes* in reducing ammonia from the wastewater. Tripathi and Shukla (1991) also found that *E. crassipes* had significant ability to remove ammonia from water. Phosphate and orthophosphate are the main form of phosphorus present in water. Aquatic macrophytes could store phosphorous in its tissue, and their growth rate is positively related to the availability of phosphorous present in water (Xie and Yu 2003). Ismail et al. (2015) reported that there is a decrease in concentration of PO_4^- in water treated with *E. crassipes* and *P. stratiotes* and found that the PO_4^- absorption ability of both treatments was highly significant. Overall, *E. crassipes* performed slightly better than *P. stratiotes*. Verma and Suthar (2014) reported that *Lemna gibba* bears great efficiency for the removal of nitrate and total phosphorus from urban wastewater without dilution. Lotus (*Nelumbo nucifera*) showed the efficient removal of TKN (Total Kjeldahl Nitrogen); NH_3–N, NO_2–N and NO_3–N; and TP (total phosphorous) from domestic wastewater (Kanabkaew and Puetpaiboon 2004) (Table 13.3).

Table 13.3 Aquatic macrophytes with potential for treating nitrate, nitrite, ammonium and phosphate contaminants from wastewater

S. no.	Macrophytes	Contaminants	Removal	References
1.	*Eichhornia crassipes*	NH_3	10.25 mg/L	Ismail et al. (2015)
		PO_4^-	2.55 mg/L	
2.	*Pistia stratiotes*	NH_3	11.7 mg/L	Ismail et al. (2015)
		PO_4^-	2.2 mg/L	
3.	*Azolla filiculoides*	NH_3	58 mg/L	Sood et al. (2011)
		PO_4^-	23.5 mg/L	
4.	*Myriophyllum aquaticum*	TNK	21.16 mg/L	Souza et al. (2013)
		TP	0.62 mg/L	
5.	*Lemna minor*	Total nitrogen	60,000 mg/L	Mkandawire and Dudel (2007)
		TP	5000 mg/L	
		K	40,000 mg/L	
		Ca	10,000 mg/L	
6.	*Spirodela polyrhiza*	N	29.8 mg/L	Mishra and Tripathi (2008)
		P	7.2 mg/L	
		Ca	2.8 mg/L	
7.	*Lemna gibba*	NO_3	9.15 mg/L	Verma and Suthar (2014)
		TP	6.90 mg/L	

13.2.2 Remediation of Organic Contaminants

A large number of organic pollutants are noticed in groundwater worldwide including nanoparticles, pesticides and their metabolites, pharmaceuticals and hormones, personal care products, industrial additives and by-products, water treatment by-products, flame/fire retardants and surfactants, caffeine and nicotine and their metabolites. Movement of these organic contaminants in water depends on solubility, charge, size, pressure and the presence of other organic contaminants. Four mechanisms are involved in phytoremediation of organic pollutants (Schnoor et al. 1995; Xia and Ma 2006):

1. Direct accumulation of contaminants and metabolism in plant tissues
2. Transpiration of VOCs (volatile organic compounds) through the green leaves
3. Release of exudates that stimulate microbial activity and biochemical transformations around the root system
4. Development of mineralization that is attributed to mycorrhizal fungi and microbial consortia associated with the root surface

13.2.2.1 Remediation of Pesticides and Persistent Organic Pollutants (POPs)

Certain chemical pollutants have the ability to persist in the environment for long duration, migrate into water and concentrate to levels that could harm human health due to their various side effects and carcinogenic nature. These chemicals are commonly called as persistent organic pollutants (POPs) (Manahan 2000; Yang et al. 2011). Pesticides are common example of POPs used in the control of pests, insects and weeds in agricultural fields. They enter the aquatic ecosystem from a number of sources such as power stations, incinerating plants as well as from transport, household furnaces, agricultural sprays and leaching from the landfills. Phytoremediation is a technique that could potentially help in the restoration of contaminated water. In few years, a number of research articles have reported the role of plants in remediating contaminated waters (Paterson et al. 1990; Shrimp et al. 1993; Schnoor et al. 1995; Simonich and Hites 1995; Watanabe 1997). Xia and Ma (2006) observed the accumulation of ethion (an organophosphorous pesticides) in *Eichhornia crassipes* and found that it decreased by 55–91% in shoots and 74–81% in roots after 1 week of incubation in culture solutions without any visible morphological changes in plants. This shows the high adoptability of *Eichhornia crassipes plant* that could decontaminate wastewater polluted with ethion. Remediations of organic contaminants by aquatic macrophytes have been presented in Table 13.4.

Table 13.4 Macrophytes with potential of organic contaminants removal from wastewater

S. no.	Plants	Organic contaminants	Accumulation	References
1.	*Pistia stratiotes*	Chlorpyrifos	0.036 mg g^{-1}	Prasertsup and Ariyakanon (2011)
2.	*Elodea canadensis*	Dimethomorph	13 ng g^{-1}	Olette et al. (2008)
		Flazasulfuron	3 ng g^{-1}	
3.	*Phragmites communis*	γ-HCH	0.97 ng g^{-1}	Guo et al. (2014)
		DDTs	0.93 ng g^{-1}	
4.	*Ipomoea aquatica*	Di-*n*-butyl phthalate	9.95 mg kg^{-1}	Cai et al. (2008)
		Polyphenols	334.89 mg g^{-1}	Deval et al. (2012)
5.	*Typha angustifolia*	Phenol	100 mg g^{-1}	Chandra and Yadav (2010)
6.	*Eichhornia crassipes*	Ethion	215.6 μg g^{-1}	Xia and Ma (2006)
7.	*Typha* spp.	Clofibric acid	20 μg g^{-1}	Dordio et al. (2009)
8.	*Hydrilla verticillata*	Chlordane	1060.95 μg g^{-1}	Chaudhry et al. (2002)
9.	*Heteranthera callifolia*	Hexane	0.123 mg g^{-1}	Denise et al. (2013)
10.	*Lemna gibba*	Alanine	23 g kg^{-1}	Mkandawire and Dudel (2007)
		Glutamic acid	36.7 g kg^{-1}	
11.	*Lemna minor*	Flazasulfuron	27 μg g^{-1}	Olette et al. (2009)
		Dimethomorph	33 μg g^{-1}	
12.	*Ceratophyllum submersum*	Aldrin	0.38 ng g^{-1}	Guo et al. (2014)
		Endosulfan	0.73 ng g^{-1}	
13.	*Azolla filiculoides*	Sulfadimethoxine	49 $μg^{-1}$	Froni et al. (2002)
		Flumequine	960 $μg^{-1}$	

13.3 Bioenergy Potential of Aquatic Macrophytes

Aquatic macrophytes are the dominant plants and play an important role in aquatic ecosystems by providing food, shelter and substantial support for many fishes and crustaceans. The aquatic macrophytes including emergent, submerged and floating macrophytes have been grown naturally and caused various environmental problems in lakes and reservoirs (Abbasi et al. 1990; Moorhead and Nordstedt 1993). Various experiments have shown that these aquatic macrophytes have potential for organic and inorganic accumulation and would be used for remediation of contaminated water. With large-scale execution of this technology, subsequent management and disposal of the huge amount of phytoremediating macrophytes with higher organic and inorganic contaminants will be a chief concern. If macrophytes will not be disposed properly, then they may create contamination in the environment. Effective and low-cost management is needed for the treatment of phytoremediating macrophytes. There are some processes for the management and disposal of aquatic macrophytes by converting them into bioenergy which is economically and environmentally viable too.

13.3.1 Lipid Production

Macrophytes have received substantial interest as a potential feedstock for biofuel production. Owing to high photosynthetic efficiency and higher biomass production, they can produce useful quantities of sugars and lipids, which are raw materials for producing biodiesel transport fuels (Tredici 2010, Slade and Bauen 2013). Different methods are used for lipid extraction; some are supercritical CO_2 extraction, subcritical water extraction and solvent extraction. Solvent extraction is one of the most widely used methods for lipid extraction, where chloroform-based extraction is effective for the laboratory and hexane based for industries because of their cost-effectiveness and convenience in handling (Halim et al. 2012). Lipid extraction by dried biomass of *Spirulina platensis* and *Chlorella pyrenoidosa* was achieved through Soxhlet extraction using hexane as a solvent (Nautiyal et al. 2014). Reddy et al. (2014) produced lipid from *Nannochloropsis salina* by analysing its fatty acid profile by direct FAME method and achieved maximum lipid extraction efficiency of 70% in conventional heating subcritical water at 220 °C and 100% in microwave-assisted subcritical water at 205 °C.

13.3.2 Bioethanol Production

Growing population and industrialization had increased energy requirement in the last century putting extra pressure on the limited natural resources. In view of this, bioethanol is considered as the most promising replacement for fossil fuels because of its renewability and 80% less greenhouse gas emission compared to gasoline (Mete et al. 2002). In recent times, research has focused on aquatic biomass to be used as raw materials including lignocelluloses, celluloses, hemicelluloses and lignin to produce bioethanol. Bioethanol may be produced from aquatic macrophytes through two different processes, i.e. hydrolysis and fermentation which require fermentable sugar that may be available in less quantity in aquatic macrophytes. A scientist follows three steps for production of bioethanol from aquatic plants, viz. (1) isolation and qualitative screening of microorganism from cow dung, municipal waste and pig and goat excreta which produce cellulose and xylem enzyme, (2) production of cellulose enzyme by addition of dried aquatic macrophyte and microorganism and (3) ethanol production through fermentation process by hydrolysis of cellulose present in aquatic plant with fermenting organism (Patel and Patel 2015; Randive et al. 2015). *E. crassipes* is found to be a prominent macrophyte for bioethanol production. Ethanol production of approximately 0.17 g/g dry weight was reported using *Saccharomyces cerevisiae* as a fermenting organism (Randive et al. 2015). Mishina et al. (2006) reported that sugar content in *Pistia* leaves is about two times higher in comparison with *E. crassipes*. This suggests that *P. stratiotes* is more suitable for ethanol production as it produces around 0.25 g/g dry biomass. Rezania et al. (2015) studied the use of malt and barley extract enhancer for ethanol production from *E. crassipes* and *P. stratiotes* and found positive results on production.

13.3.3 Biogas Production

Biogas is an eco-friendly clean fuel produced by the anaerobic digestion of organic wastes such as vegetable wastes, cow dung, industrial waste and municipal solid waste (Burke 2001; Dohanyos and Zabranska 2001). It is progressively becoming more important in household and industry due to its low costs and cleanliness. The main component of the gas is methane (50–75%), carbon dioxide (25–50%), hydrogen (0–1%) and nitrogen (0–10%) (FAO/CMS 1996). The aquatic macrophytes have immense potential in biogas production due to its huge quantity, high carbon-nitrogen ratio and high content of fermentable matter. Several aquatic macrophytes such as *Eichhornia*, *Trapa*, *Typha latifolia* and *Pistia* can be degraded easily and produce high biogas yield (Gunnerson and Stuckey 1986; Sudhakar et al. 2013; Mahmoud et al. 2015). The basic method for biogas production from macrophytes involves two stages: In the first stage, moist biomass is converted into slurry and is loaded into anaerobic digestion reactor in several stages. However, uses of single-stage reactors can cause problem by clogging with spent biomass in comparison with two-stage reactors. Therefore, reactors of stage two are more flexible for managing spent biomass by avoiding clogging. Few researchers suggest that multistage reactor increases the efficiency of digestion and production when aquatic macrophyte is mixed with organic manure. In the survey, Abbasi et al. (1990) studied the bioenergy potential of eight common aquatic macrophytes, i.e. *Azolla pinnata*, *Hydrilla verticillata*, *Nymphaea stellata*, *Salvinia molesta*, *Ceratopteris* spp., *Cyperus* spp., *Scirpus* spp. and *Utricularia reticulata*, and found that among the studied macrophytes, *S. molesta* is the highest and *Cyperus* spp. are the lowest biogas-producing plants. Production of biogas from different aquatic macrophytes and even the same spp. grown in different conditions can be variable. Taheruzzaman and Kushari (1988) reported that *Eichhornia* produce higher amount of biogas in comparison with *Hydrilla* and *Azolla* growing in Ganga River. Kayama et al. (2014) investigated chemical composition of five different submerged macrophyte species and produced methane gas with anaerobic digestion. They reported that total methane yield collected from 161.2 to 360.8 mg g^{-1} depended on species (*Elodea nuttallii* > *Egeria densa* > *Potamogeton maackianus* > *Ceratophyllum demersum* > *Potamogeton malaianus*). They recommended the delignification of the macrophytes prior to anaerobic digestion.

13.3.4 Bio-Hydrogen Production

Hydrogen is found on earth in many resources—water, hydrocarbon fuels, inorganic substances, etc. Different technologies have been developed to produce hydrogen gas: thermochemical hydrogen production (Abbasi et al. 1990), water electrolysis (Afrous et al. 2011) and biological hydrogen production (Arber 1920). Biological hydrogen production exhibits great efficiency for producing recyclable, clean and efficient energy by microbial activity on organic substance and biomass-derived sugar under anaerobic condition. Biological hydrogen production can be classified into five different groups:

1. *Direct biophotolysis*: It is a biological process using microalgae photosynthetic systems to convert solar energy into chemical energy in the form of hydrogen ($H_2O = H_2 + O_2$) (Bridgwater 1999).
2. *Indirect biophotolysis*: This involves the use of processed cyanobacteria to convert glucose into hydrogen by aerobic dark fermentation process (Gaudernack 1998).
3. *Biological water-gas shift reaction:* In this process photoheterotrophic bacteria such as *Rhodospirillum rubrum* use carbon monoxide as carbon source to generate ATP by oxidation of carbon monoxide to the reduction of H^+ to H (Gaudernack 1998).
4. *Photo-fermentation* involves photosynthetic microorganism to produce hydrogen through the action of their nitrogenase using solar energy and biomass.
5. *Dark fermentation* is a process where fermentation occurs by anaerobic microorganism such as green algae and cyanobacteria on glucose-rich substrates and can produce hydrogen at 30–80 °C chiefly in a dark condition (Lin and Jo 2003; Levin et al. 2004).

Mostly energy production from aquatic macrophyte focused on biogas and bioethanol, but few researches showed that the aquatic biomass is also used for biohydrogen gas production.

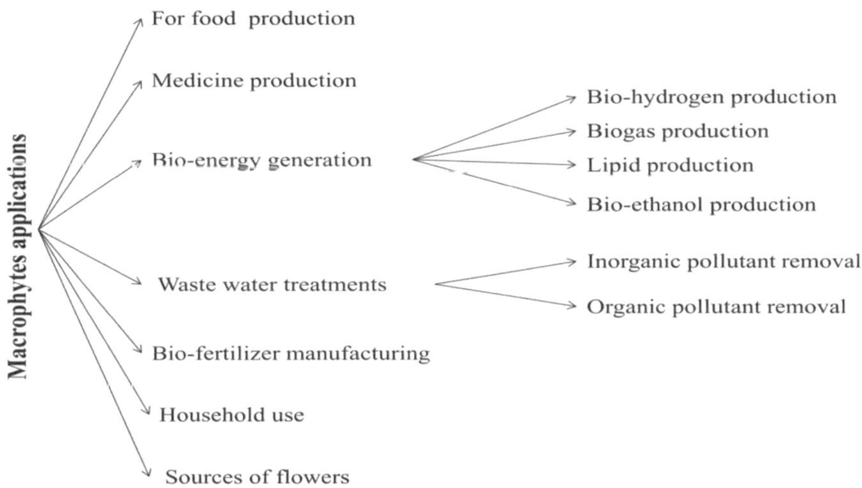

Fig. 13.3 Various industrial applications and economic importance of aquatic macrophytes

13.4 Industrial Application and Economic Importance of Aquatic Macrophytes

In addition, macrophytes have several other uses such as recreational, household, flowers, fodder, fertilizers, mulch, etc. (Fig. 13.3)

13.5 Carbon Sequestration and Other Environmental Services of Aquatic Macrophytes

Carbon sequestration refers to both natural and anthropogenic processes by which CO_2 is either removed from the atmosphere or emitted from sources and remains stored in the ocean, terrestrial environments (vegetation, soils and sediments) and geologic formations. Controlling increasing concentration of CO_2 will require mitigation with reducing emissions and increasing storage. Aquatic macrophytes play an important role in regulating exchanges of greenhouse gases, carbon dioxide, nitrous oxide and methane including water vapour. Macrophytes are capable for sequestering and storing carbon through photosynthesis and accumulation of organic matter in sediments and plant biomass. Water logging of wetland limits oxygen diffusion into sediment and creates anaerobic conditions. This condition slows down the decomposition rate and leads to storing large amounts of organic carbon in wetland sediments. Wang et al. (2011) showed carbon sequestration efficiency of macrophytes in three different months, March (3450 kg), June (4554 kg) and September (6185 kg), and found that in September (before typhoon) macrophytes contain a large amount of carbon. Hence, macrophytes can play a crucial role in managing the problem of global warming and subsequently climate change.

13.6 Conclusion

Phytoremediation is a cost-effective method for remediating water contaminated with inorganic pollutant like heavy metals, nitrate, nitrite, ammonia, phosphate and other organic pollutants like pesticides, POPs, oils, etc. The use of macrophytes is beneficial because they have tremendous capacity of absorbing nutrients (phosphate and nitrate) and removes heavy metals, pesticides and POPs from water and can easily be harvested and managed. Hence, it reduces pollution load on water. Plentifully available source of aquatic biomass and its excessive growth can be used for bioenergy production by conversion of aquatic biomass into bioenergy such as biogas, bioethanol, bio-hydrogen and lipid production. Macrophytes also offer various important uses like food, recreation, fertilizer, compost, medicine, fodder, mulch, etc. They also play a crucial role in carbon sequestration, thereby addressing the problem of global warming and climate change.

References

Abbasi SA, Nipaney PC, Schaumberg GD (1990) Bioenergy potential of eight common aquatic weed. Biol Waste 134:359–366

Afrous A, Manshouri M, Liaghat A, Pazira E, Sedghi H (2011) Mercury and arsenic accumulation by three species of aquatic plants in Dezful, Iran. Afr J Agric Res 6(24):5391–5397

Arber A (1920) A study of aquatic angiosperms. Cambridge University Press, Cambridge, 436 pp

Arora A, Sood A, Singh PK (2004) Hyperaccumulation of cadmium and nickel by *Azolla* species. Indian J Plant Physiol 3:302–304

Arora A, Saxena S, Sharma DK (2006) Tolerance and phytoaccumulation of chromium by three *Azolla* species. World J Microbiol Biotechnol 22:97–100

Bennicelli R, Stezpniewska Z, Banach A, Szajnocha K, Strowski JO (2004) The ability of *Azolla caroliniana* to remove heavy metals (Hg(II), Cr(III), Cr(VI)) from municipal waste water. Chemosphere 55:141–146

Blaylock MJ, Huang JW (2000) Phytoextraction of metals. In: Raskin I, Ensley BD (eds) Phytoremediation of toxic metals using plants to clean up the environment. Wiley, New York, pp 53–70

Boule MK, Vicentea AF, Nabaisa C, Prasad MNV, Freitas H (2009) Ecophysiological tolerance of duckweeds exposed to copper. Aquat Toxicol 91:1–9

Bridgwater AV (1999) Principles and practice of biomass fast pyrolysis processes for liquids. J Anal Appl Pyrolysis 51:3

Burke D (2001) Dairy waste anaerobic digestion handbook. Options for Gy Company, Olympia

Cai QY, Mo CH, Zenga QY, Q Ta W, Ferard JF, Ladislao BA (2008) Potential of *Ipomoea aquatic* cultivars in phytoremediation of soils contaminated with di-*n*-butyl phthalate. Environ Exp Bot 62:205–211

Chambers PA, Lacoul P, Murphy KJ, Thomaz SM (2008) Global diversity of aquatic macrophytes in freshwater. Hydrobiologia 595:9–26

Chandra R, Yadav S (2010) Potential of *Typha angustifolia* for phytoremediation of heavy metals from aqueous solution of phenol and melanoidin. Ecol Eng 36:1277–1284

Chandra R, Yadav S (2011) Phytoremediation of Cd, Cr, Cu, Mn, Fe, Ni, Pb and Zn from aqueous solution using *Phragmites cummunis, Typha angustifolia* and *Cyperus esculentus*. Int J Phytoremed 13:580–591

Chaudhry Q, Schrder P, Reichhart DW, Grajek W, Marecik R (2002) Prospects and limitations of phytoremediation for the removal of persistent pesticides in the environment. Environ Sci Pollut Res 9(1):4–17

Delgado M, Bigeriego M, Guardiola E (1993) Uptake of Zn, Cr and Cd by water hyacinth. Water Res 27:269

Denise EM, Akhere MA, Elsie Uand Ruth O (2013) Phytoextraction of total petroleum hydrocarbon in polluted environment using an aquatic macrophyte *Heteranthera callifolia* Rchb. Ex Kunth. Int J Eng Sci (IJES) 2(11):37–41

Deval CG, Mane AV, Joshi NP, Saratale GD (2012) Phytoremediation potential of aquatic macrophyte *Azolla caroliniana* with references to zinc plating effluent. Emir J Food Agric 24(3):208–223

Dhir B (2009) *Salvinia*: an aquatic fern with potential use in phytoremediation. Envion We Int J Sci Tech 4:23–27

Dhir B, Sharmila P, Pardha Saradhi P (2009) Potential of aquatic macrophytes for removing contaminants from the environment. Crit Rev Environ Sci Technol 39:754–781

Dierberg FE, DeBusk TA, Goulet NA Jr (1987) Removal of copper and lead using a thin film technique. In: Reddy KB, Smith WH (eds) Aquatic plants for water treatment and resource recovery. Magnolia Publishing Inc., Orlando, pp 497–504

Dixit A, Dixit S, Goswami S (2011) Process and plants for wastewater remediation: a review. Sci Revs Chem Commun 1(1):71–77

Dohanyos M, Zabranska J (2001) Chapter 13: Anaerobic digestion. In: Spinosa L, Vesilind PA (eds) Sludge into biosolids. IWA Publishing, London

Dordio AV, Duarte C, Barreiros M, Palace A, Carvalho J, Pinto AP, Costa CT (2009) Toxicity and removal efficiency of pharmaceutical metabolite clofibric acid by Typha spp. – potential use for phytoremediation? Bioresour Technol 100:1156–1161

Elhaak MA, Mohsen AA, Hamada EA, El-Gebaly FE (2015) Biofuel production from *phragmites australis* (cav.) and *typha domingensis* (pers.) plants of burullus lake. Egypt J Exp Biol 11(2):237–243

FAO/CMS (1996) A system approach to biogas technology. Biogas technology: a training manual for extension. website: http://www.fao.org

Forni C, Cascone A, Fiori M, Migliore L (2002) Sulfadimethoxine and *Azolla filiculoides* Lam. A model for drug remediation. Water Res 36:3398–3403

Guo W, Zhang H, Huo S (2014) Organochlorine pesticides in aquatic hydrophyte tissues and surrounding sediments in Baiyangdian wetland, China. Ecol Eng 67:150–155

Gaudernack B (1998) Photoproduction of hydrogen. IEA agreement on the production and utilization of hydrogen annual report, IEA/H2/AR-98

Gunnerson CG, Stuckey DC (1986) Anaerobic digestion: principles and practices for biogas system. World Bank technical paper 49. Washington, DC, World Bank

Greger M (1999) Metal availability and bioconcentration in plants. In: Prasad MNV, Hagemeyer J (eds) Heavy metal stress in plants: from molecule to ecosystems. Springer, Berlin/Heidelberg/Germany

Greipsson S (2011) Phytoremediation. Nat Educ Knowl 2:7

Halim R, Danquah MK, Webley PA (2012) Extraction of oil from microalgae for biodiesel production: a review. Biotechnol Adv 30:709–732

Hu C, Zhang L, Hamilton D, Zhou W, Yang T, Zhu D (2007) Physiological responses induced by copper bioaccumulation in *Eichhornia crassipes* (Mart.). Hydrobiologia 579:211–218

Ismail Z, Othman SZ, Law KH, Sulaiman AH, Hashim R (2015) Comparative performance of water hyacinth (*Eichhornia crassipes*) and water lettuce (*Pistia stratiotes*) in preventing nutrients Build-up in municipal wastewater. CLEAN Soil Air Water 43:521–531

Kalve S, Sarangi BK, Pandey RA, Chakrabarti T (2011) Arsenic and chromium hyperaccumulation by an ecotype of Pteris vittata-prospective for phytoextraction from contaminated water and soil. Curr Sci 100:888–894

Kamel AK (2013) Phytoremediation potentiality of aquatic macrophytes in heavy metal contaminated water of El-Temsah Lake, Ismailia, Egypt. Middle-East J Sci Res 14(12):1555–1568

Kanabkaew T, Puetpaiboon U (2004) Aquatic plants for domestic wastewater treatment: lotus (*Nelumbo nucifera*) and hydrilla (*Hydrilla verticillata*) systems. Songklanakarin. J Sci Technol 26(5):749–756

Kay SH, Haller WT, Garrad LA (1984) Effect of heavy metals on water hyacinth (*Eichhornia crassipes* (Mart.) Solms.). Aquat Toxicol 5:117–128
Koyamaa M, Yamamotoa S, Ishikawab K, Banc S, Toda T (2014) Anaerobic digestion of submerged macrophytes: chemical composition and anaerobic digestibility. Ecol Eng 69:304–309
Kumar N, Bauddh K, Barman SC, Singh DP (2012) Accumulation of metals in selected macrophytes grown in mixture of drain water and tannery effluent and their phytoremediation potential. J Environ Biol 33:323–327
Kumar N, Bauddh K, Kumar S, Dwivedi N, Singh DP, Barman SC (2013a) Extractability and phytotoxicity of heavy metals present in petrochemical industry sludge. Clean Techn Environ Policy 15:1033–1039
Kumar N, Bauddh K, Kumar S, Dwivedi N, Singh DP, Barman SC (2013b) Heavy metal uptake by plants naturally grown on industrially contaminated soil and their phytoremediation potential. Ecol Eng 61:491–495
Leblebici Z, Aksoy A (2011) Growth and lead accumulation capacity of *Lemna minor* and *Spirodela polyrhiza* (Lemnaceae): interactions with nutrient enrichment. Water Air Soil Pollut 214:175–184
Levin DB, Pitt L, Love M (2004) Biohydrogen production: prospects and limitations to practical application. Int J Hydrog Energy 29:173
Lin CY, Jo CH (2003) Hydrogen production from sucrose using an anaerobic sequencing batch reactor process. J Chem Technol Biotechnol 78:678
Low KS, Lee CK, Tai CH (1994) Biosorption of copper by water hyacinth roots. J Environ Sci Health A 29(1):171
Manahan S (2000) Environmental chemistry, 7th edn. Lewis Publishers/CRC Press LLC, London
Mete AM, Ulgen KO, Kirdar B, Lsen OZ, Oliver SG (2002) Improvement of ethanol production from starch by recombinant yeast through manipulation of environmental factors. Enzyme Microb Technol 31:640–647
Mishina D, Tateda M, Ike M, Fujita M (2006) Comparative study on chemical pretreatment to accelerate enzymatic hydrolysis of aquatic macrophyte biomass used in water purification process. Bioresour Technol 97:2166–2217
Mishra VK, Tripathi BD (2008) Concurrent removal and accumulation of heavy metals by the three aquatic macrophytes. Bioresour Technol 99(15):7091–7097
Mkandawire M, Taubert B, Dude EG (2004) Capacity of *Lemna gibba* L. (Duckweed) for uranium and arsenic phytoremediation in mine tailing waters. Int J Phytoremed 6(4):347–362
Mkandawire M, Dudel EG (2007) Are Lemna spp. effective phytoremediation agents. Biorem Biodiv Bioavail 1:56–71
Mo SC, Choi DS, Robinson JW (1989) Uptake of mercury from aqueous solutions by duckweed: The effect of pH, copper and humic acid. J Environ Sci Health 24:135–146
Molisani MM, Rocha R, Machado W, Barreto RC, Lacerda ID (2006) Mercury contents in aquatic macrophytes from two reservoirs in the paraiba do sul: Guandu river system Se Brazil. Braz J Biol:66–101
Moorhead KK, Nordstedr RA (1993) Anaerobic digestion of water hyacinth: effects of particles size, plants nitrogen content and inoculums volume. Bioresour Technol 44:71–76
Nautiyal P, Subramanian KA, Dastidar MG (2014) Production and characterization of biodiesel from algae. Fuel Process Technol 120:79–88
Nguyen TTT, Davy FB, Rimmer M, De Silva S (2009) Use and exchange of genetic resources of emerging species for aquaculture and other purposes. FAO/NACA expert meeting on the use and exchange of aquatic genetic resources relevant for food and agriculture, Chonburi, Thailand, 31 March–02 April 2009
Nichols PB, Couch JD, Al-Hamdani SH (2000) Selected physiological responses of *Salvinia minima* to different chromium concentrations. Aquat Bot 68:313–319
Odjegba VJ, Fasidi IO (2004) Accumulation of trace elements by *Pistia stratiotes*: implications for phytoremediation. Ecotoxicology 13:637–646

Olette R, Couderchet M, Biagianti S, Eullaffroy P (2008) Toxicity and removal of pesticides by selected aquatic plants. Chemosphere 70:1414–1421
Olette R, Couderchet M, Eullaffroy P (2009) Phytoremediation of fungicides by aquatic macrophytes: toxicity and removal rate. Ecotoxicol Environ Saf 72:2096–2101
Olguın EJ, Sánchez-Galván G, Pérez-Pérez T, Pérez-Orozco A (2005) Surface adsorption, intracellular accumulation and compartmentalization of Pb (II) in batch-operated lagoons with *Salvinia minima* as affected by environmental conditions, EDTA and nutrients. J Ind Microbiol Biotechnol 32:577–586
Patel SI, Patel NG (2015) Production of bioethanol using water hyacinth, an aquatic weed, as a substrate. J Environ Soc Sci 2(1):108
Paterson S, Mackay D, Tam D, Shiu WY (1990) Uptake of organic chemicals by plants: a review of processes, correlations and models. Chemosphere 21:297–331
Prasertsup P, Ariyakanon N (2011) Removal of chlorpyrifos by water lettuce (*Pistia stratiotes* L.) and duckweed (*Lemna minor* L.). Int J Phytoremed 13(4):383–395
Rai PK (2007a) Phytoremediation of Pb and Ni from industrial effluents using *Lemna minor:* an eco-sustainable approach. ull. Bioscience 5(1):67–73
Rai PK (2007b) Wastewater management through biomass of *Azolla pinnata*: an ecosustainable approach. Ambio 36(5):426–428
Rai PK (2008) Phytoremediation of Hg and Cd from industrial effluents using an aquatic free floating macrophyte *Azolla pinnata*. Int J Phytoremed 10:430–439
Rai PK (2009) Heavy metal phytoremediation from aquatic ecosystems with special reference to macrophytes. Environ Sci Technol 39:697–753
Rai PK (2010) Microcosm investigation of phytoremediation of Cr using *Azolla pinnata*. Int J Phytoremed 12:96–104
Rai PK, Tripathi BD (2009) Comparative assessment of *Azolla pinnata* and *Vallisneria spiralis* in Hg removal from G.B. Pant Sagar of Singrauli Industrial region, India. Environ Monit Assess 148:75–84
Randive V, Belhekar S, Paigude S (2015) Production of bioethanol from *Eichhornia crassipes* (Water Hyacinth). Int J Curr Microbiol Appl Sci 2:399–406
Reddy HK, Muppaneni T, Sun Y, Li Y, Ponnusamy S, Patil PD, Dailey P, Schaub T, Holguin FO, Dungan B, Cooke P, Lammers P, Voorhies W, Lu X, Deng S (2014) Subcritical water extraction of lipids from wet algae for biodiesel production. Fuel 133:73–81
Rezania S, Ponraj M, Talaiekhozani A, Mohamad SE, Din MFM, Taib SM (2015) Perspectives of phytoremediation using water hyacinth for removal of heavy metals, organic and inorganic pollutants in wastewater. J Environ Manag 163:125–133
Sánchez-Galván G, Monroy O, Gómez G, Olguín EJ (2008) Assessment of the hyperaccumulating lead capacity of *Salvinia minima* using bioadsorption and intracellular accumulation factors. Water Air Soil Pollut 194:77–90
Sarma H (2011) Metal hyperaccumulation in plants: a review focusing on phytoremediation technology. J Environ Sci Technol 4:118–138
Sasmaz A, Obek E (2009) The accumulation of arsenic, uranium, and boron in *Lemna gibba* L. exposed to secondary effluents. Ecol Eng 35:1564–1567
Sasmaz A, Obekb E, Hasarb H (2008) The accumulation of heavy metals in *Typha latifolia* L. grown in a stream carrying secondary effluent. Ecol Eng 33:278–284
Schnoor JL, Licht LA, McCutcheon SC, Wolfe NL, Carreira LH (1995) Phytoremediation of organic and nutrient contaminants. Environ Sci Technol 29:318A–323A
Sculthorpe CD (1967) The biology of aquatic vascular plants. St. Martin's Press, New York, 610 pp
Sharma SS, Gaur JP (1995) Potential of *Lemna polyrhiza* for removal of heavy metals. Ecol Eng 4:37–43
Shrimp JF, Tracy JC, Davies LC, Lee E, Huang W, Erikson LE, Schooner JL (1993) Beneficial effects of plants in the remediation of soils and ground water contaminated with organic materials. Crit Rev Environ Sci Technol 23(1):41–77

Simonich SL, Hites RA (1995) Organic pollutant accumulation in vegetation. Environ Sci Technol 29:2905–2914
Singh A, Prasad SM (2011) Reduction of heavy metal load in food chain: technology assessment. Rev Environ Sci Biotechnol 10:199–214
Skinner K, Wright N, Porter-Goff E (2007) Mercury uptake and accumulation by four species of aquatic plants. Environ Pollut 145:234–237
Slade R, Bauen A (2013) Micro-algae cultivation for biofuels: cost, energy balance, environmental impacts and future prospects. Biomass Bioenergy 53:29–38
Sood A, Pabbi S, Uniyal PL (2011) Effect of paraquat on lipid peroxidation and antioxidant enzymes in aquatic fern *Azolla microphylla* Kual. Russ J Plant Physiol 58:667–673
Souza FA, Dziedzic M, Cubas SA, Maranho LT (2013) Restoration of polluted waters by phytoremediation using *Myriophyllum aquaticum* (Vell.) Verdc., Haloragaceae. J Environ Manag 120:5–9
Srivastava S, Mishra S, Dwivedi S, Tripathi R (2010) Role of thiol-metabolism in arsenic detoxification in *Hydrilla verticillata* (L.f.) Royle. Water Air Soil Pollut 212:155–165
Srivastava S, Srivastava M, Suprasanna S, D'Souza F (2011) Phytofiltration of arsenic from simulated contaminated water using *Hydrilla verticillata* in field conditions. Ecol Eng 37:1937–1941
Sudhakar K, Ananthakrishnan R, Goyal AV (2013) Biogas production from a mixture of water hyacinth, water chestnut and cow dung. Int J Sci Eng Technol Res 2:1
Sune N, Sanchez G, Caffaratti S, Maine MA (2007) Cadmium and chromium, removal kinetics from solution by two aquatic macrophytes. Environ Pollut 145:467–473
Sweta BK, Singh R, Singh RP (2015) The suitability of *Trapa natans* for phytoremediation of inorganic contaminants from the aquatic ecosystems. Ecol Eng 83:39–42
Taheruzzaman Q, Kushari DP (1988) The effect of sewage enriched River Ganga water on the biomass production of Azolla-Anabaena complex. Hydrobiol Bull (Amsterdam) 22:173–181
Thayaparan M, Iqbal SS, Chathuranga PKD, Iqbal MCM (2013) Rhizofiltration of Pb by *Azolla pinnata*. Int J Environ Sci 3:6
Tredici MR (2010) Photobiology of microalgae mass cultures: understanding the tools for the next green revolution. Future Sci 1:143–162
Tripathi BD, Shukla SC (1991) Biological treatment of wastewater by selected aquatic plants. Environ Pollut 69:69–78
Verma R, Suthar S (2014) Synchronized urban wastewater treatment and biomass production using duckweed *Lemna gibba* L. Ecol Eng 64:337–343
Vesely T, Tlustos P, Szakova J (2011) The use of water lettuce (*Pistia stratiotes)* for rhizofiltration of a highly polluted solution by cadmium and lead. Int J Phytoremed 13:859–872
Vithanage M, Dabrowska BB, Mukherjee B, Sandhi A, Bhattacharya P (2012) Arsenic uptake by plants and possible phytoremediation applications: a brief overview. Environ Chem Lett 10:217–224
Wang Z, Zhang Z, Zhang J, Zhang Y, Liu H, Yan S (2011) Large-scale utilization of water hyacinth for nutrient removal in Lake Dianchi in China: the effects on the water quality, macrozoobenthos and zooplankton. Chemosphere 89:1255–1261
Watanabe ME (1997) Phytoremediation on the brink of commercialization. Envion Sci Technol 31:182A–186A
Xia H, Ma X (2006) Phytoremediation of ethion by water hyacinth (*Eichhornia crassipes*) from water. Bioresour Technol 97:1050–1054
Xie Y, Yu D (2003) The significance of lateral roots in phosphorus (P) acquisition of water hyacinth (Eichhornia crassipes). Aquat Bot 75:311–321
Xu QS, Ji WD, Yang HY, Wang HX, Xu Y, Zhao J, Shi GX (2009) Cadmium accumulation and phytotoxicity in an aquatic fern, *Salvinia natans* (Linn.). Acta Ecol Sincia 29:3019–3027
Yang D, Lauridsen H, Buels K, Chi LH, La Du J, Bruun DA et al (2011) Chlorpyrifos-oxon disrupts zebrafish axonal growth and motor behavior. Toxicol Sci 121(1):146–159

Zhang X, Lin AJ, Zhao FJ, Xu GZ, Duan GL, Zhu YG (2008) Arsenic accumulation by aquatic fern Azolla: comparison of arsenate uptake, speciation and efflux by *A. caroliniana* and *A. filiculoides*. Environ Pollut 156:1149–1155
Zhang X, Zhao FJ, Huang Q, Williams PN, Sun GX, Zhu YG (2009) Arsenic uptake and speciation in the rootless duckweed *Wolffia globosa*. New Phytol 182:421–428
Zhu YL, Zayed AM, Quian JH, Desouza M, Terry N (1999) Phytoaccumulation of trace elements by wetland plants, II: water hyacinth. J Environ Qual 28:339–444

Efficiency of Bioenergy Plant in Phytoremediation of Saline and Sodic Soil

14

Priyanka Bharti, Bhaskar Singh, Kuldeep Bauddh, R.K. Dey, and John Korstad

Abstract

Saline and sodic soils are distributed all over the world and are continuously increasing with a rapid rate and hence considered as one of the serious problems of land degradation. Land degradation is directly affecting the agricultural production. Due to limited availability of agricultural land/soil and poor soil physical and chemical characteristics, there is scarcity of food supply for the increasing population. Hence, the sodic and saline soil can be considered as an important land resource and can be utilized for economic development of the country. Several methods have been applied to restore the saline and sodic land. Chemical methods, such as using gypsum cause dissolution of calcium ion by replacing Na^+ ion through cation exchange processes. This process works efficiently but is cost intensive and not feasible for farmers as well as natural ecosystems. There is a need of sustainable and cost-effective process/technology that can help in reclamation of saline and sodic soil. In this respect, phytoremediation has emerged as a versatile technology towards the reclamation of degraded land. The purpose of phytoremediation using bioenergy crops is to obtain resources that can sustain the increasing population and simultaneously can be used for oil production. Adopting phytoremediation using energy crops also sequesters car-

P. Bharti • B. Singh • K. Bauddh
Centre for Environmental Sciences, Central University of Jharkhand, Ranchi-835205, Jharkhand, India

R.K. Dey
Centre for Applied Chemistry, Central University of Jharkhand, Ranchi-835205, Jharkhand, India

J. Korstad (✉)
Department of Biology and Renewable Energy, Oral Roberts University, 7777 South Lewis, Avenue, Tulsa, OK 74171, USA
e-mail: jkorstad@oru.edu

K. Bauddh et al. (eds.), *Phytoremediation Potential of Bioenergy Plants*,
DOI 10.1007/978-981-10-3084-0_14

bon, fixes atmospheric nitrogen in the soil, and produces oil and biomass that can be utilized as feedstock for biofuels.

Keywords
Saline soil • Sodic soil • Microbial fuel cell • Bioenergy crops • Phytoremediation

14.1 Introduction

Soil salinity and sodicity is a serious issue of land degradation worldwide and is predicted to become more of a problem in the future (Wong et al. 2008). Soil salinity refers to the high salt concentration in the soil, and sodicity is the presence of high concentration of just sodium ions (Na^+) among all of the other cations present in the soil such as magnesium (Mg^{2+}), calcium (Ca^{2+}), etc. (Bernstein 1975). Saline soil is one of the important soil resources in the world, and coastal saline soil is one of the main types of saline soil. Sandy loam and silt soil are examples of soil types present in coastal saline region (Li et al. 2016). Salt-affected soil is distributed in 831 million hectares of land worldwide including 397 and 434 million hectares of saline and sodic soil, respectively (FAO 2000). The characteristic features and principles involved in the identification, reclamation, and management of salt-affected soil are similar throughout the world. Some of the factors vary regionally such as soil characteristics, climatic condition, water availability, farm management efficiency, available resources, and economic status that lead to differences in methods and reclamation potential (Abrol et al. 1988). Various methods have been established for reclamation purposes but are associated with certain limitations. In this regard, phytoremediation can be seen as an effective, low-cost, and environmentally safe technology. It is a plant-based technology which enhances soil quality and productivity potential, thereby reducing pollutants or contaminants responsible for impaired plant growth. Some plants have more capacity to remediate degraded sodic or saline soil. It is important that the plants that are selected possess the highest potential of phytoremediation. If those plants also have the potential for high biomass production, they can be used for bioenergy generation. Bioenergy is a renewable source of energy from biological materials that produces heat, electricity, and fuel and their coproducts (Yuan et al. 2008). Among various energy sources, bioenergy is the most abundant and versatile renewable energy in the world (Zhuang et al. 2011; Edrisi and Abhilash 2016). Bioenergy is termed as the conversion of biomass into energy (McKendry 2002a). Biomass is the typical form of renewable energy that has been widely utilized as source of energy for domestic purposes since quite long ago (McKendry 2002b). Biomass can be produced by growing dedicated energy crops such as short rotation coppice (SRC), perennial grasses, forest residues, sludge from organic industrial wastes, and organic domestic wastes (McKendry 2002a). According to Ni et al. (2006), resources produced from biomass used to convert into energy have been classified into four categories:

- Energy crops: herbaceous energy crops, woody energy crops, industrial crops, agricultural crops, and aquatic crops.
- Agricultural residues and end products: crops waste and animal-produced waste.
- Forests wastes and leftover: mill wood, logging residues, trees, and shrubs residues.
- Industrial and municipal wastes: municipal solid waste, sewage sludge, and industrial effluent waste.

Biomass is used for the purpose of heating, cooling, and producing electricity and liquid biofuels. Burning fossil fuels, deforestation, and human activities have led to the emission of greenhouse gases into the atmosphere. The usage of biomass for biofuel reduces greenhouse gas emissions, making it carbon neutral (Kraxner et al. 2013). Energy produced from biomass has been categorized into two groups: biomass produced from food crops such as corn grain, sugarcane, soybean, oil seed, etc.; and biomass produced from cellulosic feedstock such as starch, sugar, fatty acid, or cellulose (Ghosh 2016). The physical quantity of biomass has enough potential for worldwide bioenergy production (Altman et al. 2015). The new renewable energy obtained through wind, solar, and biofuel is growing fast continuously and contributing to global renewable energy supply. Bioenergy is one of the alternative sources for fossil fuel, particularly for those used in transportation. Presently, commercially available biofuels are produced from starch or sugar-rich crops for bioethanol and from oilseeds for biodiesel production (Popp et al. 2014).

There are numerous plant species that are capable of cleaning up the soil. These plants are also used to obtain useful by-products such as biofuel (biodiesel or bioethanol), fiber, wood, charcoal, alkaloid, bioplastic, etc. (Tripathi et al. 2016). In India, the 1970 oil crisis has led to the establishment of bioenergy promotion (Rabindranath et al. 2010; Edrisi and Abhilash 2016).

14.1.1 Saline and Sodic Soil: Origin, Characteristics, Distribution, and Parameters for Salinity/Sodicity Measurement

14.1.1.1 Origin of Saline and Sodic Soil

Salinity is caused by natural weathering of parent material, deposition of sea salt carried by wind and rain, inundation of coastal land by tidal water, and anthropogenic activities such as excessive irrigation by underground water resulting in a rise of the water table, irrigation by salt-containing water, poor drainage, etc. (Munns 2005; Manchanda and Garg 2008; Hasanuzzaman et al. 2013; Hasanuzzaman et al. 2014). Salts present in the upper surface of the soil profile undergo hydration, hydrolysis, oxidation, solubilization, and carbonization through chemical weathering. The salts solubilize and are transported away from the origin sites through soil surfaces or groundwater. Salts in the groundwater are gradually concentrated when the water moves to more arid areas (Abrol et al. 1988).

14.1.1.2 Characteristics

Saline and sodic soil bears several features which make them unsuitable for agricultural practices. Saline and sodic soil possesses poor physical properties and fertility problem that adversely affect the growth and yield of various crops (Sumner 1993; Naidu and Rengasamy 1993; Qadir and Schubert 2002; Qadir et al. 2005). It has been reported that saline soil has electrical conductivity with value ranging from 2 to more than 32 dS/m (Richards 1954; Farifteh et al. 2008). Saline soil consists of many ions like chlorides, sulfates, nitrates and bicarbonates of sodium, calcium, magnesium, and potassium (Bul 2013). Most of saline soil also consists of some proportion of gypsum ($CaSO_4$) (Abrol et al. 1988). Sodic soil consists of sodium carbonate, sodium bicarbonate, and sodium chloride as dominating components. The physicochemical characteristics of saline and sodic soil has been depicted in Table 14.1.

14.1.1.3 Distribution of Saline and Sodic Soils

According to Szabolcs (1974), salt-affected soils can be found in North America, Mexico, Central America, South America, Africa, Southern Asia, North and Central Asia, Southeast Asia, Australia, and Europe. Several states of India such as Uttar Pradesh, West Bengal, and Gujarat are largely salt affected (Edrisi and Abhilash 2016).

Table 14.1 Physicochemical characteristic of saline and sodic soil

Parameter	Value	References
pH	>8.5	Bul (2013)
Electrical conductivity (dS/m)	>4.0	US Salinity Laboratory (1969) and Bul (2013)
Sodium absorption ratio	~13	Qadir et al. (2007)
Exchangeable sodium percentage	>15	Qadir et al. (2007)
Calcium carbonate (%)	0.80–1.05	Garg (2000), Tripathi and Singh (2005), and Singh et al. (2016)
Total soluble salts (%)	0.14–0.22	Garg (2000), Tripathi and Singh (2005), and Singh et al. (2016)
Sand (%)	43	Garg (2000); Tripathi and Singh (2005), and Singh et al. (2016)
Clay (%)	27	Garg (2000); Tripathi and Singh (2005), and Singh et al. (2016)
Silt (%)	30	Garg (2000); Tripathi and Singh (2005), and Singh et al. (2016)
Water holding capacity (%)	32–35	Garg (2000); Tripathi and Singh (2005), and Singh et al. (2016)
Cation exchange capacity ($cmole_{(+)}kg^{-1}$)	47.9	Gharaibeh et al. (2011)
Organic matter (%)	<0.1	Singh et al. (2016)

14.1.1.4 Parameters for Salinity/Sodicity Measurement

Sodium absorption ratio (SAR) is the ratio of soluble sodium to the sum of the square root of divalent cations, usually calcium (Ca^{++}) and magnesium (Mg^{++}), divided by 2:

$$SAR = \frac{Na}{\sqrt{(Ca + Mg)/2}} \text{ (Harron et al. 1983)} \tag{14.1}$$

An equivalent proportion of sodium remaining present in the cation exchange complex when expressed in terms of percentage is referred to as exchangeable sodium percentage (ESP) (Mau and Porporato 2015):

$$ESP = \frac{100(F_{Na})}{CEC} \text{ (Qadır et al. 2007)} \tag{14.2}$$

Cation exchange capacity (CEC) is the quantity of adsorbed cations on the unit mass of soil (Sposito 2008). Cation exchange capacity of a saturated saline soil paste can be analyzed by the sodium acetate method (Harron et al. 1983).

14.2 Methods for Bioenergy Generation

Biomass can be converted into bioenergy by different processes associated with various sources of biomass, conversion processes, their application, and infrastructure used (Mckendry 2002a). Biomass can be converted into bioenergy by means of producing three types of products: electrical/heat energy, transportation fuel, and chemical feedstock (Mckendry 2002b).

Biomass-based energy production processes are divided into two categories (Ni et al. 2006): thermochemical conversion and biochemical conversion. A third technology for bioenergy generation is mechanical extraction (with esterification) that produces biodiesel (McKendry 2002a).

There are three thermochemical processes (Ni et al. 2006):

- *Combustion*: Biomass burnt in air is combustion. Equipment includes stoves, furnaces, boilers, steam turbines, and turbogenerators used for the conversion of chemical energy stored in biomass into heat, mechanical power, or electricity. At a temperature range from 800 to 1000 °C, hot gases are produced by combustion of biomass. Bioenergy production efficiency by power plant is 20–40% produced mainly by pyrolysis (Verma et al. 2011). Temperature ranging from 650 to 800 K is used to convert biomass heated in the absence of air at a pressure of 0.1–0.5 Pa into biofuel such as liquid oil, charcoal, and gaseous compounds (Ni et al. 2006).
- *Liquefaction*: Biomass is converted to liquid hydrocarbon under low temperature and at higher hydrogen pressure (Warren Spring Laboratory 1993).
- *Gasification*: The gasification process is suitable for producing fuel and electricity using gas engines by gasification of biomass. This is operated either by simple technology based on a fixed-bed gasifier or fluidized bed technology.

Biochemical conversion processes are of two main types (Mckendry 2002a):

- *Anaerobic digestion*: This is the direct conversion of biomass into gas through anaerobic digestion. Biogas products are methane and carbon dioxide, along with a few other gases in smaller quantities such as hydrogen sulfide (EU 1999). Biogas is used in spark-ignition gas engines and gas turbines and can be upgraded to finer quality gases by removing carbon dioxide (Mckendry 2002a).
- *Fermentation*: In the process of fermentation, ethanol is produced from sugar crops such as sugarcane, sugar beet, and starch crops (maize and wheat). The material biomass is crushed, and starch is converted to sugar by enzymatic activities using yeast, and finally sugar is converted to ethanol (Mckendry 2002a).

Microbial fuel cell is another emerging technology where microorganisms are placed in an electro-biochemical chamber without air to oxidize organic matter and release electrons and protons, thus producing electricity (Pant et al. 2010; Hernandez-Fernandez et al. 2015) (Fig. 14.1).

14.3 Phytoremediation Potential of Energy Crops

Jatropha curcas has the potential to phytoremediate lindane and fly ash-contaminated sites by accumulating these contaminants in root followed by stem and leaf (Abhilash et al. 2013; Jamil et al. 2009). There are some species and their hybrids such as *Populus* and *Salix* known for their phytoremediation potential of contaminated sites (Zalesny et al. 2007). *Populus* is known to remediate the landfill sites, petroleum sludge, salts, heavy metals, pesticides, solvent, explosives, and radionucleotides (Burken 2001; Erdman and Christenson 2000; Gordon et al. 1997; Neill and Gordon 1994; Thompson et al. 1998; Zalesny et al. 2007). *Salix* is known to phytoremediate dairy effluent, wastewater sludge, municipal wastes, and cadmium from the contaminated sites. Plants such as *Pistacia chinensis*, *Sapium sebiferum*, and *Xanthoceras sorbifolium* are distributed in different parts of China and considered as oil-yielding plant cultivated under different range of environmental conditions (Shaoa and Chu 2008). In Greece, experimentation with *Eucalyptus* was done for wood and biomass production (Panetsos and Alizoti 1996). The result indicated that six species of *Eucalyptus*, namely, *E. bicostata*, *E. cladocalyx*, *E. viminalis*, *E. saligna*, *E. camaldulensis*, and *E. dalrympleana* are considered as dedicated energy crops (Panetsos et al. 1981). *Pinus taeda* has rapid growth on soil that is poorly or moderately drained (Coyle et al. 2008). *Liquidambar styraciflua* has potential to remediate soil contaminated with uranium (U) and thorium (Th) (Saritz 2005). *Glycine max*, *Panicum virgatum*, and *Helianthus annuus* are considered as biofuel crops grown on marginal soils such as brownfield sites (Smith et al. 2013). A study on legumes and trees for fuelwood production on sodic wasteland has been reported by Goel and Behl (2001). The legumes referred were *Acacia auriculiformis*,

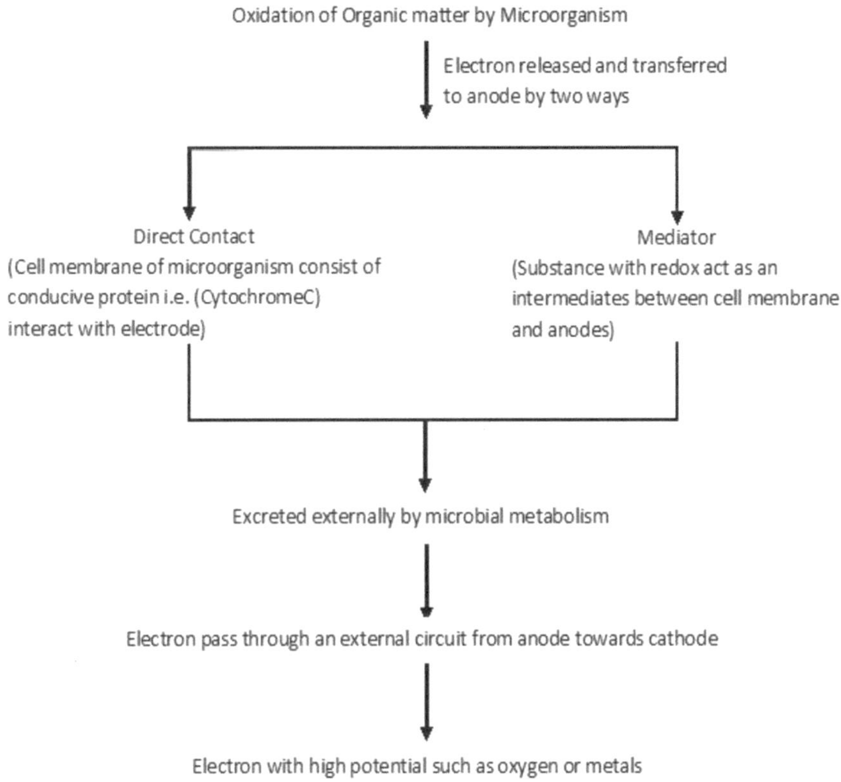

Fig. 14.1 Procedure for conversion of chemical energy of substrate into electrical energy by bacteria through microbial fuel cells (Hernandez-Fernandez et al. 2015)

A. nilotica, *Albizia lebbeck*, *A. procera*, *Dalbergia sissoo*, *Leucaena leucocephala*, *Pongamia pinnata*, *Prosopis juliflora*, and *Pithecellobium dulce*. The trees studied were *Azadirachta indica*, *Eucalyptus tereticornis*, and *Terminalia arjuna*. Among the species studied, *P. juliflora* ranked first as the most promising species for biomass production, and *Acacia nilotica* ranked second most promising species for biomass production on degraded sodic land (Goel and Behl 2001). Table 14.2 depicts the potential of pollution remediation and bioenergy production.

14.4 Reclamation of Saline and Sodic Soil

Reclamation of sodic and saline soils requires removal of most of the exchangeable cations and its replacement by Ca^{2+} ions and solubilized salts from the root zone, which can be done by various methods (Abrol et al. 1988). Before amelioration of a specific site, important factors should be considered such as soil depth to be ameliorated, presence of a dense solid layer in the subsoil, salt constituent in the soil,

Table 14.2 Pollution remediation and bioenergy potential of energy crop

Species	Remediation of particular pollutants by the plant species	Product(s)	Reference(s)
Jatropha curcas	Lindane, Fe, Al, Cr, Cu, and Mn	Biodiesel and biofertilizer	Abhilash et al. (2013) and Jamil et al. (2009)
Populus deltoids, *P. trichocarpa*, *P. nigra*, *P. maximowiczii*, *P. tremula*, *P. tremuloides*, *P. deltoids*, *P. nigra*	Explosive nitrate esters and nitro aromatics	Biomass, biogas, plywood, charcoal	Fortier et al. (2010) and Doty et al. (2007)
Salix alba, *S. viminalis*, *S. schwerinii*, *S. viminalis*	Zn, Cd, Cu, Hg, Pb, Cd	Biomass, biogas, plywood, charcoal	Delplanque et al. (2013) and Mleczek et al. (2010)
Xanthoceras sorbifolium	Not reported	Biomass, biodiesel, charcoal	Shaoa and Chu (2008)
Sapium sebiferum	Not reported	Biomass, charcoal	Shaoa and Chu (2008)
Pistacia chinensis	Not reported	Biomass, charcoal	Shaoa and Chu (2008)
Eucalyptus grandis, *E. bicostata*, *E. dalrympleana*, *E. viminalis*	PO_4^{3-}	Biomass, biogas, plywood, charcoal	Aravanopoulos (2010) and Panetsos (1988)
Pinus taeda L.	PO_4^{3-}	Biomass, biogas, plywood, charcoal	Kline and Coleman (2010) and Panetsos (1988)
Liquidambar styraciflua	Not Reported	Bioenergy, paper and pulp	Kline and Coleman (2010)
Glycine max	Cd, Cr, Ni, As, Fe, poly-aromatic hydrocarbon, atrazine	Bioenergy, bioethanol, charcoal	Smith et al. (2013) and Cutright et al. (2010)
Panicum virgatum	Cd, Cr, Ni, As, Fe, poly-aromatic hydrocarbon (PAH), atrazine	Bioethanol	Fairley (2011) and Graham Rowe (2011)
Helianthus annuus	Cd, Cr, Ni, As, Fe, poly-aromatic hydrocarbon (PAH), atrazine	Bioenergy, bioethanol, charcoal	Smith et al. (2013) and Cutright et al. (2010)
Miscanthus sinensis	Nutrients, Zn, Cd, and Pb	Bioethanol, biogas	St. Clair et al. (2008) and Zhao et al. (2012)

(continued)

Table 14.2 (continued)

Species	Remediation of particular pollutants by the plant species	Product(s)	Reference(s)
Madhuca indica	Dye removal from wastewater, fly ash	Biodiesel, biomass, charcoal	Ghadge and Raheman (2005)
Prosopis juliflora	Fly ash	Biomass, charcoal	Goel and Behl (2001)
Acacia nilotica	Not reported	Biomass, charcoal	Goel and Behl (2001)
Ricinus communis	Cd, DDT	High biomass	Huang et al. (2011)
Camelina sativa	Poly-aromatic hydrocarbon (PAH), atrazine	Biofuel	Fairley (2011) and Graham Rowe (2011)
Phragmites australis	Lindane, monochlorobenzene (MCB), dichlorobenzene (DCB), trichlorobenzene (TCB)	Bioethanol, charcoal	Sathitsuksanoh et al. (2009)
Pongamia pinnata/ Pongamia glabra	Cr, Mn, Fe, Ni, Cu, Zn, Pb, Rb, Sr, Ti, Co	Biomass, biodiesel, charcoal	Reddy et al. (2008) and Ravikumar et al. (2013)
Azadirachta indica	Cr, Mn, Fe, Ni, Cu, Zn, and Pb	Biomass, biodiesel, charcoal	Reddy et al. (2008) and Ravikumar et al. (2013)
Arundo donax	Nutrients, Cd, As, and Ni	Bioethanol, charcoal	Liu et al. (2012)
Pennisetum purpureum	Nutrients	Bioethanol	Liu et al. (2012)

availability of water for leaching, nature and depth of groundwater, topography of land, type of crops to be grown after amelioration, and climatic condition of the region (Qadir et al. 2000). Chemical amelioration of sodic land for their reclamation can be divided into three categories: gypsum and calcium chloride as a soluble calcium salt; acid-forming compounds such as sulfuric acid, iron sulfate, aluminum sulfate, lime sulfur, and pyrite; and less soluble calcium salts such as limestone (Abrol et al. 1988). Chemical amelioration using gypsum provides a source of Ca^{2+} ions directly to the soil that replaces excess Na^{+} ions while dissoluting calcite ($CaCO_3$) in the soil (Shainberg et al. 1989; Gupta and Abrol 1990; Oster et al. 1999; Qadir and Oster 2002 and Qadir et al. 2002). Methods to ameliorate saline/sodic soils include leaching salts from the upper surface of the soil and transporting them to lower depths, flushing of salts from the salt crusts at the surface and also in the shallow water table, etc. Biological amelioration involves sequestration of salts by the aerial (i.e., completely exposed in air) or shallow depth parts of plants that can be harvested and thus removes the salts from the soil (Qadir et al. 2000). Plants can improve chemical properties of soils, decrease soil pH, add organic matter, and dissolute lime (Ilyas et al. 1997). Remediation of saline soil is a critical global issue that requires multidisciplinary ways to remediate salt-affected land including agricultural practices, varieties of salt-tolerant crops, and phytoremediation.

14.5 Phytoremediation of Saline and Sodic Soil by Energy Crops

Phytoremediation is considered a cost-effective and environmentally safe technology for saline soil remediation (Hasanuzzaman et al. 2014). Phytoremediation involves various processes such as phytoextraction, phytodegradation, rhizofiltration, phytostabilization, and phytovolatilization (Fletcher 2006). According to Qadir et al. (2005), the two main advantages of phytoremediation are: i. no financial outlay needed for purchase of chemicals (for chemical amendment), and ii. salt-resistant crops generate high-value by-products. Roots of plants maintain soil structure and enhance drainage through the formation of macropores (pores greater than 0.08 mm in diameter) at deeper depths (Czarnes et al. 2000). Phytoremediation utilizing bioenergy crops/plants is one of the best technologies for remediation of saline and sodic soil because the harvested biomass can be used to produce biofuel or other commercial by-products while ameliorating the soil. The best bioenergy crops for soil amelioration should have high biomass production, be cost effective, have low contaminant content, have less nutrient and water requirements, be carbon neutral for the whole life cycle, and do not lead to the "food versus fuel" issue (Singh and Singh 2016).

Lal and Pimentel (2007) reported that several species of plant can produce abundant, good quality forage during summer, including warm season grasses such as switchgrass (*Panicum virgatum*), big bluestem (*Andropogon gerardii vitman*), and Indian grass (*Sorghastrum nutans*). Salt-tolerant grasses include Guinea grass (*Panicum maximum*), elephant grass (*Pennisetum purpureum*), and Kallar (also called Karnal) grass (*Leptochloa fusca*). Some of the short rotation woody perennials such as poplar (*Populus* spp.), willow (*Salix* spp.), and black locust (*Robinia pseudoacacia* L.) produce 10–20 tons of dry weight of biomass per hectare. Some important halophytes that grow in brackish water containing salt concentrations up to 30,000 ppm include pickle weed (*Salicornia bigelovii*), salt grass (*Distichlis palmeri*), salt brushes (*Atriplex* spp.), and few algae (e.g., *Spirulina geitleri*). Some non-edible oil-yielding plants include *Jatropha* (*Jatropha curcas*), *Pongamia* (*Millettia pinnata*), and *Madhua* (*Madhuca longifolia*). There are few energy crops in the world (*Miscanthus*, *Ricinus*, *Jatropha*, *and Populus*) that possess phytoremediation potential and act as carbon sinks, thus contributing profit through carbon tax credits (Bauddh and Singh 2012a, b; Bauddh and Singh 2015a, b; Bauddh et al. 2015a, b; 2016a, b; Pandey et al. 2016). Vetiver (*Chrysopogon zizanioides*) and lemongrass can tolerate and grow in saline soil. Live and dry biomass of vetiver and lemongrass has economic importance through their oil production (Maiti and Kumar 2016). India is the largest producer of lemongrass oil with production of 300–350 tons year^{-1}, of which 80% is exported to developed countries of the world (Lal et al. 2013). Mesquite (*Prosopis juliflora*) grows in Africa, Argentina, Australia, Brazil, Cameroon, Caribbean, Central America, Egypt, Ethiopia, Hawaii, India, Kenya, Nigeria, Pakistan, Paraguay, Peru, Ecuador, Portugal, Senegal, Spain, Sri Lanka, Sudan, Uganda, the United States, and Yemen. Mesquite is known to rehabilitate degraded saline and sodic land and concomitantly increases soil fertility by adding soil organic carbon, nitrogen, and phosphorus and decreases exchangeable Na^+ levels, pH, and electrical conductivity (Prasad and Tewari 2016). It can also be used for the production of charcoal, bioethanol, timber, fuelwood, and antibiotics (Prasad and Tewari 2016). It is estimated that crop residues and lignocellulosic residues from cereals can produce 4 billion Kg and 3 billion Kg of ethanol per year, respectively. One Mg (megagram) of corn stover can produce 280 L of ethanol, and 1 Mg of corn grains can produce 400 L of ethanol. One Mg of biomass produces about 18.5 GJ of energy. Three billion Mg of residue can produce 840 billion L of ethanol or about 56×10^9 GJ of energy (Lal 2008). Presently, global bioenergy consumption is 50 EJ yr.$^{-1}$ and is expected to reach to 80–160 EJ yr.$^{-1}$ (Pandey et al. 2016). The total biofuel production in India increased from 27.3 million L in 2007 to 46.4 million L in 2011, including 34.8 million L for bioethanol and 11.6 million L for biodiesel (Edrisi and Abhilash 2016). Marrison and Larson (1996) estimated that total bioenergy production from Africa by 2025 will be 18 EJ per year on the basis of planting crops on 10% of available land except forest, agricultural, and wilderness areas. Halophytes are being considered as potential new agricultural crops to reclaim salt-affected land. A species of halophyte (*Salicornia bigelovii*) can withstand high salinity and produce biomass and seeds of 2 tons hectare^{-1}, yielding 28% oil, 31% protein, 5% fiber, and 5% ash (Glenn et al. 1999).

Gharaibeh et al. (2011) assessed the potential of *Atriplex halimus* in reclamation of calcareous saline sodic soil. Cultivation of this plant significantly increased soil properties. The electrical conductivity (ECe) was reduced from 5.8 to 3.7 dSm^{-1} (Fig. 14.2); however, ESP was found to be decreased. ECe value after plantation of *Atriplex halimus* in salt-contaminated areas was the indication of removal of Na^+ ions from the soil.

Plantations of some plant species on degraded sodic land (D-SL) with rehabilitated *Terminalia arjuna* (R-TA), rehabilitated *Prosopis juliflora* (R-PJ), reference *Tectona grandis* (Ref-TG), rehabilitated mixed forest (R-MF), and reference mixed forest (Ref-MF) improved soil physicochemical characteristics and soil particle distribution, exchangeable sodium percentage, and microbial enzyme concentration (Singh et al. 2012). Bulk density for D-SL was 1.62 g cm^{-3}. After rehabilitation with R-TA, the bulk density reduced to 1.24 g cm^{-3} (29% decrease). With rehabilitated *Prosopis juliflora* (R-PJ), it showed a 24% decrease, and with R-MF, a 21% decrease in bulk density has been reported. Water holding capacity

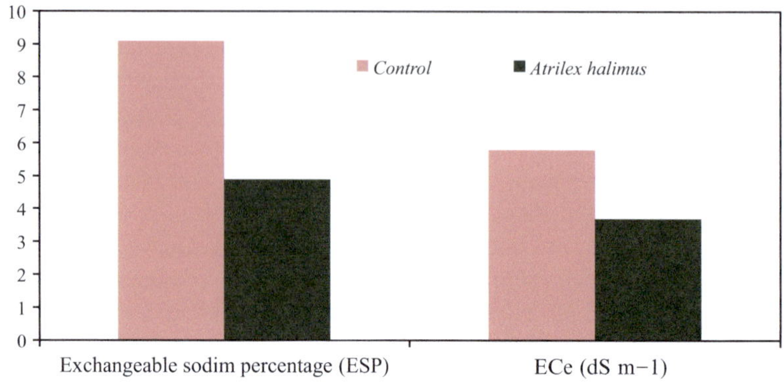

Fig. 14.2 Effect of plantation of *Atriplex halimus* in salt-contaminated areas (Gharaibeh et al. 2011)

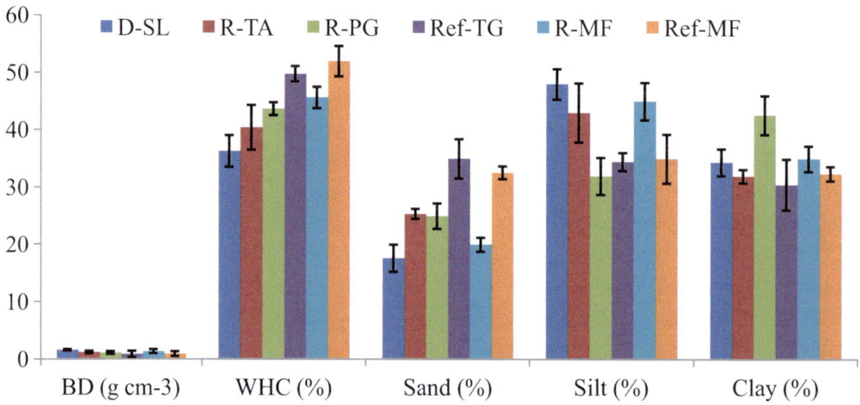

Fig. 14.3 Effect of plantation on bulk density (BD), water holding capacity (WHC), and soil particle distribution of degraded sodic land, rehabilitated land uses, and reference plantation and forest (Singh et al. 2012)

(WHC) percentage increased to 52.00 ± 2.75% in Ref-MF as compared to D-SL having WHC of 36.33±2.65%. Na^+ ion concentration decreased in the following order: R-TA (3.45 cmol kg^{-1}) <R-PJ (3.43 cmol kg^{-1}) <Ref-MF (1.96 cmol kg^{-1}) <R-MF (1.47 cmol kg^{-1}) <Ref-TG (0.80 cmol kg^{-1}). The mean values of K^+, Ca^{2+} and Mg^{2+} ion increased to 2.48, 20.35, and 5.50 cmol kg^{-1} respectively in Ref-MF. Figure 14.3 depicts the role of afforestation in improvement of physico-chemical characteristics of soil.

14.6 Conclusion

Saline and sodic soils are distributed in around 831 million hectares of land worldwide and possess various adverse features such as high pH, high exchangeable sodium percentage, high sodium adsorption ratio, and low cation exchange capacity that make a soil infertile. At the same time, fuel and energy needs are increasing globally, leading to greater emissions of greenhouse gases which results changes in climate. Chemical treatment using gypsum shows rapid amelioration of saline and sodic soil, but is a costly and non-eco-friendly approach. The emerging technology of phytoremediation using specific plants that reclaim sodic and saline land may be used to produce energy. Some of the energy crops like *Miscanthus*, *Ricinus*, *Jatropha*, and *Populus* are extensively used worldwide. Reclamation of saline and sodic soils utilizing such plants is preferable because of its applicability and sustainability. Phytoremediation using bioenergy crops could be adopted as a better approach for mitigation of major environmental concerns like land degradation, pollution, energy crisis, and climate change.

References

Abhilash PC, Singh B, Srivastava P, Schaeffer A, Singh N (2013) Remediation of lindane by *Jatropha curcas* L: utilization of multipurpose species for rhizoremediation. Biomass Bioenergy 51:160–170

Abrol IP, Yadav JSP, Massoud FI (1988) Salt-affected soils and their management, FAO soils bulletin 39. Food & Agriculture Organization, Rome

Altman I, Bergtold J, Sanders D, Johnson T (2015) Willingness to supply biomass for bioenergy production: a random parameter truncated analysis. Energy Econ 47:1–10

Aravanopoulos F (2010) Breeding of fast growing forest tree species for biomass production in Greece. Biomass Bioenergy 34:1531–1537

Bauddh K, Singh RP (2012a) Cadmium tolerance and its phytoremediation by two oil yielding plants *Ricinus communis* (L) and *Brassica juncea* (L) from the contaminated soil. Int J Phytoremed 14:772–785

Bauddh K, Singh RP (2012b) Growth, tolerance efficiency and phytoremediation potential of *Ricinus communis* (L) and *Brassica juncea* (L) in salinity and drought affected cadmium contaminated soil. Ecotoxicol Environ Saf 85:13–22

Bauddh K, Singh RP (2015a) Effects of organic and inorganic amendments on bio-accumulation and partitioning of Cd in *Brassica juncea* and *Ricinus communis*. Ecol Eng 74:93–100

Bauddh K, Singh RP (2015b) Assessment of metal uptake capacity of castor bean and mustard for phytoremediation of nickel from contaminated soil. Bioremed J 19(2):124–138

Bauddh K, Singh K, Singh B, Singh RP (2015a) *Ricinus communis*: a robust plant for bio-energy and phytoremediation of toxic metals from contaminated soil. Ecol Eng 84:640–652

Bauddh K, Singh K, Singh RP (2015b) *Ricinus communis* L a value added crop for remediation of cadmium contaminated soil. Bull Environ Contam Toxicol 96:265–269. doi:10.1007/s00128-015-1669-3

Bauddh K, Kumar A, Srivastava S, Singh RP, Tripathi RD (2016a) A study on the effect of cadmium on the antioxidative defense system and alteration in different functional groups in castor bean and Indian mustard. Arch Agron Soil Sci 62(6):877–891

Bauddh K, Singh B, Singh RP (2016b) *Ricinus communis* L. as a value added alternative for restoration of cadmium contaminated and degraded agricultural ecosystem. Bull Environ Contam Toxicol 96(2):265–269

Bernstein L (1975) Effects of salinity and sodicity on plant growth. Annu Rev Phytopathol 13(1):295–312
Bul EN (2013) Soil salinity: a neglected factor in plant ecology and biogeography. J Arid Environ 92:14–25
Burken, Joel G. (2001) Uptake and volatilization of chlorinated solvents by poplars at field-scale. In: Sixth international in situ and on site bioremediation symposium
Clair SS, Hillier J, Smith P (2008) Estimating the pre-harvest greenhouse gas costs of energy crop production. Biomass Bioenergy 32(5):442–452
Coyle DR, Coleman MD, Aubrey D (2008) Above and below ground biomass accumulation, production, and distribution of Sweetgum and Loblolly pine grown with irrigation and fertilization. Can J Res 38:1335–1348
Cutright T, Gunda N, Kurt F (2010) Simultaneous hyperaccumulation of multiple heavy metals by *Helianthus annus* grown in a contaminated sandy loam soil. Int J Phytoremed 12(6):562–573
Delplanque M, Collet S, Gratta FD, Schnuriger B, Gaucher R, Robinson B, Bert V (2013) Combustion of *Salix* used for phytoextraction: the fate of metals and viability of the processes. Biomass Bioenerg 49:160–170
Doty SL, Shang TQ, Wilson AM, Tangen J, Westergreen AD, Newman AL, Stuart SE, Gordon MP (2007) Enhanced metabolism of halogenated hydrocarbons in transgenic plants contain mammalian P450 2E.1. Proc Natl Acad Sci 97:6287–6291
Edrisi SA, Abhilash PC (2016) Exploring marginal and degraded lands for biomass and bioenergy production: an Indian scenario. Renew Sust Energ Rev 54:1537–1551
Erdman JA, Christenson S (2000) Elements in cottonwood trees as an indicator of ground water contaminated by landfill leachate. Ground Water Monit Remediat 20:120–126
EU (1999) Biomass conversion technologies. EU 18029 EN. ISBN:92-828-5363-3
Farifteh J, Van Der Meer F, Van Der Meijde M, Atzberger C (2008) Spectral characteristics of salt-affected soils: a laboratory experiment. Geoderma 145:196–206
FAO (2000) Global network as integrated soil management for sustainable use of salt-affected soils. http://www.fao.org/ag/agl/agll/spush
Fairley P (2011) Next generation biofuels. Nature 474:S2–S5
Fletcher J (2006) Introduction. In: Mackova M, Dowling D, Macek T (eds) Phytoremediation Rhizoremediation. Springer, Dordrecht
Fortier J, Gagnon D, Truax B, Lambert F (2010) Biomass and volume yield after 6 years in multiclonal hybrid poplar riparian buffer strips. Biomass Bioenerg 34(7):1028–1040
Garg VK (2000) Bioreclamation of sodic waste land - a case study. Land Degrad Dev 11:487–498
Ghadge SV, Raheman H (2005) Biodiesel production from mahua (*Madhuca indica*) oil having high free fatty acids. Biomass Bioenerg 28(6):601–605
Gharaibeh MA, Eltaif NI, Albalasmeh AA (2011) Reclamation of highly calcareous saline sodic soil using *Atriplex halimus* and by-product gypsum. Int J Phytoremed 13(9):873–883
Ghosh SK (2016) Biomass & biowaste supply chain sustainability for bioenergy and bio – fuel production. Procedia Environ Sci 3:31–39
Glenn EP, Jed Brown J, Blumwald E (1999) Salt tolerance and Crop potential of halophytes. Crit Rev Plant Sci 18(2):227–255
Goel VL, Behl HM (2001) Genetic selection and improvement of hard wood tree species for fuelwood production on sodic soil with particular reference to *Prosopis juliflora*. Biomass Bioenergy 20:9–15
Gordon M, Choe N, Duffy J, Ekuan G, Heilman P, Muiznieks I, Newman L, Ruszaj M, Shurtleff BB, Strand S, Wilmouth J (1997) Phytoremediation of trichloroethylene with hybrid poplars. In: Kruger EL, Anderson TA, Coats JR (eds) Phytoremediation of soil and water contaminants. American Chemical Society, Washington DC, pp 177–185
Graham Rowe D (2011) Beyond food versus fuel. Nature 474:S6–S8
Harron WRA, Webster GR, Cairns RR (1983) Relationship between exchangeable sodium and sodium absorbtion ratio in a solonetzic soil association. Can J Soil Sci 63(3):461–467

Hasanuzzaman M, Nahar K, Fujita M (2013) Plant response to salt stress and role of exogenous protectants to mitigate salt induced damages. In: Ahmad P, Azooz MM, MNV P (eds) Ecophysiology and responses of plants under salt stress. Springer, New York, pp 25–87

Hasanuzzaman M, Nahar K, Bhowmik PC, Ahmad Hossain M, Rahman MM, Prasad MNV, Ozturk M and Fujita M (2014) Potential use of halophytes to remediate saline soils. Biomed Res Int. http://dx.doi.org/10.1155/2014/589341

Hernandez-Fernandez FJ, Perez de los Rios A, Salar-Garcia MJ, Ortiz-Martinez VM, Lozano-Blanco LJ, Godinez C, Tomas-Alonso F, Quesada-Medina J (2015) Recent progress and perspectives in microbial fuel cells for bioenergy generation and waste water treatment. Fuel Process Technol 138:284–297

Huang H, Yu N, Wang L, Gupta DK, He Z, Wang K, Zhu Z, Yan X, Li T, Yang X (2011) The phytoremediation potential of bioenergy crop Ricinus communis for DDTs and cadmium co-contaminated soil. Bioresour Technol 102:11034–11038

Ilyas M, Qureshi RH, Qadir MA (1997) Chemical changes in a saline-sodic soil after gypsum application and cropping. Soil Technol 10:247–260

Jamil S, Abhilash PC, Sharma PN (2009) *Jatropha curcas*: a potential crop for phytoremediation of coal fly ash. J Hazard Mater 172(1):269–275

Kline KL, Coleman MD (2010) Woody energy crops in the southeastern United States: two centuries of practitioner experience. Biomass Bioenerg 34(12):1655–1666

Kraxner F, Nordstrom EM, Havlik P, Gusti M, Mosnier A, Frank S, Valin H, Fritz S, Fuss S, Kindermann G, McCallum I, Khabarov N, Bottcher H, See L, Aoki K, Schmid E, Mathe L, Obersteinner M (2013) Global bioenergy scenarios – future forest development, land – use implications, and trade offs. Biomass Bioenerg 57:86–96

Lal R, Pimentel D (2007) Biofuels from crop residues. Soil Tillage Res 93(2):237–238

Lal R (2008) Crop residues as soil amendments and feedstock for bioethanol production. Waste Manage 28:747–758

Lal K, Yadav RK, Kaur R, Bundela DS, Khan Inayat M, Chaudhary M, Meena RL, Dar SR, Singh G (2013) Productivity, essential oil yield, and heavy metal accumulation in lemon grass (*Cymbopogon flexuosus*) under varied wastewater – groundwater irrigation regimes. Ind Crop Prod 45:270–278

Li X, Kang Y, Wan S, Chen X, Liu S, Xu J (2016) Response of a salt-sensitive plant to processes of soil reclamation in two saline–sodic, coastal soils using drip irrigation with saline water. Agric Water Manag 164:223–234

Liu B, Wu X, Chang J, Gu B, Min Y, Ge Y, Shi Y, Xue H, Peng C, Wu J (2012) Constructed wetland as biofuel production systems. Nat Clim Chang 2:190–194

Maiti SK, Kumar A (2016) Energy plantation, medicinal and aromatic plants in contaminated soil. In: Bioremediation & bioeconomy. Elsevier, Amsterdam, pp 29–40

Manchanda G, Garg N (2008) Salinity and its effects on the functional biology of legumes. Acta Physiol Plant Manchanda G (eds) Springer 30(5):595–618

Marrison CI, Larson ED (1996) A preliminary analysis of the biomass energy production potential in Africa in 2025 considering projected land needs for food production. Biomass Bioenergy 10:337–351

Mau Y, Porporato A (2015) A dynamic system approach to soil salinity and sodicity. Adv Water Resour 83:68–76

McKendry P (2002a) Energy production from biomass (part 2): conversion technologies. Bioresour Technol 83:47–54

McKendry P (2002b) Energy production from biomass (part 1): overview of biomass. Bioresour Technol 83:37–46

Mleczek M, Rutkowski P, Rissmann I, Kaczmarek Z, Golinski P, Szentner K, Strazynska K, Stachowiak A (2010) Biomass productivity and phytoremediation potential of Salix alba and *Salix viminalis*. Biomass Bioenergy 34(9):1410–1418

Munns R (2005) Genes and salt tolerance: bringing them together. New Phytol 167(3):645–663

Naidu R, Rengasamy P (1993) Ion interactions and constraints to plant nutrition in Australian sodic soils. Aust J Soil Res 31:801–819
Neill OGJ, Gordon AM (1994) The nitrogen filtering capability of Carolina poplar in an artificial riparian zone. J Environ Qual 23:1218–1223
Ni M, Leung YC, Leung MKH, Sumathy K (2006) An overview of hydrogen production from biomass. Fuel Process Technol 87(5):461–472
Oster JD, Shainberg I, Abrol IP (1999) Reclamation of salt affected soils. In: Skaggs RW, van Schilfgaarde J (eds) Agricultural drainage. ASA-CSSASSSA, Madison
Pandey VC, Bajpai O, Singh N (2016) Energy crops in sustainable phytoremediation. Renew Sust Energ Rev 54:58–73
Panetsos KP, Alizoti PG (1996) Biomass production from monoclonal and mixed clonal plantations. In: Proceedings 5th national renewable energy conference, vol 3. National Research Centre Democritus, Athens, pp 201–204
Panetsos KP (1981) Biomass yield of short rotation Platanus species in Greece. In: Commercialization of production of Mediterranean climatic regions. W. Junk Publications, The Hague, pp 235–238
Panetsos KP (1988) Biomass production from fast growing forest tree species. In: Proceedings 3rd national conference of renewable energy, vol 2. Greek Productivity Centre, Thessaloniki, pp 29–36
Pant D, Van Bogaert G, Diels L, Vanbroekhoven K (2010) A review of the substrates used in microbial fuel cells (MFCs) for sustainable energy production. Bioresour Technol 101:1533–1543
Popp J, Lakner Z, Rakos MH (2014) The effect of Bioenergy expansion: food, energy and environment. Renew Sust Energ Rev 32:559–578
Prasad MNV, Tewari JC (2016) *Prosopis juliflora* (Sw) DC: potential for bioremediation and bioeconomy. In: MNV P (ed) Bioremediation & bioeconomy. Elsevier, Amsterdam, pp 49–68
Qadir M, Ghafoor A, Murtaza G (2000) Amelioration strategies for saline soils: a review. Land Degrad Dev 11:501–521
Qadir M, Schubert S (2002) Degradation processes and nutrient constraints in sodic soils. Land Degrad Dev 13:275–294
Qadir M, Oster JD (2002) Vegetative bioremediation of calcareous sodic soils: history, mechanisms, and evaluation. Irrig Sci 21:91–101
Qadir M, Qureshi RH, Ahmad N (2002) Amelioration of calcareous saline sodic soils through phytoremediation and chemical strategies. Soil Use Manag 18:381–385
Qadir M, Noble AD, Oster JD, Schubert S, Ghafoor A (2005) Driving forces for sodium removal during phytoremediation of calcareous sodic and saline–sodic. Soil Use Manag 21:173–180
Qadir M, Oster JD, Schubert S, Noble AD, Sahrawat KL (2007) Phytoremediation of sodic and saline sodic soil. Adv Agron 96:197–247
Ravikumar M, Sarita P, Naga Raju GJ, Bhuloka Reddy S (2013) Trace element accumulation in the leaves of *Azadirachta indica* and *Pongamia glabra* collected from different environmental sites. J Environ Res Dev 7(3):S410–S414
Rabindranath D, Shroff S, Rao N (2010) Bioenergy in India: barriers and policy options. United Nations Development Programme, New Delhi
Reddy BVS, Ramesh S, Kumar AA, Wani SP, Ortiz R, Ceballos H, Sreedevi TK (2008) Biofuel crops research for energy security and rural development in developing countries. Bioenergy Resour 1:248–258
Richards LA (1954) Diagnosis and improvement of saline and alkali soils. Soil Sci 78(2):154
Saritz R (2005) Phytoextraction of uranium and thorium by native trees in a contaminated wetland. J Radioanal Nucl Chem 264:417–422
Sathitsuksanoh N, Zhu Z, Templeton N, Rollin JA, Harvey SP, Zhang Y-H (2009) Saccharification of a potential bioenergy crop, *Phragmites australis* (Common Reed), by lignocellulose fractionation followed by enzymatic activities at decreased cellulase loading. Ind Eng Chem Res 48:6441–6447

Shainberg I, Sumner ME, Miller WP, Farina MPW, Pavan MA, Fey MV (1989) Use of gypsum on soils: a review. Adv Soil Sci 9:1–111

Shaoa H, Chu L (2008) Resource evaluation of typical energy plants and possible functional zone planning in China. Biomass Bioenergy 32:283–288

Singh K, Singh B, Singh RR (2012) Changes in physico-chemical, microbial and enzymatic activities during restoration of degraded sodic land: ecological suitability of mixed forest over monoculture plantation. Catena 96:57–67

Singh SP, Singh V (2016) Addressing rural decline by valuing agricultural ecosystem services and treating food production as a social and treating food production as a social contribution. Trop Ecol 57(3):381–392

Singh YP, Mishra VK, Singh S, Sharma DK, Singh D, Singh US, Singh RK, Haefele SM, Ismail AM (2016) Productivity of sodic soils can be enhanced through the use of salt tolerant rice varieties and proper agronomic practices. Field Crop Res 190:82–90

Smith L, Thelen D, MacDonald J (2013) Yield and quality analyses of bioenergy crops grown on a regulatory brownfield. Biomass Bioenergy 49:123–130

Sposito G (2008) The chemistry of soils. Oxford University Press, New York

Sumner ME (1993) Sodic soils: new perspectives. Aust J Soil Res 31(6):683–750

Szabolcs I (1974) Salt affected soils in Europe. Martinus Nijhoff, The Hague, p 63

Thompson PL, Ramer LA, Schnoor JL (1998) Uptake and transformation of TNT by hybrid poplar trees. Environ Sci Technol 32(7):975–980

Tripathi KP, Singh B (2005) The role of revegetation for rehabilitation of sodic soils in semiarid subtropical forest, India. Restor Ecol 13:29–38

Tripathi V, Edrisi SA, Abhilash PC (2016) Towards the coupling of phytoremediation with bioenergy production. Renew Sust Energ Rev 57:1386–1389

United States Salinity Laboratory Staff (1969) Diagnosis and improvement of saline and alkali soils, Agriculture handbook no 60, Revised edn. United States Department of Agriculture, Washington DC

Verma M, Godbout S, Brar SK, Solomatnikova O, Lemnay SP, Larouche JP (2011). Biofuels production from biomass by thermochemical conversion technologies. Int J Chem Eng 2012:1–8 http://dx.doi.org/10.1155/2012/542426

Warren Spring Laboratory (1993) Fundamental research on the thermal treatment of wastes and biomass: literature review of part research on thermal treatment of biomass and waste ETSU, B/T1/00208/Rep/1

Wong L, Dalal C, Greene B (2008) Salinity and sodicity effects on respiration and microbial biomass of soil. Biol Fertil Soils 44:943–953

Yuan JS, Tiller KH, Ahmad HA, Stewart NR, Stewart CN Jr (2008) Plants to power: bioenergy to fuel the future. Trends Plant Sci 13(8):421–429

Zalesny RS, Bauer EO, Hall RB, Zalesny JA, Kunzman J, Rog CJ, Riemenschneider DE (2007) Clonal variation in survival and growth of hybrid poplar and willow in an IN SITU Trialon soils heavily contaminated with petroleum hydrocarbons. Int J Phytoremed 7(3):177–197

Zhao F, Yang W, Zeng Z, Li H, Yang X, He Z, Gu B, Rafiq MT, Peng H (2012) Nutrient removal efficiency and biomass production of different bioenergy plants in hypereutropic water. Biomass Bioenergy 42:212–218

Zhuang D, Jiang D, Liu L, Huang Y (2011) Assessment of bioenergy potential on marginal land in China. Renew Sust Energ Rev 15(2):1050–1056

Managing Waste Dumpsites Through Energy Plantations

15

Vimal Chandra Pandey

Abstract

Increase in the number and area of waste dumpsites is a major concern worldwide due to modern urbanization, industrial revolution and overpopulation growth. There is a wide range of waste dumpsites such as fly ash dumps, mine spoils, red mud deposits, sewage sludge, etc. These waste dumps are a source of heavy metal pollution, the degradation of soil and water system and serious dust pollution to atmosphere. These are also recognized as ecological, economic and social challenges. Thus, the management of waste dumpsites in a safe and economic way is still a worldwide challenge. In this direction, the phytomanagement of waste dumpsites through energy plantation is a holistic approach for ecological, economic and social sustainability.

Keywords

Energy plantations • Waste dumpsites • Sustainable environment • Nonedible crops

15.1 Introduction

Continuous increase in the chain of a wide range of waste dumps, i.e. coal mine spoils, fly ash landfills, red mud deposits, chromite–asbestos mine waste dumps and other waste dumps, is a critical challenge for remediation and management at the global level (Pandey and Singh 2012; Fuller et al. 1982; Kumar and Maiti 2015). It has been proved that these abandoned man-made waste dumps affect fauna health

V.C. Pandey (✉)
Plant Ecology and Environmental Science Division, CSIR-National Botanical Research Institute, Lucknow 226001, India
e-mail: vimalcpandey@gmail.com

K. Bauddh et al. (eds.), *Phytoremediation Potential of Bioenergy Plants*,
DOI 10.1007/978-981-10-3084-0_15

nearby areas, deteriorate the environmental quality and create severe disturbances of the cultural landscape (Bryan Jr. et al. 2012; Walker and Willig 1999). The key reason of waste dumps hazards is the presence of numerous potentially toxic metals, i.e. Cd, Cr, As, Pb, Hg, Ni, Se, etc. with high concentration (Scotti et al. 1999; Pandey et al. 2011). These waste dumps create air, soil and water pollution resulting in several diseases such as cancer, cardiac disease, genetic disorders, mesothelioma, pneumoconiosis, skin irritations and other respiratory problems in humans residing close to dumpsites (Pandey et al. 2009, 2011; USEPA 2007). Hence, there is a strong public pressure and need of present and future for remediation and management of a wide variety of waste dumpsites in a safe and economic way.

In this direction, phytoremediation is most accepted, cost-effective, eco-friendly, applicable to large areas and sustainable approach, which is used to remove the contaminants from the soil and water systems (Pandey et al. 2015a). There are some limiting factors that influence phytoremediation efficiency such as the plant ability to uptake contaminants, survivability in polluted lands and low bioavailability of the contaminants (Chirakkara and Reddy 2015). Plenty of economic plant species (including trees, grasses, shrubs and aquatic plants) are subjected to characterization of their potential for sustainable phytoremediation practices. A number of economically and ecologically important plants have the ability to tolerate and remediate contaminants from the polluted lands and therefore are categorized for commercial phytoremediation (Pandey et al. 2015a). In order to fully utilize the waste dumpsites and to overcome the disadvantage of phytoremediation, a new strategy of linking phytoremediation with energy crop cultivation, with a view to achieving low-cost remediation of dumpsites through the biofuel production (Pandey et al. 2016a). Therefore, the growing energy plantations on waste dumpsites to meet biofuel as a renewable energy without affecting agricultural lands are a pressing priority. Energy plantations on waste dumpsites are one of the most economic and eco-friendly ways of harnessing solar energy through the photosynthetic process. In addition to making biofuel energy availability from waste dumpsites, it can also improve the fertility of waste dumpsites. India and other countries have thousands of hectares of waste dumpsites which could be effectively used to grow energy plantations.

Therefore, the use of energy plants to remediate hazardous waste dumpsites is considered as a highly promising and holistic approach for improving the environmental quality of the waste dumpsites (Pandey et al. 2016a; Keene and Skousen 2010; Prasad 2015). There is an increasing interest in searching of energy plants growing on a wide range of waste dumpsites. However, the use of native energy plants for the phytoremediation of waste dumpsites and biofuel production can be a useful opportunity, since native plants are genetically tolerant to high metal concentrations of waste dumpsites and have an excellent adaptation to the environmental conditions of the region than the plants introduced from other environments (Pandey and Singh 2011; Pandey et al. 2015a, b; 2016a). The screening and conservation of the genetic resource of energy plant-based phytoremediators are of vast importance for the future development of effective plans for the determination of phytoremediation of waste dumpsites and biofuel production.

15.2 The Problems Associated to Waste Dumpsites

During opencast mining of coal, the overlying soil is removed, and the fragmented rock is heaped in the form of overburden dumps. Thus, the opencast coal mining generally releases a huge amount of mining wastes to the upper part of the land surface as overburden dump materials known as coal mine spoils (inevitable by-product in the mining process) (Ghosh 2002). These overburden dumps occupy a large amount of land, which loses its original use and generally gets soil qualities degraded (Barapanda et al. 2001). As these dumping sites are generally loose, its fine particles become highly susceptible to blowing by wind. These get spread over the surrounding vegetation and disturb their natural quality, growth of fresh leaves and plant yield. Coal mine spoil also raises a number of environmental challenges, including soil erosion, dust pollution, water pollution, loss of nutrient qualities and microbial activities of the soil system and ultimately impacts on local floral diversity. Generally, coal mine spoils contain elevated concentrations of potentially harmful elements such as As, Cd, Cu, Pb, Zn, etc. which provide usually a hostile substrate for plant growth because of low organic matter, low nutrient content, high concentrations of trace elements, as well as physical disturbance such as reduced soil depth, bad soil structure and low water availability. Furthermore, the physicochemical properties of overburden dump materials are site specific and differ from one dump to another due to different geological deposits of rocks/coal (Lovesan et al. 1998). Thus, these are the reasons why most of the overburden dumps do not support plantation.

Fly ash (FA) is a coal combustion residue of thermal power plants. Generally, it is alkaline in nature, but its pH can vary from 4.5 to 12.0 depending mainly on the S content of the parent coal (Plank and Martens 1974). FA consists of many small, glass-like particles ranging in size from 0.01 to 100 mm (Davison et al. 1974) and contains a number of toxic elements like Cr, Pb, Hg, As and Cd along with many essential elements such as S, B, Ca, Na, Fe, Zn, Mn and P (Pandey et al. 2009). FA particles are very fine and light in nature and have potential to get airborne and contaminate the environment near the dumping sites (Pandey et al. 2009, 2011). Being airborne for a long period, it causes irritation to the skin, eyes, nose, throat and respiratory tract to local people. Repeated inhalation of FA dust containing crystalline silica is a main reason of bronchitis, silicosis (scarring of the lung) and lung cancer of local residents near thermal power plants. Furthermore, its higher inhalation may result in asthma-like symptoms (Davison et al. 1974; Pandey et al. 2009). Besides being a health threat, FA corrodes exposed metallic structure in its vicinity (Davison et al. 1974). The leaching of toxic elements of FA may contaminate tube well and ground water that depends on the physicochemical characteristics of ash and hydrogeologic and climatic conditions of the disposal site (Pandey et al. 2009). It is reported that Zn, Cu and Pb were found in high concentration in tube well water located in vicinity of ash pond in Orissa, whereas Cu, Mn, Pb and Zn were the main pollutants in groundwater (Praharaj et al. 2002). Thus, FA has been well recognized as an environmental hazard worldwide. Its amount is increasing continuously to encroach thousands of hectare land nearby thermal power plants all over the world.

Red mud (RM) is one of the main waste products of the alumina industry. The characteristic of RM is an alkaline (pH>11), insoluble and iron-rich (usually 10–30%) waste product (Das and Thakur 1995). It is also known by other names as bauxite residue, red slime and red sludge. About ninety million tons of RM/annum are produced worldwide (Kumar et al. 2005). Huge dumping of RM is the most important concern of environmental hazards adjacent to alumina industries. The abandoned RM deposits are damaging the environment by various ways, thus causing deleterious effect on human health and livestock. Finally, the management of various waste dumpsites, i.e. coal mine spoils, fly ash landfills, red mud deposits, chromite–asbestos mine waste dumps and other waste dumps, is an urgent need to mitigate air and water pollution in vicinity of waste dumping sites from the mine operators, coal thermal power stations, alumina industries and planners. There is also a need to assess and report the status on the management of pollutants and efficiency of current mitigation practices for improving the quality of environment of waste dumpsites.

15.3 Strategies for Minimizing the Harsh Conditions of Dumpsites to Plant Growth

The key challenges of waste dumpsites' remediation are harsh conditions (i.e. high or low pH, poor nutrient status, water scarcity, higher temperature, toxic elements, etc.) which may not be favourable for the growth and establishment of plant communities. Thus, the remediation of waste dumpsite is a most challenging task for the scientific society and practitioners. The challenge of waste dumps' remediation could be solved by implementing "sustainable phytoremediation" technology. In recent years, sustainable phytoremediation is popularized (Witters et al. 2012; Rungwa et al. 2013; Pandey et al. 2015a) for remediation and management of environmental hazards. Sustainable development is a strategy that meets the needs of the present requirements without compromising the resources of future generations and fulfils their own needs. The sustainable production covers the topics such as eco-efficiency, waste management, use of natural resources and sustainable consumption by human beings (Blok et al. 2015). The scientific community from interdisciplinary field have made sufficient efforts for solving the problems of waste dumps' remediation (Juwarkar and Jambhulkar 2008; Rau et al. 2009; Srivastava et al. 2014; Krzaklewski et al. 2012; Pandey et al. 2009; Pandey and Singh 2012). Rau et al. (2009) assessed the functional diversity of rhizobacterial taxa of a wild grass *Saccharum ravennae* colonizing abandoned fly ash dumps. Sharma et al. (2011) identified and characterized plant growth-promoting rhizobacteria associated with naturally grown *Saccharum munja* on mine spoil that can potentially increase its rate of colonization. Babu and Reddy (2011) studied about the dual inoculation of arbuscular mycorrhizal and phosphate-solubilizing fungi which contributes in sustainable maintenance of plant health in FA dumps. Pandey et al.

(2012a) suggested that *S. munja* is a wild native perennial grass which naturally colonizes extreme habitat of FA dumps. It would be potentially useful to revegetate the FA dumps more efficiently if it is introduced properly on fresh dumps to convert barren FA deposits into ecologically and socio-economically productive habitats. It is used by local people for making ropes, baskets, mats, huts, etc. to support their livelihood. Recently, Pandey et al. (2015c) also assessed the suitability of *Saccharum spontaneum* for the restoration and stabilization of bare FA dumps. The highest importance value index, visual observations and analytical results revealed that *S. spontaneum* has great ability to grow and establish on bare FA dumps and can be used as a valuable genetic resource in restoration of vast tracts of FA dumps across the world. Thus, the visual observations and field results showed that *S. spontaneum* is a promising and potential tall grass for the restoration of FA dumps.

However, some important new approaches based on previous experiments, proposed for the remediation of waste dumpsites, are (1) selection of the suitable energy plant species having capability to establish on these sites, (2) addition of organic manure as a soil ameliorant to enhance the soil nutrient status and (3) addition of native and viable microbial agents/inoculants to promote the plant growth and establishment. The selected plants for sustainable phytoremediation of waste dumpsite must be perennial in nature, unpalatable to livestock, high economic value, tolerant to adverse conditions, easy to harvest and ecologically beneficial (Pandey and Singh 2011; Pandey et al. 2015a). Incorporation of locally available blending organic amendments such as pressmud, farm yard manure, vermicomposting, composted cow dung, etc. in waste dumpsites may be cost-effective which can provide favourable and improved soil conditions for establishment of plant communities, microbes and plant–microbe interactions naturally. The native microbial communities well adapted to adverse conditions of waste sites can be selected and developed into potential inoculants that can be exploited for restoration of unused waste dumpsites into ecologically and socio-economically productive habitats.

15.4 Why Energy Plantation for the Restoration of Waste Dumpsites

Energy plantation is one that grows plant materials (e.g. biodiesel from seed or methanol from plant biomass) for their fuel value. The energy plantation on waste dumpsites is a good idea for the solution of two universal problems such as utilization of waste dumpsites and biofuel production. In other words, the energy plantation is also helpful in the restoration of waste dumpsites. The large areas that are needed for energy plantations may be obtain by utilization of waste dumpsites that are easily available in most of the countries. Thus, this type of energy plantations may not compete with food or other crops for land. Energy plantation on various waste dumpsites with different perspectives is well described in Table 15.1.

In the discussion of energy plantation on waste dumpsites, a proper goal must be focused with advantages and disadvantages. Technologically, energy plantation is a renewable and sustainable source of energy in usable forms. Ecologically, it is much

Table 15.1 Energy plantation on various waste dumpsites

SN	Energy plants	Waste dumpsites	Remarks	References
1	*Panicum virgatum* (switchgrass)	Coal mine spoil	Biofuel production	Keene and Skousen (2010)
2	*Helianthus annuus* (sunflower)	Coal mine spoil	Biofuel production	Harris et al. (2015)
3	*Jatropha curcas*	Fly ash	Phytoremediation	Jamil et al. (2009)
4	*Ricinus communis*	Fly ash-contaminated sites	Phytoremediation and energy perspective	Pandey (2013)
5	*Jatropha curcas*	Fly ash-amended soil	Growth performance	Bagchi (2013)
6	*Pongamia pinnata*	Coal mine spoil	Bio-reclamation	Chaubey et al. (2012) and Arshi (2015)
7	*Azadirachta indica*	Coal mine spoil	Bio-reclamation	Chaubey et al. (2012) and Arshi (2015)
8	*Pongamia pinnata*	Fly ash-amended soil	Growth performance and restoration potential	Singh (2013)
9	*Pongamia pinnata*	Fly ash dumps	Revegetation	Pandey et al. (2016b)
10	*Pennisetum purpureum*	Phosphatic clay	Energy production	Stricker et al. (1998)
11	*Saccharum* species	Phosphatic clay	Energy production	Stricker et al. (1998)
12	*Erianthus arundinaceum*	Phosphatic clay	Energy production	Stricker et al. (1998)
13	*Sorghum bicolor*	Phosphatic clay	Energy production	Stricker et al. (1998)
14	*Prosopis juliflora*	Fly ash dumps	Revegetation	Pandey et al. (2016b)
15	*Pongamia pinnata*	Coal mine spoil	Ecological restoration	Singh and Singh (2006)
16	*Azadirachta indica*	Coal mine spoil	Ecological restoration	Singh and Singh (2006)
17	*Populus* species	Wight industrial wastes	Remediation	Giacchetti and Sebastiani (2006)
18	*Jatropha curcas*	Coal mine overburden dump	Bio-reclamation	Arshi (2015)
19	*Saccharum spontaneum*	Fly ash dumps	Revegetation and multiple use	Pandey et al. (2015c)
20	*Saccharum spontaneum*	Coal mine overburden dump	Reclamation	Arshi (2015)
21	*Madhuca latifolia*	Coal mine overburden dump	Reclamation	Arshi (2015)

(continued)

Table 15.1 (continued)

SN	Energy plants	Waste dumpsites	Remarks	References
22	*Miscanthus giganteus*	Fly ash deposits	Biomass production	Lixandru et al. (2013)
23	*Panicum virgatum* (switchgrass)	Fly ash + poultry litter-amended soil	Biofuel feedstock production	Dzantor et al. (2015)
24	*Tripsacum dactyloides (eastern gama grass)*	Fly ash + poultry litter-amended soil	Biofuel feedstock production	Dzantor et al. (2015)

sounder than the conventional ones (fossil fuels). In fact, energy plantations will lead to long-term benefits by way of remediation and utilization of waste lands, in addition to providing local employment. Ecologically speaking, the major advantages of energy plantations on waste dumpsites are biodiversity enrichment, substrate fertility improvement, carbon sequestration, remediation of pollutants, etc. In short, energy plantations on waste dumpsites have a significant role in keeping the environment clean while supporting the livelihood (Fig. 15.1).

With the developing of energy plantation on a wide range of waste dumps sites under afforestation programme while protecting the natural forests, the Government of India should recruit policy-oriented afforestation schemes. To support this programme, energy plantations on waste dumpsites should be link with various development schemes such as Rural Landless Employment Guarantee Scheme (RLEGS), National Rural Employment Programme (NREP), Employment Guarantee Scheme (EGS), Drought Prone Areas Programme (DPAP), Western Ghats Development Scheme, etc. The objective of energy plantation on waste dumpsites should be to increase the supply of biofuel, small timbers, utilization of waste dumps, remediation of pollutants and clean environments while generating rural employment and maintaining environmental stability.

Energy crops are specifically grown as an energy plantation to produce some form of energy. Energy plantations are generally divided into two types: herbaceous and woody plantation.

Perennial grasses such as switchgrass, *Miscanthus*, etc. can be used as a potential energy crops for herbaceous energy plantation. These perennial energy grasses have the capacity to regenerate from their roots or buds at the base of the plant, offer the greatest potential for energy plantation on waste dumpsites and therefore do not require replanting for long periods of time. The second type of energy plantations is short rotation woody plantation. It includes *Salix*, poplar, etc. A number of energy crops have been considered on the waste dumpsites of phosphatic clay (a by-product of phosphate mining) through demonstration project. These crops are elephant grass (*Pennisetum purpureum*), energy cane (*Saccharum* sp.), sugarcane (*Saccharum* sp.), *Erianthus* (*Erianthus arundinaceum*), sweet sorghum (*Sorghum bicolor*) and *Leucaena* (*Leucaena leucocephala*) (Stricker et al. 1998). Lixandru et al. (2013) assessed the adaptation process of the species *Miscanthus giganteus* under the conditions of FA deposit from Utvin, Timis County, and obtained best germination

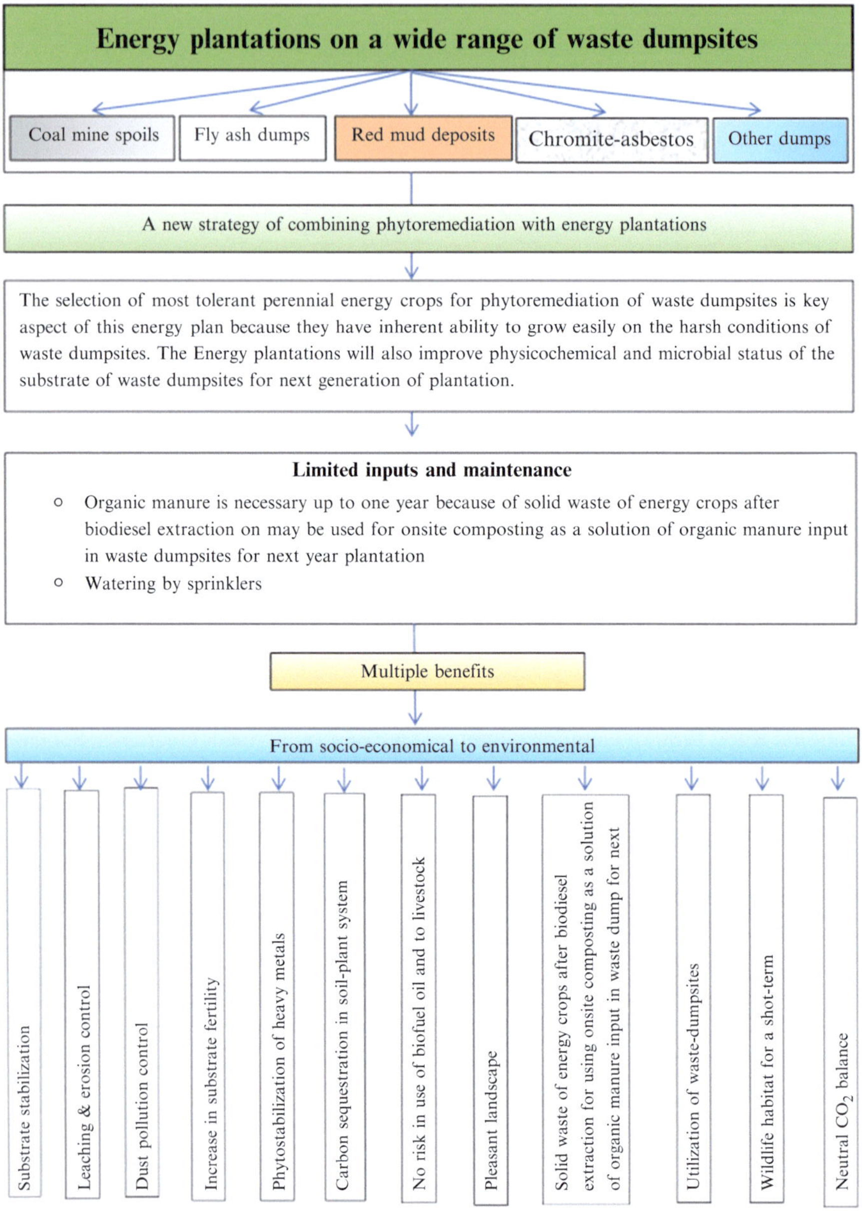

Fig. 15.1 Energy plantations on a wide range of waste dumpsites for multiple benefits from socio-economical to environmental

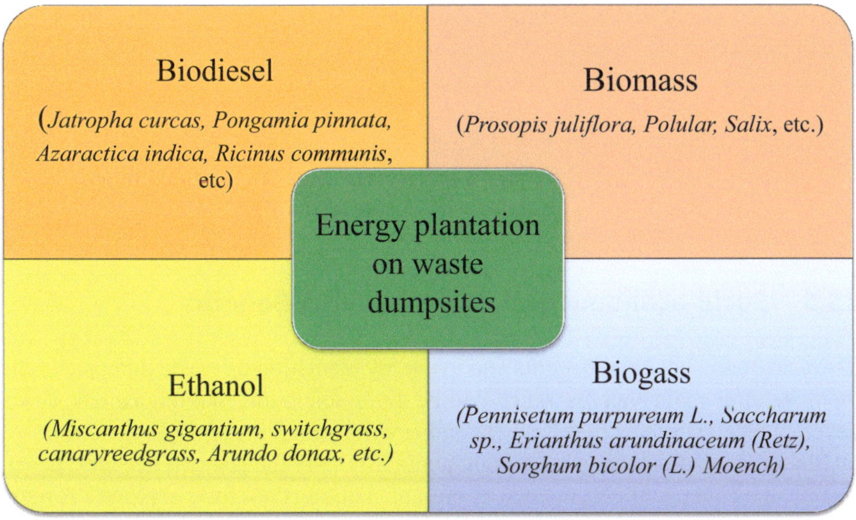

Fig. 15.2 Various categories of bioenergy source obtained from potential energy crops

percentage with sewage sludge and cattle manure. A good growth rate and higher biomass production were also recorded. Therefore, *M. giganteus* has good adaptation potential to arid conditions of the FA deposits and can be used for revegetation/phytoremediation of FA deposits with bioenergy feedstock production.

The waste dumpsites cause environmental impacts including leachate generation, toxic element pollution, dust pollution, greenhouse gas emissions, etc. Generally, revegetation is the traditional practice on waste dumpsites to lessen environmental pollution and their degradation. It can provide a source of biomass for energy production. Produced biomass from the revegetation of waste dumpsites can be converted to bioenergy through biochemical and thermochemical processes. Thus, the selection of energy plants for the revegetation of waste dumpsites will be more beneficial than traditional revegetation. Energy plants may be biomass or seed-producing plants, but it should be high-yielding crops (Fig. 15.2).

Biofuel and remediation benefits are the main considerations for selecting energy crops for energy plantations on waste dumpsites. Energy plantations on waste dumpsites are also helpful for beautification, to improve the microclimate, arrest soil erosion and many other functions. Energy crops provide an excellent ambience to the site, either residential or work areas. Energy crops absorb carbon dioxide, thereby reducing the harmful effects of air pollution. Thus, energy crops act as lungs to purify air and keep the surroundings cool, healthy and beautiful. The selection of suitable energy crops according to waste dumpsites will enhance the aesthetic value and beauty of the surroundings.

For the revegetation of waste dumpsites and bioenergy production, the energy plants should be some desirable characteristics such as high-yielding crops, high growth rate, extensive root system, perennial nature, pest and disease resistant, unpalatable, regeneration potential and tolerant to adverse conditions. Furthermore, energy plant species should also be based on the quality of waste dumpsites, availability of moisture, suitability of climate and gestation period. Therefore, energy plants having these characters should be recommended for revegetation of waste dumpsites and biofuel production. This chapter provides an overview of the managing waste dumpsites through energy plantations. The next section will discuss how energy crops can be used to generate socio-economic and environmental benefits.

15.5 Socio-economic and Environmental Benefits

There are many potential benefits due to energy plantation on waste dumpsites. But there are three main benefits which will be discussed in this section, namely, social development, economic development and environmental benefits (Table 15.2).

Phytoremediation has gained much popularity over the last two decades; nevertheless, the concept of commercial phytoremediation is not fully explored. A number of companies have started phytoremediation business in the USA, Canada and Europe. They are recognized as phytoremediation industries, namely, Applied Natural Sciences, Earth Care, Ecolotree, Phytokinetics, Phytotech, PhytoWorks, Thomas Consultants and Verdant Technologies (Glass 1999). Several projects

Table 15.2 Potential benefits of energy plantation on waste dumpsites can be social, economic, as well as environmental

SN	Social aspects	Economic aspects	Environmental aspects
1	Creation of employments and livelihoods	Phytoremediation cost-effective approach	Reduced air, soil and water pollution
2	Support for local communities	Carbon credits	Reduced soil erosion
3	Decreasing our dependency on foreign oil	Bioenergy market	Increased terrestrial carbon sinks
4	Help in basic services (pumped water, etc.	Plantation reduces input and maintenance of waste dumpsites	Enhanced biodiversity
5	Enhanced societal status	Enhance local economy by cycling of goods and services within local community	Reduced pressure on finite natural resources
6	Pride and freedom	In form of less effort (labour) in beatification of waste dumping sites	Improved substrate quality of waste dumpsites
7	Improved social unity	Income from developed park on waste dumpsites though local people	Reduced GHG emissions by using bioenergy
8	No competition between food and bioenergy for land	Developed park on waste dumpsites may attract tourism	Providing habitat for local species

Based on IEA 2005

related to phytoremediation have been successfully completed all over the world. These phytoremediation industries will increase in the future, after the development of commercial phytoremediation. It may be possible if we integrate phytoremediation with energy crops. These phytoremediation companies will generate employment for our society through nursery raising by seedling/cutting of energy crops, energy plantation, collection of seeds, de-shelling, oil extraction, oil storage, etc. Energy plantation on waste dumpsites may offer income-generating opportunities for land owners and local people, as well as promote smallholder participation in bioenergy crop production (Pandey et al. 2012b; 2016a). Energy security through the revegetation of waste dumpsites by energy crops is another important consideration or additional benefits of renewable energy. The energy plantations on waste dump sites create jobs in local areas, most of them for unskilled workers. The energy plantation on waste dumpsites could increase our energy security by decreasing our dependency on foreign oil.

Environmental benefits from the energy plantation on waste dumpsites include water quality improvements and remediation of pollutants, i.e. metal(oid)s, pesticides, radioactive elements, soil erosion, aesthetic environments, etc. (Pandey et al. 2012b; 2016a). The environmental benefits of energy plantation on waste dumpsites are given in Table 15.2. Energy crops act as filter systems, removing pollutants from irrigated landfill leachate before it pollutes groundwater or rivers. Due to filtering capabilities, energy crops are being considered as a supporting crop to be planted with traditional crops for pollution control (Shrive et al. 1994; Giacchetti and Sebastiani 2006; Zalesny et al. 2007; Aryal and Reinhold 2015). Furthermore, the gradual improvements in substrate quality after energy plantation on coal mine spoil were due to the control of soil erosion and buildup of organic matter and humus with increasing the age of energy plantations (Chaubey et al. 2012). Similar outcomes were obtained by Jain et al. (2009), Juwarkar and Jambhulkar (2009) and Nath (2009). While expanding the bioenergy programs, energy plantation on waste dumpsites has considerable potential without affecting food crop production (disturbing food security) because agricultural land is not used for energy plantation. The most important point of this bioenergy programs is no competition between food and bioenergy for land. Energy crops can be planted on a wide range of underutilized waste dumpsites such as coal fly ash dumpsites, coal mine spoils, sewage sludge, red mud deposits, phosphatic clay, etc. for revegetation programs with the objective of bioenergy production. Therefore, it should be estimated by major countries that are involved in dumping of waste. It may be accounted as thousand million acres of waste dumpsites are potentially suitable for energy crop plantation all over the world.

Another environmental benefit of using energy crops is a decrease in CO_2 emissions in comparison to fossil fuels. The energy plants grown for biomass absorb the amount of CO_2 released during their combustion for bioenergy (WRBEP 1995). Therefore, there is no net CO_2 generated because the emitted amount of CO_2 in its combustion has been earlier absorbed during its growth. The emitted amount of CO_2 by using biomass of energy crops is considerably less in comparison to coal, petroleum and natural gas, which is represented in Fig. 15.3 (MBEP 2002).

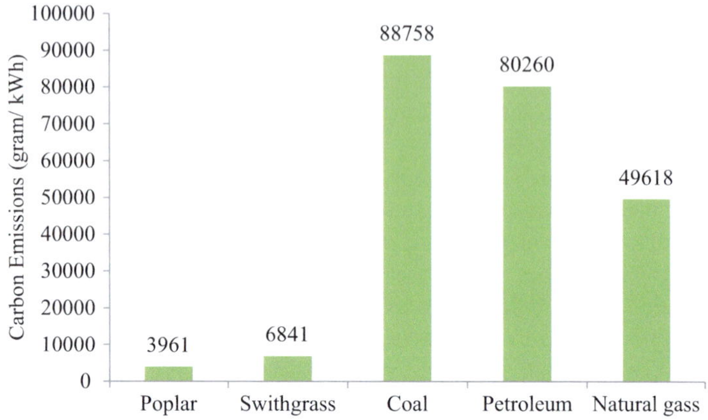

Fig. 15.3 Carbon emissions by different energy sources in energy production (Note: Carbon emissions includes production, transportation and conversion processes) (Source: MBEP 2002)

Furthermore, it has been estimated that the growing of energy crops on degraded lands facilitates soil carbon sequestration and later the improvement of soil quality (Mi et al. 2014). Therefore, bioenergy production on waste dumpsites holds a great potential for reducing toxic element pollution, sequestering atmospheric CO_2 in soil system and increasing economy as green energy. It has been increasingly recognized that planting energy crops on waste dumpsites provides multiple benefits towards social, economic and environmental (Pandey et al. 2016a).

15.6 Promotion of Energy Plantation on Waste Dumpsites

There are several benefits that could be realized by growing energy plantation on waste dump sites. However, the acceptance of energy plantation on waste dumpsites can be encouraged through additional funding, awareness and policy. Extra funding should be provided for the research of energy plantation on waste dumpsites. As a model of demonstration sites of energy plantation on waste dumpsites is needed, so owners, practitioners and researchers can see the technology at work and gain experience. Improved planting and harvesting method of energy plantation on waste dumpsites should be investigated. Focus should be intended towards generating awareness for energy plantation on waste dumpsites. Developing educational programs for responsible land owners and practitioners is an urgent need to make this energy programs on waste dumpsites. Policies could be structured to "value" the benefits of using bioenergy produced from waste dumpsites. To encourage the activities on these waste dumpsites, subsidies should be provided.

15.7 Conclusions and Future Prospects

Currently, a number of energy crops are being investigated for phytoremediation and bioenergy production, but those included (*Jatropha curcas*, *Pongamia pinnata*, *Azadirachta indica*, *Ricinus communis*, *Miscanthus species*, *Panicum virgatum* (switchgrass), *Phalaris arundinacea* (reed canary grass), *Arundo donax*, *Populus* species, *Salix*) have been hyped as the potential energy crops for linking phytoremediation with bioenergy production as well as the most widespread possibilities. Other energy crops may be better suited for a specific area but are not as appropriate for extensive use. A number of factors (i.e. fast-growing nature, high biomass-producing capacity, profuse root system, tolerant to adverse condition, unpalatable, regeneration potential, pest and disease resistant) should be considered into account when selecting an energy crop for phytoremediation of waste dumpsites. *J. curcas*, *R. communis*, *Populus* sp., *A. donax* and *Miscanthus* sp. are the potential energy crops for phytoremediation of waste dumpsites and bioenergy production due to its pure hardiness and stress-tolerant ability. These five energy crops grow fast with little maintenance on waste dumpsites as it requires minimal inputs for its establishment and may be recommended as energy plantations on waste dumpsites in India and abroad. Therefore, the revegetation of waste dumpsites with energy crops offers an eco-friendly and cost-effective approach for phytoremediating waste dumpsites with energy security. Further research as demonstration projects is needed to establish what energy crops can be successfully grown and used for energy generation as well as phytoremediation of waste dumpsites.

While promoting the energy plantation on waste dumpsites, the social, economic and environmental perspectives should be considered for sustainability issues. During energy plantation on waste dumpsites, phytoremediation and bioenergy production should be the main goal, while beautification, carbon sequestration and improving the substrate quality of waste dumpsites are the other consideration. For the success of energy plantation programs on waste dumpsites, income being the most important consideration, social and environmental aspects is also another important consideration. Cultivation and harvesting practices may vary for different energy crops that may affect biomass or their yields. Environmental conditions as well as harsh conditions of waste dumpsites must be assessed prior to farm-scale application. Furthermore, the application of PGPMs and suitable agronomic practices is the main strategy for enhancing energy crop yield during phytoremediation. Furthermore, the revegetation programs to promote energy plantation on waste dumpsites should be based on well-tested technical and economic data to guide the land owners, practitioners and general public in the right direction.

Acknowledgments Financial assistance given by the Science and Engineering Research Board (No.SR/FTP/ES-96/2012), Government of India, is gratefully acknowledged. Authors are also thankful to the Director and Head of the Plant Ecology and Environmental Science Division, CSIR-National Botanical Research Institute, Lucknow, for his kind support.

References

Arshi A (2015) Reclamation of coalmine overburden dump through environmental friendly method. Saudi J Biol Sci. doi:10.1016/j.sjbs.2015.09.009

Aryal N, Reinhold DM (2015) Reduction of metal leaching by poplars during soil treatment of wastewaters: small-scale proof of concept studies. Ecol Eng 78:53–61

Babu AG, Reddy MS (2011) Dual inoculation of arbuscular mycorrhizal and phosphate solubilizing fungi contributes in sustainable maintenance of plant health in fly ash ponds. Water Air Soil Pollut 219:3–10

Bagchi SS (2013) Study of *Jatropha Curcas* growth in fly ash amended soil. Univers J Environ Res Technol 3(3):364–374

Barapanda P, Singh SK, Pal BK (2001) Utilization of coal mining wastes: an overview. In: National seminar on environmental issues and waste management in mining and allied industries. Regional Engineering College, Rourkela, pp 177–182

Blok V, Wesselink R, Studynka O, Kemp R (2015) Encouraging sustainability in the workplace: a survey on the pro-environmental behaviour of university employees. J Clean Prod 106:55–67

Bryan AL Jr, Hopkins WA, Parikh JH, Jackson BP, Unrine JM (2012) Coal fly ash basins as an attractive nuisance to birds: parental provisioning exposes nestlings to harmful trace elements. Environ Pollut 161:170–177

Chaubey OP, Bohre P, Singhal PK (2012) Impact of bio-reclamation of coal mine spoil on nutritional and microbial characteristics – a case study. Int J Bio-Sci Bio-Technol 4(3):60–80

Chirakkara R, Reddy K (2015) Plant species identification for phytoremediation of mixed contaminated soils. J Hazard Toxic Radioact Waste 19(4):04015004. doi:10.1061/(ASCE)HZ.2153-5515.0000282

Das SN, Thakur RS (1995) International series on environment, red mud analysis and utilization. Publication and information Directorate, New Delhi

Davison RL, Natusch DFS, Wallace JR, Evans CA Jr (1974) Trace elements in fly ash: dependence of concentration on particle size. Environ Sci Technol 8:1107–1113

Dzantor EK, Adeleke E, Kankarla V, Ogunmayowa O, Hui D (2015) Using coal fly ash in agriculture: combination of fly ash and poultry litter as soil amendments for bioenergy feedstock production. CCGP 7:33–39

Fuller R, Nelson E, Richardson C (1982) Reclamation of red mud (bauxite residues) using alkaline-tolerant grasses with organic amendments. J Environ Qual 11:533–539

Ghosh R (2002) Land use in mining areas of India. Envis Monograph No. 9 by CME, ISSN: 0972 4656

Giacchetti G, Sebastiani L (2006) Metal accumulation in poplar plant grown Wight industrial wastes. Chemosphere 64:446–454

Glass DJ (1999) Current market trends in phytoremediation. Int J Phytoremediation 1(1):1–8. doi:10.1080/15226519908500001

Harris TM, Zaimes GG, Khanna V, Landis AE (2015) Sunflower cultivation on coal mine refuse piles in Appalachia for diesel biofuel production from a life-cycle perspective. Procedia Eng 118:869–878. Conference paper, International Conference on Sustainable Design, Engineering and Construction, ICSDEC 2015; Hyatt Regency McCormick Place Chicago; United States; 10 May 2015 through 13 May 2015

IEA (2005) Benefits of bioenergy. International Energy Agency, http://www.ieabioenergy.com/library/179_Benefits of Bioenergy.PDF

Jain A, Bhowmik AK, Nath V, Banerjee SK (2009) Impact of plantation on ecosystem development in drastically disturbed coal mine overburden spoils. In: Chaubey OP, Bahadur V, Shukla PK (eds) Sustainable rehabilitation of degraded ecosystems. Aavishkar Publishers Distributors, Jaipur, pp 90–206

Jamil S, Abhilash PC, Singh N, Sharma PN (2009) *Jatropha curcas*: a potential crop for phytoremediation of coal fly ash. J Hazard Mater 172:269–275

Juwarkar AA, Jambhulkar HP (2008) Restoration of fly ash dumps through biological interventions. Environ Monit Assess 139:355–365
Juwarkar AA, Jambhulkar PH (2009) Reclamation of mine spoil dump using integrated biotechnological approcach at Sasti coal mine, Maharashtra. In: Chaubey OP, Bahadur V, Shukla PK (eds) Sustainable rehabilitation of degraded ecosystems. Aavishkar Publishers Distributors, Jaipur, pp 92–108
Keene T, Skousen J (2010) Mine soil reclamation with switchgrass for biofuel production. Joint Mining Reclamation Conference on 2010 – 27th Meeting of the ASMR, 12th Pennsylvania Abandoned Mine Reclamation Conference and 4th Appalachian Regional Reforestation Initiative Mined Land Reforestation Conference, volume 1, 2010, pp 489–503
Krzaklewski W, Pietrzykowski M, Wos B (2012) Survival and growth of alders (*Alnus glutinosa* (L.) Gaertn and *Alnus incana* (L.) Moench) on fly ash technosols at different substrate improvement. Ecol Eng 49:35–40
Kumar A, Maiti SK (2015) Effect of organic manures on the growth of *Cymbopogon citrates* and *Chrysopogon zizanioides* for the phytoremediation of chromite-asbestos mine waste: a pot scale experiment. Int J Phytoremediation 17(5):437–447. doi:10.1080/15226514.2014.910174
Kumar R, Alex TC, Khan ZH, Mahapatra SP, Mehrotra SP (2005) Mechanical activation of bauxite – potential and prospects in the Bayer process. In: Kvande H (ed) Lightmetals. The Minerals, Metals and Materials Society, Warendale, pp 77–79
Lixandru B, Dragomir N, Morariu F, Popa M, Coman A, Savescu N, Mâsu S, Popescu D (2013) Researches regarding the adaptation process of the species *Miscanthus Giganteus* under the conditions of fly ash deposit from Utvin, Timis County. *Scientific Papers:* Anim Sci Biotechnol 46(1):199–203
Lovesan VJ, Kumar N, Singh TN (1998) Effect of the bulk density on the growth and biomass of the selected grasses over overburden dumps around coal mining areas. In: Proceedings of the 7th national symposium on environment, Dhanbad. pp 182–185
MBEP (Michigan Biomass Energy Program) (2002) Energy crops and their potential development in Michigan. A report of the Michigan Biomass Energy Program August, 2002
Mi J, Liu W, Yang W, Yan J, Li J, Sang T (2014) Carbon sequestration by *Miscanthus* energy crops plantations in a broad range semi-arid marginal land in China. Sci Total Environ 496:373–380
Nath S (2009) Ecosystem approach for mined land rehabilitation and present rehabilitation scenario in Jharkhand coal mines. In: Chaubey OP, Bahadur V, Shukla PK (eds) Sustainable rehabilitation of degraded ecosystems. Aavishkar Publishers Distributors, Jaipur, pp 46–66
Pandey VC (2013) Suitability of *Ricinus communis L.* cultivation for phytoremediation of fly ash disposal sites. Ecol Eng 57:336–341
Pandey VC, Singh K (2011) Is *Vigna radiata* suitable for the revegetation of fly ash landfills? Ecol Eng 37:2105–2106
Pandey VC, Singh B (2012) Rehabilitation of coal fly ash basins: current need to use ecological engineering. Ecol Eng 49:190–192
Pandey VC, Abhilash PC, Singh N (2009) The Indian perspective of utilizing fly ash in phytoremediation, phytomanagement and biomass production. J Environ Manag 90:2943–2958
Pandey VC, Singh JS, Singh RP, Singh N, Yunus M (2011) Arsenic hazards in coal fly ash and its fate in Indian scenario. Resour Conserv Recycl 55:819–835
Pandey VC, Singh K, Singh RP, Singh B (2012a) Naturally growing *Saccharum munja* on the fly ash lagoons: a potential ecological engineer for the revegetation and stabilization. Ecol Eng 40:95–99
Pandey VC, Singh K, Singh JS, Kumar A, Singh B, Singh RP (2012b) *Jatropha curcas*: a potential biofuel plant for sustainable environmental development. Renew Sust Energ Rev 16:2870–2883
Pandey VC, Pandey DN, Singh N (2015a) Sustainable phytoremediation based on naturally colonizing and economically valuable plants. J Clean Prod 86:37–39

Pandey VC, Prakash P, Bajpai O, Kumar A, Singh N (2015b) Phytodiversity on fly ash deposits: evaluation of naturally colonized species for sustainable phytorestoration. Environ Sci Pollut Res 22:2776–2787
Pandey VC, Bajpai O, Pandey DN, Singh N (2015c) *Saccharum spontaneum*: an underutilized tall grass for revegetation and restoration programs. Genet Resour Crop Evol 62(3):443–450
Pandey VC, Bajpai O, Singh N (2016a) Energy crops in sustainable phytoremediation. Renew Sust Energ Rev 54:58–73
Pandey VC, Bajpai O, Singh N (2016b) Plant regeneration potential in fly ash ecosystem. Urban For Urban Green 15:40–44
Plank CO, Martens DC (1974) Boron availability as influenced by application of fly ash to soil. Soil Sci Soc Am Proc 39:974–977
Praharaj T, Swain SP, Powell MA, Hart BR, Tripathi S (2002) Delineation of ground water contamination around an ash pond geochemical and GIS approach. Environ Int 27:631–638
Prasad MNV (2015) Phytoremediation crops and biofuels. Sust Agric Rev 17:159–261
Rau N, Mishra V, Sharma M, Das MK, Ahaluwalia K, Sharma RS (2009) Evaluation of functional diversity in rhizobacterial taxa of a wild grass (*Saccharum ravennae*) colonizing abandoned fly ash dumps in Delhi urban ecosystem. Soil Biol Biochem 41:813–821
Rungwa S, Arpa G, Sakulas H, Harakuwe A, Timi D (2013) Phytoremediation – an eco-friendly and sustainable method of heavy metal removal from closed mine environments in Papua New Guinea. Procedia Earth Planet Sci 6:269–277
Scotti A, Silva S, Botteschi G (1999) Effect of fly ash on the availability of Zn, Cu, Ni and Cd to chicory. Agric Ecosyst Environ 72:159–163
Sharma M, Mishra V, Rau N, Sharma RS (2011) Functionally diverse rhizobacteria of *Saccharum munja* (a native wild grass) colonizing abandoned morrum mine in Aravalli hills (Delhi). Plant Soil 341:447–459
Shrive SC, McBride RA, Gordon AM (1994) Photosynthetic and growth responses of two broadleaf tree species to irrigation with municipal landfill leachate. J Environ Qual 23:534–542
Singh A (2013) Comparative performance and restoration potential of *Jatropha curcas* and *Pongamia pinnata* on degraded soil of North India. Thesis submitted to University of Lucknow for the degree of Ph.D.
Singh AN, Singh JS (2006) Experiments on ecological restoration of coal mine spoil using native trees in a dry tropical environment, India: a synthesis. New For 31:25–39
Srivastava NK, Ram LC, Masto RE (2014) Reclamation of overburden and lowland in coal mining area with fly ash and selective plantation: a sustainable ecological approach. Ecol Eng 71:479–489
Stricker JA, Prine GM, Anderson DL, Shibles DB, Riddle TC (1998) Energy from crops: production and management of biomass/energy crops on phosphatic clay in Central Florida. CIR 1084, one of a series of the Agronomy Department, UF/IFAS Extension. Original publication date October 1998. Revised June 2004.Reviewed January 2015. Visit the EDIS website at http://edis.ifas.ufl.edu
US EPA (2007) Human and ecological risk assessment of coal combustion wastes. US EPA (draft)
Walker LR, Willig MR (1999) An introduction to terrestrial disturbances. In: Walker L (ed) Ecosystems of disturbed ground. Elsevier, Amsterdam, pp 1–16
Witters N, Mendelsohn R, Van Slycken S, Weyens N, Schreurs E, Meers E, Tack F, Carleer R, Vangronsveld J (2012) Phytoremediation, a sustainable remediation technology? Conclusions from a case study. I: energy production and carbon dioxide abatement. Biomass Bioenergy 39:454–469
WRBEP (Western Regional Biomass Energy Program) (1995) Environmental issues. Biomass Digest 4(2):3
Zalesny JA, Zalesny RS Jr, Coyle DR, Hall RB (2007) Growth and biomass of Populus irrigated with landfill leachate. For Ecol Manag 248:143–152

Biotechnological Intervention to Enhance the Potential Ability of Bioenergy Plants for Phytoremediation

16

Gulshan Singh, Poonam Singh, Abhishek Guldhe, Thor Axel Stenström, Faizal Bux, and Sheena Kumari

Abstract

Phytoremediation has emerged as an attractive idea for the utilization of plants and the microbial communities for the environmental cleanup. Despite of advances made in this research area, there are still blockades for its implication on a wide scale. These hurdles include limited metabolic rate of plants and failure to break down numerous complex compounds or to tolerate/accumulate heavy metals. The in-depth knowledge of the factors affecting contaminant's translocation, volatilization, uptake, bioavailability and degradation is required, in order to increase the phytoremediation potential of bioenergy crops. The use of bioenergy crops has offered a viable option for phytoremediation that can aid to the energy supply and showed a pivotal role in meeting the targeted goals for the use of renewable energy sources. Moreover, genetic engineering has opened new avenues in this research area, by offering the chance for the direct gene transfer to enhance bioenergy crops/plant capabilities for environmental cleanup. The advanced "omics" methods will increase our understanding towards integrated activity patterns between plants and associated microorganism and harness it to the growth, structural organization of microbial communities, accumulation, tolerance and detoxification to increase phytoremediation capability of the plant. In this chapter, the mechanism of phytoremediation and new high-throughput biotechnological strategies adopted to enhance the ability of phytoremediation potential of bioenergy crops have been described with challenges harnessing plants in phytoremediation.

G. Singh • P. Singh • A. Guldhe • T.A. Stenström • F. Bux • S. Kumari (✉)
Institute for Water and Wastewater Technology, Durban University of Technology, P. O Box 1334, Durban 4000, South Africa
e-mail: sheenas@dut.ac.za

K. Bauddh et al. (eds.), *Phytoremediation Potential of Bioenergy Plants*,
DOI 10.1007/978-981-10-3084-0_16

Keywords
Bioenergy • Genetic manipulation • Glutathione • Metagenomics • Phytoaccumulation

16.1 Introduction

The idea of utilizing plants for remediation emerged when the plant's potential to accumulate toxic metals and capability of metabolizing organic compounds was explored. Phytoremediation capability depends upon the synergistic plant-environment interactions predominantly microorganisms (Iban˜ez et al. 2016). It exploits the ability of selected plants to remediate pollutants from the contaminated sites. The advantages associated with this approach are numerous; the plant biomass after harvesting can be used as a substrate for biofuel; (1) the plants can act as pioneer species for the particular ecosystem if allowed to grow on-site continuously, (2) will increase the biodiversity of the targeted area and will contribute greatly to fix atmospheric CO_2 and (3) help in rebuilding the disturb soil layer.

Moreover, phytoremediation will further enable restoration of contaminated environment with comparable low costs and little collateral impacts. This is in contrast to the traditional methods used for soil remediation that consists of combination of different chemical, biological and physical practices like soil erosion, electrokinetic treatment, chemical oxidation or reduction, pollutant stabilization, excavation and incineration. These conventional methods often destroy natural habitats and are ineffective as well as expensive. The utilization of bioenergy crops (especially the short rotation crops; SRC) in phytoremediation of contaminated soil is gaining a lot of attention (Bell et al. 2014, Gomes 2012, Mleczek et al. 2010, Rowe et al. 2009). The important features required for the bioenergy plant to qualify for the phytoremediation are (1) high pollutant tolerance level, (2) rapid biomass production and (3) high growth rate.

The above requirements have motivated the development of genetically engineered plants with increased contaminant uptake abilities and enhanced tolerance to contaminants (Bell et al. 2014, Abhilash et al. 2012; Dotty 2008). Phytoremediation offers an environmental cleanup of contaminants like heavy metals (Abhilash et al. 2009; Pandey et al. 2012), metalloids (Ye et al. 2011), many organic substances including persistent organic pollutants (Passatore et al. 2014) and radioactive elements (Fulekar et al. 2010; Cerne et al. 2011). In brief, plants can uptake, accumulate and metabolize organic and inorganic contaminants and facilitate the survival and growth of microorganism capable of degrading pollutants (Iban˜ez et al. 2016).

This chapter describes and provides an overview of mechanisms of phytoremediation and new high-throughput biotechnological strategies adopted to enhance the potential ability of phytoremediation in plants as well as the challenges to harness bioenergy plants in phytoremediation. The chapter focuses primarily on plant bioremediation of metals and other strategies for contaminants like organic chemicals are also briefly mentioned.

16.2 Bioenergy Crops/Plants Used for Phytoremediation

The dedicated energy crops, characterized by high plant densities and short crop rotation, are favoured for bioenergy production as well as phytoremediation (Pandey et al. 2016). Bioenergy crops include both annuals and perennials. Examples of annuals include sorghum crops like sweet sorghum (*Sorghum bicolour* L. Moench), fibre sorghum (*S. Bicolour, L. Moench*), kenaf (*Hibiscus cannabinus* L.), rapeseed (*Brassica napus* L.) and *Brassica carinata.* Examples of perennials include:

1- Agricultural: wheat and sugar beet and cardoon (*Cynaracardunculus*)
2- Grass-like crops: reeds (*Arundo donax* L.), miscanthus (*Miscanthus giganteus*), switchgrass (*Panicum virgatum*) and canary reed grass (*Phalaris arundinacea*)
3 Forest: willows (*Salix* sp.), poplars (*Populus* sp.), eucalyptus (*Eucalyptus camaldulensis* Dehnh. and *E. Globules* Labill.) and black locust (*Robinia pseudoacacia*) (Simpson et al. 2009; Zabaniotou et al. 2008). Some examples of bioenergy crops used for phytoremediation are mentioned in Table 16.1.

16.3 Mechanism of Phytoremediation by Bioenergy Crops

The efficiency and mechanism of phytoremediation of plants depend on the type of plant species, contaminant and its bioavailability and also the soil characteristics (Jabeen et al. 2009).There are different mechanisms utilized in phytoremediation process which relates to stabilization in the rhizosphere or root uptake followed by translocation, accumulation, volatilization or degradation (Jutsz and Gnida 2015, Etim 2012, Jabeen et al. 2009). This occurs in parallel with the plant absorption of water and accumulation of essential nutrients along with the contaminants. The possible mechanisms by which bioenergy crops can remediate contaminated soil, sediments and water are briefly presented in the following sections, and pros and cons of each mechanism are summarized in Table 16.2.

16.3.1 Phytoextraction/Phytoaccumulation

In this method, plants remove contaminants by uptake and subsequent translocation of the pollutants from the soil through roots into the air exposed portions of the plants (Etim 2012). Three important characteristics are a prerequisite for a plant to qualify for phytoextraction remediation: the plants must (1) possess high metal absorbing potential, (2) must possess the capability to translocate for heavy metal into the biomass above surface and (3) must be able to contribute to increased plant biomass. The plant must also naturally have a high tolerance for metal concentrations that accumulated in their upper biomass. Phytoextraction is suitable for elimination of metals such as zinc, nickel and copper, and it has been shown that a broad variety of plants can tolerate and absorb large amounts of metals (Etim 2012).

Table 16.1 Examples of utilization of bioenergy crops for phytoremediation

Bioenergy crop	Family	Bioenergy potential	Phytoremediation potential[a]	References
Sorghum bicolour L.	Poaceae	High biomass production	Pb, Ni and Cu	Oh et al. (2015)
Zea mays L.	Poaceae	High biomass production	Pb, Cu and Cd	Oh et al. (2013)
Miscanthus	Poaceae	*Bioethanol production*	Al and oxidative stress	Ezaki et al. (2008)
Miscanthus spp.	Poaceae	*Bioethanol production*	Zn	Barbosa et al. (2015)
Jatropha curcas	Euphorbiaceae	Biodiesel production from seed oil	Cd, Cr, Ni and Zn	Chang et al. (2014)
Ricinus communis	Euphorbiaceae	Biodiesel production from seed oil	Cd and DDTs	Huang et al. (2011)
Populus	Salicaceae	Bioethanol production	Cd, Cu and Zn	Guerra et al. (2011)
Salix	Salicaceae	Bioethanol production from biomass	Cd, PbHg, Zn, Cu	Mleczek et al. (2010)
Arundo donax L.	Poaceae	Bioenergy production from biomass	Cd	Sabeen et al. (2013)
Panicum virgatum L.	Poaceae	Bioenergy production from biomass	Cr	Li et al. (2011)
Eucalyptus globulus	Myrtaceae	Bioenergy production from biomass	Fe, Cr, Mn, Ni, Cd, Pb, Zn, Cu	Luo et al. (2015)
Cannabis sativa L.	Cannabaceae	Bioenergy production from biomass	Cu, Cd, Ni	Ahmed et al. (2015)
Linum usitatissimum	Linaceae	Bioenergy production from biomass	Ni	Griga and Bjelková (2013)
Hibiscus cannabinus L.	Malvaceae	Bioenergy production from biomass	Cd, Zn	Arbaoui et al. (2013)
Cynara cardunculus L.	Compositae	Bioenergy production from biomass	Pb, Zn, Cd	Llugany et al. (2012)
Phalaris arundinacea L.	Poaceae	Bioenergy production from biomass	Heavy metal and trace metals	Polechońska and Klink (2014)

[a]*Pb* lead, *Ni* nickel, *Cu* copper, *Al* aluminium, *Zn* zinc, *Fe* iron, *Cr* chromium, *Mn* manganese, *Cd* cadmium

Table 16.2 Phytoremediation mechanisms for treatment of contaminated soils and water

Mechanisms	Metals[a]	Advantages	Disadvantages
Phytoextraction/ phytoaccumulation	Ni, Zn, Cu	Cost efficient	Remediation time is longer than other strategies and plants can't survive in highly polluted sites
		Complete removal of contaminants/ pollutants	
		Pollutants can be recycled from the biomass of plants	
		Amount of waste materials to be disposed is decreased	Metals may have a phytotoxic effect
Rhizofiltration	Ni, Pb, Zn, Cu, Cr, Cd	Ability to use both terrestrial and aquatic plants for either in situ or ex situ applications	Horizontal wetland's design must be well engineered due to harvesting and disposal of plant
			A technical knowledge of the chemical speciation/interactions is required
Phytovolatilization	Hg	Volatile organic compounds and the contaminant,	The contaminant released into the atmosphere is possible to be redeposited and accumulated in vegetation and be passed on in later products
		mercuric ion, may be converted into a less hazardous substance (i.e. elemental Hg)	
Phytostabilization	Cu, Pb, Cr As, Zn, Cd	The disposal of contaminants and soil removal is not necessary	Contaminant remain in soil
			Soil amendments and extensive use of fertilizers
		Useful for preserving surface and ground water through quick immobilization	Vegetation and soil may require long-term maintenance to prevent rerelease of the contaminants and future leaching
		Less disruptive and cost efficient	Long-term maintenance is required for soil and vegetation to avoid further release and leaching of the contaminants
			Monitoring on a regular basis is required

(continued)

Table 16.2 (continued)

Mechanisms	Metals[a]	Advantages	Disadvantages
Phytodegradation/ phytotrans-formation	Chlorinated solvents, herbicides and munitions	Remediate contaminants in soil, sediment or groundwater	Plant uptake of contaminants depends upon solubility and hydrophobicity fall into a permissible range

[a]*Pb* lead, *Ni* nickel, *Cd* cadmium, *Cu* copper, *Al* aluminium, *Zn* zinc, *Fe* iron, *Cr* chromium, *Mn* manganese, Ni nickel, *PAHs* polycyclic aromatic hydrocarbons, *PCBs* polychlorinated biphenyls

16.3.2 Rhizofiltration

Rhizofiltration relies on the capability of plant roots to uptake and confiscate metal contaminants or nutrients in excess from the wastewater streams and nutrient recycling systems (Jabeen et al. 2009). Rhizofiltration is generally used to remove metals like Pb, Cd, Cu, Ni, Zn, Cr and radionuclides (U, Cs, Sr) (USEPA 2000). Phytoextraction and rhizofiltration shares more or less similar mechanisms, but the latter is more commonly used for remediating groundwater than contaminated soil.

Longer and fibrous roots of terrestrial plants were found to be appropriate for rhizofiltration because of their larger surface area, helpful for metal sorption (Cherian and Oliveira 2005).

16.3.3 Phytovolatilization

This mechanism as its name indicate refers to (1) plant intake of contaminants/pollutants from the soil or water, followed by (2) transformation of contaminants into volatile forms and (3) transpiration of these contaminants into the atmosphere (EPA 2000). Phytovolatilization can remediate contaminants present in water sediment and soil. Mercury is the prime example of a metal contaminant remediated by phytovolatilization, but the process can also be applicable for volatile organic compounds as well as inorganic chemicals in volatile forms, such as arsenic and selenium (EPA 2000; Etim 2012; Laghlimi et al. 2015).

16.3.4 Phytostabilization

Phytostabilization is the use of particular plant species for on-site inactivation of pollutants in the groundwater and soil. The process involves different steps of sorption, precipitation, complexation or metal valence reduction (USEPA 2000) to reduce the dissemination of the contaminant. It also prevents percolation of the hazardous compounds to the groundwater and minimizes its bioavailability in the food chain. By this mechanism, plant species tolerant to metals can be used to re-establish vegetation at contaminated sites and have been shown to be very beneficial in restricting the migration of hazardous compounds through wind erosion and leachate to the groundwater (Etim 2012; Laghlimi et al. 2015).

16.3.5 Phytodegradation

Phytodegradation or phytotransformation is mainly directed towards organic contaminants. It involves the breakdown of complex organic molecules or the integration organic contaminants into plant tissues (Trap et al. 2005). Phytodegradation has been proven to be beneficial for remediating organic pollutants, for example, herbicides, chlorinated solvents and munitions (Laghlimi et al. 2015).

16.4 Genetic Engineering of Bioenergy Crops to Enhance Phytoremediation

Genetic engineering of bioenergy crops for phytoremediation has primarily focused on trait modification such plant uptake or contaminant tolerance. The bioenergy crops promote growth of associated microbial population that can be beneficial for green remediation but plant stimuli and different environmental factors also play a pivotal role.

The information on important factors for phytoremediation can provide a base for development of genetically engineered plants with enhanced remediation potential. In this context, the knowledge of unique property of metal hyperaccumulators has paved the way for the plants to clean up the contaminated soil and water (Jabeen et al. 2009).The overexpression or transfer of genes responsible for uptake, metabolism or transport of specific contaminants in transgenic plants provides a direct method to enhance the phytoremediation potential (Cherian and Oliveira 2005). Metal chelators synthesis will also increase the metal uptake ability of plants (Karenlampi et al. 2000; Cherian and Oliveira 2005). The most common groups of genes which could be targeted to enhance the phytoremediation capacity of plants are shown in Fig. 16.1.

16.4.1 Targeted Genes to Improve Phytoremediation

Genetic engineering has provided an opportunity to enhance the phytoremediation potential of plants altering their accumulation, tolerance and detoxification characteristics. The selection of target genes for developing a metal tolerant transgenic plant involves genes responsible for metal in take followed by translocation and confiscation in higher plants (Cherian and Oliveira 2005). The important gene categories selected are (1) transporters, (2) metal chelation proteins and (3) detoxification metabolic enzymes. The overexpression of any combinations of aforesaid genes is a possible stratagem for genetic engineering. Metallothioneins (MTs), phytochelatins (PCs) and glutathione (GSH) are the three classes of peptides employed

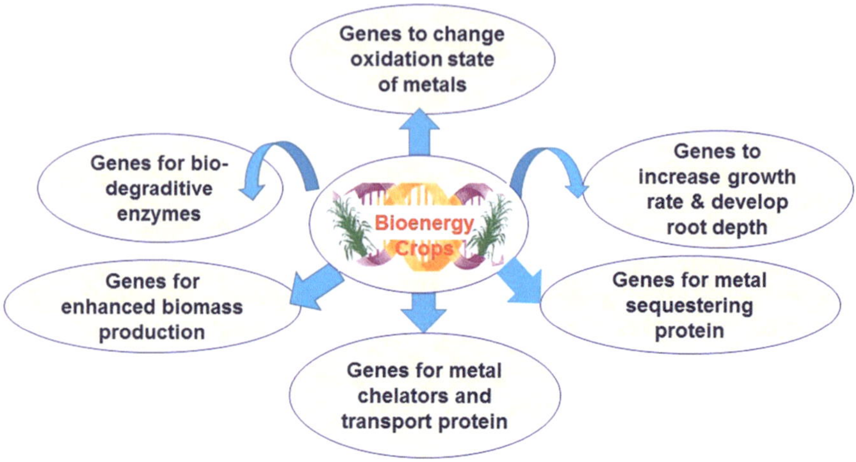

Fig. 16.1 Genes targeted for genetic engineering to enhance phytoremediation potential of bioenergy crops

in phytoremediation strategy (Cherian and Oliveira 2005). Examples of studies showing overexpression of genes in transgenic plants for phytoremediation of metals are shown in Table 16.3.

16.4.2 Alteration of Metal Uptake Mechanism

16.4.2.1 Metallothioneins

Plant metal endurance and accumulation have been affected by overproduction of different metal chelators. Metallothioneins are cysteine-rich, low molecular weight, cytoplasmic metal-binding proteins conferring heavy-metal tolerance. This class of proteins showed high affinity to Cu, Zn and Cd. The overexpression of metal chelators synthesis genes involved may result in enhanced metal internalization, metal translocation or sequestration. For instance, the accumulation of mercury and phytoremediation capability of metallothioneins gene (*mt1*) was for the first time reported in chloroplasts by Ruiz et al. (2011). In order to determine the role of metallothioneins genes in metal tolerance, the effect of cadmium on transgenic *Nicotiana tabacum* harbouring yeast metallothioneins gene was analysed (Krystofova et al. 2011). They observed higher cadmium accumulation in transgenic *Nicotiana tabacum* using yeast metallothioneins as compared to non-transgenic plants. The study concluded that the metallothioneins have enhanced the accumulation of cadmium in the roots of genetically modified plants.

16.4.2.2 Phytochelatins (PCs)

Phytochelatins are cysteine-rich metal-chelating peptides that play an important role in heavy-metal detoxification in plants and fungi by heavy-metal chelation and

Table 16.3 Selected examples of genes overexpressed in transgenic plants for enhancing metal tolerance

Genes	Gene product	Targeted species	Phytoremediation performance	References
ScMT2–1-3	Metallothionein	Chloroplast transgenic plant	Hg (mercury)	Ruiz et al. (2011)
Mt1	Metallothionein	Wild-type tobacco	Cd (cadmium)	Daghan et al. (2013)
ScMT2–1-3, (metallothionein)	Metallothionein	Sugarcane	Cu (copper)	Guo et al. (2013)
CUP1 (metallothionein)	Metallothionein	*Nicotiana tabacum*	Cd(cadmium)	Krystofova et al. (2012)
gshII	Glutathione synthetase	*Brassica juncea* L.	Cd (cadmium)	Zhu et al. (1999)
gshI	ç-Glu-Cys synthetase	*Brassica juncea* L.	Cd (cadmium)	Zhu et al. (1999)
SAT	Serine acetyl transferase	*Sedum alfredii*	Cd (cadmium)	Guo et al. (2009)
APS1	ATP sulfurylase	*Stanleya pinnata* and *Brassica juncea*	Se (selenium)	Schiavon et al. (2015)
TaPCS1	Phytochelatin synthase	*Nicotiana glauca* R. Graham	Pb (lead) and Cd (cadmium)	Gisbert et al. (2003)
AtPCS1	Phytochelatin synthase	*Arabidopsis* plants	Cd (cadmium)	Brunetti et al. (2011)
GR	Glutathione reductase	*Brassica juncea* L.	Cd (cadmium)	Pilon-Smits et al. (2000)
OASTL	Cysteine synthase	Legume crops	As (arsenic), Cd (cadmium) and Pb (lead)	Pajuelo et al. (2007)
SMT	Selenocysteine methyltransferase	*Astragalus chrysochlorus*	Se (selenium)	Çakir et al. (2016)
ATPS	ATP sulfurylase	*Brassica juncea* L.	Se (selenium)	LeDuc et al. (2006)
CGS	Cystathionine-gamma-synthase	Tobacco plants	Methionine and threonine	Hacham et al. (2008)
ArsC	Arsenate reductase	*Arabidopsis thaliana*	As (arsenic)	Mohan et al. (2016)
NtCBP4	Cation channel	Tobacco	Pb^{2+} (lead)	Sunkar et al. (2000)
AtCAX2	Vacuolar transporter	Tobacco plants	More metal ions in the shoots and higher root tonoplast transport	Hirschi et al. (2000)

(continued)

Table 16.3 (continued)

Genes	Gene product	Targeted species	Phytoremediation performance	References
ZntA	Heavy-metal transporter	*Arabidopsis* plants	Cd (cadmium)	Lee et al. (2003)
YCF1	Transport protein	*Arabidopsis*, *poplar*	Pb (lead)	Song et al. (2004)
FRE1 and FRE2	Ferric reductase	Tobacco plants	Fe (iron)	Samuelsen et al. (1998)

minimizing their bioavailability (Cherian and Oliveira 2005). Apart from their role in metal detoxification, a recent study indicated that phytochelatins have the ability to undergo long-distance transport (from root to shoot) in genetically engineered *Arabidopsis thaliana.* This was achieved by expressing the wheat (*Triticum aestivum*) PC synthase (*TaPCS1*) in an *Arabidopsis* PC-deficient mutant (Chen et al. 2006). The study depicted the role of phytochelatins in grafting technique for analysing long-distance transportation and signalling in higher plants. The role of phytochelatins (PC) in essential metal homeostasis was reported by Tennstedt et al. (2009). The *cad1-3* (a PC-deficient mutant of *Arabidopsis thaliana*) and *cad1-6* (a newly isolated second strong allele) in context to zinc (Zn) homeostasis were studied. It was found that the PC-deficient mutants showed significant Zn^{2+} hypersensitivity in the medium and leads to significant reduction in Zn accumulation in root. The above study concluded that PC formation is vital for Zn^{2+} tolerance and accumulation of Zn.

In another study, *AtPCS1* and *CePCS* genes in tobacco (*Nicotiana tabacum* var. Xanthi) were overexpressed, and their effects were compared (Wojas et al. 2008). It was found that in comparison to the wild-type and *CePCS* mutants, *AtPCS1*-overexpressed plants were found to be Cd-hypersensitive with no significant difference in cadmium accumulation. The phytochelatin synthase (PCS) activity in *AtPCS1* transformants was found to be approximately fivefold higher than in *CePCS* and wild-type plants. *CePCS* transformants showed a smaller reduction in glutathione level and less significant change in γ-glutamylcysteine concentration in comparison to *AtPCS1* expressing plants, where γ-glutamylcysteine accumulation was high with significant deletion of glutathione. The study concluded that the difference in results could be due to species-dependent differences in the activity of phytochelatin synthase.

In a study by Brunetti et al. (2015), it was reported that the *Arabidopsis thaliana* seedlings defective in the ABCC transporter AtABCC3 (*abcc3*) showed an increased sensitivity to different Cd concentrations, and increased Cd tolerance was found in seedlings overexpressing *AtABCC3*.

16.4.2.3 Glutathione (GSH) Enzymes Involved in Sulphate Assimilation Pathway

GSH enzymes, synthesized by the sulphur (S) assimilation pathway, are termed as direct precursor of metal-binding peptides like phytochelatins which play a vital role in detoxification of heavy metals (Cherian and Oliveira 2005). In a recent study,

Streptococcus thermophilus γ-glutamylcysteinesynthetase-glutathione synthetase (StGCS-GS), responsible for the synthesis of glutathione (GSH), was overexpressed in *Beta vulgaris* L., (Liu et al. 2015). The transgenic sugar beets were found to be different from their respective wild type in terms of accumulation of Cu, Zn and Cd ions in shoots and showed high levels of GSH and phytochelatin (PC) when exposed to different concentrations heavy metals. The study concluded that the enhanced metal accumulation and tolerance were probably due to the high expression of StGCS-GS followed by overproduction of phytochelatins and GSH.

In order to enhance the phytoremediation potential of plant, *AsPCS1* and *GSH1* genes were overexpressed in *Arabidopsis thaliana* (Guo et al. 2008).The total PC production in transgenic *Arabidopsis* was found to be elevated due to simultaneous overexpression of *AsPCS1* and *GSH1*. The study indicated that the dual modification of genes was capable of enhancing As and Cd accumulation and tolerance and served as a promising new strategy in phytoremediation process.

The effect of GSH homeostasis on Cd stress was explored by using the wild-type *Arabidopsis* and a double mutant *sultr1;1-sultr1;2* defective in two different high-affinity sulphate transporter (Liu et al. 2016). The double mutant was found to be more responsive to Cd concentrations in comparison to the wild type. The study reported that the GSH homeostasis and imbalance between antioxidative defence and oxidative damage were the driving force to hyperresponsiveness of double mutant to different concentrations of Cd under restricted sulphate supply.

16.4.2.4 Metal Transporters

Metal transporters have a vital role in development and growth of plants, signal transduction and metal detoxification by facilitating mobilization of metals and alkalizations across plant plasma membrane and organelle membranes. The modification of transporters such as FRE1, ZAT, FRE2 CAX2 and NtCBP4 led to the development of plant with improved metal accumulation and tolerance capability. In a recent study, isolation of *SaCAX2h* and *SaCAX2n* genes encoding CAX2-transporter-like proteins was carried out from Zn/Cd hyper- and non-hyperaccumulator ecotype of *Sedum alfredii*. It was found that the heterologous expression of both genes in the Δzrc1 yeast mutant enhances the cadmium (Cd) concentration in yeast cells. The expression profiles of these two genes varied in different Cd treatments. The study indicated that the gene *SaCAX2h* facilitates Mn and Ca sequestration into vacuoles, and its overexpression in the transgenic tobacco increased the Cd accumulation. Three Ca^{2+} transporters located in tonoplast, *ACA4*, *CAX1* (Ca^{2+}/H^{+}-antiporter) and *ACA11* (Ca^{2+}-ATPases), were found to express predominantly in Ca-rich mesophyll through the use of microarray and quantitative PCR (Conn et al. 2011).The study reported the functional role of *CAX1*, as a key regulator of apoplastic [Ca^{2+}] through compartmentation (a mechanism necessary for the optimal plant metabolism) into mesophyll vacuoles.

16.4.3 Alteration of Metabolic Pathways and Oxidative Stress Mechanisms

The phytoremediation potential of the plant can also be increased by introducing new metabolic pathways into plants for hyperaccumulation or phytovolatilization. It has been reported that bacteria has capability to convert heavy metals to less toxic states. For example, in gram-negative bacteria, mercury resistance is encoded by *merA* (mercuric ion reductase) operon, which codes for soluble NADPH-dependent FAD-containing disulphide oxidoreductase and *merB* gene which codes for organ mercurial lyase gene. *merA* enzyme reduces toxic Hg^{2+} to the less harmful metallic mercury (Hg^0) (Rugh et al. 1996). The transgenic *Arabidopsis* plants were developed by introducing *merA* and *merB* genes in the plants (Rugh et al. 1996). The resulting *merA* transgenics exhibits significantly higher endurance to Hg^{+2} and Hg in volatile form (Rugh et al. 1996). Their tolerance to methyl mercury was unaffected. Transgenic plants expressing *merB* were found to be tolerant to methyl mercury and other organomercurials in comparison to their respective wild type. The *merB* plants efficiently reduced the highly toxic methyl mercury to Hg^{+2}, which is ~100 times less hazardous to plants (Fig. 16.2).

In a recent study, *Escherichia coli* was genetically modified to exhibit mercury resistance operon genes (*merRTPAB*), in order to remediate organic and inorganic forms of mercury (Kane et al. 2016). *E. coli* exhibiting *mer operon* was further encapsulated in silica beads to develop a biological filtration material retaining the ability to degrade methyl mercury. Selenium volatilization is another area of immense interest for its potential to develop transgenic plants with improved selenium detoxification capacity. A gene (BoCOQ5-2 methyltransferase) involved in the ubiquinone biosynthetic pathway was isolated from broccoli (Zhou and Li 2010). It was found that the BoCOQ5-2 methyltransferase facilitates selenium volatilization in bacteria as well as implants and further illustrated that the different metabolic processes outside of the selenium/sulphur metabolic pathway are responsible for regulation of selenium metabolism opening new line of approaches to enhance selenium phytoremediation.

The alterations of enzymes conferring oxidative stress may also lead to increased metal uptake and tolerance. In a study conducted by Chen et al. (2015), the mechanism of antioxidatives in *Moso bamboo* in the presence of copper (Cu) stress in hydroponics and soil environment was investigated. The study demonstrated the effectiveness of plant for detoxification of Cu in soils and the role of the antioxidative defence system reactions under Cu stress. Glutathione-S-transferase (GST) enzymes have pivotal roles in detoxification of reactive oxygen or nitrogen species. In a recent study, a promoter region of 1023 *bp* length belonging to tau class

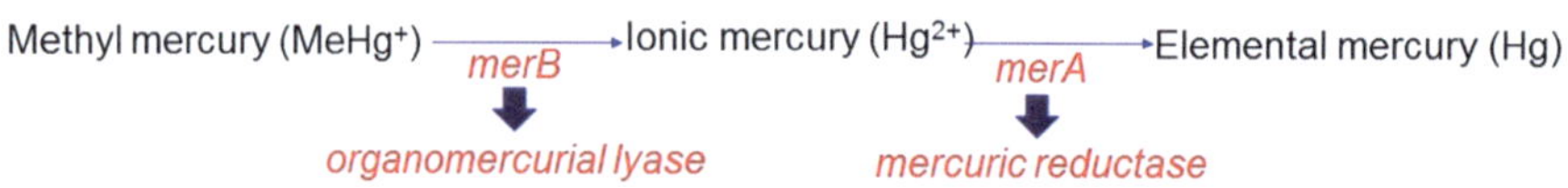

Fig. 16.2 Mercury volatilization pathway in bacteria and transgenic plants

glutathione-S-transferase (GST) enzymes from *Salicornia brachiata* was cloned and characterized in tobacco using the transgenic approach (Tiwari et al. 2016). The enhanced plant growth in transgenic plant alleviated the reactive oxygen species (ROS) buildup, maintained the ion homeostasis and enhanced the physiological status of the plant under salt and osmotic stresses. This was the result of the overexpression of the *SbUSP* gene (*Salicornia brachiate* universal stress protein) in tobacco plant (Udawat et al. 2016). The study concluded that the ectopic expression of the gene reduces salt or osmotic stress through ROS scavenging and modulation of the plant physiological state.

16.5 Genetic Manipulation of Microbial Community to Improve Phytoremediation

The emerging integrative "omics" approaches has accelerated the research to develop new strategies for increasing phytoremediation capacity of plants especially bioenergy crops (Amrani et al. 2015). The potential application of "omics" strategies like metagenomics, metatranscriptomics, metaproteomics and metabolomics in development of transgenic plant for phytoremediation will be discussed in the following sections emphasizing how "omics" bring new insight into decipher molecular mechanism of soil microbial communities and plant metabolism. Further, the coupling omics strategies and new bioinformatics methods will present detailed pictures of plant-microbe integrated activity patterns (Bell et al. 2014). These approaches will further determine how their application will magnify the growth and assembly of microbial communities and eventually improve the phytoremediation ability of the plants (Fig. 16.3) (Bell et al. 2014).

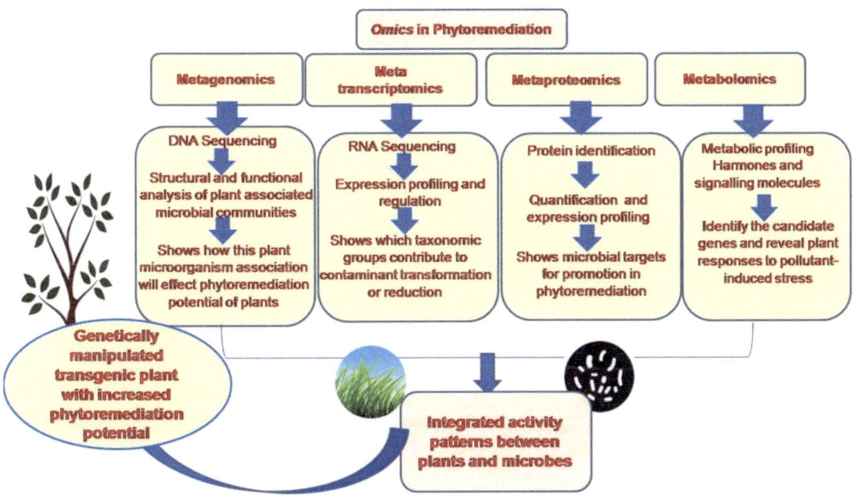

Fig. 16.3 The omics approaches employed to generate transgenic plants with improved phytoremediation potential

16.5.1 Metagenomics Approach

The term "metagenome" represents total DNA extracted from lysed cells of mixed microbial populations. This strategy has been used to determine the plant-associated microbial communities (rhizosphere, endosphere or phyllosphere) utilizing amplicon-targeting of specific genes and high-throughput 16S rDNA bacterial sequencing or shotgun sequencing (Bulgarelli et al. 2012, Peuke and Rennenberg 2005).The structure of microbial population depends on the following factors: (1) the type of plant as well as soil and somewhat on structural features of generic plant and (2) environmental factors.

In order to understand the microbial variability associated with plant environment, recent techniques such as high-throughput sequencing of bacterial 16S rDNA were adopted by various researchers (Bulgarelli et al. 2012; Lundberg et al. 2012). In a study, Yergeau et al. (2015) have used Illumina MiSeq platform to study the microbial communities of a highly petroleum-contaminated soils targeting the 16S rRNA gene's short hypervariable regions of archaea/bacteria and the ITS region of fungi. The study concluded that when soil microbiome was subjected to gamma irradiation, there were a short-term change in microbial communities and long-lasting effects on plant growth characteristics. The study again indicated the potential of modifying target plant phytoremediation characteristics through the plant-associated microbiome manipulation. In another study, high-throughput sequencing approach was used to determine the structural organization of *Arbuscular mycorrhizal* fungi (AMF) within the rhizosphere of willow cultivars introduced to hydrocarbon-contaminated soils (Hassan et al. 2014). The study provided deep knowledge of the alterations that occur as a result of plant introduction and hydrocarbon contamination. The plant-microbe integrated pattern may affect the phytoremediation property of plants which also depends upon the concentration of soil contaminants. In a study by Bell et al. (2014), targeted amplicon sequencing was used to determine the concentration of contaminants in soil and plant phylogeny. It has been reported that the horizontal gene transfer (HGT) is possible in nutritional rich hotspots like the phyllosphere, rhizosphere, decaying animal and plant tissues and manure-applied soil rather than bulk soil (Aminov 2011). In highly contaminated sites, HGT, mainly by plasmid exchange, allow microorganism to adapt to stress environments (Sentchilo et al. 2013).It is now evident from many published reports that these plasmids had an important function in shaping microbial community and hence effecting phytoremediation potential of plants by introducing contaminant degrading genes in plants that apparently lacked such genes (Bell et al. 2014, Doty et al. 2008). Another approach coupled with the metagenome is use of PhyloChip, reported by Mendes et al. (2011). Metagenomics coupled with different analysis techniques can give more information about plant microbial interaction that will help to generate an efficient phytoremediation system.

16.5.2 Metatranscriptomics Approach

Metatranscriptomics have been used to investigate the alterations in expression profiles at the messenger RNA in plant tissues, in response to specific contaminants. Transcriptomics allow the identification of functional genes and has capability to analyse thousands of genes in parallel.

The quantitative predictive capabilities of metatranscriptomics were shown in a study by Helbling et al. 2012, in which a strong association was found between relative mRNA level and atrazine-degrading activity. In spite of challenges faced in mRNA enrichment (Turner et al. 2013), advancement and reduced costs of sequencing make metatranscriptome capable of sequencing active taxonomic plant groups and associated contaminant degrading genes. In a study reported by Gonzalez et al. (2015), de novo assembled transcriptome was compared between *Salix purpurea*'s organs and impartially annotated without a priori constraint to any organism. The metatranscriptomic pattern of expression suggested a cross-tolerance mechanism, and improved biotic resistance was found in case of abiotic stress resistance systems. These findings highlighted a valuable but complex biotic and abiotic stress response explaining to some extent the high productivity of willow crop. A metatranscriptomic comparison was made to compare the alterations in microbial expression of rhizosphere in case of willow introduction in uncontaminated as well as contaminated soils (Yergeau et al. 2014).

The growth of ten *Salix* cultivars on contaminated soils was used as an example for non-model and versatile crop response for the analysis of gene expression. Independent transcriptomes (ten in number) and a global transcriptome were created by assembling de novo sequence reads and mapped against the reference genome of *Salix purpurea* 94,006 (Brereton and Gonzalez (2016). The study suggested that the pattern of *Salix* gene expression revealed the need for alteration in cultivation of specific photosynthetic apparatus and prevent the antenna complexes from oxidation stress. Furthermore, the study emphasizes the exploitation of temporal metatranscriptomic studies in order to understand the crucial microbial activities shift, associated with plants, over time.

16.5.3 Metaproteomics Approach

In comparison to other omics approaches, "metaproteomics" has been used less extensively to determine plant-microbe interaction for phytoremediation. Metaproteomics is specifically useful in environments with low diversity, for example, the phyllosphere. To identify the proteins synthesized by bacteria that make use of the fluorene and naphthalene, this technique was coupled to stable isotope probing of these organic pollutants (Herbst et al. 2013). This omics tool could also differentiate the microbes that can metabolize plant exudates from ones that can degrade organic contaminants and could further help in identification of microbial targets for enhancing phytoremediation.

The microbial communities attached with the leaves of clover soybean and *Arabidopsis thaliana* plants were assessed using metaproteogenomic approach by Delmotte et al. (2009).In their study a high consistency in bacterial communities (predominantly the alpha proteobacterial, genera (Sphingomonas and Methylo bacterium) was observed for all three selected plants species. The study provided a base for the identification of unique properties of phyllosphere bacteria and defined the plant-microbe and microbe-microbe interactions. Knief et al. (2012) have employed metaproteogenomic approach to analyse the physiology of rice-associated bacteria and archaea. The protein characterization of phyllosphere and rhizosphere microbiota associated with rice revealed that the rhizosphere was found to be enriched with proteins involved in methanogenesis and methanotrophy, whereas methanol-based methylotrophy was found in phyllosphere microbiota.

16.5.4 Metabolomics Approach

Proteomics and metabolomics approaches have been used in combination, to identify and characterize microbial agents to improve phytoremediation (Sergeant et al. 2012, Gómez Ariza et al. (2013). For instance, proteomics was used to identify 100 proteins of rice, under the short- or long-term Cd stress and differentially regulated by Silicon (Si) (Ma et al. 2016). The downregulated proteins (70%) indicated that protein use efficiency can be improved by Si, maintaining cells in the normal physiological status. The study provided protein-level information for understanding the Si-mediated Cd detoxification in rice single cells.

A comparative proteomics study was carried out to explore the potential mechanism of *Eichhornia crassipes* tolerance to Cd stress at the protein level (Li et al. 2015).The study reported that the metabolic processes and physiological activities of *E. crassipes* were affected when exposed to different concentrations of Cd stress, based on differential protein expression. The use of combinations of different omics techniques will decipher a more global vision of functional importance of genetic signatures, protein and metabolites, relationship with plants and different environmental factors.

16.6 Challenges, Concluding Remark and Future Perspectives

Phytoremediation offers enormous advantages as a green remediation cleanup approach; however, there are roadblocks to gain traction as a potential viable remediation technology due to its variable effectiveness. The successful translation of phytoremediation strategies to the contaminated site and its ability to clean up different types of contaminants is a major impediment in its application. The use of bioenergy crops and especially short rotation crops in phytoremediation of contaminated soil have paved a way to contribute to the energy supply and can thus play a vital role in meeting the targets for use of renewable energy sources. However, the main hurdle regarding the use of biomass for bioenergy is the issue of dissemination of pollutants in the biomass. The contaminants bioaccumulated in the crop may cause hindrance at later stages of biofuel production, leading the decision on whether crop uptake should be promoted or not to an open discussion forum. In-depth knowledge is needed to manage the risks of contaminant uptake by various crops, combustion/emissions and final contents in biofuel in order to gain confidence to use end-products of phytoremediation for meeting the demands of energy with minimal environmental impact. Recent development of transgenic plants and bacteria that enhance phytodegradation of contaminants could also be an exciting opportunity for bioenergy production through phytoremediation. Moreover, with new directions harnessed with biotechnological interventions especially omics approaches, phytoremediation can be presented as an efficient remediation alternative. The combination of omics approaches will definitely help in shaping microbial communities associated with plants and further alters plants response towards contaminants.

Acknowledgements Authors hereby acknowledge the support from South African Research Chairs Initiative, Department of Science and Technology and National Research Foundation of South Africa and Durban University of Technology.

References

Abhilash PC, Jamil S, Singh N (2009) Transgenic plants for enhanced biodegradation and phytoremediation of organic xenobiotics. Biotechnol Adv 27(4):474–488. doi:10.1016/j.biotechadv.2009.04.002

Abhilash PC, Powell JR, Singh HB, Singh BK (2012) Plant-microbe interactions: novel applications for exploitation in multipurpose remediation technologies. Trends Biotechnol 30(8):416–420. doi:10.1016/j.tibtech.2012.04.004

Ahmed SA, Ross SA, Slade D, Radwan MM, Khan IA, ElSohly MA (2015) Minor oxygenated cannabinoids from high potency *Cannabis sativa* L. Phytochemistry 117:194–199. doi:10.1016/j.phytochem.2015.04.007

Aminov RI (2011) Horizontal gene exchange in environmental microbiota. Front Microbiol 2:158. doi:10.3389/fmicb.2011.00158

Arbaoui S, Evlard A, Mhamdi Mel W, Campanella B, Paul R, Bettaieb T (2013) Potential of kenaf (*Hibiscus cannabinus* L.) and corn (*Zea mays* L.) for phytoremediation of dredging sludge contaminated by trace metals. Biodegradation 24(4):563–567. doi:10.1007/s10532-013-9626-5

Barbosa B, Boléo S, Sidella S, Costa J, Duarte MP, Mendes B, Cosentino SL, Fernando AL (2015) Phytoremediation of heavy metal-contaminated soils using the perennial energy crops Miscanthus spp. and Arundodonax L. BioEnergy Res 8(4):1500–1511. doi:10.1007/s12155-015-9688-9

Bell TH, Joly S, Pitre FE, Yergeau E (2014) Increasing phytoremediation efficiency and reliability using novel omics approaches. Trends Biotechnol 32(5):271–280. doi:10.1016/j.tibtech.2014.02.008

Brereton NJ, Gonzalez E (2016) Comparative transcriptomic approaches exploring contamination stress tolerance in Salix sp. Reveal the importance for a metaorganismal de novo assembly approach for nonmodel plants. Plant Physiol 171(1):3–24. doi:10.1104/pp.16.00090

Brunetti P, Zanella L, Proia A, De Paolis A, Falasca G, Altamura MM, Sanità di Toppi L, Costantino P, Cardarelli M (2011) Cadmium tolerance and phytochelatin content of Arabidopsis seedlings over-expressing the phytochelatin synthase gene AtPCS1. J Exp Bot 62:5509–5519

Brunetti P, Zanella L, De Paolis A, Di Litta D, Cecchetti V, Falasca G, Barbieri M, Altamura MM, Costantino P, Cardarelli M (2015) Cadmium-inducible expression of the ABC-type transporter AtABCC3 increases phytochelatin-mediated cadmium tolerance in Arabidopsis. J Exp Bot 66(13):3815–3829. doi:10.1093/jxb/erv185

Bulgarelli D, Rott M, Schlaeppi K, Ver Loren van Themaat E, Ahmadinejad N, Assenza F, Rauf P, Huettel B, Reinhardt R, Schmelzer E, Peplies J, Gloeckner FO, Amann R, Eickhorst T, Schulze-Lefert P (2012) Revealing structure and assembly cues for Arabidopsis root-inhabiting bacterial microbiota. Nature 488(7409):91–95. doi:10.1038/nature11336

Çakir Ö, Turgut-Kara N, Ari Ş (2016) Selenium induced selenocysteine methyltransferase gene expression and antioxidant enzyme activities in *Astragalus chrysochlorus*. Acta Bot Croat 75:11–16. doi:10.1515/botcro-2016-0015

Černe M, Smodiš B, Štrok M (2011) Uptake of radionuclides by a common reed (Phragmitesaustralis (Cav.) Trin. ex Steud.) grown in the vicinity of the former uranium mine at Žirovskivrh. Nucl Eng Des 241(4):1282–1286 doi:http://dx.doi.org/10.1016/j.nucengdes.2010.04.003

Chang F-C, Ko C-H, Tsai M-J, Wang Y-N, Chung C-Y (2014) Phytoremediation of heavy metal contaminated soil by *Jatropha curcas*. Ecotoxicology 23(10):1969–1978. doi:10.1007/s10646-014-1343-2

Chen A, Komives EA, Schroeder JI (2006) An improved grafting technique for mature Arabidopsis plants demonstrates long-distance shoot-to-root transport of phytochelatins in Arabidopsis. Plant Physiol 141(1):108–120. doi:10.1104/pp.105.072637

Chen J, Shafi M, Li S, Wang Y, Wu J, Ye Z, Peng D, Yan W, Liu D (2015) Copper induced oxidative stresses, antioxidant responses and phytoremediation potential of *Moso bamboo* (Phyllostachyspubescens). Sci Rep 5:13554. doi:10.1038/srep13554

Cherian S, Oliveira MM (2005) Transgenic plants in phytoremediation: recent advances and new possibilities. Environ Sci Technol 39(24):9377–9390. doi:10.1021/es051134l

Conn SJ, Gilliham M, Athman A, Schreiber AW, Baumann U, Moller I, Cheng NH, Stancombe MA, Hirschi KD, Webb AA, Burton R, Kaiser BN, Tyerman SD, Leigh RA (2011) Cell-specific vacuolar calcium storage mediated by CAX1 regulates apoplastic calcium concentration, gas exchange, and plant productivity in Arabidopsis. Plant Cell 23(1):240–257. doi:10.1105/tpc.109.072769

Daghan H, Arslan M, Uygur V, Koleli N (2013) Transformation of tobacco with ScMTII gene-enhanced cadmium and zinc accumulation. CLEAN Soil Air Water 41(5):503–509. doi:10.1002/clen.201200298

Delmotte N, Knief C, Chaffron S, Innerebner G, Roschitzki B, Schlapbach R, von Mering C, Vorholt JA (2009) Community proteogenomics reveals insights into the physiology of phyllosphere bacteria. Proc Natl Acad Sci 106(38):16428–16433. doi:10.1073/pnas.0905240106

Doty SL (2008) Enhancing phytoremediation through the use of transgenics and endophytes. N Phytologist 179(2):318–333. doi:10.1111/j.1469-8137.2008.02446.x

El Amrani A, Dumas AS, Wick LY, Yergeau E, Berthome R (2015) "Omics" insights into PAH degradation toward improved green remediation biotechnologies. Environ Sci Technol 49(19):11281–11291

EPA (2000) A citizen's guide to phytoremediation. EPA 542-F-98-011. United States Environmental Protection Agency. doi: http//www.bugsatwork.com/XYCLONYX/EPA_GUIDES

Etim EE (2012) Phytoremediation and its mechanisms: a review. Int J Environ Bioenergy 2(3):120–136

Ezaki B, Nagao E, Yamamoto Y, Nakashima S, Enomoto T (2008) Wild plants, *Andropogon virginicus* L. and *Miscanthus sinensis* Anders, are tolerant to multiple stresses including aluminium, heavy metals and oxidative stresses. Plant Cell Rep 27(5):951–961. doi:10.1007/s00299-007-0503-8

Fulekar MH, Singh A, Thorat V, Eapen S (2010) Phytoremediation of 137 Cs from low level nuclear waste using *Catharanthus roseus*. Ind J Pure Appl Phys 48(7):516–519

Gisbert C, Ros R, De Haro A, Walker DJ, Pilar Bernal M, Serrano R, Navarro-Aviñó J (2003) A plant genetically modified that accumulates Pb is especially promising for phytoremediation. Biochem Biophys Res Commun 303(2):440–445 doi:http://dx.doi.org/10.1016/S0006-291X(03)00349-8

Gomes HI (2012) Phytoremediation for bioenergy: challenges and opportunities. Environ Technol Rev 1(1):59–66. doi:10.1080/09593330.2012.696715

Gómez Ariza JL, García-Barrera T, García-Sevillano MA, González-Fernández M, Gómez-Jacinto V (2013) Metallomics and metabolomics of plants under environmental stress caused by metals. In: Gupta KD, Corpas JF, Palma MJ (eds) Heavy metal stress in plants. Springer, Berlin/Heidelberg, pp 173–201. doi:10.1007/978-3-642-38469-1_10

Gonzalez E, Brereton NJ, Marleau J, GuidiNissim W, Labrecque M, Pitre FE, Joly S (2015) Meta-transcriptomics indicates biotic cross-tolerance in willow trees cultivated on petroleum hydrocarbon contaminated soil. BMC Plant Biol 15:246. doi:10.1186/s12870-015-0636-9

Griga M, Bjelková M (2013) Flax (*Linum usitatissimum* L.) and Hemp (*Cannabis sativa* L.) as fibre crops for phytoextraction of heavy metals: biological, agro-technological and economical point of view. In: Gupta KD (ed) Plant-based remediation processes. Springer, Berlin/Heidelberg, pp 199–237. doi:10.1007/978-3-642-35564-6_11

Guerra F, Gainza F, Pérez R, Zamudio F (2011) Phytoremediation of heavy metals using poplars (Populus Spp.): a glimpse of the plant responses to Copper, Cadmium and Zinc stress. In: Handbook of phytoremediation, Edition: I,. Nova Science Publisher, New York, pp 387–414

Guo J, Dai X, Xu W, Ma M (2008) Over expressing GSH1 and AsPCS1 simultaneously increases the tolerance and accumulation of cadmium and arsenic in Arabidopsis thaliana. Chemosphere 72(7):1020–1026. doi:10.1016/j.chemosphere.2008.04.018

Guo W-d, Liang J, Yang X-e, Y-e C, Feng Y (2009) Response of ATP sulfurylase and serine acetyltransferase towards cadmium in hyperaccumulator Sedum alfredii Hance. J Zhejiang Univ Sci B 10(4):251–257. doi:10.1631/jzus.B0820169

Guo J, Xu L, Su Y, Wang H, Gao S, Xu J, Que Y (2013) ScMT2-1-3, a metallothionein gene of sugarcane, plays an important role in the regulation of heavy metal tolerance/accumulation. Biomed Res Int 2013:904769. doi:10.1155/2013/904769

Hacham Y, Matityahu I, Schuster G, Amir R (2008) Overexpression of mutated forms of aspartate kinase and cystathionine gamma-synthase in tobacco leaves resulted in the high accumulation of methionine and threonine. Plant J 54(2):260–271. doi:10.1111/j.1365-313X.2008.03415.x

Hassan SE-D, Bell TH, Stefani FOP, Denis D, Hijri M, St-Arnaud M (2014) Contrasting the community structure of Arbuscular Mycorrhizal fungi from hydrocarbon-contaminated and uncontaminated soils following willow (<italic>Salix</italic> spp. L.) planting. PLoS One 9(7):e102838. doi:10.1371/journal.pone.0102838

Helbling DE, Ackermann M, Fenner K, Kohler H-PE, Johnson DR (2012) The activity level of a microbial community function can be predicted from its metatranscriptome. ISME J 6(4):902–904. doi:10.1038/ismej.2011.158

Herbst FA, Bahr A, Duarte M, Pieper DH, Richnow HH, von Bergen M, Seifert J, Bombach P (2013) Elucidation of in situ polycyclic aromatic hydrocarbon degradation by functional metaproteomics (protein-SIP). Proteomics 13(18–19):2910–2920. doi:10.1002/pmic.201200569

Hirschi KD, Korenkov VD, Wilganowski NL, Wagner GJ (2000) Expression of Arabidopsis CAX2 in Tobacco. Altered metal accumulation and increased manganese tolerance. Plant Physiol 124(1):125–134

Huang H, Yu N, Wang L, Gupta DK, He Z, Wang K, Zhu Z, Yan X, Li T, Yang XE (2011) The phytoremediation potential of bioenergy crop *Ricinus communis* for DDTs and cadmium co-contaminated soil. Bioresour Technol 102(23):11034–11038. doi:10.1016/j.biortech.2011.09.067

Ibañez S, Talano M, Ontañon O, Suman J, Medina MI, Macek T, Agostini E (2016) Transgenic plants and hairy roots: exploiting the potential of plant species to remediate contaminants. New Biotechnol 33(5 Pt B):625–635 doi: http://dx.doi.org/ 10.1016/ j.nbt.2015.11.008

Jabeen R, Ahmad A, Iqbal M (2009) Phytoremediation of heavy metals: physiological and molecular mechanisms. Bot Rev 75(4):339–364. doi:10.1007/s12229-009-9036-x

Jutsz Anna M, Gnida A (2015) Mechanisms of stress avoidance and tolerance by plants used in phytoremediation of heavy metals. Arch Environ Protect 41:104–114. doi:10.1515/aep-2015-0045

Kane AL, Al-Shayeb B, Holec PV, Rajan S, Le Mieux NE, Heinsch SC, Psarska S, Aukema KG, Sarkar CA, Nater EA, Gralnick JA (2016) Toward bioremediation of methylmercury using silica encapsulated *Escherichia coli* harboring the *mer*Operon. PLoS One 11(1):e0147036. doi:10.1371/journal.pone.0147036

Kärenlampi S, Schat H, Vangronsveld J, Verkleij JAC, van der Lelie D, Mergeay M, Tervahauta AI (2000) Genetic engineering in the improvement of plants for phytoremediation of metal polluted soils. Environ Pollut 107(2):225–231 doi:http://dx.doi.org/10.1016/S0269-7491(99)00141-4

Knief C, Delmotte N, Chaffron S, Stark M, Innerebner G, Wassmann R, von Mering C, Vorholt JA (2012) Metaproteogenomic analysis of microbial communities in the phyllosphere and rhizosphere of rice. ISME J 6(7):1378–1390

Krystofova O, Zitka O, Krizkova S, Hynek D, Shestivska V, Adam V, Hubalek J, Mackova M, Macek T, Zehnalek J, Babula P, Havel L, Kizek R (2012) Accumulation of Cadmium by transgenic tobacco plants (Nicotiana tabacum L.) carrying Yeast metallothionein gene revealed by electrochemistry. Int J Electrochem Sci 7:886–907

Laghlimi M, Baghdad B, Hadi H, Bouabdli A (2015) Phytoremediation mechanisms of heavy metal contaminated soils: a Review. Open J Ecol 5:375–388

LeDuc DL, AbdelSamie M, Montes-Bayon M, Wu CP, Reisinger SJ, Terry N (2006) Over expressing both ATP sulfurylase and selenocysteinemethyltransferase enhances selenium phytoremediation traits in Indian mustard. Environ Pollut 144(1):70–76. doi:10.1016/j.envpol.2006.01.008

Lee J, Bae H, Jeong J, Lee J-Y, Yang Y-Y, Hwang I, Martinoia E, Lee Y (2003) Functional expression of a bacterial heavy metal transporter in Arabidopsis enhances resistance to and decreases uptake of heavy metals. Plant Physiol 133(2):589–596. doi:10.1104/pp.103.021972

Li X, Zhou Y, Yang Y, Yang S, Sun X, Yang Y (2015) Physiological and proteomics analyses reveal the mechanism of eichhorniacrassipes tolerance to high-concentration Cadmium stress compared with pistiastratiotes. PLoS One 10(4):e0124304. doi:10.1371/journal.pone.0124304

Li X, Zhou Y, Yang Y, Yang S, Sun X, Yang Y (2015) Physiological and proteomics analyses reveal the mechanism of Eichhornia crassipes tolerance to high-concentration cadmium. PLoS One 10(4), e0124304

Liu D, An Z, Mao Z, Ma L, Lu Z (2015) Enhanced heavy metal tolerance and accumulation by transgenic sugar beets expressing Streptococcus thermophilusStGCS-GS in the presence of Cd, Zn and Cu alone or in combination. PLoS One 10(6):e0128824. doi:10.1371/journal.pone.0128824

Liu X, Wu FH, Li JX, Chen J, Wang GH, Wang WH, Hu WJ, Gao LJ, Wang ZL, Chen JH, Simon M, Zheng HL (2016) Glutathione homeostasis and Cd tolerance in the Arabidopsis sultr1;1-sultr1;2 double mutant with limiting sulfate supply. Plant Cell Rep 35(2):397–413. doi:10.1007/s00299-015-1892-8

Llugany M, Miralles R, Corrales I, Barceló J, Poschenrieder C (2012) *Cynara cardunculus* a potentially useful plant for remediation of soils polluted with cadmium or arsenic. J Geochem Explor 123:122–127 doi:http://dx.doi.org/10.1016/j.gexplo.2012.06.016

Lundberg DS, Lebeis SL, Paredes SH, Yourstone S, Gehring J, Malfatti S, Tremblay J, Engelbrektson A, Kunin V, TGd R, Edgar RC, Eickhorst T, Ley RE, Hugenholtz P, Tringe SG, Dangl JL (2012) Defining the core Arabidopsis thaliana root microbiome. Nature 488(7409):86–90 doi:http://www.nature.com/ nature/journal/ v488/n7409/abs /nature11237

Luo J, Qi S, Peng L, Xie X (2015) Phytoremediation potential of cadmium-contaminated soil by Eucalyptus globulus under different coppice systems. Bull Environ Contam Toxicol 94(3):321–325. doi:10.1007/s00128-014-1450-z

Ma J, Sheng H, Li X, Wang L (2016) iTRAQ-based proteomic analysis reveals the mechanisms of silicon-mediated cadmium tolerance in rice (*Oryza sativa*) cells. Plant Physiol Biochem 104:71–80. doi:10.1016/j.plaphy.2016.03.024

Mendes R, Kruijt M, de Bruijn I, Dekkers E, van der Voort M, Schneider JHM, Piceno YM, DeSantis TZ, Andersen GL, Bakker PAHM, Raaijmakers JM (2011) Deciphering the rhizosphere microbiome for disease-suppressive bacteria. Science 332(6033):1097–1100. doi:10.1126/science.1203980

Mleczek M, Rutkowski P, Rissmann I, Kaczmarek Z, Golinski P, Szentner K, Strażyńska K, Stachowiak A (2010) Biomass productivity and phytoremediation potential of *Salix alba* and

Salix viminalis. Biomass Bioenergy 34 (9):1410–1418. doi:http://dx.doi.org/10.1016/j.biombioe.2010.04.012

Mohan TC, Castrillo G, Navarro C, Zarco-Fernandez S, Ramireddy E, Mateo C, Zamarreno AM, Paz-Ares J, Munoz R, Garcia-Mina JM, Hernandez LE, Schmulling T, Leyva A (2016) Cytokinin determines thiol-mediated arsenic tolerance and accumulation in Arabidopsis thaliana. Plant Physiol. doi:10.1104/pp.16.00372

Oh K, Li T, Cheng HY, Hu XF, Lin Q, Xie YH (2013) A primary study on assessment of phytoremediation potential of biofuel crops in heavy metal contaminated soil. Appl Mech Mater 295–298:1135–1138

Oh K, Cao T, Cheng H, Liang X, Hu X, Yan L, Yonemochi S, Takahi S (2015) Phytoremediation potential of sorghum as a biofuel crop and the enhancement effects with microbe inoculation in heavy metal contaminated soil. J Biosci Med 3(6):9–14

Pajuelo E, Carrasco JA, Romero LC, Chamber MA, Gotor C (2007) Evaluation of the metal phytoextraction potential of crop legumes. Regulation of the expression of O-acetylserine (thiol) lyase under metal stress. Plant Biol 9(5):672–681. doi:10.1055/s-2007-965439

Pandey VC, Singh K, Singh JS, Kumar A, Singh B, Singh RP (2012) *Jatropha curcas*: a potential biofuel plant for sustainable environmental development. Renew Sust Energ Rev 16(5):2870–2883 doi:http://dx.doi.org/10.1016/j.rser.2012.02.004

Pandey VC, Bajpai O, Singh N (2016) Energy crops in sustainable phytoremediation. Renew Sust Energ Rev 54:58–73. doi:http://dx.doi.org/10.1016/j.rser.2015.09.078

Passatore L, Rossetti S, Juwarkar AA, Massacci A (2014) Phytoremediation and bioremediation of polychlorinated biphenyls (PCBs): state of knowledge and research perspectives. J Hazard Mater 278:189–202. doi:http://dx.doi.org/10.1016/j.jhazmat.2014.05.051

Peuke AD, Rennenberg H (2005) Phytoremediation. EMBO Rep 6(6):497–501. doi:10.1038/sj.embor.7400445

Pilon-Smits EAH, Zhu YL, Sears T, Terry N (2000) Overexpression of glutathione reductase in Brassica juncea: effects on cadmium accumulation and tolerance. Physiologia Plantarum 110(4):455–460. doi:10.1111/j.1399-3054.2000.1100405.x

Polechońska L, Klink A (2014) Accumulation and distribution of macroelements in the organs of *Phalaris arundinacea* L.: implication for phytoremediation. J Environ Sci Health A 49(12):1385–1391. doi:10.1080/ 10934529.2014.928494

Rowe RL, Street NR, Taylor G (2009) Identifying potential environmental impacts of large-scale deployment of dedicated bioenergy crops in the UK. Renew Sust Energ Rev 13(1):271–290 doi:http://dx.doi.org/10.1016/j.rser.2007.07.008

Rugh CL, Wilde HD, Stack NM, Thompson DM, Summers AO, Meagher RB (1996) Mercuric ion reduction and resistance in transgenic Arabidopsis thaliana plants expressing a modified bacterial merA gene. Proc Natl Acad Sci U S A 93(8):3182–3187

Ruiz ON, Alvarez D, Torres C, Roman L, Daniell H (2011) Metallothionein expression in chloroplasts enhances mercury accumulation and phytoremediation capability. Plant Biotechnol J 9(5):609–617. doi:10.1111/j.1467-7652.2011.00616.x

Sabeen M, Mahmood Q, Irshad M, Fareed I, Khan A, Ullah F, Hussain J, Hayat Y, Tabassum S (2013) Cadmium phytoremediation by *Arundo donax* L. from contaminated soil and water. Biomed Res Int 2013:9. doi:10.1155/2013/324830

Samuelsen AI, Martin RC, Mok DWS, Mok MC (1998) Expression of the Yeast FRE genes in transgenic tobacco. Plant Physiol 118(1):51–58

Schiavon M, Pilon M, Malagoli M, Pilon-Smits EA (2015) Exploring the importance of sulfate transporters and ATP sulphurylases for selenium hyperaccumulation-a comparison of *Stanleya pinnata* and *Brassica juncea* (Brassicaceae). Front Plant Sci 6:2. doi:10.3389/fpls.2015.00002

Sentchilo V, Mayer AP, Guy L, Miyazaki R, Green Tringe S, Barry K, Malfatti S, Goessmann A, Robinson-Rechavi M, van der Meer JR (2013) Community-wide plasmid gene mobilization and selection. ISME J 7(6):1173–1186. doi:10.1038/ismej.2013.13

Sergeant K, Renaut J, Hausman J-F (2012) Proteomics as a toolbox to study the metabolic adjustment of trees during exposure to metal trace elements. In: Gupta KD, Sandalio ML (eds) Metal

toxicity in plants: perception, signaling and remediation. Springer, Berlin/Heidelberg, pp 143–164. doi:10.1007/978-3-642-22081-4_7
Simpson JA, Picchi G, Gordon AM, Thevathasan NV, Stanturf J, Nicholas I (2009) Environmental benefits associated with short-rotation woody crops. IEA Bioenergy, doi:http://www.shortrotationcrops.org/taskreports.htm
Song WY, Sohn EJ, Martinoia E, Lee YJ, Yang YY, Jasinski M, Forestier C, Hwang I, Lee Y (2004) Engineering tolerance endophytic bacteria improve phytoremediation of water soluble volatile, organic pollutants. Nat Biotechnol 22(583–588)
Sunkar R, Kaplan B, Bouche N, Arazi T, Dolev D, Talke IN, Maathuis FJ, Sanders D, Bouchez D, Fromm H (2000) Expression of a truncated tobacco NtCBP4 channel in transgenic plants and disruption of the homologous Arabidopsis CNGC1 gene confer Pb2+ tolerance. Plant J 24(4):533–542
Tennstedt P, Peisker D, Böttcher C, Trampczynska A, Clemens S (2009) Phytochelatin synthesis is essential for the detoxification of excess zinc and contributes significantly to the accumulation of zinc. Plant Physiol 149(2):938–948
Tiwari V, Patel MK, Chaturvedi AK, Mishra A, Jha B (2016) Functional characterization of the Tau Class Glutathione-S-Transferases gene (SbGSTU) promoter of *Salicornia brachiate* under salinity and osmotic stress. PLoS One 11(2):e0148494. doi:10.1371/journal.pone.0148494
Trap S, Kohler A, Larsen LC, Zambrano KC, Karlson U (2005) Phytotoxicity of fresh and weathered diesel and gasoline to willow and poplar trees. J Soils Sediments 1:71–76
Turner TR, Ramakrishnan K, Walshaw J, Heavens D, Alston M, Swarbreck D, Osbourn A, Grant A, Poole PS (2013) Comparative metatranscriptomics reveals kingdom level changes in the rhizosphere microbiome of plants. ISME J 7(12):2248–2258. doi:10.1038/ismej.2013.119
Udawat P, Jha RK, Sinha D, Mishra A, Jha B (2016) Overexpression of a cytosolic abiotic stress responsive universal stress protein (SbUSP) mitigates salt and osmotic stress in transgenic tobacco plants. Front Plant Sci 7:518. doi:10.3389/fpls.2016.00518
USEPA (2000) Introduction to phytoremediation. EPA 600/R-99/107, U.S. Environmental Protection Agency, Office of Research and Development, Cincinnati, OH
Wojas S, Clemens S, Hennig J, Sklodowska A, Kopera E, Schat H, Bal W, Antosiewicz DM (2008) Overexpression of phytochelatin synthase in tobacco: distinctive effects of AtPCS1 and CePCS genes on plant response to cadmium. J Exp Bot 59(8):2205–2219. doi:10.1093/jxb/ern092
Ye WL, Khan MA, McGrath SP, Zhao FJ (2011) Phytoremediation of arsenic contaminated paddy soils with Pterisvittata markedly reduces arsenic uptake by rice. Environ Pollut 159(12):3739–3743. doi:10.1016/j.envpol.2011.07.024
Yergeau E, Sanschagrin S, Maynard C, St-Arnaud M, Greer CW (2014) Microbial expression profiles in the rhizosphere of willows depend on soil contamination. ISME J 8(2):344–358. doi:10.1038/ismej.2013.163
Yergeau E, Bell T, Champagne J, Maynard C, Tardif S, Tremblay J, Greer C (2015) Transplanting soil microbiomes leads to lasting effects on willow growth, but not on the rhizosphere microbiome. Front Microbiol 6:1436. doi:10.3389/fmicb.2015.01436
Zabaniotou AA, Kantarelis EK, Theodoropoulos DC (2008) Sunflower shells utilization for energetic purposes in an integrated approach of energy crops: laboratory study pyrolysis and kinetics. Bioresour Technol 99(8):3174–3181. doi:10.1016/j.biortech.2007.05.060
Zhou X, Li L (2010) Think outside the box: selenium volatilization altered by a broccoli gene in the ubiquinone biosynthetic pathway. Plant Signal Behav 5(1):76–77
Zhu YL, Pilon-Smits EAH, Jouanin L, Terry T (1999) Overexpression of glutathione synthetase in Brassica juncea enhances cadmium accumulation and tolerance. Plant Physiol 119:73–79. Stress compared with Pistia stratiotes. PLOS ONE 10(4):e0124304

Sustainability of Oil Seed-Bearing Bioenergy Plants in India (Jatropha, Karanja, and Castor) for Phytoremediation: A Meta-analysis Study

17

Dipesh Kumar, Nitesh Bhatia, and Bhaskar Singh

Abstract

In the present era when several countries are facing dual challenges of energy insecurity and environmental pollution, growing plants having bioenergy potential for reclamation of contaminated sites seems to offer a rather holistic approach to tackle both the problems simultaneously. Combining both the technologies (bioenergy production and phytoremediation of contaminated sites) apparently improves the overall environmental sustainability and economic feasibility of the individual techniques. Here, we discuss the ecological sustainability and economic feasibility of an integrated approach toward bioenergy production and decontamination of polluted sites using *Jatropha curcas*, *Millettia pinnata*, and *Ricinus communis*. This review paper attempts to provide a comparative snapshot approach toward the phytoremediation dimension of the three bioenergy plants taken for study in Indian scenario.

Keywords

Bioenergy • Castor • Jatropha • Karanja • Meta-analysis • Phytoremediation • Sustainability

D. Kumar • B. Singh
Centre for Environmental Sciences, Central University of Jharkhand, Ranchi-835205, Jharkhand, India

N. Bhatia (✉)
Centre for Business Administration, Central University of Jharkhand, Ranchi-835205, Jharkhand, India
e-mail: nitesh.bhatia@cuj.ac.in

K. Bauddh et al. (eds.), *Phytoremediation Potential of Bioenergy Plants*,
DOI 10.1007/978-981-10-3084-0_17

17.1 Introduction

The worldwide demand for energy is growing day by day in line with increase in population and industrialization. Fossil fuels have historically remained the single largest source of energy for sustaining our socioeconomic structure. Since the availability of fossil fuels is finite and its demand is increasing steadily, there has been a tremendous surge in exploration and development of other alternative sources of energy. Increase in interest for alternatives is also attributed to the emission of greenhouse gases and several other noxious pollutants when fossil fuels are used for energy production. India is currently the fastest growing major economy in the world at 7.5% per annum (IMF 2016). In order to support, sustain, and supplement India's economic growth, there will be an unprecedented increase in demand for energy (Govt. of India 2009). India needs to substantially augment its installed energy capacity for meeting this increase in demand. India cannot afford to depend entirely on fossil fuels for meeting its energy needs as a significant proportion of oil (80%), coal (18%), and gas (23%) that is currently being used is imported and the share of import (particularly for petroleum and gas) is only likely to increase over time as their demand is growing steadily, while the local reserves remain insignificant. Although the share of fossil fuels in primary energy mix is likely to fall in the coming years due to the adoption of clean energy policy in which high share of renewable in energy mix is proposed, coal is highly likely to remain our single largest source of energy for atleast 2 decades. India recently through its Intended Nationally Determined Contribution (INDC) has proposed 33–35% reduction in emission intensity of its GDP (avoided emission equivalent to 3.59 billion tonne of CO_2 over business as usual) by 2030 relative to 2005 levels. This can be achieved by improvement of energy conversion efficiency by minimization of losses, by implementing energy conservation measures, by using cleaner forms of energy, and by increasing the carbon sinks or any combination thereof. For achieving the INDC targets, India has proposed to augment the share of renewable in its energy mix (40% share by 2030) and creation of carbon sinks for up to 2.5–3 billion tonnes (CO_2 equivalent) by creation of forest and tree cover by 2030. Among other renewables, bioenergy can be neutral (conditional) in terms of overall emission of CO_2 when biomass or biomass-based products are used for production of energy (Zanchi et al. 2012). The Government of India through INDC has proposed to install 10 GW of power capacity by utilizing biomass as energy source, and this can be coupled to creation of carbon sinks by employing plants having bioenergy potential. Although an extensive bioenergy policy has not been formulated, the government through its national biofuel policy 2009 can consider the integration of enhancement of carbon sink (biomass) and bioenergy production. The national biofuel policy envisages blending of 20% of bioethanol and biodiesel in gasoline and high-speed diesel, respectively, by 2017 (Govt. of India 2009). The policy focuses on utilization of marginal and non-arable lands for cultivation of non-edible biodiesel feedstock plants. The focus of the policy is mostly centered on *Jatropha curcas* (commonly known as *Jatropha*) and *Millettia pinnata* (commonly known as Karanja) as biodiesel feedstock. Biodiesel can be used in place of or in addition to mineral diesel, and the importance of diesel alternatives for India can be understood by the fact that use of diesel in India is Ca. 4.5 times higher than petrol and only an insignificant proportion of total demand is met by locally extracted petroleum which makes us highly vulnerable to supply crisis besides inducing heavy pressure on exchequer (Gopinathan and Sudhakaran 2011). Biodiesel has several advantages over mineral diesel such as higher cetane number, higher flash point and lubricity, cleaner emission profile, biodegradability, renewability, nontoxicity, and local production opportunity (Ma and Hanna 1999). Non-edible biodiesel feedstock producing oil-rich seed such as *J. curcas*, *M. pinnata*, and *R. communis* have been extensively researched for the production of biodiesel.

Anthropogenic activities such as mining of minerals, use of synthetic pesticides, discharge of untreated industrial effluents, etc. have led to the contamination of soil (Table 17.1) which is one the most important natural resources. Contaminants arising from these activities have altered the fertility, integrity, and physiochemical characteristics of soil. Among these contaminants, inorganic contaminants such as heavy metals (Hg, Cd, Cr, Co, Pb, etc.) and organics such as pesticides and other recalcitrant compounds are of especial environmental concern as they can be hazardous to most biota even at trace levels (Mattina et al. 2003). The Central Pollution Control Board (CPCB) being the statutory nodal body of India entrusted with the responsibility for monitoring and control of pollution has identified 43 critically polluted areas in 16 states having comprehensive environmental pollution index (CEPI) rating greater than 70, and several of these areas are contaminated with hazardous metals such as Cd, Cr, Hg, Pb, etc. However, a comprehensive study on distribution and areal extent of contamination in India is still awaited. The anthropogenic contamination of environment by such hazardous metals is likely to be concentrated around the industrial clusters and mining areas. The presence of heavy

Table 17.1 Major sources of hazardous metals and metalloids

Metals	Major sources
Chromium (Cr)	Mining industry, industrial coolants, leather tanning, manufacturing of chromium salts, etc.
Lead (Pb)	e-waste, lead acid batteries, paints, smelting operations, bangle industry, coal-based thermal power plants, ceramics, etc.
Arsenic (As)	Geogenic sources, smelting operations, thermal power plants, combustion of As-containing fuels, etc.
Mercury (Hg)	Combustion of coal, chlor-alkali plants, fluorescent lamps, hospital waste (damaged thermometers, sphygmomanometers, barometers), electrical appliances, etc.
Nickel (Ni)	Thermal power plants, smelting operations, battery industry, etc.
Copper (Cu)	Sulfuric acid plant, mining, smelting operations, vanadium spent catalyst, electroplating, etc.
Cadmium (Cd)	Waste batteries, zinc smelting, e-waste, paint sludge, incinerations and fuel combustion, etc.
Zinc (Zn)	Electroplating, smelting, etc.

Lone et al. (2008)

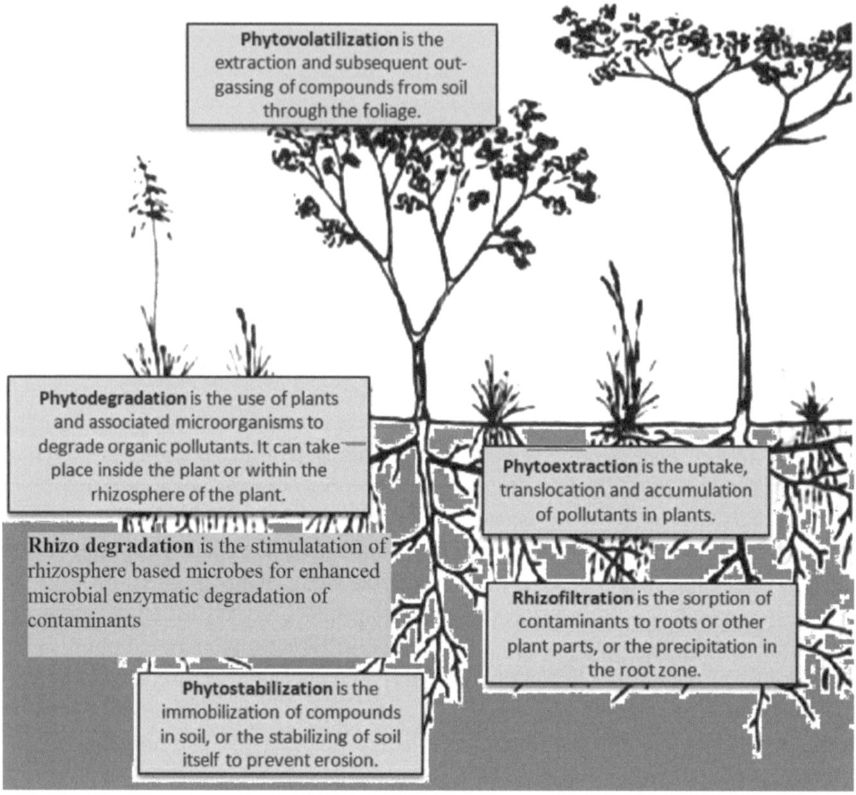

Fig. 17.1 Mechanism of phytoremediation (Adapted from Gomes 2012)

metals in soil or groundwater is potentially hazardous due to their inherent toxicity to biota in elemental and combined form (Rascio and Navari-Izzo 2011). Some of the heavy metals such as Fe, Mn, Co, Cu, Mo, Zn, and Ni are essential for normal metabolism and growth of plants, while others such as Se, As, Cd, Pb, or Hg have no known physiological role in plants. The former metals cause toxicity when their concentration exceeds supraoptimal levels, while the latter elements are inherently toxic to plants (Rascio and Navari-Izzo 2011). Techniques such as excavation followed by dumping at landfill sites and stabilization of contaminants by addition of chemical additives have been practiced, but these techniques are merely a waste containment technique, are expensive, and may involve addition of synthetic chemicals to the soil. Certain plants have shown rather appealing characteristics for bioremediation of these contaminants by utilizing a set of mechanisms (Fig. 17.1) collectively known as phytoremediation for the uptake, accumulation, stabilization, and/or detoxification of these contaminants. There are certain plants (hyperaccumulators) which can accumulate exceedingly high levels of contaminants without suffering from phytotoxicity (Tangahu et al. 2011). Unfortunately, phytoremediation is a rather slow affair (Khan 2005), and therefore the contaminated site under consideration cannot be put to any economic use for some time unless the plants used have some sort of agronomic value. Valorization of biomass produced during the period of phytoremediation by some means is therefore invaluable (Ghosh and Singh 2005). Among other valorization systems, utilization of biomass for production of bioenergy appears to offer a sustainable and integrated approach for phytoremediation and bioenergy production by using plants having bioenergy potential for phytoremediating contaminated sites (Pandey et al. 2016). Edible crops cannot be used for the phytoremediation of contaminated sites as it might lead to exposure of biota present at higher trophic levels to magnified levels of hazardous heavy metals. Certain plants have developed traits for accumulation of exceedingly high levels of heavy metals from the growth media, and high concentration of hazardous heavy metals might play a protective role for plants against herbivory (Jhee et al. 2006). Cultivation of plants for bioenergy production is perhaps the most attractive valorization opportunity for the metal-infested biomass (Ghosh and Singh 2005).

Other advantages of growing plants for decontamination can be in terms of reduction in soil erosion, improvement in the availability of soil moisture and groundwater, wasteland reclamation, and provision for wild life habitat. Thus phytoremediation appears to offer multiple environmental advantages such as it is a relatively cheap decontamination technique, and when combined with bioenergy production, the overall environmental sustainability and economic feasibility of both the technologies can be greatly improved. Therefore a critical examination of plants having bioenergy potential for their capacity to bioremediate environmental contaminants such as heavy metals and other hazardous chemicals is urgently required.

The use of edible crops for bioenergy production in India is largely unsustainable as we do not have enough produce to meet even our entire food demand, and the use of edible crops can lead to the infamous food versus fuel dilemma associated with first-generation biofuel feedstock (edible crops). The use of edible crops for bioenergy production is only feasible for nations having excess of produce. For India, second-generation biofuel feedstock (non-edible plants) are more suitable, and the national biofuel policy envisages 20% blending of biodiesel produced from non-edible feedstock in high-speed diesel by 2017 (Govt. of India 2009). The focus of the policy is on the cultivation of *J. curcas* and *M. pinnata* on waste/marginal lands. Besides these two species, other potential nonedible biodiesel feedstocks have also been analyzed by several researchers. *Ricinus communis* (commonly known as Castor) belonging to the flowering plant family Euphorbiaceae is an indigenous, fast-growing perennial shrub which produces seeds highly rich in oil (40–60%), and India is currently the largest exporter of caster bean oil. Although bioenergy plants have not been explored extensively for assessment of their phytoremediation traits, some comparative preliminary studies have highlighted the ability of few bioenergy plants for their potential use in bioremediation of contaminants. Most of the work on hyperaccumulating species capable of producing quality oil for production of biodiesel has revolved around the species of Brassicaceae (Rascio and Navari-Izzo 2011). Among the members of Brassicaceae, *Brassica napus* (rapeseed/canola), *Brassica rapa* (field mustard), and *Brassica juncea* (Indian mustard) have shown

Table 17.2 Major heavy metal-contaminated sites in India

Cr	Pb	Hg	As	Cu
Ranipet (Tamil Nadu)	Ratlam (Madhya Pradesh)	Kodaikanal (Tamil Nadu)	Tuticorin (Tamil Nadu)	Tuticorin (Tamil Nadu)
Kanpur (Uttar Pradesh)	Bandalamottu Mines (Andhra Pradesh)	Ganjam (Odisha)	Ballia and other districts (Uttar Pradesh)	Singhbhum Mines (Jharkhand)
Vadodara (Gujarat)	Vadodara (Gujarat)	Singrauli (Madhya Pradesh)	–	Malanjkhand (Madhya Pradesh)
Talcher (Odisha)	Korba (Chattisgarh)	–	–	–

CPCB (2009)

Fig. 17.2. (**a**) *Jatropha* plant, (**b**) Karanja plant, (**c**) Castor plant, (**d**) seeds of *Jatropha* plant, (**e**) seeds of Karanja plant, (**f**) seeds of castor plant

good metal accumulation traits (Ebbs and Kochian 1997), but since these species produce edible oils, they are not fit for production of biodiesel in India. Table 17.2 lists some of the major industrial clusters polluted with heavy metals and metalloids.

In this review we are reporting the suitability and sustainability of three oil seed-bearing plants, namely, *J. curcas*, *M. pinnata*, and *R. communis*, for phytoremediation of contaminated sites as well as on the potential of biomass valorization for the production of bioenergy (biodiesel) in India. Figure 17.2 illustrates *Jatropha*,

Karanja, and castor plant and their seeds. Although several routes for lignocellulosic biomass to bioenergy conversion are available and the concept of lignocellulosic biorefinery is increasingly gaining scientific attention, we will mostly deal with production of biodiesel (via transesterification) as other conversion routes are beyond the scope of this chapter.

17.2 Basics of Phytoremediation

In order to be considered as a potential phytoremediating species, a plant must have (1) effective tolerance to excessively high levels of metal(s); (2) capability to uptake, accumulate, transform, stabilize, and/or detoxify contaminants without suffering from phytotoxic damage; and (3) a fast growth rate. Studies have shown high degree of tolerance to metals and their accumulations by certain plants which naturally grow on metalliferous soil (Bert et al. 2000). Interestingly, some of the plants can also tolerate, grow, and reproduce on metal-infested sites arising due to anthropogenic activities (Rascio and Navari-Izzo 2011). Tolerance in such species (hypertolerant plants) has been explained by two different mechanisms: (1) *excluders* who retain and detoxify metals in root tissues and have shown minimum translocation of metals to aboveground biomass (Hall 2002) and (2) *hyperaccumulators* who actively take up very high levels of metals from soil with a relatively high translocation of metals to aboveground biomass. Several plant species have been recognized as hyperaccumulators (plants having ability to uptake exceedingly high levels of metals than other congener plants without showing any signs of phytotoxicity) (Reeves 2006), but only few species have been assessed for their potential application in phytoremediation, and most of the recognized hyperaccumulators have a very slow rate of growth (Cunningham et al. 1995). Several hypotheses for explaning the observed hyperaccumulation trait in certain plants have been proposed. Currently, the most accepted hypothesis envisages a protective role for the presence and excessive levels of heavy metals in aboveground biomass of hyperaccumulators against grazers, browsers, and pests (*elemental defense hypothesis*) (Noret et al. 2007). Several research-based findings have supported the *elemental defense hypothesis* including those on Ni (Jhee et al. 2006), As (Rathinasabapathi et al. 2007), Cd (Jiang et al. 2005), Se (Galeas et al. 2008), and Zn hyperaccumulators (Behmer et al. 2005). Others have proposed the existence of a joint role of secondary metabolites (phenol-based organics) and inorganic metals in defense against animal attack on plants (Boyd 2007). When the concentration of contaminants in growth substrate exceeds the threshold toxicity levels, phytotoxic effects may arise in non-tolerant species via altercation of several physiological processes operating at cellular or molecular levels (Rascio and Navari-Izzo 2011). This may arise by causing inactivation of enzymes, blocking or deactivation of metabolically important function groups, substitution or displacement of essential elements, disruption of membrane integrity, or by excessive production of reactive oxygen species and free radicals (Pagliano et al. 2006). Plants exposed to excessive levels of heavy metals resort to a variety of synergistic strategies to limit the entry of metals into cytoplasm by complexing them with several exudated organic acids or the anionic groups present on the cell wall and their subsequent storage in apoplastic regions (Watanabe and Osaki 2002). Most of the metals that enter the plant are stored in the vacuole of root cells in detoxified form by forming a complex with metal-binding peptides, amino acids, and other organic acids (Hall 2002). This limits the translocation of metals to metabolically active photosynthetic cells and thus prevents the damage and altercation of photosynthesis. Plants are also known to enhance their antioxidant system in response to oxidative stress. Few comparative studies on uptake on heavy metals by hyperaccumulating species and congener non-hyperaccumulators have revealed that most of the key processes in hyperaccumulation do not involve novel genes and depend on genes that are common to both (Verbruggen et al. 2009). These genes are differently regulated and expressed in hyperaccumulators and congener non-hyperaccumulators. Phytoremediation involves a set of mechanisms for the uptake, accumulation, storage, stabilization, and/or detoxification of contaminants present in the growth substrate. These include (1) phyto-stabilization, (2) phytoextraction, (3) rhizo-filtration, (4) phyto-volatilization, (5) rhizo-degradation, and (6) phyto-degradation (Fig. 17.1). A different set of mechanisms operates in tandem for the removal of inorganic and organic contaminants. Out of the six mechanisms mentioned above, a set of four mechanisms, namely, phytoextraction, phyto-stabilization, phyto-volatilization, and rhizo-filtration, are recognized in the phytoremediation of hazardous metals (Tangahu et al. 2011), while a different set of mechanisms operates for the phytoremediation of hazardous organics. Analysis of comparative effectiveness of plants for phytoremediation is a complicated affair attributed to the differences in the genetic makeup of plant species, presence of different levels and combinations of contaminants in growth substrates, differences in agroclimatic conditions and physicochemical characteristics of growth substrate, etc. Laboratory-based experimental setup has also been investigated, but perfect simulation of variability existing in field conditions can rarely be achieved. These issues have hampered the comparability of different studies. For making phytoremediation a profit-making affair and also for increasing the ecological sustainability and economic feasibility of the process, valorization of the produced biomass by some means is invaluable. Valorization of biomass by bioenergy production appears to be a rather promising opportunity, but for its practical application, a fast-growing plant species is required.

17.3 Sustainability of Bioenergy Plants *Jatropha*, Karanja, and Castor for Phytoremediation of Contaminated Sites

Bioenergy plants *Jatropha* and Castor have been extensively investigated for their phytoremediation potential, but research on phytoremediation efficiency of Karanja is sparse. All the three plants have high content of oil (30–60%) in their seed, and their bioenergy potential has been widely explored (Murugesan et al. 2009). Since India as of today cannot afford to employ edible crops for bioenergy production, the emphasis has been on the utilization of non-edible plants particularly for the production of biodiesel. Comparative fuel properties of *Jatropha*, Karanja, and castor

Table 17.3 Properties of *Jatropha*, Karanja, and castor oil along with their respective biodiesel and no. 2 mineral diesel

Properties	*Jatropha* oil	*Jatropha* biodiesel	Karanja oil	Karanja biodiesel	Castor oil	Castor biodiesel	Mineral diesel (no. 2)
Viscosity @ 40 °C (cSt)	24.5[a]	4.80[a]	26.88[e]	5.44[d]	240[f]	15.98[g]	1.3–4.1
Acid value (mg KOH g^{-1} oil)	28[a]	0.40[a]	19.88[d]	0.44[d]	0.36[f]	0.11[g]	–
Major fatty acids	18:0 (9.7%)[b]		16:0 (10.6%)[b]		18:1-OH (92%)[g]		–
	18:1(40.8%)		18:1(49.4%)				
	18:2 (32.1%)		18:2 (19.0%)				
Iodine value (g I_2 100g^{-1} oil)	93[b]		80.9[b]		100[f]		–
Saponification number	202.6[b]		196.7[b]		185[f]		–
Cloud point (°C)	–	8[a]	–	5[d]		−23[g]	−35 to +5
Pour point (°C)	4[a]	2[c]	–	−2[d]		−45[g]	−35 to −15
Flash point (°C)	225[a]	135[c]	–	158[d]	230[f]	190.7[g]	60 to 80
Cetane number	57[a]	57[a]	–	57[d]		50[g]	40 to 55

16:0, palmitic acid; 18:0, stearic acid; 18:1, oleic acid; 18:1-OH, ricinoleic acid; 18:2, linoleic acid
[a]Tiwari et al. (2007), [b]Bringi (1987), [c]Vyas et al. (2009), [d]Sharma et al. (2010), [e]Sharma et al. (2009), [f]Ingle and Nandedkar (2013), [g]Omari et al. (2015)

oil along with their respective biodiesel and no. 2 mineral diesel are listed in Table 17.3. Bioethanol on the other hand is mostly being produced from sugarcane juice (Ghosh and Ghose 2003). Sugarcane being an edible crop needs arable land for its cultivation besides having intensive water demand. Since the gap between demand and local availability of fertilizers has increased over time, we cannot afford to divert fertilizers from agricultural fields for the cultivation of biofuel feedstocks. Fortunately, biofuel feedstocks having very minimal demand are known not just in terms of fertilizers but also in terms of freshwater demand, and these plants have opened the door for sustainable production of biofuels in India as well as around the globe. Table 17.4 lists some of the important characters of *Jatropha*, Karanja, and castor. *Jatropha*, Karanja, and castor are known as hardy species capable of withstanding drought, varying degrees of waterlogging, and salinity. These plants can even be grown on wastelands and have the potential to bring about their reclamation (Murugesan et al. 2009). This can also facilitate creation of carbon sinks which can offset the emission of CO_2 from other sources and help India earn carbon credits. Thus these plants can serve as an ideal feedstock for the production of biodiesel, and if they can be used for the phytoremediation of contaminants, this can add another dimension of environmental sustainability. In comparison to petroleum-based fuels, the production of biodiesel is currently expensive, and the cost of the feedstock

Table 17.4 Important characteristics of *Jatropha curcas*, *Millettia pinnata*, and *Ricinus communis*

Plant	*Jatropha curcas*	*Millettia pinnata*	*Ricinus communis*
Habit	Semi-evergreen shrub or small tree	Medium-sized tree	Fast-growing perennial shrub
Oil content (wt %)	25–40[a]	30–35[d]	40–62[g]
Yield (Kg ha^{-1})	2500[b]	5499[b]	1250[h]
Gestation period (Yrs)	2[c]	4–7[e]	1[l]
N_2 fixation	No	Yes[f]	No
Moisture and nutrient requirements	Low	Low	Low
Other usage	Some industrial and medicinal usage	Used for making soap and has medicinal usage	Several industrial and medicinal usage
Cost of oil (₹Kg^{-1})	26[i](in 2012)	29.71[j] (in 2010)	52.23 [k](in 2010)
Cost of biodiesel (₹Kg^{-1})	28.64[i] (Ethyl ester)*	17.09 [j, **]	70.4 [k, ***]

[a](Openshaw 2000), [b](Azam et al. 2005), [c](Singh 2013), [d](Karmee and Chadha 2005), [e](Kesari and Rangan 2010), [f](Samuel et al. 2013), [g](Scholz and da Silva 2008), [h](Shrirame et al. 2011), [i](Nevase et al. 2012), [j](Nagarhalli et al. 2010), [k](Santana et al. 2010), [l](Greenwood and Bewley 1982)
*Without valorization of by-products by economic means
**After valorization of by-products (glycerol at 0.122$ Kg^{-1} biodiesel and seed cake at 0.1$ Kg^{-1} biodiesel produced)
***Without valorization of by-products but includes cost of glycerol separation

(vegetable oil or animal fat) is the single largest contributor (70–75% of the total cost) to the overall cost of biodiesel (Table 17.4). Thus in order to compete with petroleum-based fuels, the cost of the biodiesel has to be lowered for making it attractive to the market. Valorization of by-products of biodiesel production such as glycerol (accounting for Ca. 10% of feedstock) as raw material for cosmetic, pharmaceutical, or other industries and seed cake as organic manure/soil amendment or as animal feed can be an attractive opportunity for cost reduction of biodiesel. In this regard the concept of biorefinery for the production of multiple products from biomass by utilization of its every fraction is increasingly gaining attention. Similarly valorization of metal-infested biomass produced during the course of decontamination for production of bioenergy or bioenergy carriers can improve the overall economics of both the processes.

17.3.1 *Jatropha curcas* and Its Phytoremediation Potential

Jatropha curcas is a semi-evergreen shrub or small tree belonging to the flowering plant family Euphorbiaceae. It is native to tropical America but has naturalized in several tropical and subtropical areas of the world and grows widely in India, Africa, and Southeast Asia. It has several desirable characteristics of an agronomic crop such as easy establishment, grows relatively fast, very few pests and diseases are known, grows under various rainfall regime (200–2000 mm), can tolerate high degree of aridity, drought resistance, leaves are not browsed by herbivores, ability to grow on most terrain including gravel, sandy and saline soils, and can even be grown on wastelands (Openshaw 2000). Seeds contain a highly toxic protein called curcin and are unfit for use as an animal feed unless it is subjected to some treatment. Several plant parts have medicinal properties, and its bark contains tannin. *Jatropha* produces seed rich in oil (25–40% by wt) which has been extensively studied for its potential as an alternative to fossil fuels. *Jatropha* oil mostly contains esters of stearic, oleic, and linoleic acids (Tiwari et al. 2007) and thus possesses suitable fatty acid profile for the production of biodiesel. It has found a central role in the national biofuel policy of India as besides having suitability as a fossil fuel substitute it can also provide employment opportunities to the rural population (Govt. of India 2009).

The potential of *Jatropha* as a phytoremediating species has been analyzed for the removal of inorganic (hazardous metals) as well as organic contaminants (pesticides, petroleum products, etc.) arising out of a multitude of activities such as mining, industrial discharge, agricultural application, oil spill, etc. Luhach and Chaudhry (2012) in their pot experiment study on assessment of growth performance of *Jatropha* cultivated using various proportions of multi-metal (Cd, Cr, Cu, Ni)-infested oil refinery sludge and control soil found *Jatropha* to effectively tolerate moderate levels of metals after 2 weeks of acclimatization period without any significant effect on growth parameters chosen for study relative to the specimens grown on control soil. They reported that the majority of accumulation was in roots of *Jatropha* plant with minimal translocation to aboveground biomass, thus suggesting *Jatropha* to behave as an excluder. In a 3-month experiment on the ability of *Jatropha* to uptake hazardous metals from sewage sludge, Ahmadpour et al. (2010) reported a decrease in concentration of Pb, Cu, Cr, Cd, and Zn by 78.3%, 75%, 77.2%, 78.5%, and 67.7%, respectively. Fly ash containing Fe, Al, Mn, Cu, and Cr was amended with fertile garden soil, and it was used as a growth substrate to analyze the ability of *Jatropha* to phytoextract the above mentioned metals. Garden soil amended with fly ash was able to produce 37% more biomass compared to specimens grown only on garden soil. Further, EDTA was used as an additive to improve the mobility of metals and also to compare the phytoextraction efficiency of metals using *Jatropha* with or without EDTA amendment. EDTA dose of 0.3 g Kg^{-1} was reportedly responsible for the enhancement of metal accumulation in root, stem, and leaves by 117%, 62%, and 86%, respectively. Based on this study, the authors (Jamil et al. 2009) suggested that *Jatropha* could be effectively used for the phytoextraction of metals from fly ash given the growth substrate be amended with fertile soil and chelating agents. Agamuthu et al. (2010) in their 180-day long experiment on studying the effectiveness of *Jatropha* for the remediation of used lubricating oil-contaminated soil (containing 2.5% and 1% *w/w* lubricating oil) with or without amendment by organic wastes (brewery spent grain, spent mushroom compost, and banana skin) reported a decrease in concentration of lubricating oil in the absence of organic amendments by 56.6% and 67.3% for soil containing 2.5% and 1% lubricating oil, respectively. The decrease in concentration of lubricating oil in the presence of organic amendments was reportedly 89.6% and 96.6% for soil containing 2.5% and 1% lubricating oil by weight. In addition, Singh et al. (2013) studied the capacity of *Jatropha* for amelioration of sodic soils and reported efficiency of *Jatropha* to be equivalent to that of *Prosopis juliflora* which is generally planted for the amelioration of sodic soils. *Jatropha* plants were able to grow and uptake lindane form soil spiked with different concentrations of lindane (5, 10, 15, and 20 mg Kg^{-1}), and *Jatropha* was able to reduce the concentration of lindane in soil by 89, 82, 77, and 72%, respectively, in the four treatments mentioned (Abhilash et al. 2013). Numerous other studies have been carried out on effectiveness of *Jatropha* for the phytoremediation of environmental contaminants and have revealed that *Jatropha* is likely to be a good candidate for the phytoremediation of moderately contaminated sites.

17.3.2 *Millettia pinnata* (Karanja) and Its Phytoremediation Potential

Millettia pinnata (L.) Pierre (previously known as *Pongamia pinnata*) is a fast-growing medium-sized tree belonging to the pea family Fabaceae. Being a legume it can fix atmospheric nitrogen (Samuel et al. 2013) and thus can improve the fertility of the soil. It is native to temperate and tropical Asia and is commonly found throughout India. It is well adapted to grow in arid conditions and can even tolerate high degree of soil salinity and waterlogging (Azam et al. 2005). Its demand for moisture and nutrient is rather low and can easily grow on marginal/wasteland. The seeds of Karanja contain about 30–35% of oil having high proportion of oleic acid (No 2011). Although the reported values of cold flow properties (cloud point and pour point) of Karanja biodiesel are relatively high, considering the type of climate we experience in India, it is not a big issue. The agronomic practices for Karanja is under research, and the hardy nature of Karanja tree has earned it a place in the national biofuel policy of India (Govt. of India 2009). The plant parts have found application in treatment of several diseases as per the traditional system of Indian medicine *Ayurveda* (Tanaka et al. 1992), and its oil is also used for making soap.

Unlike *Jatropha* and castor, the research on phytoremediation potential of Karanja is sparse, but few preliminary comparative studies have provided valuable insights.

M. pinnata was grown on landfill site waste of municipal origin containing various heavy metals, and except Fe it was able to substantially reduce the concentration of all the metals in waste soil within 60 days of plantation (Shirbhate and Malode 2012). Production of high biomass at a rapid rate was observed when plants were grown on waste soil. Besides, the concentration of chlorophyll (*a* and *b*) in plant leaves and that of sodium and potassium ions in plant tissues were relatively higher when plants were grown on waste soil in comparison to growth on control. Transverse section of *M. pinnata* grown on control and waste soil was also examined for comparison, and several differences were reported. The accumulation of pollutants by plants grown on waste soil was reportedly in the form of blackish deposits in vascular and cortical tissues (Shirbhate and Malode 2012). Kumar et al. (2009) in their in vitro study on the effect of Cu and Cr on germination of

M. pinnata seeds, its growth, metal tolerance, and uptake reported that the seedlings of *M. pinnata* can tolerate the presence of Cu (50–400 μM) and Cr (100–800 μM) in the growth medium. Cu was found to have greater toxicity to M. pinnata than Cr, restricted accumulation of both the the metals in the stem was observed, and seed coats showed an excellent property to hold metals and therefore it can be employed as an absorbent for decontamination of metal infested liquid waste. Overall the presence of heavy metals in the growth medium did not seemingly affect the growth characteristics of *M. pinnata* seedling. The growth performance of 1-year-old seedlings of *M. pinnata* irrigated with industrial effluent (in various proportions of 0, 25, 50, 75, and 100% with tap water) containing heavy metals was evaluated (Manzoor et al. 2015). They reported an increase of 7.4% in shoot length and 33% increase in number of leaves compared to control when plantations were grown on 10% effluent. The leaf area and number of branches reportedly grew by 8.5 and 3%, respectively, when plantations were irrigated with 50% effluent. However the value of all the estimated parameters for growth and tolerance analysis decreased with any further increase in proportion of effluent. Three tree species (*V. parviflora*, *S. saman*, and *M. pinnata)* were compared for their growth and phytoremediation potential when grown on Cu-infested soil amended with or without VAM and/or zeolite for 170 days, and the most efficient combination was found to be that of *M. pinnata* grown on Cu-infested soil amended by both VAM and zeolite. When *M. pinnata* was grown on copper-contaminated soil (amended by addition of VAM and zeolite), it was able to accumulate 1219.75 and 26 μg g^{-1} dry matter as copper in its root and shoot, respectively (Tulod et al. 2012).

A decrease in concentration of metals (Cd, Zn, Fe, Lb, Ni, Cr, Mn, and Cu) in coal mine spoil dump by 50–60% within 3 years of plantation was reported (Singh and Juwarkar 2014). Plant species when supplemented with biofertilizers and sugar industry sludge were able to tolerate high metal content in spoil dump and produced biomass at a rapid rate (Singh and Juwarkar 2014). In a study on remediation of coal overburden (coal mine spoil dump) using ten phytoremediator species (including *M. pinnata*) assisted by addition of biofertilizers (VAM and nitrogen-fixing bacteria) and sugar industry sludge, Singh and Juwarkar (2014) reported an economic return (including ecological benefits) to the tune of 36,562 US $ 20 years after plantation. The total investment required was reported to be 1250 US $ per hectare, and the reported payback period was only 2.5 years.

17.3.3 *Ricinus communis* (Castor) and Its Phytoremediation Potential

Ricinus communis (castor) is a fast-growing perennial shrub native to India, Eastern Africa, and Southeastern Mediterranean basin, but it has widespread distribution in the tropics. It belongs to the spurge family (Euphorbiaceae), and it is currently the only member of the genus *Ricinus*. In areas with suitable climate and little anthropogenic interference, it can easily become invasive and can even grow on wastelands. Some of the characteristic features of invasive species such as easy establishment, fast growth rate, short gestation period, ability to outcompete other plants growing nearby, drought tolerance, ability to grow on marginal/wasteland, absence of animals which keep their population in check, etc. are desirable characteristics for bioenergy plants. But it can also lead to several ecological disasters if the population remains unchecked. Castor seeds contain about 40–60% of its mass as oil having ricinoleic acid (18:1-OH) as the commonest fatty acid (Ca. 90%). India is currently the world's largest producer of castor oil producing nearly 0.73 Mt. of oil annually accounting for about 60% of the global production. Ricinoleic acid (12-hydroxy-(cis)-9-octadecenoic acid) contains a hydroxyl group at 12th carbon position, and the presence of hydroxyl group gives rise to several unique properties to castor oil and biodiesel produced from castor oil. These include: high viscosity, high polarity (and hence better solubility in transesterifying alcohol), better cold flow properties, etc. The viscosity of biodiesel produced from castor oil is often very high, and hence it cannot be used in neat form. This problem can be overcome by using lower blends of biodiesel in mineral diesel or by mixing castor biodiesel with biodiesel derived from other feedstocks such as Karanja and *Jatropha* as their viscosity is relatively low but has poor cold flow properties. The seeds of castor plant contain a highly toxic compound called ricin, and castor oil has several industrial and medicinal usages.

Ricinus communis has been reported as a metal-tolerant bioenergy plant having the capability to thrive in heavily polluted soils and has recognition as a potential phytoremediator for a variety of contaminants (Rajkumar and Freitas 2008 and Huang et al. 2011). Huang et al. (2011) carried out a comparative analysis of 23 castor genotypes for their ability to remediate DDT and Cd from soil samples spiked with 1.7 and 2.8 mg Kg^{-1} of DDT and Cd, respectively. After 2 months the plants were harvested, and a significant ($p < 0.05$) difference among genotypes in aboveground and belowground biomass was observed. Having a comparatively massive root system among herbaceous plants, castor is able to penetrate several meters of soil. The accumulation of DDT and Cd was largely restricted to roots across all the genotypes. The initial concentration of Cd in the soil was 2.8 mg Kg^{-1}, while mean concentration of Cd in the root, stem, and leaves was found to be 37.63, 2.27, and 1.22 mg Kg^{-1} (dry weight), respectively. The initial levels of DDT in the soil was 2.8 mg Kg^{-1} soil, and its average concentration in the root, stem, and leaves was reportedly 70.51, 0.43, and 0.37 mg Kg^{-1} (dry weight), respectively. These results indicate the exceptional ability of castor plant to accumulate very high concentrations of DDT and Cd particularly in its massive root system. Similarly, Shi and Cai (2009) have reported excellent Cd tolerance capacity of castor plant when they were grown in pots containing soil spiked with 50–200 mg Cd Kg^{-1} soil. Thus these studies suggest high degree of tolerance of castor to Cd, and since the accumulation is mostly restricted to the root system, castor plants probably behave as excluder. Olivares et al. (2013) conducted a study on the ability of castor to uptake metals when grown on multi-metal (Zn, Cu, Pb, Mn, and Cd)-infested mine tailings. The degree of uptake of various metals varied, but the accumulation was largely restricted to belowground biomass, and the translocation factors for different metals were low. The concentration of metals in areal biomass was found to be below threshold toxicity levels for individual metals. Although the oil yield was found to correlate negatively with the metal concentration in the mine tailings, no correlation was observed between the yields and concentration of bioavailable metals, and the yield varied between 41 and 64%. Castor seedlings were grown hydroponically in nutrient media spiked with various concentrations of Pb acetate, and their growth performance and hyperaccumulation traits were measured after 28 days. Minor decrease in growth was observed in all the plants grown on Pb-spiked nutrient media across all the treatments (100, 200, and 400 mmol Pb L^{-1} of nutrient media), and the concentration of Pb in all the treatments was >1.0 g Pb Kg^{-1} dw (Romeiro et al. 2006). Therefore according to the criteria given by Raskin et al. (1994), castor possesses hyperaccumulator traits for Pb. A comparision of different sustainability dimensions of bioenergy plants Jatropha, Karanja and Castor are listed in Table 17.5.

17.4 Discussion

In order to sustain and support its economic growth rate, India will have to significantly boost up its energy supplies, and at the same time India is expected to cut down carbon footprint of its GDP. Therefore, the Government of India has announced installation of indigenous renewable energy to the tune of 40% of its primary energy mix by 2050. Bioenergy is increasingly gaining attention for it can provide carbon neutral energy. The Government of India through its national biofuel policy 2009 has announced 20% blending of biodiesel and bioethanol in high-speed diesel and petrol, respectively, by 2017. This envisages utilization of non-edible feedstocks (*Jatropha curcas* and *Millettia pinnata*) grown on marginal land or wastelands for the production of biodiesel. These plants have several interesting properties such as drought resistance, high yield, few diseases and pests, minimal requirement of nutrients, etc., besides having desirable oil composition for production of biodiesel. India being the largest producer of castor oil can also envisage its utilization for the production of biodiesel. Castor oil biodiesel when blended with mineral diesel or biodiesel from other feedstocks yields a superior quality of mineral diesel substitute or supplement. These plants (*Jatropha*, Karanja, and castor) have also been investigated for their phytoremediation potential. Several studies have shown effectiveness of these plants for the decontamination of soil and groundwater infested with hazardous metals and detrimental organics. For improving the environmental sustainability and economic feasibility of phytoremediation, valorization of biomass by some means is invaluable, and utilization of biomass for production of bioenergy appears to be a new but rather appealing opportunity. Thus, the Government of India can consider the integration of its biofuel policy with phytoremediation of contaminants.

Several important research works on phytoremediation potential of *Jatropha*, Karanja, and castor are listed in Table 17.6.

Table 17.5 Important sustainability dimensions of bioenergy plants *Jatropha*, Karanja, and castor

Sustainability dimension	Plants		
	Jatropha	Karanja	Castor
Drought resistance	Yes[a]	Yes[d]	Yes[k]
Waterlogging Tolerance	No[a]	Yes[e]	To some extent[n]
Nitrogen fixation	No[h]	Yes[f]	No[k]
Wasteland reclamation Potential	Yes[c]	Yes[d]	Yes[k]
Frost tolerance	Withstands light frost[b]	No[d]	No[k]
Soil quality	Grows on saline, sandy, gravelly stony, and saline soils[b]	Grows on sandy, stony, and clayey soils and can tolerate moderate degree of soil salinity[p]	Tolerates slight acidity (pH 5.0–6.5)[m]
Completion with local flora (invasiveness)	Yes[g]	Low[q]	Yes[g]
Nutrient demand	Low[b]	Low[d]	Reasonable demand for N, P, K, Ca, and Mg[m]
Diseases and pest	Very few[b]	Very few[d]	Very few[l]
Agronomic practices	Extensive research ongoing[i]	Ongoing[d]	Ongoing[m]
Employment opportunity	Yes[a]	Yes[d]	Yes[m]
Other usage	Used in making candles, soap, and cosmetics; used in traditional medicine, as insecticide, green manure, and fertilizer; used as fuelwood[b]	Used in making soap and fuel for lighting lamps, has several medicinal properties j	Used in the manufacturing of several industrial chemicals, pharmaceuticals, cosmetics[l]

[a](Takeda 1982), [b](Kumar and Sharma 2008), [c](Islam et al. 2011), [d](Scott et al. 2008), [e](Arpiwi et al. 2013), [f](Samuel et al. 2013), [g](Gordon et al. 2011), [h](Openshaw 2000), [i](Behera et al. 2010), [j](Singh and Pandey 1996), [k](Nass et al. 2007), [l](Sankpal and Naikwade 2013), [m](Escobar et al. 2009), [n](Nielsen et al. 2011), [o](Tomar and Gupta 1985), [p](Kesari and Rangan 2010), [q](Stanley and Ross 1989)

Table 17.6 Important studies on phytoremediation potential of *Jatropha*, Karanja, and castor

Plant	Heavy metal uptake	References
Jatropha	Al	Pandey et al. (2012)
	Ar	Pandey et al. (2012) and Nanda and Abraham (2011)
	Cd	Bauddh et al. (2015), Marques and do Nascimento (2013) and Pandey et al. (2012)
	Cr	Pandey et al. (2012), Nanda and Abraham (2011), and Mangkoedihardjo et al. (2008)
	Cu	Nanda and Abraham (2011)
	Fe	Pandey et al. (2012)
	Mg	Nanda and Abraham (2011)
	Mn	Pandey et al. (2012)
	Pb	Pandey et al. (2012) and Mangkoedihardjo (2008)
	Zn	Pandey et al. (2012)
Castor	As	Melo et al. (2009)
	Ba	Abreu et al. (2012) and Coscione and Berton (2009)
	Cd	Pandey (2013), Olivares et al. (2013), Costa et al. (2012), Bauddh and Singh (2012), Huang et al. (2011), Coscione and Berton (2009) and Lu and He (2005)
	Cr	Coscione and Berton (2009) and Pandey (2013)
	Cu	Coscione and Berton (2009) and Pandey (2013)
	Fe	Coscione and Berton (2009), and Pandey (2013)
	Mn	Pandey (2013), Coscione and Berton (2009) and Vwioko et al. (2006)
	Ni	Malarkodi et al. (2008), Vwioko et al. (2006), and Giordani et al. (2005)
	Pb	Pandey (2013), Olivares et al. 2013, Costa et al. (2012), Coscione and Berton (2009), Vwioko et al. (2006) and Romeiro et al. (2006)
	V	Vwioko et al. (2006)
	Zn	Olivares et al. (2013), Pandey (2013), and (Coscione and Berton (2009)
Karanja	Cd	Singh and Juwarkar (2014) and Manzoor et al. (2015)
	Cr	Singh and Juwarkar (2014), Shirbhate and Malode (2012) and Kumar et al. (2009)
	Cu	Singh and Juwarkar (2014), Shirbhate and Malode (2012), Kumar et al. (2009) and Tulod et al. (2012)
	Mn	Singh and Juwarkar (2014) and Shirbhate and Malode (2012)
	Ni	Singh and Juwarkar (2014) and Shirbhate and Malode (2012)
	Pb	Singh and Juwarkar (2014)
	Zn	Singh and Juwarkar (2014)

17.5 Conclusion

Facing the dual challenges of energy insecurity and environmental pollution, it is imperative for India to look for economically feasible and environmentally sustainable opportunities for achieving energy independence and also for the decontamination of environment. Combining the prospects of bioenergy production in India with phytoremediation of contaminants can provide the solutions to both the challenges. *Jatropha*, Karanja, and castor as biodiesel feedstock and phytoremediating species can be explored as their bioenergy potential is well developed, and several studies on their phytoremediation potential have substantiated their role in phytoremediation of environmental contaminants.

References

Abhilash PC, Singh B, Srivastava P, Schaeffer A, Singh N (2013) Remediation of lindane by *Jatropha curcas* L: utilization of multipurpose species for rhizoremediation. Biomass Bioenergy 51:189–193

Abreu CA, Cantoni M, Coscione AR, Paz-Ferreiro J (2012) Organic matter and barium absorption by plant species grown in an area polluted with scrap metal residue. Appl Environ Soil Sci 2012

Agamuthu P, Abioye OP, Aziz AA (2010) Phytoremediation of soil contaminated with used lubricating oil using *Jatropha curcas*. J Hazard Mater 179(1):891–894

Ahmadpour P, Nawi AM, Abdu A, Abdul-Hamid H, Singh DK, Hassan A, Majid NM, Jusop S (2010) Uptake of heavy metals by *Jatropha curcas* L. planted in soils containing sewage sludge. Am J Appl Sci 7(10):1291–1299

Arpiwi NL, Yan G, Barbour EL, Plummer JA (2013) Genetic diversity, seed traits and salinity tolerance of *Millettia pinnata* (L.) Panigrahi, a biodiesel tree. Genet Resour Crop Evol 60(2):677–692

Azam MM, Waris A, Nahar NM (2005) Prospects and potential of fatty acid methyl esters of some non-traditional seed oils for use as biodiesel in India. Biomass Bioenergy 29(4):293–302

Bauddh K, Singh RP (2012) Growth, tolerance efficiency and phytoremediation potential of *Ricinus communis* (L.) and *Brassica juncea* (L.) in salinity and drought affected cadmium contaminated soil. Ecotoxicol Environ Saf 85:13–22

Bauddh K, Singh K, Singh B, Singh RP (2015) *Ricinus communis*: a robust plant for bio-energy and phytoremediation of toxic metals from contaminated soil. Ecol Eng 84:640–652

Behera SK, Srivastava P, Tripathi R, Singh JP, Singh N (2010) Evaluation of plant performance of *Jatropha curcas* L. under different agro-practices for optimizing biomass–a case study. Biomass Bioenergy 34(1):30–41

Behmer ST, Lloyd CM, Raubenheimer D, Stewart-Clark J, Knight J, Leighton RS, Harper FA, Smith JAC (2005) Metal hyperaccumulation in plants: mechanisms of defence against insect herbivores. Funct Ecol 19(1):55–66

Bert V, Macnair MR, De Laguerie P, Saumitou-Laprade P, Petit D (2000) Zinc tolerance and accumulation in metallicolous and nonmetallicolous populations of *Arabidopsis halleri* (Brassicaceae). New Phytol 146:225–233

Boyd RS (2007) The defence hypothesis of elemental hyperaccumulation: status, challenges and new directions. Plant Soil 293:153–176

Bringi NV (1987) Non-traditional oilseeds and oils of India. New Delhi: Oxford & IBH Publishing Co. Pvt. Ltd

Coscione AR, Berton RS (2009) Barium extraction potential by mustard, sunflower and castor bean. Sci Agric 66(1):59–63

Costa ES, Guilherme L, Melo É, Ribeiro B, Inácio ESB, Severiano EC, Faquin V, Hale B (2012) Assessing the tolerance of castor bean to Cd and Pb for phytoremediation purposes. Biol Trace Elem Res 1(145):93–100
CPCB (2009) Comprehensive Environmental assessment of industrial clusters. Ecological Impact Assessment Series: EIAS/5/2009–2010. Central Pollution Control Board, Ministry of Environment and Forests, Govt. of India, New Delhi
Cunningham SD, Berti WR, Huang JW (1995) Phytoremediation of contaminated soils. Trends Biotechnol 13:393–397
Ebbs SD, Kochian LV (1997) Toxicity of zinc and copper to Brassica species: implications for phytoremediation. J Environ Qual 26(3):776–781
Escobar JC, Lora ES, Venturini OJ, Yáñez EE, Castillo EF, Almazan O (2009) Biofuels: environment, technology and food security. Renew Sust Energ Rev 13(6):1275–1287
Galeas ML, Klamper EM, Bennett LE, Freeman JL, Kondratieff BC, Quinn CF, Pilon-Smits EA (2008) Selenium hyperaccumulation reduces plant arthropod loads in the field. New Phytol 177(3):715–724
Ghosh P, Ghose TK (2003) Bioethanol in India: recent past and emerging future. In Biotechnology in India II. Springer, Berlin/Heidelberg, pp. 1–27
Ghosh M, Singh SP (2005) A review on phytoremediation of heavy metals and utilization of it's by products. Asian J Energy Environ 6(4):18
Giordani C, Cecchi S, Zanchi C (2005) Phytoremediation of soil polluted by nickel using agricultural crops. Environ Manag 36(5):675–681
Gomes HI (2012) Phytoremediation for bioenergy: challenges and opportunities. Environ Tech Rev 1(1):59–66
Gopinathan MC, Sudhakaran R (2011) Biofuels: opportunities and challenges in India. In: Biofuels. Springer, New York, pp 173–209
Gordon DR, Tancig KJ, Onderdonk DA, Gantz CA (2011) Assessing the invasive potential of biofuel species proposed for Florida and the United States using the Australian Weed Risk Assessment. Biomass Bioenergy 35(1):74–79
Greenwood JS, Bewley JD (1982) Seed development in *Ricinus communis* (castor bean). I. Descriptive morphology. Can J Bot 60(9):1751–1760
Hall JL (2002) Cellular mechanisms for heavy metal detoxification and tolerance. J Exp Bot 53:1–11
Huang H, Yu N, Wang L, Gupta DK, He Z, Wang K, Zhu Z, Yan X, Li T, Yang XE (2011) The phytoremediation potential of bioenergy crop *Ricinus communis* for DDTs and cadmium co-contaminated soil. Bioresour Technol 102(23):11034–11038
Ingle SS, Nandedkar VM (2013) Castor oil Biodiesel an alternative fuel for diesel in Compression Ignition Engine. IOSR-JMCE, ISSN:2278–1684
Islam AKMA, Yaakob Z, Anuar N (2011) Jatropha: a multipurpose plant with considerable potential for the tropics. Sci Res Essays 6(13):2597–2605
Jamil S, Abhilash PC, Singh N, Sharma PN (2009) *Jatropha curcas*: a potential crop for phytoremediation of coal fly ash. J Hazard Mater 172(1):269–275
Jhee EM, Boyd RS, Eubanks MD, Davis MA (2006) Nickel hyperaccumulation by *Streptanthus polygaloides* protects against the folivore *Plutella xylostella* (Lepidoptera: Plutellidae). Plant Ecol 183:91–104
Jiang RF, Ma DY, Zhao FJ, McGrath SP (2005) Cadmium hyperaccumulation protects *Thlaspi caerulescens* from leaf feeding damage by thrips (*Frankliniella occidentalis*). New Phytol 167:805–814
Karmee SK, Chadha A (2005) Preparation of biodiesel from crude oil of *Pongamia pinnata*. Bioresour Technol 96(13):1425–1429
Kesari V, Rangan L (2010) Development of *Pongamia pinnata* as an alternative biofuel crop—current status and scope of plantations in India. J Crop Sci Biotechnol 13(3):127–137
Khan AG (2005) Role of soil microbes in the rhizospheres of plants growing on trace metal contaminated soils in phytoremediation. J Trace Elem Med Biol 18(4):355–364

Kumar A, Sharma S (2008) An evaluation of multipurpose oil seed crop for industrial uses (*Jatropha curcas* L.): a review. Ind Crop Prod 28(1):1–10

Kumar S, Mehta JU, Hazra S (2009) In Vitro studies on chromium and copper accumulation potential of *Pongamia pinnata* (L.) Pierre Seedlings. Biorem Biodiv Bioavail 3(1):43–48

Lone MI, He Z, Stoffella PJ, Yang X-e (2008) Phytoremediation of heavy metal polluted soils and water: progresses and perspective. J Zhejiang Univ Sci B 9:210–220

Lu XY, He CQ (2005) Tolerance, uptake and accumulation of cadmium by *Ricinus communis*. J Agro Environ Sci 24(4):674–677

Luhach J, Chaudhry S (2012) Phytoremediation potential of *Jatropha curcas* for removal of heavy metals from refinery sludge. Int J Sci Eng Res 3(10):1–5

Ma F, Hanna MA (1999) Biodiesel production: a review. Bioresour Technol 70(1):1–15

Malarkodi M, Krishnasamy R, Chitdeshwari T (2008) Phytoextraction of nickel contaminated soil using castor phytoextractor. J Plant Nutr 31(2):219–229

Mangkoedihardjo S (2008) *Jatropha curcas* L. for phytoremediation of lead and cadmium polluted soil. World Appl Sci J 4(4):519–522

Mangkoedihardjo S, Ratnawati R, Alfianti N (2008) Phytoremediation of hexavalent chromium polluted soil using *Pterocarpus indicus* and *Jatropha curcas* L. World Appl Sci J 4(3):338–342

Manzoor SA, Mirza SN, Zubair M, Nouman W, Hussain SB, Mehmood S, Irshad A, Sarwar N, Ammar A, Iqbal MF, Asim A (2015) Estimating genetic potential of biofuel forest hardwoods to withstand metal toxicity in industrial effluent under dry tropical conditions. Genet Mol Res 14(3):9543–9554

Marques MC, do Nascimento CWA (2013) Analysis of chlorophyll fluorescence spectra for the monitoring of Cd toxicity in a bio-energy crop (*Jatropha curcas*). J Photochem Photobiol B 127:88–93

Mattina MI, Lannucci-Berger W, Musante C, White JC (2003) Concurrent plant uptake of heavy metals and persistent organic pollutants from soil. Environ Pollut 124(3):375–378

Melo EEC, Costa ETS, Guilherme LRG, Faquin V, Nascimento CWA (2009) Accumulation of arsenic and nutrients by castor bean plants grown on an As-enriched nutrient solution. J Hazard Mater 168(1):479–483

Murugesan A, Umarani C, Chinnusamy TR, Krishnan M, Subramanian R, Neduzchezhain N (2009) Production and analysis of bio-diesel from non-edible oils—a review. Renew Sust Energ Rev 13(4):825–834

Nagarhalli MV, Nandedkar VM, Mohite KC (2010) Emission and performance characteristics of karanja biodiesel and its blends in a CI engine and it's economics. ARPN J Eng Appl Sci 5(2):52–56

Nanda S, Abraham J (2011) Impact of heavy metals on the rhizosphere microflora of *Jatropha multifida* and their effective remediation. Afr J Biotechnol 10(56):11948–11955

Nass LL, Pereira PAA, Ellis D (2007) Biofuels in Brazil: an overview. Crop Sci 47(6):2228–2237

National Policy on Biofuels (2009) Ministry of New and Renewable Energy, Government of India

Nevase SS, Gadge SR, Dubey AK, Kadu B (2012) Economics of biodiesel production from Jatropha oil. J AGR Tech 8(2):657–662

Nielsen F, Hill B, de Jongh J (2011) Castor (*Ricinus communis*). Potential of castor for bio-fuel production. FACT-Arrakis. FACT Foundation. FACT Project no, 146

No SY (2011) Inedible vegetable oils and their derivatives for alternative diesel fuels in CI engines: a review. Renew Sust Energ Rev 15(1):131–149

Noret N, Meerts P, Vanhaelen M, Dos Santos A, Escarré J (2007) Do metal-rich deter herbivores? A field test of the defence hypothesis. Oecologia 152:92–100

Olivares AR, Carrillo-González R, González-Chávez MDCA, Hernández RMS (2013) Potential of castor bean (*Ricinus communis* L.) for phytoremediation of mine tailings and oil production. J Environ Manag 114:316–323

Omari A, Mgani QA, Mubofu EB (2015) Fatty acid profile and physico-chemical parameters of Castor oils in Tanzania. Green Sustainable Chem 5(04):154

Openshaw K (2000) A review of *Jatropha curcas*: an oil plant of unfulfilled promise. Biomass Bioenergy 19(1):1–15
Pagliano C, Raviolo M, Dalla Vecchia F, Gabbrielli R, Gonnelli C, Rascio N, Barbato R, La Rocca N (2006) Evidence for PSII-donor-side damage and photoinhibition induced by cadmium treatment on rice (*Oryza sativa* L.). J Photochem Photobiol B Biol 84(2006):70–78
Pandey VC (2013) Suitability of *Ricinus communis* L. cultivation for phytoremediation of fly ash disposal sites. Ecol Eng 57:336–341
Pandey VC, Singh K, Singh JS, Kumar A, Singh B, Singh RP (2012) *Jatropha curcas*: a potential biofuel plant for sustainable environmental development. Renew Sust Energ Rev 16(5):2870–2883
Pandey VC, Bajpai O, Singh N (2016) Energy crops in sustainable phytoremediation. Renew Sust Energ Rev 54:58–73
Rajkumar M, Freitas H (2008) Influence of metal resistant-plant growth promoting bacteria on the growth of *Ricinus communis* in soil contaminated with heavy metals. Chemosphere 71(5):834–842
Rascio N, Navari-Izzo F (2011) Heavy metal hyperaccumulating plants: how and why do they do it? And what makes them so interesting? Plant Sci 180(2):169–181
Raskin I, Kumar PN, Dushenkov JR, Salt DE (1994) Bioconcentration of heavy metals by plants. Curr Opin Biotechnol 5:285–290
Rathinasabapathi B, Rangasamy M, Froeba J, Cherry RH, McAuslane HJ, Capinera JL, Srivastava M, Ma LQ (2007) Arsenic hyperaccumulation in the Chinese brake fern (*Pteris vittata*) deters grasshopper (*Schistocerca americana*) herbivory. New Phytol 175(2):363–369
Reeves RD (2006) Hyperaccumulation of trace elements by plants. Phytorem Metal-contaminated Soils 68:68–25
Romeiro S, Lagôa AM, Furlani PR, Abreu CAD, Abreu MFD, Erismann NM (2006) Lead uptake and tolerance of *Ricinus communis* L. Braz J Plant Physiol 18(4):483–489
Samuel S, Scott PT, Gresshoff PM (2013) Nodulation in the legume biofuel feedstock tree *Pongamia pinnata*. Agric Res 2(3):207–214
Sankpal ST, Naikwade PV (2013) Important bio-fuel crops: advantages and disadvantages. IJSER 4(12)
Santana GCS, Martins PF, Da Silva NDL, Batistella CB, Maciel Filho R, Maciel MW (2010) Simulation and cost estimate for biodiesel production using castor oil. Chem Eng Res Des 88(5):626–632
Scholz V, da Silva JN (2008) Prospects and risks of the use of castor oil as a fuel. Biomass Bioenergy 32(2):95–100
Scott PT, Pregelj L, Chen N, Hadler JS, Djordjevic MA, Gresshoff PM (2008) *Pongamia pinnata*: an untapped resource for the biofuels industry of the future. Bioenergy Res 1(1):2–11
Sharma YC, Singh B, Korstad J (2009) High yield and conversion of biodiesel from a nonedible feedstock (*Pongamia pinnata*). J Agric Food Chem 58(1):242–247
Sharma YC, Singh B, Korstad J (2010) Application of an efficient nonconventional heterogeneous catalyst for biodiesel synthesis from *Pongamia pinnata* oil. Energy Fuel 24(5):3223–3231
Shi G, Cai Q (2009) Cadmium tolerance and accumulation in eight potential energy crops. Biotechnol Adv 27(5):555–561
Shirbhate N, Malode SN (2012) Heavy metals phytoremediation By *Pongamia pinnata* (L) growing in contaminated soil from municipal solid waste landfills and compost Sukali Depot, Amravati (Ms). Int J Adv Biotechnol Res 2(1):147–152
Shrirame HY, Panwar NL, Bamniya BR (2011) Bio diesel from castor oil–a green energy option. LCE 2(01):1
Singh A (2013) Recent advances in bioenergy research Volume I Editors Kumar S and Sarma A.K
Singh SK, Juwarkar AA (2014) Sustainable reclamation of coal mine spoil dump using microbe assisted phytoremediation technology. Int J Env Sci Tox 2(3):43–54
Singh RK, Pandey BL (1996) Anti-inflammatory activity of seed extracts of *Pongamia Pinnata* in Rat. Indian J Physiol Pharmacol 40:355–358

Singh K, Singh B, Tuli R (2013) Sodic soil reclamation potential of *Jatropha curcas*: a long-term study. Ecol Eng 58:434–440

Stanley TD, Ross EM (1989) Flora of south-eastern Queensland, volume 3. Queensland Department of Primary Industries

Takeda Y (1982). Development study on *Jatropha curcas* (sabu dum) oil as a substitute for diesel oil in Thailand. Interim Report of the Ministry of Agriculture, Thailand

Tanaka T, Iinuma M, Yuki K, Fujii Y, Mizuno M (1992) Flavonoids in root bark of *Pongamia pinnata*. Phytochemistry 31(3):993–998

Tangahu BV, Sheikh Abdullah SR, Basri H, Idris M, Anuar N, Mukhlisin M (2011) A review on heavy metals (As, Pb, and Hg) uptake by plants through phytoremediation. Int J Chem Eng 2011

Tiwari AK, Kumar A, Raheman H (2007) Biodiesel production from Jatropha oil (*Jatropha curcas*) with high free fatty acids: an optimized process. Biomass Bioenergy 31(8):569–575

Tomar OS, Gupta RK (1985) Performance of some forest tree species in saline soils under shallow and saline water-table conditions. Plant Soil 87:329–335

Tulod AM, Castillo AS, Carandang WM, Pampolina NM (2012) Growth performance and phytoremediation potential of *Pongamia pinnata* (L.) Pierre, *Samanea saman* (Jacq.) Merr. and *Vitex parviflora* Juss. in copper-contaminated soil amended with zeolite and VAM. Asia Life Sci 21(2):499–522

Verbruggen N, Hermans C, Schat H (2009) Molecular mechanisms of metal hyperaccumulation in plants. New Phytol 181:759–776

Vwioko DE, Anoliefo GO, Fashemi SD (2006) Metal concentration in plant tissues of *Ricinus communis* L. (castor oil) grown in soil contaminated with spent lubricating oil. JASEM 10(3):127–134

Vyas AP, Subrahmanyam N, Patel PA (2009) Production of biodiesel through transesterification of Jatropha oil using KNO_3/Al_2O_3 solid catalyst. Fuel 88:625–628

Watanabe T, Osaki M (2002) Mechanism of adaptation to high aluminium condition in native plant species growing in acid soils: a review. Commun Soil Sci Plant Anal 33:1247–1260

World Economic Outlook (2016) International Monetary Fund, April

Zanchi G, Pena N, Bird N (2012) Is woody bioenergy carbon neutral? A comparative assessment of emissions from consumption of woody bioenergy and fossil fuel. GCB Bioenergy 4(6):761–772

Phycoremediation: An Eco-friendly Algal Technology for Bioremediation and Bioenergy Production

18

Sanjay Kumar Gupta, Amritanshu Sriwastav, Faiz Ahmad Ansari, Mahmoud Nasr, and Arvind Kumar Nema

Abstract

Substantial amount of the refractory organics; inorganic nutrients, mainly nitrogen and phosphorus; heavy metals; etc. is discharged in conventional wastewater treatments. The concentration of such contaminants in the discharged wastewater depends on the performance and maintenance of the wastewater treatment plants (WWTPs). Though further reduction in such contaminants is possible with an aid of some of the advance technologies and skilled manpower, it makes wastewater treatment more expensive. More importantly, the running and maintenance of WWTPs are uncommon in economically weaker countries especially in the rural areas. This leads to the hunt of economically viable and environmentally sustainable alternative wastewater treatments. The truism nowadays is to recognize the emergence of phycoremediation as an alternative. Algae-based bioremediation has been found excellent for the nutrient, organic, pathogen, heavy metal, etc. removal from various types of wastewater. Green microalgae possess the unique potential of high photosynthetic activity compared to food crops and terrestrial plants. Therefore, such systems are capable of high biomass

S.K. Gupta (✉) • A.K. Nema
Environmental Engineering, Department of Civil Engineering, Indian Institute of Technology, Delhi, India
e-mail: sanjuenv@gmail.com

A. Sriwastav
Environmental Engineering, Department of Civil Engineering, Indian Institute of Technology, Kanpur, India

F.A. Ansari
Institute for Water and Wastewater Technologies, Durban University of Technology, Durban, South Africa

M. Nasr
Sanitary Engineering Department, Faculty of Engineering, Alexandria University, Alexandria, Egypt

K. Bauddh et al. (eds.), *Phytoremediation Potential of Bioenergy Plants*,
DOI 10.1007/978-981-10-3084-0_18

production through CO_2 sequestration from the air and nutrient and organic sequestration from water. The microalgal cells contain comparatively high lipid contents; thus, algal biomass serves as an excellent feedstock for biofuels. Therefore, the choice of algal species possessing excellent phytoremediation potential as well as capable of producing high biomass is important to consider while designing the phycoremediation-based treatment systems. In this chapter, a concise partial overview of the potential and uniqueness of phycoremediation in treating various types of water and production of algal biomass for biofuels has been discussed. The environmental sustainability and economic viability aspects of phycoremediation, factors influencing the wastewater treatment, and the limitations of such technologies are covered briefly.

Keywords
Bioremediation • Biofuels • Heavy metals • Microalgae • Nutrient removal • Phycoremediation

18.1 Introduction

18.1.1 Water Crisis

One of the major sources of pollution of natural water bodies is the release of wastewater (Nasr and Ismail 2015; Gupta et al. 2014a, b). Inadequately treated wastewater contains high concentrations of various organics as well as inorganic nutrients such as nitrogen (N) and phosphorus (P) which leads to eutrophication (Schindler et al. 2008). Water crisis is being experienced in several areas throughout the world, even though the sufficient water and land resources are available (CA 2007). If the appropriate and adequate policy measure will not be enforced and practiced, the existing land and water use practices will surely elevate the water crisis in the agricultural sector (Seckler et al. 1998; Vörösmarty et al. 2005; Lobell et al. 2008; De Fraiture and Berndes 2009). Other than water demands for the agriculture, Berndes (2002) indicated the worsening of the situation as the additional water demand for the pilot-scale production of energy crops has not been considered in such studies. Such observations clearly raise the risk of additional water demand over the water resources which are already overstressed (Varis 2007; National Academy of Sciences 2008) (Fig. 18.1).

The cultivation of crops and biomass for food, fiber, or energy requires vast amounts of water (Molden et al. 2007). In present scenario, the major proportion of the water is used in the agricultural sector, and water demand for biofuel is modest; however, with increasing demand of bioenergy, it may increase by severalfold (Molden et al. 2007). As per estimates, an average annual global evaporation is around 6800 km^3 and 7130 km^3 from the agricultural sector (Postel 1998; Rockström et al.

Fig. 18.1 A general overview of algal technologies for wastewater treatment and biomass production for biofuels

1999; Molden et al. 2007) which is much higher than the water required for human consumption. The rainfall fulfills approximately 80% of the agricultural water requirement, and the remaining 20% is required for the consumption of mankind, which is extracted from the lakes and rivers (Fraiture and Berndes 2009). As per estimated by 2050, the present cereal demand is expected to increase from 55 to 80%, and about 70–155% increase is expected in meat demand. Based on these speculations, by 2050, an increase of 70–110% is expected in crop water if the present land and water use practices will remain (Fraiture and Berndes 2009) (Tables 18.1 and 18.2).

18.1.2 Major Limitations of Conventional Wastewater Treatment Technologies

Efforts are being made over the years in the designing of various kinds of wastewater treatment systems based on the specific need and type of wastewater (Arceivala and Asolekar 2007). Various types of WWTPs exist nowadays, which may vary in capacity and design, depending upon the requirement of the specific geographical region. The design and capacity of these WWTPs are based on the geography of the locations and the amount and the type of wastewater produced. The conventional

Table 18.1 Comprehensive summary of various studies using algal technologies for wastewater treatment-coupled biofuel production

Type of wastewater (WW)	Algae species	Nutrient removal (%)	Organic removal (%)	Pathogen removal	Heavy metals (%)	Biomass yield (g, dry wt.)	Lipid yield	References
Domestic wastewater	*Chlorella sorokiniana*	N, 86.93; P, 68.24	COD, 69	> 99% TC and FC	–		22.74% w/w	Gupta et al. (2016)
Domestic wastewater	*Scenedesmus obliquus*	N, 98; P, 97	COD, 76	> 99% TC and FC	–		23% w/w	Gupta et al. (2016)
Dairy WW	*Chlamydomonas polypyrenoideum*	NO_3, 90; NO_2, 74; NH_4, 90; PO_4, 70	TS, 55; TDS, 55; COD, 35	–	–	3.8	42%	Kothari et al. (2013)
Brewery effluent	*Scenedesmus obliquus*	N, 21	COD, 57	57% TC	–	0.9	–	Mata et al. (2012)
Olive–oil mill wastewater	*Arthrospira pcrtenix*	N, 87–95.61;	COD, 29–73;					Markou et al. (2012)
		P, 15–100	phenols, 42–100					
Dairy WW	*Chlorella zofingiensis*	N, 51.7; P, 97.5	–	–	–	0.175 g	29.0% SFA 67.5% USFA	Huo et al. (2012)
Artificial wastewater	*Chlorella vulgaris*	N, 97; P, 96	COD, 86	–	–	0.89 g/L	42%	Feng et al. (2011)
	Chlorella pyrenoidosa	NO_3, 60; NO_2, 42;PO_4, 87	TS, 33; TSS, 66	–	–	6.8 g/L	51%	Kothari et al. (2012)
Wastewater	*Chlorella pyrenoidosa*	NH4–N, 74–82; TN, 68–82; P, 83–90	COD, 50–83	–	Al (65–70), Fe and Mn (98–100), Zn (56–81)	–	–	Wang et al. (2010)
Carpet industry effluents	*Botryococcus braunii*	N, 99.7–99.8; P, 98.8–99.1	–	–	–	0.034 g L^{-1} d^{-1}	13.2%	Chinnasamy et al. (2010)

Carpet industry effluents	*Chlorella saccharophila*	N, 99.7–99.8; P, 98.8–99.1	–	–	–	0.023 ± 0.004	18.10 ± 1.27	Chinnasamy et al. (2010)
Carpet industry effluents	*Dunaliella tertiolecta*	N, 99.7–99.8; P, 98.8–99.1	–	–	–	0.028 ± 0.005	15.20 ± 2.43	Chinnasamy et al. (2010)
Carpet industry effluents	*Pleurochrysis carterae*	N, 99.7–99.8; P, 98.8–99.1	–	–	–	0.033 ± 0.005	12.00 ± 0.80	Chinnasamy et al. (2010)
Swine WW	*Spirulina platensis*	N, 67– 93; P–70–93	COD, 80– 90	–	–	1.52 dry–g/L	–	Kim et al. (2000)
Livestock WW	*Botryococcus*	N, 88; P, 98	–	–	–	2.543 g L–1	19.8%	Shen et al.(2008)
Primary- and tertiary-treated WW	*Chlorella minutissima*	N, 70–80; P, 60–70; K, 45–50	BOD, 60; COD, 75				0.171 and 0.132	Malla et al.(2015)
Tannery WW	*Scenedesmus* species	NO3 >44.3; PO4 >95.	TDS, 59; TS, 61; BOD, 65; COD, 62		Cr (81–96), Cu (73.2–98), Pb (75–98)			Ajayan et al. (2015)
Industrial WW	*Chlamydomonas* sp. TAI–2	N, 100; P, 33				1.34 g L^{-1}	18.4	Wu et al. (2012)

TC total coliforms, *FC* fecal coliforms

Table 18.2 Nutrient removal efficiencies and growth rates for various algal strains immobilized under different conditions and experimental designs

Type of wastewater (WW)	Algae	Immobilization matrix	Type of bioreactor	Nutrient removal	HRT	References
Synthetic WW	*S. obliquus*	κ-Carrageenan	BBR	NH_4, 100	5 h	Chevalier and de la Noüe (1985)
Secondary WW	*P. aggregates*	Chitosan aggregates	CFBR	DIN, 98	24 h	Noüe and Proulx (1988)
Domestic WW	*P. laminosum*	Polyvinyl foam	CFBR	NO_3, 85	48 h	Garbisu et al. (1991)
Synthetic wastewater	*C. vulgaris* and *S. bijugatus*	Calcium alginate beads	PBC-BR	NH_3, 78–79	24 h	Megharaj et al.(1992)
Domestic WW	*C. vulgaris*	Calcium alginate beads		NH_4, 100	7 days	Tam et al. (1994)
Drinking water	*S. obliquus*	Polyvinyl-adsorbed cells	CFBR	NO_3, 90	3 h	Urrutia et al. (1995)
Wastewater	*C. vulgaris*	k-Carrageenan beads	BBR	NO_3, 100	48 h	Lau et al. (1998a)
Secondary WW	*C. vulgaris* and *A. brasilense*	Co-immobilized algae with bacteria	CFBR	NH_4, 91	48 h	De–Bashan et al.(2002)
Primary WW	*C. vulgaris*	Calcium alginate beads	BBR	NH_4, 100; NO_3, 96.4	48 h	Hameed (2007)
Secondary WW	*C. vulgaris* and *S. rubescens*	Immobilized	FTTL-BR	NH_4, 90	9 days	Shi et al. (2007)
Secondary wastewater	*Phormidium*	Chitosan	CFBR	NO_3, 95 PO_4, 92	24	Noue and Proulx(1988)
Polluted water	*P. laminosum*	Polyvinyl foams	CFBR	NO_3, 90; NO_2, 85; PO_4, 24	14 h	Garbisu et al. (1994)
Synthetic wastewater	*P. uncinatum*	Polyvinyl foams	CFBR	NO_3, 90	3–4 h	Gil and Serra (1993)
Natural wastewater	*S. bicellularis*	Alginate	BBR	NH_4, 100; PO_4, 88–100	2 h	Kaya et al. (1995)
Wastewater	*P. laminosum*	Hollow fibers	TPBR	NO_3, 48.7; PO_4, 99.7	48 h	Sawayama et al.(1998)
Primary wastewater	*C. vulgaris*	Carrageenan	CCMO	NH_4, 95; PO_4, 90	3 days	Lau et al. (1998a, b)
Synthetic wastewater	*C. vulgaris* and *A. brasilense*	Alginate beads	BBR	NH_4 and P	6 days	De–Bashan et al.(2002)

Modified from Fillipino et al. (2015) and Olguı´ (2003)
BBR batch bioreactor, *CFBR* continuous flow bioreactor, *PBC-BR* packed-bed column bioreactor, *FTTL-BR* flow-through twin-layer bioreactor, *TPBR* tubular photobioreactor, *CCMO* continuous contact mode operation

wastewater treatment plants (CWWTPs) mainly include the following treatment process: preliminary, primary, secondary, and tertiary followed by disinfection and disposal of solid sludge generated during various stages of the treatment. The preliminary treatments are basically simple screening of waters for the removal of solid and bigger objects which can clog or damage the further treatment processes, whereas oil and greasy substances as well as primary sludge are in the primary treatment using clarifiers and sedimentation tanks. The secondary treatment is mainly based on biological treatment, and most of the organic contents are removed due to microbial degradation followed by advance stages of treatment at tertiary level. This step comprises various treatment processes such as coagulation, flocculation, filtration and reverse osmosis, and extended biological nutrient removal and stabilization of oxygen-demanding substances. The treated wastewater is finally disinfected by chlorination, ozonation, ultraviolet (UV) lights, etc., whereas the process concentrates (biological sludge) generated during the treatment are subjected first for the dewatering and used as fertilizer (if not toxic) or subjected to further treatment or disposal which is a big bottleneck of CWWTP technologies.

The installation of WWTPs as well as the cost of wastewater treatment is considerably higher (McCarty et al. 2011) as the conventional WWTPs require high energy inputs. Moreover, the maintenance and operations need technical expertise which also incurs additional costs. Other challenges include frequent breakdowns of WWTPs due to mechanical failure or mishandling. The biological wastewater treatment is very efficient in the nutrient removal; therefore, these systems are extremely effective; however, it adds approximately 3–4 times more costs in overall treatment processes (Abdel–Raouf et al. 2012). Additionally, the unfavorable environmental conditions also lead to the failure of such systems as a major proportion of the treatment is based on the biological systems. A slight modification or variation in the physical–chemical, water, or geo-environmental conditions leads to death and decay of microflora involved in the biological nutrient removal processes, whereas excessive growth of filamentous bacteria leads to bulking and foaming (Khin and Annachhatre 2004; de–Bashan and Bashan 2004). Therefore, virtually, such systems are not affordable for the economically weaker societies hence have limited applicability.

The wastewater treatment could be ideal only if the energy and cost input is minimal. In this regards, microalgal wastewater systems are unique and comprehensively very cost effective. Such systems are being employed for decades for the treatment of various types of wastewater. As these systems are based on the algae, a phytoplankton, which requires adequate temperature and sunlight, therefore these systems are getting due considerations in the temperate regions having suitable light intensity and duration throughout the year.

18.1.3 Energy Crisis and Need of Biofuels

The existence of mankind on earth can only sustain if the natural resources, especially the nonrenewable resources, will be used rationally and in an environmentally sustainable manner. Since the industrial era, the nonrenewable resources have been used and abused as well, for the uplifting the quality of life, even without caring the nature. A limited stock of almost all the natural nonrenewable resources such as oil, gas, and coal is available now, and if the unrestrained exploitation will remain continued, the dwindling energy resources will be disbursed and diminished within the next 100–125 years (Shafiee and Topal 2009). The available stock of nonrenewable energy resources is not sufficient to cope with ever increasing energy demand of burgeoning population; therefore, energy crisis is being faced in all the corners of the globe.

The nature is signaling the mankind through various alarms in the form of natural disasters which occur due to the devastating effects of rapid industrialization and urbanization. Therefore, the urgent need of the day is to restructure natural resource conservation planning. Moreover, there is an immense need to explore eco-friendly and renewable energy resources (Ramachandra et al. 2009; Wijffels and Barbosa 2010; Demirbas 2009). That is why the hunt for alternative biofuels has gained substantial attention in the past few decades. Various types of biofuels have been identified, and the hunt is on for more economically and environmentally sustainable biofuels.

From the past few decades, the development of carbon-neutral biofuels received much attention due to rapid exploitation of fossil fuels (e.g., Sørenson 1991; IPCC 2008). The limited stock of fossil fuels as well as other nonrenewable energy resources requires the production of cleaner and sustainable fuels. The first-generation biofuels were based on the energy crops, and in the following years, the bioenergy shifted to second-generation biofuels derived from the biomass. The first- and second-generation biofuels were debated due to the use of arable land and water for the production of biomass. Various studies have shown extensive exploitation of water resources for the production of biomass (e.g., Berndes 2002; De Fraiture et al. 2008; Varis 2007; Hoekstra and Chapagain 2008). Therefore, the production of biomass for the bioenergy has always been a subject of its environmental sustainability versus economic developments.

18.2 The Uniqueness of Phycoremediation Technologies

The term phycoremediation is used specifically to denote the wastewater treatment potential of algae. Such treatment may include removal of organic compounds, inorganic nutrients, metals, pathogens, and even the emerging contaminants as well. For phycoremediation purposes, majorly those algal species used possesses high photosynthetic growth rates and potential for scavenging various types of contaminants. Various studies have demonstrated excellent phycoremediation potential of various green microalgae species (Gupta et al. 2016 and the reference there in the chapter). The phycoremediation potential actually lies and proportional to the growth rate of a particular algal species. Being photosynthetic phytoplankton, the algae require mainly sunlight for the photosynthetic activity and carbon and inorganic nutrients (N, P, Mg, and silicates) and various other micronutrients for its growth. However, the amount of solar radiation, organics, and inorganics is required for the optimal growth of algae and varies species to species. In controlled cultivation, such requirements are optimized and provided in the form of growth medium.

18.2.1 Algal Wastewater Treatment

Wastewater presents a complex matrix of various types of organic and inorganic chemicals, and with slight or no modification, it could serve as an excellent growth medium for various algal species (Abdel–Raouf et al. 2012; Cai et al. 2013, Rawat et al. 2011, 2016; Gupta et al. 2016, Olguı´ 2003). Phycoremediation has been found very effective for the treatment of various types of wastewaters, i.e., domestic sewage, agricultural waste substrates, agro-industrial wastewater, livestock wastewater, piggery wastewater, food processing effluents, as well as industrial wastes (Abdel–Raouf et al. 2012; Cai et al. 2013; Gupta et al. 2016). The success of the growth of algae over the wide range of wastewaters makes it one of the most suitable contenders for the wastewater treatment purposes. The algal treatment systems are simple and do not require major sources of energy and frequent maintenance. Moreover, the ease of the process and minimal operational requirements make it one of the best in situ bioremediation technologies.

Approximately 30,000 species of algae have been identified out of an estimated number of 350,000 to 1,000,000 species present on earth (Richmond 2008). The microalgae are very diverse and can grow autotrophically, mixotrophically, or heterotrophically in various natural habitats (Lee 2004); however, the majority of the reported algal species is photosynthetic. Microalgae are tiny self-sustained cell factories which have excellent ability to harness solar radiation and produce biomass. Microalgae cells are negatively charged cells, and its sizes range from 3 to 25 μm. The algal biomass can be used for various purposes such as food, fodder, fertilizer, and fuel (Chisti 2008; Banerjee et al. 2013; Melis 2002). The uniqueness of algae lies in its exceptionally high growth rates (approximately 20 times faster), higher productivity over the limited resources, and excellent potential to accumulate a significantly high amount of fatty acids (20–40% cellular lipids) in the cells in comparison to the oil-based crops.

Various studies have demonstrated excellent phycoremediation potential of numerous algal species for different types of wastewater having varying nutrients, pathogens, and organic loading (Gupta et al. 2016, Shriwastav et al. 2014). The most common and widely studied freshwater microalgal species are *Chlorella* and *Scenedesmus* which have demonstrated comparatively higher tolerance for varying physiological stresses and nutrients as well as organic loading. The excellent phycoremediation potential of these species for secondary or tertiary effluents coupled with a high biomass yield has been already established (Gupta et al. 2016). Notable removal of organics, nutrients, pathogens, and heavy metals along with comparable lipid yield even for primary-treated wastewaters such as domestic sewage has also been reported for both the species. In an independent study, Gupta et al. (2016) presented the comprehensive report on the use and application of *C. sorokiniana* and *S. obliquus* for the phycoremediation of untreated domestic wastewater and production of biomass for biofuels.

Most of the microalgae species are good sequester for the nutrients; therefore, microalgae-based treatment systems present several economic and environmental advantages over the conventional wastewater treatment. The nutrient-rich wastewater serves as an appropriate growth medium for the microalgae, assimilating the organic and inorganic compounds for their growth.

The microalgae cultivation requires huge amount of water as well as nutrient-rich culture media for their growth. The cost of nutrient media makes the algal biomass production costlier. The cost of fertilizers such as potash urea is about $420 per tonne and diammonium phosphate is approximately $480 per tonne, and at this the cost of potash is approximately $400 per tonne algal biomass production (Bain 2013). All of these fertilizers are commonly used as nutritional supplements. Therefore, instead of synthetic nutrient medium, the use of nutrient-rich wastewater substantially reduces the nutrient requirement for culturing of the microalgae. It has been observed that in the conventional primarily or secondary treatments, more than 50% nutrients such as nitrogen and phosphorus along with organics remain in the finally discharged wastewater. Such type of wastewaters could serve as nutrient media for sustaining the growth of algae during cultivation. In such type of arrangements, phycoremediation tremendously cut down the input cost for nutrient supplements and substantially reduces the need of freshwater for algal cultivation. Therefore, the carbon and water footprints associated with conventional water treatment are reduced substantially (Rodolfi et al. 2009, Chinnasamy et al. 2010; Samori et al. 2013; Wang et al. 2010).

Several studies have demonstrated excellent wastewater treatment by various microalgal species (Gupta et al. 2015a, b, 2016; Rawat et al. 2016; Shriwastav et al. 2014; Kim et al. 2013a, b; Ogbonna et al. 2000) along with substantial biomass production and lipid yields (Qiao and Wang 2009; Zheng et al. 2013). Several species of microalgae have been identified for its unique potential of high lipid accumulations. In a study, Griffiths and Harrison (2009) reported higher lipid yield in *C. sorokiniana*, out of 55 algal species surveyed. In one of our previous study (Gupta et al. 2016), we did an evaluation of two microalgal species *Chlorella sorokiniana* and *Scenedesmus obliquus* for their potential of domestic wastewater treatment. The phycoremediation potential of both species was evaluated with specific reference to the removal of organic carbon, nutrient, and pathogens. Our findings clearly demonstrated excellent phycoremediation potential of both species. We observed COD removal up to 76% along with 98% removal of nitrogen and phosphorus, whereas the pathogen removal was more than 99% for *S. obliquus* which was comparatively higher than the *C. sorokiniana*. In both microalgae species, *C. sorokiniana* was found comparatively better for the wastewater treatment as well as the lipid accumulation which was more than 22%. Such findings show why the selection of algal species should be based on the stress resistance potential of algal species to extend their applicability for wastewater treatment as well as production of biomass for biofuels.

Phycoremediation comprises several advantages such as excellent removal of nutrients, organics, various trace elements, heavy metals, and emerging environmental contaminant pathogens. The huge amount of sludge is produced in the conventional wastewater treatment systems, which is produced due to the use of various types of inorganic and synthetic flocculants. The disposal of sludge cost economically as well as its proper disposal is of great environmental concern. Whereas, very less or no sludge is generated in almost all types of algae-based water treatment technologies, which substantially reduces the economic as well as environmental burden of sludge disposal, another advantage of algae-based wastewater treatments is the lesser requirement of energy. In the design of the photoautotrophic algal ponds, high-rate raceways are very simple. These are either open circular shallow ponds or oval-shaped shallow pond with paddle wheels; therefore, the energy as well as operational cost is minimal. Moreover, unlike conventional WWTPs, the running and maintenance of such systems require limited skilled labor (Aziz and Ng 1992). The CO_2 is of serious environmental concern, but being photoautotrophic system, this can be used for the production of algal biomass. Along with this, most of the algal species present unique characteristics of CO_2 sequestration, so phycoremediation can also be used for reduction of carbon footprint. Even several studies have shown comparatively high production of algal biomass by purging pure CO_2 or flue gases in high-rate algal raceways. Therefore, such systems are unique with respect to the utilization of CO_2 for biomass production and release of oxygen.

18.2.2 Heavy Metal Removal by Algae

Recently, progressive industrialization has led to increase the concentrations of heavy metals such as cadmium, copper, lead, nickel, and zinc in the aquatic life (Fawzy et al. 2016a; Gupta et al. 2014a, b). Prolonged exposure to heavy metals constitutes a severe health hazard and a number of environmental problems (Nasr et al. 2015; Gupta et al. 2015a, b). The negative impact of heavy metals on the environment can be attributed to their non-degradability, toxicity, accumulation, and magnification throughout the food chain (Fawzy et al. 2016b; Gupta et al. 2014a; Chabukdhara et al. 2016). In this context, the ability of wastewater treatment systems to remove toxicity is considered a crucial important objective.

Marine and freshwater algae have been reported to eliminate heavy metals selectively from aqueous media (Lourie and Gjengedal 2011). Absorption and bioaccumulation of heavy metals within the cells of algae may create viable methods for the remediation of industrial wastewater (Tastan et al. 2012). These methods are basically followed by separation of the metal-saturated algae from the medium, resulting in high-quality reusable effluent water (Sari and Tuzen 2008).

Numerous species of algae, being living or nonliving biomass, are capable of sequestering considerable quantities of toxic heavy metal ions from aqueous solutions. There have been several reports for the accumulation of heavy metal ions Cu^{2+}, Pb^{2+}, Cr^{3+}, Ni^{2+}, Cd^{2+}, Co^{2+}, Fe^{2+}, and Mn^{2+} by algae. For example, Chen et al. (2008) found that the maximum total uptake by treated algal biomass *Undaria pinnatifida* was 24.71 mg g^{-1} for nickel (at pH 4.7) and 38.82 mg g^{-1} for copper (at pH 4.0) within a contact time of 100 min. Additionally, Gupta and Rastogi (2008) showed that Pb(II) could be removed from aqueous solutions and the adsorption capacities recorded 145.0 mg g^{-1} for *Oedogonium* sp. and 93.5 mg g^{-1} for *Nostoc* sp. In another study, Sari and Tuzen (2008) indicated that the monolayer biosorption capacity of green algal biomass (*Ulva lactuca*) for Pb(II) and Cd(II) ions was found to be 34.7 mg g^{-1} and 29.2 mg g^{-1}, respectively. Similarly, Tastan et al. (2012) studied the use *Chlorella* sp. for the removal of boron from water, and they achieved maximum removal efficiency of 32.95% at boron concentration of 5.45 mg L^{-1}. Moreover, Lourie and Gjengedal (2011) investigated the metal sorption by peat and algae-treated peat, and they found that the affinity of heavy metals to peat was Pb > Cu > Ni > Cd > Zn > Co, while for peat treated with microalga, it was Pb > Cu > Ni > Cd ≈ Co ≈ Zn.

18.2.3 Major Limitations of Microalgal Wastewater Treatments

The effectiveness of algal technologies for all types of wastewater treatment coupled with biomass production depends on the selection of the algal species, set of mechanical arrangements, and various environmental factors. Several species have been identified with great potential of lipid yield and greater biomass production in controlled conditions. However, it cannot be guaranteed that those algal species are able to produce high lipid and biomass and will be good for phycoremediation as well and vice versa. The wastewater treatment using algal systems presents a completely different set of complexities. The failure of wastewater treatment using algae occurs mainly if the selected species could not resist the varying nutrient loadings. In such cases, algal cell death occurs at the initial stages of the treatment. Moreover, physicochemical properties of various types of wastewaters vary greatly and depend upon the sources of the wastewater. Therefore, varying nutrient loadings exert physiological shock to algal species being employed for wastewater treatment. Similarly, the set of environmental variables such as varying temperatures, light intensity, and duration as well as the type of mixing, photobioreactor/raceway design, etc. also greatly affect the physiology of the algae, which directly corresponds to the survival of the particular species and thus directly or indirectly the wastewater treatment as well as biomass yield. The basic requirement is that the algal species must have great resistance to the various physiological stresses such as fluctuating nutrient as well as organic loading, having a great potential for lipid accumulation in varying nutrient loadings with physiological and environmental conditions (Gupta et al. 2016; Srivastava et al. 2014).

18.3 Phycoremediation-Coupled Algal Biomass Production for Biofuels

Rapid depletion of conventional energy resources has led to a global hunt for renewable sources of fuels, as presently almost 80% of the energy is derived from fossil fuel sources (Demirbas 2011). From the past few decades, biofuels have attracted tremendous attention of the scientific community due to limited stock of fossil fuels, continuously increasing GHGs, and climate change. The development of carbon-neutral biofuels is generally based on the two prime concerns, i.e., environmental sustainability and economic viability. The ever increasing demands of transport biofuels cannot be met with traditional feedstocks. Microalgae are receiving increasing attention for its various types of applications (Brennan and Owende 2010; Zheng et al. 2011). The recent findings claim that only algal biodiesel has the potential to fulfill the global requirement of biofuels for transport (Mata et al. 2010, Chisti 2007, 2008). The algal biomass can be used for various purposes such as biofuels, bio-fertilizers, biopolymers, bio-plastics, lubricants, paints, dyes, colorants, etc. (Cai et al. 2013). As per estimates, approximately 40,000 microalgal species have been identified so far and most of which can accumulate substantially high lipid contents in its cells. It has been found that some of the microalgae species can accumulate up to 70% cellular lipids under specific growth conditions (Chisti 2007; Buddolla et al. 2010; Brennan and Owende 2010). All types of phycoremediation technologies are given due consideration for its ability to treat wastewater and production of biomass along with the treatment. This serves dual advantages. first is the wastewater treatment and second is the production of microalgal biomass. It has already been established that algal species can accumulate 30–70% lipid contents in its cells, and the lipid yield varies with species to species as well as culture conditions. Therefore, such biomass, produced during wastewater treatment, can be used for the production of a variety of biofuels such as crude oil, biodiesel, biomethane, ethanol, hydrogen, etc. (Li et al. 2011; Gupta et al. 2016). With respect to the lipid yield, the algae have advantages over the oil crops as they do not require the arable lands for cultivation purposes.

Generally most of the microalgae species contain up to 20–40% cellular lipids; it has been evidenced that under specific growth conditions, some of the microalgae species can accumulate up to 70–80% cellular lipids (Ma and Hanna 1999). Therefore, for biofuel production purposes, the algae are considered as a potential contender (Chisti 2007; Antoni et al. 2007). Though various types of biofuels are being produced from algal biomass, however, the algal biodiesel is uniquely similar to the mineral diesel. The production of biodiesel from microalgae is being seen as a panacea for rapidly depleting conventional energy resources (Rawat et al. 2011; Singh et al. 2014). It has been established that biodiesel produced from algal lipids is similar to conventional biodiesel; therefore, with slight manipulation of its chemical characteristics, it can be used for the transport purposes (Du et al. 2008).

The cellular lipids of microalgae vary and depend on several factors such as species, growth conditions, availability of nutrients, geographical conditions, availability of nutrients and lights, etc.; however, various studies have shown that the lipid contents as well as the biomass productivity can be enhanced by manyfold by manipulation in the culture conditions (Chisti 2007; Brennan and Owende 2010 and references within). Various types of modulations such as light (Chisti 2008), temperature (Li et al. 2007; Wen and Chen 2003), cell concentrations (Miao and Wu 2006), as well as nutrient limitations, specifically nitrogen (Li et al. 2006; Xu et al. 2006), have been found very effective for the increase of the lipid yield in various microalgae species. High lipid yield by nitrogen starvation was reported in *Chlorella sorokiniana* by Zheng et al. (2013). Similarly, Ramanna et al. (2014) also reported an increase of lipid contents in the same species grown in wastewater by varying nitrogen supplementation.

Other than modulation of specific limiting factors, growth conditions such as autotrophic, heterotrophic, and mixotrophic cultivation conditions also affect the biomass and lipid yields in microalgae. Numerous studies have demonstrated comparatively higher lipid yields under heterotrophic culture conditions than the photoautotrophic cultivation (Miao and Wu 2006; Li et al. 2007, Mohamed et al. 2011), whereas higher growth rates have been observed under mixotrophic growth conditions (Miao and Wu 2006; Mohamed et al. 2011). For photoautotrophic growth, the basic requirements are the sunlight and carbon sources such as CO_2 and various types of cultivation systems such as open circular ponds, raceways, as well as photobioreactor that are used for such purposes (Muller–Feuga, 2004). Whereas for heterotrophic growth, sunlight and CO_2 both are not required, as the algae acquire all the required energy for its growth, only from carbon sources such as sugars or organic acids (Muller–Feuga 2004). Chen and Chen (2000) obtained up to 60 g biomass yield in optimal heterotrophic growth conditions, which was more than tenfold higher than the photoautotrophic yields (5 g) per liter of culture volume. Similarly, Li et al. (2007) reported comparatively very high lipid yields (up to 55.2%) in *Chlorella* sp. under heterotrophic conditions, compared to the autotrophic lipid yields (up to 14.5%).

The culture media and nutrient broths conventionally used for the algae cultivation are highly rich in nutrients, which support the continual growth of microalgae for prolonged periods. The microalgal biomass productivity in such type of growth media is comparatively high; however, the lipid yields of such biomass are lesser. Whereas the concentration of nutrients in wastewater is comparatively very lower than the nutrient broths used for the algae cultivation. It is well established that most of the microalgae accumulate higher lipid contents in nutrient stress conditions. Therefore, in case of wastewater, the microalgae get stressed due to the lower nutrient concentration; in turn this results in higher lipid accumulation. Several studies have demonstrated comparatively high to very high lipid yield in algal biomass grown in wastewaters in comparison to the nutrient broths (Singh et al. 2014; Xin et al. 2010; Zhang et al. 2013; Ramanna et al. 2014). In one of our previous study, we also recorded 27.68% and 28.36% lipid yield in the harvested biomass of *C. sorokiniana* and *S. obliquus* grown in the diluted domestic sewage. We noted that *S. obliquus* performed better with respect to the nutrient stress and lipid yield in comparison to the *C. sorokiniana*. Guldhe et al. (2014a) and Ramanna et al. (2014) also reported higher lipid accumulation under the nutrient stress in both of the species. Such findings clearly demonstrate the importance of species selection for the specific purposes as well as the potential of microalgae for higher accumulation of lipids and biomass yields in wastewater compared to the conventional nutrient broths.

18.4 Factors Influencing Phycoremediation, Algal Biomass, and Lipid Yields

There are various physical, chemical, and biological factors influencing the nutrient assimilation, growth, and lipid yields of algae; hence, such factors directly or indirectly affect the phycoremediation potential as well. Being autotroph, the growth of all the green algae depends upon their photosynthetic potential which is the major limiting factor for all autotrophic systems. The nutrient uptake and biomass yield both depend on the photosynthetic activity (Grobbelaar 2010; Wilhelm and Jakob 2011). All the physicochemical and biological phenomena, such as light intensity, absorption of light and exposure duration, mixing rate, etc. which are directly or indirectly affecting the photosynthesis, play a significant role in the performance of microalgae for both phycoremediation and biomass yield (Rawat et al. 2011; Grobbelaar 2000, 2009; González–Fernández et al. 2011). The light, temperature, and physical turbulence are the most important physical factors, whereas pH, salinity, and nutrient concentrations are the major chemical factors directly influencing the growth of algae. The interspecies competition and contamination of culture by other microalgae species, bacteria, viruses, etc. also play as the limiting biological factors (Grobbelaar 2000; González–Fernández et al. 2011).

The light penetration and utilization depend upon the depth of the culture ponds; therefore for the optimum growth, optimization of euphotic zone is very important. In case of HRAPs, biomass concentration and pond depth significantly affect the degree of light attenuation, and the light exposure to the individual algal cells depends upon the turbulence pattern (Sutherland et al. 2014). Mostert and Grobbelaar (1987) found that nutrient uptake is affected by the mixing/turbulence pattern. Therefore, the optimized speed and frequency of mixing prevent the settling of the algal cells and ensure equal exposure of light (Sutherland et al. 2014). Sutherland et al. (2014) studied the effects of temporal variation on the nutrient removal and biomass yield of algae in a full-scale HRAP. They reported that the nutrient removal and growth of microalgae are greatly influenced by several factors. Among various other physical parameters, temporal variation also affects the nutrient removal and biomass yields of microalgae. Sutherland et al. (2014) also observed reduction of nutrient removal and reduced biomass yield in wastewater HRAP during the late spring/summer season.

In phycoremediation, the algal biomass yield is not directly proportional to the nutrient removal and neither to the lipid yields. While studying the wastewater treatment potential of three microalgal species (*Scenedesmus rubescens*, *Neochloris vigensis*, and *Chlorococcum* sp.), Aravantinou et al. (2013) observed the highest growth rate for *S. rubescens* which showed the least, only 11% p, removal, whereas *N. vigensis* exhibited 53% phosphorus removal which was the highest among all three algal species. The highest lipid yield (19.29%) was also recorded for *N. vigensis* in comparison to the *S. rubescens* which showed 14.91% and the least, i.e., 6.93%, in *Chlorococcum* sp. In a 20-day period, the findings of Aravantinou et al. (2013) clearly demonstrated that the increasing biomass yield was not consistent with either the removal of the nutrients or the lipid yields in all the studied algal species.

18.5 Environmental Sustainability of Phycoremediation-Based Wastewater Treatment and Algal Biofuels

18.5.1 Reduction in Water Footprint (WF)

The ecological footprints are the best indicator of sustainability of biofuels. The most studied and important ecological footprints are carbon and water footprints. Groundwater is a scarce commodity, whereas agriculture alone endorses approximately 86% of global WF (Singh et al. 2015). An increase of 5.5% in the global blue biofuels is expected by 2030 which will significantly impose huge burden over the blue-water resources (Gerbens–Leenes et al. 2012). The first- and second-generation biofuels are based on energy crops and lignocellulosic biomass, whereas algal biofuels are considered as third-generation biofuels. The production of all types of biofuels requires huge amount of water, which could be a limiting factor. In comparison to the fossil fuels, the WF of biomass energy is approximately 70–700 times larger; therefore, the WF must be given due consideration while assessing the sustainability of biofuels (Gerbens–Leenes et al. 2012). The concept of water footprint (WF) was first introduced by Hoekstra (2003) which refers to the direct or indirect requirements and use of total water for an entire supply chain (Hoekstra et al., 2011). The WF serves as an excellent tool for the mapping of complete water consumption for any specific activity or process. Berndes (2002) first time assessed the link between biomass production and respective water uses and reported that water use is not considered in the estimation of future biomass production and use. Berndes (2002) concluded that in 2100, the global evapotranspiration will be approximately doubled in comparison with 1990 with large-scale bioenergy production. Fraiture and Berndes (2009) summarized an overview of water demand for production of both food and biofuels and explored its fate and effects on ecosystem services. Gerbens–Leenes et al. (2012) evaluated the water footprints of biofuels adopted for road transport for ten largest biofuel-consuming countries and reported an increase of about tenfold in the global annual biofuel WF. Approximately, it will reach up to 970 km3/year in 2030 which was around 90 km3/year earlier in 2005 for the global transport sector under the IEA APS scenario. The study revealed that approximately 50% of the global biofuel WF will be contributed by three counties only, i.e., the USA, China, and Brazil, even without consideration of import of biofuel, which may add additional WF. At present, the global international trade of biofuels is approximately only 6–9%, which accounts, comparatively, a shorter WF; however, it will substantially add WF with the increase of import of biofuels (IPCC 2011). For detailed report, reading of Gerbens–Leenes et al. (2012) is recommended. The study also advocated the due consideration of WF for the various types of biofuels to be used for the energy demand in the transport sector.

The precipitated water is known as green water, whereas the ground and surface water, used for cultivation purposes, is blue water. In water footprint concept, the consumptive and nonconsumptive use as well as soil moisture evaporation and evapotranspiration loss of water of these two types of water is considered separately as green and blue WF, respectively (Gerbens–Leenes et al. 2012). The requirement of water for the production of feedstock biomass used for the production of biofuels depends on the geographical as well as climatic conditions that is why the WF varies considerably in specific geographic regions. Numerous strategies are being applied to reduce the WF of biofuels. The first- and second-generation biofuels were produced from energy crops and lignocellulosic biomass, and the requirement of water is considerably high for such feedstock, thus the higher WF. The third-generation biofuels are based on algal biomass, which also requires the huge amount of water; however, production of algal biomass by utilization of wastewater substantially reduces the WF. Moreover, the WF for algal biofuels can further be reduced by biorefinery concepts (Singh et al. 2015). The phytoremediation coupled with biorefinery concept is based on several principles and ensures treatment of wastewater, production of algal biomass, and optimum utilization of biomass for the production of biofuels and various other low-value products. Batan et al. (2013) estimated the WF of a closed photobioreactor-based biofuel and assessed the WF on the basis of blue, green, and lifecycle WF.

The opportunities and challenges both are closely associated with the pilot-scale production of biofuels (Fraiture and Berndes 2009). There are several limiting factors such as the type and productivity of feedstock, geographical locations and climatic conditions, land use pattern and water management practices, etc. which regulate the prices, water and carbon footprints, as well as the environmental fate and effects of the production of biofuels (Fraiture and Berndes 2009). As per estimates, the production of bioenergy requires much more water (24–146 m^3 per gigajoule) than the fossil energy for which it is approximately 1 m^3 ET per gigajoule of energy (Gerbens–Leenes et al. 2008). The environmental sustainability in biofuel production is possible only if the WF is reduced. This can be achieved through multiple approaches such as appropriate land and water management practices for the production of feedstock biomass, development of stress- and drought-resistant high-yield energy crops by biotechnological tools, advancement of energy production processes, development of energy-efficient energy production tools and techniques, etc. In this regards, the phytoremediation technologies offer duel advantages, i.e., treatment of various types of varying-strength wastewater and production of algal biomass, which can be used for various types of energy production. Several microalgae species are excellent candidates for biomass production, whereas several species possess the excellent phycoremediation potential. Numerous studies have identified potential algal species such as *Chlorella* sp., *Scenedesmus*, etc., which have excellent potential for wastewater treatment as well as biomass production (Gupta et al. 2015a, b; Shriwastav et al. 2014 and references within). Therefore, species selection is very important for comprehensive fulfillment of both the objectives. Singh et al.(2015) also reported that the algal biofuel production in a biorefinery concept is imperative for the economic viability and environmental sustainability.

18.5.2 Reduction in Carbon Footprint

The scarcity of water, diminishing nonrenewable energy resources, and global warming due to increasing CO_2 and all other greenhouse gas (OGHGs) discharges in the atmosphere are the major global issues of the twenty-first century (Roman–Leshkov et al. 2007; Kondili and Kaldellis 2007). The major objective of the Kyoto Protocol, 1997, was the reduction of OGHGs by 5.2%, which was endorsed by more than 170 countries. As per estimates, burning of fossil fuels accounts more than two third of the total atmospheric CO_2 (Ramachandra et al. 2013).

For CO_2 mitigation, various strategies are in practice, and, broadly, all of these can be clustered into two types which are chemical and mitigation. The chemical CO_2 mitigation strategies are typically based on three different principles which are sequestration, separation, and transportation. For example, per tonne of CO_2, the transportation for 100 km was estimated approximately 1–3 USD, whereas the separation and compression costs approximately 30–50 USD (Gupta and Fan 2002; Shi and Shen 2003). All of such procedures are energy intensive and substantially expensive (Lin et al. 2003; Resnik et al. 2004). Moreover, waste generated during chemical separation and sequestration poses another environmental problem of disposal (Bonenfant et al. 2003; Yeh et al. 2001). An excellent review article on microalgal bio-mitigation of CO_2 by Wang et al. (2008) is recommended for further studies.

Biological CO_2 mitigation strategies are comparatively economical and environmentally sustainable. All the photosynthetic plants and microorganisms own unique potential of CO_2 fixation through photosynthesis and produce biomass (Kondili and Kaldellis 2007; Ragauskas et al. 2006; de Morais and Costa 2007). The share of agricultural plants in CO_2 capture is very less which is approximately 3–6% only due to slow growth rate. Out of total fossil fuel emissions, the plants can sequester only 3–6% of it (Skjanes et al. 2007). The CO_2 sequestration potential of autotrophic microalgae is 10–50 times higher in comparison to the terrestrial plants, which is due to its exceptionally high growth rates (Li et al. 2008; Usui and Ikenouchi 1997). Microalgae are capable of utilizing CO_2 in various forms for its growth from the water and atmosphere. Industrial flue gases have also been used in various proportions for the culturing and production of algal biomass. Reduction in greenhouse gases is accomplished by the biological binding of carbon dioxide obtained from organic carbon (which would have been volatilized by bacterial respiration) or carbon dioxide from the atmosphere. Energy demand is lowered as there is no need to oxygenate the reactor (Rawat et al. 2011). While studying the microalgal wastewater treatment in a full-scale HRAP, Sutherland et al.(2014) observed that along with CO_2 addition, several other factors such as the design of the pond, operational conditions, turbulence frequency, euphotic zone, etc. need to be controlled and optimized for the enhanced nutrient removal and biomass production using wastewater in a full-scale HRAP system. Therefore, phycoremediation coupled with CO_2 sequestration for biomass production elaborates its promising potentials for the consideration of green algae in the planning and execution of current CO_2 mitigation strategies as well as for biofuels.

Similar to the fossil fuels, measurement of CF of the biofuels is also recommended (Johnson and Tschudi 2012). As per recent assessments of Hammond and Seth (2013), by 2019, CF of biofuels will be approximately 0.449 bn global hectares which is substantially higher than the estimated biofuel CF of 2010 (0.248 bn global hectares). In the case of the biofuels produced from the oil seed, the CO_2 and other OGHGs emissions are comparatively very low, i.e., 0.53 g CO_2–equivalent per g dry seed, but considerably higher emissions, i.e., 0.99 g CO_2, were observed per g of oil extraction and 1.72 g CO_2 in the production of 1 g of biodiesel (Fahd et al. 2012). However, Singh et al. 2014 reported that CF of algal biofuels can be reduced substantially by following the biorefinery concepts. See Singh et al. (2014), "Sustainable Production of Biofuels from Microalgae." Using a biorefinery approach, the phycoremediation coupled with algal biofuel production in a biorefinery concept will comprise numerous advantages. The biofuel production based on the algal biomass grown in wastewater substantially reduces the use of freshwater and the use of chemical nutrients for the growth of algae. Moreover, the use of flue gases for the cultivation also plays significant role in the overall reduction of carbon and water footprints.

18.6 Economic Sustainability of Phycoremediation-Based Wastewater Treatments and Biofuels

Algal treatment systems are favored over conventional treatment processes for decentralized wastewater treatment due to the significantly lower cost of construction and operation and ease of operation without the requirement for skilled labor. The major impediments of the pilot-scale biofuel production from microalgal biomass are (i) the limited availability of freshwater for the cultivation, (ii) inefficient and energy-intensive harvesting, and (iii) solvent-based extraction of microalgal oils. These all together severely affect the lipid extraction, thus the economics of the commercial-scale algal biofuel production. Moreover the use of solvents limits the environmental sustainability of algal biofuels. As per estimates, 3726 kg of freshwater is required for the production of 1 kg of algal biomass without the recycling of water. This water footprint could be reduced if the water is recycled; however, the algae release some exudates (chemical) which are potential inhibitors; therefore, such types of water could be recycled only for few times. Furthermore, in an open pond cultivation system, the rate of evaporative loss of water is approximately 10 L. m^{-2}. day^{-1} which accounts evaporative loss 410 kg of water in the production of 1 kg algal biomass. Production of algal biomass concurrently with wastewater treatment is the basic fundamental of phycoremediation. So, even if evaporative losses were excluded, the phytoremediation could reduce more than 90% of water footprint, and if it is coupled with the use of recycling of treated water, the WF could be reduced further. The requirement of chemical nutrients and media accounts around 30% of the algal biomass production cost, and as per estimates, the cost of one tonne of algal biomass production is approximately 2500–3000 USD even in the pilot-scale production.

The wastewater comprises various types of organics and nutrients as well as trace elements which fulfill the basic growth requirement of the algae. In conventional wastewater treatment, the biological removal of 1 kg of phosphorus costs around 3.05 USD, whereas the same volume of nitrogen removal costs approximately 4.4 USD. Therefore, the use of wastewater for the cultivation of algae reduces almost 100% of freshwater and nutrient requirement which significantly reduces the water and carbon footprint as well as substantially reduces the production cost. Unlike conventional wastewater treatments which produce a huge amount of the sludge due to the use of various types of coagulants, algal wastewater treatments do not generate any sort of sludge which further reduces the economic and environmental burden of further treatment of sludge, its handling, and disposal. Moreover, almost complete wastewater treatment is achieved without any additional cost in a sustainable manner. The ease of operation and low maintenance do not require skilled labor, which further reduces the input costs. In view of environmental consciousness, phycoremediation appears as a technology perceiving wastewater as a valuable resource, zero-waste concept with lowest water, and carbon footprints.

References

Abdel–Raouf N, Al–Homaidan AA, Ibraheem IBM (2012) Microalgae and wastewater treatment. Saudi J Biol Sci 19(3):257–275

Ajayan KV, Selvaraju M, Unnikannan P, Sruthi P (2015) Phycoremediation of tannery wastewater using Microalgae Scenedesmus species. Int J Phytorem 17(10):907–916

Antoni D, Zverlov VV, Schwarz WH (2007) Biofuels from microbes. Appl Microbiol Biotechnol 77(1):23–35

Aravantinou AF, Theodorakopoulos MA, Manariotis ID (2013) Selection of microalgae for wastewater treatment and potential lipids production. Bioresour Technol 147:130–134

Arceivala SJ, Asolekar SR (2007) Wastewater treatment for pollution control and reuse. Tata McGraw Hill Education Private Limited, New Delhi

Aziz MA, Ng WJ (1992) Feasibility of wastewater treatment using the activated–algae process. Bioresour Technol 40(3):205–208

Bain B (2013) Fertilizer outlook. In: Nomura global chemical industry leaders conference. FERTECON Limited, Venice

Banerjee C, Ghosh S, Sen G, Mishra S, Shukla P, Bandopadhyay R (2013) Study of algal biomass harvesting using cationic guar gum from the natural plant source as flocculant. Carbohydr Polym 92(1):675–681

Batan L, Quinn JC, Bradley TH (2013) Analysis of water footprint of a photobioreactor microalgae biofuel production system from blue, green and lifecycle perspectives. Algal Res 2:196–203

Berndes G (2002) Bioenergy and water—the implications of large–scale bioenergy production for water use and supply. Glob Environ Chang 12(4):253–271

Bonenfant D, Mimeault M, Hausler R (2003) Determination of the structural features of distinct amines important for the absorption of CO_2 and regeneration in aqueous solution. Ind Eng Chem Res 42:3179–3184

Brennan L, Owende P (2010) Biofuels from microalgae—a review of technologies for production, processing, and extractions of biofuels and co–products. Renew Sust Energ Rev 14(2):557–577

Buddolla V, Mutanda T, White S, Bux F (2010) The Microalgae–A Future Source of Biodiesel. Dyn Biochem Process Biotechnol Mol Biol 4(1):37–47

CA (Comprehensive Assessment of Water Management in Agriculture) (2007) Water for food, water for life: a comprehensive assessment of water management in agriculture. Earthscan/International Water Management Institute, Colombo/London

Cai T, Park SY, Li Y (2013) Nutrient recovery from wastewater streams by microalgae: status and prospects. Renew Sust Energ Rev 19:360–369

Chabukdhara M, Munjal A, Nema AK, Gupta SK, Kaushal R (2016) Heavy metal contamination in vegetables grown around peri–urban and urban–industrial clusters in Ghaziabad, India. Hum Ecol Risk Assess 22(3):736–752

Chen GQ, Chen F (2006) Growing phototrophic cells without light. Biotechnol Lett 28(9):607–616

Chen Z, Ma W, Han M (2008) Biosorption of nickel and copper onto treated alga (Undaria pinnatifida): application of isotherm and kinetic models. J Hazard Mater 155:327–333

Chevalier P, de la Noüe J (1985) Wastewater nutrient removal with microalgae immobilized in carrageenan. Enzym Microb Technol 7(12):621–624

Chinnasamy S, Bhatnagar A, Hunt RW, Das KC (2010) Microalgae cultivation in a wastewater dominated by carpet mill effluents for biofuel applications. Bioresour Technol 101(9):3097–3105

Chisti Y (2007) Biodiesel from microalgae. Biotechnol Adv 25:294–306

Chisti Y (2008) Biodiesel from microalgae beats bioethanol. Trends Biotechnol 26(3):126–131

De Fraiture C, Berndes G (2009) Biofuels and water. In: Howarth RW, Bringezu S (eds) Biofuels: environmental consequences and interactions with changing land use. Proceedings of the Scientific Committee on Problems of the Environment (SCOPE) international biofuels project rapid assessment, 22–25 September 2008, Gummersbach, Germany. Cornell University, Ithaca, pp 139–153

De Fraiture C, Giordano M, Liao Y (2008) Biofuels and implications for agricultural water use: blue impacts of green energy. Water Policy 10(S1):67–81

de la Noüe J, Proulx D (1988) Biological tertiary treatment of urban wastewaters with chitosan–immobilized Phormidium. Appl Microbiol Biotechnol 29(2–3):292–297

De–Bashan LE, Bashan Y (2004) Recent advances in removing phosphorus from wastewater and its future use as fertilizer (1997–2003). Water Res 38:4222–4246

De–Bashan LE, Moreno M, Hernandez JP, Bashan Y (2002) Removal of ammonium and phosphorus ions from synthetic wastewater by the microalgae *Chlorella vulgaris* co-immobilized in alginate beads with the microalgae growth–promoting bacterium *Azospirillum brasilense*. Water Res 36(12):2941–2948

Demirbas A (2009) Progress and recent trends in biodiesel fuels. Energy Convers Manag 50(1):14–34

Demirbas MF (2011) Biofuels from algae for sustainable development. Appl Energy 88(10):3473–3480

deMorais MG, Costa JAV (2007) Biofixation of carbon dioxide by Spirulina sp. and Scenedesmus obliquus cultivated in a three stage serial tubular photobioreactor. J Biotechnol 129:439–445

Du W, Li W, Sun T, Chen X, Liu D (2008) Perspectives for biotechnological production of biodiesel and impacts. Appl Microbiol Biotechnol 79(3):331–337

Fahd S, Fiorentino G, Mellino S, Ulgiati S (2012) Cropping bioenergy and biomaterials in marginal land: the added value of the biorefinery concept. Energy 37:79–93

Fawzy M, Nasr M, Abdel–Gaber A, Fadly S (2016a) Biosorption of Cr (VI) from aqueous solution using agricultural wastes, with artificial intelligence approach. Sep Sci Technol 51(3):416–426

Fawzy M, Nasr M, Helmi S, Nagy H (2016b) Experimental and theoretical approaches for Cd(II) biosorption from aqueous solution using *Oryza sativa* biomass. Int J Phytorem. doi:10.1080/15226514.2016.1183580

Feng Y, Li C, Zhang D (2011) Lipid production of Chlorella vulgaris cultured in artificial wastewater medium. Bioresour Technol 102:101–105

Filippino KC, Mulholland MR, Bott CB (2015) Phycoremediation strategies for rapid tertiary nutrient removal in a waste stream. Algal Res 11:125–133

Garbisu C, Gil JM, Bazin MJ, Hall DO, Serra JL (1991) Removal of nitrate from water by foamimmobilized Phormidium laminosum in batch and continuous-flow bioreactors. J Appl Phycol 3(3):221–234

Garbisu C, Hall DO, Llama MJ, Serra JL (1994) Inorganic nitrogen and phosphate removal from water by free–living and polyvinyl–immobilized Phormidium laminosum in batch and continuous–flow bioreactors. Enzym Microb Technol 16(5):395–401

Gerbens–Leenes PW, Hoekstra AY, Van der Meer TH (2008) The water footprint of bioenergy: global water use for bio–ethanol, biodiesel, heat and electricity, Value of water research report series no. 34. UNESCO–IHE, Delft

Gerbens–Leenes PW, Van Lienden AR, Hoekstra AY, Van der Meer TH (2012) Biofuel scenarios in a water perspective: the global blue and green water footprint of road transport in 2030. Glob Environ Chang 22(3):764–775

Gil JM, Serra JL (1993) Nitrate removal by immobilized cells of *Phormidium uncinatum* in batch culture and a continuous–flow photobioreactor. Appl Microbiol Biotechnol 39(6):782–787

González–Fernández C, Sialve B, Bernet N, Steyer JP (2011) Impact of microalgae characteristics on their conversion to biofuel. Part 1: Focus on cultivation and biofuel production. Biofuels Bioprod Biorefin 6:105–113

Griffiths M, Harrison SL (2009) Lipid productivity as a key characteristic for choosing algal species for biodiesel production. J Appl Phycol 21(5):493–507

Grobbelaar JU (2000) Physiological and technological considerations for optimizing mass algal cultures. J Appl Phycol 12:201–206

Grobbelaar JU (2009) Upper limits of photosynthetic productivity and problems of scaling. J Appl Phycol 21:519–522

Grobbelaar JU (2010) Microalgal biomass production: challenges and realities. Photosynth Res 106:135–144

Guldhe A, Singh B, Rawat I, Bux F (2014a) Synthesis of biodiesel from Scenedesmus sp. by microwave and ultrasound assisted in situ transesterification using tungstated zirconia as a solid acid catalyst. Chem Eng Res Des 92:1503–1511

Gupta H, Fan LS (2002) Carbonation–calcination cycle using high reactivity calcium oxide for carbon dioxide separation from flue gas. Ind Eng Chem Res 41:4035–4042

Gupta VK, Rastogi A (2008) Biosorption of lead (II) from aqueous solution by non–living algal biomass Oedogonium sp. and Nostoc sp. A comparative study. Colloids Surf B: Biointerfaces 64:170–178

Gupta SK, Chabukdhara M, Pandey PK, Singh J, Bux F (2014a) Evaluation of potential ecological risk of metal contamination in the Gomti River: a biomonitoring approach. Ecotoxicol Environ Saf 110:49–55

Gupta SK, Kumar NM, Guldhe A, Ansari FA, Rawat I, Kanney K, Bux F (2014b) Design and development of polyamine polymer for harvesting microalgae for biofuels. Energy Convers Manag 84:537–544

Gupta SK, Chabukdhara M, Singh J, Bux F (2015a) Evaluation and potential health hazard of selected metals in water, sediments, and fish from the Gomti River. Hum Ecol Risk Assess 21(1):227–240

Gupta SK, Shrivastav A, Kumari S, Ansari FA, Malik A, Bux F (2015b) Phycoremediation of emerging contaminants. In Singh B, Bauddh K, Bux F (eds) Algae: role in sustainable energy production and pollution remediation. Springer International Publishing AG, Cham. pp 129–146. doi:10.1007/978–18–322–2641–34. ISBN: 978–81–322–2639–0. Publication date November 2015, Chapter 11

Gupta SK, Ansari FA, Shriwastav A, Sahoo NK, Rawat I, Bux F (2016) Dual role of Chlorella sorokiniana and Scenedesmus obliquus for comprehensive wastewater treatment and biomass production for bio-fuels. J Clean Prod 115:255–264

Hameed MA (2007) Effect of algal density in bead, bead size and bead concentrations on wastewater nutrient removal. Afr J Biotechnol 6(10)

Hammond GP, Seth SM (2013) Carbon and environmental footprinting of global biofuel production. Appl Energy 112:547–559

Hoekstra AY (2003) Virtual water trade. In: Proceedings of the international expert meeting on virtual water trade value of Wat Res Report Series No.12 IHE, Delft, The Netherlands, 12–13 December 2002

Hoekstra AY, Chapagain AK (2008) Globalisation of water. Sharing the planet's freshwater resources. Blackwell Publishing, Oxford

Hoekstra AY, Chapagain AK, Aldaya MM, Mekonnen MM (2011) The water footprint assessment manual setting the global standard. Earthscan, London

Huo S, Wang Z, Zhu S, Zhou W, Dong R, Yuan Z (2012) Cultivation of *Chlorella zofingiensis* in bench–scale outdoor ponds by regulation of pH using dairy wastewater in winter, South China. Bioresour Technol 121:76–82

IPCC (2008) Intergovernmental Panel on Climate Change. In: Proceedings of IPCC scoping meeting on renewable energy sources. IPCC Working Group III. Geneva

IPCC (2011) Intergovernmental Panel on Climate Change special report on renewable energy sources and climate change mitigation. IPCC Working Group III, Geneva, pp 28

Johnson E, Tschudi D (2012) Baseline effects on carbon footprints of biofuels: the case of wood. Environ Impact Assess Rev 37:12–17

Kaya VM, de la Noüe J, Picard G (1995) A comparative study of four systems for tertiary wastewater treatment by *Scenedesmus bicellularis*: new technology for immobilization. J Appl Phycol 7(1):85–95

Khin T, Annachhatre AP (2004) Nitrogen removal in a fluidized bed bioreactor by using mixed culture under oxygen–limited conditions. Water Sci Technol 50(6):313–320

Kim MH, Chung WT, Lee MK, Lee JY, Ohh SJ, Lee JH, Lee HY (2000) Kinetics of removing nitrogenous and phosphorus compounds from swine waste by growth of microalga, *Spirulina platensis*. J Microbiol Biotechnol 10(4):455–461

Kim S, Lee Y, Hwang SJ (2013a) Removal of nitrogen and phosphorus by *Chlorella sorokiniana* cultured heterotrophically in ammonia and nitrate. Int Biodeterior Biodegrad 85:511–516

Kim S, Park JE, Cho YB, Hwang SJ (2013b) Growth rate, organic carbon and nutrient removal rates of Chlorella sorokiniana in autotrophic, heterotrophic and mixotrophic conditions. Bioresour Technol 144:8–13

Kondili EM, Kaldellis JK (2007) Biofuel implementation in East Europe: current status and future prospects. Renew Sust Energ Rev 11:2137–2151

Kothari R, Pathak VV, Kumar V, Singh DP (2012) Experimental study for growth potential of unicellular alga *Chlorella pyrenoidosa* on dairy waste water: an integrated approach for treatment and biofuel production. Bioresour Technol 116:466–470

Kothari R, Prasad R, Kumar V, Singh DP (2013) Production of biodiesel from microalgae *Chlamydomonas polypyrenoideum* grown on dairy industry wastewater. Bioresour Technol 144:499–503

Lau PS, Tam NFY, Wong YS (1998a) Operational optimization of batchwise nutrient removal from wastewater by carrageenan immobilized Chlorella vulgaris. Water Sci Technol 38(1):185–192

Lau PS, Tam NFY, Wong YS (1998b) Effect of carrageenan immobilization on the physiological activities of Chlorella vulgaris. Bioresour Technol 63(2):115–121

Lee YK (2004) Algal nutrition: heterotrophic carbon nutrition. In: Richmond A (ed) Handbook of microalgal culture: biotechnology and applied phycology. Blackwell Science, Oxford, pp 116–124

Li M, Gong R, RaoX LZ, Wang X (2006) Effects of nitrate concentration on growth and fatty acid composition of the marine microalga *Pavlova viridis* (Prymnesiophyceae). Ann Microbiol 55(1):51–55

Li X, Xu H, Wu Q (2007) Large–scale biodiesel production from microalga Chlorella protothecoides through heterotrophic cultivation in bioreactors. Biotechnol Bioeng 98(4):764–771

Li Y, Horsman M, Wu N, Lan CQ, Dubois-Calero N (2008) Biofuels from microalgae. Biotechnol Prog 24(4):815–820

Li Y, Zhou W, Hu B, Min M, Chen P, Ruan RR (2011) Integration of algae cultivation as biodiesel production feedstock with municipal wastewater treatment: strains screening and significance evaluation of environmental factors. Bioresour Technol 102(23):10861–10867

Lin CC, Liu WT, Tan CS (2003) Removal of carbon dioxide by absorption in a rotating packed bed. Ind Eng Chem Res 42:2381–2386

Lobell D, Burke M, Tebaldi C, Mastrandrea M, Falcon W, Naylor R (2008) Prioritizing climate change adaptation needs for food security in 2030. Science 319:607–610

Lourie E, Gjengedal E (2011) Metal sorption by peat and algae treated peat: kinetics and factors affecting the process. Chemosphere 85:759–764

Ma F, Hanna MA (1999) Biodiesel production: a review. Bioresour Technol 70(1):1–15

Malla FA, Khan SA, Sharma GK, Gupta N, Abraham G (2015) Phycoremediation potential of *Chlorella minutissima* on primary and tertiary treated wastewater for nutrient removal and biodiesel production. Ecol Eng 75:343–349

Markou G, Chatzipavlidis I, Georgakakis D (2012) Cultivation of Arthrospira (*Spirulina platensis*) in olive–oil mill wastewater treated with sodium hypochlorite. Bioresour Technol 112:234–241

Mata TM, Martins AA, Caetano NS (2010) Microalgae for biodiesel production and other applications: a review. Renew Sust Energ Rev 14:217–232

Mata TM, Melo AC, Simões M, Caetano NS (2012) Parametric study of a brewery effluent treatment by microalgae *Scenedesmus obliquus*. Bioresour Technol 107:151–158

McCarty PL, Bae J, Kim J (2011) Domestic wastewater treatment as a net energy producer–can this be achieved? Environ Sci Technol 45(17):7100–7106

Megharaj M, Pearson HW, Venkateswarlu K (1992) Removal of nitrogen and phosphorus by immobilized cells of *Chlorella vulgaris* and *Scenedesmus bijugatus* isolated from soil. Enzym Microb Technol 14(8):656–658

Melis A (2002) Green alga hydrogen production: progress, challenges and prospects. Int J Hydrog Energy 27(11):1217–1228

Miao X, Wu Q (2006) Biodiesel production from heterotrophic microalgal oil. Bioresour Technol 97(6):841–846

Mohamed MS, Wei LZ, Ariff AB (2011) Heterotrophic cultivation of microalgae for production of biodiesel. Recent Pat Biotechnol 5(2):95–107

Mostert ES, Grobbelaar JU (1987) The influence of nitrogen and phosphorus on algal growth and quality in outdoor mass algal cultures. Biomass 13(4):219–233

Molden D, Frenken K, Barker R, Fraiture CD, Mati B, Svendsen M, Sadoff C, Finlayson CM, Attapatu S, Giordano M, Inocencio A (2007) Trends in water and agricultural development. In: Molden D (ed) Water for food, water for life: a comprehensive assessment of water management in agriculture. Earthscan/International Water Management Institute, London/Colombo, pp 57–89

Muller–Feuga A (2004) Microalgae for aquaculture: the current global situation and future trends. In: Richmond A (ed) Handbook of Microalgal culture. Blackwell Science, Chicester, pp 352–364

Nasr M, Ismail S (2015) Performance evaluation of sedimentation followed by constructed wetlands for drainage water treatment. Sustain Environ Res 25(3):141–150

Nasr M, Mahmoud AED, Fawzy M, Radwan A (2015) Artificial intelligence modeling of cadmium(II) biosorption using rice straw. Appl Water Sci 1–9. doi:10.1007/s13201-015-0295-x

National Academy of Sciences (2008) Water implications of biofuels production in the United States. Committee on Water Implications of Biofuel Production. National Research Council. http://www.nap.edu/catalog/12039.html

Ogbonna JC, Yoshizawa H, Tanaka H (2000) Treatment of high strength organic wastewater by a mixed culture of photosynthetic microorganisms. J Appl Phycol 12(3–5):277–284

Olguı´ EJ (2003) Phycoremediation: key issues for cost–effective nutrient removal processes. Biotechnol Adv 22(1):81–91

Postel SL (1998) Water for food production: will there be enough in 2025. Bioscience 48(8):629–637

Qiao H, Wang G (2009) Effect of carbon source on growth and lipid accumulation in Chlorella sorokiniana GXNN01. Chin J Oceanol Limnol 27(4):762–768

Ragauskas AJ, Williams CK, Davison BH, Britovsek G, Cairney J, Eckert CA, Frederick WJ Jr, Hallett JP, Leak DJ, Liotta CL, Mielenz JR (2006) The path forward for biofuels and biomaterials. Science 311(5760):484–489

Ramachandra TV, Mahapatra DM, Gordon R (2009) Milking diatoms for sustainable energy: biochemical engineering versus gasoline–secreting diatom solar panels. Ind Eng Chem Res 48(19):8769–8788
Ramachandra TV, Madhab MD, Shilpi S, Joshi NV (2013) Algal biofuel from urban wastewater in India: scope and challenges. Renew Sust Energ Rev 21:767–777
Ramanna L, Guldhe A, Rawat I, Bux F (2014) The optimization of biomass and lipid yields of Chlorella sorokiniana when using wastewater supplemented with different nitrogen sources. Bioresour Technol 168:127–135
Rawat I, Kumar RR, Mutanda T, Bux F (2011) Dual role of microalgae: phycoremediation of domestic wastewater and biomass production for sustainable biofuels production. Appl Energy 88(10):3411–3424
Rawat I, Gupta SK, Srivastav A, Singh P, Kumari S, Bux F (2016) Microalgae applications in wastewater treatment. In: Bux F, Chisti Y (eds) Algae biotechnology: products and processes. Springer International Publishing, Cham, pp 249–268. doi:10.1007/978–3–319–12334–9_13
Resnik KP, Yeh JT, Pennline HW (2004) Aqua ammonia process for simultaneous removal of CO_2, SO_2 and NOx. Int J Environ Technol Manag 4(1–2):89–104
Richmond A (ed) (2008) Handbook of microalgal culture: biotechnology and applied phycology. Wiley, Chicester
Rockström J, Gordon L, Folke C, Falkenmark M, Engwall M (1999) Linkages among water vapor flows, food production, and terrestrial ecosystem services. Conserv Ecol 3(2):5
Rodolfi L, Chini Zittelli G, Bassi N, Padovani G, Biondi N, Bonini G, Tredici MR (2009) Microalgae for oil: Strain selection, induction of lipid synthesis and outdoor mass cultivation in a low-cost photobioreactor. Biotechnol Bioeng 102(1):100–112
Roman–Leshkov Y, Barrett CJ, Liu ZY, Dumesic JA (2007) Production of dimethylfuran for liquid fuels from biomass derived carbohydrates. Nature 447:982–985
Samori G, Samori C, Guerrini F, Pistocchi R (2013) Growth and nitrogen removal capacity of *Desmodesmus communis* and of a natural microalgae consortium in a batch culture system in view of urban wastewater treatment: part I. Water Res 47(2):791–801
Sari A, Tuzen M (2008) Biosorption of Pb (II) and Cd (II) from aqueous solution using green alga (*Ulva lactuca*) biomass. J Hazard Mater 152:302–208
Sawayama S, Rao KK, Hall DO (1998) Nitrate and phosphate ion removal from water by Phormidium laminosum immobilized on hollow fibres in a photobioreactor. Appl Microbiol Biotechnol 49(4):463–468
Schindler DW, Hecky RE, Findlay DL, Stainton MP, Parker BR, Paterson MJ, Beaty KG, Lyng M, Kasian SEM (2008) Eutrophication of lakes cannot be controlled by reducing nitrogen input: results of a 37–year whole–ecosystem experiment. Proc Natl Acad Sci U S A 105(32):11254–11258
Seckler D, Amarasinghe U, Molden D, de Silva R, Barker R (1998) World water and demand and supply, 1990 to 2025: scenarios and issues, Research report 19. International Water Management Institute, Colombo
Shafiee S, Topal E (2009) When will fossil fuel reserves be diminished. Energ Policy 37(1):181–189
Shen Y, Yuan W, Pei Z, Mao E (2008) Culture of microalga Botryococcus in livestock waste water. T ASABE 51(4):1395–1400
Shi M, Shen YM (2003) Recent progresses on the fixation of carbon dioxide. Curr Org Chem 7(8):737–745
Shi J, Podola B, Melkonian M (2007) Removal of nitrogen and phosphorus from wastewater using microalgae immobilized on twin layers: an experimental study. J Appl Phycol 19(5):417–423
Shriwastav A, Gupta SK, Ansari FA, Rawat I, Bux F (2014) Adaptability of growth and nutrient uptake potential of Chlorella sorokiniana with variable nutrient loading. Bioresour Technol 174:60–66
Singh B, Guldhe A, Rawat I, Bux F (2014) Towards a sustainable approach for development of biodiesel from plant and microalgae. Renew Sust Energ Rev 29:216–245

Singh B, Guldhe A, Singh P, Singh A, Rawat I, Bux F (2015) Sustainable production of biofuels from microalgae using a biorefinary approach. In: Applied environmental biotechnology: present scenario and future trends. Springer, New Delhi, pp 115–128

Skjånes K, Lindblad P, Muller J (2007) BioCO 2–a multidisciplinary, biological approach using solar energy to capture CO 2 while producing H 2 and high value products. Biomol Eng 24(4):405–413

Sørensen B (1991) A history of renewable energy technology. Energ Policy 19(1):8–12

Sutherland DL, Howard–Williams C, Turnbull MH, Broady PA, Craggs RJ (2014) Seasonal variation in light utilisation, biomass production and nutrient removal by wastewater microalgae in a full–scale high–rate algal pond. J Appl Phycol 26(3):1317–1329

Tam NFY, Lau PS, Wong YS (1994) Wastewater inorganic N and P removal by immobilized Chlorella vulgaris. Water Sci Technol 30(6):369–374

Tastan BE, Duygu E, Donmez G (2012) Boron bioremoval by a newly isolated Chlorella sp. and its stimulation by growth stimulators. Water Res 46:167–175

Urrutia I, Serra JL, Llama MJ (1995) Nitrate removal from water by Scenedesmus obliquus immobilized in polymeric foams. Enzyme Microb Technol 17(3):200–205

Usui N, Ikenouchi M (1997) The biological CO_2 fixation and utilization project by RITE(1): highly–effective photobioreactor system. Energy Convers Manag 38(Suppl 1):S487–S492

Varis O (2007) Water demands for bioenergy production. Water Resour Dev 23(3):519–535

Vörösmarty CJ, Lévêque C, Revenga C (2005) Fresh water. In: Hassan R, Scholes R, Ash N (eds) Millenium ecosystem assessment. Ecosystems and human well–being, vol 1. Current status and trends. Island Press, Washington, DC

Wang B, Li Y, Wu N, Lan CQ (2008) CO2 bio-mitigation using microalgae. Appl Microbiol Biotechnol 79(5):707–718

Wang L, Min M, Li Y, Chen P, Chen Y, Liu Y, Wang Y, Ruan R (2010) Cultivation of green algae Chlorella sp. in different wastewaters from municipal wastewater treatment plant. Appl Biochem Biotechnol 162(4):1174–1186

Wen ZY, Chen F (2003) Heterotrophic production of eicosapentaenoic acid by microalgae. Biotechnol Adv 21(4):273–294

Wijffels RH, Barbosa MJ (2010) An outlook on microalgal biofuels. Science 329(5993):796–799

Wilhelm C, Jakob T (2011) From photons to biomass and biofuels: evaluation of different strategies for the improvement of algal biotechnology based on comparative energy balances. Appl Microbiol Biotechnol 92(5):909–919

Wu LF, Chen PC, Huang AP, Lee CM (2012) The feasibility of biodiesel production by microalgae using industrial wastewater. Bioresour Technol 113:14–18

Xin L, Hong–ying H, Ke G, Ying–xue S (2010) Effects of different nitrogen and phosphorus concentrations on the growth, nutrient uptake, and lipid accumulation of a freshwater microalga Scenedesmus sp. Bioresour Technol 101(14):5494–5500

Xu H, Miao X, Wu Q (2006) High quality biodiesel production from a microalga Chlorella protothecoides by heterotrophic growth in fermenters. J Biotechnol 126(4):499–507

Yeh JT, Pennline HW, Resnik KP (2001) Study of CO_2 absorption and desorption in a packed column. Energy Fuel 15:274–278

Zheng HL, Yin JL, Gao Z, Huang H, Ji XJ, Dou C (2011) Disruption of Chlorella vulgaris cells for the release of biodiesel–producing lipids: a comparison of grinding, ultrasonication, bead milling, enzymatic lysis and microwaves. Appl Biochem Biotechnol 164:1215–1224

Zheng Y, Li T, Yu X, Bates PD, Dong T, Chen S (2013) High–density fed–batch culture of a thermotolerant microalga Chlorella sorokiniana for biofuel production. Appl Energy 108:281–287

Coupling Phytoremediation Appositeness with Bioenergy Plants: A Sociolegal Perspective

19

Rashwet Shrinkhal

Abstract

The humanity has already started to experience the wrath of anthropogenic pollution of environment, primarily due to mindless development based upon non-renewable energy resources. Realising the need to have an environmentally friendly development along with the adoption of natural methods for corrective action to revitalise the environment, scientists are constantly engaged in developing new technologies. Biofuel production and phytoremediation techniques are practical manifestation of such an effort. However, these technologies do bear certain signs of aberration when not applied judiciously. This chapter briefly discusses some of the significant sociolegal concerns connected with the implementation of biofuel programmes and phytoremediation techniques in practical world.

Keywords

Bioenergy • Biofuel • Phytoremediation • Indigenous peoples • Food vs. fuel • ILUC • Invasive species

19.1 Introduction

Calvin Coolidge once mentioned 'The business of America is business' (Speck 1970–1971, 200) reflects the fundamental mind-set of a capitalist world which hooks on the meta-value of making money that is followed by the policy of monetisation for all goods and services for trade through market mechanism (Sinha 2014). And what remains the greatest peril of such mindless 'commodification of

R. Shrinkhal (✉)
Centre for Tribal and Customary Law, Central University of Jharkhand, Ranchi 835205, India
e-mail: rashwetshrinkhal@gmail.com

K. Bauddh et al. (eds.), *Phytoremediation Potential of Bioenergy Plants*,
DOI 10.1007/978-981-10-3084-0_19

everything' attitude? Answer is nothing but impoverished, polluted and bleeding Earth (Klein 2014). Lately realising this fact, better late than never, many now rejects the idea underlying Coolidge statement. In this process, a strong global sentiment that seems to be recognised is that the success or failure of our society cannot be determined exclusively in respect of economic growth or gross domestic product, but instead it must be determined on such scales as 'ecological balance' and 'quality of life' (Speck 1970–1971; Chowdhry 1988). There prevails the stress, a battle between two sets of values: one the outcome of a strong belief that aggressive industrialisation and economic growth symbolises the pillars of prosperity and the other the resultant of a shocking realisation that prosperity endeavours little or no happiness when the soils are contaminated, when the air is unfit to breathe and when the water is unhealthy for human consumption (Speck 1970–1971; Gadgil and Guha 1995; Bretschger and Vinogradova 2016).

This notion of pitching 'economy' against 'ecology' (Faramelli 1971) and mapping them as 'anti' or 'parallel' to each other may appear superfluous while tracking the etymology of both words. They both share the same genesis—'eco'—from the Greek *oikos*, which signifies 'house'. 'Economics' is associated with '[m]anagement of household goods', and 'ecology' corresponds to '[t]he study of the relationships in the house' (Hart 2008). Finding it necessary for the 'common good', Hart argues that 'economics' and 'ecology' need to be on the same page and establishes the link between the two:

> The common good of the integrated and interdependent biotic (and human social) community, and the common good of Earth as biota's shared home and household, find common ground in interrelated forms of justice and in interrelated form of injustice. Economic liberation and ecological liberation are linked. (Hart 2008)

In this background, the humanity started to acknowledge the truth that the complications of environment are impossible to separate from the ones of the prosperity and typically from the economic development (Gherasim and Tānase 2012). With this understanding, international community started to think on a new form of development, known as sustainable development. In time, the formulation of this notion had a continuous evolution to such an extent that today it has met a lot of criticism for its vagueness and ambiguity (Applegate and Aman, Jr. 2001). Amidst all controversy on the concept, it was meant, as expressed by the World Commission on Environment and Development (WCED), to '[e]nsure that [humanity] meets the needs of the present without comprising the ability of future generation to meet their own needs' (WCED 1987). Sustainable development is inferred as a new brand of economic development, organic and unalike to the present type, which has domineering effects on many sections of the globe during nineteenth and twentieth centuries, restricted to 'the use of natural resources of the planet, of the conventional and non-conventional form of energy, concomitantly with the protection and preservation of the environment' (Gherasim and Tānase 2012). The discourse of dominant predatory economic development, inter alia, has the following relevant characteristics (evils): (a) depletion and over-exploitation of natural resources especially fossil fuels and (b) continuous degradation of the environment (Sachs 1974). Sustainable development is also aimed to find alternative approaches having rectifying propensity against above-mentioned evils.

Working on the path of sustainable development, science and technology—the one who brought the peril—can provide solution to resource or conservation problems. The necessity to marshal suitable science and technology for sustainable development has been recognised long ago. Initial study on the sustainable yield management of renewable resources laid the basis for the International Union for Conservation of Nature's ground-breaking *World Conservation Strategy*, pronounced in 1980. The conditions for disposition of relevant experimentation as an essential module of sustainable development schemes were developed by a number of transnational technical bodies during the mid-1980s, patronised by the Brundtland Commission report *Our Common Future* in 1987 and canonised in the Agenda 21 action plan that resulted from the United Nation Conference on Environment and Development in 1992. In the following decades to come, deliberation continued to intensify on how science and technology could deliver more efficaciously to attain sustainability, involving various scientists, experts, academicians and development associations from around the world (Clark and Dickson 2003).

This book is also an attempt towards development of scientific knowledge in connection with collaboration of two scientific processes for maximum optimisation. The aim of this chapter is to throw light on some of the concerns attached with both the processes so that any policy intended to promote technologies involved should also take care of the concerns attached. This may lead to evolution of more robust normative structure dealing with the regulation of technologies discussed. In doing so, the chapter initially discusses, in brief, the major themes of the book. Subsequently, an attempt has been made to show certain sociolegal concerns. Sincere efforts have been maintained to upheld scientific overtone in writing this chapter. However, the author expresses his limitation of not having formal training in science.

19.2 Understanding the Themes Involved

19.2.1 Bioenergy: Biomass to Biofuel

Epstein in his recent work *The Moral Case for Fossil Fuels* (2014) argues that energy derived from fossil fuels has been the major propellant of human prosperity and continues to be the lone source of viable power having capability to guarantee our quality life rolling headway and lead to the pursuit of happiness (Epstein 2014, 69, 84, 135–136). Furthermore, fossil fuel energy should be persistently used as a preferable option for the time to come because its gains tremendously prevail over its costs. He further contends that there is adequately sufficient resource available and shall not be cleaned out of fossil fuels by any chance in the near future (Epstein 2014).

There are two propositions that need severance: the assertion in relation to the past and the one referring to the future. It is absolutely conceivable to come to an understanding that fossil fuel energy has brought colossal social welfare historically. To be sure, lots of persons, as well as most conservationists, would straightaway accept that fossil energy has been the supreme agent of economic growth. However, some of his claims on the future of fossil fuel is contestable, misleading or demonstrably wrong (Freeman 2015). Relevant to the theme of this Chapter, Epstein misconstrues the fact that fossil fuel is 'so plentiful as to be unlimited' (Freeman 2015). Many scientists are of the view that there is restricted amount of fossil fuel reserves both in terms of physical availability and economic viability, thus reducing them to the category of limited and non-renewable natural reserves. This arises due to the simple reason that it requires millions of years for fossil fuels to accrue, whereas the reserves are exploited with rapid pace, making it impractical for the scale of production to match with rate of exploitation. As a whole, if the extraction rate is higher than formation time, the reserves shall be limited in all probability that it will ultimately be sapped (HööK and Tang 2013).

In the past, the scientific community had shown concerns over fossil fuel depletion or over exploitation. M. K. Hubbert, a Shell scientist, predicted in 1956 that both the United States and the world production rates of oil would peak and at a later time go on decreasing. Hubbert anticipated that household production of oil would peak between 1965 and 1970 and that world production would peak near 2000 (Hubbert 1956). Although the calculation about the prospect of world oil production proved to be incorrect, there is hardly any dispute on his forecast regarding domestic oil production (McAllister 2008–2009). Much before Hubbert, it was Thomas Malthus in the year 1978, in his classical work on the 'Principles of Population', laid stress on the general limits of the earth and found resource scarcity innate in the finiteness of the globe (Malthus 1798). Similarly, the Belgian mathematician Pierre-Francois Verhulst proposed a logistic growth curve; it described the self-limiting property of population due to limited environmental resources (Verhulst 1838; Bacaër 2011). Post-Hubbert theory, it was a report entitled 'Limits to Growth' (1972), published by a group of scientists from the Massachusetts Institute of Technology, that presented an exceedingly cynical future of the world. It described that the upcoming world population level, food production and industrialisation would initially multiply exponentially but then suddenly fall down in the course of ensuing century. The crumble happens for the reason that the global economy will stretch to its physical limits in terms of non-renewable resources, agrarian production and immoderate pollution (Meadows et al. 1972; Tahvonen 2000).

As is now clear, these predictions about the exact limitation period for the exhaustion of natural resources failed to be true. Nonetheless, humanity cannot sit and relax as these predictions failed not because of any increase in net quantity of natural resources; rather it was new technological innovations which gave fresh lease to the availability of resources, in the form of increased production from the existing resources (David and Wright 1997). It is imperative to take pause and judiciously scrutinise the existing situation of the Earth. Even if we accept the plea of certain scientists, including Epstein, there is plentiful supply of fossil fuel. But does that necessarily imply that the supply must be first and foremost fossil fuel energy in eternity? Those who debate in favour of shifting focus towards renewable energy are not 'working to save planet for its own sake but rather to avoid disastrous long-term consequences to *humans*' (Freeman 2015). Many of the scientists predict or argue that there exists certain 'hidden cost' of fossil fuel combustion. Overuse of fossil fuel is often claimed to be the chief 'pathogen' for many environmental damages such as global warming, acid rain and air pollution (Liu et al. 2007). It is predicted that carbon dioxide and other greenhouse gases released after combustion of fossil fuel may have substantial damage to climate system (Balat et al. 2007). Deleterious impact of fossil fuel is also seen in the form of acid rains which is caused by sulphur dioxide (Smith 1983).

In their efforts to reduce the dependency on fossil fuels and avoid long-term harmful effect on environment, scientists have generated great interest in one of the propitious form of alternative energy: *Bioenergy*. This curiosity has been significantly moved by the petroleum price escalation between 2005 and mid-2008, as it occurred all along and even after the late 1970 energy crisis. Champions of this interest deduced bioenergy as a way to: (a) shield against mounting fossil fuel costs; (b) boost energy security by decreasing dependence on imported petroleum; (c) ration, in the near future, trustworthy domestic supplies mitigating against lessened or higher-cost supplies because of bloating worldwide demand and a decreasing international supply of conventional petroleum with increased dependence on more costly sources; (d) shape up the balance of payments by decreasing efflux of money to pay for more imported oil; and (e) decrease combustion of fossil fuel, as greenhouse gases are being emitted in this process, and GHG emissions are often linked for precipitating climate change (McCarl and Boadu 2009).

The term 'bioenergy' refers to a large renewable energy source and has been '[a] primary source of energy for most of human history' (Jennings 2007). Bioenergy is derived from resources that can be redeveloped and cannot be exhausted. Bioenergy is regarded as a renewable resource for the reason that it uses resources that can be promptly renewed, and are easily available, distinguished from fossil fuels (coal, oil and natural gas) which take thousands of years to regenerate. Thus to satisfy bioenergy requirement, energy seekers must incessantly secure ample and reliable stock of biomass at prices letting them to manoeuvre profitably (Dahiya 2014). It is imperative to understand that bioenergy is derived from *biomass*, which is 'any form of organic matter' (Hartman 2013). It is a broad term and includes numerous materials, which are essentially renewable (Uhland 2008). In reference to the production of energy, biomass can be exploited in any of the following ways: (a) ablaze directly as a fuel, either independently or mixed with fossil fuels, (b) treated to convert into a synthetic fuel that can burned to produce heat and energy or (c) breakdown of methane so that energy can be produced (Hartman 2013). Further, it is important to understand the exact meaning of certain terminology, often used as interchangeable, in bioenergy discourse. The terms like 'biofuels', 'biodiesel' and 'bioethanol' are often used as synonym of 'bioenergy'. However, these terms do carry separate meaning whose clarification will not be out of context.

'Biofuel' refers to 'the bioethanol and biodiesel and other fuels made from biomass and primary used for motive, thermal power generation' having 'certain quality specification' (S.2(f), Biofuels Act of 2006). Biofuels derived from oils, sugars and starches stocked in edible crops are referred as *first-generation biofuels*. These are obtained through comparatively modest and conventional technologies. Changeover technologies still under progress expedite the genesis of *second-generation biofuels*, also referred as advanced biofuels, which are derived from inedible crops, such as seasonal grasses and xyloid materials, and the inedible sections of food crops. *Third-generation biofuels* are derived from algae (Goh and Lee 2010; Lee and Lavoie 2013; Dahiya 2014).

'Bioethanol' is a liquid fuel prepared out of plant source and recycled elements of the food chain. Bioethanol is processed from sustainable sources of feedstock, usually plants such as wheat, sugar beet, corn, straw and wood. At present, bioethanol is derived from large-scale yeast fermentation of sugars that are obtained or made from crops followed by separation of the ethanol by distillation (Demirbas 2007). On the other hand, 'biodiesel' relates to a range of ester-based oxygenated fuels from renewable biotic descent. The same can be manufactured from processed biological oils and fats. Technically, biodiesel is defined as the 'monoalkyl esters of long-chain fatty acids derived from renewable biolipids' (Demirbas 2008a, 114). Biodiesel is usually manufactured through 'reaction of a vegetable oil or animal fat with methanol or ethanol in presence of a catalyst to yield methyl or ethyl esters and glycerine' (Demirbas 2008b)

19.2.2 Phytoremediation

In reaction to an increasing demand to focus on environmental contamination, several remediation technologies have been invented to ameliorate the contaminated soil (Riser-Roberts 1992) that are chiefly mechanical or physiochemical-based remediation techniques. These include soil washing, soil vapour extraction, soil flushing, solidification, etc. (Hamby 1996; Khan et al. 2004; Marques et al. 2009). However, these technologies are ordinarily costly and soil disturbing, from time to time making the land inadequate as a medium for new endeavours such as plant growth. As a result, a biology-based emerging technology is successful in securing the responses of both soil remediation scientists and the common people—phytoremediation.

Phytoremediation is a generic nomenclature related to technologies that use plants to remediate contaminated soil and groundwater by piling up the contaminants, generating environments contributing to the degradation of contaminants by means of natural biochemical process, volatilising contaminants to the atmosphere or by functioning as pumping mechanism to control groundwater flow (Rieger and Scott 1998). Phytoremediation consists of four mechanisms: phytovolatilisation, phytoextraction, phytodegradation and enhanced rhizosphere biodegradation. These mechanisms can be used individually, in any combination, or in association with other remedial alternatives (Dolan and Glynn 1997).

19.2.3 Coupling the Two Technologies

The capitalisation of bioenergy plants for the remediating polluted land and groundwater is an innovative strategy to derive additional benefits out of bioenergy crops (Tripathi et al. 2016). An important question, which also seems to be the aim of the present book, which needs to be understood, is why there was a need felt to couple phytoremediation technology with bioenergy plants? Possibly the answer lies in the fact that despite phytoremediation technology being well known since three decades, there seems less practical applicability on a large-scale basis (Meagher 2000; Wong 2003; Oh et al. 2013). One of the basic reasons for restricted practical applicability of phytoremediation is its 'cost'. Usually under conventional approach, only special plants, referred as hyperaccumulator, were used for phytoremediation. After being harvested, these special plants are treated with incineration due to the fact that they are loaded with high metal content. Generally such plants are scarce, having slow growth rate with no or less economic value (Oh et al. 2013). Moreover these special plants are subject to strong patent regime which makes their seeds and seedling expensive and low on accessibility (Oh et al. 2013). Coupling phytoremediation with biofuel plants could be economically valorised in the form of bioenergy, '[c] ontributing to an important environmental co-benefit' (Kathi 2016).

19.3 Emerging Issues Related with the Application of Concerned Technologies

Much of the optimism surrounding the benefits of these technologies is based on the belief that they might offer significant solution to environmental pollution and degradation, but they do come with their own sets of challenges and risks. The present section of the chapter shall attempt to thematically present the emerging issues and concerns associated with the use of these technologies.

19.3.1 Food vs. Fuel

One of the serious controversies attached with the first-generation biofuels that utilises merchandise such as corn, wheat, soybeans and vegetable oil is that the use of these merchandise knuckles with their readiness as food or animal feed. It is argued and debated by several scholars that the use of arable land for bioenergy can lead to food deficit and escalating food prices (Runge and Senauer 2007; Mitchell 2008; Tenebaum 2008; Ottinger and Miller 2010, 25; Stiegler 2014). Evidence of apprehensions are noticeable, as on 14 April 2008, the online news agency—African Energy News Review—penned down that food riots have resulted in death of five people in Haiti. It observed that 'the diversion of food crops to biofuel production was significant factor contributing to global food price rocketing by 83% in the last year, and causing violent conflicts in Haiti and other parts of the world' (Ottinger and Miller 2010).

In the year 2007, the United Nations Food and Agriculture Organisation (UNFAO) came with an estimate that there was soar in world food prices by 40% in the preceding one year, and the rise in price had adversely affected all major biofuel feedstocks comprising sugarcane, corn, rapeseed oil, palm oil and soybeans (Tenebaum 2008). In the same year, UN Special Rapporteur on the right to food labelled resulting food insecurity due to production of biofuel as a 'crime against humanity' (Lederer 2007). Similarly in the year 2008, a World Bank paper spilled to *The Guardian* estimated that biofuel production was guilty for escalation in food prices by 75% from the year 2002 to 2008 (Molony and Smith 2010).

Amidst these apprehensions and concern from a section of scholars opposed by the necessity to switch towards an alternative fuel, the key question is: do we currently have any normative framework for both production and governance of biofuel?

Possibly, at the global level, biofuel production is yet to be a subject of any legal framework involving binding multilateral treaty that especially addresses its environment and socioeconomic effects. Nonetheless the *Rio Declaration on Environment and Development* 1992 (*Rio Declaration*) and the UN good governance principles tender a comprehensive source for governance in the areas of environment and development (Lima 2009). *Rio Declaration* demands that all undertakings, including biofuel production, take place in agreement with environmental conservation. Untenable designs of production shall be scaled back, bring to an end and substituted by techniques that preserve natural resources and ecosystems' purity for the use of present and future generation (Principle 3, 4, 8 and 15 Rio Declaration 1992; Lima 2009). The *Rio Declaration* also demands that biofuel production donates effectually to hammer away inequality and cater the requirements of the poor and marginalised, who shall be given special priority. The dominant forces of the world should not take advantage of developing countries and their populations. Care should be taken that demands and vulnerability of developing countries should not be exploited to an extent that it suffers from environment or human health-related issues (Principle 5, 6 and 14 Rio Declaration 1992; Lima 2009).

The *Rio Declaration* also laid down principles for in what manner the procedure of production and development should be administered. It necessitates collaboration between people and States at the global level, with extra responsibility on the part of developed nations. Sustainability concerns should be governed with a strong policy framework through open meetings and in a general agreement fashion rather than one-sided decision. The procedure should be clear, and all engaging, and it should include accountability mechanisms (Principle 7, 10, 12, 13 and 27; Lima 2009).

In addition to the *Rio Principles*, the United Nations Development Programme *Governance and Human Development* (1997) developed certain principles: rule of law, transparency, cooperation and consensus orientation, accountability, inclusiveness and participation, responsiveness, effectiveness and fairness, biophysical sustainability and efficiency. These principles could play pivotal role for international biofuel governance. Lima has argued that how good governance principles coupled with *Rio Declaration* principles can be utilised to develop an international legal framework for biofuel governance (Table 19.1).

Table 19.1 Assessment framework for biofuel production and governance requirements

Legal principles RP (Rio principles) GGP (good governance principles)	GGP 7	RP 3; RP 4; RP 8; RP 15; GGP 7	RP 5; RP 6; RP 14; GGP 7	RP 5; RP 6; GGP 4	RP 10; GGP 1; GGP 6	RP 13; GGP 8	RP 7; RP 12; RP 27; GGP 5	RP 10; GGP 3	RP 27; GGP 2
Requirements	*Efficiency*	*Biophysical sustainability*	*Effectiveness and fairness*	*Responsiveness*	*Inclusiveness and participation*	*Accountability*	*Cooperation and consensus orientation*	*Transparency*	Rule of law
	Best use of the resource available	No negative impact on ecosystem integrity (e.g. air, water, soil, etc.)	It helps to meet people's needs, prioritising poverty irradiation, and does not exploit power or other vulnerabilities of countries or populations	It responds to the needs and interests of all, giving priority to the populations and countries who are most in need	Due involvement of all concerned stakeholders, especially those who will be most affected by decisions	Liability and compensation mechanisms are present to hold actors accountable for their decisions and actions	Multiple views are on board, interests are balanced, and decisions are taken through cooperation and consensus	Accessible and understandable information for actors; a decision-making process open to observation and scrutiny	There exists a clear and structured legal framework, which is internally consistent and enforceable
Issue	*Sustainability*		*Equity*			*Legitimacy and legality*			
Dimension	*Production*		*Governance*						

Source: Lima (2009)

19.3.2 Indirect Land Use Change

The increase in demand for ethanol and biodiesel around the world has led to risk of unintended consequences with a potential negative impact on environment. As discussed above, when requirement for biofuels escalates and food and feed crops are encouraged to be utilised for biofuels, the dearth in food production may be redressed by fresh food production on hitherto nonagricultural areas elsewhere, such as forest or grasslands (Valin et al. 2015). Thus to '[b]ring new land under cultivation— if and when happen in response to biofuel demand', it is referred as 'indirect land use change' (ILUC) ((Malins et al. 2014). Akrill and Kay in their work, *The Growth of Biofuels in the 21st Century*, explain in detail what is ILUC with a hypothetical example of German farmers. In this they portray that if a great deal of farmers from Germany decides to switch out of wheat production in favour of biofuel crops, it could escalate the price of wheat in international market. Consequently, farmers of Ukraine may be forced to plough pasture land in order to grow wheat (Akrill and Kay 2014). This is a clear-cut example of ILUC. An obvious question which comes into mind is: Why ILUC is so important? It is relevant to understand that claims in favour of biofuel production are also made on the ground that they have 'potential to mitigate climate change by reducing greenhouse gas (GHG) emissions compared to using fossil fuels' (Overmars et al. 2011). However, a complex condition stand up when extensive space of rainforest, peat bogs, grasslands and other natural lands are reduced into agricultural land as a response to biofuel production in some other place (ILUC). The complexity with ILUC is derived from the fact that, first, when jungles are in scope or length, grassland and peat bogs are ploughed, GHGs are on the loose into the atmosphere. Second, removing these stout natural sinks considerably reduces their natural capacity to absorb greenhouse gases, especially carbon (Ottinger and Miller 2010). Thus, the net result could be more GHG emission than what one reduced by preferring biofuels over fossil fuels.

A number of studies have attempted to calculate or otherwise evaluate the scale of the carbon consequences of the ILUC. The first peer-reviewed endeavour to calculate the possible extent of ILUC emission was made by Searchinger (2008). Timothy Searchinger's group applied 'the partial economic model of world agriculture' conceived by the Food and Agriculture Policy Research Institute (FAPRI) to determine how much land use change might be contemplated in corresponding the US corn ethanol mandates. It then allocated possible carbon emissions to this amount of land use change by applying 'average land carbon content estimates' from the Woods Hole Oceanographic Institution. The modelling indicated that the carbon emission intensity of land use change would be 104 grammes of carbon dioxide equivalent per megajoule (gCO_2/MJ), higher than any possible decrease from truncated fossil fuel use, and therefore, the research group contended that any likely carbon savings from truncated fossil fuel would increase rather than decrease carbon emission. Post Searchinger study, there were many attempts to analyse the ILUC adversarial impact, if any, and to quantify the same (Gallagher Review 2008; Overmars et al. 2011; Edwards et al. 2010). From the various studies done, it would be safe to argue that a consensus has developed in the scientific and regulatory community that ILUC is real. However, still confusion remains to the magnitude of the ILUC emissions.

The point which I want to make is that any policy on biofuel production should incorporate sustainability principles. Even the United Nations Framework Convention on Climate Change (UNFCCC) has acknowledged that land use change makes a far reaching contribution to global CO_2 emissions. Consequently, a crucial part of its efforts to mitigate climate change is limitation of such emissions through the development of REDD+ (reducing emission from deforestation and forest degradation, conservation of forest carbon, sustainable management of forest and enhancement of forest carbon stocks), a mechanism designed to incentivise efforts to retain and enhance forest carbon stocks in developing countries. Moreover, the European Commission also came with new rules in the year 2015, which intended to reduce the potential risk of ILUC and amended the Renewable Energy Directive (RED) (2009/28/EC 2009) and Fuel Quality Directive (2009/30/EC, 2009).

19.3.3 Concerns of Tribal and Indigenous Peoples

Besides the problems of food security and environmental concerns, biofuel is criticised for adverse social consequences it brings in the life of tribal and indigenous communities. Expanding the biofuel industry encourages governments and private investors to go to developing countries to buy or lease lands and grow biofuel crops (Chen 2016).

Nicola Colbran in her peer-reviewed work highlights the plights of local tribal communities of Indonesia. She argues that the private companies in collaboration with the Indonesian government have institutionalised oil palm production for bioenergy purpose. This in turn has led to the massive deforestation and land acquisition of tribal communities. Land alienation of tribal/indigenous peoples has direct adversarial relationship with the indigenous culture. Further, she argues that monoculture cropping pattern either in the form of oil palm or *Jatropha* poses serious threats to biodiversity. Additionally, oil palms have seriously depleted the water level creating serious water crisis among tribal areas (Colbran 2011).

Similarly, Montefrio and Sonnenfeld (2013) in their study focused on the situation of indigenous farmers in the Philippines, which portrays a murky picture of officially launched '*Jatropha* biofuel programme' in the year 2006. *Jatropha* was perceived by the governments as universal panacea for issues ranging from poverty alleviation and reduction, energy sustainability, reforestation of land denuded of trees and reinforcing food security. The justification generally forwarded by office bearers was that *Jatropha*, being a robust non-food crop, could be cultivated on land inappropriate for food production and used for restoration of discarded lands. Also, private and public firms involved in *Jatropha* cultivation encouraged the crop as 'sustainable' (Montefrio and Sonnenfeld 2013).

Though the official aim of the biofuel production programme was to achieve self-reliance in energy needs of the Philippines, the 'global and local' biofuel actors had an altogether different intention of exporting biodiesel to international market. The study further revealed that how the parastatal organisations (companies owned or controlled by governments) and private companies in between 2007 and 2010 involved several local actors including indigenous peoples and enable *Jatropha* growth through contracting farming. These companies aggressively sought for large indigenous lands on the pretext that several large tracts of indigenous lands are nothing but a waste. However, the ulterior motive was to have a contiguous biofuel cropping pattern in order to increase productivity. This whole idea of involving large tracts of indigenous land was itself contrary to the initial understanding of the biofuel crop plantation would be limited to smallholder farmers (Montefrio and Sonnenfeld 2013).

The entire discourse on biofuel production in the Philippines had some complex problems. First, the argument advanced by the company members that indigenous lands are nothing but a waste seems similar to an erstwhile doctrine of *terra nullius*, by which indigenous peoples' land and territory was referred by European invaders as 'territory of nobody', and hence it could be available for subjugation. European colonialism thrived on such an obnoxious doctrines (Colbran 2011; Anghie 1999). Second, parastatal and private companies forced contracting indigenous peoples who wanted to practise traditional swidden agriculture—a form of shifting cultivation—to grow continuously the crop of *Jatropha*, even in some cases against their wish. Third, many of these indigenous farmers used forest land for *Jatropha* farming, and the resulting deforestation deprived many forest-based indigenous population and communities. This raised the tension among the local communities itself (Montefrio and Sonnenfeld 2013). Moreover, these problems have direct bearing on the subdued cultural rights of indigenous peoples.

19.3.4 Invasive and Exotic Plants

Phytoremediation and biofuel production have a common concern regarding invasive and exotic plants. Most peoples have little or no knowledge about the difference between an invasive exotic species and a native species and will go about their life blissfully unaware of this threat (Stuart 2012). Invasive species is defined as 'an alien species whose introduction does or is likely to cause economic or environmental harm or harm to human health… Alien species means, with respect to a particular ecosystem, any species … that is not native to that ecosystem' (Executive Order No. 13112, 1999). There had been serious deleterious effect on the biodiversity by certain invasive species which includes: kudzu (*Pueraria montana*), a semi-woody perennial plant belonging to the Fabaceae or legume family (Forseth and Finnis 2004); eucalyptus plants in India, introduced for fuel and timber, turned out to be water intensive (Shiva and Bandyopadhya 1983).

Arundo donax, a plant which had both capacity of bioenergy production and phytoremediation, had altered the hydrology and eliminated native species of Florida, a glaring example which forces us to have precautionary approach while choosing plants for bioenergy or phytoremediation purposes (Vanheusden et al. 2012; Herrera and Dudley 2003).

Invasive species are imported either deliberately (generally under the presumption that introduction will revamp the environment or boost local industries) or inadvertently, as stowaways in conveyances, containers or the good themselves. Under any situation, international law has a significant role to play in the control and regulation of invasive species. In 1951, the *International Plant Protection Convention* (IPPC) created the first international regime with an objective to prevent the introduction and spread of nonindigenous plant pests via packing materials, storage places and transportation facilities. Following an obligation stipulated in the IPPC, the European and Mediterranean Plant Protection Organisation was founded in 1951. One of the main objectives of the organisation is to provide a database of the invasive plants. A major breakthrough in the control of invasive species came with the adoption of *Convention on Biological Diversity* (CBD) in the year 1993. The CBD made it obligatory on contracting states to '[p]revent the introduction of, control or eradicate to those alien species which threaten ecosystems, habitats or species' (Article 8(h), CBD 1993). The *Conference of Parties* (COP) specially referred to the potential risk of 'biocontrol agents' as invasive species. It requested member states, governmental organisations and other relevant institutions to 'take appropriate measures' in the form of 'codes of practice regarding trade and use of biocontrol agents'. It emphasised on having regulatory framework on invasive species at all three levels: domestic, regional and global (para. 55, COP 8 Decision VIII/27 2006).

19.4 Conclusion

The arguments and concerns pictured in this chapter should not be misconstrued as an attempt to thwart the development of progressive techniques of combining biofuel production with phytoremediation in spite of the fact that both the components exhibit certain deleterious effects. However, these concerns can be addressed by a robust regulatory mechanism governing biofuel production and management of phytoremediation crops. Unlike energy through fossil fuels, bioenergy is sustainable and cleaner because of less harmful emissions. Similarly phytoremediation is a more eco-friendly approach for soil and groundwater remediation. If we carefully assess the social adversity attached with the use of these technologies especially food crisis, the ILUC and impact on indigenous communities, it may be not be unwarranted to claim that most of these problems are due to policy paralysis. As regards to the problem of invasiveness of a plant species, a thorough and unbiased in situ research is required before large-scale introduction of such plant outside its place of origin. The key mantra for successful practical application of the concerned technologies is adherence to the principle of sustainable development wherein environment and human welfare constitute the core concern.

References

Ackrill R, Kay A (2014) The growth of biofuels in 21st century: policy drivers and market challenge. Palgrave MacMillan, Hampshire

Anghie A (1999) Finding the peripheries: sovereignty and colonialism in nineteenth century international law. Harv Int J 40(1):1–80

Applegate JS, Aman AC (2001) Introduction—syncopated sustainable development (sustainable development, agriculture, and the challenge of genetically modified organism symposium). Indiana J Glob Leg Stud 9(1):1–12

Bacaër N (2011) A short history of mathematical population dynamics. Springer, London

Balat M et al (2007) Influence of fossil fuel energy applications on environmental pollution. Energy Source, Part B: Econ Plan Policy 2(3):213–226

Biofuels Act (2006) Republic of the Philippines http://www.doe.gov.ph/issuances/republic-act/614-ra-9367. Accessed May 2016

Bretschger L, Vinogradova A (2016) Human development at risk: economic growth with pollution-induced health shocks. https://www.ethz.ch/content/dam/ethz/special-interest/mtec/cer-eth/resource-econ-dam/documents/people/lbretschger/Health_25.01.2016_new.pdf. Accessed April 2016

Chen Y (2016) Trade, food security, and human rights: the rules for international trade in agricultural products and the evolving world food crisis. Routledge, New York

Chowdhry K (1988) Growth or survival? India Int Cent Q 15(4):191–200

Clark WC, Dickson NM (2003) Sustainability science: the emerging research programme. http://www.pnas.org/content/100/14/8059.full.pdf. Accessed Apr 2016

Colbran N (2011) Indigenous peoples in Indonesia: at risk of disappearing as distinct peoples in the rush for biofuel. Int J Minor Gr Right 18:63–92

COP 8 Decision VIII/27 (2006) Alien species that threaten ecosystems, habitats, or species (Article 8(h)): further considerations of gaps and inconsistencies in international regulatory framework. https://www.cbd.int/decision/cop/default.shtml?id=11041. Accessed May 2016

Dahiya A (2014) Bioenergy—biomass to biofuels: an overview. In: Dahiya A (ed) Bioenergy: biomass to biofuels. Elsevier, Amsterdam, pp 1–36

David AP, Wright G (1997) Increasing returns and the genesis of American resource abundance. Ind Corp Chang 6(2):203–245

Demirbas A (2007) Producing and using bioethanol as an automotive fuel. Energy Sources, Part B: Econ Plan Policy 2(4):391–401

Demirbas A (2008a) Biofuels: securing the planet's future energy needs. Springer, Heidelberg

Demirbas A (2008b) The importance of bioethanol and biodiesel from biomass. Energy Sources, Part B: Econ Plan Policy 3(2):177–185

Directive 2009/28/EC (2009) Directive 2009/28/EC of the European Parliament and of the council on the promotion of the use of energy from renewable sources and amending and subsequently repealing directives 2001/77//EC and 2003/30/EC. Off J Eur Union L(140):16–62

Directive 2009/30/EC (2009) Fuel quality Directive. Off J Eur Union L(140):88–113

Dolan TC, Glynn BA (1997) New treatment technologies and some significant regulatory changes. Environ Claim J 10(1):113–122

Edwards EJ et al (2010) The origins of C4 grasslands: integrating evolutionary and ecosystem science. Science 328:587–591

Epstein A (2014) The moral case for fossil fuels. Penguin, New York

Faramelli NJ (1971–1972) Perilous links between economic growth, justice and ecology: a challenge for economic planners. Environ Aff 1(1):218–227

Forseth IN, Finnis AF (2004) Kudzu (*Pueraria montana*): history, physiology, and ecology combine to make major ecosystem threat. Crit Rev Plant Sci 23(5):401–413

Freeman J (2015) A critical look at "the moral case for fossil fuels". Energy Law Rev 36(2):327–354

Gadgil M, Guha R (1995) Ecology and equity. Penguin, Gurgaon

Gallagher Ed (2008) The Gallagher review of the indirect effects of biofuels production. https://www.unido.org/fileadmin/user_media/UNIDO_Header_Site/Subsites/Green_Industry_Asia_Conference__Maanila_/GC13/Gallagher_Report.pdf. Accessed May 2016
Gherasim ME, Tānase G (2012) Fundamentals of sustainable development. Contemp Read Law Soc Justice 4(1):446–456
Goh CS, Lee KT (2010) A visionary and conceptual marcoalgae-based third-generation bioethanol (TGB) biorefinery in Sabah, Malaysia as an underlay for renewable and sustainable development. Renew Sust Energ Rev 14:842–848
Hamby DM (1996) Site remediation techniques supporting environmental restoration activities – a review. Sci Total Environ 191(3):203–224
Hart J (2008) The poor of the planet and the planet of the poor: ecological ethics and economic liberation. Univ St Thomas Law J 5(1):144–182
Hartman BJ (2013) Biomass-based diesel: fueling the renewable fuel standard. LSU J Energy Law Resour 1(2):123–150
Herrera AM, Dudley TL (2003) Reduction of riparian arthropod abundance and diversity as a consequence of Gian Reed (*Arundo donax*) invasion. Biol Invasions 5(3):167–177
HööK M, Tang X (2013) Depletion of fossil fuel and anthropogenic climate change —a review. Energy Policy 52:797–809
Hubbert MK (1956) Nuclear energy and fossil fuels: API conference, San Antonio, TX, (March 7–9 1956). Subsequently published as Publication no. 95, Shell Development Co. (June 1956)
Jennings MJ (2007) Bioenergy: fueling the future. Drake J Agric Law 12(1):205–230
Kathi S (2016) Bioenergy from phytoremediated phytomass of aquatic plants via gasification. In: Prasad MNV (ed) Bioremediation and bioeconomy. Elsevier, Amsterdam, pp 111–128
Khan FI et al (2004) An overview of site remediation technologies. J Environ Manag 71:95–122
Klein N (2014) This changes everything: capitalism vs. the climate. Penguin Random House, Toronto
Lederer EM (2007) UN expert calls biofuels a crime against humanity. Live Science. Accessed Apr 2016
Lee RA, Lavoie JM (2013) From first-to-third generation biofuels: challenges of producing a commodity from a biomass of increasing complexity. Anim Front 3(2):6–11
Lima MGB (2009) Biofuel governance and international legal principles: is it equitable and sustainable. Melbourne J Int Law 10(2):470–492
Liu L et al (2007) Mitigating environmental pollution and impacts from fossil fuels: the role of alternative fuels. Energy Sources, Part A: Recovery Util Environ Eff 29(12):1069–1080
Malins C et al. (2014) A guide for the perplexed to the indirect effects of biofuels production. http://www.theicct.org/sites/default/files/publications/ICCT_A-Guide-for-the-Perplexed_Sept2014.pdf.. Accessed May 2016
Malthus TR (1798) An essay on the principles of pollution. J. Johnson, in St. Paul's Church-yard, London
Marques APGC et al (2009) Remediation of heavy metal contaminated soils: phytoremediation as a potentially promising clean-up technology. Crit Rev Environ Sci Technol 39:622–654
McAllister L (2008) America is dying and the hospital's power is shut off: the health care industry's debilitating reliance on nonrenewable energy. Quinnipaic Health Law J 12(1):69–114
McCarl BA, Boadu FO (2009) Bioenergy and U.S. renewable fuels standard: law, economic, policy/climate change and implementation concerns. Drake J Agric Law 14(1):43–74
Meadows DH et al (1972) The limits to growth. Universe Books, New York
Meagher RB (2000) Phytoremediation of toxic elemental and organic pollutants. Curr Opin Plant Biol 3(2):153–162
Mitchell D (2008) A note on rising food prices. Work bank policy research working paper no. 4682 http://papers.ssrn.com/sol3/papers.cfm?abstract_id=1233058. Accessed April 2016
Molony T, Smith J (2010) Biofuels, food security, and Africa. Afr Aff 109(436):489–498
Montefrio MJF, Sonnenfeld DA (2013) Global-local tensions in contract farming of biofuel crops involving indigenous communities in the Philippines. Soc Nat Resour 26:239–253

Oh K et al (2013) Development of profitable phytoremediation of contaminated soils with biofuel crops. J Environ Prot 4:58–64
Ottinger RL, Miller SE (2010) Bioenergy in developing countries. Renew Energy Law Policy Rev 2010(1):23–32
Overmars KP et al (2011) Indirect land use change emission related to EU biofuel consumption: an analysis based on historical data. Environ Sci Pol 14(3):248–257
Rieger GE, Scott TE (1998) Emerging technical developments-phytoremediation. Environmental Claims Journal 10(2):171–180
Riser-Roberts E (1992) Bioremediation of petroleum contaminated sites. CRC Press, Boca Raton
Runge CF, Senauer B (2007) How biofuels could starve the poor. https://www.foreignaffairs.com/articles/2007-05-01/how-biofuels-could-starve-poor. Accessed Apr 2016
Sachs I (1974) Environment and styles of development. Econ Polit Wkly 9(21):828–837
Searchinger TRH (2008) Use of U.S. croplands for biofuel increases greenhouse gases through emissions from land-use change. Science 39(5867):1238–1240
Shiva V, Bandyopadhya J (1983) Eucalyptus—a disastrous tree for India. Ecologist 13(5):184–187
Sinha JBP (2014) Psycho-social analysis of the Indian mindset. Springer, New Delhi
Smith GP II (1983) Acid rain: transnational perspective. NY School J Int Comp Law 4(3):459–502
Speck AF (1970) Environment law: the price of pollution. UMKC Law Rev 39(2):200–212
Stiegler L (2014) Avoiding the Catch-22: reforming the renewable fuel standard to protect freshwater resources and promote energy independence. Univ Richmond Law Rev 48(3):1063–1109
Stuart M (2012) Saving our ecosystems: the threat of invasive exotic species and the need to act now. Environ Claim J 24(3):220–244
Tahvonen O (2000) Economic sustainability and scarcity of natural resources: a brief historical review. http://www.rff.org/files/sharepoint/WorkImages/Download/RFF-IB-00-tahvonen.pdf. Accessed April 2016
Tenebaum DJ (2008) Food vs. Fuel: diversion of crops could cause more hunger. Environ Health Perspect 116(6):A254–A257
Tripathi V et al (2016) Towards coupling of phytoremediation with bioenergy production. Renew Sust Energ Rev 57:1386–1389
Uhland AM (2008) Improving regulations for biomass-based electrical generating facilities. Nat Resour Environ 23(1):15–18
UNCED (United Nations Conference on Environment and Development) (1992) Rio declaration on environment and development. WCED, Rio de Janerio
Valin H et al (2015) The land use change impact of biofuels consumed in the EU: quantification of area and greenhouse gas impacts. https://ec.europa.eu/energy/sites/ener/files/documents/Final%20Report_GLOBIOM_publication.pdf. Accessed Apr 2016
Vanheusden B et al (2012) Legal and economic aspects of crops selection for phytoremediation purposes and the production of biofuel. http://projects.swedgeo.se/r2/wp-content/uploads/2012/04/REJUVENATE-2-Final-report-Legal-and-economic-aspects.pdf. Accessed Apr 2016
Verhulst PF (1838) A note on the law of population growth. In: Smith D, Keyfitz N (eds) Mathematical demography. Springer, New York, pp 333–339
WECD (World Commission on Environment and Development) (1987) The Brundtland report: our common future. Oxford University Press, Oxford
Wong MH (2003) Ecological restoration of mine degradation soils, with emphasis on metal contaminated soils. Chemosphere 50(6):775–780

MIX
Papier aus verantwortungsvollen Quellen
Paper from responsible sources
FSC® C105338

If you have any concerns about our products,
you can contact us on
ProductSafety@springernature.com

In case Publisher is established outside the EU,
the EU authorized representative is:
Springer Nature Customer Service Center GmbH
Europaplatz 3, 69115 Heidelberg, Germany

Printed by Libri Plureos GmbH
in Hamburg, Germany

Springer Series on

Wave Phenomena 18

Edited by H. A. Haus

Springer Series on

Wave Phenomena

Editors: L. M. Brekhovskikh L. B. Felsen H. A. Haus
Managing Editor: H. K. V. Lotsch

Volume 1
Mechanics of Continua and Wave Dynamics 2nd Edition
By L. M. Brekhovskikh, V. Goncharov

Volume 2
Rayleigh-Wave Theory and Application
Editors: E. A. Ash, E. G. S. Paige

Volume 3
Electromagnetic Surface Excitations
Editors: R. F. Wallis, G. I. Stegeman

Volume 4
Short-Wavelength Diffraction Theory
Asymptotic Methods
By V. M. Babič, V. S. Buldyrev

Volume 5
Acoustics of Layered Media I
Plane and Quasi-Plane Waves
By L. M. Brekhovskikh, O. A. Godin

Volume 6
Geometrical Optics of Inhomogeneous Media
By Yu. A. Kravtsov, Yu. I. Orlov

Volume 7
Recent Developments in Surface Acoustic Waves
Editors: D. F. Parker, G. A. Maugin

Volume 8
Fundamentals of Ocean Acoustics
2nd Edition
By L. M. Brekhovskikh, Yu. P. Lysanov

Volume 9
Nonlinear Optics in Solids
Editor: O. Keller

Volume 10
Acoustics of Layered Media II
Point Sources and Bounded Beams
By L. M. Brekhovskikh, O. A. Godin

Volume 11
Resonance Acoustic Spectroscopy
By N. Veksler

Volume 12
Scalar Wave Theory
Green's Functions and Applications
By J. A. Desanto

Volume 13
Modern Problems in Radar Target Imaging
Editors: W.-M. Boerner, F. Molinet, H. Überall

Volume 14
Random Media and Boundaries
Unified Theory, Two-Scale Method, and Applications
By K. Furutsu

Volume 15
Caustics, Catastrophes, and Wave Fields
By Yu. A. Kravtsov, Yu. I. Orlov

Volume 16
Electromagnetic Pulse Propagation in Causal Dielectrics
By K. E. Oughstun, G. C. Sherman

Volume 17
Wave Scattering from Rough Surfaces
By A. S. Voronovich

Volume 18
Electromagnetic Wave Propagation in Turbulence
Evaluation and Application of Mellin Transforms
By R. J. Sasiela

Richard J. Sasiela

Electromagnetic Wave Propagation in Turbulence

Evaluation and Application of Mellin Transforms

With 80 Figures

Springer-Verlag
Berlin Heidelberg New York
London Paris Tokyo
Hong Kong Barcelona
Budapest

Dr. Richard J. Sasiela
Massachusetts Institute of Technology, Lincoln Laboratory,
Lexington, MA 02173-0073, USA

Series Editors:

Professor Leonid M. Brekhovskikh, Academician
Russian Academy of Sciences, P. P. Shirshov Institute of Oceanology,
23 Krasikowa St., 117218 Moscow, Russia

Professor Leopold B. Felsen, Ph.D.
Polytechnic University, Route 110, Farmingdale, NY 11735, USA

Professor Hermann A. Haus
Department of Electrical Engineering & Computer Sciences, MIT,
Cambridge, MA 02139, USA

Managing Editor: Dr.-Ing. Helmut K.V. Lotsch

Springer-Verlag, Tiergartenstrasse 17,
D-69121 Heidelberg, Germany

ISBN-13: 978-3-642-85072-1 e-ISBN-13: 978-3-642-85070-7
DOI: 10.1007/978-3-642-85070-7

Library of Congress Cataloging-in-Publication Data. Sasiela, R.J. (Richard J.), 1940 – Electromagnetic wave propagation in turbulence: evaluation and application of mellin transforms / R.J. Sasiela. p. cm. – (Springer series on wave phenomena ; 18) Includes bibliographical references and index. ISBN 3-540-57619-3 (Berlin : acid-free paper). – ISBN 0-387-57619-3 (New York : acid-free paper) 1. Electromagnetic waves – Transmission. 2. Atmospheric turbulence. 3. Mellin transform. 4. Numerical calculations. I. Title. II. Series. QC665.T7S27 1994 539.2--dc20 93-46744

Softcover reprint of the hardcover 1st edition 1994

Typesetting: Camera ready copy from the author

SPIN: 10122896 54/3140-5 4 3 2 1 0 – Printed on acid-free paper

To my parents Joseph and Helen whose encouragement,
support, and love made many things possible

Preface

This book is directed at two audiences: those interested in problems of electromagnetic wave propagation in turbulence and those interested in evaluating integrals. For the first group, the text provides a systematic way to obtain analytic answers to problems in which the scintillation is small and there are no nonlinear effects due to high optical powers. For those interested in evaluating integrals, the integration method is explained in separate chapters. In the chapters containing examples of wave propagation in turbulence, the problem is quickly reduced to one of evaluating an integral, and can be viewed as examples of the integration technique.

To address these two audiences, this book develops a systematic way of expressing solutions to problems of electromagnetic wave propagation in turbulence in integral form. It also develops Mellin transform techniques that are used to evaluate these integrals. This technique has three major advantages over others: 1) it is applicable to a wide range of problems; 2) the application of the technique is straightforward; and 3) the answers are expressed in analytic form. Mellin transform and hypergeometric functions have been a scientific backwater and are used regularly by only a few people. That is a shame for several reasons. Mellin transforms allow a deeper understanding of infinite series. Knowing the Mellin transform of a function is tantamount to knowing a convenient infinite series. Mellin transform techniques, which require an understanding of hypergeometric functions, enable one to deepen his or her understanding of elementary and transcendental functions. In addition to this pedantic usefulness, it is the natural way to solve several types of problems that have wide applicability. For instance, Mellin transforms permit one to perform integrations that are very difficult to perform by other means. They enable one to solve boundary value problems in spherical and cylindrical coordinates with the same ease that Fourier transform techniques afford in solving differential equations with constant coefficients. The self-similar characteristic of Mellin transforms leads to applications in image, radar and acoustic processing, and chaos and fractal theory.

The major part of this book develops and applies a method for evaluating integrals analytically and expressing the result either as infinite series or as a sum of generalized hypergeometric functions. At first look, the method to evaluate integrals is formidable, and the final results look very complicated. It has been suggested that results from a numerical integration can be obtained more quickly. One has to overcome these prejudices. It is true that the formalism is difficult to learn because it uses mathematical techniques that are generally unfamiliar to most scientists, but that was not a valid excuse for not learning other difficult techniques, which are part of a standard scientific education. If the formalism produces results that are useful, one should be willing to overcome the initial difficulty.

This technique has indeed proved to be very useful! The expressions for the power series, although lengthy to write down, are easily and quickly evaluated using the recursion relation for gamma functions. Recently several computer algebra programs have acquired the ability to evaluate generalized hypergeometric functions, to which the power series are equivalent. In this context the results are no more difficult to evaluate and plot for specific cases than results expressed in terms of more familiar functions such as sinusoids and exponentials. The analytic form of the answer uncovers the natural parameters of a problem and gives one insight into how important a parameter is — an insight that is difficult to develop with numerical techniques. Integrands that contain the difference of two almost equal quantities, a condition that leads to difficulties in numerical integration, are handled in the complex plane by simply deforming an integration path past a pole. Because the technique is algorithmically based, one can develop a computer algebra program that automatically evaluates these integrals, in which case the user would not need to learn the details of the technique to get an analytic answer and to generate curves for specific ranges of parameter values. The development of such a program is being investigated. Just as scientific calculators made tables of trigonometric functions and logarithms obsolete, such a computer program would do the same to most material in integral tables. It would also allow one to evaluate many integrals that are not in the integral tables.

This technique was originally developed to evaluate integrals one encounters when solving problems of electromagnetic wave propagation in turbulence. The technique enables one to solve problems in terms of integrals that are generated with filter functions that multiply the turbulence spectrum. Problems that take days to solve when one starts from first principles can often be solved in less than an hour with appropriate filter functions and Mellin transform techniques.

The techniques given in this book were developed over several years in the high-energy beam-control and propagation group at MIT Lincoln Laboratory. I would like to thank MIT Lincoln Laboratory for providing the opportunity to work on challenging problems for which this technique was developed, for the freedom to pursue research in this area, and for the chance to interact with people interested in helping to develop this technique. This work was sponsored by the Strategic Defense Initiative through the Department of the Air Force under Contract No. F19628-90-C-0002.

I chose to solve many of the problems to illustate the method developed in this book. I did not do a comprehensive literature search to see if these problems were previously solved. I apologize if I have left out relevant references.

Several people provided ideas that enabled me to develop the technique. Lee Bradley first suggested the possible usefulness of Mellin transform techniques and the existence of the Marichev text. He also suggested the use of Gegenbauer polynomials in addressing the anisoplanatism problem. The technique of evaluating integrals in several complex planes was developed in collaboration with John Shelton.

Many people had a hand in suggesting what material to include, and how it should be organized. I would like to especially thank Jonathan Schonfeld for carefully reading the first draft and suggesting many organizational changes. I have had many suggestions from Robert Kramer, Hernan Praddaude, Ronald Parenti, and Charles Primmerman. Fred Knight was very helpful in converting the text into LaTeX. Jim Eggert was particularly helpful since he was willing to read the text during several stages of the evolution of this book and made many useful suggestions. The series editor Professor Hermann Haus and the Springer-Verlag editor Helmut Lotsch made helpful suggestions on how the material should be presented. I want to thank Bill Breen, Ed Sullivan, Dave Tuells, Kevin Walsh, and their staff for producing the figures, converting them into Postscript, and printing the final copy. Sue Richardson and Katharine Krozel provided useful editorial help.

A book like this that contains so much new material and has so many complicated equations is very difficult to make error free. I would appreciate hearing any comments you have on the material or errors you have found in the text. My E-mail address on Internet is Sasiela@ll.mit.edu.

Writing a book takes a tremendous investment in time and energy that is no longer available for home life. I thank my wife Joan for being so understanding during this period.

October, 1993 Richard Sasiela
Lexington, Massachusetts

Table of Contents

Glossary

$(a)_k$	Pochhammer symbol
$A(\boldsymbol{\kappa}, z)$	Amplitude of second wave
$B_\phi(d)$	Phase correlation function
c	normalized wavenumber, speed of light
C_n^2	Turbulence strength
$C_p^\lambda(x)$	Gegenbauer polynomial
D	Aperture diameter
D_i	Inner diameter of annulus
D_s	Diameter of a finite source
$\mathcal{D}(\mathbf{d}, \boldsymbol{\rho})$	Structure function of total turbulence
D_c	Characteristic source diameter for scintillation averaging
$\mathcal{D}_v(\mathbf{r})$	Structure function of velocity
$\mathcal{D}_n(\mathbf{r})$	Structure function of atmospheric density
$\mathcal{D}_\phi(\mathbf{d}, \boldsymbol{\rho})$	Structure function of phase
$\mathcal{D}_\chi(\mathbf{d}, \boldsymbol{\rho})$	Structure function of log-amplitude
$E_n(\boldsymbol{\kappa}, z)$	Normalized axial correlation function of turbulence
$f(\kappa)$	Normalized turbulence spectrum
$F(\Omega_1, \ldots, \Omega_k)$	$(n_1!)^{\Omega_1} \cdots (n_k!)^{\Omega_k}$
$F(\gamma\boldsymbol{\kappa})$	Aperture filter function
F_N	Fresnel number
${}_pF_q[(a);(b);z]$	Generalized hypergeometric function
$g(k)$	Hill Spectra
$g(\boldsymbol{\rho})$	Aperture weighting function
$G(\gamma\boldsymbol{\kappa})$	Complex aperture filter function
$G_{p,q}^{m,n}\left[z\middle\vert\begin{matrix}a_1,\ldots,a_p\\b_1,\ldots,b_q\end{matrix}\right]$	Meijer's G-function
H	End altitude of propagation path
$H_{p,q}^{m,n}\left[z\middle\vert\begin{matrix}(a_1,\alpha_1),\ldots,(a_p,\alpha_p)\\(b_1,\beta_1),\ldots,(b_q,\beta_q)\end{matrix}\right]$	Fox's H-function
$J_\nu(x)$	Bessel function
$\mathbf{k}$	Three-dimensional propagation vector

k_0	$2\pi/\lambda$, wavenumber
k_z	Propagation vector along propagation direction
$K(\alpha)$	Modulation transfer function of a circular aperture
$K_\nu(x)$	Modified Bessel function of the third kind
L	End point of propagation path
L_i	Inner scale with definition $L_i = 2\pi/\kappa_i$
L_{in}	Inner scale with definition $L_{in} = 5.92/\kappa_i$
L_o	Outer scale
$n(\mathbf{r})$	Variation of air density with position
$P(\gamma, \kappa, z)$	Diffraction parameter
r_o	Coherence diameter of a plane wave
r_{os}	Coherence diameter of a spherical wave
R_o	Radius of curvature of beam wave at the source
$s,\ s_n,\ t$	Complex variables
$S(p_1, \ldots, p_k)$	$\sum_{n_1=p_1}^{\infty} \cdots \sum_{n_k=p_k}^{\infty} \frac{(-1)^{n_1+\cdots+n_k}}{n_1!\cdots n_k!}$
$S_\phi(\omega)$	Power spectral density of phase
$S_\chi(\omega)$	Power spectral density of log-amplitude
SR	Strehl ratio
t	Time or second complex variable
$\mathbf{u}_z$	Unit vector in propagation direction
$U(x)$	Heaviside unit step function
v_n	n^{th} velocity moment of turbulence
v_g	Wind ground speed
$v(h)$	Wind speed as a function of height
W_o	$1/e^2$ radius of beam wave
$Z(m, n)$	Zernike polynomial
$\alpha(h)$	Normalized atmospheric density versus altitude
γ	Propagation parameter
$\Gamma[x]$	Gamma function
$\delta(x-a)$	Dirac delta function
Δn_0	Difference of refractive index between two colors
θ_o	Isoplanatic angle
θ_χ	Characteristic source angle for scintillation reduction
θ_s	Angle subtended by a finite source
κ	Transverse wavenumber
κ_i	Wavenumber of inner scale
κ_o	Wavenumber of outer scale

λ	Wavelength
μ_m	m^{th} turbulence moment
μ_m^+	m^{th} upper turbulence moment
μ_m^-	m^{th} lower turbulence moment
$\nu(\boldsymbol{\kappa}, z)$	Fourier transform of refractive index variations
$\boldsymbol{\rho}$	Two-dimensional transverse spatial vector
τ	Time delay between measurement and correction
$\varphi(\boldsymbol{\rho}, z)$	Phase variation due to turbulence
$\varphi R(\boldsymbol{\rho}, z)$	Phase related quantity
$\Phi_1(\boldsymbol{\rho}, z)$	Total variation of the turbulence fluctuations
ψ	Psi or digamma function
$\chi(\boldsymbol{\rho}, z)$	Log amplitude due to turbulence
$\chi R(\boldsymbol{\rho}, z)$	Log-amplitude related quantity
ω	Radian frequency of turbulence variation
$\zeta(s)$	Riemann zeta function

1. Introduction

Solving problems of wave propagation in turbulence is a field that occupies the services of a small group of researchers. The methods used in this community and the results obtained are not generally known by researchers in other communities. The main reason is that the field is considered difficult, and if there is not an obvious need to investigate the effects of turbulence, they are neglected. The difficulty arises from the need to solve stochastic differential equations. Advances made by Tatarski and Rytov reduce problems to multiple integrals. These integrals are often difficult to evaluate since fractional exponents of functions appear in integrands. The final step in most cases is to evaluate these integrals numerically and to present the results as parametric curves. Many cases are run to develop some insight into how a quantity of interest varies with parameters. Becoming an expert in this field requires a great deal of time to become familiar with these graphical results so that one has some insight into various effects.

As pointed out above, there is a formalism for reducing a problem to quadratures. This process is lengthy, and there are several ways of doing it. Different workers use different methods to get at the same result. This makes it difficult for the novice to understand the literature and to realize that there is some underlying order. This discourages a person with only a casual interest from developing a facility in this field. It was to make the solution of these problems more algorithmic that the methods expounded in this book were developed.

In this book I use the Rytov approximation to reduce a very general problem to a triple integral. I develop techniques that allow one to evaluate these integrals analytically.

The integrals that one encounters contain products of functions of which one or more is a Bessel function. Workers in the field look for these integrals in integral tables, and if unsuccessful, resort to numerical analysis. Even numerically, some of these integrals are difficult to evaluate. The integrand is often either the product of a function that goes to infinity multiplied by one that goes to zero at one of the integration limits or the difference of two functions that each lead to a divergent integral. Great care must be exercised in evaluating these integrals.

The techniques developed in this book provide a recipe for obtaining analytical solutions. There is a saying, "You don't get something for nothing." Indeed, there is a price to be paid for being able to solve these problems more easily: The method uses mathematical techniques that are not generally taught to physicists and engineers, and one has to be willing to learn these techniques. One has to become familiar with Mellin transforms, gamma functions, generalized hypergeometric functions, asymptotic techniques, and pole-residue integration in one or more complex planes. Once these techniques are mastered, one can quickly solve complicated problems of wave propagation in turbulence. In addition, learning these techniques has some important ancillary benefits that are also discussed in this book. The techniques can be applied to most integrals in tables. In addition, one can easily evaluate integrals that are not in the tables. Since complicated integrals appear in many fields, the technique has wide applicability. Also Mellin transform techniques can be applied to solve differential and integral equations and have found many other applications as well.

An objection voiced against this method is that expressing an integral in terms of generalized hypergeometric functions is not very useful. The argument against the use of these functions is rapidly fading with the increasing accessibility of personal computers. These functions can be evaluated easily with recursion relations. These functions have become even easier to use since the introduction of computer algebra programs on personal computers that manipulate these functions just as easily as trigonometric functions. It has become easy to evaluate expressions in which these functions appear, and there is no reason to eschew them any more than Bessel functions.

1.1 Book Plan

In this book I systematize the solution of many practical problems in wave propagation through turbulence. In Sect. 2 of this chapter the Mellin transform and the Mellin convolution integrals are introduced. A short table of Mellin transforms that are used to solve propagation problems in turbulence is also given.

In Chap. 2 the Rytov approximation for wave propagation is introduced as is the concept of the turbulent spatial spectrum. I describe a fairly general problem of wave propagation in turbulence that is solved with the Rytov approximation. This problem involves the difference between the log-intensity or phase of two waves that can be propagating in different directions or be displaced from each other and have different focal distances. I show that solutions to problems of this kind can be reduced to triple integrals; one integral along the propagation direction and two along the spatial transform wavenumbers of the wave. Most problems encountered in practice are special cases of this general problem.

The expression for the variance of the phase or log-amplitude of a wave propagating from $z = 0$ to $z = L$ is of the form

$$\sigma^2 = 0.2073\, k_0^2 \int_0^L dz\, C_n^2(z) \int d\boldsymbol{\kappa}\, f(\kappa)\, g\,(\boldsymbol{\kappa}, z)\,, \tag{1.1}$$

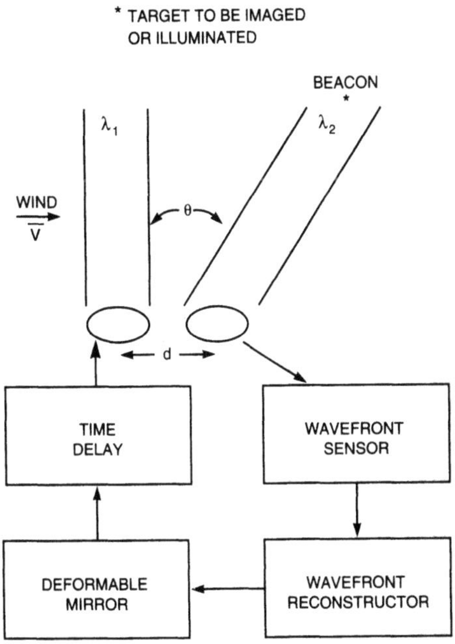

Fig. 1.1. An adaptive-optics system. A beacon sends a beam through the atmosphere, and the turbulence impresses a phase aberration on it. This phase front is sampled by a wavefront sensor, which typically responds to local wavefront tilts in sections of the aperture referred to as subapertures. Phase is reconstructed from tilts in a wavefront reconstructor, and its conjugate is impressed on a deformable mirror. This mirror is in the path of the imaging or beam projection system, with the net result that turbulence phase aberrations are corrected.

where $k_0 = 2\pi/\lambda$ with λ the propagation wavelength, $C_n^2(z)$ is the turbulence strength, $f(\kappa)$ is the turbulence spectrum in transverse spatial transform space, and $g(\boldsymbol{\kappa}, z)$ is a filter function that depends on the particular problem. The κ integration is over the entire two-dimensional space. Different problems simply necessitate the use of different filter functions. There are characteristic categories of problems. The characteristic filter functions are derived in Chap. 3. For example, the filter functions for Zernike modes, such as piston and tilt on filled and annular apertures, are derived. I show that relative displacement of two waves (anisoplanatism) can be represented as a filter function. Filter functions corresponding to a finite size source or a finite size receiver are also derived. The difference between the phase of a focused and a collimated wave affects the performance of many interesting systems, and filter functions for this problem are also evaluated.

Many applications discussed in this book are for adaptive-optics systems. An adaptive-optics system tries to correct for the effects of turbulence in an object's image or in the projection of a laser beam, and its effect on the turbulence also can be represented by filter functions. A generic system is shown in Fig. 1.1.

As I have said, this formalism reduces problems to a triple integral. The evaluation of the double integral in wavenumber space is discussed in detail in this book and has applications in many fields. The evaluation of the one remaining integral over the propagation path can be done analytically for some turbulence distribution models and is performed numerically for others. There are no numerical difficulties associated with this last integral.

The evaluation of these integrals is considered in Chap. 4 for the simplest problems in which the solution is obtained by table lookup of Mellin transforms. These integrals are of the form

$$\int_0^\infty dx\, h(x)\, x^a, \tag{1.2}$$

where $h(x)$ is typically a trigonometric or Bessel function or the square of one, or it is an algebraic function. In general the integral can be performed analytically when the function is any Meijer G-function. These integrals lead to important results for the effect of turbulence on wave propagation; for example, tilt jitter or other Zernike modes, residual phase variance due to a finite-bandwidth adaptive-optics system, and scintillation.

In order to solve more difficult problems, integrals whose integrand contains the product of two functions must be evaluated. These integrals contain one free parameter and are evaluated by using the Mellin convolution theorem to convert the integral in real space into one in the complex plane. These integrals are of the form

$$\int_0^\infty dx\, h(x)\, g\,(x/y)\, x^a, \tag{1.3}$$

where y is the free parameter, and the functions are of the same type as in (1.2). The method for evaluating this integral, which uses pole-residue integration, is developed in Chap. 5 when the parameter is positive real. The case of non-positive and complex parameters is developed in Chap. 9. Asymptotic solutions are developed. Some computer algebra programs such as Mathematica Version 2 from Wolfram Research, Inc., evaluate these integrals for a large but restricted number of functions. The functions useful in turbulence theory are included. These programs, however, do not provide the asymptotic results.

In Chap. 6 I discuss the application of these techniques to other fields. The use of Mellin transforms in the general theory of convolutional transforms is discussed. The solution of Laplace's equation in cylindrical and spherical coordinates is considered. The solution of nonlinear time-varying problems is discussed. Mellin transforms are shown to be applicable to certain classes of integral equations. Some applications to multiple integrals, to asymptotic analysis, probability theory, and to image and Doppler processing are briefly considered.

In Chap. 7 I return to the main theme of the book by taking up a host of important problems in wave propagation through turbulence in which the integral is of the form in (1.3). The following results are obtained: It is shown that the finite outer scale of turbulence can have a significant effect on tilt jitter. The effect of inner scale is shown to limit the maximum tilt that can be measured on a small aperture. It is also shown that the inner-scale size can significantly reduce the scintillation. Tilt on an annulus is considered, and it is shown that tilt jitter does not change much from that of a filled aperture until the central obscuration becomes very large. The scintillation measured from a finite size source or by a finite size receiver is shown to be considerably reduced from that of a point source or point receiver. Expressions for characteristic sizes of the source or receiver that significantly reduce scintillation are developed. The power spectral density of turbulence caused by wind transport or receiver motion is calculated. I calculate how scintillation is affected by an adaptive-optics system. The correlation function of focus is evaluated. The residual phase is calculated when one corrects the turbulence of a collimated beam by using a focused source as a reference. The residual piston and tilt for distributed and offset sources are calculated.

In addition to calculating wave variances, one can use these techniques to find the Strehl ratio as discussed in Chap. 8. Analytical expressions for the Strehl ratio cannot be calculated for every case of interest. For that reason an approximation is derived for the Strehl ratio with anisoplanatism that is much better than the Maréchal approximation. This formula is applied to anisoplanatism caused by parallel displacement, angular displacement, time displacement, chromatic displacement, and the combination of several such effects. Beam shapes for these cases are calculated in Chap. 13.

Propagation of a Gaussian wave results in integrals that have complex or negative parameters. The theory to evaluate these integrals is developed in Chap. 9 and applied in Chap. 10 to the scintillation of a Gaussian beam.

For more complicated problems there can be two or more parameters in the integral. These integrals are of the form

$$\int_0^\infty dx\, h(x)\, g_1\left(x/y_1\right) \ldots g_N\left(x/y_N\right) x^a, \tag{1.4}$$

where the functions h and g are of the same type as in (1.3). These integrals can be converted into integrals in several complex planes with the Mellin convolution theorem. Pole-residue techniques apply to this class of problems. There are complications in trying to decide which poles in the multi-dimensional space one should include when evaluating residues. A procedure is given in Chap. 11 that makes the process straightforward. I also develop a technique to evaluate both the series and steepest-descent contributions to an asymptotic solution.

For these more complicated problems, the series solutions sometimes do not converge for all parameter ranges. The source of this difficulty is illustrated in Chap. 11 by an integral whose integrand is a product of exponentials. A series of examples is then considered in Chap. 12. Several sample integrals in two or more complex planes are evaluated, and the results are shown to agree with integral tables. I then turn to four problems in wave propagation through turbulence that cover all parameter regions: Tilt anisoplanatism with outer scale is evaluated, and the effect of outer scale is shown to be not as large as its effect on pure tilt.

The method of solution shows clearly where this difference arises. Tilt with both inner and outer scale present is calculated. The solution is shown to agree with earlier work when either effect is eliminated. Next, the low-frequency regime of the tilt power spectrum is shown to be the most greatly affected by outer scale. Finally, the phase structure and correlation functions with non-zero inner and outer scales are calculated.

1.2 Introduction to Mellin Transforms

In this section I introduce the Mellin transform, and discuss a few of its properties. The Mellin transform $H(s)$ of the function $h(x)$ is defined as

$$h(x) \to H(s) \equiv \mathcal{M}\left[h\left(x\right)\right] \equiv \int_0^\infty \frac{dx}{x} h(x)\, x^s. \tag{1.5}$$

The inverse transform is

$$h(x) = \frac{1}{2\pi i} \int_C ds H(s)\, x^{-s}. \tag{1.6}$$

The integration path is a straight line extending from $\eta - i\infty$ to $\eta + i\infty$ and crossing the real axis at a value of s for which the integral in (1.5) converges. The allowed values of η typically lie in a finite range and are specified in the Mellin transform tables. The Mellin transform tables may be augmented with the following three properties that are derivable from the Mellin transform definition:

$$h(a\,x) \to a^{-s} H(s), \quad a > 0, \tag{1.7}$$

$$x^a h(x) \to H(s+a), \text{ and} \tag{1.8}$$

$$h\left(x^p\right) \to H(s/p)/\left|p\right|, \quad p \neq 0. \tag{1.9}$$

These relations are repeatedly used throughout the book to extend Mellin transforms from the tables to new functions. For instance, the Mellin transform of a Gaussian function is found from that of the exponential given in (1.45) by the use of (1.9) with $p = 2$ to obtain

$$\mathcal{M}\left[\exp\left(-x^2\right)\right] = 0.5\,\Gamma[s/2],$$

and (1.7) with $a = 1/b$ to obtain

$$\mathcal{M}\left\{\exp\left[-\left(x/b\right)^2\right]\right\} = 0.5\,b^s\Gamma[s/2]. \tag{1.10}$$

The Mellin transform is the Laplace transform in the logarithm of the variable x. This is why the Mellin transform is sometimes referred to as the *log-polar transform*. The transform of an object that is scaled in size by a factor b is related to the unscaled transform by

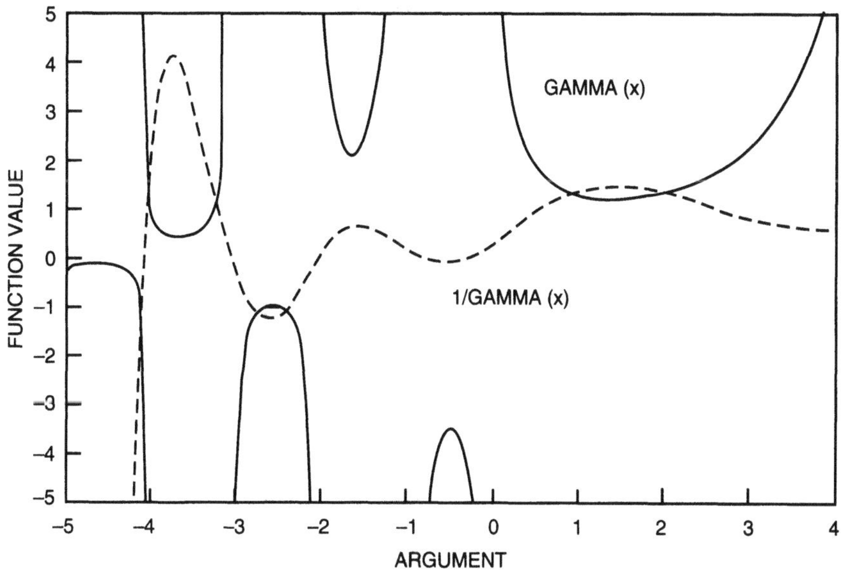

Fig. 1.2. Gamma function and its reciprocal versus its argument.

$$H(s) = \int_0^\infty \frac{dx}{x} h(bx)\, x^s = b^{-s} \int_0^\infty \frac{dy}{y}\, h(y)\, y^s. \tag{1.11}$$

This scaling factor has magnitude unity if s is confined to the imaginary axis. This is analogous to the way a time shift results in the multiplication of a Fourier transform by a complex scaling factor. This scaling property makes Mellin transforms convenient for the analysis of chaos and of acoustical, radar, and visual signals.

Gamma functions appear repeatedly in this work, and I highlight a few of their properties. The gamma function is defined as

$$\Gamma\,[s] = \int_0^\infty dx\, \exp(-x)\, x^{s-1} = \sum_{n=0}^{\infty} \frac{(-1)^n}{n!} \frac{1}{s+n} + \int_1^\infty dx\, \exp(-x)\, x^{s-1}. \tag{1.12}$$

The argument s can be complex. The last integral is entire and the only singularities of the gamma function are simple poles at the negative integers $-n$ with residue $(-1)^n/n!$. The reciprocal of the gamma function can be shown to be an entire function; therefore, the only singularities of the ratio of gamma functions come from the numerator. *This property is central to allowing one to easily evaluate integrals in the complex plane whose integrand is the ratio of gamma functions.* Plots of the gamma function and its reciprocal are shown in Fig. 1.2. The recursion relation for gamma functions is

$$\Gamma[a+1] = a\Gamma[a]. \tag{1.13}$$

With the relation $\Gamma[1] = 1$, for integer a this gives $\Gamma[a+1] = a!$. With this relation the gamma function at noninteger values can be evaluated on hand calculators such as the Hewlett-Packard 15.

To change a gamma function argument with an integer multiplier to a gamma function with a unity factor, one uses the Gauss-Legendre multiplication formula

$$\Gamma\left[my\right] = m^{my-1/2}(2\pi)^{(1-m)/2} \prod_{k=0}^{m-1} \Gamma\left[y + k/m\right]. \tag{1.14}$$

The common case of $m = 2$ is

$$\Gamma\left[2\,y\right] = 2^{2y-1}\Gamma\left[y\right]\Gamma\left[y+1/2\right]/\sqrt{\pi}. \tag{1.15}$$

To convert an argument with negative s to one with positive s, one uses the duplication formula

$$\Gamma\left[1-s\right]\,\Gamma\left[s\right] = \pi/\sin(\pi\, s). \tag{1.16}$$

For large arguments, the gamma function can be approximated by Stirling's formula

$$\Gamma[s] \sim \sqrt{2\pi}\, s^{s-1/2} \exp(-s)\left(1 + 1/12\,s + 1/288\,s^2 + \cdots\right), \quad |\arg\{s\}| < \pi. \tag{1.17}$$

Table 1.1 is a list of Mellin transforms that are used to evaluate integrals encountered in wave propagation through turbulence. Additional Mellin transforms are given in Appendix A. These are only a small fraction of the 1200 transforms available in closed form. The notation of *Marichev* (1983) and *Slater* (1966) for the ratio of gamma functions used throughout the book is

$$\Gamma\begin{bmatrix}\alpha_1,\ldots,\alpha_m\\ \beta_1,\ldots,\beta_n\end{bmatrix} \equiv \frac{\Gamma\left[\alpha_1\right]\,\Gamma\left[\alpha_2\right]\ldots\Gamma\left[\alpha_m\right]}{\Gamma\left[\beta_1\right]\,\Gamma\left[\beta_2\right]\ldots\Gamma\left[\beta_n\right]}. \tag{1.18}$$

A new notation is introduced in (1.48) in the Mellin transform table: An asterisk after a term in the gamma function signifies that the integration path passes between the first and second poles of that gamma function. The notation $\left(*^N\right)$ indicates that the integration path passes between the Nth and $N+1^{st}$ poles of the gamma function. No asterisk indicates that all poles are on one side of the integration path. The table contains the Mellin transform of the Heaviside unit step function, defined by

$$\begin{aligned} U(x) &= 1 \text{ for } x > 0, \text{ and} \\ U(x) &= 0 \text{ for } x \leq 0. \end{aligned} \tag{1.19}$$

This function can be used to convert integrals with finite limits into ones with infinite limits that can be evaluated with the techniques developed in this book.

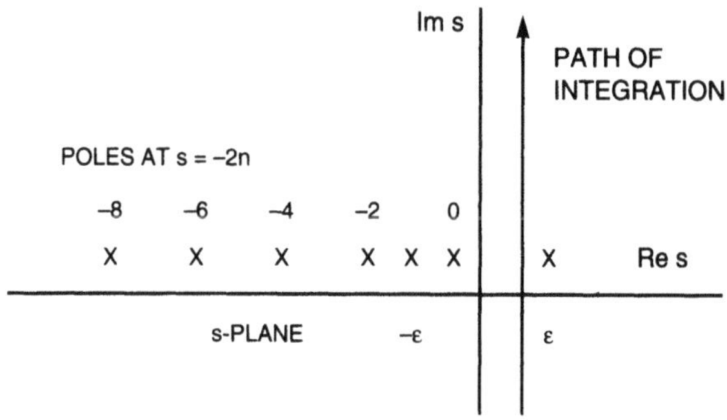

Fig. 1.3. Pole locations and integration path for the Bessel function integration. For clarity pole locations are displaced from the real axis in this and subsequent figures.

Mellin transforms are used in three ways to solve turbulence problems:

1. To evaluate integrals by table lookup.
2. To find a series representation for a function.
3. To evaluate complicated integrals.

The first application of Mellin transforms is simple yet important. Some integrals are easy to evaluate directly from the Mellin transforms. For instance, from (1.46) one finds that

$$\int_0^\infty dx\, x^a \sin(x) = 2^a \sqrt{\pi}\, \Gamma \left[\begin{matrix} a/2+1 \\ 1/2-a/2 \end{matrix} \right], \qquad \mathrm{Re}\{a+1\} < 1. \tag{1.20}$$

This technique is used to evaluate all the integrals in Chap. 4.

For some problems, i.e., those concerning anisoplanatic effects, one has to obtain the Mellin transform of the difference between a function and the first term of its power series. It is easy to show such a Mellin transform is that of the original function, except the integration path has moved over one pole. This is the analytic continuation of the integral. To illustrate, consider the transforms given in (1.49) and (1.55)

$$\mathcal{M}[J_0(x)-1] = \lim_{\varepsilon\to 0} \left\{ 2^{s-1} \Gamma \left[\begin{matrix} s/2 \\ -s/2+1 \end{matrix} \right] - \frac{1}{s+\varepsilon} + \frac{1}{s-\varepsilon} \right\}, \quad 0 < \mathrm{Re}\{s\} < \varepsilon.$$

The integration path and pole location are shown in Fig. 1.3. The poles continue periodically as $\mathrm{Re}\{s\}$ goes to minus infinity. For clarity in this figure and all subsequent ones the pole locations are slightly displaced from their true positions on the real axis.

In the limit, the poles at $s = 0$ and $s = -\varepsilon$ additively cancel. The net result is that the integration path can now cross the real axis anywhere between the poles at $s = 0$ and at $s = -2$ without changing the value of the integral. Therefore, the Mellin transform is equal to

$$\mathcal{M}\left[J_0(x)-1\right]=2^{s-1}\Gamma\left[\begin{matrix}s/2*\\ -s/2+1\end{matrix}\right], \qquad -2<\mathrm{Re}\{s\}<0. \tag{1.21}$$

In this case with the asterisk notation, the strip in complex space in which the transform is valid is completely defined, and the explicit statement of the restrictions on Re $\{s\}$ is redundant. This expression can be viewed as the analytic continuation of the integral of the Bessel function.

It is obvious how to extend this result to that in which the first m terms of the power series are subtracted from the function. In that case, the integration path moves past m poles of the original function, and this is indicated by $(*)^m$.

As a more difficult example, consider the integral of a power of the radial coordinate times a function that is related to the optical transfer function for a circular aperture obtained by *Fried* (1967). This integral occurs in the evaluation of many turbulence problems, and by using the unit step function one can put it in the form of a Mellin transform as

$$\begin{aligned}I(q)&=\int_0^1 d\alpha\,\alpha^{q+1}K(\alpha)\\ &=\int_0^\infty d\alpha\,\alpha^{q+1}\frac{16}{\pi}\left[\arccos(\alpha)\,U(1-\alpha)-\alpha\left(1-\alpha^2\right)^{1/2}U(1-\alpha)\right].\end{aligned}$$

The Mellin transform for the first function is found from (1.56) if one replaces s by $q+2$. This changes the condition for convergence to Re $\{q\}>-2$. The Mellin transform of the second integral is found from (1.53) if one replaces s by $q+3$ and also uses (1.9). The condition for convergence of this term is Re $\{q\}>-3$. For convergence of the entire integral, the more restrictive condition on the first function applies. Since $\Gamma[1/2]=\sqrt{\pi}$, the integral is equal to

$$I(q)=\frac{4}{\sqrt{\pi}}\left\{\Gamma\left[\begin{matrix}q/2+\frac{3}{2},\ -q/2-1\\ q/2+2,\ -q/2\end{matrix}\right]-\Gamma\left[\begin{matrix}q/2+\frac{3}{2}\\ q/2+3\end{matrix}\right]\right\}, \qquad \mathrm{Re}\{q\}>-2.$$

Using the gamma function recursion relation given in (1.13), one obtains

$$I(q)=\frac{8}{\sqrt{\pi}}\frac{1}{q+2}\Gamma\left[\begin{matrix}q/2+\frac{3}{2}\\ q/2+3\end{matrix}\right], \qquad \mathrm{Re}\{q\}>-2. \tag{1.22}$$

The second application of Mellin transforms is to find an infinite series for a function from its inverse Mellin transform. Consider the exponential function, whose Mellin transform is given by (1.45) as the gamma function. In Chap. 5 it is shown that the integration path for the inverse transform of the exponential function can be closed in the left-half plane without changing the value of the integral. Residues at the poles at $s=-n$ for $n=0, 1, 2, \ldots$ contribute to the integral to obtain

$$\exp(-x)=\frac{1}{2\pi i}\int_C ds\,\Gamma[s]\,x^{-s}=\sum_{n=0}^{\infty}\frac{(-1)^n}{n!}x^n. \tag{1.23}$$

This is the standard power series for the exponential function.

As a second more difficult example of obtaining a power series, consider $K_\nu(x)$, which is the modified Bessel function of the third kind with imaginary argument, and is sometimes referred to as a Macdonald function. When ν is not an integer, the poles are simple, and the integration path can be closed in the left-half plane where there are two sets of poles at $s = -\nu - 2n$ and $s = \nu - 2n$ for $n = 0, 1, 2, \ldots$. The residues at these poles give

$$\begin{aligned} K_\nu(x) &= \frac{1}{4}\frac{1}{2\pi i}\int_C ds\, \Gamma\left[s/2+\nu/2,\ s/2-\nu/2\right](x/2)^{-s} \\ &= \frac{1}{2}\sum_{n=0}^{\infty}\frac{(-1)^n}{n!}\left\{\left(\frac{x}{2}\right)^{2n+\nu}\Gamma\left[-n-\nu\right]+\left(\frac{x}{2}\right)^{2n-\nu}\Gamma\left[-n+\nu\right]\right\}. \end{aligned} \tag{1.24}$$

This result is used in Chap. 2. The Mellin transform of $K_\nu(x)$ has double poles when ν is an integer. The residues at these poles can be calculated with the technique in Sect. 5.3 to give

$$\begin{aligned} K_\nu(x) &= \frac{1}{4}\frac{1}{2\pi i}\int_C ds\, \Gamma\left[s/2+\nu/2,\ s/2-\nu/2\right](x/2)^{-s} \\ &= \sum_{n=0}^{\infty}\frac{(-1)^{\nu+1}}{n!\,(n+\nu)!}\left(\frac{x}{2}\right)^{2n+\nu}\left[\ln(x/2)-\psi(n+1)/2-\psi(\nu+n+1)/2\right] \\ &+\frac{1}{2}\sum_{n=0}^{\nu-1}\frac{(-1)^n}{n!}\left(\frac{x}{2}\right)^{2n-\nu}\Gamma\left[-n+\nu\right], \end{aligned} \tag{1.25}$$

where $\psi(x)$ is the logarithmic derivative of the gamma function.

The third and most important application of Mellin transforms is the evaluation of more complicated integrals via the Mellin convolution theorem. Consider the Mellin transform of

$$h(x) = \int_0^\infty \frac{dy}{y}\, h_0(y)h_1(x/y)\,. \tag{1.26}$$

Use the Mellin transform definition, and interchange the integration order to obtain

$$\mathcal{M}\left[h(x)\right] = \int_0^\infty \frac{dy}{y}\, h_0(y)\int_0^\infty \frac{dx}{x}\, h_1(x/y)\, x^s.$$

Letting $w = x/y$, one obtains

$$\mathcal{M}\left[h(x)\right] = \int_0^\infty \frac{dy}{y}\, h_0(y)\, y^s\int_0^\infty \frac{dw}{w}\, h_1(w)\, w^s = H_0(s)\, H_1(s)\,.$$

Therefore, the value of the original integral obtained from the inverse Mellin transform relation is

$$h(x) = \frac{1}{2\pi i} \int_C ds\, H_0(s)\, H_1(s)\, x^{-s}. \tag{1.27}$$

The Mellin convolution theorem can be applied if the Mellin transform of each function in the integrand exists. Actually, the theorem applies even if this condition is not true for one function if the second function decays fast enough. This is discussed in Chap. 9.

The convolution theorem can be generalized to integrals that contain the product of $N+1$ functions with N parameters. Consider such an integral

$$h(x_1, x_2, \ldots, x_N) = \int_0^\infty \frac{dy}{y} h_0(y) h_1(x_1/y) \cdots h_N(x_N/y). \tag{1.28}$$

Take the Mellin transform in each of the N parameters, and interchange the integration order to obtain

$$\begin{aligned} &\mathcal{M}\left[h(x_1, x_2, \ldots, x_N)\right] \\ &= \int_0^\infty \frac{dy}{y} h_0(y) \int_0^\infty \frac{dx_1}{x_1} h_1(x_1/y)\, x_1^{s_1} \ldots \int_0^\infty \frac{dx_N}{x_N} h_N(x_N/y)\, x_N^{s_N}. \end{aligned}$$

Make the substitutions $w_1 = x_1/y, \ldots, w_N = x_N/y$ to obtain

$$\begin{aligned} &\mathcal{M}\left[h(x_1, x_2, \ldots, x_N)\right] \equiv H(s_1, s_2, \ldots, s_N) \\ &= \int_0^\infty \frac{dy}{y} h_0(y)\, y^{s_1+s_2+\cdots+s_N} \int_0^\infty \frac{dw_1}{w_1} h_1(w_1)\, w_1^{s_1} \ldots \int_0^\infty \frac{dw_N}{w_N} h_N(w_N)\, w_N^{s_N}. \end{aligned}$$

This last expression is equivalent to

$$H(s_1, s_2, \ldots, s_N) = H_0(s_1 + s_2 + \cdots + s_N)\, H_1(s_1) \cdots H_N(s_N). \tag{1.29}$$

The value of the original integral is equal to the inverse Mellin transform in the N complex variables given by

$$h(x_1, \ldots, x_N) = \frac{1}{(2\pi i)^N} \int_{C_1} \cdots \int_{C_N} ds_1 \ldots ds_N\, H(s_1, \ldots, s_N)\, x_1^{-s_1} \ldots x_N^{-s_N}. \tag{1.30}$$

Methods of evaluating this integral are considered in Chap. 11. A considerable part of this book is devoted to obtaining convergent and asymptotic series for the integrals in (1.27) and (1.30) and applying these techniques to a host of problems in Chaps. 6, 7, 8, 10, 12, and 13.

1.3 Higher Transcendental Functions

Many answers to problems considered in this book are expressed as sums of generalized hypergeometric functions. Some are expressed as a Meijer G function and as a Fox H-function. The answers are abstracted from infinite sums; rules to convert these sums into generalized hypergeometric functions are given here. The definition of the generalized hypergeometric function is

$$ {}_pF_q\left[(a);(b);z\right] \equiv {}_pF_q\left[\begin{matrix}(a);z\\(b)\end{matrix}\right] \equiv \sum_{k=0}^{\infty} \frac{(a_1)_k\,(a_2)_k\cdots(a_p)_k}{(b_1)_k\,(b_2)_k\cdots(b_q)_k}\frac{z^k}{k!}, \tag{1.31}$$

where $(a) = a_1, a_2, \ldots, a_p$, $(b) = b_1, b_2, \ldots, b_q$, and the Pochhammer symbol is defined as

$$(a)_k = \frac{\Gamma[a+k]}{\Gamma[a]},\ (a)_0 = 1. \tag{1.32}$$

This generalized hypergeometric function is equal to unity when $z = 0$. For $p = q+1$, this series always converges if $|z| < 1$, and it also converges for $z = 1$ if $\mathrm{Re}\left\{\sum_{i=1}^{p} a_i - \sum_{i=1}^{q} b_i\right\} < 0$. If $p < q+1$, it converges for all z. If $p > q+1$, it does not converge, yet it is useful as an asymptotic series.

The generalized hypergeometric functions is a generalization of the Gauss hypergeometric function ${}_2F_1[\alpha,\beta;\gamma;z]$, which is defined as

$$ {}_2F_1[\alpha,\beta;\gamma;\ z] \equiv F[\alpha,\beta;\gamma;z] \equiv \sum_{k=0}^{\infty}\frac{(\alpha)_k\,(\beta)_k}{(\gamma)_k}\frac{z^k}{k}. \tag{1.33}$$

Specific convergence criteria on the unit circle can be given for this function:

1. If $0 < \mathrm{Re}\,\{\alpha+\beta-\gamma\} < 1$, the series converges on the unit circle except at $z = 1$.

2. If $\mathrm{Re}\,\{\alpha+\beta-\gamma\} < 0$, the series converges absolutely for $z \le 1$.

3. If $\mathrm{Re}\,\{\alpha+\beta-\gamma\} \ge 1$, the series diverges on the entire unit circle.

Most functions encountered in physics are special cases of Gauss's hypergeometric function. Examples are given in Table 1.2. Additional functions that are represented as hypergeometric functions are given in Appendix B.

The generalized hypergeometric function can be evaluated with a recursion relation. The function is expressed as

$$ {}_pF_q\left[(a);(b);z\right] = \sum_{k=0}^{\infty} c_k, \tag{1.34}$$

with

$$c_0 = 1,\ \text{and}\ c_{k+1} = \frac{(a_1+k)\cdots(a_p+k)}{(b_1+k)\cdots(b_q+k)}\frac{z}{(k+1)}c_k. \tag{1.35}$$

The generalized hypergeometric function can be expressed as a Mellin-Barnes integral in the form

$$\begin{aligned} &{}_pF_q\left[(a);(b);z\right] \\ &= \Gamma\left[\begin{matrix}b_1,b_2,\ldots,b_q\\a_1,a_2,\ldots,a_p\end{matrix}\right]\frac{1}{2\pi i}\int_{\eta-i\infty}^{\eta+i\infty} ds\,(-z)^s\,\Gamma\left[\begin{matrix}a_1+s,\ldots,a_p+s,\ -s\\b_1+s,\ldots,b_q+s\end{matrix}\right], \end{aligned} \tag{1.36}$$

where the integration path lies to the left of the poles of $\Gamma[-s]$ and to the right of all poles of the remaining numerator gamma functions. *From this relation, it is evident that the Mellin transform of any function that can be represented as a generalized hypergeometric function is expressible as a ratio of gamma functions.*

A further generalization of the generalized hypergeometric function is the Fox H-function defined as

$$H_{p,q}^{m,n}\left[z\left|\begin{matrix}(a_1,\alpha_1),\ldots,(a_p,\alpha_p)\\(b_1,\beta_1),\ldots,(b_q,\beta_q)\end{matrix}\right.\right]$$
$$=\frac{1}{2\pi i}\int_C ds\, z^s\frac{\prod_{i=1}^{n}\Gamma[1-a_i+\alpha_i s]\prod_{j=1}^{m}\Gamma[b_j-\beta_j s]}{\prod_{k=m+1}^{q}\Gamma[1-b_k+\beta_k s]\prod_{r=n+1}^{p}\Gamma[a_r-\alpha_r s]}, \tag{1.37}$$

where all the poles of an individual gamma function are on one side of the integration path. If the coefficients of s are rational, the solution can be expressed as a Meijer G-function defined as

$$G_{p,q}^{m,n}\left[z\left|\begin{matrix}a_1,\ldots,a_p\\b_1,\ldots,b_q\end{matrix}\right.\right]=\frac{1}{2\pi i}\int_C ds\, z^s\frac{\prod_{i=1}^{n}\Gamma[1-a_i+s]\prod_{j=1}^{m}\Gamma[b_j-s]}{\prod_{k=m+1}^{q}\Gamma[1-b_k+s]\prod_{r=n+1}^{p}\Gamma[a_r-s]}, \tag{1.38}$$

where all poles of an individual gamma function are again on one side of the integration path.

For some problems the answer is expressed as a generalized hypergeometric function minus unity. This can be expressed simply as a different generalized hypergeometric function. Consider

$${}_pF_q[(a);(b);z]-1=\sum_{k=1}^{\infty}\frac{(a_1)_k\,(a_2)_k\cdots(a_p)_k}{(b_1)_k\,(b_2)_k\cdots(b_q)_k}\frac{z^k}{k}. \tag{1.39}$$

Let $k\to k+1$ and note that from the definitions

$$(a)_{k+1}=\frac{\Gamma[a+1+k]}{\Gamma[a]}=(a+1)_k\frac{\Gamma[a+1]}{\Gamma[a]}=a\,(a+1)_k\,,\quad\text{and}$$

$$(k+1)!=k!\,\frac{(2)_k}{(1)_k}.$$

Therefore,

$${}_pF_q[(a);(b);z]-1=z\frac{\prod_{n=1}^{p}a_n}{\prod_{n=1}^{q}b_n}\,{}_{p+1}F_{q+1}[(a+1),1;(b+1),2;z]\,. \tag{1.40}$$

This relation is used in Chaps. 7 and 8.

The result of evaluating the Mellin convolution integral with pole-residue integration for problems in propagation through turbulence is generally one or more summations that contains ratios of gamma functions. As an example, consider a typical summation

$$\sum_{n=0}^{\infty} \frac{(-1)^n}{n!} \Gamma \left[\begin{matrix} n+w, & -n+r \\ n+h, & n+p \end{matrix} \right] P^{2n}. \tag{1.41}$$

This infinite series can be converted into a generalized hypergeometric function by using the function definition, and the duplication formula for gamma functions given in (1.16). I will express such a summation as $K\ {}_pF_q\left[(a)\,;(b)\,;z\right]$. The rules for obtaining this representation are:

1. The coefficient K is equal to the value of the $n = 0$ term. In this case

$$K = \Gamma \left[\begin{matrix} w, & r \\ h, & p \end{matrix} \right]. \tag{1.42}$$

2. The arguments in the gamma function for which the sign of the index n is positive, i.e., $n+w,\ n+h$, and $n+p$, lend their n-independent constants to arguments of the hypergeometric function. Those in the numerator appear before the first semicolon (they are the a_i terms), and those in the denominator appear after it (those are the b_j terms).

3. The arguments in the gamma function for which the sign of the summation index n is negative lend minus their n-independent constants, plus one, to arguments of the hypergeometric function. (The argument of this type in the example is $-n+r$. The argument to insert in the hypergeometric series is $-r+1$.) If the argument was in the numerator, it appears after the first semicolon. Those in the denominator appear before the first semicolon.

4. The magnitude of the parameter z is the quantity in the summation raised to the index power n. (In (1.41) the parameter is P^2.) The sign of the parameter is positive if the number of gamma function arguments with negative n is odd. It is negative if the number is even. The sign is positive in the example since there is one argument with negative index.

5. The subscripts before and after the function symbol are referred to as the order of the generalized hypergeometric function. The quantity before the F symbol is the number of terms before the first semicolon. The quantity after the F symbol is the number of terms between the first and second semicolons. The summation in (1.41) is equal to

$$\sum_{n=0}^{\infty} \frac{(-1)^n}{n!} \Gamma \left[\begin{matrix} n+w, & -n+r \\ n+h, & n+p \end{matrix} \right] P^{2n} = \Gamma \left[\begin{matrix} w, & r \\ h, & p \end{matrix} \right] {}_1F_3 \left[w; -r+1, h, p; P^2 \right]. \tag{1.43}$$

The third volume of the three-volume reference work by *Prudnikov, Brychkov*, and *Marichev* (1990) contains an extensive list of equivalences between generalized hypergeometric functions and more familiar functions.

Table 1.1. Mellin Transforms that are useful for turbulence problems.

$$h(x) \quad \to \mathcal{M}\left[h(x)\right] \equiv H(s) = \int_0^\infty \frac{dx}{x}\, h(x)\, x^s \tag{1.44}$$

$$\exp(-x) \quad \to \Gamma\left[s\right], \quad \mathrm{Re}\left\{s\right\} > 0 \tag{1.45}$$

$$\sin(x) \quad \to 2^{s-1}\sqrt{\pi}\,\Gamma\begin{bmatrix} 1/2+s/2 \\ 1-s/2 \end{bmatrix}, \qquad |\,\mathrm{Re}\left\{s\right\}| < 1 \tag{1.46}$$

$$\cos(x) \quad \to 2^{s-1}\sqrt{\pi}\,\Gamma\begin{bmatrix} s/2 \\ 1/2-s/2 \end{bmatrix}, \qquad 0 < \mathrm{Re}\left\{s\right\} < 1 \tag{1.47}$$

$$\sin^2\left(x^2\right) \quad \to -\frac{\sqrt{\pi}}{8}\,\Gamma\begin{bmatrix} s/4* \\ 1/2-s/4 \end{bmatrix}, \qquad -4 < \mathrm{Re}\left\{s\right\} < 0 \tag{1.48}$$

$$J_\nu(x) \quad \to 2^{s-1}\Gamma\begin{bmatrix} s/2+\nu/2 \\ \nu/2+1-s/2 \end{bmatrix}, \qquad -\mathrm{Re}\left\{\nu\right\} < \mathrm{Re}\left\{s\right\} < 3/2 \tag{1.49}$$

$$J_\nu^2(x) \quad \to \frac{1}{2\sqrt{\pi}}\,\Gamma\begin{bmatrix} s/2+\nu,\ 1/2-s/2 \\ \nu+1-s/2,\ 1-s/2 \end{bmatrix}, \quad -2\mathrm{Re}\left\{\nu\right\} < \mathrm{Re}\left\{s\right\} < 1 \tag{1.50}$$

$$J_\nu(x)\, J_{\nu+1}(x) \quad \to \quad \frac{1}{2\sqrt{\pi}}\,\Gamma\begin{bmatrix} s/2+\nu+1/2,\ 1-s/2 \\ \nu+3/2-s/2,\ 3/2-s/2 \end{bmatrix}, \qquad -1-2\mathrm{Re}\left\{\nu\right\} < \mathrm{Re}\left\{s\right\} < 2 \tag{1.51}$$

$$(1+x)^{-p} \quad \to \frac{\Gamma[s,\, p-s]}{\Gamma[p]}, \qquad 0 < \mathrm{Re}\left\{s\right\} < \mathrm{Re}\left\{p\right\} \tag{1.52}$$

$$(1-x)^{a-1}U(1-x) \quad \rightarrow \Gamma[a]\,\Gamma\begin{bmatrix} s \\ s+a \end{bmatrix}, \quad \mathrm{Re}\,\{a\} > 0,\ \mathrm{Re}\,\{s\} > 0 \tag{1.53}$$

$$(x-1)^{a-1}\,U(x-1) \quad \rightarrow \quad \Gamma[a]\Gamma\begin{bmatrix} 1-a-s \\ 1-s \end{bmatrix},$$
$$\mathrm{Re}\,\{a\} > 0,\ \mathrm{Re}\,\{a+s\} < 1 \tag{1.54}$$

$$1 \quad \rightarrow \lim_{\varepsilon\to 0}\left[\frac{1}{s+\varepsilon} - \frac{1}{s-\varepsilon}\right], \qquad -\varepsilon < \mathrm{Re}\,\{s\} < \varepsilon \tag{1.55}$$

$$\arccos(x)\,U(1-x) \rightarrow -\frac{\sqrt{\pi}}{4}\,\Gamma\begin{bmatrix} s/2+1/2,\ -s/2 \\ s/2+1,\ 1-s/2 \end{bmatrix}, \qquad \mathrm{Re}\,\{s\} > 0 \tag{1.56}$$

$$U(1-x) \quad \rightarrow \Gamma\begin{bmatrix} s \\ s+1 \end{bmatrix}, \qquad \mathrm{Re}\,\{s\} > 0 \tag{1.57}$$

$$U(x-1) \quad \rightarrow \Gamma\begin{bmatrix} -s \\ 1-s \end{bmatrix}, \qquad \mathrm{Re}\,\{s\} < 0 \tag{1.58}$$

$$K_\nu(x) \qquad \rightarrow 2^{s-2}\,\Gamma\left[s/2+\nu/2,\ s/2-\nu/2\right], \quad \mathrm{Re}\,\{s\} > |\mathrm{Re}\,\{\nu\}\,| \tag{1.59}$$

$$J_\mu(x)\,J_\lambda(x) \quad \rightarrow \quad \frac{1}{2\sqrt{\pi}}\,\Gamma\begin{bmatrix} s/2+\frac{\mu+\lambda}{2},\ \frac{1}{2}-s/2,\ 1-s/2 \\ 1-s/2+\frac{\mu+\lambda}{2},\ 1-s/2+\frac{\mu-\lambda}{2},\ 1-s/2+\frac{\lambda-\mu}{2} \end{bmatrix},$$
$$-\mathrm{Re}\ \{\mu+\lambda\} < \mathrm{Re}\,\{s\} < 1 \tag{1.60}$$

$$\delta(x-a) \quad \rightarrow a^{s-1} \tag{1.61}$$

$$K\,(x) \rightarrow \qquad \frac{8}{\sqrt{\pi}}\frac{1}{s}\Gamma\begin{bmatrix} s/2+\frac{1}{2} \\ s/2+2 \end{bmatrix}, \qquad \mathrm{Re}\,\{s\} > 0 \tag{1.62}$$

$U\,(x)$ is the Heaviside unit step function; $K_\nu\,(x)$ is the Bessel function of third kind; $\delta\,(x)$ is the Dirac delta function; and $K(x)$, defined in (1.22), is a constant times the optical transfer function on a circular aperture. An asterisk next to a gamma function entry indicates that the integration path passes between the first and second poles of that gamma function.

Table 1.2. Examples of the Gauss hypergeometric function.

$$\ln[1+z] = z\,{}_2F_1[1,1;2;-z] \tag{1.63}$$

$$\left(1+\sqrt{1-z}\right)^{1-2a} = 2^{1-2a}{}_2F_1[a,a-1/2;2a;z] \tag{1.64}$$

$$(1-z)^{-a} = {}_1F_0[a;z] \tag{1.65}$$

$$\exp(z) = {}_0F_0[z] \tag{1.66}$$

$$\sin(z) = z\,{}_0F_1\left[\tfrac{3}{2};-z^2/4\right] \tag{1.67}$$

$$\cos(z) = {}_0F_1\left[\tfrac{1}{2};-z^2/4\right] \tag{1.68}$$

$$\arcsin(z) = z\,{}_2F_1\left[1/2,1/2;3/2;z^2\right] \tag{1.69}$$

$$\arctan(z) = z\,{}_2F_1\left[1/2,1;3/2;-z^2\right] \tag{1.70}$$

$$\cos(v\arcsin z) = {}_2F_1\left[v/2,-v/2;1/2;z^2\right] \tag{1.71}$$

$$\sinh(z) = z\,{}_0F_1\left[\tfrac{3}{2};z^2/4\right] \tag{1.72}$$

$$\cosh(z) = {}_0F_1\left[\tfrac{1}{2};z^2/4\right] \tag{1.73}$$

$$J_\nu(z) = \left(\frac{z}{2}\right)^\nu \frac{1}{\Gamma[\nu+1]}\,{}_0F_1\left[\nu+1;-z^2/4\right] \tag{1.74}$$

2. Basic Equations for Wave Propagation in Turbulence

Turbulence has a large effect on electromagnetic wave propagation that can be described by stochastic differential equations. In this chapter these equations are solved with the Rytov approximation, and the main result is given in (2.85), which is the starting point for all turbulence problems considered in this book. This equation is used to find phase and log-amplitude variances. These expressions are modified to obtain expressions for the power spectral density, given in (2.101), beam profile given in (2.109), and Strehl ratio, which is the ratio of on-axis intensity to that of a diffraction-limited wave, in (2.111).

Turbulence effects arise from atmospheric refractive index variations that cause both large and small scale variations of electromagnetic-wave propagation. There are both a steady decrease of air density with altitude that causes refraction of the waves and local inhomogeneities that have a more complicated effect. The steady part of the refractive index variation $n(\mathbf{r})$ can be estimated from

$$n(\mathbf{r}) - 1 = 77.6 \times 10^{-6} \left(1 + 7.52 \times 10^{-3}/\lambda^2\right) P(\mathbf{r})/T(\mathbf{r}), \tag{2.1}$$

where λ is the propagation wavelength in μm, P is pressure in millibars, T is temperature in degrees Kelvin, and $\mathbf{r}$ is 3-dimensional spatial position. At sea level $n(\mathbf{r}) - 1$ is approximately equal to 3×10^{-4}.

In addition to this steady density variation, there are stochastic changes. Stochastic temperature variations in the air are the source of quivering images that we sometimes see. Since the atmosphere has constant pressure locally, these temperature variations cause density variations that in turn lead to refractive index variations, causing the bending and scattering of light rays. Most often these temperature variations are generated by a temperature differential between the earth and the adjoining atmosphere. In daytime, the earth is hotter than the air; this causes air closer to the ground to be hotter than that above. This negative temperature gradient bends light rays that are parallel to the earth upwards and, if strong enough, results in inverted images of the sky. This *mirage* is a common occurrence in deserts or along heated asphalt on sunny days. Stronger temperature gradients can result in the appearance of additional images.

During nighttime, temperature gradients can be positive, resulting in downward bending of light rays enabling one to see objects below the horizon and elevating images of objects that are on the horizon. (This phenomenon is called *looming*.)

2.1 Turbulence Spectra

Mirages and looming are due to large scale temperature gradients that can persist for some time. Much more commonly, temperature inversions lead to smaller scale effects that fluctuate more rapidly. For typical sizes of turbulence blobs (turbules), the Reynolds number is much greater than 1, and laminar air flow is unstable, which results in the breakup of stratified air layers into smaller blobs. Often on a hot day one can see the swirling of heated air as eddies cascade down in size. This process produces smaller and smaller eddy sizes until dissipation effects become important at a scale called the inner scale, at which point rotational energy is rapidly dissipated and flow becomes laminar. This causes a rapid decrease in spectral energy at small scales.

Different methods must be used to analyze the effects on wave propagation of large- and small-scale atmospheric temperature variations. A gradual decrease in air density with height and the dispersive characteristics of air can separate white light into spectra so that starlight at low elevation angles sometimes appears to have distinct colors. (This dispersive effect also causes the green part of the sun's spectrum to set after the red part, producing the "green flash.") These density variations last for extended periods of time, and their effect on wave propagation can be analyzed with ray optics in which density along the propagation path can be assumed to remain constant with time.

Mirages are caused by variations in air density close to the earth. Air in this region has some turbulence characteristics that cause variations in the mirage shapes; however, major features of mirages can also be determined with ray-optics propagation in a deterministic atmosphere.

Small scale fluctuations also affect wave propagation in the atmosphere and produce detectable effects. With the unaided eye, one can see scintillations in starlight as well as in street lights at some distance. Turbulence has a profound effect on the ability of astronomers to separate images of two objects that are close together. In a vacuum, resolving power of a mirror increases linearly with diameter. In the atmosphere, resolution increases until a certain diameter and then remains almost the same with increasing diameter. Thus we are led to the important concept of atmospheric coherence diameter. At sea level this diameter is about 2–15 cm at visible wavelengths. At higher elevations, at certain sites, this diameter can be tens of centimeters and is the main reason that large astronomical telescopes are located on mountain tops. Making telescope diameters larger than the coherence diameter increases the amount of light collected but hardly increases resolution for objects whose recording takes more than a fraction of a second. For smaller times images are speckled, and they contain information up to the diffraction limit of the full diameter. In some circumstances, by using elaborate processing of short term images, it is possible to resolve objects that are separated by the telescope's vacuum resolution (*Goodman* 1985).

Turbulence effects are stochastic in nature, and to characterize them in detail, one must solve the wave equation with a stochastic refractive index. This is treated in this book.

I now turn to the standard phenomenology of atmospheric turbulence. Eddy distributions at the largest sizes, where energy enters the cascade process, is highly dependent on local topography and absorption characteristics of the earth's surface, and on distribution of winds and temperature aloft. Once eddies are formed and break up into smaller eddies, the eddy size distribution follows universal laws. *Kolmogorov* (1941) and *Obukhov* (1941) were first to derive the distribution law for the relative amount of turbulence at various scale sizes in the region that is referred to as the inertial subrange. This range encompasses scale sizes from, typically, 1 millimeter to 10 meters. One can obtain this law from dimensional analysis.

Turbulence theory makes heavy use of the mean square value of the difference of a quantity at two points, i.e., the structure function. For the turbulent air speed v this is defined to be

$$\mathcal{D}_v(\mathbf{r}) = \left\langle [v(\mathbf{a}) - v(\mathbf{a}+\mathbf{r})]^2 \right\rangle, \qquad (2.2)$$

where angle brackets indicate an ensemble average.

Evolution of turbulence can be analyzed with the Navier-Stokes hydrodynamic differential equations. In the inertial subrange, where dissipation is not important, Kolmogorov and Obukhov made the key assumption that the structure function of air speed is isotropic and a function solely of separation $\mathbf{r}$ and rate of energy deposition per unit mass ε, i.e.,

$$\mathcal{D}_v(\mathbf{r}) = \mathcal{D}_v(r) = f(r, \varepsilon). \qquad (2.3)$$

The units of ε are $m^2/\sec^3$, and by Buckingham's Pi theorem (*Langhaar* 1980) the only form the solution can assume is the combination of ε and r that have units of velocity squared. Therefore, the isotropic structure function must be

$$\mathcal{D}_v(r) = e^2(\varepsilon r)^{2/3}, \qquad r >> L_i, \qquad (2.4)$$

where e^2 is a constant, and L_i is the "inner scale." This is the "two-thirds power law."

For very small separations the Reynolds number is small and friction dominates, causing rotational velocity to be dissipated, thus resulting in laminar flow. For laminar flow, the average velocity can be expanded into a Taylor series about any point. The constant terms cancel, and because velocity is assumed to be isotropic, the linear terms also cancel; therefore, the leading term is

$$\mathcal{D}_v(r) = b^2 r^2, \qquad r << L_i. \qquad (2.5)$$

In the region of greatest interest, the inertial subrange, the two-thirds power law contains all information on turbulence of practical importance. Even though this relation is simple, it is not easy to exploit for several reasons. First, it is not straightforward to incorporate this relation into Maxwell's equations. The Rytov approximation given below overcomes this difficulty.

Second, the steps necessary to obtain quantities of interest such as second moments or Strehl ratios are lengthy and not intuitive. I show in the next section that many specific problems are special cases of a general problem that is set up and reduced to integrals, thereby, making this step straightforward.

Third, many integrals that typically arise diverge because this form of the structure function is unphysical both at small and large turbulence scales. The structure function is modified below to overcome these difficulties.

Fourth and last, these integrals are difficult to evaluate analytically, and numerical methods are often used. This reduces one's physical insight into the behavior of the solution as a function of the size of the parameters. Modifications of the structure function to remove infinities results in integrals that are even more difficult to evaluate. Mellin transform techniques that are developed in this book allow one to analytically evaluate these integrals, thus enabling one to see how solutions depend on parameters.

To solve stochastic propagation problems it is convenient to analyze problems in Fourier space. For example, the expression for the phase and log-amplitude variance in Sect. 2.3 requires the correlation function of the transverse Fourier transform of density fluctuations. Let us now derive statistics of density fluctuations in Fourier space. In position space the structure function of density is given by

$$\mathcal{D}_n(\mathbf{r}) = \left\langle [n(\mathbf{a}) - n(\mathbf{a}+\mathbf{r})]^2 \right\rangle . \tag{2.6}$$

It can be shown that with the concept of conservative passive additives (*Tatarski* 1971) that this structure function has the same variation with distance as the structure function for velocity, and is expressed as

$$\mathcal{D}_n(r) = C_n^2 r^{2/3}, \qquad r >> L_i, \text{ and} \tag{2.7}$$

$$\mathcal{D}_n(r) = C_n^2 \frac{b^2}{e^2} r^2, \qquad r << L_i, \tag{2.8}$$

where C_n^2 measures turbulence strength.

A relation between the turbulence spatial spectrum and the structure function of density is found below. First, I develop some general Fourier relations in spherical coordinates. The spatial spectrum of a quantity is found from the 3-dimensional Fourier transform

$$\Psi(\mathbf{k}) = \frac{1}{(2\pi)^3} \iiint d\mathbf{r} \cos(\mathbf{k}\cdot\mathbf{r})\, \psi(\mathbf{r}), \tag{2.9}$$

where integration is over all space, and $\mathbf{k}$ is the total propagation vector, which consists of a two-dimensional transverse component $\boldsymbol{\kappa}$ and a component along the propagation vector k_z. Thus,

$$\mathbf{k} = \boldsymbol{\kappa} + \mathbf{u}_z k_z, \tag{2.10}$$

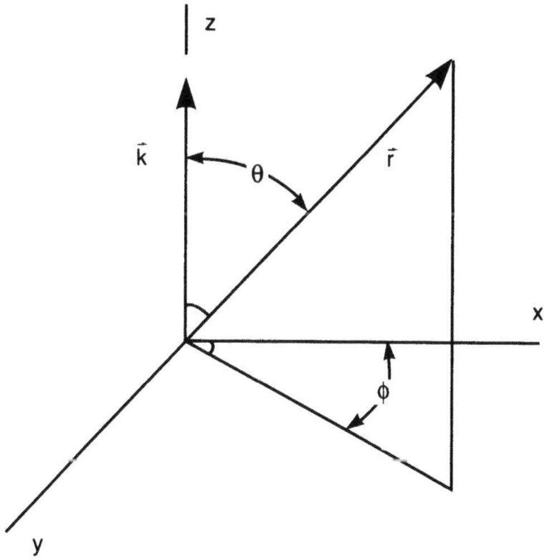

Fig. 2.1. Spherical coordinate system.

where $\mathbf{u}_z$ is a unit vector in the axial direction. The three-dimensional spatial vector is also expressed as a transverse component $\boldsymbol{\rho}$ and an axial component z as

$$\mathbf{r} = \boldsymbol{\rho} + \mathbf{u}_z \, z. \tag{2.11}$$

The exponential $\exp(i\mathbf{k} \cdot \mathbf{r})$ is replaced by $\cos(\mathbf{k} \cdot \mathbf{r})$ in (2.9) because $\Psi(\mathbf{k})$ is an even function of $\mathbf{k}$.

The inverse Fourier relation is

$$\psi(\mathbf{r}) = \iiint d\mathbf{k} \, \cos(\mathbf{k} \cdot \mathbf{r}) \, \Psi(\mathbf{k}) \, . \tag{2.12}$$

If this function is assumed to be isotropic, then this relation can be reduced to a simpler form. The integral can be expressed in spherical coordinates as

$$\Psi(k) = \frac{1}{(2\pi)^3} \int_0^{\pi} \int_0^{2\pi} \int_0^{\infty} r^2 \sin(\theta) \, dr \, d\theta \, d\phi \, \cos(\mathbf{k} \cdot \mathbf{r}) \, \psi(r) \, .$$

The $\mathbf{k}$ vector is considered to be along the z-axis; therefore, $\mathbf{k} \cdot \mathbf{r} = kr\cos(\theta)$, as is shown in Fig. 2.1. Using the relation

$$\frac{d}{d\theta} \sin[kr\cos(\theta)] = -\cos[kr\cos(\theta)] \; kr\sin(\theta) \, ,$$

one can perform the two angle integrations to give

$$\Psi(k) = \frac{1}{2\pi^2 k} \int_0^{\infty} r \, dr \, \sin(kr) \, \psi(r) \, . \tag{2.13}$$

Similarly, the angle integrations can be performed in the inverse transform to give

$$\psi(r) = \frac{4\pi}{r} \int_0^\infty k\,dk\,\sin(kr)\,\Psi(k)\,. \tag{2.14}$$

To obtain a relation between the structure function and the Fourier transform of the density distribution $n_1(\mathbf{r})$, set $\psi(\mathbf{r})$ to the correlation function of density $\langle n_1(\mathbf{a})\,n_1(\mathbf{a}+\mathbf{r})\rangle$. Since a displacement in real space corresponds to a phase shift in Fourier space, one can express the structure function of density as

$$\mathcal{D}_n(\mathbf{r}) = \iiint d\mathbf{k}\,|1-\exp(i\mathbf{k}\cdot\mathbf{r})|^2\Psi(\mathbf{k}) = 2\iiint d\mathbf{k}\,[1-\cos(\mathbf{k}\cdot\mathbf{r})]\,\Psi(\mathbf{k})\,. \tag{2.15}$$

We have assumed that the structure function is isotropic and only a function of the separation. In general $\mathcal{D}_n(\mathbf{r})$ is also a function of $\mathbf{a}$. For a stochastic process with stationary increments (that is one in which the difference in values of the process at two positions is stationary) the dependence on $\mathbf{a}$ is through a slowly varying multiplicative factor. As long as the scale length of this variation is much longer than the correlation distance of turbulence, the error in this derivation is small.

Assuming isotropic turbulence, one can express this as

$$\mathcal{D}_n(r) = 8\pi \int_0^\infty k^2\,dk \left[1 - \frac{\sin(kr)}{kr}\right]\Psi(k)\,. \tag{2.16}$$

To obtain the inverse Fourier relation, note that

$$\frac{\partial}{\partial r} r^2 \frac{\partial \mathcal{D}_n(r)}{\partial r} = 8\pi r \int_0^\infty dk\,k\,\sin(kr)\,\Psi(k)\,.$$

The inverse sine transform gives the desired relation between the turbulence spectrum and structure function as

$$\Psi(k) = \frac{1}{(2\pi k)^2} \int_0^\infty dr\,\frac{\sin(kr)}{kr}\frac{\partial}{\partial r} r^2 \frac{\partial \mathcal{D}_n(r)}{\partial r}\,. \tag{2.17}$$

If the structure function with two-thirds power law given in (2.7) is inserted into this integral and evaluated with (1.20), then one obtains the 3-dimensional Kolmogorov turbulence spectrum

$$\Psi(k) = \frac{5\,\Gamma[5/6]}{2^{4/3}\,\pi^{3/2}\,9\,\Gamma[2/3]} C_n^2\,k^{-11/3} = 0.0330054\,C_n^2\,k^{-11/3}\,. \tag{2.18}$$

As discussed above, the derivation is strictly valid only if C_n^2 does not depend on position. Later I will relax this restriction by allowing C_n^2 to vary slowly along the propagation path.

There are no convergence difficulties when one performs calculations with the structure function and its spatial spectrum; however, it will be seen that the phase spectrum of a propagating wave turns out to be infinite. The infinity comes at very small spatial wavelengths. In a physical situation, phases and intensities are uncorrelated at large separations; therefore, the structure function must approach a constant at large separations. There are many ways of achieving this, and a convenient one is to write the turbulence spectrum as

$$\Psi(k) = 0.033\, C_n^2 \left(k^2 + \kappa_o^2\right)^{-11/6}, \tag{2.19}$$

where

$$\kappa_o = 2\pi / L_o. \tag{2.20}$$

L_o is the outer-scale size. (Some authors, e.g., *Fante* (1975), omit the 2π.) This spectrum with outer scale included is called the von Kármán spectrum. If the outer scale is allowed to grow without bound the spectrum reduces to (2.18).

The structure function is found by inserting this spectrum into (2.16). To evaluate the integral the integrand is broken into two parts. The first part is evaluated with the Mellin transform in (1.52) and the relation in (1.9) to give

$$I_1 = 0.264\,\pi\, C_n^2\, \kappa_o^{-11/3} \int_0^\infty \frac{dk}{k}\, k^3 \left[\left(\frac{k}{\kappa_o}\right)^2 + 1\right]^{-11/6} = 1.0468\, C_n^2\, \kappa_o^{-2/3}.$$

The second integral is evaluated with the Mellin convolution theorem given in (1.27) and the Mellin transforms given in (1.46) and (1.52) as

$$\begin{aligned} I_2 &= -0.264\,\pi\, C_n^2\, \kappa_o^{-11/3} \int_0^\infty \frac{dk}{k}\, k^2 \sin(kr) \left[\left(\frac{k}{\kappa_o}\right)^2 + 1\right]^{-11/6} \\ &= -0.264\,\pi\, C_n^2\, \kappa_o^{-11/3} \frac{\sqrt{\pi}}{r^2\, \Gamma[11/6]} \frac{1}{2\pi i} \int_C ds \left(\frac{\kappa_o r}{2}\right)^{-s} \Gamma\left[s/2 + \tfrac{3}{2}, s/2 + \tfrac{11}{6}\right]. \end{aligned}$$

Make the replacement $s \to s - 10/3$ and recognize the integral as the inverse Mellin transform of $K_{1/3}(\kappa_o r)$ given in (1.59), where $K_{1/3}(x)$ is the modified Bessel function of the third kind with imaginary argument. The structure function of density is

$$\mathcal{D}_n(r) = \kappa_o^{-2/3} C_n^2 \left[1.0468 - 0.62029\, (\kappa_o r)^{1/3}\, K_{1/3}(\kappa_o r)\right]. \tag{2.21}$$

This structure function is plotted in Fig. 2.2.

This form of the structure function reduces to the two-thirds power law given in (2.7) for small separations. This is shown with the small argument approximation of the power series given in (1.24). The first two terms of the expansion are

$$K_{1/3}(x) \approx \frac{x^{-1/3}}{2^{2/3}} \Gamma\left[\tfrac{1}{3}\right] + \frac{x^{1/3}}{2^{4/3}} \Gamma\left[-\tfrac{1}{3}\right], \qquad x << 1.$$

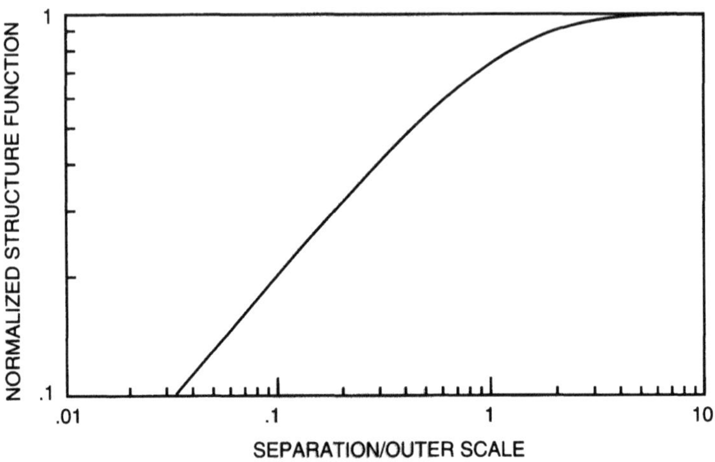

Fig. 2.2. Normalized structure function with outer scale versus separation.

The value of $\Gamma\,[1/3]$ is 2.6789, and the value of $\Gamma\,[-1/3]$ is -4.0624. With these values one obtains (2.7).

For very large separations the modified Bessel function is vanishingly small, and the structure function reduces to the constant

$$\mathcal{D}_n\,(r) = 1.0468\,\kappa_o^{-2/3} C_n^2, \qquad \kappa_o\, r >> 1. \tag{2.22}$$

The shape of the spectral region of the von Kármán spectrum in which outer scale has a significant effect has no universal theoretical justification. Measurements made over a wheat field by *Kamai* et al. (1972) support this spectral form. *Greenwood* and *Tarazano* (1974) made some measurements of turbulence spectra in areas not as topographically smooth, and their data more closely supports a model in which the turbulence spectrum is

$$\Psi(k) = 0.033\, C_n^2 \left(k^2 + k\kappa_o\right)^{-11/6}. \tag{2.23}$$

In this book the von Kármán spectrum is used except when tilt with outer scale is found for both the von Kármán and Greenwood spectra. Qualitative behavior is the same in the two cases.

There are also difficulties at large wavenumbers. From the hydrodynamic equations it can be shown that the above spectra require infinite energy dissipation due to the rapid increase in dissipation with decreasing scale size. Physically, below a certain scale size dissipative effects force the velocity to be laminar. This leads to a rapid decrease in the spectrum with decreasing scale size. Several ad hoc models are used for the spectrum in this inner scale region; the Gaussian form used in this book for some problems is referred to as the Tatarski spectrum.

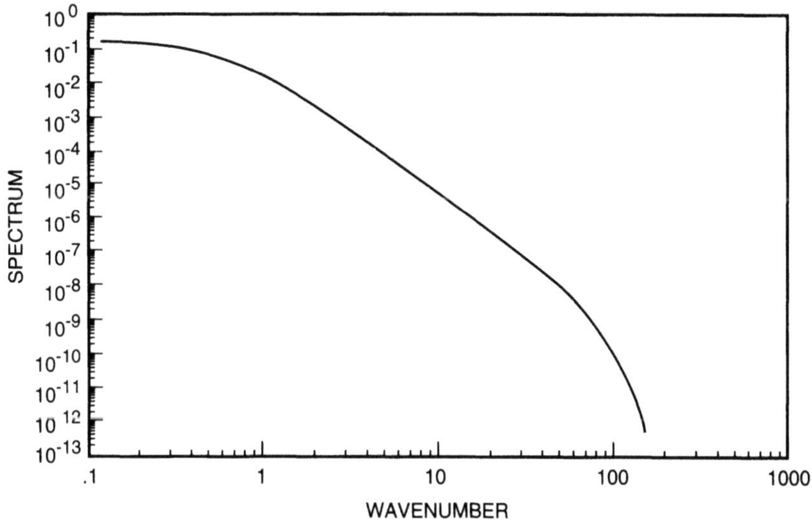

Fig. 2.3. Von Kármán spectrum with inner and outer scale present. The value of κ_i equal to $2\pi/0.1$, and the value of κ_o equal to $2\pi/10$. The presence of the outer scale causes the spectrum to approach a constant al low wavenumbers. The inner scale causes the spectrum to decay faster than Kolmogorov turbulence at high wavenumbers. The spectrum has an $-11/3$ power law behavior in the inertial subrange at intermediate wavenumbers.

A spectrum with both inner scale and von Kármán outer scale — the modified von Kármán spectrum — is

$$\Psi(k) = 0.033\, C_n^2 \left(k^2 + \kappa_o^2\right)^{-11/6} \exp\left(-k^2/\kappa_i^2\right), \tag{2.24}$$

where

$$\kappa_i = 2\pi/L_i = 5.91/L_{in}. \tag{2.25}$$

L_{in} is the commonly used inner scale. I occasionally use L_i to simplify equations. The two are related by $L_{in} = 0.94\, L_i$. The spectrum is shown in Fig. 2.3.

Greenwood and *Tarazano* (1974) and *Gurvich* et al. (1974) noted that the turbulence does not decay as rapidly as that predicted by the Kolmogorov spectrum in the region just before it starts its rapid decay due to inner scale effects. *Hill* (1978) did a hydrodynamic analysis to derive a spectrum in the region where inner scale is significant, and found a slight peak before an exponential decay. *Frehlich* (1992) has modeled the spectrum as

$$\Psi(k) = 0.033\, C_n^2 \left(k^2 + \kappa_o^2\right)^{-11/6} g\,(k)\,, \tag{2.26}$$

where a 4-term approximation for $g\,(k)$ is

$$g\,(k) = \exp\left(-\delta\, |k|\, L_{in}\right) \sum_{n=0}^{4} a_n \left(|k|\, L_{in}\right)^n. \tag{2.27}$$

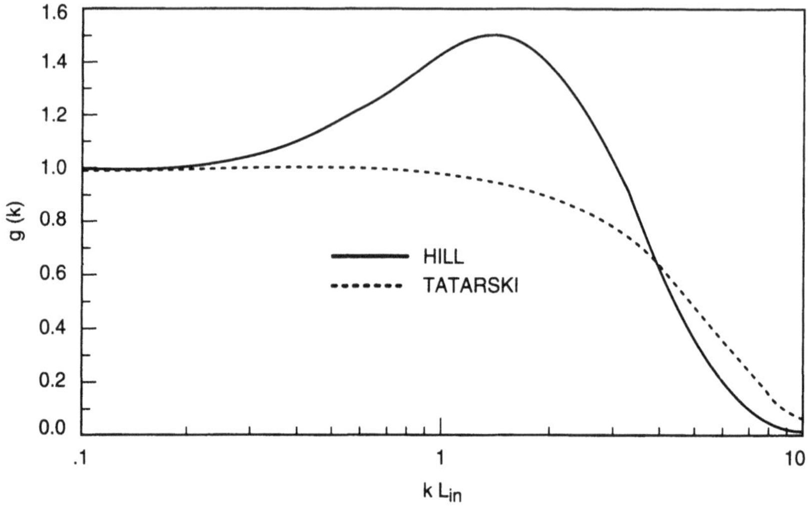

Fig. 2.4. Factor that multiplies the Kolmogorov turbulence spectrum in the Tatarski and Hill models of inner scale.

The constants are given by $\delta = 1.1090$, $a_0 = 1$, $a_1 = 0.70937$, $a_2 = 2.8235$, $a_3 = -0.28086$, and $a_4 = -0.08277$. This function is plotted versus the Tatarski function $\exp\left[-(kL_{in}/5.92)^2\right]$ in Fig. 2.4.

The spectrum given in (2.24) is used in this book for tilt calculations because it is not physically unreasonable and is mathematically convenient. The spectrum given in (2.26) can produce a significantly different scintillation at short ranges than the modified von Kármán spectrum, and it should be used in those calculations. The difference between the scintillation with these two spectra is considered in Sect. 7.9.

If the Tatarski spectrum is inserted into the integral for structure function in (2.16), then the expression can be evaluated with the Taylor series of the sinusoid and a term by term integration. Because each series term produces an integral that is absolutely convergent, this method is valid. For very small separations the first term of the expansion gives the most significant contribution as $1 - \sin(kr)/kr \approx (kr)^2/6$, for $r << L_i$. This is the only factor in the integrand that depends on the separation; therefore, for very small separations the structure function has the required quadratic dependence on separation as given in (2.5). The exponential decay is not unique in this regard; any function with rapid decay produces the correct form of structure function for small separations.

In the analysis performed in this book the 2-dimensional Fourier transform is used heavily. It is convenient to use the space coordinate for the direction of wave propagation and to use the Fourier transform of the transverse coordinates. The relationship between the 3- and 2-dimensional spectra is found by taking the Fourier transform of the propagation coordinate, and by using the evenness of the spectrum to obtain

$$\Psi(k) = \frac{C_n^2(z)}{2\pi} \int_0^\infty dz\, E_n(\boldsymbol{\kappa}, z) \cos(k_z z), \tag{2.28}$$

where $E_n(\boldsymbol{\kappa}, z)$ is the normalized axial correlation function. Notice that the turbulence strength is now allowed to vary along the propagation path. The inverse Fourier relation is

$$C_n^2(z) E(\boldsymbol{\kappa}, z) = \frac{1}{(2\pi)^2} \int d\boldsymbol{\rho}'' \exp(-i\boldsymbol{\kappa} \cdot \boldsymbol{\rho}'') \langle n_1(0,0)\, n_1(\boldsymbol{\rho}'', z) \rangle. \tag{2.29}$$

A normalized turbulence spectrum is

$$\Psi(k) \equiv 0.033\, C_n^2\, \tilde{f}(\kappa, k_z). \tag{2.30}$$

The two-dimensional spectrum is found by setting $k_z = 0$ to obtain

$$\Psi(k)|_{k_z=0} \equiv 0.033\, C_n^2(z)\, f(\kappa) = \frac{C_n^2(z)}{2\pi} \int_0^\infty dz_-\, E_n(\boldsymbol{\kappa}, z_-). \tag{2.31}$$

The normalized spectrum $f(\kappa)$ is used in the book. For the modified von Kármán spectrum it is

$$f(\kappa) = \left(\kappa^2 + \kappa_o^2\right)^{-11/6} \exp\left(-\kappa^2/\kappa_i^2\right). \tag{2.32}$$

When inner and outer scale are negligible the spectrum is

$$f(\kappa) = \kappa^{-11/3}. \tag{2.33}$$

In Sect. 2.3 the correlation function of the transverse Fourier transform of refractive index fluctuations $d\nu(\boldsymbol{\kappa}, z)$ is required. This is evaluated with the results given above. The Fourier-Stieltjes relation between the refractive index and its transform is given by

$$n_1(\boldsymbol{\rho}, z) = \int d\nu(\boldsymbol{\kappa}, z) \exp(i\boldsymbol{\kappa} \cdot \boldsymbol{\rho}). \tag{2.34}$$

The inverse transform is

$$d\nu(\boldsymbol{\kappa}, z) = \frac{d\boldsymbol{\kappa}}{(2\pi)^2} \int d\boldsymbol{\rho}\, n_1(\boldsymbol{\rho}, z) \exp(-i\boldsymbol{\kappa} \cdot \boldsymbol{\rho}).$$

The correlation function of the transverse Fourier transform of refractive index fluctuations is

$$\begin{aligned} &\langle d\nu(\boldsymbol{\kappa}, z')\, d\nu^*(\boldsymbol{\kappa}', z'') \rangle \\ &= \frac{d\boldsymbol{\kappa}\, d\boldsymbol{\kappa}'}{(2\pi)^4} \int d\boldsymbol{\rho}\, d\boldsymbol{\rho}' \exp\left[-i(\boldsymbol{\kappa} \cdot \boldsymbol{\rho} - \boldsymbol{\kappa}' \cdot \boldsymbol{\rho}')\right] \langle n_1(\boldsymbol{\rho}, z')\, n_1(\boldsymbol{\rho}', z'') \rangle. \end{aligned}$$

Make the change in variables $\boldsymbol{\rho} - \boldsymbol{\rho}' = \boldsymbol{\rho}''$ and $\boldsymbol{\rho}' = \boldsymbol{\rho}'$ to give

$$\begin{aligned}&\langle d\nu\,(\boldsymbol{\kappa}, z')\; d\nu^*\,(\boldsymbol{\kappa}', z'')\rangle \\ &= \frac{d\boldsymbol{\kappa}\, d\boldsymbol{\kappa}'}{(2\pi)^4}\int d\boldsymbol{\rho}'' d\boldsymbol{\rho}' \exp\{-i\,[\boldsymbol{\kappa}\cdot\boldsymbol{\rho}'' - (\boldsymbol{\kappa}-\boldsymbol{\kappa}')\cdot\boldsymbol{\rho}']\}\,\langle n_1(\boldsymbol{\rho}'+\boldsymbol{\rho}'', z')\, n_1(\boldsymbol{\rho}', z'')\rangle\,.\end{aligned} \tag{2.35}$$

Assume the correlation function is a function solely of the coordinate difference times a slowly varying function of position as

$$\langle n_1(\boldsymbol{\rho}'+\boldsymbol{\rho}'', z')\, n_1\,(\boldsymbol{\rho}', z'')\rangle = C_n^2\,[(z'+z'')\,/2]\,\langle n_n\,(0,0)\, n_n\,(\boldsymbol{\rho}'', z'-z'')\rangle\,,$$

where $n_n\,(\boldsymbol{\rho}, z)$ is a normalized density fluctuation. Insert this relation into (2.35). The integration over $\boldsymbol{\rho}'$ gives a delta function. Use the relation in (2.29) to find that the expected value is

$$\begin{aligned}&(2\pi)^4\,\langle d\nu\,(\boldsymbol{\kappa}, z')\; d\nu^*\,(\boldsymbol{\kappa}', z'')\rangle \\ &= E_n\,(\boldsymbol{\kappa}, |z'-z''|)\, C_n^2\,[(z'+z'')\,/2]\;\delta\,(\boldsymbol{\kappa}-\boldsymbol{\kappa}')\, d\boldsymbol{\kappa}\, d\boldsymbol{\kappa}'.\end{aligned} \tag{2.36}$$

This relation is used to find the expected value of second moments.

2.2 Rytov Approximation

Determining the turbulence effect on wave propagation requires one to solve Maxwell's equations with a refractive index that is stochastic. This can not be done exactly. An approach that uses geometric optics was first introduced by *Chernov* (1960) to solve this problem. It was soon realized that this approximation requires $L < L_i^2/\lambda$ where L is the propagation distance and λ is the propagation wavelength. This has a very limited range of applicability. It was Rytov who proposed an approximation that included diffraction effects; this resulted in a wider range of validity. This approximation could be used to obtain second moments of quantities. *Tatarski* in two seminal books (1961, 1971) codified this method and obtained solutions to a large variety of problems. His results agreed well with experiments.

Since Tatarski's books were written, many problems have been solved as interest in wave propagation in turbulence expanded because of new technologies and areas of application. These areas include sending information on optical beams close to the ground, remote sensing of the atmosphere on earth and planets, correcting atmospheric distortion to allow better resolution in optical images, and sending laser beams efficiently through the atmosphere. The last two applications require correction of atmospheric distortion by adaptive-optics systems. To correct distortion one must measure phase along a path that is close in angle to the actual path along which the laser beam propagates. How small this angle must be to obtain a good correction leads to the important concept of the isoplanatic angle. In this book anisoplanatism refers to any effect that causes a difference in the propagation paths of two waves.

When propagating laser beams parallel to the ground, one found that scintillation increased with increasing turbulence as the Rytov theory predicted until a certain level was reached at which point the measured scintillation saturated. To solve problems in the saturation region a more complicated propagation theory was developed that allowed multiple scattering of the wave as discussed by *Strohbehn* (1978). Fortunately, even when the scintillation is saturated Rytov theory typically gives the correct answer for phase disturbances, thus allowing one to treat many problems of practical interest. Intensity fluctuations in the saturation region are not considered in this book.

As the power in the laser beam increases, the air column along the propagation path heats up from energy absorption by the air. The lateral wind or beam slewing moves this heated air out of the beam in a time referred to as the wind clearing time. The resulting nonuniform temperature distribution across the beam diameter can be shown to be equivalent to a negative lens that causes the beam to expand. This "thermal blooming" is predicted by the nonlinear wave equation and has been extensively studied. Computer simulations are typically used to model the nonlinear interaction. Instabilities can set in when the laser power is high. The theory of this instability is described by *Shonfeld* (1992), and laboratory confirmation of the instability is described by *Johnson* and *Primmerman* (1989). Thermal blooming is not considered in this book. Additional information on saturated turbulence and thermal blooming is in *Strohbehn* (1978).

The average results obtained here are long-term averages. In the short term, the second moments can vary, an effect referred to as intermittency. Its effect on wave propagation is discussed in *Tatarski* and *Zavorotnyi* (1985).

To solve for effects of turbulence in the low-power regime, one starts with Maxwell's equations

$$\nabla \times \boldsymbol{\mathcal{E}}(\mathbf{r},t) = -\frac{\partial \boldsymbol{\mathcal{B}}(\mathbf{r},t)}{\partial t}, \tag{2.37}$$

$$\nabla \cdot \boldsymbol{\mathcal{E}}(\mathbf{r},t) = q(\mathbf{r},t)/\varepsilon(\mathbf{r},t), \tag{2.38}$$

$$\nabla \times \boldsymbol{\mathcal{B}}(\mathbf{r},t) = \mu_o \boldsymbol{\mathcal{J}}(\mathbf{r},t) + \mu_o \varepsilon(\mathbf{r},t)\frac{\partial \boldsymbol{\mathcal{E}}(\mathbf{r},t)}{\partial t}, \quad \text{and} \tag{2.39}$$

$$\nabla \cdot \boldsymbol{\mathcal{B}}(\mathbf{r},t) = 0, \tag{2.40}$$

where $\boldsymbol{\mathcal{E}}(\mathbf{r},t)$ is the electric field, $\boldsymbol{\mathcal{B}}(\mathbf{r},t)$ is the magnetic field, $\varepsilon(\mathbf{r},t)$ is the permittivity, μ_o is the magnetic permeability, $\boldsymbol{\mathcal{J}}(\mathbf{r},t)$ is the current density, and $q(\mathbf{r},t)$ is the charge density. Assume there is a time-harmonic variation of the fields

$$\boldsymbol{\mathcal{E}}(\mathbf{r},t) = \exp(-i\omega t)\, \mathbf{E}(\mathbf{r}). \tag{2.41}$$

$\mathbf{E}(\mathbf{r})$ can also vary with time but the time scale of variation is much longer than the harmonic variation. The air density affects the refractive index as

$$\varepsilon(\mathbf{r}) = \varepsilon_0 n^2(\mathbf{r}), \tag{2.42}$$

where the refractive index $n(\mathbf{r})$ for air is given in (2.1). The wave equation describing the electric field derived from the above relations is

$$\nabla^2 \mathbf{E}(\mathbf{r}) + k_0^2 n^2(\mathbf{r})\,\mathbf{E}(\mathbf{r}) - 2i\frac{k_0}{c}\frac{\partial\left[n^2(\mathbf{r})\,\mathbf{E}(\mathbf{r})\right]}{\partial t} + 2\nabla\left\{\mathbf{E}(\mathbf{r})\cdot\nabla\ln\left[n(\mathbf{r})\right]\right\} = 0,$$

where the free space wavenumber is given by $k_0 = \omega/c = 2\pi/\lambda$, and c is the speed of light in vacuum. The non-sinusoidal variation of the field with time is suppressed. If the rate of change of $\mathbf{E}(\mathbf{r})$ with time is much less than the sinusoidal variation $\exp(-i\omega t)$, then the third term is negligible compared to the first two. For optical frequencies this condition is satisfied. In addition, if the propagation wavelength λ is much less than the inner scale, then the last term that produces coupling between the different polarization components of the electric field can be neglected. For visible wavelengths, typical inner-scale sizes of a millimeter satisfy this condition, in which case one obtains a scalar equation for each of the electric field components separately. The equation for one component is

$$\nabla^2 u(\mathbf{r}) + k_0^2 n^2(\mathbf{r})\,u(\mathbf{r}) = 0. \tag{2.43}$$

In the Rytov method the solution is expressed as

$$u(\mathbf{r}) = \exp\left[\Phi(\mathbf{r})\right]. \tag{2.44}$$

This leads to the nonlinear Riccati equation

$$\nabla^2\Phi(\mathbf{r}) + \left[\nabla\Phi(\mathbf{r})\right]^2 = -k_0^2 n^2(\mathbf{r}). \tag{2.45}$$

If it is assumed that the magnitude of the air-density inhomogeneity is small, the refractive index is

$$n(\mathbf{r}) = 1 + \delta n_1(\mathbf{r}), \tag{2.46}$$

where $\delta n_1(\mathbf{r}) << 1$, and δ was inserted into the expression to show smallness. This Riccati equation can be solved by a perturbation method in which it is assumed that the solution can be written as a power series in δ,

$$\Phi(\mathbf{r},\delta) = \Phi_0(\mathbf{r}) + \delta\,\Phi_1(\mathbf{r}) + \delta^2\,\Phi_2(\mathbf{r}) + \cdots. \tag{2.47}$$

By inserting this into (2.45) and by separating the equations based on the power of δ, one obtains a system of equations that has been explored by *Schmeltzer* (1967).

$$\nabla^2\Phi_0(\mathbf{r}) + \nabla\Phi_0(\mathbf{r})\cdot\nabla\Phi_0(\mathbf{r}) = -k_0^2, \tag{2.48}$$

$$\nabla^2\Phi_1(\mathbf{r}) + 2\,\nabla\Phi_0(\mathbf{r})\cdot\nabla\Phi_1(\mathbf{r}) = -2\,k_0^2\,n_1(\mathbf{r}), \tag{2.49}$$

$$\nabla^2\Phi_2(\mathbf{r}) + 2\,\nabla\Phi_0(\mathbf{r})\cdot\nabla\Phi_2(\mathbf{r}) = -k_0^2\,n_1^2(\mathbf{r}) - \nabla\Phi_1(\mathbf{r})\cdot\nabla\Phi_1(\mathbf{r}),\ \text{and} \tag{2.50}$$

$$\nabla^2\Phi_m(\mathbf{r}) + 2\,\nabla\Phi_0(\mathbf{r})\cdot\nabla\Phi_m(\mathbf{r}) = -\sum_{p=1}^{m-1}\nabla\Phi_p(\mathbf{r})\cdot\nabla\Phi_{m-p}(\mathbf{r}). \tag{2.51}$$

For these equations to be valid one requires

$$|\nabla \Phi_{n+1}(\mathbf{r})| << |\nabla \Phi_n(\mathbf{r})| . \tag{2.52}$$

This condition gives this method the alternate name of "the method of smooth perturbations." I retain only the first two terms. Express the lowest-order term in the form

$$u_0(\mathbf{r}) = \exp[\Phi_0(\mathbf{r})] . \tag{2.53}$$

Equation (2.48) is equivalent to one for a wave undisturbed by turbulence

$$\nabla^2 u_0(\mathbf{r}) + k_0^2 u_0(\mathbf{r}) = 0. \tag{2.54}$$

If I set

$$\Phi_1(\mathbf{r}) = W_1(\mathbf{r}) / u_0(\mathbf{r}) , \tag{2.55}$$

the solution to second order is

$$u(\mathbf{r}) = u_0(\mathbf{r}) \exp[\Phi_1(\mathbf{r})] . \tag{2.56}$$

To obtain $\Phi_1(\mathbf{r})$, given in (2.55), one must obtain the solution of

$$\nabla^2 W_1(\mathbf{r}) + k_0^2 W_1(\mathbf{r}) = -2 k_0^2 n_1(\mathbf{r})\, u_0(\mathbf{r}) . \tag{2.57}$$

The solution to this linear differential equation with constant coefficients is obtained with the Green's function for free space propagation as

$$\Phi_1(\mathbf{r}) = \frac{2 k_0^2}{u_0(\mathbf{r})} \int dV'\, u_0(\mathbf{r}')\, n_1(\mathbf{r}')\, G(|\mathbf{r} - \mathbf{r}'|) ,$$

where the integration is over the source volume denoted by primed coordinates. The Green's function in spherical coordinates is

$$G(r) = \frac{1}{4\pi r} \exp(i k_0 r) . \tag{2.58}$$

The perturbed field is

$$\Phi_1(\mathbf{r}) = \frac{k_0^2}{2\pi u_0(\mathbf{r})} \int dV'\, u_0(\mathbf{r}')\, n_1(\mathbf{r}') \frac{\exp(i k_0 |\mathbf{r} - \mathbf{r}'|)}{|\mathbf{r} - \mathbf{r}'|} . \tag{2.59}$$

The requirement that the second term of the expansion is much greater than the sum of the remaining terms leads to the requirement that

$$k_0^{7/6} \int_0^L dz\, C_n^2(z)\, z^{5/6} < 1. \tag{2.60}$$

As shown in Sect. 4.7 this is the same as requiring log-amplitude fluctuations to be small. It has generally been found that the Rytov approximation gives a good approximation to the log-amplitude variance χ^2 in the regime

$$\chi^2 < 0.2 - 0.5. \tag{2.61}$$

Even though the solution for log-amplitude is not valid when this inequality is not satisfied, the solution for the phase *for collimated beams* has been shown to be valid.

2.3 Phase and Log-Amplitude Variances

In this section I will develop general relations for phase and log-amplitude variances that are used throughout the book to solve turbulence problems. To cover the cases of collimated and spherical waves, variances of the log-amplitude and phase are derived for a Gaussian wave with the approach of *Ishimaru* (1969). Once these quantities are found in the general case, the remainder of the derivation will be performed for infinite waves to reduce the complexity of the analysis and to obtain formulas specific for the infinite wave case, which is typically of most interest. The general case is considered in Chap. 10. The free-space propagation of a Gaussian beam has been considered in many places, and the appropriate expression for the field is

$$u_0(\boldsymbol{\rho}, z) = \frac{A}{1 + i\alpha z} \exp\left[i k_0 z - \frac{k_0 \alpha \rho^2}{2(1 + i\alpha z)} \right], \tag{2.62}$$

where $\mathbf{r} = \boldsymbol{\rho} + \mathbf{u}_z z$, and

$$\alpha = \alpha_r + i\,\alpha_i = \lambda/\pi W_o^2 + i/R_o. \tag{2.63}$$

The parameter R_o is the radius of beam curvature at the source. It is equal to infinity for a collimated beam. At a radius equal to W_o at the source, the intensity has dropped to $1/e^2$ of its value on axis. I use the paraxial assumption: which for the source located at $z = 0$ and the observation point located at $z = L$, the term in the exponential of the Green's function is expanded to obtain

$$\frac{\exp(ik_0\,|\mathbf{R} - \mathbf{r}'|)}{|\mathbf{R} - \mathbf{r}'|} \approx \frac{1}{(L - z')} \exp\left\{ ik_0(L - z')\left[1 + \frac{|\boldsymbol{\rho} - \boldsymbol{\rho}'|^2}{2\,(L - z')} - \frac{|\boldsymbol{\rho} - \boldsymbol{\rho}'|^4}{8\,(L - z')^4} \right] \right\}. \tag{2.64}$$

If one assumes that the distance to the source is much larger than the transverse coordinate, which requires $|\boldsymbol{\rho} - \boldsymbol{\rho}'|^4 << (L - z')^3\lambda$, the last term in braces can be neglected. This approximation can also be used to find an additional restriction on the applicability of the turbulence solution. The smallest turbulence sizes are the order of the inner scale; therefore, the maximum scattering angle is λ/L_i, and the maximum deviation of the beam from boresight is $\lambda\, L/L_i$. Therefore, the paraxial assumption gives the requirement that $L << L_i^4/\lambda^3$. For inner scale sizes of a millimeter or more, the range over which the assumption is valid is very large, and the condition is satisfied for most problems.

After substituting (2.64) and (2.62) into (2.59), and performing some algebraic manipulation, one finds that the turbulence-induced disturbance is

$$\Phi_1(\boldsymbol{\rho}, L) = \int_0^L dz' \int d\boldsymbol{\rho}'\, h(\boldsymbol{\rho}' - \gamma\,\boldsymbol{\rho}, L, z')\, n_1(\boldsymbol{\rho}', z'), \tag{2.65}$$

$$h(\boldsymbol{\rho}' - \gamma\,\boldsymbol{\rho}, L, z') = \frac{k_0^2}{2\pi} \frac{\exp\left\{ ik_0 \left[(\boldsymbol{\rho}' - \gamma\,\boldsymbol{\rho})^2 / 2\gamma\,(L - z') \right] \right\}}{\gamma\,(L - z')}, \text{ and} \tag{2.66}$$

$$\gamma = \frac{1 + i\alpha\, z'}{1 + i\alpha\, L}. \tag{2.67}$$

The value of γ is equal to unity for plane waves, and equal to z/L for spherical waves (corresponding to a point source at $z = 0$.) In these cases the imaginary part of γ is zero.

If the source were located at $z = L$ and the wave propagated towards $z = 0$, one obtains the appropriate relations by using the substitution $z' \rightarrow (L - z')$ in the above equations, to obtain

$$\gamma = \frac{1 + i\alpha\,(L - z')}{1 + i\alpha\, L}. \tag{2.68}$$

The value of propagation parameter γ is equal to unity for plane waves, and equal to $(L - z)/L$ for spherical waves (corresponding to a point source at $z = L$.) Notice that when the imaginary part is non-zero, then it is less than zero for both propagation directions. To find variances, the method of spectral expansion is used, in which quantities are expressed as Fourier-Stieltjes integrals of the transverse coordinates. Refractive-index fluctuations are represented in (2.34) as

$$n_1(\boldsymbol{\rho}, z) = \int d\nu\,(\boldsymbol{\kappa}, z)\exp\left(i\boldsymbol{\kappa}\cdot\boldsymbol{\rho}\right), \tag{2.69}$$

where the integral is over the two-dimensional transverse wavenumber space. The requirement that the refractive index is real leads to

$$d\nu\,(\boldsymbol{\kappa}, z) = d\nu^*\,(-\boldsymbol{\kappa}, z)\,. \tag{2.70}$$

If (2.69) is substituted into the expression for turbulence-induced fluctuations given in (2.65), and the substitution $\boldsymbol{\rho}' - \gamma\,\boldsymbol{\rho} \rightarrow -\boldsymbol{\rho}'$ is made, one obtains

$$\begin{aligned}\Phi_1(\boldsymbol{\rho}, L) &= \int_0^L dz' \int d\nu\,(\boldsymbol{\kappa}, z') \int d\boldsymbol{\rho}'\, h\,(\boldsymbol{\rho}' - \gamma\,\boldsymbol{\rho}, L, z')\,\exp\left(i\boldsymbol{\kappa}\cdot\boldsymbol{\rho}'\right)\\ &= \int_0^L dz' \int d\nu\,(\boldsymbol{\kappa}, z')\,H\,(\boldsymbol{\kappa}, L, z')\,\exp\left(i\gamma\,\boldsymbol{\kappa}\cdot\boldsymbol{\rho}\right),\end{aligned} \tag{2.71}$$

where $H\,(\boldsymbol{\kappa}, L, z)$ is the Fourier transform of $h\,(\boldsymbol{\rho}, L, z)$ given by

$$H(\boldsymbol{\kappa}, L, z) = \int d\boldsymbol{\rho}'\, h\,(\boldsymbol{\rho}', L, z)\exp\left(-i\boldsymbol{\kappa}\cdot\boldsymbol{\rho}'\right) = ik_0\exp\left[-iP\,(\gamma, \kappa, z)\right]. \tag{2.72}$$

For propagation from $z = 0$ to $z = L$, the diffraction parameter is equal to

$$P\,(\gamma, \kappa, z) = \frac{\gamma\kappa^2(L - z)}{2\,k_0}, \tag{2.73}$$

and for propagation from $z = L$ to $z = 0$

$$P\,(\gamma, \kappa, z) = \frac{\gamma\kappa^2 z}{2\,k_0}. \tag{2.74}$$

In this formulation, as pointed out by Ishimaru, the effective wavenumber is $\gamma\kappa$ rather than κ as can readily be seen in the last part of (2.71). In general, the effective wavenumber is complex. In comparing results obtained with this method to others that do not use this approach, one must make the substitution of κ for $\gamma\kappa$ in this solution to get the same form.

Notice that $H(\boldsymbol{\kappa}, L, z)$ is an even function of kappa. Also, the real part of the exponent on the far right in (2.71) is not greater than zero, which causes the integrand in (2.71) to go to zero at infinity for any angular orientation even though the complex exponential diverges for some angles.

Log-amplitude $\chi(\boldsymbol{\rho}, z)$ and phase $\varphi(\boldsymbol{\rho}, z)$ fluctuations are real quantities defined by the relation

$$\Phi_1(\boldsymbol{\rho}, z) \equiv \chi(\boldsymbol{\rho}, z) + i\varphi(\boldsymbol{\rho}, z) . \tag{2.75}$$

Therefore,

$$\chi(\boldsymbol{\rho}, z) = \frac{1}{2}\left[\Phi_1(\boldsymbol{\rho}, z) + \Phi_1^*(\boldsymbol{\rho}, z)\right], \text{ and} \tag{2.76}$$

$$\phi(\boldsymbol{\rho}, z) = \frac{1}{2i}\left[\Phi_1(\boldsymbol{\rho}, z) - \Phi_1^*(\boldsymbol{\rho}, z)\right]. \tag{2.77}$$

Let us now specialize to the infinite beam case for which γ is real; this leads to simpler expressions. Derivations for complex waves, given in Chap. 10, differ at this point. By substituting (2.71) into (2.76) and (2.77) and by using the evenness with respect to kappa, one finds

$$\chi(\boldsymbol{\rho}, L) = k_0 \int_0^L dz' \int d\nu(\boldsymbol{\kappa}, z') \sin\left[P(\gamma, \kappa, z')\right] \exp(i\gamma\boldsymbol{\kappa}\cdot\boldsymbol{\rho}), \text{ and} \tag{2.78}$$

$$\phi(\boldsymbol{\rho}, L) = k_0 \int_0^L dz' \int d\nu(\boldsymbol{\kappa}, z') \cos\left[P(\gamma, \kappa, z')\right] \exp(i\gamma\boldsymbol{\kappa}\cdot\boldsymbol{\rho}). \tag{2.79}$$

I shall only give the derivation of the variance of a phase-related quantity explicitly. Since the only difference between the phase- and log-amplitude related expressions is in the trigonometric factor, the expression for the log-amplitude variance will be written at the end by inspection.

The phase expression above applies to a single point in space. Often an aperture average with a normalized weighting function $g(\boldsymbol{\rho})$ times the phase is required, as when one finds the Zernike modes of a wave. Define the Fourier transform of the aperture function by

$$G(\gamma\boldsymbol{\kappa}) = \frac{1}{(2\pi)^2}\int d\boldsymbol{\rho}'\, g(\boldsymbol{\rho}') \exp(i\gamma\boldsymbol{\kappa}\cdot\boldsymbol{\rho}') . \tag{2.80}$$

We assume that the aperture function is normalized so that

$$\frac{4}{\pi D^2}\int d\boldsymbol{\rho}'\, g^2(\boldsymbol{\rho}') = 1. \tag{2.81}$$

In this case the phase-related quantity is

$$\phi R(L) = k_0 \int d\boldsymbol{\rho}' \int_0^L dz' \int d\nu\left(\boldsymbol{\kappa}, z'\right) \cos\left[P\left(\gamma, \kappa, z'\right)\right] g\left(\boldsymbol{\rho}'\right) \exp\left(i\gamma\boldsymbol{\kappa}\cdot\boldsymbol{\rho}'\right).$$

Using (2.80) one obtains

$$\phi R(L) = (2\pi)^2 k_0 \int_0^L dz'\, d\nu\left(\boldsymbol{\kappa}, z'\right) G\left(\gamma\,\boldsymbol{\kappa}\right) \cos\left[P\left(\gamma, \kappa, z'\right)\right]. \tag{2.82}$$

Specific aperture functions are constructed in Chap. 3.

Up to now only a single wave has been considered. Generalize this to the case of finding the difference between the phase of two waves that can have different values of γ. The analysis of two beams applies, for example, to adaptive-optics systems. This will allow one to consider the difference between displaced waves, counterpropagating waves, and also the difference between spherical and plane waves. Let the second wave be multiplied by a quantity $A\left(\boldsymbol{\kappa}, z\right)$ that can be a function of the axial coordinate and transverse wavenumber. If this multiplier is zero, one is back to the single wave case, and if it is an exponential with exponent linear in separation distance, this corresponds, by definition, to anisoplanatism. The two wave case can also express the output of a measurement system being applied to an outgoing beam. Finally, let the aperture weighting factor be different for each wave. A typical geometry for this situation is shown in Fig. 2.5. The result is

$$\begin{aligned}\phi R(L) &= (2\pi)^2 k_0 \int_0^L dz' \int d\nu\left(\boldsymbol{\kappa}, z'\right) \\ &\times\left\{G_1\left(\gamma_1\boldsymbol{\kappa}\right) \cos\left[P_1\left(\gamma_1, \kappa, z'\right)\right] - A\left(\boldsymbol{\kappa}, z'\right) G_2\left(\gamma_2\boldsymbol{\kappa}\right) \cos\left[P_2\left(\gamma_2, \kappa, z'\right)\right]\right\}.\end{aligned}$$

The second moment of $\phi R(L)$ is found by multiplying the phase or log-amplitude by its complex conjugate and calculating the expected value. The only stochastic quantity on the right is the refractive index; therefore, the second moment of the phase-related quantity is

$$\begin{aligned}\sigma_{\phi R}^2 &= \left\langle\phi_R(L)\,\phi_R^*(L)\right\rangle = (2\pi)^4 k_0^2 \int_0^L dz' \int_0^L dz'' \iint \left\langle d\nu\left(\boldsymbol{\kappa}, z'\right)\, d\nu^*\left(\boldsymbol{\kappa}', z''\right)\right\rangle \\ &\times\left\{G_1\left(\gamma_1\boldsymbol{\kappa}\right) \cos\left[P_1\left(\gamma_1, \kappa, z'\right)\right] - A\left(\boldsymbol{\kappa}, z'\right) G_2\left(\gamma_2\boldsymbol{\kappa}\right) \cos\left[P_2\left(\gamma_2, \kappa, z'\right)\right]\right\} \\ &\times\left\{G_1^*\left(\gamma_1\boldsymbol{\kappa}'\right) \cos\left[P_1^*\left(\gamma_1, \kappa', z''\right)\right] - A^*\left(\boldsymbol{\kappa}', z''\right) G_2^*\left(\gamma_2\boldsymbol{\kappa}'\right) \cos\left[P_2^*\left(\gamma_2, \kappa', z''\right)\right]\right\}.\end{aligned} \tag{2.83}$$

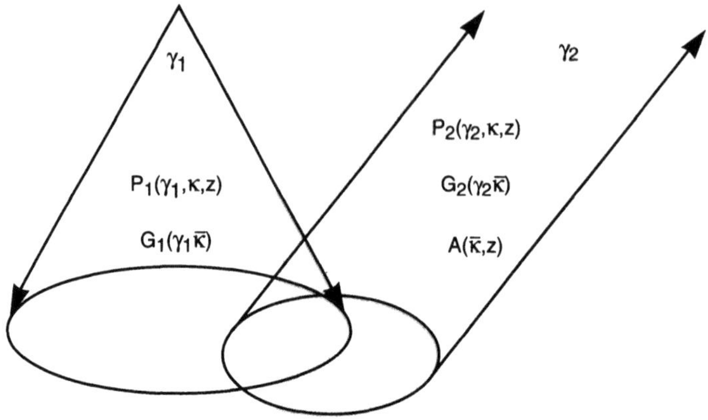

Fig. 2.5. Geometry of focused and collimated beams of different diameters propagating in different directions with a separation that depends on the axial coordinate.

The ensemble average can be simplified with the relation given in (2.36) as

$$(2\pi)^4 \langle d\nu(\boldsymbol{\kappa}, z')\, d\nu^*(\boldsymbol{\kappa}', z'')\rangle = E_n(\boldsymbol{\kappa}, |z'-z''|)\, C_n^2[(z'+z'')/2]\, \delta(\boldsymbol{\kappa}-\boldsymbol{\kappa}')\, d\boldsymbol{\kappa}\, d\boldsymbol{\kappa}', \quad (2.84)$$

where $C_n^2(z)$ is a normalizing function that is proportional to the strength of turbulence, and $E_n(\boldsymbol{\kappa}, |z'-z''|)$ is the normalized correlation function of density fluctuations.

The presence of the delta function in (2.84) allows the integration over one transverse spectral coordinate to be performed directly. To eliminate an axial integral, change the axial coordinates into sum and difference coordinates as $z_- = z' - z''$, and $z = (z'+z'')/2$. This changes the integration region from that of Fig. 2.6a to the diamond shaped region of Fig 2.6b. Perform the integration over $\boldsymbol{\kappa}'$ in (2.83) to give

$$\sigma_{\phi R}^2 = k_0^2 \underset{Diamond}{\int\int} dz_-\, dz\, C_n^2(z) \int d\boldsymbol{\kappa}\, E_n(\boldsymbol{\kappa}, |z_-|)$$
$$\times \left\{ G_1(\gamma_1 \boldsymbol{\kappa}) \cos(1,1,+) - A\left(\boldsymbol{\kappa}, z+\frac{z_-}{2}\right) G_2(\gamma_2 \boldsymbol{\kappa}) \cos(2,2,+) \right\}$$
$$\times \left\{ G_1^*(\gamma_1 \boldsymbol{\kappa}) \cos(1,1,-) - A^*\left(\boldsymbol{\kappa}, z-\frac{z_-}{2}\right) G_2^*(\gamma_2 \boldsymbol{\kappa}) \cos(2,2,-) \right\},$$

where the first argument of the cosine indicates the subscript of P, the second argument the subscript of γ, and the sign refers to the sign of $z_-/2$ in $\cos[P_a(\gamma_b, \kappa, z \pm z_-/2)]$. *Tatarski* (1971) discusses in detail how to eliminate one axial integration for the single wave case. His arguments extend to the two wave case. The integration over z_- is similar to that given in (2.29) except for two apparent differences; the integration in (2.31) extends to infinity,

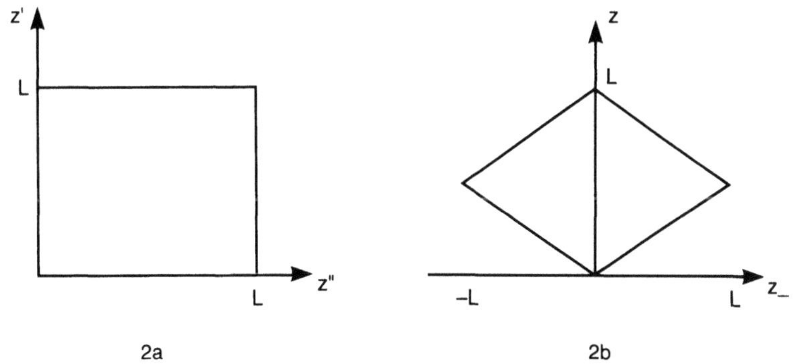

Fig. 2.6. The original and the transformed regions of integration.

while the integration above has finite limits, and there are other functions in the integrand that depend on z_-. These differences can be circumvented. The correlation function of real turbulence only has significant values over a range of tens of meters, a distance that is typically very small compared to the entire integration path. *Tatarski* (1971) shows that as long as the inner scale is much larger than the wavelength, then the variation of the trigonometric term with z_- is not significant. To eliminate the dependence on z_- in the other integrand terms, one can expand the functions in Taylor series about the point where the difference coordinate is zero. The relative magnitude of the first two terms of a Taylor series is on the order of z_-/S, where S is the distance over which the function being expanded significantly changes its magnitude. For the functions in the integrand, the distance S is either the length of the propagation path or the Fresnel distance. As long as this distance is much larger than the correlation length, the first term of the expansion is small compared to the zeroth order term. With that approximation, which is good for most practical problems, one can neglect the difference coordinate when it is added or subtracted from the sum coordinate. This assumption makes the correlation function the only one that depends on z_-. I have checked this assumption for several problems, and the error is typically on the order of one part in a million. For any specific problem one should determine if the above assumptions are violated.

The difficulty with the finite integration limits are now resolved with standard arguments: Since the correlation function is even in z_-, the integral is twice the value obtained with only the right-hand side of the diamond. The only part of the diamond-shaped area that contributes significantly to the integral is a thin strip close to the vertical axis, since the correlation function falls off very rapidly with distance along the horizontal axis. Therefore, the diamond-shaped area can be extended horizontally to infinity without significantly affecting the value of the integral. This allows the limit of the z_- integration to be changed to infinity, and the integral can be evaluated with (2.31). The same derivation can be performed for log-amplitude related quantities. *The resulting expression is the central relation of this chapter.*

$$\begin{bmatrix} \sigma_{\phi R}^2 \\ \sigma_{\chi R}^2 \end{bmatrix} = 0.2073\, k_0^2 \int_0^L dz\, C_n^2(z) \int d\boldsymbol{\kappa}\, f(\kappa)$$
$$\times \left\{ G_1(\gamma_1 \boldsymbol{\kappa}) \begin{bmatrix} \cos(1,1) \\ \sin(1,1) \end{bmatrix} - A(\boldsymbol{\kappa}, z)\, G_2(\gamma_2 \boldsymbol{\kappa}) \begin{bmatrix} \cos(2,2) \\ \sin(2,2) \end{bmatrix} \right\}$$
$$\times \left\{ G_1^*(\gamma_1 \boldsymbol{\kappa}) \begin{bmatrix} \cos(1,1) \\ \sin(1,1) \end{bmatrix} - A^*(\boldsymbol{\kappa}, z)\, G_2^*(\gamma_2 \boldsymbol{\kappa}) \begin{bmatrix} \cos(2,2) \\ \sin(2,2) \end{bmatrix} \right\}, \quad (2.85)$$

where the first argument of the cosine indicates the subscript of P and the second argument the subscript of γ in $\cos[P_a(\gamma_b, \kappa, z)]$.

This expression is the starting point for solving many turbulence problems. One can consider the variance of the difference of either the phase or log-amplitude of two waves, each of which can be propagating up or down, have different diameters, be displaced from each other with a displacement that changes along the propagation direction, have different amplitudes, and have different values of γ. The entire quantity multiplying the turbulence spectrum can be looked upon as a filter function of the spectrum. This concept is central to the formulation of solutions to different problems. *Note that often one is interested in calculating phase-related quantities in which the turbulence effects are all in the very near field, for which case, one can set the cosine terms equal to unity, thereby, simplifying the evaluation of the integral.* The phase and log-amplitude distributions are Gaussian in the region where the Rytov approximation is valid. The distribution of the amplitude is therefore log-normal. Since the intensity is the square of the amplitude, the variance of log-intensity is 4 times the variance of log-amplitude.

Specific cases of this general formula that are of interest in different problems are found by setting various terms equal to zero. These specific formulas are given in (2.86)–(2.94). For instance, a case of great interest is when there is only one beam. For this case the weighting function $A(\boldsymbol{\kappa}, z)$ is zero, and the general formula reduces to

$$\begin{bmatrix} \sigma_{\phi R}^2 \\ \sigma_{\chi R}^2 \end{bmatrix} = 0.2073\, k_0^2 \int_0^L dz\, C_n^2(z) \int d\boldsymbol{\kappa}\, f(\kappa) \begin{bmatrix} \cos^2[P(\gamma, \kappa, z)] \\ \sin^2[P(\gamma, \kappa, z)] \end{bmatrix} F(\gamma \boldsymbol{\kappa}), \quad (2.86)$$

where

$$F(\gamma \boldsymbol{\kappa}) = G(\gamma \boldsymbol{\kappa})\, G^*(\gamma \boldsymbol{\kappa}). \quad (2.87)$$

If there are two beams with weighting on the phase or log-amplitude differing by a real multiplier, i.e., $A(\boldsymbol{\kappa}, z)$ is real, and whose values of γ are the same, and of which the same aperture-average is wanted, then

$$\begin{bmatrix} \sigma_{\phi R}^2 \\ \sigma_{\chi R}^2 \end{bmatrix} = 0.2073\, k_0^2 \int_0^L dz\, C_n^2(z) \int d\boldsymbol{\kappa}\, f(\kappa)$$
$$\times \begin{bmatrix} \{\cos[P_1(\gamma, \kappa, z)] - A(\boldsymbol{\kappa}, z)\cos[P_2(\gamma, \kappa, z)]\}^2 \\ \{\sin[P_1(\gamma, \kappa, z)] - A(\boldsymbol{\kappa}, z)\sin[P_2(\gamma, \kappa, z)]\}^2 \end{bmatrix} F(\gamma \boldsymbol{\kappa}). \quad (2.88)$$

If there are two beams whose centers coincide along the path with equal beam weighting, whose values of γ are equal, but where different aperture modes are wanted, then

$$\begin{bmatrix} \sigma^2_{\phi R} \\ \sigma^2_{\chi R} \end{bmatrix} = 0.2073\, k_0^2 \int_0^L dz\, C_n^2(z) \int d\boldsymbol{\kappa}\; f(\kappa) \times \begin{bmatrix} \cos^2 [P(\gamma,\kappa,z)] \\ \sin^2 [P(\gamma,\kappa,z)] \end{bmatrix} |G_1(\gamma\boldsymbol{\kappa}) - G_2(\gamma\boldsymbol{\kappa})|^2 . \tag{2.89}$$

If there are two beams whose centers coincide along the path with equal weighting of phase and log-amplitude, whose values of γ differ, but where different aperture modes are wanted *and diffraction can be neglected,* then

$$\sigma^2_{\phi R} = 0.2073\, k_0^2 \int_0^L dz\, C_n^2(z) \int d\boldsymbol{\kappa}\, f(\kappa)\, |G_1(\gamma_1 \boldsymbol{\kappa}) - G_2(\gamma_2 \boldsymbol{\kappa})|^2 . \tag{2.90}$$

Consider the difference between two beams of equal amplitude that are displaced from each other. A displacement in real space $\mathbf{d}$ is equivalent to a phase shift in transform space; therefore

$$A(\boldsymbol{\kappa}, z) = \exp(i\gamma\, \boldsymbol{\kappa} \cdot \mathbf{d}) . \tag{2.91}$$

If there are two beams displaced from each other by a distance $\mathbf{d}$ that have the same propagation constant, and of which the same aperture mode is wanted, one obtains

$$\begin{bmatrix} \sigma^2_{\phi R} \\ \sigma^2_{\chi R} \\ \sigma^2_{R} \end{bmatrix} = \begin{bmatrix} \mathcal{D}_\phi(\mathbf{d},\boldsymbol{\rho}) \\ \mathcal{D}_\chi(\mathbf{d},\boldsymbol{\rho}) \\ \mathcal{D}(\mathbf{d},\boldsymbol{\rho}) \end{bmatrix} = 0.2073\, k_0^2 \int_0^L dz\, C_n^2(z) \int d\boldsymbol{\kappa}\, f(\kappa) \times \begin{bmatrix} \cos^2 [P(\gamma,\kappa,z)] \\ \sin^2 [P(\gamma,\kappa,z)] \\ 1 \end{bmatrix} F(\gamma\boldsymbol{\kappa})\, 2\left[1 - \cos(\gamma\boldsymbol{\kappa}\cdot\mathbf{d})\right] , \tag{2.92}$$

where σ^2_R is the sum of the phase and log-amplitude variances, and $\mathcal{D}_\phi(\mathbf{d},\boldsymbol{\rho})$, $\mathcal{D}_\chi(\mathbf{d},\boldsymbol{\rho})$, and $\mathcal{D}(\mathbf{d},\boldsymbol{\rho})$ are the structure functions for phase, log-amplitude, and their sum respectively.

Even though the filter function given in (2.85) is fairly general, it does not cover all problems of interest. Using the same techniques, one can easily construct new filter functions. For instance, in order to find the structure function of the anisoplanatic difference of a phase or log-amplitude related quantity, one must calculate the variance of the difference of the quantity evaluated at two places separated by $\mathbf{s}$. It is easy to see that the variances are equal to

$$\begin{bmatrix} \sigma^2_{\phi R} \\ \sigma^2_{\chi R} \end{bmatrix} = 0.8292\, k_0^2 \int_0^L dz\, C_n^2(z) \int d\boldsymbol{\kappa}\, f(\kappa)$$

$$\times \begin{bmatrix} \cos^2\left[P\left(\gamma, \kappa, z\right)\right] \\ \sin^2\left[P\left(\gamma, \kappa, z\right)\right] \end{bmatrix} F\left(\gamma \boldsymbol{\kappa}\right)\left[1 - \cos\left(\gamma \boldsymbol{\kappa} \cdot \mathbf{d}\right)\right]\left[1 - \cos\left(\gamma \boldsymbol{\kappa} \cdot \mathbf{s}\right)\right]. \quad (2.93)$$

The same approach used to find anisoplanatic effects can be used to find the correlation functions for phase and log-amplitude related quantities as

$$\begin{bmatrix} B_{\phi R} \\ B_{\chi R} \end{bmatrix} = \left\langle \begin{matrix} \phi R\left(\boldsymbol{\rho} + \mathbf{d}\right)\ \phi R\left(\boldsymbol{\rho}\right) \\ \chi R\left(\boldsymbol{\rho} + \mathbf{d}\right)\ \chi R\left(\boldsymbol{\rho}\right) \end{matrix} \right\rangle = 0.2073\, k_0^2 \int_0^L dz\, C_n^2(z) \int d\boldsymbol{\kappa}\, f(\kappa)$$
$$\times \left[G_1\left(\gamma_1 \boldsymbol{\kappa}\right) \alpha - A\left(\boldsymbol{\kappa}, z\right) G_2\left(\gamma_2 \boldsymbol{\kappa}\right) \beta\right]$$
$$\times \left[G_1^*\left(\gamma_1 \boldsymbol{\kappa}\right) \exp(i\gamma_1 \boldsymbol{\kappa} \cdot \mathbf{d}) \,\alpha - A^*\left(\boldsymbol{\kappa}, z\right) G_2^*\left(\gamma_2 \boldsymbol{\kappa}\right) \exp(i\gamma_2 \boldsymbol{\kappa} \cdot \mathbf{d}) \,\beta\right], \quad (2.94)$$

where

$$\alpha = \begin{bmatrix} \cos\left[P_1\left(\gamma_1, \kappa, z\right)\right] \\ \sin\left[P_1\left(\gamma_1, \kappa, z\right)\right] \end{bmatrix}, \text{and } \beta = \begin{bmatrix} \cos\left[P_2\left(\gamma_2, \kappa, z\right)\right] \\ \sin\left[P_2\left(\gamma_2, \kappa, z\right)\right] \end{bmatrix}. \quad (2.95)$$

In a similar fashion the equivalent of (2.85)–(2.93) can be found for the correlation function.

The same method of analysis as used above can be applied to find the second moment of the product of phase and log-amplitude related quantities. For a single wave the relation is

$$\sigma^2_{\phi\chi R} = 0.1037\, k_0^2 \int_0^L dz\, C_n^2(z) \int d\boldsymbol{\kappa}\, f(\kappa) \sin\left[2\, P\left(\gamma, \kappa, z\right)\right] F\left(\gamma \boldsymbol{\kappa}\right). \quad (2.96)$$

2.4 Power Spectral Density

Deleterious effects of turbulence (such as beam jitter) are ameliorated by designing suitable beam-control systems. For example, to design a tracking system with a certain level of performance requires one to know the frequency spectrum of jitter. Taylor's frozen-turbulence assumption is used in such calculations. It assumes that atmospheric-turbulence disturbances do not change in typical time scales of interest, but are transported frozen across the aperture by wind or telescope slewing. With this assumption one can transform the spatial-spectrum variables given in (2.85) into temporal-frequency variables. In particular, let

$$\omega = \boldsymbol{\kappa} \cdot \mathbf{v}. \quad (2.97)$$

Set up the local coordinate system so that the wind velocity, whose magnitude is given by $v(z)$, is in the x direction. Make the additional change of variables

$$c^2 = \frac{\kappa_y^2 v^2(z)}{\omega^2} + 1, \tag{2.98}$$

from which follows

$$\kappa = \frac{\omega c}{v(z)}, \quad \text{and} \quad d\kappa_y = \frac{\omega}{v(z)} \frac{c dc}{\sqrt{c^2 - 1}}. \tag{2.99}$$

The last change of variables was selected to produce a unity Jacobian and an integrand for which a Mellin transform exists. Express the relation for variance given in (2.85) as functions of these two new variables. Interchange the order of integration so that the omega integration is performed last, and express the integrals over omega and c, whose limits go from $-\infty$ to $+\infty$, as each double the value of an integral from 0 to $+\infty$. After these steps, the general formula for variance can be expressed as an integral over ω. Power spectral density $S(\omega)$ is related to variance σ^2 as

$$\begin{bmatrix} \sigma_{\phi R}^2 \\ \sigma_{\chi R}^2 \end{bmatrix} = \frac{1}{\pi} \int_0^\infty d\omega \begin{bmatrix} S_\phi(\omega) \\ S_\chi(\omega) \end{bmatrix}. \tag{2.100}$$

As an example, consider the power spectral density for a single wave that is given in (2.86) for a wave propagating from $z = 0$ to $z = L$. If the filter function depends on the magnitude of κ and filtering $F_f(\omega)$ due to a servo system is also included, then the power spectral density is

$$\begin{bmatrix} S_\phi(\omega) \\ S_\chi(\omega) \end{bmatrix} = 2.606\, k_0^2\, \omega\, F_f(\omega) \int_0^L dz \frac{C_n^2(z)}{v^2(z)} \int_0^\infty dc \frac{c\, U(c-1)}{\sqrt{c^2-1}}$$
$$\times f\left[\frac{\omega c}{v(z)}\right] \begin{bmatrix} \cos^2\left[\gamma\omega^2 c^2 (L-z)/2v^2(z)\, k_0\right] \\ \sin^2\left[\gamma\omega^2 c^2 (L-z)/2v^2(z)\, k_0\right] \end{bmatrix} F\left[\frac{\gamma\omega c}{v(z)}\right]. \tag{2.101}$$

The servo filter function $F_f(\omega)$ is the modulus squared of the frequency filter function of a servo system $F_f(\omega)$. For instance, in a tracking system $F_f(\omega)$ is the frequency response of the tracking-mirror servo system. The Heaviside step function is defined in (1.19). The c integration is easily evaluated with Mellin transform techniques. Examples are considered in Sects. 4.8, 7.7, and 12.7.

2.5 Beam Shape and Strehl Ratio

Second moments are sufficient for many problems. For others the analysis must be carried further. In this section, expressions for beam shape and Strehl ratio are found in terms of phase and amplitude structure functions for an unobscured circular aperture with a uniform amplitude distribution.

We use the extended Huygens-Fresnel approximation to beam propagation described by *Fante* (1985) that applies to collimated beams. It results in the following formula for a component of electric field at a distance z from a source

$$E(\boldsymbol{\rho}) = \frac{1}{i\lambda z}\int d\boldsymbol{\rho}' E(\boldsymbol{\rho}') \exp\left[\frac{i\pi}{\lambda z}(\boldsymbol{\rho}-\boldsymbol{\rho}')^2 + \chi(\boldsymbol{\rho},\boldsymbol{\rho}',z) + i\phi(\boldsymbol{\rho},\boldsymbol{\rho}',z)\right], \tag{2.102}$$

where the dependence of both the log-amplitude $\chi(\boldsymbol{\rho},\boldsymbol{\rho}',z)$ and phase fluctuations $\phi(\boldsymbol{\rho},\boldsymbol{\rho}',z)$ on the source plane at $\boldsymbol{\rho}'$ and measurement plane at $\boldsymbol{\rho}$ (see Fig. 3.1) are explicitly shown. The axial dependence of electric field is not explicitly shown. Intensity is found by multiplying the field by its complex conjugate. In the paraxial approximation that applies to waves that are confined to a small distance about the propagation direction — a condition that holds for laser-beam propagation — second-order terms in $\boldsymbol{\rho}$ and $\boldsymbol{\rho}'$ are negligible. Let the source distribution in the aperture be uniform, i.e., $E(\boldsymbol{\rho}) = E\,W(\boldsymbol{\rho})$, where

$$W(\boldsymbol{\rho}) = U(D/2 - |\boldsymbol{\rho}|)\,. \tag{2.103}$$

With this, the intensity is given by

$$\begin{aligned} E(\boldsymbol{\rho})\,E^*(\boldsymbol{\rho}) &= \frac{E^2}{(\lambda z)^2}\int d\boldsymbol{\rho}'' \int d\boldsymbol{\rho}'\, W(\boldsymbol{\rho}'')\, W(\boldsymbol{\rho}') \\ &\quad \times \exp\left\{\frac{i2\pi}{\lambda z}\left[\boldsymbol{\rho}\cdot(\boldsymbol{\rho}''-\boldsymbol{\rho}')\right] + \alpha\right\}, \end{aligned}$$

where $\alpha = \chi(\boldsymbol{\rho},\boldsymbol{\rho}'',z) + \chi(\boldsymbol{\rho},\boldsymbol{\rho}',z) + i\left[\phi(\boldsymbol{\rho},\boldsymbol{\rho}'',z) - \phi(\boldsymbol{\rho},\boldsymbol{\rho}',z)\right]$. I am interested in the ensemble average of this intensity. The turbulence fluctuations are Gaussian in character, and the log-amplitude and phase fluctuations are uncorrelated. For this distribution I use a general result for Gaussian statistics

$$\langle \exp(\beta A)\rangle = \exp\left[\frac{\beta^2}{2}\left\langle (A - \langle A\rangle)^2\right\rangle + \beta\langle A\rangle\right]. \tag{2.104}$$

Angle brackets denote ensemble average. Turbulence does not change the average energy in the beam; it just redistributes it. This leads to the requirement that the average log-intensity is reduced by the log-amplitude variance

$$\langle A\rangle = -\left\langle A^2\right\rangle. \tag{2.105}$$

If one assumes that phase and amplitude variations are uncorrelated, then use of (2.104) allows the average intensity to be expressed as

$$\begin{aligned} \langle E(\boldsymbol{\rho})\,E^*(\boldsymbol{\rho})\rangle &= \frac{E^2}{(\lambda z)^2}\int d\boldsymbol{\rho}'' \int d\boldsymbol{\rho}'\, W(\boldsymbol{\rho}'')\, W(\boldsymbol{\rho}') \\ &\quad \times \exp\left\{\frac{i2\pi}{\lambda z}\left[\mathbf{r}\cdot(\boldsymbol{\rho}''-\boldsymbol{\rho}')\right] - \mathcal{D}(\boldsymbol{\rho}''-\boldsymbol{\rho}',\boldsymbol{\rho}')\,/2\right\}, \end{aligned}$$

where the structure function is defined in (2.92).

To simplify this expression, change integration variables to normalized sum and difference coordinates as $\boldsymbol{\alpha}' = (\boldsymbol{\rho}''+\boldsymbol{\rho}')/2D$ and $\boldsymbol{\alpha} = (\boldsymbol{\rho}''-\boldsymbol{\rho}')/D$. For cases in which variance is independent of position, the structure function can be represented by a relation that only depends on the difference in position of two points, and can be written as

$$\begin{bmatrix} \mathcal{D}_\phi(\boldsymbol{\alpha}\mathcal{D},\boldsymbol{\rho}') \\ \mathcal{D}_\chi(\boldsymbol{\alpha}\mathcal{D},\boldsymbol{\rho}') \end{bmatrix} = \begin{bmatrix} D_\phi(\boldsymbol{\alpha}D) \\ D_\chi(\boldsymbol{\alpha}D) \end{bmatrix} = 2\begin{bmatrix} \langle\varphi^2(0)\rangle - \langle\varphi(\boldsymbol{\alpha}D)\,\varphi(0)\rangle \\ \langle\chi^2(0)\rangle - \langle\chi(\boldsymbol{\alpha}D)\,\chi(0)\rangle \end{bmatrix}. \tag{2.106}$$

The assumption that the structure function is a function only of the difference of positions is not true for some cases of interest. For instance, if waveform tilt on an aperture is subtracted from the turbulence-induced phase, it can be shown that average phase variance is higher in the aperture's center than at its edge. A second example occurs for the anisoplanatic effect produced when a point source that is placed over the aperture center is used as a beacon in an adaptive-optics system that is transmitting a collimated beam. In this case, the variance of the difference of a point source's and a collimated source's phase is higher at the aperture's edge than at the center.

For a uniform field distribution over a circular aperture, the integral over the sum coordinate can be performed analytically in the following manner. Express the original coordinates in terms of new ones as $\boldsymbol{\rho}'' = \boldsymbol{\alpha}'D + \boldsymbol{\alpha}D/2$, and $\boldsymbol{\rho}' = \boldsymbol{\alpha}'D - \boldsymbol{\alpha}D/2$. Expected value of the intensity divided by the intensity with no turbulence can be written as

$$I(\boldsymbol{\rho}) = \int d\boldsymbol{\alpha}\,\exp\left[\frac{i2\pi}{\lambda z}(\boldsymbol{\rho}\cdot\boldsymbol{\alpha}D) - \mathcal{D}(\boldsymbol{\alpha})/2\right] \times \int d\boldsymbol{\alpha}'\,W(\boldsymbol{\alpha}'D + \boldsymbol{\alpha}D/2)\,W(\boldsymbol{\alpha}'D - \boldsymbol{\alpha}D/2). \tag{2.107}$$

The integral over the two overlapping circles can easily be performed as shown by *Fried* (1967). The result is

$$\int d\boldsymbol{\alpha}'\,W(\boldsymbol{\alpha}'D + \boldsymbol{\alpha}D/2)\,W(\boldsymbol{\alpha}'D - \boldsymbol{\alpha}D/2) = K(\alpha) = \frac{16}{\pi}\left[\cos^{-1}(\alpha) - \alpha\left(1-\alpha^2\right)^{1/2}\right]U(1-\alpha), \tag{2.108}$$

where $U(x)$ is the Heaviside unit step function defined in (1.19). With this result the normalized intensity is given by

$$I(\boldsymbol{\rho}) = \frac{1}{2\pi}\int d\boldsymbol{\alpha}\,K(\alpha)\exp\left[ik_0 D\,\boldsymbol{\rho}\cdot\boldsymbol{\alpha}/z - \mathcal{D}(\boldsymbol{\alpha}D)/2\right]. \tag{2.109}$$

If the structure function is isotropic, the angle integration in the aperture can be performed to give

$$I(\rho) = \int_0^1 d\alpha\,\alpha\,K(\alpha)\,J_0\!\left(\frac{k_0\,\rho D\alpha}{z}\right)\exp\left[-\frac{\mathcal{D}(\alpha D)}{2}\right]. \tag{2.110}$$

The Strehl ratio, the most common description of the performance of an optical system, is given by the value of normalized intensity evaluated at the origin

$$SR = \frac{1}{2\pi}\int d\boldsymbol{\alpha}\,K(\alpha)\exp\left[-\frac{\mathcal{D}(\boldsymbol{\alpha}D)}{2}\right]. \tag{2.111}$$

When the structure function is isotropic, the integration over angle in the aperture can be performed to yield

$$SR = \int_0^1 d\alpha\, \alpha\, K(\alpha) \exp\left[-\frac{\mathcal{D}(\alpha D)}{2}\right]. \tag{2.112}$$

This can be written as

$$SR = \exp\left(-\sigma^2\right) \int_0^1 d\alpha\, \alpha\, K(\alpha) \exp\left[\langle \varphi\left(\boldsymbol{\alpha} D\right) \varphi(0)\rangle + \langle \chi\left(\boldsymbol{\alpha} D\right) \chi(0)\rangle\right], \tag{2.113}$$

where the total variance is

$$\sigma^2 = \left\langle \varphi^2(z) \right\rangle + \left\langle \chi^2(z) \right\rangle. \tag{2.114}$$

If phase and log-amplitude have a short correlation distance, then the exponential term in the integrand can be replaced by unity to give

$$SR = \exp\left(-\sigma^2\right). \tag{2.115}$$

If amplitude fluctuations are negligible, as they are for many situations, this becomes the extended Maréchal approximation

$$SR \approx \exp\left(-\sigma_\varphi^2\right), \tag{2.116}$$

where σ_φ^2 is the phase variance. If phase variance is small compared to unity and is not strongly correlated at different points, one obtains the Maréchal approximation

$$SR \approx 1 - \sigma_\varphi^2. \tag{2.117}$$

For many problems the correlation distance is not small, and the above approximation is accurate only for variances that are less than 0.1.

3. Aperture Filter Functions

In the last chapter expressions for phase and log-amplitude variances were derived. These expressions contained an aperture filter function. In this chapter explicit filter functions are derived for many cases of interest. First, those needed to calculate variances of any Zernike mode on an unobscured circular aperture are found. Next, expressions to find piston and tilt on an annular aperture are derived.

Filter functions required to determine the effect of either a finite-size source or receiver on the scintillation are derived. The filter function to be used when both the receiver and source are finite in size is also given.

Often in adaptive-optics systems the phase produced by a focused beam of finite size is used to correct an outgoing collimated beam. The source of the focused beam may not be coaxial with the outgoing beam. Filter functions that apply for each of these cases are derived.

All filter functions are given assuming the wave is collimated and infinite in extent. These filter functions can be applied to focused or finite beams by changing κ, to $\gamma\kappa$, where γ is the propagation parameter defined in (2.67) for a source at $z = 0$, and in (2.68) for a source at $z = L$.

New filter functions can be created from existing complex filter functions by taking the absolute value squared of their difference. This technique is used to find the filter function that applies to the performance of an adaptive-optics system. This technique is used mainly in the next chapter.

Filter functions derived in this chapter are for circular apertures, which are the most commonly encountered. Filter functions for other shapes can be derived with similar methods.

To evaluate various integrals, I shall need some Bessel function relations. The recurrence relations for Bessel functions are

$$\frac{d\left[r^p J_p(ar)\right]}{dr} = ar^p\, J_{p-1}(ar), \text{ and} \tag{3.1}$$

$$\frac{d}{dr}\left[\frac{J_p(ar)}{r^p}\right] = -\frac{a}{r^p}\, J_{p+1}(ar). \tag{3.2}$$

Two important special cases are

$$\frac{d\left[r\, J_1(ar)\right]}{dr} = ar\, J_0(ar), \text{ and} \tag{3.3}$$

$$\frac{d}{dr}\left[J_0(ar)\right] = -a\, J_1(ar). \tag{3.4}$$

I also use the following integral

$$J_n(r) = \frac{1}{2\pi}\int_0^{2\pi} d\varphi\, \cos(n\varphi)\exp\left[i\, r\cos(\varphi)\right] = \frac{1}{2\pi}\int_0^{2\pi} d\varphi \cos\left[r\cos(\varphi) + n\varphi\right]. \tag{3.5}$$

3.1 Circular Aperture Modes

One often is interested in obtaining weighted averages of the phase or log-amplitude over a receive aperture. The weights typically are polynomials, normally referred to as "modes." Commonly, the receive aperture is an unobscured circle or an annulus. In this section unobscured apertures are considered. Filter functions for an annulus are found in a similar fashion in the next section.

Consider the Zernike modes that were defined in a very useful way by *Noll* (1976). His expressions are for an aperture of unit diameter, and they are modified here to apply to an aperture of diameter D. The definition of the x and y components of the Zernike polynomials are

$$Z(m,n)_x = \sqrt{n+1}\, R_n^m(2\rho/D)\,\sqrt{2}\cos(m\theta), \text{ and} \tag{3.6}$$

$$Z(m,n)_y = \sqrt{n+1}\, R_n^m(2\rho/D)\,\sqrt{2}\sin(m\theta), \tag{3.7}$$

for $m \neq 0$. For $m = 0$ one uses

$$Z(0,n) = \sqrt{n+1}\, R_n^0(2\rho/D). \tag{3.8}$$

The radial function is given by

$$R_n^m(2\rho/D) = \sum_{q=0}^{(n-m)/2} \frac{(-1)^q}{q!} \frac{(n-q)!\,(2\rho/D)^{n-2q}}{[(n+m)/2-q]!\,[(n-m)/2-q]!}. \tag{3.9}$$

There are requirements that $m \leq n$, and that $n - m$ be even.

These polynomials are of interest because they correspond to the common optical aberrations. $Z(0,0)$ is piston, $Z(1,1)_x$ and $Z(1,1)_y$ are the two components of tilt, $Z(0,2)$ is focus, and $Z(2,2)_x$ and $Z(2,2)_y$ are the two components of astigmatism, $Z(1,3)$ is coma, and $Z(0,4)$ is third-order spherical distortion.

The Zernike polynomials are orthogonal over the aperture, i.e.,

$$\frac{4}{\pi D^2}\int d\boldsymbol{\rho}\, Z(m,n) Z(m',n')\, W(\boldsymbol{\rho}) = \delta_{mm'}\,\delta_{nn'}, \tag{3.10}$$

where the integral is over all space. The aperture function $W(\boldsymbol{\rho})$, defined in (2.103), is equal to unity inside the aperture and zero outside. Expressed in terms of the unit step, it is

$$W(\boldsymbol{\rho}) = U(D/2 - |\boldsymbol{\rho}|) . \tag{3.11}$$

Noll shows that the Fourier transforms of Zernike polynomials defined by

$$G(\boldsymbol{\kappa}) = \frac{4}{\pi D^2} \int d\boldsymbol{\rho} \, W(\boldsymbol{\rho}) \, Z(m,n) \exp(i\boldsymbol{\kappa} \cdot \boldsymbol{\rho}) , \tag{3.12}$$

are equal to

$$\left.\begin{array}{l} G_{m,n}(\boldsymbol{\kappa})_x \\ G_{m,n}(\boldsymbol{\kappa})_y \\ G_{0,n}(\boldsymbol{\kappa}) \end{array}\right\} = \sqrt{n+1}\frac{2J_{n+1}(\kappa D/2)}{\kappa D/2} \left\{\begin{array}{l} (-1)^{(n-m)/2} i^n \sqrt{2}\cos(m\varphi) , \\ (-1)^{(n-m)/2} i^n \sqrt{2}\sin(m\varphi) , \\ (-1)^{n/2} (m=0). \end{array}\right. \tag{3.13}$$

These are the *complex filter functions* for extracting Zernike modes. The absolute value squared of these complex filter functions, which are referred to simply as *filter functions*, are

$$\left.\begin{array}{l} F_{m,n}(\boldsymbol{\kappa})_x \\ F_{m,n}(\boldsymbol{\kappa})_y \\ F_{0,n}(\boldsymbol{\kappa}) \end{array}\right\} = (n+1)\left[\frac{2J_{n+1}(\kappa D/2)}{\kappa D/2}\right]^2 \left\{\begin{array}{l} 2\cos^2(m\varphi) , \\ 2\sin^2(m\varphi) , \\ 1\,(m=0). \end{array}\right. \tag{3.14}$$

Piston and tilt filter functions are of particular interest. Two representations of these filter functions are given. For different problems, one or the other representation is appropriate. For piston, the first representation gives the filter function that extracts phase variance due to piston. In the second representation, the filter function extracts variance of the stroke of a flat mirror that produces this piston by reflection. For tilt, the first filter function extracts phase variance due to tilt, and the second extracts the variance of the corresponding angular displacement in real space. The condition $n = m = 0$ gives *the piston phase-variance filter function*

$$F(\boldsymbol{\kappa}) = \left[\frac{2J_1(\kappa D/2)}{\kappa D/2}\right]^2 . \tag{3.15}$$

To obtain the representation that extracts the mean-square piston stroke, the filter function has to be divided by the wavenumber squared. *The filter function to find piston stroke variance is thus*

$$F(\boldsymbol{\kappa}) = \left(\frac{1}{k_0}\right)^2 \left[\frac{2J_1(\kappa D/2)}{\kappa D/2}\right]^2 . \tag{3.16}$$

This filter function can be used to calculate effects such as aperture averaging of scintillation, which affects the twinkling of stars and planets, because piston is basically a simple aperture average.

The terms $n = m = 1$ gives *the filter function to determine tilt phase variance*

$$\left.\begin{array}{l} F_x(\boldsymbol{\kappa}) \\ F_y(\boldsymbol{\kappa}) \\ F(\boldsymbol{\kappa}) \end{array}\right\} = \left[\frac{4J_2(\kappa D/2)}{\kappa D/2}\right]^2 \left\{\begin{array}{l} \cos^2(\varphi) , \\ \sin^2(\varphi) , \\ 1. \end{array}\right. \tag{3.17}$$

The three terms are the x, y, and total tilt phase variance.

The filter function for variance of tilt angle in real space is a factor of $(4/k_0\,D)^2$ times the phase-variance filter function. *The filter function to determine tilt angle variance is*

$$\left.\begin{array}{l} F_x(\boldsymbol{\kappa}) \\ F_y(\boldsymbol{\kappa}) \\ F(\boldsymbol{\kappa}) \end{array}\right\} = \left(\frac{16}{k_0 D}\right)^2 \left[\frac{J_2(\kappa D/2)}{\kappa D/2}\right]^2 \left\{\begin{array}{l} \cos^2(\varphi)\ , \\ \sin^2(\varphi)\ , \\ 1\ . \end{array}\right. \tag{3.18}$$

In some problems one needs phase variance with some Zernike modes removed. This is easily done by subtracting the phase variance of these components from the total variance. For instance, the filter function to find the variance with piston and tilt removed is

$$F(\boldsymbol{\kappa}) = 1 - \left[\frac{2J_1(\kappa D/2)}{\kappa D/2}\right]^2 - \left[\frac{4J_2(\kappa D/2)}{\kappa D/2}\right]^2 . \tag{3.19}$$

Implicit in the expression above is the requirement that cross product terms of piston and tilt average in some way to zero. This follows from the orthogonality of the Zernike polynomials over the aperture. This is not true in calculating the structure function with some Zernike modes removed because the modes are not integrated over the aperture. In that case, in order to get an orthogonal decomposition one must use Karhunen-Loève polynomials which are *statistically* independent. The lowest order Karhunen-Loève polynomials are close in shape to the Zernike polynomials. *Wang* and *Markey* (1978) show that assuming Zernike modes are statistically independent results in errors of as much as 10%.

The Zernike component of tilt is also referred to as Z-tilt. Some tilt sensors respond to this quantity, while others respond to the centroid position of the focused aperture distribution, a quantity that is closer to gradient tilt. The definition of G-tilt is

$$\text{G-tilt} = \frac{4}{\pi\, k_0^2 D^2} \int d\boldsymbol{\rho}\, \nabla_t \Phi(\boldsymbol{\rho})\, , \tag{3.20}$$

where the integration is over the circular aperture. G-tilt, in a geometric sense, is equal to the average ray direction, corresponding to the average phase gradient over the aperture. One can calculate this in transform space by multiplying the piston component of phase by $i\kappa/k_0$. *The filter function to find gradient-tilt angle variance is thus*

$$\left.\begin{array}{l} F_x(\boldsymbol{\kappa}) \\ F_y(\boldsymbol{\kappa}) \\ F(\boldsymbol{\kappa}) \end{array}\right\} = \left(\frac{4}{k_0 D}\right)^2 J_1^2(\kappa D/2) \left\{\begin{array}{l} \cos^2(\varphi)\ , \\ \sin^2(\varphi)\ , \\ 1\ . \end{array}\right. \tag{3.21}$$

The filter function to calculate phase variance of G-tilt is obtained by using the same arguments as before. *The filter function to find the phase gradient due to gradient tilt is thus*

$$\left.\begin{array}{l} F_x(\boldsymbol{\kappa}) \\ F_y(\boldsymbol{\kappa}) \\ F(\boldsymbol{\kappa}) \end{array}\right\} = J_1^2(\kappa D/2) \left\{\begin{array}{l} \cos^2(\varphi)\ , \\ \sin^2(\varphi)\ . \\ 1\ . \end{array}\right. \tag{3.22}$$

3.2 Piston and Tilt on an Annulus

As with a filled-in circular aperture, the complex filter function for Zernike tilt on an annulus is the Fourier transform of the aperture function for tilt. For tilt in the x-direction, the filter function is

$$G_x(\kappa) = a \int d\boldsymbol{\rho}\, x \exp\left(i\boldsymbol{\kappa}\cdot\boldsymbol{\rho}\right), \tag{3.23}$$

where a is a normalizing factor, and the integration is over an annulus with inner diameter D_i and outer diameter D. The angular and radial integrations can be performed with (3.5) and (3.1) to give

$$G_x\left(\kappa\right) = \frac{\pi a \cos\left(\phi\right)}{4}\left[\frac{D^3 J_2\left(\kappa D/2\right)}{\kappa D/2} - \frac{D_i^3 J_2\left(\kappa D_i/2\right)}{\kappa D_i/2}\right], \tag{3.24}$$

where ϕ is the angle in κ space. To get the correct constant assume that a pure x-tilt is applied, i.e., let the phase be

$$\phi\left(\boldsymbol{\rho}\right) = b\, x.$$

The tilt angle in real space is b/k_0. Therefore,

$$b/k_0 = a\, b \int d\boldsymbol{\rho}\, x^2 = \frac{\pi a\, b\left(D^4 - D_i^4\right)}{64}.$$

This gives $a = 64/\pi\, k_0\left(D^4 - D_i^4\right)$.

The filter function is the absolute magnitude squared of the complex filter function. *The filter function to find Zernike-tilt angle variance on an annulus is given by*

$$F_x\left(\kappa\right) = \left[\frac{16/k_0 D}{1-\left(D_i/D\right)^4}\right]^2 \left[\frac{J_2\left(\kappa D/2\right)}{\kappa D/2} - \left(\frac{D_i}{D}\right)^3 \frac{J_2\left(\kappa D_i/2\right)}{\kappa D_i/2}\right]^2 \cos^2\left(\phi\right). \tag{3.25}$$

The complex filter function for the gradient tilt on an annular aperture can be found in exactly the same way to obtain

$$G_x\left(\kappa\right) = \frac{1}{\left[1-\left(D_i/D\right)^2\right]}\left[\frac{2J_1\left(\kappa D/2\right)}{\kappa D/2} - \frac{2J_1\left(\kappa D_i/2\right)}{\kappa D_i/2}\right]\frac{i\kappa}{k_0}\cos\left(\phi\right). \tag{3.26}$$

The filter function of G-tilt angle in the x-direction is

$$F_x\left(\kappa\right) = \left\{\frac{4}{k_0 D\left[1-\left(D_i/D\right)^2\right]}\right\}^2 \left[J_1\left(\kappa D/2\right) - \frac{D_i}{D} J_1\left(\kappa D_i/2\right)\right]^2 \cos^2\left(\phi\right). \tag{3.27}$$

The y component of tilt has the cosine replaced by a sine. The total tilt has the cosine replaced by unity.

The filter function for tilt variance is obtained from the filter function for tilt. Given the tilt T on an annulus, the phase variance (now referring to averaging over the aperture) due to this tilt, which can be arbitrarily taken to be in the x direction, is

$$\sigma_T^2 = \frac{1}{A}\int d\boldsymbol{\rho}\,(k_0 \mathbf{T}\cdot\boldsymbol{\rho})^2 = \frac{k_0^2 T^2}{A}\int_{D_i/2}^{D/2} d\rho\,\rho^3 \int_0^{2\pi} d\theta\,\cos^2(\theta) = \frac{T^2 k_0^2}{16}\left(D^2 + D_i^2\right). \tag{3.28}$$

Therefore, *the filter function for total tilt phase variance on an annulus is*

$$F(\kappa) = \left[\frac{16}{1+(D_i/D)^2}\right]\left[\frac{J_2(\kappa D/2)}{\kappa D/2} - \left(\frac{D_i}{D}\right)^3 \frac{J_2(\kappa D_i/2)}{\kappa D_i/2}\right]^2. \tag{3.29}$$

As before, one obtains the $x\,(y)$ component by multiplying this expression by $\cos^2(\phi)\,\left(\sin^2(\phi)\right)$. The piston filter functions can be found in the same manner. The Zernike piston on an annulus is the Fourier transform of the aperture function, i.e.,

$$G_s(\kappa) = a\int d\boldsymbol{\rho}\,\exp\left(i\boldsymbol{\kappa}\cdot\boldsymbol{\rho}\right),$$

where a is a normalizing factor, and the integration is over the annulus. The angular and radial integrations can be performed with (3.5) and (3.3) to give

$$G(\kappa) = \pi a\left[D\,J_1(\kappa D/2) - D_i\,J_1(\kappa D_i/2)\right]. \tag{3.30}$$

To get the normalizing constant a assume that a pure piston is applied, i.e., let the phase be $\phi(\boldsymbol{\rho}) = b$. The piston in mirror stroke is b/k_0, so

$$b/k_0 = a\,b\int d\boldsymbol{\rho} = \pi a\,b\left(D^2 - D_i^2\right)/4.$$

Therefore, $a = 4/\pi k_0\,(D^2 - D_i^2)$. The complex filter function is

$$G(\kappa) = \frac{2}{k_0\left[1-(D_i/D)^2\right]}\left[\frac{J_1(\kappa D/2)}{\kappa D/2} - \left(\frac{D_i}{D}\right)^2 \frac{J_1(\kappa D_i/2)}{\kappa D_i/2}\right]. \tag{3.31}$$

The filter function is the absolute magnitude squared of the complex filter function. *The filter function for piston stroke variance on an annulus is given by*

$$F(\kappa) = \left\{\frac{2}{k_0\left[1-(D_i/D)^2\right]}\right\}^2 \left[\frac{J_1(\kappa D/2)}{\kappa D/2} - \left(\frac{D_i}{D}\right)^2 \frac{J_1(\kappa D_i/2)}{\kappa D_i/2}\right]^2. \tag{3.32}$$

Given the piston P on the aperture, the average phase variance due to piston is

$$\sigma_P^2 = \frac{1}{A}\int d\boldsymbol{\rho}\,(k_0\,P)^2 = k_0^2\,P^2,$$

where A is the aperture area. Therefore, one finds the filter function to obtain the piston phase variance by multiplying the filter function for the piston by k_0^2. *The filter function for piston phase variance on an annulus is given by*

$$F(\kappa) = \left[\frac{2}{1-(D_i/D)^2}\right]^2 \left[\frac{J_1(\kappa D/2)}{\kappa D/2} - \left(\frac{D_i}{D}\right)^2 \frac{J_1(\kappa D_i/2)}{\kappa D_i/2}\right]^2. \tag{3.33}$$

3.3 Finite Apertures and Focal Anisoplanatism

I derive filter functions for two types of problems that are analyzable in a similar fashion. The analysis is a generalization of the two-wave problem considered in Chap. 2. Here the source is allowed to be finite in size.

The first type of filter function is for either a distributed source or a finite-size receive aperture that produces phase and scintillation levels in a receiver that are very different from those of a point source or point receiver. For instance, the finite angular size of a planet causes it to have much lower scintillation than a star.

The second type of filter function applies to adaptive-optics systems. Often in the operation of an adaptive-optics system, the beacon source or reflector is either a point or a distributed source, and the corrected signal is a collimated beam. The important question that arises in this case is: How much does path difference between a beacon and corrected beam affect phase variance on the corrected beam? Variances due to this effect vary with position within the outgoing aperture. For that reason a structure function that varies solely with the difference between aperture positions does not exist. Nevertheless, one can still calculate average variances over the outgoing aperture to derive a filter function that gives the average phase variances over the aperture.

First, I consider a distributed source and finite receive aperture as shown in Fig. 3.1. This geometry applies to the physically interesting cases of reflection from a corner-cube array or a diffuse plate, or return from a planet. Let us direct our attention to problems in which the turbulence-induced effects on rays going from various source to receiver points is the most important effect, rather than problems in which the source phase is important. This assumption applies to aperture averaging of scintillation or to estimating the average wavefront slope from a distributed source.

For this case the distance between a ray parallel to the z-axis (direction of propagation) that passes through $\boldsymbol{\rho}$ on the receive aperture and a ray from aperture point $\boldsymbol{\rho}$ to the source point $\boldsymbol{\rho}'$ is

$$\mathbf{d} = \frac{z}{L}\left(\boldsymbol{\rho} - \boldsymbol{\rho}'\right). \tag{3.34}$$

This anisoplanatic shift from one point in the source plane to one point in the receive plane can be inserted into expressions for phase and log-amplitude in (2.78)–(2.79) to obtain

$$\begin{bmatrix} \chi R(L) \\ \phi R(L) \end{bmatrix} = k_0 \int_0^L dz \int d\nu(\boldsymbol{\kappa}, z) \begin{bmatrix} \sin\left[P\left(\gamma, \kappa, z\right)\right] \\ \cos\left[P\left(\gamma, \kappa, z\right)\right] \end{bmatrix} \exp(i\boldsymbol{\kappa}\cdot\boldsymbol{\rho}) \exp\!\left[i\boldsymbol{\kappa}\cdot\left(\boldsymbol{\rho} - \boldsymbol{\rho}'\right)\frac{z}{L}\right]. \tag{3.35}$$

Notice that because of the assumptions above, the initial phase is not present.

Let us consider a finite-size source where intensity $S\left(\boldsymbol{\rho}'\right)$ can vary with position, and a finite-size receiver for a problem in which the signals from each point are amplitude weighted and coherently added. The resultant expressions

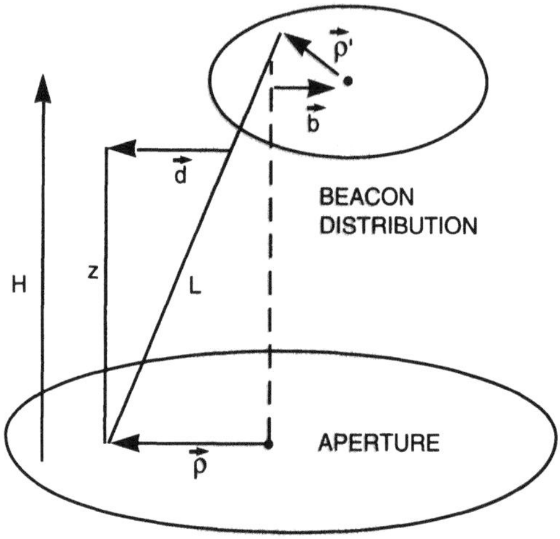

Fig. 3.1. Geometry of a distributed source and finite size aperture.

one obtains by integrating over the source and aperture and normalizing to the total source intensity are

$$\begin{bmatrix} \chi R(L) \\ \phi R(L) \end{bmatrix} = k_0 \int_0^L dz \int d\nu\,(\boldsymbol{\kappa}, z) \begin{bmatrix} \sin\left[P\left(\gamma, \kappa, z\right)\right] \\ \cos\left[P\left(\gamma, \kappa, z\right)\right] \end{bmatrix}$$

$$\times \frac{\int d\boldsymbol{\rho} \int d\boldsymbol{\rho}'\, S\left(\boldsymbol{\rho}'\right) \exp\left[i\boldsymbol{\kappa} \cdot \boldsymbol{\rho}\left(1 - \frac{z}{L}\right)\right] \exp\left(i\boldsymbol{\kappa} \cdot \boldsymbol{\rho}' \frac{z}{L}\right)}{A_A \int d\boldsymbol{\rho}'\, S\left(\boldsymbol{\rho}'\right)}, \tag{3.36}$$

where A_A is the aperture area. When the integration over the aperture is performed, one obtains by factorization *the filter function for a finite size receiver and a finite size source with nonuniform intensity*

$$F\left(\boldsymbol{\kappa}\right) = \left\{2 \frac{J_1\left[\kappa D\left(1 - z/L\right)/2\right]}{\kappa D\left(1 - z/L\right)/2}\right\}^2 \frac{\left|\int d\boldsymbol{\rho}'\, S\left(\boldsymbol{\rho}'\right) \exp\left(-i\boldsymbol{\kappa} \cdot \boldsymbol{\rho}' \frac{z}{L}\right)\right|^2}{\left|\int d\boldsymbol{\rho}'\, S\left(\boldsymbol{\rho}'\right)\right|^2}, \tag{3.37}$$

where D is the aperture diameter. If source intensity is constant, then the integrations in (3.37) can be performed yielding *the filter function for a finite size receiver and uniform circular source as*

$$F\left(\boldsymbol{\kappa}\right) = \left\{2 \frac{J_1\left[\kappa D\left(1 - z/L\right)/2\right]}{\kappa D\left(1 - z/L\right)/2}\right\}^2 \left[2 \frac{J_1\left(\kappa D_s z/2L\right)}{\kappa D_s z/2L}\right]^2, \tag{3.38}$$

where D_s is the source diameter. From this relation special cases can be found. *The filter function for a uniform, circular source located at a distance L, and a point receiver is*

$$F\left(\boldsymbol{\kappa}\right) = \left[2 \frac{J_1\left(\kappa D_s z/2L\right)}{\kappa D_s z/2L}\right]^2. \tag{3.39}$$

The filter function for a finite size receive aperture with a point source located at a distance L is

$$F(\boldsymbol{\kappa}) = \left[2\frac{J_1(\kappa D(1 - z/L)/2)}{\kappa D(1 - z/L)/2}\right]^2 . \tag{3.40}$$

If the point source is located at infinity, then the filter function for a finite size receiver is

$$F(\boldsymbol{\kappa}) = \left[2\frac{J_1(\kappa D/2)}{\kappa D/2}\right]^2 . \tag{3.41}$$

These filter functions are used in Sect. 7.8.

In an adaptive-optics system the beacon source is often small, and the transmitted beam is collimated. Rays from the small source do not sample exactly the same turbulence as that of the collimated beam; this results in a distortion on the collimated beam after it traverses the atmosphere. This effect is referred to as *focal anisoplanatism.* The phase difference arises from an offset of collimated and focused rays in the aperture. Consider the path difference between a ray parallel to the propagation direction that passes through the receiver at $\boldsymbol{\rho}$ and a ray that also passes through $\boldsymbol{\rho}$ that originates from a point in the distributed source whose center is offset from the receive aperture center by $\mathbf{b}$. The isoplanatic difference is

$$\exp(i\boldsymbol{\kappa}\cdot\boldsymbol{\rho})\exp\{i\boldsymbol{\kappa}\cdot[\boldsymbol{\rho}(1 - z/L) + (\boldsymbol{\rho}' + \mathbf{b})z/L]\} . \tag{3.42}$$

To find the phase at any point in the receive aperture, the phase expression must be integrated over the normalized source distribution. Thus, the anisoplanatic difference phase is

$$\begin{aligned}\phi R(L) &= k_0 \int_0^L dz \int d\nu(\boldsymbol{\kappa}, z)\cos[P(\gamma, \kappa, z)] \\ &\times \left[\exp(i\boldsymbol{\kappa}\cdot\boldsymbol{\rho}) - \frac{\int d\boldsymbol{\rho}' S(\boldsymbol{\rho}')\exp\{i\boldsymbol{\kappa}\cdot[\boldsymbol{\rho}(1 - z/L) + (\boldsymbol{\rho}' + \mathbf{b})z/L]\}}{\int d\boldsymbol{\rho}' S(\boldsymbol{\rho}')}\right] .\end{aligned} \tag{3.43}$$

The filter function is the absolute value squared of the term after the cosine

$$F(\boldsymbol{\kappa}) = \left|1 - \frac{\int d\boldsymbol{\rho}' S(\boldsymbol{\rho}')\exp\{i\boldsymbol{\kappa}\cdot[-\boldsymbol{\rho}z/L + (\boldsymbol{\rho}' + \mathbf{b})z/L]\}}{\int d\boldsymbol{\rho}' S(\boldsymbol{\rho}')}\right|^2 . \tag{3.44}$$

If the beacon is a uniform circular source, then the integrations can be carried out early to give

$$F(\boldsymbol{\kappa}) = 1 - 2\frac{2J_1(D_s h)}{D_s h}\cos[\boldsymbol{\kappa}\cdot z(\mathbf{b} - \boldsymbol{\rho})/L] + \left[2\frac{J_1(D_s h)}{D_s h}\right]^2 , \tag{3.45}$$

where $h = \kappa z/2L$.

The phase variance is not the same everywhere in the receive aperture. To find the average phase variance the filter function is averaged over the aperture with (3.5) to give

$$F(\boldsymbol{\kappa}) = 1 - 2\,\frac{2\,J_1(D_s\,h)}{D_s\,h}\frac{2\,J_1(D\,h)}{D\,h}\cos\left(\boldsymbol{\kappa}\cdot\frac{z\,\mathbf{b}}{L}\right) + \left[2\frac{J_1(D_s\,h)}{D_s\,h}\right]^2. \tag{3.46}$$

If there are no other filter functions in the problem that depend on the angle in κ-space, then the integral over angle can be performed; this leads to *the filter function for focal anisoplanatism with a uniform circular source offset* **b** *from boresight*

$$F(\boldsymbol{\kappa}) = 1 - 2\,\frac{2\,J_1(D_s\,h)}{D_s\,h}\frac{2\,J_1(D\,h)}{D\,h}J_0\left(\frac{\kappa b z}{L}\right) + \left[2\frac{J_1(D_s\,h)}{D_s\,h}\right]^2. \tag{3.47}$$

The filter function for focal anisoplanatism with a uniform circular source on boresight is

$$F(\boldsymbol{\kappa}) = 1 - 2\,\frac{2\,J_1(D_s\,h)}{D_s\,h}\frac{2\,J_1(D\,h)}{D\,h} + \left[2\frac{J_1(D_s\,h)}{D_s\,h}\right]^2. \tag{3.48}$$

If the source diameter is zero, then one obtains *the filter function for focal anisoplanatism with circular symmetry for a point source with no offset as*

$$F(\boldsymbol{\kappa}) = 2\left[1 - 2\,\frac{J_1(D\,h)}{D\,h}\right]. \tag{3.49}$$

This filter function for average phase variance is not derivable from (2.85) because here I want the average of the phase variance on the aperture, and previously I found the variance of a phase quantity on the aperture.

To find the piston or tilt component of focal anisoplanatism when the beacon has finite size, one calculates these components for a single source point and integrates over the source distribution. Thus the isoplanatic term in (3.42) is multiplied by a piston or tilt aperture function and integrated over the source aperture to obtain

$$\phi R(L) = k_0\int_0^L dz'\int d\nu(\boldsymbol{\kappa}, z)\cos\left[P(\gamma,\kappa,z)\right]$$
$$\times\int d\boldsymbol{\rho}'\int d\boldsymbol{\rho}\,g(\boldsymbol{\rho})\left[\exp(i\boldsymbol{\kappa}\cdot\boldsymbol{\rho}) - \frac{S(\boldsymbol{\rho}')\exp\left\{i\boldsymbol{\kappa}\cdot\left[\boldsymbol{\rho}\left(1-\frac{z}{L}\right)+(\boldsymbol{\rho}'+\mathbf{b})\frac{z}{L}\right]\right\}}{\int d\boldsymbol{\rho}'\,S(\boldsymbol{\rho}')}\right].$$

The aperture integration is evaluated to obtain an expression for a single source point at $\boldsymbol{\rho}'$ as

$$F(\boldsymbol{\kappa}) = 4\left|\frac{\nu\,J_\nu(a)}{a} - \exp\left(i\boldsymbol{\kappa}\cdot\mathbf{b}\frac{z}{L}\right)\frac{\nu\,J_\nu(e)}{e}\frac{\int d\boldsymbol{\rho}'\,S(\boldsymbol{\rho}')\exp(i\boldsymbol{\kappa}\cdot\boldsymbol{\rho}'\,z/L)}{\int d\boldsymbol{\rho}'\,S(\boldsymbol{\rho}')}\right|^2, \tag{3.50}$$

where $a = \kappa D/2$, $e = \kappa D(1 - z/L)/2$, $\nu = 1$ for piston, and $\nu = 1$ for tilt. If the source distribution is uniform, then integrations over $\boldsymbol{\rho}'$ are readily evaluated to obtain the total piston or tilt. *Thus, the filter function to find the piston or tilt component of focal anisoplanatism with a uniform circular source offset* **b** *from boresight is*

$$F(\boldsymbol{\kappa}) = \left| \frac{2\nu\, J_\nu(a)}{a} - \exp\left(i\boldsymbol{\kappa}\cdot\mathbf{b}\frac{z}{L}\right) \frac{2\nu\, J_\nu(e)}{e} \frac{2\, J_1(D_s\, h)}{(D_s\, h)} \right|^2 . \tag{3.51}$$

The filter function to find the piston or tilt component of focal anisoplanatism with a uniform circular source on boresight is

$$F(\boldsymbol{\kappa}) = \left[\frac{2\nu\, J_\nu(a)}{a} - \frac{2\nu\, J_\nu(e)}{e} \frac{2\, J_1(D_s\, h)}{(D_s\, h)} \right]^2 . \tag{3.52}$$

If the source size goes to zero, the filter function for the Zernike modes reduces to (2.90).

3.4 Adaptive-Optics Systems

In an adaptive-optics system a beacon beam is propagated downward through the atmosphere, and the phase impressed on the beam by atmospheric turbulence is measured by a phase sensor. The conjugate of the measured phase is applied at the origin to an outgoing beam along a propagation path that is displaced by a distance **d** from the beacon path. To calculate system performance, one has to calculate the residual phase or log-amplitude on the wave after correction. One finds the filter function by calculating the behavior of the system for a single phase screen at $z = z'$ and integrating over the turbulence distribution.

Consider a single phase screen at z' of strength $d\nu(\boldsymbol{\kappa}, z')\, dz'$. The phase-related quantity measured at the origin from the downward propagating beacon with the diffraction parameter in (2.73) is

$$d\nu(\boldsymbol{\kappa}, z') \cos\left(\kappa^2 z'/2k_0\right) dz'.$$

An adaptive-optics system applies the negative of this phase to a scoring beam. The phase or log-amplitude at $z = L$ is made up of two phase components. The first is due to the measured phase that is impressed at the origin, which is propagated a distance L. The second is caused by the phase screen through which the wave passes. This second component propagates a distance $L - z$, and can have an isoplanatic component. The scintillation on the beam above the turbulence is also caused by two phase components: the first due to the phase screen and the second due to the phase at the origin produced by the adaptive-optics system.

In this case the formula for variance is a modification of the standard formula because the effect of the sinusoid and cosinusoid is included in the filter function. The formula for variance is

$$\begin{bmatrix} \sigma_{\phi R}^2 \\ \sigma_{\chi R}^2 \\ \sigma_R^2 \end{bmatrix} = 0.2073\, k_0^2 \int_0^L dz\, C_n^2(z) \int d\boldsymbol{\kappa}\, f(\kappa) \begin{bmatrix} F_\phi'(\gamma\boldsymbol{\kappa}) \\ F_\chi'(\gamma\boldsymbol{\kappa}) \\ F_t'(\gamma\boldsymbol{\kappa}) \end{bmatrix} F(\gamma\boldsymbol{\kappa})\,. \tag{3.53}$$

The complex filter function associated with this process is

$$\begin{bmatrix} G_\phi'(\boldsymbol{\kappa}) \\ G_\chi'(\boldsymbol{\kappa}) \end{bmatrix} = \begin{bmatrix} \cos(aL)\cos(az) - \exp(i\,\boldsymbol{\kappa}\cdot\mathbf{d})\cos[a(L-z)] \\ \sin(aL)\cos(az) - \exp(i\,\boldsymbol{\kappa}\cdot\mathbf{d})\sin[a(L-z)] \end{bmatrix}, \tag{3.54}$$

where $a = \kappa^2/2k_0$. The filter function to obtain the phase and log-amplitude variances is the absolute value squared of the complex filter function given by

$$\begin{bmatrix} F_\phi'(\boldsymbol{\kappa}) \\ F_\chi'(\boldsymbol{\kappa}) \\ F_t'(\boldsymbol{\kappa}) \end{bmatrix} = \begin{bmatrix} |\cos(aL)\cos(az) - \exp(i\,\boldsymbol{\kappa}\cdot\mathbf{d})\cos[a(L-z)]|^2 \\ |\sin(aL)\cos(az) - \exp(i\,\boldsymbol{\kappa}\cdot\mathbf{d})\sin[a(L-z)]|^2 \\ \sin^2(aL) + 2\cos^2(aL)\,[1-\cos(\boldsymbol{\kappa}\cdot\mathbf{d})] \end{bmatrix}, \tag{3.55}$$

where the total variance is the sum of the phase and the log-amplitude variances given by $F_t'(\boldsymbol{\kappa})$ is

$$F_t'(\boldsymbol{\kappa}) = F_\phi'(\boldsymbol{\kappa}) + F_\chi'(\boldsymbol{\kappa})\,. \tag{3.56}$$

The total variance was obtained with the trigonometric identities

$$\cos(A+B) = \cos(A)\cos(B) - \sin(A)\sin(B)\,, \text{ and}$$

$$\sin(A+B) = \sin(A)\cos(B) + \cos(A)\sin(B)\,.$$

If the offset is zero, the filter function reduces to

$$\begin{bmatrix} F_\phi'(\boldsymbol{\kappa}) \\ F_\chi'(\boldsymbol{\kappa}) \\ F_t'(\boldsymbol{\kappa}) \end{bmatrix} = \begin{bmatrix} \sin^2(\kappa^2 L/2k_0)\,\sin^2(\kappa^2 z/2k_0) \\ \cos^2(\kappa^2 L/2k_0)\,\sin^2(\kappa^2 z/2k_0) \\ \sin^2(\kappa^2 z/2k_0) \end{bmatrix}. \tag{3.57}$$

One obtains the aperture-related variances by inserting this expression into (3.53).

4. Zero-Parameter Problems

In the second chapter I derived general formulas that enable one to write phase and log-amplitude variances, power spectral densities, and Strehl ratios as integrals. Different cases differ only in the appearance of different filter functions. In the last chapter explicit filter functions were derived for a variety of problems. With the use of these filter functions, integrands contain the following functions: $\sin(x)$, $\cos(x)$, $\sin^2(x)$, $\cos^2(x)$, $J_\nu(x)$, $J_\nu(x)\,J_\mu(x)$, $J_\nu^2(x)$, $\exp(x)$, $(1+x)^{-p}$, $(x-1)^{-p}\,U(x-1)$, and $K(x)$.

The Mellin transforms of these functions are given in Table 1.1. (Integrals containing $\cos^2(x)$ are evaluated by substituting $1-\sin^2(x)$.) *Therefore,* **all** *problems in which the phase or log-amplitude variances are wanted and in which the filter functions in Chap. 3 can be used are solvable by Mellin transform techniques.*

In this chapter problems of electromagnetic-wave propagation in turbulence that can be solved by table lookup are addressed. The following expressions are found: 1) variances of the Zernike tilt of collimated and focused waves, 2) gradient tilt variance, 3) variance of the difference between Zernike and gradient tilt, 4) beam movement at a target due to tilt, 5) angle-of-arrival jitter of a tracked target, 6) scintillation of collimated and focused waves, 7) phase variance of a system with temporal filtering, 8) total variance with a phase-only adaptive-optics system with and without a beacon offset. These examples illustrate how easily solutions are obtained once the technique described in this book is mastered.

The Mellin transform technique is particularly well suited to evaluate integrals obtained in wave propagation in turbulence, because the integrand term x^{-s} matches the form of the turbulence spectrum in the most important region — the inertial subrange. Results are expressed as moments of the turbulence distribution in altitude. After I define the full and partial moments, I express their values analytically for the Hufnagel-Valley turbulence model. Other frequently used turbulence models are also defined.

4.1 Turbulence Models and Moments

Several different models are commonly used to represent effects of atmospheric turbulence. The Hufnagel-Valley model is based on one proposed by *Hufnagel*

(1974) from measurements made by *Bufton* et al. (1972), and modified by *Valley* (1980) to include ground level turbulence. It is

$$C_n^2(h) = 0.00594\left(\frac{W}{27}\right)^2\left(10^{-5}\times h\right)^{10}\exp\left(-\frac{h}{1000}\right)$$
$$+2.7\times 10^{-16}\exp\left(-\frac{h}{1500}\right)+A\exp\left(-\frac{h}{100}\right), \tag{4.1}$$

where W is the pseudo-wind, and A is a parameter that is usually set equal to 1.7×10^{-14}. The turbulence is usually assumed to be zero above 30 km. The turbulence strength is usually changed by a variation of the W term. For instance, the HV-21 model has the above value for A, and W is equal to 21. This model is sometimes referred to as the $HV_{5/7}$ model since the coherence diameter is about 5 cm, and the isoplanatic angle is about 7 μrad for a wavelength of 0.5 μm.

The full moments, which are defined for propagation from the ground to space, are

$$\mu_m \equiv \int_0^\infty dz\, C_n^2(z)\, z^m = \sec^{m+1}(\xi)\int_0^\infty dh\, C_n^2(h)\, h^m, \tag{4.2}$$

where ξ is the zenith angle. *Notice that the zenith dependence is contained in the turbulence moments.* The integrations can be performed for the Hufnagel-Valley model to give

$$\mu_m = \int_0^\infty dz\, C_n^2(z)\, z^m = \sec^{m+1}(\xi)\left\{5.94\times 10^{-20+3m}\left(W/27\right)^2\Gamma\left[m+11\right]\right.$$
$$\left.+4.05\times 10^{-13}\Gamma\left[m+1\right](1500)^n + A\times 100^{m+1}\Gamma\left[m+1\right]\right\}. \tag{4.3}$$

For propagation to a distance L, which is at a height H, so that

$$L = \sec(\xi)\, H, \tag{4.4}$$

define the partial moments in the following way

$$\mu_m^+(L) \equiv \int_L^\infty dz\, C_n^2(z)\, z^m = \sec^{m+1}(\xi)\int_H^\infty dh\, C_n^2(h)\, h^m, \quad \text{and} \tag{4.5}$$

$$\mu_m^-(L) \equiv \int_0^L dz\, C_n^2(z)\, z^m = \sec^{m+1}(\xi)\int_0^H dh\, C_n^2(h)\, h^m. \tag{4.6}$$

Using the lower- and upper-incomplete gamma functions, which are

$$\gamma\left[b+1,x\right] = \int_0^x dy\, y^b\exp(-y), \quad \text{and} \quad \Gamma\left[b+1,x\right] = \int_x^\infty dy\, y^b\exp(-y), \tag{4.7}$$

one can find the partial moments of turbulence as

$$\mu_m^+(L) = \int_L^\infty dz\, C_n^2(z)\, z^m = \sec^{m+1}(\xi) \left\{ \frac{5.94}{10^{20-3m}} \left(\frac{W}{27}\right)^2 \Gamma\left[m+11, \frac{H}{1000}\right] \right.$$
$$\left. +4.05 \times 10^{-13} \Gamma\left[m+1, \frac{H}{1500}\right] (1500)^m + A \times 100^{m+1} \Gamma\left[m+1, \frac{H}{100}\right] \right\}, \tag{4.8}$$

and

$$\mu_m^-(L) = \int_0^L dz\, C_n^2(z)\, z^m = \sec^{m+1}(\xi) \left\{ \frac{5.94}{10^{20-3m}} \left(\frac{W}{27}\right)^2 \gamma\left[m+11, \frac{H}{1000}\right] \right.$$
$$\left. +4.05 \times 10^{-13} \gamma\left[m+1, \frac{H}{1500}\right] (1500)^m + A \times 100^{m+1} \gamma\left[m+1, \frac{H}{100}\right] \right\}. \tag{4.9}$$

It can be useful to express results in terms of coherence diameter and isoplanatic angle. Define the coherence diameter r_o as

$$r_o^{-5/3} = 0.423\, k_0^2\, \mu_0, \tag{4.10}$$

and isoplanatic angle
θ_o as

$$\theta_o^{-5/3} = 2.91\, k_0^2\, \mu_{5/3}. \tag{4.11}$$

Both the coherence diameter and isoplanatic angle vary as the six-fifths power of wavelength. As shown in Sect. 8.1 this definition of coherence diameter comes naturally from consideration of the beam size. As shown in Sect. 8.4.1 the isoplanatic angle is the characteristic scale for the effect of an offset angle on the Strehl ratio.

In addition to the Hufnagel-Valley model some other models are regularly used. The SLC Day model, also referred to as SLCSAT Day, is based on the day-time *Miller-Zieske* profile (1979), which can be approximated by a set of power-law segments given below:

$$\begin{aligned} C_n^2(z) &= 3.96 \times 10^{-13}/z^{1.05}, & 18.5 \le z \le 232 \\ &= 1.3 \times 10^{-15}, & 232 \le z \le 880 \\ &= 8.87 \times 10^{-7}/z^3, & 880 \le z \le 7220 \\ &= 2.0 \times 10^{-16}/z^{0.5}, & 7220 \le z \le 20500. \end{aligned} \tag{4.12}$$

The units of $C_n^2(z)$ are $m^{-2/3}$ and the altitude is in meters. The SLC Night model differs from the SLC Night model at lower altitudes

$$\begin{aligned} C_n^2(z) &= 5.0 \times 10^{-15}, & z \le 18.5 \\ &= 2.875 \times 10^{-12}/z^2, & 18.5 \le z \le 110 \\ &= 2.5 \times 10^{-16}, & 110 \le z \le 1500 \\ &= 8.87 \times 10^{-7}/z^3, & 850 \le z \le 7000 \\ &= 2.0 \times 10^{-16}/z^{0.5}, & 7000 \le z \le 20500. \end{aligned} \tag{4.13}$$

Table 4.1. Coherence diameter and isoplanatic angle for a wavelength of 0.5 μm and moments of Hufnagel-Valley turbulence models. The turbulence moments are in mks units, and the units of the nth moment are $m^{n+1/3}$.

	HV-21	HV-35	HV-54	HV-72
r_o(cm)	4.96	4.67	4.18	3.70
θ_o(μrad)	6.90	3.95	2.40	1.71
μ_0	2.23×10^{-12}	2.47×10^{-12}	2.97×10^{-12}	3.64×10^{-12}
$\mu_{5/6}$	5.45×10^{-10}	1.08×10^{-9}	2.24×10^{-9}	3.80×10^{-9}
$\mu_{5/3}$	8.97×10^{-7}	2.20×10^{-6}	5.06×10^{-6}	8.91×10^{-6}
μ_2	1.91×10^{-5}	4.97×10^{-5}	1.16×10^{-4}	2.04×10^{-4}
μ_3	2.32×10^{-1}	6.30×10^{-1}	1.49	2.64
$\mu_{11/3}$	1.32×10^{2}	3.62×10^{2}	8.58×10^{2}	1.52×10^{3}
μ_4	3.18×10^{3}	8.75×10^{3}	2.08×10^{4}	3.69×10^{4}
$\mu_{14/3}$	1.91×10^{6}	5.29×10^{6}	1.25×10^{7}	2.23×10^{7}
μ_5	4.74×10^{7}	1.31×10^{8}	3.11×10^{8}	5.53×10^{8}
μ_6	7.55×10^{11}	2.09×10^{12}	4.97×10^{12}	8.84×10^{12}

The turbulence is usually assumed to be zero above 20.5 km for both models. Table 4.1 contains various useful moments for several turbulence models in mks units. Values of coherence diameters and isoplanatic angles in the table are for a wavelength of 0.5 μm. The HV-21 and SLCSAT Day models are plotted versus altitude in Fig. 4.1.

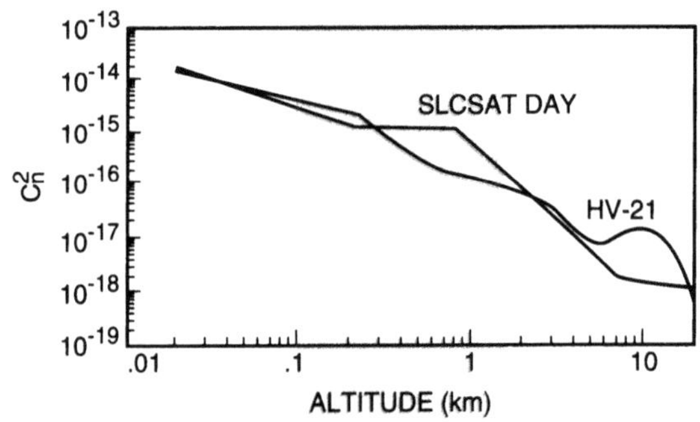

Fig. 4.1. Plots of the Hufnagel-Valley 21 and SLCSAT DAY models of turbulence versus altitude. The units of turbulence strength are $m^{-2/3}$.

4.2 Zernike Tilt and Piston for Collimated and Focused Beams

Tilt jitter is one of the most significant effects of turbulence. Tilt of a mirror is described by a Zernike tilt, and that measured by a centroid sensor is closely approximated by gradient tilt. The calculation of Zernike tilt is considered first. Since this is a single wave problem, (2.86) applies. In calculations of the tilt here and in the next two sections it will be assumed that diffraction effects are not important, thereby allowing the cosine term to be replaced by unity. The effect of diffraction on tilt is considered in Sect. 7.5. I show that as long as one is well within the Fresnel distance, effects of diffraction are negligible. If inner- and outer-scale effects are neglected, the Kolmogorov spectrum given in (2.33) applies. With the filter function for two-axis Zernike tilt given in (3.18) one obtains for the tilt angle variance

$$T_Z^2 = 0.2073\, k_0^2 \int_0^L dz\; C_n^2(z) \int d\boldsymbol{\kappa}\; \kappa^{-11/3} \left(\frac{16}{k_0 D}\right)^2 \left[\frac{J_2(\gamma\kappa D/2)}{\gamma\kappa D/2}\right]^2 . \tag{4.14}$$

The integrand in the second integral does not depend on the angle in kappa space. Integration over this angle, and substitution of $x = \gamma\kappa D/2$ yields

$$\begin{aligned} T_Z^2 &= \frac{105.1}{D^{1/3}} \int_0^L dz\, C_n^2(z)\, \gamma^{5/3} \int_0^\infty \frac{dx}{x}\, x^{-11/3} J_2^2(x) \\ &= \frac{105.1}{2\sqrt{\pi} D^{1/3}} \int_0^L dz\, C_n^2(z)\, \gamma^{5/3}\, \Gamma\left[\begin{matrix} s/2+2, & -s/2+\frac{1}{2} \\ -s/2+3, & -s/2+1 \end{matrix}\right]\Bigg|_{s=-\frac{11}{3}} \\ &= \frac{105.1}{2\sqrt{\pi} D^{1/3}} \int_0^L dz\, C_n^2(z)\, \gamma^{5/3}\, \Gamma\left[\begin{matrix} \frac{1}{6}, & \frac{7}{3} \\ \frac{29}{6}, & \frac{17}{6} \end{matrix}\right] . \end{aligned}$$

The x integration is equal to the Mellin transform given in (1.50) evaluated at $s = -11/3$. The composite gamma function defined in (1.18) is equal to the ratio of four individual gamma functions. The composite gamma function above is equal to 0.2052. For an infinite collimated beam, γ is equal to unity. For this case the integration over z can be performed with the turbulence moments given in (4.2). *The 2-axis Z-tilt angle jitter variance is*

$$T_Z^2 = \frac{6.08\, \mu_0}{D^{1/3}} = 0.3641 \left(\frac{D}{r_o}\right)^{5/3} \left(\frac{\lambda}{D}\right)^2 , \tag{4.15}$$

where the coherence diameter is defined in (4.10). The separate x and y tilt variances are half this value. Because the variance is proportional to the zeroth moment of turbulence, it varies as the secant of the zenith angle, sec (ξ). Notice that tilt goes to infinity as the diameter goes to zero. This singularity is removed in Sect. 5.3 by consideration of a finite inner scale.

For a 0.6-m-diameter aperture and HV-21 turbulence, the rms tilt is 4 μrad at zenith. This is much larger than the diffraction-limited beam size of 1 μrad at 0.5 μmet, and it causes considerable smearing of a star image.

If the turbulence is constant along the path of length L, the two-axis tilt variance is

$$T_Z^2 = \frac{6.08\, C_n^2 L}{D^{1/3}}. \tag{4.16}$$

If the wave originated as a point source propagating from the ground to space, the value of γ would be equal to z/L. *The tilt angle variance of a point source on the ground as viewed from space is*

$$T_Z^2 = \frac{6.08\, \mu_{5/3}}{D^{1/3} L^{5/3}} = \frac{0.169\, \lambda^2}{D^{1/3} L^{5/3} \theta_o^{5/3}}. \tag{4.17}$$

Therefore, a tilt sensor in space can measure the isoplanatic angle. The variance varies with zenith angle as $\sec^{8/3}(\xi)$.

This tilt also determines the resolution of an imaging system in space. Since the tilt is correlated from one point to another, the resolution can be improved by the use of short-exposure images or by tracking the image during the exposure.

By dividing the tilt variance value by the factor $(4/k_0\, D)^2$ given after (3.17) and using the relation in (4.10), one can obtain the phase variance. *The tilt phase variance is*

$$\sigma_T^2 = 0.380\, \mu_0\, k_0^2\, D^{5/3} = 0.899 \left(\frac{D}{r_o}\right)^{5/3}. \tag{4.18}$$

Total and piston phase variances are infinite for Kolmogorov turbulence. Their difference, which is piston-removed phase variance, is finite, and it is found with the relation for phase variance given in (2.86) and the filter function for piston variance given in (3.15). For a collimated beam it is

$$\sigma_{PR}^2 = 0.2073\, k_0^2 \int_0^L dz\ C_n^2(z) \int d\boldsymbol{\kappa}\ \kappa^{-11/3} \left(\frac{1}{k_0}\right)^2 \left\{1 - \left[\frac{2\, J_1(\kappa D/2)}{\kappa D/2}\right]^2\right\}.$$

The angle and axial integrations can be evaluated, and when the substitution $x = \kappa D/2$ is made one obtains

$$\sigma_{PR}^2 = 1.64\, D^{5/3} \mu_0 \int_0^\infty \frac{dx}{x}\, x^{-11/3} \left[\frac{x^2}{4} - J_1^2(x)\right] = -\frac{1.64}{2\sqrt{\pi}}\, \mu_0\, D^{5/3}\, \Gamma\left[\begin{matrix} -\frac{5}{6}, \frac{7}{3} \\ \frac{23}{6}, \frac{17}{6} \end{matrix}\right].$$

The integral of individual functions in the integrand does not converge if the exponent of the power of x that multiplies dx/x is less than -2. However, because $x^2/4$ is the first term of a power series of $J_1^2(x)$, the integral can be analytically continued and converges for the exponent in the range -2 to -4, since the residues at two poles cancel as in the derivation of (1.21). Thus the integral is simply the Mellin transform of the squared Bessel function in (1.50) evaluated at $s = -11/3$. The value of the composite gamma function is -0.9438. *Thus, the phase variance with piston removed is*

$$\sigma_{PR}^2 = 1.033 \left(\frac{D}{r_o}\right)^{5/3} . \tag{4.19}$$

Since piston and tilt are orthogonal on a circular aperture, *the phase variance with piston and tilt removed is the difference of (4.19) and (4.18), i.e.,*

$$\sigma_{PTR}^2 = 0.134 \left(\frac{D}{r_o}\right)^{5/3} . \tag{4.20}$$

These piston- and tilt-variance relationships are often used to get a first cut at system performance. Tilt variance is 87% of the piston-removed phase variance. For this reason adaptive-optics systems are usually designed to have a separate tilt mirror to limit the dynamic range of deformable mirror motion that responds only to the tilt-removed phase. If turbulence to be corrected can be so large that one can have $D/r_o = 10$, then the rms piston- and tilt-removed phase variance is 2.49 rads or 0.397 waves. This value can be used to estimate the dynamic range of actuator motion required in the deformable mirror.

4.3 Gradient Tilt

In an exactly analogous fashion, the G-tilt integral is found with the filter function given in (3.21) and the general formula in (2.86), to obtain

$$T_G^2 = 0.2073\, k_0^2 \int_0^L dz\, C_n^2(z) \int d\boldsymbol{\kappa}\; \kappa^{-11/3} \left(\frac{4}{k_0 D}\right)^2 J_1^2\left(\gamma \kappa D/2\right). \tag{4.21}$$

After the same integrations, substitutions, and Mellin transforms as above, one obtains for a collimated beam

$$T_G^2 = \frac{6.564\, \mu_0}{D^{1/3} 2\sqrt{\pi}}\, \Gamma\left[\begin{matrix} \frac{1}{6}, \frac{4}{3} \\ \frac{17}{6}, \frac{11}{6} \end{matrix}\right] = \frac{5.675\, \mu_0}{D^{1/3}} = 0.3399 \left(\frac{D}{r_o}\right)^{5/3} \left(\frac{\lambda}{D}\right)^2 . \tag{4.22}$$

This gives a tilt that is 3.5% lower than Z-tilt. This is the same result obtained by *Ellerbroek* (1984) and *Tatarski* (1971).

4.4 Difference Between Gradient and Zernike Tilt

Sometimes a target position is measured with a centroid sensor that responds essentially to G-tilt. The pointing system uses tilt from this sensor to direct a beam at a target by tilting a mirror that changes the Z-tilt of the wave. It is of interest to know what residual tilt jitter is expected at the target even if this process is performed perfectly. This problem was analyzed by *Yura* and

Tavis (1985) and given the name *centroid anisoplanatism*. The variance of the difference between G-tilt and Z-tilt is found with the filter functions for G- and Z-tilt given by the square roots of the expressions in (3.18) and (3.21). Insert these into (2.90), which determines the variance of two waves that have different aperture functions, to obtain

$$T_D^2 = 0.2073\, k_0^2 \int_0^L dz\, C_n^2(z) \int d\boldsymbol{\kappa}\, \kappa^{-11/3} \left(\frac{4}{k_0 D}\right)^2 \left[\frac{4 J_2(\kappa D/2)}{\kappa D/2} - J_1(\kappa D/2)\right]^2 . \quad (4.23)$$

If the bracket is expanded, one obtains three terms. Two are the sum of Zernike and Gradient tilts. The third can be evaluated with the Mellin transform in (1.51), to obtain

$$T_D^2 = \frac{6.564\, \mu_0}{2\sqrt{\pi} D^{1/3}} \left(16\, \Gamma\left[\begin{matrix} \frac{1}{6}, \frac{7}{3} \\ \frac{29}{6}, \frac{17}{6} \end{matrix}\right] + \Gamma\left[\begin{matrix} \frac{1}{6}, \frac{4}{3} \\ \frac{17}{6}, \frac{11}{6} \end{matrix}\right] - 8\, \Gamma\left[\begin{matrix} \frac{1}{6}, \frac{7}{3} \\ \frac{23}{6}, \frac{17}{6} \end{matrix}\right]\right) .$$

If the gamma functions are evaluated directly, then one requires five-place accuracy to achieve three-place accuracy in the final answer, since the terms in braces almost cancel. Rather than doing that, one uses the recurrence relation for gamma functions to show that

$$T_D^2 = \frac{6.564\, \mu_0}{2\sqrt{\pi} D^{1/3}}\, \Gamma\left[\begin{matrix} \frac{1}{6}, \frac{4}{3} \\ \frac{17}{6}, \frac{11}{6} \end{matrix}\right] \left(\frac{4608}{4301} + 1 - \frac{384}{187}\right) .$$

This is equal to

$$T_D^2 = \frac{0.102\, \mu_0}{D^{1/3}} = 0.0061 \left(\frac{D}{r_o}\right)^{5/3} \left(\frac{\lambda}{D}\right)^2 . \quad (4.24)$$

Beamwidth of a diffraction-limited system is $1.2\,\lambda/D$; therefore, there is a one-third of a beamwidth jitter when the aperture diameter is 6 times the coherence diameter. This tilt jitter would be observable only if the image or outgoing beam were significantly corrected for other effects of turbulence. It was shown by *Herrmann* (1981) that the major difference between the two tilts is due to coma distortion. If an adaptive-optics system corrected most of this distortion, centroid anisoplanatism would be greatly reduced.

4.5 Beam Movement at a Target

Turbulence causes the beam boresight to move at a target. This movement is found here for a source on the ground projected into space, and for a source in space projected to the ground. Tilt causes a beam to change position on a target as shown in Fig. 4.2. In this problem one finds the filter function by finding an expression for the beam movement on target for a given tilt realization, and then taking the absolute value squared. It is assumed that scintillations are small, and the beam is not broken up. If there are significant scintillations, these

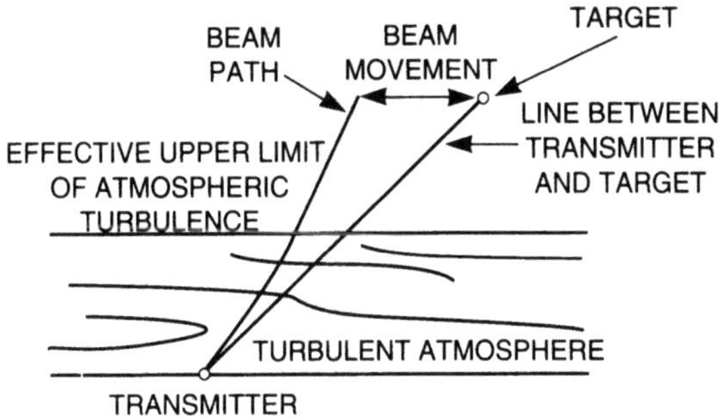

Fig. 4.2. Beam movement at a target board.

will increase the movement calculated here. It is also assumed that diffraction is unimportant; therefore, a ray-optics calculation is adequate. With these assumptions, the amount of beam jitter at a target at L is equal to the tilt from turbulence at z times the distance over which it acts, $L - z$. Therefore, the tilt filter function for this problem is

$$F(\boldsymbol{\kappa}) = (L - z)^2 \left(\frac{16}{k_0 D}\right)^2 \left[\frac{J_2(\gamma \kappa D/2)}{\gamma \kappa D/2}\right]^2 . \tag{4.25}$$

The variance of beam movement for one wave in (2.90) with Kolmogorov turbulence is

$$X^2 = 0.2073\, k_0^2 \int_0^L dz\, C_n^2(z) \int d\boldsymbol{\kappa}\, \kappa^{-11/3} (L - z)^2 \left(\frac{16}{k_0 D}\right)^2 \left[\frac{J_2(\gamma \kappa D/2)}{\gamma \kappa D/2}\right]^2 . \tag{4.26}$$

This expression can be evaluated for the collimated case, for which case $\gamma = 1$, to give

$$X^2 = \frac{6.08}{D^{1/3}} \left(L^2 \mu_0 - 2\, L\, \mu_1 + \mu_2\right) . \tag{4.27}$$

For distances at which the target is well above the turbulence, the first term in brackets is dominant; this gives the physically reasonable result that rms movement is rms tilt times distance. For a target at 100 km with HV-21 turbulence and a 60-cm aperture, the rms beam movement is 40 cm.

If turbulence is constant along the path, one finds

$$X^2 = \frac{2.03\, L^3 C_n^2}{D^{1/3}} . \tag{4.28}$$

This functional dependence with constant turbulence is the same as that reported in *Fante* (1980) where the fourth moment of the field was used to calculate the beam displacement. The coefficient found by Fante was 1.92 in that case. A Gaussian beam, as calculated by *Prokhorov* et al. (1975), gives the same dependence with the coefficient equal to 1.6. The advantage of the approach here is that the answer is arrived at in a straightforward manner, and the expression given in (4.27) applies for turbulence that can vary in space. For the beam focused at the target board, for which case $\gamma = (L - z)/L$, the movement variance is

$$X^2 = \frac{6.08}{D^{1/3}} \int_0^L dz\, C_n^2(z) \frac{(L-z)^{11/3}}{L^{5/3}}. \tag{4.29}$$

For constant turbulence along the path, this can be integrated to give

$$X^2 = \frac{2.03 L^3 C_n^2}{D^{1/3}} \left(\frac{9}{14}\right). \tag{4.30}$$

The movement variance is 64% that of a collimated beam, and the rms movement is 80% that of a collimated beam.

For a source in space, the same argument as above gives a filter function

$$F(\boldsymbol{\kappa}) = z^2 \left(\frac{16}{k_0 D}\right)^2 \left[\frac{J_2(\gamma \kappa D/2)}{\gamma \kappa D/2}\right]^2. \tag{4.31}$$

Variance of the beam movement for a collimated beam propagated from space to ground is

$$X^2 = \frac{6.08\, \mu_2}{D^{1/3}}. \tag{4.32}$$

This movement is equal to 1.2 cm for the HV-21 model with a 0.6-m aperture. Thus it is much easier to hit a target on the ground from space with a laser beam than vice versa.

4.6 Angle-of-Arrival Jitter

Suppose a tracking system is doing a perfect job of keeping a laser beam centered on a target. There will still be an angle-of-arrival jitter of the laser beam at the target because the beam has to traverse the atmosphere differently as turbulence changes. This is illustrated in Fig 4.3. This jitter causes movement of a beam relayed from a mirror at the space target to another object. The statement that a target is being perfectly tracked means that, at each turbulence realization, the distance the beam moves due to the tilted tracking mirror is exactly canceled out by tilt caused by traversing the atmospheric turbulence. The angle-of-arrival tilt will be calculated from a turbulence screen at z. In the final expression, all these differential tilts are summed by integration through the atmosphere. If T_o is the mirror tilt and dT is the turbulence-induced tilt, one obtains for perfect tracking

$$L\, dT_o - (L - z)\, dT = 0; \tag{4.33}$$

therefore,

$$dT_o = (1 - z/L)\, dT. \tag{4.34}$$

The residual tilt at the target dT_t is the mirror tilt minus tilt through turbulence. This results in the angle-of-arrival jitter

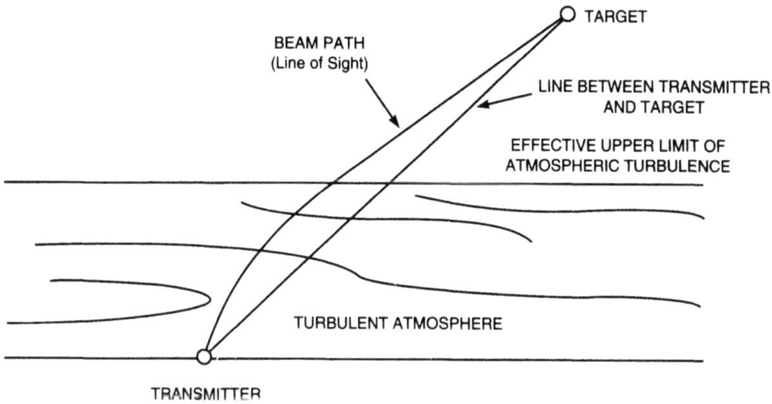

Fig. 4.3. Angle-of-arrival jitter of a beam at a target that is being perfectly tracked by a ground system.

$$dT_t = dT - dT_o = z\,dT/L. \tag{4.35}$$

Therefore, the filter function for angle-of-arrival jitter is

$$F(\boldsymbol{\kappa}) = \left(\frac{z}{L}\right)^2 \left(\frac{16}{k_0\,D}\right)^2 \left[\frac{J_2(\gamma\,\kappa D/2)}{\gamma\,\kappa D/2}\right]^2. \tag{4.36}$$

Remember that γ is equal to unity for a collimated beam, γ is equal to $1 - z/L$ for a beam focused at the target, and γ is equal to z/L for a point source on the ground. Substitution of this filter function into (2.86) gives

$$T_t^2 = 0.2073\,k_0^2 \int_0^L dz\,C_n^2(z) \left(\frac{z}{L}\right)^2 \int d\boldsymbol{\kappa}\,\kappa^{-11/3} \left(\frac{16}{k_0\,D}\right)^2 \left[\frac{J_2(\gamma\,\kappa D/2)}{\gamma\,\kappa D/2}\right]^2. \tag{4.37}$$

Evaluating this in exactly the same way as the variance of Z-tilt, one obtains for a collimated wave

$$T_t^2 = 6.08\,\frac{\mu_2}{D^{1/3}\,L^2}. \tag{4.38}$$

For a 0.6-m system with a target at 300-km range and HV-21 turbulence, the rms tilt is 39 nrads.

For a beam focused at the object, the tilt variance is

$$T_t^2 = \frac{6.08}{D^{1/3}\,L^2} \int_0^L dz\,C_n^2(z)\,z^2\,(1 - z/L)^{5/3}. \tag{4.39}$$

If the target is well above the turbulence, then the angle-of-arrival tilt jitter is almost equal to that of a collimated beam.

For a point source on the ground, γ is equal to z/L; the angle-of-arrival variance at a target in space is

$$T_t^2 = 6.08 \frac{\mu_{11/3}}{D^{1/3} L^{11/3}}. \tag{4.40}$$

For a 0.6-m system with a target at 300-km range and HV-21 turbulence, the rms tilt is 2.8 nrads.

This problem was also analyzed by *Tyler* (1985). His results are in the form of curves that are derived from a numerical integration.

4.7 Scintillation for Collimated and Focused Beams

Expressions that are derived for scintillation are only valid when the log-amplitude fluctuations are not saturated, which occurs for log-amplitude variance values that are 0.5 or less. The variance of log-intensity is four times the value of variance of log-amplitude. To calculate scintillation of a wave that propagates from 0 to L with inner and outer scale neglected, again use (2.86) with the propagation parameter given in (2.68), to obtain

$$\sigma_\chi^2 = 0.2073\, k_0^2 \int_0^L dz\, C_n^2(z) \int d\boldsymbol{\kappa}\; \kappa^{-11/3} \sin^2\left[\frac{\gamma \kappa^2 (L-z)}{2k_0}\right]. \tag{4.41}$$

As explained in Sect. 2.3, to compare this form with expressions obtained by others such as *Hufnagel* (1978), one must make the substitution $\kappa \to \kappa/\gamma$.

Integrating over angle and making the substitution $x^2 = \gamma\kappa^2(L-z)/2k_0$, one obtains

$$\sigma_\chi^2 = 0.731\, k_0^{7/6} \int_0^L dz\, C_n^2(z)(L-z)^{5/6}\gamma^{5/6} \int_0^\infty \frac{dx}{x}\, x^{-5/3} \sin^2\left(x^2\right).$$

Using the Mellin transform in (1.48) evaluated at $s = -5/3$, one obtains

$$\sigma_\chi^2 = 0.5631\, k_0^{7/6} \int_0^L dz\, C_n^2(z)(L-z)^{5/6}\gamma^{5/6}. \tag{4.42}$$

This must be integrated numerically in general. For a receiver well above the turbulence, the approximate value of σ_χ^2 for a plane wave, for which $\gamma = 1$, is

$$\sigma_\chi^2 \approx 0.5631\, k_0^{7/6} \mu_0\, L^{5/6}. \tag{4.43}$$

This varies as sec (ξ). The log-amplitude variance is greater than unity for ranges larger than 2.5 km for a propagation wavelength of 0.5 μm and for a typical turbulence strength of $10^{-14} m^{-2/3}$. Therefore, scintillation of a beam projected a modest distance along the ground or to space will be saturated.

For a point source on the ground for which $\gamma = z/L$, the scintillation observed in space is approximately equal to

$$\sigma_\chi^2 \approx 0.5631\, k_0^{7/6} \mu_{5/6}. \tag{4.44}$$

This varies as $\sec^{11/6}(\xi)$. If propagation were from L to 0, then $L-z$ should be replaced by z in (4.42). For a plane wave, for which $\gamma = 1$, I obtain the standard result

$$\sigma_\chi^2 = 0.5631\, k_0^{7/6} \mu_{5/6}. \tag{4.45}$$

This is the same scintillation level seen in space from a point source on the ground that was derived above. It is the scintillation observed from a star and is equal to 0.059 for the HV-21 model at 0.5 μm wavelength and zero zenith angle. Even though this value is small, it is significant. The variance of log-intensity is four times that of the log-amplitude; its value is 0.236. Therefore, the variance of the intensity is about 60

Similarly, for a point source in space propagating towards the ground, one obtains

$$\sigma_\chi^2 = 0.5631\, k_0^{7/6} \int_0^L dz\, C_n^2(z) \frac{(L-z)^{5/6} z^{5/6}}{L^{5/6}}. \tag{4.46}$$

For a point source well above turbulence, the scintillation is the same as that for a plane wave propagating from space, as would be expected. This result is given in (4.45).

4.8 Phase Variance with Finite Servo Bandwidth

Greenwood (1977) derived the phase variance of an adaptive-optics system with finite temporal servo bandwidth with either a one-pole filter or an infinitely sharp filter. All other errors are neglected. Residual rms-phase error from a finite frequency response is calculated here with the same servo filter treated by Greenwood. Consider the following filter

$$F_f(\omega) = \left|1 - \frac{1}{1 + i\,(\omega/\omega_{3dB})^n}\right|^2. \tag{4.47}$$

The single-pole filter case corresponds to $n = 1$, and the sharp cutoff case corresponds to $n = \infty$. Insert this filter into (2.101) with diffraction neglected. Putting the result into (2.100). and considering propagation from L_1 to L_2, one obtains

$$\sigma_\phi^2 = 0.8295\, k_0^2 \int_{L_1}^{L_2} dz\, C_n^2(z) \frac{v^{5/3}(z)}{\omega_{3dB}^{5/3}} \int_0^\infty \frac{dx}{x} \frac{x^{2n-5/3}}{1+x^{2n}} \int_0^\infty \frac{dc}{c} \frac{U(c-1)c^{-5/3}}{\sqrt{c^2-1}},$$

where $x = \omega/\omega_{3dB} = f/f_{3dB}$.

Using the Mellin transform in (1.52) with $p = 1$, the relation in (1.9) with $p = 2n$, and the identity $\Gamma[s]\,\Gamma[1-s] = \pi/\sin(\pi s)$, one obtains

$$\int_0^\infty \frac{dx}{x} \frac{x^{2n-5/3}}{1+x^{2n}} = \frac{\pi}{2n\,\sin(5\pi/6n)}.$$

Using the Mellin transform in (1.54) with $a = 1/2$, and the relation in (1.18) with $p = 2$, one obtains

$$\int_0^\infty \frac{dc}{c} \frac{U(c-1)c^{-5/3}}{\sqrt{c^2-1}} = \frac{\sqrt{\pi}}{2} \Gamma\left[\begin{matrix}\frac{4}{3}\\ \frac{11}{6}\end{matrix}\right] = 0.8413.$$

With these results, the phase variance is

$$\sigma_\phi^2 = \left(\frac{f_G}{f_{3dB}}\right)^{5/3} = \frac{0.051\, k_0^2\, v_{5/3}}{f_{3dB}^{5/3}\, n \,\sin\left(5\pi/6n\right)}, \tag{4.48}$$

where the velocity moment is defined by

$$v_m = \int_{L_1}^{L_1} dz\, C_n^2(z)\, v^m(z) = \sec\left(\xi\right) \int_{H_1}^{H_2} dh\, C_n^2(h)\, v^m(h). \tag{4.49}$$

For propagation between any two points, the integration is over the propagation path, and the velocity is the component transverse to the path. For a single-pole filter ($n=1$) one finds the characteristic frequency, sometimes referred to as the Greenwood frequency, is

$$f_G = 0.254\, k_0^{6/5}\, v_{5/3}^{3/5}. \tag{4.50}$$

For a sharp cutoff filter, the limit of the phase variance as n gets very large is found from L'Hospital's rule. This gives the same form for the Greenwood frequency; however, the coefficient is lower. In this case

$$f_G = 0.0945\, k_0^{6/5}\, v_{5/3}^{3/5}. \tag{4.51}$$

These are the same results obtained by *Greenwood* (1977).

4.9 Variances for Beams Corrected by Adaptive Optics

If an adaptive-optics system applies the conjugate in both amplitude and phase of the beacon, then an outgoing wave propagating in a direction exactly opposite to the beacon would be perfectly corrected. This result can be derived from the reciprocity relations for wave propagation. Most adaptive-optics systems apply only the conjugate of phase, and the direction of the outgoing beam may be offset from that of the beacon. For that case the filter function given in (3.57) applies, and the residual log-amplitude and phase variances, after correction by an adaptive-optics system, are given by (3.53) with $F(\gamma\,\kappa) = 1$ as

$$\begin{aligned}\begin{bmatrix}\sigma_{\phi c}^2\\ \sigma_{\chi c}^2\end{bmatrix} &= 0.2073\, k_0^2 \int_0^L dz\, C_n^2(z) \int d\boldsymbol{\kappa}\, \kappa^{-11/3}\\ &\times \begin{bmatrix}\left|\cos\left(a L\right)\cos\left(a z\right) - \exp\left(i\,\boldsymbol{\kappa}\cdot\mathbf{d}\right)\cos\left[a(L-z)\right]\right|^2\\ \left|\sin\left(a L\right)\cos\left(a z\right) - \exp\left(i\,\boldsymbol{\kappa}\cdot\mathbf{d}\right)\sin\left[a(L-z)\right]\right|^2\end{bmatrix},\end{aligned} \tag{4.52}$$

where $a = \kappa^2/2k_0$. Let us first consider the case in which the offset between the outgoing and beacon beams $\mathbf{d}$ is zero. The variance in this case is

$$\begin{bmatrix} \sigma_{\phi c}^2 \\ \sigma_{\chi c}^2 \end{bmatrix} = 0.2073\, k_0^2 \int_0^L dz\, C_n^2(z) \int d\boldsymbol{\kappa}\, \kappa^{-11/3} \begin{bmatrix} \sin^2(a\,z)\sin^2(a\,L) \\ \sin^2(a\,z)\cos^2(a\,L) \end{bmatrix}. \tag{4.53}$$

A simple relation can be found by the addition of the two variances to obtain

$$\sigma_{\phi c}^2 + \sigma_{\chi c}^2 = 0.2073\, k_0^2 \int_0^L dz\, C_n^2(z) \int d\boldsymbol{\kappa}\, \kappa^{-11/3} \sin^2\left(\frac{\kappa^2 z}{2k_0}\right). \tag{4.54}$$

This is the same expression one obtains for scintillation of a point source propagating towards the ground that was analyzed in Sect. 4.4, and the result is

$$\sigma_{\phi c}^2 + \sigma_{\chi c}^2 = 0.5631\, k_0^{7/6}\, \mu_{5/6} = \sigma_\chi^2. \tag{4.55}$$

Therefore, total variance is equal to the variance of the scintillation on the beacon. This expression includes the effect of diffraction, and should be valid for distributed turbulence.

The corrected beam in a phase-only adaptive-optics system can be distorted by several causes. Here the distortion in a system that has an anisoplanatic error produced by an offset of the outgoing beam by θ from the beacon is considered. The filter function for the total beam distortion is given in (3.55) as

$$F_t'(\boldsymbol{\kappa}) = \sin^2\left(\frac{\kappa^2 z}{2\,k_0}\right) + 2\cos^2\left(\frac{\kappa^2 z}{2\,k_0}\right)\left[1 - \cos\left(\boldsymbol{\kappa}\cdot\mathbf{d}\right)\right]. \tag{4.56}$$

The first term is the same one encountered with correctly aligned beams, and its variance is given in (4.55). It will be assumed that diffraction is negligible, which enables the first cosine to be replaced by unity. To find the total distortion with an angular offset θ for which $d = \theta\, z$, use the expression for the variances in (3.53) with outer and inner scale effects neglected. Sum the phase and log-amplitude variances and integrate over angle to obtain for the total variance of a corrected offset beam

$$\sigma_{co}^2 = \sigma_\chi^2 + 0.2073\,(4\,\pi)\,k_0^2 \int_0^L dz\, C_n^2(z) \int_0^\infty \frac{d\kappa}{\kappa}\, \kappa^{-5/3}\left[1 - J_0(\kappa\theta\, z)\right]. \tag{4.57}$$

The integral can be evaluated with the Mellin transform in (1.49) and the definition of isoplanatic angle in (4.11) to obtain

$$\sigma_{co}^2 = \sigma_\chi^2 + (\theta/\theta_o)^{5/3}. \tag{4.58}$$

The angle-dependent variance is equal to the variance of angular anisoplanatism that is derived in Sect. 8.4.2.

5. Integral Evaluation with Mellin Transforms

In this chapter I continue the development of techniques for evaluating the wavenumber integrals in turbulence propagation theory. With appropriate normalization, the wavenumber (κ) integration can be expressed in a standard form depending only on zero, one, or more parameters. If no parameters are present, the integration is performed simply by table lookup as was done in the last chapter. The one parameter case requires a transformation of the integral into the complex plane that is subsequently evaluated by pole-residue techniques. Details of this process are considered in this chapter. Integrals with more than one parameter are evaluated by extensions of this technique and are considered in Chap. 11. Tatarski considered the evaluation of one-parameter integrals in which inner scale was finite. He expanded the function that multiplies the decaying exponential, which contains the inner scale, into a Taylor series, and integrated term by term. Since the integral over each term of the power series converges absolutely, this method is valid. Tatarski expressed the resulting power series as a hypergeometric function. His approach does not always work with zero inner scale and restricts the range of problems that can be solved. This limitation does not apply to the technique discussed here. The integrals evaluated here are of the form

$$I = \int_0^\infty dx\, x^a\, f\left(b\, x^c\right)\, g\left(d\, x^e\right), \tag{5.1}$$

where $f(x)$ and $g(x)$ are Meijer's G-functions, a special case of which are generalized hypergeometric functions. The only restriction on a, b, c, d, and e is that the integral converges. Sometimes it will be required that $d^c/b^e \neq 1$.

A powerful method for evaluating these integrals is described by *Marichev* (1983). His method applies to all integrals whose integrand is the product of two generalized hypergeometric functions and a power of the integration variable. He shows that the integral, which can be transformed into a Mellin-Barnes integral in the complex plane, can be expressed as a finite sum of generalized hypergeometric functions or, equivalently, as a Meijer G-function.

He puts integrals into a standard form to which *Slater's* (1966) theorem applies. This is convenient for some problems and is also used here. However, for other problems this approach is not always convenient. The calculation of the Strehl ratio in turbulence in Chap. 8 is an example. The Strehl ratio for uncorrected turbulence can be written as the sum of six generalized hypergeometric functions of the form ${}_6F_{11}[\,]$. This form is not convenient for either obtaining physical insight or in numerical evaluation. For these cases, I short circuit the step of transforming the integrand into standard form, since I can obtain the answer as a more convenient infinite series.

The parameters in the integrand can be less than or greater than unity. For small parameter values the Taylor series is useful, and for large values an asymptotic solution is sometimes applicable. Each of these solutions is obtained in a straightforward manner. The Taylor series can be expressed as generalized hypergeometric functions. For the problems considered in this book, the solutions will be given in terms of both generalized hypergeometric and power series.

In this chapter the theory of using Mellin transforms to obtain both Taylor series and asymptotic solutions is given. The formulas are a generalization of those given by *Marichev* (1983) and apply to cases where coefficients of the complex variable in gamma functions need not be unity. I discuss examples with repeated poles in the complex plane, which give rise to characteristic logarithmic terms.

5.1 Integral Evaluation with One Parameter

The Mellin transform pair is given by

$$h(x) \to H(s) \equiv M\left[h\left(x\right)\right] \equiv \int_0^\infty \frac{dx}{x}\, h(x)\, x^{\,s}, \quad \text{and} \tag{5.2}$$

$$h(x) = \frac{1}{2\pi i} \int_C ds\, H(s)\, x^{-s}. \tag{5.3}$$

The integral in (5.2) only converges when the value of s is within certain limits. The integration path in the inverse transform goes from $\eta - i\infty$ to $\eta + i\infty$, and the value of the real part of s along the integration path is determined by convergence properties of the function.

The Mellin transform of any function that can be expressed as a generalized hypergeometric function times a power of the integration variable is a ratio of gamma functions as discussed in Sect. 1.3. This category includes most common functions such as algebraic, exponential, trigonometric, inverse trigonometric, hyperbolic, logarithmic, complete elliptic, sine and cosine integrals, error functions, incomplete gamma, parabolic cylinder, Gegenbauer polynomials, Hermite, Laguerre, Whittaker, Legendre, Bessel, and other orthogonal functions. The Mellin transform of these functions and others are given in Appendix A.

Marichev (1983) lists 1200 Mellin transforms that are expressed as the ratio of gamma functions. There is also a similar list in *Prudnikov* et al. (1990). *Oberhettinger* (1974) also has an extensive list of Mellin transforms, but they are not all expressed as ratios of gamma functions. Table 1.1 contains Mellin transforms of functions that are used for problems of wave propagation in turbulence. Other transforms are in Appendix A. The range of s for which the integral converges is also given.

The Mellin transform definition and a list of Mellin transforms is all that is necessary to solve many simple problems as in Chap. 4. More complicated problems are solved with the Mellin convolution integral, which is the parallel to the Fourier convolution integral. This was derived in Sect. 1.2 as

$$h(x) = \int_0^\infty \frac{dy}{y}\, h_0(y)\, h_1(x/y) \to \mathcal{M}\left[h\left(x\right)\right] \equiv H(s) = H_0(s)H_1(s). \tag{5.4}$$

Notice the appearance of the multiplier $1/y$, and the appearance of $1/y$ in the argument of the second function. Obtaining the Mellin transform of a function in which the variable is the reciprocal of the usual variable, as required above, is trivially obtained by the replacement of s by $-s$ in the function's Mellin transform, as in (1.9). The single parameter in (5.4) is x or its inverse.

Using this formula one can transform the general integral given in the introduction to this chapter into an integral in the complex plane as

$$\begin{aligned} I &= \int_0^\infty dw\, w^a\, f\left(b\, w^c\right)\, g\left(d\, w^e\right) \\ &= \frac{b^{-(1+a)/c}}{e\, c} \frac{1}{2\pi i} \int_C ds\, \left(d^{1/e}\, b^{-1/c}\right)^s\, F\left(\frac{s+a+1}{c}\right)\, G\left(-\frac{s}{e}\right) \end{aligned} \tag{5.5}$$

for $c > 0$. When $c < 0$, the sign of the result changes. F and G are the Mellin transforms of f and g. The integration path in the complex plane has to be consistent with the restrictions on s in the individual Mellin transforms. The integral in the complex plane can always be performed with the methods to be discussed when $d\, b^{-e/c}$ is not equal to unity.

In the single parameter case, one evaluates the integral in the complex plane by closing the integration path in a clockwise or counterclockwise direction depending on the integrand. By Cauchy's residue theorem the value of the integral is $2\pi i$ times the sum of residues of the enclosed poles.

I now develop a method for evaluating the integral as a Taylor or asymptotic series that gives an accurate answer with the fewest number of terms for the case of a a single parameter. Since each Mellin transform in (5.5) is the ratio of gamma functions, the general form of the integral in the complex plane is a Mellin-Barnes integral given by

$$h(x) = \frac{1}{2\pi i} \int_{C'} ds\, x^{-s}\, \frac{\prod_{i=1}^{A} \Gamma\left[a_i + \alpha_i\, s\right] \prod_{j=1}^{B} \Gamma\left[b_j - \beta_j\, s\right]}{\prod_{k=1}^{C} \Gamma\left[c_k + \gamma_k\, s\right] \prod_{m=1}^{D} \Gamma\left[d_m - \delta_m\, s\right]}. \tag{5.6}$$

The terms are arranged so that all coefficients of s are positive. The poles closest to the coordinate axis and integration path for a typical integral are shown in

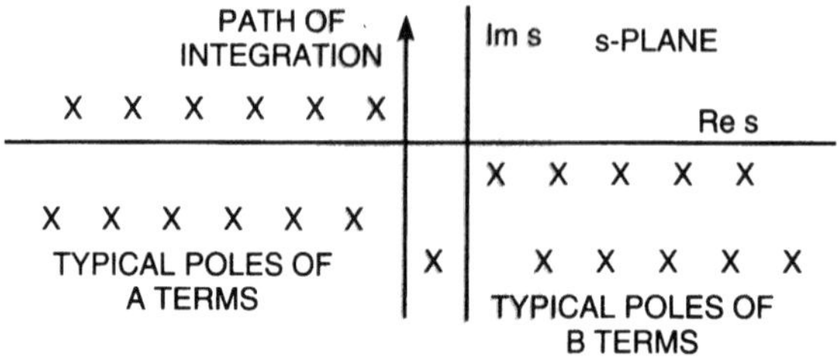

Fig. 5.1. Typical integration path and pole locations for the general integral. The poles extend to infinity in both directions.

Fig. 5.1. In general, x can be negative or complex, as it is in the beam-wave case. For infinite waves x is positive and real. This is the case considered in this chapter. The general case of complex x is treated in Chap. 9. The integration path can have all poles of a gamma function on one side of the integration path or can split its poles. In all the turbulence problems considered so far, the integration path had all the poles or all the poles but the first and rarely the second on one side of the integration path. The methods developed here allow one to evaluate the integral regardless of the pole locations.

The general integral in (5.90) for all the poles of an individual gamma function to one side of the integration path can be expressed as a Fox H-function, defined in (1.37), as

$$h(x) = H^{B,\,A}_{A+D,\,B+C}\left[\frac{1}{x}\,\middle|\,\begin{matrix}(1-a_1,\alpha_1),\ldots,(1-a_A,\ \alpha_A),(d_1,\delta_1),\ldots,(d_D,\delta_D)\\(b_1,\beta_1),\ldots,(b_B,\beta_B),(1-c_1,\gamma_1),\ldots,(1-c_C,\gamma_C)\end{matrix}\right]. \tag{5.7}$$

Since this function is not readily evaluated with current software, I generally do not express the integral in this form but as generalized hypergeometric functions. If software becomes available to evaluate the Fox H-function, this form, which is found by inspection, would become the form of choice.

To apply Cauchy's residue theorem, the integration path must be closed at infinity. Marichev considered the conditions that allow closing the path along the infinite semicircle when the coefficients of s in the gamma functions in (5.90) are unity. In this form it is easy to reduce the integrals to sums of generalized hypergeometric functions. One can obtain unity coefficients when all the coefficients of s are rational by the substitution $s = a\,y$, where a is the least common denominator of the coefficients of s. This substitution makes all the coefficients of y integers, and application of the Gauss-Legendre multiplication formula given in (1.14), results in unity coefficients of y. The number of gamma functions is equal to the sum of the coefficients of y. Using that procedure, one arrives at final answers that are sometimes the sum of many high-order generalized hypergeometric functions. For some problems this form is not only lengthy to evaluate, but also it provides little physical insight. In such cases, a different procedure can be used. The coefficients of s are left as they are, and the final answer is obtained as a rapidly converging series in powers of the parameters. This method works even if the coefficients are irrational. Separate power series are obtained for large and small parameter values, while under certain conditions for large parameter values, an asymptotic solution can be obtained. For many problems one can reduce the integral to sums of a few generalized hypergeometric functions.

Pole-residue integration is the method used to evaluate these integrals. To apply this method, the integration path must be closed in the complex plane, and the value of the integral is $2\pi i$ times the sum of the residues at the enclosed poles. The only singularities of $\Gamma\left[a_i + \alpha_i\, s\right]$ are simple poles that occur when $s = -\left(n + a_i\right)/\alpha_i$ with residues equal to $(-1)^n/\alpha_i\, n!$. The resulting residue sum is a power series of the parameter that can be expressed as

$$h(x) = \sum_n x^{-s_n}\, G\left(s_n\right), \tag{5.8}$$

where $G\left(s_n\right)$ is the residue at the pole occurring at s_n. The summation is over all poles enclosed within the integration path. To close the path on the infinite semicircle without affecting the value of the integral, the integrand must decrease faster than $1/s$ for large values of s. Conditions for convergence along the semicircle in the left half plane are now obtained.

To determine whether the integration path can be closed in the left half plane, the asymptotic behavior of the integrand must be examined. The behavior of the gamma function for large arguments is given by Stirling's formula

$$\Gamma\left[s\right]\ \sqrt{2\pi}\, s^{s-1\ 2} \exp(-s)\left(1 + \frac{s}{12} + \frac{s^2}{288} + \cdots\right), \quad \left|\arg\{s\}\right| < \pi. \tag{5.9}$$

This can be used to find the asymptotic form for the gamma functions with negative coefficients of s; however, it is not valid in the entire left half plane when the gamma function argument is positive. To obtain a valid expression, the duplication formula

$$\frac{1}{\Gamma\left[s\right]} = \frac{\sin(\pi\, s)}{\pi}\Gamma\left[1 - s\right] \tag{5.10}$$

is used to eliminate the gamma functions with positive coefficients of s; this results in an integrand of the following form

$$In = x^{-s}\pi^{A-C}\, \frac{\prod\limits_{j=1}^{B} \Gamma\left[b_j - \beta_j\, s\right] \prod\limits_{k=1}^{C} \Gamma\left[1 - c_k - \gamma_k\, s\right]\ \sin\left[\left(c_k + \gamma_k\, s\right)\pi\right]}{\prod\limits_{m=1}^{D} \Gamma\left[d_m - \delta_m\, s\right] \prod\limits_{i=1}^{A} \Gamma\left[1 - a_i - \alpha_i\, s\right]\ \sin\left[\left(a_i + \alpha_i\, s\right)\pi\right]}. \tag{5.11}$$

All constants are assumed to be real, and in addition, the coefficients of s are assumed to be non-negative. To examine behavior at negative infinity, the following two relations are necessary

$$\lim_{s \to -\infty} \sin\left[(a + \alpha\, s)\pi\right] = O\left[\exp\left(\alpha\pi\ \left|\mathrm{Im}\,\{s\}\right|\right)\right], \quad \text{and} \tag{5.12}$$

$$\lim_{s \to -\infty} \Gamma[s] = O\{\exp[(s-1/2)\ln(s) - s]\}. \tag{5.13}$$

The symbol O is used to denote the quantity's order of magnitude. Using these relations in the integrand, rearranging terms, and using the definitions in Table 5.1, one shows the integrand is

$$In = O\left[\exp\left(\Delta \mathrm{Re}\{s\}\ln|s| + [-\ln|s| - \Delta + \Delta']\,\mathrm{Re}\{s\} + [\nu - \Xi/2]\ln|s|\right.\right.$$
$$\left.\left. + |\mathrm{Im}\{s\}|\,[\Xi \arg\{-s\} + \pi(C' - A')] + \mathrm{Im}\{s\}\arg\{-s\}\right)\right]. \tag{5.14}$$

It is assumed that the original integral converged; therefore, the sign of the exponent as the integration path goes to $\pm\infty$ along the imaginary axis does not have to be considered. The sign of the exponent must be negative for the integral along the semicircle to be negligible. There is a hierarchy of terms that determine the properties at infinity. The terms in decreasing order of importance are $\mathrm{Re}\{s\}\ln|s|$, $\mathrm{Re}\{s\}$, and $\ln|s|$. *The sign of these terms determines the direction of path closure.*

The dominant term that determines the sign of the exponent at infinity is $\Delta\mathrm{Re}\{s\}\ln|s|$. Δ is the sum of the coefficients of s in the numerator gamma functions minus those in the denominator. If $\Delta > 0$, the integral can be closed in the left half plane. A similar analysis shows that when $\Delta < 0$ the integral can be closed in the right half plane without changing the value of the integral. For both cases, a single power series that applies to all parameter values is obtained, and it converges quickly for small values of the parameter. For large parameter values the series converges slowly, and there can be numerical difficulties when one evaluates a summation composed of large terms that may alternate in sign. In this parameter regime, an asymptotic solution is appropriate, and the method to derive this solution is discussed in the next section.

If $\Delta = 0$, the term with $[-\ln|x| + \Delta']\,\mathrm{Re}\{s\}$ is the most important. Let

$$\Pi = -\ln|x| + \Delta'. \tag{5.15}$$

For $\Pi < 0$, the path can be closed in the right half plane. For $\Pi > 0$, the path can be closed in the left half plane. Typically, if $\Delta = 0$, then also $\Delta' = 0$, and in that case, for $|x| > 1$ the path can be closed in the right half plane, while for $|x| < 1$ the path can be closed in the left half plane. Here, separate power series for large and small values of x are obtained, and both converge rapidly, except possibly close to x equal to unity.

If both $\Delta = 0$ and $\Pi = 0$, the integrand behaves as

$$I = O\left[s^{(\nu - \Xi/2)}\right]. \tag{5.16}$$

For convergence on the infinite semicircle, the integrand has to decrease faster than $1/s$, and this gives the condition

$$\Omega = \nu - \Xi/2 + 1 < 0. \tag{5.17}$$

If this condition is satisfied, the integration path can be closed in either direction. If this condition is not satisfied, the path cannot be closed at infinity, and pole-residue integration cannot be used.

To illustrate the technique consider the integral

$$I = \int_0^\infty \frac{dy}{y}\, y^\mu \exp(-ay)\, J_\nu(by)\,. \tag{5.18}$$

After normalizing by letting $x = ay$, one can convert the integral into one in the complex plane with the Mellin convolution integral, the Mellin transforms in (1.45) and (1.49), and the relation in (1.8), to give

$$I = \frac{1}{2\,a^\mu}\frac{1}{2\pi i}\int_C ds \left(\frac{2a}{b}\right)^{-s} \Gamma\left[\begin{matrix} s+\mu,\ -s/2+\nu/2 \\ s/2+\nu/2+1 \end{matrix}\right]. \tag{5.19}$$

With the use of (1.37) the solution can be written by inspection as a Fox H-function

$$I = \frac{1}{2\,a^\mu} H_{1,2}^{1,1}\left[\left(\frac{2\,a}{b}\right) \middle| \begin{matrix} (1-\mu,\ 1) \\ (\nu/2,\ 1/2),\ (-\nu/2,\ 1/2) \end{matrix}\right]. \tag{5.20}$$

To express this as a Meijer G-function, make the substitution $s \to 2\,s$ in (5.19) and use the multiplication formula in (1.15) to obtain

$$I = \frac{1}{2\sqrt{\pi}}\left(\frac{2}{a}\right)^\mu \frac{1}{2\pi i}\int_C ds \left(\frac{b}{a}\right)^{2s} \Gamma\left[\begin{matrix} s+\mu/2,\ s+\mu/2+\frac{1}{2},\ -s+\nu/2 \\ s+\nu/2+1 \end{matrix}\right]. \tag{5.21}$$

From (1.38) the solution can be written as

$$I = \frac{1}{2\sqrt{\pi}}\left(\frac{2}{a}\right)^\mu G_{2,2}^{1,2}\left[\left(\frac{b}{a}\right)^2 \middle| \begin{matrix} 1-\mu/2,\ 1/2-\mu/2 \\ \nu/2,\ -\nu/2 \end{matrix}\right]. \tag{5.22}$$

The solution can also be expressed as a generalized hypergeometric function. Since $\Delta = 0$ in (5.21), for $b/a > 1$ close the integration path to the right and obtain pole contributions at $s = \nu/2 + n$ for $n = 0, 1, \ldots$. The value of the integral is

$$I = \frac{2^{\mu-1} b^\nu}{\sqrt{\pi}\, a^{\nu+\mu}} \sum_{n=0}^\infty \frac{(-1)^n}{n!}\left(\frac{b}{a}\right)^{2n} \Gamma\left[\begin{matrix} n+\frac{\mu+\nu}{2},\ n+\frac{\mu+\nu+1}{2} \\ n+\nu+1 \end{matrix}\right]. \tag{5.23}$$

Use the rules in Sect. 1.3 to convert this into a generalized hypergeometric function, and use the multiplication theorem to simplify the expression to

$$I = \left(\frac{b}{2\,a}\right)^\nu a^{-\mu}\Gamma\left[\begin{matrix}\nu+\mu \\ \nu+1\end{matrix}\right] {}_2F_1\left[\frac{\mu+\nu}{2}, \frac{\mu+\nu+1}{2}; \nu+1; -\left(\frac{b}{a}\right)^2\right], \quad b > a. \tag{5.24}$$

This is the same formula as in *Gradshteyn* and *Ryzhik* (1980, Eq. 6.621 #1).

When $a/b > 1$ the integration path is closed to the left, and one obtains pole contributions at $s = -n - \mu/2$, and $s = -n - \mu/2 - 1/2$ for $n = 0, 1, \ldots$. The value of the integral expressed as generalized hypergeometric functions is

$$I = \frac{1}{2}\left(\frac{2}{b}\right)^{\mu} \Gamma\left[\begin{matrix}\frac{\nu+\mu}{2} \\ \frac{\nu-\mu+2}{2}\end{matrix}\right] {}_2F_1\left[\frac{\mu+\nu}{2}, \frac{\mu-\nu}{2}; \frac{1}{2}; -\left(\frac{a}{b}\right)^2\right]$$
$$+\frac{a}{b}\left(\frac{2}{b}\right)^{\mu} \Gamma\left[\begin{matrix}\frac{\nu+\mu+1}{2} \\ \frac{\nu-\mu+1}{2}\end{matrix}\right] {}_2F_1\left[\frac{\mu+\nu+1}{2}, \frac{\mu-\nu+1}{2}; \frac{3}{2}; -\left(\frac{a}{b}\right)^2\right], \quad a > b. \quad (5.25)$$

I now examine the convergence properties when $a = b$, in which case the integral is

$$I = \int_0^\infty \frac{dy}{y} y^\mu \exp(-y) J_\nu(y). \quad (5.26)$$

Equation (5.17) states that the solution in (5.24) converges when $\mu < 1/2$. This is equivalent to the third condition given after (1.33). The integral converges even if this requirement is not satisfied; however, Mellin transform techniques do not apply. It will be shown in Chap. 11 that the single point at which the method does not apply expands to a finite volume when there is more than one parameter.

An integral can also be evaluated easily when an integrand factor has the first term of its power series subtracted. Consider, e.g.,

$$I = \int_0^\infty \frac{dy}{y} y^\mu [\exp(-ay) - 1] J_\nu(by). \quad (5.27)$$

With this change (5.21) becomes

$$I = \frac{2^{\mu-1}}{\sqrt{\pi}\, a^\mu} \frac{1}{2\pi i} \int_C ds \left(\frac{b}{a}\right)^s \Gamma\left[\begin{matrix} s+\mu/2*,\ s+\mu/2+\frac{1}{2},\ -s+\nu/2 \\ s+\nu/2+1\end{matrix}\right], \quad (5.28)$$

where the asterisk indicates that the integration path falls between the first and second poles of a gamma function. The solution in this case is simply the solution above with one series term moved from the expression valid for $a > b$ to the one for $a < b$ to give

$$I = \left(\frac{b}{2a}\right)^{\nu} a^{-\mu} \Gamma\left[\begin{matrix}\nu+\mu \\ \nu+1\end{matrix}\right] {}_2F_1\left[\frac{\mu+\nu}{2}, \frac{\mu+\nu+1}{2}; \nu+1; -\left(\frac{b}{a}\right)^2\right]$$
$$-\frac{1}{2}\left(\frac{2}{b}\right)^{\mu} \Gamma\left[\begin{matrix}\frac{\nu+\mu}{2} \\ \frac{\nu-\mu+2}{2}\end{matrix}\right], \quad b > a, \text{ and} \quad (5.29)$$

$$I = \frac{1}{2}\left(\frac{2}{b}\right)^{\mu} \Gamma\left[\begin{matrix}\frac{\nu+\mu}{2} \\ \frac{\nu-\mu+2}{2}\end{matrix}\right] \left\{ {}_2F_1\left[\frac{\mu+\nu}{2}, \frac{\mu-\nu}{2}; \frac{1}{2}; -\left(\frac{a}{b}\right)^2\right] - 1\right\}$$
$$+\frac{a}{b}\left(\frac{2}{b}\right)^{\mu} \Gamma\left[\begin{matrix}\frac{\nu+\mu+1}{2} \\ \frac{\nu-\mu+1}{2}\end{matrix}\right] {}_2F_1\left[\frac{\mu+\nu+1}{2}, \frac{\mu-\nu+1}{2}; \frac{3}{2}; -\left(\frac{a}{b}\right)^2\right], \quad a > b. \quad (5.30)$$

The first function on the right of the last equation can be expressed as another generalized hypergeometric function with the relation in (1.40) to obtain

$$I = \left(\frac{a}{b}\right)^2 \left(\frac{2}{b}\right)^{\mu} \Gamma\left[\begin{matrix}\frac{\nu+\mu+2}{2} \\ \frac{\nu-\mu}{2}\end{matrix}\right] {}_3F_2\left[\frac{\mu+\nu+2}{2}, \frac{\mu-\nu+2}{2}, 1; \frac{3}{2}, 2; -\left(\frac{a}{b}\right)^2\right]$$
$$+\frac{a}{b}\left(\frac{2}{b}\right)^{\mu} \Gamma\left[\begin{matrix}\frac{\nu+\mu+1}{2} \\ \frac{\nu-\mu+1}{2}\end{matrix}\right] {}_2F_1\left[\frac{\mu+\nu+1}{2}, \frac{\mu-\nu+1}{2}; \frac{3}{2}; -\left(\frac{a}{b}\right)^2\right], \quad a > b. \quad (5.31)$$

Similarly, the integral can be evaluated when the first term of the power series is subtracted from the second factor in the integrand in (5.18)

$$I = \int_0^{\infty} \frac{dy}{y}\, y^{\mu} \exp(-ay) \left[J_{\nu}(by) - \frac{1}{\Gamma[\nu+1]}\left(\frac{by}{2}\right)^{\nu}\right]. \quad (5.32)$$

In this case the asterisk is after the third term in the gamma function in (5.21). The integral can be evaluated to give

$$I = \frac{1}{2}\left(\frac{2}{b}\right)^{\mu} \Gamma\left[\begin{matrix}\frac{\nu+\mu}{2} \\ \frac{\nu-\mu+2}{2}\end{matrix}\right] {}_2F_1\left[\frac{\mu+\nu}{2}, \frac{\mu-\nu}{2}; \frac{1}{2}; -\left(\frac{a}{b}\right)^2\right] - \left(\frac{b}{2a}\right)^{\nu} a^{-\mu} \Gamma\left[\begin{matrix}\nu+\mu \\ \nu+1\end{matrix}\right]$$
$$+\frac{a}{b}\left(\frac{2}{b}\right)^{\mu} \Gamma\left[\begin{matrix}\frac{\nu+\mu+1}{2} \\ \frac{\nu-\mu+1}{2}\end{matrix}\right] {}_2F_1\left[\frac{\mu+\nu+1}{2}, \frac{\mu-\nu+1}{2}; \frac{3}{2}; -\left(\frac{a}{b}\right)^2\right], \quad a > b, \text{ and} \quad (5.33)$$

$$I = \left(\frac{b}{2a}\right)^{\nu} a^{-\mu} \Gamma\left[\begin{matrix}\nu+\mu \\ \nu+1\end{matrix}\right] \left\{ {}_2F_1\left[\frac{\mu+\nu}{2}, \frac{\mu+\nu+1}{2}; \nu+1; -\left(\frac{b}{a}\right)^2\right] - 1\right\}, \quad b > a. \quad (5.34)$$

Using (1.40) one can show that the last expression is

$$I = -\left(\frac{b}{2a}\right)^{\nu+2} a^{-\mu} \Gamma\left[\begin{matrix}\nu+\mu+2 \\ \nu+2\end{matrix}\right]$$
$$\times {}_3F_2\left[\frac{\mu+\nu+2}{2}, \frac{\mu+\nu+3}{2}, 1; \nu+2, 2; -\left(\frac{b}{a}\right)^2\right], \quad b > a. \quad (5.35)$$

To summarize, if $\Delta \neq 0$, a solution is obtained for any parameter value. If $\Delta = 0$, a solution is found for all parameter values except for certain conditions, when the solution is not found for the parameter equal to unity.

5.2 Asymptotic Solutions

Asymptotic solutions are often used to obtain insight into the behavior of a solution for large parameter values and to ease the computations. There are many techniques to obtain an asymptotic solution. The Mellin transform technique is applicable to a wide variety of problems, including all turbulence problems—but by no means to all. *Wong* (1989) discusses several ways to find asymptotic solutions. He has a small section on Mellin transforms that is not as general as that presented here. Many of the examples that he solves by other techniques could be solved by the Mellin transform method. In general, Mellin transform techniques are underutilized.

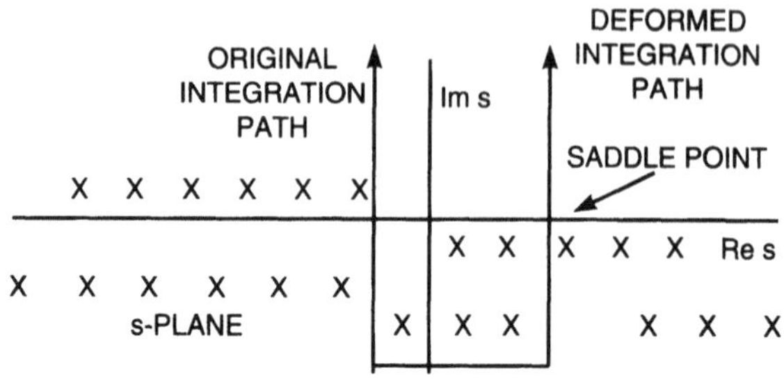

Fig. 5.2. Original and deformed integration paths used to evaluate the asymptotic series.

When $\Delta = 0$, separate power series are obtained for small and large parameter values. Each Taylor series has good convergence properties except possibly when the parameter is equal to one, which allows them to be readily evaluated without numerical difficulties. The solution for large parameters is a convergent Taylor series. I do not refer to such a solution as an asymptotic solution. Wong calls these asymptotic series. I reserve the term asymptotic series for non-convergent series.

When $\Delta \neq 0$, a single power series is obtained, which converges slowly for large parameter values. To overcome numerical difficulties, for large parameter values an asymptotic solution is preferred. One finds the asymptotic solution not by closing the integration path at infinity, but by deforming the integration path to go through a saddle point in the direction of steepest descent as shown in Fig. 5.2. The asymptotic solution is composed of the two contributions from poles crossed in the path deformation and from the steepest-descent integration. The steepest-descent contribution can be found for complex parameters; however, this is more complicated than that for real values. The case when x is positive real is the only one considered in this chapter. The general case is considered in Chap. 9.

The value of the integral along the new integration path is found by a modification of the method of steepest descent. In this method the path is deformed from the original path to a path through a saddle point such that the function's value decreases most rapidly along the path away from the saddle point. Along this new path, the value of the integrand is a maximum at the saddle point and decreases away from it. For large parameter values, the saddle point is located at a large value of s in the complex plane. Near the saddle point one writes

$$I = \int_C ds \, \exp\left[f(s)\right] . \tag{5.36}$$

Write the argument of the exponent as a Taylor series given by

$$f(s) \approx f\left(s_o\right) + \frac{df\left(s_o\right)}{ds}\left(s - s_o\right) + \frac{1}{2}\frac{d^2 f\left(s_o\right)}{ds^2}\left(s - s_o\right)^2 . \tag{5.37}$$

It may be shown that if the parameter is very large, then only these three terms are necessary to express the value of the integral with sufficient accuracy. At a saddle point, the value of the first derivative is zero. This condition gives equations to determine the real and imaginary parts of s_o at the saddle point. It will be seen that the contribution of derivative terms higher than the second to the integral are small compared to the second and can be normally neglected. The integral along this path is approximately that of a Gaussian function with infinite limits that is easily evaluated to give

$$I \approx \int\limits_C ds \exp\left[f(s_o) + \frac{1}{2}\frac{d^2 f(s_o)}{ds^2}(s - s_o)^2\right] = \sqrt{-\frac{2\pi}{d^2 f(s_o)/ds^2}} \exp\left[f(s_o)\right]. \tag{5.38}$$

This is the first term of the asymptotic expansion. Additional terms can be found if desired by the inclusion of more terms in the expansion of $f(s)$ in (5.37). For problems examined to date, the first term of the asymptotic series gives sufficient accuracy to get a seamless fit between the Taylor and asymptotic series. For that reason, it is the only one evaluated here.

The asymptotic value of the integral has contributions from poles crossed in the path deformation, in addition to the steepest-descent contribution. Under certain conditions, one of these two terms dominates, and the other can be neglected. If there are no poles to the right of the original integration path, the steepest-descent contribution must always be included. The pole contribution decays algebraically with the parameter. I show below that the steepest-descent contribution can vary sinusoidally, can have exponential decay, or can have exponential growth. For the sinusoidal variation, both pole and steepest-descent contributions are important, and both must be retained. When the steepest-descent contribution decays exponentially, it can be neglected if there are any pole contributions. If the steepest-descent contribution increases exponentially, pole contributions are negligible. Conditions that allow the steepest-descent contribution to be neglected will be considered later.

The asymptotic value of the integral in (5.90) is now found. Only the case $\Delta > 0$ needs to be considered. In this case, the integration path can be closed in the left half plane to obtain the convergent power series. If $\Delta < 0$, the substitution $s \to -s$ transforms the integral to one in which $\Delta > 0$, and the results derived below again apply. For the asymptotic series, the behavior of the integrand in the right half plane must be examined. As a simple case, the asymptotic series for $B = D = 0$ in (5.90) will be derived first. (This case also encompasses the situation in which $B = D$, but all the poles of the B gamma functions are cancelled by poles of the denominator gamma functions. In this case the ratio of gamma functions can be replaced by a polynomial.) When there are no poles in the right-half plane, the asymptotic series does not have a branch cut in the right half plane, and the evaluation of the steepest-descent contribution is straightforward. The behavior of the integrand for large s is needed. Unlike the order of magnitude calculation previously made to determine whether the integral converges on an infinite circle, more accurate integrand estimates are required to obtain the actual value of the integral. The asymptotic form of a gamma function is

$$\Gamma\left[a_i + \alpha_i\, s\right] \sim \sqrt{2\pi}\, \exp\left[(a_i + \alpha_i\, s - 1/2)\, \ln\left(a_i + \alpha_i\, s\right) - a_i - \alpha_i\, s\right]. \quad (5.39)$$

The following approximation for the logarithm is used

$$\ln\left(a_i + \alpha_i\, s\right) = \ln\left[\alpha_i\, s\, \left(1 + a_i/\alpha_i\, s\right)\right]\; \ln\left(\alpha_i\right) + \ln\left(s\right) + a_i/\alpha_i\, s. \quad (5.40)$$

If these are inserted into (5.90) and a similar expansion is performed for the other gamma functions, one obtains

$$h\left(x\right) = \frac{(2\pi)^{\Xi/2}}{2\pi\, i} \int\limits_C ds \exp\left\{\Delta'' - \Delta'''/2 \right.$$
$$\left. + s\left[\Delta' - \Delta - \ln\left(x\right)\right] + \ln(s)\left(\nu - \Xi/2\right) + \Delta s \ln(s)\right\}. \quad (5.41)$$

The new symbols introduced above are defined in Table 5.1.

If the exponent is designated by $f(s)$, the saddle point is located at

$$\frac{df(s)}{ds} = 0 = \left[\Delta' - \Delta - \ln\left(x\right)\right] + \frac{\nu - \Xi/2}{s} + \Delta\, \left[1 + \ln\left(s\right)\right]. \quad (5.42)$$

Since $|s|$ is large (I will check this below by consistency), its value can be approximated by

$$\ln\left(s_o\right) \approx \left[\ln\left(x\right) - \Delta'\right]/\Delta, \quad (5.43)$$

which yields

$$s_o = x^{1/\Delta} \exp\left(-\Delta'/\Delta\right). \quad (5.44)$$

The value of s_o is large and positive since x is large; this agrees with the assumption that the integration path was to be moved far into the right half plane.

The second derivative is

$$\frac{d^2 f(s_o)}{ds^2} = -\frac{\nu - \Xi/2}{s_o^2} + \frac{\Delta}{s_o} \approx \frac{\Delta}{s_o}. \quad (5.45)$$

Since this is positive in the right half plane, the steepest-descent path must be in the imaginary direction as shown in Fig. 5.2 to make the quadratic term in (5.37) negative. Each higher-order derivative has an increasingly higher power of s in the denominator. Therefore, the higher-order derivatives are small compared to the second, and the assumption that $f(s)$ can be accurately expressed as a constant term plus a quadratic term is valid. If these values are substituted into (5.38), one obtains for the steepest-descent contribution

$$E(x) = \frac{(2\pi)^{(\Xi-1)/2}}{\sqrt{\Delta}} x^{\rho} \exp\left[-\Delta\, x^{1/\Delta} \exp\left(-\Delta'/\Delta\right) + \Delta'' - \Delta'''/2 - \rho\Delta'\right]. \quad (5.46)$$

For most problems encountered in propagation through turbulence one has the condition, $\Delta' = \Delta'' = \Delta''' = 0$. For that case one obtains the expression used by *Marichev* (1983, Eq. 4.40)

$$E(x) = \frac{(2\pi)^{(\Xi-1)/2}}{\sqrt{\Delta}} x^{\rho} \exp\left(-\Delta\, x^{1/\Delta}\right). \tag{5.47}$$

This solution has an exponential decay and can be neglected if there are pole contributions.

The more general case has B and D not equal to zero in (5.11). This case is more complicated since one must deal with the multi-valued nature of the asymptotic behavior of the functions. *Luke* (1969) derives expressions for this case, and a similar approach can be used to obtain the steepest-descent contribution for the case where the coefficients of s in the gamma functions are not unity. The details of this derivation are considered in Chap. 9 where x is allowed to be complex. The result for $\Lambda' \geq 0$ is

$$\begin{aligned} E(x) &= \frac{2(2\pi)^{(\Xi-1)/2}}{\sqrt{\Delta}} x^{\rho} \\ &\times \exp\left[-\Delta x^{1/\Delta} \exp\left(-\Delta'/\Delta\right) \cos\left(\pi\Lambda'/\Delta\right) + \Delta'' - \Delta'''/2 - \rho\Delta'\right] \\ &\times \cos\left[\Delta x^{1/\Delta} \exp\left(-\Delta'/\Delta\right) \sin\left(\pi\Lambda'/\Delta\right) + \pi\left(-\rho\Lambda' - \Lambda'' + \Lambda/2\right)\right]. \end{aligned} \tag{5.48}$$

Notice the factor of 2 that multiplies this expression that is not present in (5.92). This is a consequence of the addition of two steepest-descent contributions that are complex conjugates of each other. In (5.93) if $B = D = 0$ or, equivalently, if the argument of the second cosine is zero, then the answer must be divided by two. If $\Lambda' < 0$, then the steepest-descent contribution has exponential decay. The decay coefficient is found with the procedure discussed in Sect. 9.2. For the simpler case that applies to many problems of interest in which

$$\Delta' = \Delta'' = \Delta''' = 0, \quad B = B', \text{ and } D = D', \tag{5.49}$$

one obtains a result equivalent to that obtained by *Marichev* (1980, Eq. 4.51)

$$\begin{aligned} E(x) &= \frac{2(2\pi)^{(\Xi-1)/2}}{\sqrt{\Delta}} x^{\rho} \exp\left[-\Delta x^{1/\Delta} \cos\left(\pi\Lambda/\Delta\right)\right] \\ &\times \cos\left\{\Delta x^{1/\Delta} \sin\left(\pi\Lambda/\Delta\right) + \pi\left[\left(1/2 - \rho\right)\Lambda - \Lambda''\right]\right\}. \end{aligned} \tag{5.50}$$

Designate the sum of the residues of all poles crossed in the integration path deformation by $W(x)$. The general asymptotic solution is

$$h(x) \sim W(x) + E(x). \tag{5.51}$$

The general equations necessary to obtain $E(x)$ are given in Table 5.1. In some cases, one of E or W is insignificant compared to the other. That will be obvious once the terms are evaluated. Rather than always evaluating the steepest-descent contribution, one can state conditions in which only the pole contributions are significant for the asymptotic solution. These are

$$\text{If } B' > 0 \text{ and } \Lambda'/\Delta \neq 1/2, \text{ then} \quad h\left(x\right) \sim W\left(x\right). \tag{5.52}$$

In this case the asymptotic solution has pole contributions, and the steepest-descent contribution has exponential decay, which makes it negligible.

As an example of obtaining an asymptotic series, consider the inverse transform of the function considered in (1.21)

$$J_0(x) - 1 = \frac{1}{2\pi i}\int_C ds\,\frac{1}{2}\left(\frac{x}{2}\right)^{-s} \Gamma\left[\begin{matrix} s/2* \\ -s/2+1 \end{matrix}\right]. \tag{5.53}$$

Using (5.94) in (5.51), one obtains the pole residue at $s = 0$, and the saddle point contribution, to get

$$J_0(x) - 1 \sim \sqrt{2/\pi x}\,\cos(x - \pi/4) - 1. \tag{5.54}$$

This is the standard result.

Asymptotic solutions have useful properties that allow them to be treated almost like Taylor series, and the ones obtained here are all of the Poincaré type (*Bleistein* and *Handelsman* 1986) that have the following characteristics:

1. The error in truncation is less than the first term neglected.
2. The asymptotic series of the sum or product of two functions is equal to the sum or product of the asymptotic series of the individual functions.
3. If the asymptotic series of the integral or derivative of a function exists, it is equal to the integral or derivative of the asymptotic series of that function.
4. The asymptotic series is unique for a given function.

As the number of terms in the asymptotic series is increased, the error in the value of the integral first decreases, and after some number of terms n_o starts to increase. The value of n_o can be found from the position of the steepest descent path for a given parameter value. In practice, one does not evaluate n_o since a few terms of the asymptotic series give an accurate result for large parameter values.

$E(x)$ is the first term of the series representation of the asymptotic series. Typically, for most problems this answer is accurate enough. If more terms of the series are required, they can be obtained with Luke's approach. This approach is complicated, and different expressions apply for different conditions on the parameters. The details of this approach will not be given here since, for all problems considered so far, the first term is sufficient to produce a region of overlap in which the values obtained from the Taylor and the asymptotic series are very close to each other.

5.2.1 Alternate Method of Integral Evaluation

The method of evaluating integrals presented in this chapter is straightforward in concept. One finds Taylor series by closing the integration path in a direction that does not change the value of the integral. For some problems an asymptotic solution is found by the addition of the residues of poles on the other side of the integration path to a steepest-descent contribution. In Chap. 11 I will consider the evaluation of integrals in multiple complex planes. A natural approach in this more complicated scenario would be to extend the technique developed in this chapter. Unfortunately, one encounters difficulties in trying to decide which pole-contributions to include in a solution. The Mellin convolution theorem leads to gamma functions that contain the sum of complex integration variables; this results in a problematic coupling of integration paths.

Fortunately, another approach does provide a conceptually straightforward generalization to several complex planes. This approach does not rely on considering how the contours must be closed. Instead all power series arising from poles on either side of the integration are written down. One obtains the value of the integral by choosing the correct power series. The Taylor series solution is simply the sum of those power series that converge for the parameter size of interest.

The asymptotic solution is obtained in the following manner: Express all power series in terms of the parameter or its inverse, whichever gives a positive exponent. The asymptotic series is the sum of those diverging series for which the parameter in the series is less than unity. A steepest-descent contribution must be added to the asymptotic series to obtain a complete solution.

If the integration path split the poles of a gamma function so that p poles are on the other side of the integration path, then that summation starts at the index equal to p. If a series that starts with the index equal to p is excluded from the solution by the criteria above, then the series with index value varying from zero to p must be added to the solution.

It is easy to see that this approach gives the same answer as that obtained with the method expounded in this chapter. In this second method one is not concerned with how contours are closed. The series that are selected from all possible power series are those with the appropriate convergence properties. This method can be applied directly to the evaluation of integrals in multiple complex planes. It is further developed in Chap. 11, and used in Chaps. 12 and 13 to solve specific problems.

5.3 Multiple Poles

I show in this section that the occurrence of multiple poles, although requiring special treatment, does not pose any conceptual difficulties when one calculates the power or asymptotic series. One finds the residue at a multiple pole by expanding all integrand functions into a Laurent series centered at the multiple-pole position, by multiplying *all* the series together, and by determining the coefficient of the resulting simple pole. This is illustrated with examples that contain double and triple poles.

In turbulence theory there are typically fractional powers in integrals. The Mellin transforms of the functions encountered in turbulence theory do not have multiple poles. The poles of the transforms of functions that multiply the fractional powers are at negative integers or halfway between the negative integers. When the fractional power is introduced in the convolution theorem, the poles of the convolved functions do not coincide. I have not encountered any problems in turbulence wave propagation that have multiple poles. However, in other contexts some of the functions in the integral can have Mellin transforms that have multiple poles. For that reason the method to evaluate integrals in which multiple poles occur is developed here. Multiple poles have been encountered in the expansion of $K_\nu(x)$ in Chap. 1, and they will be encountered again in the evaluation of the potential surrounding a uniformly charged cylinder in Sect. 6.2.

A function that has multiple poles and occurs in elementary particle physics is the polylogarithm. The polylogarithm function of order n, $L^n(z)$, arises from an integral with a multiple pole of order n. The definition is

$$L^n(z) = \frac{1}{2\pi i}\int\limits_C ds\,\frac{\Gamma[s,\,-s]}{(-s)^{n-1}}\,(-z)^{-s}, \quad n = 0,\ 1,\ 2,\ \ldots. \tag{5.55}$$

This function is discussed extensively in Marichev. Its Mellin transform is given in Appendix A.

5.3.1 Expansion of Integrand Functions

To calculate the residue in the presence of a multiple pole, the functions in the integrand are expanded in power series at the pole, and the coefficient of the simple pole term is calculated. To facilitate this process, in this section two expansions of the gamma function are derived: one when the singular point is not at a pole of the gamma function; the second when it is. The function z^{-s} is also expanded in a power series.

The expansion of the gamma functions is normally expressed in terms of the logarithmic derivative of the gamma function, defined by

$$\psi(z) \equiv \frac{d\ln(\Gamma[z])}{dz} = \frac{\Gamma'[z]}{\Gamma[z]}. \tag{5.56}$$

This psi or digamma function is given in *Gradshteyn* and *Ryzhik* (1980, Eq. 8.362 #1) as

$$\psi(z) = -\mathbf{C} + \sum_{n=0}^{\infty}\left(\frac{1}{n+1} - \frac{1}{z+n}\right), \tag{5.57}$$

where $\mathbf{C}$ is Euler's constant, which is equal to 0.577216... . The derivative of the last expression gives the trigamma function

$$\psi'(z) = \sum_{m=0}^{\infty}\frac{1}{(z+m)^2}. \tag{5.58}$$

The higher order derivatives are given by

$$\frac{d^n \ln\left(\Gamma\left[z\right]\right)}{dz^n} \equiv \psi^{(n)}\left(z\right) = \left(-1\right)^{n+1} n!\sum_{m=0}^{\infty} \frac{1}{\left(z+m\right)^{n+1}}. \tag{5.59}$$

These expressions can be summed numerically to find the values, which are most often needed: $\psi\left(1\right) = -\mathbf{C} = -0.577216$; $\psi\left(0.5\right) = -\mathbf{C} - 2\ln\left(2\right) = -1.964$; $\psi'\left(-0.5\right) = 4 - \pi^2/2 = 8.935$; $\psi\left(-0.5\right) = 2 - \mathbf{C} - 2\ln\left(2\right) = -0.03649$; $\psi'\left(0.5\right) = \pi^2/2 = 4.935$; and $\psi'\left(1\right) = \pi^2/5 = 1.645$.

For positive z one expands

$$\Gamma\left[z+\varepsilon\right] = \Gamma\left[z\right]\left\{1 + \varepsilon\,\psi(z) + \varepsilon^2\left[\psi'(z) + \psi^2(z)\right]/2 + \cdots\right\}. \tag{5.60}$$

The Gamma function $\Gamma\left[s\right]$ has a pole at $s = -N$. To find the Laurent expansion about a pole use the identity

$$\Gamma\left[-N+\varepsilon\right] = \frac{\left(-1\right)^N}{\varepsilon}\frac{\Gamma\left[1+\varepsilon\right]\Gamma\left[1-\varepsilon\right]}{\Gamma\left[1+N+\varepsilon\right]}. \tag{5.61}$$

This identity can be proved with the recursion relation in (1.13) and the identity $\Gamma\left[s\right]\,\Gamma\left[1-s\right] = \pi/\sin\left(\pi s\right)$. If each of the three gamma functions on the right is expanded with (5.61), one gets the series representation

$$\Gamma\left[-N+\varepsilon\right] = \frac{\left(-1\right)^N}{\varepsilon N!}\sum_{n=0}^{\infty} a_n(N)\,\varepsilon^n. \tag{5.62}$$

The first few coefficients of this expansion are

$$a_0(N) = 1, \quad a_1(N) = \psi(1+N), \quad \text{and} \tag{5.63}$$

$$a_2(N) = \psi'(1) + [\psi^2(1+N) - \psi'(1+N)]/2. \tag{5.64}$$

Finally the Taylor series expansion of z^{-s} about the point $s = k$ is

$$z^{-s}\Big|_{s=k+\varepsilon} = z^{-k}\sum_{n=0}^{\infty}\frac{(-1)^n}{n!}\left[\ln(z)\right]^n \varepsilon^n ! = z^{-k}\left\{1 - \varepsilon\ln\left(z\right) + \frac{\varepsilon^2}{2}\left[\ln\left(z\right)\right]^2 + \cdots\right\}. \tag{5.65}$$

The integrals considered in the following examples are of the form

$$I = \frac{1}{2\pi i}\int_C ds\, F(s), \tag{5.66}$$

where the integration path and integrand are defined in each particular problem.

5.3.2 Example 1

Consider the following integrand with a double pole

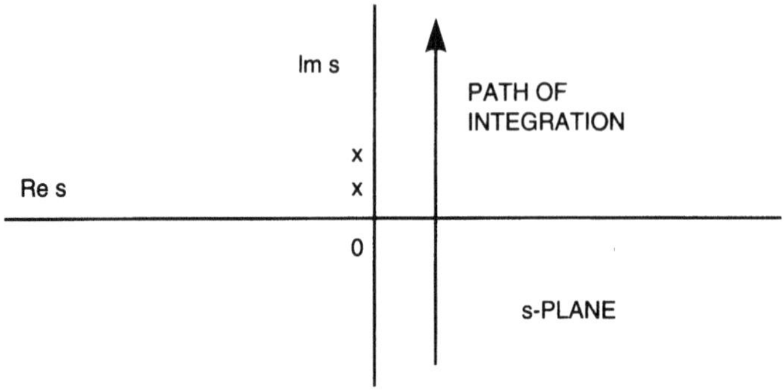

Fig. 5.3. The integration path and pole location for Examples 1 and 2.

$$F(s) = \frac{1}{s^2} = \Gamma\left[\begin{matrix} s,\ s \\ s+1,\ s+1 \end{matrix}\right]. \tag{5.67}$$

I seek to evaluate

$$I = \frac{1}{2\pi i} \int\limits_C ds\, \Gamma\left[\begin{matrix} s,\ s \\ s+1,\ s+1 \end{matrix}\right]. \tag{5.68}$$

The integration path and pole location are shown in Fig. 5.3. It is obvious from the first term to the right of the equality in (5.67) that there is no term that varies as $1/s$; therefore, the integral is zero. This is not obvious when the function is expressed as the ratio of gamma functions. If the Gamma functions in the numerator and denominator are expanded with (5.62) and (5.60), the terms with $1/\varepsilon$ cancel, and no simple pole exists. Therefore, the integral is zero.

This integration in the complex plane corresponds to an integral in real space. Using the Mellin convolution integral, one finds that an integral that would give this integral in the complex plane is

$$I = \int\limits_0^\infty \frac{dx}{x} U(1-x)\, U\left(1 - z/x\right), \quad z = 1. \tag{5.69}$$

This integral is easily evaluated to give

$$I = \int\limits_z^1 \frac{dx}{x} = -\ln(z)\, U\left(1 - z\right), \tag{5.70}$$

which is indeed equal to zero for $z = 1$.

5.3.3 Example 2

Consider the slightly more complicated integrand

$$F(s) = \frac{z^{-s}}{s^2} = z^{-s}\Gamma\left[\begin{matrix} s,\ s \\ s+1,\ s+1 \end{matrix}\right]. \tag{5.71}$$

The integration path and pole location are the same as in Fig. 5.3 . Here $\Delta = 0$, and the direction of path closure depends on the magnitude of z. For $z > 1$ the integration path can be closed in the right half plane, and the value of the integral is zero. For $z < 1$ the integration path can be closed in the left half plane. If the expansions in (5.62), (5.60), and (5.65) are used, the $1/\varepsilon$ term comes from the second term of the first equation and the second terms of the second equation. One obtains for the value of the integral

$$I = -\ln(z)\,U(1-z). \tag{5.72}$$

The integral in real space corresponding to this complex integration is the same one given in (5.69).

5.3.4 Example 3

Consider the integrand

$$F(s) = \frac{\Gamma[s]}{s} z^{-s} = \Gamma\left[\begin{matrix} s*,\ s* \\ s+1 \end{matrix}\right] z^{-s}. \tag{5.73}$$

The asterisks appear after both numerator gamma functions because the chosen integration path passes between the first and second poles of the gamma functions. The integration path and pole location are shown in Fig. 5.4. This corresponds to the real space convolution integral

$$I = \int_0^\infty \frac{dx}{x} \left[\exp(-x) - 1\right]\left[1 - U(1 - z/x)\right] = \int_0^z \frac{dx}{x} \left[\exp(-x) - 1\right]. \tag{5.74}$$

The poles go to infinity to the left just as they do in the succeeding examples. For this example, $\Delta = 1$, and the integration path can be closed to the left for all z. All the enclosed poles are simple since the only double pole at $z = 0$ is on the right side of the integration path. The integral is equal to

$$I = -\sum_{n=1}^{\infty} \frac{(-1)^n}{n!} \frac{z^n}{n}. \tag{5.75}$$

For large z an asymptotic series can be found. Since $B = 0$, the steepest-descent contribution has exponential decay as given in (5.92). With (5.63) and (5.65) for the numerator and (5.60) for the denominator, the asymptotic series, which only has 3 terms in this case, is

$$I \sim \exp(-z)/z + \ln(z) + \mathbf{C}, \quad z >> 1. \tag{5.76}$$

The decaying exponential term is negligible.

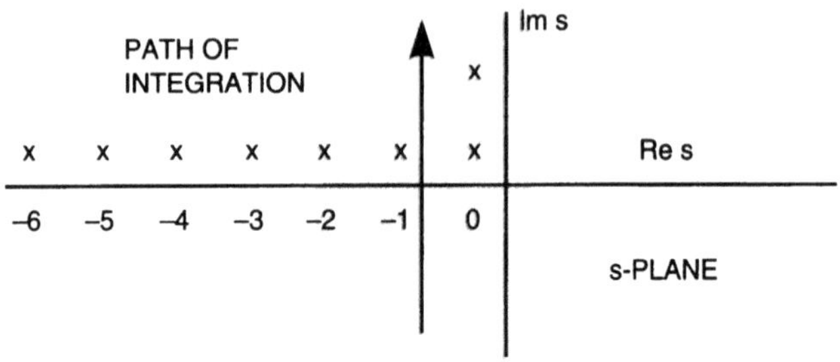

Fig. 5.4. The integration path and pole location for Example 3.

5.3.5 Example 4

Consider the integrand

$$F(s) = \Gamma\,[s*,\,-s]\; z^{-s}. \tag{5.77}$$

The integration path and pole location are shown in Fig. 5.5. This corresponds to the real space convolution integral

$$I = \int_0^\infty \frac{dx}{x}\,[\exp(-x) - 1]\,\exp(-x/z). \tag{5.78}$$

Since $\Delta = \Delta' = 0$, the integration path can be closed in the left half plane when $z < 1$, and in the right half plane when $z > 1$. There is a double pole to the right at $z = 0$ that can be evaluated with (5.62), (5.60), and (5.65). This yields

$$I = \sum_{n=1}^{\infty} \frac{(-z)^n}{n} = -\ln(z+1), \quad z < 1, \quad \text{and} \tag{5.79}$$

$$I = \sum_{n=1}^{\infty} \frac{(-z)^{-n}}{n} + \ln(z) = -\ln(z+1) + \ln(z) = -\ln(1 - 1/z), \quad z > 1. \tag{5.80}$$

5.3.6 Example 5

Consider the integrand

$$F(s) = \Gamma\left[s*,\,-s - N\left(*^N\right)\right] z^{-s}. \tag{5.81}$$

The $\left(*^N\right)$ term indicates that the first N terms of the power series are subtracted from the function whose Mellin transform results in that gamma function. The integration path and pole location are shown in Fig. 5.6 for the case $N = 4$. For general N this corresponds to the real space convolution integral

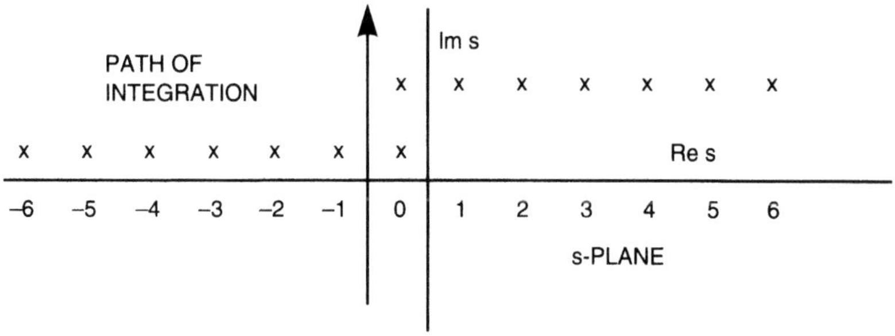

Fig. 5.5. The integration path and pole location for Example 4.

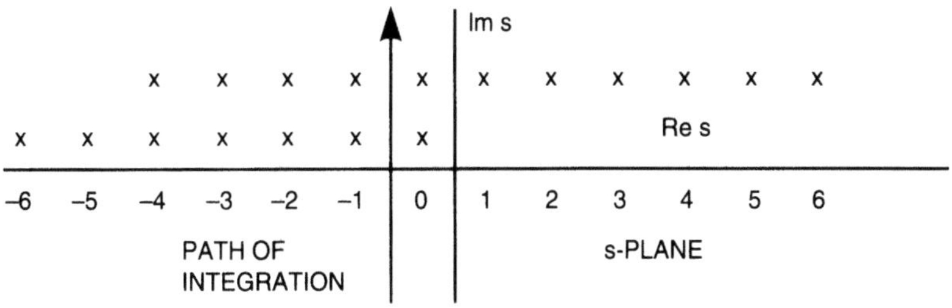

Fig. 5.6. The integration path and pole location for Example 5.

$$I = \int_0^\infty \frac{dx}{x} \left[\exp(-x) - 1\right] \left(\frac{z}{x}\right)^N \left[\exp\left(-x/z\right) - \sum_{n=0}^{N-1} \frac{1}{n\,!} \left(-\frac{x}{z}\right)^n\right]. \tag{5.82}$$

The value of $\Delta = \Delta' = 0$, and the direction of path closure depends on the magnitude of z. There is one double pole to the right of the integration path, N double poles to the left, and simple poles on both sides of the integration path. The result is

$$\begin{aligned} I = (-1)^N \sum_{n=1}^{N} & \frac{\ln(z) + \psi(1+N-n) - \psi(1+n)}{n\,!\,(N-n)\,!} \\ & + (-z)^{N+1} \sum_{n=0}^{\infty} \frac{n!\,(-z)^n}{(N+n+1)!}, \qquad z < 1, \quad \text{and} \end{aligned} \tag{5.83}$$

$$I = (-1)^N \frac{\psi(1) - \psi(1+N) - \ln(z)}{N!} + (-1)^N \sum_{n=0}^{\infty} \frac{n\,!\,(-z)^{-n-1}}{(N+n+1)!}, \qquad z > 1. \tag{5.84}$$

5.3.7 Example 6

Consider the following integrand that has triple poles

$$F(s) = \Gamma\left[s*,\ s,\ -N - s\left(*^N\right)\right] z^{-s}. \tag{5.85}$$

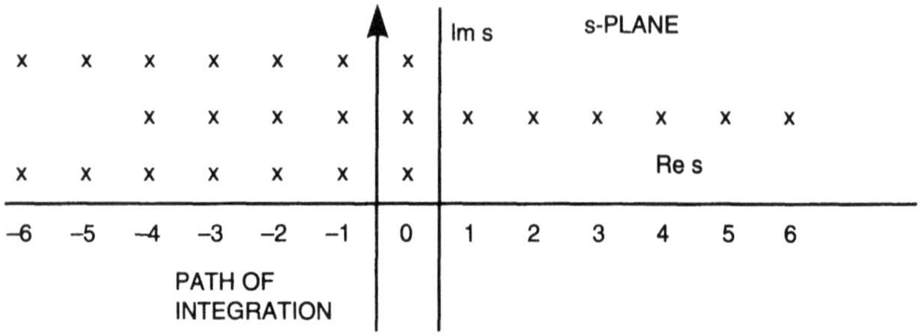

Fig. 5.7. The integration path and pole location for Example 6.

The integration path and pole location are shown in Fig. 5.7 for the case $N = 4$. For this case $\Delta = 1$, and there is one triple pole to the right of the integration path, and N triple poles and an infinity of double poles to the left. Three terms in the Laurent expansions of each factor must be retained to get the complete residue at each of the triple poles. Since $\Delta = 1$, the integration path can be closed in the left half plane. The residue at a double pole that occurs at $s = -k$, $k = N+1$, $N+2\ldots$ is referred to as $R2\,(k)$ and is given by

$$R2\,(k) = \Gamma\,[k-N]\,[2\psi(1+k) - \psi(k-N) - \ln(z)]\,\frac{z^k}{(k\,!)^2}. \tag{5.86}$$

The residue at a triple pole that occurs at $s = -1, -2, \ldots, N$ is referred to as $R3\,(k)$ and is given by

$$\begin{aligned} R3\,(k) = &\Big\{\ln^2(z)\,/2 + \ln(z)\ \left[\psi\,(1+N-k) - 2\psi\,(1+k)\right] \\ &+2\psi^2(1+k) + 3\psi'(1) + \Big[\psi^2(1+N-k) - \psi'(1+N-k)\Big]\,/2 \\ &-\psi'(1+k) - 2\psi(1+N-k)\,\psi(1+k)\Big\}\,\frac{(-1)^{N-k-1}z^k}{(k!)^2(N-k)!}. \end{aligned} \tag{5.87}$$

The value of the integral is thus

$$I = \sum_{k=N+1}^{\infty} R2\,(k) + \sum_{k=1}^{N} R3\,(k). \tag{5.88}$$

For large values of z an asymptotic series is needed. Since the condition in (5.52) applies, the asymptotic series is equal to the residues of the poles to the right of the integration path. This contains one triple pole, and the sum of the single poles at $s = n - N$, for $n = N+1, N+2\ldots$. The asymptotic series is thus

$$\begin{aligned} I \sim &\frac{(-1)^{N+1}}{z}\sum_{n=0}^{\infty}\frac{(-1)^n\,(n!)^2}{(N+n+1)!}\,z^{-n} + \frac{z^N}{(N!)^2}\Big\{\ln(z)\,[\psi(1) - 2\psi(1+N)] + \frac{\ln^2(z)}{2} \\ &+2\psi^2(1+N) + \frac{5\psi'(1)}{2} + \frac{\psi^2(1)}{2} - \psi'(1+N) - 2\psi(1)\,\psi(1+N)\Big\}. \end{aligned} \tag{5.89}$$

Table 5.1 Equations to obtain the steepest-descent term $E(x)$.

Equation whose asymptotic value is wanted for large values of z.

$$h(x) = \frac{1}{2\pi i} \int_{C'} ds\, x^{-s} \; \frac{\prod_{i=1}^{A} \Gamma\,[a_i + \alpha_i\, s]\; \prod_{j=1}^{B} \Gamma\,[b_j - \beta_j\, s]}{\prod_{k=1}^{C} \Gamma\,[c_k + \gamma_k\, s]\; \prod_{m=1}^{D} \Gamma\,[d_m - \delta_m\, s]} \tag{5.90}$$

$$\Xi' = A + D - B - C, \quad \Xi = A + B - C - D$$

$$\Delta = \sum_{i=1}^{A} \alpha_i + \sum_{m=1}^{D} \delta_m - \sum_{j=1}^{B} \beta_j - \sum_{k=1}^{C} \gamma_k$$

$$\Delta' = \sum_{i=1}^{A} \alpha_i \ln(\alpha_i) + \sum_{m=1}^{D} \delta_m \ln(\delta_m) - \sum_{j=1}^{B} \beta_j \ln(\beta_j) - \sum_{k=1}^{C} \gamma_k \ln(\gamma_k)$$

$$\nu = \sum_{i=1}^{A} a_i + \sum_{j=1}^{B} b_j - \sum_{k=1}^{C} c_k - \sum_{m=1}^{D} d_m$$

$$A' = \sum_{i=1}^{A} \alpha_i, \quad B' = \sum_{j=1}^{B} \beta_j, \quad C' = \sum_{k=1}^{C} \gamma_k, \quad D' = \sum_{m=1}^{D} \delta_m$$

$$\rho = [\nu + (1 - \Xi)/2]\,/\Delta, \quad B'' = \sum_{j=1}^{B} b_j, \quad D'' = \sum_{m=1}^{D} d_m$$

$$\Lambda = D - B, \quad \Lambda' = D' - B', \quad \Lambda'' = D'' - B''$$

$$\Delta'' = \sum_{i=1}^{A} a_i \ln(\alpha_i) + \sum_{j=1}^{B} b_j \ln(\beta_j) - \sum_{k=1}^{C} c_k \ln(\gamma_k) - \sum_{m=1}^{D} d_m \ln(\delta_m)$$

$$\Delta''' = \sum_{i=1}^{A} \ln(\alpha_i) + \sum_{j=1}^{B} \ln(\beta_j) - \sum_{k=1}^{C} \ln(\gamma_k) - \sum_{m=1}^{D} \ln(\delta_m)$$

If $B = D = 0$ and $\Lambda' > 0$, then

$$E(x) = \frac{(2\pi)^{(\Xi-1)/2}}{\sqrt{\Delta}} x^{\rho} \exp\left[-\Delta\, x^{1/\Delta} \exp\left(-\Delta'/\Delta\right) + \Delta'' - \Delta'''/2 - \rho\Delta'\right]. \quad (5.91)$$

If $\Delta' = \Delta'' = \Delta''' = 0$, $\Lambda' > 0$ and $B = D = 0$, then

$$E(x) = \frac{(2\pi)^{(\Xi-1)/2}}{\sqrt{\Delta}} x^{\rho} \exp\left(-\Delta\, x^{1/\Delta}\right). \quad (5.92)$$

If B or $D \neq 0$ and $\Lambda' > 0$, then

$$E(x) = \frac{2(2\pi)^{(\Xi-1)/2}}{\sqrt{\Delta}} x^{\rho}$$

$$\times \exp\left[-\Delta x^{1/\Delta} \exp\left(-\Delta'/\Delta\right) \cos\left(\pi\Lambda'/\Delta\right) + \Delta'' - \Delta'''/2 - \rho\Delta'\right]$$

$$\times \cos\left[\Delta x^{1/\Delta} \exp\left(-\Delta'/\Delta\right) \sin\left(\pi\Lambda'/\Delta\right) + \pi\left(-\rho\Lambda' - \Lambda'' + \Lambda/2\right)\right]. \quad (5.93)$$

If B or $D \neq 0$ and $\Lambda' > 0$, and $\Delta' = \Delta'' = \Delta''' = 0$, $B = B'$, $D = D'$, then

$$E(x) = \frac{2(2\pi)^{(\Xi-1)/2}}{\sqrt{\Delta}} x^{\rho} \exp\left[-\Delta x^{1/\Delta} \cos\left(\pi\Lambda/\Delta\right)\right]$$

$$\times \cos\left\{\Delta x^{1/\Delta} \sin\left(\pi\Lambda/\Delta\right) + \pi\left[\left(1/2 - \rho\right)\Lambda - \Lambda''\right]\right\}. \quad (5.94)$$

If $\Lambda' \leq 0$, then $E(x)$ has exponential decay with specific values discussed in Sect. 9.2.

6. Other Applications of Mellin Transforms

In addition to helping one to evaluate integrals, Mellin transform techniques are useful in many fields. They were used to find the magnetic field of a spherical magnet (*Lipinski* et al. 1975), for Doppler signal processing (*Casasent* and *Psaltis* 1976), in transient propagation (*Lee* 1987), to analyze mammalian hearing (*Altes* 1978), for quantum field theory (*Cicuti* and *Montaldi* 1975; *Birmingham* and *Sen* 1987), for nonlinear systems (*Banai* 1970), in chaos theory (*Bessis* et al. 1987), in digital signal processing (*Oppenheim* and *Schafer* 1975), in summing series and in probability theory (*Andrews* and *Shivamoggi* 1988), and by using the scale-invariant property of Mellin transforms to do optical processing of images (*Casasent* and *Psaltis* 1976a, 1976b; *Zwicke* and *Kiss* 1983). Some of these applications are considered in this chapter.

I will use the Mellin transform to solve Laplace's equation in spherical coordinates and also in cylindrical coordinates when there is no axial dependence. I will also apply the Mellin transform to a circuit problem with time-varying elements. This capability is part of a general theory that reduces differential equations to algebraic ones: Fourier transforms are useful for differential equations with constant coefficients since their use leads to solutions proportional to the transform function $\exp(i\omega t)$. Similarly, the Hankel transform converts differential equations whose solutions can be expressed as Bessel functions into algebraic equations. Likewise, the Mellin transform converts differential equations whose solutions can be expressed in simple powers into algebraic equations.

To solve differential equations the Mellin transform is useful because of the relation between the differential of a quantity and its Mellin transform. This can be expressed in two ways as

$$\left(x\frac{d}{dx}\right)^n h(x) \to (-s)^n H(s)\,, \text{ and} \tag{6.1}$$

$$\left(\frac{d}{dx}x\right)^n h(x) \to (1-s)^n H(s)\,. \tag{6.2}$$

Differential equations in which the derivative can be put in either form are easily solvable with Mellin transforms.

In solving differential equations with Mellin transforms, one is faced with the task of deciding where the integration contour can cross the real axis. The restrictions on the integration path arise mainly from the physical requirements, e.g., the quantities calculated must be finite at the origin and at infinity. The physical requirement that a function decay at infinity means the integration path must cross the real axis to the right of any poles in the left half plane. The requirement that the function must be finite at the origin means the integration path must cross the real axis to the left of any poles in the right half plane. Sometimes this requirement has to be relaxed, as in Sect. 6.2 where the potential surrounding a uniformly charged cylinder is calculated.

One might think that other restrictions on the integration path would be found by looking more carefully at the requirements on s for the above equations to be valid. These equations are derived from the definition of the Mellin transform by an integration by parts. This integration produces

$$\mathcal{M}\left[\left(x\frac{d}{dx}\right)^2 h(x)\right] = \left[x^{s+1}\frac{\partial h\left(x\right)}{\partial x} - s\,x^s\,h\left(x\right)\right]\Bigg|_0^\infty + s^2 H\left(s\right). \tag{6.3}$$

Therefore, for (6.1) to be valid, the first term on the right side of the equation must be zero. If $h\left(x\right)$ behaves as $x^{-\lambda}$ as $x \to 0$, and as $x^{-\mu}$ as $x \to \infty$, then the requirement on the integration path is that $\lambda < \operatorname{Re}\{s\} < \mu$. Before the problem is solved one does not know how the solution will behave at these limits. However, since $\Delta = \Delta' = 0$ for all problems discussed in this chapter, the direction of path closure is determined by the magnitude of the normalized parameter x. The path is closed to the left for small x and closed to the right for large x. The behavior of the solution in these limits is determined by poles closest to the integration path. A little thought will convince the reader that no matter where the integration path is placed, the solution must satisfy the requirement $\lambda < \operatorname{Re}\{s\} < \mu$. Therefore, the requirement that the boundary term vanish imposes no additional restrictions on the integration path.

6.1 Convolutional Transforms

Fourier, Bessel, and Mellin transforms are special cases of a general "convolutional" transform. These transforms are important because a convolution of functions with appropriate factors can be written as the inverse transform of the product of the functions' transforms. Functions that satisfy the convolution theorem are called Fourier kernels. In general, a necessary and sufficient condition for a function $f\left(x\right)$ to be Fourier kernel is given in *Sneddon* (1972) as

$$F\left(s\right)\,F\left(1-s\right) = 1, \tag{6.4}$$

where $F\left(s\right)$ is the Mellin transform of $f\left(x\right)$. Using (1.46), (1.47), and (1.49) one can see that $\sqrt{2/\pi}\,\sin\left(x\right)$, $\sqrt{2/\pi}\,\cos\left(x\right)$, and $\sqrt{x}\,J_\nu\left(x\right)$ are Fourier kernels.

A more general transform that encompasses the ones above has a Meijer G-function as the Fourier kernel. Remember that the Meijer G-function is a generalization of the generalized hypergeometric function and, therefore, includes most familiar analytic functions. For this transform one can write a transform pair as

$$g(u) = \int_0^\infty dx\, k(ux) f(x), \text{ and} \tag{6.5}$$

$$f(x) = \int_0^\infty du\, h(ux) g(u), \tag{6.6}$$

where $k(x) = 2\gamma x^{\gamma-1/2} G^{m,\,p}_{p+q,\,m+n}\left[x^{2\gamma}\left| \begin{matrix} a_p, & b_q \\ c_m, & d_n \end{matrix} \right.\right]$, and (6.7)

$$h(x) = 2\gamma x^{\gamma-1/2} G^{n,\,q}_{p+q,\,m+n}\left[x^{2\gamma}\left| \begin{matrix} -b_q, & -a_p \\ -d_n, & -c_m \end{matrix} \right.\right]. \tag{6.8}$$

This kernel is symmetric, i.e., $h(x) = k(x)$, when $p = q,\ m = n,\ a_j + b_j = 0$ for $j = 0,\ 1,\ 2,\ \ldots, p$ and $c_k + d_k = 0$, for $k = 0,\ 1,\ 2, \ldots, m$.

6.2 Laplace's Equation in Cylindrical Coordinates

Laplace's equation in cylindrical coordinates for the electric potential V is

$$\left(\rho\frac{\partial}{\partial\rho}\right)^2 V + \frac{\partial^2 V}{\partial\phi^2} + \rho^2\frac{\partial^2 V}{\partial z^2} = -q\rho^2/\varepsilon, \tag{6.9}$$

where q is the charge density, and ε is the permittivity. If there is axial variation, then the radial dependence can be expressed in terms of Bessel functions. If there is no axial variation, then Mellin transform techniques can be applied.

Two problems are solved in this section. In the first problem the potential is found outside a cylinder of radius a, on which a potential is impressed that varies with respect to infinity as the cosine of the angular position on the cylinder. In the second problem the potential is calculated for a uniform charge on the cylinder.

Consider the simple case in which the ϕ dependence is $\cos(\nu\phi)$. Use separation of variables to write the potential and charge as the product of a term that depends on the radial coordinate and a term that depends on the angular coordinate as

$$q = q_\rho \cos(\nu\phi)\,, \text{ and} \tag{6.10}$$

$$V = V_\rho \cos(\nu\phi)\,. \tag{6.11}$$

The requirement that the solution applies to real space forces the angular function to repeat every 2π; therefore, $\nu = 2\pi n$. Inserting (6.10) and (6.11) into the differential equation, one obtains

$$\left(\rho\frac{\partial}{\partial\rho}\right)^2 V_\rho - \nu^2 V_\rho = -q_\rho\rho^2/\varepsilon.$$

Take the Mellin transform of the equation to obtain

$$\hat{V}_\rho(s) = \frac{\mathcal{M}\left(q_\rho \rho^2/\varepsilon\right)}{(s+\nu)(s-\nu)}, \tag{6.12}$$

where the circumflex indicates the Mellin transform of the function. The potential is given by the inverse Mellin transform as

$$V_\rho = \frac{1}{2\pi i}\int_C ds\, \rho^{-s} \frac{\mathcal{M}\left(q_\rho \rho^2/\varepsilon\right)}{(s+\nu)(s-\nu)}. \tag{6.13}$$

Let us evaluate the integral for the case in which the radial distribution of charge is a delta function at radius a,

$$q_\rho = -V_o \frac{2\nu\varepsilon}{a}\delta(\rho - a).$$

The Mellin transform of the delta function is given in (1.61).

First, consider the case $\nu \neq 0$. In this case there are two simple poles. The integration path passes between the two poles at $s = \pm\nu$ as discussed in the introduction to this chapter. Equation (6.13) with the assumed charge distribution is

$$V_\rho = -V_o \frac{1}{2\pi i}\int_C ds \left(\frac{\rho}{a}\right)^{-s} \frac{2\nu}{(s+\nu)(s-\nu)}. \tag{6.14}$$

For $\rho > a$ the integration path must be closed to the right to obtain the pole contribution at $s = \nu$, and for $\rho < a$ the integration path must be closed to the left to obtain the pole contribution at $s = -\nu$. Therefore,

$$\begin{aligned} V &= V_o \left(\frac{a}{\rho}\right)^{\nu} \cos(\nu\phi), \qquad \rho > a, \text{ and} \\ V &= V_o \left(\frac{\rho}{a}\right)^{\nu} \cos(\nu\phi), \quad \rho < a. \end{aligned} \tag{6.15}$$

Let us consider a second case in which there is no azimuthal dependence, i.e. $\nu = 0$. The requirement that the potential must be finite at the origin and at infinity cannot be satisfied for this case. Since there is no source at the origin, the requirement that the potential is finite at the origin is kept. This requires the integration path to pass to the left of both poles. In this case both poles coalesce, and one must find the residue from the double pole. Let the charge distribution in this case q'_ρ have the same radial distribution as in the previous problem except that $q'_\rho = q_\rho/2\nu$, then the potential is

$$V = -\lim_{\nu \to 0} \frac{1}{2\pi i}\int_C ds\, \rho^{-s} \frac{\mathcal{M}\left(q'_\rho \rho^2/\varepsilon\right)}{(s+\nu)(s-\nu)}. \tag{6.16}$$

As described in Sect. 5.3, one solves problems with multiple poles by expanding all functions about singularities and by calculating residues of the simple poles. The potential is given by

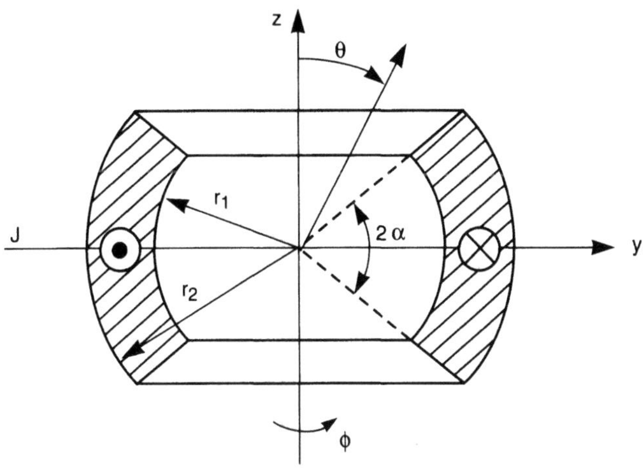

Fig. 6.1. Spherical coil geometry used to determine the vector potential.

$$V = -V_o \lim_{\nu \to 0} \frac{1}{2\pi i} \int_C ds \, \frac{[1 - s \ln(\rho/a)]}{(s+\nu)(s-\nu)}.$$

For $\rho > a$ the integration path must be closed to the right to obtain the pole contribution at the double pole, and for $\rho < a$ the integration path must be closed to the left where there are no poles. The potential is thus

$$V = V_o \ln(a/\rho)\, U(\rho - a). \tag{6.17}$$

This is the standard result for a uniformly charged cylinder.

6.3 Laplace's Equation in Spherical Coordinates

As another example of the use of the Mellin transform to reduce a differential equation to an algebraic one, let us derive the vector potential of a spherical magnetic coil, as first considered by *Lipinski* et al. (1975). The geometry of the problem is shown in Fig. 6.1.

The vector potential **A** satisfies

$$\nabla^2 \mathbf{A} = -\mu_o \mathbf{J}, \tag{6.18}$$

where μ_o is magnetic permeability, and **J** is the current density. In spherical coordinates the ϕ component of **A**, A_ϕ, satisfies

$$\left(r\frac{\partial}{\partial r}\right)^2 A_\phi + \left(r\frac{\partial}{\partial r}\right) A_\phi - \frac{A_\phi}{\sin^2(\theta)} + \frac{1}{\sin(\theta)}\frac{\partial}{\partial \theta}\sin(\theta)\frac{\partial A_\phi}{\partial \theta} + \frac{1}{\sin^2(\theta)}\frac{\partial^2 A_\phi}{\partial \phi^2} = -\mu_o J_\phi r^2. \tag{6.19}$$

The Mellin transform of this equation is

$$\left(s^2 - s\right) \hat{A}_\phi - \frac{\hat{A}_\phi}{\sin^2(\theta)} + \frac{1}{\sin(\theta)} \frac{\partial}{\partial \theta} \sin(\theta) \frac{\partial \hat{A}_\phi}{\partial \theta} + \frac{1}{\sin^2(\theta)} \frac{\partial^2 \hat{A}_\phi}{\partial \phi^2} = -\mu_o \mathcal{M}\left(J_\phi r^2\right), \tag{6.20}$$

where the circumflex above the quantity indicates its Mellin transform.

For the problem under consideration there is no ϕ dependence; therefore, the equation reduces to

$$\left(s^2 - s\right) \hat{A}_\phi - \frac{\hat{A}_\phi}{\sin^2(\theta)} + \frac{1}{\sin(\theta)} \frac{\partial}{\partial \theta} \sin(\theta) \frac{\partial \hat{A}_\phi}{\partial \theta} = -\mu_o \mathcal{M}\left(J_\phi r^2\right). \tag{6.21}$$

Separate variables by expressing the θ dependence in terms of associated Legendre functions $P_n^1(\cos\theta)$. Thus the vector potential is

$$A_\phi = \sum_{n=0}^{\infty} b_n P_n^1(\cos\theta)\, a(r). \tag{6.22}$$

Assume the current density can be expressed in the same form

$$J_\phi = J\, f(\theta)\, g(r), \tag{6.23}$$

where J is a constant, and the θ dependence of current is expressed as

$$f(\theta) = \sum_{n=0}^{\infty} G_n P_n^1(\cos\theta), \tag{6.24}$$

where

$$G_n = \frac{2n+1}{2n(n+1)} \int_{-1}^{1} d(\cos\theta)\, P_n^1(\cos\theta)\, f(\theta). \tag{6.25}$$

Because of the orthogonality properties of associated Legendre functions, (6.21) becomes

$$\left(s^2 - s\right) b_n - n(n+1)\, b_n = -\mu_o J G_n \mathcal{M}\left[g(r)\, r^2\right]. \tag{6.26}$$

If the current is uniform between radii r_1 and r_2, then $g(r)$ is expressed in terms of the unit step function as

$$g(r) = U(r_2 - r) - U(r_1 - r). \tag{6.27}$$

The Mellin transform in (1.57) gives

$$\mathcal{M}\left(g(r)\, r^2\right) = -\Gamma\begin{bmatrix} s+2 \\ s+3 \end{bmatrix} \left(r_1^{s+2} - r_2^{s+2}\right). \tag{6.28}$$

Inserting this into (6.26), solving for b_n, and inserting the result into (6.22), one finds

$$A_\phi = \mu_o J \sum_{n=0}^{\infty} G_n P_n^1(\cos\theta) \frac{1}{2\pi i} \int_C ds \left[r_1^2 \left(\frac{r}{r_1}\right)^{-s} - r_2^2 \left(\frac{r}{r_2}\right)^{-s} \right] \times \frac{1}{(s+2)(s+n)(s-n-1)}. \tag{6.29}$$

The pole locations and integration path that guarantees physical behavior at $r = 0$ and infinity are shown in Fig. 6.2. For this case $\Delta = 0$, so the direction of path closure is determined by the parameter size. For $r < r_1$ and $r < r_2$, the integration path must be closed to the left, and there are poles at $s = -n$ and -2. The residue at $s = -2$ is zero because the radial factor has a first-order zero there. This gives

$$A_\phi = \mu_o J \sum_{n=0}^{\infty} \frac{G_n P_n^1(\cos\theta)}{(n-2)(2n+1)} \left[r_1^2 \left(\frac{r}{r_1}\right)^n - r_2^2 \left(\frac{r}{r_2}\right)^n \right]. \tag{6.30}$$

For $r > r_1$ and $r < r_2$ the integration path must be closed to the right for the first term in brackets and closed to the left for the second, in which case one obtains a residue at the pole located at $s = 1 + n$. The result is

$$A_\phi = \mu_o J \sum_{n=0}^{\infty} G_n P_n^1(\cos\theta) \times \left[-\frac{r_1^2}{(n+3)(2n+1)} \left(\frac{r_1}{r}\right)^{n+1} - \frac{r_2^2}{(n-2)(2n+1)} \left(\frac{r}{r_2}\right)^n + \frac{r^2}{(n-2)(n+3)} \right]. \tag{6.31}$$

For $r > r_1$ and $r > r_2$ both integration paths must be closed to the right to obtain

$$A_\phi = \mu_o J \sum_{n=0}^{\infty} \frac{G_n P_n^1(\cos\theta)}{(n+3)(2n+1)} \left[r_2^2 \left(\frac{r_2}{r}\right)^{n+1} - r_1^2 \left(\frac{r_1}{r}\right)^{n+1} \right]. \tag{6.32}$$

This agrees with the results obtained by *Lipinski* et al. (1975).

6.4 Time Varying Problems

Mellin transform techniques are useful for certain time-varying problems in which the derivatives can be put in the form $(t\, d/dt)^n$. To illustrate this with a concrete example, consider the circuit problem depicted in Fig. 6.3 that was considered by *Zemanian* (1987), in which inductance increases linearly with time. The differential equation describing the relation between driving voltage and current is

$$v(t) = \frac{d\, L(t)\, i(t)}{dt} + R\, i(t) = \frac{Lt\, d\, i(t)}{dt} + (R + L)\, i(t). \tag{6.33}$$

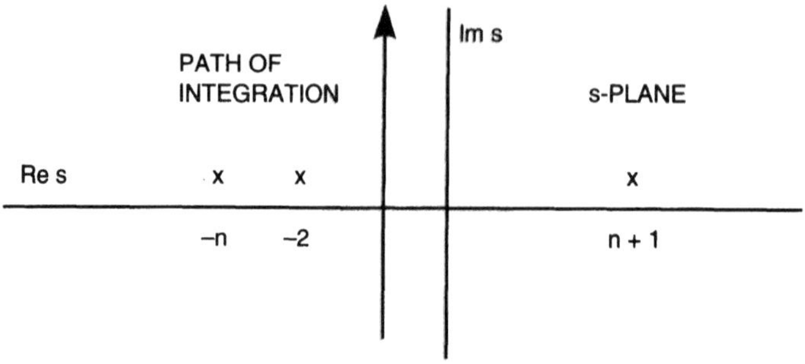

Fig. 6.2. Pole location and integration path for the spherical coil problem.

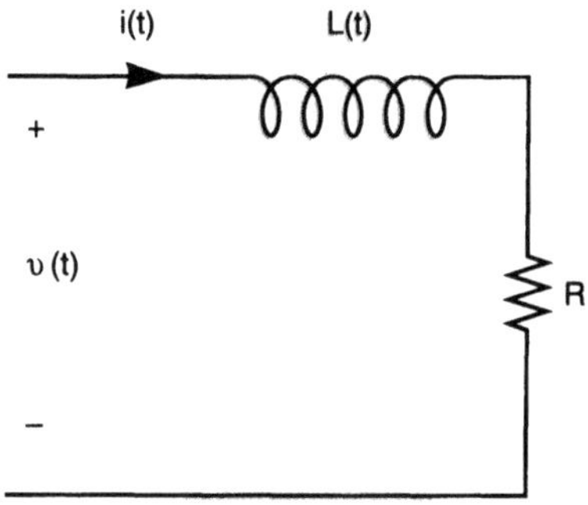

Fig. 6.3. Circuit with time varying inductor.

Taking the Mellin transform of this equation, one obtains

$$I(s) = \frac{V(s)}{R + L - Ls}. \tag{6.34}$$

The current is the inverse Mellin transform of $I(s)$, given by

$$i(t) = \frac{1}{L}\frac{1}{2\pi i}\int_C ds\, t^{-s}\frac{V(s)}{-s + 1 + R/L}. \tag{6.35}$$

Consider a specific problem in which an increasing voltage is turned on at time $t = T$. The voltage is

$$v(t) = t^{\nu} U(t - T). \tag{6.36}$$

The current in this case is

$$i(t) = -\frac{1}{2\pi i}\frac{1}{L}\int_C ds \left(\frac{t}{T}\right)^{-s}\frac{1}{(-s + 1 + R/L)(s + \nu)}. \tag{6.37}$$

There are poles at $s = 1 + R/L$ and $-\nu$. The pole location and integration path are shown in Fig. 6.4. For $t < T$ close the integration path to the left and obtain

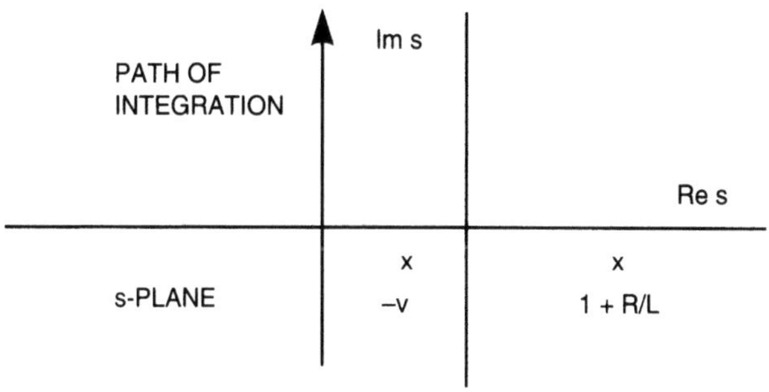

Fig. 6.4. Pole location and integration path for the circuit problem.

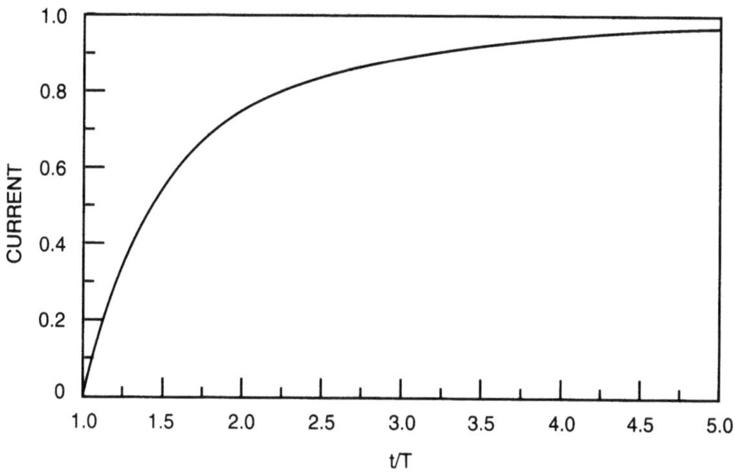

Fig. 6.5. Response of the circuit to a step in voltage.

zero. For $t > T$ the integration path must be closed in the right half plane. The residues of the two poles contribute to give

$$i\,(t) = \frac{1}{(L + R + L\nu)}\left[\left(\frac{t}{T}\right)^{\nu} - \left(\frac{t}{T}\right)^{-1-L/R}\right] U\,(t - T)\,. \qquad (6.38)$$

The specific case of $\nu = 0$, and $L = R = 1/2$ corresponds to the application of a step in voltage, and the time response of the circuit is

$$i\,(t) = \left[1 - \left(\frac{t}{T}\right)^{-2}\right], \quad t > T. \qquad (6.39)$$

This response is shown in Fig. 6.5.

6.5 Integral Equations

Mellin transforms provide a direct means to enable one to solve certain types of integral equations. Two examples are given here. As a basis to enable the solution of these equations, first consider the Mellin transform of the convolution integral

$$I = \mathcal{M}\left[\int_0^\infty dx\, f(x)\, k(xw)\right] = \int_0^\infty dw\, w^{s-1} \int_0^\infty dx\, f(x)\, k(xw)\,.$$

Let $y = xw$, and interchange the integration order to obtain

$$I = \int_0^\infty dx\, f(x)\, x^{-s} \int_0^\infty dy\, y^{s-1}\, k(y) = F(1-s)\, K(s)\,. \tag{6.40}$$

For the first problem consider the equation in which the functions k and g are given, and one has to find f.

$$\int_0^\infty dx\, f(x)\, k(xw) = g(w)\,.$$

Take the Mellin transform to obtain

$$F(1-s)\, K(s) = G(s)\,.$$

Let $s \to 1-s$ to obtain

$$F(s) = \frac{G(1-s)}{K(1-s)}. \tag{6.41}$$

Formally, the solution is

$$f(x) = \frac{1}{2\pi i}\int_C ds\, x^{-s}\, \frac{G(1-s)}{K(1-s)}. \tag{6.42}$$

As a more difficult example, consider

$$g(w) + \int_0^\infty dx\, f(x)\, k(xw) = f(w)\,. \tag{6.43}$$

Take the Mellin transform to obtain

$$G(s) + F(1-s)\, K(s) = F(s)\,.$$

Let $s \to 1-s$ to obtain a second equation. Eliminate $F(1-s)$ to obtain

$$F(s) = \frac{G(s) + G(1-s)\, K(s)}{1 - K(1-s)\, K(s)}.$$

Formally, then

$$f(x) = \frac{1}{2\pi i}\int_C ds\, x^{-s}\, \frac{G(s) + G(1-s)\, K(s)}{1 - K(1-s)\, K(s)}. \tag{6.44}$$

6.6 Multiple Integrals and Asymptotic Series from Power Series

It is often possible for one to evaluate a double integral by representing the integrand by a power series, and integrating term by term. In some cases it may be difficult to numerically evaluate the resultant series for large values of the expansion parameter. In such cases Mellin transform techniques can sometimes be used to obtain an asymptotic series.

To illustrate this technique, the double integral of an incomplete gamma function will be considered. The power series of the incomplete gamma function is

$$\gamma(a,x) = \sum_{n=0}^{\infty} \frac{(-1)^n}{n!} \frac{x^{a+n}}{(a+n)}. \tag{6.45}$$

Suppose the following integral is to be evaluated

$$P(a,x) = \int_0^x dw \int_0^w dy\, \gamma(a,\ y) = x^{a+2} \sum_{n=0}^{\infty} \frac{(-1)^n}{n!} \frac{x^n}{(n+a)(n+a+1)(n+a+2)}. \tag{6.46}$$

The power series was integrated term by term, a method that is allowed as long as the resultant series converges, which it does in this case.

Sometimes when the expansion parameter is large, however, the power series converges slowly, and one can encounter numerical problems in summing large terms that alternate in sign. In such cases, the techniques that have been developed here can be used to find an asymptotic series. When the original summation has coefficients that can be represented by gamma functions in the numerator and denominator, it can easily be represented by a Mellin-Barnes integral. Once this integral is obtained, an asymptotic series can be found easily.

For example, consider the series in (6.46). It is not difficult to see with the recursion relation for gamma functions, $\Gamma[n+1] = n\,\Gamma[n]$, that the series can be represented as

$$\begin{aligned} P(a,x) &= x^{a+2} \sum_{n=0}^{\infty} \frac{(-1)^n}{n!} \frac{\Gamma[n+a]\, x^n}{\Gamma[n+a+3]} = x^{a+2}\, \Gamma\left[\begin{matrix} a \\ a+3 \end{matrix}\right] {}_1F_1\left[a; a+3; -x\right] \\ &= \frac{x^{a+2}}{2\pi i} \int_C ds\, x^{-s}\, \Gamma\left[\begin{matrix} s,\ -s+a \\ -s+a+3 \end{matrix}\right]. \end{aligned} \tag{6.47}$$

The integration path passes between the poles of the first gamma function and the three poles of the second gamma function that are not canceled out by the gamma function in the denominator. Since $\Delta = 1$, the integration path should be closed in the left half plane, and one obtains the summation given in (6.46). To obtain the asymptotic series, note that since the condition given in (5.52) applies, one has only to sum the contributions of the three poles to the right of the integration path to give

$$P(a,x) \sim \frac{\Gamma[a]\, x^2}{2} \left[1 - \frac{2a}{x} + \frac{(a+1)\,a}{x^2}\right]. \tag{6.48}$$

6.7 Miscellaneous Additional Applications

Automatic recognition of an object in a cluttered environment is difficult to achieve, and it is an area of active research. One technique is to correlate a picture of the object with a scene that is being analyzed. If this is done literally, there is little chance of success since the reference object and the object in the scene under analysis can be displaced from each other, rotated with respect to each other, and be scaled differently. If the Fourier transform of the object is found, a translation in space corresponds to a phase shift in Fourier space. Therefore, the correlation of the magnitude of the Fourier transform of a picture with the same picture shifted in space will be high. Similarly, one can handle a rotation by obtaining the Fourier transform in radial coordinates, so that a rotation corresponds to a phase shift. In an analogous fashion the self-similar properties of Mellin transforms can be used to compare objects at two different scalings as considered by *Casasent* and *Psaltis* (1976a, 1976b). This was discussed in Sect. 1.2.

If either an acoustic wave (*Altes* 1978) or a radar wave (*Zwicke* and *Kiss* 1983) that has time dependence $\exp(i\omega t)$ is bounced off a moving object, there is a Doppler shift that is proportional to frequency; this results in a return that varies as $\exp(i\omega\alpha t)$. This corresponds to a time scaling of the return waveform; therefore, the magnitudes of the Mellin transforms of the transmitted and received waveforms are the same. This property can be used in a correlator to obtain the time delay and hence, the range of the reflecting object.

Self-similarity is also characteristic of chaos and fractals. Mellin transforms techniques have also been applied to problems in this area by *Bessis* et. al (1984, 1987).

Andrews and *Shivamoggi* (1988) considered the application of Mellin transforms to sum certain series and to find probability distribution functions. A brief review of these topics is considered here. The sum of the series $f(nx)$ can be easily shown to be equal to

$$\sum_{n=1}^{\infty} f(nx) = \mathcal{M}^{-1}[\zeta(s)\, F(s)], \tag{6.49}$$

where $\zeta(s)$ is the Riemann zeta function defined by

$$\zeta(s) = \sum_{n=1}^{\infty} \frac{1}{n^s}, \quad \mathrm{Re}\{s\} > 1. \tag{6.50}$$

Since the only singularity of the zeta function is a simple pole at $s = 1$ with a residue of 1 (*Whittaker* and *Watson* 1927), the inverse Mellin transform in (6.49) can be easily evaluated.

Andrews and *Shivamoggi* (1988) also showed that if

$$x = \prod_{i=1}^{N} y_i, \tag{6.51}$$

where y_i are independent, positive, random variables with probability densities p_i, that the probability density function of x is

$$p_x = \mathcal{M}^{-1}\left[\prod_{i=1}^{N} P_i(s)\right]. \tag{6.52}$$

The last example that I present illustrating the use of the Mellin transform is in the field of digital signal processing. Often modern electronic systems digitally sample the electronic signals and manipulate these signals in many ways to achieve the desired output. One of the most useful manipulations is linear filtering. In this application a continuous time signal $x(t)$ is sampled to produce a discrete time series $x(n)$ that is manipulated with the z-transform (*Oppenheim* and *Schafer* 1975). The z-transformed time series $X(z)$ is related to the original time series by the relation

$$X(z) = \sum_{n=-\infty}^{\infty} x(n)\, z^{-n}. \tag{6.53}$$

The inverse transform is easily found to be

$$x(n) = \frac{1}{2\pi i}\int_C dz\, X(z)\, z^{n-1} = \mathcal{M}[X(z)], \tag{6.54}$$

where the integration path is chosen to insure convergence. Here n serves as the transform variable rather than s. To obtain the original time series from its z-transform, one has to find the Mellin transform of the transformed time series. One can find the z-transform of a product of two time series with the convolution theorem just as one did in evaluating integrals that were the product of two functions.

The z-transform is also useful in that it enables one to eliminate noise and emphasize various characteristics of optical images. One can digitally filter these images by finding the two-dimensional z-transform of the digitally-sampled original image and by applying various linear filters in transform space. One obtains the filtered image by evaluating the Mellin transform of the transformed and filtered image. The Mellin transform in this case requires an integration in two complex planes. The theory of the evaluation of these integrals, which is developed latter in this book, applies in this case. The theory also applies to signals that have more than two dimensions.

7. Examples with a Single Positive Parameter

Several interesting problems relating to wave propagation in turbulence are solved in this chapter. The Zernike tilt jitter is calculated for finite outer scale with both von Kármán and Greenwood models for outer scale effects. It is shown that outer scale significantly affects tilt even if the diameter is much smaller than the outer scale. It is shown that inner scale limits the maximum tilt that can be measured on an aperture. Tilt is calculated on an annulus, and it is shown that this tilt does not differ much from that of a filled aperture unless most of the aperture is obscured. The effect of diffraction on tilt is calculated. It is shown that tilt variance goes to half of its near-field value in the far field. The amount of tilt difference between two displaced sources is calculated. The power spectral density of tilt is found. An asymptotic series describes the behavior at high frequencies. I show that finite size sources and apertures can significantly reduce scintillation. Characteristic sizes for the source and aperture at which the reduction is significant are found. Inner scale is found to reduce the scintillation for typical inner-scale sizes. Scintillation of a beam corrected for turbulence by the use of a beacon that is displaced from the corrected beam is found. The scintillation difference between two sources in the sky is found. The phase variance of focus is found, and the correlation function of focus is calculated. The problem of correcting turbulence with artificial beacons is addressed, and the phase variances with a point, distributed, and offset beacon are calculated.

The first few problems are solved in detail, and subsequent discussions leave out some intermediate steps that are repeated from problem to problem. This chapter illustrates how readily solutions are obtained with the algorithmic approach developed in this book. Many of these problems are difficult to solve by other means.

The solutions to many problems contain a single turbulence moment; however, for those problems in which the integration parameter is a function of the propagation coordinate, the solution is the sum of terms with different turbulence moments. For most problems considered here this does not pose a problem; however, there are cases in which the individual terms are infinite. In that case an infinite sum of moments cannot be used, and the function in the axial integral must be evaluated at each position and multiplied by the turbulence strength.

An example of an integral in which this occurs is $\int_0^L dz\, C_n^2(z)\, \exp(-a/z)$. This integral is finite; however, if the exponential is expanded in a Taylor series, for the case in which the turbulence is non-zero at the origin, only the integral of the first term is finite. As examples of cases where this occurs in practice are the aperture-averaged scintillation considered in Sect. 7.8; the evaluation of scintillation with inner scale as treated in Sect. 7.3, if one were to consider the case of turbulence that varies along the path; and the calculation of tilt power spectral density as discussed in Sect. 7.7, if the velocity were slew dominated, for which case it is proportional to z.

7.1 Tilt for the von Kármán Spectrum

The significant effect of outer scale on tilt jitter was pointed out by *Valley* (1979) and a solution for all Zernike modes is given by *Winker* (1991). To set up this problem, the general expression given in (2.86) with diffraction and inner scale neglected is used with the filter function for Zernike tilt given in (3.17), for a plane wave ($\gamma = 1$). The von Kármán turbulence spectrum given in (2.24) with zero inner-scale size is used to obtain for two-axis tilt variance,

$$T_o^2 = 0.2073\, k_0^2 \int_0^L dz\, C_n^2(z) \int d\boldsymbol{\kappa} \left(\kappa^2 + \kappa_o^2\right)^{-11/6} \left(\frac{16}{k_0 D}\right)^2 \left[\frac{J_2(\kappa D/2)}{\kappa D/2}\right]^2 . \tag{7.1}$$

It is assumed that outer scale does not depend on z. The integrations over angle and z can be performed easily. Make the substitution $x = \kappa D/2$ to obtain

$$T_o^2 = \frac{1334\, \mu_0\, \kappa_o^{-11/3}}{D^4} \int_0^\infty \frac{dx}{x} J_2^2(x) \left[\left(\frac{x}{\kappa_o D/2}\right)^2 + 1\right]^{-11/6} . \tag{7.2}$$

This varies with zenith angle as $\sec(\xi)$. This expression can be converted into an integral in the complex plane with the use of the Mellin convolution integral. Define $h_0(x) = J_2^2(x)$, and $h_1(a/x) = \left[(a/x)^{-2} + 1\right]^{-11/6}$. Use the Mellin transform in (1.50) for the first function. For the second, use the Mellin transform in (1.52) with $q = 11/6$, and use (1.9) with $p = 2$ to obtain

$$T_o^2 = \frac{1334\, \mu_0\, \kappa_o^{-11/3}}{4\sqrt{\pi}\, \Gamma\left[\frac{11}{6}\right] D^4} \frac{1}{2\pi i} \int_C ds \left(\frac{\kappa_o D}{2}\right)^{-s}$$
$$\times \Gamma\left[\begin{matrix} s/2+2,\ -s/2+\frac{1}{2},\ -s/2,\ s/2+\frac{11}{6} \\ -s/2+3,\ -s/2+1 \end{matrix}\right], \tag{7.3}$$

where the integration path does not split any poles of an individual gamma function. After the substitution $s \longrightarrow 2s$, one obtains

$$T_o^2 = \frac{400\, \mu_0\, \kappa_o^{-11/3}}{D^4} \frac{1}{2\pi i} \int_C ds \left(\frac{\kappa_o D}{2}\right)^{-2s} \Gamma\left[\begin{matrix} s+2,\ -s+\frac{1}{2},\ -s,\ s+\frac{11}{6} \\ -s+3,\ -s+1 \end{matrix}\right] . \tag{7.4}$$

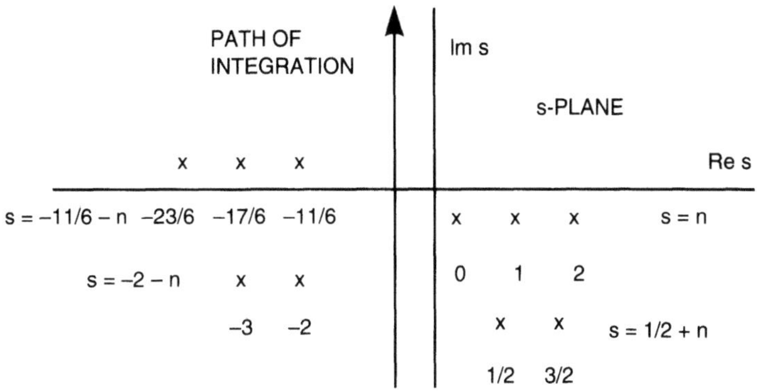

Fig. 7.1. Location of the poles and integration path for determining the tilt with finite outer scale for the von Kármán turbulence spectrum.

The only singularities of the integrand are associated with the numerator gamma functions that have poles where the argument of the gamma function is a negative integer. The poles of $\Gamma[a\,s]$ occur at $s = -n/a$ and have a residue $(-1)^n\,/an!$ for $n = 0,\ 1,\ 2,\ \ldots$. The locations of the poles closest to the origin are shown in Fig. 7.1.

As discussed in Chapter 5, the quantity Δ that determines the direction of path closure is the sum of the coefficients of s of the numerator gamma functions minus the sum of the coefficients of s in the denominator gamma functions. For this integral $\Delta = 2$, and the integration path can be closed in the left half plane. There are pole contributions at $s = -11/6 - n$, and $s = -2 - n$ for $n = 0,\ 1,\ 2,\ \ldots$. Use the definition of outer scale given in (2.20) to obtain

$$T_o^2 = \frac{6.08\,\mu_0}{D^{1/3}}\left[5.18\sum_{n=0}^{\infty}\frac{(-1)^n}{n!}\left\{\left(\frac{\pi D}{L_o}\right)^{2n}\Gamma\left[\begin{matrix}-n+\frac{1}{6},\ n+\frac{7}{3},\ n+\frac{11}{6}\\ n+\frac{29}{6},\ n+\frac{17}{6}\end{matrix}\right]\right.\right.$$
$$\left.\left.+\left(\frac{\pi D}{L_o}\right)^{2n+1/3}\Gamma\left[\begin{matrix}n+\frac{5}{2},\ n+2,\ -n-\frac{1}{6}\\ n+5,\ n+3\end{matrix}\right]\right\}\right]. \tag{7.5}$$

This expression can be converted into the sum of two generalized hypergeometric functions with the rules in Sect. 1.3 to give

$$T_o^2 = \frac{6.08\,\mu_0}{D^{1/3}}\left\{{}_2F_3\left[\frac{11}{6},\frac{7}{3};\frac{5}{6},\frac{29}{6},\frac{17}{6};\left(\frac{\pi D}{L_o}\right)^2\right]\right.$$
$$\left.-1.4234\left(\frac{D}{L_o}\right)^{1/3}{}_2F_3\left[2,\frac{5}{2};\frac{7}{6},5,3;\left(\frac{\pi D}{L_o}\right)^2\right]\right\}. \tag{7.6}$$

The lowest order terms in (7.5) for $D/L_o << 1$ are

$$T_o^2 \approx \frac{6.08\,\mu_0}{D^{1/3}}\left(1 - 1.42\,e^{1/3} + 3.70\,e^2 - 4.01\,e^{7/3} + 4.21\,e^4 - 4.00\,e^{13/3}\right), \tag{7.7}$$

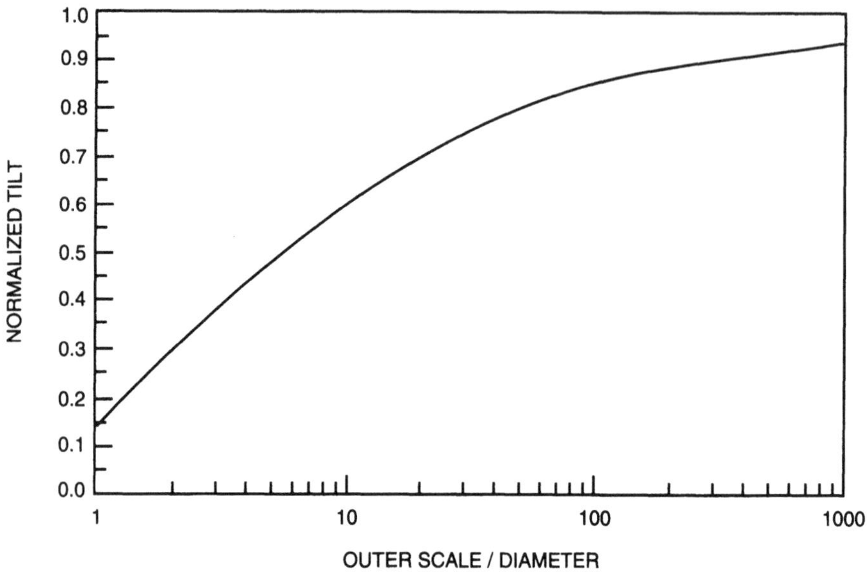

Fig. 7.2. Tilt variance for the von Kármán turbulence spectrum with outer scale, normalized to tilt variance with infinite outer scale. Notice that there is a significant reduction of tilt even if outer scale is 100 times the diameter. This curve is independent of the turbulence distribution along the propagation path.

where $e = D/L_o$. Tilt jitter is affected by outer scale, and the fractional decrease from the value with infinite outer scale is shown in Fig. 7.2. If outer-scale size is 100 times the aperture diameter, then the rms tilt jitter is still decreased by 15%. This decrease occurs because tilt jitter is caused mainly by long wavelength turbulence. A finite outer scale decreases turbulence at long wavelengths and, thereby, decreases tilt variance.

The first two terms in braces give a good approximation to the tilt variance if outer scale is larger than several times the aperture diameter. This is an example of how to write down the answer as a few terms of a power series. If outer-scale size were much smaller than diameter, then the series would converge slowly, and an asymptotic series would be appropriate.

7.2 Tilt for the Greenwood Spectrum

The equation to find tilt variance with the Greenwood outer-scale model is the same as the one in the last section except that the spectrum in (2.23) is used,

$$T_o^2 = 0.2073\, k_0^2 \int_0^L dz\, C_n^2(z) \int d\boldsymbol{\kappa}\, \left(\kappa^2 + \kappa\kappa_o\right)^{-11/6} \left(\frac{16}{k_0 D}\right)^2 \left[\frac{J_2(\kappa D/2)}{\kappa D/2}\right]^2 . \tag{7.8}$$

It is assumed again that outer scale does not depend on z. The integrations over angle and z can be performed easily. Make the substitution $x = \kappa D/2$ to obtain

$$T_o^2 = \frac{1334\,\mu_0 \kappa_o^{-11/6}}{D^4}\left(\frac{D}{2}\right)^{11/6}\int_0^\infty \frac{dx}{x}\,x^{-11/6}J_2^2(x)\left(\frac{x}{\kappa_o D/2}+1\right)^{-11/6}. \tag{7.9}$$

This varies with zenith angle as $\sec(\xi)$. Define $h_0(x) = x^{-11/6}J_2^2(x)$, and $h_1(a/x) = \left[(a/x)^{-1}+1\right]^{-11/6}$. Use the Mellin transform in (1.50) for the first function, exploiting (1.8) with $a = -11/6$. For the second function, start with (1.52) with $q = 11/6$, and use (1.9) with $p = -1$. After the substitution $s \to 2s$, one obtains

$$T_o^2 = \frac{800\,\mu_0\,\kappa_o^{-11/6}}{D^4}\left(\frac{D}{2}\right)^{11/6}\frac{1}{2\pi i}\int_C ds\left(\frac{\kappa_o D}{2}\right)^{-2s}$$
$$\times\Gamma\left[\begin{matrix} s+\frac{13}{12},\ -s+\frac{17}{12},\ -2s,\ 2s+\frac{11}{6} \\ s+\frac{47}{12},\ -s+\frac{23}{12}\end{matrix}\right], \tag{7.10}$$

where the integration path does not split any poles of an individual gamma function. Since once again $\Delta = 2$, the integration path can be closed in the left half plane, and there are pole contributions at $s = -11/12 - n/2$, and $-13/12 - n$ for $n = 0, 1, 2, \ldots$. Use the definition of outer scale given after (2.19) to obtain

$$T_o^2 = \frac{6.08\,\mu_0}{D^{1/3}}\left\{\sum_{n=0}^\infty \frac{(-1)^n}{n!}\left[5.18\left(\frac{\pi D}{L_o}\right)^n \Gamma\left[\begin{matrix} -n/2+\frac{1}{6},\ n/2+\frac{7}{3},\ n+\frac{11}{6} \\ n/2+\frac{29}{6},\ n/2+\frac{17}{6}\end{matrix}\right]\right.\right.$$
$$\left.\left. +15.18\left(\frac{D}{L_o}\right)^{1/3}\Gamma\left[\begin{matrix} n+\frac{5}{2},\ 2n+\frac{13}{6},\ -2n-\frac{1}{3} \\ n+5,\ n+3\end{matrix}\right]\left(\frac{\pi D}{L_o}\right)^{2n}\right]\right\}. \tag{7.11}$$

If the coefficients of s in the gamma functions in (7.10) are made unity with the multiplication theorem for gamma functions given in (1.15), the solution can be expressed as

$$T_o^2 = \frac{6.08\,\mu_0}{D^{1/3}}\left\{{}_3F_4\left[\frac{11}{12}, \frac{7}{3}, \frac{17}{12}; \frac{5}{6}, \frac{29}{6}, \frac{17}{6}, \frac{1}{2}; -\left(\frac{\pi D}{L_o}\right)^2\right]\right.$$
$$+1.764\left(\frac{D}{L_o}\right){}_3F_4\left[\frac{17}{12}, \frac{23}{12}, \frac{17}{6}; \frac{7}{6}, \frac{4}{3}, \frac{3}{2}\,\frac{16}{3}, \frac{10}{3}, -\left(\frac{\pi D}{L_o}\right)^2\right]$$
$$\left.-1.848\left(\frac{D}{L_o}\right)^{1/3}{}_3F_4\left[\frac{5}{2}, \frac{13}{12}, \frac{19}{12}; \frac{7}{6}, 5, 3\,\frac{7}{6}, \frac{2}{3}, -\left(\frac{\pi D}{L_o}\right)^2\right]\right\}. \tag{7.12}$$

This is plotted in Fig. 7.3.

The lowest-order terms of the summations in (7.11) for $D/L_o << 1$ give

$$T_o^2 \approx \frac{6.08\,\mu_0}{D^{1/3}}\left(1 - 1.85\,e^{1/3} + 1.76\,e - 5.24\,e^2 + 6.70\,e^{7/3} - 3.77\,e^3\right), \tag{7.13}$$

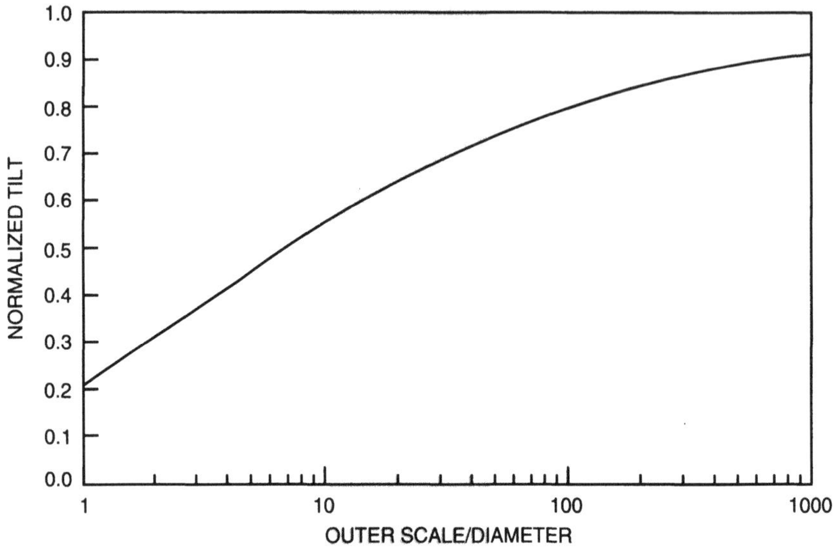

Fig. 7.3. Tilt variance for the Greenwood turbulence spectrum with outer scale, normalized to tilt variance with infinite outer scale. Notice that there is a significant reduction of tilt even if outer scale is 100 times the diameter. This curve is independent of the turbulence distribution along the propagation path.

where $e = D/L_o$. Tilt jitter is also significantly affected by outer-scale effects for the Greenwood spectrum. The leading correction term has the 1/3 power of D/L_o in both spectra considered here; this results in a significant effect on tilt jitter even when outer scale is much larger than aperture diameter.

7.3 Tilt with Finite Inner Scale

The expression for Zernike tilt variance given in (4.15) goes to infinity when the diameter goes to zero. This obviously is an incorrect physical limit; to determine the correct tilt for very small apertures, the inner scale must be included. To find this tilt, the turbulence spectrum with Tatarski's model for inner scale is used in (4.15) with diffraction neglected to obtain

$$T_i^2 = \frac{105.1\,\mu_0}{D^{1/3}} \int_0^\infty \frac{dx}{x}\, x^{-11/3} J_2^2(x) \exp\left[-(x/b)^2\right], \tag{7.14}$$

where $b = \kappa_i D/2$, and it is assumed that the inner scale does not depend on z. This varies with zenith angle as $\sec(\xi)$. Mellin transforms in (1.50) and (1.45) can be used in the convolution integral, and after the substitution $s \to 2s$, one obtains

$$T_i^2 = \frac{29.64\,\mu_0}{D^{1/3}} \frac{1}{2\pi i} \int_C ds\, \Gamma\begin{bmatrix} s+\frac{1}{6}, & -s+\frac{7}{3}, & -s \\ -s+\frac{29}{6}, & -s+\frac{17}{6} \end{bmatrix} b^{-2s}. \tag{7.15}$$

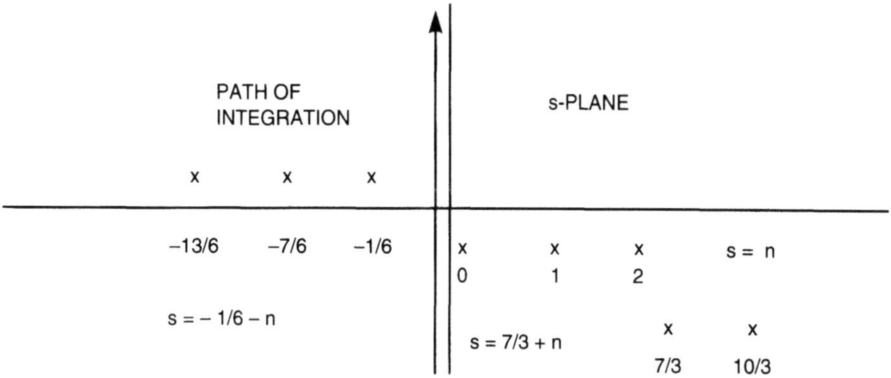

Fig. 7.4. Pole location and integration path for determining the tilt with finite inner scale.

The pole locations are shown in Fig. 7.4.

Since $\Delta = 1$, the integration path can be closed in the left half plane. There are poles at $s = -1/6 - n$, for $n = 0, 1, 2, \ldots$. The tilt variance is

$$T_i^2 = \frac{29.64\,\mu_0}{D^{1/3}} \sum_{n=0}^{\infty} \frac{(-1)^n}{n!} \Gamma \begin{bmatrix} n+\frac{5}{2},\, n+\frac{1}{6} \\ n+5,\, n+3 \end{bmatrix} \left(\frac{\pi D}{L_i}\right)^{2n+1/3} . \tag{7.16}$$

With the rules in Sect. 1.3 the solution is

$$T_i^2 = \frac{6.69\,\mu_0}{L_i^{1/3}}\, {}_2F_2\left[\tfrac{5}{2}, \tfrac{1}{6}; 5, 3; -\left(\frac{\pi D}{L_i}\right)^2\right] . \tag{7.17}$$

If inner scale is larger than diameter, then tilt variance determined only from the leading series term is

$$T_i^2 \approx \frac{6.08\,\mu_0}{(0.75 L_i)^{1/3}}, \quad L_i >> D. \tag{7.18}$$

From this solution it is seen that for inner scale large with respect to D, the tilt formula is the same as that with zero inner scale except that the diameter to place in the formula is 0.75 times the inner scale. If the inner scale is 1 millimeter, the rms tilt one would measure with a very small aperture is 12 μrads with the HV-21 turbulence distribution.

For inner scale much smaller than diameter, a case that normally occurs, the above power series converges slowly. For this case, an asymptotic solution is appropriate. The conditions given in (5.52) are satisfied, which means that the asymptotic solution is simply found from the pole residues in the right half plane that are located at $s = m + 7/3$, and for $m = 0, 1, 2, \ldots$. This gives

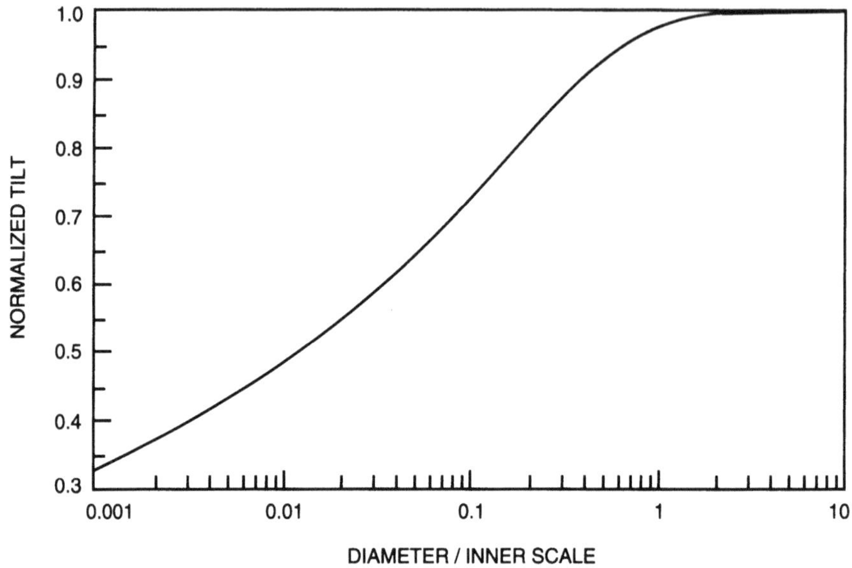

Fig. 7.5. Effect of inner scale on the tilt, normalized to that with zero inner scale. This curve is independent of the turbulence distribution along the propagation path. Tilt is close to that of tilt with zero inner scale once the diameter is larger than inner scale.

$$T_i^2 = \frac{6.08\,\mu_0}{D^{1/3}} \sum_{m=0}^{n_o} \frac{(-1)^m}{m!} 4.875 \left\{ \Gamma \begin{bmatrix} m+\frac{1}{6},\ -m+\frac{7}{3} \\ -m+\frac{29}{6},\ -m+\frac{17}{6} \end{bmatrix} \left(\frac{L_i}{\pi D}\right)^{2m} \right.$$
$$\left. + 0.00479 \left(\frac{L_i}{D}\right)^{14/3} \left(\frac{L_i}{\pi D}\right)^{2m} \Gamma \begin{bmatrix} m+\frac{5}{2},\ -m-\frac{7}{3} \\ -m+\frac{5}{2},\ -m+\frac{1}{2} \end{bmatrix} \right\}. \tag{7.19}$$

The value to use for n_o was discussed in Sect. 5.2.

The two most significant pole contributions are at $s = 0$ and 1 in the first summation. The approximate solution from those poles is

$$T_i^2 \approx \frac{6.08\,\mu_0}{D^{1/3}} \left[1 - 0.089 \left(\frac{L_i}{D}\right)^2 \right], \quad D >> L_i. \tag{7.20}$$

The first term corresponds to the result given in (4.15) where the inner scale is zero. Tilt with inner scale is plotted in Fig. 7.5.

7.4 Piston- and Tilt-Removed Phase Variance on an Annulus

Most apertures for astronomical telescopes and beam directors have a central obscuration whose diameter is 25% or less of the aperture diameter. In most analyses the effect of this obscuration is ignored because it is assumed to be small. Here it is shown explicitly that the central obscuration has little effect on piston-removed phase and tilt.

To obtain piston-removed phase variance σ_{PR}^2 on an annulus, the filter function that removes piston phase variance given in (3.32) must be subtracted from unity in the standard formula in (2.86) for phase variance of a single wave. This yields

$$\sigma_{PR}^2 = 0.2073\, k_0^2 \int_0^L dz\, C_n^2(z) \int d\boldsymbol{\kappa}\, \kappa^{-11/3} \times \left\{ 1 - \frac{4}{\left[1-(D_i/D)^2\right]^2} \left[\frac{J_1(\kappa D/2)}{\kappa D/2} - \left(\frac{D_i}{D}\right)^2 \frac{J_1(\kappa D_i/2)}{\kappa D_i/2} \right]^2 \right\}. \tag{7.21}$$

For vanishingly small wavenumbers the filter function for piston approaches unity, thereby the unity term in the braces is cancelled, thus the apparent singularity at the origin in wavenumber space is cancelled. Expand the expression in braces, and rearrange terms to obtain three integrals that each converge to give

$$\sigma_{PR}^2 = \frac{5.212\, k_0^2\, \mu_0}{\left[1-(D_i/D)^2\right]^2} \int_0^\infty \frac{d\kappa}{\kappa} \kappa^{-5/3} \left(- \left\{ \left[\frac{J_1(\kappa D/2)}{\kappa D/2} \right]^2 - \frac{1}{4} \right\} - \left\{ \left[\left(\frac{D_i}{D}\right)^2 \frac{J_1(\kappa D_i/2)}{\kappa D_i/2} \right]^2 - \frac{1}{4} \left(\frac{D_i}{D}\right)^4 \right\} + \left[2 \frac{J_1(\kappa D/2)}{\kappa D/2} \left(\frac{D_i}{D}\right)^2 \frac{J_1(\kappa D_i/2)}{\kappa D_i/2} - \frac{1}{2} \left(\frac{D_i}{D}\right)^2 \right] \right). \tag{7.22}$$

The first term in braces is the only term remaining when the central obscuration goes to zero, and produces the standard result in (4.19). The second term in braces gives the same result with a multiplier.

Let $x = \kappa D/2$, and $\beta = D_i/D$ to obtain

$$\sigma_{PR}^2 = 1.033 \left(\frac{D}{r_o}\right)^{5/3} \frac{\left(1+\beta^{17/3}\right)}{\left(1-\beta^2\right)^2} + 1.94 \left(\frac{D}{r_o}\right)^{5/3} \frac{4}{\left(1-\beta^2\right)^2} \beta I. \tag{7.23}$$

The remaining integral to be evaluated is

$$I = \int_0^\infty \frac{dx}{x} x^{-11/3} \left[J_1(x)\, J_1(\beta x) - \frac{x}{2} \frac{\beta x}{2} \right]. \tag{7.24}$$

The form of this integral is not one that has been considered before. It is the product of two functions minus the product of the first terms of their series expansions. If A and B are the two functions, and a and b are the first terms of the series expansions, then the following identity will express the integral in a form that can be evaluated with the techniques that have been developed.

$$AB - ab = (A - a)(B - b) + a(B - b) + b(A - a). \tag{7.25}$$

The original integral I can now be broken up into the sum of three separate integrals that are

$$I_1 = \int_0^\infty \frac{dx}{x} x^{-11/3} \left[J_1(x) - \frac{x}{2}\right] \left[J_1(\beta x) - \frac{\beta x}{2}\right], \tag{7.26}$$

$$I_2 = 0.5\,\beta^{8/3} \int_0^\infty \frac{dy}{y} y^{-8/3} \left[J_1(y) - \frac{y}{2}\right] = -0.1525\,\beta^{8/3}, \text{ and} \tag{7.27}$$

$$I_3 = 0.5\,\beta \int_0^\infty \frac{dy}{y} y^{-8/3} \left[J_1(y) - \frac{y}{2}\right] = -0.1525\,\beta. \tag{7.28}$$

The last two integrals were evaluated with the Mellin transform in (1.49).

Use the convolution theorem to evaluate I_1. By using (1.49) and letting $s \to 2s$, one can transform I_1 into the complex plane as follows

$$I_1 = \frac{2^{-14/3}}{2\pi i} \int_C ds\, \beta^{2s}\, \Gamma\left[\begin{matrix} s - \frac{4}{3}*,\ -s + \frac{1}{2}* \\ -s + \frac{10}{3},\ s + \frac{3}{2} \end{matrix}\right], \tag{7.29}$$

where the asterisk indicates that the integration path passes between the first and second poles of the gamma function.

Since $\Delta = \Delta' = 0$, and $\beta < 1$, the integration path is closed to the right. The contributions of the poles at $s = 4/3$, and $s = 1/2 + n$ for $n = 1, 2, \ldots$, give

$$I_1 = 2^{-14/3} \left\{ 3.873\,\beta^{8/3} + \beta \sum_{n=1}^{\infty} \frac{(-1)^n}{n!} \beta^{2n}\, \Gamma\left[\begin{matrix} n - \frac{5}{6} \\ -n + \frac{17}{6},\ n + 2 \end{matrix}\right] \right\}. \tag{7.30}$$

From the rules in Sect. 1.3 this is equivalent to

$$I_1 = 0.1525\,\beta^{8/3} - 0.1525\,\beta \left({}_2F_1\left[-\tfrac{5}{6}, -\tfrac{11}{6}; 2; \beta^2\right] - 1\right). \tag{7.31}$$

Therefore, the sum of the three integrals is

$$I = I_1 + I_2 + I_3 = -0.1525\,\beta\, {}_2F_1\left[-\tfrac{5}{6}, -\tfrac{11}{6}; 2; \beta^2\right]. \tag{7.32}$$

The piston-removed phase variance is thus

$$\sigma_{PR}^2 = \frac{1.033}{(1 - \beta^2)^2} \left(\frac{D}{r_o}\right)^{5/3} \left(1 + \beta^{17/3} - 1.146\,\beta^2\, {}_2F_1\left[-\tfrac{5}{6}, -\tfrac{11}{6}; 2; \beta^2\right]\right). \tag{7.33}$$

This is plotted in Fig. 7.6.

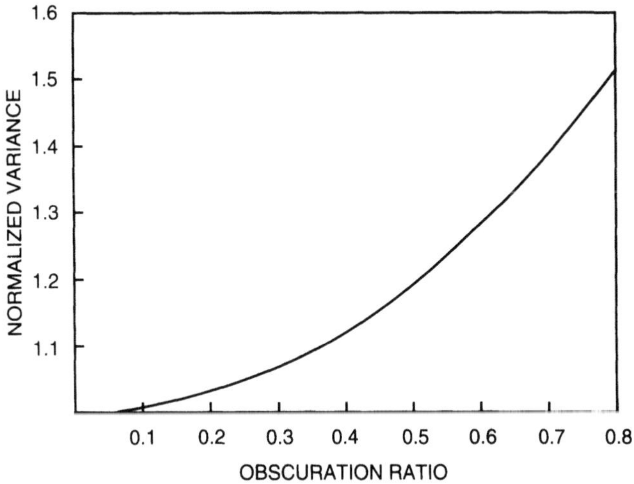

Fig. 7.6. Normalized piston-removed phase variance on an annulus versus the obscuration ratio. For a typical obscuration ratio of 25% there is little change in tilt from that of a filled aperture. This curve is independent of the turbulence distribution along the propagation path.

The expression for Zernike tilt phase variance on an annulus is obtained by inserting the filter function in (3.25) into the standard formula (2.86) for the phase variance of a single wave with diffraction neglected. The tilt variance in this case is thus

$$T_{az}^2 = 0.2073\, k_0^2 \int_0^L dz\, C_n^2(z) \int d\boldsymbol{\kappa} \left[\frac{16}{k_0 D\,(1-\beta^4)}\right]^2 \times \left[\frac{J_2\,(\kappa D/2)}{\kappa D/2} - \beta^3 \frac{J_2\,(\kappa \beta D/2)}{\kappa \beta D/2}\right]^2 . \tag{7.34}$$

The angular and axial integrations can be performed easily, and if the square of the sum in brackets is expanded, the first term gives the Zernike tilt of a full aperture, and the second term squared is the same as the first term except it is multiplied by a constant. Making the substitution $x = \kappa D/2$, one can thus express the tilt variance as

$$T_{az}^2 = \frac{6.08\,\mu_0}{D^{1/3}\,(1-\beta^4)^2} \left[1 + \beta^{23/3} - 34.55\,\beta^2 \int_0^\infty \frac{dx}{x}\, x^{-11/3} J_2(x)\, J_2\,(\beta x)\right] . \tag{7.35}$$

Using the Mellin convolution theorem with the Mellin transform in (1.49) and making the substitution $s \to 2s$, one obtains for the integral in (7.35)

$$I = \frac{2^{-14/3}}{2\pi i} \int_C ds\, \beta^{2s}\, \Gamma \left[\begin{matrix} s - \frac{5}{6},\ -s+1 \\ -s + \frac{23}{6},\ s+2 \end{matrix}\right] . \tag{7.36}$$

Since $\Delta = \Delta' = 0$, and the obscuration ratio is always less than 1, the integration path is closed in the right half plane. The integral is equal to the residues at $s = 1 + n$, for $n = 0, 1, 2, \ldots$. Tilt variance is thus equal to

$$T_{az}^2 = \frac{6.08\,\mu_0}{D^{1/3}\left(1-\beta^4\right)^2}\left(1+\beta^{23/3} - 1.36\sum_{n=0}^{\infty}\frac{(-1)^n}{n!}\beta^{4+2n}\,\Gamma\left[\begin{matrix} n+\frac{1}{6} \\ -n+\frac{17}{6},\ n+3\end{matrix}\right]\right). \tag{7.37}$$

This can be expressed as a generalized hypergeometric function with the rules in Sect. 1.3,

$$T_{az}^2 = \frac{6.08\,\mu_0}{D^{1/3}\left(1-\beta^4\right)^2}\left(1+\beta^{23/3} - 2.196\,\beta^4\,{}_2F_1\left[\tfrac{1}{6}, -\tfrac{11}{6}; 3; \beta^2\right]\right). \tag{7.38}$$

The first few terms of the solution, which give an answer accurate to better than 1% for $\beta < 0.5$, are

$$T_{az}^2 \approx 6.08\,\mu_0 \frac{\left(1+\beta^{23/3} - 2.196\,\beta^4 + 0.2237\,\beta^6 - 0.02718\,\beta^8 - 0.0006544\,\beta^{10}\right)}{D^{1/3}\left(1-\beta^4\right)^2}. \tag{7.39}$$

Therefore, using the constant in (3.28) to convert tilt angle variance to tilt phase variance, one obtains

$$\sigma_T^2 = 0.899\left(\frac{D}{r_o}\right)^{5/3}\frac{\left(1+\beta^2\right)}{\left(1-\beta^4\right)^2}\left(1+\beta^{23/3} - 2.196\,\beta^4\,{}_2F_1\left[\tfrac{1}{6}, -\tfrac{11}{6}; 3; \beta^2\right]\right). \tag{7.40}$$

Piston and tilt modes are orthogonal on an annulus; therefore, the phase variance with piston and tilt removed σ_{PTR}^2 is

$$\sigma_{PTR}^2 = \sigma_{PR}^2 - \sigma_T^2. \tag{7.41}$$

This is equal to

$$\begin{aligned}\sigma_{PTR}^2 = \left(\frac{D}{r_o}\right)^{5/3}&\left\{\frac{1.033}{\left(1-\beta^2\right)^2}\left(1+\beta^{17/3} - 1.146\,\beta^2\,{}_2F_1\left[-\tfrac{5}{6}, -\tfrac{11}{6}; 2; \beta^2\right]\right)\right.\\ &\left.-0.899\left[\frac{1+\beta^2}{\left(1-\beta^4\right)^2}\right]\left(1+\beta^{23/3} - 2.196\,\beta^4\,{}_2F_1\left[\tfrac{1}{6}, -\tfrac{11}{6}; 3; \beta^2\right]\right)\right\}.\end{aligned} \tag{7.42}$$

The piston- and tilt-removed phase variance is plotted in Fig. 7.7. For large obscuration ratios the terms in brackets almost cancel. To obtain an accurate answer for obscuration ratios approaching unity, the numerical constants must be replaced by their exact values, which are $1.146 \to 2^{-8/3}\sqrt{\pi}\,\Gamma\left[\frac{23}{6}\right]/\Gamma\left[\frac{7}{3}\right]$, and $2.196 \to 2^{-11/3}\sqrt{\pi}\,\Gamma\left[\frac{29}{6}\right]/\Gamma\left[\frac{7}{3}\right]$.

In a similar way gradient tilt variance can be calculated with (2.86), and the filter function for gradient tilt on an annulus given in (3.27), to obtain

$$T_{ag}^2 = \frac{3.317}{\left[D\left(1-\beta^2\right)\right]^2}\int_0^L dz\,C_n^2(z)\int d\boldsymbol{\kappa}\,\left[J_1\left(\kappa D/2\right) - \beta J_1\left(\kappa D_i/2\right)\right]^2. \tag{7.43}$$

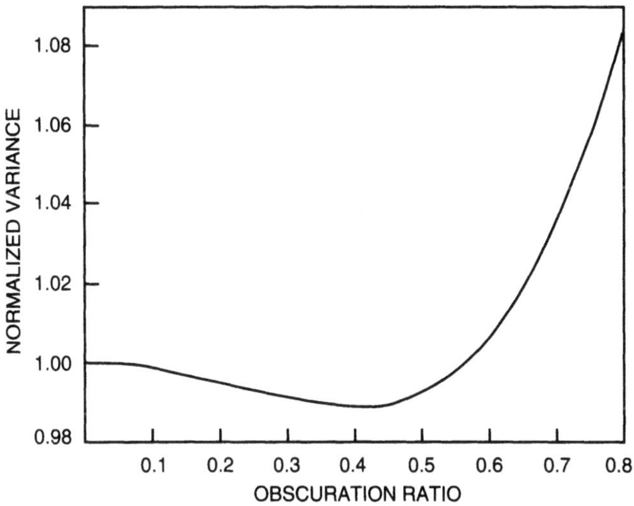

Fig. 7.7. Normalized piston- and tilt-removed phase variance on an annulus versus the obscuration ratio. For a typical obscuration ratio of 25% there is little change in variance from that of a filled aperture. This curve is independent of the turbulence distribution along the propagation path.

This is evaluated in the now standard way to obtain

$$T_{ag}^2 = \frac{5.675\,\mu_0}{D^{1/3}\left(1-\beta^2\right)^2}\left(1+\beta^{11/3}-2.156\,\beta^2\,{}_2F_1\left[\tfrac{1}{6},-\tfrac{5}{6};2;\beta^2\right]\right). \tag{7.44}$$

The first few terms of the solution, which result in an answer accurate to better than 1% for $\beta < 0.5$, are

$$\begin{aligned} T_{ag}^2 &\approx 5.675\,\mu_0 \\ &\times \frac{\left(1+\beta^{11/3}-2.156\,\beta^2+0.1497\,\beta^4+0.00485\,\beta^6+0.00102\,\beta^8\right)}{D^{1/3}\left(1-\beta^2\right)^2}. \end{aligned} \tag{7.45}$$

Normalized Zernike and gradient tilt variances are plotted in Fig. 7.8.

7.5 Effect of Diffraction on Tilt

As a beam propagates, it eventually spreads due to diffraction, and tilt can be affected. The effect of diffraction on Zernike tilt is obtained by not setting the cosine term equal to 1 as has been done until now. The formula for tilt of an infinite plane wave that propagates from ground to space is obtained from the standard formula for variance of a single wave given in (2.86), with the filter function for Zernike tilt given in (3.18), to obtain

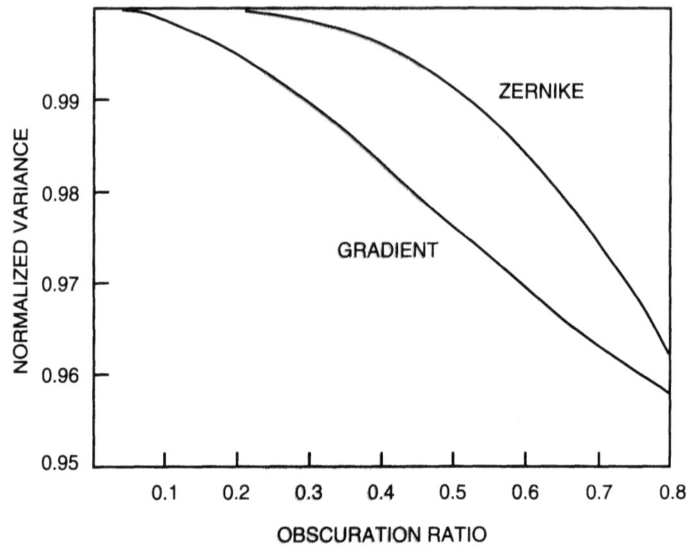

Fig. 7.8. Normalized Zernike and gradient tilt variances on an annulus versus the obscuration ratio. For a typical obscuration ratio of 25% there is little change in variance from that of a filled aperture. These curves are independent of the turbulence distribution along the propagation path.

$$T_d^2 = \frac{53.07}{D^2} \int_0^L dz\, C_n^2(z) \int d\boldsymbol{\kappa}\, \kappa^{-11/3} \left[\frac{J_2(\kappa D/2)}{\kappa D/2}\right]^2 \cos^2\left[\frac{\kappa^2(L-z)}{2k_0}\right]. \quad (7.46)$$

Integrate over angle, let $x = \kappa D/2$, and $y = D\sqrt{k_0/2(L-z)}$ to obtain

$$T_d^2 = \frac{105}{D^{1/3}} \int_0^L dz\, C_n^2(z) \int_0^\infty \frac{dx}{x} x^{-11/3} J_2^2(x) \left[1 - \sin^2(x/y)^2\right]. \quad (7.47)$$

The first term in brackets corresponds to Zernike tilt without diffraction. With the Mellin convolution theorem, the Mellin transforms given in (1.50) and (1.48), and the substitution $s \to 4s$, one obtains

$$T_d^2 = \frac{6.08\,\mu_0}{D^{1/3}}$$
$$\times \left(1 + \frac{4.318}{\mu_0} \int_0^L dz\, C_n^2(z) \frac{1}{2\pi i} \int_C ds\, y^{-4s}\, \Gamma\left[\begin{matrix} 2s+\frac{1}{6},\ -2s+\frac{7}{3},\ -s* \\ -2s+\frac{29}{6},\ -2s+\frac{17}{6},\ s+\frac{1}{2} \end{matrix}\right]\right).$$

With the multiplication formula for gamma functions in (1.15) one obtains

$$I = 2^{-25/6} \frac{1}{2\pi i} \int_C ds\, y^{-4s}\, \Gamma\left[\begin{matrix} s+\frac{1}{12},\ s+\frac{7}{12},\ -s+\frac{7}{6},\ -s+\frac{5}{3},\ -s* \\ -s+\frac{29}{12},\ -s+\frac{35}{12},\ -s+\frac{17}{12},\ -s+\frac{23}{12},\ s+\frac{1}{2} \end{matrix}\right]. \quad (7.48)$$

Since $\Delta = 2$, the integration path should be closed in the left half plane. There are poles located at $s = 0$, and $s = -1/12-n$, and $-7/12-n$ for $n = 0, 1, 2, \ldots$. The pole at zero gives a term that is minus one-half the diffraction-free term. The full result is

$$T_d^2 = \frac{6.08\,\mu_0}{D^{1/3}}\left[0.5 + \frac{2.159}{\mu_0}\int_0^L dz\, C_n^2(z)\right.$$
$$\times\left(\sum_{n=0}^{\infty}\frac{(-1)^n}{n!}\left\{\left[\frac{k_0 D^2}{2(L-z)}\right]^{n+1/12}\Gamma\left[\begin{matrix}-n+\frac{1}{2},\ n+\frac{5}{4},\ n+\frac{7}{3},\ n+\frac{1}{12}\\ n+3,\ n+2,\ -n+\frac{5}{12},\ n+\frac{5}{2},\ n+\frac{3}{2}\end{matrix}\right]\right.\right.$$
$$\left.\left.\left.+\left[\frac{k_0 D^2}{2(L-z)}\right]^{n+7/12}\Gamma\left[\begin{matrix}-n-\frac{1}{2},\ n+\frac{7}{4},\ n+\frac{9}{4},\ n+\frac{7}{12}\\ n+3,\ n+2,\ -n-\frac{1}{12},\ n+\frac{5}{2},\ n+\frac{14}{3}\end{matrix}\right]\right\}\right)\right]. \quad (7.49)$$

For very long ranges the first term dominates, and one finds that tilt variance is one-half that of the tilt in the near field. When the observation plane is well above the turbulence, $L-z$ can be replaced by L in the above expression. Use the rules in Sect. 1.3 to obtain in that limit

$$T_d^2 = \frac{6.08\,\mu_0}{D^{1/3}}$$
$$\times\left\{0.5 - 0.07341\left(\frac{k_0 D^2}{L}\right)^{1/12} {}_4F_5\left[\frac{5}{4},\frac{7}{3},\frac{1}{2},-\frac{3}{2};2,3,\frac{1}{2},\frac{3}{2},\frac{5}{2};\left(\frac{k_0 D^2}{2L}\right)^2\right]\right.$$
$$\left.+0.01095\left(\frac{k_0 D^2}{L}\right)^{7/12} {}_4F_5\left[\frac{7}{4},\frac{9}{4},\frac{7}{12},\frac{13}{12};2,3,\frac{14}{3},\frac{3}{2},\frac{5}{2};\left(\frac{k_0 D^2}{2L}\right)^2\right]\right\}. \quad (7.50)$$

Tilt is independent of wavelength at very long and short ranges. For intermediate ranges tilt can depend on wavelength.

Define the Fresnel number as

$$F_N = W_o^2/\lambda L, \quad (7.51)$$

where W_o is the beamwidth defined in (2.63). In terms of F_N, the first few terms of the series solution are

$$T_d^2 \approx \frac{6.08\,\mu_0}{D^{1/3}}$$
$$\times\left(0.5 + 0.3911\,F_N^{1/6} + 0.004572\,F_N^{7/6} - 0.002281\,F_N^{13/6} - 8.912\times10^{-5}\,F_N^{19/6}\right). \quad (7.52)$$

This solution converges rapidly for Fresnel numbers comparable or smaller than unity. The factor of one-half in the tilt for large ranges is due to conversion of half the phase variance into log-amplitude variance. There is a conversion back and forth between amplitude and phase scintillations. The conversion rate varies with spatial wavelength, this results in an equipartition of variance at long ranges. For large distances the first two terms in parenthesis in (7.52) provide a good estimate of tilt variance. The one-sixth power law dependence of the second term causes the variance to approach the infinite-range value slowly.

For small ranges an asymptotic solution is appropriate, and the simple poles in the right half plane that are located at $s = 7/6 + n$, and $s = 5/3 + n$ for $n = 0, 1, 2, \ldots$, and at $s = n$, for $n = 1, 2, 3, \ldots$ give an asymptotic series that is the sum of these terms plus the $E(x)$ term. Again assuming that the target is much higher than the top of the turbulence, one finds

$$\begin{aligned} &T_d^2 \, \frac{6.08\,\mu_0}{D^{1/3}} \\ &\times \left(1 + 2.159 \left\{ \sum_{n=0}^{n_o} \frac{(-1)^n}{n\,!} \left(\frac{2L}{k_0 D^2}\right)^{n+7/3} \Gamma\left[\begin{matrix} n+\frac{5}{2}, -n/2-\frac{7}{6} \\ -n+\frac{5}{2}, -n+\frac{1}{2}, n/2+\frac{5}{3} \end{matrix}\right] \right.\right. \\ &\left.\left. +2 \sum_{n=1}^{n_o} \frac{(-1)^n}{n\,!} \left(\frac{2L}{k_0 D^2}\right)^{2n} \Gamma\left[\begin{matrix} 2n+\frac{1}{6}, -2n+\frac{7}{3} \\ -2n+\frac{29}{6}, -2n+\frac{17}{6}, n+\frac{1}{2} \end{matrix}\right] + E\left(\frac{2L}{k_0 D^2}\right) \right\}\right) . \end{aligned} \tag{7.53}$$

The first few terms of the series, plus the E term, for $F_N >> 1$ give

$$T_d^2 \approx \frac{6.08\,\mu_0}{D^{1/3}} \left[1 - \frac{0.7353}{F_N^2} + \frac{0.5419}{F_N^{7/3}} - \frac{0.112}{F_N^{10/3}} - \frac{0.2363}{F_N^{31/6}} \cos\left(\frac{\pi F_N}{2} + \frac{\pi}{4}\right)\right] . \tag{7.54}$$

Tilt variance normalized to the value at very short distances is plotted in Fig. 7.9 and Fig. 7.10 on linear and logarithmic scales. From this result one sees that the Fresnel number has to be very small before tilt variance drops to the far-field value. For accurate tracking one wants small deviations from the near field value; tilt variance has dropped by 5% when the Fresnel number is 3.

7.6 Tilt Anisoplanatism

Tilt anisoplanatism is the differential tilt jitter between two sources. The size of tilt anisoplanatism is of interest in many important applications. If the position of a source were measured, and a laser beam were directed at the source, there could be a jitter if the transmit aperture were not exactly aligned with the receive aperture. If a satellite were tracked, and a laser beam sent to the satellite, so that the laser beam would have to lead the perceived position of the target by an angle determined by light velocity and satellite velocity (about 50 μrads for low-altitude satellites), then this difference between the receiver and transmitter pointing angles leads to jitter of the laser beam at the target. Tilt anisoplanatism also accounts for the apparent jitter between two stars simultaneously viewed through the atmosphere. In this section I calculate Z-tilt anisoplanatism for two apertures displaced in distance or displaced in angle. In the second case, physical displacement is a function of the distance from the source. G-tilt anisoplanatism has been calculated by *Ellerbroek* and *Roberts* (1984), and one can obtain their

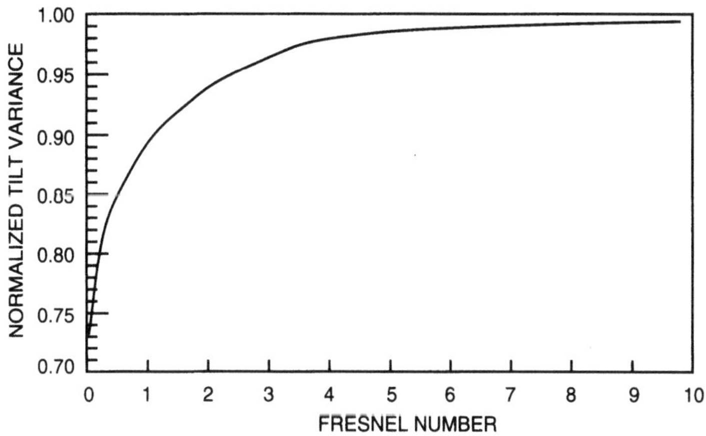

Fig. 7.9. Tilt versus the Fresnel number on a linear scale. If the range is such that the Fresnel number is greater than 3, then tilt variance is within 5% of the very near field tilt variance. This curve is independent of the turbulence distribution along the propagation path.

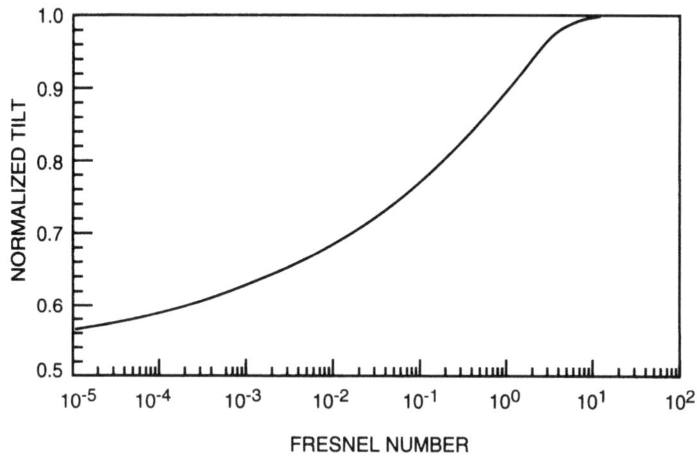

Fig. 7.10. Tilt versus Fresnel number on a logarithmic scale. This plot more clearly shows tilt at low Fresnel numbers.

results by using the approach illustrated below with a G-tilt rather than a Z-tilt filter function.

Tilt anisoplanatism is often much less than a beamwidth for scenarios considered in adaptive optics. When the beam is corrected, the tilt measured by sensors is closer to the Zernike tilt even for a centroid sensor, which for uncorrected beams measures a value closer to the gradient tilt (*Herrmann* 1981). For corrected beams the results obtained here are closer to values that would be measured than the G-tilt values.

To find the total value of tilt anisoplanatism and components both parallel and perpendicular to the displacement direction, (2.92) with diffraction neglected is used with the filter function for Zernike tilt given in (3.18), to obtain

$$\begin{bmatrix} \sigma_{\parallel}^2 \\ \sigma_{\perp}^2 \\ \sigma^2 \end{bmatrix} = 0.2073 \int_0^L dz\, C_n^2(z) \int d\boldsymbol{\kappa} \begin{bmatrix} \cos^2(\varphi) \\ \sin^2(\varphi) \\ 1 \end{bmatrix} \kappa^{-11/3} \times \left(\frac{16}{D}\right)^2 \left[\frac{J_2(\kappa D/2)}{\kappa D/2}\right]^2 2\left\{1 - \cos\left[\kappa d \cos(\varphi)\right]\right\}, \quad (7.55)$$

where σ^2 is the total variance, which is equal to the sum of the parallel and perpendicular variances.

The angular integration can be performed with *Gradshteyn* and *Ryzhik* (1980, Eq. 8.4.11 #7)

$$\int_0^{\pi} dx \exp\left[i\beta\cos(x)\right]\sin^{2\nu}(x) = \sqrt{\pi}\left(\frac{2}{\beta}\right)^{\nu} \Gamma\left[\nu + \tfrac{1}{2}\right] J_\nu(\beta), \quad \mathrm{Re}\{\nu\} > -\tfrac{1}{2}. \quad (7.56)$$

Use the trigonometric identity $\cos^2(\varphi) = 1 - \sin^2(\varphi)$ to express the first integral in (7.55) in the form of (7.56). Using (3.5) one can perform the φ integration to obtain

$$\begin{bmatrix} \sigma_{\parallel}^2 \\ \sigma_{\perp}^2 \\ \sigma^2 \end{bmatrix} = \frac{667}{D^2} \int_0^L dz\, C_n^2(z) \begin{bmatrix} I_T - I_1 \\ I_1 \\ I_T \end{bmatrix}. \quad (7.57)$$

The two integrals that allows one to evaluate all three variances in kappa-space are

$$I_1 = -\int_0^{\infty} d\kappa\, \kappa^{-8/3} \left[\frac{J_2(\kappa D/2)}{\kappa D/2}\right]^2 \left[\frac{J_1(\kappa d)}{\kappa d} - \frac{1}{2}\right], \text{ and} \quad (7.58)$$

$$I_T = -\int_0^{\infty} d\kappa\, \kappa^{-8/3} \left[\frac{J_2(\kappa D/2)}{\kappa D/2}\right]^2 \left[J_0(\kappa d) - 1\right]. \quad (7.59)$$

Set $y = \kappa d$, and $x = 2d/D$, to obtain

$$\begin{bmatrix} I_1 \\ I_T \end{bmatrix} = -\frac{4d^{11/3}}{D^2} \int_0^{\infty} \frac{dy}{y} \begin{bmatrix} y^{-14/3}\left[J_1(y) - \frac{y}{2}\right] \\ y^{-11/3}\left[J_0(y) - 1\right] \end{bmatrix} J_2^2(y/x). \quad (7.60)$$

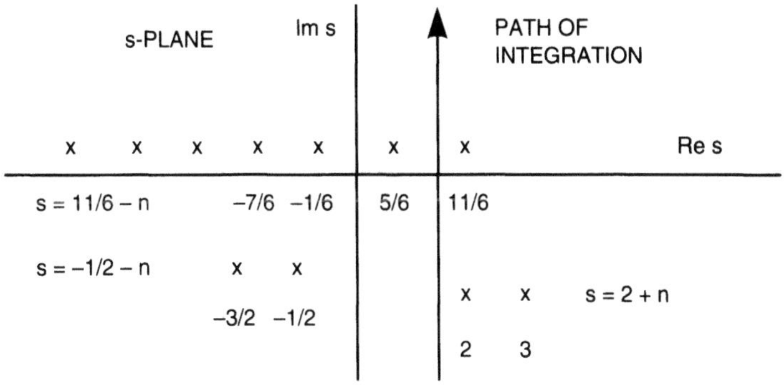

Fig. 7.11. Pole location and integration path for tilt anisoplanatism.

The Mellin transforms of the functions in brackets can be found from (1.49) and (1.8) with $a = -14/3$ and $-11/3$. The Mellin transform of J_2^2 is given in (1.50). After the substitution $s \to 2s$, the integrals in (7.60) can thus be transformed into

$$\begin{bmatrix} I_1 \\ I_T \end{bmatrix} = -\frac{0.0889\, d^{11/3}}{D^2} \frac{1}{2\pi i} \int_C ds \left(\frac{d}{D}\right)^{-2s} \times \Gamma \begin{bmatrix} s - \frac{11}{6}*,\ -s+2,\ s+\frac{1}{2} \\ s+3,\ s+1 \end{bmatrix} \begin{bmatrix} 1/2\Gamma\left[-s+\frac{23}{6}\right] \\ 1/\Gamma\left[-s+\frac{17}{6}\right] \end{bmatrix}. \tag{7.61}$$

The integration path and pole locations are the same for both integrals and are shown in Fig. 7.11.

Since $\Delta = \Delta' = 0$, the direction of path closure is determined by the parameter d/D in the integral. Different Taylor's expansions are obtained for small and large displacements. The point in the atmosphere at which the character of the solution changes from one series to another is where the two beams just barely overlap. For $d/D < 1$, the integration path is closed to the left, and one obtains the residues of poles at $s = -1/2 - n$ for $n = 0, 1, 2, \ldots$, and $s = 11/6 - n$, for $n = 1, 2, 3, \ldots$. For $d/D > 1$, the path is closed to the right, and one obtains the residues of poles at $s = 2 + n$ for $n = 0, 1, 2, \ldots$, and the single pole at $s = 11/6$. The resultant values of the integrals for small displacements are

$$\begin{bmatrix} I_1 \\ I_T \end{bmatrix}_L = -\frac{0.0889\, d^{11/3}}{D^2} \times \left\{ \sum_{n=1}^{\infty} \frac{(-1)^n}{n!} \left(\frac{d}{D}\right)^{-11/3+2n} \Gamma \begin{bmatrix} n+\frac{1}{6},\ -n+\frac{7}{3} \\ n+1,\ -n+\frac{29}{6},\ -n+\frac{17}{6} \end{bmatrix} \begin{bmatrix} \frac{0.5}{n+1} \\ 1 \end{bmatrix} + \sum_{n=0}^{\infty} \frac{(-1)^n}{n!} \left(\frac{d}{D}\right)^{1+2n} \Gamma \begin{bmatrix} -n-\frac{7}{3},\ n+\frac{5}{2} \\ n+\frac{10}{3},\ -n+\frac{5}{2},\ -n+\frac{1}{2} \end{bmatrix} \begin{bmatrix} \frac{0.5}{n+\frac{10}{3}} \\ 1 \end{bmatrix} \right\}, \quad d < D. \tag{7.62}$$

For large displacements the integrals are equal to

$$\begin{bmatrix} I_1 \\ I_T \end{bmatrix}_U = \frac{0.0889\, d^{11/3}}{D^2} \left\{ \left(\frac{d}{D}\right)^{-11/3} \Gamma\begin{bmatrix} \frac{1}{6}, \frac{7}{3} \\ \frac{29}{6}, \frac{17}{6} \end{bmatrix} \begin{bmatrix} 0.5 \\ 1 \end{bmatrix} \right.$$
$$\left. - \sum_{n=0}^{\infty} \frac{(-1)^n}{n!} \left(\frac{d}{D}\right)^{-4-2n} \Gamma\begin{bmatrix} n+\frac{1}{6},\ n+\frac{5}{2} \\ -n+\frac{5}{6},\ n+5,\ n+3 \end{bmatrix} \begin{bmatrix} \frac{0.5}{-n+\frac{5}{6}} \\ 1 \end{bmatrix} \right\}, \quad d > D. \tag{7.63}$$

Substituting the last two results into (7.60) and performing the integration over the axial coordinate, one finds that tilt variances for small displacements are

$$\begin{bmatrix} \sigma_{\parallel}^2 \\ \sigma_{\perp}^2 \end{bmatrix} = -\frac{29.6\, \mu_0}{D^{1/3}}$$
$$\times \left\{ \sum_{n=1}^{\infty} \frac{(-1)^n}{n!} \left(\frac{d}{D}\right)^{2n} \Gamma\begin{bmatrix} n+\frac{1}{6},\ -n+\frac{7}{3} \\ n+2,\ -n+\frac{29}{6},\ -n+\frac{17}{6} \end{bmatrix} \begin{bmatrix} 2n+1 \\ 1 \end{bmatrix} \right.$$
$$\left. + \sum_{n=0}^{\infty} \frac{(-1)^n}{n!} \left(\frac{d}{D}\right)^{2n+14/3} \Gamma\begin{bmatrix} -n-\frac{7}{3},\ n+\frac{5}{2} \\ n+\frac{13}{3},\ -n+\frac{5}{2},\ -n+\frac{1}{2} \end{bmatrix} \begin{bmatrix} 2n+\frac{17}{3} \\ 1 \end{bmatrix} \right\}, \quad d < D. \tag{7.64}$$

For large displacements one obtains

$$\begin{bmatrix} \sigma_{\parallel}^2 \\ \sigma_{\perp}^2 \end{bmatrix} = \frac{6.08\, \mu_0}{D^{1/3}} \left\{ \begin{bmatrix} 1 \\ 1 \end{bmatrix} \right.$$
$$\left. -4.86 \sum_{n=0}^{\infty} \frac{(-1)^n}{n!} \left(\frac{D}{d}\right)^{1/3+2n} \Gamma\begin{bmatrix} n-\frac{5}{6},\ n+\frac{5}{2} \\ -n+\frac{5}{6},\ n+5,\ n+3 \end{bmatrix} \begin{bmatrix} 2\left(n-\frac{1}{3}\right) \\ -1 \end{bmatrix} \right\}, \quad d > D. \tag{7.65}$$

For very large displacements the first term is the only significant one. Tilt variance for each component for very large separations is equal to the total tilt variance of a single wave given in (4.15). This is what one would expect since two uncorrelated tilts are being subtracted.

For small displacements, the first term in (7.64) is the only significant term and gives

$$\begin{bmatrix} \sigma_{\parallel}^2 \\ \sigma_{\perp}^2 \end{bmatrix} \approx \frac{2.67\, \mu_0}{D^{1/3}} \left(\frac{d}{D}\right)^2 \begin{bmatrix} 3 \\ 1 \end{bmatrix}. \tag{7.66}$$

The parallel component is three times the perpendicular component of tilt variance. The value of total rms tilt is

$$T \approx \frac{3.27\, \mu_0^{1/2}}{D^{1/6}} \frac{d}{D}. \tag{7.67}$$

In some systems that propagate laser beams, a tracking system looks through a certain part of the main aperture, and the laser beam may propagate out of a different part of the aperture. This misregistration can cause a relative jitter on the laser beam if the tracked signal is used to direct it. For a 60-cm aperture with a displacement of 5 cm, the rms jitter is 442 nanoradians for an HV-21 turbulence model at zenith.

Change the summations to generalized hypergeometric functions, and convert the expression of one minus such a function into a different generalized hypergeometric function with (1.40), to obtain

$$\begin{bmatrix}\sigma_\parallel^2\\ \sigma_\perp^2\end{bmatrix} = \frac{6.08\,\mu_0}{D^{1/3}}\left\{\begin{bmatrix}1.318\left(\frac{d}{D}\right)^2 {}_5F_4\left[\frac{7}{6},-\frac{17}{6},-\frac{5}{6},\frac{5}{2},1;-\frac{1}{3},3,\frac{3}{2},2;\left(\frac{d}{D}\right)^2\right]\\ 0.4392\left(\frac{d}{D}\right)^2 {}_4F_3\left[\frac{7}{6},-\frac{17}{6},-\frac{5}{6},1;-\frac{1}{3},3,2;\left(\frac{d}{D}\right)^2\right]\end{bmatrix}\right.$$

$$\left.+\begin{bmatrix}2.19\left(\frac{d}{D}\right)^{14/3} {}_4F_3\left[\frac{5}{2},-\frac{3}{2},\frac{1}{2},\frac{23}{6};\frac{17}{6},\frac{13}{3},\frac{10}{3};\left(\frac{d}{D}\right)^2\right]\\ 0.0388\left(\frac{d}{D}\right)^{14/3} {}_3F_2\left[\frac{5}{2},-\frac{3}{2},\frac{1}{2};\frac{13}{3},\frac{10}{3};\left(\frac{d}{D}\right)^2\right]\end{bmatrix}\right\},\quad d<D, \tag{7.68}$$

and

$$\begin{bmatrix}\sigma_\parallel^2\\ \sigma_\perp^2\end{bmatrix} = \frac{6.08\,\mu_0}{D^{1/3}}\left\{\begin{bmatrix}1\\ 1\end{bmatrix} - \begin{bmatrix}0.532\left(\frac{D}{d}\right)^{1/3} {}_4F_3\left[-\frac{5}{6},\frac{5}{2},\frac{1}{6},\frac{2}{3};5,3,-\frac{1}{3};\left(\frac{D}{d}\right)^2\right]\\ 0.798\left(\frac{D}{d}\right)^{1/3} {}_3F_2\left[-\frac{5}{6},\frac{5}{2},\frac{1}{6};5,3;\left(\frac{D}{d}\right)^2\right]\end{bmatrix}\right\},$$
$$d>D. \tag{7.69}$$

These variances vary with zenith angle as $\sec(\xi)$. They are plotted in Fig. 7.12 versus displacement.

If one is interested in calculating tilt anisoplanatism for two beams pointing in different directions, then displacement is a function of distance, and the axial integration results in the summation that contains various different moments of the turbulence distribution. For the Hufnagel-Valley-turbulence model these can be evaluated analytically; therefore, it is very easy to make calculations for specific turbulence levels, aperture sizes, and zenith angles.

If the displacement between the two beams is an angle θ, then the expression for tilt anisoplanatism is obtained with the substitution $d = \theta z$ in the above equations. The displacement is a small distance low in the atmosphere, however, it can exceed the diameter at higher altitudes. Therefore, the solution for small displacement applies at low altitudes and that for large displacements applies at higher altitudes. The solution is a sum of a lower and upper contribution,

$$\begin{bmatrix}\sigma_\parallel^2\\ \sigma_\perp^2\end{bmatrix} = \begin{bmatrix}\sigma_\parallel^2\\ \sigma_\perp^2\end{bmatrix}_L + \begin{bmatrix}\sigma_\parallel^2\\ \sigma_\perp^2\end{bmatrix}_U . \tag{7.70}$$

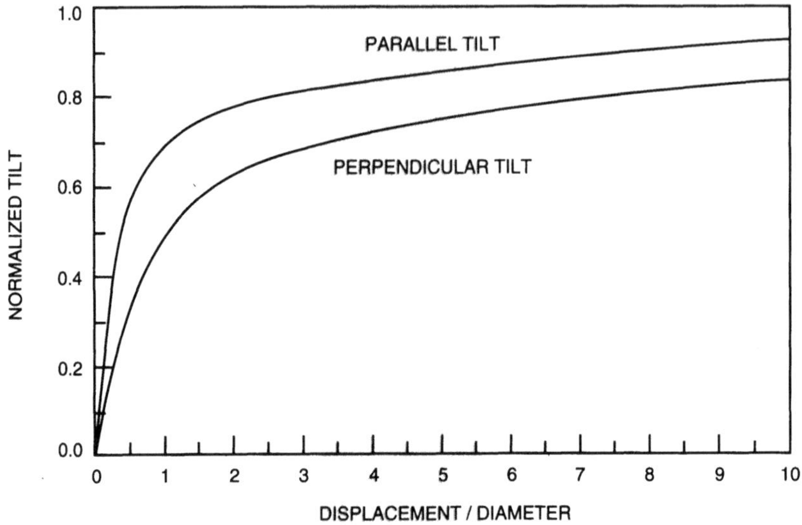

Fig. 7.12. Tilt anisoplanatism versus displacement, normalized to aperture tilt. The tilt component parallel to the displacement is 1.73 times that of the perpendicular component, for small displacements. For very large displacements these components are equal. The tilt variance increases rapidly with displacement so that when the displacement is equal to the diameter, tilt variance is one-half that of a single wave. This curve is independent of the turbulence distribution along the propagation path.

The lower contribution, with $H_c = D\cos(\xi)/\theta$, is

$$\begin{bmatrix}\sigma_{\|}^2 \\ \sigma_{\perp}^2\end{bmatrix}_L = \frac{6.08}{D^{1/3}} \int_0^{H_c} dz\, C_n^2(z)$$

$$\times \left\{ \begin{bmatrix} 1.318\left(\frac{\theta z}{D}\right)^2 {}_5F_4\left[\frac{7}{6}, -\frac{17}{6}, -\frac{5}{6}, \frac{5}{2}, 1; -\frac{1}{3}, 3, \frac{3}{2}, 2; \left(\frac{\theta z}{D}\right)^2\right] \\ 0.4392\left(\frac{\theta z}{D}\right)^2 {}_4F_3\left[\frac{7}{6}, -\frac{17}{6}, -\frac{5}{6}, 1; -\frac{1}{3}, 3, 2; \left(\frac{\theta z}{D}\right)^2\right] \end{bmatrix} \right.$$

$$\left. + \begin{bmatrix} 2.19\left(\frac{\theta z}{D}\right)^{14/3} {}_4F_3\left[\frac{5}{2}, -\frac{3}{2}, \frac{1}{2}, \frac{23}{6}; \frac{17}{6}, \frac{13}{3}, \frac{10}{3}; \left(\frac{\theta z}{D}\right)^2\right] \\ 0.388\left(\frac{\theta z}{D}\right)^{14/3} {}_3F_2\left[\frac{5}{2}, -\frac{3}{2}, \frac{1}{2}; \frac{13}{3}, \frac{10}{3}; \left(\frac{\theta z}{D}\right)^2\right] \end{bmatrix} \right\}. \tag{7.71}$$

The upper contribution is

$$\begin{bmatrix}\sigma_{\|}^2 \\ \sigma_{\perp}^2\end{bmatrix}_U = \frac{6.08}{D^{1/3}}$$

$$\times \left\{ \mu_0^+(H_c) - \int_{H_c}^{L} dz\, C_n^2(z) \begin{bmatrix} 0.532\left(\frac{D}{\theta z}\right)^{1/3} {}_4F_3\left[-\frac{5}{6}, \frac{5}{2}, \frac{1}{6}, \frac{2}{3}; 5, 3, -\frac{1}{3}; \left(\frac{D}{\theta z}\right)^2\right] \\ 0.798\left(\frac{D}{\theta z}\right)^{1/3} {}_3F_2\left[-\frac{5}{6}, \frac{5}{2}, \frac{1}{6}; 5, 3; \left(\frac{D}{\theta z}\right)^2\right] \end{bmatrix} \right\}. \tag{7.72}$$

The upper and lower partial turbulence moments, μ_m^+ and μ_m^-, are defined in Sect. 4.1.

The turbulence above 30 km is small and is generally negligible. For small displacement angles such that H_c is higher than the uppermost turbulence, the anisoplanatic tilt variance is given solely by the lower contribution, and the leading terms of the series are

$$\begin{bmatrix}\sigma_{\|}^2\\ \sigma_{\perp}^2\end{bmatrix} = \frac{2.67\,\mu_2^-(H_c)}{D^{1/3}}\left(\frac{\theta}{D}\right)^2\begin{bmatrix}3\\1\end{bmatrix} - \frac{3.68\,\mu_4^-(H_c)}{D^{1/3}}\left(\frac{\theta}{D}\right)^4\begin{bmatrix}5\\1\end{bmatrix} + \frac{2.35\,\mu_{14/3}^-(H_c)}{D^{1/3}}\left(\frac{\theta}{D}\right)^{14/3}\begin{bmatrix}\frac{17}{3}\\1\end{bmatrix} + \cdots . \quad (7.73)$$

The total anisoplanatic tilt multiplied by the diameter is plotted versus aperture diameter for various angular separations in Fig. 7.13 for the HV-21 turbulence model. As one can observe in the figure, the plotted values decrease as the one-sixth power of the diameter for most regions of the plots. In those regions, the tilt is given with small error solely by the first term of the series. The one-term approximation applies when $\theta < D/40,000$. In this region for the HV-21 turbulence model, the total rms tilt anisoplanatism σ_T is given by

$$\sigma_T(\text{nrad}) \approx 14.2\,\theta\,(\mu\text{rad})\,\sec^{1.5}(\xi)\;/D^{7/6}.$$

Define a characteristic tilt anisoplanatic angle as the angle between two sources for which the tilt jitter is equal to one-half the diffraction-limited beam size. At this angle the jitter radius is equal to that of the perfectly corrected spot. For these small angles the first term of the expression above gives a reasonable approximation of the angle jitter, and the partial moment becomes a full moment. This characteristic tilt anisoplanatic angle θ_{Tc} is

$$\theta_{Tc} = \frac{0.184\,\lambda\,D^{1/6}}{\mu_2^{1/2}}. \quad (7.74)$$

For a HV-21 turbulence model and a 1-m diameter aperture at 0.5 μm operating wavelength the characteristic tilt anisoplanatic angle is 21 μrad.

For large angles in which the transition altitude is close to the ground, tilt variance is given solely by the upper contribution as

$$\begin{bmatrix}\sigma_{\|}^2\\ \sigma_{\perp}^2\end{bmatrix} \approx \frac{1}{D^{1/3}}\left\{6.08\,\mu_0^+(H_c)\begin{bmatrix}1\\1\end{bmatrix} - 4.85\,\mu_{-1/3}^+(H_c)\left(\frac{D}{\theta}\right)^{1/3}\begin{bmatrix}\frac{2}{3}\\1\end{bmatrix} + 0.0205\,\mu_{-7/3}^+(H_c)\left(\frac{D}{\theta}\right)^{7/3}\begin{bmatrix}-\frac{4}{3}\\1\end{bmatrix} + \cdots . \right. \quad (7.75)$$

The leading term is twice the tilt variance of a single wave. The angular separation of the beams has to be very large before this asymptotic limit is approached.

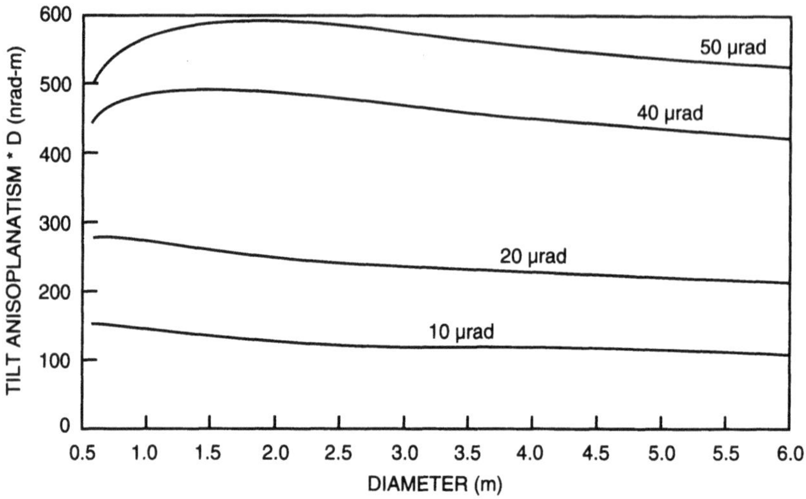

Fig. 7.13. Tilt anisoplanatism at zenith times aperture diameter versus aperture diameter with the HV-21 turbulence model. Plots are given for source separations of 10, 20, 40, and 50 μrad.

7.7 Power Spectral Density of Tilt

In this section, the power spectral density for tilt of a collimated wave is derived. Plane-wave results have been discussed briefly in *Tatarski* (1971), and presented in more detail in *Greenwood* and *Fried* (1976) who made a simplifying assumption in order to obtain simple, analytic results. *Fields* (1983) coined the term "parallel approximation" for this simplification. One consequence of the parallel approximation is that the rate at which spectra decay at high frequencies is underestimated. *Tyler* (1986) subsequently analyzed plane-wave tilt spectra without making this approximation. However, his results remain in integral form, which has an integral over a dummy variable related to spatial frequency and an integral over altitude. In a subsequent report, *Vaughn* (1986) provided numerical techniques to solve the integrals presented by Tyler, but no solutions have been published that do not rely heavily on numerical integration. A similar approach has been used in tilt spectra associated with a point source (spherical wave analysis). This leads to integral expressions given in *Hogge* and *Butts* (1976) and *Butts* (1980) that also must be evaluated numerically.

Power spectral density of tilt is found here with the use of the general expression for power spectral density given in (2.101) and the filter function for Zernike tilt in (3.18). It is assumed that one is in the near field so that the cosine term can be replaced by unity. This assumption breaks down at sufficiently high frequencies, where the exact equation must be used. The effect of including the cosine term is to lower the high-frequency spectrum. The spectral density is

$$S_T(\omega) = 2.606\, k_0^2 \omega \int_0^L dz\, \frac{C_n^2(z)}{v^2(z)}$$

$$\times \int_0^\infty dc\, \frac{c\, U(c-1)}{\sqrt{c^2-1}} \left[\frac{\omega c}{v(z)}\right]^{-11/3} \left(\frac{16}{k_0 D}\right)^2 \left\{\frac{J_2\left[\omega c D/2v(z)\right]}{\omega c D/2v(z)}\right\}^2 . \quad (7.76)$$

Let $x = 2v(z)/\omega D$, to give

$$S_T(\omega) = \frac{2668}{D^4 \omega^{14/3}} \int_0^L dz\, C_n^2(z)\, v^{11/3}(z) \int_0^\infty \frac{dc}{c} \frac{U(c-1)}{\sqrt{c^2-1}} J_2^2\left(\frac{c}{x}\right) c^{-11/3}. \quad (7.77)$$

To evaluate the integral over c, define the functions $h_1(x/c) = J_2^2(c/x)$, and $h_2(c) = U(c-1)/\sqrt{c^2-1}\, c^{-11/3}$. The integral over c can be expressed as a Mellin convolution integral with the Mellin transforms in (1.50) and (1.54). Let $s \to 2s$ to obtain for the c integral

$$I = \frac{1}{2}\frac{1}{2\pi i} \int_C ds\, \Gamma\left[\begin{matrix} -s+\frac{1}{2},\; s+2,\; s+\frac{7}{3} \\ -s+1,\; -s+3,\; s+\frac{17}{6} \end{matrix}\right] \left[\frac{\omega D}{2v(z)}\right]^{-2s} . \quad (7.78)$$

Since $\Delta = 2$, the integration path is closed in the left-half plane, and the contributions at the two sets of poles at $s = -2-n$, and $s = -7/3 - n$ for $n = 0, 1, 2, \ldots$ give

$$I = \frac{1}{2}\sum_{n=0}^\infty \frac{(-1)^n}{n!} \left\{ \Gamma\left[\begin{matrix} n+\frac{5}{2},\; -n+\frac{1}{3} \\ n+3,\; n+5,\; -n+\frac{5}{6} \end{matrix}\right] \left[\omega D/2v(z)\right]^{4+2n} \right.$$

$$\left. + \Gamma\left[\begin{matrix} n+\frac{17}{6},\; -n-\frac{1}{3} \\ n+\frac{10}{3},\; n+\frac{16}{3},\; -n+\frac{1}{2} \end{matrix}\right] \left[\omega D/2v(z)\right]^{14/3+2n} \right\} . \quad (7.79)$$

To obtain the spectral density, the velocity profile must be inserted into the above series, and the integrations along the path performed. Various velocity moments, defined in (4.49), have to be calculated. The final result is a series solution for the power spectral density,

$$S_T(\omega) = 52.53 \left(\frac{2}{\omega}\right)^{2/3} \sum_{n=0}^\infty \frac{(-1)^n}{n!} \left\{ \Gamma\left[\begin{matrix} n+\frac{5}{2},\; -n+\frac{1}{3} \\ n+3,\; n+5,\; -n+\frac{5}{6} \end{matrix}\right] \left(\frac{\omega D}{2}\right)^{2n} v_{-2n-1/3} \right.$$

$$\left. + \Gamma\left[\begin{matrix} n+\frac{17}{6},\; -n-\frac{1}{3} \\ n+\frac{10}{3},\; n+\frac{16}{3},\; -n+\frac{1}{2} \end{matrix}\right] \left(\frac{\omega D}{2}\right)^{2n+2/3} v_{-2n-1} \right\} . \quad (7.80)$$

For the slew-dominated case in which the velocity is proportional to axial position, the velocity moments in the above expression are infinite. To obtain a solution for this problem, one must evaluate the power series in (7.79), which is finite, and then perform the axial integration.

The low-frequency asymptote comes from the first term of the first series, and the dependence is $\omega^{-2/3}$. For large frequencies, one obtains an asymptotic series by deforming the integration path into the right half plane, thus obtaining the pole contributions at $s = -1/2 - n$ for $n = 0, 1, 2 \ldots, n_o$. The $E(x)$ term is significant in this case. It is found with (5.94). Inserting this into (5.51), one finds the asymptotic solution is

$$I = \frac{1}{2\sqrt{\pi}}\left[\frac{2v(z)}{\omega D}\right]^{3/2}\cos\left\{\left[\frac{\omega D}{v(z)}\right]-\frac{\pi}{4}\right\}$$
$$+\frac{1}{2}\sum_{n=0}^{n_o}\frac{(-1)^n}{n!}\Gamma\left[\begin{matrix} n+\frac{5}{2},\, n+\frac{17}{6} \\ n+\frac{10}{3},\, -n+\frac{1}{2},\, -n+\frac{5}{2}\end{matrix}\right]\left[\frac{2v(z)}{\omega D}\right]^{1+2n}. \quad (7.81)$$

The power spectral density at high frequencies is thus

$$S_T(\omega) \sim 52.53\, D^{2/3}\left(\frac{2}{\omega D}\right)^{17/3}$$
$$\times\left\{\int_0^L dz\, C_n^2(z)\frac{1}{\sqrt{\pi}}\left[\frac{2v(z)}{\omega D}\right]^{1/2} v^{14/3}(z)\cos\left\{\left[\frac{\omega D}{v(z)}\right]-\frac{\pi}{4}\right\}\right.$$
$$\left.+\sum_{n=0}^{n_o}\frac{(-1)^n}{n!}\Gamma\left[\begin{matrix} n+\frac{5}{2},\, n+\frac{17}{6} \\ n+\frac{10}{3},\, -n+\frac{1}{2},\, -n+\frac{5}{2}\end{matrix}\right]\left(\frac{2}{\omega D}\right)^{2n} v_{2n+14/3}\right\}. \quad (7.82)$$

The high-frequency asymptote comes from the first series term, and it varies as $\omega^{-17/3}$.

If velocity is constant along the path, spectral density for small values of $\omega D/2v$ is found with the rules in Sect. 1.3 as

$$S_T(\omega) = 3.453\,\frac{\mu_0\, D^{2/3}}{v}\left(\frac{2v}{\omega D}\right)^{2/3}\left\{{}_2F_3\left[\frac{5}{2},\frac{1}{6};3,5,-\frac{2}{3};-\left(\frac{\omega D}{2v}\right)^2\right]\right.$$
$$\left.-0.5394\left(\frac{\omega D}{2v}\right)^{2/3}{}_2F_3\left[\frac{17}{6},\frac{1}{2};\frac{10}{3},\frac{16}{3},\frac{4}{3};-\left(\frac{\omega D}{2v}\right)^2\right]\right\}. \quad (7.83)$$

The spectral density for large values of $\omega D/2v$ is

$$S_T(\omega) = 52.53\frac{\mu_0\, D^{2/3}}{v}\left(\frac{2v}{\omega D}\right)^{17/3}\left\{\frac{1}{\sqrt{\pi}}\left(\frac{2v}{\omega D}\right)^{1/2}\cos\left[\left(\frac{\omega D}{v}\right)-\frac{\pi}{4}\right]\right.$$
$$\left.+\sum_{n=0}^{n_o}\frac{(-1)^n}{n!}\Gamma\left[\begin{matrix} n+\frac{5}{2},\, n+\frac{17}{6} \\ n+\frac{10}{3},\, -n+\frac{1}{2},\, -n+\frac{5}{2}\end{matrix}\right]\left(\frac{2v}{\omega D}\right)^{2n}\right\}, \quad fD/v > 1. \quad (7.84)$$

Power spectral density is plotted in Fig. 7.14. The spectrum is normalized to give a value of unity when fD/v is unity. In the figure both Taylor series (first 40 terms) and asymptotic solution (first 5 terms) are plotted. For $fD/v < 1$, the asymptotic series was set equal to unity since the series is very inaccurate in that region, and above 5 the Taylor series was discarded because it becomes inaccurate with the 40-term approximation. The two series give virtually the

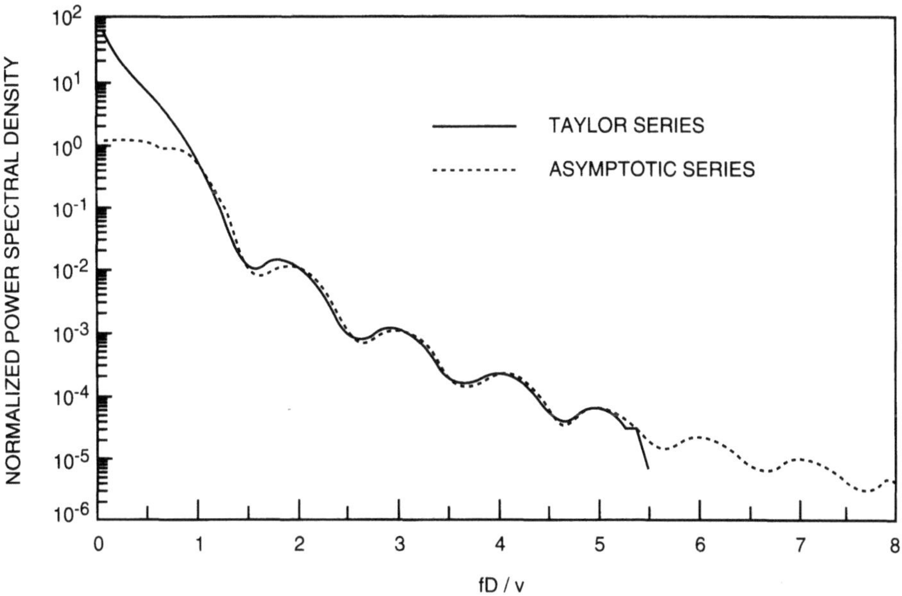

Fig. 7.14. Log-linear plot of the power spectral density of tilt. The velocity and turbulence are constant along the path. Notice the agreement of asymptotic and Taylor series over a considerable range. This curve is independent of the turbulence distribution along the propagation path.

same answer in the parameter range between 1 and 5. Notice the strong influence of the cosine term on the spectrum. If velocity is allowed to vary along the path, various cosine terms for different velocities tend to average to zero, and the spectrum does not exhibit nearly as strong an oscillatory behavior.

7.8 Scintillation with Finite Apertures and Sources

Scintillation in a receiver is considerably reduced from that measured with a very small aperture if the aperture is a few inches in diameter. Also, the scintillation from a finite size object such as a planet is considerably lower than that from a point source such as a star. Both of these averaging effects can be calculated with the filter function approach. In these problems one is interested in the average scintillation over an aperture or at a point receiver from a finite source. One has to average the intensity but the statistical expressions are given in terms of the log intensity. The Rytov approximation, upon which this work is based, is valid for low levels of scintillation, and in that approximation the normalized intensity is well approximated by the log intensity. Therefore, the average of the log intensity approximates the average of the intensity. The scintillation from stars is low, and the approximation holds for that practical problem.

The exact solution was found by *Fried* (1967) by using a numerical integration. *Andrews* (1992) used the low scintillation approximation to obtain asymptotic results for constant turbulence with finite inner scale. His results agree with those given here.

The scintillation averaged over an aperture is obtained with the filter function in (3.41) for a source at infinity that is inserted into the formula for variance given in (2.86). The aperture-averaged scintillation is

$$\sigma_{\chi A}^2 \approx 0.2073\, k_0^2 \int_0^L dz\, C_n^2(z) \int d\boldsymbol{\kappa}\, f(\kappa) \sin^2\left[P(\gamma,\kappa,z)\right] \left[\frac{2J_1(\gamma\kappa D/2)}{\gamma\kappa D/2}\right]^2 . \tag{7.85}$$

The diffraction parameter is $P(\gamma,\kappa,z) = \gamma\kappa^2 z/2k_0$ for propagation towards the receiver. Assuming Kolmogorov turbulence and performing the angular integration, one obtains

$$\sigma_{\chi A}^2 = 1.303\, k_0^2 \int_0^L dz\, C_n^2(z) \int_0^\infty \frac{d\kappa}{\kappa}\, \kappa^{-5/3} \sin^2\left(\frac{\gamma\kappa^2 z}{2k_0}\right) \left[\frac{2J_1(\gamma\kappa D/2)}{\gamma\kappa D/2}\right]^2 . \tag{7.86}$$

Let $w^2 = \gamma\kappa^2 z/2k_0$, and $x = \sqrt{2z/\gamma D^2 k_0}$, to obtain

$$\sigma_{\chi A}^2 = 2.925\, k_0^{7/6} \int_0^L dz\, C_n^2(z)\, (z\gamma)^{5/6} x^2 \int_0^\infty \frac{dw}{w}\, w^{-11/3} \sin^2\left(w^2\right) J_1^2(w/x) . \tag{7.87}$$

The last integral can be evaluated with the Mellin transforms in (1.48) and (1.50). After the substitution $s \to -2s$ it is equal to

$$\begin{aligned} I &= \int_0^\infty \frac{dw}{w}\, w^{-11/3} \sin^2\left(w^2\right) J_1^2(w/x) \\ &= -\frac{1}{8}\frac{1}{2\pi i} \int_C ds\, x^{2s}\, \Gamma\left[\begin{matrix} -s/2 - \frac{11}{12}*,\ s+1,\ -s+\frac{1}{2} \\ s/2 + \frac{17}{12},\ -s+2,\ -s+1 \end{matrix}\right] . \end{aligned} \tag{7.88}$$

Because both the sign of the integration variable and the direction of the integration path are changed, the sign of the expression remains unchanged. Since $\Delta = 1$, the integration path is closed to the left, and one obtains residues of poles at $s = -11/6$, and at $s = -n-1$ for $n = 0, 1, 2, \ldots$, which gives

$$I = -\frac{1}{8}\sum_{n=0}^\infty \frac{(-1)^n}{n!} x^{-2n-2}\, \Gamma\left[\begin{matrix} n/2 - \frac{5}{12},\ n+\frac{3}{2} \\ -n/2 + \frac{11}{12},\ n+3,\ n+2 \end{matrix}\right] - 0.1331\, x^{-11/3} . \tag{7.89}$$

With this result the scintillation is equal to

$$\begin{aligned} \sigma_{\chi A}^2 = -0.3656\, k_0^{7/6} \int_0^L dz\, C_n^2(z)\, (z\gamma)^{5/6} \Bigg\{ 0.1331 \left(\frac{\gamma D^2 k_0}{2z}\right)^{5/6} \\ + \sum_{n=0}^\infty \frac{(-1)^n}{n!} \left(\frac{\gamma D^2 k_0}{2z}\right)^n \Gamma\left[\begin{matrix} n/2 - \frac{5}{12},\ n+\frac{3}{2} \\ -n/2 + \frac{11}{12},\ n+3,\ n+2 \end{matrix}\right] \Bigg\} . \end{aligned} \tag{7.90}$$

For very small diameters the first series term dominates, and one obtains the point-source result.

One cannot evaluate this expression by using turbulence moments, since the negative moments are equal to infinity. In this case one obtains the scintillation by first evaluating the power series in (7.90), and then performing the axial integration.

Very often the parameter in the previous equation is very large, and the answer can be approximated by the first term of the asymptotic series. The steepest-descent contribution is negligible since it has exponential decay. The pole at $s = 1/6$ gives this contribution as

$$\sigma_{\chi A}^2 \approx \frac{4.34}{D^{7/3}} \int_0^L dz\, C_n^2(z)\, \frac{z^2}{\gamma^{1/3}}. \tag{7.91}$$

For propagation from space to ground $\gamma = (L - z)/L$. If the source is well above the atmosphere, then this is close to unity. In this case the scintillation is equal to

$$\sigma_{\chi A}^2 \approx \frac{4.34\,\mu_2}{D^{7/3}}. \tag{7.92}$$

The aperture-averaged scintillation does not depend on the wavelength. That is the reason that no scintillation-induced color is evident in a star image as seen in a telescope. The aperture-averaged scintillation depends on the second moment of turbulence, which is close to the five-thirds moment that is needed to calculate the isoplanatic angle. By modifying the aperture function using masks in their isoplanometer, *Walters* et al. (1979) were able to closely approximate the aperture-averaged scintillation by the five-third moment of turbulence. In this manner the instrument is used to estimate isoplanatic angle.

If turbulence is constant along the path, for plane-wave propagation the variance is

$$\sigma_{\chi A}^2 \approx \frac{1.45\,C_n^2\,L^3}{D^{7/3}}. \tag{7.93}$$

Scintillation with no averaging is given in (4.44) as

$$\sigma_\chi^2 = 0.5631\,k_0^{7/6}\,\mu_{5/6}. \tag{7.94}$$

The scintillation reduction ratio is

$$\frac{\sigma_{\chi A}^2}{\sigma_\chi^2} \approx 7.7\,\frac{\mu_2}{\mu_{5/6}\,D^{7/3}k_0^{7/6}} = 0.902\,\frac{\mu_2}{\mu_{5/6}}\left(\frac{\lambda}{D^2}\right)^{7/6} = \left(\frac{D_c}{D}\right)^{7/3}, \quad D >> D_c, \tag{7.95}$$

where D_c is a characteristic diameter for scintillation averaging. It is given by

$$D_c = 0.957\left(\frac{\mu_2}{\mu_{5/6}}\right)^{3/7}\sqrt{\lambda}. \tag{7.96}$$

For the HV-21 model of turbulence, the relevant turbulence moments are $\mu_2 = 1.91 \times 10^{-5}\mathrm{m}^{7/3}$ and $\mu_{5/6} = 5.45 \times 10^{-10}\mathrm{m}^{7/6}$. At a wavelength of 0.5 μm the characteristic diameter is 6 cm.

If turbulence is constant along the path, (7.95) becomes

$$\frac{\sigma_{\chi A}^2}{\sigma_\chi^2} \approx 0.109 \left(\frac{\sqrt{\lambda L}}{D/2} \right)^{7/3}, \quad D >> \sqrt{\lambda L}. \tag{7.97}$$

This formula agrees with the power-law dependence given in *Tatarski* (1971). For this case the characteristic diameter is

$$D_c = 0.774 \sqrt{\lambda L}. \tag{7.98}$$

Next, the case of a source of finite angular extent viewed by a point receiver is considered. In this case the filter function for a source of diameter D_s given in (3.39) is rewritten in terms of the angle subtended by the source θ_s, where $\theta_s = D_s/L$. The filter function is thus

$$F(\gamma \boldsymbol{\kappa}) = \left[\frac{2 J_1 (\gamma \kappa \theta_s z/2)}{\gamma \kappa \theta_s z/2} \right]^2 . \tag{7.99}$$

The analysis proceeds in the same manner as above, and with the definitions $w^2 = \gamma \kappa^2 z / 2 k_0$, and $x = \sqrt{2 / z \theta_s^2 k_0 \gamma}$, one obtains (7.87), which is evaluated in (7.88) to (7.90). If the value of x is inserted into (7.90), one obtains

$$\sigma_{\chi A}^2 = -0.3663\, k_0^{7/6} \int_0^L dz\, C_n^2(z)\, (z\gamma)^{5/6} \left\{ -0.1331 \left(\frac{\gamma z \theta_s^2 k_0}{2} \right)^{-5/6} \right.$$
$$\left. + \sum_{n=0}^{\infty} \frac{(-1)^n}{n!} \left(\frac{\gamma z \theta_s^2 k_0}{2} \right)^n \Gamma \left[\begin{matrix} n/2 - \frac{5}{12},\ n + \frac{3}{2} \\ -n/2 + \frac{11}{12},\ n+3,\ n+2 \end{matrix} \right] \right\} . \tag{7.100}$$

The leading term of the asymptotic series is

$$\sigma_{\chi A}^2 \approx \frac{4.34}{\theta_s^{7/3}} \int_0^L dz\, C_n^2(z)\, (\gamma z)^{-1/3} . \tag{7.101}$$

The negative moment of turbulence is finite. The leading term for scintillation from a finite source, just as for aperture averaging, does not depend on the wavelength. For propagation from space to the ground $\gamma = (L - z)/L$. If the source is well above the atmosphere, then this is close to unity. For large source sizes the scintillation reduction ratio is

$$\frac{\sigma_{\chi A}^2}{\sigma_\chi^2} \approx 7.7 \frac{\mu_{-1/3}}{\mu_{5/6}\, \theta_s^{7/3}\, k_0^{7/6}} = 0.902 \frac{\mu_{-1/3}}{\mu_{5/6}} \left(\frac{\lambda}{\theta_s^2} \right)^{7/6} = \left(\frac{\theta_c}{\theta_s} \right)^{7/3}, \quad \theta_s >> \theta_\chi, \tag{7.102}$$

where θ_χ, the characteristic angle of the source for scintillation reduction, is

$$\theta_\chi = 0.957 \left(\frac{\mu_{-1/3}}{\mu_{5/6}} \right)^{3/7} \sqrt{\lambda}. \tag{7.103}$$

For the HV-21 model the characteristic angle is equal to 35.2 μrad at a wavelength of 0.5 μm.

If turbulence is constant along the path, the scintillation reduction factor is

$$\frac{\sigma_{\chi A}^2}{\sigma_\chi^2} \approx 2.48 \left(\frac{\lambda}{L\theta_s^2}\right)^{7/6}, \quad \theta_s >> \sqrt{\lambda/L}. \tag{7.104}$$

The characteristic angle in this case is

$$\theta_\chi = 1.48 \sqrt{\lambda/L}. \tag{7.105}$$

7.9 Scintillation with Finite Inner Scale

The phase disturbances that results from turbulence bend the ray paths of the propagating waves. These bent rays after propagating some distance have regions of bunching and expansion, thus leading to scintillation. For close distances it is the shorter scale turbulence that affects the scintillation most significantly. Therefore, it is at short ranges that inner scale affects the scintillation most strongly, and measurements in this regime have confirmed the presence of the bump in the Hill turbulence spectrum. Most of these measurements are performed over a horizontal path for which the turbulence can be considered to be constant. I find the scintillation versus propagation distance on a collimated beam for both the Hill and Tatarski inner-scale spectra for the case in which the turbulence is constant along the propagation direction. I show the difference between their scintillation and the one with zero inner scale that for constant turbulence in (4.45) is

$$\sigma_\chi^2 = 0.3071\, k_0^{7/6}\, L^{11/6}\, C_n^2. \tag{7.106}$$

Hill and *Clifford* (1978) found the scintillation for the Hill spectrum by using a numerical integration.

To obtain analytical results, use the formula for scintillation given in (2.86), with the turbulence spectrum given in (2.32), which has the Tatarski inner-scale spectrum. Set the outer-scale size to infinity to obtain

$$\chi_T^2 = 0.2073\, k_0^2 \int_0^L dz\, C_n^2(z) \int d\boldsymbol{\kappa}\, \kappa^{-11/3} \exp\left[-\left(\frac{\kappa^2}{\kappa_i^2}\right)\right] \sin^2\left(\frac{\kappa^2 z}{2k_0}\right). \tag{7.107}$$

The plane wave is propagated from $z = 0$ to $z = L$. Assume that the turbulence strength is constant, integrate over angle in kappa-space and along the propagation direction, to obtain

$$\chi_T^2 = -0.6513\, k_0^3 C_n^2 \int_0^\infty \frac{d\kappa}{\kappa}\, \kappa^{-11/3} \exp\left[-\left(\frac{\kappa^2}{\kappa_i^2}\right)\right] \left[\sin\left(\frac{\kappa^2 L}{k_0}\right) - \frac{\kappa^2 L}{k_0}\right]. \tag{7.108}$$

Let $y = \kappa\, (L/k_0)^{1/2}$, and $x = \kappa_i\, (L/k_0)^{1/2}$. Use the Mellin convolution theorem, the substitution $s \to 4s$, and the Legendre multiplication theorem in (1.14) to obtain

$$\chi_T^2 = -0.04571\, k_0^{7/6}\, L^{11/6}\, C_n^2 \frac{1}{2\pi i} \int_C ds\, x^{-4s}\, \Gamma\left[\begin{matrix} s - \frac{5}{12}*, \ -s, \ -s + \frac{1}{2} \\ \frac{23}{12} - s \end{matrix}\right]. \quad (7.109)$$

Since $\Delta = 0$, the direction of path closure is determined by the parameter size. Evaluating this integral in the standard way, one obtains

$$\chi_T^2 = 0.3072\, k_0^{7/6}\, L^{11/6}\, C_n^2 \left\{ {}_2F_1\left[-\tfrac{5}{12}, -\tfrac{11}{12}; \tfrac{1}{2}; -x^{-4}\right]\right.$$

$$\left. +6.841\, {}_2F_1\left[\tfrac{1}{12}, -\tfrac{5}{12}; \tfrac{3}{2}; -x^{-4}\right]/x - 7.086\, x^{-5/6}\right\}, \quad x > 1, \text{ and} \quad (7.110)$$

$$\chi_T^2 = 2.177\, k_0^{7/6}\, L^{11/6}\, C_n^2\, x^{-5/3}\, {}_2F_1\left[\tfrac{1}{12}, -\tfrac{5}{12}; \tfrac{3}{2}; -x^4\right], \quad x < 1. \quad (7.111)$$

This is equivalent to a result given by the computer algebra program Mathematica that applies over the entire parameter range as

$$\chi_T^2 = 0.1977\, k_0^{7/6}\, L^{11/6}\, C_n^2 \frac{\left\{-11\, x^2 + 6\, \left(1 + x^4\right)^{11/12} \sin\left[11 \arctan\left(x^2\right)/6\right]\right\}}{x^2 \mid (x^{5/3}) \mid}. \quad (7.112)$$

I will now find the scintillation for the Hill spectra given in (2.27). The scintillation in this case is

$$\chi_H^2 = 0.2073\, k_0^2 \int_0^L dz\, C_n^2(z)$$

$$\times \sum_{n=0}^{4} a_n \int d\boldsymbol{\kappa}\, \kappa^{-11/3} \left(\kappa\, L_{in}\right)^n \exp\left(-\delta\, |\kappa|\, L_{in}\right) \sin^2\left(\frac{\kappa^2 z}{2k_0}\right). \quad (7.113)$$

The constants are given by $\delta = 1.1090$, $a_0 = 1$, $a_1 = 0.70937$, $a_2 = 2.8235$, $a_3 = -0.28086$, and $a_4 = -0.08277$.

Again assume the turbulence is constant along the path, integrate over angle and propagation direction, and make the substitutions $y = \kappa\, (L/k_0)^{1/2}$, and $x = \sqrt{L/k_0}\,/\,(L_{in}\,\delta)$ to obtain

$$\chi_H^2 = -0.6513\, k_0^{7/6}\, L^{11/6}\, C_n^2$$

$$\times \sum_{n=0}^{4} \frac{a_n}{(\delta\, x)^n} \int_0^\infty \frac{dy}{y}\, y^{-11/3+n} \exp\left(-y/x\right) \left[\sin\left(y^2\right) - y^2\right]. \quad (7.114)$$

Use the Mellin convolution theorem and the substitution $s \to 4s$ to get

$$\chi_H^2 = -0.324\, k_0^{7/6}\, L^{11/6}\, C_n^2 \sum_{n=0}^{4} \frac{a_n}{(\delta\, x)^n} \frac{1}{2\pi i} \int_C ds \left(\frac{x^2}{2}\right)^{2s} \Gamma\left[\begin{matrix} -s - \frac{5}{12} + \frac{n}{4}*, \ 4s \\ \frac{23}{12} - \frac{n}{4} + s \end{matrix}\right]. \quad (7.115)$$

The multiplication theorem can be used to obtain unity coefficients in the gamma functions, in which case one obtains a solution which is the sum of four ${}_2F_1\,[\,]$ functions and the residue at the single pole. Rather than doing that, I give the single series solution. Since $\Delta = 2$, the integration path is closed to

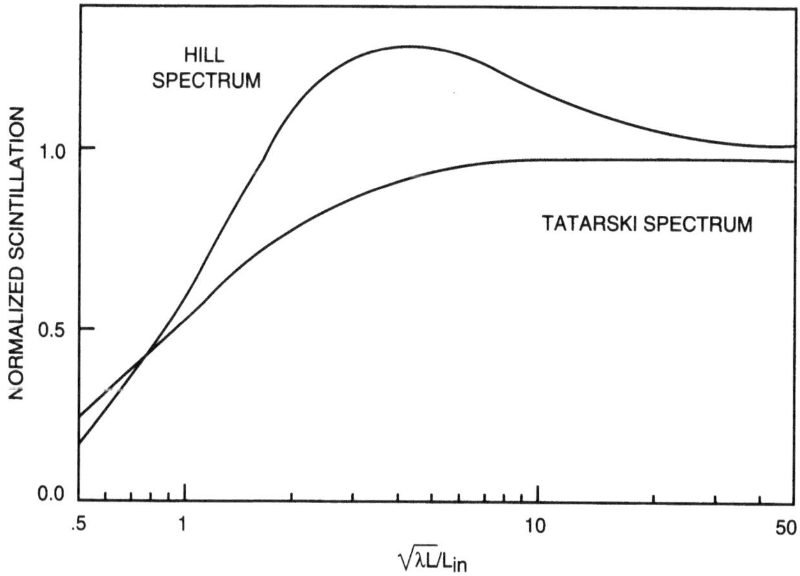

Fig. 7.15. Scintillation for a collimated beam propagating over a region of constant turbulence normalized to the scintillation with zero inner scale for the Tatarski and Hill spectra of inner scale. This is plotted versus the normalized propagation distance.

the left, and one obtains the contributions from the poles at $s = -5/12 + n/4$ and at $s = -m/4$ for $m = 0, 1, 2, \ldots$ to obtain

$$\chi_H^2 = 0.081\, k_0^{7/6} L^{11/6} C_n^2 \sum_{n=0}^{4} \frac{a_n}{\delta^n} \left\{ 4.514\, \Gamma\,[n - 5/3] \left(L_{in}\, \delta \sqrt{\frac{2\, k_0}{L}} \right)^{5/3} \right.$$
$$\left. - \sum_{m=0}^{\infty} \frac{(-1)^m}{m!} \left(L_{in}\, \delta \sqrt{\frac{2\, k_0}{L}} \right)^{m+n} \Gamma \left[\begin{array}{c} (n+m)/4 - \frac{5}{12} \\ -(n+m)/4 + \frac{23}{12} \end{array} \right] \right\}. \quad (7.116)$$

The scintillation for the Tatarski and Hill spectra versus propagation distance are plotted in Fig. 7.15. Notice the substantial difference in the scintillation, which can be used to verify the actual spectrum shape.

For small distances one can obtain an asymptotic series for the Hill spectrum. The steepest-descent contribution is exponentially small since $\Lambda' = -1$ and can be neglected. The most significant contribution is from the first pole on the left located at $s = 7/12$. The normalized scintillation for small propagation distances is

$$\chi_H^2 / \sigma_\chi^2 \approx 0.90 \left(\sqrt{\lambda\, L} / L_{in} \right)^{7/3}, \quad \sqrt{\lambda\, L} / L_{in} << 1. \quad (7.117)$$

The first term of the relation in (7.111) gives the Tatarski scintillation for small distances as

$$\chi_T^2/\sigma_\chi^2 \approx 1.22\left(\sqrt{\lambda L}/L_{in}\right)^{7/3}, \quad \sqrt{\lambda L}/L_{in} << 1. \tag{7.118}$$

Both spectra produce the same functional form in the asymptotic limit although the coefficients differ.

7.10 Scintillation Anisoplanatism

Here, differential scintillation between two uncorrected beams that originate exoatmospherically and are displaced from each other is calculated. One application of this result is to find the scintillation difference between the components of a double star system. The second application is to the reconstruction of a diffraction-limited image of a target outside the atmosphere. In this reconstruction, which was applied by *Lynds* et al. (1976) to obtain a resolved image of Betelgeuse, the short exposure images of an object are added to each other after the brightest speckles are aligned. This procedure works only if speckles from different parts of the image are strongly correlated.

The variance of log-amplitude difference in (2.92) for two offset infinite waves that propagate from $z = 0$ to $z = L$ is

$$\chi^2 = 0.2073\, k_0^2 \int_0^L dz\, C_n^2(z) \int d\boldsymbol{\kappa}\, \kappa^{-11/3} \sin^2\left(\frac{\kappa^2 z}{2k_0}\right) 2\left[1 - \cos\left(\boldsymbol{\kappa}\cdot\mathbf{d}\right)\right]. \tag{7.119}$$

Perform the angle integration and let $y = \kappa d$, and $x^2 = 2\, k_0\, d^2/z$ to obtain

$$\chi^2 = -2.606\, k_0^2 \int_0^\infty dz\, C_n^2(z) d^{5/3} \int_0^\infty \frac{dy}{y}\, y^{-5/3}\left[J_0(y) - 1\right] \sin^2\left[\left(\frac{y}{x}\right)^2\right]. \tag{7.120}$$

The last integral can be converted into one in the complex plane with the Mellin transforms in (1.49) and (1.48), the substitution $s \to 4s$, and the multiplication theorem for gamma functions, to give

$$I = -\frac{\sqrt{\pi}}{2^{19/3}}\frac{1}{2\pi i}\int_C ds \left(\frac{x}{4}\right)^{-4s} \Gamma\left[\begin{matrix} s - \frac{5}{12}*,\ s + \frac{1}{12},\ -s* \\ \frac{11}{12} - s,\ \frac{17}{12} - s,\ s + \frac{1}{2}\end{matrix}\right]. \tag{7.121}$$

Since $\Delta = 2$, close the integration path in the left half plane, and pick up the pole contributions at $s = 0,\ 5/12 - n$ for $n = 1, 2, 3, \ldots$, and $-1/12 - n$ for $n = 0, 1, 2, \ldots$. Therefore, the scintillation difference, after the use of the rules in Sect. 1.3 to change the summation into a generalized hypergeometric function, and (1.40) to reduce one minus that function to another generalized hypergeometric function, is

$$\chi^2 = k_0^2 \int_0^\infty dz\, C_n^2(z)\, d^{5/3}$$

$$\left\{1.455 - 0.08754 \left(\frac{k_0\, d^2}{z}\right)^{1/6} {}_2F_3\left[\tfrac{1}{12}, \tfrac{7}{12}; \tfrac{3}{2}, \tfrac{3}{2}, 1; \left(\frac{k_0\, d^2}{8\, z}\right)^2\right]\right.$$

$$\left. +0.00244 \left(\frac{k_0\, d^2}{z}\right)^{7/6} {}_3F_4\left[\tfrac{7}{12}, \tfrac{13}{12}, 1; \tfrac{3}{2}, \tfrac{3}{2}, 2, 2; \left(\frac{k_0\, d^2}{8\, z}\right)^2\right]\right\}. \quad (7.122)$$

For very large separations, the first pole on the right side of the integration path located at $s = 5/12$ is the main contributor. Evaluating the residue at this pole, one obtains for large path separations

$$\chi^2 \approx 2 \times 0.5631\, k_0^{7/6} \mu_{5/6}. \quad (7.123)$$

This is twice the scintillation for a point source, and it is what is expected since scintillations from the two sources are uncorrelated for large separations. The z integration in (7.122) cannot be performed term by term, since after the first few terms the integral diverges at z equal to zero. Either the hypergeometric functions must be evaluated or the asymptotic solution must be used for small values of z. The integral over z of the first two terms of the general solution converge, and for two paths separated by a small angle the scintillation is

$$\chi^2 \approx 0.50\, (\theta/\theta_o)^{5/3} - 0.08754\, k_0^{13/6}\, \theta^2\, \mu_{11/6}, \quad \theta << \theta_o. \quad (7.124)$$

The above approximation is valid only for angles smaller than the isoplanatic angle. The scintillation increases rapidly with angle and is substantial when separation is equal to the isoplanatic angle. This is to be contrasted with tilt anisoplanatism in which tilt difference is small compared to tilt from a single object even for angles many times the isoplanatic angle. The reason for this difference is that tilt is caused by turbulence with long wavelengths, and there has to be a substantial difference in path separations to get a sizeable effect. In contrast, small wavelength disturbances contribute most strongly to scintillation, and a small separation between paths produces significant effects.

One cannot practically use the relative scintillation difference between stars to get a direct measurement of turbulence distribution since, for reasonable star separations, several terms of the series solution are needed to get the correct scintillation, and each term contains a different turbulence moment.

7.11 Focus Anisoplanatism

It has been proposed to determine range to an airborne vehicle by using the focus component of a wavefront return from the vehicle. The concept requires that focus be measured with an error of less than 1/1000 of a wave. There are stable systematic errors such as optical aberrations of the measurement system that, hopefully, can be taken out with calibrations. There are also systematic errors that change with time such as the static focus term of the aerodynamic boundary layer that varies with angle-of-attack, altitude, and aircraft speed. There are also stochastic noise sources such as focal changes caused by the vibrations of optics and turbulence in the path. These last errors can be reduced by averaging. To calculate the effect of averaging individual focus measurements as the aircraft moves lateral to the line of sight, the focus correlation distance must be found.

I calculate the focus variance and focus correlation distance. The focus variance is found with the filter function for focus, which is the $Z(0,2)$ Zernike mode, and is given in (3.14) as

$$F(\boldsymbol{\kappa}) = 3\left[\frac{2\,J_3(\kappa D/2)}{\kappa D/2}\right]^2. \tag{7.125}$$

Insert this into (2.86) for single-wave variance with diffraction neglected to get

$$\sigma_f^2 = 0.2073\,k_0^2 \int_0^L dz\, C_n^2(z) \int d\boldsymbol{\kappa}\, \kappa^{-11/3}\, 3\left[\frac{2J_3(\kappa D/2)}{\kappa D/2}\right]^2. \tag{7.126}$$

Integrate over axial distance and angle, and make the substitution $x = \kappa D/2$, to obtain

$$\sigma_f^2 = 4.923\,k_0^2\,\mu_0\,D^{5/3} \int_0^\infty \frac{dx}{x}\, x^{-11/3}\, J_3^2(x)\,, \tag{7.127}$$

where μ_0 is the zeroth moment of turbulence. The value of the last integral is found with the Mellin transform in (1.50) that is evaluated at $s = -11/3$, and is equal to 1.996×10^{-3}. Since the coherence diameter is $r_o^{-5/3} = 0.423\,k_0^2\,\mu_0$, the phase variance due to focus is

$$\sigma_f^2 = 0.023\,(D/r_o)^{5/3}. \tag{7.128}$$

The focus correlation function is the variance of focus difference between two displaced apertures and is given by (2.92) with the focus filter function as

$$C_f(\mathbf{d}) = 1.244\,k_0^2 \int_0^L dz\, C_n^2(z) \int d\boldsymbol{\kappa}\, \kappa^{-11/3} \left[\frac{2J_3(\kappa D/2)}{\kappa D/2}\right]^2 [1 - \cos(\boldsymbol{\kappa}\cdot\mathbf{d})]. \tag{7.129}$$

Assume that the displacement is constant and perpendicular to the axial direction. Perform the axial and angular integrations and make the substitutions $x = \kappa D/2$, and $y = D/2d$, to obtain

$$C_f(d) = 23.28\,(D/r_o)^{5/3} \int_0^\infty \frac{dx}{x}\, x^{-11/3} J_3^2(x)\,[1 - J_0(x/y)]. \tag{7.130}$$

Form the ratio of this quantity with the focus phase variance given in (7.128) to obtain

$$R(d) = \frac{\sigma_{fd}^2}{\sigma_f^2} = 1002 \int_0^\infty \frac{dx}{x}\, x^{-11/3} J_3^2(x) \left[1 - J_0\left(x/y\right)\right]. \tag{7.131}$$

Notice that this ratio does not depend on the turbulence strength along the path.

Inserting the transforms in (1.50) and (1.49) into the Mellin convolution integral given in (1.27), and making the substitution $s \rightarrow 2s$, one obtains

$$R\left(d\right) = -282.7 \frac{1}{2\pi i} \int_C ds \left(\frac{D}{d}\right)^{-2s} \Gamma\left[\begin{matrix} s+\frac{7}{6},\ -s+\frac{7}{3},\ -s* \\ -s+\frac{35}{6},\ -s+\frac{17}{6},\ s+1 \end{matrix}\right]. \tag{7.132}$$

Since $\Delta = \Delta' = 0$, the integration path is closed to the right for separations less than the diameter, and closed to the left for larger separations. Evaluating the pole residues at $s = 7/3 + n$ for $n = 0,\ 1,\ 2,\ \ldots$, and $s = n$ for $n = 1,\ 2,\ 3\ \ldots$, one obtains for small displacements

$$R(d) = -282.7 \left\{ \sum_{n=1}^{\infty} \frac{(-1)^n}{n!} \left(\frac{d}{D}\right)^{2n} \Gamma\left[\begin{matrix} n+\frac{7}{6},\ -n+\frac{7}{3} \\ -n+\frac{35}{6},\ -n+\frac{17}{6},\ n+1 \end{matrix}\right] \right.$$
$$\left. + \sum_{n=0}^{\infty} \frac{(-1)^n}{n!} \left(\frac{d}{D}\right)^{2n+14/3} \Gamma\left[\begin{matrix} n+\frac{7}{2},\ -n-\frac{7}{3} \\ -n+\frac{7}{2},\ -n+\frac{1}{2},\ n+\frac{10}{3} \end{matrix}\right] \right\}, \quad d < D. \tag{7.133}$$

For large displacements one uses the pole-residues at $s = -7/6 - n$ for $n = 0,\ 1,\ 2,\ \ldots$ and the single pole at $s = 0$. The result is

$$R(d) = 2 - 282.7 \sum_{n=0}^{\infty} \frac{(-1)^n}{n!} \left(\frac{D}{d}\right)^{2n+7/3} \Gamma\left[\begin{matrix} n+\frac{7}{2},\ n+\frac{7}{6} \\ n+7,\ n+4,\ -n-\frac{1}{6} \end{matrix}\right], \quad d > D. \tag{7.134}$$

Notice that as the separation gets very large, the ratio becomes 2, which is to be expected since one is subtracting two quantities that are uncorrelated. Expressed as generalized hypergeometric functions with the rules in Sect. 1.3 and (1.40), one obtains

$$R(d) = 15.51 \left(\frac{d}{D}\right)^2 {}_4F_3\left[\frac{13}{6}, -\frac{23}{6}, -\frac{5}{6}, 1; -\frac{1}{3}, 2, 2; \left(\frac{d}{D}\right)^2\right]$$
$$+ 74.76 \left(\frac{d}{D}\right)^{14/3} {}_3F_2\left[\frac{7}{2}, -\frac{5}{2}, \frac{1}{2}; \frac{10}{3}, \frac{10}{3}; \left(\frac{d}{D}\right)^2\right], \quad d < D, \tag{7.135}$$

and $R(d) = 2 + 0.03008\, (D/d)^{7/3}\, {}_3F_2\left[\frac{7}{2}, \frac{7}{6}, \frac{7}{6}; 7, 4; (D/d)^2\right], \quad d > D.$ (7.136)

This function is plotted in Fig. 7.16. It shows that the focus term is essentially uncorrelated when the separation is one-half the aperture diameter.

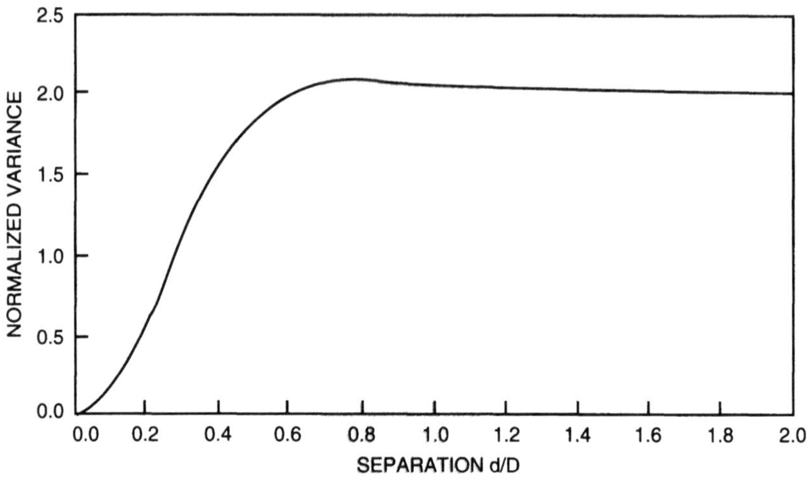

Fig. 7.16. Normalized difference of the focus variance versus separation of two apertures. When the path separation is one-half the aperture diameter the focus components on the two apertures are essentially uncorrelated. This curve is independent of the turbulence distribution along the propagation path.

7.12 Focal Anisoplanatism for Point Sources

Focal anisoplanatism refers to errors made by using a phase disturbance measured with a focused beam to correct for turbulence effects on a collimated beam. If a point source is very far away, the error is very small. For closer sources, such as artificial beacons that are used in the 'guidestar' concept [*Gardner* et al. (1990); *Murphy* et al. (1991); *Primmerman* et al. (1991); *Fugate* et al. (1991); *Welsh* and *Gardner* (1989); *Welsh* and *Thompson* (1991)], the effect can be significant and may result in a low Strehl ratio. The phase variance from this effect is calculated in this section. First, an expression for variance is obtained that has both piston and tilt components present. Next, expressions for piston and tilt variances are obtained. If the last variances are subtracted from the total variance, then one obtains the piston- and tilt-removed phase variances due to focal anisoplanatism.

The technique to find the variance is the same one that has been used on previous problems. The filter function for a distributed, circular, uniform beacon of diameter D_s and offset $\mathbf{b}$ that is at a range L and altitude H as given in (3.47) is inserted into the formula for phase variance of a single wave given in (2.86). Assume one is operating in the near field, thereby allowing the cosine term to be replaced by 1.

After doing the angle integration, one obtains for the aperture-averaged phase variance

$$\sigma_{-d}^2 = 1.303\, k_0^2 \int_0^L dz\, C_n^2(z) \int_0^\infty d\kappa\, \kappa^{-8/3}$$
$$\times \left\{ 1 - \frac{2\, J_1\,(a\, D_s)}{a\, D_s} \frac{2\, J_1\,(a\, D)}{a\, D} J_0 \left(\frac{\kappa b z}{L} \right) + \left[2 \frac{J_1\,(a\, D_s)}{a\, D_s} \right]^2 \right\}, \quad (7.137)$$

where $a = \kappa\, z/2L$. The subscript d on the variance refers to a distributed source. The minus subscript means that it is the component from the turbulence below the beacon. The component above the beacon will be added to this to obtain the total variance. If the d subscript is missing, it means a point source is being considered. Piston and tilt are present in σ_d^2.

This equation is the starting point for the evaluation of the total phase variance for all cases considered in this and the next two section. The problem is solved in steps, since the results for a point source, a distributed source, and a displaced source are of interest for different situations.

The phase variance of a point source on the aperture center with piston and tilt present can be found from (7.137) by setting the source diameter and displacement equal to zero to obtain

$$\sigma_-^2 = 2.606\, k_0^2 \int_0^L dz\, C_n^2(z) \int_0^\infty d\kappa\, \kappa^{-8/3} \left[1 - \frac{2 J_1\,(a D)}{a D} \right]. \quad (7.138)$$

Let $x = \kappa\, D\, z/2\, L$ to obtain

$$\sigma_-^2 = -5.212\, k_0^2 \int_0^L dz\, C_n^2(z) \left(\frac{D\, z}{2\, H} \right)^{5/3} \int_0^\infty dx\, x^{-8/3-1} \left[J_1(x) - \frac{x}{2} \right]. \quad (7.139)$$

The last integral converges by analytical continuation as discussed in evaluating (5.5), and its value of 0.305 is just the Mellin transform of the Bessel function given in (1.49) evaluated at $s = -8/3$. The phase variance is equal to

$$\sigma_-^2 = 0.5\, k_0^2\, \mu_{5/3}^- \left(\frac{D}{L} \right)^{5/3} = \left(\frac{0.348\, D}{L\, \theta_o^-} \right)^{5/3}, \quad (7.140)$$

where the partial isoplanatic angle is defined as

$$\left(\theta_o^- \right)^{-5/3} = 2.91\, k_0^2\, \mu_{5/3}^-. \quad (7.141)$$

The phase variance results from an angular offset of the collimated and focused rays. The phase variance varies with radius; however, one can consider the average phase variance to be due to an angular offset equal to that of the ray that emanates from the point that is about 0.7 of the radius from the center. This angle is $0.348\, D/L$.

This phase variance with piston included is finite unlike the result for unfiltered turbulence, which is infinite. The infinite result comes from the zero spatial wavelength term. This infinity cancels out in the subtraction of the collimated beam phase from that of a focused beam. Even though the above result is finite, it might be possible that the major component of this variance is due to the piston, which is no practical interest. It can be shown that at satellite altitudes the piston contribution is less than 10% of the variance. At 10 km it can be as large as 20%. Therefore, the above simple expression is a reasonable zero-order approximation to phase variance. A 60-cm aperture pointed at a point source at 300 km would have the same phase variance as two beams with angular offset of 0.7 μrads. For typical isoplanatic angles, this will produce a very small variance.

Often one is interested in phase variance with piston or piston and tilt removed. The phase variance due to piston and tilt can be calculated separately and subtracted from the total phase variance to obtain these. Piston and tilt phase variances are found by using (2.90) with filter functions given in (3.51)

$$\sigma^2_{-Z} = 0.2073\, k_0^2 \int_0^L dz\, C_n^2(z)\, 4\,\nu^2 \int d\boldsymbol{\kappa}\, \kappa^{-11/3}$$
$$\times \left\{ \frac{J_\nu\,(\kappa D/2)}{\kappa D/2} - \exp\left(i\,\boldsymbol{\kappa}\cdot\mathbf{b}\frac{z}{L}\right) \frac{J_\nu\,[\kappa D(1-z/L)/2]}{\kappa D(1-z/L)/2} \frac{2\,J_1\,(\kappa\,D_s\,z/2\,L)}{\kappa\,D_s\,z/2\,L} \right\}^2, \tag{7.142}$$

where D_s is the source diameter, and $\mathbf{b}$ is displacement of the source from boresight. The value of ν is 1 for piston and 2 for tilt. The subscript Z is either P or T.

First consider the case of zero source size and no offset. Perform the angular integration. The evaluation of the integral in kappa space poses some difficulty. The two terms in brackets almost cancel especially for very high beacon altitudes. If the bracket is expanded, three terms result. Two of these terms can be evaluated by table lookup. The term that is the product of the two terms must be evaluated with the Mellin convolution integral. The final answer is expanded in a power series in the parameter z/L to enable the answer to be expressed as a sum of turbulence moments. Let $\kappa D/2 = x$, and $y = 1/(1-z/L)$. The variances are equal to

$$\sigma^2_{-Z} = 1.642\, k_0^2\, \nu^2\, D^{5/3} \int_0^L dz\, C_n^2(z)\, I, \quad \text{where} \tag{7.143}$$

$$I = \int_0^\infty dx\, x^{-14/3}$$
$$\times \left\{ \left[J_\nu^2\,(x) - \frac{a\,x^2}{4} \right] + \left[y^2\, J_\nu^2\,(x/y) - \frac{a\,x^2}{4} \right] - 2\left[J_\nu\,(x)\; y\, J_\nu\,(x/y) - \frac{a\,x^2}{4} \right] \right\}. \tag{7.144}$$

The value of a is unity for piston and zero for tilt. Add the constant terms, whose sum is zero, to obtain three integrals that converge. Express the integral as the sum of three integrals

$$I = I_1 + I_2 + I_3, \tag{7.145}$$

to obtain for the first two terms

$$I_1 + I_2 = \left\{ \frac{\left(1 + y^{-5/3}\right)}{2\sqrt{\pi}} \Gamma \begin{bmatrix} -\frac{11}{6} + \nu, \frac{7}{3} \\ \nu + \frac{17}{6}, \frac{17}{6} \end{bmatrix} \right\}. \tag{7.146}$$

For piston the third integral is

$$I_3 = -2\, y \int_0^\infty \frac{de}{e} e^{-11/3} \left[J_1(e) J_1\left(\frac{e}{y}\right) - \frac{e}{2} \frac{e}{2y} \right]. \tag{7.147}$$

If $y = 1/\beta$, this is the same integral that was evaluated in (7.24). Its value is

$$I_3 = -y\, 2^{-11/3} \Gamma \begin{bmatrix} -\frac{5}{6} \\ \frac{17}{6} \end{bmatrix} {}_2F_1\left[-\tfrac{5}{6}, -\tfrac{11}{6}; 2; (1 - z/L)^2\right]. \tag{7.148}$$

The piston phase variance due to turbulence below the beacon is

$$\sigma_{-P}^2 = 0.5\, k_0^2\, D^{5/3} \int_0^L dz\, C_n^2(z)$$
$$\times \left\{ {}_2F_1\left[-\tfrac{5}{6}, -\tfrac{11}{6}; 2; (1 - z/L)^2\right] - \frac{\left[1 + (1 - z/L)^{5/3}\right]}{\sqrt{\pi}\, 2^{-8/3}} \Gamma \begin{bmatrix} \frac{7}{3} \\ \frac{23}{6} \end{bmatrix} \right\}. \tag{7.149}$$

This can be integrated numerically; however, to express this as turbulence moments, the expression in braces is expanded in a Taylor series about the point $z/L = 0$. Some care must be taken in expanding this series to obtain a short series that is a good approximation to the exact calculation. It has been found that a good approximation to the exact result for typical turbulence models and aperture diameters is

$$\sigma_{-P}^2 \approx 0.0833\, k_0^2\, D^{5/3} \frac{\mu_2^-(H)}{L^2}. \tag{7.150}$$

The integrations can be performed for tilt to obtain

$$\sigma_{-T}^2 = 0.8345\, k_0^2\, D^{5/3} \int_0^L dz\, C_n^2(z)$$
$$\times \left\{ \frac{\left[1 + (1 - z/L)^{5/3}\right]}{\sqrt{\pi}\, 2^{-11/3}} \Gamma \begin{bmatrix} \frac{7}{3} \\ \frac{29}{6} \end{bmatrix} - (1 - z/L)\, {}_2F_1\left[\tfrac{1}{6}, -\tfrac{11}{6}; 3; (1 - z/L)^2\right] \right\}. \tag{7.151}$$

After expansion of this Taylor series in the same way as was done for piston, the tilt variance is closely approximated by

$$\sigma^2_{-T} \approx 0.368\, k_0^2\, D^{5/3}\, \frac{\mu_2^-(H)}{L^2}. \tag{7.152}$$

The piston-removed variance is

$$\sigma^2_{-PR} = \left(\frac{0.348\, D}{H\theta_o^-}\right)^{5/3} - \sigma^2_{-P}. \tag{7.153}$$

The piston- and tilt-removed variance is

$$\sigma^2_{-PTR} = \left(\frac{0.348 D}{H\,\theta_o^-}\right)^{5/3} - \sigma^2_{-P} - \sigma^2_{-T}. \tag{7.154}$$

With the above results, a good approximation to the piston-removed phase is

$$\sigma^2_{-PR} = \left(\frac{0.348\, D}{H\theta_o^-}\right)^{5/3} - \sigma^2_{-P} \approx \frac{k_0^2\, D^{5/3}}{2}\left[\frac{\mu_{5/3}^-(H)}{L^{5/3}} - 0.167\,\frac{\mu_2^-(H)}{L^2}\right]. \tag{7.155}$$

Similarly, a good approximation to the piston- and tilt-removed phase is

$$\sigma^2_{-PTR} = \left(\frac{0.348 D}{H\,\theta_o^-}\right)^{5/3} - \sigma^2_{-P} - \sigma^2_{-T} \approx \frac{k_0^2\, D^{5/3}}{2}\left[\frac{\mu_{5/3}^-(H)}{L^{5/3}} - 0.903\frac{\mu_2^-(H)}{L^2}\right]. \tag{7.156}$$

For changes in zenith angle, when the beacon is kept at the same altitude, it is clear from the variation of each term in brackets that the zenith dependence is simply $\sec(\xi)$.

The effect of turbulence above the beacon is obtained by modifying the results in Sect. 4.2. The phase variance with piston removed is

$$\sigma^2_{+PR} = 1.033\left(D/r_o^+\right)^{5/3}, \tag{7.157}$$

where the coherence diameter looking up from the beacon altitude is

$$\left(r_o^+\right)^{-5/3} = 0.423\, k_0^2\, \mu_0^+. \tag{7.158}$$

The expression for phase variance from turbulence above the beacon with piston and tilt removed is

$$\sigma^2_{+PTR} = 0.134\left(D/r_o^+\right)^{5/3}. \tag{7.159}$$

The contribution due to the unsensed turbulence increases with increasing aperture size, and one will find that the beacon altitude must be increased for larger diameters in order to keep this component manageable.

The total phase variance with piston removed is

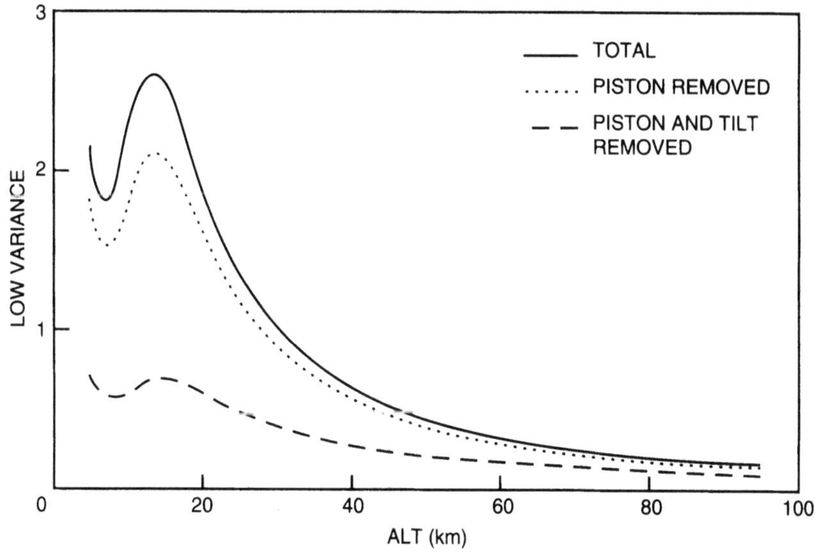

Fig. 7.17. Phase variance of turbulence below the beacon. The total, piston-removed, and tilt- and piston-removed phase variances are plotted.

$$\sigma_{PR}^2 = \sigma_{-PR}^2 + \sigma_{+PR}^2. \tag{7.160}$$

The total phase variance with piston and tilt removed is

$$\sigma_{PTR}^2 = \sigma_{-PTR}^2 + \sigma_{+PTR}^2. \tag{7.161}$$

In Fig. 7.17 the total, piston-removed, and piston and tilt-removed phase variances due to turbulence below the beacon are plotted for a single beacon operating with a 60-cm diameter aperture with the HV-21 turbulence model. In Fig. 7.18 the piston-removed, and piston and tilt-removed phase variances of the entire atmosphere are plotted for the same conditions.

7.13 Focal Anisoplanatism for Distributed Sources

The artificial beacon source is adequately approximated by a point source for very distant beacons. This, however, is a poor approximation for some closer beacons. In trying to use resonant sodium backscatter from the sodium layer at about 90 km, one wants to use as large a beacon as possible that does not significantly increase the phase variance over that of a point source. A large beacon is desirable because the return from a given area saturates at a certain power. To increase the return signal, not only must the beacon power be increased, but also, the focused spot size must be increased.

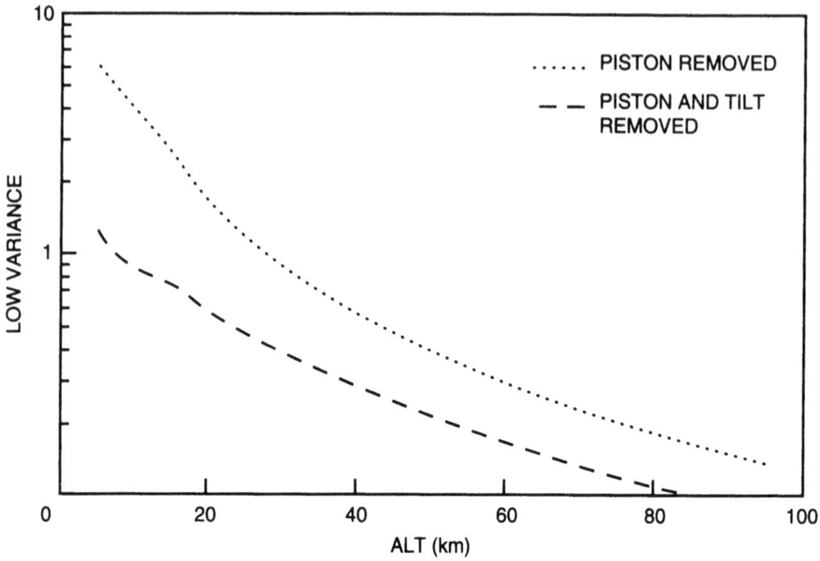

Fig. 7.18. Piston-removed phase variance and piston- and tilt-removed phase variance caused by turbulence of the entire atmosphere.

The expression for the phase variance for a distributed source centered on the aperture with piston and tilt present is found from (7.137) by setting **b** equal to zero to obtain

$$\sigma^2_{-d} = 1.303\, k_0^2 \int_0^L dz\, C_n^2(z) \int_0^\infty d\kappa\, \kappa^{-8/3} \times \left\{ 1 - 8 \frac{J_1(aD_s)}{aD_s} \frac{J_1(aD)}{aD} + \left[\frac{2J_1(aD_s)}{aD_s} \right]^2 \right\}, \tag{7.162}$$

where $a = \kappa z/2L$. To evaluate this integral, it is written as the sum of two integrals.

Let $w = \kappa D\, z/2\,L$, $y = \kappa D_s\, z/2\,L$, and $x = D/D_s$, to obtain

$$\sigma^2_{-d} = 6.55 \left(\frac{0.348\, D}{L\, \theta_o^-} \right)^{5/3} (-I + J). \tag{7.163}$$

The J term is easily evaluated with (1.50) to give

$$J = \frac{1}{2} \left(\frac{D_s}{D} \right)^{5/3} \int_0^\infty dy\, y^{-11/3-1} \left[J_1^2(y) - \left(\frac{y}{2} \right)^2 \right] = -0.1331 \left(\frac{D_s}{D} \right)^{5/3}. \tag{7.164}$$

The first integral is

$$I = x \int_0^\infty \frac{dw}{w} w^{-11/3} \left[J_1(w) J_1 \left(\frac{w}{x} \right) - \frac{w}{2} \frac{w}{2x} \right]. \tag{7.165}$$

The value of I depends on x, which is not a function of z. J is also not a function of z. Therefore, the phase variance due to turbulence below the beacon depends solely on the 5/3 moment of turbulence.

The expression for I is exactly the same expression as that in (7.147). Use the results of the evaluation of that integral to obtain for $D_s/D < 1$

$$\sigma^2_{-d} = \left(\frac{0.348\, D}{L\,\theta_o^-}\right)^{5/3} \left\{ {}_2F_1\left[-\tfrac{5}{6}, -\tfrac{11}{6}; 2; \left(\frac{D_s}{D}\right)^2\right] - 0.872 \left(\frac{D_s}{D}\right)^{5/3} \right\}, \quad D_s/D < 1. \tag{7.166}$$

The first few terms of this solution are

$$\sigma^2_{-d} = \left(\frac{0.348\, D}{L\,\theta_o^-}\right)^{5/3} \left[1 - 0.872 \left(\frac{D_s}{D}\right)^{5/3} + 0.764 \left(\frac{D_s}{D}\right)^2 - 0.0177 \left(\frac{D_s}{D}\right)^4 - 0.000287 \left(\frac{D_s}{D}\right)^6 + \cdots, \quad D_s/D << 1. \right. \tag{7.167}$$

The first three terms in brackets give a good approximation to the phase variance when $D_s/D < 1$.

When $D_s/D > 1$, the variance is

$$\sigma^2_{-d} = \left(\frac{0.348\, D_s}{H\,\theta_o^-}\right)^{5/3} \left\{ -0.128 + {}_2F_1\left[-\tfrac{5}{6}, -\tfrac{11}{6}; 2; \left(\frac{D}{D_s}\right)^2\right] \right\}, \quad D_s/D > 1. \tag{7.168}$$

The first few terms of this solution are

$$\sigma^2_{-d} = \left(\frac{0.348\, D_s}{L\,\theta_o^-}\right)^{5/3} \left[0.872 + 0.764 \left(\frac{D}{D_s}\right)^2 - 0.0177 \left(\frac{D}{D_s}\right)^4 - 0.000287 \left(\frac{D}{D_s}\right)^6 + \cdots, \quad D_s/D >> 1. \right. \tag{7.169}$$

The normalized variance is plotted in Fig. 7.19. Initially, the phase variance caused by the turbulence below the beacon decreases as source size increases, then it increases.

7.14 Focal Anisoplanatism for Offset Sources

The beacon may not be placed in the exact location that the correction is wanted; this results in a decreased Strehl ratio. This error can be caused by errors both in positioning the beacon and in knowing which direction to point the beacon. If a laser beam is to be projected in the point-ahead direction, which is typically about 50 μrad, there is an error made in pointing the mirrors caused by tilt anisoplanatism. This error can be as much as a microradian.

The phase variance with a point source offset from the origin is obtained by setting the source diameter equal to zero in (7.137). After performing the angle integration, one obtains

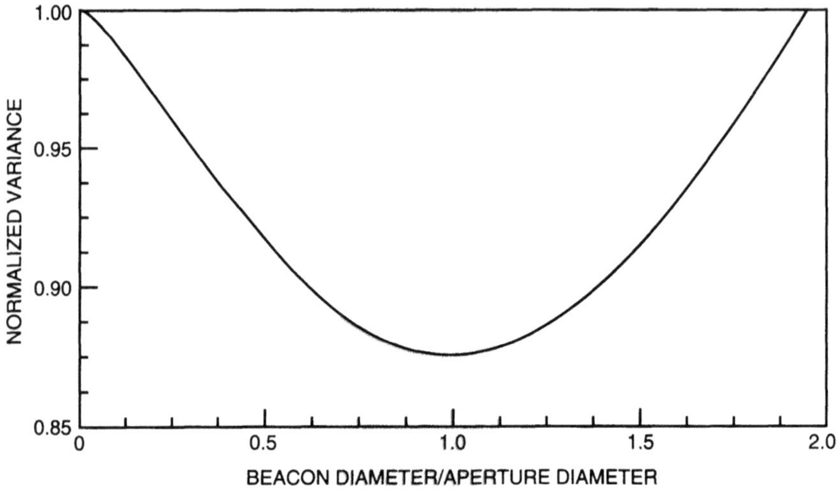

Fig. 7.19. Phase variance below the beacon. Effect of beacon diameter on the phase variance due to focal anisoplanatism normalized to the phase variance due to focal anisoplanatism from a point source with piston and tilt present.

$$\sigma_{-o}^2 = 1.303\, k_0^2 \int_0^L dz\, C_n^2(z) \int_0^\infty d\kappa\, \kappa^{-11/3} 2 \left[1 - \frac{2J_1(\kappa Dz/2L)}{\kappa Dz/2L} J_0\left(\frac{\kappa bz}{L}\right)\right]. \tag{7.170}$$

The o subscript refers to an offset source. Make the substitutions $y = \kappa Dz/2L$, and $x = D/2b$, to obtain

$$\sigma_{-o}^2 = 2.606\, k_0^2 \left(\frac{D}{2L}\right)^{5/3} \int_0^L dz\, C_n^2(z)\, z^{5/3} \int_0^\infty \frac{dy}{y}\, y^{-5/3} \left[1 - \frac{2\,J_1\,(y)}{y} J_0\,(y/x)\right]. \tag{7.171}$$

Since the parameter in the last integral does not depend on the axial coordinate, the axial integration can be performed to give

$$\sigma_{-o}^2 = 5.212\, k_0^2\, \mu_{5/3}^- \left(\frac{D}{2L}\right)^{5/3} \int_0^\infty \frac{dy}{y}\, y^{-8/3} \left[\frac{y}{2} - J_1\,(y)\; J_0\,(y/x)\right]. \tag{7.172}$$

The expression in brackets can be manipulated into two integrable terms as

$$1 - \frac{2J_1\,(y)}{y} J_0\,(y/x) = \left[1 - J_0\,(y/x)\right] + \frac{2}{y}\left[\frac{y}{2} - J_1\,(y)\right] J_0\,(y/x)\,. \tag{7.173}$$

The integral of the first term is evaluated by table lookup and the analytical continuation arguments used in evaluating (1.21) to give

$$\sigma_1^2 = 1.118\, k_0^2\, \mu_{5/3}^- \left(\frac{2b}{L}\right)^{5/3}. \tag{7.174}$$

The last integral can be converted into an integral in the complex plane as

$$I = -\frac{2^{-11/3}}{2\pi i} \int_C ds \left(\frac{D}{2b}\right)^{-2s} \Gamma \left[\begin{matrix} s - \frac{5}{6}*, \ -s \\ -s + \frac{17}{6}, \ s + 1 \end{matrix}\right]. \tag{7.175}$$

Since $\Delta = 0$, the direction of path closure depends on the parameter magnitude. When the displacement is less than aperture radius, the integration path is closed to the right, and there are poles at $s = n$ for $n = 0, 1, 2, \ldots$ and at $s = 5/6$. The last term cancels the contribution of σ_1^2. For this case, phase variance from turbulence below the beacon is

$$\sigma_{-o}^2 = -0.1293\, k_0^2\, \mu_{5/3}^- \left(\frac{D}{L}\right)^{5/3} \times \sum_{n=0}^{\infty} \frac{(-1)^n}{n!} \left(\frac{2b}{D}\right)^{2n} \Gamma \left[\begin{matrix} n - \frac{5}{6} \\ -n + \frac{17}{6}, \ n + 1 \end{matrix}\right], \quad 2b/D < 1. \tag{7.176}$$

This can also be written as

$$\sigma_{-o}^2 = 0.5\, k_0^2 \mu_{5/3}^- \left(\frac{D}{L}\right)^{5/3} {}_2F_1 \left[-\tfrac{5}{6}, -\tfrac{11}{6}; 1; \left(\frac{2b}{D}\right)^2\right], \quad 2b/D < 1. \tag{7.177}$$

The variance normalized to that of an on-axis source is

$$\begin{aligned} \frac{\sigma_{-o}^2}{\sigma_{-o}^2\,(b=0)} &= {}_2F_1 \left[-\tfrac{5}{6}, \ -\tfrac{11}{6}; \ 1; \ \left(\frac{2b}{D}\right)^2\right] \\ &= 1 + 6.12 \left(\frac{b}{D}\right)^2 - 0.848 \left(\frac{b}{D}\right)^4 + \cdots, \quad 2b/D < 1. \end{aligned} \tag{7.178}$$

When the displacement is zero, the hypergeometric function is equal to unity, and one obtains the previous result

$$\sigma_{-o}^2\,(b=0) = 0.5\, k_0^2\, \mu_{5/3}^- \left(\frac{D}{L}\right)^{5/3}. \tag{7.179}$$

When the displacement is larger than the radius, the integration path is closed to the left, and there are poles at $s = -n + 5/6$ for $n = 1, 2, \ldots$, thus one obtains

$$\sigma_{-o}^2 = 1.293\, k_0^2\, \mu_{5/3}^- \left(\frac{D}{L}\right)^{5/3} \sum_{n=0}^{\infty} \frac{(-1)^n}{n!} \left(\frac{D}{2b}\right)^{2n-5/3} \Gamma \left[\begin{matrix} n - \frac{5}{6} \\ -n + \frac{11}{6}, \ n + 2 \end{matrix}\right], \quad 2b/D > 1. \tag{7.180}$$

The variance normalized to that of an on-axis source is

$$\frac{\sigma_{-o}^2}{\sigma_{-o}^2\,(b=0)} = 1.833 \left(\frac{2b}{D}\right)^{5/3} {}_2F_1 \left[-\tfrac{5}{6}, -\tfrac{5}{6}; 2, \left(\frac{D}{2b}\right)^2\right], \quad 2b/D > 1. \tag{7.181}$$

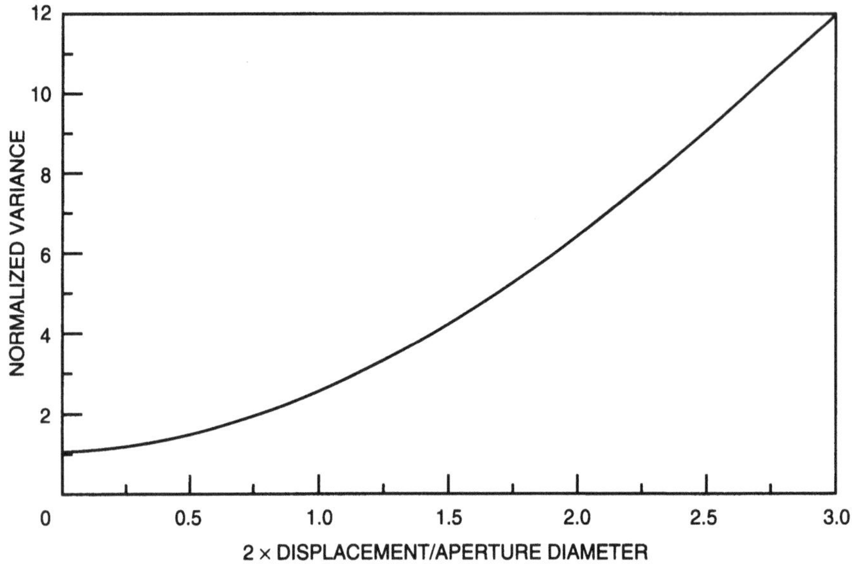

Fig. 7.20. Total phase variance below the beacon for focal anisoplanatism with a displaced point source. The variance is normalized to that of a source on axis. Piston and tilt are present. It is plotted versus twice the displacement divided by the aperture diameter.

The variance normalized to that with no offset is plotted in Fig. 7.20.

For larger source displacements than those plotted, one can use the approximation

$$\frac{\sigma^2_{-o}}{\sigma^2_{-o}\,(b=0)} \approx 1.833 \left(\frac{2b}{D}\right)^{5/3}, \quad 2b/D > 3. \tag{7.182}$$

The phase variances in this and the last section can be used to determine the importance of beacon size and placement; however, the piston and tilt terms are included. The evaluation of the integrals to determine the phase variance with these components removed can be done by the techniques discussed in Chap. 11. The analysis is complicated and lengthy and is not given here.

8. Strehl Ratio

Strehl ratio is an important performance measure for adaptive-optics systems. Its calculation is more complicated than that of variances. Nevertheless, it is shown below that one can obtain exact expressions for the Strehl ratio for propagation through uncorrected turbulence, and through turbulence for which all modes except tilt have been corrected. Asymptotic expressions are also obtained that apply when the distortion is high.

The effect of anisoplanatism on the Strehl is especially important. Here I consider isoplanatic effects of aperture displacement in space and angle, time delay, and chromatic offsets. A formula is obtained that is a good approximation to the Strehl ratio when it is above 0.2. Fortunately, this is the region of greatest interest in adaptive-optics system design.

8.1 Strehl Ratio for Propagation Through Turbulence

The problem of determining the on-axis intensity of a plane wave propagating through atmospheric turbulence was analyzed by *Fried* (1966) and the resultant integral was evaluated numerically. Here a compact analytic expression for the Strehl ratio is obtained. A second form expressible as an infinite series is also obtained. This second solution is more convenient, since a few series terms give an accurate answer for small levels of turbulence. An asymptotic series is obtained that is applicable when turbulence is more severe. It is shown that the first two terms of the series give an accurate result in the regime in which most astronomical telescopes operate.

Strehl ratio is the most common description of optical system performance and is given by normalized intensity evaluated at the beam center. Its general expression was derived in Sect. 2.5, and for isotropic turbulence is given in (2.112) as

$$SR = \int_0^1 d\alpha\, \alpha\, K(\alpha)\, \exp\left[-\mathcal{D}(\alpha D)/2\right]. \tag{8.1}$$

The integral is over a circular aperture of unit radius, $\mathcal{D}(\alpha D)$ is the structure function, $\boldsymbol{\alpha}$ is the normalized radius that is equal to unity at the aperture edge, and $K(\alpha)$ is a constant times the optical transfer function, given by

$$K(\alpha) = \frac{16}{\pi}\left[\cos^{-1}(\alpha) - \alpha\left(1-\alpha^2\right)^{1/2}\right] U(1-\alpha), \tag{8.2}$$

where $U(x)$ is the Heaviside unit step function defined in (1.19).

The structure function is given in (2.92), and for the Kolmogorov turbulence spectrum with inner and outer scale neglected is

$$\mathcal{D}(\alpha D) = 0.4146\, k_0^2 \int_0^\infty dz\, C_n^2(z) \int d\boldsymbol{\kappa}\, \kappa^{-11/3}\left[1 - \cos\left(\gamma \boldsymbol{\kappa} \cdot \boldsymbol{\alpha} D\right)\right]. \tag{8.3}$$

Integrate over angle in kappa space with (3.5), and perform the last integration with the relation in (1.21), to obtain

$$\begin{aligned}\mathcal{D}(\alpha D) &= 2.605\, k_0^2 \int_0^\infty dz\, C_n^2(z)\, \gamma^{5/3} \int_0^\infty d\kappa\, \kappa^{-5/3-1}\left[1 - J_0\left(\kappa \alpha D\right)\right] \\ &= 2.605\, k_0^2 (\alpha D)^{5/3} \int_0^\infty dz\, C_n^2(z)\, \gamma^{5/3} \mathcal{M}\left[1 - J_0(x)\right]\Big|_{s=-5/3}.\end{aligned} \tag{8.4}$$

Write the structure function in the form

$$\mathcal{D}(\alpha D) = 6.88\, (\alpha D/r_o)^{5/3}. \tag{8.5}$$

For plane waves $\gamma = 1$; this makes the coherence diameter r_o equal to

$$r_o = \left(0.423\, k_0^2\, \mu_0\right)^{-3/5}. \tag{8.6}$$

This reproduces the standard definition of r_o. For spherical waves with a focus at L, $\gamma = z/L$, and one obtains the spherical-wave coherence diameter

$$r_{os} = L\left(0.423\, k_0^2\, \mu_{5/3}\right)^{-3/5} = 3.18\, L\theta_o, \tag{8.7}$$

where L is the distance to the beam focus, and θ_o is the isoplanatic angle defined in (4.11).

The coherence diameter for a plane wave corresponds to the aperture size that in a vacuum gives the same resolution as the best average resolution of a ground-based aperture. The point-source coherence diameter can be used to get the average resolution from space when one is viewing the earth. For the HV-21 turbulence model the plane-wave coherence diameter is about 5 cm for a system that operates at 0.5 μm. For a system looking down from 100 km above the earth the coherence diameter is 2.2 m at this wavelength, which is a factor of 44 times the resolution looking up. This is why the resolution of an imaging system looking down toward the earth is much greater than that of an imaging system on the ground looking at an object in space.

Another type of coherence diameter ρ_o is sometimes used, and it is related to r_o by $r_o = 3.18\, \rho_o$. This coherence diameter results in a unity coefficient when inserted in (8.5).

Insert the structure function into the integral for Strehl ratio to obtain

$$SR = \int_0^\infty d\alpha\, \alpha\, K(\alpha) \exp\left[-3.44\, (\alpha D/r_o)^{5/3}\right]. \tag{8.8}$$

This is the integral that was evaluated numerically by *Fried* (1966). It is analytically evaluated here. Use the Mellin convolution integral given in (1.27) and the Mellin transforms in (1.62) and (1.45), and (1.9) with $p = 5/3$, to express the Strehl ratio as

$$SR = \frac{24}{5\sqrt{\pi}} \frac{1}{2\pi i} \int_C ds\, \Gamma\left[\begin{matrix} s/2 + \frac{3}{2},\ -3s/5 \\ s/2 + 3 \end{matrix}\right] \left(\frac{1}{s+2}\right) \left[3.44\, (D/r_o)^{5/3}\right]^{-3s/5}. \tag{8.9}$$

Let $s \to -5s/3$, and replace the fraction by a ratio of gamma functions with the recursion relation $-s + 6/5 = \Gamma\left[-s + \frac{11}{5}\right] / \Gamma\left[-s + \frac{6}{5}\right]$ to obtain

$$SR = \frac{24}{5\sqrt{\pi}} \frac{1}{2\pi i} \int_C ds\, \Gamma\left[\begin{matrix} s,\ -5s/6 + \frac{3}{2},\ -s + \frac{6}{5} \\ -5s/6 + 3,\ -s + \frac{11}{5} \end{matrix}\right] \left[3.44 \left(\frac{D}{r_o}\right)^{5/3}\right]^{-s}. \tag{8.10}$$

The coefficients of s in the gamma functions can be made equal to unity by the substitution $s \to 6s$ and with the Gauss-Legendre multiplication theorem in (1.14) to obtain

$$SR = \frac{\sqrt{0.6}}{25\pi^3} \frac{1}{2\pi i} \int_C ds \left[\left(\frac{6}{3.44}\right)^6 \left(\frac{r_o}{D}\right)^{10}\right]^s \Gamma\left[\begin{matrix} s,\ s + \frac{1}{6},\ s + \frac{1}{3},\ s + \frac{1}{2} \\ -s + \frac{6}{5},\ -s + \frac{3}{5},\ -s + \frac{4}{5} \end{matrix}\right]$$
$$\times \Gamma\left[\begin{matrix} s + \frac{2}{3},\ s + \frac{5}{6},\ -s + \frac{1}{5},\ -s + \frac{3}{10},\ -s + \frac{1}{2},\ -s + \frac{7}{10},\ -s + \frac{9}{10},\ -s + \frac{11}{10} \\ -s + 1,\ -s + \frac{6}{5},\ -s + \frac{7}{5} \end{matrix}\right]. \tag{8.11}$$

The result can be written by inspection as a Meijer G-function, which is defined in (1.38), or as a Fox H-function, which is defined in (1.37), as

$$= \frac{\sqrt{0.6}}{25\pi^3} G_{12,6}^{6,6}\left[\left(\frac{6}{3.44}\right)^6 \left(\frac{r_o}{D}\right)^{10} \middle| \begin{matrix} 1,\ \frac{5}{6},\ \frac{2}{3},\ \frac{1}{2},\ \frac{1}{3},\ \frac{1}{6},\ \frac{6}{5},\ \frac{3}{5},\ \frac{4}{5},\ 1,\ \frac{6}{5},\ \frac{7}{5} \\ \frac{1}{5},\ \frac{3}{10},\ \frac{1}{2},\ \frac{7}{10},\ \frac{9}{10},\ \frac{11}{10} \end{matrix}\right]$$

$$= \frac{24}{5\sqrt{\pi}} H_{1,4}^{2,1}\left[\frac{1}{3.44}\left(\frac{r_o}{D}\right)^{5/3} \middle| \begin{matrix} (1,\ 1),\ \left(3,\ \frac{5}{6}\right),\ \left(\frac{11}{5},\ 1\right) \\ \left(\frac{3}{2},\ \frac{5}{6}\right),\ \left(\frac{6}{5},\ 1\right) \end{matrix}\right]. \tag{8.12}$$

An equivalent solution, which provides more physical insight, expresses the Strehl ratio as an infinite sum in which coefficients of the integers in the arguments of the gamma functions are fractions. Since $\Delta = 1$ in (8.10), the integral can be closed in the left-half plane. Encircled poles are located at $s = -n$, for $n = 0, 1, 2, \ldots$. Residues at the enclosed poles give the following convergent series

$$SR = \frac{24}{5\sqrt{\pi}} \sum_{n=0}^{\infty} \frac{(-1)^n}{n!} \Gamma \begin{bmatrix} \frac{3}{2} + 5n/6, \ n + \frac{6}{5} \\ 3 + 5n/6, \ n + \frac{11}{5} \end{bmatrix} \left[3.44 \left(\frac{D}{r_o} \right)^{5/3} \right]^n . \tag{8.13}$$

The first few terms of this series are

$$SR = 1 - 1.032 \left(\frac{D}{r_o} \right)^{5/3} + 0.7982 \left(\frac{D}{r_o} \right)^{10/3} - 0.5015 \left(\frac{D}{r_o} \right)^{5} + \cdots . \tag{8.14}$$

For large diameters this series converges slowly, and an asymptotic series is more convenient. For this case one can show with (5.93) that the steepest-descent contribution has exponential decay; therefore, the asymptotic solution is equal to the sum of pole residues on the left side of the integration path. The poles are at $s = 6/5$, and $s = 6n/5 + 9/5$ for $n = 0, 1, 2, \ldots$. (Remember to multiply the second set of residues by 6/5 because the coefficient of s in the gamma function is not unity.) The gamma function that produces the pole at $s = 6/5$ only has a single contributing pole because the residues at the other poles of the gamma function are zero. The asymptotic series is

$$SR = \left(\frac{r_o}{D} \right)^2 - \frac{24}{5\sqrt{\pi}} 3.44^{-9/5} \sum_{n=0}^{n_o} \frac{(-1)^n}{n! \left(n + \frac{1}{2} \right)} \Gamma \begin{bmatrix} 6n/5 + \frac{9}{5} \\ -n + \frac{3}{2} \end{bmatrix} 3.44^{-6n/5} \left(\frac{r_o}{D} \right)^{3+2n} , \tag{8.15}$$

where n_o is a number of terms discussed in Sect. 5.2. The first few terms of this series are

$$SR = \left(\frac{r_o}{D} \right)^2 - 0.6159 \left(\frac{r_o}{D} \right)^3 + 0.05 \left(\frac{r_o}{D} \right)^5 + 0.00661 \left(\frac{r_o}{D} \right)^7 + \cdots . \tag{8.16}$$

It is this relation that was the basis for the definition of coherence diameter. For very large apertures the field on axis has the same value as would be produced by a transmitter with diameter r_o that propagates through free space. The Strehl ratio is plotted in Fig. 8.1.

If one calculates the Strehl ratio using solely the first term of the asymptotic series for $D/r_o = 2$, 5, and 10, then one obtains errors of 32%, 13%, and 6%. The use of the first two terms produces errors of 0.8%, 0.04%, and 0.005%, respectively. Therefore, the first two terms of the asymptotic expansion give a value of Strehl ratio when $D/r_o > 2$ that has an error of less than 1%.

8.2 Strehl Ratio with Beam Jitter

Typical adaptive-optics systems correct higher-order terms; tilt is corrected by a separate tracking system and mirror. Any jitter associated with this tracking system reduces intensity on axis, and its effect on system resolution must be known during system design. For propagation through turbulence, jitter can be expressed as a function of turbulence strength. *Yura* and *Tavis* (1985) solved for the structure function of the difference between Zernike and gradient tilts. Many tracking systems measure gradient tilt, which is sensitive to higher-order

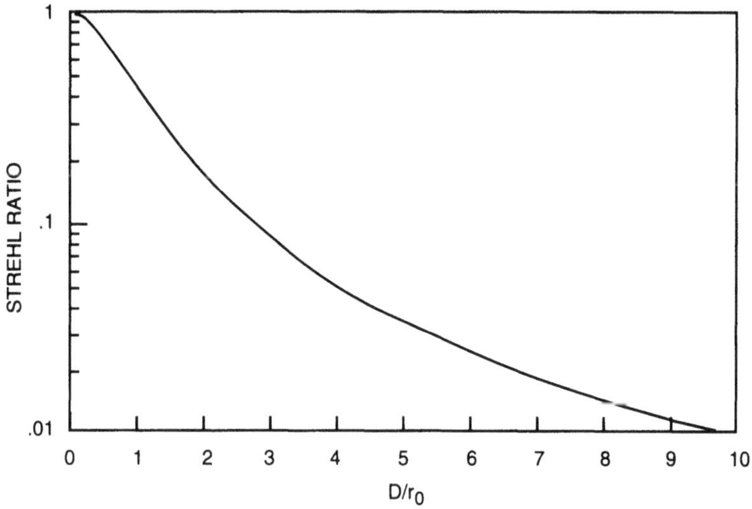

Fig. 8.1. Strehl ratio for uncorrected turbulence versus D/r_o.

Zernike aberrations. The difference between gradient and Zernike tilt represents an inherent source of tilt jitter in such a system and is referred to as centroid anisoplanatism. An adaptive-optics system reduces this error as explained in Sect. 4.4. Yura and Tavis presented numerical results for the Strehl ratio. They also presented a simple expression that is a reasonable approximation to the Strehl ratio. Here an analytic solution is derived for the Strehl ratio for centroid anisoplanatism and also for the Zernike and gradient tilts.

The structure function of tilt-induced phase is equal to the expected value of the square of the difference between tilt-induced phase at two points in the aperture, and is expressed as

$$\mathcal{D}(\boldsymbol{\rho}_1 - \boldsymbol{\rho}_2) = \left\langle \left[\phi\left(\boldsymbol{\rho}_1\right) - \phi\left(\boldsymbol{\rho}_2\right)\right]^2 \right\rangle = k_0^2 \left\langle \left[\mathbf{T} \cdot \left(\boldsymbol{\rho}_1 - \boldsymbol{\rho}_2\right)\right]^2 \right\rangle = \frac{k_0^2 T^2}{2} \left|\boldsymbol{\rho}_1 - \boldsymbol{\rho}_2\right|^2, \tag{8.17}$$

where $\mathbf{T}$ is the two-axis tilt of a wave that is randomly distributed in angle. The structure function is solely a function of the position difference. This tilt can be due solely to tracking jitter or can be considered to be due to atmospheric turbulence. If it is due to turbulence, then it is assumed that there is an adaptive-optics system that corrects all higher order modes of wavefront distortion, which leaves tilt as the only remaining distortion. The same expression with different constants is valid for gradient and Zernike tilt and for their difference. Two-axis tilt variances for atmospheric turbulence, which are listed in (4.15), (4.22), and (4.24), are given by

$$T^2 = \frac{1}{\pi^2} \left(\frac{D}{M r_o}\right)^{5/3} \left(\frac{\lambda}{D}\right)^2 , \tag{8.18}$$

where M is equal to 0.4642, 0.4838, and 5.394 for Zernike tilt (Z-tilt), gradient tilt (G-tilt), and their difference respectively. Therefore,

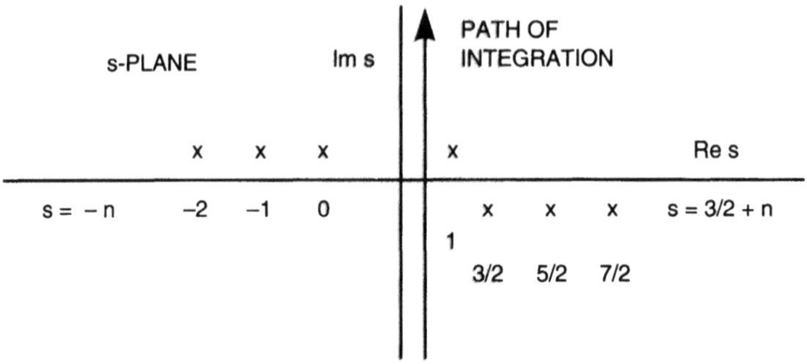

Fig. 8.2. Pole location and integration path in the complex plane for the problem to determine the Strehl ratio with tilt jitter.

$$\mathcal{D}\left(\boldsymbol{\rho}_1-\boldsymbol{\rho}_2\right)/2=D\left(\alpha D\right)/2=\alpha^2\left(\frac{D}{Mr_o}\right)^{5/3}=\alpha^2\sigma^2, \tag{8.19}$$

where $\sigma^2=(D/Mr_o)^{5/3}$. This form leads to one plot that applies to all three tilts. With this relation one can write the Strehl ratio as

$$SR=\int_0^1 \alpha\, d\alpha\, K(\alpha)\exp\left(-\sigma^2\alpha^2\right). \tag{8.20}$$

To evaluate this integral use the Mellin transforms in (1.62) and (1.10) in the Mellin convolution integral, and let $s\to 2s$ to obtain

$$SR=-\frac{4}{\sqrt{\pi}}\frac{1}{2\pi i}\int_C ds\left(\frac{1}{s-1}\right)\Gamma\left[\begin{matrix} s,\ -s+\frac{3}{2}\\ -s+3\end{matrix}\right]\left(\sigma^2\right)^{-s}. \tag{8.21}$$

The pole location and integration path are shown in Fig. 8.2. The recursion relation for gamma functions, $\Gamma[s]=(s-1)\,\Gamma[s-1]$, is used to simplify the integrand. The substitution $s\to s+1$ gives

$$SR=-\frac{4}{\sigma^2\sqrt{\pi}}\frac{1}{2\pi i}\int_C ds\,\Gamma\left[\begin{matrix} s*,\ -s+\frac{1}{2}\\ -s+2\end{matrix}\right]\left(\sigma^2\right)^{-s}, \tag{8.22}$$

where the asterisk signifies that the integration path passes between the first and second poles of that gamma function.

For this case $\Delta=1$, and the integration path can be closed in the left-half plane. The enclosed poles are located at $s=-n$, $n=1,2,3,\ldots$. The Strehl ratio given by the sum of the residues at the enclosed poles is

$$\begin{aligned} SR&=\frac{4}{\sqrt{\pi}}\sum_{n=0}^{\infty}\frac{(-1)^n}{(n+1)!}\,\Gamma\left[\begin{matrix} n+\frac{3}{2}\\ n+3\end{matrix}\right]\left(\frac{D}{Mr_o}\right)^{5n/3}\\ &={}_2F_2\left[\frac{3}{2},1;3,2;-\left(\frac{D}{Mr_o}\right)^{5/3}\right]. \end{aligned} \tag{8.23}$$

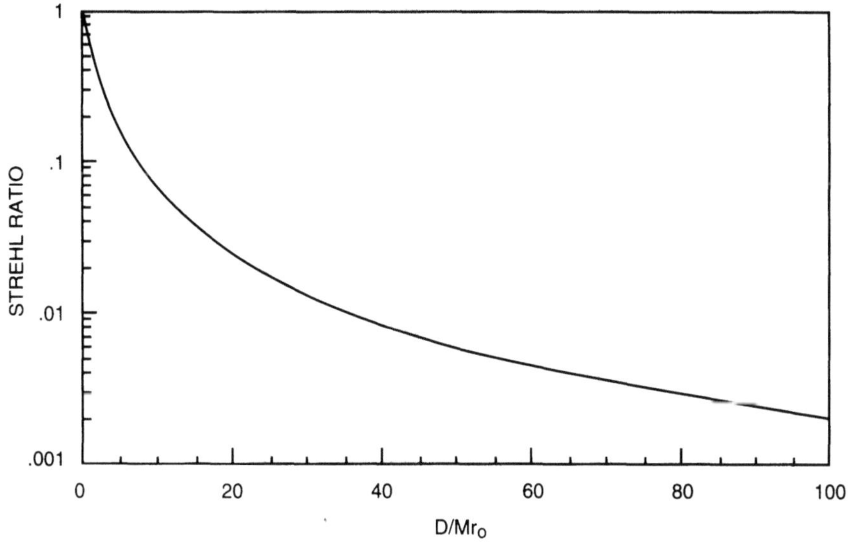

Fig. 8.3. General Strehl ratio for any type of tilt. M is equal to 0.4642, 0.4838, and 5.394 for Z-tilt, G-tilt, and their difference, respectively.

This can also be expressed as a confluent hypergeometric function as

$$SR = 4\left(\frac{Mr_o}{D}\right)^{5/3}\left\{1 - {}_1F_1\left[\tfrac{1}{2}; 2; -\left(\frac{D}{Mr_o}\right)^{5/3}\right]\right\}. \tag{8.24}$$

Strehl ratio is plotted for the general case in Fig. 8.3. The first few terms of this series are

$$SR = 1 - 0.25\left(\frac{D}{Mr_o}\right)^{5/3} + 0.0521\left(\frac{D}{Mr_o}\right)^{10/3} - 0.00911\left(\frac{D}{Mr_o}\right)^{5} + \cdots. \tag{8.25}$$

For large parameter values an asymptotic solution is needed to get an approximation that is accurate with only a few terms. One could use the results in *Abramowitz* and *Stegun* (1964) for confluent hypergeometric functions or, alternatively, one can easily calculate the series. The steepest-descent contribution given by (5.93) decays exponentially in this case. The asymptotic series is equal to the residues of the poles on the right side of the integration path located at $s = 0$, and $s = n + 1/2$ for $n = 0, 1, 2, \ldots$. The asymptotic series is

$$SR = 4\left(\frac{Mr_o}{D}\right)^{5/3} - \frac{4}{\sqrt{\pi}}\sum_{n=0}^{n_o}\frac{(-1)^n}{n!}\Gamma\left[\begin{matrix} n+\frac{1}{2} \\ -n+\frac{3}{2}\end{matrix}\right]\left(\frac{Mr_o}{D}\right)^{5n/3+5/2}. \tag{8.26}$$

The first few terms of this series are

$$SR = 4\left(\frac{Mr_o}{D}\right)^{5/3} - 4.51\left(\frac{Mr_o}{D}\right)^{5/2} + 1.13\left(\frac{Mr_o}{D}\right)^{25/6} + 0.423\left(\frac{Mr_o}{D}\right)^{35/6} + \cdots. \tag{8.27}$$

Yura gives an approximation

$$SR \approx \left[1 + \frac{1}{4}\left(\frac{D}{M r_o}\right)^{5/3}\right]^{-1} = \left[1 + \left(\frac{\pi T D}{2\lambda}\right)^2\right]^{-1} . \tag{8.28}$$

One can obtain this expression by dividing unity by the sum of unity and the inverse of the first term of the asymptotic series. For very large or small diameters this approximation gives the same result as the exact calculation. For intermediate diameters this approximation gives a Strehl ratio that is too large, but not by more than 10%.

8.3 Strehl Ratio with Anisoplanatism

In adaptive-optics systems the paths of the beacon signal and the imaging or projected laser should be the same. In general, this is not possible to achieve in practice, and there is a degradation in performance caused by time delays, displacement of the two paths by translation and angle, and differences in wavelength of the beacon and measurement or projecting systems. The effect of displacement, angular mispointing, time delay, and atmospheric dispersion can each be treated as an anisoplanatic effect. In fact, if all displacements are present simultaneously, they can be combined to get a total offset of the measurement from the imaging paths. In this section, the effect of that displacement on the Strehl ratio is determined. To find the Strehl ratio, the structure function must first be determined. Use the expression for the structure function in (2.92) with Kolmogorov turbulence to get

$$\mathcal{D}(\alpha D) = 0.4146\, k_0^2 \int_0^\infty dz\, C_n^2(z) \int d\boldsymbol{\kappa}\, \kappa^{-11/3} \times \left[1 - \cos\left(\boldsymbol{\kappa} \cdot \boldsymbol{\alpha} D\right)\right] 2\left\{1 - \cos\left[\boldsymbol{\kappa} \cdot \mathbf{d}(z)\right]\right\} . \tag{8.29}$$

The integrand can be expanded and the trigonometric identity for the product of cosines used to get

$$I = 2 \int d\boldsymbol{\kappa}\, \kappa^{-11/3} \left(1 - \cos\left(\boldsymbol{\kappa} \cdot \boldsymbol{\alpha} D\right) - \cos\left[\boldsymbol{\kappa} \cdot \mathbf{d}(z)\right] + \cos\left\{\boldsymbol{\kappa} \cdot \left[\boldsymbol{\alpha} D + \mathbf{d}(z)\right]\right\}/2 + \cos\left\{\boldsymbol{\kappa} \cdot \left[\boldsymbol{\alpha} D - \mathbf{d}(z)\right]\right\}/2\right) . \tag{8.30}$$

This expression is integrated over angle with (3.5), and after rearranging terms, one obtains

$$I = 4\pi \int_0^\infty d\kappa\, \kappa^{-8/3} \left(\left\{1 - J_0\left(\kappa \alpha D\right\}\right] + \left\{1 - J_0\left[\kappa d(z)\right]\right\} - \left\{1 - J_0\left[\kappa \left|\boldsymbol{\alpha} D + \mathbf{d}(z)\right|\right]\right\}/2 - \left\{1 - J_0\left[\kappa \left|\boldsymbol{\alpha} D - \mathbf{d}(z)\right|\right]\right\}/2\right) . \tag{8.31}$$

Using (1.21) for each of the four terms, one obtains for the structure function

$$\mathcal{D}(\alpha D) = 2(2.91)\, k_0^2 \int_0^\infty dz\, C_n^2(z)$$
$$\times \left[(\alpha D)^{5/3} + d^{5/3}(z) - \tfrac{1}{2} \left|\boldsymbol{\alpha} D + \mathbf{d}(z)\right|^{5/3} - \tfrac{1}{2} \left|\boldsymbol{\alpha} D - \mathbf{d}(z)\right|^{5/3}\right]. \quad (8.32)$$

The sum of terms in brackets almost cancel, which causes difficulties if one tries to evaluate this integral numerically. The terms in the absolute value sign are equal to

$$\left|\boldsymbol{\alpha} D \pm \mathbf{d}(z)\right|^{5/3} = \left[(\alpha D)^2 \pm 2\alpha D\, d(z) \cos(\varphi) + d^2(z)\right]^{5/6}, \quad (8.33)$$

where φ is the angle between $\boldsymbol{\alpha}$ and $\mathbf{d}$.

This expression can be simplified and numerical difficulties eliminated with the use of Gegenbauer polynomials $C_p^\lambda(x)$. Their generating function is

$$\left(1 - 2ax + a^2\right)^{-\lambda} = \sum_{p=0}^{\infty} C_p^\lambda(x)\, a^p. \quad (8.34)$$

These functions are sometimes referred to as ultraspherical functions because they are a generalization of the Legendre polynomials $P_n(t)$ whose generating function is

$$\left(1 - 2ax + a^2\right)^{-1/2} = \sum_{p=0}^{\infty} P_p(x)\, a^p. \quad (8.35)$$

The Gegenbauer polynomials can be represented as

$$C_p^\lambda\left[\cos(\varphi)\right] = \sum_{m=0}^{p} \frac{\Gamma[\lambda + m]\, \Gamma[\lambda + p - m] \cos[(p - 2m)\varphi]}{m!\,(p - m)!\, \{\Gamma[\lambda]\}^2}. \quad (8.36)$$

A Gegenbauer polynomial that is useful in this analysis is

$$C_2^{-5/6}\left[\cos(\varphi)\right] = \tfrac{5}{6}\left[1 - \tfrac{1}{3}\cos^2(\varphi)\right]. \quad (8.37)$$

For $\alpha D > d(z)$ the terms in the structure function can be expanded in Gegenbauer polynomials. The zeroth and all odd order terms cancel. When the summation index is changed by the substitution $m \rightarrow 2m$, the result is

$$\mathcal{D}(\alpha D) = 2(2.91)\, k_0^2 \int_0^\infty dz\, C_n^2(z)$$
$$\times \left\{ d^{5/3}(z) - (\alpha D)^{5/3} \sum_{p=1}^{\infty} C_{2p}^{-5/6}[\cos(\varphi)] \left[\frac{d(z)}{\alpha D}\right]^{2p} \right\}. \quad (8.38)$$

It is this canceling of the first two terms of the power series that would cause numerical difficulties if the original integral were evaluated numerically. Define

$$d_m \equiv 2.91\, k_0^2 \int_0^\infty dz\, C_n^2(z)\, d^m(z), \text{ and} \quad (8.39)$$

$$\sigma_\varphi^2 = d_{5/3}. \tag{8.40}$$

Unlike the calculation for the Strehl ratio for uncorrected turbulence and for corrected turbulence with tilt jitter, an exact analytical solution cannot be found for anisoplanatism. Fortunately for adaptive-optics systems, the Strehl ratio should be fairly high by design, which requires the structure function to be small. This assumption allows one to retain only the first term of the Gegenbauer expansion to give

$$\mathcal{D}(\alpha D) = 2\sigma_\varphi^2 - 2x, \tag{8.41}$$

where

$$x = d_2\left[1 - \tfrac{1}{3}\cos^2(\varphi)\right]\tfrac{5}{6}(\alpha D)^{-1/3}. \tag{8.42}$$

I justify this single-term approximation below by showing that its value is close to the exact result. The assumption that $\alpha D > d(z)$ is not true close to the center of the aperture, but it is typically true over most of the aperture. There will be an error, which increases with each term of the series, caused by the integration of this approximate expression over the entire aperture. In fact because of this assumption, if the exponential is expanded in a power series, only the integral of the first six terms converges. An expansion in Gegenbauer polynomials that applies close to the aperture center can be performed, and different expansions can be used for different parts of the aperture. These integrals cannot be evaluated in simple form, which is why it is not done here. The Strehl ratio with the six-term approximation is

$$SR \approx \frac{\exp\left(-\sigma_\varphi^2\right)}{2\pi}\int d\boldsymbol{\alpha}\, K(\alpha)\left(1 + x + \frac{x^2}{2} + \frac{x^3}{6} + \frac{x^4}{24} + \frac{x^5}{120}\right). \tag{8.43}$$

If just the first term in the last parentheses is retained, the result is equivalent to the extended Maréchal approximation. It is shown below that the six-term approximation is best for aperture sizes that are normally encountered. The integrals over angle and aperture coordinate are performed analytically. The angle integral for the nth term, after the use of the binomial theorem, is proportional to

$$\Phi(n) = \frac{1}{2\pi}\int_0^{2\pi} d\varphi\left[1 - \tfrac{1}{3}\cos^2(\varphi)\right]^n = \frac{1}{2\pi}\sum_{m=0}^{n}\binom{n}{n-m}3^{-m}\int_0^{2\pi} d\varphi\,\cos^{2m}(\varphi), \tag{8.44}$$

where $$\binom{n}{n-m} = \frac{n!}{(n-m)!\,m!}. \tag{8.45}$$

Equation (4.641 #4) in *Gradshteyn* and *Ryzhik* (1980) is

$$\int_0^{\pi/2} d\varphi\,\cos^{2m}(\varphi) = \frac{\pi(2m-1)!!}{2(2m)!!}, \quad \text{where} \tag{8.46}$$

$(2m-1)!! = (2m-1)(2m-3)\ldots(3)(1)$, and $(2m)!! = (2m)(2m-2)\ldots(4)(2)$. With these relations, the angle integral is equal to

$$\Phi(n) = 1 - \sum_{m=1}^{n} \binom{n}{n-m} 3^{-m} \frac{(2m-1)!!}{(2m)!!}. \tag{8.47}$$

The needed values are $\Phi(0) = 1$, $\Phi(1) = 0.8333$, $\Phi(2) = 0.7083$, $\Phi(3) = 0.6134$, $\Phi(4) = 0.5404$, and $\Phi(5) = 0.4836$. The aperture integration for the nth term is proportional to

$$Y(n) = \int_0^1 d\alpha\, \alpha^{1-n/3} K(\alpha). \tag{8.48}$$

From (1.62), the needed values are $Y(0) = 1$, $Y(1) = 1.402$, $Y(2) = 2.087$, $Y(3) = 3.396$, $Y(4) = 6.419$, and $Y(5) = 16.94$. If these values are used, then the approximation to the Strehl ratio is

$$SR = \exp\left(-\sigma_\varphi^2\right) \times \left(1 + 0.9736E + 0.5133E^2 + 0.2009E^3 + 0.06970E^4 + 0.02744E^5\right), \tag{8.49}$$

where $E = d_2/D^{1/3}$.

As mentioned above the Gegenbauer polynomial expansion is not valid near the center of the aperture. There is an error made in the use of this approximation for the central part of the aperture that increases with each term in the approximation. One has to determine whether this error is less or greater than the increased accuracy achieved over the remainder of the aperture by the inclusion of additional series terms. To resolve these uncertainties, I compared the Strehl ratio using various numbers of terms with the exact calculations.

I calculated the Strehl ratio numerically for the case in which the displacement does not vary with propagation distance. In Fig. 8.4 are plotted the exact Strehl ratio versus displacement for the Hufnagel-Valley 21 model of turbulence, and the Strehl ratio from (8.49) for $D/r_o = 1$ with only the unity term in brackets (Maréchal approximation), and with different numbers of terms in the bracket. This case of low turbulence is a region in which the approximation fares most poorly. The curves are plotted past their region of applicability to show how they begin to deviate from the correct values. The approximation that includes three or more terms is good out to a displacement that is one-third of the diameter, and the extended Maréchal approximation is always worse than that obtained with additional terms. The three-, four-, five-, and six-term approximations are the best.

In Fig. 8.5 and Fig. 8.6 are plotted the comparisons of the exact Strehl ratio for $D/r_o = 5$ and 10 with the extended Maréchal approximation and the approximation derived here. The six-term approximation is the best, and it gives accurate Strehl ratio values down to values of 0.2. The six-term approximation is always best when $D/r_o > 5$. The extended Maréchal approximation is poor. The six-term approximation gives results accurate to 1% over the first half of the range of plotted values of d/D for $D/r_o = 5$ and 10. The error approaches

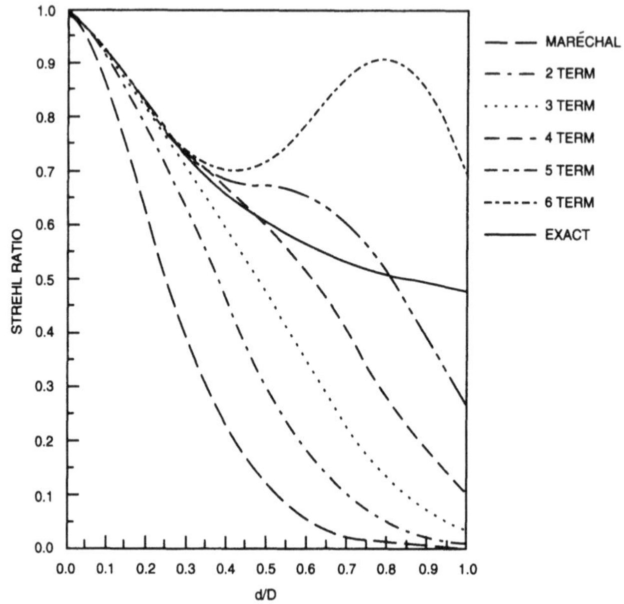

Fig. 8.4. Comparison of the Maréchal and two- to six-term approximations with the exact value of Strehl ratio, for an anisoplanatic displacement, for D/r_o equal to 1.

10% at the maximum values of d/D that are plotted. As the ratio of coherence diameter to aperture diameter gets smaller, the approximation improves. The close agreement between the exact results and those obtained with this approximation shows that the use of the one-term expansion in Gegenbauer polynomials was justified.

I now apply this approximate expression to calculate the Strehl ratio for various types of specific anisoplanatic effects.

8.4 Strehl Ratio for Various Anisoplanatic Effects

The Strehl ratio for various types of anisoplanatic effects is found with the expression given in (8.49), and the three terms

$$E = d_2/D^{1/3}, \tag{8.50}$$

$$\sigma_{\varphi}^2 = d_{5/3}, \text{ and} \tag{8.51}$$

$$d_m \equiv 2.91\, k_0^2 \int_0^{\infty} dz\, C_n^2(z)\, d^m(z). \tag{8.52}$$

8.4.1 Displacement Anisoplanatism

In this simplest case of displacement anisoplanatism, which was treated in the Sect. 8.3, the displacement is constant along the propagation direction.

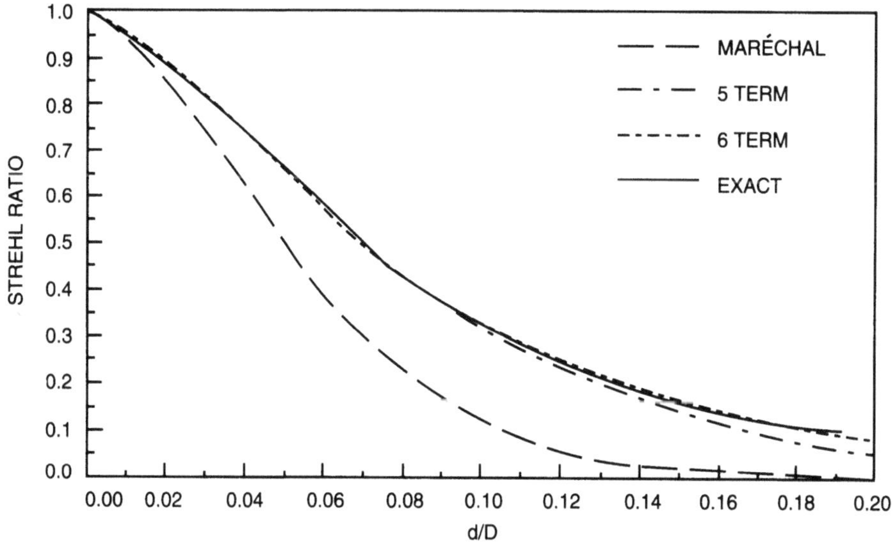

Fig. 8.5. Comparison of Maréchal and five- and six-term approximations with the exact value of the Strehl ratio, with an anisoplanatic displacement that is constant along the propagation path, for D/r_o equal to 5.

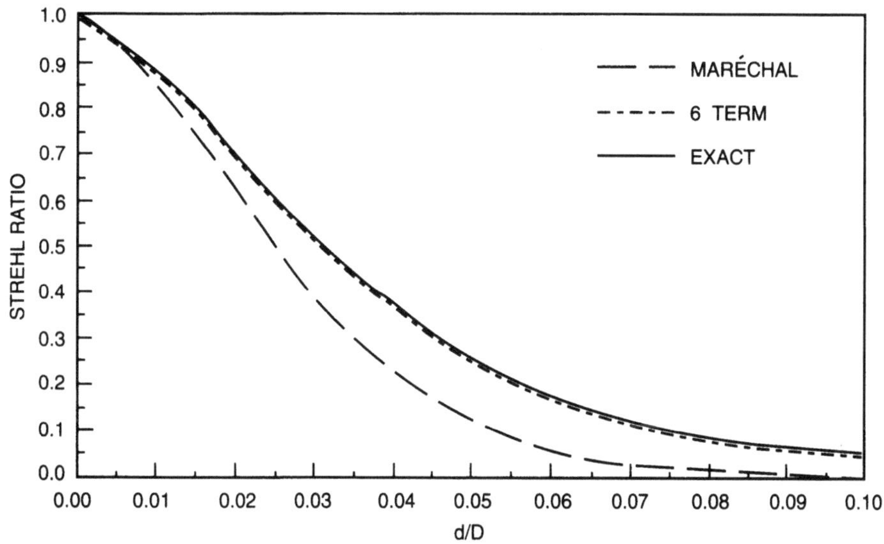

Fig. 8.6. Comparison of Maréchal and six-term approximations with the exact value of the Strehl ratio, with an anisoplanatic displacement, for D/r_o equal to 10.

Terms needed to find the Strehl ratio are $d(z) = d$, $d_2 = 2.91\,k_0^2\,\mu_0\,d^2$, $E = 6.88\,(d/D)^2\,(D/r_o)^{5/3}$, and $\sigma_\varphi^2 = 2.91\,k_0^2\,\mu_0\,d^{5/3} = 6.88\,(d/r_o)^{5/3}$, where the turbulence moments are defined in (4.2), and the coherence diameter is given by $r_o^{-5/3} = 0.423\,k_0^2\,\mu_0$. The Strehl ratios are plotted in Fig. 8.5 and Fig. 8.6.

8.4.2 Angular Anisoplanatism

When the propagation beam is offset by a constant angle from the direction along which turbulence is measured, the effect is called angular anisoplanatism (*Fried* 1982). It arises naturally when one is tracking a satellite target and directing a laser beam at it. Because of the finite speed of light, the laser beam has to lead the tracking direction; this results in an angular difference between the direction along which the target is tracked and the one along which the laser beam is directed. This error can be eliminated if the target has a reflector for the beacon that extends a suitable distance in the point-ahead direction. For the case of an angular error θ, one obtains

$$d(z) = \theta\,z, \tag{8.53}$$

$$d_2 = 2.91\,k_0^2\,\mu_2\,\theta^2, \tag{8.54}$$

$$E = 6.88\,\frac{\mu_2}{\mu_0}\left(\frac{\theta}{D}\right)^2\left(\frac{D}{r_o}\right)^{5/3}, \text{ and} \tag{8.55}$$

$$\sigma_\varphi^2 = 2.91\,k_0^2\,\theta^{5/3}\int_0^L dz\,C_n^2(z)\,z^{5/3} = (\theta/\theta_o)^{5/3}, \tag{8.56}$$

where the isoplanatic angle is given by $\theta_o = \left(2.91\,k_0^2\,\mu_{5/3}\right)^{-3/5}$. The relation in (8.56) defines the isoplanatic angle. The Strehl ratio versus offset angle for various turbulence models is plotted in Fig. 8.7 for a beam directed vertically and in Fig. 8.8 for a beam directed at 30 degrees off zenith.

8.4.3 Time-Delay Anisoplanatism

If there is a time delay τ between when turbulence is measured and a correction is applied to the deformable mirror, there is a degradation in performance as noted by *Tyler* (1984). This effect is not often thought of as an anisoplanatic effect; however, it can be treated as such. The Strehl ratio decays because the wind velocity $v(z)$ carries the turbules across the field of view. Using Taylor's frozen turbulence assumption discussed in Sect. 2.4, one obtains

$$d(z) = v(z)\,\tau, \tag{8.57}$$

$$d_2 = 2.91\,k_0^2\int_0^L dz\,C_n^2(z)\,v^2(z)\,\tau^2 = (\tau/\tau_2)^2, \tag{8.58}$$

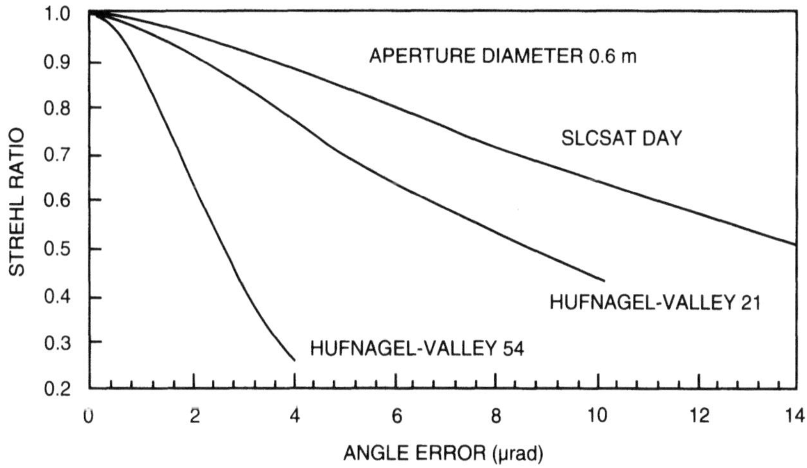

Fig. 8.7. Strehl ratio for angular anisoplanatic error at zenith, for various turbulence models, versus separation angle for a 0.6-m system. Upper-altitude turbulence has a strong effect on Strehl ratio.

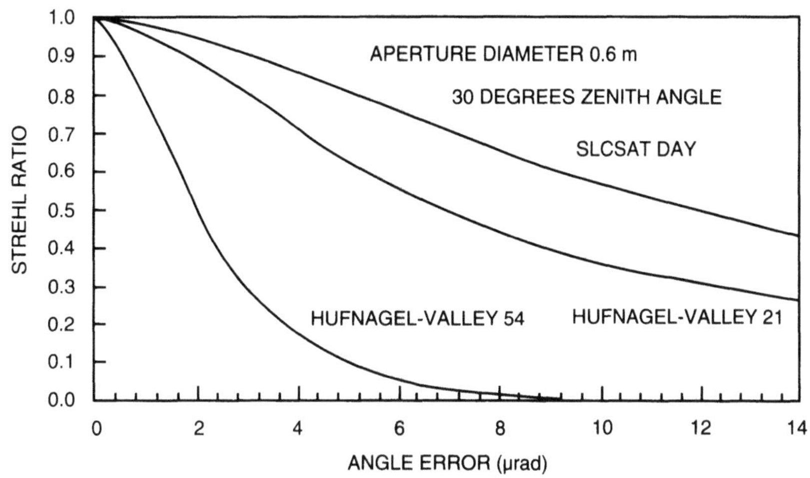

Fig. 8.8. Strehl ratio for angular anisoplanatism at 30 degrees from zenith for a 0.6-m system.

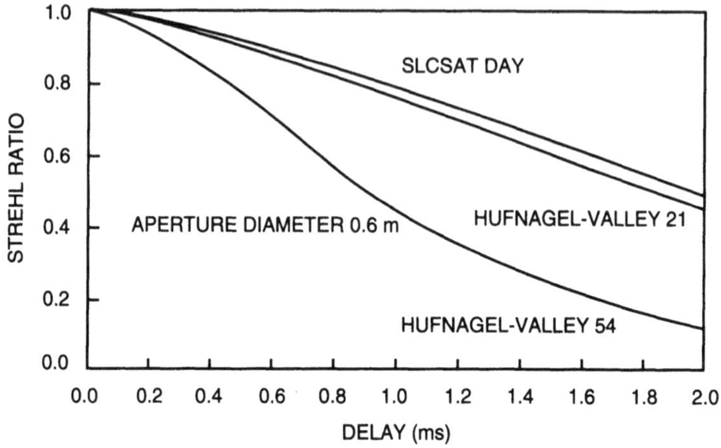

Fig. 8.9. Strehl ratio versus time delay at zenith for 0.6-m system.

$$E = \frac{\tau^2}{\tau_2^2 D^{1/3}}, \quad \text{and} \tag{8.59}$$

$$\sigma_\varphi^2 = 2.91\, k_0^2 \int_0^L dz\, C_n^2(z)\, v^{5/3}(z)\, \tau^{5/3} = \left(\tau/\tau_{5/3}\right)^{5/3}. \tag{8.60}$$

The temporal moment is defined as

$$\tau_m = \left[2.91\, k_0^2 \int_0^L dz\, C_n^2(z)\, v^m(z)\right]^{-3/5}. \tag{8.61}$$

The time $\tau_{5/3}$ is the characteristic time of the problem. It is related to the Greenwood frequency, defined in Sect. 4.8, by

$$f_G\, \tau_{5/3} = \left[\frac{0.0175}{n \sin(5\pi/6n)}\right]^{3/5}, \tag{8.62}$$

where n is the order of the filter in the servo system. For a single-pole filter $f_G\, \tau_{5/3} = 0.134$ as noted by *Fried* (1990).

The Strehl ratio is plotted in Fig. 8.9 for various turbulence models at zenith for a Bufton wind model with a ground speed of 5 m/s, and it is plotted at 30 degrees off zenith in Fig. 8.10. The Bufton wind model gives the wind versus altitude as

$$v(h) = v_g + 30 \exp\left[-\left(\frac{h - 9400}{4800}\right)^2\right], \tag{8.63}$$

where v_g is the ground speed.

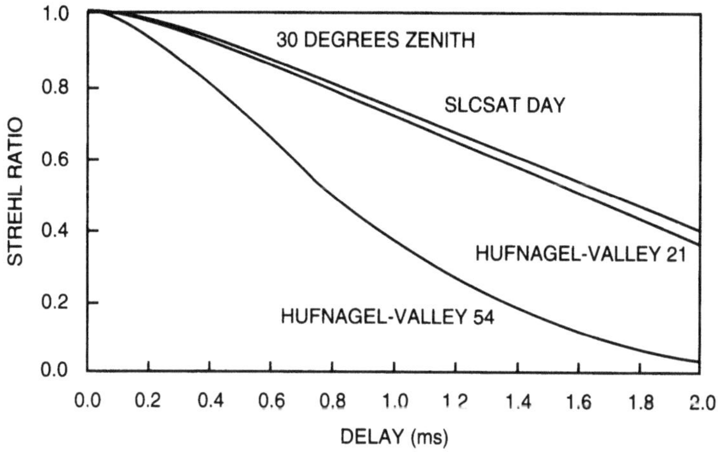

Fig. 8.10. Strehl ratio versus time delay for 0.6-m system at 30 degrees zenith angle.

8.4.4 Chromatic Anisoplanatism

If the beacon beam that senses the turbulence has a wavelength different from that of the laser beam that is sent out, then the two beams will follow different paths through the atmosphere because of the atmosphere's dispersive properties. The analysis given here parallels that given by *Belsher* and *Fried* (1981). Geometric optics is used to obtain the path displacement for chromatic anisoplanatism. If $n(h)$ is the refractive index seen by the first wave and $n(h) + \Delta n(h)$ is that seen by the second and if $\xi(h)$ is the zenith angle of the first and $\xi(h) + \Delta\xi(h)$ is that of the second, Snell's law, which states that $n \sin(\xi)$ is a constant, gives

$$\Delta\xi(h) = -\Delta n(h) \tan[\xi(h)] . \tag{8.64}$$

This can be integrated along the path to yield

$$\mathbf{d}_c(z) = \int_0^z dz\, \Delta\boldsymbol{\xi}(h) = -\frac{\boldsymbol{\xi} \sin(\xi)\, \Delta n_0}{\xi \cos^2(\xi)} \int_0^z dz'\, \alpha(z') , \tag{8.65}$$

where Δn_0 is the difference in refractive index between the two colors at wavelengths λ_1 and λ_2 and $\alpha(z)$ is the air density normalized to the value at sea level. The change of refractive index with wavelength has been given by *Allen* (1963) as

$$\Delta n_0 = \left(\lambda_1^2 - \lambda_2^2\right) \left[\frac{29498.1}{(146\lambda_2^2 - 1)(146\lambda_1^2 - 1)} + \frac{255.4}{(41\lambda_2^2 - 1)(41\lambda_1^2 - 1)}\right] 10^{-6}, \tag{8.66}$$

when the wavelengths are given in μm. The atmospheric density versus altitude is given in *Cole* (1965). The ratio of the atmospheric density at any altitude normalized to that at sea level was approximated by Belsher and Fried as

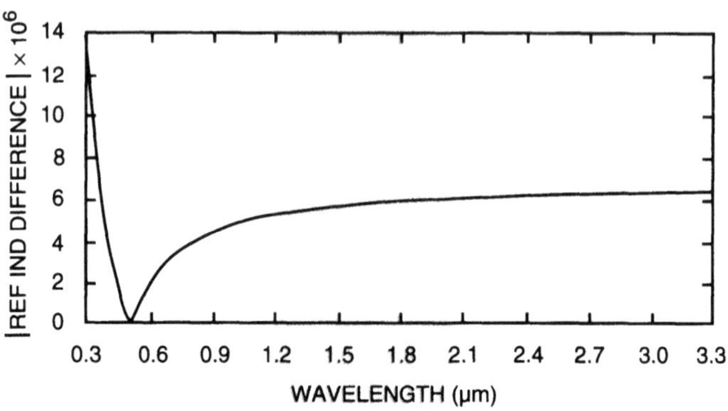

Fig. 8.11. A million times the difference in refractive index between 0.5 μm and another wavelength.

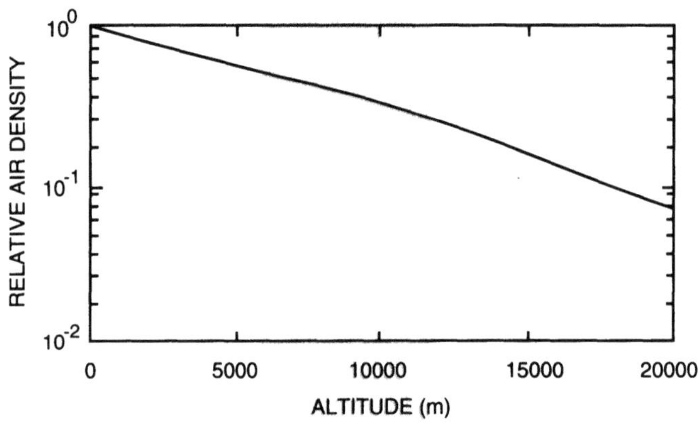

Fig. 8.12. Normalized air density versus altitude.

$$\alpha(h) = \exp\left(-1.11 \times 10^{-4} h\right), \quad h < 10\,\text{km, and}$$

$$\alpha(h) = 1.6 \exp\left(-1.57 \times 10^{-4} h\right), \quad h > 10\,\text{km}. \tag{8.67}$$

A plot of the absolute value difference in the refractive index between waves at 0.5 μm and other wavelengths is shown in Fig. 8.11. The normalized air density versus height is plotted in Fig. 8.12.

To calculate the Strehl ratio, the displacement as a function of range is found with (8.64). At the target range R, the displacement is $d_c(L)$, and to hit the target squarely one must change the launching angle by $d_c(L)/L$. Thus the beam displacement along the path is

$$\mathbf{d}_c(z) = -\frac{\boldsymbol{\xi} \sin(\xi)\, \Delta n_0}{\xi \cos^2(\xi)} \left[\int_0^z dz'\, \alpha(z') - \frac{z}{L} \int_0^L dz'\, \alpha(z') \right]. \tag{8.68}$$

Table 8.1. Values of T_2 and $T_{5/3}$ for various turbulence models. The units of T_2 are $m^{1/3}$, and $T_{5/3}$ is dimensionless.

Model	T_2	$T_{5/3}$
SLCDAY	2.706×10^{-6}	2.004×10^{-7}
SLCNIGHT	2.258×10^{-6}	1.512×10^{-7}
HV-21	6.160×10^{-6}	3.596×10^{-7}
HV-54	3.399×10^{-5}	1.867×10^{-6}
HV-72	5.949×10^{-5}	3.247×10^{-6}

Define the integral of air density as

$$I(z) = \int_0^z dz' \, \alpha(z') . \tag{8.69}$$

This is evaluated as

$$\begin{aligned} I(z) &= 9010 \left[1 - \exp\left(-1.11 \times 10^{-4} z\right)\right], \quad z < 10\,\text{km, and} \\ I(z) &= 8161 - 10\,190 \exp\left(-1.57 \times 10^{-4} z\right), \quad z > 10\,\text{km}. \end{aligned} \tag{8.70}$$

The moments of this displacement are

$$d_m = \left[\frac{\sin(\xi)\, \Delta n_0}{\cos^2(\xi)}\right]^m T_m, \tag{8.71}$$

where

$$T_m = 2.91\, k_0^2 \sec(\xi) \int_0^H dh\, C_n^2(h) \left[I(h) - \frac{h \sec(\xi)}{L} I(L)\right]^m . \tag{8.72}$$

H is the altitude of the target. The last term in brackets goes to zero as the range becomes infinite. A Strehl ratio comparison of a target at 300 km and one at infinity is plotted in Fig. 8.13. There is very little difference in the two results. For infinite range, the above equation reduces to

$$T_m = 2.91\, k_0^2 \sec(\xi) \int_0^H dh\, C_n^2(h)\, I^m(h). \tag{8.73}$$

The result for finite range entails a numerical calculation for each range. The infinite-range results can be calculated once for each turbulence model and used for different zenith angles. Table 8.1 contains values of T_2 and $T_{5/3}$ for various turbulence models.

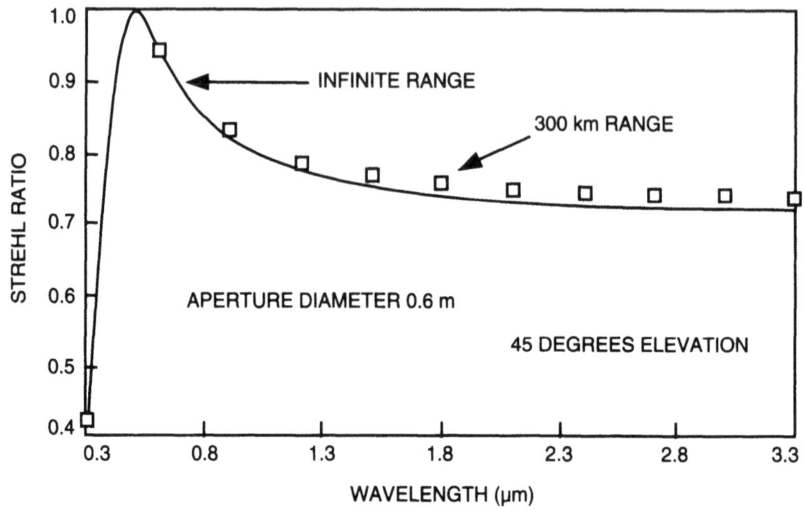

Fig. 8.13. Comparison of the Strehl ratio at infinite and 300 km range versus beacon wavelength. The two results are close in value. This allows a simpler calculation for finite range problems.

The relevant quantities to calculate Strehl ratio are

$$E = \left[\frac{\sin(\xi)\,\Delta n_0}{\cos^2(\xi)}\right]^2 \frac{T_2}{D^{1/3}}, \text{ and} \tag{8.74}$$

$$\sigma_\varphi^2 = \left[\frac{\sin(\xi)\,\Delta n_0}{\cos^2(\xi)}\right]^{5/3} T_{5/3}. \tag{8.75}$$

As a specific example, Fig. 8.14 contains plots of Strehl ratio for the SLCSAT day model for various elevation angles for an infinite-range target, when one beam is at 0.5-μm wavelength.

8.4.5 Combined Displacement Anisoplanatism

If there are several anisoplanatic effects present, with each not decreasing the Strehl ratio much, it is a common practice to multiply the Strehl ratios for the individual effects to get a combined Strehl ratio. The validity of this assumption is now examined. The total displacement that is due to a translation, an angular offset, a time delay, and a chromatic offset is

$$\mathbf{d}_t(z) = \mathbf{d} + \boldsymbol{\theta} z + \mathbf{v}(z)\tau + \mathbf{d}_c(z), \tag{8.76}$$

where chromatic displacement is given in (8.69). The two terms necessary to calculate the Strehl ratio are $E = d_2/D^{1/3}$, and $\sigma_\varphi^2 = d_{5/3}$, where

$$d_m = 2.91\, k_0^2 \int_0^\infty dz\, C_n^2(z)\, d_t^m(z).$$

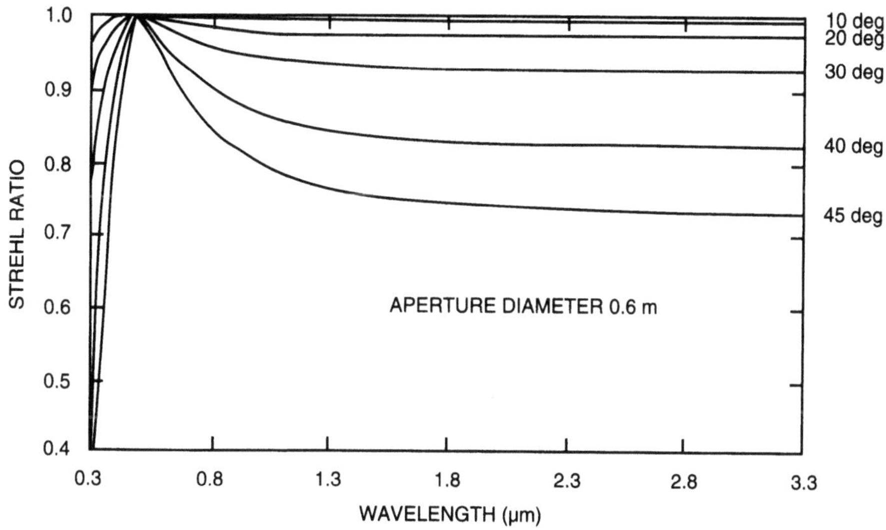

Fig. 8.14. Strehl ratio for SLCSAT day turbulence with the scoring beam at 0.5 μm for a 0.6-m system and zenith angles between 10 and 45 degrees. Notice that above 45 degrees elevation angle the Strehl ratio exceeds 0.7 for the first beam at 0.5 μm and the second at any longer wavelength.

Let us examine the case in which the effects of displacement, angular offset, and temporal delay are comparable. Chromatic anisoplanatism is assumed to be zero. To obtain comparable effects from the remaining three types of anisoplanatism, the angular offset in microradians is 850 times the displacement in meters. The time delay in milliseconds is 130 times the displacement in meters for the Bufton wind model with a ground speed of 5 m/sec.

In Fig. 8.15 the Strehl ratios are plotted for these effects present individually and combined for the SLCSAT Day turbulence model with a 0.6-m diameter system. It was assumed that the three vector displacements either were all in the same direction or were separated by 120 degrees. Also plotted is the product of the Strehl ratios from the individual effects. Note that the Strehl ratio for all the effects combined can be less than or greater than the Strehl ratio obtained by the common practice of multiplying individual Strehl ratios. The relative direction of the displacements is crucial. It is possible to get some cancellation if the displacements cancel. *Tyler* et al. (1982) took advantage of the vector nature of the displacement almost to eliminate the effect of chromatic anisoplanatism on an adaptive-optics system by choosing an optimal offset angle of a beacon from the propagation direction.

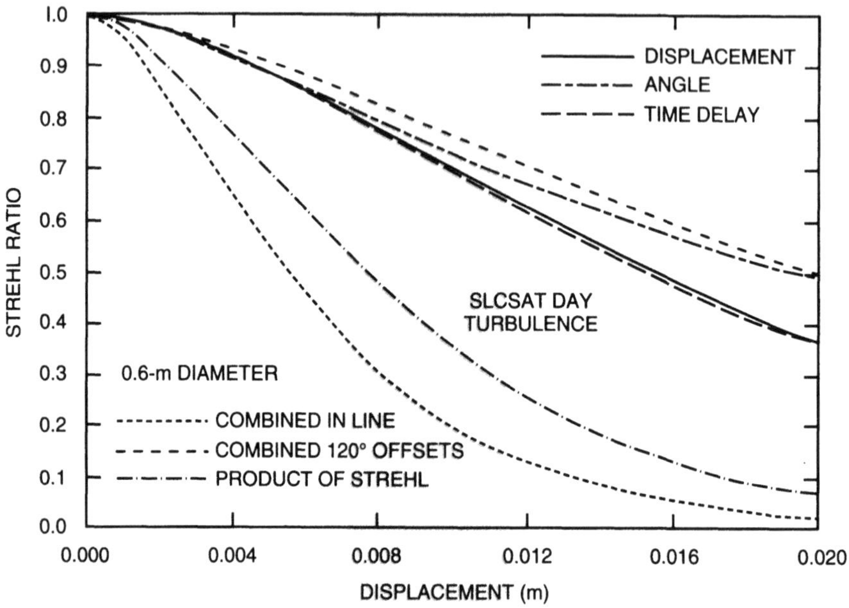

Fig. 8.15. Strehl ratios for displacement, angle, and time delay anisoplanatism presented separately and combined for the SLCSAT-Day turbulence model. The displacement is constant along the propagation path. The angle offset in microradians is 850 times the displacement in meters. The time delay in milliseconds is 130 times the displacement in meters. The combined effects are for all displacements in line and at 120 degree to each other.

9. Mellin Transforms with a Complex Parameter

When one analyzes beam-wave propagation through turbulence, the propagation parameter γ is complex. This causes complications when one evaluates the steepest-descent contribution to an asymptotic solution. In some problems, the spatial filter functions contain a decaying exponential times a function that grows exponentially when γ is complex. Consequently, the Mellin convolution relation must be generalized to allow for convergent integrals that contain integrands that are the product of functions that exhibit exponential growth and exponential decay. In addition, methods developed in Chap. 5 must be extended to include asymptotic results for a non-positive parameter. This requires an extension of saddle-point results to complex-parameter cases. These extensions are discussed in this chapter.

The theory of asymptotic expressions for general integrals has been considered by *Luke* (1969). *Marichev* (1983) states asymptotic results for integrals with complex parameters that apply specifically to Mellin-Barnes integrals with unity coefficients of the complex variable in the gamma functions. These results are generalized here to allow asymptotic series to be written for Mellin-Barnes integrals in which the coefficients of the complex variable are not unity. Also, the results are given in a form that is directly applicable to the evaluation of integrals encountered in turbulence problems.

9.1 Mellin-Barnes Integrals with Complex Parameters

In beam-wave analysis, the integral expression that must be evaluated is

$$h(z) = \int_0^\infty \frac{dy}{y}\, h_0(y)\, h_1\left(z/y\right), \tag{9.1}$$

where the parameter z in general is complex. If $z = x$, x real, the convolution integral in (1.27) may be used to obtain the Mellin transform of the integral

$$h(x) = \int_0^\infty \frac{dy}{y}\, h_0(y) h_1\left(x/y\right) \rightarrow \mathcal{M}\left[h(x)\right] = H(s) = H_0(s)\, H_1(s). \tag{9.2}$$

The function $h(x)$ is written as an inverse Mellin transform that is evaluated with pole-residue methods. In the case of a complex parameter, however, the Mellin transform of one integrand function may be undefined because it exhibits exponential growth. Let that function be $h_1(x)$. It is not obvious how to apply the convolution integral in this case. *Marichev* (1983, Theorem 23) gives a generalized Parseval's equality that is applicable here. Before the statement of the theorem, it is necessary to show how to express a function with a complex parameter as a Mellin-Barnes integral.

The general integral of interest is given by (5.90)

$$h(z) = \frac{1}{2\pi i} \int_{C'} ds\, z^{-s} \frac{\prod_{i=1}^{A} \Gamma\left[a_i + \alpha_i\, s\right] \prod_{j=1}^{B} \Gamma\left[b_j - \beta_j\, s\right]}{\prod_{k=1}^{C} \Gamma\left[c_k + \gamma_k\, s\right] \prod_{m=1}^{D} \Gamma\left[d_m - \delta_m\, s\right]}. \tag{9.3}$$

The parameter z in the above equation may, in general, be complex. In this discussion if the parameter is real, it will be represented by x. The contour C' along which the integral is normally evaluated is given by the vertical line from $\eta - i\infty$ to $\eta + i\infty$, where η is a real constant chosen to insure that the integration path lies within the strips of analyticity of the integrand functions. When the parameter is complex, however, convergence conditions may require that the integration path be closed either in the left-half-plane or the right-half-plane. This issue is discussed further by *Luke* (1969) and *Marichev* (1983). An example of this condition is provided by the Bessel function of the first kind $J_\nu(z)$. If $z = x$ is real, the Mellin-Barnes integral representation of $J_\nu(z)$ is given by

$$J_v(x) = \frac{1}{4\pi i} \int_{\eta - i\infty}^{\eta + i\infty} ds\, \Gamma\left[\begin{matrix} s/ + \nu/2 \\ -s/2 + 1 + \nu/2 \end{matrix}\right] \left(\frac{x}{2}\right)^{-s}, \tag{9.4}$$

where $x > 0$, and $-\mathrm{Re}\{v\} < \eta < 3/2$. If $z = \beta x$ is complex, a similar integral expression may be written,

$$J_v(\beta x) = \frac{1}{4\pi i} \int_{C_{-\infty}} ds\, \Gamma\left[\begin{matrix} s/2 + \nu/2 \\ -s/2 + 1 + \nu/2 \end{matrix}\right] \left(\frac{\beta x}{2}\right)^{-s}, \tag{9.5}$$

where the new integration path $C_{-\infty}$ is the left loop surrounding all poles of $\Gamma\left[(v + s)/2\right]$. These poles occur at $s = -v - 2n$ for $n = 0, 1, 2, \ldots$.

In problems treated in this book and encountered in analyzing the effects of turbulence, the parameters a_i, α_1, b_j, β_j, c_k, γ_k, d_m, and δ_m are all real. Therefore, all the integrand's poles occur along the real axis. This allows us to paraphrase *Marichev's* theorem (1983) as below:

Theorem 9.1 *If the following three conditions hold:*

a) for certain $x > 0$ and η, the integrals

$$\int_0^\infty \frac{dy}{y}\, h_0(y)\, h_1(x/y), \quad \frac{1}{2\pi i} \int_{\eta - i\infty}^{\eta + i\infty} ds\, H_0(s)\, H_1(s)\, x^{-s} \tag{9.6}$$

exist;

b) $H_0(s)$ *is an analytic function of s without singularities for* $Re\{s\} < \eta$ (*or* $Re\{s\} > \eta$)*; and*

c) $h_1(y)$ *is a function of hypergeometric type (a hypergeometric function with a power of the argument multiplying it, or such that some terms of its power series are subtracted from it) such that* $H_1(s)$ *exists with*

$$h_1(y) = \frac{1}{2\pi i} \int\limits_{L\mp\infty} ds\, H_1(s)\, y^{-s}, \tag{9.7}$$

where the integration is over the appropriate loop $L_{-\infty}$ (*or* $L_{+\infty}$) *such that all the poles of* $H_1(s)$ *are enclosed, then*

$$\int\limits_0^\infty \frac{dy}{y}\, h_0\,(y)\; h_1(z/y) = \frac{1}{2\pi i} \int\limits_{\eta-i\infty}^{\eta+i\infty} ds\, H_0(s)\, H_1(s)\, z^{-s}. \tag{9.8}$$

The Bessel function satisfies the conditions required for $h_1(y)$, and for complex z can have exponential growth. In practice, when conditions for the theorem are satisfied one substitutes the Mellin transforms of the functions with real parameters into the convolution integral and ignores the fact that the individual Mellin transforms may not exist for a complex parameter.

Use of this result is illustrated in Sect. 9.3 where the Mellin transform of $\exp(-\alpha x^2)\, J_v(\beta x)$ is derived for β complex.

9.2 Asymptotic Results with a Complex Parameter

As discussed in Chap. 5, when the parameter associated with a Mellin convolution is large, the series result obtained by pole-residue techniques may converge slowly. Consequently, asymptotic results are useful when accurate approximations are desired. Asymptotic expressions were obtained in Chap. 5 when the parameter was assumed to be positive real (complex argument has zero imaginary part). These results are now generalized to allow complex or negative parameters.

The general form of the Mellin-Barnes integral that must be evaluated is given in (9.3). The definitions in the table at the end of this chapter are used in our analysis. One finds the asymptotic form of (9.1) by deforming the original integration path so that the new path goes through all saddle points in the direction of steepest descent. The asymptotic approximation of the integral includes the pole residues picked up through path deformation. One calculates the asymptotic contribution of the deformed path by computing a steepest-descent contribution, and one finds the pole-residue contributions by changing the normal direction of contour closure as described in Chap. 5.

To find the steepest-descent contribution to the integral, one uses the asymptotic form of the integrand. The identity

$$\Gamma[s]\,\Gamma[1-s] = \pi/\sin(\pi s) \tag{9.9}$$

allows (9.1) to be expressed so that the coefficients of s in all gamma functions are positive as in (5.18). The asymptotic expressions for the gamma functions,

$$\Gamma[e+\varepsilon s] \sim \sqrt{2\pi}\,\exp\left[(\varepsilon s + e - 1/2)\ln(\varepsilon s) - \varepsilon s\right], \tag{9.10}$$

where e and ε are arbitrary numbers, and the sine functions

$$\sin(\pi s) \sim \tfrac{1}{2}\exp\left[-i\pi\left(s-\tfrac{1}{2}\right)\right]$$

when $\mathrm{Im}\{s\} > 0$, and

$$\sin(\pi s) \sim \tfrac{1}{2}\exp\left[i\pi\left(s-\tfrac{1}{2}\right)\right]$$

when $\mathrm{Im}\{s\} < 0$, may be used to obtain the asymptotic form for the integral,

$$h(z) \sim -i(2\pi)^{\Xi/2-1}\int ds\,\exp\left[f(s)\right]. \tag{9.11}$$

The function in the exponent is given by

$$\begin{aligned} f(s) &= \Delta'' - \Delta'''/2 + s\left[\Delta' - \Delta - \ln(z)\right] \\ &+ \ln(s)\left(\nu - \frac{\Xi}{2} + \Delta s\right) - i\lambda\pi\left(-\Lambda'' + \Lambda/2 + s\Lambda'\right). \end{aligned} \tag{9.12}$$

This expression is valid for $-\pi < \arg\{s\} < \pi$. When $\mathrm{Im}\{s\} > 0$, then $\lambda = 1$, and when $\mathrm{Im}\{s\} < 0$, then $\lambda = -1$.

To find the steepest-descent contribution, the saddle points of $f(s)$ must be found. These are the points at which

$$\frac{df(s)}{ds} = 0 = \Delta\ln(s) + \Delta'\ln(z) - i\lambda\pi\Lambda'. \tag{9.13}$$

Terms which depend upon $1/s$ have been dropped from this expression because at the saddle point $|s|$ is large for large $|z|$. Two cases must be considered. If a saddle point exists in the upper half plane, then the solution of (9.13) for s requires $\lambda = 1$. Similarly lower-half-plane saddle points require $\lambda = -1$.

The upper-half-plane saddle point is located at

$$s_o = z^{1/\Delta}\exp\left[-\frac{1}{\Delta}\left(\Delta' - i\pi\Lambda'\right)\right], \tag{9.14}$$

and the requirement on $\arg\{z\}$ for the existence of the saddle point in the upper half-plane is

$$-\pi\Lambda' < \arg\{z\} < \pi\left(\Delta - \Lambda'\right). \tag{9.15}$$

Similarly, the lower-half-plane saddle point is located at

$$s_o = z^{1/\Delta}\exp\left[-\frac{1}{\Delta}\left(\Delta' + i\pi\Lambda'\right)\right], \tag{9.16}$$

and the corresponding restriction on $\arg\{z\}$ is

$$-\pi(\Delta - \Lambda') < \arg\{z\} < \pi\Lambda'. \tag{9.17}$$

If $\Lambda' = 0$, then for positive-real z the saddle point is located on the real axis. If $-1 < \Lambda'/\Delta < 0$, there is a contradiction. If one assumes the saddle point is in the upper half plane, one finds that the saddle point specified by (9.14) is located in the lower half plane. A contradiction also results if one assumes that the saddle point is located in the lower half plane. The only other possibility is to assume that the saddle point is located on the real axis. For this case the sinusoid cannot be represented by their asymptotic equivalents. A simple case arises when $B = D = 0$. For this case one should set $\lambda = 0$ in (9.12), and indeed the single saddle point is located on the real axis at

$$s_o = z^{1/\Delta}\exp(-\Delta'/\Delta). \tag{9.18}$$

One obtains the correct value for the steepest-descent contribution in this case from the final answer given in (9.25) and (9.26) by setting $\Lambda'' = \Lambda' = \Lambda = 0$ and dividing the answer by two.

For the more general case of $\Lambda' < 0$, (9.12) becomes

$$\begin{aligned} f(s) &= \Delta'' - \Delta'''/2 + s\left[\Delta' - \Delta - \ln(z)\right] \\ &+ \ln(s)\left(\nu - \frac{\Xi}{2} + \Delta s\right) - i\lambda\pi\left(-\Lambda'' + \Lambda/2 + s\Lambda'\right). \end{aligned} \tag{9.19}$$

The steepest-descent point in this case is located at the solution of the equation

$$\begin{aligned} s_o &= z^{1/\Delta}\exp\left(-\frac{\Delta'}{\Delta}\right) \\ &\times \exp\left(\sum_{m=1}^{D}\pi\delta_m\cot\left[\pi\left(\delta_m s_o + 1 - d_m\right)\right] - \sum_{j=1}^{B}\pi\beta_j\cot\left[\pi\left(\beta_j s_o + 1 - b_j\right)\right]\right). \end{aligned} \tag{9.20}$$

The right-hand side of the equation is always positive for positive-real z; therefore, the saddle point is located in the right half plane. The equation has multiple solutions but the one closest to the origin gives the largest contribution. For this case the saddle point has exponential decay. Often knowledge of this fact allows the steepest-descent contribution to the value of the integral to be neglected compared to pole contributions. If there are no pole contributions in the right-half plane, then (9.20) must be solved to find the steepest-descent contribution.

For the case in which the saddle point is on the real axis, it is located in the right half plane. Therefore, in deforming the path to pass through the saddle point. one picks up poles that were not included in the Taylor series solution. When there are two complex saddle points, they are also located in the right half plane. To prove that, first note that all Mellin transforms have fewer or the same number of denominator gamma functions than numerator gamma functions (This stems from the requirement that the inverse transform integrand must go to zero at $\pm\infty$.) Also note that for this case $\Delta' > 0$ and $\Delta > 0$. The direction of the upper-complex saddle point is determined by

$$0 < \frac{\Lambda'}{\Delta} = \frac{1}{2} - \frac{1}{2}\left(\frac{A+B-C-D}{\Delta}\right) \leq \frac{1}{2}. \tag{9.21}$$

Therefore, the saddle point specified by (9.14) is located in the upper-right complex plane for positive-real z.

One finds the line integral contribution to the asymptotic form of the original integral by evaluating the steepest-descent contribution at upper- and lower-half-plane saddle points. The exponent in (9.12) is expanded in a Taylor series given by

$$f(s) \approx f(s_o) + \frac{df(s_o)}{ds}(s-s_o) + \frac{1}{2}\frac{d^2f(s_o)}{ds^2}(s-s_o)^2. \tag{9.22}$$

The magnitudes of second and higher-order derivatives at the saddle point are proportional to s_o^{1-n}, where n is the order of the derivative. Since $|s_o|$ is large, higher-order terms may be neglected. At the saddle point

$$f(s) \approx f(s_o) + \frac{1}{2}\frac{d^2f(s_o)}{ds^2}(s-s_o)^2. \tag{9.23}$$

With this approximation and an appropriate change of variables, the integral in (9.11) becomes the integral of a Gaussian function that, when evaluated, gives the following results

$$h(z) \sim E(z) = E_1(z) + E_2(z), \tag{9.24}$$

where $E_1(z)$ and $E_2(z)$ are the upper- and lower-half-plane contributions, respectively. $E_1(z)$ is given by

$$E_1(z) = \frac{(2\pi)^{(\Xi-1)/2}}{\sqrt{\Delta}} z^{\rho} \exp\Bigg[-\rho\left(\Delta' - i\pi\Lambda'\right) - \Delta z^{1/\Delta} \exp\left(\frac{-\Delta' + i\pi\Lambda'}{\Delta}\right) + \Delta'' - \frac{\Delta'''}{2} - i\pi\left(-\Lambda'' + \frac{\Lambda}{2}\right)\Bigg] \tag{9.25}$$

when $-\pi\Lambda' < \arg\{z\} < \pi(\Delta - \Lambda')$, and $E_1(z) = 0$ otherwise. $E_2(z)$ is given by

$$E_2(z) = \frac{(2\pi)^{(\Xi-1)/2}}{\sqrt{\Delta}} z^{\rho} \exp\Bigg[-\rho\left(\Delta' + i\pi\Lambda'\right) - \Delta z^{1/\Delta} \exp\left(\frac{-\Delta' + i\pi\Lambda'}{\Delta}\right) + \Delta'' - \frac{\Delta'''}{2} + i\pi\left(-\Lambda'' + \frac{\Lambda}{2}\right)\Bigg] \tag{9.26}$$

when $-\pi(\Delta - \Lambda') < \arg\{z\} < \pi\Lambda'$, and $E_2(z) = 0$ otherwise.

I have actually underestimated the range of $\arg\{z\}$ over which the approximations in (9.25) or (9.26) are valid. It may be extended by using asymptotic approximations for the gamma function $\Gamma[\alpha]$, that are valid for other ranges of $\arg\{a\}$. When this is done, results from steepest-descent methods reported here match those obtained by *Luke* (1969) that were derived from the properties of the Meijer G function. *Marichev* (1983) summarizes these results. With this extension, (9.25) can be used for $-\pi\Lambda' < \arg\{z\} < \pi(\Delta - \Lambda' + \varsigma)$, and (9.26) can be used for $-\pi(\Delta - \Lambda' + \varsigma) < \arg\{z\} < \pi\Lambda'$ where $\varsigma = 1/2$ when $\Delta = 1$, and $\varsigma = 1$ when $\Delta > 1$.

To determine which terms dominate the asymptotic solution, the steepest-descent and the pole-residue portions of the asymptote must be examined. If $E(z)$ exhibits exponential growth, it will dominate the asymptote. If $E(z)$ exhibits exponential decay, and there is a pole-residue contribution to the asymptote, then the pole-residue portion is dominant. An example that illustrates the use of these techniques is given in the next section.

When the parameter z is positive real, these results reduce to the saddle-point expression given in (5.93).

9.3 The Mellin Transform of an Exponential Times a Bessel Function

In this section the technique that was developed above is used to evaluate an integral with a complex parameter to illustrate how to find the pole and steepest-descent contributions in this case.

The integral expression for the Mellin transform of $\exp(-\alpha x^2)\, J_v(\beta x)$ is given by

$$Q(s) = \mathcal{M}\left[\exp\left(-\alpha x^2\right) J_\nu(\beta x)\right] = \int_0^\infty \frac{dx}{x}\, x^s \exp\left(-\alpha x^2\right) J_\nu(\beta x), \tag{9.27}$$

where b is complex, and a is real and greater than zero. This integral is of interest not only because its solution demonstrates the techniques described in this chapter, but also because it occurs in the log-amplitude and phase variance expressions for propagation in the presence of Kolmogorov turbulence through complex cascaded optical systems, referred to as ABCD systems (*Yura* and *Hanson* 1989), and in the evaluation of the scintillation in the next chapter. In addition, one can use this transform as an intermediate step in finding the Mellin transform of $\exp(-\alpha x^2)\, J_v^2(\beta x)$, which is encountered in beam-wave tilt analysis.

I use the extension to the Mellin convolution integral discussed in Sect. 9.2, with the expression for the Mellin-Barnes representation of the Bessel function given in (9.4), to evaluate the integral given above. Note that Mellin transforms of Bessel functions with complex arguments do not converge although they do satisfy condition (c) of the theorem in Sect. 9.2. The exponential function that appears in the integrand satisfies condition (b) of the theorem and balances the exponential growth of the Bessel function (when β is complex) with exponential decay.

As discussed in Sect. 9.1, the Mellin transform of the Bessel function to be inserted into the convolution integral is given by

$$\mathcal{M}\left[J_\nu(\beta x)\right] = \frac{1}{2}\Gamma\left[\begin{matrix} s/2+\nu/2 \\ -s/2+\nu/2+1 \end{matrix}\right]\left(\frac{\beta}{2}\right)^{-s} . \tag{9.28}$$

With these results the Mellin transform in (1.45), and the substitution $u \to 2u$, (9.27) is equal to

$$Q(s) = \left(\frac{\alpha^{-s/2}}{2}\right)\frac{1}{2\pi i}\int_{\eta-i\infty}^{\eta+i\infty} du\, \Gamma\left[\begin{matrix} -u+\frac{s}{2},\ u+\frac{\nu}{2} \\ -u+\frac{\nu}{2}+1 \end{matrix}\right]\left(\frac{\beta^2}{4\alpha}\right)^{-u} , \tag{9.29}$$

where $\mathrm{Re}\,\{-\nu/2\} < \eta < \mathrm{Re}\,\{-s/2\}$. In this integral $\Delta = 1$; therefore, in order to obtain a converging series solution for the integral, the integration path is closed to the left, and one obtains the pole contributions at $u = -n - \nu/2$ for $n = 0, 1, 2, \ldots$. Pole-residue techniques applied to the integral give

$$\exp\left(-\alpha x^2\right) J_\nu(\beta x) \to \frac{1}{2\alpha^{s/2}}\left(\frac{\beta^2}{4\alpha}\right)^{\nu/2}\sum_{n=0}^{\infty}\frac{(-1)^n}{n!}\Gamma\left[\begin{matrix} s/2+\nu/2+n \\ \nu+1+n \end{matrix}\right]\left(\frac{\beta^2}{4\alpha}\right)^n$$

$$= \frac{1}{2\alpha^{s/2}}\left(\frac{\beta^2}{4\alpha}\right)^{\nu/2}\Gamma\left[\begin{matrix} \frac{\nu+s}{2} \\ \nu+1 \end{matrix}\right]{}_1F_1\left[\frac{\nu}{2}+\frac{s}{2};\nu+1;-\frac{\beta^2}{4\alpha}\right], \tag{9.30}$$

for $\mathrm{Re}\,\{\nu+s\} > 0$, and $\alpha > 0$.

The natural parameter that results from this analysis is $\beta^2/4\alpha$. When the magnitude of this parameter is small, the series given in (9.30) converges quickly, thereby only a few terms are needed to accurately approximate the transform. If the parameter magnitude is large, and an approximation for the integral is needed, the asymptotic solution for (9.29) is more appropriate. The integral is closed to the right in order to pick up the pole-residue contributions. One obtains path-integral contributions to the asymptotic form by evaluating the integral with the steepest-descent methods discussed in Sect. 9.2.

The poles which contribute to the asymptotic form for the Mellin transform occur at the points $u = n + s/2$ for $n = 0, 1, 2, \ldots$. This leads to a pole-residue contribution given by

$$W = \frac{1}{2}\left(\frac{2}{\beta}\right)^s\sum_{n=0}^{n_o}\frac{(-1)^n}{n!}\Gamma\left[\begin{matrix} n+s/2+\nu/2 \\ -n-s/2+\nu/2+1 \end{matrix}\right]\left(\frac{\beta^2}{4\alpha}\right)^{-n} . \tag{9.31}$$

n_o is selected to minimize the error in the asymptotic approximation as discussed in Sect. 5.2.

Added to W are the steepest-descent components of the integral. Equations (9.25) and (9.26) are evaluated, and it is found that

$$E_1 = \frac{\alpha^{1-s}}{2}\left(-\frac{\beta^2}{4}\right)^{s/2-1} \exp\left(-\frac{\beta^2}{4\alpha}+\frac{i\pi\nu}{2}\right)$$

$$\text{for } 0 < \arg\left\{\beta^2/4\alpha\right\} < 3\pi/2, \text{ and} \tag{9.32}$$

$$E_1 = \frac{\alpha^{1-s}}{2}\left(-\frac{\beta^2}{4}\right)^{s/2-1} \exp\left(-\frac{\beta^2}{4\alpha}-\frac{i\pi\nu}{2}\right)$$

$$\text{for } -3\pi/2 < \arg\left\{\beta^2/4\alpha\right\} < 0. \tag{9.33}$$

These terms are dominant only when Re $\{\beta^2/4\alpha\} < 0$ in which case E_1 or E_2 exhibit exponential growth. When E_1 or E_2 have sinusoidal behavior (when Re $\{\beta^2/4\alpha\} = 0$), both W and E components must be included in the asymptotic solution. Combining these results one obtains the following asymptotic form for the Mellin transform of $\exp(-\alpha x^2)\ J_v(\beta x)$:

$$Q(s) \sim E_1 \text{ when } \frac{\pi}{2} < \arg\left\{\beta^2/4\alpha\right\} < \frac{3\pi}{2}, \tag{9.34}$$

$$Q(s) \sim E_1 + W \text{ when } \arg\left\{\beta^2/4\alpha\right\} = \frac{\pi}{2}, \tag{9.35}$$

$$Q(s) \sim W \text{ when } -\frac{\pi}{2} < \arg\left\{\beta^2/4\alpha\right\} < \frac{\pi}{2}, \tag{9.36}$$

$$Q(s) \sim E_2 + W \text{ when } \arg\left\{\beta^2/4\alpha\right\} = -\frac{\pi}{2}, \text{ and} \tag{9.37}$$

$$Q(s) \sim E_2 \text{ when } -\frac{3\pi}{2} < \arg\left\{\beta^2/4\alpha\right\} < -\frac{\pi}{2}. \tag{9.38}$$

As a check on this result, examine the case when $s = 1$, for which results are available in the literature. Replace β by $i\delta$ (δ is still complex.) Note the identity that relates Bessel and modified Bessel functions,

$$I_\nu(z) = \exp\left(-\frac{i\pi\nu}{2}\right) J_\nu(iz), \quad \left[-\pi < \arg\{z\} \le \tfrac{\pi}{2}\right], \tag{9.39}$$

and the relation in *Gradshteyn* and *Ryzhik* (1980, Eq. 9.238 #2)

$$I_\nu(x) = \frac{2^{-\nu}}{\Gamma[\nu+1]} x^\nu \exp(-x)\ {}_1F_1\left[\frac{1}{2}+\nu; 1+2\nu; 2x\right]. \tag{9.40}$$

Substitute these results in (9.30) to give

$$\int_0^\infty \frac{dx}{x} x \exp\left(-\alpha x^2\right) I_\nu(\delta x) = \frac{1}{2}\sqrt{\frac{\pi}{\alpha}} \exp\left(\frac{\delta^2}{8\alpha}\right) I_{\nu/2}\left(\frac{\delta^2}{8\alpha}\right) \tag{9.41}$$

for $\mathrm{Re}\{\nu\} > -1$, and $\alpha > 0$. This agrees with the result given in *Gradshteyn* and *Ryzhik* (1980, Eq. 6.618 #4).

If δ is real, *Gradshteyn* and *Ryzhik* (1980, Eq. 8.451 #5) gives the asymptotic form for the modified Bessel function

$$I_\nu(z) \sim \frac{\exp(z)}{\sqrt{2\pi z}}. \tag{9.42}$$

This leads to the asymptotic form for the integral given by

$$\begin{aligned} Q(1) &= \exp\left(\frac{i\pi\nu}{2}\right) \int_0^\infty \frac{dx}{x}\, x \exp\left(-\alpha x^2\right) I_\nu(\delta x) \\ &\sim \frac{1}{\delta} \exp\left(\frac{\delta^2}{4\alpha}\right) \exp\left(\frac{i\pi\nu}{2}\right), \quad \left|\frac{\delta^2}{\alpha}\right| >> 1. \end{aligned} \tag{9.43}$$

With these methods I can obtain asymptotic results for arbitrary values of s. For the present case $\arg\{\beta^2/4\alpha\} = \pi$, and the steepest-descent contribution dominates. One finds this by inserting (9.32) into (9.34) to obtain

$$Q(s) \sim \frac{\alpha^{1-s}}{2} \left(-\frac{\beta^2}{4}\right)^{s/2-1} \exp\left(-\frac{\beta^2}{4\alpha}\right) \exp\left(\frac{i\pi\nu}{2}\right), \quad \left|\frac{\beta^2}{\alpha}\right| >> 1. \tag{9.44}$$

For $s = 1$ this result is identical to (9.43). The case of arbitrary s is not generally available.

10. Examples with a Single Complex Parameter

In Chapter 2 I introduced a method for easily expressing in integral form the second moment of phase and log-amplitude related quantities for infinite collimated and focused waves propagating through turbulence. It was shown how these integrals could be evaluated using Mellin transform techniques. I extend these methods here to solve for the propagation of beam waves with Gaussian shape. Earlier results are a special case of this more general case. To express results as integrals, the approach in Chap. 2 is followed to the point where the propagation parameter γ was assumed to be real. Here γ must be complex to handle beams of finite extent. I derive the filter function for two counterpropagating or copropagating waves.

The filter function for the general case has a complex propagation parameter, which results in functions with exponential growth in the integrand. The Mellin convolution integral was generalized to this contingency in the last chapter by applying Parseval's equality. The Taylor series obtained using pole-residue integration are not more difficult to derive in this case. For large parameters I derive asymptotic expressions that have a more complicated dependence than for the infinite-wave case, because the functional form can change depending on the argument of the propagation constant in the complex plane. The method is applied to find the scintillation on a beam wave.

10.1 Phase and Log-Amplitude Variances of Beam Waves

The derivation of the second moment expressions for beam-wave problems is the same as in Chap. 2 up to (2.77). At that point the propagation parameter γ is allowed to be complex, which requires that the succeeding equations be modified. The relevant equations from Chap. 2 are repeated here as (10.1) to (10.9) for convenience. The phase and log-amplitude are given by

$$\begin{bmatrix} i\phi\left(\boldsymbol{\rho}, z\right) \\ \chi\left(\boldsymbol{\rho}, z\right) \end{bmatrix} = \tfrac{1}{2}\left[\Psi_1(\boldsymbol{\rho}, z) \mp \Psi_1^*(\boldsymbol{\rho}, z)\right], \tag{10.1}$$

where

$$\Psi_1(\boldsymbol{\rho}, L) = \int_0^L dz' \int d\nu\,(\boldsymbol{\kappa}, z')\, H(\boldsymbol{\kappa}, L, z') \exp\left(i\gamma\boldsymbol{\kappa}\cdot\boldsymbol{\rho}\right), \tag{10.2}$$

$$H(\boldsymbol{\kappa}, L, z) = i k_0 \exp\left[-iP\left(\gamma, \kappa, z\right)\right], \tag{10.3}$$

$k_0 = 2\pi/\lambda$, and λ is the wavelength in the propagating medium. The requirement that the refractive index is real gives

$$d\nu(\boldsymbol{\kappa}, z) = d\nu^*(-\boldsymbol{\kappa}, z). \tag{10.4}$$

For propagation from $z = 0$ to $z = L$ the diffraction parameter is given by

$$P\left(\gamma, \kappa, z\right) = \frac{\gamma\kappa^2(L - z)}{2k_0}, \tag{10.5}$$

and for propagation from $z = L$ to $z = 0$, the diffraction parameter is equal to

$$P\left(\gamma, \kappa, z\right) = \frac{\gamma\kappa^2 z}{2k_0}. \tag{10.6}$$

The value of γ is

$$\gamma = \frac{1 + i\alpha\, z}{1 + i\alpha\, L} \tag{10.7}$$

for a wave propagating from $z = 0$ to $z = L$, and

$$\gamma = \frac{1 + i\alpha\,(L - z)}{1 + i\alpha\, L} \tag{10.8}$$

for a wave propagating from $z = L$ to $z = 0$. The value of α is

$$\alpha = \alpha_r + i\,\alpha_i = \lambda/\pi\, W_o^2 + i/R_o. \tag{10.9}$$

The parameter R_o is the radius of curvature of the beam at the source. It is equal to infinity for a collimated beam. At the radius equal to W_o, the intensity has dropped to $1/e^2$ of its value on axis.

Change the sign of κ in (10.2), take the complex conjugate, use (10.4) and the evenness of $H\left(\boldsymbol{\kappa}, L, z\right)$ with respect to κ. Substitute the result and (10.2) into (10.1) to obtain

$$\begin{bmatrix} i\phi\left(\boldsymbol{\rho}, L\right) \\ \chi\left(\boldsymbol{\rho}, L\right) \end{bmatrix} = \int_0^L dz' \int \frac{d\nu\left(\boldsymbol{\kappa}, z'\right)}{2} \left[H\left(\boldsymbol{\kappa}, L, z'\right) \exp\left(i\gamma\boldsymbol{\kappa}\cdot\boldsymbol{\rho}\right) \mp H^*\left(\boldsymbol{\kappa}, L, z'\right) \exp\left(i\gamma^*\boldsymbol{\kappa}\cdot\boldsymbol{\rho}\right)\right]. \tag{10.10}$$

These expressions apply to a single point in space. Often the aperture average of the phase times a weighting function $g\left(\boldsymbol{\rho}\right)$ is required, as in the case of tilt. For the beam wave case, the aperture averaging function does not have to be centered on the beam axis but can be displaced by a distance $\boldsymbol{\rho}$. In that case phase and log-amplitude related quantities are

$$\begin{bmatrix} i\phi R(\boldsymbol{\rho}, L) \\ \chi R(\boldsymbol{\rho}, L) \end{bmatrix} = \frac{1}{2} \int d\boldsymbol{\rho}' \int_0^L dz' \int d\nu\,(\boldsymbol{\kappa}, z')\; g\,(\boldsymbol{\rho}' - \boldsymbol{\rho})$$
$$\times \left[H(\boldsymbol{\kappa}, L, z') \exp\left(i\gamma \boldsymbol{\kappa} \cdot \boldsymbol{\rho}'\right) \mp H^*(\boldsymbol{\kappa}, L, z') \exp\left(i\gamma^* \boldsymbol{\kappa} \cdot \boldsymbol{\rho}'\right)\right]. \quad (10.11)$$

Now interchange the integration order. The integral over a finite aperture converges and is recognized as the two-dimensional Fourier transform of the aperture function

$$G\,(\gamma\boldsymbol{\kappa}) = \int d\boldsymbol{\rho}'\, g\,(\boldsymbol{\rho}') \exp\left(i\gamma \boldsymbol{\kappa} \cdot \boldsymbol{\rho}'\right) \quad (10.12)$$

with anisoplanatic phase shift $\exp\left(i\gamma \boldsymbol{\kappa} \cdot \boldsymbol{\rho}\right)$. This gives

$$\begin{bmatrix} i\phi R\,(\boldsymbol{\rho}, L) \\ \chi R\,(\boldsymbol{\rho}, L) \end{bmatrix} = \frac{1}{2} \int_0^L dz' \int d\nu\,(\boldsymbol{\kappa}, z')$$
$$\times \left[H\,(\boldsymbol{\kappa}, L, z')\, G(\gamma\boldsymbol{\kappa}) \exp\left(i\gamma \boldsymbol{\kappa} \cdot \boldsymbol{\rho}\right) \mp H^*\,(\boldsymbol{\kappa}, L, z')\, G\,(\gamma^*\boldsymbol{\kappa}) \exp\left(i\gamma^* \boldsymbol{\kappa} \cdot \boldsymbol{\rho}\right)\right]. \quad (10.13)$$

Up to now only a single wave has been considered. Let us generalize this to the case of finding the difference between the phase and log-amplitude of two waves that can have different propagation constants. This will allow us to consider the difference between displaced counterpropagating and copropagating waves that are focused or collimated. Let the second wave be multiplied by a quantity $A\,(\boldsymbol{\kappa}, z)$ that can be a function of the axial coordinate and transverse wavenumber. If this multiplier is zero, one is back to the single-wave case, and if it is a complex exponential, it corresponds to anisoplanatism. Finally, let the aperture weighting factor be different for each wave. The result is

$$\begin{bmatrix} i\phi R\,(\boldsymbol{\rho}, L) \\ \chi R\,(\boldsymbol{\rho}, L) \end{bmatrix} = \frac{1}{2} \int_0^L dz' \int d\nu\,(\boldsymbol{\kappa}, z')$$
$$\times \left\{\left[H_1\,(\boldsymbol{\kappa}, L, z')\, G_1\,(\gamma_1 \boldsymbol{\kappa}) \exp\,(a_1) \mp H_1^*\,(\boldsymbol{\kappa}, L, z')\, G_1\,(\gamma_1^* \boldsymbol{\kappa}) \exp\,(a_1^*)\right]\right.$$
$$\left. -A\,(\boldsymbol{\kappa}, z')\left[H_2(\boldsymbol{\kappa}, L, z')\, G_2(\gamma_2 \boldsymbol{\kappa}) \exp\,(a_2) \mp H_2^*(\boldsymbol{\kappa}, L, z')\, G_2(\gamma_2^* \boldsymbol{\kappa}) \exp\,(a_2^*)\right]\right\}, \quad (10.14)$$

where $a_n = i\gamma_n \boldsymbol{\kappa} \cdot \boldsymbol{\rho}$. The second moment of this quantity is found by multiplying it by its complex conjugate and taking the expected value. The same manipulations as between (2.83) and (2.85) are applied to eliminate integrations over κ and the axial coordinate, to obtain

$$\begin{bmatrix}\sigma^2_{\phi R}(\boldsymbol{\rho})\\ \sigma^2_{\chi R}(\boldsymbol{\rho})\end{bmatrix} = \frac{0.2073}{4}\int_0^L dz\, C_n^2(z)\int d\boldsymbol{\kappa}\, f(\kappa)$$

$$\times\{[H_1(\boldsymbol{\kappa},L,z)\,G_1(\gamma_1\boldsymbol{\kappa})\exp(a_1)\mp H_1^*(\boldsymbol{\kappa},L,z)\,G_1(\gamma_1^*\boldsymbol{\kappa})\exp(a_1^*)]$$

$$-A(\boldsymbol{\kappa},z)\,[H_2(\boldsymbol{\kappa},\ L,\ z)\,G_2(\gamma_2\boldsymbol{\kappa})\exp(a_2)\mp H_2^*(\boldsymbol{\kappa},L,z)\,G_2(\gamma_2^*\boldsymbol{\kappa})\exp(a_2^*)]\}$$

$$\times\{[H_1^*(\boldsymbol{\kappa},L,z)\,G_1^*(\gamma_1\boldsymbol{\kappa})\exp(-a_1^*)\mp H_1(\boldsymbol{\kappa},L,z)\,G_1^*(\gamma_1^*\boldsymbol{\kappa})\exp(-a_1)]$$

$$-A^*(\boldsymbol{\kappa},z)\,[H_2^*(\boldsymbol{\kappa},L,z)\,G_2^*(\gamma_2\boldsymbol{\kappa})\exp(-a_2^*)\mp H_2(\boldsymbol{\kappa},L,z)\,G_2^*(\gamma_2^*\boldsymbol{\kappa})\exp(-a_2)]\}. \tag{10.15}$$

This is the central formula for beam-wave problems and is the analog of (2.85). This expression does not simplify as nicely as its analog did for the special cases considered in Chap. 2. Consequently, only the expression for a single beam will be treated explicitly. If there is only one beam, $A(\boldsymbol{\kappa},z)$ is zero. The above expression, which has two exponentials with opposite signs, can be simplified by using the evenness of $H(\boldsymbol{\kappa},L,z)$ and $f(\kappa)$ with respect to κ and the relation $G(\gamma\boldsymbol{\kappa}) = G^*(-\gamma^*\boldsymbol{\kappa})$, which comes from the Fourier-transform definition in (10.12). If in addition $G(\gamma\boldsymbol{\kappa})$ is even in κ, which it is for all filter functions considered in this book, then Eq. (10.15) reduces to

$$\begin{bmatrix}\sigma^2_{\phi R}(\boldsymbol{\rho})\\ \sigma^2_{\chi R}(\boldsymbol{\rho})\end{bmatrix} = \frac{0.2073}{2}\int_0^L dz\, C_n^2(z)\int d\boldsymbol{\kappa}\, f(\kappa)$$

$$\times\Big[|H(\boldsymbol{\kappa},L,z)\,G(\gamma\boldsymbol{\kappa})|^2\exp(-2\gamma_i\boldsymbol{\kappa}\cdot\boldsymbol{\rho})\mp \mathrm{Re}\Big\{[H(\boldsymbol{\kappa},L,z)]^2 G(\gamma\boldsymbol{\kappa})\,G(-\gamma\boldsymbol{\kappa})\Big\}\Big]. \tag{10.16}$$

For the Zernike modes defined in (3.6) to (3.9), $G(-\gamma\boldsymbol{\kappa}) = (-1)^m G(\gamma\boldsymbol{\kappa})$, where m is the order of the azimuthal mode.

From (10.3) and (10.5) for a wave going from $z = 0$ to L, one obtains

$$|H(\boldsymbol{\kappa},L,z)|^2 = k_0^2\exp\left[\gamma_i\kappa^2(L-z)/k_0\right],\ \text{and} \tag{10.17}$$

$$\mathrm{Re}\left\{H(\boldsymbol{\kappa},L,z)^2\right\} = \mathrm{Re}\left\{k_0^2\exp\left[-i\gamma\kappa^2(L-z)/k_0\right]\right\}, \tag{10.18}$$

where γ is expressed in real and imaginary parts as

$$\gamma = \gamma_r + i\gamma_i. \tag{10.19}$$

Insert these relations into (10.16) to obtain

$$\begin{bmatrix} \sigma^2_{\phi R}(\boldsymbol{\rho}) \\ \sigma^2_{\chi R}(\boldsymbol{\rho}) \end{bmatrix} = \frac{0.2073\, k_0^2}{2} \int_0^L dz\, C_n^2(z) \int d\boldsymbol{\kappa}\, f(\kappa)$$
$$\times \left[F(\gamma\boldsymbol{\kappa}) \exp(\gamma_i b) \exp(-2\gamma_i \boldsymbol{\kappa}\cdot\boldsymbol{\rho}) \pm (-1)^m \operatorname{Re}\left\{ \exp(-i\gamma_i b)\, G^2(\gamma\boldsymbol{\kappa}) \right\} \right], \tag{10.20}$$

where $b = \kappa^2(L-z)/k_0$, and $F(\gamma\boldsymbol{\kappa}) = G(\gamma\boldsymbol{\kappa})\, G^*(\gamma\boldsymbol{\kappa})$. For γ real this reduces to the result in (2.90).

The real expression in the last term can be expressed as the exponential of the real part times the cosine of the imaginary part of the argument. This form is closer to the form in (2.85) when γ is real. Although it is tempting to do this, I will not, since it results in an integral with an additional paramc ter, thereby complicating evaluation. If $G(\gamma\boldsymbol{\kappa})$ does not depend on angle, the angular integration can be performed readily to give

$$\begin{bmatrix} \sigma^2_{\phi R}(\boldsymbol{\rho}) \\ \sigma^2_{\chi R}(\boldsymbol{\rho}) \end{bmatrix} = \frac{1.303\, k_0^2}{2} \int_0^L dz\, C_n^2(z) \int_0^\infty d\kappa\, \kappa\, f(\kappa)$$
$$\times \left[F(\gamma\kappa) \exp(\gamma_i)\, J_0(2i\gamma_i \kappa\rho) \pm (-1)^m \operatorname{Re}\left\{ \exp(-i\gamma)\, G^2(\gamma\boldsymbol{\kappa}) \right\} \right]. \tag{10.21}$$

In the general case one finds the equivalent of the correlation function given in (2.94) by applying an anisoplanatic shift to the second terms in braces in (10.15). In a similar fashion a formula can be derived for the structure function.

10.2 Power Spectral Density of Beam Waves

The derivation of power spectral density follows the same procedure that was used in Chap. 2. The arguments that lead up to (2.101) also apply here, to obtain for the power spectral density of a single wave,

$$\begin{bmatrix} S_\phi(\omega) \\ S_\chi(\omega) \end{bmatrix} = 1.303\, k_0^2\, \omega \int_0^L dz\, \frac{C_n^2(z)}{v^2(z)} \int_0^\infty dc\, \frac{c\, U(c-1)}{\sqrt{c^2-1}}\, f\left[\frac{\omega\, c}{v(z)}\right] F_f(\omega)$$
$$\times \left(\exp\left[\frac{\gamma_i \omega^2 c^2 (L-z)}{v^2(z) k_0}\right] F\left[\frac{\gamma\,\omega\, c}{v(z)}\right] J_0\left[\frac{2i\gamma_i \omega c \rho}{v(z)}\right] \right.$$
$$\left. \pm (-1)^m \operatorname{Re}\left\{ \exp\left[-i\frac{\gamma\omega^2 c^2 (z-L)}{v^2(z) k_0}\right] G^2\left[\frac{\gamma\,\omega\, c}{v(z)}\right] \right\} \right), \tag{10.22}$$

where $F_f(\omega)$ is the response of a servo system.

This reduces to the earlier results for collimated and spherical waves when the appropriate values of gamma are inserted. I shall not consider further the power spectral density of a beam wave.

10.3 Scintillation on Beam Waves

An adaptive-optics system requires a reference wave that propagates along the same path that is to be imaged or along which a laser beam is to be propagated. The reference wave provides information used to deform a mirror to correct for turbulence. It can be generated by an active source on the target or by a laser signal reflected from a retroreflector on the target. A minimum return energy is necessary to drive the wavefront sensor that measures the integrated turbulence distribution. For good performance the signal must also not drop out too frequently due to scintillation fades. To properly design a beacon transmitter one needs to know the intensity and scintillation for various beacon-transmitter diameters and wavelengths. Knowledge of the variation of average intensity and scintillation as one moves off the optical axis is necessary to specify the accuracy of the reference-wave tracking system. This set of problems is analyzed for collimated beams in this section.

The analysis of this problem is complicated since different effects occur as the transmitter diameter varies. For very small diameters, scintillation approaches that of a point source, and for very large diameters, the scintillation approaches that of an infinite plane wave. Scintillation of the plane wave is several orders of magnitude larger than that of a point source for typical target ranges. In the intermediate range the behavior is complicated by two factors. As the diameter gets larger the beamwidth gets smaller than the turbulence-induced-tilt jitter. In this regime there is a rapid increase in scintillation with increasing beam diameter. For smaller apertures one finds that scintillation can decrease with increasing diameter. For intermediate size apertures, the scintillation of a beam wave on axis can be less than that of a point source.

Scintillation also varies with the distance off axis. It will be shown that the scintillation can grow rapidly as one moves away from beam boresight.

This problem has been partially analyzed by others. *Ishimaru* (1969, 1978a, 1978b) develops an expression for scintillation of a beam wave. The formula applies to the general case in which turbulence is allowed to vary along the propagation path; however, he evaluates the expression only for constant turbulence strength along the beam path. *Kinoshita* et al. (1968) also finds the scintillation for constant turbulence, but with a Gaussian correlation function. The scintillation for propagation with constant turbulence strength was found by *Fried* and *Siedman* (1967), and for ground-to space-propagation where the turbulence is not constant by *Fried* (1967). The Kolmogorov turbulence spectrum is used in these calculations. In *Fried* (1967), the approximation is made that the target is well into the far field of the beam, which implies that the diameter cannot be large, yet the analysis is carried out to large beams, for which it is found that the scintillation decreases as the $-7/3$ power of the Gaussian beam size. It is not clear to what range of diameters this solution applies.

To obtain an expression for scintillation of a wave that propagates from ground to space, use (10.21) with the aperture filter function set equal to unity to obtain

$$\sigma_\chi^2(\boldsymbol{\rho}) = 0.2073\,\pi\,k_0^2 \int_0^L dz\, C_n^2(z) \int_0^\infty d\kappa\,\kappa\, f(\kappa)$$
$$\times \left(J_0\left(2i\gamma_i\,\kappa\,\rho\right) \exp\left[\frac{\gamma_i\,\kappa^2(L-z)}{k_0}\right] - \operatorname{Re}\left\{ \exp\left[i\frac{\gamma\,\kappa^2(L-z)}{k_0}\right]\right\}\right). \tag{10.23}$$

The propagation parameter is given in terms of its real and imaginary parts as

$$\gamma_r = \frac{1+\alpha^2 z\,L}{1+(\alpha\,L)^2} = \frac{F_N + F_N^{-1}\,z/L}{F_N + F_N^{-1}}, \quad \text{and} \tag{10.24}$$

$$\gamma_i = -\frac{\alpha(L-z)}{1+(\alpha\,L)^2} = -\frac{1-z/L}{F_N+F_N^{-1}}, \tag{10.25}$$

and the Fresnel number is defined to be

$$F_N = 1/\alpha\,L = \pi\,W_o^2/\lambda\,L. \tag{10.26}$$

For a collimated beam $\alpha = \lambda/\pi W_o^2$. The inner scale of turbulence with the Tatarski spectrum can be included without an increase in complexity of the expression. The spectrum given in (2.32) is used with infinite outer scale. After the use of these relations, the log-amplitude variance is equal to

$$\sigma_\chi^2(\rho) = 0.2073\,\pi\,k_0^2 \int_0^L dz\, C_n^2(z) \int_0^\infty \frac{d\kappa}{\kappa}\,\kappa^{-5/3}$$
$$\times \left(J_0\left(2i\gamma_i\,\kappa\,\rho\right) \exp\left[\frac{\gamma_i\,\kappa^2(L-z)}{k_0} - \frac{\kappa^2}{\kappa_i^2}\right] - \operatorname{Re}\left\{ \exp\left[-i\frac{\gamma\,\kappa^2(L-z)}{k_0} - \frac{\kappa^2}{\kappa_i^2}\right]\right\}\right). \tag{10.27}$$

The Bessel function can be left with an imaginary argument because the exponential that multiplies it causes the integrand to go to zero at infinity. As shown in the previous chapter, one can obtain the Mellin transform of this function. Make the changes of variables $A^2 = -\gamma_i(L-z)/k_0 + 1/\kappa_i^2$, $B^2 = i\gamma(L-z)/k_0 + 1/\kappa_i^2$, and $\kappa A = x$ in the first integral, and $\kappa B = x$ in the second. and $y = A/2i\gamma_i\,\rho$. The integral is then equal to

$$\sigma_\chi^2(\rho) = 0.2073\,\pi\,k_0^2 \int_0^L dz\, C_n^2(z)$$
$$\times \left(A^{5/3} \int_0^\infty \frac{dx}{x}\, x^{-5/3} \exp\left(-x^2\right) J_0\left(x/y\right) - \operatorname{Re}\left\{ B^{5/3} \int_0^\infty \frac{dx}{x}\, x^{-5/3} \exp\left(-x^2\right)\right\}\right)$$
$$= 0.1037\,\pi\,k_0^2\,\Gamma\left[-\tfrac{5}{6}\right] \int_0^L dz\, C_n^2(z)\,\left(I_1 - I_2\right). \tag{10.28}$$

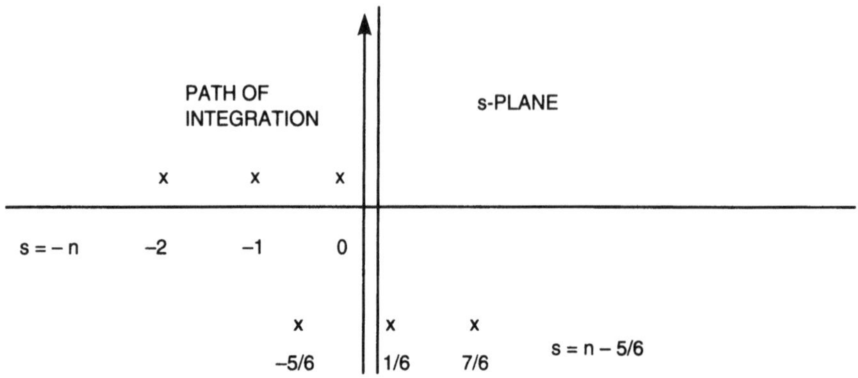

Fig. 10.1. Pole locations and integration path for the beam-wave scintillation problem.

Each integral diverges due to a simple pole at the origin. The residues at these two poles cancel, so the region of validity of the integral can be analytically continued, which enables these integrals to be evaluated with Mellin transforms.

The second integral in braces, with the pole contribution neglected, is equal to

$$I_2^* = \frac{2}{\Gamma\left[-\frac{5}{6}\right]} \mathrm{Re}\left\{\frac{B^{5/3}}{2}\,\Gamma[s/2]|_{s=-5/3}\right\} = \mathrm{Re}\left\{B^{5/3}\right\}. \tag{10.29}$$

The Mellin convolution theorem and the substitution $s \to -2s$ in the first integral gives

$$I_1 = \frac{A^{5/3}}{\Gamma\left[-\frac{5}{6}\right]} \frac{1}{2\pi i} \int\limits_C ds \left(-\frac{\gamma_i^2 \rho^2}{A^2}\right)^{-s} \Gamma\left[\begin{matrix} -s-\frac{5}{6}, s \\ -s+1 \end{matrix}\right]. \tag{10.30}$$

The pole locations and integration path are shown in Fig. 10.1.

Since $\Delta = 1$, close the integration path to the left, and obtain pole contributions at $s = -n$ for $n = 0, 1, 2, \ldots$. As explained above, the residue at $s = -5/6$ is cancelled by an opposite contribution from the second integral so this pole contribution can be discarded. The integral is thus effectively equal to

$$I_1^* = \frac{A^{5/3}}{\Gamma\left[-\frac{5}{6}\right]} \sum_{n=0}^{\infty} \frac{1}{n!} \left(\frac{\gamma_i^2 \rho^2}{A^2}\right)^n \Gamma\left[\begin{matrix} n-\frac{5}{6} \\ n+1 \end{matrix}\right] = A^{5/3}{}_1F_1\left[-\tfrac{5}{6}; 1; \frac{\gamma_i^2 \rho^2}{A^2}\right]. \tag{10.31}$$

When the parameter is large, an asymptotic series is appropriate. The steepest-descent contribution given in (5.94) grows exponentially; therefore, the pole contributions are negligible, and the asymptotic series is

$$I_1 \sim \frac{A^{5/3}}{\Gamma\left[-\frac{5}{6}\right]} \left(\frac{A^2}{\gamma_i^2 \rho^2}\right)^{11/6} \exp\left(\frac{\gamma_i^2 \rho^2}{A^2}\right). \tag{10.32}$$

Using these results, one obtains for the Taylor series solution

$$\sigma_\chi^2(\rho) = \frac{2.175\, k_0^{7/6} L^{5/6}}{\left(F_N + F_N^{-1}\right)^{5/6}} \int_0^L dz\, C_n^2(z)\, (1 - z/L)^{5/3}$$

$$\times \mathrm{Re}\left\{ \left[i\frac{F_N + F_N^{-1} z/L}{(1 - z/L)} + 1 + \zeta \right]^{5/6} - (1+\zeta)^{5/6} {}_1F_1\left[-\tfrac{5}{6}; 1; \frac{k_0\, \rho^2/L}{\left(F_N + F_N^{-1}\right)(1+\zeta)} \right] \right\} \tag{10.33}$$

where

$$\zeta = \frac{k_0 \left(F_N + F_N^{-1}\right)}{L \kappa_i^2 \left(1 - z/L\right)^2}. \tag{10.34}$$

This expression is valid out to radii for which the paraxial assumption holds, which requires that $\rho^4 << L^3 \lambda$ as discussed in Sect. 2.3.

If the turbulence is constant along the propagation path, this result reduces to that obtained by Ishimaru. This also reduces to the standard results for a collimated or focused infinite wave. For a collimated infinite wave the Fresnel number is infinite, and the scintillation variance with zero inner-scale is

$$\sigma_\chi^2(\rho) = \lim_{F_N \to \infty} \frac{2.175\, k_0^{7/6}}{\left(F_N + F_N^{-1}\right)^{5/6}} \mathrm{Re}\left\{ \int_0^L dz\, C_n^2(z) \left[i\left(F_N + F_N^{-1} z/L\right) \right]^{5/6} (L - z)^{5/6} \right\}$$

$$= 2.175\, k_0^{7/6} \mathrm{Re}\left\{ i^{5/6} \right\} \int_0^L dz\, C_n^2(z)\, (L - z)^{5/6} = 0.5631\, k_0^{7/6} \int_0^L dz\, C_n^2(z)\, (L - z)^{5/6}, \tag{10.35}$$

since $\mathrm{Re}\left\{ i^{5/6} \right\} = \cos(5\pi/12) = 0.2588$. For a focused infinite beam the Fresnel number goes to zero at the source and one obtains

$$\sigma_\chi^2(\rho) = \lim_{F_N \to 0} \frac{2.175\, k_0^{7/6}}{\left(F_N + F_N^{-1}\right)^{5/6}} \mathrm{Re}\left\{ \int_0^L dz\, C_n^2(z) \left[i\left(F_N + F_N^{-1} z/L\right) \right]^{5/6} (L - z)^{5/6} \right\}$$

$$= 0.5631\, k_0^{7/6} \int_0^L dz\ C_n^2(z) (L - z)^{5/6} \left(z/L\right)^{5/6}. \tag{10.36}$$

The limiting expressions in (10.35) and (10.36) are the same as in (4.42) and (4.46).

Fig. 10.2 shows the log-amplitude variance versus Fresnel number for various target distances, normalized to scintillation of a point source given by (10.36). The scintillation approaches that of a point source for small aperture sizes. As the aperture size increases, the scintillation decreases initially and then increases to approach the value for an infinite diameter source. The results plotted are correct only when $\sigma_\chi^2 < 0.5$ because above this value the Rytov approximation breaks down, and scintillation is less than that calculated above. For a system that operates at 0.5-μm wavelength with the Hufnagel-Valley 21 turbulence

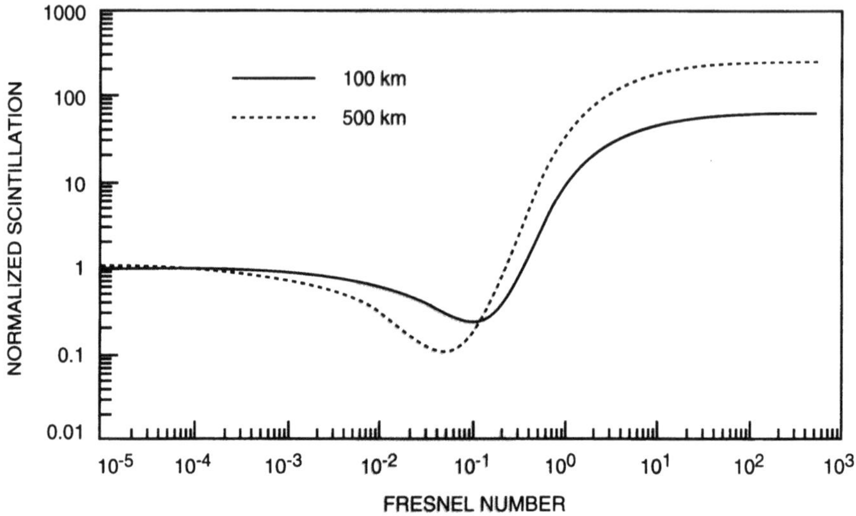

Fig. 10.2. Scintillation on axis versus Fresnel number for a target at 100 and 500 km range for HV-21 turbulence. The Fresnel number is defined in (10.26). The scintillation is normalized by that of a point source. Notice that it is possible to have less scintillation than from a point source.

model, the point source scintillation is 0.059; therefore, the results when the normalized scintillation is above 4 are too high.

Scintillation with inner-scale sizes of 10 mm and zero at a target distance of 500 km are plotted versus the Fresnel number in Fig. 10.3. For larger Fresnel numbers the two curves are identical. An inner-scale size of 1 mm, a value that is often quoted as typical, produces scintillation that is virtually identical to that with zero inner scale.

Scintillation as a function of distance off axis is considered for several specific cases that are close to the minimum of the scintillation curve in Fig. 10.2. Consider a target at 500 km, with a system that operates in the visible at 0.5 μm wavelength. The value of $\sqrt[4]{L^3\lambda}$ is 500 meters; therefore, the paraxial assumption is valid to about 50 meters off axis. For beam radii of 0.625, 1.25, and 2.5 cm, the beam waist at 500 km is 12, 6, and 3 meters, respectively. Fig. 10.4 shows the scintillation out to those radii for these three cases. Notice that even though scintillation on axis is smaller for a 2.5-cm beam, it increases more rapidly away from beam center than for beams that started off with smaller beam waists at the transmitter. These results can be used to do a tradeoff in a beacon illumination system between tracking accuracy versus laser power required to get an intensity above a certain level at a space target.

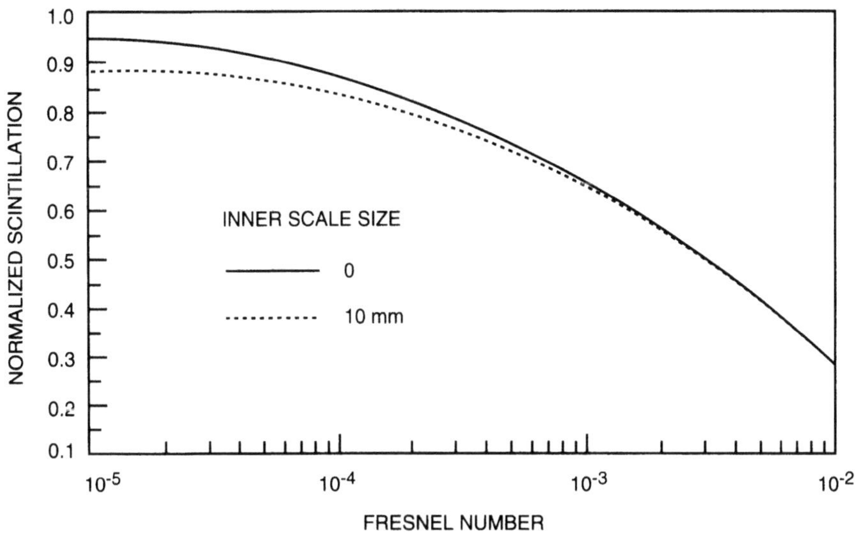

Fig. 10.3. Scintillation on axis versus Fresnel number for a target at 500 km with HV-21 turbulence and inner scale of zero and 10 mm.

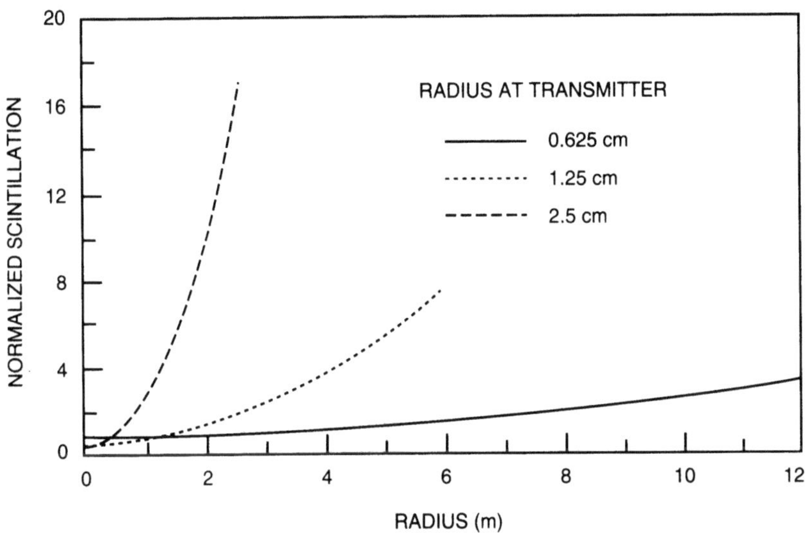

Fig. 10.4. Normalized scintillation versus distance from the beam center at a 500 km target for various transmitted beam sizes, for the HV-21 turbulence model. The increase of scintillation with distance off axis is a strong function of beam size.

11. Mellin Transforms in N Complex Planes

In Chaps. 4 and 7 Mellin transform techniques were used to solve problems of wave propagation through turbulence in which there was zero or one parameter. Here, that technique is generalized to allow one to solve problems with any number of parameters. Analytical solutions that are easily evaluated are obtained for problems that have previously been considered analytically intractable and for which only numerical results are available. The method to obtain the solution in integral form by the insertion of the transverse spatial filter functions into the standard formula for variances is exactly the same as that discussed in Chap. 2. This process produces a three-dimensional integral over the transverse spatial-transform coordinates and the propagation path. The integration over angle in transform space can usually be performed analytically. The integration over the magnitude of the spatial transform coordinate in which there are two or more parameters in the integrand is addressed in this chapter. I show that it can be evaluated to give a series solution. The remaining integration over the propagation path can be performed term by term in most cases. For some cases the infinite series terms after the axial integration are infinite, in which cases one must evaluate the integration along the propagation direction first.

The integration over the magnitude of the transform coordinate is the one that is difficult. For some problems more than one parameter remains in the integrand after a normalization of the variables. For N parameters, this integral can be transformed into an integral in N complex planes with the Mellin convolution integral extended to N functions given in (1.29) as

$$\begin{aligned} h\left(x_1, \ldots, x_N\right) &= \int\limits_0^\infty \frac{dy}{y} h_0(y) \prod_{j=1}^N h_j\left(x_j/y\right) \rightarrow M\left[h\left(x_1, \ldots, x_N\right)\right] \\ &= H_0\left(s_1 + s_2 + \cdots + s_N\right) \prod_{j=1}^N H_j\left(s_j\right) = H\left(s_1, s_2, \ldots, s_N\right). \end{aligned} \quad (11.1)$$

The original integral is equal to an integration in N complex planes given by

$$h\left(x_1, \ldots, x_N\right) = \frac{1}{(2\pi i)^N} \int\limits_{C_1} \cdots \int\limits_{C_N} ds_1 \cdots ds_N \, H\left(s_1, s_2, \ldots, s_N\right) x_1^{-s_1} \ldots x_N^{-s_N}. \quad (11.2)$$

For the problems that are considered here, the function $H\left(s_1, s_2, \ldots, s_N\right)$ is the product of ratios of gamma functions. The evaluation of integrals of this type is an area of mathematics that is not well developed. Pole-residue integration cannot easily be applied in problems; also, the steepest-descent contributions for the asymptotic series have been evaluated only for limited conditions. Because the complex variables appear either alone or in sums, however, *Shelton* and *I* (1993) have been able to extend pole-residue integration into this new area. I show how to evaluate these integrals and develop Taylor series and asymptotic solutions. The technique to select which power series apply for different parameter regimes is new.

If a function in the integrand of the original integral has the first term of its power series subtracted from it, the integration path moves over one pole as it did in the single-parameter case. This property eliminates difficulties one encounters in evaluating some integrals numerically. Deciding which pole residues contribute is more difficult than in the single-parameter case; however, procedures are given to make those decisions.

The solution for multiparameter problems is reduced to a technique that is straightforward, in principle, but may be algebraically lengthy. The types of problems it is possible to solve with this technique often are very difficult to solve with standard methods.

We begin by introducing the necessary background. First, the convergence properties of multiple infinite series are examined. Next, I show that Cauchy's residue theorem can be extended into N complex planes for integrands that contain the product of ratios of gamma functions. The value of the integral depends on the residues at points at which there is a confluence of N poles. After all the possible power series are written down, which corresponds to all ways of closing contours, those that are pertinent to the parameter range of interest are selected with a simple procedure that relies on the convergence properties of the series. The justification for this approach is given in the next section where it is shown that if a power series converged, then the integration path could have been closed to enclose the poles that lead to that power series.

This procedure is demonstrated on a "toy" problem in which the integrand is a product of three unit step functions. A second "toy" problem in which the integrand is the product of three exponentials is subsequently solved. In this example one clearly sees why the power series solution sometimes is valid in a parameter region that is smaller than the convergence region of the integral. This example is also used to illustrate how to evaluate integrals in which the integrand functions have terms of their power series subtracted from them.

It is straightforward to write down the solution as convergent power series. This is a complete solution to the problem. Sometimes these solutions converge slowly and require great accuracy for individual series terms; however, computer algebra programs in which one can specify the accuracy of calculation are suitable for this purpose. For faster convergence and to obtain insight into how the solution varies with the parameters, an asymptotic solution is appropriate. It is easy to find the asymptotic series. The technique to obtain the steepest-descent contribution is more complicated than that for a single complex plane (*Bleistein* and *Handelman* 1986; *Dingle* 1973). Methods to find the asymptotic series and saddle-point contributions are developed here. Restricting the class of integrals to Mellin-Barnes integrals with certain restrictions on functions allows one to obtain readily evaluated expressions for the saddle-point contribution to the asymptotic form of the integral. Results on saddle-point contributions available in the literature are not directly applicable here because the integrals are generally not in the standard form assumed in traditional saddle-point treatments. In addition, in problems considered here, saddle-point contributions are not only necessary when all parameters in the original integral are large, but may also be required when one or more of the parameters are small. The reason behind this phenomenon is discussed in Sect. 11.4.

The form of the solutions, which is composed of the sum of multiple infinite summations, looks complicated. However, because typically few terms are necessary to get an answer that is accurate to 1%, these solutions are easy to evaluate. The examples in this book were evaluated on a personal computer with short, easily-written Basic programs.

11.1 Convergence of Multi-Parameter Series

The convergence properties of power series obtained by pole-residue integration are used in Sect. 11.3 in criteria to decide whether to include a particular series in the solution. For that reason these properties are important. The conditions for convergence of multiple series are more complex than those for a single series. First, consider the double series

$$F(x, y)=\sum_{n=0}^{\infty} \sum_{m=0}^{\infty} A_{m,n} x^{m} y^{n}. \tag{11.3}$$

The series coefficients, as found by pole-residue techniques, are of the form

$$A_{m,n}=\prod_{i} \Gamma\left[a_{i}+u_{i} m+v_{i} n\right] / \Gamma\left[a_{i}\right], \tag{11.4}$$

where the coefficients u_i and v_i are typically either 1, -1, or 0. One achieves this form from the normal form by moving gamma functions in the denominator into the numerator with the identity $1/\Gamma[1-s]=\Gamma[s]\sin(\pi s)/\pi$. Some solutions can have the sign of $A_{m,n}$ alternate in one or both indices. If the series converges absolutely, this alternating series also converges.

The form of the coefficients in (11.4) is a special case of the general function with arbitrary integer coefficients studied by *Horn* (1889). This function can be generalized to N variables, in which case the coefficients of the power series are of the form

$$A_{n_{1}, \ldots, n_{N}}=\prod_{i} \Gamma\left[a_{i}+c_{i 1} n_{1}+\cdots+c_{i N} n_{N}\right] / \Gamma\left[a_{i}\right]. \tag{11.5}$$

This is also the form of the coefficients of the residues for the typical N-parameter case. Horn derived general convergence criteria for multiple series, which are discussed for double series by *Erdélyi* (1981, Page 227) and for multiple series in *Exton* (1978, Page 30). I give here the criteria for a double series, followed by the general criteria.

The criteria for convergence give two regions of parameter space, and the region of convergence is the space within the intersection of the two regions. The first region comes from the requirement that the series must converge when one of the indices is zero. The second region results from a rearrangement of the series terms so that one obtains an infinite sum of a finite sum of terms for which the sum of the indices is the same. The two series that result are

$$F(x,y) = \sum_{n=0}^{\infty}\sum_{p=0}^{n} A_{n-p,p}x^{n-p}y^{p} = \sum_{m=0}^{\infty}\sum_{p=0}^{m} A_{p,m-p}x^{p}y^{m-p}. \tag{11.6}$$

The ratio test, which requires each successive term to decrease in the limit as the index goes to infinity, is applied to each series. In the series in (11.6) it is applied to each term of the second series. To express the result in a convenient form define the functions

$$\frac{A_{m+1,n}}{A_{m,n}} = P_1(m,n), \quad \text{and} \quad \frac{A_{m,n+1}}{A_{m,n}} = P_2(m,n), \tag{11.7}$$

where it is required that these be algebraic functions, a condition that is satisfied if the coefficients of the indices (m and n) in the gamma functions are integers. Also, define

$$\Phi_1(m,n) = \lim_{\varepsilon\to\infty} P_1(\varepsilon m, \varepsilon n), \quad \text{and} \quad \eta = |y| = 1/\left|\Phi_2(0,1)\right|. \tag{11.8}$$

The first region in the $x-y$ plane is a rectangle in the quadrant in which both variables are positive. Two sides of the rectangle are formed from the x and y axes and the origin. The other two sides are the two lines

$$\xi = |x| = 1/\left|\Phi_1(1,0)\right|, \quad \text{and} \quad \eta = |y| = 1/\left|\Phi_2(0,1)\right|. \tag{11.9}$$

The second region is a surface that includes the origin and the x and y axes and is bounded by the surface that is defined by the parametric equations

$$\xi = |x| = 1/\Phi_1(m,n), \quad \text{and} \quad \eta = |y| = 1/\Phi_2(m,n). \tag{11.10}$$

This second surface may be identical to the rectangle defined above or may be a subregion of the rectangle.

For problems solved by Mellin transform techniques, the region defined by (11.9) is a unit square. The parametric equations are always of the form

$$\xi = |x| = \frac{(m+n)^k}{m^k}, \quad \text{and} \quad \eta = |y| = \frac{(m+n)^k}{n^k}, \tag{11.11}$$

where k is an integer that can be positive or negative. The equation of the bounding surface is

$$|x|^{1/k} + |y|^{1/k} = 1. \tag{11.12}$$

For positive k this region lies within the unit square; therefore, it determines the region of convergence. For parameter values that lie between this curve and the unit square no analytical solutions can be found with this method.

For negative k the unit square lies within this region; therefore, the series converges if the magnitude of both parameters is less than unity. Solutions for all parameter values except possibly those on the boundary of the unit square are obtained in this case.

To summarize the convergence criteria for two variables, define $\Omega_2(\Omega_1)$ as the sum of all coefficients of $n(m)$ in the numerator gamma functions minus the sum of those in the denominator. A factorial function is equivalent to a gamma function in this definition. I refer to $\Omega_2(\Omega_1)$ as the "net factorial power" of $n(m)$. From the criteria above, the following statements are evident. If $\Omega_2(\Omega_1)$ is negative the series converges for all values of $x(y)$. If $\Omega_2(\Omega_1)$ is negative, and $\Omega_1(\Omega_2)$ is zero, then the series converges for all values of $x(y)$ and for $y(x)$ less than unity. All these series are Taylor series. If either Ω_2 or Ω_1 is positive, then the series does not converge for any finite values of x or y. These series are asymptotic series. The only complicated case occurs when $\Omega_1 = \Omega_2 = 0$. In this case the convergence criteria vary from problem to problem. Examples of this case are considered in the next section and in Chap. 12.

In a similar manner as in the single complex plane case, one can define quantities Δ_{s_1} and Δ_{s_2} that are the sum of coefficients of s_1 and s_2 in the numerator gamma functions minus the sum of coefficients of the denominator gamma functions in an integral. The condition $\Delta_{s_1} = \Delta_{s_2} = 0$ corresponds to the case $\Omega_1 = \Omega_2 = 0$.

For the general case, a summation with p variables is

$$F(x_1, \ldots, x_p) = \sum_{n_1=0}^{\infty} \cdots \sum_{n_p=0}^{\infty} A_{n_1,\ldots,n_p} x_1^{n_1} \cdots x_p^{n_p}. \tag{11.13}$$

Define

$$\frac{A_{n_1,\ldots,n_{q-1},(n_q+1),n_{q+1},\ldots,n_p}}{A_{n_1,\ldots,n_q,\ldots,n_p}} = P_q(n_1,\ldots,n_p), \text{ and} \tag{11.14}$$

$$\Phi_q(n_1,\ldots,n_q,\ldots,n_p) = \lim_{\varepsilon\to\infty} P_q(\varepsilon n_1,\ldots,\varepsilon n_p), \quad q = 1, 2, \ldots, p. \tag{11.15}$$

The space within the intersection of two regions is again the region of convergence. The first region is a hyper-rectangle formed by the origin, the positive part of the coordinate axes, and a bounding hyperplane given by

$$\xi_q = |x_q| = 1/\left|\Phi_q(0,\ldots,1,\ldots,0)\right|, \quad q = 1, 2, \ldots, p. \tag{11.16}$$

Only the qth term in parenthesis is not equal to zero.

The second region is a hypersurface that includes the origin, the coordinate axes, and a bounding surface that is defined by the parametric equations

$$\xi_q = |x_q| = 1/\Phi_q(n_1,\ldots,n_p), \; q = 1, 2, \ldots, p. \tag{11.17}$$

The region of convergence equivalent to the one in (11.12) is

$$|x_1|^{1/k} + |x_2|^{1/k} + \cdots + |x_p|^{1/k} = 1. \tag{11.18}$$

These regions are found for a particular problem in Sect. 12.3.

Criteria for convergence of functions that are not Horn functions, a case that does not occur in the examples in this book, can be developed with the approach developed by *Slater* (1966) that is used to find the convergence regions of the four Appell functions.

The results on convergence can be used to show that if a power series converges, then the integration path can be closed to include the poles that lead to that power series. This result is proved for two complex planes. The proof is easily generalized to N complex planes.

11.2 Path Closure at Infinity

In this section I show that if the power series converges, then the integration path could have been closed at infinity to encircle the poles that generate the power series. Consider the general form of the integral in two complex planes

$$I = \frac{1}{(2\pi i)^2} \int_C ds \int_C dt\, x^{-s} y^{-t} F(s,t)\,, \tag{11.19}$$

where s and t are the two complex variables. To be able to close the contour at negative infinity for the two complex variables, one requires the contribution of the line integral at infinity to be zero, i.e.,

$$\lim_{\substack{s \to -\infty \\ t \to -\infty}} x^{-s} y^{-t} F(s,t)\; s\, t = 0. \tag{11.20}$$

If $F(s,t)$ has more gamma functions in both complex variables in the denominator than in the numerator, then it is easy to show that (11.20) is true. If $F(s,t)$ has more gamma functions in the t complex variable in the denominator than in the numerator and the same number for s, then the series converges if $x < 1$. This is shown with l'Hospital's rule, which gives

$$\lim_{s \to -\infty} x^{-s}\, s = \lim_{s \to -\infty} \frac{1}{\ln(x)\, x^s} = 0. \tag{11.21}$$

Since the limit in this equation is zero, one can close the integration path along an infinite semicircle in the left half plane about the poles that contribute to the series.

The only complicated case is the one in which there are the same number of gamma functions in each variable in the numerator and denominator. Consider the case when the infinite series results from poles in the left half plane for both complex variables. The integrand for large values of the complex variables, for the case of the integrand composed of the ratio of gamma functions, is given by

$$\lim_{\substack{s \to -\infty \\ t \to -\infty}} x^{-s} y^{-t} F(s,t)\; s\, t = \lim_{\substack{m \to \infty \\ n \to \infty}} x^m y^n \left\{ \Gamma \begin{bmatrix} m+n \\ m,\, n \end{bmatrix} \right\}^k m\, n = 0, \tag{11.22}$$

where k is an integer that can be positive or negative. For convergence, the mildest restrictions on x and y are that they are both less than unity. There can be constants added to the arguments of the gamma functions. These constants do not affect the limits; therefore, for notational convenience they are not explicitly displayed. For large index values one obtains from Stirling's approximation to the gamma function

$$\Gamma\begin{bmatrix} m+n \\ m,n \end{bmatrix} \sim A\frac{(m+n)^{m+n}}{m^m n^n}, \tag{11.23}$$

where A is a factor that does not affect convergence at infinity.

There are two cases to consider, one in which one index goes to ∞ with the other remaining finite, and one in which both indices go to ∞. Suppose n goes to ∞ and m remains finite, then

$$\lim_{n\to\infty} x^m y^n \left\{\Gamma\begin{bmatrix} m+n \\ m,\, n \end{bmatrix}\right\}^k m\,n = A\frac{\exp(mk)\,x^m}{m^{mk-1}} \lim_{n\to\infty} y^n n^{mk+1}. \tag{11.24}$$

I made use of the relation

$$\lim_{n\to\infty} (1+m/n)^n = \exp(m)\,. \tag{11.25}$$

If k is negative, then the limit in (11.24) is obviously zero for $|y| < 1$. For k positive, one has

$$\begin{aligned} \lim_{n\to\infty} y^n n^{mk+1} &= \lim_{n\to\infty} \frac{n^{mk+1}}{y^{-n}} = \lim_{n\to\infty} \frac{(mk+1)\,n^{mk}}{-\ln(y)\; y^{-n}} \\ &= \ldots = \lim_{n\to\infty} \frac{(mk+1)\,!}{(-1)^{mk+1}\ln^{mk+1}(y)\; y^{-n}} = 0. \end{aligned} \tag{11.26}$$

Consider the case in which both indices go to ∞. I will let $m = n$. With the form given in (11.22) series convergence requires

$$y^{1/k} = 1 - x^{1/k} - \delta, \tag{11.27}$$

where δ is a small positive number. One obtains

$$\lim_{\substack{s\to-\infty \\ t\to-\infty}} x^{-s} y^{-t}\, F(s,t)\; s\,t = A \lim_{n\to\infty} n^2 \left[4^k x \left(1 - x^{1/k} - \delta\right)^k\right]^n. \tag{11.28}$$

The maximum value of the expression, which one finds by setting the derivative of the term in braces equal to zero, occurs at $x_{Max} = (1/2)^k$. Inserting this into the above expression, one finds

$$\lim_{\substack{s\to-\infty \\ t\to-\infty}} x^{-s} y^{-t}\, F(s,t)\; s\,t = A \lim_{n\to\infty} n^2 (1-2\,\delta)^{k\,n} = \lim_{n\to\infty} \frac{n^2}{\exp(2\,\delta\,k\,n)} = 0. \tag{11.29}$$

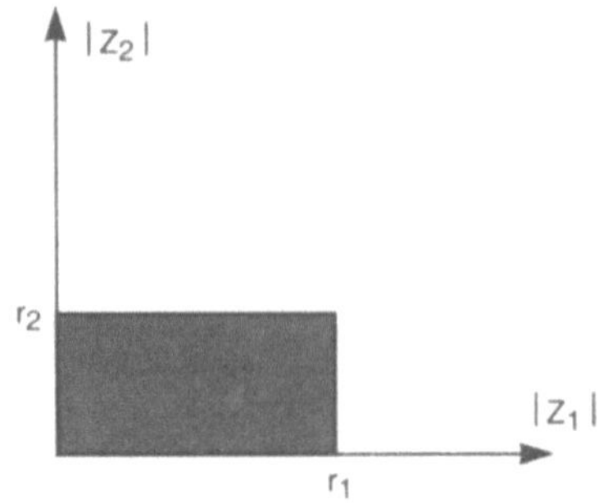

Fig. 11.1. Representation of a polydisk in the space of the absolute value of each complex variable.

Therefore, the contributions along the infinite paths in the two complex variables are zero, and one could close the integration path at infinity without affecting the value of the integral.

The infinite series coefficients are the values of the residues of the integrand at the pole locations. It may turn out that many of those residues are zero, which is the case when a gamma function in the denominator is of the form $\Gamma[1-n-m]$. For that case only the $n = m = 0$ term contributes. Notice that the proof above depended only on the indices in the gamma functions and not on any constants added to them. Therefore, when calculating whether to include a series one must determine if the series is of a convergent form rather than whether it actually converges. A series of convergent form is defined to be one in which the parameters lie within the convergent region defined in (11.9) for the two complex-plane case and (11.16)–(11.17) for the general case.

11.3 Integration in Multiple Complex Planes

The space of several complex variables can be very complicated geometrically, and much of the recent work has been concerned with the validity of basic concepts in complex geometries. For problems in which the Mellin transform theorem applies, the only singularities of the integrand are poles of finite order and branch cuts. To evaluate these integrals only the simplest concepts are necessary. One can obtain the residues at the poles with Cauchy's integral formula on polydisks. A polydisk $P(\mathbf{a}, \mathbf{r})$ is the space in the N-dimensional complex plane $\mathbb{C}^N$ that is analogous to the unit circle in one complex plane. It has a multiradius $\mathbf{r} = (r_1, \ldots, r_N)$ centered at $\mathbf{a}$ and is defined by the condition

$$P(\mathbf{a}, \mathbf{r}) = \left\{ z \in \mathbb{C}^N, \quad |z_j - a_j| < r_j, \quad 1 \le j \le N \right\}. \tag{11.30}$$

An example of a polydisk in two complex planes is represented in Fig. 11.1.

The extension of the Cauchy pole-residue integral theorem given in *Range* (1986, Page 8) to the polydisk is

$$f(z_1, z_2, \ldots, z_N) = \frac{1}{(2\pi i)^N} \int\limits_{C_1} \ldots \int\limits_{C_N} ds_1 \ldots ds_N \frac{f(s_1, s_2, \ldots, s_N)}{(s_1 - z_1)\ldots(s_N - z_N)}, \quad z \in P, \tag{11.31}$$

where the contour encloses all poles, and the function $f(s_1, s_2, \ldots, s_N)$ is holomorphic (has no singularities) in the integration region. For Mellin transform problems there are poles that depend on the individual complex variables as above, and poles that have the sum of the variables. The method to evaluate these is to expand the integrand in partial fractions in which there are N independent poles in the denominator of each term. The terms with the sum of the variables is converted into the above form by a change of a single variable

$$s_{N+1} = s_1 + s_2 + \cdots + s_N. \tag{11.32}$$

Since the Wronskian of the transformation is unity, this new variable is treated just like the others.

The method to evaluate an integral in which the integrand is a ratio of gamma functions multiplied by a power of the integration variable is first considered for two complex planes. The generalization of this result to more complex planes is obvious. The partial fraction expansion is equivalent to one choosing from the set of all possible poles the pairs of poles in which the variables are independent. A gamma function has an infinite number of simple poles that occur at the points where the gamma function's argument has a negative integer value. For integrands that contain the product of gamma functions, one must consider all sets of pole pairs (2-poles) formed by gamma functions whose arguments are linearly independent in the two complex variables. The complete solution is the sum of all pole residues for these sets of 2-poles that are contained in the closed integration path. To determine which poles are enclosed in the integration path is not obvious for the multiple complex-plane geometry.

I will generally evaluate the complex-contour integration by simultaneously performing the integration over N-complex planes. As an example I will first evaluate the integral traditionally and then with the method espoused here. As shown above, if one writes down the power series resulting from the evaluation at the multiple poles, then the series that should be included are the ones that converge, since the contour path could be closed at infinity thereby encircling those poles without affecting the value of the integral. As a "toy" problem, consider the integration of a product of unit step functions given by

$$\begin{aligned} I &= d \int\limits_0^\infty \frac{dx}{x} x^d U(a-x)\, U(x-b)\, U(x-c) \\ &= 0 \text{ if } a < b \text{ or } a < c, \\ &= a^d - c^d \text{ if } a > b, \ a > c, \text{ and } c > b, \\ &= a^d - b^d \text{ if } a > b, \ a > c, \text{ and } b > c. \end{aligned} \tag{11.33}$$

$U(x)$ is unity for $x > 0$ and zero for $x < 0$. Convert this into an integral in two complex planes with the convolution integral in (11.1), the inverse transform relation in (11.2) and the Mellin transforms in (1.57) and (1.58), to obtain

$$I=\frac{da^d}{(2\pi i)^2}\int_{C_1}\int_{C_2} ds_1\, ds_2 \left(\frac{b}{a}\right)^{-s_1}\left(\frac{c}{a}\right)^{-s_2}\Gamma\left[\begin{matrix} s_1+s_2+d,\ s_1,\ s_2 \\ s_1+s_2+d+1,\ s_1+1,\ s_2+1\end{matrix}\right]$$

$$=\frac{d\,a^d}{(2\pi i)^2}\int_{C_1}\int_{C_2} ds_1\, ds_2 \left(\frac{b}{a}\right)^{-s_1}\left(\frac{c}{a}\right)^{-s_2}\frac{1}{(s_1+s_2+d)\ s_1\, s_2}$$

$$=\frac{a^d}{(2\pi i)^2}\int_{C_1}\int_{C_2} ds_1\, ds_2 \left(\frac{b}{a}\right)^{-s_1}\left(\frac{c}{a}\right)^{-s_2}$$

$$\times\left[\frac{1}{s_1\, s_2}-\frac{1}{(s_1+s_2+d)\ s_1}-\frac{1}{(s_1+s_2+d)\ s_2}\right]. \qquad (11.34)$$

The last form of the integrand is used in the traditional evaluation of the integral. The integration path as determined from the individual Mellin transforms is such that if one closed the path in the negative half plane in each complex plane, then one would include all poles. If $a < b$ or $a < c$, at least one of the integration paths must be closed in the right half plane of s_1 or s_2, and one obtains zero. If $a > b$ or $a > c$, perform the s_2 integration first, and obtain contributions from one pole in each of the first two terms and from two poles of the third term. For two of the terms the s_1 integration can be subsequently performed since $a > b$ and $a > c$ to give

$$I = a^d - b^d + \frac{c^d}{2\pi i}\int_{C_1} ds_1 \left(\frac{c}{b}\right)^{s_1}\left[\frac{1}{s_1+d}-\frac{1}{s_1}\right]. \qquad (11.35)$$

If $c < b$, the integration path is closed to the right, and the integral does not have any contribution. If $c > b$, the integration path is closed to the left, and both poles contribute. The net result is that one obtains the result in (11.33).

Next, the solution is found by the method developed in this book by listing the three possible power series that are solutions of the complex integration that appear in the first line of (11.34) and by choosing those that converge. Since information is required on the convergence of the series, the full power series are written down even if all the terms but the first have a coefficient of zero. The locations of the 2-poles that contribute are:

(1) $-s_1 = -n;\ -s_2 = -m;$ therefore, $s_1 = n;\ s_2 = m,$
(2) $s_1 + s_2 + d = -m;\ -s_1 = -n;$ therefore, $s_1 = n;\ s_2 = -n - m - d,$ and
(3) $s_1 + s_2 + d = -n;\ -s_2 = -m;$ therefore, $s_1 = -n - m - d;\ s_2 = m.$
(11.36)

This leads to the following three double series:

$$I_1 = d\,a^d\, S(0,0)\left(\frac{b}{a}\right)^n\left(\frac{c}{a}\right)^m\Gamma\left[\begin{matrix} -m-n-d \\ 1-m,\ 1-n,\ 1-n-m-d\end{matrix}\right] = a^d, \quad (11.37)$$

$$I_2 = d\,a^d\,S\,(0,0)\left(\frac{b}{c}\right)^n\left(\frac{c}{a}\right)^{m+d}\Gamma\left[\begin{matrix} m-n-d \\ 1-m,\,1-n,\,1-m+n-d\end{matrix}\right] = -c^d, \tag{11.38}$$

$$I_3 = d\,a^d\,S\,(0,0)\left(\frac{b}{a}\right)^{n+d}\left(\frac{c}{b}\right)^m\Gamma\left[\begin{matrix} m-n-d \\ 1-m,\,1-n,\,1-n+m-d\end{matrix}\right] = -b^d. \tag{11.39}$$

I have introduced a notation to keep the equation size manageable. In the above equation and many subsequent equations I use

$$S\,(p_1,\ldots,p_k) = \sum_{n_1=p_1}^{\infty}\cdots\sum_{n_k=p_k}^{\infty}\frac{(-1)^{n_1+\cdots+n_k}}{n_1!\cdots n_k!}. \tag{11.40}$$

If the series is asymptotic in any variable, then an a is inserted next to the number that represents the lower limit.

As explained at the end of Sect. 11.2, even if a series is truncated because terms of the series are zero, one must still apply the convergence criteria given in Sect. 11.1 to determine if a given series should be included in the solution for particular parameter values. The convergence criterion in Sect. 11.1 requires $a > b$ and $a > c$ for I_1 to be included, $c > b$ and $a > c$ for I_2 to be included, and $a > b$ and $b > c$ for I_3 to be included. Include only those series that converge in the region of convergence to obtain the same result as in (11.33).

The method discussed here replaces considerations of contour closure necessary in the usual treatment of complex-contour integrals with a criterion based upon series convergence. In the physical problems this results in a tremendous simplification. When the Mellin transform of products of more than three hypergeometric functions are sought, the resulting complex contour integral may contain 2 or more complex variables of integration. The geometric considerations necessary to determine the direction of contour closure rapidly become very cumbersome especially as three or more complex variables are introduced. Replacing this often intractable requirement with an algebraic basis for determining the proper solution provides a method for easily solving these otherwise difficult integrals. In addition, this technique may be used to find the asymptotic series for an integral of Mellin type. The asymptotic forms are enumerated as part of the solution process for convergent series solutions; this requires only the occasional addition of a steepest-descent contribution.

To apply this method to find the convergent and asymptotic series, the series that result from pole-residue evaluation are written such that the exponents that appear in the power series are positive. By this I mean, for example, that if I have double series in indices m and n, the parameters in the power series are raised to the powers m and n rather than $-m$ and $-n$. *All possible infinite series are enumerated and the Taylor series solution is the sum of those series that converge for the parameter range of interest. Parameters that appear in these power series may have magnitudes greater than 1. When an approximation is desired that contains only a few terms, in these cases an asymptotic approximation should be used. One obtains this by examining all possible series and retaining those that contain parameters with all magnitudes less than 1.*

The asymptotic approximation appropriate for a particular parameter range is the sum of those series that contain series parameters that are less than unity. A steepest-descent contribution must be added when any Taylor series are excluded because their parameters have magnitudes greater than 1. This leads to a straightforward way to evaluate the integral that no longer requires explicit consideration of the direction of contour closure in a $2N$-dimensional space.

Using this method to evaluate the integral given by (11.34), one must write down all combinations of two independent poles, which are called 2-poles, evaluate the residue at each of these 2-poles, and sum the contributions of the enclosed 2-poles. Let the integral be a product of factors in which there are n_1 poles that contain only the first variable, n_2 poles with only the second, and n_3 poles with the sum of the two variables. The number of potential infinite series in the solution is equal to the total number of ways that two poles can be picked from the total number of poles—$A = (n_1 + n_2 + n_3)(n_1 + n_2 + n_3 - 1)$, minus the number of ways in which two poles could be chosen solely from each of the three categories. Therefore, the total number of 2-poles for the two complex-plane case is

$$\begin{aligned} N_t &= 0.5\left[A - n_1(n_1 - 1) - n_2(n_2 - 1) - n_3(n_3 - 1)\right] \\ &= n_1 n_2 + n_1 n_3 + n_3 n_2. \end{aligned} \tag{11.41}$$

The last form of the equation, in which there is the product of every combination of the number of each complex variable with one deleted, also applies to cases where there are more than two complex planes.

In the discussion that follows, a Taylor series is defined to be a series that converges. An asymptotic series is a series that diverges as the number of terms in the series gets arbitrarily large. For some problems the solution contains two Taylor series, one of which applies when the parameter is less than unity and the other when the parameter is greater than unity. Some authors, i.e., *Wong* (1989), call this second series an asymptotic series. In this book the term asymptotic series is only applied to a divergent series.

A second "toy" problem is solved to illustrate two important concepts. First I show that the restricted region of validity of the solution as compared to the region of convergence of the integral results from the power series form of the solution. The same integral is then evaluated with terms of the power series subtracted from the integrand functions to illustrate that these integrals pose no conceptual difficulties. Consider an integral that contains the product of three exponentials

$$I = \int_0^\infty du \exp(-u) \exp(-xu) \exp(-yu). \tag{11.42}$$

For $1 + x + y > 0$ this is easily solved to give

$$I = \frac{1}{1 + x + y}. \tag{11.43}$$

To evaluate this with Mellin transform techniques, use the Mellin convolution theorem with the Mellin transform in (1.45) to obtain

$$I = \frac{1}{(2\pi i)^2} \int_{C_1} \int_{C_2} ds_1\, ds_2\, \Gamma[s_1 + s_2 + 1,\ -s_1,\ -s_2]\, x^{s_1} y^{s_2}. \tag{11.44}$$

Since $n = m = p = 1$, (11.41) states that there are three sets of 2-poles in the integrand. These three 2-poles are:

$$\begin{aligned} &(1)\quad -s_1 = -n;\ -s_2 = -m, \\ &(2)\quad s_1 + s_2 + 1 = -m;\ -s_1 = -n,\ \text{and} \\ &(3)\quad s_1 + s_2 + 1 = -n;\ -s_2 = -m. \end{aligned} \tag{11.45}$$

These 2-poles lead to the following powers of the parameters and powers of the factorial of the summation indices in the summation:

$$\begin{aligned} &(1)\quad x^n y^m (n!)^0 (m!)^0, \\ &(2)\quad x^n y^{-n-m-1} (n!)^0 (m!)^0 = y^{-m-1} \left(\frac{x}{y}\right)^n F(0,0),\ \text{and} \\ &(3)\quad x^{-n-m-1} y^m (n!)^0 (m!)^0 = x^{-n-1} \left(\frac{y}{x}\right)^m F(0,0). \end{aligned} \tag{11.46}$$

New parameters were formed in the second and third terms that are combinations of the original parameters. The constant terms added to the exponent do not affect series convergence and will be suppressed in subsequent problems. The powers of the factorials represent the behavior of the terms of the summation as the index goes to infinity. The powers of these factorials were previously defined as Ω. I have introduced the notation

$$F(\Omega_1, \ldots, \Omega_k) = (n_1!)^{\Omega_1} \ldots (n_k!)^{\Omega_k}. \tag{11.47}$$

The three double sums that result from the evaluation of the integral at these three 2-poles are:

$$I_1 = S(0,0)\ \Gamma[1 + n + m]\, x^n y^m, \tag{11.48}$$

$$I_2 = S(0,0)\ \Gamma[1 + n + m] \left(\frac{x}{y}\right)^n y^{-m},\ \text{and} \tag{11.49}$$

$$I_3 = S(0,0)\ \Gamma[1 + n + m]\, x^{-n} \left(\frac{y}{x}\right)^m. \tag{11.50}$$

The definition of $S(0,0)$ was introduced in the last example. In this example $\Omega_1 = \Omega_2 = 0$ for all three series; therefore, the three series are Taylor series. Their convergence depends on the sizes of the parameters. The parameters x and y can be either less than or greater than unity. To be specific, assume that $1 > x > y$. To find the region of convergence of I_1 note that

$$\Phi_1(m,n) = \frac{n+m}{m}, \quad \text{and} \quad \Phi_2(m,n) = \frac{n+m}{n}. \tag{11.51}$$

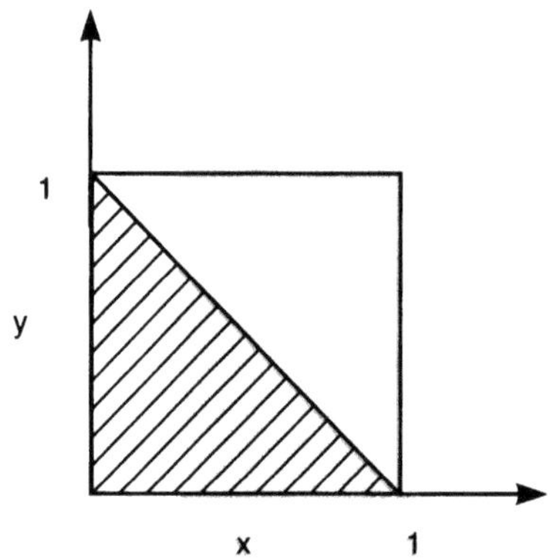

Fig. 11.2. Hatched area is the parameter region in which the power series converges.

Therefore, the first criterion for the region of convergence gives the unit square. The parametric equations obtained from the second criterion for the region of convergence are

$$\xi = \frac{m}{m+n}, \quad \text{and} \quad \eta = \frac{n}{m+n}. \tag{11.52}$$

The variable m and n can be eliminated from these two equations, which leads to the condition $\eta + \xi < 1$. Therefore, this double series converges when $x + y < 1$. This is sometimes true for the assumed relation between the parameters as shown in Fig. 11.2. From the same criteria one finds that I_2 converges when $1 + x < y$, and I_3 converges when $1 + y < x$. These last two conditions are never satisfied with the assumed relations between the parameters. Since these series do not converge, they are not included in the solution.

The value of the integral is

$$I = \int_0^\infty du \, \exp(-u) \exp(-xu) \exp(-yu) = S\,(0,0) \; \Gamma[1+n+m] \, x^n y^m, \quad x+y < 1. \tag{11.53}$$

If the ordering of the magnitudes of 1, x, and y were chosen differently, then one of the other power series would have been selected.

To show the equivalence of the direct result in (11.43) with this double infinite series, note that for $x + y < 1$, one obtains

$$I = \frac{1}{1+x+y} = \sum_{n=0}^{\infty} (-1)^n \, (x+y)^n \,. \tag{11.54}$$

This can be expressed as a double series by using the binomial theorem when $x > y$ to obtain

$$I = \sum_{n=0}^{\infty} \sum_{m=0}^{n} (-1)^n x^n \left(\frac{y}{x}\right)^m \frac{n!}{(n-m)!\, m!}. \tag{11.55}$$

If one changes variables to $m \to m'$ and $n \to n' + m'$ with $m \leq n$, then the summation reduces to the result given in (11.53) that was obtained with Mellin transform techniques. Therefore, the Mellin transform method produces the correct result although it was expressed as an infinite series rather than a simple function as in the direct integration. In general, one would not recognize the double series that was obtained as being equivalent to the simple function. The conditions on the parameters are more restrictive than the requirement that $1 + x + y > 0$ as obtained by the direct integration. This is a direct result of the requirement that the solution be expressible as a double series, and results from the restriction on the expansion to obtain the series in (11.54).

From this example, one learns that in solving problems of this type, the conditions obtained to get an answer, even though they are correct, may be too restrictive. An answer exists for other parameter values, but it cannot be expressed as a double series. To find a solution in those regions, specialized techniques that are peculiar to individual integrals are used (*Gervois* and *Navelet* 1985, 1986).

The fact that an answer is not in the most compact form and expressible in terms of simple functions is not a major deficiency because more complicated problems cannot, in general, be expressed as elementary functions. If the original function that is represented by the double summation can be found, then the region in which the solution applies can be extended. To do this one must invert two power series in succession. I am unaware of any general techniques to do this.

Up to now it has been assumed that the integration path has all poles of each gamma function on one side. For this case all summations start at the index equal to zero. There is an additional complication that arises when a summation does not start at zero. In the single complex plane case if there is a term of the form $\sum_{n=b}^{\infty} D_n x^n$ that applies when x is less than unity, then $\sum_{n=0}^{b-1} D_n x^n$ must be added to the series that applies when x is greater than unity. This is easily generalizable to the case of two complex planes. Consider first the case where one series starts at zero, and the other starts at the index equal to unity. This series can be represented as

$$B_1 = \sum_{n=1}^{\infty} \sum_{m=0}^{\infty} A_{mn} x^n y^m \tag{11.56}$$

and applies when x and y are less than unity. In parallel to the single complex plane case, when x is greater than unity and y is less than unity, the term

$$B_2 = \sum_{m=0}^{\infty} A_{m0} y^m \tag{11.57}$$

must be added to the solution. Similarly, for the other index starting at unity, assume the series

$$C_1 = \sum_{n=0}^{\infty} \sum_{m=1}^{\infty} A_{mn} x^n y^m \tag{11.58}$$

Table 11.1. Expressions to include for different parameter ranges when the indices m and n do not start at zero.

x	y	Terms to include
<1	<1	$A_1\ B_1\ C_1$
>1	<1	$A_2\ B_2$
<1	>1	$A_3\ C_3$
>1	>1	A_4

applies when x and y are less than unity. When x is less than unity and y is greater than unity, then

$$C_3 = \sum_{n=0}^{\infty} A_{0n} x^n \tag{11.59}$$

must be added to the solution.

In the most general case both indices can start at a non-zero value. Assume the expression that applies when x and y are less than unity is

$$A_1 = \sum_{n=a}^{\infty} \sum_{m=b}^{\infty} A_{mn} x^n y^m. \tag{11.60}$$

When x is greater than unity and y is less than unity, then

$$A_2 = \sum_{n=0}^{a-1} \sum_{m=b}^{\infty} A_{mn} x^n y^m \tag{11.61}$$

must be added to the solution. When x is less than unity and y is greater than unity, then

$$A_3 = \sum_{n=a}^{\infty} \sum_{m=0}^{b-1} A_{mn} x^n y^m \tag{11.62}$$

must be added to the solution. When x and y are greater than unity, then

$$A_4 = \sum_{n=0}^{a-1} \sum_{m=0}^{b-1} A_{mn} x^n y^m \tag{11.63}$$

must be added to the solution. All possible solutions in which the summations do not start at zero are summarized in Table 11.3.1. For each set of parameter values for a given pole set, only one of the tabulated series is included in the solution. If the indices m and n start at zero, then none of these additional expressions is included.

To illustrate this, consider a modified version of the integral considered in the previous example

$$I = \int_0^{\infty} du \exp(-u) \left[\exp(-xu) - 1\right] \exp(-yu). \tag{11.64}$$

This integral transforms into

$$I = \frac{1}{(2\pi i)^2} \int\limits_{C_1} \int\limits_{C_2} ds_1\, ds_2\, \Gamma\left[s_1 + s_2 + 1,\ -s_1*,\ -s_2\right] x^{s_1} y^{s_2}, \tag{11.65}$$

where * means that the integration path passes between the first and second poles of that gamma function. The three 2-poles are:

$$\begin{aligned} &(1) \quad -s_1 = -n*;\ -s_2 = -m, \\ &(2) \quad s_1 + s_2 + 1 = -m;\ -s_1 = -n*, \text{ and} \\ &(3) \quad s_1 + s_2 + 1 = -n;\ -s_2 = -m. \end{aligned} \tag{11.66}$$

The asterisk indicates that the summation index starts at 1. The solution contains the same summation as in the first example except that the n summation starts at 1. There is a potential to pick up one term of the second summation; however, that is added to the solution when $y > 1$ and $y/x > 1$. Since that is not the case being considered, this term is not added to the solution. Therefore, the value of this integral is

$$I = S\,(1,0)\ \Gamma[1 + n + m]\, x^n y^m = \frac{1}{1 + x + y} - \frac{1}{1 + y}, \quad x + y < 1. \tag{11.67}$$

This is the same result that is obtained from direct integration.

If unity is also subtracted from the last exponential in (11.64), then the answer is obtained in a similar manner. Consider the following integral

$$I = \int\limits_0^\infty du\ \exp(-u)\left[\exp(-xu) - 1\right]\left[\exp(-yu) - 1\right]. \tag{11.68}$$

This transforms into

$$I = \frac{1}{(2\pi i)^2} \int\limits_{C_1} \int\limits_{C_2} ds_1\, ds_2\, \Gamma\left[s_1 + s_2 + 1,\ -s_1*,\ -s_2*\right] x^{s_1} y^{s_2}. \tag{11.69}$$

The three 2-poles are:

$$\begin{aligned} &(1) \quad -s_1 = -n*;\ -s_2 = -m*, \\ &(2) \quad s_1 + s_2 + 1 = -m;\ -s_1 = -n*, \text{ and} \\ &(3) \quad s_1 + s_2 + 1 = -n;\ -s_2 = -m*. \end{aligned} \tag{11.70}$$

The solution contains the same summation as in the first example except that the n and m summations both start at 1. Again there is a potential to pick up one term of the second and third summations; however, the first term of the second is added to the solution when $x > 1$ and $y/x > 1$, and the first term of the third is added to the solution when $y > 1$ and $y/x > 1$. Since these are not the cases being considered, these terms are not added to the solution. The value of this integral is thus

$$I = S(1,1)\,\Gamma[1+n+m]\,x^n y^m = \frac{1}{1+x+y} - \frac{1}{1+y} - \frac{1}{1+x} + 1, \quad x+y<1\,. \tag{11.71}$$

This is the same result that is obtained from direct integration. In these examples no terms from the other series appeared in the solution.

The problem of being unable to represent the solution to an integral for all parameter ranges surfaces in various fields. For instance, one can show (*Jackson* 1962) that the electric potential in space caused by an infinitely thin conducting disc of radius a with a charge q is

$$V(\rho, z) = q \int_0^\infty dk \frac{\sin(ka)}{ka} \exp(-kz)\, J_0(k\rho)\,. \tag{11.72}$$

One can evaluate this integral with the methods developed in the chapter to obtain a solution as a double series. This series does not converge for some ratios of the parameters close to unity. For this case a solution that applies for all space is available in *Gradshteyn* and *Ryzhik* (1980, 6.752 #1) as

$$V(\rho, z) = \frac{q}{a} \sin^{-1}\left(\frac{2a}{\sqrt{(\rho-a)^2+z^2} + \sqrt{(\rho+a)^2+z^2}}\right). \tag{11.73}$$

The above result was obtained with (3.5) to convert the Bessel function into an integral over angle, the evaluation of the integral over k, followed by the evaluation of the integral over angle. This technique does not work if the index of the Bessel function is different than unity or if the power of the integration variable in the integrand is other than -1. I know of no general procedures to convert double power series into such complicated functional relations.

11.3.1 Evaluation Procedure for N Complex Planes

The procedure to obtain Taylor series solutions in more than 2 complex planes is easily extended from the procedure given above. After all possible power series are determined, choose only those series that converge for the parameter range of interest. With this procedure it is straightforward to obtain Taylor series for very complicated integrals. Examples of the application this procedure to 3 and N complex planes are given in Sect. 12.3.

11.4 Asymptotic Solution in Two or More Complex Planes

In general, different Taylor (convergent) series may be readily obtained for various parameter regimes. For many problems this solution is adequate. For other problems the Taylor series converge slowly for some parameter regimes, and it is desirable to obtain asymptotic solutions in these regimes.

The procedure to obtain the series that converge most rapidly is rather simple; write down the series terms that correspond to every pole set in the entire complex plane in a form such that the parameters have positive exponents, and choose only those series for which the parameter magnitude is less than unity. If a Taylor series convergent for the parameter magnitudes over the range of interest is excluded because of this procedure, then an asymptotic solution that consists of an asymptotic series and a steepest-descent contribution must be included. Normally this will be obvious because the procedure will automatically require the inclusion of an asymptotic series. In some cases, however, no asymptotic series will be called for because the coefficients in the asymptotic series are all zero. In these cases the steepest-descent contribution by itself is the asymptotic form for the excluded Taylor series.

It is possible to have the sum of Taylor and asymptotic solutions for the same range of parameter values. This is a consequence of the fact that the integrand's decomposition into the partial-fraction expansion as given in (11.34) results in a sum of functions. Some of the functions may contribute a Taylor series, and some may contribute an asymptotic form to the final solution. This procedure is applied in Sects. 12.4 to 12.7.

The technique to obtain asymptotic expressions for an integral in a single complex plane discussed in Chaps. 5 and 9 is extended here to obtain steepest-descent contributions for integrals in two or more complex planes. A simple case to evaluate occurs in the parameter regime where the power series is convergent in one parameter and asymptotic in the other. In that case the integration in the complex variable that leads to the convergent series is performed first, followed by the evaluation of the asymptotic solution for the second integral. The result is a sum of power series that are the ones that would be chosen from all possible power series by the application the rules of the last section. In addition, one needs to obtain the steepest-descent contribution.

The situation is more complicated when the parameter regime is such that both power series are asymptotic and a change of variables cannot be make to make one integral result in a convergent series. I show that when the complex variables are not strongly coupled that one can still perform the integrations serially.

Let us discuss the general case. As for a single complex plane, the integration path of the complex variables is deformed so that all saddle points of the integrand are crossed in the directions of steepest descent. The asymptotic expression for the integral is given by the sum of the contributions of poles crossed as a result of the path deformation, and the contribution of the path integral along the steepest-descent path. The asymptotic series are found with the procedure discussed in the last section. In this section the method to find the steepest-descent contribution is discussed.

For a single-complex-plane case the steepest-descent contribution was found by a modification of the normal method in which one is concerned with evaluating integrals of the form

$$I(\lambda) = \int_C ds\, g(s) \exp[-\lambda f(s)] , \tag{11.74}$$

where λ is large. The difference in the application of this technique for Mellin transform problems is that there is no large quantity that is immediately evident. It was shown in Chap. 5 when the parameter was large that the second derivative of $f(s)$ was much larger than the higher order derivatives at the saddle point with the result that the procedure essentially reduced to the normal steepest-descent procedure. For more than one complex plane a similar procedure is attempted. It is found that it does not work in general; however, for a class of problems one can find the steepest-descent contribution by the simple process of evaluating integrals in each complex plane in sequence. This special case occurs often in practice, and it applies to every turbulence problem discussed in the next chapter. The only case for which a general method to obtain the asymptotic solution for two complex planes is not available is when there is a steepest-descent contribution that must be retained for the first integration no matter which integral is performed first.

To develop this method, I will discuss the standard problem of steepest descent in two variables, then I will modify those results to apply to Mellin transform problems. The integral to be evaluated with the normal method is

$$I(\lambda) = \int_{C_2} ds_2 \int_{C_1} ds_1\, g(s_1, s_2) \exp[-\lambda f(s_1, s_2)] . \tag{11.75}$$

We will recast the Mellin convolution integral into this form. The general form for the inverse Mellin transform integral in two complex planes is

$$h(x_1, x_2) = \frac{1}{(2\pi i)^2} \int_{C_2} \int_{C_1} ds_1\, ds_2\, H_0(s_1 + s_2)\ H_1(s_1)\, H_2(s_2)\, x_1^{-s_1}\, x_2^{-s_2}, \tag{11.76}$$

where the paths C_1 and C_2 represent a sheet in 4-space that must be contained within the hyper-volumes where $H_0(s_1 + s_2)$, $H_1(s_1)$, and $H_2(s_2)$ are analytic. In the one complex-plane case, asymptotic results are needed only when the parameter magnitude $|x|$ is large. In two complex planes, asymptotic analysis is more complicated for the following reason: Expansion of the integral in (11.76) into a form that allows the use of pole-residue methods was discussed in Sect. 11.2. Pole-residue integration produces terms that each contain the product of a holomorphic function and a function with one simple pole for each complex variable. The resultant power series may contain a factor $x_1^{n+m} x_2^{-m}$ that reduces to $(x_1/x_2)^n x_1^m$. The ratio $|x_1/x_2|$ may be large even though $|x_1|$ and $|x_2|$ are small. This type of possibility demonstrates the potential need for asymptotic analysis even when both $|x_1|$ and $|x_2|$ are small. Three distinct cases are possible: $|x_1|$ and $|x_2|$ are small, $|x_1|$ is small (large) and $|x_2|$ is large (small), and $|x_1|$ and $|x_2|$ are large.

In problems discussed in this book, the integrand may be expressed as the ratio of products of gamma functions

$$H_0(s_1+s_2)\,H_1(s_1)\,H_2(s_2) = \left(\frac{\prod_{j=1}^{A_1}\Gamma\left[a_{1,j}+s_1\right]\ \prod_{j=1}^{B_1}\Gamma\left[b_{1,j}-s_1\right]}{\prod_{j=1}^{C_1}\Gamma\left[c_{1,j}+s_1\right]\ \prod_{j=1}^{D_1}\Gamma\left[d_{1,j}-s_1\right]}\right)$$
$$\times\left(\frac{\prod_{j=1}^{A_2}\Gamma\left[a_{2,j}+s_2\right]\ \prod_{j=1}^{B_2}\Gamma\left[b_{2,j}-s_2\right]}{\prod_{j=1}^{C_2}\Gamma\left[c_{2,j}+s_2\right]\ \prod_{j=1}^{D_2}\Gamma\left[d_{2,j}-s_2\right]}\right)$$
$$\times\left(\frac{\prod_{j=1}^{A_3}\Gamma\left[a_{3,j}+s_1+s_2\right]\ \prod_{j=1}^{B_3}\Gamma\left[b_{3,j}-s_1-s_2\right]}{\prod_{j=1}^{C_3}\Gamma\left[c_{3,j}+s_1+s_2\right]\ \prod_{j=1}^{D_3}\Gamma\left[d_{3,j}-s_1-s_2\right]}\right). \tag{11.77}$$

The coefficients of s_1, s_2, and s_1+s_2 are unity in the above expression, unlike the more general case considered in the single-parameter case. If integrals that contain noninteger, rational coefficients must be evaluated, make the substitution $s \to u\,s$ where u is the least common denominator of the coefficients, and use the Gauss-Legendre multiplication theorem to expand the gamma functions until all the coefficients are unity.

Define the following quantities:

$$\Delta_n = A_n + D_n - B_n - C_n \quad \text{for } n = 1,\ 2,\ 3, \text{ and} \tag{11.78}$$

$$\Delta_{s_n} = \Delta_n + \Delta_3 \quad \text{for } n = 1,\ 2. \tag{11.79}$$

These Δ's play an important role in the determination of the asymptotic properties of the integral. If $\Delta_{s_1} = \Delta_{s_2} = 0$, no asymptotic form for $h(x_1, x_2)$ is necessary. As in the 1-dimensional case, path closure is determined by the magnitude of the parameters x_1 and x_2, and a rapidly converging series always results in the parameter regime in which the series converges. If either Δ_{s_1} or Δ_{s_2} is less than zero, then a change of variables $s_1 \to -s_1$ and / or $s_2 \to -s_2$ should be used to rewrite the integral so that these Δ's are greater than or equal to zero.

The identity $\Gamma[s]\,\Gamma[1-s] = \pi\csc(\pi s)$ is used to rewrite gamma functions that contain $-s_1$, $-s_2$, and $-(s_1+s_2)$ in terms of gamma functions that contain s_1, s_2, and s_1+s_2. This allows the use of an asymptotic expansion for all gamma functions that is valid in the sector where the phase of the argument of the gamma function is greater than $-\pi$ and less than π. Because the integrand's behavior at large values of s_1 and s_2 is of interest, the gamma functions are replaced with the asymptotic equivalent given as

$$\Gamma[a+s] \sim \sqrt{2\pi}\exp\left[(s+a-1/2)\ln(s) - s\right]. \tag{11.80}$$

The asymptotic form for the sine terms obtained through application of the previously discussed identity is also used. One must use different representations for the sine terms depending upon the sign of the imaginary part of the sine's argument.

Table 11.2. Conditions on the imaginary parts of s_1, s_2, and $s_1 + s_2$ that require separate treatment.

Case 1	$\mathrm{Im}\{s_1\}>0$	$\mathrm{Im}\{s_2\}>0$	$\mathrm{Im}\{s_1+s_2\}>0$	$\lambda_1=1$	$\lambda_2=1$	$\lambda_3=1$
Case 2	$\mathrm{Im}\{s_1\}<0$	$\mathrm{Im}\{s_2\}<0$	$\mathrm{Im}\{s_1+s_2\}<0$	$\lambda_1=-1$	$\lambda_2=-1$	$\lambda_3=-1$
Case 3a	$\mathrm{Im}\{s_1\}>0$	$\mathrm{Im}\{s_2\}<0$	$\mathrm{Im}\{s_1+s_2\}>0$	$\lambda_1=1$	$\lambda_2=-1$	$\lambda_3=1$
Case 3b	$\mathrm{Im}\{s_1\}>0$	$\mathrm{Im}\{s_2\}<0$	$\mathrm{Im}\{s_1+s_2\}<0$	$\lambda_1=1$	$\lambda_2=-1$	$\lambda_3=-1$
Case 4a	$\mathrm{Im}\{s_1\}<0$	$\mathrm{Im}\{s_2\}>0$	$\mathrm{Im}\{s_1+s_2\}>0$	$\lambda_1=-1$	$\lambda_2=1$	$\lambda_3=1$
Case 4b	$\mathrm{Im}\{s_1\}<0$	$\mathrm{Im}\{s_2\}>0$	$\mathrm{Im}\{s_1+s_2\}<0$	$\lambda_1=-1$	$\lambda_2=1$	$\lambda_3=-1$

If $\mathrm{Im}\{\pi s\} > 0$, then

$$\sin(\pi s) \sim \tfrac{1}{2} \exp\left[-i\pi\left(s-\tfrac{1}{2}\right)\right]. \tag{11.81}$$

Similarly, if $\mathrm{Im}\{\pi s\} < 0$, then

$$\sin(\pi s) \sim \tfrac{1}{2} \exp\left[i\pi\left(s-\tfrac{1}{2}\right)\right]. \tag{11.82}$$

Incorporate the sign of $i\pi$ into quantities λ as in Sect. 9.1, one for each s_1, s_2, and $s_1 + s_2$. This leads to six conditions that require separate treatment as tabulated in Table 11.2.

Cases 1, 2, 3, and 4 correspond to different quadrants in 4-space. The distinctions in Case 3 (a and b) and in Case 4 (a and b) arise because of the possibility that Im $(s_1 + s_2)$ may be positive or negative in each instance.

Combining the exponential form of the gamma functions with the logarithm of the parameters x_1 and x_2 one obtains

$$h(x_1, x_2) = (2\pi)^{([\Xi_1+\Xi_2+\Xi_3]/2-2)} \int_{C_1}\int_{C_2} ds_1\, ds_2\, \exp\left[f(s_1, s_2)\right], \tag{11.83}$$

where

$$\begin{aligned} f(s_1, s_2) = &\sum_{j=1}^{2} \Big\{ (\nu_j - \Xi_j/2 + \Delta_j s_j) \ln(s_j) - s_j \ln(x_j) - \Delta_j s_j \\ &- i\pi\lambda_j \left[\Lambda''_j + \Lambda_j\left(s_j + \tfrac{1}{2}\right)\right]\Big\} + \left[\nu_3 - \Xi_3/2 + \Delta_3(s_1+s_2)\right] \ln(s_1+s_2) \\ &- \Delta_3(s_1+s_2) - i\pi\lambda_3\left[\Lambda''_3 + \Lambda_3\left(s_1+s_2+\tfrac{1}{2}\right)\right], \end{aligned} \tag{11.84}$$

$$\Xi_n = A_n + B_n - C_n - D_n, \tag{11.85}$$

$$\nu_n = \sum_{j=1}^{A_n} a_{n,j} + \sum_{j=1}^{B_n} b_{n,j} - \sum_{j=1}^{C_n} c_{n,j} - \sum_{j=1}^{D_n} d_{n,j}, \tag{11.86}$$

$$\Lambda_n = B_n - D_n, \quad \text{and} \tag{11.87}$$

$$\Lambda_n'' = \sum_{j=1}^{B_n} b_{n,j} - \sum_{j=1}^{D_n} d_{n,j}, \tag{11.88}$$

for $n = 1$ to 3. The contributions of the sine terms, that depend upon the quadrant of the saddle point, are contained in the terms dependent upon λ_n.

The first step in one's calculating the integral's steepest-descent value is to locate the saddle points of $f(s_1, s_2)$. These points occur at $(\tilde{s}_1, \tilde{s}_2)$ where

$$0 = \left.\frac{\partial f(s_1, s_2)}{\partial s_1}\right|_{(s_1,s_2)=(\tilde{s}_1,\tilde{s}_2)} \quad \text{and} \quad 0 = \left.\frac{\partial f(s_1, s_2)}{\partial s_2}\right|_{(s_1,s_2)=(\tilde{s}_1,\tilde{s}_2)}. \tag{11.89}$$

This gives the two equations

$$\begin{aligned}\frac{\partial f(s_1, s_2)}{\partial s_1} \approx\ & -\ln(x_1) + \Delta_1 \ln(s_1) + \Delta_3 \ln(s_1 + s_2) \\ & + i\pi\left[\lambda_1 (B_1 - D_1) + \lambda_3 (B_3 - D_3)\right] \\ & + \frac{\nu_1 - \Xi_1/2}{s_1} + \frac{\nu_3 - \Xi_3/2}{s_1 + s_2}, \quad \text{and}\end{aligned} \tag{11.90}$$

$$\begin{aligned}\frac{\partial f(s_1, s_2)}{\partial s_2} \approx\ & -\ln(x_2) + \Delta_2 \ln(s_2) + \Delta_3 \ln(s_1 + s_2) \\ & + i\pi\left[\lambda_2 (B_2 - D_2) + \lambda_3 (B_3 - D_3)\right] \\ & + \frac{\nu_2 - \Xi_2/2}{s_2} + \frac{\nu_3 - \Xi_3/2}{s_1 + s_2}.\end{aligned} \tag{11.91}$$

Another set of partial derivatives that is necessary to evaluate saddle-point contributions are given by

$$\frac{\partial^2 f(s_1, s_2)}{\partial^2 s_1} = \alpha(s_1, s_2) \approx \frac{\Delta_1}{s_1} + \frac{\Delta_3}{s_1 + s_2} - \frac{\nu_1 - \Xi_1/2}{s_1^2} - \frac{\nu_3 - \Xi_3/2}{(s_1 + s_2)^2}, \tag{11.92}$$

$$\frac{\partial^2 f(s_1, s_2)}{\partial^2 s_2} = \beta(s_1, s_2) \approx \frac{\Delta_2}{s_2} + \frac{\Delta_3}{s_1 + s_2} - \frac{\nu_2 - \Xi_2/2}{s_2^2} - \frac{\nu_3 - \Xi_3/2}{(s_1 + s_2)^2}, \tag{11.93}$$

$$\text{and} \quad \frac{\partial^2 f(s_1, s_2)}{\partial s_1 \partial s_2} = \gamma(s_1, s_2) \approx \frac{\Delta_3}{s_1 + s_2} - \frac{\nu_3 - \Xi_3/2}{(s_1 + s_2)^2}. \tag{11.94}$$

At the saddle point, Taylor's theorem may be used to show that $f(s_1, s_2)$ is given by

$$\begin{aligned} f(s_1, s_2) \approx\ & f(\tilde{s}_1, \tilde{s}_2) + \frac{(s_1 - \tilde{s}_1)^2}{2} \alpha(\tilde{s}_1, \tilde{s}_2) \\ & + \frac{(s_2 - \tilde{s}_2)^2}{2} \beta(\tilde{s}_1, \tilde{s}_2) + (s_1 - \tilde{s}_1)(s_2 - \tilde{s}_2)\, \gamma(\tilde{s}_1, \tilde{s}_2). \end{aligned} \tag{11.95}$$

Only terms up to quadratic order in s_1 and s_2 are kept. As in the case of a single complex variable, higher-order terms will generally not contribute significantly to the integral. The first term in the expansion given above does not depend upon the integration variables and may be moved outside the integral. The remaining integral is evaluated along the paths of steepest descent, i.e., along the paths where the imaginary part of $f(s_1, s_2)$ is constant. The result is a 2-dimensional analog of the Gaussian integral that is obtained by the application of steepest-descent techniques to an integral over a single complex variable.

First let us restrict ourselves to the case when Δ_1, Δ_2, and Δ_3 are all positive. In that case when $|x_1|$ and $|x_2|$ are large, the saddle points occur where $|s_1|$, $|s_2|$, and $|s_1 + s_2|$ are large. This fact allows one to neglect terms with the square of the complex variables in the denominators of (11.92)–(11.94). The final result of this manipulation is

$$h(x_1, x_2) \sim (2\pi)^{[(\Delta_1+\Delta_2+\Delta_3)/2-1]} \frac{\exp[f(\tilde{s}_1, \tilde{s}_2)]}{\sqrt{\alpha(\tilde{s}_1, \tilde{s}_2)\,\beta(\tilde{s}_1, \tilde{s}_2) - \gamma^2(\tilde{s}_1, \tilde{s}_2)}}. \tag{11.96}$$

The term under the square root can be expressed as

$$\alpha(\tilde{s}_1, \tilde{s}_2)\,\beta(\tilde{s}_1, \tilde{s}_2) - \gamma^2(\tilde{s}_1, \tilde{s}_2) = \frac{\Delta_1\Delta_2}{\tilde{s}_1\,\tilde{s}_2} + \frac{\Delta_1\Delta_3}{\tilde{s}_1\,(\tilde{s}_1 + \tilde{s}_2)} + \frac{\Delta_2\Delta_3}{(\tilde{s}_1 + \tilde{s}_2)\,\tilde{s}_2}. \tag{11.97}$$

This term is referred to as Hesse's determinant, and can be generalized to N complex variables (Ishimaru 1991).

If Δ_1, Δ_2, and Δ_3 are all positive, then one can find the saddle points. The second derivatives are large, and the steepest-descent contribution can be found. This typically is a difficult procedure because the equations to be solved are nonlinear.

For many integrals to be evaluated and all the problems in turbulence that I have considered, a simpler procedure applies. Looking at the Mellin transform tables, one sees that the sums of the coefficients of s in the numerator minus those in the denominator are either positive or zero. One could possibly obtain a negative sum by finding the Mellin transform of the inverse of the variable. However, for turbulence problems and many integrals, the functions do not have the inverse of the argument, and the quantities Δ_1, Δ_2, and Δ_3 are not all positive. First, assume that they are all nonzero and arrange the functions so that Δ_3 is negative. Because of the difference in sign, a large parameter size does not guarantee that the second derivatives given in (11.92)–(11.94) are large or even positive. It is not clear that there is a steep saddle point in this case that allows the line integral to be approximated by an integration along a short segment. I will not discuss this case further.

A case that often occurs in practice is the one in which one of the Δ's is zero. The signs of the complex variables can be changed if necessary so that the other two Δ's are positive. The functions can be arranged in the integral so that the Δ that is zero is Δ_3. In this case the saddle point is approximately located at

$$s_1 \approx x_1^{1/\Delta_1}, \quad \text{and} \quad s_2 \approx x_2^{1/\Delta_2}. \tag{11.98}$$

In this case if x_1 is large, then s_1 is large, and if x_2 is large, then s_2 is large. The derivatives higher than second order are much smaller than the second order ones. The steepest-descent results in (11.96) apply. Notice that in (11.97) only one of the three terms on the right is non-zero. The denominator term has the product of the two non-zero Δ's. This suggests that the double integral might be broken into the product of two single integrals. Thus examine $f(s_1, s_2)$ in (11.84). With $\Delta_3 = 0$, the only term that couples the two complex variables is $\ln(s_1 + s_2)$. As in the single complex plane case, the dominant term in this exponential function are terms of the form $s\ \ln(s)$. There are two terms of that form due to Δ_1 and Δ_2, and the coupling term is small compared to these terms. Since I am only interested in calculating the first term of the steepest-descent contribution, this coupling is neglected, and the double integral is performed as a cascade of two single integrals.

When one of the Δ's is zero, the three cases of relative parameter magnitude are all treated similarly to obtain the steepest-descent contribution. In the first case where the magnitudes of both parameters are small, the integral over one of the two complex variables is evaluated to obtain the pole contributions. One or more of the resulting power series depend on the ratio of the parameters. For only those series in which this ratio can be large is the second integral examined to obtain solely the steepest-descent contributions. For the second case when one parameter is large and the other small, the integration over the complex variable associated with the large variable is performed first to obtain only the steepest-descent contribution. The remaining integral is performed with pole-residue techniques. Doing the integration in this order reduces the amount of algebra. In the final case both parameters are large. One obtains the steepest-descent contribution to the double integral by first evaluating the asymptotic solution to the first integral. This includes both the asymptotic series and the steepest-descent contribution. The second integral is then evaluated to obtain only the steepest-descent contribution from each series, and the steepest-descent contribution and asymptotic series for the steepest-descent result of the first integration.

In actuality the functions do not have to be arranged so that when one of the Δ's is zero it is Δ_3. Suppose Δ_2 is zero, by making the substitutions $s_1 + s_2 \to s_2$ and $s_2 \to s_1 + s_2$ one can make the same arguments as above. Therefore, the integrals can be evaluated one at a time as long as any of the Δ's is zero. *In fact, even though it was assumed that the coefficients of s in the gamma functions were unity in the derivation of (11.96), for the case in which the integration is performed sequentially, the coefficients of s can be arbitrary.* This procedure is illustrated in Sects. 12.4 to 12.7.

A similar analysis applies for more than two complex planes when one of the Δ's is zero. Arrange the functions in the integrand so that the Δ associated with the sum of the N complex variables is zero. The function in the exponential, $f(s_1, s_2, \dots s_N)$, can again be approximated by the sum of terms that each depend on a single complex variable, thereby allowing one to perform the integration as a cascade of integrals.

If two of the Δ's are zero, that implies that either Δ_{s_1} or Δ_{s_2} is zero. In this case the series in one of the variables is a convergent series with a net factorial of zero. The asymptotic series has to be found in only one variable. One of Poincaré's theorems on asymptotic series given in Sect. 5.2 states that the integral of asymptotic series is equal to the asymptotic series of the integral. Therefore, the asymptotic solution can be found for the integral whose complex variable has the non-zero value of Δ_s, and the remaining integral can be evaluated with pole-residue techniques. This applies even if Δ_1, Δ_2, and Δ_3 are all positive.

Above I showed that if Δ_1, Δ_2, and Δ_3 are all positive and both parameters are large, then it is not possible in general to obtain a steepest-descent contribution to the integral. If only one of the parameters is large, then an asymptotic solution can be found for one variable and pole-residue integration performed for the second. This can be generalized to the case of N parameters. Define Δ_1 to Δ_{N+1} as in (11.78). If all these Δ's are positive and only one parameter is large, then one evaluates the integral by obtaining the steepest-descent contribution of the complex variable associated with the large parameter followed by evaluating the remaining multiple integral by the method described in this chapter.

The results of this chapter are now summarized. To evaluate an integral with N parameters, one must first determine the locations of the N-poles and the power series associated with them. The Taylor series solution is simply the sum of those series that converge in the parameter regime of interest. Conditions for convergence are given in this chapter. An exact convergent solution is always obtained except for certain parameter regimes when more than one of the Δ's is zero. In those parameter regimes no valid solution is obtained.

If terms of the power series are subtracted from some integrand functions, then the summations do not start at zero. Criteria for the inclusion of additional partial series in the solution were developed.

If some of the series converge slowly, it is often possible to obtain an asymptotic solution for the integral. This solution is composed of a sum of series plus a steepest-descent contribution. To find the series solution, express all power series such that the parameters in the series are raised to a positive exponent. The asymptotic series part of the solution is simply the sum of all series that have all parameters in the summation less than unity for the parameter regime of interest.

The steepest-descent contribution can be found for those cases in which the integrals can be arranged so that either the result of the first integration is a rapidly convergent series for the parameter range of interest or one of the Δ's is zero. For these cases one finds the steepest-descent part of the solution by performing the integrals sequentially. There are several ways in which the first integral can result in a convergent series:

1. We have seen that the result of the first integration is a rapidly convergent series regardless of the size of the parameter when either Δ_{s_1} or Δ_{s_2} is zero. In this case separate power series are obtained for small and large parameters.

2. If neither Δ_{s_1} nor Δ_{s_2} is zero, then the order of the integrals sometimes may be arranged so that the single power series that results from the first integration converges rapidly for the parameter range of interest. In this case the steepest-descent contribution is found when the second integral is subsequently evaluated.

3. If neither Δ_{s_1} nor Δ_{s_2} is zero but $\Delta_1 = \pm\Delta_2$, then by making the substitutions $s \to \mp s$, $t \to s + t$, and $s + t \to t$ one obtains $\Delta_{s_1} = 0$. This has now been reduced to the first case.

For cases where a steepest-descent contribution cannot be found by sequential evaluation of the integrals in the complex plane and all the Δ's are positive, one can obtain that contribution by solving the nonlinear equations given in this section. For the remaining cases it is not clear that an asymptotic solution with a single-term, steepest-descent contribution can be found that gives an accurate approximation to the integral.

12. Examples with N Parameters

The mathematical technique developed in the last chapter is used to find analytical solutions to seven problems. The first three are purely mathematical problems, and the last four are problems related to turbulence.

The first three examples have solutions that are in integral tables. The same solution and range of applicability is obtained with the Mellin transform method. The second problem, which contains a product of three Bessel functions, has applicability in the calculation of the correlation coefficients of various modes on a circular aperture. The third example has an integrand that contains a sinusoid and the product of Bessel functions. Two cases are considered, in the first there are three Bessel functions and in the second there are N. This example is the only one given in this book that illustrates how to use the technique in more than two complex planes.

For the first four problems asymptotic solutions are not required. In the last three problems they are required. It may not be necessary to use asymptotic solutions for certain applications since if enough Taylor series terms are used they can provide a sufficiently accurate answer. In the calculation of the power spectral density in Sect. 7.7 a 40-term Taylor series was sufficient to calculate the spectrum to large frequencies. The asymptotic solution does have one advantage over the Taylor series in that the dependence on parameters is clearly evident.

Four problems in wave propagation in turbulence are solved. These problems were chosen to illustrate the various techniques one must use to obtain asymptotic solutions under different conditions. The first three are generalizations of problems solved earlier in the book. It is shown that these solutions approach the former solutions at the proper limits. These three solutions determine the effect of outer scale on tilt anisoplanatism, the effect on tilt variance with both inner-scale and outer-scale effects present, and the tilt power spectral density with outer-scale effects. The fourth solution gives phase structure and correlation functions with inner and outer scale effects. These solutions demonstrate one of the most valuable assets of this technique. From the first term that couples the two effects, one can determine when the coupling starts to affect the result. From this term one can determine when coupling can be neglected. The specific form of the coupling is difficult to ascertain when numerical methods are used to evaluate the integral.

For notational convenience, in two-complex-plane problems the two complex variables are designated s and t rather than s_1 and s_2. The notation introduced in the last chapter is used extensively in this chapter. For convenience it is repeated here.

$$S(p_1, \dots, p_k) = \sum_{n_1=p_1}^{\infty} \cdots \sum_{n_k=p_k}^{\infty} \frac{(-1)^{n_1+\cdots+n_k}}{n_1! \cdots n_k!}. \tag{12.1}$$

If an a is placed next to the limit, it indicates that the series is asymptotic and only a finite number of terms should be summed. The argument of $S(p_1, \dots, p_k)$ has the n limit first followed by the m limit. Sometimes when there are multiple series, the summations are separated by other terms. In that case there will be two terms of the form $S(0)$, and I will add either an n or an m subscript to make it clear which summation is first.

The other function gives the factorial behavior of series terms as the summation index goes to infinity.

$$F(\Omega_1, \dots, \Omega_k) = (n_1!)^{\Omega_1} \cdots (n_k!)^{\Omega_k}. \tag{12.2}$$

12.1 An Integral with Two Bessel Functions and a Sinusoid

This first example is chosen because the value of the integral is a particularly simple expression. Consider the integral in *Gradshteyn* and *Ryzhik* (1980, Eq. 6.711 # 3)

$$I = \int_0^\infty du\, u^{\nu-\mu-2} J_\mu(au)\, J_\nu(bu) \sin(cu) = \frac{2^{\nu-\mu-1} a^\mu b^{-\nu} c\, \Gamma[\nu]}{\Gamma[\mu+1]}, \tag{12.3}$$

with $a > 0$, $b > 0$, $b - a > c > 0$, and $0 < \mathrm{Re}\{n\} < \mathrm{Re}\{m\} + 3$. The last condition results from the requirement that the integral must converge at the upper and lower limits.

Without loss of generality, the functions can be arranged so that $a < b$. This integral is put into the standard form by the following transformations $au = y$, $y = a/b$, and $z = a/c$, to give

$$I = a^{-\nu+\mu+1} \int_0^\infty \frac{dx}{x} x^{\nu-\mu-1} J_\mu(x)\, J_\nu\left(\frac{x}{y}\right) \sin\left(\frac{x}{z}\right).$$

This can be converted into an integral in two complex planes with the Mellin transforms in (1.49) and (1.46) and the Mellin convolution integral given in (1.30). After the substitutions $s \to 2s$ and $t \to 2t$, one obtains

$$I = \frac{\sqrt{\pi}}{2} \left(\frac{a}{2}\right)^{1-\nu+\mu} \frac{1}{(2\pi i)^2} \int_{C_1} \int_{C_2} ds\, dt \left(\frac{a}{b}\right)^{-2s} \left(\frac{a}{c}\right)^{-2t}$$

$$\times \Gamma\left[\begin{matrix} s+t+\nu/2-1/2,\ -s+\nu/2,\ -t+1/2 \\ -s-t-\nu/2+\mu+3/2,\ s+1+\nu/2,\ t+1 \end{matrix}\right].$$

Table 12.1. Range of applicability of various series.

2-Pole Set	Solution Type	Parameter Range	
1	Taylor	$b < a$	$c < a$
2	Taylor	$b < c$	$a < c$
3	Taylor	$a < b$	$c < b$

Since $\Delta_s = \Delta_t = 0$, there is no asymptotic series.

The technique discussed in Chap. 11 is used to evaluate this integral. The locations of the 2-poles are:

$$\begin{array}{ll} (1) & -t + 1/2 = -m;\ -s + \nu/2 = -n, \\ (2) & -s + \nu/2 = -n;\ s + t + \nu/2 - 1/2 = -m, \text{ and} \\ (3) & -t + 1/2 = -m;\ s + t + \nu/2 - 1/2 = -n. \end{array} \tag{12.4}$$

To determine which 2-pole contributions must be retained, the above relations are used to find the power-law dependence of the parameters for large values of m and n. If the exponents have m and n indices in the same exponent, one must separate the terms by defining new parameters. This statement is made clearer when one performs this process for the three sets of two-poles to give:

$$\begin{array}{ll} (1) & \left(\frac{b}{a}\right)^{2n} \left(\frac{c}{a}\right)^{2m} F(0,0), \\ (2) & \left(\frac{b}{c}\right)^{2n} \left(\frac{a}{c}\right)^{2m} F(0,0), \text{ and} \\ (3) & \left(\frac{a}{b}\right)^{2n} \left(\frac{c}{b}\right)^{2m} F(0,0). \end{array} \tag{12.5}$$

As discussed in Chap. 11 the values of Ω_1 and Ω_2 in $F(\Omega_1, \Omega_2)$ are the values of the exponent of the factorial function. These are referred to as the "net factorial power", and they determine the series type. If both values are non-positive, then the series is a Taylor series, otherwise it is an asymptotic series. The parameters with a single index in the exponent are the natural parameters of the problems, and are independent of initial substitutions made to put the integral into the canonical form. The constant terms in the exponents do not affect series convergence and, therefore, the direction of path closure, and have not been displayed explicitly. The three power series are Taylor series that apply to different parameter ranges. The range of applicability of the various series is given in Table 12.1.

From the condition $b - a > c > 0$ the first two series diverge, in which case only the third series is convergent, and the integral is equal to

$$I = \sqrt{\pi}\frac{a^{\mu}\, c}{b^{\nu}\, 2^{2-\nu+\mu}} S\,(0,0) \left(\frac{a}{b}\right)^{2n} \left(\frac{c}{b}\right)^{2m} \Gamma\left[\begin{matrix} m+n+\nu \\ n+\mu+1, -m-n+1,\, m+3/2 \end{matrix}\right]. \tag{12.6}$$

The term $\Gamma[-m-n+1]$ in the denominator is infinite at all n and m except $m = n = 0$. Therefore, all residues vanish except those at $m = n = 0$. The single series term gives the result in (12.3). The particular value of the power of x produces this simple form for the result. The expression is more complicated for a general exponent in the original integral, and this method can be used to get a double-series solution for this case. This will not be pursued here since this type of series is illustrated in the next example.

One must examine the convergence region of this solution. Since the net factorial power for both summation indices is zero, one has to use the criteria for convergence of the power series given in Sect. 11.2. Even though terms other than the first are zero, in order to get the first term, one had to close the integration path at infinity to encompass all terms of this series; therefore, the convergence criteria must be applied as if all series terms are non-zero. Equation (11.8) gives $\Phi_1(m,n) = (m+n)^2/m^2$, and $\Phi_2(m,n) = (m+n)^2/n^2$. From (11.9) one finds the first of the two regions is a rectangle with sides of unity length. The parametric equations given by (11.10) are $\xi = |x| = m^2/(m+n)^2$, and $\eta = |y| = n^2/(m+n)^2$. These equations satisfy the relation $\sqrt{x} + \sqrt{y} < 1$.

Since $x = (a/b)^2$ and $y = (c/b)^2$, one requires that $a/b + c/b < 1$. The area in the x–y plane defined by this condition is contained within the unit square; therefore, this more restrictive condition defines the region of convergence. From this, one obtains the same condition as that in the integral tables given by $b - a > c$.

12.2 An Integral with Three Bessel Functions

Consider the following integral whose value is needed to evaluate the Zernike covariance matrix

$$I = \int_0^{\infty} \frac{d\omega}{\omega}\, \omega^{d} J_{\mu}(a\omega)\, J_{\nu}(b\omega)\, J_{\rho}(c\omega). \tag{12.7}$$

This integral is tabulated in *Gradshteyn and Ryzhik* (1980, Eq. 6.578 # 1). The condition for convergence at the lower limit is $\mu+\nu+\rho+d > 0$. For convergence at the upper limit one requires $\mathrm{Re}\,\{d\} < 5/2$. Without loss of generality, one can arrange the Bessel functions so that $c > b > a$. With this last condition the only region in parameter space for which a solution is required is the hatched region of Fig. 12.1.

Watson (1934) considered this integral but was unable to evaluate it. *Bailey* (1936) found its value when $a + b < c$, which is in the integral table, and is the result derived below.

Evaluate the integral by letting $x = c\omega$ to give

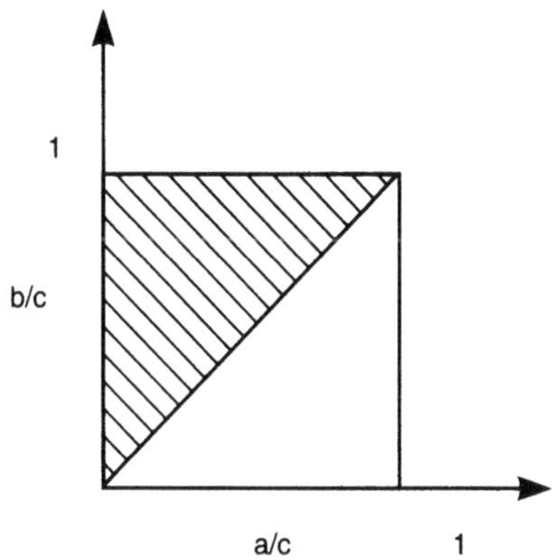

Fig. 12.1. Hatched area is the parameter region in which a solution is required for the integral of 3 Bessel functions.

$$I = c^{-d} \int_0^\infty \frac{dx}{x}\, x^d\, J_\rho(x)\, J_\mu\left[x/\left(c/a\right)\right]\, J_\nu\left[x/\left(c/b\right)\right].$$

Use the generalized Mellin convolution theorem and the Mellin transform in (1.49), and make the substitutions $s \to 2s$, and $t \to 2t$, to write the integral in the s and t complex planes as

$$I = \frac{2^{d-1}}{c^d} \frac{1}{(2\pi i)^2} \int_{C_1} \int_{C_2} ds\, dt \left(\frac{c}{b}\right)^{-2s} \left(\frac{c}{a}\right)^{-2t}$$

$$\times \Gamma\left[\begin{matrix} s+t+\left(d+\rho\right)/2,\ -s+\nu/2,\ -t+\mu/2 \\ -s-t+\left(\rho-d\right)/2+1,\ s+\nu/2+1,\ t+1+\mu/2 \end{matrix}\right].$$

Since $\Delta_s = \Delta_t = 0$, there is no asymptotic series. The 2-poles are located at:

1) $s+t+\left(d+\rho\right)/2 = -n\,;\ -s+\nu/2 = -m,$
2) $s+t+\left(d+\rho\right)/2 = -m\,;\ -t+\mu/2 = -n,$ and
3) $-s+\nu/2 = -n\,;\ -t+\mu/2 = -m.$

These 2-poles lead to terms of the type:

$$(1)\quad \left(\frac{c}{a}\right)^{2n} \left(\frac{b}{a}\right)^{2m} F\left(0,0\right),$$

$$(2)\quad \left(\frac{a}{b}\right)^{2n} \left(\frac{c}{b}\right)^{2m} F\left(0,0\right),\ \text{and}$$

$$(3)\quad \left(\frac{b}{c}\right)^{2n} \left(\frac{a}{c}\right)^{2m} F\left(0,0\right).$$

These three sets of 2-poles give three Taylor series that are applicable for different parameter ranges, given in Table 12.2.

Table 12.2. Range of applicability of various series for the problem in which the product of three Bessel functions is integrated.

2–Pole Set	Solution Type	Parameter Range	
1	Taylor	$c < a$	$b < a$
2	Taylor	$a < b$	$c < b$
3	Taylor	$b < c$	$a < c$

For the case being considered in which $c > b > a$, the power series that corresponds to the third pole set is the only one applicable, and the solution is

$$I = \frac{2^{d-1}a^{\mu}b^{\nu}}{c^{d+\mu+\nu}} S(0,0) \left(\frac{b}{c}\right)^{2n} \left(\frac{a}{c}\right)^{2m}$$
$$\times \Gamma \left[\begin{matrix} m+n+(\nu+\rho+\mu+d)/2 \\ -m-n+(\rho-\nu-\mu-d)/2+1,\ m+1+\mu,\ n+1+\nu \end{matrix} \right] .$$

The summation can be written as an Appell function of the fourth kind to get exactly the same result as in the integral tables

$$I = \frac{2^{d-1}a^{\mu}b^{\nu}}{c^{d+\mu+\nu}} \Gamma \left[\begin{matrix} (\nu+\rho+\mu+d)/2 \\ 1-(\nu-\rho+\mu+d)/2,\ 1+\nu,\ 1+\mu \end{matrix} \right]$$
$$\times F_4 \left[\frac{(\nu+\rho+\mu+d)}{2}, \frac{(\nu-\rho+\mu+d)}{2}, 1+\mu, 1+\nu; \left(\frac{a}{c}\right)^2, \left(\frac{b}{c}\right)^2 \right] . \quad (12.8)$$

Since the net factorial power for both summation indices is zero, one has to examine the convergence criteria of the power series in this case. The analysis to find the region of convergence proceeds exactly as in the last section, and the parametric equations are identical. Since in this problem $x = (b/c)^2$, and $y = (a/c)^2$, one obtains the same condition as in the integral tables $a + b < c$. The two potential regions of convergence represented by the above inequality and the unit square are shown in Fig. 12.2. Both hatched regions are ones in which the power series does not apply even though the parameters are less than one. The horizontally hatched region is half the area of interest as depicted in Fig. 12.1. In that region $1/2 < c/2b < 1$. For parameter values in this region, the integration path cannot be closed around the poles, and the above technique does not give an answer even though the integral converges. *Gervois* and *Navelet* (1985, 1986) recently gave results in the horizontally hatched region for special values of the order of the Bessel function and exponent in the integration. The analytic value of the integral for arbitrary parameter values is not known.

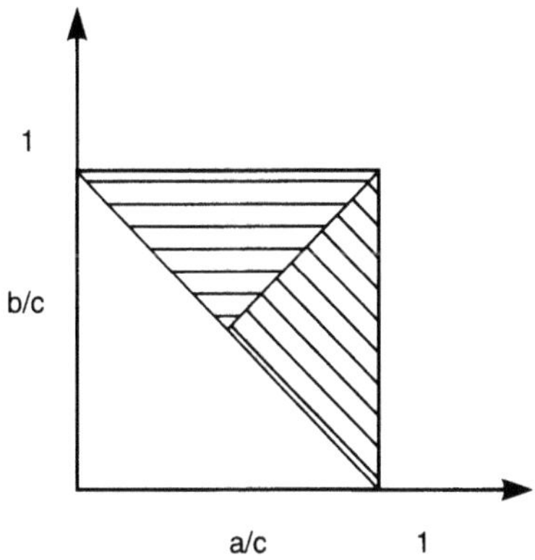

Fig. 12.2. Regions in which the power series has different convergence properties for the integral of 3 Bessel functions.

12.3 Example in Three and N Complex Planes

As an example of the technique in three complex planes, an integral similar to the first example, whose value has a particularly simple form, is worked out. By examining the steps in obtaining this solution, one can obtain by inspection the result of the integration in N complex planes. Consider the following integral

$$\begin{aligned} I &= \int_0^\infty du\, u^{\nu-\mu-\lambda-2}\, J_\nu(bu)\, \sin(cu)\, J_\mu(au)\, J_\lambda(du) \\ &= \frac{2^{\nu-\mu-\lambda-1}\, b^{-\nu}\, \Gamma[\nu]\, c\, a^\mu\, d^\lambda}{\Gamma[\mu+1]\, \Gamma[\lambda+1]}, \end{aligned} \tag{12.9}$$

with $a > 0$, $b > 0$, $d > 0$, $b > a + c + d > 0$, and $0 < \text{Re}\,\{\nu\} < \text{Re}\,\{\mu + \lambda\} + 3.5$. The last condition results from the convergence requirement at the upper and lower limits.

Without loss of generality, functions can be arranged so that $a < b$ and $d < b$. This integral is converted into standard form by the following transformations $au = x$, $y = a/b$, $z = a/c$, and $w = a/d$, to give

$$I = a^{-\nu+\mu+\lambda+1} \int_0^\infty \frac{dx}{x}\, x^{\nu-\mu-\lambda-1} J_\mu(x)\, J_\nu\left(\frac{x}{y}\right) \sin\left(\frac{x}{z}\right) J_\lambda\left(\frac{x}{w}\right).$$

This can be converted into an integral in three complex planes with the Mellin transforms in (1.49) and (1.46) and the Mellin convolution integral given in (1.30). The third complex variable is r.

After the substitutions $s \to 2s$, $r \to 2r$, and $t \to 2t$, one obtains

$$I = \sqrt{\pi} a^{-\nu+\mu+\lambda+1} 2^{-2+\nu-\mu-\lambda} \frac{1}{(2\pi i)^3} \int_{C_1} \int_{C_2} \int_{C_3} ds\, dt\, dr \left(\frac{a}{b}\right)^{-2s} \left(\frac{a}{c}\right)^{-2t} \left(\frac{a}{d}\right)^{-2r}$$

$$\times \Gamma \begin{bmatrix} s+t+r+\nu/2-\lambda/2-1/2,\ -s+\nu/2,\ -t+1/2,\ -r+\lambda/2 \\ -s-t-r-\nu/2+\lambda/2+\mu+3/2,\ s+1+\nu/2,\ t+1,\ r+1+\lambda/2 \end{bmatrix}.$$

The technique discussed in the last chapter is used to evaluate this integral. The locations of the 3-poles are:

$$\text{(1)}\quad -t+\frac{1}{2} = -m;\ -s+\frac{\nu}{2} = -n;\ -r+\frac{\lambda}{2} = -p,$$

$$\text{(2)}\quad -s+\frac{\nu}{2} = -n;\ s+t+r+\nu/2-\frac{\lambda}{2}-\frac{1}{2} = -m;\ -r+\frac{\lambda}{2} = -p,$$

$$\text{(3)}\quad -t+\frac{1}{2} = -m;\ s+t+r+\frac{\nu}{2}-\frac{\lambda}{2}-\frac{1}{2} = -n;\ -r+\lambda/2 = -p, \text{ and}$$

$$\text{(4)}\quad s+t+r+\frac{\nu}{2}-\frac{\lambda}{2}-\frac{1}{2} = -p;\ -s+\frac{\nu}{2} = -n;\ -t+\frac{1}{2} = -m.$$

To determine which 3-pole contributions must be retained, the above relations are used to find the power-law dependence of the variables for large values of m, n, and p, which are:

$$\text{(1)}\quad \left(\frac{b}{a}\right)^{2n} \left(\frac{c}{a}\right)^{2m} \left(\frac{d}{a}\right)^{2p} F(0,0,0),$$

$$\text{(2)}\quad \left(\frac{b}{c}\right)^{2n} \left(\frac{a}{c}\right)^{2m} \left(\frac{d}{c}\right)^{2p} F(0,0,0),$$

$$\text{(3)}\quad \left(\frac{a}{b}\right)^{2n} \left(\frac{c}{b}\right)^{2m} \left(\frac{d}{b}\right)^{2p} F(0,0,0), \text{ and}$$

$$\text{(4)}\quad \left(\frac{b}{d}\right)^{2n} \left(\frac{c}{d}\right)^{2m} \left(\frac{a}{d}\right)^{2p} F(0,0,0).$$

All four lead to Taylor series that converge for different parameter ranges. It will be required that $c < b$, and since $a < b$ and $d < b$, only the third Taylor series converges so that the integral is equal to

$$I = \sqrt{\pi} a^{\mu} c\, 2^{-2+\nu-\mu-\lambda} b^{-\nu} d^{\lambda} S(0,0,0) \left(\frac{a}{b}\right)^{2n} \left(\frac{c}{b}\right)^{2m} \left(\frac{d}{b}\right)^{2p}$$

$$\times \Gamma \begin{bmatrix} m+n+p+\nu \\ n+\mu+1,\ -m-n-p+1,\ m+3/2,\ p+\lambda+1 \end{bmatrix}. \qquad (12.10)$$

The term $\Gamma[-m-n-p+1]$ in the denominator is infinite at all n, m, and p except for $m = n = p = 0$. Therefore, all poles have zero residue except the

3-pole at $m = n = p = 0$. The one term that remains gives the same result as that in (12.9). The particular form of the power of x produces the simple form for the result, and the expression is more complicated for a general exponent.

As before, because the net factorial power for the three summation indices is zero, one has to examine the criteria for convergence of Sect. 11.2. Equation (11.15) gives $\Phi_1(m,n,p) = (m+n+p)^2/m^2$, $\Phi_2(m,n,p) = (m+n+p)^2/n^2$, and $\Phi_3(m,n,p) = (m+n+p)^2/p^2$. From (11.17) one finds that the hyper-rectangle of possible convergence has sides of unity length. The parametric equations to determine the second convergence region are $\xi = |x| = n^2/(m+n+p)^2$, $\eta = |y| = m^2/(m+n+p)^2$, and $\phi = |z| = p^2/(m+n+p)^2$.These conditions require $\sqrt{x} + \sqrt{y} + \sqrt{z} < 1$.

Since $x = (a/b)^2$, $y = (c/b)^2$, and $z = (d/b)^2$, then $b > c + a + d$. The volume in the x–y–z plane defined by this condition is contained within the hyper-rectangle; therefore, this more restrictive condition defines the convergence region. The above result agrees with the special case of the integral in *Prudnikov* et al. (1990, Eq. 7 of Sect. 2.12.44) when one notes that

$$J_{1/2}(cx) = \sqrt{\frac{2}{\pi cx}} \sin(cx). \tag{12.11}$$

The extension of this technique to evaluate integrals with more Bessel functions, which require more than three complex planes, is obvious. The result is

$$\begin{aligned} I &= \int_0^\infty du\, u^{\left(\nu-2-\sum_{i=1}^{N} a_i\right)} J_\nu(bu) \sin(cu) \prod_{i=1}^{N} J_{a_i}(d_i u) \\ &= 2^{\left(\nu-1-\sum_{i=1}^{N} a_i\right)} \frac{c}{b^\nu} \Gamma[\nu] \prod_{i=1}^{N} \frac{d_i^{a_i}}{\Gamma[1+a_i]}. \end{aligned} \tag{12.12}$$

For convergence at the lower and upper limits, one requires

$$0 < \mathrm{Re}\{\nu\} < \mathrm{Re}\left\{\sum_{i=1}^{N} a_i\right\} + \frac{N+5}{2}. \tag{12.13}$$

The convergence criteria are easily extended into N-complex planes to obtain $b > c + \sum_{i=1}^{N} d_i$. This integration formula also agrees with the special case of the above-mentioned integral in *Prudnikov* et al. (1990).

12.4 Effect of Outer Scale on Tilt Anisoplanatism

Tilt anisoplanatism with an infinite outer scale was considered in Sect. 7.6. Here this problem is solved with a finite outer turbulence scale for the von Kármán spectrum. I will show that outer scale's effect is not as significant here as its effect on tilt in Sect. 7.1. The equivalent of (7.55) with outer-scale effects included is

$$\begin{bmatrix} \sigma_{\parallel}^2 \\ \sigma_{\perp}^2 \\ \sigma^2 \end{bmatrix} = 0.4146 \left(\frac{16}{D}\right)^2 \int_0^L dz\, C_n^2(z) \int d\boldsymbol{\kappa} \begin{bmatrix} \cos^2(\varphi) \\ \sin^2(\varphi) \\ 1 \end{bmatrix}$$

$$\times \left(\kappa^2 + \kappa_o^2\right)^{-11/6} \left[\frac{J_2\left(\kappa D/2\right)}{\kappa D/2}\right]^2 \left\{1 - \cos\left[\kappa d \cos\left(\varphi\right)\right]\right\}. \qquad (12.14)$$

Let $w = \kappa d$, $y = 2\pi d/L_o$, and $x = 2d/D$, and integrate over angle. The equivalent of (7.61) is

$$\begin{bmatrix} I_1 \\ I_T \end{bmatrix} = -\frac{4\,d^{11/3}}{D^2} \int_0^\infty \frac{dw}{w} \begin{bmatrix} w^{-14/3}\left[J_1(w) - \frac{w}{2}\right] \\ w^{-11/3}\left[J_0(w) - 1\right] \end{bmatrix} J_2^2\left(w/x\right) \left[1 + \left(y/w\right)^2\right]^{-11/6}.$$

Using the Mellin convolution theorem with the Mellin transforms in (1.49), (1.50), and (1.52), and making the substitutions $s \to 2s$ and $t \to 2t$, one can convert this into an integral in two complex planes as

$$\begin{bmatrix} I_1 \\ I_T \end{bmatrix} = \frac{-0.0945\,d^{11/3}}{D^2} \frac{1}{(2\pi i)^2} \int_{C_1}\int_{C_2} ds\,dt \left(\frac{\pi d}{L_o}\right)^{-2s} \left(\frac{d}{D}\right)^{-2t}$$

$$\times \Gamma\left[\begin{matrix} s + t - \frac{11}{6}*,\ -t + 2,\ t + \frac{1}{2},\ s,\ -s + \frac{11}{6} \\ t + 3,\ t + 1 \end{matrix}\right] \begin{bmatrix} \dfrac{1}{2\Gamma\left[-s-t+\frac{23}{6}\right]} \\ \dfrac{1}{\Gamma\left[-s-t+\frac{17}{6}\right]} \end{bmatrix}.$$

The asterisk indicates that the integration path passes between the first and second poles of that gamma function.

The list of 2-poles are:

(1) $-s + \frac{11}{6} = -m;\ t + \frac{1}{2} = -n,$
(2) $-s + \frac{11}{6} = -m;\ -t + 2 = -n,$
(3) $-s + \frac{11}{6} = -m;\ s + t - \frac{11}{6} = -n*,$
(4) $s = -m;\ \frac{1}{2} + t = -n,$
(5) $s = -m;\ -t + 2 = -n,$
(6) $s = -m;\ s + t - \frac{11}{6} = -n*$
(7) $t + \frac{1}{2} = -n;\ s + t - \frac{11}{6} = -m*$, and
(8) $-t + 2 = -n;\ s + t - \frac{11}{6} = -m*.$

The values of m and n are integers that vary between 0 and ∞. The asterisk after a term means that the index varies between 1 and ∞ on one side of the integration path, and is equal to 0 on the other. The general terms of the various possible series are:

$$(1) \quad \left(\frac{L_o}{\pi d}\right)^{2m} \left(\frac{d}{D}\right)^{2n} F\left(0, 2\right),$$

$$(2) \quad \left(\frac{L_o}{\pi d}\right)^{2m} \left(\frac{d}{D}\right)^{2n} F\left(0, 2\right),$$

Table 12.3. Series ranges of applicability for most rapid convergence, for the problem where the effect of outer scale on tilt anisoplanatism is found.

2-Pole Set	Solution Type	Parameter	Range
1	Asymp	$\pi D/L_o > 1$	$d < D$
2	Asymp	$\pi d/L_o > 1$	$d < D$
3	Asymp	$\pi D/L_o > 1$	$d < D$
4	Taylor	$\pi d/L_o < 1$	$d < D$
5	Taylor	$\pi d/L_o < 1$	$d > D$
6	Taylor	$\pi D/L_o < 1$	$d < D$
7	Asymp	$\pi D/L_o > 1$	$\pi d/L_o < 1$
8	Taylor	$\pi D/L_o < 1$	$\pi d/L_o < 1$

$$(3) \quad \left(\frac{L_o}{\pi D}\right)^{2m} \left(\frac{d}{D}\right)^{2n*} F(0,2),$$

$$(4) \quad \left(\frac{\pi d}{L_o}\right)^{2m} \left(\frac{d}{D}\right)^{2n} F(0,-2),$$

$$(5) \quad \left(\frac{\pi d}{L_o}\right)^{2m} \left(\frac{D}{d}\right)^{2n} F(0,-2),$$

$$(6) \quad \left(\frac{d}{D}\right)^{2n*} \left(\frac{\pi D}{L_o}\right)^{2m} F(0,-2),$$

$$(7) \quad \left(\frac{L_o}{\pi D}\right)^{2n} \left(\frac{\pi d}{L_o}\right)^{2m*} F(2,-2), \text{ and}$$

$$(8) \quad \left(\frac{\pi D}{L_o}\right)^{2n} \left(\frac{\pi d}{L_o}\right)^{2m*} F(-2,-2).$$

This process produces the parameters that are relevant to the problem. The new parameter introduced above is $\pi D/L_o$. In this analysis, this ratio as well as $\pi d/L_o$ is considered to be less than 1. To determine which pole residues must be included, the rule given in the last chapter must be applied. These results are summarized in Table 12.3. Series 1, 2, 3, and 7 are asymptotic series that apply when the displacement or diameter is much greater than the outer-scale size. I am not interested in this parameter range; therefore, these series are discarded. For $d/D < 1$, Series 4 for all m and n, and Series 6 for all m and n greater than 0 contribute. For $d/D > 1$, Series 5 for all n and m, and Series 6 for $n = 0$ and all m contribute. Series 8 contributes for all ranges of d/D for all n and for m greater than 0. Series 6 corresponds to a series of type B_1 in Table 11.1 and the term with $n = 0$ must be added to the solution when $d/D > 1$.

Since the only relevant series are Taylor series, steepest-descent contributions do not have to be added to the solution. With these properties the value of the integral for small displacements, $d/D < 1$, can be expressed as

$$\begin{bmatrix} I_1 \\ I_T \end{bmatrix} = \frac{-0.0945\, d^{11/3}}{D^2}$$
$$\times\left\{S(0,0)\left(\frac{\pi d}{L_o}\right)^{2m}\left(\frac{d}{D}\right)^{2n+1}\Gamma\begin{bmatrix} -m-n-\frac{7}{3},\ n+\frac{5}{2},\ m+\frac{11}{6} \\ -n+\frac{5}{2},\ -n+\frac{1}{2},\ n+m+\frac{10}{3} \end{bmatrix}\begin{bmatrix} \frac{0.5}{n+m+\frac{10}{3}} \\ 1 \end{bmatrix}\right.$$
$$+S(1,0)\left(\frac{\pi D}{L_o}\right)^{2m}\left(\frac{d}{D}\right)^{-\frac{11}{3}+2n}\Gamma\begin{bmatrix} n-m+\frac{1}{6},\ m-n+\frac{7}{3},\ m+\frac{11}{6} \\ -n+m+\frac{29}{6}, -n+m+\frac{17}{6},\ n+1 \end{bmatrix}\begin{bmatrix} \frac{0.5}{n+1} \\ 1 \end{bmatrix}$$
$$\left.+\left(\frac{D}{d}\right)^4 S(0,1)\left(\frac{\pi d}{L_o}\right)^{2m+\frac{1}{3}}\left(\frac{\pi D}{L_o}\right)^{2n}\Gamma\begin{bmatrix} n+\frac{5}{2}, -m-n-\frac{1}{6},\ m+n+2 \\ n+5,\ n+3,\ m+1 \end{bmatrix}\begin{bmatrix} \frac{0.5}{m+1} \\ 1 \end{bmatrix}\right\}. \tag{12.15}$$

For large displacements, $d/D > 1$, the integral is equal to

$$\begin{bmatrix} I_1 \\ I_T \end{bmatrix} = \frac{-0.0945\, d^{11/3}}{D^2}\left\{\left(\frac{D}{d}\right)^4 S(0,1)\left(\frac{\pi d}{L_o}\right)^{2m+1/3}\left(\frac{\pi D}{L_o}\right)^{2n}\right.$$
$$\times\Gamma\begin{bmatrix} n+\frac{5}{2},\ -m-n-\frac{1}{6},\ m+n+2 \\ n+5,\ n+3,\ m+1 \end{bmatrix}\begin{bmatrix} \frac{0.5}{m+1} \\ 1 \end{bmatrix}$$
$$+S(0,0)\left(\frac{\pi d}{L_o}\right)^{2m}\left(\frac{d}{D}\right)^{-4-2n}\Gamma\begin{bmatrix} -m+n+\frac{1}{6},\ n+\frac{5}{2},\ m+\frac{11}{6} \\ n+5,\ n+3,\ m-n+\frac{5}{6} \end{bmatrix}\begin{bmatrix} \frac{0.5}{m-n+\frac{5}{6}} \\ 1 \end{bmatrix}$$
$$\left.-\left(\frac{D}{d}\right)^{11/3} S(0)\left(\frac{\pi D}{L_o}\right)^{2m}\Gamma\begin{bmatrix} -m+\frac{1}{6},\ m+\frac{7}{3},\ m+\frac{11}{6} \\ m+\frac{29}{6},\ m+\frac{17}{6} \end{bmatrix}\begin{bmatrix} 0.5 \\ 1 \end{bmatrix}\right\}. \tag{12.16}$$

If the outer scale goes to infinity, only terms with $m = 0$ in the first two summations for large or small displacements contribute, and these expressions reduce to the ones in which the outer scale was neglected in (7.63) and (7.64).

For relatively small displacements, $d/D < 1$, the tilt variance is

$$\begin{bmatrix} \sigma_{\parallel}^2 \\ \sigma_{\perp}^2 \end{bmatrix} = \frac{-31.5\,\mu_0}{d^{1/3}}$$
$$\times\left\{S(0,0)\left(\frac{\pi d}{L_o}\right)^{2m}\left(\frac{d}{D}\right)^{2n+5}\Gamma\begin{bmatrix} -m-n-\frac{7}{3},\ n+\frac{5}{2},\ m+\frac{11}{6} \\ -n+\frac{5}{2}, -n+\frac{1}{2}, n+m+\frac{13}{3} \end{bmatrix}\begin{bmatrix} 2n+2m+\frac{17}{3} \\ 1 \end{bmatrix}\right.$$
$$+S(1,0)\left(\frac{\pi D}{L_o}\right)^{2m}\left(\frac{d}{D}\right)^{2n+1/3}\Gamma\begin{bmatrix} n-m+\frac{1}{6},\ m-n+\frac{7}{3},\ m+\frac{11}{6} \\ -n+m+\frac{29}{6}, -n+m+\frac{17}{6}, n+2 \end{bmatrix}\begin{bmatrix} 2n+1 \\ 1 \end{bmatrix}$$
$$\left.+S(0,1)\left(\frac{\pi d}{L_o}\right)^{2m+1/3}\left(\frac{\pi D}{L_o}\right)^{2n}\Gamma\begin{bmatrix} n+\frac{5}{2}, -m-n-\frac{1}{6}, m+n+2 \\ n+5,\ n+3,\ m+2 \end{bmatrix}\begin{bmatrix} 2m+1 \\ 1 \end{bmatrix}\right\}. \tag{12.17}$$

The most significant terms for small displacements are the $n = 1$, $m = 0$, and the $n = 1$, $m = 1$ terms of the second summation, which give

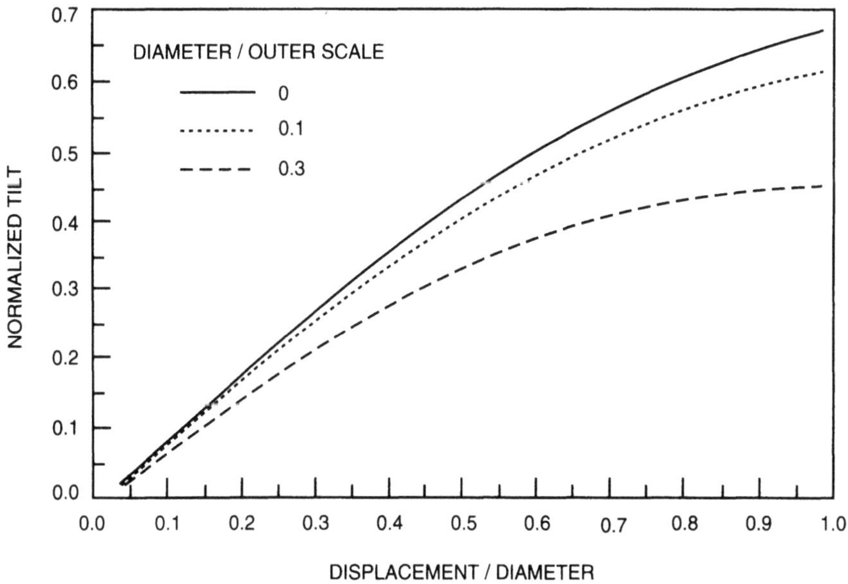

Fig. 12.3. Tilt parallel to the displacement, versus displacement, for various outer-scale sizes, normalized to the tilt on the aperture.

$$\begin{bmatrix} \sigma_{\parallel}^2 \\ \sigma_{\perp}^2 \end{bmatrix} \approx \frac{2.67\mu_0}{D^{1/3}} \left(\frac{d}{D}\right)^2 \begin{bmatrix} 3 \\ 1 \end{bmatrix} \left[1 - 20.6 \left(\frac{D}{L_o}\right)^2 + 27.4 \left(\frac{D}{L_o}\right)^{7/3}\right]. \tag{12.18}$$

The first term is the same as that in (7.66) in which outer scale was neglected. In Sect. 7.1 where tilt with outer scale was considered, outer scale had a significant effect on tilt because the leading term had the ratio of diameter to outer scale raised to the inverse one-third power. Here, the effect is small if the outer-scale size is significantly greater than the diameter because the leading term is the inverse second power of that ratio. The physical reason this occurs is that for small relative aperture displacements the two apertures see the same tilt from long-wavelength turbulence, which cancels when the tilts are subtracted from each other. Mathematically, this subtraction results from the requirement that the third summation starts at $m = 1$ rather than 0. The term that was eliminated has the one-third power-law dependence on outer scale.

Plots of tilt for various ratios of outer scale to diameter are given in Fig. 12.3 and Fig. 12.4 for the parallel and perpendicular components of tilt anisoplanatism, respectively. When outer scale is 100 times the diameter, the curve is virtually identical to that for infinite outer scale, and is not plotted on the graph. This is in contrast to the result for tilt with outer scale as seen in Fig.7.2, for which tilt is reduced by 15% for the same ratio of outer scale to diameter.

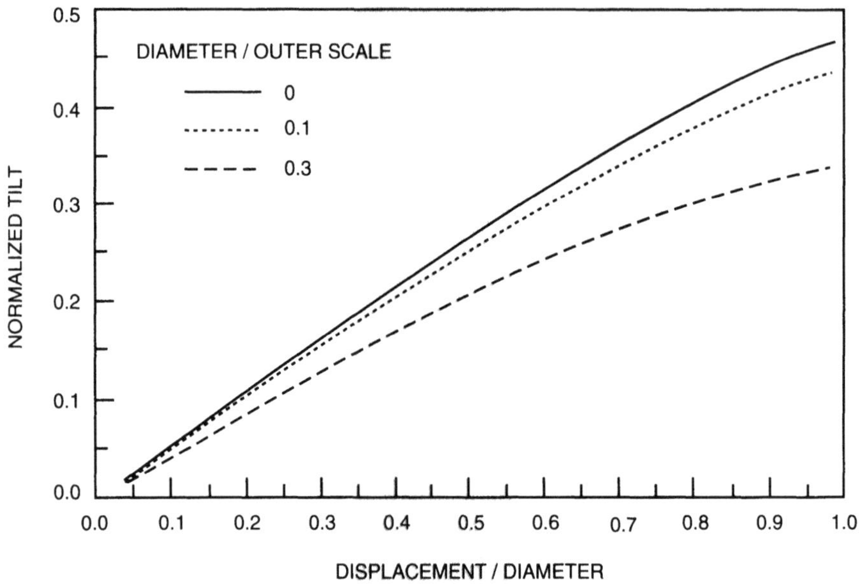

Fig. 12.4. Tilt perpendicular to the displacement, versus displacement, for various outer-scale sizes, normalized to the tilt on the aperture.

12.5 Tilt with Inner and Outer Scale

Tilt with outer scale was found in Sect. 7.1 and with inner scale in Sect. 7.3. Here tilt with both effects present is found. The integral expression for tilt with non-zero inner and finite outer scale can be found from (7.1) by the insertion of the Tatarski inner-scale term to obtain

$$\begin{aligned} T_o^2 &= 0.2073\, k_0^2 \int_0^L dz\, C_n^2(z) \int d\boldsymbol{\kappa}\, \left(\kappa^2 + \kappa_o^2\right)^{-11/6} \left(\frac{16}{k_0 D}\right)^2 \\ &\quad \times \left[\frac{J_2(\kappa D/2)}{\kappa D/2}\right]^2 \exp\left(-\kappa^2/\kappa_i^2\right). \end{aligned} \tag{12.19}$$

The integrations over angle and z can easily be performed. The substitution $x = \kappa D/2$ gives

$$T_o^2 = \frac{1334\, \mu_0}{D^4\, \kappa_o^{11/3}} \int_0^\infty \frac{dx}{x}\, J_2^2(x) \left[\left(\frac{x}{\kappa_o D/2}\right)^2 + 1\right]^{-11/6} \exp\left\{-\left[x/\left(\kappa_i D/2\right)\right]^2\right\}.$$

Using the Mellin convolution integral and the Mellin transforms given by (1.50), (1.52), and (1.45), and making the substitutions $s \to 2s$ and $t \to 2t$, one obtains

$$T_o^2 = \frac{400\mu_0 \kappa_o^{-11/3}}{D^4} \frac{1}{(2\pi i)^2} \int_{C_1}\int_{C_2} ds\, dt \left(\frac{\kappa_o D}{2}\right)^{-2s} \left(\frac{\kappa_i D}{2}\right)^{-2t}$$

$$\times \Gamma\left[\begin{matrix} s+t+2,\ -s-t+\frac{1}{2},\ -s,\ s+\frac{11}{6},\ -t \\ -s-t+3,\ -s-t+1 \end{matrix}\right]. \qquad (12.20)$$

From (11.77) one finds that $\Delta_s = 2$, and $\Delta_t = 1$. There are 8 sets of 2-poles that are given by:

$$\begin{array}{ll}
(1) & s+t+2=-n;\ -s=-m, \\
(2) & s+t+2=-n;\ s+\frac{11}{6}=-m, \\
(3) & s+t+2=-n;\ -t=-m, \\
(4) & -s-t+\frac{1}{2}=-n;\ -s=-m, \\
(5) & -s-t+\frac{1}{2}=-n;\ s+\frac{11}{6}=-m, \\
(6) & -s-t-\frac{1}{2}=-n;\ -t=-m, \\
(7) & -s=-n;\ -t=-m, \text{ and} \\
(8) & s+\frac{11}{6}=-n;\ -t=-m.
\end{array}$$

This leads to terms of the type:

$$\begin{array}{ll}
(1) & \left(\frac{L_o}{L_i}\right)^{2m} \left(\frac{\pi D}{L_i}\right)^{2n} F(-1,1), \\
(2) & \left(\frac{L_i}{L_o}\right)^{2m} \left(\frac{\pi D}{L_i}\right)^{2n} F(-1,-1), \\
(3) & \left(\frac{L_i}{L_o}\right)^{2m} \left(\frac{\pi D}{L_o}\right)^{2n} F(-2,-1), \\
(4) & \left(\frac{L_o}{L_i}\right)^{2m} \left(\frac{L_i}{\pi D}\right)^{2n} F(1,1), \\
(5) & \left(\frac{L_i}{L_o}\right)^{2m} \left(\frac{L_i}{\pi D}\right)^{2n} F(1,-1), \\
(6) & \left(\frac{L_i}{L_o}\right)^{2m} \left(\frac{L_o}{\pi D}\right)^{2n} F(2,-1), \\
(7) & \left(\frac{L_o}{\pi D}\right)^{2n} \left(\frac{L_i}{\pi D}\right)^{2m} F(2,1), \text{ and} \\
(8) & \left(\frac{\pi D}{L_o}\right)^{2n} \left(\frac{L_i}{\pi D}\right)^{2m} F(-2,1).
\end{array}$$

The Taylor series solution is the sum of the second and third series. This converges slowly for normal parameter values, and the solution is not expressed in this form. Instead, a solution that contains asymptotic series that gives an accurate answer with the fewest number of terms is found.

Table 12.4. Series ranges of applicability for most rapid convergence, for the problem in which the effect of inner and outer scale on the tilt is found.

2–Pole Set	Solution Type	Parameter	Range
1	Asymp	$L_i > L_o$	$L_i/\pi D > 1$
2	Taylor	$L_i < L_o$	$L_i/\pi D > 1$
3	Taylor	$L_i < L_o$	$L_o/\pi D > 1$
4	Asymp	$L_i > L_o$	$L_i/\pi D < 1$
5	Asymp	$L_i < L_o$	$L_i/\pi D < 1$
6	Asymp	$L_i < L_o$	$L_o/\pi D < 1$
7	Asymp	$L_i/\pi D < 1$	$L_o/\pi D < 1$
8	Asymp	$L_i/\pi D < 1$	$L_o/\pi D > 1$

Table 12.4 contains the range of applicability of the various series. Terms 1, 2, 4, 6, and 7 apply only when the outer scale is less than the diameter or the inner scale, or if the inner scale is larger than the diameter. Therefore, these are not included in the solution of interest. The only terms that are appropriate in the case in which outer scale is larger than the diameter and also inner scale is less than diameter are 3, 5, and 8.

These asymptotic and Taylor series lead to

$$
\begin{aligned}
T_o^2 &= \frac{6.08\,\mu_0}{D^{1/3}} \\
&\times\Bigg\{5.18\,S\,(0,0)\left(\frac{\pi D}{L_o}\right)^{2n+1/3}\left(\frac{L_i}{L_o}\right)^{2m}\Gamma\left[\begin{matrix} n+\frac{5}{2},\ n+m+2,\ -n-m-\frac{1}{6} \\ n+5,\ n+3 \end{matrix}\right] \\
&+5.18\,S\,(0,0a)\left(\frac{\pi D}{L_o}\right)^{2n}\left(\frac{L_i}{\pi D}\right)^{2m}\Gamma\left[\begin{matrix} m-n+\frac{1}{6},\ n-m+\frac{7}{3},\ n+\frac{11}{6} \\ n-m+\frac{29}{6},\ n-m+\frac{17}{6} \end{matrix}\right] \\
&+0.0248\,S\,(0a,0)\left(\frac{L_i}{D}\right)^{14/3}\left(\frac{L_i}{L_o}\right)^{2m}\left(\frac{L_i}{\pi D}\right)^{2n} \\
&\qquad\times\Gamma\left[\begin{matrix} n+\frac{5}{2},\ m+\frac{11}{6},\ -n-m-\frac{7}{3} \\ -n+\frac{5}{2},\ -n+\frac{1}{2} \end{matrix}\right]\Bigg\}.
\end{aligned}
\tag{12.21}
$$

The steepest-descent contribution must be added to these asymptotic series. The technique discussed in Sect. 1.4 is used to determine this quantity. As seen from (12.20), the parameter associated with the complex variable s is small, and the parameter $\pi D/L_o$ associated with the complex variable t is large. Since $\Delta_1 = 0$, the complex integral is evaluated sequentially with the evaluation of the steepest-descent contribution over the variable t and normal, pole-residue evaluation over the variable s. Equation (12.20) is rewritten as

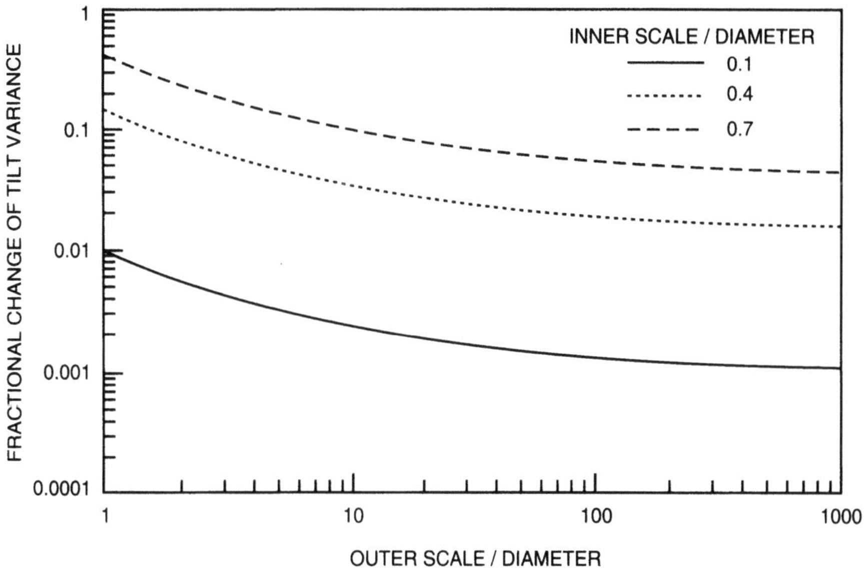

Fig. 12.5. Fractional change of the tilt variance versus the ratio of the outer-scale size to the diameter, with various inner-scale sizes. Notice that the effect of inner scale is to change the scaling of the curves. There is little change in shape. This occurs because inner and outer scale operate at opposite ends of the turbulence spectrum and their effects are almost independent.

$$T_o^2 = \frac{400\,\mu_0 \kappa_o^{-11/3}}{D^4}\,\frac{1}{2\pi i}\int_{C_1} ds\,\left(\frac{\kappa_o D}{2}\right)^{-2s}\,\Gamma\left[-s\right]$$

$$\times \frac{1}{2\pi i}\int_{C_2} dt\,\left(\frac{\kappa_i D}{2}\right)^{-2t}\,\Gamma\left[\begin{matrix} s+t+2,\ -s-t+\frac{1}{2},\ -t,\ s+\frac{11}{6} \\ -s-t+3,\ -s-t+1 \end{matrix}\right]. \quad (12.22)$$

Using (5.94) to evaluate the steepest-descent contribution, one finds that it has exponential decay and can be neglected compared to the pole contributions. This exponential decay was the same behavior found in Sect. 7.3 when the inner scale was much smaller than the diameter. Therefore, the solution in (12.21) is the complete solution.

If inner scale goes to zero, the only terms left are the $m=0$ terms of the first and last summations, which reduce to the series obtained in (7.5). If the outer scale goes to infinity, the only terms left are the $n = 0$ term of the first series, and the $m=0$ term of the third series, which reduce to the series obtained in (7.19). A plot of the change in variance, divided by the variance with zero inner-scale size, is given in Fig. 12.5.

12.6 Power Spectrum of Tilt with Outer Scale

The temporal power spectrum of Zernike tilt with infinite outer-scale size was found in Sect. 7.7. When finite outer scale is considered, a two-dimensional complex integral must be evaluated to compute the tilt power spectral density. The integral that must be evaluated has the general form given in (7.76) except that the spectrum has outer scale included and is equal to

$$S_T(\omega) = 2.606\, k_0^2 \omega \int_0^L dz\, \frac{C_n^2(z)}{v^2(z)}$$
$$\times \int_0^\infty c\, dc\, \frac{U(c-1)}{\sqrt{c^2-1}} \cos^2\left[\frac{\gamma\,\omega^2 c^2 z}{2 v^2(z)\, k_0}\right] f\left[\frac{\omega\, c}{v(z)}\right] F\left[\frac{\gamma\,\omega\, c}{v(z)}, z\right]. \tag{12.23}$$

One neglects diffraction by setting the cosine term equal to unity. The collimated beam case is considered for which $\gamma = 1$. The filter function for tilt is given in (3.17). Change variables by letting $x = 2\, v(z)/\omega\, D$, and $y = v(z)\, \kappa_o/\omega$. The tilt spectrum is expressed as

$$S_T(\omega) = \frac{2668}{\omega\, D^4\, \kappa_o^{11/3}} \int_0^L dz\, C_n^2(z) \int_0^\infty \frac{dc}{c} \frac{U(c-1)}{\sqrt{c^2-1}}\, J_2^2\left(\frac{c}{x}\right) \left[1 + \left(\frac{c}{y}\right)^2\right]^{-11/6}. \tag{12.24}$$

With the generalized Mellin convolution integral and the Mellin transforms in (1.54), (1.50), and (1.52), and after the substitution $s \to -2s$ and $t \to -2t$, the integral over c transforms into

$$I = \frac{1}{2\Gamma[11/6]}\, \frac{1}{(2\pi i)^2} \int_{C_1}\int_{C_2} ds\, dt\, \Gamma\left[\begin{matrix} s+2,\ -s+\frac{1}{2},\ t,\ -t+\frac{11}{6},\ s+t+\frac{1}{2} \\ -s+1,\ -s+3,\ s+t+1 \end{matrix}\right]$$
$$\times \left[\frac{\omega D}{2 v(z)}\right]^{-2s} \left[\frac{L_o\, \omega}{2\pi\, v(z)}\right]^{-2t}.$$

From the definition in (11.79), $\Delta_s = 2$, and $\Delta_t = 0$. There are eight sets of 2-poles that can contribute to I:

(1) $s+2 = -m;\ t = -n,$
(2) $-s+1/2 = -m;\ t = -n,$
(3) $s+2 = -m;\ -t+11/6 = -n,$
(4) $-s+1/2 = -m;\ -t+11/6 = -n,$
(5) $s+t+1/2 = -m;\ t = -n,$
(6) $s+t+1/2 = -m;\ -t+11/6 = -n,$
(7) $s+2 = -m;\ s+t+1/2 = -n,$ and
(8) $-s+1/2 = -m;\ s+t+1/2 = -n.$

For each of these cases, the 2-poles lead to power series that contain terms which vary as:

Table 12.5. Properties of series solutions and ranges of applicability for most rapid series convergence in the problem in which the temporal power spectrum of tilt with outer scale is found.

2 Pole Set	Solution Type	Parameter	Range
1	Taylor	$2\pi v(z)/\omega L_o > 1$	$2v(z)/\omega D > 1$
2	Asymp	$2\pi v(z)/\omega L_o > 1$	$2v(z)/\omega D < 1$
3	Taylor	$2\pi v(z)/\omega L_o < 1$	$2v(z)/\omega D > 1$
4	Asymp	$2\pi v(z)/\omega L_o < 1$	$2v(z)/\omega D < 1$
5	Asymp	$\pi D/L_o > 1$	$2v(z)/\omega D > 1$
6	Taylor	$\pi D/L_o < 1$	$2v(z)/\omega D > 1$
7	Taylor	$2\pi v(z)/\omega L_o > 1$	$\pi D/L_o < 1$
8	Asymp	$2\pi v(z)/\omega L_o > 1$	$\pi D/L_o > 1$

$$(1)\quad \left[\frac{\omega D}{2v(z)}\right]^{2m}\left[\frac{L_o\,\omega}{2\pi\, v(z)}\right]^{2n} F(0,-2),$$

$$(2)\quad \left[\frac{2v(z)}{\omega D}\right]^{2m}\left[\frac{L_o\,\omega}{2\pi\, v(z)}\right]^{2n} F(0,2),$$

$$(3)\quad \left[\frac{\omega D}{2v(z)}\right]^{2m}\left[\frac{2\pi\, v(z)}{L_o\,\omega}\right]^{2n} F(0,-2),$$

$$(4)\quad \left[\frac{2v(z)}{\omega D}\right]^{2m}\left[\frac{2\pi\, v(z)}{L_o\,\omega}\right]^{2n} F(0,2),$$

$$(5)\quad \left[\frac{\omega D}{2v(z)}\right]^{2m}\left(\frac{L_o}{\pi D}\right)^{2n} F(2,-2),$$

$$(6)\quad \left[\frac{\omega D}{2v(z)}\right]^{2m}\left(\frac{\pi D}{L_o}\right)^{2n} F(-2,-2),$$

$$(7)\quad \left(\frac{\pi D}{L_o}\right)^{2m}\left[\frac{L_o\,\omega}{2\pi\, v(z)}\right]^{2n} F(0,-2),\quad \text{and}$$

$$(8)\quad \left(\frac{L_o}{\pi D}\right)^{2m}\left[\frac{L_o\,\omega}{2\pi\, v(z)}\right]^{2n} F(0,2).$$

Table 12.5 summarizes the series type and the parameter range over which the series should be used to obtain a solution that requires the least number of series terms to obtain an accurate answer.

Here I assume that $D/L_o << 1$, which results in eliminating the contribution of asymptotic series 2, 5, and 8. The remaining 5 series fall into three parameter regimes:

1. For low frequencies, $\omega < 2\pi v(z)/L_o$, terms 1, 6, and 7 provide Taylor series. The natural parameters in (12.24) are given by $\omega D/2v(z)$ and $L_o\omega/2\pi v(z)$.

2. At mid-frequencies, $2\pi v(z)/L_o < \omega < 2v(z)/D$, terms 3 and 6 provide Taylor series.

3. At high frequencies, $\omega > 2v(z)/D$, terms 3 and 6 continue to provide the Taylor series solution for the spectrum although they converge slowly. The asymptotic series from term 4 and the steepest-descent contribution may be used in this regime rather than the slowly converging Taylor series to approximate the spectrum with only a few series terms.

The steepest-descent contribution to the asymptotic form of I is obtained with the use of the procedure explained in Sect. 11.4. Since $\Delta_2 = \Delta_3 = 0$, the integral is evaluated sequentially. Accordingly, the integral in (12.24) is rewritten as

$$E = \frac{1}{2\Gamma[11/6]}\frac{1}{2\pi i}\int_{C_2} dt\,\Gamma\left[t,\ -t+\tfrac{11}{6}\right]\left[\frac{L_o\omega}{2\pi v(z)}\right]^{-2t} I_2, \tag{12.25}$$

where

$$I_2 = \frac{1}{2\pi i}\int_{C_1} ds\,\Gamma\left[\begin{matrix} s+2,\ -s+\frac{1}{2},\ s+t+\frac{1}{2} \\ -s+1,\ -s+3,\ s+t+1\end{matrix}\right]\left[\frac{\omega D}{2v(z)}\right]^{-2s}. \tag{12.26}$$

Equation (5.94) is used to show that

$$I_2 \sim \frac{1}{\sqrt{\pi}}\left[\frac{2v(z)}{\omega D}\right]^{3/2}\cos\left[\frac{\omega D}{v(z)} - \frac{\pi}{4}\right]. \tag{12.27}$$

In this problem I_2 does not depend upon t, and the evaluation of (12.25) is trivial. Because $\Delta_t = 0$ and $L_o\omega/2\pi\,v(z) > 1$, the contour in (12.25) is closed to the right, and pole residues are evaluated at the points $t = m + 11/6$. This gives a series result that can be expressed as a generalized hypergeometric function with the rules in Sect. 1.3 as

$$\begin{aligned} E &\sim \frac{\kappa_o^{11/3}/\sqrt{2/\pi}}{D^{3/2}\Gamma(11/6)}\left[\frac{v(z)}{\omega}\right]^{31/6}\cos\left[\frac{\omega D}{v(z)} - \frac{\pi}{4}\right] S(0)\,\Gamma\left[m+\tfrac{11}{6}\right]\left[\frac{2\pi\,v(z)}{L_o\omega}\right]^{2m} \\ &\sim \frac{\kappa_o^{11/3}}{D^{3/2}}\sqrt{\frac{2}{\pi}}\left[\frac{v(z)}{\omega}\right]^{31/6}\cos\left[\frac{\omega D}{v(z)} - \frac{\pi}{4}\right]{}_1F_0\left[\frac{11}{6};-\left[\frac{2\pi v(z)}{L_o\omega}\right]^2\right]. \end{aligned} \tag{12.28}$$

With the identity given in (1.65), this becomes

$$E \sim \frac{\kappa_o^{11/3}}{D^{3/2}}\sqrt{\frac{2}{\pi}}\left[\frac{v(z)}{\omega}\right]^{31/6}\left\{1+\left[\frac{2\pi v(z)}{L_o\omega}\right]^2\right\}^{-11/6}\cos\left[\frac{\omega D}{v(z)} - \frac{\pi}{4}\right]. \tag{12.29}$$

Using pole residues to evaluate series 1, 3, 4, 6, and 7 combined with the result given above leads to the expressions that follow for the power spectral density of tilt with finite outer scale.

For $\omega < 2\pi v(z)/L_o$, after the interchange of m and n in the first summation,

$$S_T(\omega) = \frac{88.64}{\kappa_o^{2/3}} S(0,0) \left(\frac{\pi D}{L_o}\right)^{2n} v_{-2m-1}$$

$$\times \left\{ \Gamma \begin{bmatrix} n+\frac{5}{2},\ m-n+\frac{1}{3},\ -m+n+\frac{3}{2} \\ -m+\frac{1}{2},\ n+3,\ n+5 \end{bmatrix} \left(\frac{L_o\omega}{2\pi}\right)^{2m} \right.$$

$$\left. + \left(\frac{\pi D}{L_o}\right)^{2/3} \Gamma \begin{bmatrix} n+\frac{11}{6},\ m+n+\frac{17}{6},\ -m-n-\frac{1}{3} \\ -m+\frac{1}{2},\ m+n+\frac{10}{3},\ m+n+\frac{16}{3} \end{bmatrix} \left(\frac{\omega D}{2}\right)^{2m} \right\}. \quad (12.30)$$

The contribution from Series 1 does not appear here because the coefficients in this series are all zero. The nth velocity moment defined in (4.49) is expressed as v_n in these results. For $2\pi v(z)/L_o < \omega < 2v(z)/D$ one obtains

$$S_T(\omega) = \frac{52.53\, D^{2/3}}{\Gamma[11/6]} \left(\frac{2}{\omega D}\right)^{2/3} S(0,0) \left(\frac{\omega D}{2}\right)^{2m}$$

$$\times \left\{ \Gamma \begin{bmatrix} m+\frac{5}{2},\ n+\frac{11}{6},\ -m+n+\frac{1}{3} \\ m+3,\ m+5,\ -m+n+\frac{5}{6} \end{bmatrix} \left(\frac{2\pi}{L_o\omega}\right)^{2n} v_{-2m+2n-1/3} \right.$$

$$\left. + \Gamma \begin{bmatrix} n+\frac{11}{6},\ m+n+\frac{17}{6},\ -m-n-\frac{1}{3} \\ -m+\frac{1}{2},\ m+n+\frac{10}{3},\ m+n+\frac{16}{3} \end{bmatrix} \left(\frac{\omega D}{2}\right)^{2/3} \left(\frac{\pi D}{L_o}\right)^{2n} v_{-2m-1} \right\}.$$

(12.31)

For $\omega > 2v(z)/D$, (12.30) provides a slowly converging Taylor series solution. The asymptotic solution for the power spectral density in this regime is given by

$$S_T(\omega) = \frac{52.53\, D^{2/3}}{\Gamma[11/6]} \left(\frac{2}{\omega D}\right)^{17/3}$$

$$\times \left\{ \int_0^L dz\, C_n^2(z) \frac{\Gamma[11/6]}{\sqrt{\pi}} \left[\frac{2\, v(z)}{\omega D}\right]^{1/2} \frac{v(z)^{14/3} \cos\left[\omega D/v(z) - \frac{\pi}{4}\right]}{\left[1+(2\pi v(z)/L_o\omega)^2\right]^{11/6}} \right.$$

$$\left. + S(0,0a) \Gamma \begin{bmatrix} m+\frac{5}{2},\ n+\frac{11}{6},\ m+n+\frac{17}{6} \\ -m+\frac{1}{2}, -m+\frac{5}{2}, m+n+\frac{10}{3} \end{bmatrix} \left(\frac{2}{\omega D}\right)^{2m} \left(\frac{2\pi}{L_o\omega}\right)^{2n} v_{2m+2n+14/3} \right\}.$$

(12.32)

At low frequencies, $\omega < 2\pi v(z)/L_o$, the tilt spectrum is flat, or stated equivalently, the spectrum is proportional to ω^0. In the mid-range of frequencies, $2\pi v(z)/L_o < \omega < 2v(z)/D$, the tilt spectrum is proportional to $\omega^{-2/3}$.

This regime corresponds to the low-frequency portion of the tilt spectrum in the absence of outer-scale effects. At high frequencies, $\omega > 2v(z)/D$, the tilt spectrum is proportional to $\omega^{-17/3}$ as is the case when outer scale is neglected. In the limit as L_o grows large and outer scale no longer influences the tilt spectrum, (12.30) and (12.31) reduce to the expressions for tilt spectrum without outer scale derived in Sect. 7.7.

For a constant velocity (independent of altitude), the tilt spectrum for the three frequency regimes is given by the following equations. For $\omega < 2\pi v(z)/L_o$,

$$S_T(\omega) = \frac{88.64\,\mu_0}{v\,\kappa_o^{2/3}} S\,(0,0) \left(\frac{\pi D}{L_o}\right)^{2n}$$

$$\times \left\{ \Gamma \begin{bmatrix} n+\frac{5}{2},\ m-n+\frac{1}{3},\ -m+n+\frac{3}{2} \\ -m+\frac{1}{2},\ n+3,\ n+5 \end{bmatrix} \left(\frac{L_o\omega}{2\pi v}\right)^{2m} \right.$$

$$\left. + \left(\frac{\pi D}{L_o}\right)^{2/3} \Gamma \begin{bmatrix} n+\frac{11}{6},\ m+n+\frac{17}{6},\ -m-n-\frac{1}{3} \\ -m+\frac{1}{2},\ m+n+\frac{10}{3},\ m+n+\frac{16}{3} \end{bmatrix} \left(\frac{\omega D}{2v}\right)^{2m} \right\}. \quad (12.33)$$

For $2\pi v/L_o < \omega < 2v/D$,

$$S_T(\omega) = \frac{52.53\,\mu_0\,D^{2/3}}{v\,\Gamma[11/6]} S\,(0,0)$$

$$\times \left\{ \Gamma \begin{bmatrix} m+\frac{5}{2},\ n+\frac{11}{6},\ -m+n+\frac{1}{3} \\ m+3,\ m+5,\ -m+n+\frac{5}{6} \end{bmatrix} \left(\frac{\omega D}{2v}\right)^{2m-2/3} \left(\frac{2\pi v}{L_o\omega}\right)^{2n} \right.$$

$$\left. +\Gamma \begin{bmatrix} n+\frac{11}{6},\ m+n+\frac{17}{6},\ -m-n-\frac{1}{3} \\ -m+\frac{1}{2},\ m+n+\frac{10}{3},\ m+n+\frac{16}{3} \end{bmatrix} \left(\frac{\omega D}{2v}\right)^{2m} \left(\frac{\pi D}{L_o}\right)^{2n} \right\}. \quad (12.34)$$

For $\omega > 2v/D$, (12.33) provides a slowly converging Taylor series solution. The asymptotic solution for the power spectral density in this regime is given by

$$S_T(\omega) = \frac{52.53\,\mu_0\,D^{2/3}}{v} \left(\frac{2v}{\omega D}\right)^{17/3} \left\{ \frac{1}{\sqrt{\pi}} \left(\frac{2v}{\omega D}\right)^{1/2} \frac{\cos\left(\omega D/v - \frac{\pi}{4}\right)}{\left[1+(2\pi v/L_o\omega)^2\right]^{11/6}} \right.$$

$$\left. + \frac{S\,(0,0a)}{\Gamma\,[11/6]} \Gamma \begin{bmatrix} m+\frac{5}{2},\ n+\frac{11}{6},\ m+n+\frac{17}{6} \\ -m+\frac{1}{2},\ -m+\frac{5}{2},\ m+n+\frac{10}{3} \end{bmatrix} \left(\frac{2v}{\omega D}\right)^{2m} \left(\frac{2\pi v}{L_o\,\omega}\right)^{2n} \right\}. \quad (12.35)$$

The power spectral density is plotted in Fig. 12.6 as a function of $f\,D/v$ for several values of the parameter D/L_o. The curve has been normalized so that the power spectral density at $f\,D/v = 1$ is unity. These spectra are similar to the spectra without outer scale given in Fig. 7.14 except that the spectrum saturates below the frequency $\omega = 2\pi v/L_o$ due to outer-scale effects. Because it was assumed that $L_o > D$, outer scale has little impact on the high-frequency portion of the tilt spectrum.

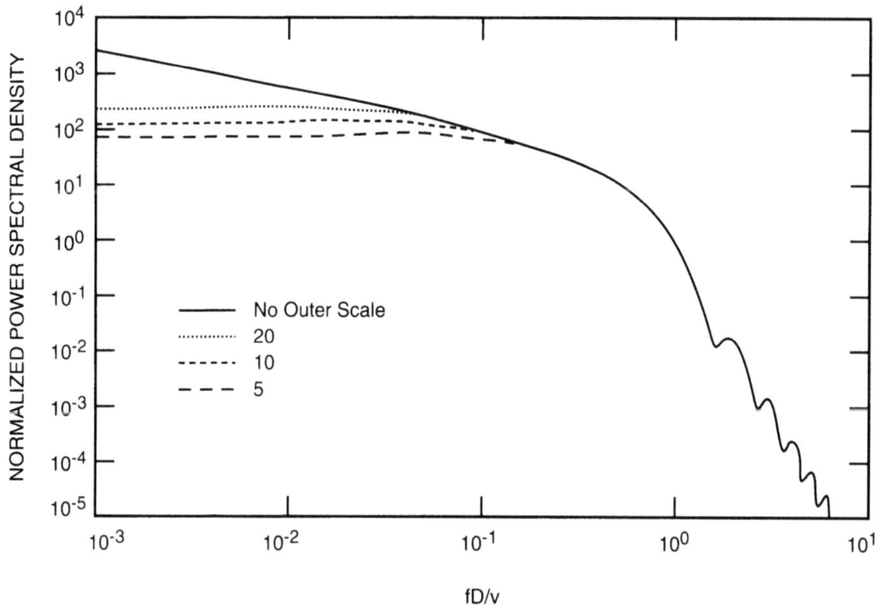

Fig. 12.6. Normalized tilt power spectral density with the assumption of a constant wind velocity and several ratios of outer-scale size to aperture diameter. Outer scale causes a leveling off of the spectral power at small frequencies.

12.7 Structure and Correlation Functions with Inner and Outer Scales

In Sect. 12.6 the impact of non-zero inner and finite outer scale on tilt variance was examined. Inner and outer scales also influence the phase variance of a wave that has propagated through turbulence. For example, if infinite outer scale in Kolmogorov turbulence is assumed, the phase variance is unbounded, and if finite outer scale in Kolmogorov turbulence is assumed, the phase variance is finite. Unfortunately, the inclusion of inner and outer scale in turbulence calculations often leads to integrals that are difficult to evaluate. This is one of the reasons to assume that the outer scale is infinite and to work around the problem of infinite phase. Structure functions are often used in calculations rather than correlation functions for this reason. As shown in this section, Mellin transform methods can be used to solve for both structure and correlation functions with inner and outer scale. First, one derives the structure function, after which solving for the phase correlation function is a simple extension.

The expression for the structure function in (2.92) with the turbulence spectrum in (2.32) is

$$\mathcal{D}(\mathbf{d}) = 0.4146\, k_0^2 \int_0^L dz\, C_n^2(z)$$
$$\times \int d\boldsymbol{\kappa}\, \left(\kappa^2+\kappa_o^2\right)^{-11/6} \exp\left(-\kappa^2/\kappa_i^2\right) \cos^2\left[\frac{\kappa^2(L-z)}{2k_0}\right] \left[1-\cos\left(\boldsymbol{\kappa}\cdot\mathbf{d}\right)\right]. \tag{12.36}$$

Furthermore, it is assumed that turbulence is in the near field of the receiver; therefore, the $\cos^2$ term may be set to unity. The integral over angle in κ space may be performed to give

$$\mathcal{D}(d) = 2.605\, k_0^2 \int_0^L dz\, C_n^2(z) \int_0^\infty d\kappa\, \kappa \left(\kappa^2+\kappa_o^2\right)^{-11/6} \exp\left(-\kappa^2/\kappa_i^2\right) \left[1-J_0(\kappa\, d)\right]. \tag{12.37}$$

The integral over κ is evaluated with Mellin transform techniques. In order to generalize the results, the exponent $-11/6$ is replaced with $-\alpha/2$. For Kolmogorov turbulence, $\alpha = 11/3$. After the change of variables $q = \kappa d$, $x = d\kappa_i$, and $y = d\kappa_o$, the integral over κ becomes

$$Q = \frac{1}{d^2\kappa_o^\alpha} \int_0^\infty \frac{dq}{q}\, q^2 \left[1-J_0(q)\right] \exp\left[-(q/x)^2\right] \left[1+(q/y)^2\right]^{-\alpha/2}.$$

With standard Mellin transform techniques, this integral is converted into a contour integral over two complex planes with the transforms in (1.45), (1.49), and (1.52), and the substitutions $s \to -2s$ and $t \to -2t$ to obtain

$$Q = \frac{-2}{d^2\Gamma[\alpha/2]\,\kappa_o^\alpha}\frac{1}{(2\pi i)^2}\int_{C_1}\int_{C_2} ds\, dt \left(\frac{x}{2}\right)^{-2s}\left(\frac{y}{2}\right)^{-2t}$$
$$\times \Gamma\left[\begin{matrix} s+t+1*,\ -s,\ -t,\ t+\alpha/2 \\ -s-t \end{matrix}\right].$$

The * indicates that the contour of integration is shifted to the left by one pole location. In this problem, $\Delta_s = 1$, and $\Delta_t = 2$. The two-poles associated with the integrand are located at:

(1) $s+t+1 = -m*;\ -s = -n,$
(2) $s+t+1 = -n*;\ -t = -m,$
(3) $s+t+1 = -n*;\ \alpha/2+t = -m,$
(4) $-s = -n;\ -t = -m,$ and
(5) $-s = -n;\ \alpha/2+t = -m.$

The range of the integers m and n are from 0 to ∞. The asterisk after an index indicates that the range of that index is 1 to ∞ for one series and 0 for the other. The relations in the last set of equations lead to terms of the form:

Table 12.6. Series ranges of applicability for most rapid convergence for structure function with inner and outer scale present.

2-Pole Set	Solution Type	Parameter	Range
1	Taylor	$L_i < L_o$	$\pi d < L_o$
2	Asymp	$L_i > L_o$	$\pi d < L_i$
3	Taylor	$L_i < L_o$	$\pi d < L_i$
4	Asymp	$\pi d > L_i$	$\pi d > L_o$
5	Asymp	$\pi d > L_i$	$\pi d < L_o$

$$
\begin{aligned}
&(1) \quad \left(\frac{\pi d}{L_o}\right)^{2m*} \left(\frac{L_i}{L_o}\right)^{2n} F(-1,-2)\,, \\
&(2) \quad \left(\frac{L_o}{L_i}\right)^{2m} \left(\frac{\pi d}{L_i}\right)^{2n*} F(-1,1)\,, \\
&(3) \quad \left(\frac{L_i}{L_o}\right)^{2m} \left(\frac{\pi d}{L_i}\right)^{2n*} F(-1,-1)\,, \\
&(4) \quad \left(\frac{L_o}{\pi d}\right)^{2m} \left(\frac{L_i}{\pi d}\right)^{2n} F(1,2)\,, \text{ and} \\
&(5) \quad \left(\frac{\pi d}{L_o}\right)^{2m} \left(\frac{L_i}{\pi d}\right)^{2n} F(1,-2)\,.
\end{aligned} \tag{12.38}
$$

The sum of Taylor series 1 and 3 gives the complete solution for all parameter values. The Taylor series solution, however, converges slowly when $L_i/\pi < d < L_o/\pi$, and even more slowly when $d > L_o/\pi$. To ease computational requirements, I seek asymptotic approximations when $d > L_i/\pi$ that allow me to approximate the structure function with only a few terms. The approximations also more clearly illustrate the functional dependence of the structure function on parameters $\pi d/L_i$, $\pi d/L_o$, and L_i/L_o in the regimes where many terms from the Taylor series are required. One can determine the parameter regimes where asymptotic approximations are useful by examining Table 12.6.

There are three regimes in this problem. In the first regime where $d < L_i/\pi$, Taylor series 1 and 3 converge rapidly. In the second regime that occurs when $L_i/\pi < d < L_o/\pi$, Taylor series 1, asymptotic series 5, and a steepest-descent component should be used to obtain accurate estimates with a few terms. In the third regime, where $d > L_o/\pi$, asymptotic series 4 and a steepest-descent component provide a good approximation to the structure function. Because the inner-scale size L_i is always less than the outer-scale size L_o, the series that corresponds to the second pole set in the table never contributes. The series that results from the evaluation of the residues at 2-poles 1, 3, 4, and 5 are given below:

$$Q_1 = \frac{-2\pi^2\kappa_o^{-\alpha}}{L_o^2\,\Gamma\,[\alpha/2]} S\,(0,1) \left(\frac{L_i}{L_o}\right)^{2n} \left(\frac{\pi\,d}{L_o}\right)^{2m} \Gamma\left[\begin{matrix} m+n+1,\ -m-n+\alpha/2-1 \\ m+1 \end{matrix}\right],$$

$$Q_3 = \frac{-2\pi^2\,(d/2)^{\alpha}}{L_i^2\,\Gamma\,[\alpha/2]} S\,(1,0) \left(\frac{L_i}{L_o}\right)^{2m} \left(\frac{\pi\,d}{L_i}\right)^{2n-\alpha} \Gamma\left[\begin{matrix} -m+n-\alpha/2+1,\ m+\alpha/2 \\ n+1 \end{matrix}\right],$$

$$Q_4 = \frac{-2\,\kappa_o^{-\alpha}}{d^2\,\Gamma\,[\alpha/2]} S\,(0a,0a) \left(\frac{L_i}{\pi\,d}\right)^{2n} \left(\frac{L_o}{\pi\,d}\right)^{2m} \Gamma\left[\begin{matrix} m+n+1,\ m+\alpha/2 \\ -m-n \end{matrix}\right] = 0,$$

and

$$Q_5 = \frac{-2\,(d/2)^{\alpha}}{d^2\,\Gamma\,[\alpha/2]} S\,(0a,0) \left(\frac{L_i}{\pi\,d}\right)^{2n} \left(\frac{\pi\,d}{L_o}\right)^{2m} \Gamma\left[\begin{matrix} -m+n-\alpha/2+1,\ m+\alpha/2 \\ m-n+\alpha/2 \end{matrix}\right].$$

Because $\Gamma[-m-n]$ is infinite for all m and n, Q_4 is equal to zero. I must also be concerned with the B_2 and C_3 terms discussed in Sect. 11.3 and defined in (11.57) and (11.59). In the second and third regimes, the third set of two-poles contributes a series of type B_2. The first set of two-poles contributes a series of type C_3 in the third regime. These terms are given by

$$B_2 = \frac{-2\pi}{\Gamma\,[\alpha/2]\cos\,(\pi\,\alpha/2)} \left(\frac{L_i}{2\pi}\right)^{\alpha} \left(\frac{\pi}{L_i}\right)^{2} \exp\left(\frac{L_i}{L_o}\right)^{2}, \quad \text{and} \tag{12.39}$$

$$C_3 = \frac{-2}{\Gamma\,[\alpha/2]} \left(\frac{L_o}{2\pi}\right)^{\alpha} \left(\frac{\pi}{L_o}\right)^{2} \sum_{n=0}^{\infty} (-1)^n\,\Gamma\,[-n+\alpha/2-1] \left(\frac{L_i}{L_o}\right)^{2n}. \tag{12.40}$$

The steepest-descent contributions to the asymptotic forms for the integral must be calculated for the second and third regimes. Since $\Delta_2 = 0$, the steepest-descent contribution is found by sequential evaluation of the two-dimensional integral.

In the second regime, x is large and y is small. The integral may be rewritten as

$$\begin{aligned} \tilde{Q} &= \frac{-2\,\kappa_o^{-\alpha}}{d^2\,\Gamma\,[\alpha/2]} \frac{1}{(2\pi\,i)^2} \int_{C_1} dt \left(\frac{y}{2}\right)^{-2t} \Gamma\,[-t,\,t+\alpha/2] \\ &\quad \times \int_{C_2} ds \left(\frac{x}{2}\right)^{-2s} \Gamma\left[\begin{matrix} s+t+1*,\ -s \\ -s-t \end{matrix}\right]. \end{aligned} \tag{12.41}$$

The tilde over Q is used to acknowledge the fact that pole contributions may be counted more than once as a result of our separating the integrations over s and t. As I noted before, multiple counting of poles residues is not a concern when the saddle-point solution is being sought since pole residues are neglected. Using (5.94) one finds that the steepest-descent contribution has exponential decay and can be neglected compared to the pole contributions in this parameter regime.

The steepest-descent contribution in the third parameter regime is now calculated. Both $x/2$ and $y/2$ are greater than one, which requires the full asymptotic solution for the first integral. There are pole contributions at $t = n$ for $n = 0, 1, 2, \ldots$ as well as a steepest-descent contribution. Exponentially decaying terms are retained since there is no pole contribution in this region for the correlation function. The result of the evaluation of the t integration is

$$\tilde{Q} = \frac{-2\,\kappa_o^{-\alpha}}{d^2\Gamma\left[\alpha/2\right]}\frac{1}{2\pi i}\int_{C_2} ds\,\left(\frac{x}{2}\right)^{-2s}\left[S(0a)\left(\frac{y}{2}\right)^{-2n}\Gamma[n+\alpha/2]\,\Gamma\left[\begin{matrix} s+n+1*, \; -s \\ -s-n \end{matrix}\right]\right.$$
$$\left. +\sqrt{\pi}\,\Gamma\left[-s\right]\left(\frac{y}{2}\right)^{2s+(\alpha+1)/2}\exp\left(-y\right)\right]. \qquad (12.42)$$

For the first term $\Delta_s = 1$, and the parameter $x/2$ is large. The steepest-descent contribution is obtained from (5.94). In the last term $\Delta_s = -1$; therefore, let $s \to -s$ to make $\Delta_s = 1$. The parameter y/x is small, and the integration path is closed to the left. There are pole contributions at $s = -m$ for $m = 0, 1, 2, \ldots$. Adding all these contributions gives the steepest-descent contribution as

$$Q \sim -\frac{\kappa_o^{-\alpha}}{\Gamma\left[\alpha/2\right]}\left\{\frac{\kappa_i^2\,S(0a)}{2\,\exp\left[\left(\pi d/L_i\right)^2\right]}\left(\frac{\pi\,L_o d}{L_i^2}\right)^{2n} + \frac{2\sqrt{\pi}\left(\pi d/L_o\right)^{(\alpha+1)/2}}{d^2\exp\left[\left(L_i/L_o\right)^2+\kappa_o d\right]}\right\}. \qquad (12.43)$$

Collecting all the results from this section and assuming that the inner- and outer-scale lengths do not change with altitude and performing the integration over z, one obtains the final expression for the structure function given by the following: Regime 1 ($d < L_i/\pi$)

$$\mathcal{D}(d) = \frac{-51.42\,k_0^2\,\mu_0}{\Gamma\left[\alpha/2\right]\kappa_o^{\alpha}}$$
$$\times\left\{\left(\frac{1}{L_o}\right)^2 S\left(0,1\right)\left(\frac{L_i}{L_o}\right)^{2n}\left(\frac{\pi d}{L_o}\right)^{2m}\Gamma\left[\begin{matrix} m+n+1, \; -m-n-1+\alpha/2 \\ m+1 \end{matrix}\right]\right.$$
$$\left. +\left(\frac{1}{L_i}\right)^2\left(\frac{L_i}{L_o}\right)^{\alpha} S\left(1,0\right)\left(\frac{L_i}{L_o}\right)^{2m}\left(\frac{\pi d}{L_i}\right)^{2n}\Gamma\left[\begin{matrix} -m+n-\alpha/2+1, m+\alpha/2 \\ n+1 \end{matrix}\right]\right\}. \qquad (12.44)$$

Regime 2 ($L_i/\pi < d < L_o/\pi$)

$$\mathcal{D}(d) = \frac{-51.42\,k_0^2\,\mu_0}{\Gamma\left[\alpha/2\right]}$$
$$\times\left\{\left(\frac{1}{L_o}\right)^2\kappa_o^{-\alpha}S\left(0,1\right)\left(\frac{L_i}{L_o}\right)^{2n}\left(\frac{\pi d}{L_o}\right)^{2m}\Gamma\left[\begin{matrix} m+n+1, \; -m-n-1+\alpha/2 \\ m+1 \end{matrix}\right]\right.$$
$$\left. +\left(\frac{1}{\pi d}\right)^2\left(\frac{d}{2}\right)^{\alpha} S(0a,0)\left(\frac{L_i}{\pi d}\right)^{2n}\left(\frac{\pi d}{L_o}\right)^{2m}\Gamma\left[\begin{matrix} -m+n-\alpha/2+1, m+\alpha/2 \\ m-n+\alpha/2 \end{matrix}\right]\right\}. \qquad (12.45)$$

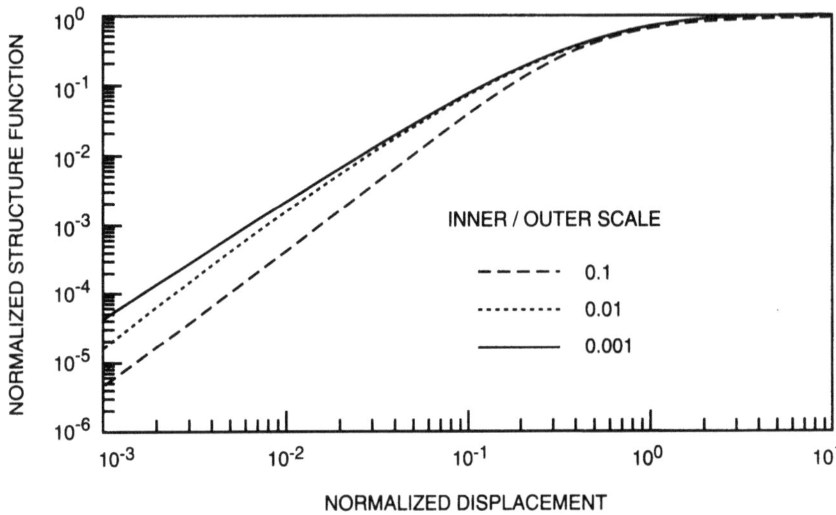

Fig. 12.7. Structure function for several values of L_i/L_o. The structure function has been normalized by twice the variance and has been plotted against the normalized displacement, $\pi d/L_o$. Notice that as the inner scale gets smaller, the structure function increases at small separations because the turbulence increases at the smaller scales.

Regime 3 ($d > L_o/\pi$)

$$\mathcal{D}(d) = \frac{k_0^2\,\mu_0}{\Gamma\left[\alpha/2\right]\,\kappa_o^\alpha}\left\{\frac{51.42}{L_o^2}S_n(0)\Gamma\left[-n-1+\alpha/2\right]\left(\frac{L_i}{L_o}\right)^{2n}\right.$$
$$\left.-1.30\,\kappa_i^2\,\exp\left[-\left(\pi d/L_i\right)^2\right]S(0a)\left(\frac{\pi\,L_o d}{L_i^2}\right)^{2n} - \frac{9.23\,\left(\pi d/L_o\right)^{(\alpha+1)/2}}{d^2\,\exp\left[\left(L_i/L_o\right)^2+\kappa_o d\right]}\right\}. \tag{12.46}$$

The first term in Regime 2 comes from the B_2 term from the third set of two-poles. In Regime 3, the terms are the B_2 and C_3 components from Series 3 and 1, respectively. The steepest-descent contributions have been neglected here since they decay exponentially with increasing d and contribute negligibly to the final solution.

It may be shown that in the limit as $L_o \to \infty$, *Tatarski's* (1961) result for the structure function with inner scale is reproduced. For this case $\alpha = 11/3$. A plot of the structure function for several values of L_i/L_o is shown in Fig. 12.7. Note that in Regime 2, which spans the range from $\pi d/L_o = L_i/L_o$ to $\pi d/L_o = 1$, the structure function has a 5/3-power-law dependence. In Regime 1, which occurs where $\pi d/L_o < L_i/L_o$, there is a power-law dependence of 2. The structure function approaches a constant equal to twice the phase variance in Regime 3.

Because a finite outer scale is included in this problem, the variance is bounded, and a phase correlation function may be computed. One can obtain this result from the previous derivation by noting that the correlation function defined in (2.94) is given by

$$\mathcal{B}_\phi(d) = 2.605\, k_0^2 \int_0^L dz\, C_n^2(z) \int_0^\infty d\kappa\, \kappa \left(\kappa^2 + \kappa_o^2\right)^{-11/6} \exp\left(-\kappa^2/\kappa_i^2\right) J_0(\kappa d)/2. \tag{12.47}$$

This differs from the expression for the structure function given in (12.38) in the factor associated with the Bessel function. As a result, the two-dimensional contour integral expression for the structure function (12.38) may be used with slight modification. Analytic continuation need not be used to get the Mellin transform of the factor that contains the Bessel function, and the asterisks in (12.38) should be eliminated. The subsequent analysis is identical except that summation indices for terms due to the first and second set of two-poles range from 0 to ∞ for both m and n, and there are no terms due to the B_2 and C_3 components.

The expression for the phase correlation function becomes:
Regime 1 ($d < L_i/\pi$)

$$\mathcal{B}_\phi(d) = \frac{25.71\, k_0^2\, \mu_0}{\Gamma[\alpha/2]\, \kappa_o^\alpha}$$
$$\times \left\{ \left(\frac{1}{L_o}\right)^2 S(0,0) \left(\frac{L_i}{L_o}\right)^{2n} \left(\frac{\pi d}{L_o}\right)^{2m} \Gamma\left[\begin{matrix} m+n+1,\ -m-n-1+\alpha/2 \\ m+1 \end{matrix}\right]\right.$$
$$\left. + \left(\frac{1}{L_i}\right)^2 \left(\frac{L_i}{L_o}\right)^\alpha S(0,0) \left(\frac{L_i}{L_o}\right)^{2m} \left(\frac{\pi d}{L_i}\right)^{2n} \Gamma\left[\begin{matrix} -m+n-\alpha/2+1, m+\alpha/2 \\ n+1 \end{matrix}\right]\right\}. \tag{12.48}$$

Regime 2 ($L_i/\pi < d < L_o/\pi$)

$$\mathcal{B}_\phi(d) = \frac{25.71\, k_0^2\, \mu_0}{\Gamma[\alpha/2]\, \kappa_o^\alpha}$$
$$\times \left\{ \left(\frac{1}{L_o}\right)^2 S(0,0) \left(\frac{L_i}{L_o}\right)^{2n} \left(\frac{\pi d}{L_o}\right)^{2m} \Gamma\left[\begin{matrix} m+n+1,\ -m-n-1+\alpha/2 \\ m+1 \end{matrix}\right]\right.$$
$$\left. + \left(\frac{1}{\pi d}\right)^2 \left(\frac{\pi d}{L_o}\right)^\alpha S(0a,0) \left(\frac{L_i}{\pi d}\right)^{2n} \left(\frac{\pi d}{L_o}\right)^{2m} \Gamma\left[\begin{matrix} -m+n-\alpha/2+1, m+\alpha/2 \\ m-n+\alpha/2 \end{matrix}\right]\right\}. \tag{12.49}$$

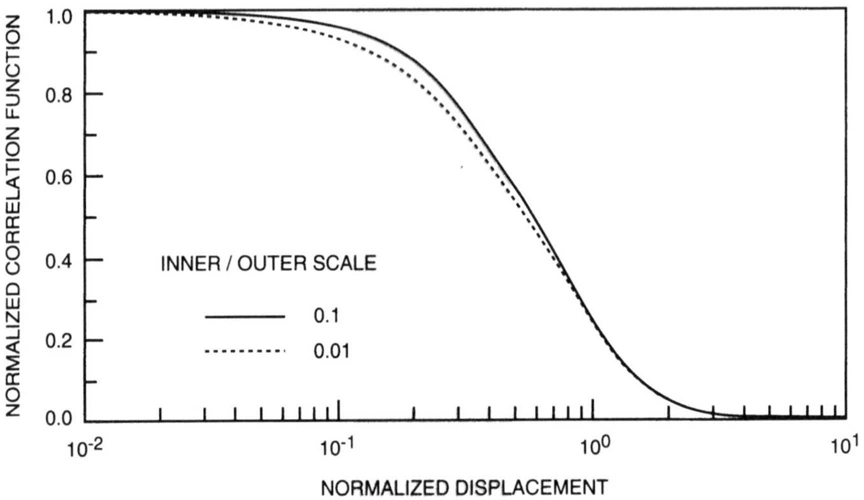

Fig. 12.8. Normalized phase correlation function for several values of L_i/L_o. The correlation function has been plotted against the normalized displacement $\pi d/L_o$. As inner scale increases, the displacement must be larger to get the same reduction in correlation.

Regime 3 ($d > L_o/\pi$)

$$\mathcal{B}_\phi(d) = -\frac{k_0^2\,\mu_0}{\Gamma\,[\alpha/2]\,\kappa_o^\alpha}$$
$$\times\left\{0.65\,\kappa_i^2 \exp\!\left[-\,(\pi\,d/L_i)^2\right] S\,(0a)\left(\frac{\pi\,L_o d}{L_i^2}\right)^{2n} + \frac{4.62\,(\pi\,d/L_o)^{(\alpha+1)/2}}{d^2 \exp\!\left[(L_i/L_o)^2 + \kappa_o d\right]}\right\}. \tag{12.50}$$

The steepest-descent contributions in Regime 2 have been discarded because of their negligible contribution. Note that in Regime 3 the only contribution comes from steepest-descent terms. A plot of the correlation function for the same parameter ranges used for the structure function in Fig. 12.7 is given in Fig. 12.8 for $\alpha = 11/3$. How L_i/L_o affects the correlation function is less obvious than its effect on the structure function.

13. Beam Shape

Propagating a wave through turbulence not only reduces the Strehl ratio, but also changes the beam shape. Mellin transform techniques can be used to calculate the beam shape using the same technique as previously applied. Determining the beam shape simply adds another parameter to an integral. The beam profile after propagating through uncorrected turbulence, or through a medium with turbulence-induced beam jitter present, or with anisoplanatic effects can all be represented as special cases of a general integral. Results from evaluating the general integral are used to find the average beam profiles for the three specific cases.

13.1 General Formula for Beam Shape

A framework for finding the beam profile for any ratio of coherence diameter to aperture diameter is developed here. The starting point for obtaining the beam profile with uncorrected turbulence is the general expression for the normalized beam shape with isotropic turbulence in (2.110) and the structure function for uncorrected turbulence given in (8.7). This gives

$$I(\rho) = \int_0^1 d\alpha\, \alpha\, K(\alpha)\, J_0\left(\frac{k_0\, \rho\, D\alpha}{z}\right) \exp\left[-3.44\left(\frac{\alpha D}{r_o}\right)^{5/3}\right]. \tag{13.1}$$

Similar integrals are required to determine the beam shape both with tilt jitter and with anisoplanatism. To evaluate the three beam shapes at the same time, define

$$P(\rho, \nu, q, g) = \int_0^\infty \frac{d\alpha}{\alpha}\, \alpha^{2+\nu} K(\alpha)\, J_0\left(\frac{k_0\, \rho\, D\alpha}{z}\right) \exp\left(-g^q \alpha^q\right), \tag{13.2}$$

where for the case of uncorrected turbulence

$$I(\rho) = P\left(\rho, 0, 5/3, \frac{3.44^{3/5} D}{r_o}\right). \tag{13.3}$$

This integral will be evaluated by using the Mellin transform techniques in two complex planes developed in Chap. 11. Let $x = z/k_0 \rho D$, and $y = 1/g$. Using the Mellin convolution integral and the Mellin transforms in (1.62), (1.49), and (1.45), one can transform the integral into

$$P(\rho,\nu,q,g) = \frac{4}{q\sqrt{\pi}}\frac{1}{(2\pi i)^2}\int_{C_1}\int_{C_2} ds\,dt\,(2x)^{-s}\,y^{-t}$$

$$\times\frac{1}{(s+t+\nu+2)}\Gamma\left[\begin{matrix}\frac{s+t+\nu}{2}+\frac{3}{2},\ -s/2,\ -t/q\\ \frac{s+t+\nu}{2}+3,\ s/2+1\end{matrix}\right].$$

Let $s\to 2s$, and $t\to 2t$, and use the recursion relation

$$\frac{1}{(s+t+\nu/2+1)} = \frac{\Gamma[s+t+\nu/2+1]}{\Gamma[s+t+\nu/2+2]}$$

to obtain

$$P(\rho,\nu,q,g) = \frac{8}{q\sqrt{\pi}}\frac{1}{(2\pi i)^2}\int_{C_1} ds\int_{C_2} dt\,(2x)^{-2s}\,y^{-2t}$$

$$\times\Gamma\left[\begin{matrix}s+t+\nu/2+\frac{3}{2},\ s+t+\nu/2+1,\ -s,\ -2t/q\\ s+t+\nu/2+3,\ s+t+\nu/2+2,\ s+1\end{matrix}\right].$$

To evaluate this integral, 2-poles must be listed, and these are:

$$\begin{aligned}
&(1)\quad s+t+\nu/2+\tfrac{3}{2} = -m;\ -s=-n,\\
&(2)\quad s+t+\nu/2+1 = -m;\ -s=-n,\\
&(3)\quad s+t+\nu/2+\tfrac{3}{2} = -m;\ -2t/q=-n,\\
&(4)\quad s+t+\nu/2+1 = -m;\ -2t/q=-n,\ \text{and}\\
&(5)\quad -s=-n;\ -2t/q=-m.
\end{aligned}$$

These lead to terms of the form:

$$(1)\quad \left(\frac{1}{g}\right)^{2m}\left(\frac{k_0\,\rho D}{2gz}\right)^{2n}(m!)^{2/q}\,(n!)^{2/q-2},$$

$$(2)\quad \left(\frac{1}{g}\right)^{2m}\left(\frac{k_0\,\rho D}{2gz}\right)^{2n}(m!)^{2/q}\,(n!)^{2/q-2},$$

$$(3)\quad \left(\frac{2z}{k_0\,\rho D}\right)^{2m}\left(\frac{2zg}{k_0\,\rho D}\right)^{qn}(m!)^{2}\,(n!)^{q-1},$$

$$(4)\quad \left(\frac{2z}{k_0\,\rho D}\right)^{2m}\left(\frac{2zg}{k_0\,\rho D}\right)^{qn}(m!)^{2}\,(n!)^{q-1},\ \text{and}$$

$$(5)\quad \left(\frac{k_0\,\rho D}{2z}\right)^{2n} g^{qm}\,(m!)^{-1}\,(n!)^{-2}.$$

The series in n can be a Taylor or an asymptotic series depending on the value of q. The ranges of applicability of the 5 double series to get the most rapid convergence are given in Table 13.1.

Table 13.1. Region in which the series have most rapid convergence for the general problem of finding the beam shape with uncorrected turbulence, tilt jitter, or anisoplanatism.

2-Pole Set	Solution Type	Parameter	Range
1	Asymp	$g > 1$	$2gz > k_0\rho D$
2	Asymp	$g > 1$	$2gz > k_0\rho D$
3	Asymp	$2gz < k_0\rho D$	$k_0\rho D > 2z$
4	Asymp	$2gz < k_0\rho D$	$k_0\rho D > 2z$
5	Taylor	$g < 1$	$k_0\rho D < 2z$

As described in Chap. 11 these series are useful for different parameter ranges. The single Taylor series gives a solution that is applicable for all parameter ranges and is a complete solution to the problem. It is given by

$$P(\rho, \nu, q, g) = \frac{4}{\sqrt{\pi}} S(0,0) \left(\frac{k_0\rho D}{2z}\right)^{2n} \frac{g^{qm}}{n!}$$
$$\times \left(\frac{1}{n + mq/2 + \nu/2 + 1}\right) \Gamma \left[\begin{matrix} n + mq/2 + \nu/2 + \frac{3}{2} \\ n + mq/2 + \nu/2 + 3 \end{matrix}\right] . \quad (13.4)$$

Because of the presence of the term $(n!)^2\, m!$ in the denominator, the above series converges rapidly even for relatively large distances off the beam axis and can be used rather than the asymptotic series. For instance, if the parameter $k_0\rho D/2z$ is equal to 10, then the error in using 30 terms of the series is one part in 10^5 of the value on axis.

For some parameter ranges a large number of terms is necessary to get an accurate answer. In these regions an asymptotic series may be more convenient. The asymptotic series given by the second and first set of 2-poles is

$$P(\rho, \nu, q, g)_{AS} \sim \frac{8}{q\sqrt{\pi}} S(0a) \left(\frac{k_0\rho D}{2zg}\right)^{2n} \left\{ \left(\frac{1}{g}\right)^{\nu+2} \sqrt{\pi}\Gamma\, [2n/q + \nu/q + 2/q] \right.$$
$$\left. + S_m(0) \frac{1}{\left(m + \frac{1}{2}\right)} \left(\frac{1}{g}\right)^{2m+\nu+3} \Gamma \left[\begin{matrix} 2n/q + 2m/q + \nu/q + 3/q \\ -m + \frac{3}{2} \end{matrix}\right] \right\} . \quad (13.5)$$

Since $\Delta_3 = 0$, the steepest-descent contribution is found by sequential evaluation of the two-dimensional integral. In the regime in which the above equation applies, integrate (13.4) first over s assuming that x is large, and then integrate over t assuming y is small. The result is equal to zero.

Table 13.2. Region in which the series have most rapid convergence for uncorrected turbulence.

2–Pole Set	Solution Type	Parameter	Range
1	Asymp	$0.4765r_o < D$	$k_0\rho r_o < 4.197z$
2	Asymp	$0.4765r_o < D$	$k_0\rho r_o < 4.197z$
3	Asymp	$k_0\rho r_o > 4.197z$	$k_0\rho D > 2z$
4	Asymp	$k_0\rho r_o > 4.197z$	$k_0\rho D > 2z$
5	Taylor	$0.4765r_o > D$	$k_0\rho D < 2z$

The asymptotic solution given by the third and fourth set of 2-poles is

$$P(\rho,\nu,q,g)_{AS} \sim \frac{4}{\sqrt{\pi}} S(0a)\left(\frac{2g\,z}{k_0\rho D}\right)^{nq}\left\{\left(\frac{2z}{k_0\,\rho\,D}\right)^{\nu+2}\sqrt{\pi}\Gamma\begin{bmatrix} nq/2+\nu/2+1 \\ -nq/2-\nu/2 \end{bmatrix}\right.$$
$$\left. -S_m(0a)\frac{1}{\left(m+\frac{1}{2}\right)}\left(\frac{2z}{k_0\,\rho\,D}\right)^{2m+\nu+3}\Gamma\begin{bmatrix} m+nq/2+\nu/2+\frac{3}{2} \\ -m+\frac{3}{2},\ -m-nq/2-\nu/2-\frac{1}{2} \end{bmatrix}\right\}. \tag{13.6}$$

The steepest-descent contribution that is found by integrating (13.4) first over t assuming that y is large, then over s assuming x is small, is

$$P(\rho,\nu,q,g)_{SD} = -\frac{32}{\pi}\left(\frac{z}{k_0\,\rho D}\right)^3 \sin\left(\frac{k_0\,\rho\,D}{z}\right)\exp\left[-\left(\frac{g}{2}\right)^q\right]. \tag{13.7}$$

13.2 Beam Shape for Uncorrected Turbulence

For the specific case of uncorrected turbulence, the beam shape is given in (13.3) as $P\left(\rho,0,5/3,3.44^{3/5}D/r_o\right)$. The range of applicability of the series is obtained by inserting the parameters into Table 13.1 to give the ranges in Table 13.2.

The Taylor series solution for this case that applies for all turbulence values and distances from the axis is

$$I(\rho) = \frac{4}{\sqrt{\pi}} S(0,0)\frac{1}{n!}\left(\frac{k_0\rho D}{2z}\right)^{2n}\left[3.44\left(\frac{D}{r_o}\right)^{5/3}\right]^m$$
$$\times\left(\frac{1}{n+5m/6+1}\right)\Gamma\begin{bmatrix} n+5m/6+\frac{3}{2} \\ n+5m/6+3 \end{bmatrix}. \tag{13.8}$$

Notice that the only way turbulence enters the expression is in the form of the coherence diameter.

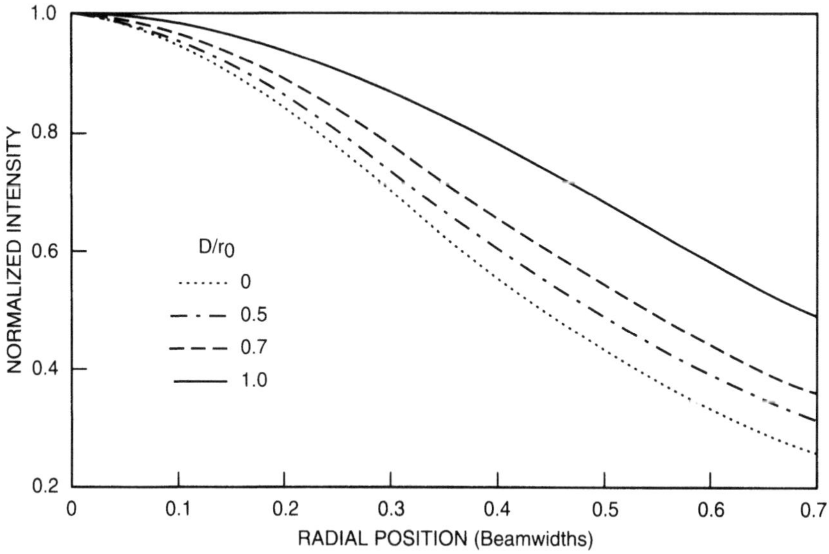

Fig. 13.1. Beam shape of a wave that propagated through turbulence with the ratio D/r_o equal to 0, 0.5, 0.7, and 1.0.

It is instructive to consider the solution for infinite r_o. That can be done by examining the solution above, or alternatively, by evaluating the integration over t in (13.4) and only evaluating the pole at $t = 5m/6$. Set $m = 0$ to obtain

$$I(\rho) = \frac{1}{2\pi i} \int_C ds \, \frac{2}{\sqrt{\pi}} \Gamma \begin{bmatrix} s/2, -s/2 + \frac{3}{2} \\ -s/2 + 2, -s/2 + 3 \end{bmatrix} \left(\frac{k_0 \rho D}{2z} \right)^s . \tag{13.9}$$

From (1.50), one sees that this integral produces an Airy pattern given by

$$I(\rho) = \left[J_1 \left(\frac{k_0 \rho D}{2z} \right) / \left(\frac{k_0 \rho D}{4z} \right) \right]^2 . \tag{13.10}$$

The beamwidth in the plots in this chapter is the value of radius that results in a zero value for the intensity. The argument is 3.8 at this radius, and the normalizing function to convert the arguments of the function into beamwidths is $1.22 \rho D / \lambda z$.

The beam profiles versus axial position of a wave that propagated through turbulence with values of D/r_o of 0, 0.5, 0.7, and 1 are plotted in Fig. 13.1. The beam shape is similar for all cases out to the beamwidth that is plotted. If the beamwidth increased by the inverse of the square root of the Strehl ratio, then the normalized half-power beamwidths would be 1, 1.15, 1.28, and 1.60 for these cases. The beamwidths are 1, 1.11, 1.23, and 1.56, which are close to the above values.

Only a few terms of the Taylor series are necessary to get an accurate answer for low turbulence and small distances from the axis. For stronger turbulence conditions or larger distances off axis, asymptotic solutions are easier to evaluate. The asymptotic series in (13.5) are applicable when the coherence diameter is much less than the aperture diameter, and the angle off-axis is less than the diffraction-limited size of an aperture of diameter r_o. These are given by

$$I(\rho) \sim \frac{24}{5\sqrt{\pi}} S(0) \frac{1}{n!} \left(\frac{0.238\, r_o\, k_0\, \rho}{z}\right)^{2n} \left\{ \left(\frac{0.4765\, r_o}{D}\right)^2 \sqrt{\pi}\Gamma\left[6n/5 + \tfrac{6}{5}\right] \right.$$

$$\left. - S_m(0) \frac{1}{\left(m + \frac{1}{2}\right)} \left(\frac{0.4765\, r_o}{D}\right)^{2m+3} \Gamma\left[\begin{matrix} 6n/5 + 6m/5 + \frac{9}{5} \\ -m + \frac{3}{2} \end{matrix}\right] \right\} . \quad (13.11)$$

For the second and fourth sets of poles the only value of m that gives a nonzero contribution is $m = 0$. The third and fourth double series in (13.4) and the steepest-descent contribution in (13.4) gives the asymptotic solution that is applicable for large distances from the axis. These expressions are

$$I(\rho) \sim \frac{4}{\sqrt{\pi}} S(0a) \left(\frac{4.197\, z}{r_o k_0 \rho}\right)^{5n/3} \left\{ \left(\frac{2z}{k_0 \rho D}\right)^2 \sqrt{\pi}\Gamma\left[\begin{matrix} 5n/6 + 1 \\ -5n/6 \end{matrix}\right] \right.$$

$$\left. - S_m(0a) \frac{1}{\left(m + \frac{1}{2}\right)} \left(\frac{2z}{k_0\, \rho\, D}\right)^{2m+3} \Gamma\left[\begin{matrix} m + 5n/6 + \frac{3}{2} \\ -m + \frac{3}{2},\ -m - 5n/6 - \frac{1}{2} \end{matrix}\right] \right\}$$

$$- \frac{32}{\pi} \left(\frac{z}{k_0\, \rho D}\right)^3 \sin\left(\frac{k_0\, \rho\, D}{z}\right) \exp\left[-\left(\frac{3.44^{3/5} D}{2\, r_o}\right)^{5/3}\right] . \quad (13.12)$$

Two curves are plotted in Fig. 13.2. One is the beam shape with turbulence; the other is the beam shape for a wave with no turbulence that was transmitted from an aperture with a diameter equal to the coherence diameter. The horizontal axis is distance normalized to the beamwidth so that a value of unity corresponds to the first zero of the Airy pattern of an unperturbed beam. Notice that the shapes of the two profiles are close to each other when the intensity is above 0.2 of the value on axis.

13.3 Beam Shape with Tilt Jitter

There are some occasions when the higher-order Zernike modes of turbulence are well corrected, yet tilt jitter remains. One can use an artificial beacon to correct for higher-order turbulence terms as discussed in Sects. 7.12–7.14. This beacon provides no information on the jitter of the source. Since tilt jitter is several microradians, it must be well corrected to prevent significant tilt smearing of the image. One technique of reducing the jitter is to track a bright source close to the object to be corrected. There is some tilt remaining that results from the angular offset of the guide star that will degrade the image. In another scenario, which is not common, one can correct for turbulence in an adaptive-optics system and track on the uncorrected image. This results in a tilt jitter

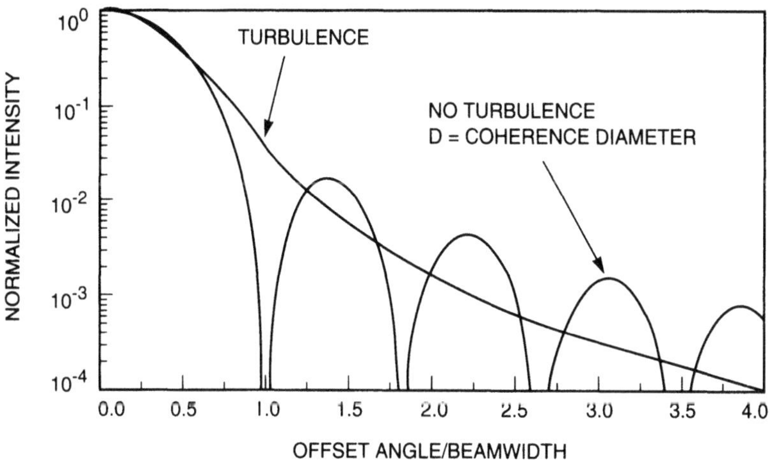

Fig. 13.2. Beam shape with and without turbulence. Notice how closely the average beam profile in turbulence is approximated down to a normalized intensity of 0.2 by a diffraction-limited beam profile transmitted from a coherence-diameter sized aperture.

due to the difference of gradient and Zernike tilt, which is referred to as centroid anisoplanatism by *Yura* and *Tavis* (1980). Tilt jitter affects the beam shape, and it is shown that it basically increases its area by a factor equal to the inverse of the reduction in Strehl ratio.

The expression for beam shape with tilt jitter is given by substituting the structure function in (8.19) into (2.110) to obtain

$$\begin{aligned} I(\rho) &= K(\alpha)\, J_0\left(\frac{k_0 \rho D \alpha}{z}\right) \exp\left[-\alpha^2 \left(\frac{D}{M r_o}\right)^{5/3}\right] \\ &= P\left(\rho, 0, 2, \left(\frac{D}{M r_o}\right)^{5/6}\right). \end{aligned} \tag{13.13}$$

M is equal to 0.4642, 0.4838, and 5.394 for Zernike tilt (Z-tilt), gradient tilt (G-tilt), and their difference respectively. The ranges of applicability of the series are obtained by inserting the parameters into Table 13.1 to give the ranges in Table 13.3.

The Taylor series good for all radii and turbulence strengths is

$$I(\rho) = \frac{4}{\sqrt{\pi}} S(0,0) \frac{1}{n!} \left(\frac{k_0 \rho D}{2z}\right)^{2n} \left(\frac{D}{M r_o}\right)^{5m/3} \left(\frac{1}{n+m+1}\right) \Gamma\left[\begin{matrix} n+m+\frac{3}{2} \\ n+m+3 \end{matrix}\right]. \tag{13.14}$$

Notice that the only way turbulence enters the formulas is in the form of the coherence diameter as was the case for uncorrected turbulence. The beam shape is plotted in Fig.13.3 for $D/M r_o$ equal to 0, 1, 2, and 3 versus the beamwidth $1.2\lambda z/D$. The normalized beamwidths are 1, 1.09, 1.30, and 1.58 for these cases.

Table 13.3. Region in which the series have most rapid convergence for beam shape with beam jitter.

2–Pole Set	Solution Type	Parameter	Range
1	Asymp	$M r_o < D$	$(M r_o)^{5/6} k_0 \rho D^{1/6} < 2z$
2	Asymp	$M r_o < D$	$(M r_o)^{5/6} k_0 \rho D^{1/6} < 2z$
3	Asymp	$k_0 \rho D > 2z$	$(M r_o)^{5/6} k_0 \rho D^{1/6} > 2z$
4	Asymp	$k_0 \rho D > 2z$	$(M r_o)^{5/6} k_0 \rho D^{1/6} > 2z$
5	Taylor	$M r_o > D$	$k_0 \rho D < 2z$

If the beamwidth varied as the inverse of the square root of the Strehl ratio, they would be 1, 1.12, 1.37, and 1.66, which are close to the actual values. This expression reduces to an Airy pattern if the coherence diameter becomes infinite. The asymptotic series that applies when the angle off-axis is much less than the diffraction-limited size of an aperture of diameter D and the coherence diameter is smaller than the aperture diameter is given in (13.5) as

$$I(\rho) \sim \frac{4}{\sqrt{\pi}} S(0) \left[\frac{D^{1/6} k_0 \rho}{2z} (M r_o)^{5/6} \right]^{2n}$$
$$\times \left\{ \left(\frac{M r_o}{D} \right)^{5/3} \sqrt{\pi} - S_m(0a) \frac{1}{n! \left(m + \frac{1}{2} \right)} \left(\frac{M r_o}{D} \right)^{5m/3+3} \Gamma \left[\begin{matrix} n + m + \frac{3}{2} \\ -m + \frac{3}{2} \end{matrix} \right] \right\}. \tag{13.15}$$

The asymptotic solution that applies when the angle off-axis is much greater than the diffraction-limited size of an aperture of diameter D is given in (13.6) and (13.7) as

$$I(\rho) \sim \frac{4}{\sqrt{\pi}} S(0a) \left[\frac{r_o k_0 \rho}{2z} \left(\frac{D}{M r_o} \right)^{5/6} \right]^{2n} \left\{ \left(\frac{2z}{k_0 \rho D} \right)^2 \sqrt{\pi} \Gamma \left[\begin{matrix} n+1 \\ -n \end{matrix} \right] \right.$$
$$\left. - S_m(0a) \frac{1}{\left(m + \frac{1}{2} \right)} \left(\frac{2z}{k_0 \rho D} \right)^{2m+3} \Gamma \left[\begin{matrix} m + n + \frac{3}{2} \\ -m + \frac{3}{2},\ -m - n - \frac{1}{2} \end{matrix} \right] \right\}$$
$$- \frac{32}{\pi} \left(\frac{z}{k_0 \rho D} \right)^3 \sin \left(\frac{k_0 \rho D}{z} \right) \exp \left[-\frac{1}{4} \left(\frac{D}{M r_o} \right)^{5/3} \right]. \tag{13.16}$$

The Taylor series expression given in (13.14) reduces to the Strehl ratio given in (8.28) if it is evaluated on axis. The ratio of beam intensity on axis with tilt jitter to that with all turbulence for very high turbulence levels is

$$\frac{I(Jitter)}{I(Turbulence)} = 4 M^{5/3} \left(\frac{D}{r_o} \right)^{1/3}, \qquad D >> r_o. \tag{13.17}$$

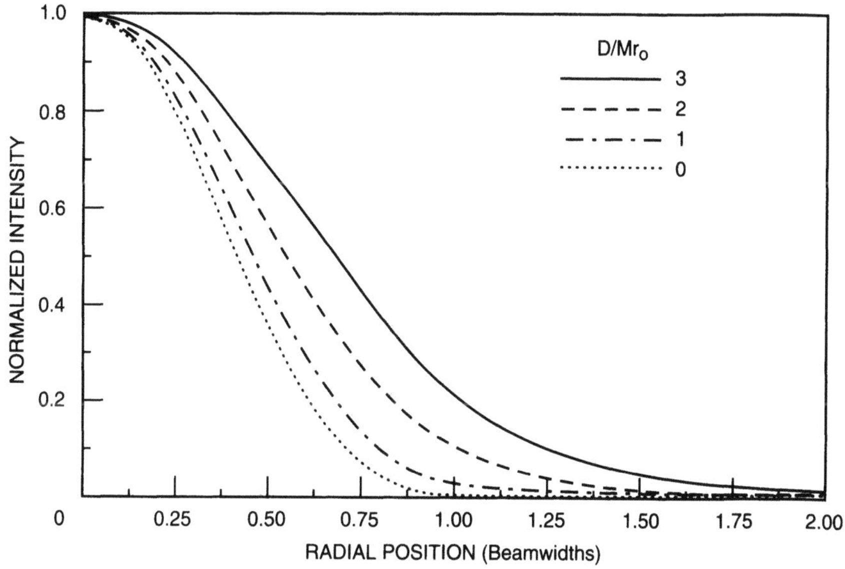

Fig. 13.3. Beam shape of a wave that propagated through turbulence with the ratio D/Mr_o equal to 0, 1, 2, and 3 that has all distortions except tilt corrected.

The same result is also obtained by taking the ratio of Strehls in (8.27) and (8.16).

13.4 Beam Shape with Anisoplanatism

Next, the more complicated problem of finding the beam shape with anisoplanatism is solved. I expand the exponential in the integral into a series of Gegenbauer polynomials as was done in the calculation of Strehl ratio. For this reason, the range of applicability of this calculation is the same as that for the calculation of Strehl ratio. The beam profile calculation is valid when the displacements at the altitudes that contribute most to the reduction in Strehl ratio are less than 0.3 of the beam diameter. Fortunately, for adaptive-optics systems, that is often the region of greatest interest. The starting point is the expression for the beam profile given in (2.109), repeated here

$$I(\boldsymbol{\rho}) = \frac{1}{2\pi} \int d\boldsymbol{\alpha}\, K(\alpha) \exp\left[i k_0 D \boldsymbol{\rho} \cdot \boldsymbol{\alpha}/z - \mathcal{D}(\boldsymbol{\alpha} D)/2\right]. \qquad (13.18)$$

The integration is over the unit circle. The structure function is the same as that used for Strehl ratio. It is again expanded into Gegenbauer polynomials, and only the first term is retained. If the exponential is expanded into a power series as was done before, the expression equivalent to that given in (8.43) is

$$I(\boldsymbol{\rho}) \approx \frac{\exp\left(-\sigma_\varphi^2\right)}{2\pi} \int d\boldsymbol{\alpha}\, K(\alpha) \exp\left(i k_0 D \boldsymbol{\rho} \cdot \boldsymbol{\alpha}/z\right) \times \left(1 + x + \frac{x^2}{2} + \frac{x^3}{6} + \frac{x^4}{24} + \frac{x^5}{120}\right), \tag{13.19}$$

where x is the same quantity as defined in (8.42) and repeated here

$$x = \tfrac{5}{6} d_2 \left[1 - \tfrac{1}{3}\cos^2(\varphi)\right] (\alpha D)^{-1/3}. \tag{13.20}$$

The integration over angle is not as straightforward as before since there is an additional term in the exponential that also depends on angle. Rather than performing a complicated calculation, at this point I make an approximation. The angular dependence of x is weak, especially for the lower-order terms. I will replace the angular expression in x by the values averaged over angle that were calculated for the Strehl ratio. This eliminates any angular dependence of the beam profile; therefore, the calculations below give the beam profile averaged over angle. The angular integration can then be evaluated to get a Bessel function for the exponential term. Define

$$Q(\nu) = \frac{1}{P(0,\nu,0,0)} \int_0^1 \frac{d\alpha}{\alpha} \alpha^{\upsilon+2} K(\alpha) J_0(k_0 \rho D \alpha/z) = \frac{P(\rho,\nu,0,0)}{P(0,\nu,0,0)}, \tag{13.21}$$

where $P(0,\nu,0,0)$ is found from (13.4) as

$$P(0,\nu,0,0) = \frac{4}{\sqrt{\pi}} \frac{1}{(\nu/2+1)} \Gamma\begin{bmatrix} \nu/2+\frac{3}{2} \\ \nu/2+3 \end{bmatrix}. \tag{13.22}$$

With these definitions, the beam profile is

$$I(\rho) \approx \exp\left(-\sigma_\varphi^2\right) \left[Q(0) + 0.9736 E Q\left(-\tfrac{1}{3}\right) + 0.5133 E^2 Q\left(-\tfrac{2}{3}\right) + 0.2009 E^3 Q(-1) + 0.06970 E^4 Q\left(-\tfrac{4}{3}\right) + 0.02744 E^5 Q\left(-\tfrac{5}{3}\right)\right]. \tag{13.23}$$

The value of $Q(\nu)$ is obtained from the evaluation $P(\rho,\nu,0,0)$. Substituting the values of the parameters into Table 13.1, one obtains Table 13.4 for the ranges of applicability of the series.

Since the conditions for Series 1 and 2 cannot be satisfied, these series do not apply for this problem. The Taylor series solution obtained from (13.4) is

$$Q(\nu) = (\nu/2+1)\Gamma\begin{bmatrix} \nu/2+3 \\ \nu/2+\frac{3}{2} \end{bmatrix} \times S(0) \frac{1}{n!\,(n+1+\nu/2)} \Gamma\begin{bmatrix} n+\frac{3}{2}+\nu/2 \\ n+3+\nu/2 \end{bmatrix} \left(\frac{k_0 \rho D}{2z}\right)^{2n} = {}_2F_3\left[\tfrac{3}{2}+\nu/2, 1+\nu/2; 3+\nu/2, 2+\nu/2, 1; -\left(\frac{k_0 \rho D}{2z}\right)^2\right]. \tag{13.24}$$

Table 13.4. Region in which the series have most rapid convergence for beam shape with anisoplanatism for which $g = 0$.

2–Pole Set	Solution Type	Parameter	Range
1	Asymp	$0 < 1$	$k_0 \rho D > g z$
2	Asymp	$0 > 1$	$k_0 \rho D > g z$
3	Asymp	$k_0 \rho D > 0$	$k_0 \rho D > 2 z$
4	Asymp	$k_0 \rho D > 0$	$k_0 \rho D > 2 z$
5	Taylor	$0 < 1$	$k_0 \rho D < 2 z$

For large radii, an asymptotic solution is needed. This series is obtained from (13.6) and (13.7) as

$$Q(\nu) \sim (\nu/2+1)\Gamma\begin{bmatrix}\nu/2+3\\ \nu/2+\frac{3}{2}\end{bmatrix}$$
$$\times\left\{S(0)\frac{-1}{\left(m+\frac{1}{2}\right)}\Gamma\begin{bmatrix}m+\frac{3}{2}+\nu/2\\ -m-\frac{1}{2}-\nu/2,\ -m+\frac{3}{2}\end{bmatrix}\left(\frac{2z}{k_0\rho D}\right)^{2m+3+\nu}\right.$$
$$\left.+\sqrt{\pi}\Gamma\begin{bmatrix}\nu/2+1\\ -\nu/2\end{bmatrix}\left(\frac{2z}{k_0\rho D}\right)^{\nu+2}-\frac{8}{\sqrt{\pi}}\left(\frac{z}{k_0\rho D}\right)^{3}\sin\left(\frac{k_0\rho D}{z}\right)\right\}. \quad (13.25)$$

The beam pattern can be plotted for various types of anisoplanatism by insertion of the appropriate values of E and σ_ϕ^2 which were derived in Sects. 8.4.1 to 8.4.5. The beam pattern is plotted for angular anisoplanatism in Figs. 13.4 and 13.5 for various values of offset angles. Notice how the dips in the Airy pattern fill in as anisoplanatism increases, and the power gets redistributed to larger radial distances. In Fig. 13.6 the curves are normalized to unity at zero radius for angular offsets of 0, 9, and 18 μrad. The intensity on axis for these three cases is 1, 0.7, and 0.45, and they reach a normalized intensity of 0.5 at 0.423, 0.435, and 0.494 beamwidths, respectively. If anisoplanatism causes the peak to spread by an amount proportional to the inverse of the square root of the Strehl ratio, then the half power points would occur at 0.423, 0.505, and 0.630 beamwidths. The beamwidth of the central core is narrower for these anisoplanatism levels that indicates that the effect of small offsets is to decrease the power on axis without a substantial affect on the near-axis beamwidth. Small angular anisoplanatic errors mainly affect the high spatial spectral components, which scatter energy out to large angles, thus they produce a distribution with a narrow peak and a low plateau of energy that extends much further out.

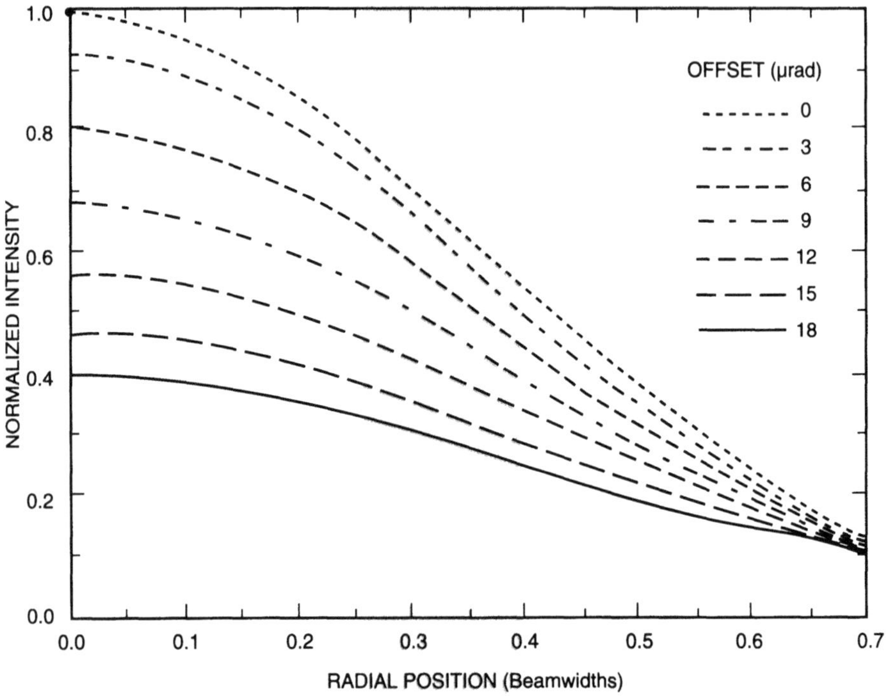

Fig. 13.4. Beam shape close to the beam axis with angular anisoplanatism for a 0.6-m aperture at zenith for the SLCSAT day model. The intensity is normalized to that with no turbulence.

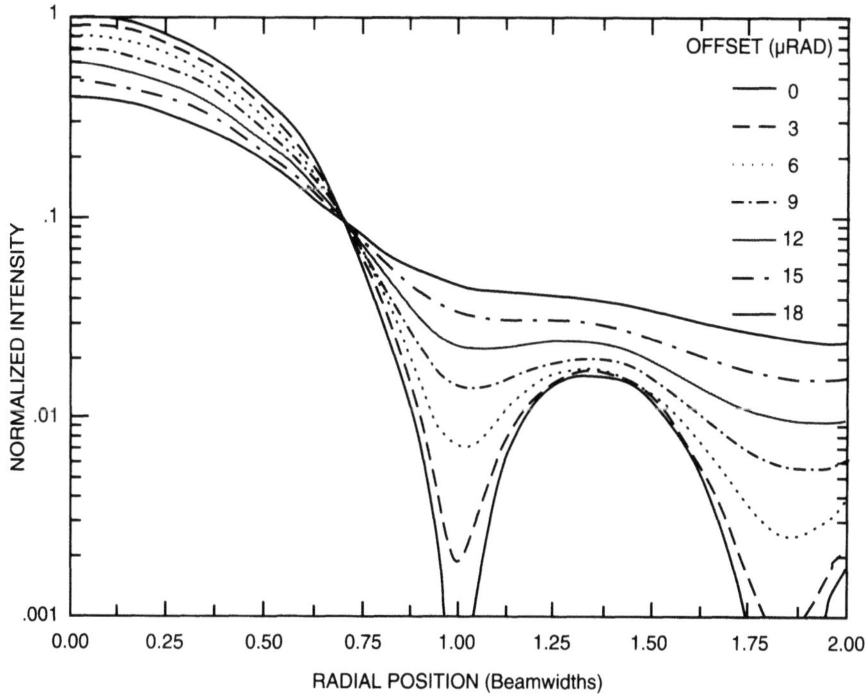

Fig. 13.5. Beam shape with angular anisoplanatism for a 0.6-m aperture at zenith for the SLCSAT day model. The intensity is normalized to that with no turbulence.

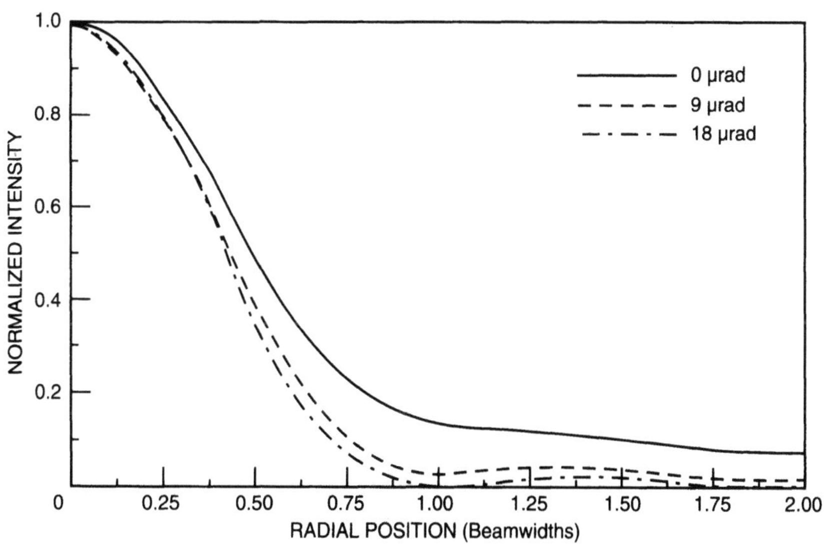

Fig. 13.6. Beam shape with angular anisoplanatism for a 0.6-m aperture at zenith for the SLCSAT day model with offset of 0, 9, and 18 μrad. The intensity is normalized to that with no turbulence.

A. Additional Mellin Transforms

This table contains Mellin transforms to supplement the ones at the end of Chap. 1. These are still a small fraction of the transforms that are listed in *Marichev* (1983). The special functions that are not commonly used are defined in Appendix B. The value of n is an integer in the transforms below.

$$h(x) \qquad \rightarrow M\left[h(x)\right] \equiv H(s) = \int_0^\infty \frac{dx}{x}\, h(x)\, x^s \tag{A.1}$$

$$\frac{1}{1-x} \rightarrow \pi\Gamma\left[\begin{matrix} s,\ 1-s \\ s+\frac{1}{2},\ \frac{1}{2}-s \end{matrix}\right], \qquad 0 < \mathrm{Re}\,\{s\} < 1 \tag{A.2}$$

$$|1-x|^{-p} \quad \rightarrow \quad \frac{\pi}{\Gamma\left[p\right]\cos\left(\pi p/2\right)}\Gamma\left[\begin{matrix} s,\ p-s \\ s+(1-p)/2,\ (1+p)/2-s \end{matrix}\right], \qquad 0 < \mathrm{Re}\,\{s\} < \mathrm{Re}\,\{p\} < 1 \tag{A.3}$$

$$\left[x^2+2x\cos\left(\pi b\right)+1\right]^{-1} \quad \rightarrow \quad -\frac{\pi}{\sin\left(\pi b\right)}\Gamma\left[\begin{matrix} s,\ 1-s \\ s\,b-b,\ 1+b-s\,b \end{matrix}\right] \qquad 0 < \mathrm{Re}\,\{s\} < 2\,,\ \ |b| < 1 \tag{A.4}$$

$$\sin\left(x+b\pi\right) \rightarrow \sqrt{\pi}\,2^{s-1}\Gamma\left[\begin{matrix} s/2,\ s/2+\frac{1}{2} \\ s/2+b,\ 1-b-s/2 \end{matrix}\right], \qquad 0 < \mathrm{Re}\,\{s\} < 1 \tag{A.5}$$

$$\cos\left(x+b\pi\right) \rightarrow \sqrt{\pi}\,2^{s-1}\Gamma\left[\begin{matrix} s/2,\ s/2+\frac{1}{2} \\ s/2+b+\frac{1}{2},\ \frac{1}{2}-b-s/2 \end{matrix}\right], \qquad 0 < \mathrm{Re}\,\{s\} < 1 \tag{A.6}$$

$$\sin^{2n-1}(x) \quad \rightarrow \quad \frac{\sqrt{\pi}}{4^{n-s/2}}\sum_{k=0}^{n-1}(-1)^{n+k-1}\frac{(2n-1)!}{k!\,(2\,n-k-1)!}\,(2\,n-2\,k-1)^{-s} \times\Gamma\left[\begin{matrix} s/2+\frac{1}{2} \\ 1-s/2 \end{matrix}\right], \qquad |\mathrm{Re}\,\{s\}| < 1 \tag{A.7}$$

$$\sin^{2n}(x) \;\rightarrow\; 2^{-2n}\sqrt{\pi}\sum_{k=0}^{n-1}(-1)^{n+k}\frac{(2n)!}{k!\,(2n-k)!}(n-k)^{-s}\Gamma\left[\begin{matrix} s/2 \\ \frac{1}{2}-s/2\end{matrix}\right],$$
$$-2<\mathrm{Re}\{s\}<0 \qquad \text{(A.8)}$$

$$\cos^{2n-1}(x) \;\rightarrow\; 4^{s/2-n}\sqrt{\pi}\sum_{k=0}^{n-1}\frac{(2n-1)!}{k!\,(2n-k-1)!}(2n-2k-1)^{-s}\Gamma\left[\begin{matrix} s/2 \\ \frac{1}{2}-s/2\end{matrix}\right],$$
$$0<\mathrm{Re}\{s\}<1 \qquad \text{(A.9)}$$

$$\cos^{2n}(x)-1 \;\rightarrow\; 2^{-2n}\sqrt{\pi}\sum_{k=0}^{n-1}\frac{(2n)!}{k!\,(2n-k)!}(n-k)^{-s}\Gamma\left[\begin{matrix} s/2 \\ \frac{1}{2}-s/2\end{matrix}\right],$$
$$-2<\mathrm{Re}\{s\}<0 \qquad \text{(A.10)}$$

$$\ln(x)\,U(1-x) \rightarrow \Gamma\left[\begin{matrix} s,\,-s \\ 1+s,\,1-s\end{matrix}\right],\quad 0<\mathrm{Re}\{s\} \qquad \text{(A.11)}$$

$$\ln(x)\,U(x-1) \rightarrow -\Gamma\left[\begin{matrix} s,\,-s \\ 1+s,\,1-s\end{matrix}\right],\quad 0>\mathrm{Re}\{s\} \qquad \text{(A.12)}$$

$$\ln(1+x) \rightarrow -\Gamma[s,\,-s],\quad -1<\mathrm{Re}\{s\}<0 \qquad \text{(A.13)}$$

$$\ln|1-x| \rightarrow -\pi\,\Gamma\left[\begin{matrix} s,\,-s \\ s+\frac{1}{2},\,\frac{1}{2}-s\end{matrix}\right],\quad -1<\mathrm{Re}\{s\}<0 \qquad \text{(A.14)}$$

$$\frac{\ln(x)}{x-1} \rightarrow \Gamma[s,\,s,\,1-s,\,1-s],\quad 0<\mathrm{Re}\{s\}<1 \qquad \text{(A.15)}$$

$$\frac{\ln(x)}{x+1} \rightarrow -\pi\,\Gamma\left[\begin{matrix} s,\,s,\,1-s,\,1-s \\ s+\frac{1}{2},\,\frac{1}{2}-s\end{matrix}\right],\quad 0<\mathrm{Re}\{s\}<1 \qquad \text{(A.16)}$$

$$L^{n}(-x) \rightarrow (-s)^{1-n}\,\Gamma[s,\,-s],\quad -1<\mathrm{Re}\{s\}<0 \qquad \text{(A.17)}$$

$$\sinh[v\,\mathrm{arcsinh}(x)] \;\rightarrow\; \frac{v\cos(\pi v/2)}{4\sqrt{\pi}}\,\Gamma\left[\begin{matrix} s/2+\frac{1}{2},\,v/2-s/2,\,-v/2-s/2 \\ 1-s/2\end{matrix}\right],$$
$$-1<\mathrm{Re}\{s\}<-|\mathrm{Re}\{v\}| \qquad \text{(A.18)}$$

$$\arcsin(x)\,U(1-x) \rightarrow \frac{\pi}{4}\Gamma\left[\begin{matrix} s/2 \\ s/2+1\end{matrix}\right]+\frac{\sqrt{\pi}}{4}\Gamma\left[\begin{matrix} s/2+\frac{1}{2},\,-s/2 \\ s/2+1,\,1-s/2\end{matrix}\right],\; -1<\mathrm{Re}\{s\}$$
$$\text{(A.19)}$$

$$\arccos(x)\,U(1-x) \to -\frac{\sqrt{\pi}}{4}\,\Gamma\left[\begin{matrix} s/2+\frac{1}{2},\ -s/2 \\ s/2+1,\ 1-s/2 \end{matrix}\right], \qquad 0<\mathrm{Re}\{s\} \tag{A.20}$$

$$\arctan(x) \to \frac{1}{4}\,\Gamma\left[\begin{matrix} s/2+\frac{1}{2},\ \frac{1}{2}-s/2,\ -s/2 \\ 1-s/2 \end{matrix}\right], \qquad -1<\mathrm{Re}\{s\}<0 \tag{A.21}$$

$$K(x) \to \frac{\pi}{4}\,\Gamma\left[\begin{matrix} s/2,\ \frac{1}{2}-s/2 \\ s/2+\frac{1}{2},\ 1-s/2 \end{matrix}\right], \qquad 0<\mathrm{Re}\{s\}<1 \tag{A.22}$$

$$K(ix) \to \frac{1}{4}\,\Gamma\left[\begin{matrix} s/2,\ \frac{1}{2}-s/2,\ \frac{1}{2}-s/2 \\ 1-s/2 \end{matrix}\right], \qquad 0<\mathrm{Re}\{s\}<1 \tag{A.23}$$

$$K\left(\sqrt{1-x}\right)U(1-x) \to \frac{\pi}{2}\,\Gamma\left[\begin{matrix} s,\ s \\ s+\frac{1}{2},\ s+\frac{1}{2} \end{matrix}\right], \qquad 0<\mathrm{Re}\{s\} \tag{A.24}$$

$$E\left(\sqrt{1-x}\right)U(1-x) \to \frac{\pi}{2}\,\Gamma\left[\begin{matrix} s,\ s+1 \\ s+\frac{1}{2},\ s+\frac{3}{2} \end{matrix}\right], \qquad 0<\mathrm{Re}\{s\} \tag{A.25}$$

$$E\left(\frac{1}{\sqrt{1+x}}\right)\frac{1}{\sqrt{1+x}} \to \Gamma\left[\begin{matrix} s,\ s+1,\ \frac{1}{2}-s \\ s+\frac{1}{2} \end{matrix}\right], \qquad 0<\mathrm{Re}\{s\}<1/2 \tag{A.26}$$

$$Ei(-x) \to -\Gamma\left[\begin{matrix} s,\ s \\ s+1 \end{matrix}\right], \qquad 0<\mathrm{Re}\{s\} \tag{A.27}$$

$$\exp(x)\,Ei(-x) \to -\Gamma[s,\ s,\ 1-s], \qquad 0<\mathrm{Re}\{s\}<1 \tag{A.28}$$

$$Si(x) \to -\frac{\sqrt{\pi}}{4}2^s\,\Gamma\left[\begin{matrix} s/2,\ s/2+\frac{1}{2} \\ s/2+1,\ 1-s/2 \end{matrix}\right], \qquad -1<\mathrm{Re}\{s\}<0 \tag{A.29}$$

$$si(x) \to -\frac{\sqrt{\pi}}{4}2^s\,\Gamma\left[\begin{matrix} s/2,\ s/2+\frac{1}{2} \\ s/2+1,\ 1-s/2 \end{matrix}\right], \qquad 0<\mathrm{Re}\{s\}<2 \tag{A.30}$$

$$Ci(x) \to \frac{\sqrt{\pi}}{4}2^s\,\Gamma\left[\begin{matrix} s/2,\ -s/2 \\ 1-s/2,\ \frac{1}{2}-s/2 \end{matrix}\right], \qquad 0<\mathrm{Re}\{s\}<2 \tag{A.31}$$

$$\mathrm{Erf}(x) \to \frac{1}{2\sqrt{\pi}}\,\Gamma\left[\begin{matrix} s/2+\frac{1}{2},\ -s/2 \\ 1-s/2 \end{matrix}\right], \qquad -1<\mathrm{Re}\{s\}<0 \tag{A.32}$$

$$\exp\left(-x^2\right)\mathrm{Erf}(ix) \to \frac{i}{2}\,\Gamma\left[\begin{matrix} s/2+\frac{1}{2},\ \frac{1}{2}-s/2 \\ 1-s/2 \end{matrix}\right], \qquad 1>|\mathrm{Re}\{s\}| \tag{A.33}$$

$$\mathrm{Erfc}(x) \to \frac{1}{2\sqrt{\pi}}\,\Gamma\left[\begin{matrix} s/2+\frac{1}{2},\ s/2 \\ s/2+1 \end{matrix}\right], \qquad 0<\mathrm{Re}\{s\} \tag{A.34}$$

$$\exp\left(x^2\right)\mathrm{Erfc}(x) \to \frac{1}{2\pi}\,\Gamma\left[s/2+\tfrac{1}{2},\ s/2,\ \tfrac{1}{2}-s/2\right], \qquad 0<\mathrm{Re}\{s\}<1 \tag{A.35}$$

$$S(x) \to -\frac{2^s}{4}\,\Gamma\left[\begin{matrix} s/2+\frac{3}{4},\ s/2 \\ s/2+1,\ \frac{3}{4}-s/2 \end{matrix}\right], \qquad -3/2<\mathrm{Re}\{s\}<0 \tag{A.36}$$

$$C(x) \to -\frac{2^s}{4}\Gamma\begin{bmatrix} s/2+\frac{1}{4},\ s/2 \\ s/2+1,\ \frac{1}{4}-s/2 \end{bmatrix}, \qquad -1/2 < \mathrm{Re}\{s\} < 0 \tag{A.37}$$

$$\gamma[a,\, x] \to \Gamma\begin{bmatrix} s+a,\ -s \\ 1-s \end{bmatrix}, \qquad -\mathrm{Re}\{a\} < \mathrm{Re}\{s\} < 0 \tag{A.38}$$

$$\Gamma[a,\, x] \to \Gamma\begin{bmatrix} s+a,\ s \\ s+1 \end{bmatrix}, \qquad \mathrm{Re}\{s\} > -\mathrm{Re}\{a\},\ 0 < \mathrm{Re}\{s\} \tag{A.39}$$

$$\exp\left(-x^2/4\right) D_\nu(x) \to 2^{s/2+\nu/2-1/2}\,\Gamma\begin{bmatrix} s/2,\ s/2+\frac{1}{2} \\ s/2+(1-\nu)/2 \end{bmatrix}, \quad 0 < \mathrm{Re}\{s\} \tag{A.40}$$

$$\begin{aligned} \exp\left(x^2/4\right) D_\nu(x) \;\to\; & \left[2^{-s/2+\nu/2+3/2}\sqrt{\pi}\,\Gamma[-\nu]\right]^{-1} \\ & \times\Gamma\left[s/2,\ s/2+\tfrac{1}{2},\ -\nu/2-s/2\right], \quad 0 < \mathrm{Re}\{s\} < -\mathrm{Re}\{\nu\} \end{aligned} \tag{A.41}$$

$$\begin{aligned} N_\nu(x) \;\to\; & \frac{2^s}{2}\Gamma\begin{bmatrix} s/2+\nu/2,\ s/2-\nu/2 \\ s/2-(1+\nu)/2,\ (3+\nu)/2-s/2 \end{bmatrix}, \\ & |\,\mathrm{Re}\{\nu\}\,| < \mathrm{Re}\{s\} < 3/2 \end{aligned} \tag{A.42}$$

$$\begin{aligned} H_\nu(x) \;\to\; & \frac{2^s}{2}\Gamma\begin{bmatrix} s/2+(1+\nu)/2,\ (1-\nu)/2-s/2 \\ 1+\nu/2-s/2,\ 1-\nu/2-s/2 \end{bmatrix}, \\ & \mathrm{Re}\{s\} < 3/2, \quad -1 < \mathrm{Re}\ \{s+\nu\} < 1 \end{aligned} \tag{A.43}$$

$$\begin{aligned} s_{u,\nu}(x) \;\to\; & 2^{s/2+u-2}\,\Gamma\left[\frac{u-\nu+1}{2},\ \frac{u+\nu+1}{2}\right] \\ & \times\Gamma\begin{bmatrix} s/2+(u+1)/2, (1-u)/2-s/2 \\ 1-\nu/2-s/2,\ 1+\nu/2-s/2 \end{bmatrix}, \\ & |\mathrm{Re}\{s+u\}| < 1, \mathrm{Re}\{s\} < 3/2 \end{aligned} \tag{A.44}$$

$$\exp(-x)\ I_\nu(x) \to \frac{2^s}{\sqrt{\pi}}\Gamma\begin{bmatrix} s+\nu,\ \frac{1}{2}-s \\ 1+\nu-s \end{bmatrix}, \qquad \mathrm{Re}\{\nu\} < \mathrm{Re}\{s\} < 1/2 \tag{A.45}$$

$$\exp(-x)\,K_\nu(x) \to 2^s\sqrt{\pi}\Gamma\begin{bmatrix} s+\nu,\ s-\nu \\ s+\frac{1}{2} \end{bmatrix}, \qquad |\,\mathrm{Re}\{\nu\}\,| < \mathrm{Re}\{s\} \tag{A.46}$$

$$\exp(x)\,K_\nu(x) \quad\rightarrow\quad \frac{2^s\cos(\pi\nu)}{\sqrt{\pi}}\varGamma\left[s+\nu,\ s-\nu,\ \tfrac{1}{2}-s\right],$$
$$|\mathrm{Re}\{\nu\}| < \mathrm{Re}\{s\} < 1/2 \qquad \text{(A.47)}$$

$$P_n(2x-1)\,U(1-x) \rightarrow \varGamma\begin{bmatrix} s,\ s \\ s+n+1,\ s-n \end{bmatrix}, \quad 0 < \mathrm{Re}\{s\} \qquad \text{(A.48)}$$

$$P_n(2x-1)\,U(x-1) \rightarrow \varGamma\begin{bmatrix} -n-s,\ 1+n-s \\ 1-s,\ 1-s \end{bmatrix}, \qquad -n > \mathrm{Re}\{s\} \qquad \text{(A.49)}$$

$$P_n(x)\,U(1-x) \rightarrow \frac{1}{2}\varGamma\begin{bmatrix} s/2,\ s/2+\frac{1}{2} \\ s/2+(1-n)/2,\ s/2+1+n/2 \end{bmatrix}, \quad 0 < \mathrm{Re}\{s\} \quad \text{(A.50)}$$

$$P_n(x)\,U(x-1) \rightarrow \frac{1}{2}\varGamma\begin{bmatrix} (1+n)/2-s/2,\ -n/2-s/2 \\ 1-s/2,\ \frac{1}{2}-s/2 \end{bmatrix}, \quad -n > \mathrm{Re}\{s\} \ \text{(A.51)}$$

$$(1-x)^{-1/2}\,U(1-x)\,T_n(2x-1) \quad\rightarrow\quad \sqrt{\pi}\,\varGamma\begin{bmatrix} s,\ s+\frac{1}{2} \\ s+\frac{1}{2}-n,\ s+\frac{1}{2}+n \end{bmatrix},$$
$$0 < \mathrm{Re}\{s\} \qquad \text{(A.52)}$$

$$(1-x)^{1/2}\,U(1-x)\,U_n(2x-1) \quad\rightarrow\quad \sqrt{\pi}\,\frac{n+1}{2}\,\varGamma\begin{bmatrix} s,\ s-\frac{1}{2} \\ s+\frac{3}{2}+n,\ s-\frac{1}{2}-n \end{bmatrix},$$
$$0 < \mathrm{Re}\{s\} \qquad \text{(A.53)}$$

$$(x-1)^{1/2}\,U(x-1)\,U_n(2x-1) \quad\rightarrow\quad \sqrt{\pi}\,\frac{n+1}{2}\,\varGamma\begin{bmatrix} \frac{3}{2}+n-s,\ -\frac{1}{2}-n-s \\ 1-s,\ \frac{3}{2}-s \end{bmatrix},$$
$$-1/2-n > \mathrm{Re}\{s\} \qquad \text{(A.54)}$$

$$H_n(x)\exp(-x/2) \rightarrow 2^{n-1}\,\varGamma\begin{bmatrix} s/2,\ s/2+\frac{1}{2} \\ s/2+(1-n)/2 \end{bmatrix}, \qquad 0 < \mathrm{Re}\{s\} \qquad \text{(A.55)}$$

$$L_n^a(x)\,\exp(-x) \rightarrow \frac{1}{n!}\varGamma\begin{bmatrix} s,\ 1+n+a-s \\ 1+a-s \end{bmatrix}, \qquad 0 < \mathrm{Re}\{s\} \qquad \text{(A.56)}$$

$$C_n^{\lambda}(2x-1)\,U(1-x)\,(1-x)^{\lambda-1/2} \;\rightarrow\; \frac{(2\lambda)_n}{n!}\,\Gamma\left[\lambda+\tfrac{1}{2}\right]$$

$$\times\Gamma\left[\begin{matrix} s,\ s-\lambda+\frac{1}{2} \\ s+\lambda+n+\frac{1}{2},\ s-\lambda-n+\frac{1}{2}\end{matrix}\right],\quad 0<\mathrm{Re}\{s\},\ \mathrm{Re}\{\lambda\}>-1/2 \qquad \text{(A.57)}$$

$$C_n^{\lambda}(2x-1)\,U(x-1)\,(x-1)^{\lambda-1/2} \;\rightarrow\; \frac{(2\lambda)_n}{n!}\,\Gamma\left[\lambda+\tfrac{1}{2}\right]$$

$$\times\Gamma\left[\begin{matrix} \frac{1}{2}-n-\lambda-s,\ \frac{1}{2}+n+\lambda-s \\ 1-s,\ \lambda+\frac{1}{2}-s\end{matrix}\right],$$

$$1/2-n-\mathrm{Re}\{\lambda\}>\mathrm{Re}\{s\},\mathrm{Re}\{\lambda\}>-1/2 \qquad \text{(A.58)}$$

$$P_n^{(a,b)}(2x-1)\,U(1-x)\,(1-x)^{a} \;\rightarrow\; \frac{\Gamma[a+n+1]}{n!}$$

$$\times\Gamma\left[\begin{matrix} s,\ s-b \\ s+a+n+1,\ s-b-n\end{matrix}\right],\qquad 0<\mathrm{Re}\{s\},\ \mathrm{Re}\{a\}>-1 \qquad \text{(A.59)}$$

$$P_n^{(a,b)}(2x-1)\,U(x-1)\,(x-1)^{a} \;\rightarrow\; \frac{\Gamma[a+n+1]}{n!}$$

$$\times\Gamma\left[\begin{matrix} 1+n+b-s,\ -a-b-s \\ 1+b-s,\ 1-s\end{matrix}\right],\mathrm{Re}\{s\}<-n-\mathrm{Re}\{a\},\mathrm{Re}\{a\}>-1 \qquad \text{(A.60)}$$

$$P_{\nu}^{u}(2x-1)\,U(1-x)\,(1-x)^{-u/2} \;\rightarrow\; \Gamma\left[\begin{matrix} s+u/2,\ s-u/2 \\ s+\nu+1-u/2,\ s-\nu-u/2\end{matrix}\right],$$

$$\mathrm{Re}\{s\}>|\mathrm{Re}\{u\}|/2,\ \mathrm{Re}\{u\}<1 \qquad \text{(A.61)}$$

$$P_{\nu}^{u}(2x-1)\,U(x-1)\,(x-1)^{-u/2} \;\rightarrow\; \Gamma\left[\begin{matrix} u/2-\nu-s,\ u/2+1+\nu-s \\ 1+u/2-s,\ 1-u/2-s\end{matrix}\right],$$

$$\mathrm{Re}\{s\}<1+\mathrm{Re}\{u/2\}+\mathrm{Re}\{\nu\},\ \mathrm{Re}\{s\}<\mathrm{Re}\{u/2\}-\mathrm{Re}\{\nu\},\ \mathrm{Re}\{u\}<1 \qquad \text{(A.62)}$$

$$(1+x)^{-u/2}\,Q_{\nu}^{u}(1+2x) \;\rightarrow\; \frac{\exp(i\pi u)}{2}$$

$$\times\Gamma\left[\begin{matrix} s+u/2,\ s-u/2,\ \nu+1+u/2-s \\ s+\nu+1-u/2\end{matrix}\right],$$

$$|\mathrm{Re}\{u\}|/2<\mathrm{Re}\{s\}<1+\mathrm{Re}\ \{\nu+u/2\} \qquad \text{(A.63)}$$

$$ {}_1F_1\left[a;c;-x\right] \to \Gamma\begin{bmatrix} c \\ a \end{bmatrix} \Gamma\begin{bmatrix} a-s,\ s \\ c-s \end{bmatrix}, \qquad 0 < \mathrm{Re}\{s\} < \mathrm{Re}\{a\} \tag{A.64} $$

$$ \exp(-x)\,{}_1F_1\left[a;c;x\right] \to \Gamma\begin{bmatrix} c \\ c-a \end{bmatrix} \Gamma\begin{bmatrix} c-a-s,\ s \\ c-s \end{bmatrix}, \quad 0 < \mathrm{Re}\{s\} < \mathrm{Re}\{c-a\} \tag{A.65} $$

$$ \Psi\left[a,c;x\right] \to \frac{1}{\Gamma\left[a,\ 1+a-c\right]}\Gamma\left[s,\ s+1-c,\ a-s\right], \quad \mathrm{Re}\{c-1\} < \mathrm{Re}\{s\} < \mathrm{Re}\{a\},\ 0 < \mathrm{Re}\{s\} \tag{A.66} $$

$$ \exp(-x)\,\Psi\left[a,c;x\right] \to \Gamma\begin{bmatrix} s,\ s+1-c \\ s+a+1-c \end{bmatrix}, \quad \mathrm{Re}\{c-1\} < \mathrm{Re}\{s\},\ 0 < \mathrm{Re}\{s\} \tag{A.67} $$

$$ \exp(-x/2)\,M_{a,u}(x) \to \Gamma\begin{bmatrix} 2u+1 \\ a+u+\frac{1}{2} \end{bmatrix} \Gamma\begin{bmatrix} s+u+\frac{1}{2},\ a-s \\ u+\frac{1}{2}-s \end{bmatrix}, \quad -\mathrm{Re}\{u-1/2\} < \mathrm{Re}\{s\} < \mathrm{Re}\{a\} \tag{A.68} $$

$$ \exp(-x/2)\,W_{a,u}(x) \to \Gamma\begin{bmatrix} s+u+\frac{1}{2},\ s+\frac{1}{2}-u \\ s+1-a \end{bmatrix}, \quad \mathrm{Re}\{s\} < 1/2 + |\mathrm{Re}\{u\}| \tag{A.69} $$

$$ \exp(x/2)\,W_{a,u}(x) \to \frac{1}{\Gamma\left[\frac{1}{2}-u-a,\ \frac{1}{2}+u-a\right]} \times\Gamma\left[s-u+\tfrac{1}{2},\ s+\tfrac{1}{2}+u,\ -a-s\right], \quad -1/4 + |\mathrm{Re}\{u\}| < \mathrm{Re}\{s\} < -|\mathrm{Re}\{a\}| \tag{A.70} $$

$$ {}_1F_1\left[a,b;c;-x\right] \to \Gamma\begin{bmatrix} c \\ a,\ b \end{bmatrix} \Gamma\begin{bmatrix} a-s,\ b-s,\ s \\ c-s \end{bmatrix}, \quad 0 < \mathrm{Re}\{s\} < \mathrm{Re}\{a\}, \mathrm{Re}\{s\} < \mathrm{Re}\{b\} \tag{A.71} $$

$$ Ai(z) \to \frac{2^{s/3}}{2\pi\, 3^{7/6}}\Gamma\left[s/3+\tfrac{1}{3},\ s/3\right], \quad \mathrm{Re}\{s\} > 0 \tag{A.72} $$

B. Trancendental Functions

Below are the definitions of functions whose Mellin transforms are given in App. A. If an expression contains gamma functions, it is not valid for arguments that give an infinite result.

Polylogarithm

$$L^n(z) = \sum_{k=1}^{\infty} \frac{z^k}{k^n} \qquad |z| < 1 \tag{B.1}$$

Complete elliptic integral of the second kind

$$E(k) = \int_0^{\pi/2} d\varphi \sqrt{1 - k^2 \sin^2(\varphi)} = \frac{\pi}{2}\, {}_2F_1\left[-\tfrac{1}{2}, \tfrac{1}{2}; 1; k^2\right] \tag{B.2}$$

Elliptic integral of the first kind

$$K(k) = \int_0^{\pi/2} \frac{d\varphi}{\sqrt{1 - k^2 \sin^2(\varphi)}} = \frac{\pi}{2}\, {}_2F_1\left[\tfrac{1}{2}, \tfrac{1}{2}; 1; k^2\right] \tag{B.3}$$

Exponential integral

$$Ei(x) = \int_{-\infty}^{x} dt\, \frac{\exp(t)}{t} = -\exp(x)\, \Psi[1, 1; -x], \quad x < 0 \tag{B.4}$$

Lower Sine integral

$$Si(x) = \int_0^x dt\, \frac{\sin(t)}{t} \tag{B.5}$$

Upper Sine integral

$$si(x) = Si(x) - \frac{\pi}{2} = -\int_x^{\infty} dt\, \frac{\sin(t)}{t} \tag{B.6}$$

Cosine integral

$$Ci(x) = -\int_x^\infty dt \, \frac{\cos(t)}{t} \tag{B.7}$$

Error function

$$\mathrm{Erf}(x) = \frac{2}{\sqrt{\pi}} \int_x^\infty dt \, \exp\left(-t^2/2\right) = \frac{2x}{\sqrt{\pi}} \, {}_1F_1\left[-\tfrac{1}{2}; \tfrac{3}{2}; -x^2\right] \tag{B.8}$$

Complementary error function

$$\mathrm{Erfc}(x) = 1 - \mathrm{Erf}(x) = \frac{\exp(-x^2)}{\sqrt{\pi}} \Psi\left[-\tfrac{1}{2}, \tfrac{1}{2}; x^2\right] \tag{B.9}$$

Fresnel sine integral

$$S(x) = \frac{1}{\sqrt{2\pi}} \int_0^x dt \, \frac{\sin(t)}{\sqrt{t}} \tag{B.10}$$

Fresnel cosine integral

$$C(x) = \frac{1}{\sqrt{2\pi}} \int_0^x dt \, \frac{\cos(t)}{\sqrt{t}} \tag{B.11}$$

Lower incomplete gamma function

$$\gamma[a, x] = \frac{x^a}{a} \, {}_1F_1[a; a+1; -x], \quad \mathrm{Re}\{a\} > 0 \tag{B.12}$$

Upper incomplete gamma function

$$\Gamma[a, x] = \exp(-x) \, \Psi[1-a, 1-a; x] \tag{B.13}$$

Parabolic cylinder function

$$D_\nu(x) = 2^{\nu/2} \exp\left(-z^2/4\right) \Psi\left[-\nu/2, \tfrac{1}{2}; z^2/2\right] \tag{B.14}$$

Bessel function of the second kind or Neumann function

$$\begin{aligned} N_\nu(z) &= -\frac{1}{\pi}\left(\frac{z}{2}\right)^{-\nu} \Gamma[\nu] \, {}_0F_1\left[1-\nu; -z^2/4\right] \\ &\quad - \frac{1}{\pi}\cos(\pi\nu)\left(\frac{z}{2}\right)^{\nu} \Gamma[-\nu] \, {}_0F_1\left[1+\nu; -z^2/4\right] \end{aligned} \tag{B.15}$$

Struve function

$$H_\nu(z) = \frac{2}{\sqrt{\pi}}\left(\frac{2}{z}\right)^{\nu+1} \frac{1}{\Gamma\left[\nu + \frac{3}{2}\right]} \, {}_1F_2\left[1; \tfrac{3}{2}, \nu + \tfrac{3}{2}; -z^2/4\right] \tag{B.16}$$

Lommel function

$$s_{u,\nu}(z) = \frac{z^{u+1}}{(u-\nu+1)(u+\nu+1)}\,{}_1F_2\left[1;(u-\nu+3)/2,(u+\nu+3)/2;-z^2/4\right] \tag{B.17}$$

Bessel function of imaginary argument

$$I_\nu(z) = \left(\frac{z}{2}\right)^\nu \frac{1}{\Gamma[\nu+1]}\,{}_0F_1\left[\nu+1;z^2/4\right] \tag{B.18}$$

Bessel Function of the Third Kind or Macdonald or Kelvin function

$$K_\nu(z) = \frac{1}{2}\left(\frac{z}{2}\right)^\nu \Gamma[-\nu]\,{}_0F_1\left[\nu+1;z^2/4\right] + \frac{1}{2}\left(\frac{z}{2}\right)^{-\nu}\Gamma[\nu]\,{}_0F_1\left[1-\nu;z^2/4\right] \tag{B.19}$$

Legendre polynomials

$$P_n(z) = \frac{2^{-n}}{n!}\frac{d^n}{dz^n}\left(z^2-1\right)^n = {}_2F_1\left[-n,1+n;,1;\frac{1-z}{2}\right] \tag{B.20}$$

Chebyshev or Tchebcheff polynomial of the first kind

$$T_n(z) = \cos[n\arccos(z)] = {}_2F_1[-n,n;1/2;(1-z)/2] \tag{B.21}$$

Chebyshev polynomial of the second kind

$$U_n(z) = \frac{\sin[(n+1)\arccos(z)]}{\sqrt{1-z^2}} = (n+1)\,{}_2F_1[-n,n+2;3/2;(1-z)/2] \tag{B.22}$$

Hermite polynomials

$$H_n(z) = (-1)^n \exp\left(z^2\right)\frac{d^n}{dz^n}\exp\left(-z^2\right) = 2^{2n}z\,{}_1F_1\left[(1-n)/2;\tfrac{3}{2};z^2\right] \tag{B.23}$$

Laguerre polynomials

$$L_n^a(z) = \frac{\exp(z)\,z^{-a}}{n!}\frac{d^n}{dz^n}\left[\exp(-z)\,z^{n+a}\right] = \Gamma\left[\begin{matrix}a+n+1\\ a+1,\ n+1\end{matrix}\right]{}_1F_1[-n;a+1;z] \tag{B.24}$$

Gegenbauer polynomials

$$C_p^\lambda(z) = \Gamma\left[\begin{matrix}2\lambda+p\\ 2\lambda,\ p+1\end{matrix}\right]{}_2F_1\left[-p,p+2\lambda;\lambda+\tfrac{1}{2};\frac{1-z}{2}\right] \tag{B.25}$$

Jacobi polynomials

$$P_n^{(a,b)}(z) = \Gamma\left[\begin{matrix}a+n+1\\ a+1,\ n+1\end{matrix}\right]{}_2F_1\left[-n,n+a+b+1;1-a;\frac{1-z}{2}\right] \tag{B.26}$$

Associated Legendre function of the first kind

$$P_\nu^u(z) = \left(\frac{z+1}{z-1}\right)^{u/2} \frac{1}{\Gamma[1-u]} {}_2F_1\left[-\nu, 1+\nu; 1-u; \frac{1-z}{2}\right], \ |\arg(z \pm 1)| < \pi \tag{B.27}$$

Associated Legendre function of the second kind

$$\begin{aligned} Q_\nu^u(z) &= 2^{-\nu-1}\Gamma\begin{bmatrix} u+\nu+1 \\ \nu+\frac{3}{2} \end{bmatrix} \exp(iu\pi)\sqrt{\pi}\, z^{-u-\nu-1}\left(z^2-1\right)^{u/2} \\ &= {}_2F_1\left[(u+\nu+1)/2, (u+\nu+2)/2; \nu+3/2; 1/z^2\right], \\ &\qquad |\arg(z\pm 1)| < \pi, \ \ |\arg z| < \pi \end{aligned} \tag{B.28}$$

Tricomi function

$$\Psi[a, c; x] = \Gamma\begin{bmatrix} 1-c \\ a-c+1 \end{bmatrix} {}_1F_1[a; c; z] + z^{1-c}\Gamma\begin{bmatrix} c-1 \\ a \end{bmatrix} {}_1F_1[a-c+1; 2-c; z] \tag{B.29}$$

Degenerate Whittaker hypergeometric function

$$M_{b,u}(z) = \exp(-z/2)\, z^{u+1/2} = {}_1F_1[1/2-b+u; 2u+1; z] \tag{B.30}$$

Airy function

$$\begin{aligned} Ai(z) &= \frac{1}{\pi}\int_0^\infty dw\, \cos\left(w^3/3 + w\, z\right) = \frac{z^{1/2}}{\pi\sqrt{3}} K_{1/3}\left(2z^{3/2}/3\right) \\ &= \frac{3^{-2/3}}{\Gamma\left[\frac{2}{3}\right]} {}_0F_1\left[\tfrac{2}{3}; z^3/9\right] - \frac{3^{-1/3}}{\Gamma\left[\frac{1}{3}\right]} z\, {}_0F_1\left[\tfrac{4}{3}; z^3/9\right] \end{aligned} \tag{B.31}$$

References

Chap. 1

Fried, D. L., Optical Heterodyne Detection of an Atmospherically Distorted Signal Wave Front, Proc. IEEE, **55**, (1967) 57–67

Marichev, O. I., *Integral Transforms of Higher Transcendental Functions*, Ellis Horwood Limited, Chichester, England, 1983

Prudnikov, A. P., Brychkov, Y. A., Marichev, O. I., *Integrals and Series*, Gordon and Breach Science Publishers, New York, 1990

Slater, L. J., *Generalized hypergeometric functions*, Cambridge University Press, London-New York, 1966

Chap. 2

Chernov, L. A., *Wave Propagation in a Random Media, Dover Publications*, Inc., New York, 1960

Fante, R., L., *Progress in Optics* (Edited by Wolf, E.) North-Holland, Amsterdam, 1985

Fante, R. L., Electromagnetic Beam Propagation in Turbulent Media, Proc. of IEEE, **63**, (1975) 1669–1692

Frehlich, R., Laser Scintillation Measurements of the Temperature Spectrum in the Atmospheric Surface Layer, Journal of the Atmospheric Sciences, **49**, (1992) 1494–1509

Fried, D. L., Optical Heterodyne Detection of an Atmospherically Distorted Signal Wave Front, Proc. IEEE, **55**, (1967) 57–67

Goodman, J. W., *Statistical Optics*, John Wiley & Sons, Inc, New York, 1985

Greenwood, D. P., Tarazano, D. O., A Proposed Form for the Atmospheric Microtemperature Spatial Spectrum in the Input Range, RADC-TR-74-19 (ADA 776294/1GI) (Rome Air Development Center, 1974)

Gurvich, A. S., Time, N. S., Turovtseva, V. F., Turchin, V. F., *Reconstruction of the temperature fluctuation spectrum of the atmosphere from optical measurements*, Izvestiya, Atmospheric and Oceanic Physics, (1974)

Hill, R. J., Clifford, S. F., Modified spectrum of atmospheric temperature fluctuations and its application to optical propagation, J. Opt. Soc. Am., **68**, (1978) 892–899

Ishimaru, A., Fluctuations of a beam wave propagating through a locally homogeneous medium, Radio Sci., **4**, (1969) 293–305

Johnson, B., Primmerman, C. A., Experimental observation of phase-conjugate instability for thermal-blooming compensation, Opt. Lett., **14**, (1989) 639–651

Kaimal, J. C., Izumi, Y., Cote, O. R., Quart. J. Roy. Meteorol. Soc., **98**, (1972) 563

Kolmogorov, A. N., Dissipation of energy in locally isotropic turbulence, Doklady Akad. Nauk. SSSR, **32**, (1941) 16

Langhaar, H. L., *Dimensional analysis and Theory of Models*, John Wiley & Sons, Inc., Huntington, New York, 1980

Obukhov, A. M., On the distribution of energy in the spectrum of turbulent flow, Doklady Akad. Nauk SSSR, **32**, (1941) 19

Schmeltzer, R. A., Means, Variances, and Covariances For Laser Beam Propagation Through A Random Medium, Quart. Appl. Math., **24**, (1967) 339–354

Shonfeld, J. F., Linearized theory of thermal-blooming phase-compensation instability with realistic adaptive-optics geometry, J. Opt. Soc. Am. B, **9**, (1992) 1803–1812

Strohbehn, J. W., *Laser Beam Propagation in the Atmosphere*, Springer-Verlag, Berlin, 1978

Tatarski, V. I., *Wave Propagation In a Turbulent Medium*, Dover Publications, Inc., New York, 1961

Tatarski, V. I., *The Effects Of The Turbulent Atmosphere On Wave Propagation*, U. S. Department Of Commerce, 1971

Tatarski, V. I., Zavorotnyi, V. U., Wave propagation in random media with fluctuating turbulent parameters, J. Opt. Soc. Am. A, **2**, (1985) 2069–2076

Chap. 3

Noll, R. J., Zernike polynomials and atmospheric turbulence, J. Opt. Soc. Am., **3**, (1976) 207–211

Wang, J. Y., Markey, J. K., Modal compensation of atmospheric turbulence phase distortion, J. Opt. Soc. Am., **68**, (1978) 78–87

Chap. 4

Bufton, J. L., Minott, P. O., Fitzmaurice, M. W., Titterton, P. J., Measurements of Turbulence Profiles in the Troposphere, J. Opt. Soc. Am., **62**, (1972) 1068–1070

Ellerbroek, B. L., The Temporal Dependence of Turbulence-Induced Line-of-Sight Jitter as Measured by a Wavefront Gradient Sensor, tOSC Report No. TR-572 (the Optical Sciences Company, Placentia, California, 1984)

Fante, R. L., Electromagnetic Beam Propagation in Turbulent Media: An Update, Proc. of IEEE, **68**, (1980) 1424–1443

Greenwood, D. P., Bandwidth specifications for adaptive optics systems, J. Opt. Soc. Am., **67**, (1977) 390–393.

Herrmann, J., Cross Coupling and aliasing in modal wave-front estimation, J. Opt. Soc. Am., **71**, (1981) 989–992

Hufnagel, R. E., *Digest of Topical Meeting on Optical Propagation Through Turbulence* (Optical Society of America, Washington, D. C., 1974)

Hufnagel, R. E., *The Infrared Handbook* (Edited by Wolfe, W.L. & Zeiss, G.J.) 6-18 (Environmental Research Institute of Michigan, Michigan, 1978)

Miller, M. G., Zieske, P. L., Turbulence Environmental Characterization, RADC-TR-79-131 (Rome Air Development Center, Griffiss Air Force Base, N. Y., 1979)

Prokhorov, A. M., Bunkin, F. V., Gochelashvily, K. S., Shishov, V. I., Laser Irradiance Propagation in Turbulent Media, Proc. of IEEE, **63**, (1975) 790–811

Tatarski, V. I., *The Effects Of The Turbulent Atmosphere On Wave Propagation*, U. S. Department Of Commerce, 1971

Tyler, G. A., The Magnitude of Angle-of-Arrival Jitter in Ground to Relay Mirror Propagation for Various Turbulence Models, tOSC No. TR-6213 (The Optical Sciences Company, 1985)

Valley, G. C., Isoplanatic degradation of tilt correction and short-term imaging system, Appl. Opt., **19**, (1980) 574–577

Yura, H. T., Tavis, M. T., Centroid anisoplanatism, J. Opt. Soc. Am. A, **2**, (1985) 765–773

Chap. 5

Bleistein, N., Hendelsman, R. A., *Asymptotic Expansions of Integrals*, Dover, New York, 1986

Luke, Y., *The Special Functions and Their Approximations*, Academic Press, New York, 1969

Marichev, O. I., *Integral Transforms of Higher Transcendental Functions*, Ellis Horwood Limited, Chichester, England, 1983

Oberhettinger, F., *Tables of Mellin Transforms*, Springer-Verlag, Berlin, Germany, 1974

Prudnikov, A. P., Brychkov, Y. A., Marichev, O. I., *Integrals and Series*, Gordon and Breach Science Publishers, New York, 1990

Slater, L. J., *Generalized hypergeometric functions*, Cambridge University Press, London-New York, 1966

Wong, R., *Asymptotic Approximations Of Integrals*, Academic Press, Inc., Boston, Ma., 1989

Chap. 6

Altes, R. A., The Fourier-Mellin Transform and Mammalian Hearing, J. Acoust. Soc. Am., **63**, (1978) 174–183

Andrews, L. C., Shivamoggi, B. K., *Integral Transforms for Engineers and Applied Mathematicians*, Macmillan Publishing Company, New York, 1988

Banai, V. S., Transient Response of Euler-Cauchy-Type Time-Varying Nonlinear Systems Using Multidimensional Mellin Transforms, Proc. IEEE, **117**, (1970) 1156–1160

Bessis, D., Fournier, J. D., Servizi, G., Turchetti, G., Vaienti, S., Mellin Transforms of Correlation Integrals and Generalized Dimensions of Strange Sets, Phys. Rev. A, **36**, (1987) 920–928

Bessis, D., Geronimo, J. S., Moussa, P., Mellin Transforms Associated with Julia Sets and Physical Applications, J. Stat. Phys., **34**, (1984) 75–110

Birmingham, D., Sen, S., A Mellin Summation Technique, J. Phys. A: Math Gen., **20**, (1987) 4557–4560

Casasent, D., Psaltis, D., Doppler signal processing: a new technique, Appl. Opt., **15**, (1976) 2015

Casasent, D., Psaltis, D., Position, Rotation, and Scale Invariant Optical Correlation, Appl. Opt., **15**, (1976) 1795–1799

Casasent, D., Psaltis, D., Scale Invariant Optical Transform, Opt. Eng., **15**, (1976) 258–261

Cicuta, G. M., Montaldi, E., Remarks on the Full Asymptotic Expansion of Feynman Parametrized Integrals, Letters to Nuovo Cimento, **13**, (1975) 310–312

Lee, H.-M., Rise and Fall of Directed Transient; Use of Mellin Transformation in Time Domain Problems, Radio Sci., **6**, (1987) 1102–1108

Lipinski, W., Rolicz, P., Sikora, R., Application of Integral Transforms to the Analysis of the Magnetic Field of a Spherical Coil, IEEE Trans. on Magnetics, **11**, (1975) 1552–1554

Oppenheim, A. V., Schafer, R. W., *Digital Signal Processing*, Prentice-Hall, Inc., New Jersey, 1975

Sneddon, I. H., *The Use of Integral Transforms*, McGraw-Hill Book Company, New York, 1972

Whittaker, E. T., Watson, G. N., *A Course of Modern Analysis*, Cambridge University Press, Cambridge, 1927

Zemanian, A. H., *Generalized Integral Transformations*, Dover Publications Inc., New York, 1987

Zwicke, P., Kiss, I., Jr., A New Implementation of the Mellin Transform and its Application to Radar Classification of Ships, IEEE Trans. on Pattern Rec. and Machine Int., **5**, (1983) 191–199

Chap. 7

Andrews, L. C., Aperture-averaging factor for optical scintillation of plane and spherical waves in the atmosphere, J. Opt. Soc. Am. A, **9**, (1992) 597–600

Butts, R. R., Spectra of Turbulence Induced Wavefront Aberrations, AFWL-TR-80-107 (Air Force Weapons Laboratory, 1980)

Ellerbroek, B. L., Roberts, P. H., Turbulence Induced Angular Separation Errors; Expected Values For the SOR-2 Experiment, tOSC Report No. TR-613 (the Optical Sciences Company, Placentia, California, 1984)

Fields, D. A., High frequency behavior of the tilt spectrum of atmospheric turbulence, Appl. Opt., **22**, (1983) 645–647

Fried, D. L., Aperture Averaging of Scintillation, J. Opt. Soc. Am., **57**, (1967) 169–175

Fugate, R. Q., Fried, D. L., Ameer, G. A., Boeke, B. R., Browne, S. L., Roberts, P. H., Ruane, R. E., Tyler, G. A., Wopat, L. M., Measurement of atmospheric distortion using scattered light from a laser guide star, Nature, **353**, (1991) 144–146

Gardner, C. S., Welsh, B. M., Thompson, L. A., Design and Performance Analysis of Adaptive Optical Telescopes Using Laser Guide Stars, Proc. IEEE, **78**, (1990) 1721–1743

Gradshteyn, I. S., Ryzhik, I. M., *Table of Integrals, Series, and Products*, Academic Press, New York, 1980

Greenwood, D. P., Fried, D. F., Power spectra requirements for wave-front-compensative systems, J. Opt. Soc. Am., **66**, (1976) 193–206

Herrmann, J., Cross Coupling and aliasing in modal wave-front estimation, J. Opt. Soc. Am., **71**, (1981) 989–992

Hill, R. J., Clifford, S. F., Modified spectrum of atmospheric temperature fluctuations and its application to optical propagation, J. Opt. Soc. Am., **68**, (1978) 892–899

Hogge, C. B., Butts, R. R., Frequency Spectra for the Geometric Representation of Wavefront Distortions Due to Atmospheric Turbulence, IEEE Trans. Antennas Propaga., **24**, (1976) 144–154

Lynds, C. R., Worden, S. P., Harvey, J. W., Digital Image Reconstruction Applied To Alpha Orionis, The Astrophysical Journal, **207**, (1976) 174–180

Murphy, D. V., Primmerman, C. A., Zollars, B. G., Barclay, H. T., Experimental Demonstration of Atmospheric Compensation Using Multiple Synthetic Beacons, Opt. Lett., **15**, (1991) 1797–1799

Primmerman, C. A., Murphy, D. V., Page, D. A., Zollars, B. G., Barclay, H. T., Compensation of Atmospheric Optical Distortion Using a Synthetic Beacon, Nature, **353** (1991) 141–143

Tatarski, V. I., *The Effects Of The Turbulent Atmosphere On Wave Propagation*, U. S. Department Of Commerce, 1971

Tyler, G. A., The Power Spectrum for G-tilt and Z-tilt, tOSC Report No. TR-700 (the Optical Sciences Company, Placentia, California, 1986)

Valley, G. C., Long- and short-term Strehl ratios for turbulence with finite inner and outer scales, Appl. Opt., **18**, (1979) 984–987

Vaughn, J. L., Calculation of the Power Spectra of Z-tilt and G-tilt, tOSC Report No. TR-710 (the Optical Sciences Company, Placentia, California, 1986)

Walters, D. L., Favier, D. L., Hines, J. R., Vertical Path Atmospheric MTF Measurements, J. Opt. Soc. Am., **69**, (1979) 828–837

Welsh, B. M., Gardner, C. S., Performance Analysis of Adaptive-Optics Systems using Laser Guide Stars and Slope Sensors, J. Opt. Soc. Am., **6**, (1989) 1913–1923

Welsh, B. M., Thompson, L. A., Effects of Turbulence-Induced Anisoplanatism on the Imaging Performance of Adaptive-Astronomical Telescopes using Laser Guide Stars, J. Opt. Soc. Am. A, **8**, (1991) 69–80

Winker, D. M., Effect of a finite outer scale on the Zernike decomposition of atmospheric optical turbulence, J. Opt. Soc. Am. A, **8**, (1991) 1568–1573

Chap. 8

Abramowitz, M., Stegun, I. A., *Handbook of Mathematical Functions*, National Bureau of Standards, Washington, 1964

Allen, C. W., *Astrophysical Quantities*, The Athone Press, University of London, 1963

Belsher, J., Fried, D., Chromatic Refraction Induced Pseudo Anisoplanatism, tOSC Report No. TR-433 (the Optical Sciences Company, Placentia, California, 1981)

Cole, A. E., Court, A., Kantor, A. J., *Handbook of Geophysics and Space Environments*, Edited by Valley, Shea L, McGraw-Hill Book Company, Inc., New York, 1965

Fried, D., Anisoplanatism in adaptive optics, J. Opt. Soc. Am., **72**, (1982) 52–61

Fried, D. L., Limiting Resolution Looking Down Through the Atmosphere, J. Opt. Soc. Am, **56**, (1966) 1380–1384

Fried, D. L., Time-delay-induced mean-square error in adaptive optics, J. Opt. Soc. Am. A, **7**, (1990) 1224–1225

Gradshteyn, I. S., Ryzhik, I. M., *Table of Integrals, Series, and Products*, Academic Press, New York, 1980

Tyler, G., Belsher, J., Fried, D., Amelioration of Chromatic Refraction Induced Pseudoanisoplanatism, tOSC Report No. TR-465 (the Optical Sciences Company, Placentia, California, 1982)

Tyler, G. A., Turbulence-induced adaptive-optics performance degradation: evaluation in the time domain, J. Opt. Soc. Am. A, **1**, (1984) 251–262

Yura, H. T., Tavis, M. T., Centroid anisoplanatism, J. Opt. Soc. Am. A, 2, (1985) 765–773

Chap. 9

Gradshteyn, I. S., Ryzhik, I. M., *Table of Integrals, Series, and Products*, Academic Press, New York, 1980

Luke, Y., *The Special Functions and Their Approximations*, Academic Press, New York, 1969

Marichev, O. I., *Integral Transforms of Higher Transcendental Functions*, Ellis Horwood Limited, Chichester, England, 1983

Yura, H. T., Hanson, S. G., Second order statistics for wave propagation through complex optical systems, J. Opt. Soc. Am. A, **6**, (1989) 564–575

Chap. 10

Fried, D., Scintillation of a Ground-to-Space Laser Illuminator, J. Opt. Soc. Am., **57**, (1967) 980–983

Fried, D., Seidman, J., Laser-Beam Scintillation in the Atmosphere, J. Opt. Soc. Am., **57**, (1967) 181–185

Ishimaru, A., Fluctuations of a beam wave propagating through a locally homogeneous medium, Radio Sci., **4**, (1969) 293–305

Ishimaru, A., *Laser Beam Propagation in the Atmosphere* (Edited by Strohbehm, J.W.) Springer-Verlag, Berlin, 1978

Ishimaru, A., *Wave Propagation and Scattering in Random Media*, Academic Press, New York, 1978

Kinoshita, Y., Asakura, T., Suzuki, M., Fluctuation Distribution of a Gaussian Beam Propagating Through a Random Medium, J. Opt. Soc. Am., **58**, (1968) 798–807

Chap. 11

Bleistein, N., Hendelsman, R. A., *Asymptotic Expansions of Integrals*, Dover, New York, 1986

Dingle, R. B., *Asymptotic Expansions: Their Derivation and Interpretation*, Academic Press, London, 1973

Erdélyi, A., *Higher Transcendental Functions*, Robert E. Kieger Publishing Company, Malabar, Florida, 1981

Exton, H., *Multiple Hypergeometric Functions and Applications*, John Wiley & Sons, Chichester, England, 1978

Gervois, A., Navelet, H., Integrals of three Bessel functions and Legendre functions. I, J. Math. Phys., **26**, (1985) 633–644

Gervois, A., Navelet, H., Integrals of three Bessel functions and Legendre functions. II, J. Math. Phys., **27**, (1986) 688–695

Gradshteyn, I. S., Ryzhik, I. M., *Table of Integrals, Series, and Products*, Academic Press, New York, 1980

Horn, J., Ueber die Convergenz der hypergeometrischen Reihen zweier und dreier veränderlichen, Math. Ann., **34**, (1889) 544–600

Ishimaru, A., *Electromagnetic Wave Propagation, Radiation, and Scattering*, Prentice Hall, Englewood Cliffs, New Jersey, 1991

Jackson, J. D., *Classical Electromagnetics*, John Wiley & Sons, Inc., New York, 1962

Range, R. M., *Holomorphic Functions and Integral Representations in Several Complex Variables*, Springer-Verlag, Berlin, Germany, 1986

Sasiela, R. J., Shelton, J. D., Mellin transform methods applied to integral evaluation: Taylor series and asymptotic approximations, J. Math. Phys., **34**, (1993) 2572–2617

Slater, L. J., *Generalized hypergeometric functions*, Cambridge University Press, London-New York, 1966
Wong, R., *Asymptotic Approximations Of Integrals*, Academic Press, Inc., Boston, Ma., 1989

Chap. 12

Bailey, W. N., Some Infinite Integrals Involving Bessel Functions, Proc. London Math. Soc., **40**, (1936) 37–48
Gervois, A., Navelet, H., Integrals of three Bessel functions and Legendre functions. I, J. Math. Phys., **26**, (1985) 633–644
Gradshteyn, I. S., Ryzhik, I. M., *Table of Integrals, Series, and Products*, Academic Press, New York, 1980
Prudnikov, A. P., Brychkov, Y. A., Marichev, O. I., *Integrals and Series*, Gordon and Breach Science Publishers, New York, 1990
Tatarski, V. I., *Wave Propagation In a Turbulent Medium*, Dover Publications, Inc., New York, 1961
Watson, G. N., An Infinite Integral Involving Bessel Functions, Journal London Math Soc., **9**, (1934) 16–22

Chap. 13

Yura, H. T., Tavis, M. T., Centroid anisoplanatism, J. Opt. Soc. Am. A, **2**, (1985) 765–773

Index

Springer-Verlag and the Environment

We at Springer-Verlag firmly believe that an international science publisher has a special obligation to the environment, and our corporate policies consistently reflect this conviction.

We also expect our business partners – paper mills, printers, packaging manufacturers, etc. – to commit themselves to using environmentally friendly materials and production processes.

The paper in this book is made from low- or no-chlorine pulp and is acid free, in conformance with international standards for paper permanency.

MIX
Papier aus verantwortungsvollen Quellen
Paper from responsible sources
FSC® C105338

If you have any concerns about our products,
you can contact us on
ProductSafety@springernature.com

In case Publisher is established outside the EU,
the EU authorized representative is:
Springer Nature Customer Service Center GmbH
Europaplatz 3, 69115 Heidelberg, Germany

Printed by Libri Plureos GmbH
in Hamburg, Germany